STUDENT SOLUTIONS MANUAL

ALGEBRA

FOR COLLEGE STUDENTS

STUDENT SOLUTIONS MANUAL
GLORIA E. LANGER

SECOND EDITION

ALGEBRA
FOR COLLEGE STUDENTS

ROBERT F. BLITZER

PRENTICE HALL, UPPER SADDLE RIVER, NJ 07458

Production Editor: *Lisa Protzmann*
Acquisitions Editor: *Melissa Acuna*
Supplement Acquisitions Editor: *Audra Walsh*
Production Coordinator: *Alan Fischer*

©1995 by Prentice-Hall, Inc.
A Simon & Schuster Company
Upper Saddle River, New Jersey 07458

Printed in the United States of America

10 9 8 7 6 5 4 3 2 1

ISBN: 0-13-359654-0

Prentice-Hall International (UK) Limited, *London*
Prentice-Hall of Australia Pty. Limited, *Sydney*
Prentice-Hall Canada Inc., *Toronto*
Prentice-Hall Hispanoamericana, S.A., *Mexico*
Prentice-Hall of India Private Limited, *New Delhi*
Prentice-Hall of Japan, Inc., *Tokyo*
Simon & Schuster Asia Pte. Ltd., *Singapore*
Editora Prentice-Hall do Brasil, Ltda., *Rio de Janeiro*

Algebra for College Students

Table of Contents

Algebra for College Students

Chapter 1   The Real Numbers

Section 1.1   The Real Numbers and Their Properties

Problem Set 1.1, pp. 15-18

1. $\{x \mid x$ is an even natural number between 14 and 20, inclusively$\}$
$\boxed{\{14, 16, 18, 20\}}$

3. $\{x \mid x$ is a prime number between 10 and 26$\}$ **5.** $\{x \mid x$ is a natural number that is divisible by 5$\}$
$\boxed{\{11, 13, 17, 19, 23\}}$ $\boxed{\{5, 10, 15, 20, \dots \}}$

7. $\{x \mid x$ is a whole number but is not a natural number$\}$
$\boxed{\{0\}}$

9. $\{a/b \mid a = 2$ or 3 and $b = 2$ or 3$\}$ **11.** $\{x \mid x$ is an integer and $x^2 = 1\}$
$\boxed{\left\{1, \dfrac{2}{3}, \dfrac{3}{2}\right\}}$ $\boxed{\{-1, 1\}}$

13. $\{x \mid x$ is a positive real number and $x^2 = 5\}$ **15.** $\{x \mid x$ is an integer but not a natural number$\}$
$\boxed{\{\sqrt{5}\}}$ $\boxed{\{\dots, -4, -3, -2, -1, 0\}}$

17. $\{x \mid \sqrt{x}$ is a natural number less than or equal to 2$\}$
$\boxed{\{1, 2\}}$

19. $\{x \mid x$ is the fractional form of $0.\overline{6}\}$ **21.**   Natural numbers: $\boxed{7, \dfrac{18}{2}, 100}$

       $\boxed{\left\{\dfrac{2}{3}\right\}}$

23.   Integers: $\boxed{-10, 0, 7, \dfrac{18}{2}, 100}$ **25.**   Irrational numbers: $\boxed{-\sqrt{2}, \sqrt{3}, \pi}$

27.   Every natural number is a whole number.   **29.** $\{1, 2, 5\} \subseteq \{1, 2, 5, 8\}$
$\boxed{\text{True}}$ $\boxed{\text{True}}$

31. $\{1, 2, 5\} \subseteq \{1, 2, 5\}$ **33.** $1 \subseteq \{1, 2, 5\}$
$\boxed{\text{True}}$ $\boxed{\text{False}}$; $1 \in \{1, 2, 5\}$ and $\{1\} \subseteq \{1, 2, 5\}$

35. $\{1\} \subseteq \{1, 2, 5\}$ **37.** $26 \notin \{1, 2, 3, 4, \dots\}$
$\boxed{\text{True}}$ $\boxed{\text{False}}$; $26 \in \{1, 2, 3, 4, \dots, 25, 26, \dots \}$

39. Every rational number is an integer.

$\boxed{\text{False}}$; $\dfrac{18}{2} = 9$ is a rational number *and* an integer, but $\dfrac{9}{2} = 4.5$ is a rational number but **not** an integer

41. Some whole numbers are not integers.

$\boxed{\text{False}}$; *All* whole numbers are integers. {Whole numbers} $\subseteq$ {Integers}

43. Some real numbers are not rational numbers

$\boxed{\text{True}}$

45. Some real numbers are irrational numbers.

$\boxed{\text{True}}$

47. π cannot be expressed as the quotient of two integers.

$\boxed{\text{True}}$

49. $\dfrac{3}{0}$ is a whole number

$\boxed{\text{False}}$; $\dfrac{3}{0}$ is undefined and *not* a whole number

51. Every prime number is an integer.

$\boxed{\text{True}}$

53. One is a prime number.

$\boxed{\text{False}}$; A prime number is greater than 1.

55. $\pi \cdot 5 = 5\pi$

$\boxed{\text{True}}$; $\pi \cdot 5 = 5 \cdot \pi = 5\pi$ by Commutative property

57. $7[5 + (-5)] = 0$

$\boxed{\text{True}}$; $7[5 + (-5)] = 7 \cdot 0 = 0$ by inverse property of addition and multiplication property of zero

59. $-(7 \cdot 4) = (-7)(-4)$

$\boxed{\text{False}}$; $(-7)(-4) = +(7 \cdot 4) \neq -(7 \cdot 4)$

61. $2(3 \cdot 8) = (2 \cdot 3) \cdot (2 \cdot 8)$

$\boxed{\text{False}}$; $(2 \cdot 3) \cdot (2 \cdot 8) = 2 \cdot [2(3 \cdot 8)] \neq 2(3 \cdot 8)$

63. $14x + x = (14 + 1)x$

$\boxed{\text{True}}$; by the distributive property

65. $4 + (3 + x) = \boxed{(4 + 3) + x = 7 + x}$

67. $(y + 5) + 9 = \boxed{y + (5 + 9) = y + 14}$

69. $7(4y) = \boxed{(7 \cdot 4)y = 28y}$

71. $\dfrac{1}{5}(5x) = \boxed{\left(\dfrac{1}{5} \cdot 5\right) x = x}$

73. $5(7 + 3x) = 5 \cdot 7 + 5 \cdot 3x = \boxed{35 + 15x}$

75. $\dfrac{1}{3}(9x + 6) = \dfrac{1}{3} \cdot 9x + \dfrac{1}{3} \cdot 6 = \boxed{3x + 2}$

77. $\dfrac{1}{3}(4x + 12) = \dfrac{1}{3} \cdot 4x + \dfrac{1}{3} \cdot 12 = \boxed{\dfrac{4}{3}x + 4}$

79. $(5x + 2)8 = 5x \cdot 8 + 2 \cdot 8 = \boxed{40x + 16}$

81. $5(7x + 3) + 10 = 5 \cdot 7x + 5 \cdot 3 + 10 = \boxed{35x + 15 + 10 = 35x + 25}$

83. $7 + (3 + 4x)5 = 7 + 3 \cdot 5 + 4x \cdot 5 = \boxed{7 + 15 + 20x = 22 + 20x}$

85. $4(3x + 2y + 6) + 25 = 4 \cdot 3x + 4 \cdot 2y + 4 \cdot 6 + 25 = \boxed{12x + 8y + 24 + 25 = 12x + 8y + 49}$

87. $7 \cdot 6 = 6 \cdot 7$

$\boxed{\text{Commutative property of multiplication}}$

89. $6 + 0 = 6$

$\boxed{\text{Identity property of addition}}$

91. $7 + (-7) = 0$

  Inverse property of addition

93. $3(8 + 4) = 3(4 + 8)$

  Commutative property of addition

95. $(2 \cdot 3) \cdot 4 = (3 \cdot 2) \cdot 4$

  Commutative property of multiplication

97. $7 = 7$

  Reflexive property of equality

99. $(x + 7) + 3 = x + (7 + 3)$

  Associative property of addition

101. $-a + 0 = -a$

  Identity property of addition

103. $5 \cdot [4 + (7 + 3)] = 5 \cdot 4 + 5 \cdot (7 + 3)$

  Distributive property

105. $5 \cdot [4 + (7 + 3)] = 5 \cdot [4 + (7 + 3)]$

  Reflexive property of equality

107. $9 \cdot [2 + (3 + 0)] = 9 \cdot [(2 + 3) + 0]$

  Associative property of addition

109. $9 \cdot [2 + (3 + 0)] = [2 + (3 + 0)] \cdot 9$

  Commutative property of multiplication

111. If $4x = 21$, then $\frac{1}{4}(4x) = \frac{1}{4}(21)$.

  Multiplication principle

113. If $-(-a) = 0 + a$ and $0 + a = a$, then $-(-a) = a$.

  Transitive property of equality, or substitution

115. $(7 + x) \cdot 0 = 0$

  Multiplicative property of zero

117. $7 + 15$ is a real number

  Closure property of addition

119.

Statement	Reason
1. $a = b$	1. Given
2. ac is a real number	2. Closure of multiplication
3. $ac = ac$	3. Reflexive property
4. $ac = bc$	4. Substitution

121.

Statement	Reason
1. $a + c = b + c$	1. Given
2. $(a + c) + (-c) = (b + c) + (-c)$	2. Addition principle
3. $(a + c) + (-c) = a$	3. Problem 120
$(b + c) + (-c) = b$	
4. $a = b$	4. Substitution

123. a. $S \circ S = S, S \circ L = L, S \circ A = A, S \circ R = R$
  $L \circ S = L, L \circ L = A, L \circ A = R, L \circ R = S$
  $A \circ S = A, A \circ L = R, A \circ A = S, A \circ R = L$
  $R \circ S = R, R \circ L = S, R \circ A = L, R \circ R = A$

  Each operation results in a unique element that is a memeber of $\{S, L, A, R\}$.

b. $(L \circ A) \circ R = R \circ R = A$
  $L \circ (A \circ R) = L \circ L = A$
  associative property

c. $L \circ R = S$
  $R \circ L = S$
  $L \circ R = R \circ L$
  commutative property

  d. $\boxed{S}$ since $S \circ S = S, S \circ L = L, S \circ A = A, S \circ R = R$, etc

   $\boxed{\text{identity}}$

  e. $S \circ \boxed{S} = S$

   $L \circ \boxed{R} = S$

   $A \circ \boxed{A} = S$

   $R \circ \boxed{L} = S$

   $\boxed{\text{inverse property}}$

125. $a \circ b = aa + bb$

    $b \circ a = bb + aa$

    $a \circ b = b \circ a$

    $\boxed{\text{Yes}}$ $\circ$ is a commutative operation

Section 1.2 The Real Number Line, Order, and Absolute Value

Problem Set 1.2, pp. 28-30

1. $\{x \mid x \le 2\}$

   $\boxed{(-\infty, 2]}$

3. $\{x \mid x > -3\}$

   $\boxed{(-3, \infty)}$

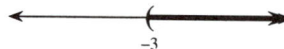

5. $\{x \mid x < -1\}$

   $\boxed{(-\infty, -1)}$

      -1

7. $\{x \mid x \le 0\}$

   $\boxed{(-\infty, 0]}$

      0

9. $\{x \mid 0 < x < 3\}$

   $\boxed{(0, 3)}$

     0 3

11. $\{x \mid -2 \le x < 1\}$

   $\boxed{[-2, 1)}$

     -2 1

13. $\{x \mid -2 \le x \le -1\}$

   $\boxed{[-2, -1]}$

     -2 -1

15. x lies between 5 and 12, excluding 5 and 12

   set-builder: $\boxed{\{x \mid 5 < x < 12\}}$

   interval: $\boxed{(5, 12)}$

17. x lies between 2 and 13, excluding 2 and including 13

   set-builder: $\boxed{\{x \mid 2 < x \le 13\}}$

   interval: $\boxed{(2, 13]}$

19. x is at most 6

set-builder: $\boxed{\{x \mid x \le 6\}}$

interval: $\boxed{(-\infty, 6]}$

21. x is at least 2 and at most 5

set-builder: $\boxed{\{x \mid 2 \le x \le 5\}}$

interval: $\boxed{[2, 5]}$

23. x is not more than 60

set-builder: $\boxed{\{x \mid x \le 60\}}$

interval: $\boxed{(-\infty, 60]}$

25. x is negative and at least -2

set-builder: $\boxed{\{x \mid -2 \le x < 0\}}$

intervale: $\boxed{[-2, 0)}$

27. $|x| < 3$

$\boxed{\{x \mid -3 < x < 3\}}$

$\boxed{(-3, 3)}$

29. $|x| \le 2$

$\boxed{\{x \mid -2 \le x \le 2\}}$

$\boxed{[-2, 2]}$

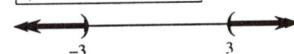

31. $|x| > 3$

$\boxed{\{x \mid x > 3 \text{ or } x < -3\}}$

$\boxed{(-\infty, -3) \text{ or } (3, \infty)}$

33. $|x| \ge 6$

$\boxed{\{x \mid x \ge 6 \text{ or } x \le -6\}}$

$\boxed{(-\infty, -6] \text{ or } [6, \infty)}$

35. $|x| < \dfrac{1}{2}$

$\boxed{\left\{ x \mid -\dfrac{1}{2} < x < \dfrac{1}{2} \right\}}$

$\boxed{\left(-\dfrac{1}{2}, \dfrac{1}{2}\right)}$

37. $|x| > \dfrac{1}{4}$

$\boxed{\left\{ x \mid x > \dfrac{1}{4} \text{ or } x < -\dfrac{1}{4} \right\}}$

$\boxed{\left(-\infty, -\dfrac{1}{4}\right) \text{ or } \left(\dfrac{1}{4}, \infty\right)}$

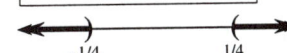

39. $\boxed{D}$ is true;   A is not true: $\{x \mid x < 3\} \rightarrow (-\infty, 3)$ *not* $[-\infty, 3)$

B is not true: $|x| > -1 \rightarrow x$ is any real number

C is not true: $|a + b| \le |a| + |b|$

D is true: None of the above is true.

41. $\boxed{C}$ is true; $-(-6) = 6 \ge |-6| = 6$

43. $\boxed{C}$ is true; $|x| = 7.1563$

$x = \pm 7.1563$

45. $\boxed{B}$ is true;   distance between x and 4 is at least 8

$|x - 4| \ge 8$

47. $|3| < |-5|$

$3 < 5$ $\boxed{\text{True}}$

49. $|6| \le |-6|$

$6 \le 6$ $\boxed{\text{True}}$

51. $|-7| > |-15|$

$7 > 15$ $\boxed{\text{False}}$ since $7 < 15$

53. $|-1| < -3 < -12$

$1 < -3 < -12$ $\boxed{\text{False}}$ since $-12 < -3 < 1$

55. $\left|-12\right| > 4(2+1)$

$\quad\quad 12 > 4(3)$

$\quad\quad 12 > 12$ $\boxed{\text{False}}$ since $12 \geq 12$

57. $\left|-13\right| = \left|13\right|$

$\quad\quad 13 = 13$ $\boxed{\text{True}}$

59. $a \geq \left|a\right|$

$\boxed{\text{False}}$ since $a \geq \left|a\right|$ only when $a \geq 0$

61. $\left|-a\right| = -\left|a\right|$

$\quad\quad a = -a$ $\boxed{\text{False}}$

63. $\left|x\right| < -1$

$\boxed{\text{No values of } x}$ since the absolute value of a number is positive.

65. $\left|x\right| < 0$

$\boxed{\text{No values of } x}$ since the absolute value of a number is positive.

Review Problems

70. $\{x \mid x$ is a prime number between 10 and 18$\}$
$\boxed{\{11, 13, 17\}}$

71. $13 + (-13) = 0$
$\boxed{\text{Inverse property of addition}}$

72. $-13 \in \{x \mid x$ is a rational number$\}$
$\boxed{\text{True}}$

Section 1.3 Operations with the Real Numbers

Problem Set 1.3, pp. 38-40

1. $(-3) + (-9) = \boxed{-12}$

3. $5 + (-17) = \boxed{-12}$

5. $9 + (-11) = \boxed{-2}$

7. $-10 + 14 = \boxed{4}$

9. $-8 - 5 = \boxed{-13}$

11. $(-5.2) + (1.7) = \boxed{-3.5}$

13. $(-7.415) + (+3.2) = \boxed{-4.215}$

15. $\frac{5}{8} + \left(-\frac{1}{3}\right) = \frac{15}{24} + \left(-\frac{8}{24}\right) = \boxed{\frac{7}{24}}$

17. $13 - (-17) = 13 + (+17) = \boxed{30}$

19. $-10 - (-7) = -10 + (+7) = \boxed{-3}$

21. $23 + (-35) = \boxed{-12}$

23. $-27 + (-23) = \boxed{-50}$

25. $14 - 24 = \boxed{-10}$

27. $-\frac{5}{8} - \left(-\frac{1}{6}\right) = -\frac{5}{8} + \left(+\frac{1}{6}\right) = -\frac{15}{24} + \left(+\frac{4}{24}\right) = \boxed{-\frac{11}{24}}$

29. $-4 + 12 + 5 = -4 + (12 + 5) = -4 + 17 = \boxed{13}$

31. $5 + (-8) + (-3) = 5 + [(-8) + (-3)] = 5 + (-11) = \boxed{-6}$

33. $11 - (-5) - (-3) = 11 + (+5) + (+3) = 16 + 3 = \boxed{19}$

35. $-8 - (-12) - (5 - 7)$
$= -8 + (+12) - (-2)$
$= -8 + [(+12) + (+2)]$
$= -8 + (+14)$
$= \boxed{6}$

37. $-8 - [2 - (-7)]$
$= -8 - [2 + (+7)]$
$= -8 - 9$
$= \boxed{-17}$

39. $17 + (-5) - (-3 + 11 - 4)$
$= 17 + (-5) - [11 + (-3 - 4)]$
$= 17 + (-5) - [11 + (-7)]$
$= 17 + (-5) - 4$
$= 17 + (-5) + (-4)$
$= 17 + (-9)$
$= \boxed{8}$

41. $-\dfrac{3}{5} - \left(\dfrac{1}{2} - \dfrac{3}{10}\right)$
$= -\dfrac{3}{5} - \left(\dfrac{5}{10} - \dfrac{3}{10}\right)$
$= -\dfrac{3}{5} - \dfrac{2}{10}$
$= -\dfrac{3}{5} - \dfrac{1}{5}$
$= \boxed{-\dfrac{4}{5}}$

43. $-\dfrac{5}{7} + \dfrac{5}{14} - \left(-\dfrac{3}{4}\right)$
$= -\dfrac{20}{28} + \dfrac{10}{28} + \left(+\dfrac{21}{28}\right)$
$= -\dfrac{20}{28} + \dfrac{31}{28}$
$= \boxed{\dfrac{11}{28}}$

45. $(1 - 3) - (6 + 4)$
$= -2 - 10$
$= \boxed{-12}$

47. $-4 - 7 - 10 + 8 + 9 - 1$
$= -11 - 10 + 8 + 9 - 1$
$= -21 + 8 + 9 - 1$
$= -13 + 9 - 1$
$= -4 - 1$
$= \boxed{-5}$

49. $17 - [13 - (9 - 19)]$
$= 17 - [13 - (-10)]$
$= 17 - (13 + 10)$
$= 17 - 23$
$= \boxed{-6}$

51. $(-13) + (-15) + [-20 - (13 - 18)]$
$= (-13) + (-15) + [-20 - (-5)]$
$= (-13) + (-15) + [-20 + (+5)]$
$= (-13) + (-15) + (-15)$
$= (-28) + (-15)$
$= \boxed{-43}$

53. $-8 + \left|-10\right| + 6$
$= -8 + 10 + 6$
$= 2 + 6$
$= \boxed{8}$

55. $13 + \left|-7\right| + \left|-6\right|$
$= 13 + 7 + 6$
$= \boxed{26}$

57. $\left|-20\right| - \left|-18\right| + 3$
$= 20 - 18 + 3$
$= 2 + 3$
$= \boxed{5}$

59. $\left|-12\right| + \left|3\right| - \left(\left|9\right| + \left|-11\right|\right)$
$= 12 + 3 - (9 + 11)$
$= 15 - 20$
$= \boxed{-5}$

61. $[7.5 + (-4)] - [8.5 + (-6)]$
$= 3.5 - 2.5$
$= \boxed{1}$

63. $(-8)(-7) = \boxed{56}$

65. $\frac{1}{5}(-75) = \boxed{-15}$

67. $-18\left(-\frac{1}{3}\right) = \boxed{6}$

69. $-\frac{2}{3}\left(\frac{15}{16}\right) = \boxed{-\frac{5}{8}}$

71. $0.8(-0.3) = \boxed{-0.24}$

73. $\frac{1}{12}(-9)(-4) = \boxed{3}$

75. $-55\left(-\frac{1}{11}\right)\left(-\frac{1}{5}\right) = \boxed{-1}$

77. $-\frac{1}{9}(-54)\left(-\frac{1}{3}\right) = \boxed{-2}$

79. $(-3)(-1)(-2)\left(-\frac{1}{2}\right)(-4) = \boxed{-12}$

81. $-\frac{1}{8}(-4)\left(-\frac{1}{2}\right)(-6) = \boxed{9}$

83. $(-0.1)(-3)(-6)(2) = \boxed{-3.6}$

85. $6^3 = 6 \cdot 6 \cdot 6 = \boxed{216}$

87. $(-5)^2 = (-5)(-5) = \boxed{25}$

89. $-5^2 = -5 \cdot 5 = \boxed{-25}$

91. $(-1)^6 = (-1)(-1)(-1)(-1)(-1)(-1) = \boxed{1}$

93. $-1^6 = -1 \cdot 1 \cdot 1 \cdot 1 \cdot 1 \cdot 1 = \boxed{-1}$

95. $\left(\frac{2}{5}\right)^4 = \frac{2}{5} \cdot \frac{2}{5} \cdot \frac{2}{5} \cdot \frac{2}{5} = \boxed{\frac{16}{625}}$

97. $\left(-\frac{2}{7}\right)^2 = \left(-\frac{2}{7}\right)\left(-\frac{2}{7}\right) = \boxed{\frac{4}{49}}$

99. $4^2 \cdot 3^3 = 16 \cdot 27 = \boxed{432}$

101. $(-2)^3(-4)^2 = (-8)(16) = \boxed{-128}$

103. $\left(-\frac{2}{3}\right)^3 \cdot 9^2 = \left(-\frac{8}{27}\right)(81) = \boxed{-24}$

105. $-\left(\frac{1}{2}\right)^5 \cdot (-2)^6 = -\left(\frac{1}{32}\right)(64) = \boxed{-2}$

107. $\left(-\frac{4}{5}\right)^2 \cdot 5^3 = \left(\frac{16}{25}\right)(125) = \boxed{80}$

109. $-(-5)^3 = -(-125) = \boxed{125}$

111. $[-(-5)]^3 = 5^3 = \boxed{125}$

113. $\frac{-120}{-12} = \boxed{10}$

115. $\frac{0}{-18} = \boxed{0}$

117. $15 \div \left(-\frac{1}{5}\right) = 15 \cdot \left(\frac{-5}{1}\right) = \boxed{-75}$

119. $-\frac{3}{8} \div \left(-\frac{2}{5}\right) = -\frac{3}{8} \cdot \left(-\frac{5}{2}\right) = \boxed{\frac{15}{16}}$

121. $\dfrac{-\dfrac{9}{26}}{\dfrac{16}{27}} = -\frac{9}{26} \cdot \left(-\frac{27}{16}\right) = \boxed{\frac{243}{416}}$

123. $0 \div \left(-\frac{17}{53}\right) = 0 \cdot \left(-\frac{53}{17}\right) = \boxed{0}$

125. $1\frac{2}{3} \div \left(-6\frac{2}{3}\right) = \frac{5}{3} \div \left(-\frac{20}{3}\right) = \frac{5}{3} \cdot \left(-\frac{3}{20}\right) = \boxed{-\frac{1}{4}}$

127. The sum of two negative numbers is $\boxed{\text{always}}$ a negative number.

129. The sum of a positive number and a negative number is $\boxed{\text{sometimes}}$ a negative number.

131. The sum of two positive numbers is $\boxed{\text{never}}$ zero.

133. The sum of zero and a negative number is $\boxed{\text{always}}$ a negative number.

135. position of the submarine relative to the surface:
$88 - 148 = -60$
$\boxed{60 \text{ ft below the surface}}$

137. $8848 - 19{,}763 = -10{,}915$
$\boxed{10{,}915 \text{ m below sea level}}$

139. $\boxed{\text{B}}$ is true; Neither division nor subtraction is commutative.

For problems 141-155, let $a = 1$, $b = 2$, and $c = 3$.

141.
$$\begin{aligned}
a(b-c) &= ab - ac \\
1(2-3) &= 1 \cdot 2 - 1 \cdot 3 \\
1(-1) &= 2 - 3 \\
-1 &= -1 \quad \boxed{\text{True}}
\end{aligned}$$

143.
$$\begin{aligned}
(a-b)-c &= a-(b-c) \\
(1-2)-3 &= 1-(2-3) \\
-1-3 &= 1-(-1) \\
-4 &= 1+1 \\
-4 &= 2 \quad \boxed{\text{False}}
\end{aligned}$$

145.
$$\begin{aligned}
|a-b| &\geq |a|-|b| \\
|1-2| &\geq |1|-|2| \\
|-1| &\geq 1-2 \\
1 &\geq -1 \quad \boxed{\text{True}}
\end{aligned}$$

147.
$$\begin{aligned}
\frac{a+b}{c} &= \frac{a}{c} + \frac{b}{c} \\
\frac{1+2}{3} &= \frac{1}{3} + \frac{2}{3} \\
\frac{3}{3} &= \frac{3}{3} \quad \boxed{\text{True}}
\end{aligned}$$

149.
$$\begin{aligned}
(a \div b) \div c &= a \div (b \div c) \\
(1 \div 2) \div 3 &= 1 \div (2 \div 3) \\
\frac{1}{2} \cdot \frac{1}{3} &= 1 \div \left(\frac{2}{3}\right) \\
\frac{1}{6} &= 1 \cdot \frac{3}{2} \\
\frac{1}{6} &= \frac{3}{2} \quad \boxed{\text{False}}
\end{aligned}$$

151.
$$\begin{aligned}
\frac{1}{a+b} &= \frac{1}{a} + \frac{1}{b} \\
\frac{1}{1+2} &= \frac{1}{1} + \frac{1}{2} \\
\frac{1}{3} &= \frac{3}{2} \quad \boxed{\text{False}}
\end{aligned}$$

153.
$$\begin{aligned}
\frac{1}{a}(b-c) &= \frac{1}{a} \cdot b - \frac{1}{a} \cdot c \\
\frac{1}{1}(2-3) &= \frac{1}{1} \cdot 2 - \frac{1}{1} \cdot 3 \\
1(-1) &= 2 - 3 \\
-1 &= -1 \quad \boxed{\text{True}}
\end{aligned}$$

155. $(-b)^n$ and $-b^n$ are sometimes equal
$(-2)^n = -2^n$ when n is an odd integer
$\boxed{\text{True}}$

157.   Statement   Reason

1. $\dfrac{a+b}{c} = (a+b) \cdot \dfrac{1}{c}$ | 1. Definition of division

2. $\quad = a \cdot \dfrac{1}{c} + b \cdot \dfrac{1}{c}$ | 2. Distributive property

3. $\quad = \dfrac{a}{c} + \dfrac{b}{c}$ | 3. Definition of division

4. $\dfrac{a+b}{c} = \dfrac{a}{c} + \dfrac{b}{c}$ | 4. Transitive property

Review Problems

161. x is positive and at most 7

set-builder: $\boxed{\{x \mid 0 < x \le 7\}}$

interval: $\boxed{(0, 7]}$

162. $\{x \mid -2 < x \le 1\}$

$\boxed{(-2, 1]}$

163.   If $3x + y = 6$ and $y = 7x - 2$, then $3x + 7x - 2 = 6$.

$\boxed{\text{Substitution}}$

Section 1.4 Order of Operations and Mathematical Models

Problem Set 1.4, pp. 47-50

1. $7 + 6 \cdot 3 = 7 + 18 = \boxed{25}$

3. $4(-5) - 6(-3) = -20 + 18 = \boxed{-2}$

5. $6 - 4(-3) - 5 = 6 + 12 - 5 = 18 - 5 = \boxed{13}$

7. $3 - 5(-4 - 2) = 3 - 5(-6) = 3 + 30 = \boxed{33}$

9. $(2 - 6)(-3 - 5) = (-4)(-8) = \boxed{32}$

11. $3(-2)^2 - 4(-3)^2 = 3(4) - 4(9) = 12 - 36 = \boxed{-24}$

13. $(2 - 6)^2 - (3 - 7)^2 = (-4)^2 - (-4)^2 = 16 - 16 = \boxed{0}$

15. $6(3 - 5)^3 - 2(1 - 3)^3$
$= 6(-2)^3 - 2(-2)^3$
$= 6(-8) - 2(-8)$
$= -48 + 16$
$= \boxed{-32}$

17. $8^2 - 16 \div 2^2 \cdot 4 - 3$
$= 64 - 16 \div 4 \cdot 4 - 3$
$= 64 - 4 \cdot 4 - 3$
$= 64 - 16 - 3$
$= 48 - 3$
$= \boxed{45}$

19. $\dfrac{4^2 + 3^3}{5^2 - (-18)}$
$= \dfrac{16 + 27}{25 + 18}$
$= \dfrac{43}{43}$
$= \boxed{1}$

21. $20 - 4 \left(\dfrac{8 - 2}{3 - 6} \right) \div \dfrac{1}{2}$
$= 20 - 4 \left(\dfrac{6}{-3} \div \dfrac{1}{2} \right)$
$= 20 - 4((-2) \cdot 2)$
$= 20 - 4(-4)$
$= 20 + 16$
$= \boxed{36}$

23. $\left(\dfrac{1}{2}\right)^2 + \left(\dfrac{6-4}{5}\right)^2 + \left(\dfrac{5+2}{10}\right)^2$

$= \left(\dfrac{1}{2}\right)^2 + \left(\dfrac{2}{5}\right)^2 + \left(\dfrac{7}{10}\right)^2$

$= \dfrac{1}{4} + \dfrac{4}{25} + \dfrac{49}{100}$

$= \dfrac{25}{100} + \dfrac{16}{100} + \dfrac{49}{100}$

$= \dfrac{90}{100}$

$= \boxed{\dfrac{9}{10}}$

25. $-3[8 + (-6)] \div [-4 - (-5)]$

$= -3[2] \div [-4 + 5]$

$= -6 \div 1$

$= \boxed{-6}$

27. $\left(\dfrac{1}{2} - \dfrac{7}{4}\right) \div \left(1 - \dfrac{3}{8}\right)$

$= \left(\dfrac{2}{4} - \dfrac{7}{4}\right) \div \left(\dfrac{8}{3} - \dfrac{3}{8}\right)$

$= -\dfrac{5}{4} \div \dfrac{5}{8}$

$= -\dfrac{5}{4} \cdot \dfrac{8}{5}$

$= \boxed{-2}$

29. $\dfrac{1}{4} - 6(2 + 8) \div \left(-\dfrac{1}{3}\right)\left(-\dfrac{1}{9}\right)$

$= \dfrac{1}{4} - 6(10) \cdot (-3)\left(-\dfrac{1}{9}\right)$

$= \dfrac{1}{4} - 60 \cdot \dfrac{1}{3}$

$= \dfrac{1}{4} - 20$

$= \dfrac{1}{4} - \dfrac{80}{4}$

$= \boxed{-\dfrac{79}{4}}$

31. $6.8 - (0.3)^2 \div 0.09$

$= 6.8 - (0.09) \div 0.09$

$= 6.8 - (0.09) \cdot \dfrac{1}{(0.09)}$

$= 6.8 - 1$

$= \boxed{5.8}$

33. $\dfrac{1}{2} - \left(\dfrac{2}{3} \cdot \dfrac{9}{5}\right) + \dfrac{3}{10}$

$= \dfrac{1}{2} - \dfrac{6}{5} + \dfrac{3}{10}$

$= \dfrac{5}{10} - \dfrac{12}{10} + \dfrac{3}{10}$

$= -\dfrac{4}{10}$

$= \boxed{-\dfrac{2}{5}}$

35. $8 - 3[-2(2 - 5) - 4(8 - 6)]$

$= 8 - 3[-2(-3) - 4(2)]$

$= 8 - 3[6 - 8]$

$= 8 - 3(-2)$

$= 8 + 6$

$= \boxed{14}$

37. $\dfrac{2(-2) - 4(-3)}{5 - 8}$

$= \dfrac{-4 + 12}{-3}$

$= \dfrac{8}{-3}$

$= \boxed{-\dfrac{8}{3}}$

39. $10 - (-8) \left[\dfrac{2(-3) - 5(6)}{7 - (-1)} \right]$

$= 10 - (-8) \left[\dfrac{-6 - 30}{7 + 1} \right]$

$= 10 + 8 \left[\dfrac{-36}{8} \right]$

$= 10 - 36$

$= \boxed{-26}$

41. $6 - (-12) \left[\dfrac{2 - 4(3 - 7)}{-4 - 5(1 - 3)} \right]$

$= 6 - (-12) \left[\dfrac{2 - 4(-4)}{-4 - 5(-2)} \right]$

$= 6 + 12 \left[\dfrac{2 + 16}{-4 + 10} \right]$

$= 6 + 12 \left[\dfrac{18}{6} \right]$

$= 6 + 12(3)$

$= 6 + 36$

$= \boxed{42}$

43. $2 \left[-5 - \dfrac{1}{3}(17 + 4) \right]$

$= 2 \left[-5 - \dfrac{1}{3}(21) \right]$

$= 2 \left[-5 - 7 \right]$

$= 2(-12)$

$= \boxed{24}$

45. $-\dfrac{4}{5} \left[8(-3) + (-50)\left(-\dfrac{1}{2}\right) \right]$

$= -\dfrac{4}{5} [-24 + 25]$

$= -\dfrac{4}{5}(1)$

$= \boxed{-\dfrac{4}{5}}$

47. $\dfrac{-7\left(\dfrac{3 - 10}{5 - (-2)}\right) - 8\left(\dfrac{-8 - 3}{7 - 6}\right)}{2\left(\dfrac{-2 - 9}{5 \cdot 3 - 2^2}\right) + 9\left(\dfrac{2 - 6 - 5}{2 - (-1)}\right)}$

$= \dfrac{-7\left(\dfrac{-7}{5 + 2}\right) - 8\left(\dfrac{-11}{1}\right)}{2\left(\dfrac{-11}{15 - 4}\right) + 9\left(\dfrac{-9}{2 + 1}\right)}$

$= \dfrac{-7\left(-\dfrac{7}{7}\right) + 88}{2\left(-\dfrac{11}{11}\right) + 9\left(-\dfrac{9}{3}\right)}$

$= \dfrac{7 + 88}{-2 - 27}$

$= \dfrac{95}{-29}$

$= \boxed{-\dfrac{95}{29}}$

49. a. $E = 155x^2 - 65x + 150$

$\quad E = 155(2)^2 - 65(2) + 150$ Substitute 2 for x.

$\quad E = 155(4) - 130 + 150$

$\quad E = 620 + 20$

$\quad E = 640$

A person walking 2 meters/second uses $\boxed{640\text{ W}}$ of energy.

b. $E = 250x + 100$
$E = 250(4) + 100$
$E = 1000 + 100$
$E = 1100$
A runner moving at 4 meters/second uses $\boxed{1100\ \text{W}}$ of energy.

51. $d = 0.042s^2 + 1.1s$
$s = 50$: $d = 0.042(50)^2 + 1.1(50)$
$\qquad\qquad d = 0.042(2500) + 55$
$\qquad\qquad d = 105 + 55$
$\qquad\qquad d = 160$
$s = 30$: $d = 0.042(30)^2 + 1.1(30)$
$\qquad\qquad d = 0.042(900) + 33$
$\qquad\qquad d = 37.8 + 33$
$\qquad\qquad d = 70.8$
difference in distance: $160 - 70.8 = 89.2$

A car going 50 mph will travel $\boxed{89.2\ \text{ft}}$ farther than a car going 30 mph.

53. $R = 206.835 - (1.015w + 0.846\,s)$
$(w = 10,\ s = 2.5)$:
$\qquad R = 206.835 - (1.015 \cdot 10 + 0.846 \cdot 2.5)$
$\qquad R = 206.835 - (10.15 + 2.115)$
$\qquad R = 206.835 - 12.265$
$\qquad \boxed{R = 194.57}$

55. $N = (14{,}400 + 120t + 100t^2) \div (144 + t^2)$
$(t = 2)$:
$\qquad N = (14{,}400 + 120 \cdot 2 + 100 \cdot 2^2) \div (144 + 2^2)$
$\qquad N = (14{,}400 + 240 + 100 \cdot 4) \div (144 + 4)$
$\qquad N = (14{,}640 + 400) \div (148)$
$\qquad N = (15{,}040) \div (148)$
$\qquad N \approx 101.62 \approx 102$
$\boxed{\text{Approximately 102 bacteria}}$ are present after 2 hours.

57. $T = 3(A - 20)^2 \div 50 + 10$
$A = 30$: $T = 3(30 - 20)^2 \div 50 + 10$
$\qquad\qquad T = 3(10)^2 \div 50 + 10$
$\qquad\qquad T = 3(100) \div 50 + 10$
$\qquad\qquad T = 300 \div 50 + 10$
$\qquad\qquad T = 6 + 10$
$\qquad\qquad T = 16$
$A = 40$: $T = 3(40 - 20)^2 \div 50 + 10$
$\qquad\qquad T = 3(20)^2 \div 50 + 10$
$\qquad\qquad T = 3(400) \div 50 + 10$
$\qquad\qquad T = 1200 \div 50 + 10$
$\qquad\qquad T = 24 + 10$
$\qquad\qquad T = 34$
percent increase $= \dfrac{34 - 16}{16} = \dfrac{18}{16} = \dfrac{9}{8} = 1.125 = \boxed{112.5\%}$

59. $C = \begin{cases} 1.22x & \text{if } 0 \le x \le 12 \\ 10x - 105.36 & \text{if } 12 < x \le 24 \\ 50x - 1065.36 & \text{if } x > 24 \end{cases}$

a. $x = 12$
$C = 1.22x$
$C = 1.22(12)$
$C = 14.64$
monthly water bill: $\boxed{\$14.64}$

b. $x = 24$
$C = 10x - 105.36$
$C = 10(24) - 105.36$
$C = 240 - 105.36$
$C = 134.64$
monthly water bill: $\boxed{\$134.64}$

c. $x = 30$
$C = 50x - 1065.36$
$C = 50(30) - 1065.36$
$C = 1500 - 1065.36$
$C = 434.64$
montly water bill: $\boxed{\$434.64}$

61. $t = \begin{cases} \dfrac{1}{24}T + \dfrac{11}{4} & \text{if } 30 \le T \le 36 \\ \dfrac{4}{3}T - \dfrac{175}{4} & \text{if } 36 < T \le 39 \end{cases}$

a. $T = 32$
$t = \dfrac{1}{24}T + \dfrac{11}{4}$
$t = \dfrac{1}{24}(32) + \dfrac{11}{4}$
$t = \dfrac{4}{3} + \dfrac{11}{4}$
$t = \dfrac{16 + 33}{12} = \dfrac{49}{12} = 4\dfrac{1}{12}$
generation time: $\boxed{4\dfrac{1}{12}\text{ hr}}$

b. $T = 39$
$t = \dfrac{4}{3}T - \dfrac{175}{4}$
$t = \dfrac{4}{3}(39) - \dfrac{175}{4}$
$t = 4(13) - \dfrac{175}{4}$
$t = \dfrac{208}{4} - \dfrac{175}{4} = \dfrac{33}{4} = 8\dfrac{1}{4}$
generation time: $\boxed{8\dfrac{1}{4}\text{ hr}}$

c. $T = 36$

$t = \dfrac{1}{24} T + \dfrac{11}{4}$

$t = \dfrac{1}{24}(36) + \dfrac{11}{4}$

$t = \dfrac{6}{4} + \dfrac{11}{4}$

$t = \dfrac{17}{4} = 4\dfrac{1}{4}$

generation time: $\boxed{4\dfrac{1}{4}\,\text{hr}}$

63. $\boxed{\text{C}}$ is true

65. $7.9[18 - 2.6(-9.3)] \approx \boxed{333.22}$

  7.9 $\boxed{\times}$ $\boxed{(}$ 18 $\boxed{-}$ 2.6 $\boxed{\times}$ 9.3 $\boxed{+/-}$ $\boxed{)}$ $\boxed{=}$
  Display: 333.222

67. $73(1072.9 - 6783.5) = \boxed{-416{,}873.8}$

  73 $\boxed{\times}$ $\boxed{(}$ 1072.9 $\boxed{-}$ 6783.5 $\boxed{)}$ $\boxed{=}$
  Display: −416873.8

69. $(-2.7)^6 - 3.1(263) \approx \boxed{-427.88}$

  $\boxed{+/-}$ 2.7 $\boxed{y^x}$ 6 $\boxed{-}$ 3.1 $\boxed{\times}$ 263 $\boxed{=}$
  Display: −427.87951

71. $17.4(-1.07)^{10} \approx \boxed{34.23}$

  17.4 $\boxed{\times}$ $\boxed{(}$ $\boxed{+/-}$ $\boxed{)}$ 1.07 $\boxed{y^x}$ 10 $\boxed{=}$
  Display: 34.228434

73. $\dfrac{432}{-1.045^8} = -\dfrac{432}{1.045^8} \approx \boxed{-303.78}$

  $\boxed{\pm}$ 432 $\boxed{\div}$ 1.045 $\boxed{y^x}$ 8 $\boxed{=}$
  Display: 303.7759749

75. $3.8(50 - 2.3^4) \div 6.1 \approx \boxed{13.71}$

  3.8 $\boxed{\times}$ $\boxed{(}$ 50 $\boxed{-}$ 2.3 $\boxed{y^x}$ 4 $\boxed{)}$ $\boxed{\div}$ 6.1 $\boxed{=}$
  Display: 13.714823

77. $\boxed{8 - 2 \cdot (3 - 4) = 10}$

79. $\boxed{\left(2 \cdot 5 - \dfrac{1}{2} \cdot 10\right) \cdot 9 = 45}$

81. $1 + 2 + 3 + 4 + \ldots + n = \dfrac{n(n + 1)}{2}$

difference between (the sum of first 100 positive mulitples of 3) and (the sum of the first 100 positive even integers):

$(3 + 6 + 9 + \ldots + 300) - (2 + 4 + 6 + \ldots + 200)$
$= 3(1 + 2 + 3 + \ldots + 100) - 2(1 + 2 + 3 + \ldots + 100)$
$= (3 - 2)(1 + 2 + 3 + \ldots + 100)$
$= 1\,\dfrac{(100)(101)}{2} = 50(101) = \boxed{5050}$

Review Problems

86. $\boxed{\text{B}}$ is true

87. $\{x \mid x \text{ is a composite number between 3 and 7}\}$

  $\boxed{\{4, 6\}}$

88. $\{x \mid |x| \geq 6\}$

$\boxed{(-\infty, -6] \text{ or } [6, \infty)}$

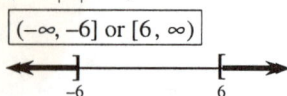

Section 1.5 Algebraic Expressions

Problem Set 1.5, pp. 53-55

1. $-3x + 7x - 6x = \boxed{-2x}$

3. $4x^2y - 8x^2y = \boxed{-4x^2y}$

5. $-2x + 7y + 9x = \boxed{7x + 7y}$

7. $5x^2y - 3xy^2 + 2x^2y = \boxed{7x^2y - 3xy^2}$

9. $-4x^2 + 5y^2 - 3x^2 - 6y^2 = \boxed{-7x^2 - y^2}$

11. $7x^3y - 3 + 6x^3y - 8 = \boxed{13x^3y - 11}$

13. $3x^2y - 2xy^2 + 5x^2y = \boxed{8x^2y - 2xy^2}$

15. $2(x + 3) + 3(x + 2) = 2x + 6 + 3x + 6 = \boxed{5x + 12}$

17. $-4(b - 3) - 2(b + 1) = -4b + 12 - 2b - 2 = \boxed{-6b + 10}$

19. $5(x^2 + 3) - 7(x^2 - 4) = 5x^2 + 15 - 7x^2 + 28 = \boxed{-2x^2 + 43}$

21. $-3(y^2 - 1) - (y^2 - 7) = -3y^2 + 3 - y^2 + 7 = \boxed{-4y^2 + 10}$

23. $3(7x - 2) + 4(5x - 6) = 21x - 6 + 20x - 24 = \boxed{41x - 30}$

25. $6(3b - 1) - 4(2b - 5) = 18b - 6 - 8b + 20 = \boxed{10b + 14}$

27. $-2(x^2 - 1) - 4(3x^2 - 5) = -2x^2 + 2 - 12x^2 + 20 = \boxed{-14x^2 + 22}$

29. $4(2x + 5y) - 6(3x - 2y) = 8x + 20y - 18x + 12y = \boxed{-10x + 32y}$

31. $2x - 5(x - 6y) = 2x - 5x + 30y = \boxed{-3x + 30y}$

33. $-3(5x - z) - (z - 4y) + 5(x + 3y - z)$
$= -15x + 3z - z + 4y + 5x + 15y - 5z$
$= \boxed{-10x + 19y - 3z}$

35. $3[x - 5(5 - 3x)]$
$= 3[x - 25 + 15x]$
$= 3[16x - 25]$
$= \boxed{48x - 75}$

37. $4[x - 2(x - 3y)]$
$= 4[x - 2x + 6y]$
$= 4(-x + 6y)$
$= \boxed{-4x + 24y}$

39. $5(3x - 2y) - 4(-4x + 5y)$
$= 15x - 10y + 16x - 20y$
$= \boxed{31x - 30y}$

41. $4 - 3(4x - 3y) - 3(-2x + 3y)$
$= 4 - 12x + 9y + 6x - 9y$
$= \boxed{-6x + 4}$

43. $\frac{1}{3}[8a - 2(a - 12) + 3]$

$= \frac{1}{3}[8a - 2a + 24 + 3]$

$= \frac{1}{3}[6a + 27]$

$= \boxed{2a + 9}$

45. $(x^3 - 2x^2 + 3x + 4) - (x^2 + 4x - 1)$

$= x^3 - 2x^2 + 3x + 4 - x^2 - 4x + 1$

$= \boxed{x^3 - 3x^2 - x + 5}$

47. $3[x^2 + 5(x^2 + 3)] - 6$
$= 3[x^2 + 5x^2 + 15] - 6$
$= 3(6x^2 + 15) - 6$
$= 18x^2 + 45 - 6$
$= \boxed{18x^2 + 39}$

49. $4x^2 - [2(x^2 - y^2) - 2(x^2 + y^2)]$
$= 4x^2 - [2x^2 - 2y^2 - 2x^2 - 2y^2]$
$= 4x^2 - (-4y^2)$
$= \boxed{4x^2 + 4y^2}$

51. $10x - 6x \cdot 3 + 15y^2 \div 5 \cdot 3$
$= 10x - 18x + 3y^2 \cdot 3$
$= \boxed{-8x + 9y^2}$

53. $9y + 6y \div 2y - 8 \cdot 3$
$= 9y + 3 - 24$
$= \boxed{9y - 21}$

55. $\left(-\frac{4}{5}\right) \div \left(-\frac{2}{5}x + 6x\right)$

$= \left(-\frac{4}{5}\right) \div \left(-\frac{2x}{5} + \frac{30x}{5}\right)$

$= \left(-\frac{4}{5}\right) \div \left(\frac{28x}{5}\right)$

$= \left(-\frac{4}{5}\right) \cdot \left(\frac{5}{28x}\right)$

$= \boxed{-\frac{1}{7x}}$

57. $4x + 8x \div 2x - 6x \cdot (-2) + 5$

$= 4x + 4 + 12x + 5$

$= \boxed{16x + 9}$

59. $a^2b^3 - ab + 2a^2b^3$
$= \boxed{3a^2b^3 - ab}$
$(a = 1,\ b = -2)$:
$= 3(1)^2(-2)^3 - (1)(-2)$
$= 3(-8) + 2$
$= -24 + 2$
$= \boxed{-22}$

61. $-4(2a^2 - b) - 5(b - a^2)$
$= -8a^2 + 4b - 5b + 5a^2$
$= \boxed{-3a^2 - b}$
$(a = -2,\ b = -1)$:
$= -3(-2)^2 - (-1)$
$= -3(4) + 1$
$= -12 + 1$
$= \boxed{-11}$

63. $2(a^2b + 3) - 3(a^2b - 1) - 4(a^2b + 5)$
$= 2a^2b + 6 - 3a^2b + 3 - 4a^2b - 20$
$= \boxed{-5a^2b - 11}$
$(a = -1,\ b = -2)$:
$= -5(-1)^2(-2) - 11$
$= -5(1)(-2) - 11$
$= 10 - 11$
$= \boxed{-1}$

65. $3[x - 2(x + 2y)]$
$= 3[x - 2x - 4y]$
$= 3(-x - 4y)$
$= \boxed{-3x - 12y}$
$\left(x = -\frac{1}{3},\ y = \frac{1}{6}\right)$:
$= -3\left(-\frac{1}{3}\right) - 12\left(\frac{1}{6}\right)$
$= 1 - 2$
$= \boxed{-1}$

67. $N = 3(x + 19x - y) + 7[y - (7x - 3y)]$
$= 3(20x - y) + 7(y - 7x + 3y)$
$= 60x - 3y + 7(-7x + 4y)$
$= 60x - 3y - 49x + 28y$
$= \boxed{11x + 25y}$

69. $a - (y - b) = \boxed{a - y + b}$

71. $3x^2 + 5x^2 = \boxed{8x^2}$

73. $7 - 3(x - 5y) = \boxed{7 - 3x + 15y}$

75. $6x + 3x \div \dfrac{1}{3} = \boxed{6x + 3x \cdot 3 = 6x + 9x = 15x}$

77. 1. x
2. $2x$
3. $2x + 9$
4. $(2x + 9) + x$
5. $[(2x + 9) + x] \div 3$
6. $[(2x + 9) + x] \div 3 + 4$
7. $[(2x + 9) + x] \div 3 + 4 - x$
$= (3x + 9) \div 3 + 4 - x$
$= x + 3 + 4 - x$
$= 7$

79. $\boxed{D}$ is true

81. $0.03(4.7x - 5.9) - 0.07(3.8x - 61)$
$= 0.141x - 0.177 - 0.266x + 4.27$
$= \boxed{-0.125x + 4.093}$

Review Problems

85. $I = \begin{cases} 200t + 600 & \text{if } 0 \le t < 5 \\ 200(1 + t) + 1300 & \text{if } 5 \le t < 10 \\ 100(1 + t)^2 & \text{if } t \ge 10 \end{cases}$

1989: $t = 9$
$I = 200(1 + 8) + 1300$
$= 200(9) + 1300$
$= 1800 + 1300$
$= 3100$

1994: $t = 13$
$I = 100(1 + 13)^2$
$= 100(14)^2$
$= 100(196)$
$= 19600$

percent increase: $\dfrac{(1994) - (1989)}{(1989)}$

$= \dfrac{19,600 - 3100}{3100}$

$= \dfrac{16,500}{3100}$

≈ 5.3226

$\approx 532\%$

The percent increase in inventory from 1989 to 1994 is $\boxed{\text{approximately } 532\%}$.

86. If $3 = x$, then $x = 3$

$\boxed{\text{Symmetric property of equality}}$

87. $6 - 4[-2(3 - 6) - 5(19 - 16)]$

$= 6 - 4[-2(-3) - 5(3)]$

$= 6 - 4[6 - 15]$

$= 6 - 4(-9)$

$= 6 + 36$

$= \boxed{42}$

Section 1.6 Properties of Integral Exponents

Problem Set 1.6, pp. 64-65

1. $x^2 \cdot x^5 = x^{2+5} = \boxed{x^7}$

3. $x^2 \cdot x^3 \cdot x^8 = x^{2+3+8} = \boxed{x^{13}}$

5. $(2y^5)(-3y^8) = -6y^{5+8} = \boxed{-6y^{13}}$

7. $(-5x^3y^2)(2xy^{17}) = -10x^{3+1}y^{2+17} = \boxed{-10x^4y^{19}}$

9. $(x^4)^8 = x^{4 \cdot 8} = \boxed{x^{32}}$

11. $(4x)^3 = 4^3x^3 = \boxed{64x^3}$

13. $(3xy)^4 = 3^4x^4y^4 = \boxed{81x^4y^4}$

15. $(2xy^2)^3 = 2^3x^3y^{2 \cdot 3} = \boxed{8x^3y^6}$

17. $(-3x^2y^5)^2 = (-3)^2x^{2 \cdot 2}y^{5 \cdot 2} = \boxed{9x^4y^{10}}$

19. $(2xy)(4x)^2 = (2xy)(4^2x^2) = (2xy)(16x^2) = \boxed{32x^3y}$

21. $(4xy)(-2x^2y) + 17x^3y^2$

$= -8x^3y^2 + 17x^3y^2$

$= \boxed{9x^3y^2}$

23. $(2x)^3(-3xy) + 25x^4y$

$= (2^3x^3)(-3xy) + 25x^4y$

$= (8x^3)(-3xy) + 25x^4y$

$= -24x^4y + 25x^4y$

$= \boxed{x^4y}$

25. $\left(\dfrac{x}{y}\right)^6 = \boxed{\dfrac{x^6}{y^6}}$

27. $\left(\dfrac{-3x}{y}\right)^4 = \dfrac{(-3)^4x^4}{y^4} = \boxed{\dfrac{81x^4}{y^4}}$

29. $\left(\dfrac{x^4}{y^2}\right)^3 = \boxed{\dfrac{x^{12}}{y^6}}$

31. $\left(\dfrac{-5x^3}{2y}\right)^3 = \dfrac{(-5)^3x^9}{2^3y^3} = \boxed{\dfrac{-125x^9}{8y^3}}$

33. $6^0 = \boxed{1}$

35. $17^0 = \boxed{1}$

37. $(6x)^0 = \boxed{1}$

39. $5^{-2} = \dfrac{1}{5^2} = \boxed{\dfrac{1}{25}}$

41. $(-4)^{-3} = \dfrac{1}{(-4)^3} = \dfrac{1}{-64} = \boxed{-\dfrac{1}{64}}$

43. $(-4)^{-2} = \dfrac{1}{(-4)^2} = \boxed{\dfrac{1}{16}}$

45. $-4^{-2} = -\dfrac{1}{4^2} = \boxed{-\dfrac{1}{16}}$

47. $\left(\dfrac{3}{4}\right)^{-2} = \dfrac{1}{\left(\dfrac{3}{4}\right)^2} = \dfrac{1}{\dfrac{9}{16}} = \boxed{\dfrac{16}{9}}$

49. $\dfrac{1}{5^{-3}} = 5^3 = \boxed{125}$

51. $\dfrac{1}{(-3)^{-4}} = (-3)^4 = \boxed{81}$

53. $\dfrac{1}{-3^{-4}} = -3^4 = \boxed{-81}$

55. $\dfrac{x^{16}}{x^8} = x^{16-8} = \boxed{x^8}$

57. $\dfrac{8x^7}{2x^4} = \left(\dfrac{8}{2}\right)\left(\dfrac{x^7}{x^4}\right) = 4x^{7-4} = \boxed{4x^3}$

59. $\dfrac{-100x^{18}}{25x^{17}} = \left(\dfrac{-100}{25}\right)\left(\dfrac{x^{18}}{x^{17}}\right) = -4x^{18-17} = \boxed{-4x}$

61. $\dfrac{20x^4y^3}{5xy^3} = \left(\dfrac{20}{5}\right)\left(\dfrac{x^4}{x}\right)\left(\dfrac{y^3}{y^3}\right) = 4x^{4-1}y^{3-3} = 4x^3y^0 = \boxed{4x^3}$

63. $\dfrac{x^3}{x^9} = x^{3-9} = x^{-6} = \boxed{\dfrac{1}{x^6}}$

65. $\dfrac{20x^3}{-5x^4} = \left(\dfrac{20}{-5}\right)\left(\dfrac{x^3}{x^4}\right) = -4x^{3-4} = -4x^{-1} = \boxed{\dfrac{-4}{x}}$

67. $\dfrac{16x^3}{8x^{10}} = \left(\dfrac{16}{8}\right)\left(\dfrac{x^3}{x^{10}}\right) = 2x^{3-10} = 2x^{-7} = \boxed{\dfrac{2}{x^7}}$

69. $\dfrac{20a^3b^8}{2ab^{13}} = \left(\dfrac{20}{2}\right)(a^{3-1})(b^{8-13}) = 10a^2b^{-5} = \boxed{\dfrac{10a^2}{b^5}}$

71. $\dfrac{1}{b^{-5}} = \boxed{b^5}$

73. $x^3 \cdot x^{-12} = x^{3-12} = x^{-9} = \boxed{\dfrac{1}{x^9}}$

75. $(2a^5)(-3a^{-7}) = -6a^{5-7} = -6a^{-2} = \boxed{\dfrac{-6}{a^2}}$

77. $(3a^7)(-2a^{-3}) = -6a^{7-3} = \boxed{-6a^4}$

79. $(x^3)^{-6} = x^{3(-6)} = x^{-18} = \boxed{\dfrac{1}{x^{18}}}$

81. $(x^{-3})^{-6} = x^{(-3)(-6)} = \boxed{x^{18}}$

83. $\dfrac{x^3}{x^{-7}} = x^{3-(-7)} = x^{3+7} = \boxed{x^{10}}$

85. $\dfrac{6y^2}{2y^{-8}} = 3y^{2-(-8)} = 3y^{2+8} = \boxed{3y^{10}}$

87. $\dfrac{4x^6}{-20x^{-11}} = \dfrac{1}{-5}x^{6-(-11)} = \boxed{\dfrac{-x^{17}}{5}}$

89. $\dfrac{x^{-7}}{x^3} = x^{-7-3} = x^{-10} = \boxed{\dfrac{1}{x^{10}}}$

91. $\dfrac{x^{-7}}{x^{-3}} = x^{-7-(-3)} = x^{-7+3} = x^{-4} = \boxed{\dfrac{1}{x^4}}$

93. $\dfrac{30x^2y^5}{-6x^8y^{-3}} = -5x^{2-8}y^{5-(-3)} = -5x^{-6}y^{5+3} = -5x^{-6}y^8 = \boxed{\dfrac{-5y^8}{x^6}}$

95. $\dfrac{25a^{-8}b^2}{-75a^{-3}b^4} = \dfrac{1}{-3}a^{-8-(-3)}b^{2-4} = \dfrac{-1}{3}a^{-8+3}b^{-2} = \dfrac{-1}{3}a^{-5}b^{-2} = \boxed{\dfrac{-1}{3a^5b^2}}$

97. $\left(\dfrac{x^3}{x^{-5}}\right)^2 = (x^{3-(-5)})^2 = (x^{3+5})^2 = (x^8)^2 = \boxed{x^{16}}$

99. $\left(\dfrac{15a^4b^2}{-5a^{10}b^{-3}}\right)^3 = (-3a^{4-10}b^{2-(-3)})^3 = (-3a^{-6}b^5)^3 = (-3)^3(a^{-6})^3(b^5)^3 = -27a^{-18}b^{15} = \boxed{\dfrac{-27b^{15}}{a^{18}}}$

101. $\left(\dfrac{10y^2}{y}\right) + \left(\dfrac{4xy^4}{xy^3}\right)$

$= 10y + 4y$

$= \boxed{14y}$

103. $P = (2x^2y)^3 - 46{,}500$

$\boxed{P = 8x^6y^3 - 46{,}500}$

$(x = 3,\ y = 2)$:

$P = 8(3)^6(2)^3 - 46{,}500$

$= 8(729)(8) - 46{,}500$

$= 46{,}656 - 46500$

$= 156$

Profit: $\boxed{\$156}$

105. $\boxed{D}$ is true; $-3^{-2} = -\dfrac{1}{3^2} = -\dfrac{1}{9}$

107. $3(0.36)^{-2} - 5(0.47)^{-3} \approx \boxed{-25.011}$

$3 \boxed{\times} 0.36 \boxed{y^x} 2 \boxed{+/-} \boxed{-} 5 \boxed{\times} 0.47 \boxed{y^x} 3 \boxed{+/-} \boxed{=}$

Display: -25.010738

109. $[6.13 + (0.38)^{-1}]^{-3} \approx \boxed{0.001}$

$\boxed{(} 6.13 \boxed{+} 0.38 \boxed{y^x} 1 \boxed{+/-} \boxed{)} \boxed{y^x} 3 \boxed{+/-} \boxed{=}$

Display: 0.0014868

For problems 111-115: Let $a = -1$, $b = -2$, $n = 2$, $m = 3$.

111. $b^n + b^m = b^{n+m}$

$(-2)^2 + (-2)^3 = (-2)^{2+3}$

$\qquad 4 - 8 = (-2)^5$

$\qquad\quad -4 = -32 \qquad$ False

Statement is *not true*.

113. $(a+b)^n = a^n + b^n$

$(-1-2)^2 = (-1)^2 + (-2)^2$

$\qquad (-3)^2 = 1 + 4$

$\qquad\quad 9 = 5 \qquad$ False

Statement is *not true*.

115. $\quad (-b)^{-n} = \dfrac{1}{b^n} \qquad n = 2$, an even integer

$[-(-2)]^{-2} = \dfrac{1}{(-2)^2}$

$\qquad 2^{-2} = \dfrac{1}{4}$

$\qquad \dfrac{1}{4} = \dfrac{1}{4} \qquad$ True

Now let $n = 3$ an odd integer.

$[-(-2)]^{-3} = \dfrac{1}{(-2)^3}$

$\qquad 2^{-3} = \dfrac{1}{-8}$

$\qquad \dfrac{1}{8} = -\dfrac{1}{8} \qquad$ False

Statement is *not true*.

117. $(5^x)^{x+y} + 5^x \cdot 5^{x+y}$

$= 5^{x(x+y)} + 5^{x+(x+y)}$

$= \boxed{5^{x^2+xy} + 5^{2x+y}}$

Review Problems

120. $[6(xy - 1) - 2xy] - 4[(2xy + 3) - 2(xy + 2)]$
$= [6xy - 6 - 2xy] - 4[2xy + 3 - 2xy - 4]$
$= (4xy - 6) - 4(-1)$
$= 4xy - 6 + 4$
$= \boxed{4xy - 2}$

121.   multiplicative inverse of $-\dfrac{3}{4}$

$= \dfrac{1}{-\dfrac{3}{4}} = \boxed{-\dfrac{4}{3}}$

122. $\dfrac{3}{7^2 + (-7)(-5)} - \left[\left(\dfrac{2}{3} + \dfrac{1}{4}\right) \cdot \dfrac{6}{7}\right]$

$= \dfrac{3}{49 + 35} - \left(\dfrac{11}{12}\right) \cdot \dfrac{6}{7}$

$= \dfrac{3}{84} - \dfrac{11}{14}$

$= \dfrac{1}{28} - \dfrac{22}{28}$

$= -\dfrac{21}{28}$

$= \boxed{-\dfrac{3}{4}}$

Section 1.7 Scientific Notation

Problem Set 1.7, pp. 71-73

1. $(0.000\ 37)(8,300,000)$
$= (3.7 \times 10^{-4})(8.3 \times 10^{6})$

$= 30.71 \times 10^2$

$= 3.071 \times 10 \times 10^2$
$= \boxed{3.071 \times 10^3}$

3. $(4,200,000,000,000) \div (14,000)$
$= (4.2 \times 10^{12}) \div (1.4 \times 10^{4})$
$= \dfrac{4.2 \times 10^{12}}{1.4 \times 10^{4}}$
$= 3 \times 10^{12 - 4}$
$= \boxed{3 \times 10^8}$

5. $(840,000)(0.000\ 000\ 000\ 000\ 007\ 1)$

$= (8.4 \times 10^{5})(7.1 \times 10^{-15})$
$= 59.64 \times 10^{-10}$
$= 5.964 \times 10 \times 10^{-10}$
$= \boxed{5.964 \times 10^{-9}}$

7. $\dfrac{480,000,000,000}{0.000\ 12}$
$= \dfrac{4.8 \times 10^{11}}{1.2 \times 10^{-4}}$
$= 4 \times 10^{11 - (-4)}$
$= \boxed{4 \times 10^{15}}$

9. $\dfrac{0.000\ 000\ 096}{16,000}$

$= \dfrac{9.6 \times 10^{-8}}{1.6 \times 10^4}$

$= 6 \times 10^{-8-4}$

$= \boxed{6 \times 10^{-12}}$

11. $\dfrac{(90,000)(0.004)}{(0.0003)(120)}$

$= \dfrac{(9 \times 10^4)(4 \times 10^{-3})}{(3 \times 10^{-4})(1.2 \times 10^2)}$

$= \dfrac{(9)(4) \times 10^{4-3}}{(3)(1.2) \times 10^{-4+2}}$

$= \dfrac{36 \times 10}{3.6 \times 10^{-2}}$

$= 10 \times 10^{1+2}$

$= 1 \times 10 \times 10^3$

$= \boxed{1 \times 10^4}$

13. $\dfrac{(0.000\ 035)(40,000)}{(14,000)(0.000\ 25)}$

$\dfrac{(3.5 \times\ ^{-5})(4 \times 10^4)}{(1.4 \times 10^4)(2.5 \times 10^{-4})}$

$= \dfrac{3(3.5)(4) \times 10^{-5+4}}{(1.4)(2.5) \times 10^{4-4}}$

$= \dfrac{14 \times 10^{-1}}{3.5 \times 10^0}$

$= \boxed{4 \times 10^{-1}}$

15. $\dfrac{\text{(average length of human foot)}}{\text{(length of hydrogen atom)}}$

$= \dfrac{200 \text{ millimeters}}{0.000\ 000\ 03 \text{ milliliters}}$

$= \dfrac{2 \times 10^2}{3 \times 10^{-8}}$

$= 0.\overline{6} \times 10^{2-(-8)}$

$= 6.\overline{6} \times 10^{-1} \times 10^{10}$

$= \boxed{6.\overline{6} \times 10^9 \text{ times as large}}$

17. $\dfrac{\text{(energy released by H-bombs)}}{\text{(energy released by cricket's chirp)}}$

$= \dfrac{10,000,000,000,000,000,000,000,000 \text{ ergs}}{9000 \text{ ergs}}$

$= \dfrac{1 \times 10^{28}}{9 \times 10^3}$

$= 0.\overline{1} \times 10^{28 \cdot 3}$

$= 1.\overline{1} \times 10^1 \times 10^{25}$

$= \boxed{1.\overline{1} \times 10^{24} \text{ times greater}}$

19. $P = \dfrac{(8.1 \times 10^7)}{(3000+t)^4}$

$(t = 700 \text{ days}):$

$P = \dfrac{8.1 \times 10^7}{(300+700)^4}$

$= \dfrac{8.1 \times 10^7}{(1000)^4}$

$= \dfrac{8.1 \times 10^7}{(1 \times 10^3)^4}$

$= \dfrac{8.1 \times 10^7}{1 \times 10^{12}}$

$= 8.1 \times 10^{7-12}$

$= \boxed{8.1 \times 10^{-5}}$

21. time for light of the sun to reach Pluto

$\approx \dfrac{4.6 \times 10^9 \text{ miles}}{1.86 \times 10^5 \text{ miles/second}}$

$\approx \left(\dfrac{4.6}{1.86}\right) \times 10^{9-5} \text{ sec}$

$\approx \boxed{2.47 \times 10^4 \text{ sec}}$

23. $\boxed{D}$ is true;　$(4 \times 10^3) + (3 \times 10^2)$

$= 4000 + 300$

$= 4300$

$= 43 \times 10^2$

25. $\dfrac{1.782 \times 10^4}{(5.2 \times 10^4)(6.84 \times 10^{-5})} \approx \boxed{5.010 \times 10^3}$

1.782 $\boxed{\text{EE}}$ 4 $\boxed{\div}$ 5.2 $\boxed{\text{EE}}$ 4 $\boxed{\div}$ 6.84 $\boxed{\text{EE}}$ 5 $\boxed{+/-}$ $\boxed{=}$

Display: 5.010   3

27. $\dfrac{(0.000\ 068\ 3)(4{,}870{,}000{,}000)}{(0.006\ 45)} \approx \boxed{5.157 \times 10^7}$

6.83 $\boxed{\text{EE}}$ 5 $\boxed{+/-}$ $\boxed{\times}$ 4.87 $\boxed{\text{EE}}$ 9 $\boxed{\div}$ 6.45 $\boxed{\text{EE}}$ 3 $\boxed{+/-}$ $\boxed{=}$

Display: 5.157   7

29. $\dfrac{(-6.34 \times 10^8)(0.76 \times 10^{-1})}{(8.2 \times 10^4)(42.2 \times 10^{-8})} \approx \boxed{-1.392 \times 10^9}$

$\boxed{+/-}$ 6.34 $\boxed{\text{EE}}$ 8 $\boxed{\times}$ 7.6 $\boxed{\text{EE}}$ 1 $\boxed{+/-}$ $\boxed{\div}$ 8.2 $\boxed{\text{EE}}$ 4 $\boxed{\div}$ 4.22 $\boxed{\text{EE}}$ 7 $\boxed{+/-}$ $\boxed{=}$

Display: −1.392   9

31. $\dfrac{(-9.3 \times 10^{-8})(4.7 \times 10^7)}{(3.5 \times 10^5)(7.65 \times 10^4)} \approx \boxed{-1.632 \times 10^{-10}}$

$\boxed{+/-}$ 9.3 $\boxed{\text{EE}}$ 8 $\boxed{+/-}$ $\boxed{\times}$ 4.7 $\boxed{\text{EE}}$ 7 $\boxed{\div}$ 3.5 $\boxed{\text{EE}}$ 5 $\boxed{\div}$ 7.65 $\boxed{\text{EE}}$ 4 $\boxed{=}$

Display: −1.632   −10

Review Problems

36. $\{x \mid -3 \le x < 5\}$

$\boxed{[-3,5)}$

37. $\{x \mid |x| < 3\}$

$\boxed{(-3, 3)}$

38. $-6 - \dfrac{20}{11 - \dfrac{8}{6 - \dfrac{14}{1 - 8}}}$

$= -6 - \dfrac{20}{11 - \dfrac{8}{6 - \dfrac{14}{-7}}}$

$= -6 - \dfrac{20}{11 - \dfrac{8}{6 + 2}}$

$= -6 - \dfrac{20}{11 - \dfrac{8}{8}}$

$= -6 - \dfrac{20}{11 - 1}$

$= -6 - \dfrac{20}{10}$

$= -6 - 2$

$= \boxed{-8}$

Chapter 1 Review Problems

Review Problems, pp. 76-78

1. $\{x \mid x$ is a natural number that is divisible by $4\}$
$\boxed{\{4, 8, 12, 16, \dots\}}$

2. $\{x \mid x$ is a whole number but not a natural number$\}$
$\boxed{\{0\}}$

3. $\{x \mid x$ is a prime factor of $90\}$
$\boxed{\{2, 3, 5\}}$ since the factors of 90 are $2 \cdot 3^2 \cdot 5$

4. $3 \in \{2, 4, 6, \dots\}$
$\boxed{\text{False}}$ since $3 \notin \{2, 4, 6, \dots\}$

5. $\{13\} \le \{1, 2, 3, \dots\}$
$\boxed{\text{True}}$

6. All irrational numbers are real numbers.
$\boxed{\text{True}}$

7. Some integers are not rational numbers.
$\boxed{\text{False}}$ since $\{$Integers$\} \subseteq \{$Rational numbers$\}$

8. $3 + \pi = \pi + 3$
$\boxed{\text{Commutative property of addition}}$

9. $(7 \cdot 4) \cdot 6 = 7 \cdot (4 \cdot 6)$
$\boxed{\text{Associative property of multiplication}}$

10. If $x = 6$, then $6 = x$.
$\boxed{\text{Symmetric property of equality}}$

11. $4 + 0 = 4$
$\boxed{\text{Identity property of addition}}$

12. $\frac{1}{3}(3) = 1$
$\boxed{\text{Inverse property of multiplication}}$

13. $5(x + y) = 5x + 5y$
$\boxed{\text{Distributive property}}$

14. If $9x + 6 = 10$, then $9x + 6 + (-6) = 10 + (-6)$
$\boxed{\text{Addition principle}}$

15. $-(-\sqrt{3}) = \sqrt{3}$
$\boxed{\text{Double negative property}}$

16. additive inverse of 8: $\boxed{-8}$

17. reciprocal of $\frac{1}{13}$: $\dfrac{1}{\frac{1}{13}} = \boxed{13}$

18. $\{x \mid x \le 1\}$
$\boxed{(-\infty, 1]}$

19. $\{x \mid x \ge -2\}$
$\boxed{[-2, \infty)}$

20. $\{x \mid -1 < x \le 2\}$
$\boxed{(-1, 2]}$

21. $\{x \mid |x| < 3\}$
$\boxed{(-3, 3)}$

22. $\{x \mid |x| \ge 1\}$
$\boxed{(-\infty, -1] \text{ or } [1, \infty)}$

23. x lies between -3 and 6, including -3 and excluding 6

set-builder: $\boxed{\{x \mid -3 \le x < 6\}}$

interval: $\boxed{[-3, 6)}$

24. x is at most 12

set-builder: $\boxed{\{x \mid x \le 12\}}$

interval: $\boxed{(-\infty, 12]}$

25. $\left|-23\right| + \left|17\right| - \left|-6\right|$

$= 23 + 17 - 6$

$= \boxed{34}$

26. $\left|x\right| = 6$

$\boxed{x = 6 \text{ or } x = -6}$

27. $3 + (-17) + (-25) = -14 - 25 = \boxed{-39}$

28. $16 - (-14) = 16 + 14 = \boxed{30}$

29. $-11 - [-17 + (-3)] = -11 - (-20) = -11 + 20 = \boxed{9}$

30. $\left|-17\right| + \left|3\right| - (\left|10\right| + \left|-13\right|)$

$= 17 + 3 - (10 + 13)$

$= 20 - 23$

$= \boxed{-3}$

31. $(-0.2)(-0.5) = \boxed{0.1}$

32. $-\dfrac{1}{2}(-16)(-3) = \boxed{-24}$

33. $-12 \div \dfrac{1}{4} = -12 \cdot 4 = \boxed{-48}$

34. $\left(-\dfrac{1}{2}\right)^3 \cdot 2^4 = \dfrac{1}{(-2)^3} \cdot 2^4 = -\dfrac{1}{2^3} \cdot 2^4 = -2^{4-3} = \boxed{-2}$

35. $-\dfrac{2}{7} \div \left(-\dfrac{3}{7}\right) = -\dfrac{2}{7} \cdot \left(-\dfrac{7}{3}\right) = \boxed{\dfrac{2}{3}}$

36. $-3[4 - (6 - 8)] = -3[4 - (-2)] = -3(4 + 2) = -3(6) = \boxed{-18}$

37. $8^2 - 36 \div 3^2 \cdot 4 - (-7)$

$= 64 - 36 \div 9 \cdot 4 + 7$

$= 64 - 4 \cdot 4 + 7$

$= 64 - 16 + 7$

$= \boxed{55}$

38. $\dfrac{(-2)^4 + (-3)^2}{2^2 - (-21)}$

$= \dfrac{16 + 9}{4 + 21}$

$= \dfrac{25}{25}$

$= \boxed{1}$

39. $25 \div \left(\dfrac{8+9}{2^3-3}\right) - 6$

$= 25 \div \left(\dfrac{17}{8-3}\right) - 6$

$= 25 \div \dfrac{17}{5} - 6$

$= 25 \cdot \dfrac{5}{17} - 6$

$= \dfrac{125}{17} - \dfrac{102}{17}$

$= \boxed{\dfrac{23}{17}}$

40. $6 - (-20)\left[\dfrac{6-1(6-10)}{14-3(6-8)}\right]$

$= 6 + 20\left[\dfrac{6-1(-4)}{14-3(-2)}\right]$

$= 6 + 20\left(\dfrac{6+4}{14+6}\right)$

$= 6 + 20\left(\dfrac{10}{20}\right)$

$= 6 + 10$

$= \boxed{16}$

41. $\left(\dfrac{1}{4} - \dfrac{3}{8}\right) \div \left(-\dfrac{3}{5} - \dfrac{1}{4}\right)$

$= \left(\dfrac{2}{8} - \dfrac{3}{8}\right) \div \left(-\dfrac{12}{20} - \dfrac{5}{20}\right)$

$= \left(-\dfrac{1}{8}\right) \div \left(-\dfrac{17}{20}\right)$

$= \left(-\dfrac{1}{8}\right) \cdot \left(-\dfrac{20}{17}\right)$

$= \boxed{\dfrac{5}{34}}$

42. $\dfrac{3}{4} - \dfrac{5}{8} - \left(-\dfrac{1}{2}\right)$

$= \dfrac{6}{8} - \dfrac{5}{8} + \dfrac{4}{8}$

$= \boxed{\dfrac{5}{8}}$

43. $\dfrac{(10-6)^2 + (-2)(-3)}{10 - (-4)(3)}$

$= \dfrac{4^2 + 6}{10 + 12}$

$= \dfrac{16 + 6}{22}$

$= \dfrac{22}{22}$

$= \boxed{1}$

44. $(2.6)(-5.4) \div (1.8) - (-5.7)$

$= -14.04 \div 1.8 + 5.7$

$= -7.8 + 5.7$

$= \boxed{-2.1}$

45. $\dfrac{9(-1)^3 - 3(-6)^2}{5 - 8}$

$= \dfrac{9(-1) - 3(36)}{-3}$

$= \dfrac{-9 - 108}{-3}$

$= \dfrac{-117}{-3}$

$= \boxed{39}$

46. $\dfrac{(7-9)^3 - (-4)^2}{2 + 2(8) \div 4}$

$= \dfrac{(-2)^3 - 16}{2 + 16 \div 4}$

$= \dfrac{-8 - 16}{2 + 4}$

$= \dfrac{-24}{6}$

$= \boxed{-4}$

47. $2^4(-1)^{50} + \left|-3\right|^3$

$= 16(1) + 3^3$

$= 16 + 27$

$= \boxed{43}$

48. $6 - [-3(-2)^3 \div (-8)]^2$

$= 6 - [-3(-8) \div (-8)]^2$

$= 6 - [+24 \div (-8)]^8$

$= 6 - (-3)^2$

$= 6 - 9$

$= \boxed{-3}$

49. $N = t^2 + 6t + 300$ t: years after 1990

(1991: $t = 1$):
$N = 1^2 + 6(1) + 300$
$N = 307$
(1995: $t = 5$)
$N = 5^2 + 6(5) + 300$
$N = 25 + 30 + 300$
$N = 355$
(1995) − (1991):
$355 - 307 = 48$
GNP: $\boxed{48 \text{ billion dollars}}$

50. $Q = -t^3 + 9t^2 + 12t$ t: hours after 8:00 A.M.

(noon: $t = 4$):
$\begin{aligned} Q &= -4^3 + 9(4^2) + 12(4) \\ &= -64 + 9(16) + 48 \\ &= -64 + 144 + 48 \\ &= 128;\ 128 \div 4 = 32 \end{aligned}$

(2:00 P.M.: $t = 6$):
$\begin{aligned} Q &= -6^3 + 9(6^2) + 12(6) \\ &= -216 + 9(36) + 72 \\ &= -216 + 324 + 72 \\ &= 180;\ 180 \div 6 = 30 \end{aligned}$

> At noon: 128 total units produced or 32 units/hr;
> At 2:00 P.M.: 180 total units produced or 30 units/hr;
> Efficiency has decreased by 2 units/hr.

51. $k = 293$
$C = k - 273$
$C = 293 - 273 = 20$
$F = \dfrac{9}{5} C + 32$
$F = \dfrac{9}{5}(20) + 32 = 36 + 32 = 68$
$\boxed{68°\text{F}}$

52. $L = 3(T \div 2 - 2)^2$ where $4 \le T \le 20$
($T = 12$):
$L = 3(12 \div 2 - 2)^2$
$L = 3(6 - 2)^2 = 3(4^2) = 3(16) = 48$
($T = 16$):
$L = 3(16 \div 2 - 2)^2$
$L = 3(8 - 2)^2 = 3(6^2) = 3(36) = 108$

difference: $108 - 48 = \boxed{60}$

53. $V = C(1 - rt)$
($C = 1200$, $r = 7\% = 0.07$, $t = 4$):
$V = 1200[1 - 0.07(4)]$
$V = 1200(1 - 0.28)$
$V = 1200(0.72)$
$V = 864$
value: $\boxed{\$864}$

54. $D = \begin{cases} 70 & \text{if } S < 15{,}000 \\ 70 + 0.02(S - 15{,}000) & \text{if } 15{,}000 \leq S < 30{,}000 \\ 160 + 0.03(S - 30{,}000) & \text{if } S \geq 30{,}000 \end{cases}$

$(S = 13{,}000)$:
$D = 70$

$(S = 18{,}000)$:
$D = 70 + 0.02(18{,}000 - 15{,}000)$
$D = 70 + 0.02(3000)$
$D = 70 + 60$
$D = 130$

$(S = 40{,}000)$:
$D = 160 + 0.03(40{,}000 - 30{,}000)$
$D = 160 + 0.03(10{,}000)$
$D = 160 + 300$
$D = 460$

total dues: $2(S = 13{,}000) + 3(S = 18{,}000) + 10(S = 40{,}000)$
$= 2(70) + 3(130) + 10(460)$
$= 140 + 390 + 4600$
$= \boxed{\$5130}$

55. $6(2x - 3) - 5(3x - 2)$
$= 12x - 18 - 15x + 10$
$= \boxed{-3x - 8}$

56. $6[b - 3(a - 6b)]$
$= 6[b - 3a + 18b]$
$= 6(19b - 3a)$
$= \boxed{114b - 18a}$

57. $3x - [y - (2x - 3y)]$
$= 3x - (y - 2x + 3y)$
$= 3x - (4y - 2x)$
$= 3x - 4y + 2x$
$= \boxed{5x - 4y}$

58. $-x^2 - [3y^2 - (2x^2 - y^2)]$
$= -x^2 - (3y^2 - 2x^2 + y^2)$
$= -x^2 - (4y^2 - 2x^2)$
$= -x^2 - 4y^2 + 2x^2$
$= \boxed{x^2 - 4y^2}$

59. $8x - 2x \cdot 5 + 6x^2 \div 3 \cdot 2$
$= 8x - 10x + 2x^2 \cdot 2$
$= \boxed{-2x + 4x^2}$

60. $[6(xy - 1) - 2xy] - [4(2xy + 3) - 2(xy + 2)]$
$= (6xy - 6 - 2xy) - (8xy + 12 - 2xy - 4)$
$= (4xy - 6) - (6xy + 8)$
$= 4xy - 6 - 6xy - 8$
$= \boxed{-2xy - 14}$

61. $-\dfrac{1}{2}(x + 4) - \dfrac{1}{4}(-3x + 8)$
$= -\dfrac{1}{2}x - 2 + \dfrac{3x}{4} - 2$
$\boxed{\dfrac{1}{4}x - 4}$

62. $(-3y^7)(-8y^6) = 24y^{7+6} = \boxed{24y^{13}}$

63. $(7x^3y)^2 = 7^2x^{3 \cdot 2}y^2 = \boxed{49x^6y^2}$

64. $(-3xy)(2x^2)^3 = (-3xy)(8x^6) = \boxed{-24x^7y}$

65. $\left(\dfrac{2}{3}\right)^{-2} = \dfrac{1}{\left(\dfrac{2}{3}\right)^2} = \dfrac{1}{\dfrac{4}{9}} = \boxed{\dfrac{9}{4}}$

66. $(-6xy)(-3x^2y) - 25x^3y^2$

$= 18x^3y^2 - 25x^3y^2$

$= \boxed{-7x^3y^2}$

67. $\dfrac{16y^3}{-2y^{10}} = -8y^{3-10} = -8y^{-7} = \boxed{\dfrac{-8}{y^7}}$

68. $(-3x^4)(-4x^{-11}) = 12x^{4-11} = 12x^{-7} = \boxed{\dfrac{12}{x^7}}$

69. $\dfrac{12x^7}{-4x^{-3}} = -3x^{7-(-3)} = -3x^{7+3} = \boxed{-3x^{10}}$

70. $\dfrac{-10a^5b^6}{20a^{-3}b^{11}} = -\dfrac{1}{2}a^{5-(-3)}b^{6-11} = -\dfrac{1}{2}a^8b^{-5} = \boxed{-\dfrac{a^8}{2b^5}}$

71. $(-2)^{-3} + 2^{-2} + \dfrac{1}{2}x^0$

$= \dfrac{1}{(-2)^3} + \dfrac{1}{2^2} + \dfrac{1}{2}(1)$

$= \dfrac{1}{-8} + \dfrac{1}{4} + \dfrac{1}{2}$

$= -\dfrac{1}{8} + \dfrac{6}{8}$

$= \boxed{\dfrac{5}{8}}$

72. $\left(\dfrac{-2}{ab}\right)^5 = \dfrac{(-2)^5}{a^5b^5} = \boxed{-\dfrac{32}{a^5b^5}}$

73. $(3x^4y^{-2})(-2x^5y^{-3}) = -6x^{4+5}y^{-2-3} = -6x^9y^{-5} = \boxed{-\dfrac{6x^9}{y^5}}$

74. $(0.000\,92)(7,400,000,000,000)$

$= (9.2 \times 10^{-4})(7.4 \times 10^{12})$

$= (9.2)(7.4) \times 10^{-4+12}$

$= 68.08 \times 10^8$

$= 6.808 \times 10 \times 10^8$

$= \boxed{6.808 \times 10^9}$

75. $\dfrac{0.000\,000\,000\,000\,33}{0.000\,66}$

$= \dfrac{3.3 \times 10^{-13}}{6.6 \times 10^{-4}}$

$= 0.5 \times 10^{-13-(-4)}$

$= 5 \times 10^{-1} \times 10^{-9}$

$= \boxed{5 \times 10^{-10}}$

76. $\dfrac{(92{,}000{,}000)(0.0036)}{(0.018)(4000)}$

$= \dfrac{(9.2 \times 10^7)(3.6 \times 10^{-3})}{(1.8 \times 10^{-2})(4 \times 10^3)}$

$= \left(\dfrac{9.2}{4}\right)\left(\dfrac{3.6}{1.8}\right) \times \dfrac{10^{7-3}}{10^{-2+3}}$

$= (2.3)(2) \times \dfrac{10^4}{10}$

$= \boxed{4.6 \times 10^3}$

77. 1 light year $= (186000 \text{ mi/sec})(60 \text{ sec/min})(60 \text{ min/hr})(24 \text{ hr/day})(365 \text{ days})$

$\approx \boxed{5.866 \times 10^{12} \text{ mi}}$

78. time $= \dfrac{\text{distance}}{\text{rate}} = \dfrac{6 \times 10^{17} \text{ mi}}{20{,}000 \text{ mi/hr}} = \dfrac{6 \times 10^{17}}{2 \times 10^4} \text{ hr} = \boxed{3 \times 10^{13} \text{ hr}}$

Algebra for College Students

Chapter 2   Linear Equations and Inequalities

Section 2.1 Solving Linear Equations

Problem Set 2.1, pp. 91-92

1.
$$3x + 5 = 2x + 13$$
$$3x + 5 - 2x = 2x + 13 - 2x$$
$$x + 5 = 13$$
$$x + 5 - 5 = 13 - 5$$
$$x = 8$$
$$\boxed{\{8\}}$$

3.
$$8x - 2 = 7x - 5$$
$$8x - 7x - 2 = 7x - 5 - 7x$$
$$x - 2 = -5$$
$$x - 2 + 2 = -5 + 2$$
$$x = -3$$
$$\boxed{\{-3\}}$$

5.
$$7x + 4 = x + 16$$
$$7x + 4 - x = x + 16 - x$$
$$6x + 4 = 16$$
$$6x + 4 - 4 = 16 - 4$$
$$6x = 12$$
$$\frac{1}{6}(6x) = \frac{1}{6}(12)$$
$$x = 2$$
$$\boxed{\{2\}}$$

7.
$$8y - 3 = 11y + 9$$
$$8y - 3 - 11y = 11y + 9 - 11y$$
$$-3y - 3 = 9$$
$$-3y - 3 + 3 = 9 + 3$$
$$-3y = 12$$
$$\left(-\frac{1}{3}\right)(-3y) = \left(-\frac{1}{3}\right)(12)$$
$$y = -4$$
$$\boxed{\{-4\}}$$

9.
$$8z + 11.7 = 9z - 15$$
$$8z + 11.7 - 9z = 9z - 15 - 9z$$
$$-z + 11.7 = -15$$
$$-z + 11.7 - 11.7 = -15 - 11.7$$
$$-z = -26.7$$
$$z = 26.7$$
$$\boxed{\{26.7\}}$$

11.
$$8x - 3.1 = 1x + 9.02$$
$$8x - 3.1 - 11x = 11x + 9.02 - 11x$$
$$-3x - 3.1 = 9.02$$
$$-3x - 3.1 + 3.1 = 9.02 + 3.1$$
$$-3x = 12.12$$
$$\left(-\frac{1}{3}\right)(-3x) = \left(-\frac{1}{3}\right)(12.12)$$
$$x = -4.04$$
$$\boxed{\{-4.04\}}$$

13.
$$\frac{1}{8}y + \frac{1}{4} = \frac{1}{2}$$
$$\frac{1}{8}y + \frac{1}{4} - \frac{1}{4} = \frac{1}{2} - \frac{1}{4}$$
$$\frac{1}{8}y = \frac{1}{4}$$
$$8\left(\frac{1}{8}y\right) = 8\left(\frac{1}{4}\right)$$
$$y = 2$$
$$\boxed{\{2\}}$$

15.
$$13x - (5x - 2) = -14$$
$$13x - 5x + 2 = -14$$
$$8x + 2 = -14$$
$$8x + 2 - 2 = -14 - 2$$
$$8x = -16$$
$$\frac{1}{8}(8x) = \frac{1}{8}(-16)$$
$$x = -2$$
$$\boxed{\{-2\}}$$

17.
$$
\begin{aligned}
2 - (7y + 5) &= 13 - 3y \\
2 - 7y - 5 &= 13 - 3y \\
-7y - 3 &= 13 - 3y \\
-7y - 3 + 3y &= 13 - 3y + 3y \\
-4y - 3 &= 13 \\
-4y - 3 + 3 &= 13 + 3 \\
-4y &= 16 \\
\left(-\tfrac{1}{4}\right)(-4y) &= \left(-\tfrac{1}{4}\right)(16) \\
y &= -4
\end{aligned}
$$

$\boxed{\{-4\}}$

19.
$$
\begin{aligned}
3(y - 3) - 2(y - 2) &= 7 \\
3y - 9 - 2y + 4 &= 7 \\
y - 5 &= 7 \\
y - 5 + 5 &= 7 + 5 \\
y &= 12
\end{aligned}
$$

$\boxed{\{12\}}$

21.
$$
\begin{aligned}
4(2y + 1) - 29 &= 3(2y - 5) \\
8y + 4 - 29 &= 6y - 15 \\
8y - 25 &= 6y - 15 \\
8y - 25 - 6y &= 6y - 15 - 6y \\
2y - 25 &= -15 \\
2y - 25 + 25 &= -15 + 25 \\
2y &= 10 \\
\tfrac{1}{2}(2y) &= \tfrac{1}{2}(10) \\
y &= 5
\end{aligned}
$$

$\boxed{\{5\}}$

23.
$$
\begin{aligned}
2(3x - 7) - (6 - x) &= 4 - (-x + 6) \\
6x - 14 - 6 + x &= 4 + x - 6 \\
7x - 20 &= x - 2 \\
7x - 20 - x &= x - 2 - x \\
6x - 20 &= -2 \\
6x - 20 + 20 &= -2 + 20 \\
6x &= 18 \\
\tfrac{1}{6}(6x) &= \tfrac{1}{6}(18) \\
x &= 3
\end{aligned}
$$

$\boxed{\{3\}}$

25.
$$
\begin{aligned}
3(2x - 1) &= 5[3x - (2 - x)] \\
6x - 3 &= 5(3x - 2 + x) \\
6x - 3 &= 5(4x - 2) \\
6x - 3 &= 20x - 10 \\
6x - 3 - 20x &= 20x - 10 - 20x \\
-14x - 3 &= -10 \\
-14x - 3 + 3 &= -10 + 3 \\
-14x &= -7 \\
-\tfrac{1}{14}(-14x) &= -\tfrac{1}{14}(-7) \\
x &= \tfrac{1}{2}
\end{aligned}
$$

$\boxed{\left\{\tfrac{1}{2}\right\}}$

27.
$$
\begin{aligned}
7[(3y - 2) - (y - 4)] &= 28 \\
7(3y - 2 - y + 4) &= 28 \\
7(2y + 2) &= 28 \\
14y + 14 &= 28 \\
14y + 14 - 14 &= 28 - 14 \\
14y &= 14 \\
\tfrac{1}{14}(14y) &= \tfrac{1}{14}(14) \\
y &= 1
\end{aligned}
$$

$\boxed{\{1\}}$

29.
$$
\begin{aligned}
-2\{7 - [4 - 2(1 - x) + 3]\} &= 10 - [4x - 2(x - 3)] \\
-2[7 - (4 - 2 + 2x + 3)] &= 10 - (4x - 2x + 6) \\
-2[7 - (2x + 5)] &= 10 - (2x + 6) \\
-2(7 - 2x - 5) &= 10 - 2x - 6 \\
-14 + 4x + 10 &= 4 - 2x \\
4x - 4 &= 4 - 2x \\
4x - 4 + 2x &= 4 - 2x + 2x \\
6x - 4 &= 4 \\
6x - 4 + 4 &= 4 + 4 \\
6x &= 8 \\
\frac{1}{6}(6x) &= \frac{1}{6}(8) \\
x &= \frac{4}{3}
\end{aligned}
$$

$$\boxed{\left\{\frac{4}{3}\right\}}$$

31.
$$
\begin{aligned}
-\frac{1}{5}(10x - 5) &= -3(-x + 1) \\
-2x + 1 &= 3x - 3 \\
-2x + 1 - 3x &= 3x - 3 - 3x \\
-5x + 1 &= -3 \\
-5x + 1 - 1 &= -3 - 1 \\
-5x &= -4 \\
\left(-\frac{1}{5}\right)(-5x) &= \left(-\frac{1}{5}\right)(-4) \\
x &= \frac{4}{5}
\end{aligned}
$$

$$\boxed{\left\{\frac{4}{5}\right\}}$$

33.
$$
\begin{aligned}
\frac{y}{4} &= 2 + \frac{y - 3}{3} \\
12\left(\frac{y}{4}\right) &= 12\left(2 + \frac{y - 3}{3}\right) \\
3y &= 24 + 4(y - 3) \\
3y &= 24 + 4y - 12 \\
3y &= 12 + 4y \\
3y - 4y &= 12 + 4y - 4y \\
-y &= 12 \\
y &= -12
\end{aligned}
$$

$$\boxed{\{-12\}}$$

35.
$$
\begin{aligned}
\frac{y - 3}{12} + \frac{y - 1}{6} &= \frac{y + 2}{9} - 1 \\
36\left(\frac{y - 3}{12} + \frac{y - 1}{6}\right) &= 36\left(\frac{y + 2}{9} - 1\right) \\
3(y - 3) + 6(y - 1) &= 4(y + 2) - 36 \\
3y - 9 + 6y - 6 &= 4y + 8 - 36 \\
9y - 15 &= 4y - 28 \\
9y - 15 - 4y &= 4y - 28 - 4y \\
5y - 15 &= -28 \\
5y - 15 + 15 &= -28 + 15 \\
5y &= -13 \\
\frac{1}{5}(5y) &= \frac{1}{5}(-13) \\
y &= -\frac{13}{5}
\end{aligned}
$$

$$\boxed{\left\{-\frac{13}{5}\right\}}$$

37.
$$
\begin{aligned}
\frac{x}{5} - 6 &= 6 + \frac{1}{5}x \\
5\left(\frac{x}{5} - 6\right) &= 5\left(6 + \frac{1}{5}x\right) \\
5\left(\frac{x}{5}\right) - 5(6) &= 5(6) + 5\left(\frac{1}{5}x\right) \\
x - 30 &= 30 + x \\
x - 30 - x &= 30 + x - x \\
-30 &= 30 \qquad \text{contradiction}
\end{aligned}
$$

no solution

$$\boxed{\varnothing}$$

39.
$$1 - \frac{5y+6}{50} = \frac{45y+43}{100} + \frac{y}{5}$$
$$100\left(1 - \frac{5y+6}{50}\right) = 100\left(\frac{45y+43}{100} + \frac{y}{5}\right)$$
$$100 - 2(5y+6) = 45y + 43 + 20y$$
$$100 - 10y - 12 = 65y + 43$$
$$88 - 10y = 65y + 43$$
$$88 - 10y - 65y = 65y + 43 - 65y$$
$$88 - 75y = 43$$
$$88 - 75y - 88 = 43 - 88$$
$$-75y = -45$$
$$\left(-\frac{1}{75}\right)(-75y) = \left(-\frac{1}{75}\right)(-45)$$
$$y = \frac{3}{5}$$
$$\boxed{\left\{\frac{3}{5}\right\}}$$

41.
$$\frac{1}{3}(6x+9) = -2(x-1)$$
$$2x + 3 = -2x + 2$$
$$2x + 3 + 2x = -2x + 2 + 2x$$
$$4x + 3 = 2$$
$$4x + 3 - 3 = 2 - 3$$
$$4x = -1$$
$$\frac{1}{4}(4x) = \frac{1}{4}(-1)$$
$$x = -\frac{1}{4}$$
$$\boxed{\left\{-\frac{1}{4}\right\}}$$

43.
$$0.02(y-100) = 62 + 0.06y$$
$$0.02y - 2 = 62 + 0.06y$$
$$0.02y - 2 - 0.06y = 62 + 0.06y - 0.06y$$
$$-2 - 0.04y = 62$$
$$-2 - 0.04y + 2 = 62 + 2$$
$$-0.04y + 2 = 64$$
$$\left(-\frac{1}{0.04}\right)(-0.04y) = \left(-\frac{1}{0.04}\right)(64)$$
$$y = -1600$$
$$\boxed{\{-1600\}}$$

45.
$$0.09(-y+5000) = 513 - 0.12y$$
$$-0.09y + 450 = 513 - 0.12y$$
$$-0.09y + 450 + 0.12y = 513 - 0.12y + 0.12y$$
$$0.03y + 450 = 513$$
$$0.03y + 450 - 450 = 513 - 450$$
$$0.03y = 63$$
$$\left(\frac{1}{0.03}\right)(0.03y) = \left(\frac{1}{0.03}\right)(63)$$
$$y = 2100$$
$$\boxed{\{2100\}}$$

47.
$$0.2z = 35 + 0.05(z-100)$$
$$0.2z = 35 + 0.05z - 5$$
$$0.2z = 30 + 0.05z$$
$$0.2z - 0.05z = 30 + 0.05z - 0.05z$$
$$0.15z = 30$$
$$\left(\frac{1}{0.15}\right)(0.15z) = \left(\frac{1}{0.15}\right)(30)$$
$$z = 200$$
$$\boxed{\{200\}}$$

49.
$$0.35y - 0.1 = 0.15y + 0.2$$
$$0.35y - 0.1 - 0.15y = 0.15y + 0.2 - 0.15y$$
$$0.20y - 0.1 = 0.2$$
$$0.20y - 0.1 + 0.1 = 0.2 + 0.1$$
$$0.2y = 0.3$$
$$\left(\frac{1}{0.2}\right)(0.2y) = \left(\frac{1}{0.2}\right)(0.3)$$
$$y = \frac{3}{2}$$
$$\boxed{\left\{\frac{3}{2}\right\}}$$

51.
$$10x - 2(4+5x) = -8$$
$$10x - 8 - 10x = 8$$
$$-8 = -8 \quad \text{True}$$
$$\boxed{\{x \mid x \in R\}}$$

53.
$$10x - 2(4+5x) = 8$$
$$10x - 8 - 10x = 8$$
$$-8 = 8 \quad \text{True}$$
$$\boxed{\{x \mid x \in R\}}$$

55.
$$\begin{aligned}
7x - 2[3(1 - x)] &= -(4 + 2 - 9x - 4x) \\
7x - 2(3 - 3x) &= -(6 - 13x) \\
7x - 6 + 6x &= -6 + 13x \\
13x - 6 &= -6 + 13x \\
13x - 6 - 13x &= -6 + 13x - 13x \\
-6 &= -6 \quad \text{True}
\end{aligned}$$

$$\boxed{\{x \mid x \in R\}}$$

57.
$$\begin{aligned}
8x - [5x + 2(x - 3) - (3x + 4)] &= 30 \\
8x - (5x + 2x - 6 - 3x - 4) &= 30 \\
8x - (4x - 10) &= 30 \\
4x + 10 &= 30 \\
4x + 10 - 10 &= 30 - 10 \\
4x &= 20 \\
\tfrac{1}{4}(4x) &= \tfrac{1}{4}(20) \\
x &= 5
\end{aligned}$$

$$\boxed{\{5\}}$$

59.
$$\begin{aligned}
12\left(\frac{y}{3} - \frac{1}{2}\right) &= 8\left(\frac{y}{2} - 1\right) \\
4y - 6 &= 4y - 8 \\
4y - 6 - 4y &= 4y - 8 - 4y \\
-6 &= -8 \quad \text{contradiction}
\end{aligned}$$

no solution

$$\boxed{\varnothing}$$

61.
$$\begin{aligned}
2(3y + 4) - 4 &= 9y + 4 - 3y \\
6y + 8 - 4 &= 6y + 4 \\
6y + 4 &= 6y + 4 \\
6y + 4 - 6y &= 6y + 4 - 6y \\
4 &= 4 \quad \text{True}
\end{aligned}$$

$$\boxed{\{y \mid y \in R\}}$$

63.
$$\begin{aligned}
-11p + 4(p - 3) + 6p &= 4p - 12 \\
-11p + 4p - 12 + 6p &= 4p - 12 \\
-p - 12 &= 4p - 12 \\
-p - 12 - 4p &= 4p - 12 - 4p \\
-5p - 12 &= -12 \\
-5p - 12 + 12 &= -12 + 12 \\
-5p &= 0 \\
p &= 0
\end{aligned}$$

$$\boxed{\{0\}}$$

65.
$$\begin{aligned}
3(7 + 4m) &= -4[6 - (-2 + 3m)] \\
21 + 12m &= -4(6 + 2 - 3m) \\
21 + 12m &= -4(8 - 3m) \\
21 + 12m &= -32 + 12m \\
21 + 12m - 12m &= -32 + 12m - 12m \\
21 &= -36 \quad \text{contradiction}
\end{aligned}$$

no solution

$$\boxed{\varnothing}$$

67. $\boxed{D}$ is true;
$$\begin{aligned}
7(y + 1) + 5(-y + 5) &= 2(y + 3) - 7 \\
7y + 7 - 5y + 25 &= 2y + 6 - 7 \\
2y + 32 &= 2y - 1 \\
2y + 32 - 2y &= 2y - 1 - 2y \\
32 &= -1 \quad \text{contradiction}
\end{aligned}$$
no solution
$$\varnothing$$

Review Problems

74. $\{x \mid |x| > 2\}$

$$\boxed{(-\infty, -2) \text{ or } (2, \infty)}$$

75. $4x^2 - [6y^2 - (3y^2 - 8x^2)]$

$$\begin{aligned}
&= 4x^2 - (6y^2 - 3y^2 + 8x^2) \\
&= 4x^2 - (3y^2 + 8x^2) \\
&= 4x^2 - 3y^2 - 8x^2 \\
&= \boxed{-4x^2 - 3y^2}
\end{aligned}$$

76. $\dfrac{-30x^2 y^8}{10x^{-4}y^{11}} = -3x^{2 - (-4)}y^{8 - 11} = -3x^6 y^{-3} = \boxed{\dfrac{-3x^6}{y^3}}$

Section 2.2 Mathematical Models

Problem Set 2.2, pp. 98-101

1.
$$W = \frac{127}{5}d + 40$$

$W = 294$ grams:
$$294 = \frac{127}{5}d + 40$$
$$5(294) = 5\left(\frac{127}{5}d + 40\right)$$
$$1470 = 127d + 200$$
$$1470 - 200 = 127d + 200 - 200$$
$$1270 = 127d$$
$$\frac{1270}{127} = d$$
$$10 = d$$

$\boxed{10 \text{ days}}$

3.
$$R = 75L + 1310$$

$R = 1535$ milliseconds:
$$1535 = 75L + 1310$$
$$225 = 75L$$
$$\frac{225}{75} = L$$
$$3 = L$$

level of semantic memory: $\boxed{3}$

5.
$$\frac{W}{4A} = P$$
$$P = 28 \text{ pounds/square inch}$$
$$A = 245 \text{ square inches}$$
$$\frac{W}{4(24)} = 28$$
$$W = 4(24)(28)$$
$$W = 2688$$

$\boxed{2688 \text{ lb}}$

7. demand: $p = -0.2q + 4$

supply: $p = 0.07q + 0.76$
$$q = 10$$
$$p = -0.2(10) + 4$$
$$p = -2 + 4$$
$$p = 2$$

Thus, consumers will purchase 10,000 pounds of grapes when the price is $\boxed{\$2/\text{pound}}$.
$$2 = 0.07q + 0.76$$
$$1.24 = 0.07q$$
$$17.714 \approx q$$

Suppliers will be willing to supply approximately $\boxed{17{,}714 \text{ pounds}}$ at \$2/lb.

Because $\boxed{\text{supply exceeds the demand}}$ at \$2/lb, prices will come $\boxed{\text{down}}$.

9.
$$p = Kd + 1$$

$p = 10.94$ atmosphere, $d = 100$ meters:
$$10.94 = 100K + 1$$
$$9.94 = 100K$$
$$\frac{9.94}{100} = K$$
$$K = \boxed{0.0994}$$
$$p = 0.0994d + 1$$
$$d = 50 \text{ meters}$$
$$p = 0.0994(50) + 1$$
$$p = 4.97 + 1$$
$$p = 5.97$$

$\boxed{5.97 \text{ atm}}$

11.
$$s = 3.514268Y - 226.2953$$
$s = 93.5\%:$
$$93.5 = 3.514268Y - 226.2953$$
$$319.7953 = 3.514268Y$$
$$Y \approx 90.000121 \approx 91$$

$\boxed{1991}$

13. For women: $h = 61.412 + 2.317f$
$h = 160$ cm:
$$160 = 61.412 + 2.317f$$
$$98.588 = 2.317f$$
$$f \approx 42.55$$

length of femur: $\boxed{\text{approximately 42.55 cm}}$

15.
$$N = \frac{GMC}{(G-M)DP}$$
$$G = 24 \text{ mpg}, M = 12 \text{ mpg}, D = 12{,}000 \text{ miles/years},$$
$$P = \$1.40/\text{gallon}, N = 5 \text{ years}$$
$$5 = \frac{24(12)C}{(24-12)(12{,}000)(1.40)}$$
$$5 = \frac{24(12)C}{(12)(12{,}000)(1.40)}$$
$$\frac{5(12{,}000)(1.40)}{24} = C$$
$$3500 = C$$

cost of new car: $\boxed{\$3500}$

17.
$$V = C\left(1 - \frac{n}{N}\right)$$
$$C = \$10{,}000, N = 20 \text{ years}, V = \$6500$$
$$6500 = 10{,}000\left(1 - \frac{n}{20}\right)$$
$$6500 = 10{,}000 - 500n$$
$$6500 + 500n = 10{,}000$$
$$500n = 3500$$
$$n = 7$$

$\boxed{\text{After 7 yr}}$ the equipment value will be \$6500.

19.
$$R = -\frac{35}{2}L + 195$$
$$R + \frac{35}{2}L = 195$$
$$\frac{35}{2}L = 195 - R$$
$$\frac{2}{35}\left(\frac{35}{2}L\right) = \frac{2}{35}(195 - R)$$
$$L = \boxed{\frac{2}{35}(195 - R)}$$
$R = 125:$
$$L = \frac{2}{35}(195 - 125)$$
$$L = \frac{2}{35}(70)$$
$$L = 4$$

wool length: $\boxed{4 \text{ cm}}$

This approach is a $\boxed{\text{more efficient method}}$ than the one used in Problem 2.

21.
$$E = mc^2$$
$$\left(\frac{1}{c^2}\right)(E) = \left(\frac{1}{c^2}\right)(mc^2)$$
$$\frac{E}{c^2} = m$$
$$m = \boxed{\frac{E}{c^2}}$$

23.
$$ax - bx = 13$$
$$x(a - b) = 13$$
$$\left(\frac{1}{a-b}\right)x(a-b) = \left(\frac{1}{a-b}\right)13$$
$$x = \boxed{\frac{13}{a-b}}$$

25.
$$ax + 13 = bx - 12$$
$$ax + 13 - bx = bx - 12 - bx$$
$$ax - bx + 13 = -12$$
$$ax - bx + 13 - 13 = -12 - 13$$
$$ax - bx = -25$$
$$x(a - b) = -25$$
$$x = -\frac{25}{a - b}$$
$$\boxed{x = -\frac{25}{a - b} \text{ or } x = \frac{25}{b - a}}$$

27.
$$D = 0.8(L - S) + 5$$
$$D = 0.8L - 0.85 + S$$
$$D = 0.8L + 0.25$$
$$D - 0.8L = 0.25$$
$$\frac{1}{0.2}(D - 0.8L) = S$$
$$\boxed{S = \frac{D}{0.2} - 4L \text{ or } S = 5D - 4L}$$

29.
$$P = C + MC$$
$$P - C = MC$$
$$\frac{P - C}{C} = M$$
$$\boxed{M = \frac{P - C}{C} \text{ or } M = \frac{P}{C} - 1}$$

31.
$$P = 2L + 2W$$
$$P - 2L = 2W$$
$$\frac{P - 2L}{2} = W$$
$$\boxed{W = \frac{P - 2L}{2} \text{ or } W = \frac{P}{2} - L}$$

33.
$$B = \frac{1}{7}ac$$
$$7B = ac$$
$$\frac{7B}{C} = a$$
$$\boxed{a = \frac{7B}{C}}$$

35.
$$a = \frac{d}{7}(B + x)$$
$$7a = d(B + x)$$
$$7a = dB + dx$$
$$7a - dx = dB$$
$$\frac{7a - dx}{d} = B$$
$$\boxed{B = \frac{7a - dx}{d}}$$

37.
$$\frac{2}{7}(c + d) = 4(c - a)$$
$$2(c + d) = 28(c - a)$$
$$2c + 2d = 28c - 28a$$
$$2d = 28c - 28a - 2c$$
$$2d = 26c - 28a$$
$$d = 13c - 14a$$
$$\boxed{d = 13c - 14a}$$

39.
$$A = \frac{1}{7}B(C + D)$$
$$7A = B(C + D)$$
$$7A = BC + BD$$
$$7A - BC = BD$$
$$\frac{7A - BC}{B} = D$$
$$\boxed{D = \frac{7A - BC}{B}}$$

41.
$$\frac{4xy}{z} = -5$$
$$4xy = -5z$$
$$-x = \frac{-5z}{4y}$$
$$\boxed{x = \frac{-5z}{4y}}$$

43.
$$\frac{1}{3}(c - dx) + 2d = \frac{1}{6}bd - x)$$
$$6\left[\frac{1}{3}(c - dx) + 2d\right] = 6\left[\frac{1}{6}(6d - x)\right]$$
$$2(c - dx) + 12d = 6d - x$$
$$2c - 2dx + 12d = 6d - x$$
$$2c - 2dx + 12d + x = 6d$$
$$-2dx + x = 6d - 2c - 12d$$
$$x(-2d + 1) = -6d - 2c$$
$$x = \frac{6d - 2c}{-2d + 1}$$
$$\boxed{x = \frac{2c + 6d}{2d - 1}}$$

45.
$$\frac{acx}{3} + \frac{4c}{5} = \frac{2acx}{3}$$
$$15\left(\frac{acx}{3} + \frac{4c}{5}\right) = 15\left(\frac{2acx}{3}\right)$$
$$5acx + 12c = 10acx$$
$$5acx + 12c - 10acx = 0$$
$$-5acx = -12c$$
$$x = \frac{-12c}{-5ac}$$
$$\boxed{x = \frac{12}{5a}}$$

47.
$$Bx = G[x - (r + t)]$$
$$Bx = G(x - r - t)$$
$$Bx = Gx - Gr - Gt$$
$$Bx - Gx = -Gr - Gt$$
$$x(B - G) = -G(r + t)$$
$$x = \frac{-G(r + t)}{B - G}$$
$$\boxed{x = \frac{G(r + t)}{G - B}}$$

49. $\boxed{D}$ is true;   None is true.

A is *not* true:
$$S = P + Prt$$
$$S = P(1 + rt)$$
$$P = \frac{S}{1 + rt}$$

B is *not* true:
$$I = Prt$$
$$t = \frac{I}{pr}$$

C is *not* true:
$$S = 2LW + 2LH + 2WH$$
Using the distributive property.
$$S - 2LW = H(2L + 2W)$$
$$H = \frac{S - 2LW}{2L + 2W}$$

Review Problems

54. x is at least 7
$$\boxed{\{x \mid x \geq 7\}}$$

55. $\boxed{C}$ is not true: $3 \in \{1, 2, 3, \ldots\}$ *not* $3 \subseteq \{1, 2, 3, \ldots\}$

56. $\{x \mid x$ is a natural number less than 17 that is divisible by 3$\}$
$$\boxed{\{3, 6, 9, 12, 15\}}$$

Section 2.3 Problem Solving

Problem Set 2.3, pp. 113-117

1. Let x = the number.
$$
\begin{aligned}
12x - 6 &= 7x + 24 \\
12x - 6 - 7x &= 24 \\
5x - 6 &= 24 \\
5x - 6 + 6 &= 24 + 6 \\
5x &= 30 \\
x &= 6
\end{aligned}
$$
The number is $\boxed{6}$.

3. Let x = the number.
$$
\begin{aligned}
3(x - 5) &= 2(1 + 2x) \\
3x - 15 &= 2 + 4x \\
3x - 4x - 15 &= 2 \\
-x - 15 &= 2 \\
-x &= 2 + 15 \\
-x &= 17 \\
x &= -17
\end{aligned}
$$
The number is $\boxed{-17}$.

5. Let
$$
\begin{aligned}
x &= \text{first integer} \\
x + 1 &= \text{second consecutive integer} \\
x + 2 &= \text{third consecutive integer}
\end{aligned}
$$

$$
\begin{aligned}
x + (x + 1) + (x + 2) &= 25 \\
3x + 3 &= 25 \\
3x &= 2 \\
x &= \frac{22}{3} \quad \text{which is not an integer}
\end{aligned}
$$

$\boxed{\text{No integers satisfy the given condition.}}$

7. Let
$$
\begin{aligned}
x &= \text{first consecutive even integer} \\
x + 2 &= \text{second consecutive even integer} \\
x + 4 &= \text{third consecutive even integer}
\end{aligned}
$$

$$
\begin{aligned}
x + 4 &= 2x - 6 \\
x + 4 - 2x &= -6 \\
-x + 4 &= -6 \\
-x &= -6 - 4 \\
x &= -10 \\
x &= 10 \\
x + 2 &= 12 \\
x + 4 &= 14
\end{aligned}
$$
The integers are $\boxed{10, 12, \text{ and } 14}$.

9. Let

$$
\begin{aligned}
x &= \text{first even integer} \\
x + 2 &= \text{second even integer} \\
x + 4 &= \text{third even integer}
\end{aligned}
$$

$$
\begin{aligned}
x + (x + 4) &= \frac{1}{2}(x + 2) + 15 \\
2x + 4 &= \frac{1}{2}x + 1 + 15 \\
2x + 4 &= \frac{1}{2}x + 16 \\
4x + 8 &= x + 32 \\
4x + 8 - x &= 32 \\
3x + 8 &= 32 \\
3x &= 32 - 8 \\
3x &= 24 \\
x &= 8 \\
x + 2 &= 10 \\
x + 10 &= 12
\end{aligned}
$$

The integers are $\boxed{8, 10, \text{and } 12}$.

11. Let

$$
\begin{aligned}
x &= \text{first consecutive even integer} \\
x + 2 &= \text{second consecutive even integer} \\
x + 4 &= \text{third consecutive even integer}
\end{aligned}
$$

$$
\begin{aligned}
2(x + 2) &= x + (x + 4) \\
2x + 4 &= 2x + 4 \\
4 &= 4 \quad \text{True}
\end{aligned}
$$

$\{x \mid x \in R\}$

$\boxed{\text{Conditions are true for any three consecutive even integers.}}$

13. Let

$$
\begin{aligned}
x &= \text{first consecutive even integer} \\
x + 2 &= \text{second consecutive even integer} \\
x + 4 &= \text{third consecutive even integer}
\end{aligned}
$$

$$
\begin{aligned}
\frac{x + (x + 2) + (x + 4)}{3} &= 20 \\
3x + 6 &= 60 \\
3x &= 54 \\
x &= 18 \\
x + 2 &= 20 \\
x + 4 &= 22
\end{aligned}
$$

The integers are $\boxed{18, 20 \text{ and } 22}$.

15. Let

$$
\begin{aligned}
x &= \text{number of atoms of oxygen} \\
2x + 1 &= \text{number of atoms of carbon} \\
2x + 1 - 1 = 2x &= \text{number of atoms of hydrogen}
\end{aligned}
$$

$$
\begin{aligned}
(\text{oxygen atoms}) + (\text{carbon atoms}) + (\text{hydrogen atoms}) &= 21 \\
x + (2x + 1) + 2x &= 21 \\
5x + 1 &= 21 \\
5x &= 20 \\
x &= 4 \\
2x + 1 &= 9
\end{aligned}
$$

number of atoms of carbon: $\boxed{9}$

17. Let

$$
\begin{aligned}
x &= \text{length of the Rhine River} \\
x + 160 &= \text{length of St. Lawrence River} \\
2(x + 160) &= \text{length of Zambezi River} \\
6x - 320 &= \text{length of Amazon River}
\end{aligned}
$$

$$
\begin{aligned}
6x - 320 &= x + (x + 160) + 2(x + 160) + 1440 \\
6x - 320 &= 4x + 1920 \\
2x - 320 &= 1920 \\
2x &= 2240 \\
x &= 1120 \\
2(x + 160) &= 2(1120 + 160) = 2(1280) = 2560
\end{aligned}
$$

The length of the Zambezi River is $\boxed{2560 \text{ km}}$.

19. Let

$$
\begin{aligned}
x &= \text{number of national parks in Canada} \\
4x - 28 &= \text{number of rational parks in the United States}
\end{aligned}
$$

$$
\begin{aligned}
x + (4x - 28) &= 106 + \frac{6}{7}(4x - 28) \\
5x - 28 &= 106 + \frac{6}{7}(4x - 28) \\
7(5x - 28) &= 7(106) + 6(4x - 28) \\
35x - 196 &= 742 + 24x - 168 \\
35x - 196 &= 574 + 24x \\
11x - 196 &= 574 + 24x \\
11x &= 770 \\
x &= 70 \\
4x - 28 &= 4(70) - 28 = 280 - 28 = 252
\end{aligned}
$$

number of national parks:

$\boxed{\text{Canada: 70 parks; USA: 252 parks}}$

21. Let

$$
\begin{aligned}
x &= \text{smaller number} \\
4x + 1 &= \text{larger number}
\end{aligned}
$$

$$
\begin{aligned}
3(4x + 1) - 2x &= 43 \\
12x + 3 - 2x &= 43 \\
10x + 3 &= 43 \\
10x &= 40 \\
x &= 4 \\
4x + 1 &= 4(4) + 1 = 16 + 1 = 17
\end{aligned}
$$

The numbers are $\boxed{4 \text{ and } 17}$.

23. Let

$$x = \text{Barry's age now}$$
$$4x = \text{Simon's age now}$$
$$x + 4 = \text{Barry's age in 4 years}$$
$$4x + 4 = \text{Simon's age in 4 years}$$

$$4x + 4 = 2(x + 4)$$
$$4x + 4 = 2x + 8$$
$$2x = 4$$
$$x = 2$$
$$4x = 4(2) = 8$$

Present ages: $\boxed{\text{Barry, 2 years old; Simon, 8 years old}}$

25. Let

$$x = \text{unknown number of years ago}$$
$$23 - x = \text{Rita's age } x \text{ years ago}$$
$$15 - x = \text{Joel's age } x \text{ years ago}$$

$$23 - x = 3(15 - x)$$
$$23 - x = 45 - 3x$$
$$23 - x + 3x = 45$$
$$2x = 45 - 23$$
$$2x = 22$$
$$x = 11$$

$\boxed{\text{11 years ago}}$

27. Let

$$x = \text{age of woman}$$
$$3x = \text{age of "uncle"}$$
$$x + 20 = \text{age of women in 20 years}$$
$$3x + 20 = \text{age of "uncle" in 20 years}$$

$$3x + 20 = 2(x + 20)$$
$$3x + 20 = 2x + 40$$
$$x = 20$$
$$3x = 60$$

$\boxed{\text{"uncle", 60 years old; woman, 20 years old}}$

29. The Bundy's will pay $\frac{1}{3}$ ($264,000) = $88,000.

The balance is $264,000 - $88,000 - $176,000.
Let

$$x = \text{the bachelor's share}$$
$$3x = \text{the Brady's share}$$
$$\frac{3x}{2} = 1.5x = \text{the couple's share}$$

$$x + 3x + 1.5x = 176,000$$
$$5.5x = 176,000$$
$$x = 32,000$$
$$3x = 3(32,000) = 96,000$$
$$1.5x = 1.5(32,000) = 48,000$$

Thus, each party pays as follows:

$\boxed{\begin{array}{l}\text{Bundy's: \$88,000; Brady's: \$96,000;} \\ \text{Bachelor: \$32,000; couple: \$48,000}\end{array}}$

31. Let $x =$ "my" age

$$x + \frac{1}{2}x + \frac{1}{3}x + 3(3) = 6(20) + 10$$

$$\frac{11}{6}x + 9 = 120 + 10$$

$$\frac{11}{6}x = 130 - 9$$

$$\frac{11}{6}x = 121$$

$$x = \frac{6}{11}(121)$$

$$x = 66$$

$\boxed{66 \text{ years old}}$

33. Let

$$x = \text{Tom's age when daughter was born}$$
$$2x - 44 = \text{age of Tom's wife when daughter was born}$$

$$x + (2x - 44) = 64$$
$$3x - 44 = 64$$
$$3x = 108$$
$$x = 36$$

Tom's daughter was born when Tom was 36 years old or in 1952 + 36 or 1988. In 2005, Tom's daughter will be 2005 − 1988 = $\boxed{17 \text{ years old}}$.

35. Let

$$x = \text{the number of A's}$$
$$x + 2 = \text{the number of B's}$$
$$3x - 4 = \text{the number of C's}$$
$$x - 4 = \text{the number of D's}$$

$$4x + 3(x + 2) + 2(3x - 4) + 1(x - 4) = 50$$
$$4x + 3x + 6 - 6x - 8 + x - 4 = 50$$
$$14x - 6 = 50$$
$$14x = 56$$
$$x = 4 \quad \text{(A's)}$$
$$x + 2 = 6 \quad \text{(B's)}$$
$$3x - 4 = 12 - 4 = 8 \quad \text{(C's)}$$
$$x - 4 = 4 - 4 = 0 \quad \text{(D's)}$$

Janelle has $\boxed{4 \text{ A's, 6 B's, 8 C's, and no D's}}$.

37. Let

$$x = \text{number of hours the tutor worked}$$
$$30 - x = \text{number of hours the tutor does not work}$$

$$4.40x = 6.60(30 - x)$$
$$4.40x = 198 - 6.60x$$
$$11x = 198$$
$$x = 18$$

$\boxed{18 \text{ hours}}$ worked

39. Let x = the number of guests.

A rice dish for every two guests	plus	A broth dish for every three guests	plus	A meat dish for every four guests	equals	65 dishes
↓	↓	↓	↓	↓	↓	↓
$\frac{x}{2}$	+	$\frac{x}{3}$	+	$\frac{x}{4}$	=	65

$$\frac{x}{2} + \frac{x}{3} + \frac{x}{4} = 65$$
$$12\left(\frac{x}{2} + \frac{x}{3} + \frac{x}{4}\right) = 12(65)$$
$$6x + 4x + 3x = 780$$
$$13x = 780$$
$$x = 60$$

Thus, there were $\boxed{60 \text{ guests}}$.

41. Let

$$x = \text{width of rectangle}$$
$$3x - 11 = \text{length of rectangle}$$

$$2x + 2(3x - 11) = 82$$
$$2x + 6x - 22 = 82$$
$$8x - 22 = 82$$
$$8x = 104$$
$$x = 13$$
$$3x - 11 = 3(13) - 11 = 39 - 11 = 28$$

$\boxed{\text{width: } 13 \text{ inches; Length: } 28 \text{ inches}}$

43. Let

$$x = \text{width of rectangular pool}$$
$$2x + 6 = \text{length of rectangular pool}$$

$$2x + 2(2x + 6) = 228$$
$$2x + 4x + 12 = 228$$
$$6x + 12 = 228$$
$$6x = 216$$
$$x = 36$$
$$2x + 6 = 2(36) + 6 = 72 + 6 = 78$$

$\boxed{\text{width: } 36 \text{ feet; length: } 78 \text{ feet}}$

45. Let

x = length of shorter side of rectangle
$x + 2$ = length of larger side

$$\left(\frac{1}{2}x + 5\right) + 2(x + 2) = 44$$
$$\frac{1}{2}x + 5 + 2x + 4 = 44$$
$$\frac{5}{2}x + 9 = 44$$
$$\frac{5x}{2} = 35$$
$$x = 14$$
$$x + 2 = 16$$

area = $x(x + 2) = 14(16) = 224$

area: $\boxed{224 \text{ cm}^2}$

47. Let

x = length of smaller base of trapezoid
$3x - 1$ = length of each of the nonparallel sides of the trapezoid
$5x + 2$ = length of longer base of trapezoid

$$x + (3x - 1) + (3x - 1) + (5x + 2) = 36$$
$$12x = 36$$
$$x = 3$$
$$3x - 1 = 3(3) - 1 = 8$$
$$5x + 2 = 5(3) + 2 = 17$$

$\boxed{\text{smaller base, 3 yards; each nonparallel side, 8 yards; longer base, 17 yards}}$

49. Let

x = width of rectangle
$x + 6$ = length of rectangle
$2(x + 6)$ = length of new rectangle
$x + 3$ = width of new rectangle

$$2[2(x + 6)] + 2(x + 3) = 26 + [2x + 2(x + 6)]$$
$$2(2x + 12) + 2x + 6 = 26 + 2x + 2x + 12$$
$$4x + 24 + 2x + 6 = 4x + 38$$
$$6x + 30 = 4x + 38$$
$$2x + 30 = 38$$
$$2x = 8$$
$$x = 4$$
$$x + 6 = 4 + 6 = 10$$

dimensions of original rectangle: $\boxed{\text{width: 4 cm; length: 10 cm}}$

51. Let

x = number of oranges in each of the original 3 groups
$3x$ = total number of oranges in original groups
$3x - 11$ = total number of oranges in new group

$$\frac{3x - 11}{2} = 17$$
$$3x - 11 = 34$$
$$3x = 45$$
$$x = 15$$

number in original group: $\boxed{15 \text{ oranges}}$

53. Let

$$
\begin{aligned}
x &= \text{the number of hours of darkness} \\
x + 6\ \text{hr}\ 6\ \text{min} &= \text{the number of hours of daylight}
\end{aligned}
$$

$$
\begin{aligned}
6\ \text{hr}\ 6\ \text{min} &= 6.1\ \text{hours} \\
x + (x + 6.1) &= 24 \\
2x &= 17.9 \\
x &= 8.95 \\
8.95\ \text{hours} &= 8\ \text{hours} + 0.95(60)\ \text{minutes} = 8\ \text{hours}\ 57\ \text{minutes}
\end{aligned}
$$

Darkness is experienced $\boxed{8\ \text{hours}\ 57\ \text{minutes}}$.

55. Let

$$
\begin{aligned}
x &= \text{number of Republicans} \\
5x &= \text{number of Democrats}
\end{aligned}
$$

$$
\begin{aligned}
5x &= 2(x + 12) \\
5x &= 2x + 24 \\
3x &= 24 \\
x &= 8 \\
5x &= 40
\end{aligned}
$$

$\boxed{40\ \text{Democrats}}$

57. Let

$$
\begin{aligned}
x &= \text{length of base of isosceles triangle} \\
x + 3 &= \text{length of each equal side of isosceles triangle}
\end{aligned}
$$

$$
\begin{aligned}
x + (x + 3) + (x + 3) &= 27 \\
3x + 6 &= 27 \\
3x &= 21 \\
x &= 7 \\
x + 3 &= 7 + 3 = 10
\end{aligned}
$$

$\boxed{\text{Base: 7 feet; each equal side: 10 feet}}$

59. Let

$$
\begin{aligned}
x &= \text{length of } BE \\
x &= \text{length of } EC \\
x + 2 &= \text{length of } AE \\
x + 1 &= \text{length of } AC \\
2x &= \text{length of } AB
\end{aligned}
$$

$$
\begin{aligned}
AB + BE + EC + AC + AE &= 45 \\
2x + x - x + (x + 1) + (x + 2) &= 45 \\
6x + 3 &= 45 \\
6x &= 42 \\
x &= 7 \\
2x &= 14
\end{aligned}
$$

length of AB: $\boxed{14\ \text{m}}$

61. Let

$$x = \text{amount of sqaure feet of wall painted}$$
$$30(10) - x = \text{amount of square feet of wall papered}$$

$$
\begin{aligned}
1.40x + 0.60(300 - x) &= 324 \\
1.40x + 180 - 0.60x &= 324 \\
0.80x + 180 &= 324 \\
0.80x &= 144 \\
x &= 180
\end{aligned}
$$

amount of wall painted: $\boxed{180 \text{ ft}^2}$

63. Let $x =$ the number of condominiums Reuben owns.

Miami: $\dfrac{1}{2}x$

Atlanta: $\dfrac{1}{8}x$

New York: $\dfrac{1}{12}x$

Chicago: $\dfrac{1}{20}x$

San Francisco: $\dfrac{1}{30}x$

Boulder: 50

$$x = \frac{1}{2}x + \frac{1}{8}x + \frac{1}{12}x + \frac{1}{20}x + \frac{1}{30}x + 50$$

Multiplying both sides of 120:

$$
\begin{aligned}
120x &= 60x + 15x + 10x + 6x + 4x + 6000 \\
120x - 95x &= 6000 \\
25x &= 6000 \\
x &= 240
\end{aligned}
$$

Reuben owns $\boxed{240 \text{ condominiums}}$.

65. Let $x =$ the amount Nathan had before finding $2.

$$
\begin{aligned}
(\text{amount with \$2}) &= 5(\text{amount would have had if lost \$2}) \\
x + 2 &= 5(x - 2) \\
x + 2 &= 5x - 10 \\
12 &= 4x \\
3 &= x
\end{aligned}
$$

Nathan had $\boxed{\$3}$ originally.

67. Let $x =$ amount of money you had originally.

amount of money spent: $\dfrac{1}{3}x$

amount left: $x - \dfrac{1}{3}x = \dfrac{2}{3}x$

amount of money lost: $\dfrac{2}{3}\left(\dfrac{2}{3}x\right) = \dfrac{4}{9}x$

amount left: $\dfrac{2}{3}x - \dfrac{4}{9}x = \dfrac{6}{9}x - \dfrac{4}{9}x = \dfrac{2}{9}x$

$$
\begin{aligned}
\frac{2}{9}x &= 12 \\
x &= 54
\end{aligned}
$$

$\boxed{\text{original amount of money, \$54}}$

69. Let x = the man's final age.

$$\left.\begin{array}{l} \text{childhood: } \dfrac{1}{6}x \\[3ex] \text{youth: } \dfrac{1}{12}x \end{array}\right\} \text{extraneous facts}$$

childless marriage: $\dfrac{1}{7}x$; 5 years

$$\frac{1}{7}x = 5$$
$$x = 35$$

$\boxed{\text{man's final age, 35 years old}}$

71. Let

$\quad x$ = the total number of votes cast

$\quad 180$ = number of votes for candidate A

$\quad 0.40x + 40$ = the number of votes for candidate B

$\quad 0.50(\text{candidate B}) - 40 = 0.50(0.40x - 40) = 0.20x + 20$ = number of votes for candidate C

$\quad 0.50(\text{candidate C}) + 20 = 0.50(0.20x + 20) + 20 = 0.10x + 10 + 20 = 0.10x + 30$ = number of votes cast

for candidate D.

$$\begin{aligned} x - (\text{candidate B} + \text{candidate C} + \text{candidate D}) &= (\text{candidate A}) \\ x - [(0.40x + 40) + (0.20x + 20) + (0.10x + 30)] &= 180 \\ x - (0.70x + 90) &= 180 \\ 0.30x - 90 &= 180 \\ 0.30x &= 270 \\ x &= 900 \end{aligned}$$

candidate A: 180

candidate B: $0.40x + 40 = 360 + 40 = 400$

candidate C: $0.20x + 20 = 180 + 20 = 200$

candidate D: $0.10x + 30 = 90 + 20 = 120$

$\boxed{\text{900 votes cast; candidate B won}}$

Review Problems

76. $P = \dfrac{100(p + q)}{pq}$

$(P = 300, q = -50)$:

$P = \dfrac{100(300 - 50)}{300(-50)}$

$ = \left(\dfrac{100}{300}\right)\left(\dfrac{250}{-50}\right)$

$ = \dfrac{1}{3}(-5)$

$ = \boxed{-\dfrac{5}{3}}$

77. $D = A(n - 1)$

$\dfrac{D}{A} = n - 1$

$\dfrac{D}{A} + 1 = n$

$\boxed{n = \dfrac{D}{A} + 1 \text{ or } n = \dfrac{D + A}{A}}$

78. $10^3 + 10^0 - 10^{-1} = 1000 + 1 - \dfrac{1}{10} = 1001 - 0.1 = \boxed{1000.9}$

Section 2.4 More Problem Solving

Problem Set 2.4, pp. 125-128

1. Let

$$x = \text{number of dimes}$$
$$20 - x = \text{number of quarters}$$

	Face value	Number of Coins	Value of coins
dimes	10¢	x	$10(x)$
quarters	25¢	$20 - x$	$25(20 - x)$

$$
\begin{aligned}
10x + 25(20 - x) &= 320 \\
10x + 500 - 25x &= 320 \\
-15x &= -180 \\
x &= 12 \\
20 - x &= 20 - 12 = 8
\end{aligned}
$$

$\boxed{12 \text{ dimes, } 8 \text{ quarters}}$

3. Let

$$x = \text{number of nickels}$$
$$x + 5 = \text{number of quarters}$$

	Face value	Number of coins	Value of coins
nickels	5¢	x	$5x$
quarters	25¢	$x - 3$	$25(x + 5)$

$$
\begin{aligned}
5x + 25(x + 5) &= 215 \\
5x + 25x + 125 &= 215 \\
30x &= 90 \\
x &= 3 \\
x + 5 &= 3 + 5 = 8
\end{aligned}
$$

$\boxed{3 \text{ nickels, } 8 \text{ quarters}}$

5. Let

$$x = \text{number of half-dollars}$$
$$x - 3 = \text{number of quarters}$$
$$3x + 2 = \text{number of dimes}$$

	Face value	Number of coins	Value of coins
half-dollar	50¢	x	$50x$
quarters	25¢	$x - 3$	$25(x - 3)$
dimes	10¢	$3x + 2$	$10(3x + 2)$

$$
\begin{aligned}
50x + 25(x - 3) + 10(3x + 2) &= 680 \\
50x + 25x - 75 + 30x + 20 &= 680 \\
105x - 55 &= 680 \\
105x &= 735 \\
x &= 7 \\
x - 3 &= 4 \\
3x + 2 &= 3(7) + 2 = 23
\end{aligned}
$$

$\boxed{7 \text{ half-dollars, } 4 \text{ quarters, } 23 \text{ dimes}}$

7. Let

$$
\begin{aligned}
x &= \text{number of \$5 bills} \\
x + 6 &= \text{number of \$10 bills} \\
4x - 3 &= \text{number of \$20 bills}
\end{aligned}
$$

	Face value	Number of bills	Value of bills
$5 bills	$5	x	$5x$
$10 bills	$10	$x + 6$	$10(x + 6)$
$20 bills	$20	$4x - 3$	$20(4x - 3)$

$$
\begin{aligned}
5x + 10(x + 6) + 20(4x - 3) &= 665 \\
5x + 10x + 60 + 80x - 60 &= 665 \\
95x &= 665 \\
x &= 7 \\
4x - 3 &= 4(7) - 3 = 28 - 3 = 25
\end{aligned}
$$

$\boxed{25}$ $20 bills

9. Let

$$
\begin{aligned}
x &= \text{amount invested at 8\%} \\
20{,}000 - x &= \text{amount invested at 5\%}
\end{aligned}
$$

	P ·	r ·	t =	Interest
8% stock	x	0.08	2	$0.08(2)x$
5% stock	$20{,}000 - x$	0.05	3	$0.05(3)(20{,}000 - x)$

$$
\begin{aligned}
0.08(2)x + 0.05(3)(20{,}000 - x) &= 3160 \\
0.16x + 3000 - 0.15x &= 3160 \\
0.01x &= 160 \\
x &= 16{,}000 \\
20{,}000 - x &= 4000
\end{aligned}
$$

$\boxed{\$16{,}000 \text{ at } 8\% \text{ for 2 years; } \$4000 \text{ at } 5\% \text{ for 3 years}}$

11. Let

$$
\begin{aligned}
x &= \text{amount invested at 9\%} \\
35{,}000 - x &= \text{amount invested at 6\%}
\end{aligned}
$$

	P ·	r ·	t =	Interest
9% stock	x	0.09	2	$0.09(2)x$
6% stock	$35{,}000 - x$	0.06	4	$0.06(4)(35{,}000 - x)$

$$
\begin{aligned}
0.09(2)x &= 0.06(4)(35{,}000 - x) \\
0.18x &= 8400 - 0.24x \\
0.42x &= 8400 \\
x &= 20{,}000 \\
35{,}000 - x &= 35{,}000 - 20{,}000 = 15{,}000
\end{aligned}
$$

$\boxed{\$20{,}000 \text{ at } 9\% \text{ for 2 years; } \$15{,}000 \text{ at } 6\% \text{ for 4 years}}$

$$
\begin{aligned}
\text{Interest:} \quad 0.09(2)(20{,}000) &= 3600 \\
0.06(4)(15{,}000) &= 3600
\end{aligned}
$$

Total interest: $3600 + $3600 = $\boxed{\$7200}$

13. Let

$$
\begin{aligned}
x &= \text{amount invested at 5\%} \\
x &= \text{amount invested at 7\%} \\
30,000 - 2x &= \text{amount invested at 9\%}
\end{aligned}
$$

$$
\begin{aligned}
0.05(2)x + 0.07(2)x + 0.09(2)(30,000 - 2x) &= 4920 \\
0.10x + 0.14x + 5400 - 0.36x &= 4920 \\
0.12x &= 480 \\
x &= 4000 \\
30,000 - 2x &= 30,000 - 8000 = 22,000
\end{aligned}
$$

$\boxed{\text{\$4000 each at 5\% and 7\%; \$22,000 at 9\%}}$

15. Let

$$
\begin{aligned}
x &= \text{amount invested at 5\%} \\
50,000 - x &= \text{amount invested at 16\%} \quad \text{(after 1 year, loses at 8\%)}
\end{aligned}
$$

$$
\begin{aligned}
-0.08(50,000 - x) + 0.05(3)x &= 5200 \\
-4000 + 0.08x + 0.15x &= 5200 \\
0.23x &= 9200 \\
x &= 40,000 \\
50,000 - x &= 10,000
\end{aligned}
$$

$\boxed{\text{\$40,000 at 5\%; \$10,000 at 16\%}}$

17. Let $x =$ number of hours when the planes are 2500 miles apart.

	Rate, R	$\cdot$	Time, t	$=$	Distance, D
faster plane	300		x		$300x$
slower plane	200		x		$200x$

$$
\begin{aligned}
300x + 200x &= 2500 \\
500x &= 2500 \\
x &= 5
\end{aligned}
$$

$\boxed{\text{5 hours}}$

19. Let

$$
\begin{aligned}
x &= \text{rate of slowre truck} \\
x + 5 &= \text{rate of faster truck}
\end{aligned}
$$

	Rate, R	$\cdot$	Time, t	$=$	Distance, D
slower truck	x		5		$5x$
faster truck	$x + 5$		5		$5(x + 5)$

$$
\begin{aligned}
5x + 5(x + 5) &= 600 \\
5x + 5x + 25 &= 600 \\
10x &= 575 \\
x &= 57.5 \\
x + 5 &= 62.5
\end{aligned}
$$

$\boxed{\text{speed of slower truck: 57.5 mph; speed of faster truck: 62.5 mph}}$

21. Let
$$x = \text{Bob's rate}$$
$$x + 5 = \text{Joe's rate}$$

	Rate, R $\cdot$	Time, t $=$	Distance, D
Bob	x	4	$4x$
Joe	$x + 5$	4	$4(x + 5)$

$$
\begin{aligned}
4x + 4(x + 5) &= 420 \\
4x + 4x + 20 &= 420 \\
8x &= 400 \\
x &= 50 \\
x + 5 &= 55
\end{aligned}
$$

$\boxed{\text{Bob, 50 mph; Joe, 55 mph}}$

23. Let
$$x = \text{time out going}$$
$$10 - x = \text{time returning}$$

	Rate, R $\cdot$	Time, t $=$	Distance, D
out going trip	20	x	$20x$
returning trip	30	$10 - x$	$30(10 - x)$

$$
\begin{aligned}
\text{distance going} &= \text{distance returning} \\
20x &= 30(10 - x) \\
20x &= 300 - 30x \\
50x &= 300 \\
x &= 6 \\
10 - x &= 4
\end{aligned}
$$

$\boxed{\text{out going: 6 hr; return: 4 hr}}$

distance $= 20x = 20(6) = 120$

$\boxed{\text{120 miles each way}}$

25. Let
$$x = \text{time for second traveler}$$
$$x + 3 = \text{time for first traveler}$$

	Rate, R $\cdot$	Time, t $=$	Distance, D
first traveler	$x + 3$	50	$50(x + 3)$
second traveler	x	60	$60x$

$$
\begin{aligned}
\text{distance of first traveler} &= \text{distance of second traveler} \\
50(x + 3) &= 60x \\
50x + 150 &= 60x \\
150 &= 10x \\
15 &= x
\end{aligned}
$$

The second traveler will overtake the first traveler in $\boxed{\text{15 hours}}$.

27. Let

$$t = \text{time jogging at 6 mph (in hours)}$$
$$2\tfrac{1}{2} - t = \text{time running at 10 mph (in hours)}$$

	Rate, R $\cdot$	Time, t $=$	Distance, D
jogging	6	t	$6t$
running	10	$2\tfrac{1}{2} - t$	$100\left(\tfrac{5}{2} - t\right)$

(distance to jog) + (distance to run) = (total distance)

$$6t + 10\left(\tfrac{5}{2} - t\right) = 23$$
$$6t + 25 - 10t = 23$$
$$-4t + 25 = 23$$
$$-4t = -2$$
$$t = \tfrac{1}{2} \quad \text{(jogging)}$$
$$2\tfrac{1}{2} - t = 2 \quad \text{(running)}$$

distance jogging: $6t = 6\left(\tfrac{1}{2}\right) = 3$

distance running: $10\left(2\tfrac{1}{2} - t\right) = 10(2) = 20$

jogging: 30 min, 3 miles; running: 2 hr, 20 miles

29. Let

$$r = \text{swimming speed}$$
$$10r = \text{cycling speed}$$
$$\tfrac{1}{2}(10r) = \text{running speed}$$

	Rate, R $\cdot$	Time, t $=$	Distance, D
swim	r	1	$1(r)$
cycled	$10r$	3	$3(10r)$
ran	$5r$	2	$2(5r)$

$$r + 30r + 10r = 82$$
$$41r = 82$$
$$r = 2 \quad \rightarrow \text{ distance: } 1(r) = 2$$
$$10r = 20 \quad \rightarrow \text{ distance: } 30r = 60$$
$$5r = 10 \quad \rightarrow \text{ distance: } 10r = 20$$

swimming: 2 mph, 2 mi;
cycling: 20 mph, 60 mi;
running: 10 mph, 20 mi

31. Let

$$x = \text{number of liters of solution A(30\% acid)}$$
$$60 - x = \text{number of liters of solution B(60\% acid)}$$

	Number of liters	·	Percentage of acid	=	Amount of acid
solution A (30% acid)	x		30% (0.30)		$0.30x$
solution B (60% acid)	$60 - x$		60% (0.60)		$0.60(60 - x)$
50% acid solution	60		50% (0.50)		$0.50(60)$

$$0.30x + 0.60(60 - x) = 0.50(60)$$
$$0.30x + 36 - 0.60x = 30$$
$$-0.30x = -6$$
$$x = 20 \quad \text{(solution A)}$$
$$60 - x = 60 - 20 = 40 \quad \text{(solution B)}$$

$\boxed{\text{20 gallon solution A; 40 gallon solution B}}$

33. Let $x =$ number of liters of 80% glucose solution.

	Number of liters	·	Percentage of glucose	=	Amount of glucose
80% solution	x		0.80		$0.80x$
40% solution	30		0.40		$0.40(30)$
50% solution	$x + 30$		0.50		$0.50(x + 30)$

$$0.80x + 0.40(30) = 0.50(x + 30)$$
$$0.80x + 12 = 0.50x + 15$$
$$0.30x = 3$$
$$x = 10 \quad \text{(80\% solution)}$$
$$x + 30 = 40 \quad \text{(mixture)}$$

$\boxed{\text{10 liters at 80\%; 40 liters in the mixture}}$

35. Let $x =$ number of cubic millimeters (at 0% alcohol)

	Number of cubic of millimeters	·	Percentage of alcohol	=	Amount of alcohol
water (0% alcohol)	x		0		$0x$
40% solution	50		0.40		$0.4(50)$
16% solution	$x + 50$		0.16		$0.16(x + 15)$

$$0x + 0.40(50) = 0.16(x + 15)$$
$$20 = 0.16x + 8$$
$$12 = 0.16x$$
$$75 = x \quad \text{(water)}$$

$\boxed{\text{75 mm}^3 \text{ of water}}$

37. Let

$$x = \text{number of gallons of pure antifreeze to be added (100\% solution)}$$
$$5 - x = \text{number of gallons of 40\% solution left after } x \text{ gallons drained}$$

	Number of gallons	·	Percentange of antifreeze	=	Amount of antifreeze
pure antifreeze (100% solution)	x		100%(1.00)		$1.00x$
40% solution	$5 - x$		40%(0.40)		$0.40(5 - x)$
50% solution	5		50%(0.50)		0.50(5)

$$
\begin{aligned}
1.00x + 0.40(5 - x) &= 0.50(5) \\
x + 2 - 0.40x &= 2.5 \\
0.6x &= 0.5 \\
x &= \frac{5}{6} \quad \text{(pure antifreeze)}
\end{aligned}
$$

$\boxed{\dfrac{5}{6}\text{ gal}}$ of pure antifreeze to be added

39. $\boxed{C}$ is true; 45 mintues $= \dfrac{3}{4}$ hour

$$
\begin{aligned}
\text{distance} &= \text{rate} \cdot \text{time} \\
&= x\left(\frac{3}{4}\right) \text{ miles} \\
&= \frac{3}{4}x \text{ miles}
\end{aligned}
$$

41. Let

$$
\begin{aligned}
r &= \text{athlete's running speed} \\
2r &= \text{athlete's biking speed} \\
0.40r &= \text{athlete's walking speed}
\end{aligned}
$$

Let "+" indicate initial direction and "−" indicate opposite direction.

	Rate, R	·	Time, t	=	Distance, D
running	r		8		$r(8)$
biking	$2r$		3		$2r(3)$ (in opposite direction)
walking	$0.40r$		1		$0.40r(1)$

$$
\begin{aligned}
8r - 3(2r) + 0.40r &= 9.6 \\
8r - 6r + 0.4r &= 9.6 \\
2.4r &= 9.6 \\
r &= 4 \quad \text{(running)} \\
2r &= 8 \quad \text{(biking)} \\
0.40r &= 0.40(4) = 1.6 \quad \text{(walking)}
\end{aligned}
$$

$\boxed{\text{running, 4 mph; biking(riding), 8 mph; walking, 1.6 mph}}$

43. Let

$$x = \text{number of questions in first part of test}$$
$$50 - x = \text{number of questions in second part of test}$$

	number of questions	$\cdot$	percent correct	$=$	number correct
first part	x		80%(0.80)		$0.80x$
second part	$50 - x$		20%(0.20)		$0.20(50 - x)$
total	50		65% − 3% = 62%		0.62(50)

$$
\begin{aligned}
0.80x + 0.20(50 - x) &= 0.62(50) \\
0.80x + 10 - 0.20x &= 31 \\
0.60x &= 21 \\
x &= 35 \quad \text{(first part)} \\
50 - x &= 50 - 35 = 15 \quad \text{(second part)}
\end{aligned}
$$

first part, 35; second part, 15

45. Let

$$x = \text{number of half-dollars}$$
$$2x = \text{number of quarters}$$
$$\tfrac{1}{2}x = \text{number of dollar bills}$$

	Face value	$\cdot$	Number of coins or bills	$=$	Value of coin or bill
half-dollar	$0.50		x		$0.50x$
quarters	$0.25		$2x$		$0.25(2x)$
dollar bills	$1.00		$\tfrac{1}{2}x$		$1.00\left(\tfrac{1}{2}x\right)$

$$
\begin{aligned}
0.50x + 0.25(2x) + 1.00\left(\tfrac{1}{2}x\right) &= 20.00 \\
0.50x + 0.50x + 0.50x &= 20.00 \\
1.50x &= 20.00 \\
x &= 13.\overline{3} \\
2x &= 26.\overline{6} \\
\tfrac{1}{2}x &= 6.\overline{6}
\end{aligned}
$$

Since the number of coins or bills must be an integer and $13.\overline{3}$ is *not* an integer, it is not possible to make change for a $20 bill with the given restrictions.

Review Problems

51.
$$
\begin{aligned}
ay &= 3r - by \\
ay + by &= 3r \\
y(a + b) &= 3r \\
y &= \frac{3r}{a + b}
\end{aligned}
$$

52. $4a + 3b = 3b + a4$

Commutative properties of both addition and multiplication

53.

$$\frac{3x-2}{4} - \frac{x-2}{3} = \frac{13}{4} - \frac{10x-8}{12}$$

$$12\left(\frac{3x-2}{4} - \frac{x-2}{3}\right) = 12\left(\frac{13}{4} - \frac{10x-8}{12}\right)$$

$$3(3x-2) - 4(x-2) = 3(13) - (10x-8)$$

$$9x - 6 - 4x + 8 = 39 - 10x + 8$$

$$5x + 2 = -10x + 47$$

$$15x = 45$$

$$x = 3$$

$$\boxed{\{3\}}$$

Section 2.5 Solving Linear Inequalities

Problem Set 2.5, pp. 135-136

1. $x + 7 > 10$
   $x > 3$
   $\boxed{\{x \mid x > 3\}}$
   $\boxed{(3, \infty)}$

3. $x - 5 \le -7$
   $x \le -2$
   $\boxed{\{x \mid x \le -2\}}$
   $\boxed{(-\infty, -2]}$

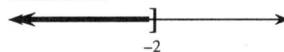

5. $3(x + 1) - 5 < 2x + 1$
   $3x + 3 - 5 < 2x + 1$
   $3x - 2 < 2x + 1$
   $x < 3$
   $\boxed{\{x \mid x < 3\}}$
   $\boxed{(-\infty, 3)}$

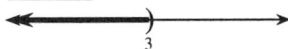

7. $8x + 3 > 3(2x + 1) + x + 5$
   $8x + 3 > 6x + 3 + x + 5$
   $8x + 3 > 7x + 8$
   $x > 5$
   $\boxed{\{x \mid x > 5\}}$
   $\boxed{(5, \infty)}$

9. $8x - 2(2x + 3) \ge 0$
   $8x - 4x - 6 \ge 0$
   $4x \ge 6$
   $x \ge \dfrac{3}{2}$
   $\boxed{\left\{ x \mid x \ge \dfrac{3}{2} \right\}}$
   $\boxed{\left[\dfrac{3}{2}, \infty\right)}$

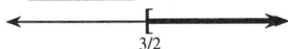

11. $12\left(\dfrac{1}{4} + \dfrac{x}{3}\right) < 15$
   $3 + 4x < 15$
   $4x < 12$
   $x < 3$
   $\boxed{\{x \mid x < 3\}}$
   $\boxed{(-\infty, 3)}$

13. $\dfrac{1}{4}(x+3) \;<\; 4x - 2(x-3)$

$12\left[\dfrac{1}{4}(x+3)\right] \;<\; 12[4x - 2(x-3)]$

$3(x+3) \;<\; 48x - 24(x-3)$

$3x + 9 \;<\; 48x - 24x + 72$

$3x + 9 \;<\; 24x + 72$

$-21x \;<\; 63$

$x \;>\; -3$

$\boxed{\{x \mid x > -3\}}$

$\boxed{(-3, \infty)}$

15. $7(y+4) - 13 \;<\; 12 + 13(3+y)$

$7y + 28 - 13 \;<\; 12 + 39 + 13y$

$7y + 15 \;<\; 13y + 51$

$-6y \;<\; 36$

$y \;>\; -6$

$\boxed{\{y \mid y > -6\}}$

$\boxed{(-6, \infty)}$

17. $3[3(y+5) + 8y + 7] + 5[3(y-6) - 2(3y-5)] \;<\; 2(4y+3)$

$3(3y + 15 + 8y + 7) + 5(3y - 18 - 6y + 10) \;<\; 8y + 6$

$3(11y + 22) + 5(-3y - 8) \;<\; 8y + 6$

$33y + 66 - 15y - 40 \;<\; 8y + 6$

$18y + 26 \;<\; 8y + 6$

$10y \;<\; -20$

$y \;<\; -2$

$\boxed{\{y \mid y < -2\}}$

$\boxed{(-\infty, -2)}$

19. $4(3y+2) - 3y \;<\; 3(1+3y) - 7$

$12y + 8 - 3y \;<\; 3 + 9y - 7$

$9y + 8 \;<\; 9y - 4$

$8 \;<\; -4 \qquad$ False

No solution

$\boxed{\varnothing}$

21. $\dfrac{5}{6}[3(y+2) - 2(y+1)] \;<\; \dfrac{2}{3}\left[\dfrac{1}{2}(y-4) - 2(y-9)\right]$

$(\times 12):\quad 10[3(y+2) - 2(y+1)] \;<\; 8\left[\dfrac{1}{2}(y-4) - 2(y-9)\right]$

$10(3y + 6 - 2y - 2) \;<\; 8\left(\dfrac{1}{2}y - 2 - 2y + 18\right)$

$10(y+4) \;<\; 8\left(-\dfrac{3}{2}y + 16\right)$

$10y + 40 \;<\; -12y + 128$

$22y \;<\; 88$

$y \;<\; 4$

$\boxed{\{y \mid y < 4\}}$

$\boxed{(-\infty, 4)}$

23. Let x = number of pounds of uranium.
$$2x - 6 \; < \; 18$$
$$2x \; < \; 24$$
$$x \; < \; 12$$
$\boxed{\text{less than 12 lb}}$

25. Let x = the number.
$$8x - 1 \; \leq \; 29 + 3x$$
$$5x \; \leq \; 30$$
$$x \; \leq \; 6$$
$\boxed{\{x \mid x \leq 6\}}$

27. Let x = the number.
$$8x \; > \; 4x + 6$$
$$4x \; > \; 6$$
$$x \; > \; \frac{3}{2}$$
$\boxed{\left\{ x \mid x > \dfrac{3}{2} \right\}}$

29. $\boxed{B}$ is true; Let
$$x \; = \; \text{Mia's age}$$
$$2x + 3 \; = \; \text{Eleanor's age}$$
$$x + (2x + 3) \; \geq \; 24$$
$$3x + 3 \; \geq \; 24$$
$$3x \; \geq \; 21$$
$$x \; \geq \; 7 \quad \text{(Mia} \geq 7)$$

31. Let x = monthly sales
$$0.30(x - 1000) \; > \; 700$$
$$0.30x - 300 \; > \; 700$$
$$0.30x \; > \; 1000$$
$$x \; > \; 3333.\overline{3}$$
$\boxed{\text{more than } \$3{,}333.34}$

33. Let x = number of kilometers.
agency A: $C = 0.15x + 4$ dollars
agency B: $C = 0.05x + 20$ dollars

$$(\text{agency A}) \; < \; (\text{agency B})$$
$$0.15x + 4 \; < \; 0.05x + 20$$
$$0.10x \; < \; 16$$
$$x \; < \; 160$$
$\boxed{\text{fewer than 160 km}}$

35. Let x = number of raffle tickets.
$$3x + 10 \; \leq \; 25$$
$$3x \; \leq \; 15$$
$$x \; \leq \; 5$$
maximum number of raffle tickets: $\boxed{5}$

37. Let x = number of days.
$$N = 400 - 8x \; < \; 328$$
$$400 - 8x \; < \; 328$$
$$-8x \; < \; -72$$
$$x \; > \; 9$$
$\boxed{\text{more than 9 days}}$

39. Let x = number of hours.
first club: $500 + 1(x)$
second club: $440 + 1.75(x)$

$$(\text{first club}) \; < \; (\text{second club})$$
$$500 + x \; < \; 440 + 1.75x$$
$$-0.75x \; < \; -60$$
$$x \; > \; 80$$
$\boxed{\text{more than 80 hr}}$

41. Let x = number of checks.
bank: $8 + 0.05x$
credit union: $2 + 0.08x$

$$(\text{bank}) \; > \; (\text{credit union})$$
$$8 + 0.05x \; > \; 2 + 0.08x$$
$$-0.03x \; > \; -6$$
$$x \; < \; 200$$
$\boxed{\text{fewer than 200 checks}}$

43. Let

$$x = \text{number of dimes}$$
$$12 - x = \text{number of quarters}$$

$$
\begin{aligned}
0.10x + 0.25(12 - x) &< 1.95 \\
0.10x + 3 - 0.25x &< 1.95 \\
-0.15x &< -1.05 \\
x &> 7 \quad \text{(dimes)}
\end{aligned}
$$

Possible combinations:

dimes	x	8	9	10	11	12
quarters	$12 - x$	4	3	2	1	0

> 8 dimes, 4 quarters;
> 9 dimes, 3 quarters;
> 10 dimes, 2 quarters;
> 11 dimes, 1 quarter;
> 12 dimes, no quarters

45. $\boxed{B}$ is true;

$$
\begin{aligned}
\frac{7 - 5x}{-2} &\geq -1 \\
7 - 5x &\leq 2 \\
-5x &\leq -5 \\
x &\geq 1 \\
4 &\in \{x \mid x \geq 1\}
\end{aligned}
$$

47. Let x = the cost of meat pie.
Each tourist had $x - 7$ and $x - 2$ dollars.
Together they had $(x - 7) + (x - 2)$ or $2x - 9$ dollars, where

$$
\begin{aligned}
2x - 9 &< x \\
x &< 9
\end{aligned}
$$

Since the first tourist needed $7 more, $x > 7$.
Since $x > 7$ and $x < 9$,

$$
\begin{aligned}
7 &< x < 9 \\
x &= 8
\end{aligned}
$$

and the meat pie costs $8.00.

> more than $7, less than $9
> cost of meat pie, $8

Review Problems

51.

$$
\begin{aligned}
y - (3 + y) &= 5 + \frac{2(y - 2)}{4} \\
y - 3 - y &= 5 + \frac{1}{2}(y - 2) \\
-3 &= 5 + \frac{1}{2}y - 1 \\
-3 &= 4 + \frac{1}{2}y \\
-7 &= \frac{1}{2}y \\
-7 &= \frac{1}{2}y \\
-14 &= y
\end{aligned}
$$

$$\boxed{\{-14\}}$$

52.
$$w = \frac{11}{2}h - 220$$
$$2w = 11h - 440$$
$$-11h = -2w - 440$$

$$\boxed{h \approx \frac{1}{11}(2w + 440) \text{ or } \frac{2}{11}(w + 220)}$$

$(w = 154:)$

$$h = \frac{2}{11}(154 + 220) = \frac{2}{11}(374) = 2(34) = 68$$

$$\boxed{68 \text{ inches}}$$

53. $-4x^2 - [6y^2 - (5x^2 - 2y^2)]$
$= -4x^2 - (6y^2 - 5x^2 + 2y^2)$
$= -4x^2 - (8y^2 - 5x^2)$
$= -4x^2 - 8y^2 + 5x^2$
$= \boxed{x^2 - 8y^2}$

Section 2.6 Compound Inequalities

Problem Set 2.6, pp. 145-147

1. $\{x \mid 5x < -20)\} \cap \{x \mid 3x > -18\}$
$$5x < -20 \qquad 3x > -18$$
$$x < -4 \qquad x > -6$$

$$\boxed{\{x \mid -6 < x < -4\}}$$

$$\boxed{(-6, -4)}$$

3. $\{y \mid y - 4 \le 2\} \cap \{y \mid 3y + 1 > -8\}$
$$y - 4 \le 2 \qquad 3y + 1 > -8$$
$$y \le 6 \qquad 3y > -9$$
$$\qquad\qquad y > -3$$

$$\boxed{\{y \mid -3 < y \le 6\}}$$

$$\boxed{(-3, 6]}$$

5. $2x > 5x - 15$ and $7x > 2x + 10$
$$-3x > -15 \qquad 5x > 10$$
$$x < 5 \qquad x > 2$$
$$\{x \mid x < 5\} \cap \{x \mid x > 2\}$$

$$\boxed{x \mid 2 < x < 5\}}$$

$$\boxed{(2, 5)}$$

7. $\{y \mid 4(1-y) < -6\} \cap \left\{ y \mid \dfrac{y-7}{5} \leq -2 \right\}$

$$4(1-y) < -6 \qquad\qquad \dfrac{y-7}{5} \leq -2$$
$$4 - 4y < -6 \qquad\qquad y - 7 \leq -10$$
$$-4y < -10 \qquad\qquad\qquad y \leq -3$$
$$y > \dfrac{5}{2}$$

$y \geq \dfrac{5}{2}$ *and* $y \leq -3$ must be satisfied.

No real numbers are less than -3 and also greater than $\dfrac{5}{2}$.

The solution is $\boxed{\varnothing}$. $\boxed{\text{No solution}}$

9. $\{x \mid x - 1 \leq 7x - 1\} \cap \{x \mid 4x - 7 < 3 - x\}$
$$x - 1 \leq 7x - 1 \qquad\qquad 4x - 7 < 3 - x$$
$$-6x \leq 0 \qquad\qquad\qquad 5x < 10$$
$$x \geq 0 \qquad\qquad\qquad\qquad x < 2$$
$$\{x \mid x \geq 0\} \cap \{x \mid x < 2\}$$

$\boxed{\{x \mid 0 \leq x < 2\}}$

$\boxed{[0, 2)}$

11. $\left\{ x \mid \dfrac{9 + 4x}{3} > -5 \right\} \cap \left\{ x \mid \dfrac{x}{3} + 4 < 3 \right\}$

$$\dfrac{9 + 4x}{3} > -5 \qquad\qquad \dfrac{x}{3} + 4 < 3$$
$$9 + 4x > -15 \qquad\qquad \dfrac{x}{3} < -1$$
$$4x > -24 \qquad\qquad\qquad x < -3$$
$$x > -6$$
$$\{x \mid x > -6\} \cap \{x \mid x < -3\}$$

$\boxed{x \mid -6 < x < -3\}}$

$\boxed{(-6, -3)}$

13. $2 < x - 5 \leq 7$
$$7 < x \leq 12$$

$\boxed{\{x \mid 7 < x \leq 12\}}$

$\boxed{(7, 12]}$

15. $-5 \leq 2x - 7 < 9$
$$2 \leq 2x < 16$$
$$1 \leq x < 8$$

$\boxed{\{x \mid 1 \leq x < 8\}}$

$\boxed{[1, 8)}$

17. $-6 \; < \; \frac{1}{2}y - 4 < -3$

  $-2 \; < \; \frac{1}{2}y < 1$

  $-4 \; < \; y < 2$

  $\boxed{\{y \mid -4 < y < 2\}}$

  $\boxed{(-4, 2)}$

19. $-2 \; < \; -3y \leq 3$

  $\frac{2}{3} \; > \; y \geq -1$

  $-1 \; \leq \; -y < \frac{2}{3}$

  $\boxed{\left\{ y \mid -1 \leq y < \frac{2}{3} \right\}}$

  $\boxed{\left[-1, \frac{2}{3} \right)}$

21. $-8 \; < \; 2 - 3y < 10$

  $-10 \; < \; -3y < 8$

  $\frac{10}{3} \; < \; y > -\frac{8}{3}$

  $-\frac{8}{3} \; < \; y < \frac{10}{3}$

  $\boxed{\left\{ y \mid -\frac{8}{3} < y < \frac{10}{3} \right\}}$

  $\boxed{\left(-\frac{8}{3}, \frac{10}{3} \right)}$

23. $-2 \; \leq \; -\frac{1}{2}y + 3 \leq 6$

  $-5 \; \leq \; -\frac{1}{2}y \leq 3$

  $10 \; \geq \; y \geq -6$

  $-6 \; \leq \; y \leq 10$

  $\boxed{\{y \mid -6 \leq y \leq 10\}}$

  $\boxed{[-6, 10]}$

25. $-7 \; \leq \; 8 - 3y \leq 20 \quad$ and $\quad -7 \; < \; 6y - 1 < 41$

  $-15 \; \leq \; -3y \leq 12 \qquad\qquad -6 \; < \; 6y < 42$

  $5 \; \geq \; y \geq -4 \qquad\qquad\quad -1 \; < \; y < 7$

  $-4 \; \leq \; y \leq 5$

  $\{y \mid -4 \leq y \leq 5\} \; \cap \; \{y \mid -1 < y < 7\}$

  $\boxed{\{y \mid -1 < y \leq 5\}}$

  $\boxed{(-1, 5]}$

27. $2y - 3 < 3y + 1 < 6y + 2$

$\{y \mid 3y + 1 > 2y - 3\} \quad \cap \quad \{y \mid 3y + 1 < 6y + 2\}$

$\quad 3y + 1 > 2y - 2 \qquad\qquad 3y + 1 < 6y + 2$

$\qquad y > -4 \qquad\qquad\qquad -3y < 1$

$\qquad\qquad\qquad\qquad\qquad\qquad\quad y > -\dfrac{1}{3}$

$\{y \mid y > -4\} \quad \cap \quad \left\{ \{y \mid y \ge -\dfrac{1}{3}\} \right\}$

$\boxed{\left\{ y \mid y > -\dfrac{1}{3} \right\}}$

$\boxed{\left(-\dfrac{1}{3}, \infty \right)}$

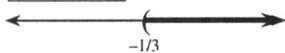

−1/3

29. $\{x \mid 3x - 12 > 0\} \quad \cup \quad \{x \mid 2x - 3 \le -6\}$

$\quad 3x - 12 > 0 \qquad\qquad\quad 2x - 3 \le -6$

$\qquad 3x > 12 \qquad\qquad\qquad 2x \le -3$

$\qquad\quad x > 4 \qquad\qquad\qquad\quad x \le -\dfrac{3}{2}$

$\{x \mid x > 4\} \quad \cup \quad \left\{ x \mid x \le -\dfrac{3}{2} \right\}$

$\boxed{\left\{ x \mid x \le -\dfrac{3}{2} \text{ or } x > 4 \right\}}$

$\boxed{\left(-\infty, -\dfrac{3}{2} \right) \cup (4, \infty)}$

−3/2   4

31. $\{x \mid 3x + 6 < 8\} \quad \cup \quad \{x \mid -2x + 3 > -2\}$

$\quad 3x + 6 < 8 \qquad\qquad\quad -2x + 3 > -2$

$\qquad 3x < 2 \qquad\qquad\qquad -2x > -5$

$\qquad\quad x < \dfrac{2}{3} \qquad\qquad\qquad x < \dfrac{5}{2}$

$\left\{ x \mid x < \dfrac{2}{3} \right\} \quad \cup \quad \left\{ x \mid x < \dfrac{5}{2} \right\}$

$\boxed{\left\{ x \mid x < \dfrac{5}{2} \right\}}$

$\boxed{\left(-\infty, \dfrac{5}{2} \right)}$

5/2

33. $\{x \mid x - 14 < 26 - 3x\} \quad \cup \quad \{x \mid 31 - 5x > 1 - 8x\}$

$\quad x - 14 < 26 - 3x \qquad\qquad 31 - 5x > 1 - 8x$

$\qquad 4x < 40 \qquad\qquad\qquad\quad 3x > -30$

$\qquad\quad x < 10 \qquad\qquad\qquad\quad x > -10$

$\{x \mid x < 10\} \cup \{x \mid x > -10\}$

$\boxed{\{x \mid x \in R\}}$

$\boxed{(-\infty, \infty)}$

0

35. $\{y \mid 12y + 6 \ > \ 4(5y - 1)\} \ \cup \ \{y \mid 9y + 4 \ \geq \ 5(3y - 4)\}$

$\quad\quad 12y + 6 \ < \ 4(5y - 1) \quad\quad\quad 9y + 4 \ \geq \ 5(3y - 4)$

$\quad\quad 12y + 6 \ < \ 20y - 4 \quad\quad\quad\quad 9y + 4 \ \geq \ 15y - 20$

$\quad\quad\quad\quad -8y \ < \ -10 \quad\quad\quad\quad\quad\quad -6y \ \geq \ -24$

$\quad\quad\quad\quad\quad\quad y \ > \ \dfrac{5}{4} \quad\quad\quad\quad\quad\quad\quad\quad y \ \leq \ 4$

$\quad \left\{ y \mid y > \dfrac{5}{4} \right\} \ \cup \ \{y \mid y \leq 4\}$

$\quad \boxed{\left\{ y \mid y > \dfrac{5}{4} \right\}}$

$\quad \boxed{(-\infty, \infty)}$

37. $\left\{ x \mid \dfrac{2x}{9} + 5 < 7 \right\} \quad \cup \quad \left\{ x \mid 3 - x > \dfrac{x - 3}{2} \right\}$

$\quad\quad \dfrac{2x}{9} + 5 \ < \ 7 \quad\quad\quad\quad 3 - x \ > \ \dfrac{x - 3}{2}$

$\quad\quad\quad \dfrac{2x}{9} \ < \ 2 \quad\quad\quad\quad\quad 6 - 2x \ > \ x - 3$

$\quad\quad\quad\quad x \ < \ 9 \quad\quad\quad\quad\quad\quad -3x \ > \ -9$

$\quad\quad\quad\quad\quad\quad\quad\quad\quad\quad\quad\quad\quad\quad x \ < \ 3$

$\quad \{x \mid x < 9\} \cup \{x \mid x < 3\}$

$\quad \boxed{\{x \mid x < 9\}}$

$\quad \boxed{(-\infty, 9)}$

39. $\{x \mid 9 - 3(x + 5) \ > \ -7x + 22\} \ \cup \ \{x \mid 2x - 8 \ > \ -2(-3x + 17)\}$

$\quad\quad 9 - 3(x + 5) \ > \ -7x + 22 \quad\quad\quad 2x - 8 \ > \ -2(-3x + 7)$

$\quad\quad\quad 9 - x - 15 \ > \ -7x + 22 \quad\quad\quad 2x - 8 \ > \ 6x - 14$

$\quad\quad\quad\quad -3x - 6 \ > \ -7x + 22 \quad\quad\quad\quad -4x \ > \ -6$

$\quad\quad\quad\quad\quad\quad 4x \ > \ 28 \quad\quad\quad\quad\quad\quad\quad x \ < \ \dfrac{3}{2}$

$\quad\quad\quad\quad\quad\quad\quad x \ > \ 7$

$\quad \{x \mid x > 7\} \cup \left\{ x \mid x < \dfrac{3}{2} \right\}$

$\quad \boxed{\left\{ x \mid x < \dfrac{3}{2} \text{ or } x > 7 \right\}}$

$\quad \boxed{\left(-\infty, \dfrac{3}{2}\right) \cup (7, \infty)}$

41. $\{x \mid 2 \leq x - 5 \leq 7\}$ $\cup$ $\{x \mid 5 \leq -2x - 7 < 9\}$

$$2 \leq x - 5 \leq 7 \qquad\qquad 5 \leq -2x - 7 < 9$$
$$7 \leq x \leq 12 \qquad\qquad 12 \leq -2x < 16$$
$$-6 \geq x > -8$$
$$-8 < x \leq -6$$

$\{x \mid 7 \leq x \leq 12\}$ $\cup$ $\{x \mid -8 < x \leq -6\}$

$\boxed{\{x \mid -8 < x \leq -6 \text{ or } 7 \leq x \leq 12\}}$

$\boxed{(-8, -6] \cup [7, 12]}$

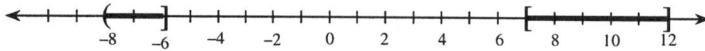

43. $C = 25x + 4500$
$$4775 \leq C \leq 4925$$
$$4775 \leq 25x + 4500 \leq 4925$$
$$275 \leq 25x \leq 425$$
$$\boxed{11 \leq x \leq 17}$$

45. $P = 20 + 17(n - 1)$
$$54 \leq P \leq 326$$
$$54 \leq 20 + 17(n - 1) \leq 326$$
$$54 \leq 20 + 17n - 17 \leq 326$$
$$54 \leq 3 + 17n \leq 326$$
$$51 \leq 17n \leq 323$$
$$\boxed{3 \leq n \leq 19}$$

47. Let
$$x = \text{number of books}$$
$$\text{cost} = 2x + 500$$
$$\text{Revenue} = 20x$$
a. Profit = Revenue – cost
$$\text{Profit} = 20x - (2x + 500)$$
$$\text{Profit} = 20x - 2x - 500$$
$$\boxed{\text{Profit} = 18x - 500}$$
b. $4090 \leq \text{Profit} \leq 13{,}990$
$$4090 \leq 18x - 500 \leq 13{,}990$$
$$4590 \leq 18x \leq 14{,}490$$
$$\boxed{255 \leq x \leq 805}$$

49. Let
$$x = \text{width of rectangle}$$
$$3x - 11 = \text{length of rectangle}$$
$$\text{perimeter} = 2x + 2(3x - 11)$$
$$= 2x + 6x - 22$$
$$= 8x - 22$$
$$10 \leq \text{Perimeter} \leq 114$$
$$10 \leq 8x - 22 \leq 114$$
$$32 \leq 8x \leq 136$$
$$4 \leq x \leq 17$$
$$\boxed{\text{Maximum width: 17 in.; Minimum width: 4 in.}}$$

51. $\text{mean} = \dfrac{90 + 70 + 82 + 80 + 90 + x}{6}$

$$= \dfrac{412 + x}{6}$$

$$80 \leq \text{mean} \leq 85$$

$$80 \leq \dfrac{412 + x}{6} \leq 85$$

$$480 \leq 412 + x \leq 510$$

$$\boxed{68 \leq x \leq 98}$$

53. Let
$$t = \text{time for cars to meet (in hours)}$$
$$\text{distance} = 40t + 50t = 90t$$
$$90 < \text{distance} < 120$$
$$90 < 90t < 120$$
$$1 < t < \frac{4}{3} = 1\frac{1}{3} \qquad \left(1\frac{1}{3} \text{ hr} = 1 \text{ hr } 20 \text{ min}\right)$$

$$\boxed{\text{Between 1 hr and } 1\frac{1}{3} \text{ hr (or 1 hr 20 min)}}$$

55.
$$
\begin{aligned}
\text{perimeter} &= 2y + 2(y+4) \\
&= 2y + 2y + 8 \\
&= 4y + 8
\end{aligned}
$$
$$
\begin{aligned}
18 &\leq \text{perimeter} \leq 22 \\
18 &\leq 4y + 8 \leq 22 \\
10 &\leq 4y \leq 14 \\
\frac{5}{2} &\leq y \leq \frac{7}{2}
\end{aligned}
$$

$\boxed{2\dfrac{1}{2} \text{ meters} \leq y \leq 3\dfrac{1}{2} \text{ meters}}$

57. "Base angles have equal measure" means the triangle is isosceles and the sides opposite the equal angles are of equal measure.
$$
\begin{aligned}
\text{Perimeter} &= (y+2) + (y+2) + y \\
&= 3y + 4
\end{aligned}
$$
$$
\begin{aligned}
18 &\leq \text{perimeter} \leq 22 \\
18 &\leq 3y + 4 \leq 22 \\
14 &\leq 3y \leq 18 \\
\frac{14}{3} &\leq y \leq 6
\end{aligned}
$$

$\boxed{4\dfrac{2}{3} \text{ meters} \leq y \leq 6 \text{ meters}}$

59. $\boxed{A}$ is true; $(-\infty, -1] \cap [-4, \infty)$
$$[-4, -1] \quad \text{True}$$

61. Let
$$
\begin{aligned}
x &= \text{cost of picture frame} \\
5 - x &= \text{cost of print} \\
2(5-x) &< \text{cost of a picture frame} < 3(5-x) \\
10 - 2x &< x < 15 - 3x
\end{aligned}
$$
$$
\begin{aligned}
&\{x \mid 10 - 2x < x\} \cap \{x \mid x < 15 - 3x\} \\
&\quad -3x < -10 \qquad\quad 4x < 15 \\
&\quad\quad x > \frac{10}{3} \qquad\qquad x < \frac{15}{4}
\end{aligned}
$$
$$
\{x \mid x > 3.3\overline{3}\} \cap \{x \mid x < 3.75\}
$$
$$
\{x \mid 3.3\overline{3} < x < 3.75\}
$$

$\boxed{\$3.33 < \text{cost of picture frame} < \$3.75}$

Review Problems

64. Let $x = $ the number.
$$
\begin{aligned}
6(x-6) &= 9(x-9) \\
6x - 36 &= 9x - 81 \\
-3x &= -45 \\
x &= 15
\end{aligned}
$$
The number is $\boxed{15}$.

65.
$$
p = \frac{5}{11}d + 15
$$
$(p = 60)$:
$$
\begin{aligned}
60 &= \frac{5}{11}d + 15 \\
45 &= \frac{5}{11}d \\
99 &= d
\end{aligned}
$$
The depth is $\boxed{99 \text{ feet}}$.

66. Let

$$
\begin{aligned}
x &= \text{Gloria's age} \\
x+7 &= \text{Mike's age} \\
x-10 &= \text{Gloria's age ten years ago} \\
(x+7)-10 &= \text{Mike's age ten years ago}
\end{aligned}
$$

$$
\begin{aligned}
(x+7)-10 &= 2(x-10)-5 \\
x-3 &= 2x-20-5 \\
x-3 &= 2x-25 \\
22 &= x \\
x &= 22 \qquad \text{(Gloria)} \\
x+7 &= 29 \qquad \text{(Mike)}
\end{aligned}
$$

$\boxed{\text{Gloria, 22 years old; Mike, 29 years old}}$

Section 2.7 Equations and Inequalities Involving Absolute Value

Problem Set 2.7, pp. 152-154

1.
$$
\begin{aligned}
|x-2| &= 7 \\
x-2 &= 7 \qquad \text{or} \qquad -(x-2) = 7 \\
x &= 9 \qquad\qquad\qquad\quad -x+2 = 7 \\
&\qquad\qquad\qquad\qquad\qquad\quad -x = 5 \\
&\qquad\qquad\qquad\qquad\qquad\quad\ x = -5
\end{aligned}
$$

$\boxed{\{-5,9\}}$

Check:

$$
\begin{aligned}
x &= 9 & x &= -5 \\
|9-2| &= 7 & |-5-2| &= 7 \\
|7| &= 7 & |-7| &= 7 \\
7 &= 7 \ \checkmark & 7 &= 7 \ \checkmark
\end{aligned}
$$

The solution checks.

3.
$$
\left|\frac{4y-2}{3}\right| = 2
$$

$$
\begin{aligned}
\frac{4y-2}{3} &= 2 & \text{or} & & -\left(\frac{4y-2}{3}\right) &= 2 \\
4y-2 &= 6 & & & 4y-2 &= -6 \\
4y &= 8 & & & 4y &= -4 \\
y &= 2 & & & y &= -1
\end{aligned}
$$

$\boxed{\{-1,2\}}$

Check:

$$y = 2$$

$$\left|\frac{4 \cdot 2 - 2}{3}\right| = 2$$

$$\left|\frac{6}{3}\right| = 2$$

$$|2| = 2$$

$$2 = 2 \sqrt{}$$

$$y = -1$$

$$\left|\frac{4(-1) - 2}{3}\right| = 2$$

$$\left|\frac{-6}{3}\right| = 2$$

$$|-2| = 2$$

$$2 = 2 \sqrt{}$$

The solution checks.

5. Let $x =$ the number.

$$|2x - 3| = 13$$

$2x - 3 = 13$	or	$2x - 3 = -13$
$2x = 16$		$2x = -10$
$x = 8$		$x = -5$

The number is $\boxed{8 \text{ or } -5}$.

7. Let $x =$ the number.

$$|x - 6| = 5$$

$x - 6 = 5$	or	$x - 6 = -5$
$x = 11$		$x = 1$

The number is $\boxed{11 \text{ or } 1}$.

9. Let $x =$ the number.

$$|7 + 5x| = 32$$

$7 + 5x = 32$	or	$7 + 5x = -32$
$5x = 25$		$5x = -39$
$x = 5$		$x = -\frac{39}{5} = -7\frac{4}{5}$

The number is $\boxed{5 \text{ or } -7\frac{4}{5}}$.

11. Let $x =$ the number.

$$\left|\frac{1}{6}(5x - 1)\right| = 4$$

$\frac{1}{6}(5x - 1) = 4$	or	$\frac{1}{6}(5x - 1) = -4$
$5x - 1 = 24$		$5x - 1 = -24$
$5x = 25$		$5x = -23$
$x = 5$		$x = -\frac{23}{5} = -4\frac{3}{5}$

The number is $\boxed{5 \text{ or } -4\frac{3}{5}}$.

13. Let

$$x = \text{the number}$$
$$5x - 3 = \text{the other number}$$
$$\left| (5x - 3) - x \right| = 11$$
$$\left| 4x - 3 \right| = 11$$

$$4x - 3 = 11 \quad \text{or} \quad 4x - 3 = -11$$
$$4x = 14 \qquad\qquad 4x = -8$$
$$x = \frac{7}{2} \qquad\qquad x = -2$$

$$5x - 3 = 5\left(\frac{7}{2}\right) - 3 = \frac{29}{2} \qquad 5x - 3 = -10 - 3 = -13$$

The numbers are $\boxed{-2 \text{ and } -13 \text{ or } \frac{7}{2} \text{ and } \frac{29}{2}}$.

15. $\left| 4x + 3 \right| = -5$

An absolute value cannot equal a negative number.

The solution set is $\boxed{\varnothing}$.

17.
$$\left| 2y + 2 \right| = \left| y + 2 \right|$$
$$2y + 2 = y + 2 \quad \text{or} \quad 2y + 2 = -(y + 2)$$
$$y = 0 \qquad\qquad 2y + 2 = -y - 2$$
$$3y = -4$$
$$y = -\frac{4}{3}$$

$$\boxed{\left\{ -\frac{4}{3}, 0 \right\}}$$

19.
$$\left| 3x - 3 \right| = \left| x + 4 \right|$$
$$3x - 3 = x + 4 \quad \text{or} \quad 3x - 3 = -(x + 4)$$
$$2x = 7 \qquad\qquad 3x - 3 = -x - 4$$
$$x = \frac{7}{2} \qquad\qquad 4x = -1$$
$$x = -\frac{1}{4}$$

$$\boxed{\left\{ -\frac{1}{4}, \frac{7}{2} \right\}}$$

21.
$$\left| 2x - 4 \right| = \left| 2x + 3 \right|$$
$$2x - 4 = 2x + 3 \quad \text{or} \quad 2x - 4 = -(2x + 3)$$
$$-4 = 3 \text{ False} \qquad 2x - 4 = -2x - 3$$
$$4x = -1$$
$$x = \frac{1}{4}$$

$$\boxed{\left\{ \frac{1}{4} \right\}}$$

23. $\left|\dfrac{2}{3}x - 2\right| = \left|\dfrac{1}{3}x + 3\right|$

$\dfrac{2}{3}x - 2 = \dfrac{1}{3}x + 3$ or $\dfrac{2}{3}x - 2 = -\left(\dfrac{1}{3}x + 3\right)$

$\dfrac{1}{3}x = 5$ $\dfrac{2}{3}x - 2 = -\dfrac{1}{3}x - 3$

$x = 15$ $x = -1$

$\boxed{\{-1, 15\}}$

25. $\left|8x + 10\right| = \left|2(4x + 5)\right|$

$8x + 10 = 2(4x + 5)$ or $8x + 10 = -2(4x + 5)$

$8x + 10 = 8x + 10$ $8x + 10 = -8x - 10$

$10 = 10$ True $16x = -20$

$\{x \mid x \in R\}$ $x = -\dfrac{5}{4}$

$\boxed{\{x \mid x \in R\}}$

27. $\left|2x + 4\right| - 2 = 8$

$\left|2x + 4\right| = 10$

$2x + 4 = 10$ or $2x + 4 = -10$

$2x = 6$ $2x = -14$

$x = 3$ $x = -7$

$\boxed{\{-7, 3\}}$

29. $\left|x - 7\right| = -(x - 7)$

$\left|x - 7\right| = -x + 7$

Since $\left|x - 7\right| \geq 0,$

then $-x + 7 \geq 0$

$-x \geq -7$

$x \leq 7$

$\boxed{\{x \mid x \leq 7\}}$

31. $\left|2y - 6\right| < 8$

$-8 < 2y - 6 < 8$

$-2 < 2y < 14$

$-1 < y < 7$

$\boxed{\{y \mid -1 < y < 7\}}$

$\boxed{(-1, 7)}$

33. $\left|2(x + 1) + 3\right| \leq 5$

$-5 \leq 2(x + 1) + 3 \leq 5$

$-5 \leq 2x + 2 + 3 \leq 5$

$-5 \leq 2x + 5 \leq 5$

$-10 \leq 2x \leq 0$

$-5 \leq x \leq 0$

$\boxed{\{x \mid -5 \leq x \leq 0\}}$

$\boxed{[-5, 0]}$

35.
$$\left| \frac{2(3x-1)}{3} \right| \leq \frac{1}{6}$$

$$-\frac{1}{6} < \frac{2(3x-1)}{3} < \frac{1}{6}$$

$$-1 < 4(3x-1) < 1$$

$$-1 < 12x - 4 < 1$$

$$3 < 12x < 5$$

$$\frac{1}{4} < x < \frac{5}{12}$$

$$\boxed{\left(\frac{1}{4}, \frac{5}{12} \right)}$$

37.
$$|3x - 1| > 13$$

$$\begin{array}{lll} 3x - 1 > 13 & \text{or} & 3x - 1 < -13 \\ 3x > 14 & & 3x < -12 \\ x > \dfrac{14}{3} & & x < -4 \end{array}$$

$$\boxed{\left\{ x \mid x < -4 \text{ or } x > \frac{14}{3} \right\}}$$

$$\boxed{(-\infty, -4) \cup \left(\frac{14}{3}, \infty \right)}$$

39.
$$|2(x-3) + 3(x+2) - 17| \geq 13$$

$$|2x - 6 + 3x + 6 - 17| \geq 13$$

$$|5x - 17| \geq 13$$

$$\begin{array}{lll} 5x - 17 \geq 13 & \text{or} & 5x - 17 \leq -13 \\ 5x \geq 30 & & 5x \leq 4 \\ x \geq 6 & & x \leq \dfrac{4}{5} \end{array}$$

$$\boxed{\left\{ x \mid x \leq \frac{4}{5} \text{ or } x \geq 6 \right\}}$$

$$\boxed{\left(-\infty, \frac{4}{5} \right] \cup [6, \infty]}$$

41. $\left| \dfrac{2y+2}{4} \right| > 2$

$\dfrac{2y+2}{4} > 2$ or $\dfrac{2y+2}{4} < -2$

$2y+2 > 8$ $\qquad\qquad$ $2y+2 < -8$

$2y > 6$ $\qquad\qquad\quad$ $2y < -10$

$y > 3$ $\qquad\qquad\quad$ $y < -5$

$\boxed{\{y \mid y < -5 \text{ or } y > 3\}}$

$\boxed{(-\infty, -5) \cup (3, \infty)}$

43. $\left| \dfrac{7x-2}{4} \right| \geq \dfrac{5}{4}$

$\dfrac{7x-2}{4} \geq \dfrac{5}{4}$ or $\dfrac{7x-2}{4} \leq -\dfrac{5}{4}$

$7x-2 \geq 5$ $\qquad\qquad$ $7x-2 \leq -5$

$7x \geq 7$ $\qquad\qquad\quad$ $7x \leq -3$

$x \geq 1$ $\qquad\qquad\quad$ $x \leq -\dfrac{3}{7}$

$\boxed{\left\{ x \mid x \leq -\dfrac{3}{7} \text{ or } x \geq 1 \right\}}$

$\boxed{\left(-\infty, -\dfrac{3}{7}\right) \cup [1, \infty)}$

45. Let x = the number.

$|2x+1| < 13$

$-13 < 2x+1 < 13$

$-14 < 2x < 12$

$-7 < x < 6$

$\boxed{\{x \mid -7 < x < 6\}}$

47. Let x = the number.

$|2x-1| - 2 < 3$

$|2x-1| < 5$

$-5 < 2x-1 < 5$

$-4 < 2x < 6$

$-2 < x < 3$

$\boxed{\{x \mid -2 < x < 3\}}$

49. Let x = the number.

$|2x+1| \leq 3$

$-3 \leq 2x+1 \leq 3$

$-4 \leq 2x \leq 2$

$-2 \leq x \leq 1$

$\boxed{\{x \mid -2 \leq x \leq 1\}}$

51. Let x = the number.

$|2x-1| > 1$

$2x-1 > 1$ or $2x-1 < -1$

$2x > 2$ $\qquad\qquad$ $2x < 0$

$x > 1$ $\qquad\qquad\quad$ $x < 0$

$\boxed{\{x \mid x < 0 \text{ or } x > 1\}}$

53. Let x = the number.

$$|6x - 3| \geq 9$$

$$6x - 3 \geq 9 \qquad \text{or} \qquad 6x - 3 \leq -9$$
$$6x \geq 12 \qquad\qquad\qquad 6x \leq -6$$
$$x \geq 2 \qquad\qquad\qquad x \leq -1$$

$$\boxed{\{x \mid x \leq -1 \text{ or } x \geq 2\}}$$

55.
$$|x - 2{,}560{,}000| \leq 135{,}000$$
$$-135{,}000 \leq x - 2{,}560{,}000 \leq 135{,}000$$
$$2{,}425{,}000 \leq x \leq 2{,}695{,}000$$

$$\boxed{\text{High: } 2{,}695{,}000 \text{ barrels; Low: } 2{,}425{,}000 \text{ barrels}}$$

57.
$$|p - 0.35\%| \leq 0.16\%$$
$$-0.16\% \leq p - 0.35\% \leq 0.16\%$$
$$0.19\% \leq p \leq 0.51\%$$
$$100{,}000p = \text{number of defective products}$$
$$100{,}000(0.0019) \leq 100{,}000p \leq 100{,}000(0.0051)$$
$$190 \leq \text{number of defective products} \leq 510$$
$$(\text{Let } R = \text{cost of refunds} = 6 \text{ (number of defective products)}):$$
$$6(190) \leq 6(\text{number of defective products}) \leq 6(150)$$
$$1140 \leq R \leq 3060$$

$$\boxed{\{R \mid \$1140 \leq R \leq \$3060\}}$$

59.
$$|2x - 1| + 8 \geq 10$$
$$|2x - 1| \geq 2$$
$$2x - 1 \geq 2 \qquad \text{or} \qquad 2x - 1 \leq -2$$
$$2x \geq 3 \qquad\qquad\qquad 2x \leq -1$$
$$x \geq \frac{3}{2} \qquad\qquad\qquad x \leq -\frac{1}{2}$$

$$\boxed{\left\{ x \mid x \leq -\frac{1}{2} \text{ or } x \geq \frac{3}{2} \right\}}$$

61. $|2x + 4| < -1$

An absolute value cannot be less than 0.
There is no solution.
The solution set is $\boxed{\varnothing}$.

63.
$$|2x - b| \leq x \quad (c > 0)$$
$$-c \leq 2x - b \leq c$$
$$-c + b \leq 2x \leq c + b$$
$$\frac{b - c}{2} \leq x \leq \frac{b + c}{2}$$

$$\boxed{\left\{ x \mid \frac{b - c}{2} \leq x \leq \frac{b + c}{2} \right\}}$$

65.
$$|10 - x| > 5$$
$$10 - x > 5 \qquad \text{or} \qquad 10 - x < -5$$
$$-x > -5 \qquad\qquad\qquad -x < -15$$
$$x < 5 \qquad\qquad\qquad x > 15$$

$$\boxed{\{x \mid x < 5 \text{ or } x > 15\}}$$

67. $\boxed{D}$ is true

69. $|x + c| = x + c$

When $x + c \geq 0$ or $x \geq -c$ when $x + c < 0$ or $x < -c$

$$\begin{array}{rcll} x + c &=& x + c & \\ c &=& c & \text{True} \\ \text{when } x &\geq& -c & \end{array} \quad \text{or} \quad \begin{array}{rcl} x + c &=& -(x + c) \\ x + c &=& -x - c \\ 2x &=& -2c \\ x &=& -c \quad \text{but } x < -c \end{array}$$

Thus, no solution or $\varnothing$

$$\boxed{\{x \mid x \geq -c\}}$$

71.
$$\begin{array}{rcl} |x| + x &=& 4 \\ |x| &=& 4 - x \end{array}$$

$$4 - x \geq 0: \quad \begin{array}{rcl} x &=& 4 - x \\ 2x &=& 4 \\ x &=& 2 \end{array} \qquad 4 - x < 0,\ x > 4: \quad \begin{array}{rcl} x &=& -(4 - x) \\ x &=& -4 + x \\ 0 &=& -4 \quad \text{False} \end{array}$$

$$\boxed{\{2\}} \qquad\qquad\qquad\qquad\qquad \varnothing$$

73.
$$\begin{array}{rcl} |x + 2| &\geq& 4x \end{array}$$

$$\begin{array}{rcl} x + 2 &\geq& 4x \\ -3x &\geq& -2 \\ x &\leq& \dfrac{2}{3} \end{array} \qquad \text{or} \qquad \begin{array}{rcl} x + 2 &\leq& -4 \\ 5x &\leq& -2 \\ x &\leq& -\dfrac{2}{5} \end{array}$$

$$\boxed{\left\{ x \mid x \leq \dfrac{2}{3} \right\}}$$

Review Problems

77. Let

$$\begin{array}{rcl} x &=& \text{number of quarters} \\ 2x &=& \text{number of dimes} \\ 2\dfrac{1}{2}(2x) = 5x &=& \text{number of nickels} \end{array}$$

$$\begin{array}{rcll} 0.25x + 0.10(2x) + 0.05(5x) &=& 11.20 & \\ 0.25x + 0.20x + 0.25x &=& 11.20 & \\ 0.70x &=& 11.20 & \\ x &=& 16 & \text{(quarters)} \\ 2x &=& 32 & \text{(dimes)} \\ 5x &=& 80 & \text{(nickels)} \end{array}$$

$$\boxed{16 \text{ quarters}, 32 \text{ dimes, and } 80 \text{ nickels}}$$

78. Let

$$t = \text{time for trip (going) at 45 mph}$$
$$6 - t = \text{time for trip (returning) at 55 mph}$$

$$\text{distance going} = \text{distance returning}$$
$$45t = 55(6 - t)$$
$$45t = 330 - 55t$$
$$100t = 330$$
$$t = 3.3\text{ h}$$
$$\text{distance} = 45(3.3) = 148.5\text{ miles}$$

$$\boxed{148.5\text{ miles}}$$

79. Let x = number of ounces of 4% alcohol.
$$0.04x + 0.40(70) = 0.18(x + 70)$$
$$0.04x + 28 = 0.18x + 12.6$$
$$-0.14x = -15.4$$
$$x = 110$$

$$\boxed{110\text{ oz}}$$

Section 2.8 Critical Thinking

Problem Set 2.8, pp. 158-161

1.
$$-3 - 5(2 - x) + 6x = -6 - 2[-1 + 4(x - 6)]$$
$$-3 - 10 + 5x + 6x = -6 - 2(-1 + 4x - 24)$$
$$11x - 13 = -6 - 2(4x - 25)$$
$$11x - 13 = -6 - 8x + 50$$
$$11x - 13 = -8x + 44$$
$$19x = 57$$
$$\boxed{x = 3}$$

a	4	8	3
2	b	c	14
1	d	6	13
12	e	11	0

column 4: $3 + 14 + 13 + 0 = 30$ (total = 30)

row 1: $a + 4 + 8 + 3 = 30$

$a + 15 = 30$

$a = 15$

row 4: $12 + e + 11 + 0 = 30$

$e + 23 = 30$

$e = 7$

column 3: $8 + c + 6 + 11 = 30$

$c + 25 = 30$

$c = 5$

row 2: $2 + b + 5 + 14 = 30$
$b + 21 = 30$
$b = 9$
row 3: $1 + d + 6 + 13 = 30$
$d + 20 = 30$
$d = 10$

15	4	8	3
2	9	5	14
1	10	6	13
12	7	11	0

3. Siccup, ⎹ Wiccup, Niccup, Ficcup, Biccup, ⎸ Hiccup

5.
$$28 \overline{)4368} \quad 156$$

```
      156
28 | 4368
     28
     156
     140
      168
      168
        0
```

7. Since the three people are ex-teenagers, using

$\boxed{2} \cdot 5 \cdot 7 \cdot \boxed{13} \cdot \underline{29}$

we see that their ages are
$2 \cdot 13 = 26, 5 \cdot 7 = 35,$ and 29
The sum of their ages is
$26 + 35 + 29 = \boxed{90}$

9. Using a systematic list like the one shown below, the answer is $\boxed{8}$.

$1^2 + 1^2 = 2$ $2^2 + 2 = 8$ $3^2 + 3^2 = 18$
$2^2 + 1^2 = 5$ $3^2 + 2^2 = 13$
$3^2 + 1^2 = 10$ $4^2 + 2^2 = 20$
$4^2 + 1^2 = 17$

11. $\boxed{\dfrac{3}{6} = \dfrac{7}{14} = \dfrac{29}{58} \text{ or } \dfrac{3}{6} = \dfrac{9}{18} = \dfrac{27}{54}}$

$a = 3, b = 6, d = 1, f = 2, h = 5$
and $c = 7, e = 4, g = 9, i = 8$
or $c = 9, e = 8, g = 7, i = 4$

13. row 6:

| 1 | 5 | 10 | 10 | 5 | 1 |

row 7:

| 1 | 6 | 15 | 20 | 15 | 6 | 1 |

15.

Input		Output
23		11
11	eleven	6
6	six	3
3	three	5
5	five	4
4	four	4
$\boxed{4}$		

17. We are the number of letters in the underlined words.

19.

n	n^2	digital root		
1	1			
2	4			
3	9			
4	16	$1 + 6 = 7$		
5	25	$2 + 5 = 7$		
6	36	$3 + 6 = 9$		
7	49	$4 + 9 = 13$	$1 + 3 = 4$	
8	64	$6 + 4 = 10$	$1 + 0 = 1$	
9	81	$8 + 1 = 9$		
10	100	$1 + 0 + 0 = 1$		
11	121	$1 + 2 + 1 = 4$		
12	144	$1 + 4 + 4 = 9$		
13	169	$1 + 6 + 9 = 16$	$1 + 6 = 7$	
14	196	$1 + 9 + 6 = 16$	$1 + 6 = 7$	
15	225	$2 + 2 + 5 = 9$		
16	256	$2 + 5 + 6 = 13$	$1 + 3 = 4$	
17	289	$2 + 8 + 9 = 19$	$1 + 9 = 10$	$1 + 0 = 1$
etc				

all possible digital roots: $\boxed{1, 4, 7, 9}$

21. Lord Elphick's son

23.

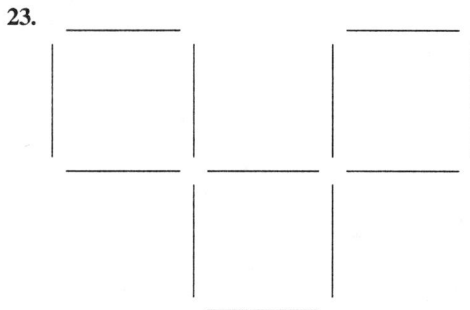

25. Three spelling errors plus the false claim that it only contains one mistake.

27. Answers will vary.

$2 \boxed{-} 9 + \boxed{+} 3 \boxed{+} 5 \boxed{+} 7 \boxed{+} 1 = 9$

or

$2 \boxed{+} 9 \boxed{-} 3 \boxed{-} 5 \boxed{+} 7 \boxed{-} 1 = 9$

29.

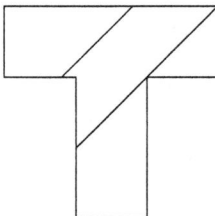

Review Problems

35. Let

$$x \ = \ \text{first consecutive odd integer}$$
$$x + 2 \ = \ \text{second consecutive odd integer}$$
$$x + 4 \ = \ \text{third consecutive odd integer}$$
$$2[x + (x + 2)] \ = \ 4(x + 4) - 12$$
$$2(2x + 2) \ = \ 4x + 16 - 12$$
$$4x + 4 \ = \ 4x + 4$$
$$4 \ = \ 4 \quad \text{True}$$

| Any three consecutive odd integers will satisfy the conditions. |

36. side of square: $2(x + 3) - 5 = 2x + 6 - 5 = 2x + 1$
side of square: $17 - (4x - 2) = 17 - 4x + 2 = 19 - 4x$

$$2x + 1 \ = \ 19 - 4x$$
$$6x \ = \ 18$$
$$x \ = \ 3$$

side of square: $2x + 1 = 2(3) + 1 = 7$

perimeter: $4(7) = \boxed{28 \text{ in.}}$

area: $7(7) = \boxed{49 \text{ in.}^2}$

37. Let x = age of philosopher.

youngster: $\dfrac{1}{4} x$

young woman: $\dfrac{1}{5} x$

adult: $\dfrac{1}{3} x$

dotage: 13

$$x = \frac{1}{4} x + \frac{1}{5} x + \frac{1}{3} x + 13$$

$$x = \frac{15}{60} x + \frac{12}{60} x + \frac{20}{60} x + 13$$

$$\frac{60x}{60} \ = \ \frac{47}{60} x + 13$$

$$\frac{13x}{60} \ = \ 13$$

$$x \ = \ 60$$

$\boxed{60 \text{ years old}}$

Chapter 2   Review Problems

Review Problems, pp. 164-167

1.
$$
\begin{aligned}
2(y - 2) &= 2[y - 5(1 - y)] \\
2y - 4 &= 2(y - 5 + 5y) \\
2y - 4 &= 2(6y - 5) \\
2y - 4 &= 12y - 10 \\
-10y &= -6 \\
y &= \frac{3}{5}
\end{aligned}
$$

$$\boxed{\left\{\frac{3}{5}\right\}}$$

2.
$$
\begin{aligned}
-3\{5 - [1 - (1 - x) + 2]\} &= 7 - [9x - 3(x + 1)] \\
-3[5 - (1 - 1 + x + 2)] &= 7 - (9x - 3x - 3) \\
-3[5 - (x + 2)] &= 7 - (6x - 3) \\
-3(5 - x - 2) &= 7 - 6x + 3 \\
-3(3 - x) &= 10 - 6x \\
-9 + 3x &= 10 - 6x \\
9x &= 19 \\
x &= \frac{19}{9}
\end{aligned}
$$

$$\boxed{\left\{\frac{19}{9}\right\}}$$

3.
$$
\begin{aligned}
3(x - 1) + 7(x - 3) &= 5(2x - 5) \\
3x - 3 + 7x - 21 &= 10x - 25 \\
10x - 24 &= 10x - 25 \\
-24 &= -25 \qquad \text{False}
\end{aligned}
$$

no solution

$$\boxed{\varnothing}$$

4.
$$
\begin{aligned}
\frac{2y}{3} &= 6 - \frac{y}{4} \\
4(2y) &= 12(6) - 3y \\
8y &= 72 - 3y \\
11y &= 72 \\
y &= \frac{72}{11}
\end{aligned}
$$

$$\boxed{\left\{\frac{72}{11}\right\}}$$

5.
$$
\begin{aligned}
\frac{3z + 1}{3} - \frac{13}{2} &= \frac{1 - z}{4} \\
4(3z + 1) - 6(13) &= 3(1 - z) \\
12z + 4 - 78 &= 3 - 3z \\
15z &= 77 \\
z &= \frac{75}{15}
\end{aligned}
$$

$$\boxed{\left\{\frac{77}{15}\right\}}$$

6.
$$
\begin{aligned}
0.07y + 0.06(1400 - y) &= 90 \\
0.07y + 84 - 0.06y &= 90 \\
0.01y &= 6 \\
y &= 600
\end{aligned}
$$

$$\boxed{\{600\}}$$

7.
$$
\begin{aligned}
2(x + 6) + 3(x + 1) &= 4x + 10 + x + 5 \\
2x + 12 + 3x + 3 &= 5x + 15 \\
5x + 15 &= 5x + 15 \\
15 &= 15 \qquad \text{True}
\end{aligned}
$$

$$\boxed{\{x \mid x \in R\}}$$

8.
$$\left|\frac{3x + 4}{2}\right| = 5$$

$$
\begin{aligned}
\frac{3x + 4}{2} &= 5 \qquad \text{or} \qquad & \frac{3x + 4}{2} &= -5 \\
3x + 4 &= 10 & 3x + 4 &= -10 \\
3x &= 6 & 3x &= -14 \\
x &= 2 & x &= -\frac{14}{3}
\end{aligned}
$$

$$\boxed{\left\{x \mid x = -\frac{14}{3}, 2\right\}}$$

9. $|4x - 3| = |7x + 9|$

$4x - 3 = 7x + 9$ or $4x - 3 = -(7x + 9)$

$-3x = 12$ $4x - 3 = -7x - 9$

$x = -4$ $11x = -6$

$x = -\dfrac{6}{11}$

$\left\{ x \mid x = -4, -\dfrac{6}{11} \right\}$

10. $3(z + 5) \leq 6(z + 1)$

$3z + 15 \leq 6z + 6$

$-3z \leq -9$

$z \geq 3$

$\{z \mid z \geq 3\}$

$[3, \infty)$

11. $6y + 5 > -2(y - 3) - 25$

$6y + 5 > -2y + 6 - 25$

$6y + 5 > -2y - 19$

$8y > -24$

$y > -3$

$\{y \mid y > -3\}$

$(-3, \infty)$

12. $3(2y - 1) - 2(y - 4) \geq 7 + 2(3 + 4y)$

$6y - 3 - 2y + 8 \geq 7 + 6 + 8y$

$4y + 5 \geq 8y + 13$

$-4y \geq 8$

$y \leq -2$

$\{y \mid y \leq -2\}$

$(-\infty, -2]$

13. $\dfrac{3}{4}(z - 2) - \dfrac{1}{3}(5 - 2z) > -2$

$(\times 12):$ $9(z - 2) - 4(5 - 2z) > -24$

$9z - 18 - 20 + 8z > -24$

$17z - 38 > -24$

$17z > 14$

$z > \dfrac{14}{17}$

$\left\{ z \mid z > \dfrac{14}{17} \right\}$

$\left(\dfrac{14}{17}, \infty \right)$

14. $|6x + 5| > 29$

$6x + 5 > 29$ or $6x + 5 < -29$

$6x > 24$ $6x < -34$

$x > 4$ $x < -\dfrac{17}{3}$

$\left\{ x \mid x < -\dfrac{17}{3} \text{ or } x > 4 \right\}$

$\left(-\infty, -\dfrac{17}{3} \right) \cup (4, \infty)$

15.
$$\left| 7x + 3 \right| \leq 4$$
$$-4 \leq 7x + 3 \leq 4$$
$$-7 \leq 7x \leq 1$$
$$-1 \leq x \leq \frac{1}{7}$$
$$\boxed{\left\{ x \mid -1 \leq x \leq \frac{1}{7} \right\}}$$
$$\boxed{\left[-1, \frac{1}{7} \right]}$$

16.
$$\left| 3 - 2x \right| \leq 1$$
$$-1 \leq 3 - 2x \leq 1$$
$$-4 \leq -2x \leq -2$$
$$2 \geq x \geq 1$$
$$1 \leq x \leq 2$$
$$\boxed{\{ x \mid 1 \leq x \leq 2 \}}$$
$$\boxed{[1, 2]}$$

17.
$$\left| -4 - 2x \right| < 2$$
$$-2 < -4 - 2x < 2$$
$$2 < -2x < 6$$
$$-1 > x > -3$$
$$-3 < x < -1$$
$$\boxed{\{ x \mid -3 < x < -1 \}}$$
$$\boxed{(-3, -1)}$$

18.
$$\frac{1}{3} - \frac{1}{2} z < 12$$
$$2 - 3z < 72$$
$$-3z < 70$$
$$z > -\frac{70}{3}$$
$$\boxed{\left\{ z \mid z > -\frac{70}{3} \right\}}$$
$$\boxed{\left(-\frac{70}{3}, \infty \right)}$$

19.
$$-1 < 4y + 2 \leq 6$$
$$-3 < 4y \leq 4$$
$$-\frac{3}{4} < y \leq 1$$
$$\boxed{\left\{ y \mid -\frac{3}{4} < y \leq 1 \right\}}$$
$$\boxed{\left(-\frac{3}{4}, 1 \right]}$$

20.
$$-13 < 3 - 4x \leq 13$$
$$-16 < -4x \leq 10$$
$$4 > x \geq -\frac{5}{2}$$
$$-\frac{5}{2} \leq x < 4$$
$$\boxed{\left\{ x \mid -\frac{5}{2} \leq x < 4 \right\}}$$
$$\boxed{\left[-\frac{5}{2}, 4 \right)}$$

21. $\left\{ y \mid 5 - y > \dfrac{y-1}{3} \right\} \quad \cap \quad \left\{ y \mid \dfrac{2y}{3} + 5 > 7 \right\}$

$$5 - y > \frac{y-1}{3} \qquad\qquad \frac{2y}{3} + 5 > 7$$
$$15 - 3y > y - 1 \qquad\qquad \frac{2y}{3} > 2$$
$$-4y > -16 \qquad\qquad y > 3$$
$$y < 4$$
$$\{ y \mid y < 4 \} \quad \cap \quad \{ y \mid y > 3 \}$$
$$\boxed{\{ y \mid 3 < y < 4 \}}$$
$$\boxed{(3, 4)}$$

22.
$$-4 \le \frac{4-x}{3} < 0$$
$$-12 \le 4 - x < 0$$
$$-16 \le -x < -4$$
$$16 \ge x > 4$$
$$4 < x \le 16$$

$$\boxed{\{x \mid 4 < x \le 16\}}$$

$$\boxed{(4, 16]}$$

23.

$5x + 3 \le 3x - 1$	and	$4 - x > 5x - 6$
$2x \le -4$		$-6x > -10$
$x \le -2$		$x < \dfrac{5}{3}$

$$\{x \mid x \le -2\} \qquad \cap \qquad \left\{ x \mid x < \frac{5}{3} \right\}$$

$$\boxed{\{x \mid x \le -2\}}$$

$$\boxed{(-\infty, -2]}$$

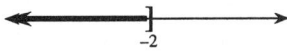

24. $\left\{ y \mid 3 - y > \dfrac{2y-6}{3} \right\} \quad \cup \quad \left\{ y \mid \dfrac{5y}{3} + 6 < 1 \right\}$

$$9 - 3y > 2y - 6 \qquad\qquad \frac{5y}{3} < -5$$
$$-5y > -15 \qquad\qquad y < -3$$
$$y < 3$$
$$\{y \mid y < 3\} \qquad \cup \qquad \{y \mid y < -3\}$$

$$\boxed{\{y \mid y < 3\}}$$

$$\boxed{(-\infty, 3)}$$

25.

$7x + 5 < 2(3x - 1)$	or	$7x + 2 \ge 3(2x - 1)$
$7x + 5 < 6x - 2$		$7x + 2 \ge 6x - 3$
$x < -7$		$x \ge -5$
$\{x \mid x < -7\}$	$\cup$	$\{x \mid x \ge -5\}$

$$\boxed{\{x \mid x < -7 \text{ or } x \ge -5\}}$$

$$\boxed{(-\infty, -7) \text{ or } [-5, \infty)}$$

26.
$$\frac{ay}{3} - \frac{by}{2} = \frac{17}{6}$$
$$2ay - 3by = 17$$
$$y(2a - 3b) = 17$$

$$\boxed{y = \frac{17}{2a - 3b}}$$

27.
$$4x - 3y = 15$$
$$4x = 3y + 15$$
$$\boxed{x = \frac{3y + 15}{4}}$$

28.
$$A = \frac{1}{2}h(B+b)$$
$$2A = h(B+b)$$
$$\frac{2A}{h} = B+b$$
$$\frac{2A}{h} - b = B$$
$$\boxed{B = \frac{2A}{h} - b}$$

29.
$$S = 2\pi r^2 + 2\pi rh$$
$$S - 2\pi r^2 = 2\pi rh$$
$$\frac{S - 2\pi r^2}{2\pi r} = h$$
$$\boxed{h = \frac{S - 2\pi r^2}{2\pi r}}$$

30.
$$A = 8.34\,FC$$
$$(A = 2001.6 \text{ lb},\ c = 30 \text{ ppm}):$$
$$2001.6 = 8.34\,F(30)$$
$$F = 8$$
$$\boxed{8 \text{ million gallons}}$$

31.
$$R = 1.63t + 9.85$$
$$(R = 32.67):$$
$$32.67 = 1.63t + 9.85$$
$$22.82 = 1.63t$$
$$14 = t$$
year: $1965 + 14 = 1979$
$$\boxed{1979}$$

32.
$$C = 400(3x - 1) + 700$$
$$R = 1250x$$
$$1250x = 400(3x - 1) + 700$$
$$1250x = 1200x - 400 + 700$$
$$50x = 300$$
$$x = 6$$
$$\boxed{\text{zero profit: } 6 \text{ units}}$$
Profit: $R > C$
$$x > 6$$
$$\boxed{\text{Profit: more than 6 units}}$$

33. $F = \frac{1}{4}C + 40$

a.
$$4F = C + 160$$
$$4F - 160 = C$$
$$\boxed{C = 4F - 160}$$

b.
$$C = 200$$
$$F = \frac{1}{4}C + 40$$
$$F = \frac{1}{4}(200) + 40$$
$$F = 50 + 40$$
$$F = 90$$
$$\boxed{\text{control temperature at } 90\,^\circ\text{F}}$$

34.
$$N = 600 - 5x$$
$$N < 505$$
$$600 - 5x < 505$$
$$95 < 5x$$
$$19 < x$$
$$\boxed{\text{after 19 days}}$$

35. Let $x =$ number of checks.
first method: $11 + 0.06x$
second method: $4 + 0.20x$
$$11 + 0.06x < 4 + 0.20x$$
$$-0.14x < -7$$
$$x > 50$$
$$\boxed{\text{more than 50 checks}}$$

36. $C = \frac{5}{9}(F - 32)$
$$15 \le C \le 35$$
$$15 \le \frac{5}{9}(F - 32) \le 35$$
$$\frac{9}{5}(15) \le F - 32 \le \frac{9}{5}(35)$$
$$27 \le F - 32 \le 63$$
$$59 \le F \le 95$$
$$\boxed{59^\circ \le F \le 95^\circ}$$

37.
$$|x - 73| \le 67$$
$$-67 \le x - 73 \le 67$$
$$6 \le x \le 140$$

5 days: $5x$
$$5(6) \le 5x \le 5(140)$$
$$30 \le 5 \text{ days} \le 700$$
$$\boxed{\text{High: 700; Low: 30}}$$

38. $\left| p - 0.3\% \right| \leq 0.2\%$

$\qquad -0.2\% \leq p - 0.3\% \leq 0.2\%$

$\qquad 0.1\% \leq p \leq 0.5\%$

$\qquad 0.001 \leq p \leq 0.005$

$\qquad 100 \leq 100{,}000p \leq 500$

(Refund cost = $5x$(each defective part)):

$\qquad 5(100) \leq$ Refund cost $\leq 5(500)$

$\qquad 500 \leq R \leq 2500$

$\boxed{\{R \mid \$500 \leq R \leq \$2500\}}$

39. Let x = the number.

$\qquad 7x - 1 = 9 + 5x$

$\qquad 2x = 10$

$\qquad x = 5$

The number is $\boxed{5}$.

40. Let

$\qquad x =$ first consecutive integer

$\qquad x + 2 =$ second consecutive integer

$\qquad x + 4 =$ third consecutive integer

$\qquad x + (x + 4) = 3(x + 2)$

$\qquad 2x + 4 = 3x + 6$

$\qquad -x = 2$

$\qquad x = -2$

$\qquad x + 2 = 0$

$\qquad x + 4 = 2$

The integers are $\boxed{-2,\ 0,\ \text{and } 2}$.

41. China: 1.2 billion = 1200 million

Russia: 149 million

(India) + (Brazil) + (U.S.) + (Indonesia) = 1480 million

India: 3(U.S.) + 115 million

Brazil: (U.S.) − 101 million

Indonesia: $\dfrac{1}{5}$(India) + 15 million

$\qquad = \dfrac{1}{5}[3(\text{U.S.}) + 115] + 15$ million

$\qquad = \dfrac{3}{5}(\text{U.S.}) + 23 + 15$ million

$\qquad = \dfrac{3}{5}(\text{U.S.}) + 38$ million

Let x = U.S. population (in millions)

$\qquad (3x + 115) + (x - 101) + x + \left(\dfrac{3}{5}x + 38 \right) = 1480$

$\qquad \dfrac{28}{5}x + 52 = 1480$

$\qquad \dfrac{28}{5}x = 1428$

$\qquad x = 225 \ \text{(U.S.)}$

$\qquad 3x - 115 = 880 \ \text{(India)}$

$\qquad x - 101 = 154 \ \text{(Brazil)}$

$\qquad \dfrac{3}{5}x + 38 = 191 \ \text{(Indonesia)}$

$\boxed{\text{India: 880 million; Brazil: 154 million; U.S.: 255 million; Indonesia: 191 million}}$

$\boxed{\text{Yes, Indonesia and Brazil rank ahead of Russia.}}$

42. Let

$$
\begin{array}{rcl rcl}
x &=& \text{Lorna's age} & x+5 &=& \text{Lorna's age in 5 years} \\
x+2 &=& \text{Lisa's age} & (x+2)+5 &=& \text{Lisa's age in 5 years} \\
x-3 &=& \text{Harry's age} & (x-3)+5 &=& \text{Harry's age in 5 years}
\end{array}
$$

$$
\begin{array}{rcl}
3(\text{Lisa's age in 5 years}) &=& 2(\text{Lorna's age in 5 years} + \text{Harry's age in 5 years}) - 7 \\
3(x+7) &=& 2[(x+5)+(x+2)] - 7 \\
3x+21 &=& 2(2x+7) - 7 \\
3x+21 &=& 4x+14-7 \\
-x &=& -14 \\
x &=& 14 \qquad \text{(Lorna's age)} \\
x+2 &=& 16 \qquad \text{(Lisa's age)} \\
x-3 &=& 11 \qquad \text{(Harry's age)}
\end{array}
$$

All three children are over 10 years of age. Our movie star need not worry about being in violation of the child labor laws.

$\boxed{\text{No}}$

43. Let

$$
\begin{array}{rcl rcl}
x &=& \text{Jill's age} & x-6 &=& \text{Jill's age 6 years ago} \\
78-x &=& \text{Phil's age} & (78-x)+6 &=& \text{Phil's age in 6 years}
\end{array}
$$

$$
\begin{array}{rcl}
(\text{Phil's age in 6 years}) &=& 2(\text{Jill's age in 6 years ago}) \\
(78-x)+6 &=& 2(x-6) \\
84-x &=& 2x-12 \\
-3x &=& -96 \\
x &=& 32
\end{array}
$$

Jill is 32 years old (less than 35) so Phil will not date Jill.

$\boxed{\text{No}}$

44. Let

$$
\begin{array}{rcl}
x &=& \text{number of miles} \\
\text{cost} &=& 19(\text{number of days}) + 0.23x
\end{array}
$$

$(\text{cost} = 154.20, \ 4 \text{ days}):$

$$
\begin{array}{rcl}
154.20 &=& 19(4) + 0.23x \\
154.20 &=& 76 + 0.23x \\
78.20 &=& 0.23x \\
340 &=& x
\end{array}
$$

$\boxed{340 \text{ miles}}$

45.

$$
\begin{array}{rcl}
2[x-(1-3x)]-1 &=& 3(x+4) \\
2(x-1+3x)-1 &=& 3x+12 \\
2(4x-1)-1 &=& 3x+12 \\
8x-2-1 &=& 3x+12 \\
8x-3 &=& 3x+12 \\
5x &=& 15 \\
x &=& 3
\end{array}
$$

side of square: $3(x+4)$ yards
$$= 3(3+4) = 3(7) = 21 \text{ yards}$$

$$
\text{Perimeter} = 4\,(\text{side of square})
$$
$$= 4(21) = \boxed{84 \text{ yards}}$$

area $= (\text{side of square})^2$
$$= (21)^2 = \boxed{441 \text{ yd}^2}$$

46. Let

$$\begin{aligned} x &= \text{length of side of the square} \\ 2x &= \text{length of the rectangle} \\ x - 1 &= \text{width of the rectangle} \end{aligned}$$

$$\begin{aligned} (\text{Perimeter of rectangle}) &= 8 + (\text{perimeter of square}) \\ [2(2x) + 2(x - 1)] &= 8 + (4x) \\ 4x + 2x - 2 &= 8 + 4x \\ 2x &= 10 \\ x &= 5 \end{aligned}$$

length of side of the square: $\boxed{5 \text{ yards}}$

47. Let

$$\begin{aligned} x &= \text{height of trapezoid} \\ x - 2 &= \text{length of each nonparallel side of trapezoid} \\ x + 2 &= \text{length of shorter base} \\ 3x - 8 &= \text{length of longer base} \end{aligned}$$

$$\begin{aligned} \text{Perimeter} &= 26 \text{ yards} \\ 2(x - 2) + (x + 2) + (3x - 8) &= 26 \\ 2x - 4 + x + 2 + 3x - 8 &= 26 \\ 6x - 10 &= 26 \\ 6x &= 36 \\ x &= 6 \qquad \text{(height)} \\ x + 2 &= 8 \qquad \text{(shorter base, } b_1) \\ 3x - 8 &= 18 - 8 = 10 \qquad \text{(longer base, } b_2) \end{aligned}$$

$$\begin{aligned} \text{area} &= \frac{1}{2} h (b_1 + b_2) \\ &= \frac{1}{2} 6(8 + 10) \\ &= 3(18) \\ &= 54 \end{aligned}$$

area of trapezoid: $\boxed{54 \text{ yards}^2}$

48. $\angle COB$ is a right angle:

$$\begin{aligned} 3x - 6 &= 90 \\ 3x &= 96 \\ x &= 32 \end{aligned}$$

$\boxed{x = 32°}$

If $\angle DOA$ is a straight angle, then

$$\begin{aligned} 2x + (3x - 6) + (x - 5) &= 180 \\ 6x - 11 &= 180 \\ 6(32) - 11 &= 180 \\ 192 - 11 &= 180 \\ 181 &= 180 \qquad \text{False} \end{aligned}$$

$\boxed{\angle DOA \text{ is } \textit{not} \text{ a straight angle}}$

49.

$$\begin{aligned}
\text{length of sheet metal} &= 16 \\
\text{width of sheet metal} &= 12 \\
\text{length of box} &= 16 - 2x \\
\text{width of box} &= 12 - 2x
\end{aligned}$$

$$\begin{aligned}
(\text{length of box}) &= 2(\text{width of box}) \\
16 - 2x &= 2(12 - 2x) \\
16 - 2x &= 24 - 4x \\
2x &= 8 \\
x &= 4
\end{aligned}$$

length of square: $\boxed{4 \text{ yards}}$

50. Let

$$\begin{aligned}
x &= \text{number of dimes} \\
2x &= \text{number of nickels} \\
x + 2 &= \text{number of quarters} \\
0.10x + 0.05(2x) + 0.25(x + 2) &= 4.10 \\
0.10x + 0.10x + 0.25x + 0.50 &= 4.10 \\
0.45x + 0.50 &= 4.10 \\
0.45x &= 3.60 \\
x &= 8 \quad \text{(dimes)} \\
2x &= 16 \quad \text{(nickels)} \\
x + 2 &= 10 \quad \text{(quarters)}
\end{aligned}$$

$\boxed{16 \text{ nickels}, 8 \text{ dimes}, 10 \text{ quarters}}$

51. Let

$$\begin{aligned}
x &= \text{number of \$5 bills} \\
2x + 1 &= \text{number of \$10 bills} \\
3x - 5 &= \text{number of \$20 bills} \\
5x + 10(2x + 1) + 20(3x - 5) &= 165 \\
5x + 20x + 10 + 60x - 100 &= 165 \\
85x - 90 &= 165 \\
85x &= 255 \\
x &= 3 \quad \text{(\$5 bills)} \\
2x + 1 &= 7 \quad \text{(\$10 bills)} \\
3x - 5 &= 4 \quad \text{(\$20 bills)}
\end{aligned}$$

$\boxed{\$5 \text{ bills}, 3; \$10 \text{ bills}, 7; \$20 \text{ bills}, 4}$

52. Let

$$\begin{aligned}
x &= \text{amount invested at 6\% for 2 years} \\
5000 - x &= \text{amount invested at 7\% for 3 years} \\
0.06(2)x + 0.07(3)(5000 - x) &= 735 \\
0.12x + 1050 - 0.21x &= 735 \\
-0.09x &= -315 \\
x &= 3500 \quad \text{(at 6\%)} \\
5000 - x &= 1500 \quad \text{(at 7\%)}
\end{aligned}$$

$\boxed{\$3500 \text{ at 6\% for 2 years}; \$1500 \text{ at 7\% for 3 years}}$

53. Let

$$
\begin{aligned}
x &= \text{amount invested at 5\% for 10 years} \\
36{,}000 - x &= \text{amount invested at 8\% for 5 years}
\end{aligned}
$$

$$
\begin{aligned}
0.05(10)x &= 0.08(5)(36{,}000 - x) \\
0.5x &= 14{,}400 - 0.4x \\
0.9x &= 14{,}400 \\
x &= 16{,}000 \quad \text{(at 5\%)} \\
36{,}000 - x &= 20{,}000 \quad \text{(at 8\%)}
\end{aligned}
$$

$\boxed{\$16{,}000 \text{ at 5\% for 10 years; } \$20{,}000 \text{ at 8\% for 5 years}}$

54. Let

$$ t = \text{number of hours when the cars will be 660 mile apart} $$

$$
\begin{aligned}
50t + 60t &= 660 \\
110t &= 660 \\
t &= 6
\end{aligned}
$$

$\boxed{6 \text{ hours}}$

55. Let

$$
\begin{aligned}
x &= \text{speed of slower train} \\
x + 20 &= \text{speed of faster train}
\end{aligned}
$$

$$
\begin{aligned}
5x + 5(x + 20) &= 500 \\
5x + 5x + 100 &= 500 \\
10x &= 400 \\
x &= 40 \quad \text{(slower train)} \\
x + 20 &= 60 \quad \text{(faster train)}
\end{aligned}
$$

$\boxed{\text{slower train: 40 mph; faster train: 60 mph}}$

56. Let $t = $ time (in hr) faster planes meet

$$
\begin{aligned}
650t + 550t &= 3000 \\
1200t &= 3000 \\
t &= 2.5
\end{aligned}
$$

The planes will meet in $\boxed{2.5 \text{ hours}}$.

57. Let

$$
\begin{aligned}
t &= \text{time (in hr) for cops to catch the robber} \\
t + 2 &= \text{time (in hr) robbers travel before caught (2 more hours than time for cops)}
\end{aligned}
$$

$$
\begin{aligned}
(\text{distance traveled by robbers}) &= (\text{distance traveled by cops}) \\
85(t + 2) &= 105t \\
85t + 170 &= 105t \\
-20t &= -170 \\
t &= 8.5 \text{ hr}
\end{aligned}
$$

time caught: 12 noon + 8.5 hr = 8:30 P.M.
distance traveled: $105t$

$$ = 105(8.5) = 892.5 \text{ miles} $$

$\boxed{8\frac{1}{2} \text{ hr; 8:30 P.M.; 892.5 miles}}$

58. Let

$$x = \text{number of liters of 90\% sulfuric acid solution}$$
$$20 - x = \text{number of liters of 75\% sulfuric acid solution}$$

	total amount	percentage of solution	=	amount of solution
90% solution	x	0.90		$0.90x$
75% solution	$20 - x$	0.75		$0.75(20 - x)$
78% solution	20	0.78		$0.78(20)$

$$0.09x + 0.75(20 - x) = 0.78(20)$$
$$0.90x + 15 - 0.75x = 15.6$$
$$0.15x = 0.6$$
$$x = 4 \quad \text{(90\% solution)}$$
$$20 - x = 16 \quad \text{(75\% solution)}$$

$\boxed{\text{4 liters at 90\%; 16 liters at 75\%}}$

59. Let

$$x = \text{number of liters of pure alcohol}$$
$$x + 12 = \text{number of liters of 90\% alcohol solution}$$

	total amount	percentage of solution	=	amount of solution
pure alcohol	x	1.00		$1.00x$
45% solution	12	0.45		$0.45(12)$
60% solution	$x + 12$	0.60		$0.60(x + 12)$

$$x + 0.45(12) = 0.60(x + 12)$$
$$x + 5.4 = 0.60x + 7.2$$
$$0.40x = 1.8$$
$$x = 4.5$$

$\boxed{\text{4.5 liters}}$

60. Let

$$x = \text{number of type A radios at \$12 each}$$
$$40 - x = \text{number of type B radios at \$10 each}$$
$$\text{cost (correct)} = 12x + 10(40 - x)$$
$$\text{cost (when error was made)} = 10x + 12(40 - x)$$

$$\text{cost (when error was made)} = \text{cost (correct)} - 20$$
$$10x + 12(40 - x) = 12x + 10(40 - x) - 20$$
$$10x + 480 - 12x = 12x + 400 - 10x - 20$$
$$-2x + 480 = 2x + 380$$
$$-4x = -100$$
$$x = 25 \quad \text{(Type A)}$$
$$40 - x = 15 \quad \text{(Type B)}$$

$\boxed{\text{25 type A; 15 type B}}$

61.

a	3	x
5	b	9
6	c	4

$$\frac{x-2}{2} - \frac{x-3}{4} = \frac{7}{4}$$

$$2(x-2) - (x-3) = 7$$

$$2x - 4 - x + 3 = 7$$

$$x - 1 = 7$$

$$\boxed{x = 8}$$

column 3: $8 + 9 + 4 = 21$ (total)

row 1: $a + 3 + 8 = 21$

$a + 11 = 20$

$a = 10$

row 2: $5 + b + 9 = 21$

$b + 14 = 21$

$b = 7$

row 3: $6 + c + 4 = 21$

$c + 10 = 21$

$c = 11$

10	3	8
5	7	9
6	11	4

62. $\boxed{\text{12 ways}}$

Possible ways:

dimes	nickels	quarters
2	1	0
2	0	5
1	3	0
1	2	5
1	1	10
1	0	15
0	5	0
0	4	5
0	3	10
0	2	15
0	1	20
0	0	25

63. Product = 96
Factors of 96 = $2 \cdot 2 \cdot 2 \cdot 2 \cdot 2 \cdot 3$
one is a teenager: $2 \cdot 2 \cdot 2 \cdot 2 = 16$
remaining factors: 2 and 3
ages of three people: $\boxed{2, 3, \text{ and } 16}$

64. $IQ = \dfrac{100M}{C}$

    a. $\quad 80 \;\le\; IQ \le 140$

$\qquad\quad 80 \;\le\; \dfrac{100M}{C} \le 140$

$\qquad\quad 80 \;\le\; \dfrac{100M}{11} \le 140$

$\qquad 880 \;\le\; 100M \le 1540$

$\qquad \boxed{8.8 \;\le\; M \le 15.4}$

Algebra for College Students

Chapter 3   Graphing and Functions

Section 3.1 The Cartesian Coordinate System and Graphing Equations

Problem Set 3.1, pp. 180-183

1.   (1, 4): $\boxed{\text{I}}$ **3.**   (−3, −5): $\boxed{\text{III}}$ **5.** $\left(-\dfrac{3}{2}, 0\right)$: $\boxed{\text{On } x\text{-axis}}$ **7.** $\left(-3, -\dfrac{3}{2}\right)$: $\boxed{\text{III}}$

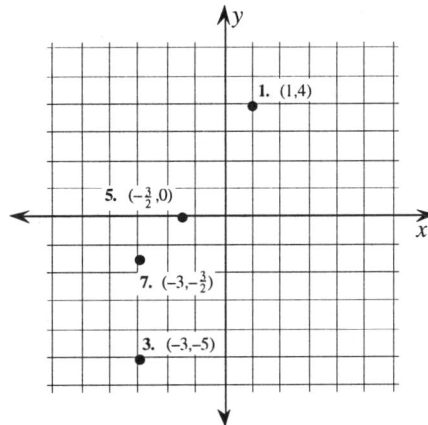

9.   A(0, 6), B(2, 4), C(−2, 1), D(2, −4), E(0, −5)
$$-2 \ \leq \ x \leq 2$$
$$-5 \ \leq \ y \leq 6$$

11. A(−2, 0), B(−1, 3), C(0, 4), D(1, 3), E(2, 0)
$$-2 \ \leq \ x \leq 2$$
$$0 \ \leq \ y \leq 4$$

13. a. $\boxed{64 \text{ ft}}$; $\boxed{\text{after 2 seconds}}$

  b. $\boxed{\text{after 4 seconds}}$

15. a.   1977: (2, 40) number of enrollees: 40
   1979: (4, 25) number of enrollees: 25
   percent decrease $= \dfrac{40 - 25}{40} = \dfrac{15}{40} = \dfrac{3}{8} = 0.375 = \boxed{37.5\%}$

  b. 1977: (2, 40) number of enrollees: 40
   1981: (6, 50) number of enrollees: 50
   percent increase $= \dfrac{50 - 40}{40} = \dfrac{10}{40} = \dfrac{1}{4} = 0.25 = \boxed{25\%}$

  c. (14, 60) number of enrollees: 60
   1975 + 14 = $\boxed{1989}$
   $\boxed{60 \text{ people}}$

17.

F	14	32	50	59	68
C	−10	0	10	15	20

F	Calculate C using $C = \dfrac{5}{9}(F - 32)$	Ordered Pair
14	$C = \dfrac{5}{9}(14 - 32) = -10$	(14, −10)
32	$C = \dfrac{5}{9}(32 - 32) = 0$	(32, 0)
50	$C = \dfrac{5}{9}(50 - 32) = 10$	(50, 10)
59	$C = \dfrac{5}{9}(59 - 32) = 15$	(59, 15)
68	$C = \dfrac{5}{9}(68 - 32) = 20$	(68, 20)

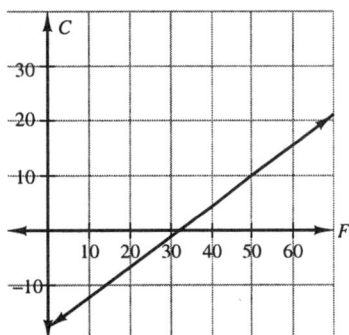

19. $p = t^3 - 12t^2 + 36t + 10$

t	0	1	2	3	4	5	6	7	8
p	10	35	42	37	26	15	10	17	42

$t = 0$: $p = 0^3 - 12(0^2) + 36(0) + 10 = 10$
$t = 1$: $p = 1^3 - 12(1^2) + 36(1) + 10 = 35$
$t = 2$: $p = 2^3 - 12(2^3) + 36(2) + 10 = 42$
$t = 3$: $p = 3^3 - 12(3^2) + 36(3) + 10 = 37$
$t = 4$: $p = 4^3 - 12(4^2) + 36(4) + 10 = 26$
$t = 5$: $p = 5^3 - 12(5^2) + 36(5) + 10 = 15$
$t = 6$: $p = 6^3 - 12(6^2) + 36(6) + 10 = 10$
$t = 7$: $p = 7^3 - 12(7^2) + 36(7) + 10 = 17$
$t = 8$: $p = 8^3 - 12(8^2) + 36(8) + 10 = 42$

a.   relative minimum ($t > 0$): (6, 10)

  6 seconds; 10 inches

b.   relative maximum: (2, 42)

  2 seconds; 42 inches

c.   decreases between relative maximum and relative minimum:

  Between 2 seconds and 6 seconds

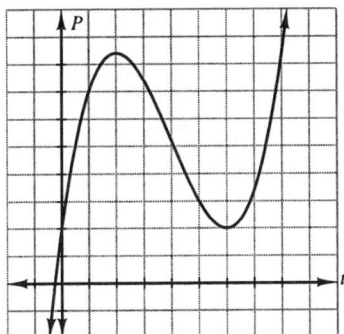

21. $y + 5x - 4 = x^2$
$y = x^2 - 5x + 4$

x	-2	-1	0	1	2
y	18	10	4	0	-2

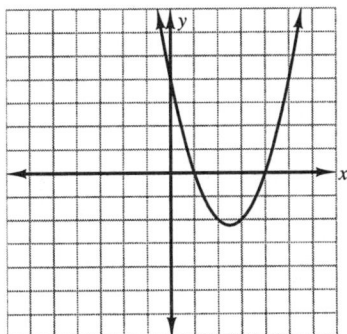

23. $y = x^2 + x - 6$

x	-2	-1	0	1	2
y	-4	-6	-6	-4	0

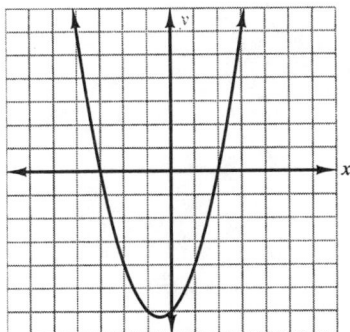

25. $y = x^2 - 3x + 1$

x	-2	-1	0	1	2
y	11	5	1	-1	1

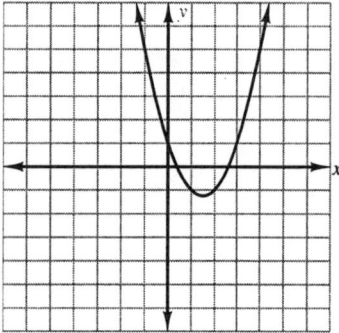

27. $y = x^2 - 4$

x	-2	-1	0	1	2
y	0	-3	-4	-3	0

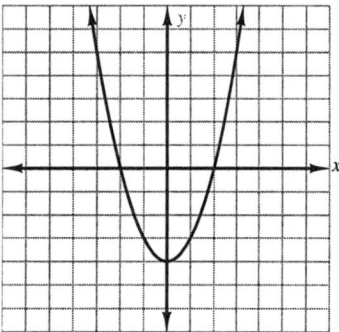

29. $y = x^2 + x$

x	-2	-1	0	1	2
y	2	0	0	2	6

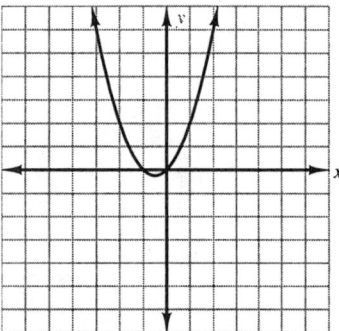

31.

$$y - x = -x^2 + 1$$
$$y = -x^2 + x + 1$$

x	-2	-1	0	1	2
y	-5	-1	1	1	-1

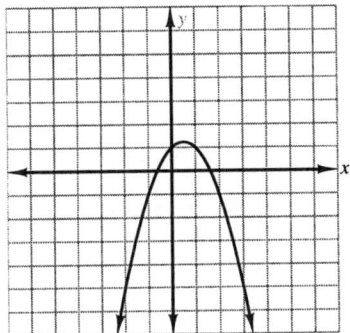

33. $y = -x^2$

x	-2	-1	0	1	2
y	-4	-1	0	-1	-4

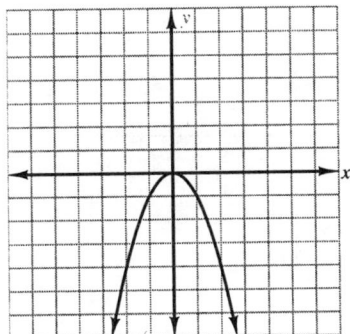

35. $y - x^3 + 2x = 0$
$y = x^3 - 2x$

x	-2	-1	0	1	2
y	-4	1	0	-1	-4

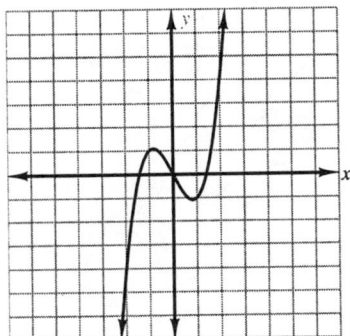

37. $y = x^3 - 3x - 1$

x	-2	-1	0	1	2
y	-3	1	-1	-3	1

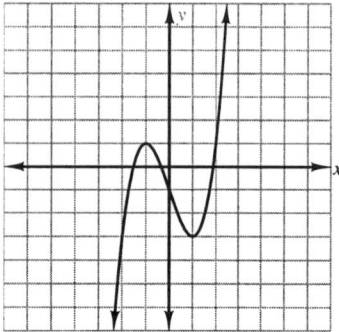

For exercises 39-77, points will vary. Sample values are given.

39. $x - 2y = 6$
Use the intercepts: $(0, -3), (6, 0)$

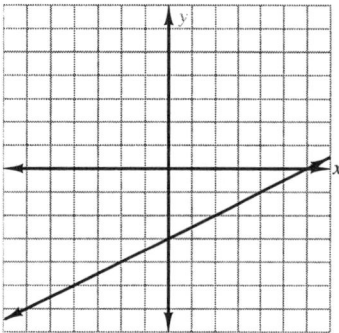

41. $x + 3y = 6$
Use the intercepts: $(0, 2), (6, 0)$

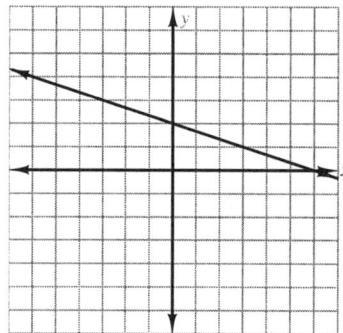

43. $5x - 4y = 20$
Use the intercepts: $(0, -5), (4, 0)$

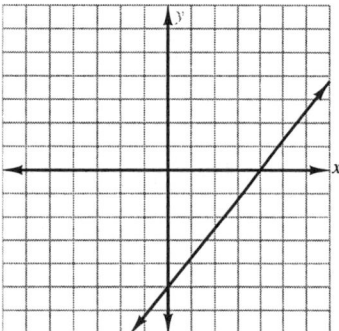

45. $2x + y = 5$
Use the points: $(0, 5), (1, 3)$

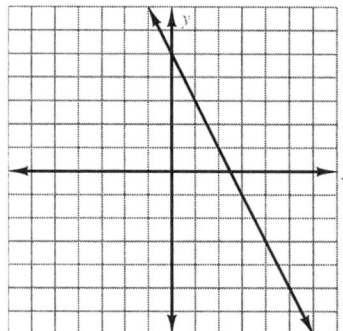

47. $2x + 3y = 5$
Use the points: $(4, -1), (-2, 3)$

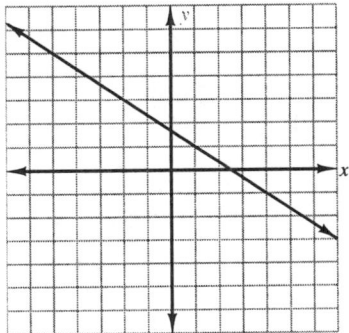

49. $y = -2x + 1$
Use the points: $(0, 1), (1, -1)$

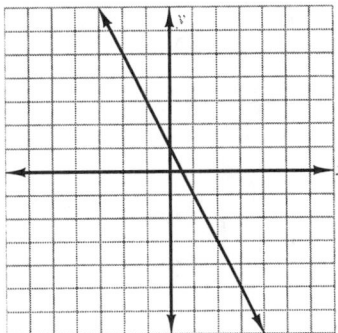

51. $y = 3x + 2$
Use the points: $(0, 2), (-1, -1)$

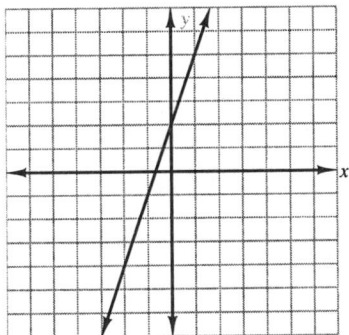

53. $y = x$
Use the points: $(0, 0), (1, 1)$

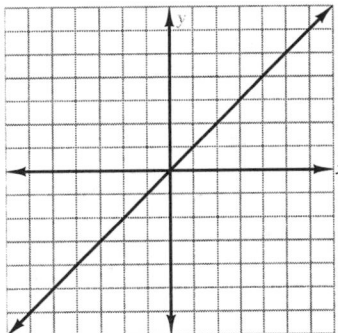

55. $y = 3x$
Use the points: $(0, 0), (1, 3)$

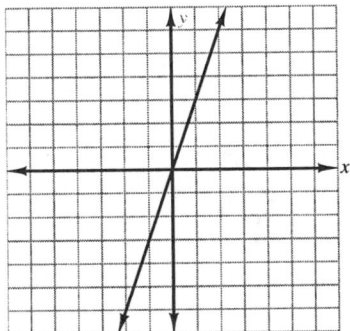

57. $x - y = 0$
$y = x$
Use the points: $(0, 0), (2, 2)$

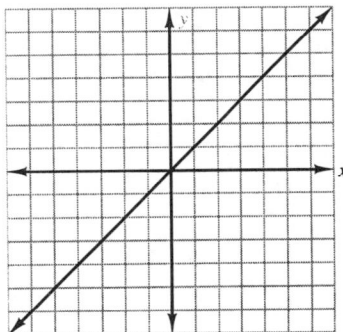

59. $y - 3 = 0$
$y = 3$ is a horizontal line through $(0, 3)$.

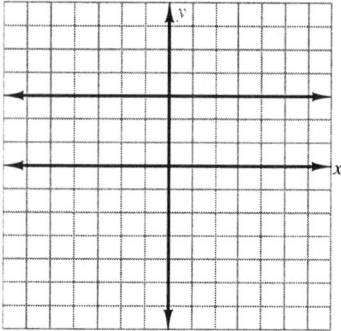

61. $y = -2$ is a horizontal line through $(0, -2)$.

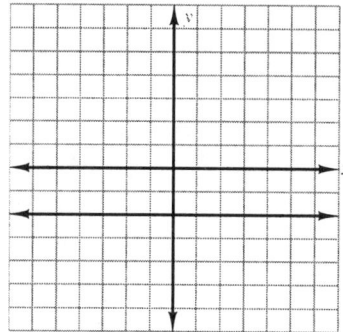

63. $y = 0$ is a horizontal line through the origin. (the x-axis)

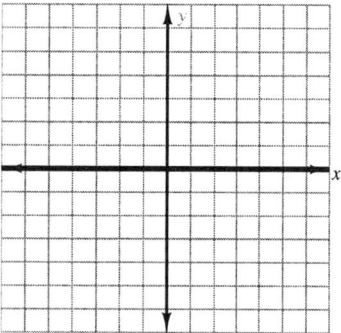

65. $y + 2 = -3$
$y = -5$ is a horizontal line through $(0, -5)$.

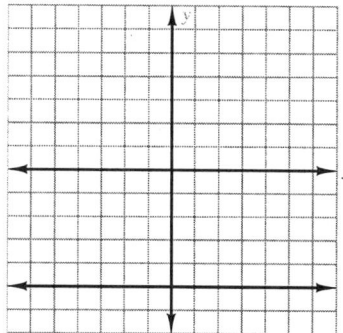

67. $x = -4$ is a vertical line through $(-4, 0)$.

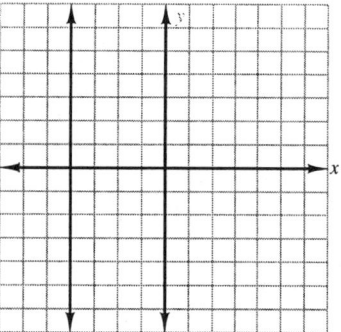

69. $x + 2 = -4$
$x = -6$ is a vertical line through $(-6, 0)$.

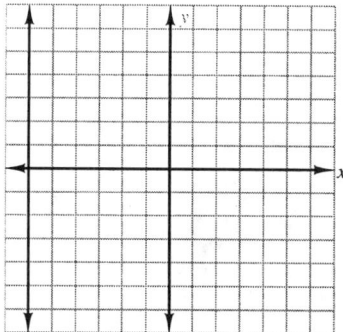

71. $x = 0$ is a vertical line through the origin.
(the y-axis).

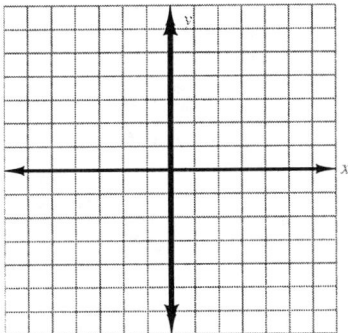

73.
$$A - 30P = -15$$
$$30P = A + 15$$
$$P = \frac{1}{30}A + \frac{1}{2}$$

Use $(15, 1)$ and $(45, 2)$

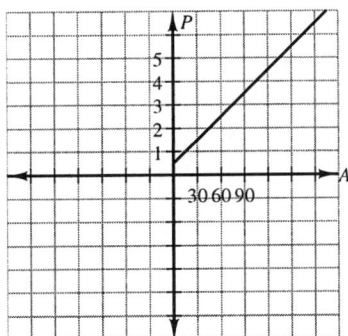

75.
$$33p + 15d = 495$$
$$15d = -33p + 495$$
$$d = -\frac{33}{15}p + 33$$
$$d = -33\left(\frac{p}{15} - 1\right)$$

Use $(0, 33)$ and $(15, 0)$.

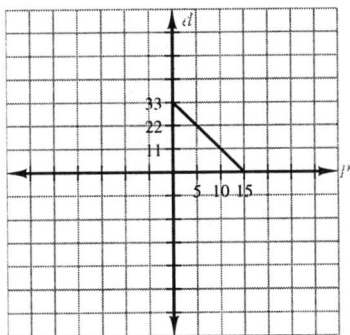

77. $E = 20I$ for $0 \leq I \leq 1$

Use $(0, 0)$, $(1, 20)$

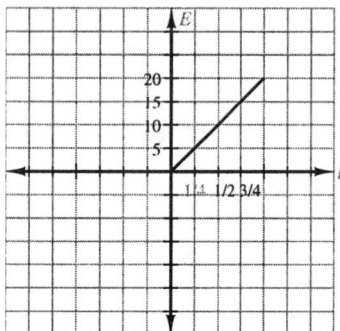

79. $\boxed{C}$ is true; $6 - y = 0$

$y = 6$ is a horizontal line

81.

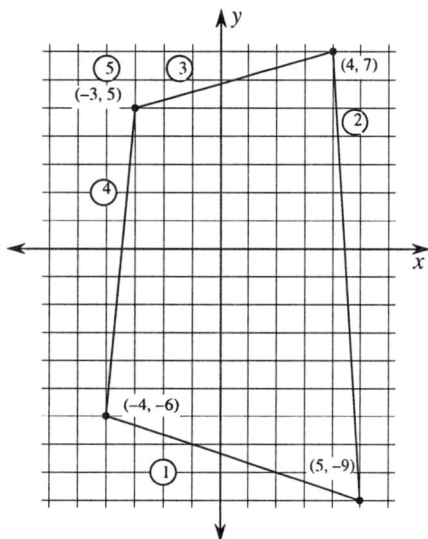

Consider the location after using fourth move, since the point is always located on the negative side of the *x*-axis.

To find the area:

Enclose the quadrilateral in the smallest possible rectangle. The corners of the rectangle will be at $(5, -9)$, $(5, 7)$, $(-4, 7)$, $(-4, -9)$. Once this is accomplished, find the area of the rectangle:

$$A_{\text{Rectangle}} = [7 - (-9)][5 - (-4)] = (7 + 9)(5 + 4) = (16)(9) = 144$$

Then subtract the pieces that are in excess of the desired quadrilateral:

1. triangle at bottom with corners: $(-4, -6)$, $(-4, -9)$, $(5, -9)$

$$A_1 = \frac{1}{2}(5 - (-4))[-6 - (-9)] = \frac{1}{2}(9)(3) = 13.5$$

2. triangle at right with corner: $(5, -9)$, $(5, 7)$, $(4, 7)$

$$A_2 = \frac{1}{2}(7 - (-9)(5 - 4) = \frac{1}{2}(16)(1) = 8$$

3.   triangle at top with corner: $(4, 7), (-3, 7), (-3, 5)$

$$A_3 = \frac{1}{2}(4 - (-3))(7 - 5) = \frac{1}{2}(7)(2) = 7$$

4.   triangle at left with corner: $(-3, 5), (-4, 5), (-4, -6)$

$$A_4 = \frac{1}{2}(5 - (-6))(-3 - (-4)) = \frac{1}{2}(11)(1) = 5.5$$

5.   rectangle at top left with corner: $(-3, 5), (-4, 5), (-4, 7), (-3, 7)$

$$A_5 = (7 - 5)(-3 - (-4)) = 2(1) = 2$$

$$
\begin{aligned}
\text{area of quadrilateral} \ &= \ A_{\text{Rectangle}} - (A_1 + A_2 + A_3 + A_4 + A_5) \\
&= \ 144 - (13.5 + 8 + 7 + 5.5 + 2) \\
&= \ 144 - 36 \\
&= \ 108
\end{aligned}
$$

$\boxed{\text{area, 108 square units}}$

83.

Move (n)	Coordinate $(0, 0)$	Pattern	
1	$(1, -1)$		$+n, -n$
2	$(3, 1)$	$(1 + 2, -1 + 2)$	$+n, +n$
3	$(0, 4)$	$(3 - 3, 1 + 3)$	$-n, +n$
4	$(-4, 0)$	$(0 - 4, 4 - 4)$	$-n, -n$
5	$(1, -5)$	$(-4 + 5, 0 - 5)$	
6	$(7, 1)$	$(1 + 6, -5 + 6$	
7	$(0, 8)$	$(7 - 7, 1 + 7)$	
8	$(-8, 0)$	$(0 - 8, 8 - 8)$	

Note the pattern for every 4 moves.

After four moves: $(-4, 0)$
After eight moves: $(-8, 0)$

*   For the x-coordinate the pattern is: add the move number, add the move number, subtract the move number, subtract the move number (at this point the x-coordinate is the negative value of the move) and then repeat the pattern.

Since $\dfrac{220}{4} = 55$ (an integer) the x-coordinate for the particle after 220 moves will be in the fourth point of the pattern at the negative value of the move number or -220

*   For the y-coordinate the fourth part of the pattern is on the x-axis.

Thus the coordinate of the particle after 220 moves: $\boxed{(-220, 0)}$

Review Problems

87.

$$\frac{2x + 3}{5} - \frac{3x - 1}{2} = \frac{4x + 7}{2}$$

$$
\begin{aligned}
2(2x + 3) - 5(3x - 1) &= 5(4x + 7) \\
4x + 6 - 15x + 5 &= 20x + 35 \\
-11x + 11 &= 20x + 35 \\
-31x &= 24 \\
x &= -\frac{24}{21}
\end{aligned}
$$

$\boxed{\left\{ \dfrac{24}{31} \right\}}$

88. Let

$$x = \text{number of gallons of 20\% acid solution}$$
$$x + 10 = \text{number of gallons of 30\% acid solution}$$
$$0.20x + 0.50(10) = 0.30(x + 10)$$
$$0.20x + 5 = 0.30x + 3$$
$$-0.10x = -2$$
$$x = 20$$

$\boxed{20 \text{ gallons}}$

89.

$4x < x - 6$	or	$6x > 2x + 8$
$3x < -6$		$4x > 8$
$x < -2$		$x > 2$

$\boxed{\{x \mid x < -2 \text{ or } x > \}}$

$\boxed{(-\infty, -2) \cup (2, \infty)}$

Section 3.2 Linear Inequalities Containing Two Variables

Problem Set 3.2, pp. 190-193

1. $x + y \geq 2$

   1. Graph $x + y = 2$.
      The graph is indicated by a solid line since equality is included. $(\geq)$
   2. Select a test point not on the line: $(0, 0)$
      $$0 + 0 \geq 2$$
      $$0 \geq 2 \quad \text{False}$$
      Shade the half-plane not containing $(0, 0)$.

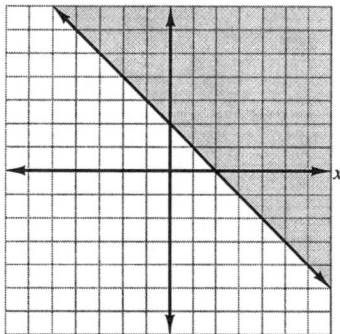

3. $3x - y \geq 6$
   1. Graph $3x - y = 6$ with a solid line ($\geq$).
   2. Test point: (0, 0)
       $$0 - 0 \; \geq \; 6$$
       $$0 \; \geq \; 6 \quad \text{False}$$
   3. Shade the half-plane not containing (0, 0).

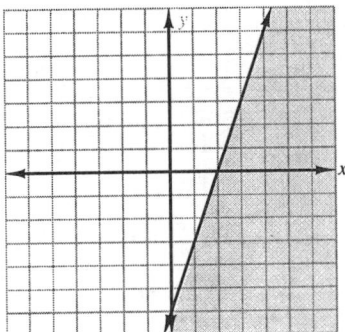

5. $2x + 3y > 12$
   1. Graph $2x + 3y = 12$ with a dashed line(>).
   2. Test point: (0, 0)
       $$0 + 0 \; > \; 12$$
       $$0 \; > \; 12 \quad \text{False}$$
   3. Shade the half-plane *not* containing (0, 0).

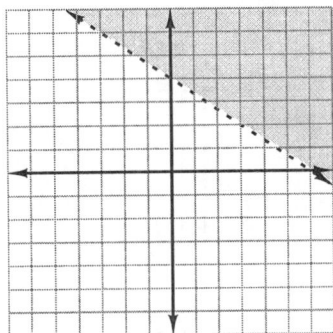

7. $y \geq 2x - 1$
   1. Graph $y = 2x - 1$ with a solid line ($\geq$).
   2. Test point: (0, 0)
       $$0 \; \geq \; 0 - 1$$
       $$0 \; \geq \; -1 \quad \text{True}$$
   3. Shade the half-plane containing (0, 0)

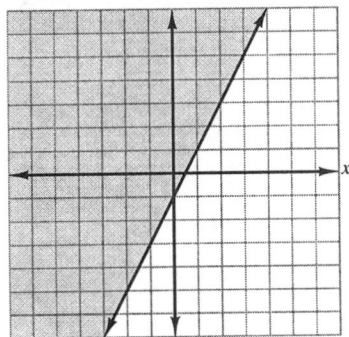

9. $y < x$
   1. Graph $y = x$ with a dashed line.
   2. Test point, (1, 0): $0 < 1$ True
   3. Shade the half-plane containing (1, 0).

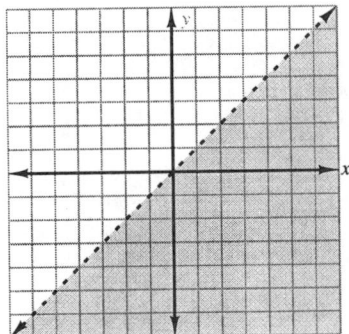

11. $2x + y \leq 0$
   1. Graph $2x + y = 0$ with solid line.
   2. Test point, (1, 0): $1 + 0 \leq 0$ False
   3. Shade half-plane *not* containing (1, 0).

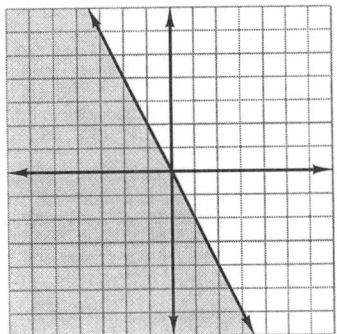

13. $x \le 1$
  1. Graph $x = 1$ with a solid line.
  2. Shade half-plane left of $x = 1$.

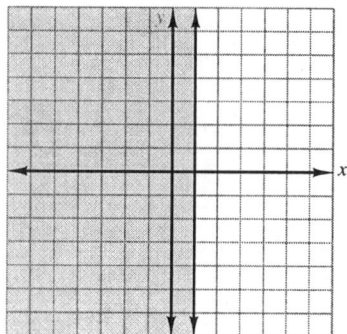

15. $y > -1$
  1. Graph $y = -1$ with a dashed line.
  2. Shade half-plane above $y = -1$.

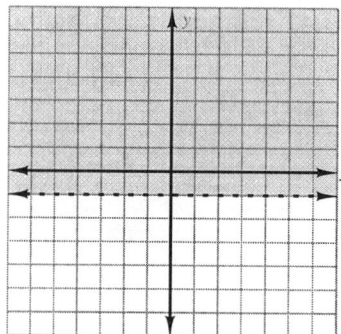

17. $x \le 1$ and $y \le 2$
  1. Graph $x = 1$ with a solid line.
     Graph $y = 2$ with a solid line.
  2. Shade half-plane left of $x = 1$.
     Shade half-plane above $y = 2$.
  3. The graph of the intersection (*and*) includes all points belonging to intersection of the two graphs. (the shaded area)

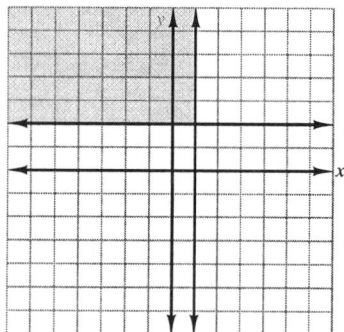

19. $x \ge 0$ and $y \le 0$
  1. Graph $x = 0$ with a solid line.
     Graph $y = 0$ with a solid line.
  2. Shade half-plane right of $x = 0$.
     Shade half-plane below $y = 0$.
  3. The graph of the intersection (*and*) includes all points belonging to the intersection of the two graphs. (the shaded area)

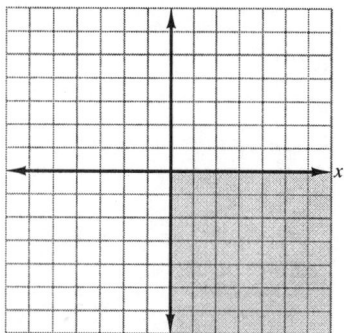

21. $3x - 2y > 9$
1.   Graph $3x - 2y = 9$ with a dashed line.
2.   Test point, $(0, 0)$: $0 - 0 > 9$ False
3.   Shade half-plane not including $(0, 0)$.

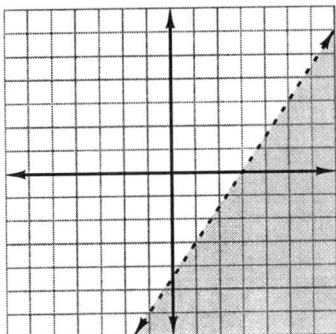

23. $2x + 3y > 6$ and $y \geq 0$
1.   Graph $2x + 3y = 6$ with a dashed line.
     Test point, $(0, 0)$: $0 + 0 > 6$ False
     Shade half-plane not including $(0, 0)$.
2.   Add $y \geq 0$ by graphing $y = 0$ with a solid line and shading half-plane above $y = 0$.
3.   The graph of the intersection (*and*) includes all points of the graph of $2x + 3y > 6$ *above* $y = 0$.
     (the x-axis)

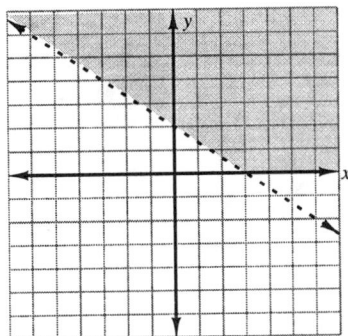

25. $x - 3y > -6$ and $x \geq 0$

   1.   Graph $x - 3y = -6$ with a dashed line.

       Test point, $(0, 0)$: $0 - 0 > -6$ True

       Shade half-plane including $(0, 0)$

   2.   Add $x \geq 0$ by graphing $x = 0$ with a solid line and shading half-plane right of $x = 0$.

   3.   The graph of the intersection (*and*) includes all points of the graph of $x - 3y > -6$ *right* of $x = 0$ (the y-axis)

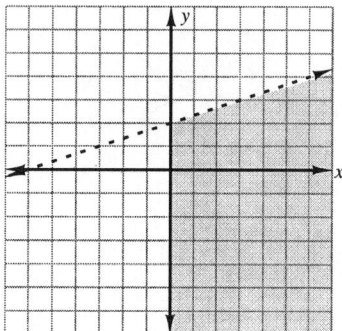

27. $x - y \geq 4$ and $x + y \leq 6$

   1.   Graph $x - y = 4$ with solid line.

       Test point, $(0, 0)$: $0 - 0 \geq 4$ False

       Shade half-plane not including $(0, 0)$

          (below $x - y = 4$)

   2.   Graph $x + y = 6$ with a solid line.

       Test point, $(0, 0)$: $0 + 0 = 6$ True

       Shade half-plane including $(0, 0)$. (below $x + y = 6$)

   3.   The graph of the intersection (*and*) includes all points belonging to the intersection of the two graphs. (below $x - y = 4$ and below $x + y = 6$)

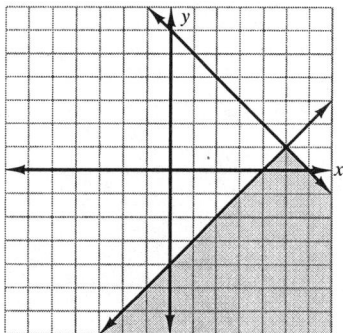

29. $x + 2y \le 16$ and $3x + y \le 18$ and $x \ge 0$ and $y \ge 0$

 1. Graph $x + 2y = 16$ with a solid line.
   Test point, $(0, 0) \le 16$ True
   Shade half-plane including $(0, 0)$ (below $x + 2y = 16$)

 2. Graph $3x + y = 18$ with a solid line
   Test point, $(0, 0)$: $0 + 0 \le 18$ True
   Shade half-plane includin $(0, 0)$ (below $3x + y = 18$)

 3. Add $x \ge 0$ and $y \ge 0$ by graphing $x = 0$ and $y = 0$ with a solid line and shading Quadrant I.

 4. The graph of the intersection of the four graphs includes the intersection of the graphs of $x + 2y \le 16$ and $3x + y \le 18$ which occur in Quadrant I.

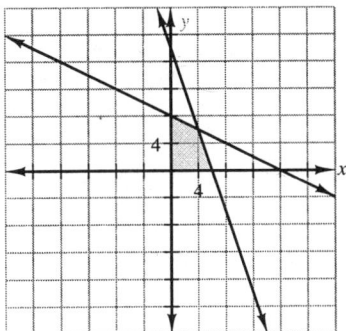

31. $2x + y \ge -2$ or $y \le 4$

 1. Graph $2x + y = -2$ with a solid line.
   Test point, $(0, 0)$: $0 + 0 \ge -2$ True
   Shade half-plane including $(0, 0)$ (to the right and above $2x + y = -2$)

 2. The graph of $y \le 4$ is the solid line $(y = 4)$ and the half-plane left of $y = 4$

 3. The graph of the union (*or*) includes all points belonging to *either* $2x + y \ge -2$ *or* $y \le 4$. (includes *both* half-plane)

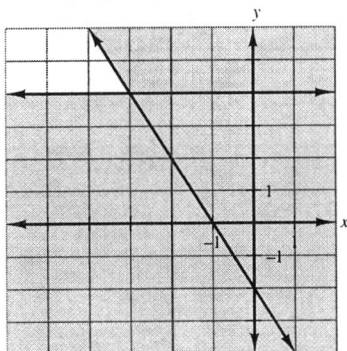

33. $x > -4 + y$ or $4x + 3y > 12$
  1.   Graph $x = -4 + y$ with a dashed line.
       Test point, $(0, 0)$: $0 > -4 + 0$ True
       Shade half-plane including $(0, 0)$ (to the right and below $x = -4 + y$)
  2.   Graph $4x + 3y = 12$ with a dashed line.
       Test point, $(0, 0)$: $(0 + 0 > 12$ False
       Shade half-plane not including $(0, 0)$ (to the right and above $4x + 3y = 12$)
  3.   The graph of the union (*or*) includes all points belonging to *either* $x > -4 + y$ *or* $4x + 3y > 12$.
       (includes *both* half-planes)

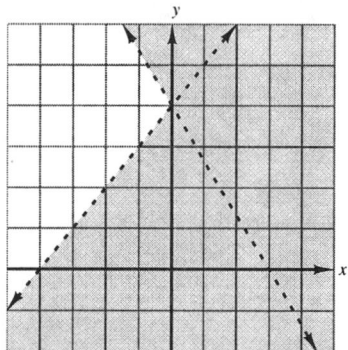

35. $3x - y < 6$ or $2y > x - 4$
  1.   Graph $3x - y = 6$ with a dashed line.
       Test point, $(0, 0)$: $0 - 0 < 6$ True
       Shade half-plane including $(0, 0)$ (to the left of $3x - y = 6$)
  2.   Graph $2y = x - 4$ with a dashed line.
       Test point, $(0, 0)$: $0 > 0 - 4$ True
       Shade half-plane including $(0, 0)$ (to the left and above $2y = x - 4$)
  3.   The graph of the union (*or*) includes all points belonging to *either* $3x - y < 6$ *or* $2y > x - 4$.
       (includes *both* half-planes)

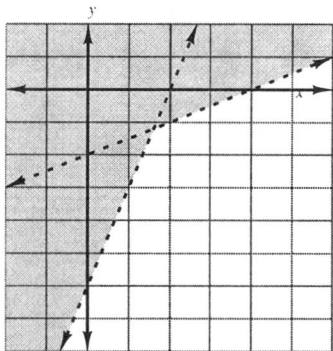

37. $60x + 120y \le 6000$ and $x \ge 0$ and $y \ge 0$

   1.   Graph $60x + 120y = 6000$ with a solid line.

   2.   Shade the half-plane below and to the left of $60x + 120y = 6000$ (including $(0, 0)$)

   3.   Since $x \ge 0$ and $y \ge 0$, the intersection only includes the portion of the half-plane in the first quadrant where $x \ge 0$ and $y \ge 0$.

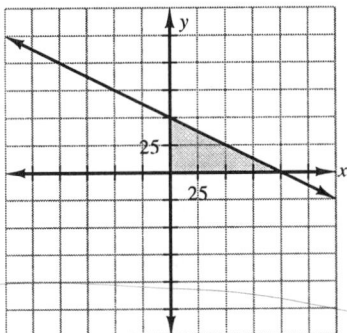

39. $\quad |x| \le 2$

$\quad\quad -2 \le x \le 2$

Graph $x = \pm 2$ with solid lines.

Shade the portion of the plane between the two vertical line $x = +2$ and $x = -2$.

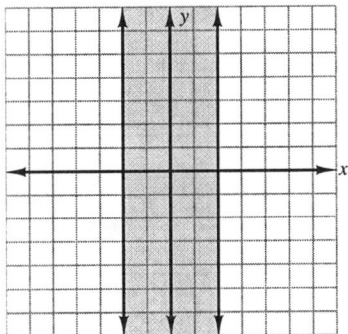

41. $\quad |y| \le 2$

$\quad\quad -2 \le y \le 2$

Graph $y = \pm 2$ with solid lines.

Shade portion of the plane between the two horizontal line $y = +2$ and $y = -2$.

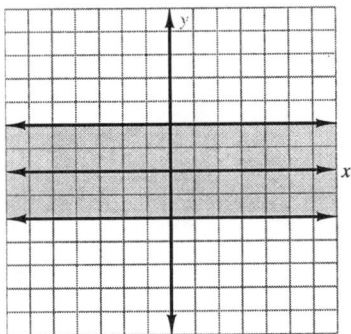

43. $\left| 2x + 1 \right| \leq 3$

$\qquad -3 \;\leq\; 2x + 1 \leq 3$

$\qquad -4 \;\leq\; 2x \leq 2$

$\qquad -2 \;\leq\; x \leq 1 \qquad$ (solid lines)

Shade between solid vertical lines $x = -2$ and $x = 1$

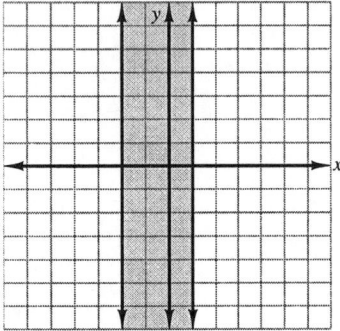

45. $\left| 3x + 1 \right| < 7$

$\qquad -7 \;<\; 3x + 1 < 7$

$\qquad -8 \;<\; 3x < 6$

$\qquad -\dfrac{8}{3} \;<\; x < 2 \qquad$ (dashed lines)

Shade between dashed vertical line $x = -\dfrac{8}{3}$ and $x = 2$.

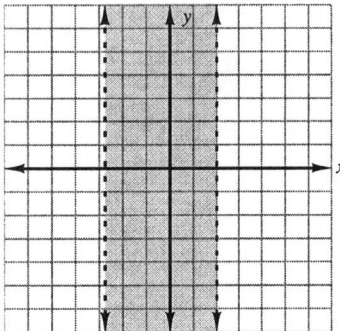

47. $\left| 2y + 1 \right| \le 3$

$-3 \le 2y + 1 \le 3$

$-4 \le 2y + 1 \le 2$

$-2 \le y \le 1$ (solid lines)

Shade between solid horizontal lines $y = -2$ and $y = 1$.

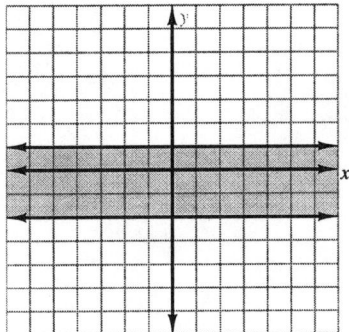

49. $\left| 3y + 1 \right| < 7$

$-7 < 3y + 1 < 7$

$-8 < 3y < 6$

$-\dfrac{8}{3} < y < 2$ (dashed lines)

Shade between dashed horizontal lines $y = -\dfrac{8}{3}$ and $y = 2$.

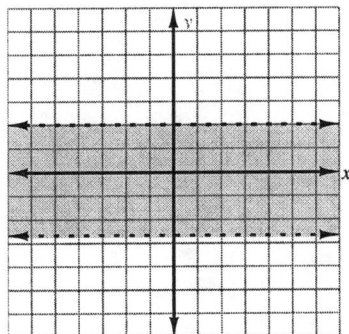

51. $\left|2x+1\right| > 3$

$$2x+1 > 3 \qquad \text{or} \qquad 2x+1 < -3$$
$$2x > 2 \qquad\qquad\qquad 2x < -4$$
$$x > 1 \qquad\qquad\qquad x < -2 \quad \text{(dashed lines)}$$

Shade right of $x = 1$ and left of $x = -2$.

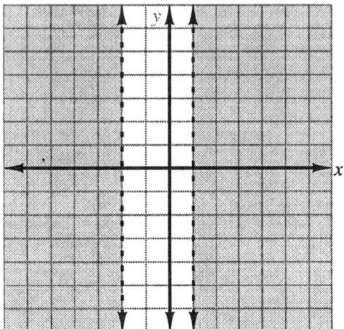

53. $\left|3x+1\right| \geq 7$

$$3x+1 \geq 7 \qquad \text{or} \qquad 3x+1 \leq -7$$
$$3x \geq 6 \qquad\qquad\qquad 3x \leq -8$$
$$x \geq 2 \qquad\qquad\qquad x \leq -\frac{8}{3} \quad \text{(solid lines)}$$

Shade right of $x = 2$ and left of $x = -\frac{8}{3}$.

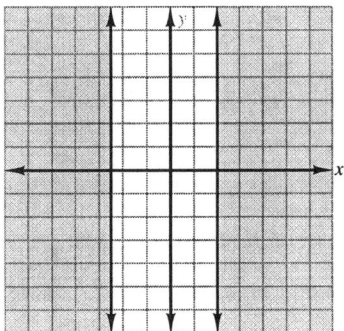

55. $\left| 2y + 1 \right| \ \geq \ 3$

$$
\begin{array}{lllll}
2y + 1 & \geq & 3 & \qquad \text{or} \qquad & 2y + 1 & \leq & -3 \\
2y & \geq & 2 & & 2y & \leq & -4 \\
y & \geq & 1 & & y & \leq & -2 \qquad \text{(solid lines)}
\end{array}
$$

Shade above $y = 1$ and below $y = -2$.

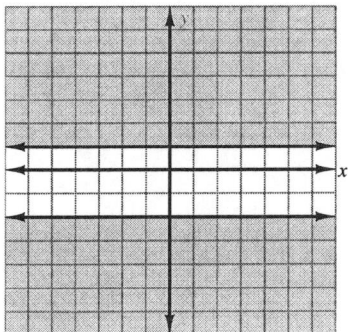

57. $\left| x \right| \leq 2$ and $\left| y \right| \leq 3$

1. Graph $-2 \leq x \leq 2$ (solid lines)
2. Graph $-3 \leq y \leq 3$ (solid lines)
3. The graph of the intersection includes all points belonging to the portion of the plane between the two vertical lines $x = +2$ and $x = -2$ and between the two horizontal lines $y = +3$ and $y = -3$.

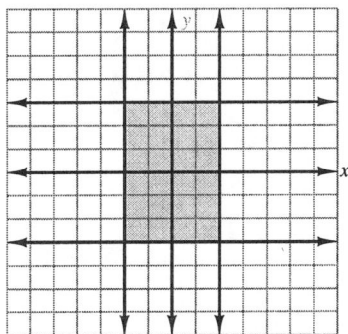

59. $\left| x \right| \geq 2$ and $\left| y \right| \geq 3$

$x \geq 2$ or $x \leq -2$ $y \geq 3$ or $y \leq -3$
              (solid lines)

The graph of the intersection includes all point belonging to the portion of the planes to the right of $x = 2$ and to the left of $x = -2$ *and* above $y = 3$ and below $y = -3$.

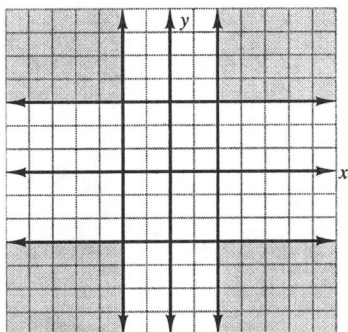

61. $\left| 2x + 1 \right| \leq 3$ and $\left| y \right| \leq 3$

$-3 \leq 2x + 1 \leq 3$ $-3 \leq y \leq 3$
$-4 \leq 2x \leq 2$
$-2 \leq x \leq 1$
(solid lines)

Shade between $x = -2$ and $x = 1$ and between $y = -3$ and $y = 3$.

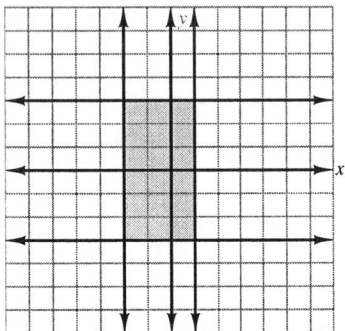

63. $\left| 2x + 4 \right| \geq 6$ or $\left| y \right| \leq 2$

$2x + 4 \geq 6$ or $2x + 4 \leq -6$ $-2 \leq y \leq 2$ (solid lines)
$2x \geq 2$ $2x \leq -10$
$x \geq 1$ $x \leq -5$ (solid lines)

The graph of the union includes the portion of the plane right of $x = 1$ and left of $x = -5$ plus between $y = -2$ and $y = 2$.

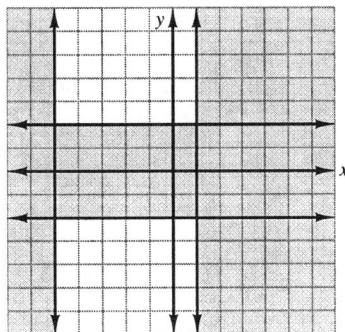

65. Let

x = number of type A pots
y = number of type B pots

$$\boxed{\begin{array}{l} x + 2y \le 6 \\ x + y \le 4 \\ x \ge 0 \\ y \ge 0 \end{array}}$$

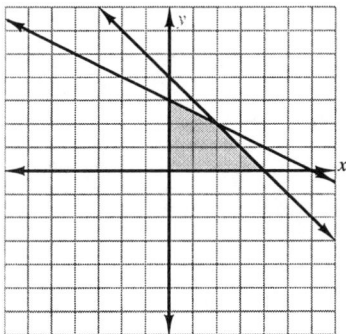

67. $I = 25x + 15y$

The maximum income must occur at one of the corner points: $(3, 0)$, $(5, 4)$, $(0, 6)$, or $(0, 3)$

At $(5, 4)$: $I = 25(5) + 15(4)$
            $= 125 + 60 = 185$ maximum
At $(3, 0)$: $I = 25(3) + 15(0)$
            $= 75$
At $(0, 6)$: $I = 25(0) + 15(6)$
            $= 90$
At $(0, 3)$: $I = 25(0) + 15(3)$
            $= 45$

The maximum income is $\boxed{\$185}$.

69. a. $\boxed{P = 6x + 20y}$

   b. $\boxed{\begin{array}{l} x + y \le 200,\ x \ge 0 \\ 10x + 25y \le 5000,\ y \ge 0 \end{array}}$

   c.

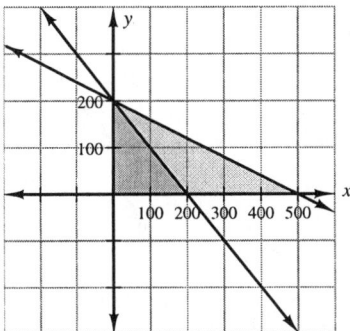

   d. Corner points: $(0, 0)$, $(200, 0)$, $(0, 200)$
       At $(0, 0)$: $P = 0 + 0 = 0$
       At $(200, 0)$: $P = 6(200) + 20(0) = 1200$
       At $(0, 20)$: $P = 6(0) + 20(200) = 4000$

   Maximum weekly profit: $\boxed{\$4000}$

71. $\boxed{C}$ is true;

$$
\begin{array}{llll}
 & 2x - 3y & \le & -6 \\
(3, 1792): & 2(3) - (1792) & \le & -6 \\
 & 6 - 1792 & \le & -6 \\
 & -1792 & \le & -6 \\
 & & \text{True} &
\end{array}
\qquad \text{or} \qquad
\begin{array}{llll}
 & x + 2y & \ge & 4 \\
 & 3 + 2(179) & \ge & 4 \\
 & 3 + 3584 & \ge & 4 \\
 & 3587 & \ge & 4 \\
 & & \text{True} &
\end{array}
$$

73. $\left| 3x - y \right| \le -6$

Absolute value is always positive and cannot be negative.
No solution: $\varnothing$

or $\left| 4x - y \right| \ge -8$

Graph
$$
\begin{array}{lll}
4x - y & = & 8 \\
y & = & 4x - 8
\end{array}
\qquad \text{and} \qquad
\begin{array}{lll}
4x - y & = & -8 \quad \text{with solid lines} \\
y & = & 4x + 8
\end{array}
$$

Since $\left| 4x - y \right|$ is always positive and always greater than -8.
Shade entire plane (both half-planes).

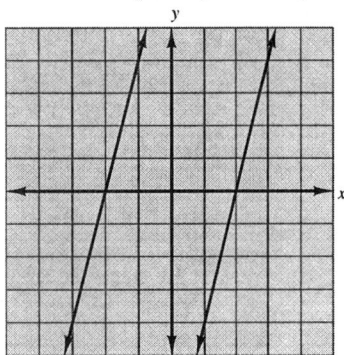

Review Problems

76.
$$
\begin{array}{llll}
7x + 5 & \le & 3 \\
7x & \le & -2 \\
x & \le & -\dfrac{2}{7}
\end{array}
\qquad \text{and} \qquad
\begin{array}{llll}
-3x & \ge & -9 \\
x & \le & 3
\end{array}
$$

$$\boxed{\left\{ x \mid x \le -\dfrac{2}{7} \right\}}$$

$$\boxed{\left(-\infty \ , \ -\dfrac{2}{7} \right]}$$

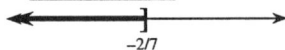

77. $|4 - 3x| \geq 10$

$$4 - 3x \geq 10 \qquad \text{or} \qquad 4 - 3x \leq -10$$
$$-3x \geq 6 \qquad\qquad\qquad -3x \leq -14$$
$$x \leq -2 \qquad\qquad\qquad\quad x \geq \frac{14}{3}$$

$$\boxed{\left\{ x \mid x \leq -2 \text{ or } x \geq \frac{14}{3} \right\}}$$

$$\boxed{(-\infty, -2] \cup x \geq \left[\frac{14}{3}, \infty \right)}$$

78. Let

$$x = \text{number of pennies}$$
$$2x + 1 = \text{number of dimes}$$
$$x - 3 = \text{number of quarters}$$
$$0.01x + 0.10(2x + 1) + 0.25(x - 3) = 3.03$$
$$0.01x + 0.20x + 0.10 + 0.25x - 0.75 = 3.03$$
$$0.46x - 0.65 = 3.03$$
$$0.46x = 3.68$$
$$x = 8 \quad \text{(pennies)}$$
$$2x + 1 = 17 \quad \text{(dimes)}$$
$$x - 3 = 5 \quad \text{(quarters)}$$

$$\boxed{8 \text{ pennies, 17 dimes, 5 quarters}}$$

Section 3.3 Introduction to Functions

Problem Set 3.3, pp. 202-206

1. $\{(1, 3), (1, 7), (1, 10)\}$

$\boxed{\text{not a function}}$ since $(1, 3)$, $(1, 7)$ and $(1, 10)$ have the same first component

$\boxed{\text{Domain} = \{ x \mid x = 1 \}}$

$\boxed{\text{Range} = \{ y \mid y = 3, 7, 10 \}}$

3. $\{(2, 3), (2, 4), (3, 3), (3, 4)\}$

$\boxed{\text{not a function}}$ since $(2, 3)$ and $(2, 4)$ have the same first component and $(3, 3)$ and $(3, 4)$ have the same first component

$\boxed{\text{Domain} = \{ x \mid x = 2, 3 \}}$

$\boxed{\text{Range} = \{ y \mid y = 3, 4 \}}$

5. $\{(-1, -1), (0, 0), (1, 1), (2, 2)\}$

$\boxed{\text{Function}}$ since each first component is assigned exactly one second component

$\boxed{\text{Domain} = \{ x \mid x = -1, 0, 1, 2 \}}$

$\boxed{\text{Range} = \{ y \mid y = -1, 0, 1, 2 \}}$

7. $\{(1, -1), (2, -2), (3, -3), (4, -4)\}$

$\boxed{\text{Function}}$ since each first component is assigned exactly one second component

$\boxed{\text{Domain} = \{ x \mid x = 1, 2, 3, 4\}}$

$\boxed{\text{Range} = \{ y \mid y = -4, -3, -2, -1\}}$

9. $f(x) = \left(\dfrac{x}{12.3}\right)^3$

$f(70) = \left(\dfrac{70}{12.3}\right)^3$

$\boxed{f(70) \approx 184}$

$\boxed{\text{Threshold weight is approximately 184 lb for a 70-inch tall man.}}$

11. $y = 0.4x + 0.88$

   a. $\boxed{f(x) = 0.4x + 0.88}$

   b. $f(2) = 0.4(2) + 0.88$

         $f(2) = 0.8 + 0.88$

         $\boxed{f(12) = 1.68}$

         $\boxed{\text{A subject isolated for 2 hours takes 1.68 minutes to get through the maze.}}$

13. $y = x^3 - 12x^2 + 36x + 10$

   a. $\boxed{f(x) = x^3 - 12x^2 + 36x + 10}$

   b. $f(5) = 5^3 - 12(5^2) + 36(5) + 10$

         $f(5) = 125 - 300 + 180 + 10$

         $f(5) = 315 - 300$

         $\boxed{f(5) = 15}$

         $\boxed{\text{After 5 seconds the particle is 15 inches from the origin.}}$

   c. $f(2) = 2^3 - 12(2^2) + 36(2) + 10$

         $f(2) = 8 - 48 + 72 + 10$

         $\boxed{f(2) = 42}$

   d. $\dfrac{f(5) - f(2)}{3} = \dfrac{15 - 42}{3} = \dfrac{-27}{3} = -9$

         $\boxed{-9 \text{ inches/second}}$

15. $y = 14x^3 - 17x^2 - 16x + 34 \quad \text{for } 1.5 \le x \le 3.5$

   a. $\boxed{f(x) = 14x^3 - 17x^2 - 16x + 34 \quad \text{for } 1.5 \le x \le 3.5}$

   b. $f(2) = 14(2^3) - 17(2^2) - 16(2) + 34$

         $f(2) = 112 - 68 - 32 + 34$

         $\boxed{f(2) = 46}$

         A moth with abdominal width of 2 mm produces 46 eggs.

   c. $\boxed{\text{no}}$; 1 is not in the specified domain of $1.5 \le x \le 3.5$

   d. $f(3.2) - f(1.83)$

17. $x + y = 2$

             $y = -x + 2$

   $\boxed{f(x) = -x + 2}$

         $f(-3) = -(-3) + 2 = 3 + 2 = 5$

   $\boxed{f(-3) = 5}$

   $\boxed{f(a) = -a + 2}$

19.
$$2x + y = 7$$
$$y = -2x + 7$$
$$\boxed{f(x) = -2x + 7}$$
$$f(-3) = -2(-3) + 7 = 6 + 7 = 13$$
$$\boxed{f(-3) = 13}$$
$$\boxed{f(a) = -2a + 7}$$

21.
$$2x + 3y = 8$$
$$3y = -2x + 8$$
$$y = -\frac{2}{3}x + \frac{8}{3}$$
$$\boxed{f(x) = -\frac{2}{3}x + \frac{8}{3} \text{ or } f(x) = \frac{-2x + 8}{3}}$$
$$f(-3) = \frac{-2(-3) + 8}{3} = \frac{6 + 8}{3} = \frac{14}{3}$$
$$\boxed{f(-3) = \frac{14}{3}}$$
$$\boxed{f(a) = \frac{-2a + 8}{3}}$$

23. $\boxed{y \text{ is not a function of } x}$
A vertical line drawn through $x = 0$ intersects the graph twice. (This occurs with vertical lines drawn through other values of x as well.)
$\boxed{\text{Domain} = \{ x \mid -5 \le x \le 5\} \text{ or } [-5, 5]}$
$\boxed{\text{Range} = \{ y \mid -2 \le y \le 2 \text{ or } [-2, 2]}$

25. $\boxed{y \text{ is a function of } x}$
Each vertical line that can be drawn intersects the graph only once. For each domain value x there is only one range value y.
$\boxed{\text{Domain} = \{ x \mid x \in R\} \text{ or } (-\infty, \infty)}$
$\boxed{\text{Range} = \{ y \mid y \in R\} \text{ or } (-\infty, \infty)\}}$

27. $\boxed{y \text{ is a function of } x}$
Each verical line that can be drawn intersects the graph only once. For each domain value x there is only one range value of y.
$\boxed{\text{Domain} = \{ x \mid x \in R\} \text{ or } (-\infty, \infty)}$
$\boxed{\text{Range} = \{ y \mid y \le 3\} \text{ or } (-\infty, 3]}$

29. $\boxed{y \text{ is not a function of } x}$
A vertical line drawn through $x = 2$ intersects the graph four times.
$\boxed{\text{Domian} = \{ x \mid x = 2\} \text{ or } [2]}$
$\boxed{\text{Range} = \{ y \mid y = -2, 1, 3, 5\} \text{ or } [-2, 1, 3, 5]}$

31. $\boxed{y \text{ is a function of } x}$
Each vertical line that can be drawn intersects the graph only one.
$\boxed{\text{Domian} = \{ x \mid x = -6, -5, -4, -3, -2, -1, 0, 1, 2, 3, 4, 5, 6\}}$
$\boxed{\text{Range} = \{ y \mid y = 0, 1, 2, 3, 4, 5, 6\} \text{ or } [0, 1, 2, 3, 4, 5, 6]}$

33. $\boxed{y \text{ is a function of } x}$
Each vertical line that can be drawn intersects the graph only once.
$\boxed{\text{Domain} = \{\, x \mid x \in R \,\} \text{ or } (-\infty, \infty)}$
$\boxed{\text{Range} = \{\, y \mid y \in R \,\} \text{ or } (-\infty, \infty)}$

35. $\boxed{y \text{ is not a function of } x}$
A vertical line drawn through $x = 0$ intersects the graph four times.
$\boxed{\text{Domain} = \{\, x \mid -2 \le x \le 2 \,\} \text{ or } [-2, 2]}$
$\boxed{\text{Range} = \{\, y \mid -5 \le y \le 6 \,\} \text{ or } [-5, 6]}$

37. a. $\boxed{\text{For each value of time there is one value of height.}}$

b. $\boxed{\text{From 3 seconds to 12 seconds}}$

c. $\boxed{\text{From 17 seconds + 30 seconds}}$

d. $\boxed{y = 0. \text{ The vulture was on the ground (not in flight).}}$

39. $f(x) = x^2 + 3x - 1$
$f(-1) = (-1)^2 + 3(-1) - 1 = 1 - 3 - 1 = \boxed{-3}$

41.
$$\begin{aligned}
f(x) &= x^2 + 3x - 1 & g(x) &= \sqrt{x} + 2x - 3 \\
f(2) &= 2^2 + 3(2) - 1 & g(9) &= \sqrt{9} + 2(9) - 3 \\
&= 4 + 6 - 1 & &= 3 + 18 - 3 \\
&= 9 & &= 18
\end{aligned}$$
$f(2) + g(9) = 9 + 18 = \boxed{27}$

43.
$$\begin{aligned}
f(x) &= x^2 + 3x - 1 & g(x) &= \sqrt{x} + 2x - 3 \\
f(0) &= 0 + 0 - 1 = -1 & g(0) &= 0 + 0 - 3 = -3
\end{aligned}$$
$f(0) \cdot g(0) = (-1)(-3) = \boxed{3}$

45. $f(x) = 3x^2 - 7x + 5$
$f(2.43) = 3(2.43)^2 - 7(2.43) + 5 \approx \boxed{5.705}$

47. $f(x) = 3x^2 - 7x + 5 \qquad\qquad g(x) = \sqrt{x} + 7$
$$\begin{aligned}
f(1.46) + g(5) &= [3(1.46)^2 - 7(1.46) + 5] + (\sqrt{5} + 7) \\
&\approx \boxed{10.411}
\end{aligned}$$

49. $f(x) = 3x^2 - 7x + 5 \qquad\qquad g(x) = \sqrt{x} + 7$
$$\begin{aligned}
f(-35) \cdot g(2.16) &= [3(-35)^2 - 7(-35) + 5](\sqrt{2.16} + 7) \\
&\approx \boxed{33{,}243.548}
\end{aligned}$$

51. $f(x) = 6 - |x - 4|$
$f(16) = 6 - |16 - 4| = 6 - |12| = 6 - 12 = -6$
$f(-1) = 6 - |-1 - 4| = 6 - |-5| = 6 - 5 = 1$
$f(16) - f(-1) = -6 - 1 = \boxed{-7}$

53.
$$f(x) = 3x - 1 \qquad\qquad g(x) = 4x^2 - 2x + 3$$
$$f(2) = 3(2) - 1 = 6 - 1 = 5 \qquad = 4(1) + 2 + 3$$
$$= 9$$

$$5f(2) + 4g(-1) = 5(5) + 4(9) = 25 + 36 = \boxed{61}$$

55. $f(x) = 3x - 1$
$$f(a) = \boxed{3a - 1}$$

57.
$$f(x) = 3x - 1 \qquad\qquad g(x) = 4x^2 - 2x + 3$$
$$f(a) = 3a - 1 \qquad\qquad g(a) = 4a^2 - 2a + 3$$
$$f(a) + g(a) = (3a - 1) + (4a^2 - 2a + 3)$$
$$= \boxed{4a^2 + a + 2}$$

59.
$$f(a) = 3a - 1 \qquad\qquad g(a) = 4a^2 - 2a + 3$$
$$5f(a) - 4g(a) = 5(3a - 1) - 4(4a^2 - 2a + 3)$$
$$= 15a - 5 - 16a^2 + 8a - 12$$
$$= \boxed{-16a^2 + 23a - 17}$$

61. $\boxed{C}$ is true; $\{(1, 7), (3, 7), (5, 7), (7, 7)\}$ is a function since each first component is assigned exactly one second component

63. a. $f(x)$ if x is even $= \boxed{-\dfrac{x}{2}}$

b. $f(x)$ if x is even $= \boxed{\dfrac{x+1}{2}}$

c. $f(20) + f(40) + f(65)$
$$= \frac{-20}{2} + \left(-\frac{40}{2}\right) + \frac{65+1}{2}$$
$$= -10 + (-20) + 33$$
$$= \boxed{3}$$

Review Problems

68. Let
$$x = \text{length of the shorter base } (b_1)$$
$$3x = \text{length of the longer base } (b_2)$$
$$\text{height } (h) = 8 \text{ cm}$$
$$\text{area } (A) = 96 \text{ cm}^2$$
$$\text{area} = \frac{1}{2} h(b_1 + b_2)$$
$$96 = \frac{1}{2}(8)(x + 3x)$$
$$96 = 4(4x)$$
$$96 = 16x$$
$$6 = x \quad \text{(shorter base)}$$
$$3x = 18 \quad \text{(longer base)}$$

The lengths of the bases are $\boxed{6 \text{ cm and } 18 \text{ cm}}$.

69. Let t = time (in hours) when the cars will be 400 miles apart

$$60t + 40t = 400$$
$$100t = 400$$
$$t = 4$$

$\boxed{4 \text{ hours}}$

70.

$$P = \frac{2}{5}w\left(1 + \frac{n}{50}\right)$$
$$P = 720$$
$$w = 1000$$
$$720 = \frac{2}{5}(1000)\left(1 + \frac{n}{50}\right)$$
$$720 = 400\left(1 + \frac{n}{50}\right)$$
$$720 = 400 + 8n$$
$$320 = 8n$$
$$40 = n$$

$\boxed{40 \text{ years}}$

Section 3.4 The Slope of a Line

Problem Set 3.4, pp. 214-216

1. (2, −2) and (4, 2)

$m = \dfrac{2 - (-2)}{4 - 2} = \dfrac{4}{2} = \boxed{2}$

3. (0, 0) and (7, 3)

$m = \dfrac{3 - 0}{7 - 0} = \boxed{\dfrac{3}{7}}$

5. (4, -3) and (−6, 2)

$m = \dfrac{2 - (-3)}{-6 - 4} = \dfrac{5}{-10} = \boxed{-\dfrac{1}{2}}$

7. (4, −1) and (2, 7)

$m = \dfrac{7 - (-1)}{2 - 4} = \dfrac{8}{-2} = \boxed{-4}$

9. (3, 5) and (4, 5)

$m = \dfrac{5 - 5}{4 - 3} = \dfrac{0}{1} = \boxed{0}$

11. (−6, 2) and (−6, −9)

$m = \dfrac{-9 - 2}{-6 - (-6)} = \dfrac{-11}{0}$ is $\boxed{\text{undefined (no slope)}}$

13. y-intercept = 4: (0, 4)

slope = $3\left(-\dfrac{3}{1}\right) = \dfrac{\text{rise}}{\text{run}}$

start at (0, 4); move 3 units up and 1 unit to the right: (1, 7)

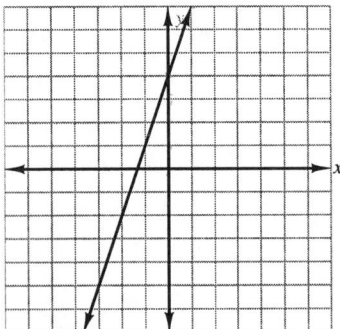

15. y-intercept = 4: (0, 4)

slope $= -3 = \left(-\dfrac{3}{1}\right)$

start at (0,) move 3 units down and 1 unit to the right: (1, 1)

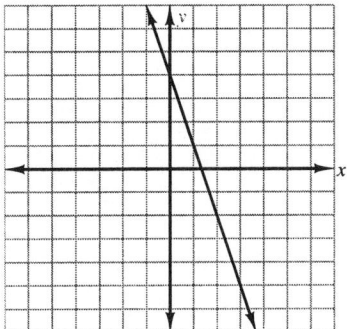

17. Passing through (−1, 3) and slope $= -\dfrac{3}{4}$

Use points (−1, 3) and (−1 + 4, 3 − 3) = (3, 0)

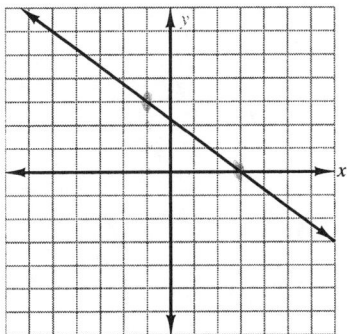

19. Passing through (0, 5) and slope = 0: A line with slope equal to zero and passing through (0, 5) is a horizontal line through (0, 5); $y = 5$.

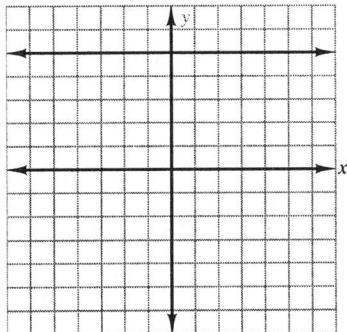

21. Passing through $(-2, -5)$ and slope $= 0$
horizontal line through $(-2, -5)$; $y = -5$.

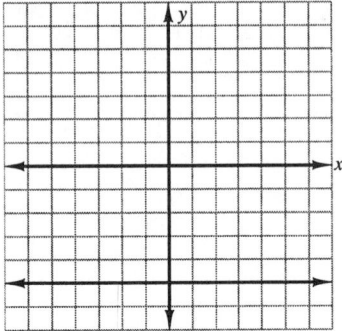

23. x-intercept of 4 and y-intercept of -2.
$(4, 0)$ and $(0, -2)$

$$m = \frac{-2 - 0}{0 - 4} = \frac{-2}{-4} = \boxed{\frac{1}{2}}$$

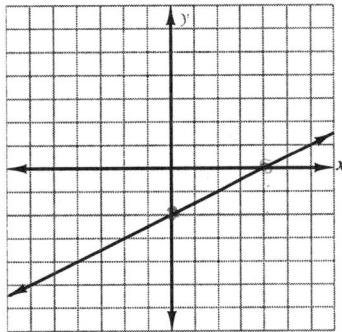

25. $2x - y = 6$
$$\boxed{y = 2x - 6}$$
$(0, -6), (3, 0)$

$$m = \frac{0 - (-6)}{3 - 0} = \frac{6}{3} = \boxed{2}$$

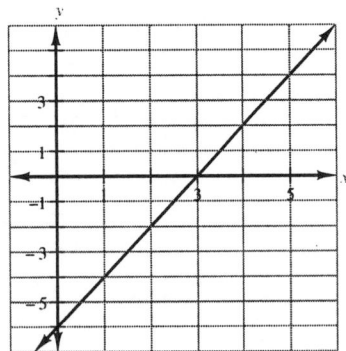

27. $-x + y = 4$

$\boxed{y = x + 4}$

$(0, 4), (-4, 0)$

$m = \dfrac{0 - 4}{-4 - 0} = \boxed{1}$

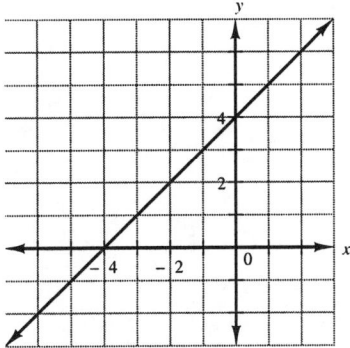

29. $x + 2y = 0$

$\boxed{y = -\dfrac{1}{2} x}$

$(0, 0), (2, -1)$

$m = \dfrac{-1 - 0}{2 - 0} = \boxed{-\dfrac{1}{2}}$

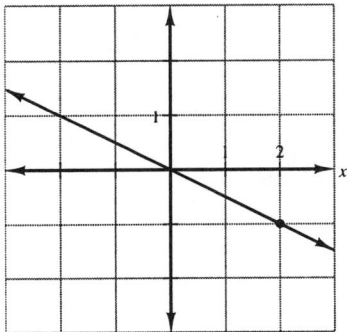

31. $\boxed{y = 3x - 1}$

$(0, -1), \left(\dfrac{1}{3}, 0\right)$

$m = \dfrac{0 - (-1)}{\dfrac{1}{3} - 0} = \dfrac{1}{\dfrac{1}{3}} = \boxed{3}$

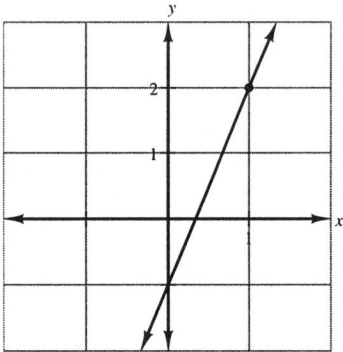

33. $\boxed{y = 5}$

horizontal line, $m = \boxed{0}$

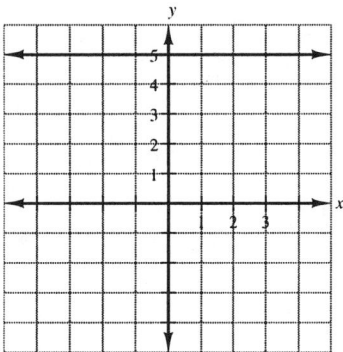

35. $\boxed{x = -3}$

vertical line, $\boxed{\text{undefined slope (no slope)}}$

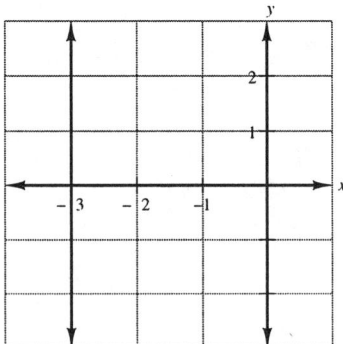

37. $6x + 2y = 10$ and $3x + y = 7$

$(0, 5),\ \left(\dfrac{5}{3}, 0\right)$ $(0, 7),\ \left(\dfrac{7}{3}, 0\right)$

$m_1 = \dfrac{0 - 5}{\dfrac{5}{3} - 0} = \dfrac{-5}{\dfrac{5}{3}} = -3$ $m_2 = \dfrac{0 - 7}{\dfrac{7}{3} - 0} = \dfrac{-7}{\dfrac{7}{3}} = -3$

$$m_1 = m_2$$

Since the slopes are equal, the lines are $\boxed{\text{parallel}}$.

39. $3x - 2y = 6$ and $2x + 3y = -6$

$(0, -3),\ (2, 0)$ $(0, -2),\ (-3, 0)$

$m_1 = \dfrac{0 - (-3)}{2 - 0} = \dfrac{3}{2}$ $m_2 = \dfrac{0 - (-2)}{-3 - 0} = -\dfrac{2}{3}$

Since $m_1 m_2 = \left(\dfrac{3}{2}\right)\left(-\dfrac{2}{3}\right) = -1$, the lines are $\boxed{\text{perpendicular}}$.

41. $2x + y = 1$ and $x - y = 2$

$(0, 1),\ \left(\dfrac{1}{2}, 0\right)$ $(0, -2),\ (2, 0)$

$m_1 = \dfrac{0 - 1}{\dfrac{1}{2} - 0} = \dfrac{-1}{\dfrac{1}{2}} = -2$ $m_2 = \dfrac{0 - (-2)}{2 - 0} = \dfrac{2}{2} = 1$

Since $m_1 \neq m_2$ and $m_1 m_2 = (-2)(1) = -2 \neq -1$, the lines are $\boxed{\text{neither}}$ parallel nor perpendicular.

43. The slope of any line parallel to the line through $(-3, -1)$ and $(1, 4)$ is the same as the slope through the points:

$m = \dfrac{4 - (-1)}{1 - (-3)} = \boxed{\dfrac{5}{4}}$

45. $(2, -3),\ (5, 2)$

$m_1 = \dfrac{2 - (-3)}{5 - 2} = \dfrac{5}{3}$

slope of line perpendicular: $m_2 = \dfrac{-1}{m_1} = \dfrac{-1}{\dfrac{5}{3}} = \boxed{-\dfrac{3}{5}}$

47. $m = \dfrac{3}{4}$ Let x = vertical change.

$\dfrac{3}{4} = \dfrac{x}{16}$

$x = \dfrac{3}{4}(16) = \boxed{12}$

49. $(x, 2),\ (1, 0)$ parallel to $(2, 3),\ (-2, 1)$

$m_1 = \dfrac{0 - 2}{1 - x} = \dfrac{-2}{1 - x}$ $m_2 = \dfrac{1 - 3}{-2 - 2} = \dfrac{-2}{-4} = \dfrac{1}{2}$

$m_1 = m_2$

$\dfrac{-2}{1 - x} = \dfrac{1}{2}$

$-4 = 1 - x$

$x = \boxed{5}$

51. $A(-3,2)$, $B(-1,-2)$, $C(3,0)$

slope of AB: $\dfrac{-2-2}{-1-(-3)} = \dfrac{-4}{2} = -2$

slope of BC: $\dfrac{0-(-2)}{3-(-1)} = \dfrac{2}{4} = \dfrac{1}{2}$

slope of $AC = \dfrac{0-2}{3-(-3)} = \dfrac{-2}{6} = \dfrac{-1}{3}$

$\boxed{AB,\, -2;\ BC,\, \dfrac{1}{2};\ AC,\, -\dfrac{1}{3}}$

53.

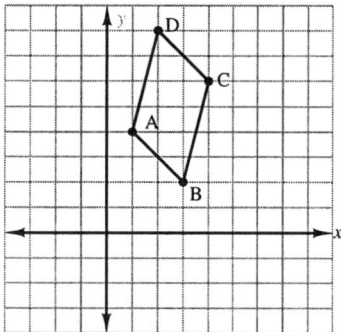

$A(1,4)$, $B(3,2)$, $C(4,6)$, $D(2,8)$

a. $m_1 = $ slope of $AB = \dfrac{2-4}{3-1} = \dfrac{-2}{2} = -1$

$m_2 = $ slope of $CD = \dfrac{8-6}{2-4} = \dfrac{2}{-2} = -1$

$m_1 = m_2$: AB is parallel to CD.

$m_3 = $ slope of $BC = \dfrac{6-2}{4-3} = \dfrac{4}{1} = 4$

$m_4 = $ slope of $AD = \dfrac{8-4}{2-1} = \dfrac{4}{1} = 4$

$m_3 = m_4$: AD is parallel to BC

Both pairs of opposite sides are parallel; thus $ABCD$ is a parallelagram.

b. $m_1 = $ slope of $AB = -1$

$m_3 = $ slope of $BC = 4$

$m_1 m_3 = -1(4) = -4 \neq -1$

The slopes are not negative reciprocals.

AB and BC are not perpendicular.

Thus, the figure is not a rectangle.

55. $A(a,b)$, $B(a+c,b)$, $c(a+c,b+c)$, $D(a,b+c)$

The diagonals are AC and BD.

$m_1 = $ slope of $AC = \dfrac{(b+c)-b}{a-(a+c)} = \dfrac{c}{c} = 1$

$m_2 = $ slope of $BD = \dfrac{(b+c)-b}{a-(a+c)} = \dfrac{c}{-c} = -1$

The slopes are negative reciprocals.

Thus, the diagonals are perpendicular.

57. pitch $= \dfrac{\text{rise}}{\text{run}} = \dfrac{8 \text{ ft}}{28 \text{ ft}} = \boxed{\dfrac{2}{7}}$

59. $D(1960, 1{,}698{,}281)$, $E(1970, 1{,}539{,}233)$

slope of $DE = \dfrac{1{,}539{,}233 - 1{,}698{,}281}{1970 - 1960}$

$= \dfrac{-159{,}048}{10}$

$= \boxed{-15{,}904.8}$

> On the average, the population is decreasing by 15,904.8 (approximately 15,905) people per year between 1960 and 1970.

61. 10 tons of pollutant, 36,000 fish population: (10, 36,000)
30 tons of pollutant, 14,000 fish population: (30, 14,000)

average rate of fish decrease for every ton of pollutant $= \dfrac{14{,}000 - 36{,}000}{30 - 10}$

$= \dfrac{-22{,}000}{20}$

$= -1100$

$\boxed{-1100 \text{ fish per ton of pollutant}}$

63. (10 seconds, 250 feet), (14 seconds, 290 feet)

Average velocity $= \dfrac{290 \text{ feet} - 250 \text{ feet}}{14 \text{ seconds} - 10 \text{ seconds}}$

$= \dfrac{40 \text{ feet}}{4 \text{ seconds}}$

$= \boxed{10 \text{ ft/sec}}$

65. $\boxed{\text{C}}$ is true; A line with zero slope is a horizontal line and represents a function.

Review Problems

69. Let $x =$ the length of x rectangle.

$\dfrac{x - 1}{4} =$ the width of rectangle

$2x + \dfrac{2(x - 1)}{4} = 62$

$4x + x - 1 = 124$

$5x = 125$

$x = 25$ (length)

$\dfrac{x - 1}{4} = \dfrac{25 - 1}{4} = \dfrac{24}{4} = 6$ (width)

length, 25 cm; width, 6 cm

$\boxed{6 \text{ cm} \times 25 \text{ cm}}$

70. $\dfrac{1}{6}(2x+4) = \dfrac{1}{4}(2x+4) - \dfrac{5}{8}x$

$4(2x+4) = 6(2x+4) - 15x$

$8x+16 = 12x+24 - 15x$

$8x+16 = -3x+24$

$11x = 8$

$x = \dfrac{8}{11}$

$$\boxed{\left\{\dfrac{8}{11}\right\}}$$

71. $\{x \mid 3(x-2) - 2x > -6\} \quad \cap \quad \{x \mid 9 - x \geq 5\}$

$3x - 6 - 2x > -6 \qquad\qquad\qquad -x \geq -4$

$x > 0 \qquad\qquad\qquad\qquad\quad x \leq 4$

$\boxed{\{x \mid 0 < x \leq 4\}}$

$\boxed{(0, 4]}$

Section 3.5 Linear Equations

Problem Set 3.5, pp. 225-229

1.   Passing through $(-3, 2)$ with slope $= 2$

point-slope form: $\boxed{y - 2 = 2(x + 3)}$

$y - 2 = 2x + 6$

slope-intercept form: $\boxed{y = 2x + 8}$

standard form: $\boxed{2x - y = -8}$

3.   Passing through $(1, 0)$ with slope $= \dfrac{1}{2}$

point-slope: $\boxed{y - 0 = \dfrac{1}{2}(x - 1)}$

slope-intercept: $\boxed{y = \dfrac{1}{2}x - \dfrac{1}{2}}$

$2y = x - 1$

standard: $\boxed{x - 2y = 1}$

5.   Passing through $(-8, 6)$ with slope $= -\dfrac{2}{3}$

point-slope: $\boxed{y - 6 = -\dfrac{2}{3}(x + 8)}$

$y - \dfrac{8}{3} = -\dfrac{2}{3}x - \dfrac{16}{3}$

slope-intercept: $\boxed{y = -\dfrac{2}{3}x + \dfrac{2}{3}}$

$3y = -2x + 2$

standard: $\boxed{2x + 3y = 2}$

7. Passing through (5, 8) and (3, 16)

$$m = \frac{16 - 8}{3 - 5} = \frac{8}{-2} = -4$$

point-slope: $\boxed{y - 8 = -4(x - 5)}$

or $\boxed{y - 16 = -4(x - 3)}$

$$y - 8 = -4x + 20$$

slope-intercept: $\boxed{y = -4x + 28}$

standard: $\boxed{4x + y = 28}$

9. Passing through (5, 6) with x-intercept $= 11$

(5, 6) and (11, 0)

$$m = \frac{0 - 6}{11 - 5} = -\frac{6}{6} = -1$$

point-slope: $\boxed{y - 6 = -1(x - 5)}$

$$y - 6 = -x + 5$$

slope-intercept: $\boxed{y = -x + 11}$

standard: $\boxed{x + y = 11}$

11. (10, 115), (30, 125)

$$m = \frac{125 - 115}{30 - 10} = \frac{10}{20} = \frac{1}{2}$$

point-slope: $\boxed{y - 115 = \frac{1}{2}(x - 10) \text{ or } y - 125 = \frac{1}{2}(x - 30)}$

$$y - 115 = \frac{1}{2}x - 5$$

$$\boxed{y = \frac{1}{2}x + 110}$$

$x = 80$: $y = \frac{1}{2}(80) + 110 = 40 + 110 = 150$

blood pressure: $\boxed{150}$

13. (35, 95), (0, 32)

$$m = \frac{32 - 95}{0 - 35} = \frac{-63}{-35} = \frac{9}{5}$$

$$\boxed{y - 32 = \frac{9}{5}(x - 0) \text{ or } y - 95 = \frac{9}{5}(x - 35)}$$

$$\boxed{y = \frac{9}{5}x + 32}$$

$x = 80$: $y = \frac{9}{5}(80) + 32 = 144 + 32 = 176$

Fahrenheit temperature: $\boxed{176°}$

15. $(2, 0.37), (8, 1.39)$

$$m = \frac{1.39 - 0.37}{8 - 2} = \frac{1.02}{6} = 0.17$$

$$y - 0.37 = 0.17(x - 2)$$
$$y - 0.37 = 0.17x - 0.34$$
$$y = 0.17x + 0.03$$

a. $x = 15$: $y = 0.17(15) + 0.03$
$$y = 2.55 + 0.03$$
$$y = 2.58$$

$\boxed{\$2.58}$

b. $x = 6$: $y = 0.17(6) + 0.03$
$$y = 1.02 + 0.03$$
$$y = 1.05$$

$\boxed{\$1.05}$

17. $y = -2x - 1$

$m = -2$; y-intercept: $(0, -1)$

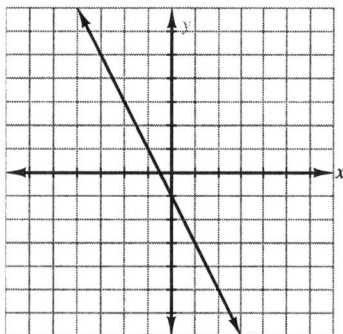

19. $y = \frac{1}{2}x + 1$

$m = \frac{1}{2}$; y-intercept: $(0, 1)$

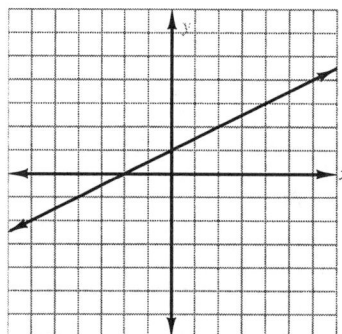

21. $y = \frac{1}{2}x - 1$

$m = \frac{1}{2}$; y-intercept: $(0, -1)$

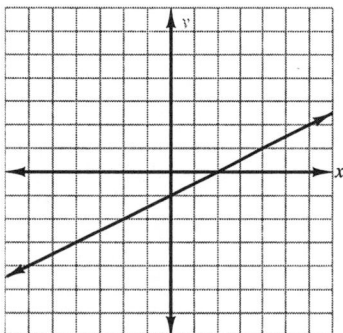

23. $y = -\frac{1}{2}x + 1$

$m = -\frac{1}{2}$; y-intercept: $(0, 1)$

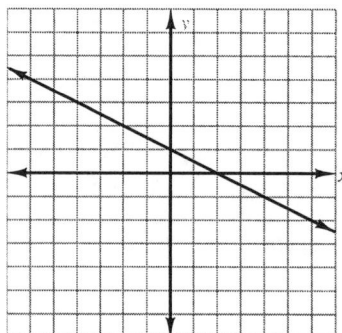

25. $y = -\dfrac{1}{2}x - 1$

$m = -\dfrac{1}{2}$; y-intercept: $(0, -1)$

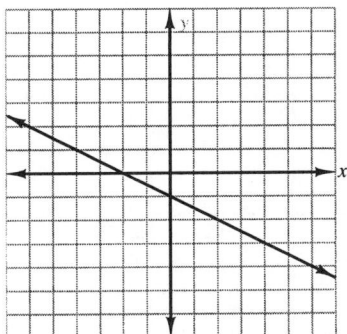

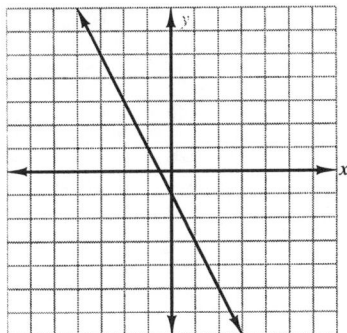

27. $x + y = 3$

$y = -x + 3$

$m = -1$; y-intercept: $(0, 3)$

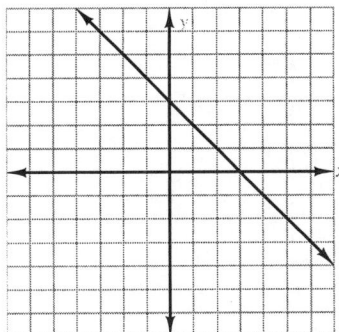

29. $2x + y = -1$

$y = -2x - 1$

$m = -2$; y-intercept: $(0, -1)$

31. $2x + 3y = 6$

$3y = -2x + 6$

$y = -\dfrac{2}{3}x + 2$

$m = -\dfrac{2}{3}$; y-intercept: $(0, 2)$

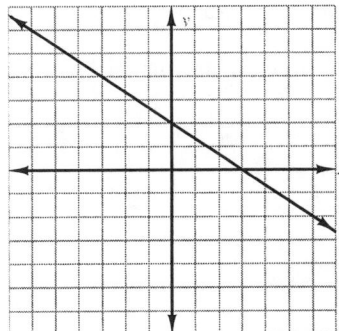

33. $-2x + 3y = 6$

$$3y = 2x + 6$$
$$y = \frac{2}{3}x + 2$$

$m = \frac{2}{3}$; y-intercept: $(0, 2)$

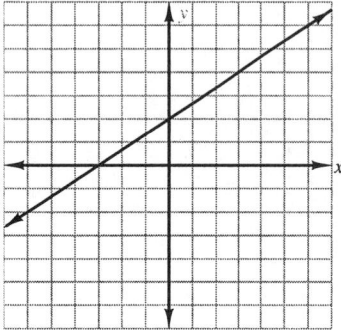

35. $\boxed{y > 2x - 1}$

Graph $y = 2x - 1$ with a dashed line

$m = 2$; y-intercept: $(0, -1)$

Shade half-plane above line since $y > 2x - 1$

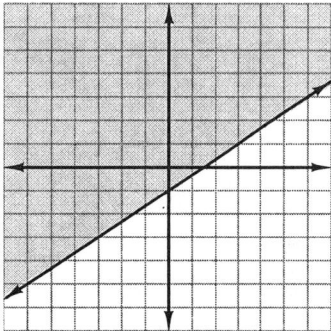

37. $\boxed{y \geq \frac{2}{3}x - 1}$

Graph $y = \frac{2}{3}x - 1$ with a solid line

$m = \frac{2}{3}$; y-intercept: $(0, -1)$

Shade half-plane above line since $y \geq \frac{2}{3}x - 1$

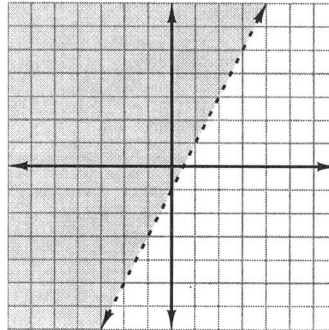

39.

$$2x + 5y > 10$$
$$5y > -2x + 10$$
$$\boxed{y > -\frac{2}{5}x + 2}$$

Graph $y = -\frac{2}{5}x + 2$ with a dashed line

$m = -\frac{2}{5}$; y-intercept: $(0, 2)$

Shade half-plane above line since $y > -\frac{2}{5}x + 2$.

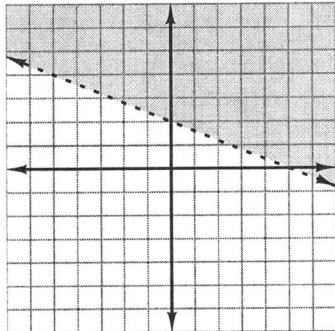

41.

$$x - 2y > 0$$
$$-2y > -x$$
$$\boxed{y < \frac{1}{2}x}$$

Graph $y = \frac{1}{2}x$ with a dashed line

$m = \frac{1}{2}$; y-intercept: $(0, 0)$

Shade half-plane below line since $y < \frac{1}{2}x$.

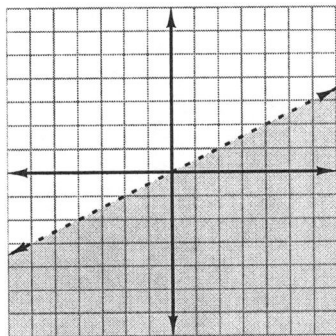

43. $4x - 3y \geq 12$

$-3y \geq -4x + 12$

$$\boxed{y \leq \frac{4}{3}x - 4}$$

Graph $y = \frac{4}{3}x - 4$ with a solid line

$m = \frac{4}{3}$; *y*-intercept: $(0, -4)$

Shade half-plane below line since $y \leq \frac{4}{3}x - 4$.

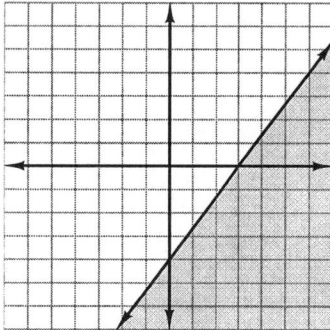

45. $-2x - 2y \leq -10$

$x + y \geq 5$

$$\boxed{y \geq -x + 5}$$

Graph $y = -x + 5$ with a solid line

$m = -1$; *y*-intercept: $(0, 5)$

Shade half-plane above line since $y \geq -x + 5$.

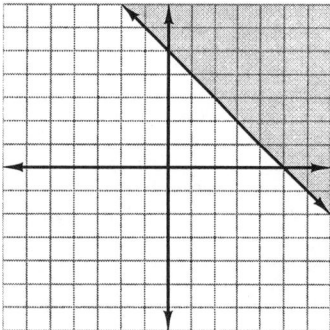

47. $y = 2x + 3$ and $y = 2x + 17$

$m_1 = 2$ $m_2 = 2$

$m_1 = m_2 = 2$

The slopes are equal.

The lines are $\boxed{\text{parallel}}$.

49. $y = 2x + 3$ and $y = -\dfrac{1}{2}x + 5$

$m_1 = 2$ $m_2 = -\dfrac{1}{2}$

$$m_1 m_2 = 2\left(-\dfrac{1}{2}\right) = -1$$

The slopes are <u>negative reciprocals</u>.

The lines are $\boxed{\text{perpendicular}}$.

51.
$$\begin{aligned}
2x + 3y &= 5 \\
3y &= -2x + 5 \\
y &= -\dfrac{2}{3}x + \dfrac{5}{3} \\
m_1 &= -\dfrac{2}{3}
\end{aligned}$$
  and
$$\begin{aligned}
2x - 3y &= 11 \\
-3y &= -2x + 11 \\
y &= \dfrac{2}{3}x - \dfrac{11}{3} \\
m_2 &= \dfrac{2}{3}
\end{aligned}$$

$$m_1 \neq m_2$$
$$m_1 m_2 = \left(-\dfrac{2}{3}\right)\left(\dfrac{2}{3}\right) = -\dfrac{4}{9} \neq -1$$

The slopes are not equal and are not negative reciprocals.
The lines are not parallel and not perpendicular.

The lines are $\boxed{\text{intersecting, but not perpendicular}}$.

53.
$$\begin{aligned}
y &= 4x \\
m_1 &= 4
\end{aligned}$$
  and
$$\begin{aligned}
y &= \dfrac{1}{4}x \\
m_2 &= \dfrac{1}{4}
\end{aligned}$$

$$m_1 \neq m_2$$
$$m_1 m_2 = 4\left(\dfrac{1}{4}\right) = 1$$

The slopes are not equal one are not negative reciprocals.
The lines are not parallel and not perpendicular.

The lines are $\boxed{\text{intersecting, but not perpendicular}}$.

55.
$$\begin{aligned}
y &= 2 \\
m_1 &= 0
\end{aligned}$$
  and
$$\begin{aligned}
y &= 4 \\
m_2 &= 0
\end{aligned}$$

The slopes are equal.

The lines are $\boxed{\text{parallel}}$.

57. $x = 0$ and $y = 0$
$x = 0$ is a vertical line (y-axis)
$y = 0$ is a horizontal line (x-axis)

The lines are $\boxed{\text{perpendicular}}$.

59. Passing through $(1, 3)$ and parallel to $y = 2x - 1$
$y = 2x - 1$: $m = 2$
slope of line parallel $= 2$
point-slope: $\boxed{y - 3 = 2(x - 1)}$
$$y - 3 = 2x - 2$$
slope-intercept: $\boxed{y = 2x + 1}$
standard: $\boxed{2x - y = -1}$

61. Passing through $(1, -3)$ and perpendicular to $y = 2x - 1$.

$y = 2x - 1$: $m = 2$

slope of line perpendicular: $m = -\dfrac{1}{2}$

point-slope: $\boxed{y + 3 = -\dfrac{1}{2}(x - 1)}$

$\qquad\qquad y + 3 = -\dfrac{1}{2}x + \dfrac{1}{2}$

slope-intercept: $\boxed{y = -\dfrac{1}{2}x - \dfrac{5}{2}}$

standard: $2y = -x - 5$

$\qquad\qquad \boxed{x + 2y = -5}$

63. Passing through $(-2, -5)$ and parallel to $2x + y = 4$.

$\qquad 2x + y = 4$

$\qquad\qquad y = -2x + 4$: $m = -2$

slope of line parallel: $m = -2$

point-slope: $\boxed{y + 5 = -2(x + 2)}$

$\qquad\qquad y + 5 = -2x - 4$

slope intercept: $\boxed{y = -2x - 9}$

standard: $\boxed{2x + y = -9}$

65. Passing through $(2, -3)$ and perpendicular to $-\dfrac{1}{3}x + y = 4$

$\qquad -\dfrac{1}{3}x + y = 4$

$\qquad\qquad y = \dfrac{1}{3}x + 4$: $m = \dfrac{1}{3}$

slope of line perpendicular: $m = -3$

point-slope: $\boxed{y + 3 = -3(x - 2)}$

$\qquad\qquad y + 3 = -3x + 6$

slope-intercept: $\boxed{y = -3x + 3}$

standard: $\boxed{3x + y = 3}$

67. Containing $(-1, 3)$ and parallel to $2x - 4y = 12$:

$\qquad 2x - 4y = 12$

$\qquad x - 2y = 6$

$\qquad -2y = -x - 6$

$\qquad\qquad y = \dfrac{1}{2}x + 3$: $m = \dfrac{1}{2}$

slope of line parallel: $m = \dfrac{1}{2}$

point-slope: $\boxed{y - 3 = \dfrac{1}{2}(x + 1)}$

$\qquad\qquad y - 3 = \dfrac{1}{2}x + \dfrac{1}{2}$

slope-intercept: $\boxed{y = \dfrac{1}{2}x + \dfrac{7}{2}}$

standard: $2y = x + 7$

$\qquad\qquad \boxed{x - 2y = -7}$

69. Containing $(1, -3)$ and perpendicular to $2x - 4y = 12$

$$2x - 4y = 12$$
$$x - 2y = 6$$
$$x = \frac{1}{2}x + 3: \ m = \frac{1}{2}$$

slope of line perpendicular: -2

point-slope: $\boxed{y + 3 = -2(x - 1)}$

$$y + 3 = -2x + 2$$

slope-intercept: $\boxed{y = -2x - 1}$

standard: $\boxed{2x + y = -1}$

71. Containing the origin and perpendicular to the line $5x + 3y = 10$:

$$5x + 3y = 10$$
$$3y = -5x + 10$$
$$y = -\frac{5}{3}x + \frac{10}{3}: \ m = -\frac{5}{3}$$

slope of line perpendicular: $\dfrac{3}{5}$

origin: $(0, 0)$

point-slope: $\boxed{y - 0 = \dfrac{3}{5}(x - 0)}$

slope-intercept: $\boxed{y = \dfrac{3}{5}x}$

standard: $\boxed{3x - 5y = 0}$

73. Containing $(2, 4)$ and parallel to the line passing through $(1, -5)$ and $(0, -6)$

$$m = \frac{-6 - (-5)}{0 - 1} = \frac{-1}{-1} = 1$$

slope of line parallel: $m = 1$

point-slope: $\boxed{y - 4 = 1(x - 2)}$

slope-intercept: $\boxed{y = x + 2}$

standard: $\boxed{x - y = -2}$

75. Containing $(-3, -7)$ and perpendicular to the line passing through $\left(8, -\dfrac{3}{2}\right)$ and $\left(0, \dfrac{5}{2}\right)$

$$m = \frac{\dfrac{5}{2} - \left(-\dfrac{3}{2}\right)}{0 - 8} = \frac{\dfrac{8}{2}}{-8} = -\frac{1}{2}$$

slope of line perpendicular: $m = -\dfrac{1}{-\dfrac{1}{2}} = 2$

point-slope: $\boxed{y + 7 = 2(x + 3)}$

$$y + 7 = 2x + 6$$

slope-intercept: $\boxed{y = 2x - 1}$

standard: $\boxed{2x - y = 1}$

77. Passing through (–2, 4) and parallel to the x-axis parallel to the x-axis: (horizontal line) $m = 0$, $y = 0$
horizontal line thorugh (–2, 4): $m = 0$, $y = 4$

point-slope: $\boxed{y - 4 = 0(x + 2)}$

slope-intercept: $\boxed{y = 4}$

standard: $\boxed{y = 4}$

79. Perpendicular to $3x - 2y = 4$ with same x-intercept

$$
\begin{aligned}
3x - 2y &= 4 \\
-2y &= -3x + 4 \\
y &= \frac{3}{2}x - 2: \ m = \frac{3}{2}; \ x\text{-intercept: } \left(\frac{4}{3}, 0\right)
\end{aligned}
$$

slope of line perpendicular: $m = -\dfrac{1}{\frac{3}{2}} = -\dfrac{2}{3}$

point-slope: $\boxed{y - 0 = -\dfrac{2}{3}\left(x - \dfrac{4}{3}\right)}$

slope-intercept: $\boxed{y = -\dfrac{2}{3}x + \dfrac{8}{9}}$

$$9y = -6x + 8$$

standard: $\boxed{6x + 9y = 8}$

81.
$$
\begin{aligned}
Ax + By &= C \\
By &= -Ax + C \\
y &= -\frac{Ax}{B} + \frac{C}{B}: \ m = -\frac{A}{B}
\end{aligned}
$$

slope of line parallel: $m = \boxed{-\dfrac{A}{B}}$

83. (c, r):
(291 micrograms/cubic meter, 103 deaths),
(582 micrograms/cubic meters, 112 deaths)

$$
\begin{aligned}
m &= \frac{112 - 103}{582 - 291} = \frac{9}{291} = \frac{3}{97} \\
r - 103 &= \frac{3}{97}(c - 291) \\
r - 103 &= \frac{3}{97}c - 9 \\
\boxed{r = \frac{3}{97}c + 94}
\end{aligned}
$$

85. $\boxed{D}$ is true;
$$
\begin{aligned}
Ax + By &= C \\
By &= C - Ax \\
y &= -\frac{Ax}{B} + \frac{C}{B}: \ \text{slope, } m = -\frac{A}{B}
\end{aligned}
$$

87. $40° \text{ E} = 25° \text{ M}$: (M, E): (25, 40)
$280° \text{ E} = 125° \text{ M}$: (125, 280)

$$m = \frac{280 - 40}{125 - 25} = \frac{240}{100} = \frac{12}{5}$$

$$E - 40 = \frac{12}{5}(M - 25)$$

$$E - 40 = \frac{12}{5}M - 60$$

$$\boxed{E = \frac{12}{5}M - 20}$$

89.
$$by = 8x - 1, \text{ slope} = -2$$

$$y = \frac{8}{b} - 1$$

$$\frac{8}{b} = -2$$

$$\frac{b}{8} = -\frac{1}{2}$$

$$b = \boxed{-4}$$

91. $Ax + By = C$

(t, u): $At + Bu = C$

(v, w): $Av + Bu = C$

$$At + Bu = Av + Bw$$

$$A(t - v) = B(w - u)$$

$$\frac{A}{B}(t - v) = w - u$$

$$\boxed{\frac{A}{B} = \frac{w - u}{t - v}}$$

93. line passing through (0, 4) and $(a - 2, 6)$ and has x-intercept of a: $(a, 0)$

$$m = \frac{6 - 4}{a - 2 - 0} = \frac{2}{a - 2}$$

$$y - 4 = \frac{2}{a - 2}(x - 0)$$

$$y = \frac{2}{a - 2}x + 4$$

x-intercept: $(a, 0) \rightarrow x = a, y = 0$

$$0 = \frac{2a}{a - 2} + 4$$

$$\frac{2a}{a - 2} = -4$$

$$2a = -4a + 8$$

$$6a = 8$$

$$a = \boxed{\frac{4}{3}}$$

95.

Number	Term		Pattern
first	17	(1, 17)	$8 \cdot 1 + 9 = 17$
second	25	(2, 25)	$8 \cdot 2 + 9 = 25$
third	33	(3, 33)	$8 \cdot 3 + 9 = 33$
fourth	41	(4, 41)	$8 \cdot 4 + 9 = 41$
fifth	49	(5, 49)	$8 \cdot 5 + 9 = 49$
n^{th}	$8n + 9$	$\boxed{(n, 8n + 9)}$	$8n + 9$

Review Problems

102. Let

$$x \quad = \quad \text{first consecutive odd integer}$$
$$x+2 \quad = \quad \text{second consecutive odd integer}$$
$$x+4 \quad = \quad \text{third consecutive odd integer}$$
$$x+2(x+2) \quad = \quad 2(x+4)+3$$
$$x+2x+4 \quad = \quad 2x+8+3$$
$$3x+4 \quad = \quad 2x+11$$
$$x \quad = \quad 7$$
$$x+2 \quad = \quad 9$$
$$x+4 \quad = \quad 11$$

The integers are $\boxed{7, 9 \text{ and } 11}$.

103. Let

$$x \quad = \quad \text{amount invested at 8\% for 3 years}$$
$$50,000-x \quad = \quad \text{amount invested at 11\% for 2 years}$$
$$0.08(3)(x)+0.11(2)(50,000-x) \quad = \quad 11400$$
$$0.24x+11,000-0.22x \quad = \quad 11,400$$
$$0.02x \quad = \quad 400$$
$$x \quad = \quad 20000 \text{ at 8\%}$$
$$50,000-x \quad = \quad 30000 \text{ at 11\%}$$

$\boxed{\$20,000 \text{ at 8\% for 3 yr; } \$30,000 \text{ at 11\% for 2 yr}}$

104. possible combinations of 13 coins:

nickels	quarters	Value	$1.65 < x < 2.65$?
13	0	0.65	No
12	1	$0.60 + 0.25 = 0.85$	No
11	2	$0.55 + 0.50 = 1.05$	No
10	3	$0.50 + 0.75 = 1.25$	No
9	4	$0.45 + 1.00 = 1.45$	No
8	5	$0.40 + 1.25 = 1.65$	No
7	6	$0.35 + 1.50 = 1.85$	Yes
6	7	$0.30 + 1.75 = 2.05$	Yes
5	8	$0.25 + 2.00 = 2.25$	Yes
4	9	$0.20 + 2.25 = 2.45$	Yes
3	10	$0.15 + 2.50 = 2.65$	No
2	11	$0.10 + 2.75 = 2.85$	No
1	12	$0.05 + 3.00 = 3.05$	No
0	13	$0.00 + 3.25 = 3.25$	No

Possible combinations are:

$\boxed{\begin{array}{l} \text{6 quarters, 7 nickels; 7 quarters, 6 nickels;} \\ \text{8 quarters, 5 nickels; 9 quarters, 4 nickels} \end{array}}$

Section 3.6 Variation

Problem Set 3.6, pp. 238-239

1. $C = kd$

Given: $d = 2$ feet, $C = 2\pi$ feet

$2\pi = k(2)$

$\pi = k$

$C = \pi d$

At $r = 8$ feet; $d = 16$ feet, $C = ?$

$C = \pi(16)$

$C = 16\pi$

The circumference is $\boxed{16\pi \text{ ft}}$.

3. $d = kr^2$

Given: $d = 200$ feet, $r = 60$ mph

$200 = k(60)^2$

$200 = k(3600)$

$\dfrac{1}{18} = k$

$d = \dfrac{1}{18} r^2$

At $r = 100$ mph; $d = ?$

$d = \dfrac{1}{18} r^2$

$d = \dfrac{1}{18} (100)^2$

$d = \dfrac{10000}{18} = 555\dfrac{5}{9}$

The stopping distance is $\boxed{555\dfrac{5}{9} \text{ feet}}$.

5. $V = \dfrac{k}{p}$

Given: $V = 32 \text{ cm}^3, p = 8$ lb

$32 = \dfrac{k}{8}$

$256 = k$

$V = \dfrac{256}{p}$

At $V = 40 \text{ cm}^3, p = ?$

$40 = \dfrac{256}{p}$

$p = \dfrac{256}{40} = 6.4$

The pressure is $\boxed{6.4 \text{ lb}}$.

7. Let

$$I = \text{intensity}$$
$$d = \text{distance}$$
$$I = \frac{k}{d^2}$$

Given: $I = 25$ ft-candles, $d = 4$ ft.

$$25 = \frac{k}{4^2}$$
$$k = 25(16) = 400$$
$$I = \frac{400}{d^2}$$

At $d = 6$ ft, $I = ?$

$$I = \frac{400}{6} = \frac{400}{36} = 11\frac{1}{9}$$

The illumination is $\boxed{11\frac{1}{9} \text{ foot candles}}$.

9. Let

$$V = \text{volume}$$
$$T = \text{temperature}$$
$$P = \text{pressure}$$
$$V = \frac{kT}{P}$$

Given: $T = 100$ Kelvin, $P = 15$ kgm/m^2, $V = 20$ m^3

Find k: $20 = \dfrac{k(100)}{15}$

$$k = 3$$
$$V = \frac{3T}{P}$$

At $T = 150$ Kelvin, $P = 30$ kgm/m^2, $V = ?$

$$V = \frac{3(150)}{50} = 15$$

The volume is $\boxed{15 \text{ m}^3}$.

11. Let

$$I = \text{simple interest}$$
$$r = \text{interest rate}$$
$$t = \text{time that money is invested}$$
$$I = krt$$

Given: $r = 12\% = 0.12$, $k = 2$ years, $I = \$280$

$$280 = k(0.12)(2)$$
$$\frac{3500}{3} = k$$
$$I = \frac{3500}{3}rt$$

At $r = 16\% = 0.16$, $t = 4$ years, $I = ?$

$$I = \frac{3500}{3}(0.16)(4) = 746.6\overline{6}$$

The investment will yield $\boxed{\text{approximately } \$746.67}$.

13. Let

F = force of attraction
m_1 = first mass
m_2 = second mass
d = distance between masses

$$F = \frac{km_1m_2}{d^2}$$

Given: m_1 = 4 units, m_2 = 2 units, d = 3 feet, F = 16 units

$$16 = \frac{k4(2)}{3^2}$$

$$18 = k$$

$$F = \frac{18m_1m_2}{d^2}$$

At m_1 = 5 units, m_2 = 3 units, d = 2 feet, F = ?

$$F = \frac{18(5)(3)}{2^2}$$

$$F = 67.5$$

The force is $\boxed{67.5 \text{ units}}$.

15. Let

H = number of hours
T = number of tasks
P = number of people

$$H = \frac{kT}{P}$$

Given: P = 2, T = 6, H = 4

$$4 = \frac{k6}{2}$$

$$\frac{4}{3} = k$$

$$H = \frac{4T}{3P}$$

At T = 18, H = 8, P = ?

$$8 = \frac{4(18)}{3P}$$

$$P = 3$$

$\boxed{3 \text{ people}}$ are required

17. Let

R = electrical resistance
L = length
D = diameter

$$R = \frac{kL}{D^2}$$

Given: L = 720 feet, $D = \frac{1}{4}$-inch, $R = 1\frac{1}{2}$ ohms

$$\frac{3}{2} = \frac{k(720)}{\left(\frac{1}{4}\right)^2}$$

$$\frac{\frac{3}{2}\left(\frac{1}{16}\right)}{720} = k$$

$$\frac{1}{7680} = k$$

$$R = \frac{L}{7680D^2}$$

At $L = 960$ feet, $D = 2\left(\frac{1}{4}\text{-inch}\right) = \frac{1}{2}\text{-inch}$, $R = ?$

$$R = \frac{960}{7680\left(\frac{1}{2}\right)^2} = \frac{1}{2} = 0.5$$

The resistance is $\boxed{0.5 \text{ ohm}}$.

19. Let

$$
\begin{aligned}
F &= \text{centrifugal force} \\
r &= \text{radius of ciruclar path} \\
M &= \text{body's mass} \\
t &= \text{time it takes to move about one full circle} \\
F &= \frac{krM}{t^2}
\end{aligned}
$$

Given: $M = 6\text{-grams}$, $r = 100$ cm, $t = 2$ seconds, $F = 6000$ dymes

$$
\begin{aligned}
6000 &= \frac{k100(6)}{2^2} \\
40 &= k \\
F &= \frac{40rM}{t^2}
\end{aligned}
$$

At $M = 18\text{-gram}$, $r = 100$ cm, $t = 3$ seconds, $F = ?$

$$
\begin{aligned}
F &= \frac{40(100)(18)}{3^2} \\
F &= 8000
\end{aligned}
$$

The centrifugal force is $\boxed{8000 \text{ dynes}}$.

21. $y = \dfrac{4}{x}$

x	-4	-2	0	2	4
$y = \dfrac{4}{x}$	-1	-2	undefined	2	1

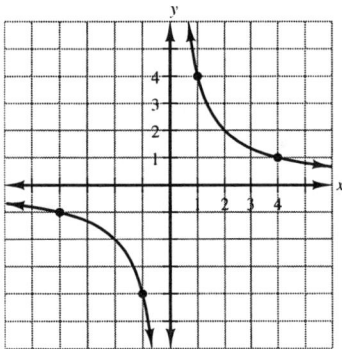

23. $y = -\dfrac{1}{x}$

x	-2	-1	$-\dfrac{1}{2}$	0	$\dfrac{1}{2}$	1	2
$y = -\dfrac{1}{x}$	$\dfrac{1}{2}$	1	2	undefined	-2	-1	$-\dfrac{1}{2}$

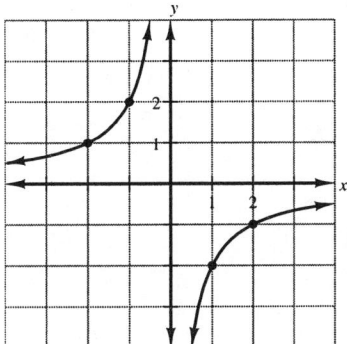

25. $y = -\dfrac{6}{x}$

x	-6	-3	-2	-1	0	1	2	3	6
$y = -\dfrac{6}{x}$	1	2	3	6	undefined	-6	-3	-2	-1

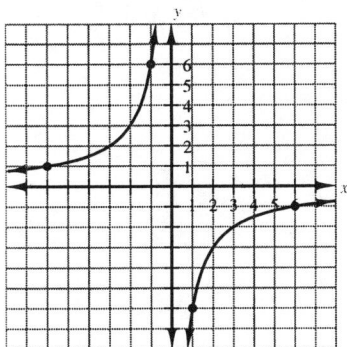

27. $\boxed{C}$ is true

29. $y = \dfrac{K_1}{x},\ x = \dfrac{K_2}{z}$

$y = \dfrac{K_1}{K_2/z} = \dfrac{K_1 z}{K_2}$

$\boxed{y \text{ is directly proportional to } z}$

constant of variation: $\boxed{\dfrac{K_1}{K_2}}$

Review Problems

32. A line passing through (–4, 5) and perpendicular to $4x - 5y = 7$.

slope of $4x - 5y = 7$:

$$-5y = -4x + 7$$

$$y = \frac{4}{3}x - \frac{7}{5}$$

$$m = \frac{4}{5}$$

slope of line perpendicular: $\dfrac{-1}{\frac{4}{5}} = -\dfrac{5}{4}$

$$y - 5 = -\frac{5}{4}(x + 4)$$

$$y - 5 = -\frac{5}{4}x - 5$$

$$y = -\frac{5}{4}x$$

$$4y = -5x$$

$$\boxed{5x + 4y = 0}$$

33.

$$S = 2LW + 2LH + 2WH$$

$$2LW + 2LH + 2WH = S$$

$$2LW + 2WH = S - 2LH$$

$$W(2L + 2H) = S - 2LH$$

$$\boxed{W = \frac{S - 2LH}{2L + 2H}}$$

34. F, C: (32,0), (212, 100)

$$m = \frac{100 - 0}{212 - 32} = \frac{100}{180} = \frac{5}{9}$$

$$C - 0 = \frac{5}{9}(F - 32)$$

$$\boxed{C = \frac{5}{9}(F - 32)}$$

$F = 122$: $C = \dfrac{5}{9}(122 - 32)$

$$C = \frac{5}{9}(90)$$

$$C = 50$$

$$\boxed{50°C}$$

Section 3.7 Graphic Solutions to Systems of Equations

Problem Set 3.7, pp. 244-246

1. From the graph: $\boxed{\{(1, 3), (1, -3), (-1, 3), (-1, -3)\}}$

(1, 3):

$$4x^2 + y^2 = 13 \qquad\qquad x^2 + y^2 = 10$$

$$4(1)^2 + 3^2 = 13 \qquad\qquad 1^2 + 3^2 = 10$$

$$4 + 9 = 13 \qquad\qquad 1 + 9 = 10$$

$$13 = 13 \;\; \surd \qquad\qquad 10 = 10 \;\; \surd$$

$(1, -3)$: $\begin{aligned} 4(1)^2 + (-3)^2 &= 13 \\ 4 + 9 &= 13 \\ 13 &= 13 \ \sqrt{} \end{aligned}$ $\begin{aligned} 1^2 + (-3)^2 &= 10 \\ 1 + 9 &= 10 \\ 10 &= 10 \ \sqrt{} \end{aligned}$

$(-1, 3)$: $\begin{aligned} 4(-1)^2 + 3^2 &= 13 \\ 4 + 9 &= 13 \\ 13 &= 13 \ \sqrt{} \end{aligned}$ $\begin{aligned} (-1)^2 + 3^2 &= 10 \\ 1 + 9 &= 10 \\ 10 &= 10 \ \sqrt{} \end{aligned}$

$(-1, -3)$ $\begin{aligned} 4(-1)^2 + (-3)^2 &= 13 \ \sqrt{} \\ 4 + 9 &= 13 \\ 13 &= 13 \ \sqrt{} \end{aligned}$ $\begin{aligned} (-1)^2 + (-3)^2 &= 10 \\ 1 + 9 &= 10 \\ 10 &= 10 \ \sqrt{} \end{aligned}$

3. From the graph $\boxed{\{(0, 3), (1, 1), (2, -1)\}}$

Check: $\begin{aligned} y &= x^3 - 3x^2 + 3 \end{aligned}$ $\begin{aligned} 2x + y &= 3 \end{aligned}$

$(0, 3)$: $\begin{aligned} 3 &= 0 - 0 + 3 \\ 3 &= 3 \ \sqrt{} \end{aligned}$ $\begin{aligned} 2(0) + 3 &= 3 \\ 0 + 3 &= 3 \\ 3 &= 3 \ \sqrt{} \end{aligned}$

$(1, 1)$: $\begin{aligned} 1 &= 1^3 - 3(1^2) + 3 \\ 1 &= 1 - 3 + 3 \\ 1 &= 1 \ \sqrt{} \end{aligned}$ $\begin{aligned} 2(1) + 1 &= 3 \\ 2 + 1 &= 3 \\ 3 &= 3 \ \sqrt{} \end{aligned}$

$(2, -1)$: $\begin{aligned} -1 &= 2^3 - 3(2^2) + 3 \\ -1 &= 8 - 12 + 3 \\ -1 &= -1 \ \sqrt{} \end{aligned}$ $\begin{aligned} 2(2) + (-1) &= 3 \\ 4 - 1 &= 3 \\ 3 &= 3 \ \sqrt{} \end{aligned}$

5. From the graph: $\boxed{\{(4, 2), (4, -2)\}}$

Check: $\begin{aligned} x - y^2 &= 0 \end{aligned}$ $\begin{aligned} x &= 4 \end{aligned}$

$(4, 2)$: $\begin{aligned} 4 - (2)^2 &= 0 \\ 4 - 4 &= 0 \\ 0 &= 0 \ \sqrt{} \end{aligned}$ $\begin{aligned} 4 &= 4 \ \sqrt{} \end{aligned}$

$(4, -2)$: $\begin{aligned} 4 - (-2)^2 &= 0 \\ 4 - 4 &= 0 \\ 0 &= 0 \ \sqrt{} \end{aligned}$ $\begin{aligned} 4 &= 4 \ \sqrt{} \end{aligned}$

7. From the graph: $\boxed{\{(4, 2), (4, -2)\}}$

Check: $\begin{aligned} 2y^2 &= x + 4 \end{aligned}$ $\begin{aligned} x &= y^2 \end{aligned}$

$(4, 2)$: $\begin{aligned} 2(2^2) &= 4 + 4 \\ 2(4) &= 8 \\ 8 &= 8 \ \sqrt{} \end{aligned}$ $\begin{aligned} 4 &= 2^2 \\ 4 &= 4 \ \sqrt{} \end{aligned}$

$(4, -2)$: $\begin{aligned} 2(-2)^2 &= 4 + 4 \\ 2(4) &= 8 \\ 8 &= 8 \ \sqrt{} \end{aligned}$ $\begin{aligned} 4 &= (-2)^2 \\ 4 &= 4 \ \sqrt{} \end{aligned}$

9. From the graph: $\boxed{\{(2, 2), (-1, -1)\}}$

Check: $\begin{aligned} y^2 &= x + 2 \end{aligned}$ $\begin{aligned} y &= x \end{aligned}$

$(2, 2)$: $\begin{aligned} 2^2 &= 2 + 2 \\ 4 &= 4 \ \sqrt{} \end{aligned}$ $\begin{aligned} 2 &= 2 \ \sqrt{} \end{aligned}$

$(-1, -1)$: $\begin{aligned} (-1)^2 &= -1 + 2 \\ 1 &= 1 \ \sqrt{} \end{aligned}$ $\begin{aligned} -1 &= -1 \ \sqrt{} \end{aligned}$

11. $x + y = 4$
$x - y = 2$

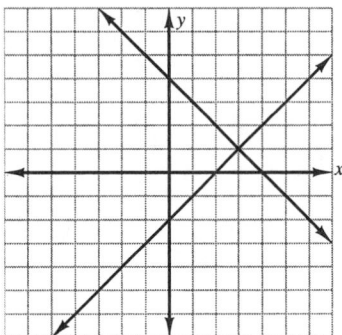

$\boxed{\{(3, 1)\}}$

13. $x - 3y = 2$
$y = 6 - x$

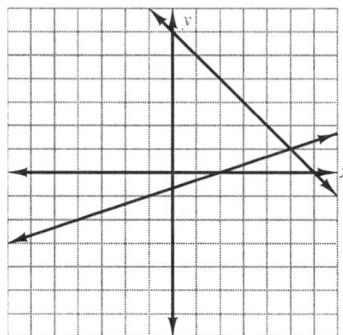

$\boxed{\{(5, 1)\}}$

15. $2x = 3y - 6$
$-3x + y = -5$

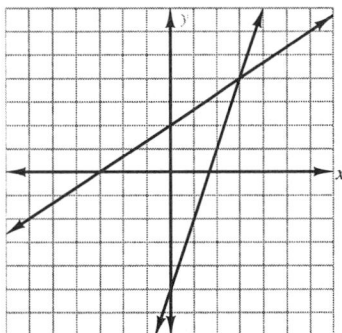

$\boxed{\{(3, 4)\}}$

17. $x + 2y = 1$
$x = 3$

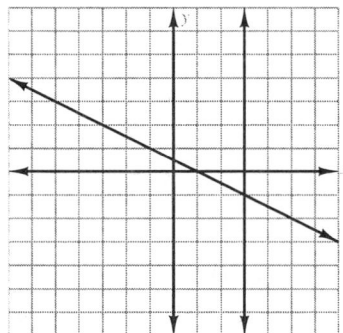

$\boxed{\{(3, -1)\}}$

19. $3x + y = 13$

$6x + 2y = 12$

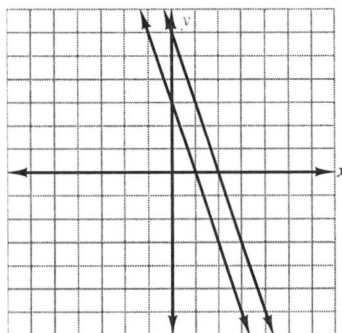

parallel lines
$\boxed{\varnothing \text{ (inconsistent)}}$

21. $\dfrac{x}{3} - \dfrac{y}{4} = 1$

$\dfrac{2x}{3} - \dfrac{y}{2} = 1$

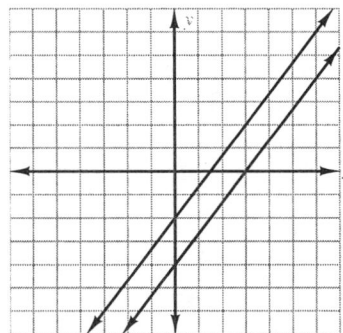

parallel lines
$\boxed{\varnothing \text{ (inconsistent)}}$

23. $2x + 2y = 1$
$4x + 4y = 2$

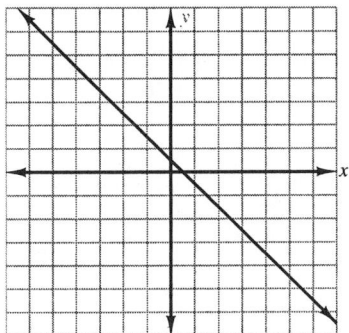

same line
Dependent

25. $3x - 3y = 6$
$2x - 2y = 4$

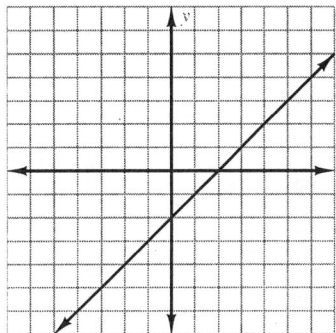

same line
Dependent

27. $y = 500x$
$y = 50x + 1000$

a.

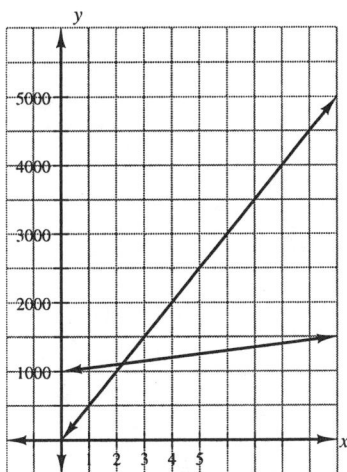

Achilles overtakes the tortoise in 2.2 minutes. This occurs approximately 110 yards from the starting point.

31. $3x - 2y = 4$

[D] $6x - 4y = 6$

$(\div 2)$ $3x - 2y = 3$ is a line parallel to $3x - 2y = 4$
Thus, the system is inconsistent.

33. $5x - 10y = 40$ $\left(\rightarrow \times \frac{2}{5} \right)$ $2x - 4y = 16$

          $2x + ky = -30$ $2x + ky = -30$
The system of linear equations is inconsistent when $k = -4$.

Review Problems

37.
$$\frac{x}{3} - \frac{x-5}{5} = 3$$

$$15\left(\frac{x}{3} - \frac{x-5}{5}\right) = 15(3)$$

$$5x - 3(x-5) = 45$$
$$5x - 3x + 15 = 45$$
$$2x + 15 = 45$$
$$2x = 30$$
$$x = 15$$

y	10	15
12	a	b
13	c	d

diagonal: $15 + a + 13 = 28 + a$ (total)
row 2: $12 + a + b$
$$12 + a + b = 28 + a$$
$$b = 16$$
column 2: $10 + a + c$
$$10 + a + c = 28 + a$$
$$c = 18$$

y	10	15
12	a	16
13	18	d

All columns, rows and diagonal have the same sum.

row 1: $y + 10 + 15 = 25 + y$
row 2: $12 + a + 16 = 28 + a$
row 3: $13 + 18 + d = 31 + d$
column 1: $y + 12 + 13 = 25 + y$
column 2: $10 + a + 18 = 28 + d$
column 3: $15 + 16 + d = 31 + d$
diagonal (top left): $y + a + d$
diagonal (top right): $13 + a + 15 = 28 + a$

Let total $= y + a + d$
Then $y + a + d = 25 + y \;\rightarrow\; a + d = 25$ 1
         $y + a + d = 28 + a \;\rightarrow\; y + d = 28$ 2
         $y + a + d = 31 + d \;\rightarrow\; y + a = 31$ 3

Subtract equation 2 from equation 3.

$$\left.\begin{array}{r} y + a = 31 \\ \underline{y + d = 28} \end{array}\right\}$$
$$a - d = 3$$

$$\begin{array}{rcl} a - d & = & 3 \\ \underline{a + d} & = & \underline{25} \\ 2a & = & 28 \\ a & = & 14 \\ a + d & = & 25 \qquad \rightarrow \qquad 14 + d = 25 \\ & & \qquad\qquad\qquad\qquad\quad\; d = 11 \\ y + a & = & 31 \qquad \rightarrow \qquad y + 14 = 31 \\ & & \qquad\qquad\qquad\qquad\quad\; y = 17 \end{array}$$

17	10	15
12	14	16
13	18	11

38. Let x = price before the increase.

$$\begin{array}{rcl} x + 0.16x & = & 3.19 \\ 1.16x & = & 3.19 \\ x & = & 2.75 \end{array}$$

$\boxed{\$2.75}$

39.
$$\left| 2x + 3 \right| = \left| 4x - 5 \right|$$

$$\begin{array}{rcl} 2x + 3 & = & 4x - 5 \\ -2x & = & -8 \\ x & = & 4 \end{array} \qquad \text{or} \qquad \begin{array}{rcl} 2x + 3 & = & -(4x - 5) \\ 2x + 3 & = & -4x + 5 \\ 6x & = & 2 \\ x & = & \dfrac{1}{3} \end{array}$$

$$\boxed{\left\{ \dfrac{1}{3}, 4 \right\}}$$

Chapter 3   Review Problems

Review Problems, pp. 248-251

1. $y = x^2 - 6x + 8$

x	0	1	2	3	4	5
y	8	3	0	-1	0	3

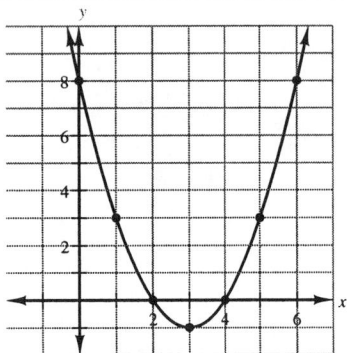

The graph defines y as a function of x, since each value of x has only one y-value.

2. $y = \dfrac{10{,}000x}{100 - x}$

x	0	20	50	60	70	80	90	95
y	100	124.75	166	199	248.5	496	991	9901

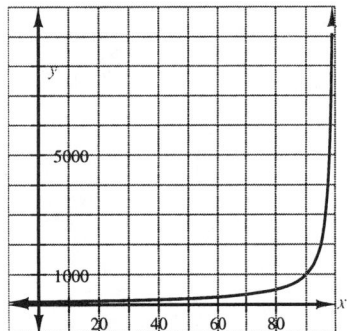

The cost of removing the pollutants from the lake increases rapidly as the percent of removed pollutants appoaches 100%.

3.   **a.**   minimum temperature (time): 5 P.M.
            minimum temperature (time): $-4°$

   b.   maximum temperature (time): 8 P.M.
            maximum temperature: $16°$

   c. x-intercepts: 4 and 6
            At 4 P.M. and 6 P.M. the temperature is $0°$

   d. y-intercept: 12
            at noon the temperature is $12°$

   e.   7 P.M.: $4°$
            8 P.M.: $16°$

            percent increase in temperature: $\dfrac{16 - 4}{4} = \dfrac{12}{4} = 3 = \boxed{300\%}$

4. $2x - 4y = -8$
   x-intercepts: $(-4, 0)$
   y-intercepts: $(0, 2)$

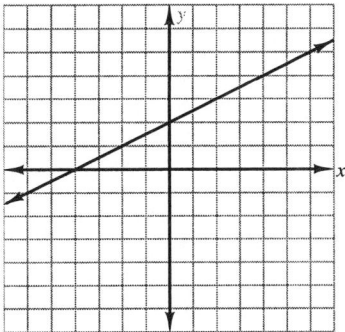

5. $x - 3y \leq 6$

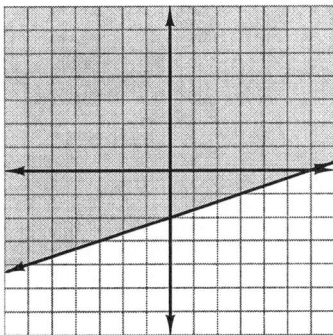

6. $x \geq 3$ and $y \leq 0$

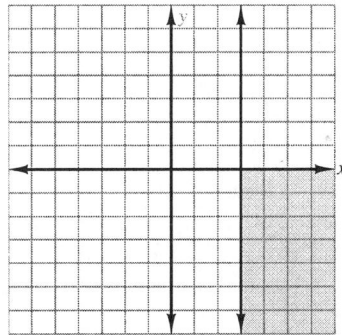

7. $2x - y > -4$ and $x \geq 0$

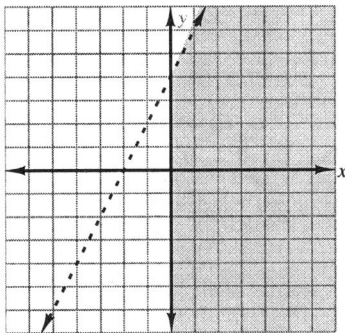

8. $3x - y \leq 6$ or $x + y \geq 2$

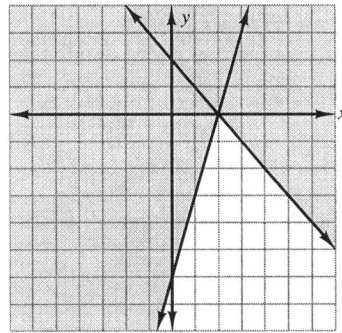

9. $\begin{aligned} |2x + 1| &\leq 3 \\ -3 \leq 2x + 1 &\leq 3 \\ -4 \leq 2x &\leq 2 \\ -2 \leq x &\leq 1 \end{aligned}$

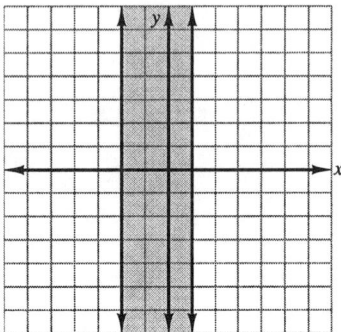

10. $\begin{aligned} |3y + 2| &> 8 \end{aligned}$

$\begin{array}{lll} 3y + 2 > 8 & \text{or} & 3y + 2 < -8 \\ 3y > 6 & & 3y < -10 \\ y > 2 & & y < -\dfrac{10}{3} \end{array}$

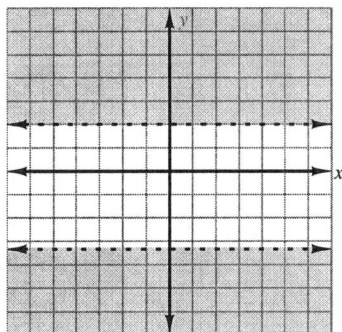

11. $\{(2, 7), (3, 9), (5, 7)\}$

  Function   since each first component is assigned exactly one second component.

  $\text{Domain} = \{\, x \mid x = 2, 3, 5\,\}$

  $\text{Range} = \{\, y \mid y = 7\,\}$

12. $\{(1, 10), (2, 500), (13, \pi)\}$

  Function   since each first component is assigned exactly one second component.

  $\text{Domain} = \{\, x \mid x = 1, 2, 13\,\}$

  $\text{Range} = \{\, y \mid y = 10, 500, \pi\,\}$

13. $\{(1, 2), (3, 4), (5, 6)\})$

  Function   since each first component is assigned exactly one second component.

  $\text{Domain} = \{\, x \mid x = 1, 3, 5\,\}$

  $\text{Range} = \{\, y \mid y = 2, 4, 6\,\}$

14. $\{(12, 13), (14, 15), (12, 19)\}$

$\boxed{\text{Not a function}}$ since $(12, 13)$ and $(12, 19)$ have the same first component.

$\boxed{\text{Domain} = \{ x \mid x = 12, 14\}}$

$\boxed{\text{Range} = \{ y \mid y = 13, 15, 19\}}$

15. $y = 4x - 148$

   a. $\boxed{f(x) = 4x - 148}$

   b. $f(60) = 4(60) - 149 = 240 - 148 = 92$

      $\boxed{f(60) = 92}$

      $\boxed{\text{At } 60° \text{ a cricket chirps 92 times/minutes.}}$

   c. range: 164

        $164 = 4x - 148$

        $312 = 4x$

         $78 = x$

      domain: $\boxed{78}$

      $\boxed{\text{If the cricket chirps 164 times/minute, the temperature is } 78°.}$

16. $s = -16t^2 + 64t + 80$

   a. $\boxed{f(t) = -16t^2 + 64t + 80}$

   b. $f(0) = -0 + 0 + 80 = 80$

      $\boxed{f(0) = 80}$

      $\boxed{\text{Initially the ball is 80 feet high.}}$

   c. $f(3) = -16(3^2) + 64(3) + 80$

      $f(3) = -144 + 192 + 80$

      $\boxed{f(3) = 128}$

      $\boxed{\text{After 3 seconds the ball is 128 feet high.}}$

17. $f(t) = t^3 + 2t^2 - 5t + 3$

   a. $f(3) = 3^3 + 2(3^2) - 5(3) + 3 = 27 + 18 - 15 + 3 = 33$

      $f(1) = 1^3 + 2(1^2) - 5(1) + 3 = 1 + 2 - 5 + 3 = 1$

      $f(3) - f(1) = 33 - 1 = \boxed{32}$

   b. The particle moves 32 inches (from 1 inch to 33 inches) on the number line between 1 and 3 seconds.

18. $4x + y = 17$

          $y = -4x + 17$

   $\boxed{f(x) = -4x + 17}$

   $f(-5) = -4(-5) + 17 = 20 + 17 = 37$

   $\boxed{f(-5) = 37}$

   $\boxed{f(a) = -4a + 17}$

19. $2x - 3y = 9$

          $-3y = -2x + 9$

            $y = \dfrac{2}{3}x - 3$

   $\boxed{f(x) = \dfrac{2}{3}x - 3}$

   $f(-5) = \dfrac{2}{3}(-5) - 3 = -\dfrac{10}{3} - \dfrac{9}{3} = -\dfrac{19}{3}$

   $\boxed{f(-5) = -\dfrac{19}{3}}$

   $\boxed{f(a) = \dfrac{2}{3}a - 3}$

20. $3y - 18 = 0$
$3y = 18$
$y = 6$

$\boxed{f(x) = 6}$

$\boxed{f(-5) = 6}$

$\boxed{f(a) = 6}$

21. $\boxed{y \text{ is not a function of } x}$

A vertical line drawn through $x = 1$ (or other values of x in the domain of x) intersects the graph twice.

$\boxed{\begin{array}{c} \text{Domain} = \{ x \mid 0 \leq x \leq 3 \} \\ [0, 3] \\ \text{Range} = \{ y \mid -2 \leq y \leq 4 \} \\ [-2, 4] \end{array}}$

22. $\boxed{y \text{ is a function of } x}$

Each vertical line that can be drawn intersects the graph only once.

$\boxed{\begin{array}{c} \text{Domain} = \{ x \mid -4 < x < 0 \text{ or } x \geq 3 \} \\ (-4, 0) \text{ or } [3, \infty) \\ \text{Range} = \{ y \mid y \leq -1 \text{ or } y = 2 \} \\ (-\infty, -1) \text{ or } [2] \end{array}}$

23. $\boxed{y \text{ is a function of } x}$

Each vertical line that can be drawn intersects the graph only once

$\boxed{\begin{array}{c} \text{Domain} = \{ x \mid x \in R \} \\ (-\infty, \infty) \\ \text{Range} = \{ y \mid 0 \leq y < 1 \} \\ [0, 1) \end{array}}$

24. $\boxed{y \text{ is a function of } x}$

Each vertical line that can be drawn intersects the graph only once.

$\boxed{\begin{array}{c} \text{Domain} = \{ x \mid -3 \leq x \leq 3 \} \\ [3, 3] \\ \text{Range} = \{ y \mid 0 \leq y \leq 3 \} \\ [0, 3] \end{array}}$

25. $\boxed{y \text{ is a function of } x}$

Each vertical line that can be drawn intersects the graph only once.

$\boxed{\begin{array}{c} \text{Domain} = \{ x \mid x \in R \} \\ (-\infty, \infty) \\ \text{Range} = \{ y \mid -1 < y < 1 \} \\ (-1, 1) \end{array}}$

26. $\boxed{y \text{ is not a function of } x}$

A vertical line drawn through $x = 3$ intersects the graph infinitely many times.

$$\begin{array}{l} \text{Domain} = \{\, x \mid x = 3 \,\} \\ \qquad\qquad [3] \\ \text{Range} = \{\, y \mid y \in R \,\} \\ \qquad\quad (-\infty, \infty) \end{array}$$

27. $\boxed{y \text{ is a function of } x}$

Each vertical line that can be drawn intersects the graph only once.

$$\begin{array}{l} \text{Domain} = \{\, x \mid x = -4, -2, 0, 2, 4 \,\} \\ \qquad\qquad [-4], [-2], [0], [2], [4] \\ \text{Range} = \{\, y \mid y = -4, -2, 0, 2, 4 \,\} \\ \qquad\qquad [-4], [-2], [0], [2], [4] \end{array}$$

28. $\boxed{y \text{ is not a function of } x}$

A vertical line drawn through $x = 2$ (and also through $x = 3$) intersects the graph twice.

$$\begin{array}{l} \text{Domain} = \{\, x \mid x = 2, 3 \,\} \\ \qquad\qquad [2], [3] \\ \text{Range} = \{\, y \mid y = 1, 3, 4, 6 \,\} \\ \qquad\quad [1], [3], [4], [6] \end{array}$$

29.

$$\begin{aligned} f(x) &= x^2 - 5x + 2 \\ f(-2) &= (-2)^2 - 5(-2) + 2 \\ &= 4 + 10 + 2 \\ &= 16 \end{aligned}$$

$$\begin{aligned} g(x) &= 3\sqrt{x} - x + 4 \\ g(16) &= 3\sqrt{16} - 16 + 4 \\ &= 3(4) - 12 \\ &= 12 - 12 \\ &= 0 \end{aligned}$$

$$f(-2) - g(16) = 16 - 0 = \boxed{16}$$

30. x-intercept of 2: $(2, 0)$

slope of $-\dfrac{2}{3}$

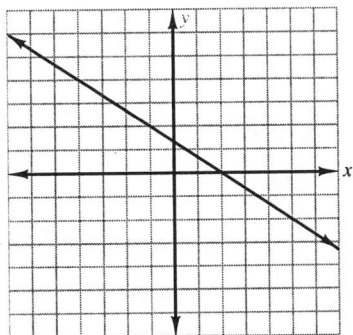

31. line passing through $(1, -3)$

slope of -2

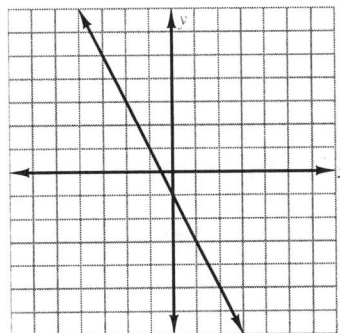

32. line passing through $(-2, 3)$ with slope $= -4$

$$\begin{aligned} y - 3 &= -4(x + 2) \\ y - 3 &= -4x - 8 \\ \boxed{y &= -4x - 5} \end{aligned}$$

33. line passing through $(1, 3)$ and $(-4, 18)$

$$\begin{aligned} m &= \frac{18 - 3}{-4 - 1} = \frac{15}{-5} = -3 \\ y - 3 &= -3(x - 1) \quad \text{or} \quad y - 18 = -3(x + 4) \\ y - 3 &= -3x + 3 \\ \boxed{y &= -3x + 6} \end{aligned}$$

34. Measurement 1: $(40, 47.2)$
Measurement 2: $(60, 54.8)$
$$m = \frac{54.8 - 47.2}{60 - 40} = \frac{7.6}{20} = 0.38$$

  a. $y - 47.2 = 0.38(x - 40)$ or $y - 54.8 = 0.38(x - 60)$
   $y - 47.2 = 0.38x - 15.2$
   $\boxed{y = 0.38x + 32}$

  b. $y = 80$
   $y = 0.38(80) + 32 = 30.4 + 32 = 62.4$
   $\boxed{62.4 \text{ bushels}}$

35. $y = \frac{3}{4}x - 2$

    $m = \frac{3}{4}$

y-intercept: $(0, -2)$

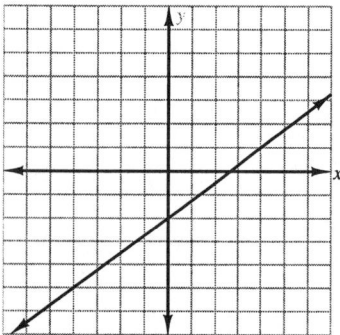

36. $3y - 2x \le 6$

$3y - 2x = 6:$ $3y = 2x + 6$

    $y = \frac{2}{3}x + 2$ $m = \frac{2}{3}; (0, 2)$

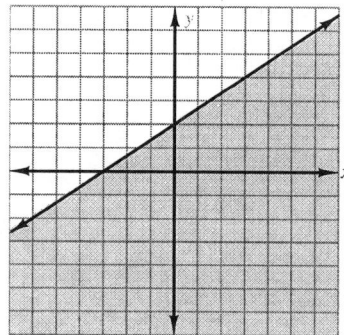

37. passing through $(-1, -4)$ and parallel to $2x - 3y = 6$
slope of line $2x - 3y = 6:$ $-3y = -2x + 6$

    $y = \frac{2}{3}x - 2$

    $m = \frac{2}{3}$

slope of line parallel: $m = \frac{2}{3}$

point-slope: $\boxed{y + 4 = \frac{2}{3}(x + 1)}$

slope-intercept: $y + 4 = \frac{2}{3}x + \frac{2}{3}$

    $\boxed{y = \frac{2}{3}x - \frac{10}{3}}$

standard: $3y = 2x - 10$
    $\boxed{2x - 3y = 10}$

38. Containing (2, –3) and perpendicular to $4y - 2x = -8$
slope of line $4y - 2x = -8$: $\begin{aligned} 4y &= 2x - 8 \\ y &= \frac{1}{2}x - 2 \\ m &= \frac{1}{2} \end{aligned}$

slope of line perpendicular: $m = -\dfrac{1}{\frac{1}{2}} = -2$

point-slope: $\boxed{y + 3 = -2(x - 2)}$
slope-intercept: $y + 3 = -2x + 4$
$\boxed{y = -2x + 1}$

standard: $\boxed{2x + y = 1}$

39. Containing the origin and parallel to the line passing through (1, 4) and (–2, 10)
$m = \dfrac{10 - 4}{-2 - 1} = \dfrac{6}{-3} = -1$
slope of line parallel: $m = -2$
origin: (0, 0)
point-slope: $\boxed{y - 0 = -2(x - 0)}$
slope-intercept: $\boxed{y = -2x}$
standard: $\boxed{2x + y = 0}$

40. passing through (–3, 2) and perpendicular to the line whose x-intercept is 4 and whose y-intercepts is –2:
x-intercept of 4: (4, 0)
y-intercept of –2: (0, –2)
slope of line: $\dfrac{-2 - 0}{0 - 4} = \dfrac{-2}{-4} = \dfrac{1}{2}$

slope of line perpendicular: $m = -\dfrac{1}{\frac{1}{2}} = -2$

point-slope: $\boxed{y - 2 = -2(x + 3)}$
slope-intercept: $y - 2 = -2x - 6$
$\boxed{y = -2x - 4}$

standard: $\boxed{2x + y = -4}$

41. $\begin{aligned} 4x + 8y &= 10 \\ 8y &= -4x + 10 \\ y &= -\frac{1}{2}x + \frac{5}{4} \\ m_1 &= -\frac{1}{2} \end{aligned}$ and $\begin{aligned} 2x &= 12 - 4y \\ 4y &= -2x + 12 \\ y &= -\frac{1}{2}x + 3 \\ m_2 &= -\frac{1}{2} \end{aligned}$

$$m_1 = m_2$$

The slopes are equal.
The lines are $\boxed{\text{parallel}}$.

42.

$$
\begin{aligned}
x - 3y &= 7 \\
-3y &= -x + 7 \\
y &= \frac{1}{3}x - \frac{7}{3} \\
m_1 &= \frac{1}{3}
\end{aligned}
\qquad \text{and} \qquad
\begin{aligned}
3x + y &= 7 \\
y &= -3x + 7 \\
m_2 &= -3
\end{aligned}
$$

$$
m_1 m_2 = \frac{1}{3}(-3) = -1
$$

The slopes are negative reciprocals.

The lines are $\boxed{\text{perpendicular}}$.

43.

$$
\begin{aligned}
3x - y &= -2 \\
-y &= -3x - 2 \\
y &= 3x + 2 \\
m_1 &= 3
\end{aligned}
\qquad \text{and} \qquad
\begin{aligned}
3x + y &= 2 \\
y &= -3x + 2 \\
m_2 &= -3
\end{aligned}
$$

$$
m_1 \neq m_2
$$
$$
m_1 m_2 = 3(-3) = -9 \neq -1
$$

The slopes are neither equal nor negative reciprocals.

The line are $\boxed{\text{intersecting but not perpendicular}}$.

44. Let

$$
\begin{aligned}
d &= \text{distance} \\
t &= \text{time the object falls} \\
d &= kt^2
\end{aligned}
$$

Given: $t = 3$ seconds, $d = 144$ feet

$$
\begin{aligned}
144 &= k\,3^2 \\
16 &= k \\
d &= 16t^2
\end{aligned}
$$

At $t = 7$ seconds, $d = ?$

$$
d = 16(7^2) = 784
$$

After 7 seconds the object will fall $\boxed{784 \text{ feet}}$.

45. Let

$$
\begin{aligned}
F &= \text{gravitational force} \\
d &= \text{distance between masses} \\
F &= \frac{k}{d^2}
\end{aligned}
$$

Given: $d = 5$ feet; $F = 100$ pounds

$$
\begin{aligned}
100 &= \frac{k}{5^2} \\
2500 &= k \\
F &= \frac{2500}{d^2}
\end{aligned}
$$

At $d = 10$ feet, $F = ?$

$$
F = \frac{2500}{10^2} = \frac{2500}{100} = 25
$$

The gravitational force is $\boxed{25 \text{ pounds}}$.

46. Let

$$
\begin{aligned}
D &= \text{number of days} \\
C &= \text{number of cars} \\
P &= \text{number of people} \\
D &= \frac{kC}{P}
\end{aligned}
$$

Given: D = 10 days, C = 100 cars, P = 7 people

$$10 = \frac{k\,100}{7}$$

$$0.7 = k$$

$$D = \frac{0.7\,C}{P}$$

At $\quad P$ = 10 people, C = 400 cars, D = ?

$$D = \frac{0.7(400)}{10} = 28$$

The number of days is $\boxed{28}$.

47. $y = -\dfrac{6}{x}$

x	-6	-3	-2	-1	0	1	2	3	6
y	1	2	3	6	undefined	-6	-3	-2	-1

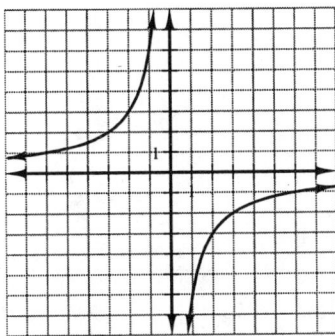

48. From the graph: $\boxed{\{(1, 1), (0, 0), (-2, -8)\}}$

Check:	$y = x^3$	$y = 2x - x^2$
(1, 1):	$1 = 1^3$	$1 = 2(1) - 1^2$
	$1 = 1 \ \checkmark$	$1 = 1 \ \checkmark$
(0, 0):	$0 = 0 \ \checkmark$	$0 = 0 - 0 = 0 \ \checkmark$
$(-2, -8)$:	$-8 = (-2)^3$	$-8 = 2(-2) - (-2)^2$
	$-8 = -8 \ \checkmark$	$-8 = -4 - 4$
		$-8 = -8 \ \checkmark$

49. $x + y = 5$
$3y - y = 3$

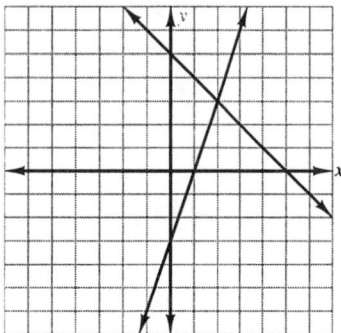

$\{(2, 3)\}$

50. $3x - 2y = 6$
$6x - 4y = 12$

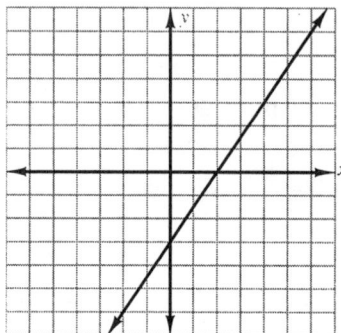

same line

dependent

51. $y = \dfrac{3}{5}x - 3$
$2x - y = -4$

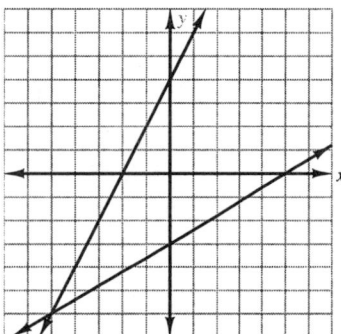

$\{(-5, -6)\}$

52. $y = -x + 4$
$3x + 3y = -6$

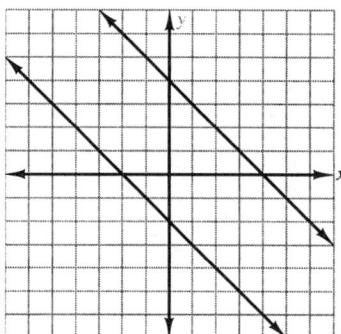

parallel lines

inconsistent

53. $|x| < 4$ or $|y| < 3$
$-4 \le x \le 4$ or $-3 < y < 3$

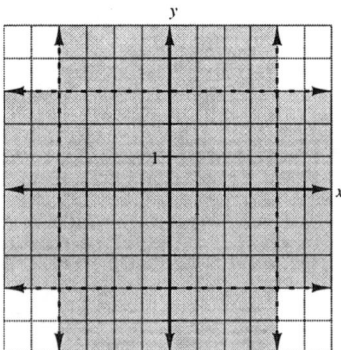

54. (x, y): (number of units demanded, price per widget)
(48,000, \$8), (36,000, \$11)

$$m = \frac{11-8}{36,000-48,000} = \frac{3}{-12,000} = \frac{-1}{4000}$$

$$y - 8 = -\frac{1}{4000}(x - 48,000)$$

$$y - 8 = -\frac{1}{4000}x + 12$$

$$\boxed{y = -\frac{1}{4000}x + 20}$$

$y = \$12$:

$$12 = -\frac{1}{4000}x + 20$$

$$\frac{1}{4000}x = 8$$

$$x = 32,000$$

$$\boxed{32,000 \text{ widgets}}$$

55. Cost: $f(x) = 1.10x + 1080$
Revenue: $g(x) = 1.3x$

a. From the graph, $\boxed{5400 \text{ gallons}}$

b. From the graph, $\boxed{x < 5400 \text{ gallons}}$

c. $\boxed{\text{Sales cannot be negative}}$

d. Profit $=$ Revenue $-$ Cost
$= 1.3x - (1.10x + 1080)$
$= 0.2x - 1080$

$x = 8000$ gallons
Profit $= 0.2(8000) - 1080$
$= 1600 - 1080$
$= 520$
$\boxed{\$520 \text{ per week}}$

Cumulative Review Problems (Chapters 1-3)

Cumulative Review, pp. 251-252

1. $\{x \mid x$ is a natural number that is divisible by 3 and a factor of 45$\}$
factors of 45: 1, 3, 5, 9, 15, 45
factors divisible by 3: 3, 9, 15, 45
$\boxed{\{3, 9, 15, 45\}}$

2. $5(2x + y) = 5(y + 2x)$

$\boxed{\text{Commutative property of addition}}$

3. $\{x \mid |x| \le 2\} = \{x \mid -2 \le x \le 2\}$

$\boxed{[-2, 2]}$

4. $\dfrac{\left(-\frac{3}{4} \cdot 12\right) - 7}{-9 - \left(-\frac{3}{4} \cdot 8\right)} = \dfrac{-9 - 7}{-9 - (-6)} = \dfrac{-16}{-9 + 6} = \dfrac{-16}{-3} = \boxed{\dfrac{16}{3}}$

5. $16 - 4\{6 - 2(x - 3y) + x\}$
$= 16 - 4(6 - 2x + 6y + x)$

$= 16 - 4(6 - x + 6y)$

$= 16 - 24 + 4x - 24y$

$= \boxed{-8 + 4x - 24y}$

6. $(-4x^2y^{-5})(-3x^{-1}y^{-4})$
$= 12x^1y^{-9}$

$= \boxed{\dfrac{12x}{y^9}}$

7. $\dfrac{(700{,}000{,}000)(0.000\,09)}{(0.000\,000\,3)(40{,}000)}$

$= \dfrac{(7 \times 10^8)(9 \times 10^{-5})}{(3 \times 10^{-7})(4 \times 10^4)}$

$= \dfrac{7(9) \times 10^3}{4(3) \times 10^{-3}}$

$= \dfrac{21}{4} \times 10^{3+3}$

$= \boxed{5.25 \times 10^6}$

8.
$$4y - 3(5 - 2y) = 6(y - 3) + 2y + 1$$
$$4y - 15 + 6y = 6y - 18 + 2y + 1$$
$$10y - 15 = 8y - 17$$
$$2y = -2$$
$$y = -1$$
$$\boxed{\{-1\}}$$

9.
$$-4 \le \dfrac{1 - 2x}{6} < 0$$
$$-24 \le 1 - 2x < 0$$
$$-25 \le -2x < -1$$
$$\dfrac{25}{2} \le x > \dfrac{1}{2}$$
$$\dfrac{1}{2} < x \le \dfrac{25}{2}$$

$$\boxed{\left\{ x \mid \dfrac{1}{2} < x \le \dfrac{25}{2} \right\}}$$

$$\boxed{\left(\dfrac{1}{2}, \dfrac{25}{2} \right]}$$

10.
$$\{x \mid 3(x + 2) - 2x < 4x\} \quad \cup$$
$$3x + 6 - 2x < 4x$$
$$x + 6 < 4x$$
$$-3x < -6$$
$$x > 2$$

$$\{x \mid 2(3x + 2) - 2x < -2(4 - 3x)\}$$
$$6x + 4 - 2x < -8 + 6x$$
$$4x + 4 < -8 + 6x$$
$$-2x < -12$$
$$x > 6$$

$$\boxed{\{x \mid x > 2\}}$$
$$\boxed{(2, \infty)}$$

11.
$$2(r - 3s) = 6t$$
$$2r - 6s = 6t$$
$$-6s = -2r + 6t$$
$$\boxed{s = \dfrac{1}{3}r - t \text{ or } s = \dfrac{r - 3t}{3}}$$

12. Let

$$
\begin{aligned}
x &= \text{number of nickels}\\
x+1 &= \text{number of dimes}\\
x+6 &= \text{number of pennies}\\
0.01(x+6)+0.05x+0.10(x+1) &= 4.80\\
0.01x+0.06+0.05x+0.10x+0.10 &= 4.80\\
0.16x+0.16 &= 4.80\\
0.16x &= 4.64\\
x &= 29 \quad \text{(nickels)}\\
x+1 &= 30 \quad \text{(dimes)}\\
x+6 &= 35 \quad \text{(nickels)}
\end{aligned}
$$

$\boxed{35 \text{ pennies}, 29 \text{ nickels}, 30 \text{ dimes}}$

13. Let

$$
\begin{aligned}
x &= \text{amount invested at 18\% profit}\\
25{,}000-x &= \text{amount invested at 11\% loss}\\
0.18x-0.11(25{,}000-x) &= 2180\\
0.18x-2750+0.11x &= 2180\\
0.29x &= 4930\\
x &= 17{,}000 \quad \text{(at 18\% profit)}\\
25{,}000-x &= 8000 \quad \text{(at 11\% loss)}
\end{aligned}
$$

$\boxed{\$17{,}000 \text{ at } 18\%; \$8000 \text{ at } 11\%}$

14. Let

$$
\begin{aligned}
x &= \text{amount of 65\% pure silver}\\
100-x &= \text{amount of 45\% pure silver}\\
0.65x+0.45(100-x) &= 50\\
0.65x+45-0.45x &= 50\\
0.20x &= 5\\
x &= 25 \quad \text{(at 65\%)}\\
100-x &= 75 \quad \text{(at 45\%)}
\end{aligned}
$$

$\boxed{25 \text{ grams at } 65\%; 75 \text{ grams at } 45\%}$

15. Let

$$
\begin{aligned}
x &= \text{first consecutive even integer}\\
x+2 &= \text{second consecutive even integer}\\
x+4 &= \text{third consecutive even integer}\\
5x &= 2[(x+2)+(x+4)]-4\\
5x &= 2(2x+6)-4\\
5x &= 4x+12-4\\
5x &= 4x+8\\
x &= 8\\
x+2 &= 10\\
x+4 &= 12
\end{aligned}
$$

The integers are $\boxed{8, 10 \text{ and } 12}$.

16. Let t = time for runner B to overtake runner A (in seconds)

	Rate, R	$\cdot$	Time, t	=	Distance, D
Runner A	8		$t + 2$		$8(t + 2)$
Runner B	10		t		$10t$

$$
\begin{aligned}
8(t + 2) &= 10t \\
8t + 16 &= 10t \\
16 &= 2t \\
8 &= t
\end{aligned}
$$

$\boxed{8 \text{ seconds}}$ for runner B to overtake runner A.

distance = $10(8) = 80$

$\boxed{80 \text{ meters}}$

17. Ana's number is odd.

José's number $=$ Ana's number $+ 1$
            which is an even number

Jud's number $=$ José's number $- 3$
            which is an odd number

$\boxed{\text{No; Jud's number is odd.}}$

18. Let x = number of dimes.

$$
\begin{aligned}
0.1x + 0.90 &= 0.25x \\
0.90 &= 0.15x \\
6 &= x
\end{aligned}
$$

$\boxed{6 \text{ dimes}}$

19. $\boxed{2 \text{ dozen}}$

20. $x - 2y \le -4$ and $y \le 0$

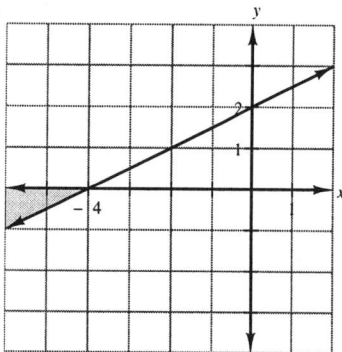

21.
$$
\begin{aligned}
2x - 4y &= 8 \\
-4y &= -2x + 8 \\
y &= \frac{1}{2}x - 2
\end{aligned}
$$

$\boxed{f(x) = \dfrac{1}{2}x - 2}$

$f(-4) = \dfrac{1}{2}(-4) - 2 = -2 = -2 = -4$

$\boxed{f(-4) = -4}$

22. $\boxed{y \text{ is not a function of } x}$

At $x = 0$, a vertical line intersects the graph twice.

$$\boxed{\begin{array}{l} \text{Domain} = \{\, x \mid x \geq -3 \,\} \\ \qquad [-3, \infty) \\ \text{Range} = \{\, y \mid y \in R \,\} \\ \qquad (-\infty, \infty) \end{array}}$$

23. $\boxed{y \text{ is a function of } x}$

$$\boxed{\begin{array}{l} \text{Domain} = \{\, x \mid x \in R \,\} \\ \qquad (-\infty, \infty) \\ \text{Range} = \{\, y \mid y \in R \,\} \\ \qquad (-\infty, \infty) \end{array}}$$

24. Passing through $(2, 4)$ and $(-5, -2)$

$$m = \frac{-2-4}{-5-2} = \frac{-6}{-7} = \frac{6}{7}$$

point-slope: $\boxed{y - 4 = \frac{6}{7}(x - 2)}$ or $\boxed{y - 2 = \frac{6}{7}(x + 5)}$

$$y - 4 = \frac{6}{7}x - \frac{12}{7}$$

slope-intercept: $y = \frac{6}{7}x - \frac{12}{7} + \frac{28}{7}$

$$\boxed{y = \frac{6}{7}x + \frac{16}{7}}$$

standard: $7y = 6x + 16$

$$\boxed{6x - 7y = -16}$$

25. Passing through $(-3, 4)$ and perpendicular to the line whose equation is $4x - 2y = 5$.

slope of line $4x - 2y = 5$:
$$\begin{aligned} -2y &= -4x + 5 \\ y &= 2x - \frac{5}{2} \\ m &= 2 \end{aligned}$$

slope of line perpendicular: $m = -\frac{1}{2}$

point-slope: $\boxed{y - 4 = -\frac{1}{2}(x + 3)}$

slope-intercept: $y - 4 = -\frac{1}{2}x - \frac{3}{2}$

$$\boxed{y = -\frac{1}{2}x + \frac{5}{2}}$$

standard: $2y = -x + 5$

$$\boxed{x + 2y = 5}$$

26. $H = \dfrac{kF}{m}$

Given: $F = 750$ degrees, $m = 200$; $H = 30$ oersteds

$3 = k\dfrac{750}{200}$

$\dfrac{4}{5} = k$

$H = \dfrac{4F}{5M}$

At $F = 500$ degrees, $m = 175$, $H = ?$

$H = \dfrac{4(500)}{5(175)} = \dfrac{16}{7} = 2\dfrac{2}{7}$

$\boxed{2\dfrac{2}{7}\text{ oersteds}}$

27. $y = \dfrac{8}{x}$

x	-4	-2	0	2	4
y	-2	-4	undefined	4	2

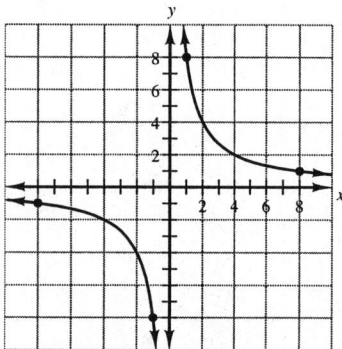

28. From the graph: $\boxed{\{(-4, 3), (-4, -3), (4, 3), (4, -3)\}}$

Check: $x^2 + y^2 = 25$ $x^2 + 4y^2 = 52$

$(\pm4, \pm3)$: $(\pm4)^2 + (\pm3)^2 = 25$ $(\pm4)^2 + 4(\pm3)^2 = 52$

$16 + 9 = 25$ $16 + 4(9) = 52$

$25 = 25$ √ $16 + 36 = 52$

$52 = 52$ √

29. $x + y = 6$
$2x + 4y = 16$

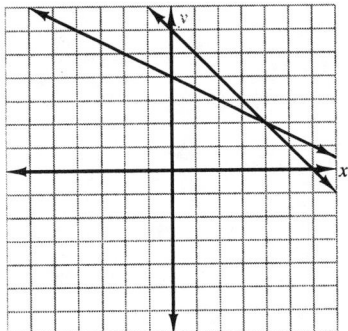

$\{(4, 2)\}$

30.

	y	$=$	$8000 - kx$
Given:	x	$=$	10 years, $y = \$500$
	500	$=$	$8000 - 10k$
	$10k$	$=$	7500
	k	$=$	750

$y = 8000 - 750x$

At $x = 5$ years, $y = ?$

$y = 8000 - 750(5)$
$y = 8000 - 3750$
$y = 4250$

Value of machine: $\$4250$

Algebra for College Students

Chapter 4   Systems of Linear Equations

Section 4.1 Linear Systems of Equations in Two Variables

Problem Set 4.1, pp. 264-267

1. $x = 2y - 5$
$\underline{x - 3y = 8}$ (substitute for x) $\rightarrow$ $(2y - 5) - 3y = 8$
$-y - 5 = 8$
$-y = 13$
$y = -13$

$x = 2y - 5$ (replace y with -13)
$x = 2(-13) - 5 = -26 - 5 = -31$

$\boxed{\{(-31, -13)\}}$

3. $4x + y = 5$ (solve for y) $\rightarrow$ $y = 5 - 4x$
$2x - 3y = 13$ (substitute for y) $\rightarrow$ $2x - 3(5 - 4x) = 13$
$2x - 15 + 12x = 13$
$14x = 28$
$x = 2$

$y = 5 - 4x$ (replace x with 2)
$y = 5 - 4(2)$
$y = 5 - 8$
$y = -3$

$\boxed{\{(2, -3)\}}$

5. $x + y = 0$ (solve for y) $\rightarrow$ $y = -x$
$3x + 2y = 5$ (substitute for y) $\rightarrow$ $3x + 2(-x) = 5$
$3x - 2x = 5$
$x = 5$

$y = -x$ (replace x with 5)
$y = -5$

$\boxed{\{(5, -5)\}}$

7. $7x - 3y = 23$
$x + 2y = 13$ (solve for x) $\rightarrow$ $x = 13 - 2y$
(substitute for x in first equation) $\rightarrow$ $7(13 - 2y) - 3y = 23$
$91 - 14y - 3y = 23$
$91 - 17y = 23$
$-17y = -68$
$y = 4$

$x = 13 - 2y$ (Replace y with 4)
$x = 13 - 2(4)$
$x = 5$

$\boxed{\{(5, 4)\}}$

9. $3x - 2y = 14$ (solve for y) $\rightarrow$

$2x + 3y = -8$

$$-2y = 14 - 3x$$
$$2y = 3x - 14$$
$$y = \frac{3}{2}x - 7$$

(substitute for y in second equation) $\rightarrow$ $2x + 3\left(\frac{3}{2}x - 7\right) = -8$

$$2x + \frac{9}{2}x - 21 = -8 \quad \text{(multiply by 2)}$$
$$4x + 9x - 42 = -16$$
$$13x = 26$$
$$x = 2$$

$y = \frac{3}{2}x - 7$ (Replace x with 2)

$y = \frac{3}{2}(2) - 7$

$y = 3 - 7$

$y = -4$

$\boxed{\{(2, -4)\}}$

11. $5x - y = 32$ (solve for y) $\rightarrow$ $y = 5x - 32$

$x - 2y - 19 = 0$ (substitute for y) $\rightarrow$ $x - 2(5x - 32) - 19 = 0$

$$x - 10x + 64 - 19 = 0$$
$$-9x = -45$$
$$x = 5$$

$y = 5x - 32$ (Replace with 5)

$y = 5(5) - 32$

$y = -7$

$\boxed{\{(5, -7)\}}$

13. $y - 3x = 2$ (substitute for x) $\rightarrow$ $y - 3\left(\frac{1}{4}y\right) = 2$

$x = \frac{1}{4}y$

$$y - \frac{3}{4}y = 2 \quad \text{(multiply by 4)}$$
$$4y - 3y = 8$$
$$y = 8$$

$x = \frac{1}{4}y$ (Replace y with 8)

$x = \frac{1}{4}(8)$

$x = 2$

$\boxed{\{(2, 8)\}}$

15. $\frac{x}{4} = 9 + \frac{y}{5}$ (substitute for y) $\rightarrow$ $\frac{x}{4} = 9 + \frac{5x}{5}$

$y = 5x$

$$\frac{x}{4} = 9 + x \quad \text{(multiply by 4)}$$
$$x = 36 + 4x$$
$$3x = -36$$
$$x = -12$$

$y = 5x$ (Replace x with -12)

$y = 5(-12)$

$y = -60$

$\boxed{\{(-12, -60)\}}$

17.

$$\begin{aligned} y - bx &= 2 \\ 3bx + 2y &= 1 \end{aligned}$$

(solve for y) $\rightarrow$

(substitute for y) $\rightarrow$

$$\begin{aligned} y &= bx + 2 \\ 3bx + 2(bx + 2) &= 1 \\ 3bx + 2(bx + 2) &= 1 \\ 5bx + 2bx + 4 &= 1 \\ x &= -\frac{3}{5b} \end{aligned}$$

$$y = bx + 2 \qquad \left(\text{Replace } x \text{ with } -\frac{3}{5b} \right)$$

$$y = b \left(-\frac{3}{5b} \right) + 2$$

$$y = -\frac{3}{5} + 2$$

$$y = \frac{7}{5}$$

$$\boxed{\left\{ \left(-\frac{3}{5b}, \frac{7}{5} \right) \right\}}$$

19.

$$\begin{aligned} x + y &= 7 \\ \underline{x - y} &= \underline{3} \qquad \text{(Add)} \\ 2x &= 10 \\ x &= 5 \qquad \text{(Replace } x \text{ with 5 in first equation)} \rightarrow \end{aligned}$$

$$\begin{aligned} 5 + y &= 7 \\ y &= 2 \end{aligned}$$

$$\boxed{\{(5, -2)\}}$$

21.

$$\begin{aligned} 12x + 3y &= 15 \\ \underline{2x - 3y} &= \underline{13} \qquad \text{(Add)} \\ 14x &= -28 \\ x &= 2 \qquad \text{(Replace } x \text{ with 2 in second equation)} \rightarrow \end{aligned}$$

$$\begin{aligned} 2(2) - 3y &= 13 \\ 4 - 3y &= 13 \\ -3y &= 9 \\ y &= -3 \end{aligned}$$

$$\boxed{\{(2, -3)\}}$$

23.

$$\begin{aligned} x - 2y &= 5 \\ 5x - y &= -2 \end{aligned}$$

(no change) $\rightarrow$

($\times$ –2) $\rightarrow$

$$\begin{aligned} x - 2y &= 5 \\ \underline{-10x + 2y} &= \underline{4} \\ -9x &= 9 \\ x &= -1 \end{aligned}$$

$$\begin{aligned} 5x - y &= -2 \\ 5(-1) - y &= -2 \\ y &= -5 + 2 \\ y &= -3 \end{aligned}$$

$$\boxed{\{(-1, -3)\}}$$

25.

$$\begin{aligned} 2x - 9y &= 5 \\ 3x - 3y &= 11 \end{aligned}$$

(no change) $\rightarrow$

($\times$ –3) $\rightarrow$

$$\begin{aligned} 2x - 9y &= 5 \\ \underline{-9x + 9y} &= \underline{-33} \\ -7x &= -28 \\ x &= 4 \end{aligned}$$

$$\begin{aligned} 3x - 3y &= 11 \\ 3(4) - 3y &= 11 \\ 12 - 3y &= 11 \\ -3y &= -1 \\ y &= \frac{1}{3} \end{aligned}$$

$$\boxed{\left\{ \left(4, \frac{1}{3} \right) \right\}}$$

27.

$$
\begin{array}{rcl}
3x - 7y &=& 1 \\
2x - 3y &=& -1
\end{array}
$$

$(\times 2) \rightarrow$
$(\times 3) \rightarrow$

$$
\begin{array}{rcl}
6x - 14y &=& 2 \\
-6x + 9y &=& 3 \\
\hline
-5y &=& 5 \\
y &=& -1
\end{array}
$$

$$
\begin{array}{rcl}
2x - 3y &=& -1 \\
2x - 3(-1) &=& -1 \\
2x + 3 &=& 1 \\
2x &=& -4 \\
x &=& -2
\end{array}
$$

$$\boxed{\{(-2, -1)\}}$$

29.

$$
\begin{array}{rcl}
4x + y &=& 2 \\
2x - 3y &=& 8
\end{array}
$$

$(\times 3) \rightarrow$
$(\text{unchanged}) \rightarrow$

$$
\begin{array}{rcl}
12x + 3y &=& 6 \\
2x - 3y &=& 8 \\
\hline
14x &=& 14 \\
x &=& 1
\end{array}
$$

$$
\begin{array}{rcl}
4x + y &=& 2 \\
4(1) + y &=& 2 \\
y &=& 2 - 4 \\
y &=& -2
\end{array}
$$

$$\boxed{\{(1, -2)\}}$$

31.

$$
\begin{array}{rcl}
2y &=& 5 - 5x \\
9x - 15 &=& -3y
\end{array}
$$

(Rearrange)

$$
\begin{array}{rcl}
5x + 2y &=& 5 \\
9x + 3y &=& 15
\end{array}
$$

$(\times 3) \rightarrow$
$(\times -2) \rightarrow$

$$
\begin{array}{rcl}
15x + 6y &=& 15 \\
-18x - 6y &=& -30 \\
\hline
-3x &=& -15 \\
x &=& 5
\end{array}
$$

$$
\begin{array}{rcl}
5x + 2y &=& 5 \\
5(5) + 2y &=& 5 \\
2y &=& -20 \\
y &=& -10
\end{array}
$$

$$\boxed{\{(5, -10)\}}$$

33.

$$
\begin{array}{rcl}
9x + \dfrac{4}{3}y &=& 5 \\[2mm]
4x - \dfrac{1}{3}y &=& 5
\end{array}
$$

$(\times 3) \rightarrow$
$(\times 12) \rightarrow$

$$
\begin{array}{rcl}
27x + 4y &=& 15 \\
48x - 4y &=& 60 \\
\hline
75x &=& 75 \\
x &=& 1
\end{array}
$$

$$
\begin{array}{rcl}
12x - y &=& 15 \\
12 - y &=& 15 \\
y &=& -3
\end{array}
$$

(multiply equation 2 by 3)
(substitute 1 for x)

$$\boxed{\{(1, -3)\}}$$

35.

$$\frac{1}{8}y + \frac{1}{5} = 5 \qquad (\times 5) \rightarrow \qquad \frac{5}{8}y + x = 25$$

$$\frac{1}{3}y + \frac{1}{2}x = 13 \qquad (\times 2) \rightarrow \qquad -\frac{2}{3}y - x = -26$$

$$\overline{\frac{5}{8}y - \frac{2}{3}y = -1} \quad \text{(multiply by 24)}$$

$$15y - 16y = -24$$
$$y = 24$$

$$\frac{1}{3}y + \frac{1}{2}x = 13 \qquad \text{(multiply by 6)}$$
$$2y + 3x = 78 \qquad \text{(replace } y \text{ with 24)}$$
$$2(24) + 3x = 78$$
$$3x = 30$$
$$x = 10$$

$$\boxed{\{(10, 24)\}}$$

37. $2x = 6 - 3y$
$9x = 8y + 4$
(Rearrange)

$$2x + 3y = 6 \qquad (\times 8) \rightarrow \qquad 16x + 24y = 48$$
$$9x - 8y = 4 \qquad (\times 3) \rightarrow \qquad \underline{27x - 24y = 12}$$
$$43x = 60$$
$$x = \frac{60}{43}$$

$$2x + 3y = 6$$
$$2\left(\frac{60}{43}\right) + 3y = 6$$
$$3y = \frac{-120}{43} + \frac{258}{43}$$
$$3y = \frac{138}{43}$$
$$y = \frac{46}{43}$$

$$\boxed{\left\{ \left(\frac{60}{43}, \frac{46}{43} \right) \right\}}$$

39.

$$\frac{x}{10} + \frac{y}{5} = \frac{1}{2} \qquad (\times 5) \rightarrow \qquad \frac{x}{2} + y = \frac{5}{2}$$

$$\frac{x}{4} - \frac{y}{3} = \frac{-5}{12} \qquad (\times 3) \rightarrow \qquad \frac{3x}{4} - y = -\frac{5}{4}$$

$$\frac{3x}{4} + \frac{x}{2} = \frac{5}{2} - \frac{5}{4}$$
$$\frac{5x}{4} = \frac{5}{4}$$
$$5x = 5$$
$$x = 1$$

$$\frac{x}{10} + \frac{y}{5} = \frac{1}{2} \qquad \text{(multiply by 10)}$$
$$x + 2y = 5 \qquad \text{(replace } x \text{ with 1)}$$
$$1 + 2y = 5$$
$$2y = 5$$
$$y = 2$$

$$\boxed{\{(1, 2)\}}$$

41.

$$\begin{array}{rcl} 3x + 3y & = & 2 \\ 2x + 2y & = & 3 \end{array}$$

$(\times 20) \to$
$(\times -3) \to$

$$\begin{array}{rcl} 6x + 6y & = & 4 \\ \underline{-6x - 6y} & = & \underline{-9} \\ 0 & = & -5 \quad \text{False} \end{array}$$

Contradiction
No solution, $\varnothing$

$\boxed{\text{inconsistent}}$

43.

$$\begin{array}{rcl} 2x - y & = & -2 \\ 4x - 2y & = & 5 \end{array}$$

$(\times -2) \to$
$(\text{unchanged}) \to$

$$\begin{array}{rcl} -4x + 2y & = & 4 \\ 4x - 2y & = & 5 \\ 0 & = & 9 \quad \text{False} \end{array}$$

Contradiction
No solution, $\varnothing$.

$\boxed{\text{inconsistent}}$

45.

$$\begin{array}{rcl} \dfrac{x}{3} + \dfrac{y}{5} & = & 15 \\ 10x + 6y & = & 5 \end{array}$$

$(\times -30) \to$
$(\text{unchanged}) \to$

$$\begin{array}{rcl} -10x - 6y & = & -450 \\ \underline{10x + 6y} & = & \underline{5} \\ 0 & = & -445 \quad \text{False} \end{array}$$

Contradiction
No solution, $\varnothing$.

$\boxed{\text{inconsistent}}$

47. $2x = 3 - 2y$
$3y = 4 - 3x$
(Rearrange)

$$\begin{array}{rcl} 2x + 2y & = & 3 \\ 3x + 3y & = & 4 \end{array}$$

$(\times 3) \to$
$(\times -2) \to$

$$\begin{array}{rcl} 6x + 6y & = & 9 \\ \underline{-6x - 6y} & = & \underline{-8} \\ 0 & = & -1 \quad \text{False} \end{array}$$

Contradiction
No solution, $\varnothing$.

$\boxed{\text{inconsistent}}$

49.

$$\begin{array}{rcl} 2x + y - 4 & = & 0 \\ \underline{2x + y - 2} & = & \underline{0} \quad \text{(Subtract)} \\ -2 & = & 0 \quad \text{False} \end{array}$$

Contradiction
No solution $\varnothing$.

$\boxed{\text{inconsistent}}$

51.

$$\begin{array}{rcl} 4x - 6y & = & 14 \\ 2x - 3y & = & 7 \end{array}$$

$(\text{unchanged}) \to$
$(\times -2) \to$

$$\begin{array}{rcl} 4x - 6y & = & 14 \\ \underline{-4x + 6y} & = & \underline{-14} \\ 0 & = & 0 \quad \text{True} \end{array}$$

Identity
Infinite number of solutions

$\boxed{\text{Dependent}}$

53.

$$3(x-3)-2y=0$$
$$2(x-y)=-x-3$$ (Rearrange and simplify)

$$3x-9-2y=0$$
$$2x-2y=-x-3$$ (Rearrange)

$$
\begin{array}{rcl}
3x-2y &=& 9\\
\underline{3x-2y} &=& \underline{-3} \quad \text{(subtract)}\\
0 &=& 6 \quad \text{False}
\end{array}
$$

Contradiction
No solution, $\varnothing$.

Inconsistent

55.

$$\dfrac{x-1}{2}=\dfrac{y-2}{8}$$ $(\times 8) \rightarrow$ $4x-4 = y-2$ (rearrange)

$$
\begin{array}{rcl}
4x-y &=& 2
\end{array}
$$

$$12x-3y = 6$$ $(\div -3) \rightarrow$
$$
\begin{array}{rcl}
\underline{-4x+y} &=& \underline{-2} \quad \text{True}\\
0 &=& 0
\end{array}
$$

Identity
Infinite number of solutions

Dependent

57.   [A] is true; $y = 4x-3$ $(\times -1)$ $-y = -4x+3$

$$y = 4x+5$$ (no change) $\underline{y = 4x+5}$

$$0 = 8 \quad \text{False}$$

Contradiction
No soluiton, $\varnothing$.

59.

$$
\begin{array}{rcl}
2y+2x &=& -2-4y\\
7y+27 &=& 3x+y
\end{array}
$$
(simplify) $\rightarrow$
$$
\begin{array}{rcl}
2x+6y &=& -2\\
-3x+6y &=& -27
\end{array}
$$
(no change) $\rightarrow$
$$
\begin{array}{rcl}
2x+6y &=& -2\\
\underline{3x-6y} &=& \underline{27}\\
5x &=& 25 \quad \text{(Add)}\\
x &=& 5
\end{array}
$$

$$
\begin{array}{rcl}
2x+6y &=& -2\\
2(5)+6y &=& -2\\
6y &=& -12\\
y &=& -2
\end{array}
$$

$\{(5, -2)\}$

61.

$$
\begin{array}{rcl}
3(x+y) &=& 6\\
3(x-y) &=& -36
\end{array}
$$
$(\div 3) \rightarrow$
$(\div 3) \rightarrow$
$$
\begin{array}{rcl}
x+y &=& 2\\
\underline{x-y} &=& \underline{-12}\\
2x &=& -10 \quad \text{(Add)}\\
x &=& -5
\end{array}
$$

$$
\begin{array}{rcl}
x+y &=& 2\\
-5+y &=& 2\\
y &=& 7
\end{array}
$$

$\{(-5, 7)\}$

63.
$$x + y = 50 \qquad (\times -6) \rightarrow \qquad -6x - 6y = -300$$
$$\frac{x}{2} + \frac{3y}{5} = 14 \qquad (\times 10) \rightarrow \qquad \underline{5x + 6y = 140}$$
$$-x = -160$$
$$x = 160$$

$$x + y = 50$$
$$160 + y = 50$$
$$y = -110$$

$$\boxed{\{(160, -110)\}}$$

65.
$$\frac{1}{2}(x - y) - \frac{1}{6}(x - 4y) = 4$$
$$\frac{1}{9}(x + y) - \frac{1}{6}(x - 2y) = \frac{11}{9}$$

$(\times 6)\quad 3(x - y) - (x - 4y) = 6(4)$
$(\times 18)\quad \underline{3(x - y) - 3(x - 2y) = 2(11)}$
(simplify):
$$3x - 3y - x + 4y = 24$$
$$\underline{2x + 2y - 3x + 6y = 22}$$
(simplify):

$$2x + y = 24 \qquad (\text{no change}) \rightarrow \qquad 2x + y = 24$$
$$-x + 8y = 22 \qquad (\times 2) \rightarrow \qquad \underline{-2x + 16y = 44}$$
$$17y = 68$$
$$y = 4$$

$$2x + y = 24$$
$$2x + 4 = 24$$
$$2x = 20$$
$$x = 10$$

$$\boxed{\{(10, 4)\}}$$

67.
$$5x - 4y - 14 = 0$$
$$\underline{2x - 7y + 3 = 0}$$
(rearrange):

$$5x - 4y = 14 \qquad (\times 7) \rightarrow \qquad 35x - 28y = 98$$
$$2x - 7y = -3 \qquad (\times -4) \rightarrow \qquad \underline{-8x + 28y = 12}$$
$$27x = 110$$
$$x = \frac{110}{27}$$

$$2x - 7y = -3$$
$$\frac{2(110)}{27} - 7y = -3$$
$$-7y = -\frac{81}{27} - \frac{220}{27}$$
$$-7y = \frac{-301}{27}$$
$$y = \frac{43}{27}$$

$$\boxed{\left\{ \left(\frac{110}{27}, \frac{43}{27} \right) \right\}}$$

69.

$$\frac{2}{x+y} = \frac{2}{3}$$

$$\frac{3}{2x-y} = \frac{1}{2}$$

$(\times 3(x+y))$	$6 = 2(x+y)$	$(\div 2) \to$	$3 = x+y$
$(\times 2(2x-y))$	$6 = 2x-y$	(no change) $\to$	$6 = 2x-y$

$$9 = 3x$$
$$3 = x$$
$$3 = x+y$$
$$3 = 3+y$$
$$0 = y$$

$\boxed{\{(3,0)\}}$

71.

$$\frac{3+y}{2} = \frac{3x}{5}$$

$$\frac{3x-2}{4} = \frac{y}{3}$$

$(\times 10(3+y))\quad 15 + 5y = 6x$
$(\times 12)\quad\quad\quad\; 9x - 6 = 4y$
(rearrange)

$-6x + 5y = -15 \qquad (\times 4) \to \qquad -24x + 20y = -60$
$\;\;9x - 4y = 6 \qquad\;\; (\times 5) \to \qquad\;\; 45x - 20y = 30$

$$21x = -30$$
$$x = -\frac{10}{7}$$

$$-6x + 5y = -15$$
$$-6\left(-\frac{10}{7}\right) + 5y = -15$$
$$\frac{60}{7} + 5y = -\frac{105}{7}$$
$$5y = -\frac{165}{7}$$
$$y = -\frac{33}{7}$$

$\boxed{\left\{\left(-\dfrac{10}{7}, -\dfrac{33}{7}\right)\right\}}$

73.

$2.5x - 3y = 7.312 \qquad\quad (\times 2.5) \to$
$-3x + 2.5y = -7.125 \qquad (\times 3) \to$

$(2.5)(2.5)x - 7.5y = (2.5)(7.3125)$
$\underline{-9x + 7.5y = 3(-7.125)}$
$2.5^2 x - 9x = 2.5(7.3125) + 3(7.3125)$
$x = [2.5(7.3125) + 3(-7.125)] \div (2.5^2 - 9)$
$x = 1.125$

$$-3x + 2.5y = -7.125$$
$$2.5y = -7.125 + 3x$$
$$y = \frac{-7.125 + 3(1.125)}{2.5}$$
$$y = -1.05$$

$\boxed{\{(1.125, -1.5)\}}$

75. Passing through the intersection of $x + y = 4$ and $x - y = 0$ with slope $= 3$.
The intersection occurs at the solution of the system of equations: $x + y = 4$ and $x - y = 0$.

$$\begin{aligned}
x + y &= 4 \\
\underline{x - y} &= \underline{0} \\
2x &= 4 \\
x &= 2
\end{aligned}$$

$$\begin{aligned}
x + y &= 4 \\
2 + y &= 4 \\
y &= 2
\end{aligned}$$

$\{(2, 2)\}$

$$\begin{aligned}
y - 2 &= 3(x - 2) \\
y - 2 &= 3x - 6
\end{aligned}$$

$$\boxed{y = 3x - 4}$$

77. Passing through the intersection of $3x + 4y = -1$ and $x + 5y = 7$ and parallel to the line whose equation is $3x - y = 5$.

$$\begin{array}{llll}
3x + 4y = -1 & \text{(no change)} \rightarrow & 3x + 4y &= -1 \\
x + 5y = 7 & (\times -3) \rightarrow & \underline{-3x - 15y} &= \underline{-21} \\
& & -11y &= -22 \\
& & y &= 2
\end{array}$$

$$\begin{aligned}
x + 5y &= 7 \\
x + 10 &= 7 \\
x &= -3
\end{aligned}$$

$\{(-3, 2)\}$
slope of line: $3x - y = 5$:

$$\begin{aligned}
y &= 3x - 5 \\
m &= 3
\end{aligned}$$

slope of line parallel: $m = 3$

$$\begin{aligned}
y - 2 &= 3(x + 3) \\
y - 2 &= 3x + 9
\end{aligned}$$

$$\boxed{y = 3x + 11}$$

79. Passing through the intersection of $x + 4y = 16$ and $2x + 3y = 17$ and perpendicular to the line whose equation is $3x - 9y = 10$.

$$\begin{array}{llll}
x + 4y = 16 & (\times -2) \rightarrow & -2x - 8y &= -32 \\
2x + 3y = 17 & \text{(no change)} \rightarrow & \underline{2x + 3y} &= \underline{17} \\
& & -5y &= -15 \\
& & y &= 3
\end{array}$$

$$\begin{aligned}
x + 4y &= 16 \\
x + 12 &= 16 \\
x &= 4
\end{aligned}$$

$\{(4, 3)\}$

slope of line $3x - 9y = 10$:

$$9y = 3x - 10$$
$$y = \frac{1}{3}x - \frac{10}{9}$$
$$m = \frac{1}{3}$$

slope of perpendicular: $-\dfrac{1}{\frac{1}{3}} = -3$

$$y - 3 = -3(x - 4)$$
$$y - 3 = -3x + 12$$
$$\boxed{y = -3x + 15}$$

81. Costs: $y = 800{,}000 + 45x$

Revenue $\quad y = 65x$

$$
\begin{aligned}
\text{cost} &= \text{Revenue} \\
800{,}000 + 45x &= 65x \\
800{,}000 &= 20x \\
40{,}000 &= x
\end{aligned}
$$

$\boxed{40{,}000 \text{ units}}$

83.

$$x + y = 10$$
$$\frac{x}{20} + \frac{y}{10} = \frac{7}{10}$$

(solve for y) $\quad y = 10 - x$

$(\times 20) \rightarrow \quad \underline{x + 2y = 14}$

(substitute for y in second equation):

$$
\begin{aligned}
x + 2(10 - x) &= 14 \\
x + 2(10 - x) &= 14 \\
x + 20 - 2x &= 14 \\
-x &= -6 \\
x &= 6
\end{aligned}
$$

$$
\begin{aligned}
y &= 10 - x \\
y &= 10 - 6 = 4
\end{aligned}
$$

$\boxed{x(\text{gold}),\ 6 \text{ lb};\ y(\text{silver}),\ 4 \text{ lb}}$

Archimedes was treated dishonestly by the jeweler.

$\boxed{\text{Yes}}$

85. $7x - 4y = 6$

$2x - 3y = 11$

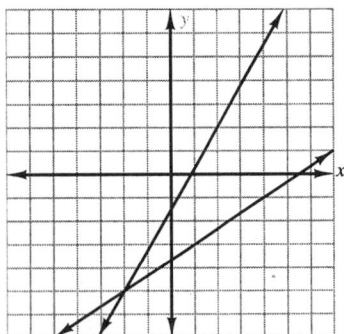

$\boxed{\{(-2, -5)\}}$

87.
$$\begin{aligned}3x + 3y &= 17 - A \qquad (\times -1) \rightarrow\\ 4x + 3y &= 18 - A\end{aligned}$$

$$\begin{aligned}-3x - 3y &= -17 + A\\ \underline{4x + 3y} &= \underline{18 - A}\\ x &= 1\\ 3x + 3y &= 17 - A\\ 3 + 3y &= 17 - A\\ 3y &= 14 - A\\ y &= \frac{14 - A}{3}\end{aligned}$$

$$\boxed{\left\{\left(1, \frac{14 - A}{3}\right)\right\}}$$

89.
$$\begin{aligned}Ax + By &= C\\ \underline{-Ax + By} &= \underline{C}\\ 2By &= 2C\\ y &= \frac{C}{B}\end{aligned}$$

$$\begin{aligned}Ax + By &= C\\ Ax + B\left(\frac{C}{B}\right) &= C\\ Ax + C &= C\\ Ax &= 0\\ x &= 0\end{aligned}$$

but
$$A = B \text{ and } C = 3A$$
$$y = \frac{C}{B} = \frac{3A}{A} = 3$$

$$\boxed{\{(0, 3)\}}$$

91.
$$\begin{aligned}ax + by &= c\\ \underline{ax - by} &= \underline{d}\\ 2ax &= c + d\\ x &= \frac{c + d}{2a}\end{aligned}$$

$$\begin{aligned}ax + by &= c\\ \frac{a(c + d)}{2a} + by &= c\\ \frac{c + d}{2} + by &= c\\ (\times 2) \quad c + d + 2by &= 2c\\ 2by &= c - d\\ y &= \frac{c - d}{2b}\end{aligned}$$

$$\boxed{\left\{\left(\frac{c + d}{2a}, \frac{c - d}{2b}\right)\right\}}$$

93.
$$\begin{aligned}6x - 9y &= 3 \qquad (\div 3) \rightarrow \qquad 2x - 3y = 1\\ 4x - 6y &= a \qquad (\div 2) \rightarrow \qquad 2x - 3y = \frac{a}{2}\end{aligned}$$

The system has an infinite number of solutions when
$$\frac{a}{2} = 1$$
or
$$\boxed{a = 2}$$

95.

$$ax + by = c$$
$$\underline{bx + ay = c}$$
$$(a+b)x + (a+b)y = 2c \qquad \text{(add)}$$

The coefficients of x and y are zero when

$$a + b = 0$$

or
$$\boxed{a = -b}$$

Thus,

$$0 + 0 = 2c$$
$$0 = 2c \qquad \text{False (contradiction)}$$

No solution.

The system is inconsistent.

97.

$$\frac{5}{x} + \frac{6}{y} = \frac{19}{6}$$

$$\frac{3}{x} + \frac{4}{y} = 2$$

$$\left(\text{Let } a = \frac{1}{x} \text{ and } b = \frac{1}{y}\right):$$

$$5a + 6b = \frac{19}{6} \qquad (\times 6) \rightarrow \qquad 30a + 36b = 19$$

$$3a + 4b = 2 \qquad (\times -9) \rightarrow \qquad \underline{-27a - 36b = -18}$$

$$3a = 1$$

$$a = \frac{1}{3}$$

$$3a + 4b = 2$$
$$3\left(\frac{1}{3}\right) + 4b = 2$$
$$1 + 4b = 2$$
$$4b = 1$$
$$b = \frac{1}{4}$$

$$x = \frac{1}{a} = \frac{1}{1/3} = 3$$
$$y = \frac{1}{b} = \frac{1}{1/4} = 4$$

$$\boxed{\{(3, 4)\}}$$

Review Problems

102. Let

$$x = \text{number of quarters}$$
$$x + 6 = \text{number of dimes}$$

$$0.25x + 0.10(x + 6) = 3.05$$
$$0.25x + 0.10x + 0.60 = 3.05$$
$$0.35x = 2.45$$
$$x = 7 \qquad \text{(quarters)}$$
$$x + 6 = 13 \qquad \text{(dimes)}$$

$$\boxed{7 \text{ quarters}, 13 \text{ dimes}}$$

103.

$$\frac{x+6}{8} - \frac{x-2}{12} = \frac{5}{6}$$

$$(\times 24) \quad 3(x+6) - 2(x-2) = 4(5)$$

$$3x + 18 - 2x + 4 = 20$$

$$x + 22 = 20$$

$$x = -2$$

$$\boxed{\{-2\}}$$

104.

$$\{x \mid 4x + 1 \leq x + 13\} \quad \cap \quad \{x \mid 2(3 - 2x) \leq 18\}$$

$$4x + 1 \leq x + 13 \qquad\qquad\qquad 6 - 4x \leq 18$$

$$3x \leq 12 \qquad\qquad\qquad\qquad -4x \leq 12$$

$$x \leq 4 \qquad\qquad\qquad\qquad\quad x \geq -3$$

$$\{x \mid x \leq 4\} \quad \cap \qquad \{x \mid x \geq -3\}$$

$$\boxed{\{x \mid -3 \leq x \leq 4\}}$$

or $\boxed{[-3, 4]}$

Section 4.2   Problem Solving Using Systems of Equations

Problem Set 4.2, pp. 275-278

1. Let

$$x = \text{first number}$$
$$y = \text{second number}$$

$$2x - y = 9 \qquad\quad (\times 3) \rightarrow \qquad 6x - 3y = 27$$
$$2x + 3y = -3 \qquad (\text{no change}) \rightarrow \qquad \underline{2x + 3y = -3}$$
$$8x = 24$$
$$x = 3$$

$$2x - y = 9$$
$$2(3) - y = 9$$
$$-y = 3$$
$$y = -3$$

The numbers are $\boxed{3 \text{ and } -3}$.

3. Let

$$t = \text{the tens' digit}$$
$$u = \text{the units' digit}$$
$$\text{original number} = 10t + u$$
$$\text{number with digits reversed} = 10u + t$$
$$t + u = 9$$
$$\underline{10u + t = (10t + u) + 45}$$

(simplify):

$t + u$	$=$	9	(no charge)	$t + u$	$=$	9
$-9t + 9u$	$=$	45	$(\div 9) \rightarrow$	$\underline{-t + u}$	$=$	$\underline{5}$
				$2u$	$=$	14
				u	$=$	7

$$t + u = 9$$
$$t + 7 = 9$$
$$t = 2$$

The number is $\boxed{27}$.

5. Let

$$t = \text{the tens' digit}$$
$$u = \text{the units' digit}$$
$$\text{original number} = 10t + u$$

$$u = 2t + 1$$
$$(10t + u) + u = 3t + 35$$

(simplify)	$u - 2t$	$=$	1	$(\times -2)$	$-2u + 4t$	$= -2$
	$2u + 7t$	$=$	35	(no change)	$\underline{2u + 7t}$	$= \underline{35}$
					$11t$	$= 33$
					t	$= 3$

$$u - 2t = 1$$
$$u - 2(3) = 1$$
$$u = 7$$

The number is $\boxed{37}$.

7. $m\angle A + m\angle B = 180$ (angles A and B are supplementary)
    $m\angle A = m\angle C$ (corresponding angles are equal in measrure)
(Substitute):

$$(4x - 2y + 4) + (12x + 6y + 12) = 180$$
$$4x - 2y + 4 = 6x - 24$$

(Simplify):

$16x + 4y$	$=$	164	$(\div 4) \rightarrow$	$4x + y$	$= 41$
$-2x - 2y$	$=$	-28	$(\div 2) \rightarrow$	$\underline{-x - y}$	$= \underline{-14}$
				$3x$	$= 27$
				x	$= 9$

$$x + y = 14$$
$$9 + y = 14$$
$$y = 5$$

$$\boxed{x = 9, y = 5}$$

$m\angle A = (4x - 2y + 4)° = (4 \cdot 9 - 2 \cdot 5 + 4)° = 30°$
$m\angle B = (12x + 6y + 12)° = (12 \cdot 9 + 6 \cdot 5 + 12)° = 150°$
$m\angle C = (6x - 24)° = (6 \cdot 9 - 24)° = 30°$

$$\boxed{m\angle A = 30°, m\angle B = 150°, m\angle C = 30°}$$

9. $[5x - (2y - 80)] + 2y = 180$ (consecutive angles are supplementary)
$5x - (2y - 80) = 3x$ (opposite angles are equal in measure)
(Simplify):

$$
\begin{aligned}
5x - 2y + 80 + 2y &= 180 \\
5x &= 100 \\
x &= 20
\end{aligned}
\qquad
\begin{aligned}
5x - 2y + 80 &= 3x \\
2x - 2y &= -80 \\
x - y &= -40 \\
20 - y &= -40 \\
-y &= -60 \\
y &= 60
\end{aligned}
$$

$(3x)° = (3 \cdot 20)° = \boxed{60°}$

$(2y)° = (2 \cdot 60)° = \boxed{120°}$

$5x - (2y - 80)° = [5 \cdot 20 - (2 \cdot 60 - 80)]° = [100 - (120 - 80)]° = (100 - 40)° = \boxed{60°}$

Since opposite angles are equal in measure, the remaining angle is $(2y)° = \boxed{120°}$

11. Let
$$
\begin{aligned}
x &= \text{length of } EB, \, x = \text{length of } BC, \, x = \text{length of } AD \\
y &= \text{length of } AE, \, y = \text{length of } EC \\
x + y &= \text{length of } AD
\end{aligned}
$$

perimeter of parallelogram $ABCD = 50$
$$
\begin{aligned}
AE + EB + BC + DC + AD &= 50 \\
y + x + x + (x + y) + x &= 50 \\
4x + 2y &= 50
\end{aligned}
$$

perimeter of trapezoid $AECD = 39$
$$
\begin{aligned}
AE + EC + DC + AD &= 39 \\
y + y + (x + y) + x &= 39 \\
2x + 3y &= 39
\end{aligned}
$$

$$
\begin{aligned}
4x + 2y &= 50 \\
2x + 3y &= 39
\end{aligned}
\qquad
\begin{aligned}
(\div -2) &\to \\
(\text{no change}) &\to
\end{aligned}
\qquad
\begin{aligned}
-2x - y &= -25 \\
2x + 3y &= 39 \\
2y &= 14 \\
y &= 7 \quad (AE)
\end{aligned}
$$

$$
\begin{aligned}
2x + y &= 25 \\
2x + 7 &= 25 \\
2x &= 18 \\
x &= 9 \quad (EB)
\end{aligned}
$$

$$
x + y = 7 + 9 = 16 \quad (DC)
$$

$\boxed{\text{length } AE, \, 7 \text{ m; length } EB, \, 9 \text{ m; length of } DC, \, 16 \text{ m}}$

13. Let

$$x \;=\; \text{hours of tutoring}$$
$$y \;=\; \text{hours of grading}$$

$$70x + 50y \;=\; 622.50 \qquad (\times -4) \to$$
$$90x + 40y \;=\; 719 \qquad (\times 5) \to$$

$$-280x - 200y \;=\; -2490$$
$$\underline{450x + 200y \;=\; 3595}$$
$$170x \;=\; 11.05$$
$$x \;=\; 6.5 \quad \text{(tutors)}$$

$$70x + 50y \;=\; 622.50$$
$$70(6.5) + 50y \;=\; 622.50$$
$$455 + 50y \;=\; 622.50$$
$$50y \;=\; 167.50$$
$$y \;=\; 3.35 \quad \text{(graders)}$$

$$\boxed{\text{tutors: } \$6.50; \text{ graders: } \$3.35}$$

15. Let

$$x \;=\; \text{Louis' daily wage}$$
$$y \;=\; \text{Lestat's daily wage}$$
$$8x + 10y \;=\; 640$$

$$12x \;=\; 9y \qquad \text{(solve for } y) \to \qquad y \;=\; \frac{4}{3}x$$

$$\text{(substitute for } y \text{ in first equation)} \qquad 8x + 10\left(\frac{4}{3}x\right) \;=\; 640$$
$$24x + 40x \;=\; 1920$$
$$64x \;=\; 1920$$
$$x \;=\; 30 \quad \text{(Louis)}$$

$$y \;=\; \frac{4}{3}x$$
$$y \;=\; \frac{4}{3}(30) = 40 \quad \text{(Lestat)}$$

$$\boxed{\text{Louis: } \$30; \text{ Lestat: } \$40}$$

17. Let

$$x \;=\; \text{cost of orange trees}$$
$$y \;=\; \text{cost of grapefruit trees}$$

$$3x + 4y \;=\; 22 \qquad (\times -3) \to$$
$$4x + 6y \;=\; 31 \qquad (\times 2) \to$$

$$-9x - 12y \;=\; -66$$
$$\underline{8x + 12y \;=\; 62}$$
$$-x \;=\; -4$$
$$x \;=\; 4 \quad \text{(orange trees)}$$

$$3x + 4y \;=\; 22$$
$$3(4) + 4y \;=\; 22$$
$$4y \;=\; 10$$
$$y \;=\; 2.50 \quad \text{(grapefruit trees)}$$

$$\boxed{\text{cost of orange tree: } \$4; \text{ cost of grapefruit tree: } \$2.50}$$

19. Let

x = speed of plane in still air
y = speed of wind

	rate, R	×	time, t (hours)	=	Distance, D (km)
with the wind	$x + y$		5		3000
against the wind	$x - y$		6		3000

$$
\begin{aligned}
5(x + y) &= 3000 & (\div 5) \rightarrow & & x + y &= 600 \\
6(x - y) &= 3000 & (\div 6) \rightarrow & & \underline{x - y} &\underline{= 500} \\
& & & & 2x &= 1100 \\
& & & & x &= 550 \quad \text{(plane)}
\end{aligned}
$$

$$
\begin{aligned}
x + y &= 600 \\
550 + y &= 600 \\
y &= 50 \quad \text{(wind)}
\end{aligned}
$$

speed of plane in still air, 550 kilometer per hour;
speed of wind, 50 kilometers per hour

21. Let

x = rate of paddling in still water
y = rate of current

	R	×	t (hours)	=	D
with the wind	$x + y$		2.5		D
against the wind	$x - y$		5		D

$$
\begin{aligned}
2.5(x + y) &= D & (\div 2.5) \rightarrow & & x + y &= \frac{D}{2.5} = \frac{2}{5} D \\
5(x - y) &= D & (\div 5) & & \underline{x - y} &\underline{= \frac{1}{5} D} \\
& & & & 2x &= \frac{3}{5} D \\
& & & & x &= \frac{3}{10} D \quad \text{(paddling in still water)}
\end{aligned}
$$

$$
\begin{aligned}
x + y &= \frac{2}{5} D \\
\frac{3}{10} D + y &= \frac{4}{10} D \\
y &= \frac{1}{10} D \quad \text{(current)}
\end{aligned}
$$

$$
\begin{aligned}
x &= 3 \left(\frac{1}{10} D \right) \\
x &= 3y \quad \text{(three times faster)}
\end{aligned}
$$

Thus, Steve's rate of paddling in still water is three times faster than the current's speed.

23. Let

$$x = \text{number of dogs}$$
$$y = \text{number of canaries}$$

$4x + 2y$	$= 72$	$(\div -2) \rightarrow$	$-2x - y$	$=$	-36
$x + y$	$= 23$	$(\text{no change}) \rightarrow$	$\underline{x + y}$	$\underline{=}$	$\underline{23}$
			$-x$	$=$	-13
			x	$=$	13 (dogs)

$$
\begin{aligned}
x + y &= 23 \\
13 + y &= 23 \\
y &= 10 \quad \text{(canaries)}
\end{aligned}
$$

$\boxed{\text{Dogs: 13; Canaries: 10}}$

25. Let

$x = \text{age of son}$		$x - 6 = \text{age of son 6 years ago}$	
$y = \text{age of mother}$		$y - 6 = \text{age of mother 6 years ago}$	

$$
\begin{aligned}
y &= 3x & (\text{no change}) \rightarrow & & y &= 3x \\
x - 6 &= \tfrac{1}{4}(y - 6) & (\text{simplify}) \rightarrow & & 4(x - 6) &= y - 6
\end{aligned}
$$

$$(\text{simplify and substitute for } y) \rightarrow \quad
\begin{aligned}
4x - 24 &= 3x - 6 \\
x &= 18 \quad \text{(son)} \\
y &= 3x = 3(18) = 54 \quad \text{(mother)}
\end{aligned}
$$

$\boxed{\text{son: 18; mother: 54}}$

27. Let

$$
\begin{aligned}
x &= \text{length of body} \\
y &= \text{length of tail} \\
\text{length of head} &= 6 \text{ inches}
\end{aligned}
$$

$$
\begin{aligned}
(\text{tail}) &= (\text{head}) + \tfrac{1}{4}(\text{body}) \\
(\text{body}) &= [(\text{head}) + (\text{tail})] + 3
\end{aligned}
$$

$$
\begin{aligned}
y &= 6 + \tfrac{1}{4}x & (\text{simplify/rearrange}) \rightarrow & & -x + 4y &= 24 \\
x &= (6 + y) + 3 & (\text{rearrange}) \rightarrow & & \underline{x - y} &\;\underline{=\;9} \\
& & & & 3y &= 33 \\
& & & & y &= 11 \quad \text{(tail)}
\end{aligned}
$$

$$x = y + 9 = 11 + 9 = 20 \quad \text{(body)}$$

$$
\begin{aligned}
\text{length of lobster} &= \text{length of head} + \text{length of body} + \text{length of tail} \\
&= 6 + 20 + 11 \\
&= 37
\end{aligned}
$$

$\boxed{\text{37 inches}}$

29. $y = mx + b$

$(8, -2)$:	$-2 = m(8) + b$	(no change) $\rightarrow$	$-2 = 8m + b$
$(-4, -8)$:	$-8 = m(-4) + b$	$(\times 2) \rightarrow$	$\underline{-16 = -8m + 2b}$

$$-18 = 3b$$
$$-6 = b$$

$$-2 = 8m + b$$
$$-2 = 8m - 6$$
$$4 = 8m$$
$$\frac{1}{2} = m$$

$$\boxed{m = \frac{1}{2}, \; b = -6}$$

$$y = \frac{1}{2}x - 6$$

31. $y = ax^2 + bx - 1$

$(1, 4)$: $4 = a(1^2) + b(-1) - 1$

$(-2, 1)$: $1 = a(-2)^2 + b(-2) - 1$

(simplify and rearrange):	$a + b = 5$	(no change) $\rightarrow$	$a + b = 5$
	$4a - 2b = 2$	$(\div 2) \rightarrow$	$\underline{2a - b = 1}$

$$3a = 6$$
$$a = 2$$

$$a + b = 5$$
$$2 + b = 5$$
$$b = 3$$

$$\boxed{a = 2, \; b = 3}$$

$$y = 2x^2 + 3x - 1$$

33. $s = S_0 + v_0 t - 16t^2$

$t = 5$ seconds; $s = 10{,}000$ feet; $(5, 10{,}000)$: $10{,}000 = S_0 + 5v_0 - 16(5^2)$

$t = 10$ seconds; $s = 8550$ feet; $(10, 8550)$: $8550 = S_0 + 10v_0 - 16(10^2)$

(simplify and rearrange):	$s_0 + 5v_0 = 10{,}400$	(no change) $\rightarrow$	$s^0 + 5v_0 = 10{,}400$
	$s_0 + 10v_0 = 10{,}150$	$(\times -1) \rightarrow$	$\underline{-s_0 - 10v_0 = -10{,}150}$

$$-5v_0 = 250$$
$$v_0 = -50 \quad \text{(initial velocity)}$$

$$s_0 + 5v_0 = 10{,}400$$
$$s_0 + 5(-50) = 10{,}400$$
$$s = 10{,}650 \quad \text{(initial height)}$$

$$\boxed{\text{Initial height: } 10{,}650 \text{ ft; Initial velocity: } -50 \text{ ft/sec}}$$

35. $y = ax^2 + bx + 32$

$x = \dfrac{1}{2}$ second, $y = 36$ feet; $\left(\dfrac{1}{2}, 36\right)$: $36 = a\left(\dfrac{1}{2}\right)^2 + 32$

$x = 2$ seconds, $y = 0$ feet; $(2, 0)$: $0 = a(2^2) + b(2) + 32$

$$\begin{array}{rcl}
\text{(simplify and rearrange):} \quad \dfrac{1}{4}a + \dfrac{1}{2}b &=& 4 \qquad (\times 4) \rightarrow \quad a + 2b = 16 \\
4a + 2b &=& -32 \qquad (\times -1) \rightarrow \quad \underline{-4a - 2b = 32} \\
&& \qquad\qquad\qquad\qquad 3a = 48 \\
&& \qquad\qquad\qquad\qquad\; a = -16
\end{array}$$

$$\begin{array}{rcl}
a + 2b &=& 16 \\
-16 + 2b &=& 16 \\
2b &=& 32 \\
b &=& 16
\end{array}$$

$\boxed{a = -16,\, b = 16}$

$y = -16x^2 + 16x + 32$

37. $cx = dy$

Ricky: 80 pounds, x = distance Ricky is from fulcrum

Fred: 100 pounds, y = distance Fred is from fulcrum

$$\begin{array}{rcl}
80x &=& 100y \\
x + y &=& 9 \qquad \text{(solve for } y) \rightarrow \qquad y = 9 - x \\
&& \qquad\qquad \text{(substitute for } y) \qquad 80x = 100(9 - x) \\
&& \qquad\qquad\qquad\qquad\qquad\qquad\quad 80x = 900 - 100x \\
&& \qquad\qquad\qquad\qquad\qquad\qquad\; 180x = 900 \\
&& \qquad\qquad\qquad\qquad\qquad\qquad\qquad x = 5 \quad \text{(Ricky)} \\
&& \qquad\qquad\qquad\qquad\quad y = 9 - x = 9 - 5 = 4 \quad \text{(Fred)}
\end{array}$$

$\boxed{\text{Ricky: 5 feet; Fred: 4 feet}}$

39. Let

$\qquad x$ = cost of mangos   (sold at profit of 20%)

$\qquad y$ = cost of avocados   (sold at loss of 2%)

$x + y = 67$ (solve for y) $\rightarrow$ $y = 67 - x$

(mangos at profit of 20%) + (avocados at loss of 2%) = (total profit of 8.56)

$$\begin{array}{rcl}
0.20x + (-0.02y) &=& 8.56 \\
0.20x - 0.02y &=& 8.56 \\
\text{(substitute for } y) && \\
0.20x - 0.02(67 - x) &=& 8.56 \\
0.20x - 1.34 + 0.02x &=& 8.56 \\
0.20x - 1.34 + 0.02x &=& 8.56 \\
0.22x &=& 9.90 \\
x &=& 45 \quad \text{(mangos)}
\end{array}$$

$$\begin{array}{rcl}
x + y &=& 67 \\
45 + y &=& 67 \\
y &=& 22 \quad \text{(avocados)}
\end{array}$$

$\boxed{\text{Mangos: \$45; Avocados: \$22}}$

41. Let

R = number of people on right
L = number of people on left

	New left	New right
Wednesday: Jane changed from left to right:	$L-1$	$R+1$
equation:	$L-1 = R+1$	

Next Wednesday: Jane moved back to left:	$L-1+1$	$R+1-1$
Ted changed from right to left:	$L-1+1+1$	$R+1-1-1$
equation:	$L+1 = 2(R-1)$	

System of equations

$$
\begin{aligned}
L-1 &= R+1 \\
L+1 &= 2(R-1)
\end{aligned}
\qquad
\begin{aligned}
\rightarrow \\
\rightarrow
\end{aligned}
\qquad
\begin{aligned}
L &= R+2 \\
L &= 2R-3
\end{aligned}
$$

$$
\begin{aligned}
2R-3 &= R+2 \\
R &= 5
\end{aligned}
$$

$L = R+2 = 5+2 = 7$

Right: 5; Left: 7

43. Let

x = the assitant's federal tax
y = the assitant's state tax
x = $0.2(9800-y)$ (These equations are translated from the given conditions.)
y = $0.1(9800-x)$

Multiply both sides of each equation by 10 and simplifying, we obtain:

$$
\begin{aligned}
10x + 2y &= 19{,}600 & \text{(no change)} \rightarrow & & 10x + 2y &= 19{,}600 \\
x + 10y &= 9800 & (\times -10) \rightarrow & & \underline{-10x - 100y} &= \underline{98{,}000} \\
& & \text{(Add):} & & -98y &= -78{,}400 \\
& & & & y &= 800
\end{aligned}
$$

$$
\begin{aligned}
\text{Since } x + 10y &= 9800 \\
x + 10(800) &= 9800 \\
x &= 1800
\end{aligned}
$$

The teacher's assistant paid $1800 in federal tax and $800 in state tax.

State: $800; Federal: $1800

45. Let

x = the number of boys
y = the number of girls

Since he has as many brothers as he has sisters, counting himself, the number of boys exceeds the number of girls by 1. Hence

$$x = y+1$$

His sister has twice as many brothers as she has sisters. Hence

$$x = 2(y-1)$$

Substituting $y + 1$ for x (from the first equation), we obtain

$$\begin{aligned} y + 1 &= 2(y - 1) \\ y + 1 &= 2y - 2 \\ 3 &= y \end{aligned}$$

Since $x = y + 1$, we obtain $x = 3 + 1 = 4$.

There are 4 boys and 3 girls in this family. (Check this against the original conditions of the problem.)

$\boxed{\text{3 girls; 4 boys}}$

47. Let

$$\begin{aligned} x &= \text{the speed of the escalator (in steps/second)} \\ y &= \text{the number of steps in the escalator} \end{aligned}$$

Walking at 2 steps/second, the top is reached after 32 steps, so it takes 16 seconds to reach the top.

Thus, $y = 16x + 32$.

Walking at 1 step/second, the top is reached after 20 steps, so it takes 20 seconds to reach the top.

Thus, $y = 20x + 20$.

Using substitution, we obtain

$$\begin{aligned} 20x + 20 &= 16x + 32 \\ 4x &= 12 \\ x &= 3 \end{aligned}$$

The speed of the escalator, x, is $\boxed{\text{3 steps per second}}$.

There are $y = 20x + 20 = 20(3) + 20$ or $\boxed{\text{80 steps}}$ in the escalator.

Review Problems

51.

$$\begin{aligned} 6x - 3y &= 15 \\ -3y &= -6x + 15 \\ y &= 2x - 5 \end{aligned}$$

$\boxed{f(x) = 2x - 5}$

$f(-5) = 2(-5) - 5) = -10 - 5 = \boxed{-15}$

$f(a) = \boxed{2a - 5}$

52. a line having x-intercept of 3 and a slope of $-\dfrac{1}{2}$:

$$(3, 0),\ m = -\frac{1}{2}$$

$$y - 0\ =\ -\frac{1}{2}(x - 3)$$

$$y\ =\ -\frac{1}{2}x + \frac{3}{2}$$

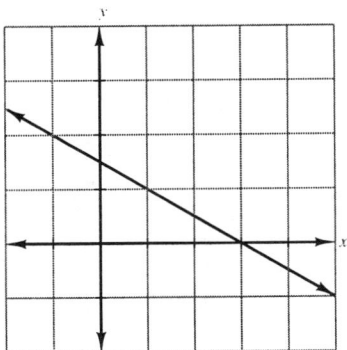

53. $\begin{aligned} x - 2y\ &\le\ -4 \\ y\ &\le\ 0 \end{aligned}$

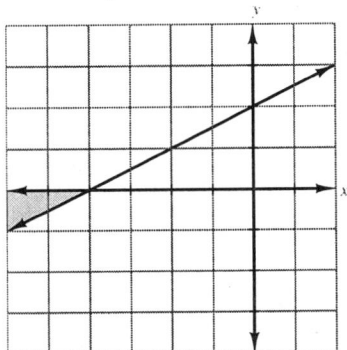

Section 4.3 Algebraic Solution to a System of Three Linear Equations in Three Variables

Problem Set 4.3, pp. 286-289

(*Note:* **For all systems of equations, even though the equations are not numbered, they will be identified as Equations (1) or (2) or Equations (1), (2) or (3) in order. Problem 1 shows the equation numbering sequence. However the remaining problems will not, except when new equations result from these equations.**)

1. $x + y + 2z = 11$
    $x + y + 3z = 14$
    $x + 2y - z = 5$

(Equations 1 and 2): $x + y + 2z\ =\ 11$
         $(\times -1)$ $\underline{-x - 3z\ =\ -14}$
              $-z\ =\ -3$
               $z\ =\ 3$

(Substitute $z = 3$ into Equations 1 and 3):

$$
\begin{array}{lll}
x + y + 2(3) = 11 & \text{(simplify)}(\times{-1}) \to & -x - y = -5 \\
x + 2y - 3 = 5 & \text{(simplify)} \to & \underline{x + 2y = 8} \\
& & y = 3
\end{array}
$$

(Substitute $z = 3$ and $y = 3$ into Equation 1):

$$
\begin{aligned}
x + 3 + 6 &= 1 \\
x &= 2
\end{aligned}
$$

$$\boxed{\{(2, 3, 3)\}}$$

3.
$$
\begin{aligned}
4x - y + 2z &= 11 \\
x + 2y - z &= -1 \\
2x + 2y - 3z &= -1
\end{aligned}
$$

(Equations 1 and 2):

$$
\begin{array}{rl}
& 4x - y + 2z = 11 \\
(\times 2) & \underline{2x + 4y - 2z = -2} \\
& 6x + 3y = 9
\end{array}
$$

(Equations 2 and 3):

$$
\begin{array}{rl}
(\times{-3}) & -3x - 6y + 3z = 3 \\
& \underline{2x + 2y - 3z = -1} \\
& -x - 4y = 2
\end{array}
$$

$$
\begin{array}{l}
6x + 3y = 9 \\
\underline{-x - 4y = 2}
\end{array}
$$

$$
\begin{array}{rl}
(\div 3) \to & 2x + y = 3 \\
(\times 2) \to & \underline{-2x - 8y = 4} \\
& -7y = 7 \\
& y = -1
\end{array}
$$

$$
\begin{aligned}
2x + y &= 3 \\
2x + (-1) &= 3 \\
2x &= 4 \\
x &= 2
\end{aligned}
$$

(Substitute $x = 2$ and $y = -1$ into Equation 2):

$$
\begin{aligned}
x + 2y - z &= -1 \\
2 + 2(-1) - z &= -1 \\
-z &= -1 \\
z &= 1
\end{aligned}
$$

$$\boxed{\{(2, -1, 1)\}}$$

5.
$$
\begin{array}{lll}
3x + 5y + 2z = 0 & (\times 4) \to & 12x + 20y + 8z = 0 \\
12x - 15y + 4z = 12 & (\times 2) \to & 24x - 30y + 8z = 24 \\
6x - 25y - 8z = 8 & \text{(no change)} \to & 6x - 25y - 8z = 8
\end{array}
$$

(Equations 1 and 3):

$$
\begin{aligned}
12x + 20y + 8z &= 0 \\
\underline{6x - 25y - 8z} &= \underline{8} \\
18x - 5y &= 8
\end{aligned}
$$

(Equations 2 and 3):

$$
\begin{aligned}
24x - 30y + 8z &= 24 \\
\underline{6x - 25y - 8z} &= \underline{8} \\
30x - 55y &= 32
\end{aligned}
$$

$$
\begin{array}{l}
18x - 5y = 8 \\
\underline{30x - 55y = 32}
\end{array}
$$

$$
\begin{array}{rl}
(\times{-11}) \to & -198x + 55y = -88 \\
& \underline{30x - 55y = 32} \\
& -168x = -56 \\
& x = \dfrac{1}{3}
\end{array}
$$

$$18x - 5y = 8$$
$$18\left(\frac{1}{3}\right) - 5y = 8 \quad \left(\text{substitute for } x = \frac{1}{3}\right)$$
$$6 - 5y = 8$$
$$-5y = 2$$
$$y = -\frac{2}{5}$$

$$3x + 5y + 2z = 0$$
$$3\left(\frac{1}{3}\right) + 5\left(-\frac{2}{5}\right) + 2z = 0 \quad \left(\text{substitute for } x = \frac{1}{3} \text{ and } y = -\frac{2}{5}\right)$$
$$1 - 2 + 2z = 0$$
$$2z = 1$$
$$x = \frac{1}{2}$$

$$\left\{ \left(\frac{1}{3}, -\frac{2}{5}, \frac{1}{2}\right) \right\}$$

7.

$2x - 4y + 3z = 17$	(no change) $\rightarrow$	$2x - 4y + 3z = 17$
$x + 2y - z = 0$	($\times 3$) $\rightarrow$	$3x + 6y - 3z = 0$
$4x - y - z = 6$	($\times 3$) $\rightarrow$	$12x - 3y - 3z = 18$

(Equations 1 and 2):
$$2x - 4y + 3z = 17$$
$$\underline{3x + 6y - 3z = 0}$$
$$5x + 2y = 17$$

(Equations 1 and 3):
$$2x - 4y + 3z = 17$$
$$\underline{12x - 3y - 3z = 18}$$
$$14x - 7y = 35$$
$$(\div 7) \rightarrow \quad 2x - y = 5$$

$5x + 2y = 17$	(no change) $\rightarrow$	$5x + 2y = 17$
$2x - y = 5$	($\times 2$) $\rightarrow$	$4x - 2y = 10$
		$9x = 27$
		$x = 3$

$$2x - y = 5$$
$$2(3) - y = 5$$
$$-y = -1$$
$$y = 1$$

$$x + 2y - z = 0$$
$$3 + 2(1) - z = 0$$
$$-z = -5$$
$$z = 5$$

$$\{(3, 1, 5)\}$$

9.

$2x + y = 2$	($\times -3$) $\rightarrow$	$-6x - 3y = -6$
$x + y - z = 4$	(add Equations 2 and 3) $\rightarrow$	$\underline{4x + 3y = 4}$
$3x + 2y + z = 0$		$-2x = -2$
		$x = 1$

$$2x + y = 2 \quad \text{(Equation 1)}$$
$$2(1) + y = 1 \quad (x = 1)$$
$$y = 0$$

$$x + y - z = 4 \quad \text{(Equation 2)}$$
$$1 + 0 - z = 4 \quad (x = 1, y = 0)$$
$$-z = 3$$
$$z = -3$$

$$\{(1, 0, -3)\}$$

11.

$$\begin{aligned}x + y &= -4\\ y - z &= 1 \qquad (\times 3) \rightarrow\\ 2x + y + 3z &= -21\end{aligned}$$

$$\begin{aligned}3y - 3z &= 3\\ \underline{2x + y + 3z} &= \underline{-21}\\ 2x + 4y &= -18 \qquad \text{(add Equations 2 and 3)}\\ (\div 2) \rightarrow \quad x + 2y &= -9\end{aligned}$$

$$\begin{aligned}x + y &= -4 \qquad (\times -2) \rightarrow\\ x + 2y &= -9\end{aligned}$$

$$\begin{aligned}-2x - 2y &= 8\\ \underline{x + 2y} &= \underline{-9}\\ -x &= -1\\ x &= 1\end{aligned}$$

$$\begin{aligned}x + y &= -4 \qquad \text{(Equation 1)}\\ 1 + y &= -4\\ y &= -5\end{aligned}$$

$$\begin{aligned}y - z &= 1 \qquad \text{(Equation 2)}\\ -5 - z &= 1\\ -z &= 6\\ z &= -6\end{aligned}$$

$\boxed{\{(1, -5, -6)\}}$

13.

$$\begin{aligned}6x - y + 3z &= 9 \qquad \rightarrow\\ \tfrac{1}{4}x - \tfrac{1}{2}y - \tfrac{1}{3}z &= -1 \qquad (\times -12) \rightarrow\\ -x + \tfrac{1}{6}y - \tfrac{2}{3}z &= 0 \qquad (\times 6) \rightarrow\end{aligned}$$

$$\begin{aligned}6x - y + 3z &= 9\\ -3x + 6y + 4z &= 12\\ -6x + y - 4z &= 0\end{aligned}$$

(Equations 1 and 3):
$$\begin{aligned}6x - y + 3z &= 9\\ \underline{-6x + y - 4z} &= \underline{0}\\ -z &= 9\\ z &= -9\end{aligned}$$

(Equation 2 and 3):
$$\begin{aligned}-3x + 6y + 4z &= 12\\ \underline{-6x + y - 4z} &= \underline{0}\\ -9x + 7y &= 12 \qquad \text{(Equation 5)}\end{aligned}$$

$$\begin{aligned}6x - y + 3z &= 9 \qquad \text{(Equation 1)}\\ 6x - y - 27 &= 9\\ 6x - y &= 36 \qquad \text{(Equation 4)}\end{aligned}$$

$$\begin{aligned}\text{(Equation 4):} \quad 6x - y &= 36 \qquad (\times 7) \rightarrow\\ \text{(Equation 5):} \quad -9x + 7y &= 12\end{aligned}$$

$$\begin{aligned}42x - 7y &= 252\\ \underline{-9x + 7y} &= \underline{12}\\ 33x &= 264\\ x &= 8\end{aligned}$$

$$\begin{aligned}6x - y &= 36 \qquad \text{(Equation 4)}\\ 6(8) - y &= 36\\ -y &= -12\\ y &= 12\end{aligned}$$

• $\boxed{\{(8, 12, -9)\}}$

15.

$$\begin{aligned}2x + y + 4z &= 4\\ x - y + z &= 6 \qquad (\times 2) \rightarrow\\ x + 2y + 3z &= 5\end{aligned}$$

$$\begin{aligned}2x - 2y + 2z &= 12\\ \underline{x + 2y + 3z} &= \underline{5}\\ 3x + 5z &= 17 \qquad \text{(Equation 4)}\end{aligned}$$

(Add Equation 1 and Equation 2):
$$3x + 5z = 10$$

(Equation 4)
$$(\times -1) \quad \underline{-3x - 5z = -17}$$
$$0 = -7 \qquad \text{False (Contradiction)}$$

No solution.

$\boxed{\varnothing}$

17.

$$\begin{array}{rl} x - 4y + z &= -5 \\ 3x - 12y + 3z &= -15 \\ 2x - 8y + 2z &= -10 \end{array}$$

$$\begin{array}{l} \text{(no change)} \rightarrow \\ (\div 3) \rightarrow \\ (\div 2) \rightarrow \end{array}$$

$$\begin{array}{rl} x - 4y + z &= -5 \\ x - 4y + z &= -5 \\ x - 4y + z &= -5 \end{array}$$

Each equation represents the same line.
Dependent system with infinitely every solutions.

$$\boxed{\{x, y, z \mid x - 4y + z = -5\}}$$

19.

$$\begin{array}{rl} x + y &= 4 \\ x + z &= 4 \\ y + z &= 4 \end{array}$$

$$\begin{array}{l} \rightarrow \\ (\times -1) \rightarrow \end{array}$$

$$\begin{array}{rl} x + z &= 4 \\ -y - z &= -4 \\ \hline x - y &= 0 \quad \text{(Equation 4)} \end{array}$$

$$\begin{array}{rll} x + y &= 4 & \text{(Equation 1)} \\ x - y &= 0 & \text{(Equation 4)} \\ \hline 2x &= 4 \\ x &= 2 \end{array}$$

$$\begin{array}{rll} x + y &= 4 & \text{(Equation 1)} \\ 2 + y &= 4 \\ y &= 2 \end{array} \qquad \begin{array}{rll} x + z &= 4 & \text{(Equation 2)} \\ 2 + z &= 4 \\ z &= 2 \end{array}$$

$$\boxed{\{(2, 2, 2)\}}$$

21.

$$\begin{array}{rl} 7z - 3 &= 2(x - 3y) \\ 5y + 3z - 7 &= 4x \\ 4 + 5z &= 3(2x - y) \end{array}$$

(Simplify and rearrange):

$$\begin{array}{rl} -2x + 6y + 7z &= 3 \\ -4x + 5y + 3z &= 7 \\ -6x + 3y + 5z &= -4 \end{array}$$

(Equations 1 and 2):

$$\begin{array}{l} (\times -2) \rightarrow \end{array} \quad \begin{array}{rl} 4x - 12y - 14z &= -6 \\ -4x + 5y + 3z &= 7 \\ \hline -7y - 11z &= 1 \quad \text{(Equation 4)} \end{array}$$

(Equations 1 and 3):

$$\begin{array}{l} (\times -3) \rightarrow \end{array} \quad \begin{array}{rl} 6x - 18y - 21z &= -9 \\ -6x + 3y + 5z &= -4 \\ \hline -15y - 16z &= -13 \quad \text{(Equation 5)} \end{array}$$

(Equations 4 and 5):

$$\begin{array}{l} (\times -15) \rightarrow \\ (\times 7) \end{array} \quad \begin{array}{rl} 105y + 165z &= -15 \\ -105y - 112z &= -91 \\ \hline 53z &= -106 \\ z &= -2 \end{array}$$

(Equation 4)

$$\begin{array}{rl} -7y - 11z &= 1 \\ -7y - 11(-2) &= 1 \\ -7y + 22 &= 1 \\ -7y &= -21 \\ y &= 3 \end{array}$$

(Equation 1)

$$\begin{array}{rl} -2x + 6y + 7z &= 3 \\ -2x + 6(3) + 7(-2) &= 3 \\ -2x + 18 - 14 &= 3 \\ -2x &= -1 \\ x &= \frac{1}{2} \end{array}$$

$$\boxed{\left\{ \left(\frac{1}{2}, 3, -2 \right) \right\}}$$

23.

$$\begin{aligned} x + y &= 3 \\ x + y + z &= 3 \\ x + y - z &= 3 \end{aligned}$$

$\rightarrow$

(Add Equations 2 and 3):

$2x + 2y = 6$ $(\div -2) \rightarrow$

$$\begin{aligned} x + y &= 3 \\ -x - y &= -3 \\ \hline 0 &= 0 \quad \text{True} \end{aligned}$$

Dependent system with infinitely many solutions.

$$\boxed{\{x, y, z \mid x + y = 3\}}$$

25. Let

$\begin{aligned} x &= \text{first number} \\ y &= \text{second number} \\ z &= \text{third number} \end{aligned}$

$$\begin{aligned} x + y + z &= 16 \\ 2x + 3y + 4z &= 46 \\ 5x - y &= 31 \end{aligned}$$

$(\times -4) \rightarrow$

$$\begin{aligned} -4x - 4y - 4z &= -64 \\ 2x + 3y + 4z &= 46 \\ \hline -2x - y &= -18 \quad \text{(Equation 4)} \end{aligned}$$

(Equations 3 and 4):

$(\times -1) \rightarrow$
$$\begin{aligned} 5x - y &= 31 \\ 2x + y &= 18 \\ \hline 7x &= 49 \\ x &= 7 \quad \text{(first number)} \end{aligned}$$

$$\begin{aligned} 5x - y &= 31 \quad \text{(Equation 3)} \\ 5(7) - y &= 31 \\ -y &= -4 \\ y &= 4 \quad \text{(second number)} \end{aligned}$$

$$\begin{aligned} x + y + z &= 16 \quad \text{(Equation 1)} \\ 7 + 4 + z &= 16 \\ z &= 5 \quad \text{(third number)} \end{aligned}$$

The numbers are $\boxed{7, 4 \text{ and } 5}$.

27. Let

$\begin{aligned} x &= \text{number of nickels} \\ y &= \text{number of dimes} \\ z &= \text{number of quarters} \end{aligned}$

$$\begin{aligned} x + y + z &= 17 \\ 0.05x + 0.10y + 0.25z &= 2 \\ x + y &= z + 9 \end{aligned}$$

$\begin{aligned} &\rightarrow \\ &\rightarrow \\ &\rightarrow \end{aligned}$

$$\begin{aligned} x + y + z &= 17 \\ 5x + 10y + 25z &= 200 \\ x + y - z &= 9 \end{aligned}$$

(Equation 1 and 3):

$$\begin{aligned} x + y + z &= 17 \\ x + y - z &= 9 \\ \hline 2x + 2y &= 26 \end{aligned}$$

$(\div 2) \quad x + y = 13 \quad \text{(Equation 4)}$

(Equation 2 and 3):

$(\div 5) \rightarrow \quad x + 2y + 5z = 40$

$(\times 5) \rightarrow \quad 5x + 5y - 5z = 45$

$$\overline{6x + 7y = 85} \quad \text{(Equation 5)}$$

$(\times -7) \quad$
$$\begin{aligned} -7x - 7y &= -91 \quad \text{(Equation 4)} \\ 6x + 7y &= 85 \quad \text{(Equation 5)} \\ \hline -x &= -6 \\ x &= 6 \quad \text{(nickels)} \end{aligned}$$

(Equation 4): $x + y = 13$
$6 + y = 13$
$y = 7$ (dimes)

(Equation 1): $x + y + z = 17$
$6 + 7 + z = 17$
$x = 4$ (quarters)

$\boxed{6 \text{ nickels, 7 dimes, 4 quarters}}$

29. $y = \dfrac{1}{2} Ax^2 + Bx + C$

$(x = 1, y = 46)$: $(1, 46)$: $46 = \dfrac{1}{2} A(1^2) + B(1) + C$

$(x = 2, y = 84)$: $(2, 84)$: $84 = \dfrac{1}{2} A(2^2) + B(2) + C$

$(x + 3, y = 114)$: $(3, 114)$: $114 = \dfrac{1}{2} A(3^2) + B(3^2) + C$

(simplify):

$$46 = \dfrac{1}{2} A + B + C \qquad (\times 2) \to \qquad A + 2B + 2C = 92$$
$$84 = 2A + 2B + C \qquad \to \qquad 2A + 2B + C = 84$$
$$114 = \dfrac{9}{2} A + 3B + C \qquad (\times 2) \to \qquad 9A + 6B + 2C = 228$$

(Equation 1 and 2):

$$\begin{array}{rrl}
& A + 2B + 2C & = 92 \\
(\times -2) & -4A - 4B - 2C & = -168 \\ \hline
& -3A - 2B & = -76 \quad \text{(Equation 4)}
\end{array}$$

(Equation 1 and 3):

$$\begin{array}{rrl}
& A + 2B + 2C & = 92 \\
(\times -1) & -9A - 6B - 2C & = -22B \\ \hline
& -8A - 4B & = -136 \\
(\div 2) & 4A + 2B & = 68 \quad \text{(Equation 5)}
\end{array}$$

(Equation 4): $-3A - 2B = -76$
(Equation 5): $\dfrac{4A + 2B = 68}{A = -8}$

(Equation 5): $4A + 2B = 68$
$4(-8) + 2B = 68$
$2B = 100$
$B = 50$

(Equation 2): $2A + 2B + C = 84$
$2(-8) + 2(50) + C = 84$
$84 + C = 84$
$C = 0$

$\boxed{A = -8, B = 50, C = 0}$

$y = \dfrac{1}{2}(-8)x^2 + 50x + 0$
$y = -4x^2 + 50x$

31. Let
$x = $ length of BD, $x = $ length of DC
$2x = $ length of AB, $2x = $ length of BC
$y = $ length of AC, $z = $ length of AD
perimeter of $\triangle ABC$: $4x + y = 80$
perimeter of $\triangle ADB$: $3x + z = 70$
perimeter of $\triangle ACD$: $x + y + z = 48$
(Equations 2 and 3):

$$\begin{array}{rrl}
& 3x + z & = 70 \\
(\times -1) & -x - y - z & = -48 \\ \hline
& 2x - y & = 2 \quad \text{(Equation 4)} \\
& 4x + y & = 80 \quad \text{(Equation 1)} \\ \hline
& 6x & = 102 \\
& x & = 17 \\
& 2x & = 34 \quad (AB)
\end{array}$$

$$4x + y = 80 \quad \text{(Equation 1)}$$
$$4(17) + y = 80$$
$$y = 12 \quad (AC)$$

$$x + y + z = 48 \quad \text{(Equation 3)}$$
$$17 + 12 + z = 48$$
$$z = 19 \quad (AD)$$

$$\boxed{AB,\ 34 \text{ ft};\ AD,\ 19 \text{ ft};\ AC,\ 12 \text{ ft}}$$

33. Let

x = number of triangles
y = number of rectangles   (each contains 2 red roses)
z = number of pentagons   (each contains 5 carnations)

$$
\begin{array}{ll}
x + y + z = 40 & \rightarrow \\
3x + 4y + 5z = 153 & (\times -1) \rightarrow
\end{array}
\qquad
\begin{array}{l}
3x + 4y + 5z = 153 \\
\underline{-2y - 5z = -72} \\
3x + 2y = 81 \quad \text{(Eq. 4)}
\end{array}
$$

(Equations 1 and 2):

$$
\begin{array}{rl}
(\times -5): & -5x - 5y - 5z = -200 \\
& \underline{3x + 4y + 5z = 153} \\
& -2x - y = -47
\end{array}
$$

$$
\begin{array}{rll}
(\times 2) & -4x - 2y = -94 & \text{(Equation 5)} \\
& \underline{3x + 2y = 81} & \text{(Equation 4)} \\
& -x = -13 & \\
& x = 13 & \text{(triangles)} \\
(\times -1) & 2x + y = 47 & \text{(Equation 5)} \\
& 2(13) + y = 47 & \\
& y = 21 & \text{(rectangles)}
\end{array}
$$

$$
\begin{array}{ll}
x + y + z = 40 & \text{(Equation 1)} \\
13 + 21 + z = 40 & \\
z = 6 & \text{(pentagons)}
\end{array}
$$

$$\boxed{13 \text{ triangles},\ 21 \text{ rectangles},\ 6 \text{ pentagons}}$$

35. Let

L = length of rectangular solid
W = width of rectangular solid
H = height of rectangular solid

$$
\begin{array}{lll}
2L + 2W = 16 & (\div 2) \rightarrow & L + W = 8 \\
2L + 2H = 18 & (\div 2) \rightarrow & L + H = 9 \\
2W + 2H = 14 & (\div 2) \rightarrow & \underline{-W - H = -7} \\
& \text{(add Equation 2 and 3)} & L - W = 2 \quad \text{(Equation 4)}
\end{array}
$$

$$
\begin{array}{ll}
\text{(Equation 1):} & L + W = 8 \\
\text{(Equation 4):} & \underline{L - W = 2} \\
& 2L = 10 \\
& L = 5
\end{array}
$$

$$
L = 5 \qquad
\begin{array}{ll}
L + W = 8 & \text{(Equation 1)} \\
5 + W = 8 & \\
W = 3 &
\end{array}
\qquad
\begin{array}{ll}
L + H = 9 & \text{(Equation 2)} \\
5 + H = 9 & \\
H = 4 &
\end{array}
$$

$$\boxed{\text{length, 5 cm; width, 3 cm; height, 4 cm}}$$

37. Let

$$x \;=\; \text{amount invested at 8\%}$$
$$y \;=\; \text{amount invested at 10\%}$$
$$z \;=\; \text{amount invested at 12\%}$$

$$x + y + z \;=\; 6700$$
$$0.08x + 0.10y + 0.12z \;=\; 716 \qquad (\times 100) \to \qquad 8x + 10y + 12z \;=\; 71600$$
$$z \;=\; (x + y) + 300$$

(Equations 1 and 3):
$$
\begin{aligned}
x + y + z &= 6700 \\
\underline{-x - y + z} &= \underline{\;\;300} \\
2z &= 7000 \\
z &= 3500 \qquad \text{(at 12\%)}
\end{aligned}
$$

(Equations 1 and 2, replace 3500 for z):
$$
\begin{array}{lll}
(\times -5) \to & x + y + 3500 = 6700 & \to \qquad -5x - 5y = -16000 \\
(\div 2) \to & 4x + 5y + 6(3500) = 35800 & \to \qquad \underline{4x + 5y = \;\;14800} \\
& & \qquad\qquad -x = -1200 \\
& & \qquad\qquad\;\; x = 1200 \qquad \text{(at 8\%)}
\end{array}
$$

(Equation 1)
$$
\begin{aligned}
x + y &= 3200 \\
1200 + y &= 3200 \\
y &= 2000 \qquad \text{(at 10\%)}
\end{aligned}
$$

$$\boxed{\text{\$1200 at 8\%; \$2000 at 10\%; \$3500 at 12\%}}$$

39. Let

$$u \;=\; \text{units' digit}$$
$$t \;=\; \text{tens' digit}$$
$$h \;=\; \text{hundreds' digit}$$

$$
\begin{array}{lll}
h + t + u = 11 & \to & h + t + u = 11 \\
h + t = u - 1 & \to & h + t - u = -1 \\
100h + 10t + u = (t + u) + 27 & \to & 100h + 9t = 227
\end{array}
$$

(Equations 1 and 2):
$$
\begin{array}{ll}
& h + t + u = 11 \\
(\times -1) & \underline{-h - t + u = \;\;1} \\
& 2u = 12 \\
& u = 6
\end{array}
$$

(Equations 1 and 2):
$$
\begin{array}{ll}
(\text{add}) & 2h + 2t = 10 \\
(\div 2) & h + t = 5 \qquad \text{(Equation 4)}
\end{array}
$$

(Equations 4 and 3):
$$
\begin{array}{ll}
(\times -9) & -9h - 9t = -45 \\
& \underline{100h + 9t = \;\;227} \\
& 91h = 182 \\
& h = 2
\end{array}
$$

(Equation 4):
$$
\begin{aligned}
h + t &= 5 \\
2 + t &= 5 \\
t &= 3
\end{aligned}
$$

$$h = 2, t = 3, u = 6$$

number: $\boxed{236}$

41. Let

$$x \;=\; \text{output for Press I} \quad \text{(in output/hour)}$$
$$y \;=\; \text{output for Press II} \quad \text{(in output/hour)}$$
$$z \;=\; \text{output for Press III} \quad \text{(in output/hour)}$$

$$
\begin{array}{lll}
8x + 4y + 7z = 1270 & \to & 8x + 4y + 7z = 1270 \\
4x + y + 7z = 730 & (\times -1) \to & \underline{-4x - y - 7z = -730} \\
2x + 5y + 0z = 550 & & 4x + 3y = 540 \qquad \text{(Equation 4)}
\end{array}
$$

(Equations 3 and 4):

$(\times -2)$
$$-4x - 10y = -1100$$
$$\underline{4x + 3y = 540}$$
$$-7y = -560$$
$$y = 80 \qquad \text{(Press II)}$$

(Equation 3)
$$2x + 5y = 550$$
$$2x + 5(80) = 550$$
$$2x = 150$$
$$x = 75 \qquad \text{(Press I)}$$

(Equation 2)
$$4x + y + 7z = 730$$
$$4(75) + 80 + 7z = 730$$
$$7z = 350$$
$$z = 50 \qquad \text{(Press III)}$$

> Press I: 75 units/hr;
> Press II: 80 units/hr;
> Press III: 50 units/hr

43. Let
$$x = \text{rate uphill}$$
$$y = \text{rate on level ground}$$
$$z = \text{rate downhill}$$

$$2x + 3y = 115$$
$$2z + 3y = 135 \qquad \rightarrow \qquad 3y + 2z = 135$$
$$\frac{x + z}{2} = y \qquad \rightarrow \qquad x - 2y + z = 0$$

(Equations 2 and 3):
$$3y + 2z = 135$$
$(\times -2)$
$$\underline{-2x + 4y - 2z = 0}$$
$$-2x + 7y = 135 \qquad \text{(Eq. 4)}$$
$$\underline{2x + 3y = 115} \qquad \text{(Eq. 1)}$$
$$10y = 250$$
$$y = 25 \qquad \text{(level ground)}$$

(Equation 1):
$$2x + 3y = 115$$
$$2x + 3(25) = 115$$
$$2x = 40$$
$$x = 20 \quad \text{(uphill)}$$

(Equation 2):
$$3y + 2z = 135$$
$$3(25) + 2z = 135$$
$$2z = 60$$
$$z = 30 \quad \text{(downhill)}$$

> Uphill: 20 kph; Level ground: 25 kph; Downhill: 30 kph

45. Let

$$u \;=\; \text{units' digit}$$
$$t \;=\; \text{tens' digit}$$
$$h \;=\; \text{hundreds' digit}$$

$$100h + 10t + u \;=\; \text{original number}$$
$$h + 10t + 100u \;=\; \text{number with digits reversed}$$

$$\begin{aligned}
h + t + u &= 6 \\
h + 10t + 100u &= (100h + 10t + u) - 99 \\
\underline{t} &= \underline{h + u} \\
h + t + u &= 6 \\
-99h + 99u &= -99 \quad \rightarrow -h + u = -1 \\
\underline{-h + t - u} &= \underline{0}
\end{aligned}$$

(Equations 1 and 3):
$$\begin{aligned}
h + t + u &= 6 \\
\underline{-h + t - u} &= \underline{0} \\
2t &= 6 \\
t &= 3
\end{aligned}$$

$(\times -1)$

$(\div 2)$

(Equations 1 and 3):
$$\begin{aligned}
h + t + u &= 6 \\
\underline{h - t + u} &= \underline{0} \\
2h + 2u &= 6 \\
h + u &= 3 \quad \text{(Eq 4)}
\end{aligned}$$

(Equations 2 and 4):
$$\begin{aligned}
-h + u &= -1 \\
\underline{h + u} &= \underline{3} \\
2u &= 2 \\
u &= 1
\end{aligned}$$

(Equations 1):
$$\begin{aligned}
h + t + u &= 6 \\
h + 3 + 1 &= 6 \\
h &= 2
\end{aligned}$$

$h = 2, \ t = 3, \ u = 1$

number: $\boxed{231}$

47.
$$\begin{aligned}
x &= Ax^2 + (B - 4A)x + (4A - 2B + C) \\
0x^2 + (1)x + 0 &= Ax^2 + (B - 4A)x + (4A - 2B + C)
\end{aligned}$$

Equate coefficients of x^2, x and x^0 on the left and right sides:
$$\begin{aligned}
0 &= A &\rightarrow& & A &= 0 \\
1 &= B - 4A &\rightarrow& & B - 4A &= 1 \\
0 &= 4A - 2B + C &\rightarrow& & 4A - 2B + C &= 0
\end{aligned}$$

(Equations 2 and 1):
$$\begin{aligned}
B - 4A &= 1 \\
B - 4(0) &= 1 \quad (A = 0) \\
B &= 1
\end{aligned}$$

(Equation 3):
$$\begin{aligned}
4A - 2B + C &= 0 \\
4(0) - 2(1) + C &= 0 \quad (A = 0, \, B = 1) \\
C &= 2
\end{aligned}$$

$\boxed{A = 0, \, B = 1, \, C = 2}$

49. $-12x^5 + 9x^3 + 10 = (6B + 4C)x^5 + (3A + 3B)x^3 + (2A - 3C)$

Equate coefficients on the left and right sides.

$$\begin{array}{rcl}
-12 & = & 6B + 4C \\
9 & = & 3A + 3B \\
10 & = & 2A - 3C
\end{array} \qquad \begin{array}{l} (\div 2) \rightarrow \\ (\div 3) \rightarrow \\ \rightarrow \end{array} \qquad \begin{array}{rcl}
3B + 2C & = & -6 \\
A + B & = & 3 \\
2A - 3C & = & 10
\end{array}$$

(Equations 2 and 3):

$$\begin{array}{rrcl}
(\times -2) & -2A - 2B & = & -6 \\
 & \underline{2A - 3C} & \underline{=} & \underline{10} \\
 & -2B - 3C & = & 4 \quad \text{(Equation 4)}
\end{array}$$

(Equations 1 and 4):

$$\begin{array}{rrcl}
(\times 3) & 9B + 6C & = & -18 \\
 & \underline{-4B - 6C} & \underline{=} & \underline{8} \\
 & 5B & = & -10 \\
 & B & = & -2
\end{array}$$

(Equation 2): $\begin{array}{rcl} A + B & = & 3 \\ A - 2 & = & 3 \\ A & = & 5 \end{array}$ (Equation 1): $\begin{array}{rcl} 3B + 2C & = & -6 \\ 3(-2) + 2C & = & -6 \\ C & = & 0 \end{array}$

$$\boxed{A = 5, B = -2, C = 0}$$

51. $\boxed{C}$ is true

53. $\begin{array}{rcl} 1.5x + y - 0.2z & = & 0.05 \\ 2x - 3.8y + 2.1z & = & 3.26 \\ 3.7x - 0.2y + 0.05z & = & 0.41 \end{array}$

(Equations 1 and 2):

$$\begin{array}{rrcl}
(\times 3.8) & 1.5(3.8)x + 3.8y - 0.2(3.8)z & = & 0.05(3.8) \\
 & \underline{2x - 3.8y + 2.1z} & \underline{=} & \underline{3.26} \\
x[1.5(3.0) + 2] + z[-0.2(3.8) + 2.1]] & = & 0.05(3.8) + 3.26 \\
 & 7.7x + 1.34z & = & 3.45 \quad \text{(Equation 4)}
\end{array}$$

(Equations 1 and 3):

$$\begin{array}{rrcl}
(\times 0.2) & 1.5(0.2)x + 0.2y - 0.2(0.2)z & = & 0.05(0.2) \\
 & \underline{3.7x - 0.2y + 0.05z} & \underline{=} & \underline{0.41} \\
x[1.5(0.2) + 3.7] + [-0.2(0.2) + 0.05] & = & 0.05(0.2) + 0.41 \\
 & 4x + 0.01z & = & 0.42 \quad \text{(Equation 5)}
\end{array}$$

(Equations 4 and 5):

$$\begin{array}{rrcl}
(\times -4) & 7.7(-4)x + 1.34(-4)z & = & 3.45(-4) \\
(\times 7.7) & \underline{7.7(4)x + 0.01(7.7)z} & \underline{=} & \underline{0.42(7.7)} \\
 & z[(1.34)(-4) + 0.01(7.7)] & = & 3.45(-4) + 0.42(7.7) \\
 & -5.283z & = & -10.566 \\
 & z & = & 2
\end{array}$$

(Equation 5): $\begin{array}{rcl} 4x + 0.01z & = & 0.42 \\ 4x + 0.02 & = & 0.42 \\ 4x & = & 0.4 \\ x & = & 0.1 \end{array}$

(Equation 1): $\begin{array}{rcl} 1.5x + y - 0.2z & = & 0.05 \\ 1.5(0.1) + y - 0.2(2) & = & 0.05 \\ y & = & 0.05 - 1.5(0.1) + 0.2(2) \\ y & = & 0.3 \end{array}$

$$\boxed{\{(0.1, 0.3, 2)\}}$$

55.
$$\begin{aligned}
x - y + 2z - 2w &= -1 \\
x - y - z + w &= -4 \\
-x + 2y - 2z - w &= -7 \\
2x + y + 3z - w &= 6
\end{aligned}$$

(Add Equations 2 and 3) $\rightarrow$ $y - 3z = -11$ (Equation 5)

(Add Equatins 2 and 4) $\rightarrow$ $3x + 2z = 2$ (Equation 6)

Eliminate w first:

(Equations 1 and 2):
$$\begin{aligned}
x - y + 2z - 2w &= -1 \\
(\times 2) \quad \underline{2x - 2y - 2z + 2w} &= \underline{-8} \\
3x - 3y &= -9 \\
(\div 3) \qquad\qquad x - y &= -3 \quad \text{(Equation 7)}
\end{aligned}$$

Now we have a system of 3 equations:
$$\begin{aligned}
y - 3z &= -11 \quad \text{(Equation 5)} \rightarrow & y - 3z &= -11 \\
3x + 2z &= 2 \quad \text{(Equation 6)} \\
x - y &= -3 \quad \text{(Equation 7)} & \underline{x - y} &= \underline{-3} \\
& & x - 3z &= -14 \quad \text{(Eq 8)}
\end{aligned}$$

System of 2 equations:
$$\begin{aligned}
3x + 2z &= 2 \quad \text{(Equation 6)} \\
(\times -3) \quad \underline{-3x + 9z} &= \underline{42} \quad \text{(Equation 8)} \\
11z &= 4x \\
z &= 4
\end{aligned}$$

$$\begin{aligned}
3x + 2z &= 2 \quad \text{(Equation 6)} \\
3x + 2(4) &= 2 \\
3x &= -6 \\
x &= -2
\end{aligned}$$

$$\begin{aligned}
x - y &= -3 \quad \text{(Equation 7)} \\
-2 - y &= -3 \\
-y &= -1 \\
y &= 1
\end{aligned}$$

$$\begin{aligned}
x - y - z + w &= -4 \quad \text{(Eq 2)} \\
-2 - 1 - 4 + w &= -4 \\
w &= 3
\end{aligned}$$

$$\boxed{\{(-2, 1, 4, 3)\}}$$

57. Let A, B, C, D, E, F, and G represent the cost of an apple tree, banana tree, carambella tree, dogwood tree, elm tree, fern, and maple tree, respectively. Thus,

$$\begin{aligned}
A + B &= 21 \\
B + C &= 24 \\
C + D &= 32 \\
D + E &= 37 \\
E + F &= 31 \\
F + G &= 25 \\
\underline{G + A} &= \underline{26}
\end{aligned}$$

(Add:) $2A + 2B + 2C + 2D + 2E + 2F + 2G = 196$

$\left(\text{multiply by } \frac{1}{2}\right):$ $A + B + C + D + E + F + G = 98$

$$\begin{array}{ccc}
\uparrow & \uparrow & \uparrow
\end{array}$$

$\begin{pmatrix}\text{substitute from} \\ \text{equations } 1, 3 \\ \text{and } 5:\end{pmatrix}$ $21 \qquad 32 \qquad 31$

$$\begin{aligned}
84 + G &= 98 \\
G &= 14
\end{aligned}$$

Substituting 14 for G in the last equation gives
$$14 + A = 26$$
$$A = 12$$

Now substitute 12 for A in the first equation and find B. Once B is known, use this value in the second equation to find C, etc.

The cost of each plant is ⎹ $12 (apple tree), $9 (banana), $15 (carambolla), $17 (dogwood), $20 (elm), $11 (fern), and $14 (maple).

59. Use $y = ax^2 + bx = c$. First substitute $(-1, -2)$:
$$a - b + c = -2$$

Now substitute $(1, 0)$:
$$a + b + c = 0$$

Finally substitute $(2, 7)$:
$$4a + 2b + c = 7$$

Solving this system gives $a = 2$, $b = 1$, and $c = -3$.

The equation $y = ax^2 + bx + c$ becomes
$$y = 2x^2 + x - 3$$

Since $(3, y)$ satisfies this equation
$$y = 2(3^2) + 3 - 3$$
$$\boxed{y = 18}$$

Review Problems

64. line passing through $(3, 5)$ and $(4, 2)$:
$$m = \frac{2 - 5}{4 - 3} = -3$$
point-slope: $\boxed{y - 5 = -3(x - 3) \text{ or } y - 2 = -3(x - 4)}$
$$y - 5 = -3x + 9$$
slope-intercept: $\boxed{y = -3x + 14}$
standard: $\boxed{3x + y - 14 = 0}$

65. $y = -2x + 4$

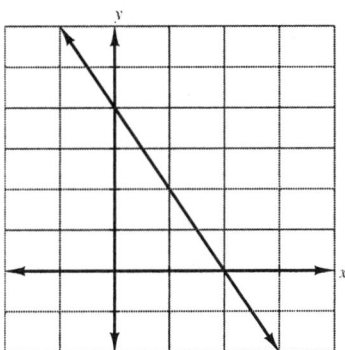

x-intercept: $(2, 0)$
y-intercept: $(0, 4)$

area $= \dfrac{1}{2} (2)(4) = \boxed{4 \text{ square units}}$

66. a. $\boxed{y \text{ is not a function of } x}$

> Domain $= \{x \mid x = -2, 1\}$
>   or $[-2], [1]$
>
> Range $= \{y \mid y = 1, 2\}$
>   or $[1], [2]$

b. $\boxed{y \text{ is a function of } x}$

> Domain $= \{x \mid -2 \le x \le 2\}$
>   or $[-2, 2]$
>
> Range $= \{y \mid 0 \le y \le 2\}\}$
>   or $[0, 2]$

c. $\boxed{y \text{ is a not a function of } x}$

> Domain $= \{x \mid -2 \le x \le 2\}$
>   or $[-2, 2]$
>
> Range $= \{y \mid -2 \le y \le 2\}\}$
>   or $[-2, 2]$

d. $\boxed{y \text{ is a function of } x}$

> Domain $= \{x \mid x \in R\}$
>   or $(-\infty, \infty)$
>
> Range $= \{y \mid y = 3\}$
>   or $[3]$

Section 4.4 Solving Linear Systems of Equations in Two Variables Using Determinants and Cramer's Rule

Problem Set 4.4, pp. 296-297

1. $\begin{vmatrix} 5 & 7 \\ 2 & 3 \end{vmatrix} = 5(3) - 2(7) = 15 - 14 = \boxed{1}$ **3.** $\begin{vmatrix} -4 & 1 \\ 5 & 6 \end{vmatrix} = -4(6) - 5(1) = -24 - 5 = \boxed{-29}$

5. $\begin{vmatrix} -7 & 14 \\ 2 & -4 \end{vmatrix} = (-7)(-4) - (2)(14) = 28 - 28 = \boxed{0}$

7. $\begin{vmatrix} -5 & -1 \\ -2 & -7 \end{vmatrix} = (-5)(-7) - (-2)(-1) = 35 - 2 = \boxed{33}$

9. $\begin{vmatrix} \frac{1}{2} & \frac{1}{2} \\ \frac{1}{8} & -\frac{3}{4} \end{vmatrix} = \left(\frac{1}{2}\right)\left(-\frac{3}{4}\right) - \left(\frac{1}{8}\right)\left(\frac{1}{2}\right) = -\frac{3}{8} - \frac{1}{16} = \boxed{-\frac{7}{16}}$

11. $x + y = 7$
$x - y = 3$

$$D = \begin{vmatrix} 1 & 1 \\ 1 & -1 \end{vmatrix} = -1 - 1 = -2$$

$$Dx = \begin{vmatrix} 7 & 1 \\ 3 & -1 \end{vmatrix} = -7 - 3 = -10$$

$$Dy = \begin{vmatrix} 1 & 7 \\ 1 & 3 \end{vmatrix} = 3 - 7 = -4$$

$$x = \frac{Dx}{D} = \frac{-10}{-2} = 5$$

$$y = \frac{Dy}{D} = \frac{-4}{-2} = 2$$

$\boxed{\{(5, 2)\}}$

13. $12x + 3y = 15$
$2x - 3y = 13$

$$D = \begin{vmatrix} 12 & 3 \\ 2 & -3 \end{vmatrix} = -36 - 6 = -42$$

$$Dx = \begin{vmatrix} 15 & 3 \\ 13 & -3 \end{vmatrix} = -45 - 39 = -84$$

$$Dy = \begin{vmatrix} 12 & 15 \\ 2 & 13 \end{vmatrix} = 156 - 30 = 126$$

$$x = \frac{Dx}{D} = \frac{-84}{-42} = 2$$

$$y = \frac{Dy}{D} = \frac{126}{-42} = -3$$

$\boxed{\{(2, -3)\}}$

15. $\quad 4x - 5y = 17$
$\qquad 2x + 3y = 3$

$$D = \begin{vmatrix} 4 & -5 \\ 2 & 3 \end{vmatrix} = 12 - (-10) = 22$$

$$Dx = \begin{vmatrix} 17 & -5 \\ 3 & 3 \end{vmatrix} = 51 - (-15) = 66$$

$$Dy = \begin{vmatrix} 4 & 17 \\ 2 & 3 \end{vmatrix} = 12 - 34 = -22$$

$$x = \frac{Dx}{D} = \frac{66}{22} = 3$$

$$y = \frac{Dy}{D} = \frac{-22}{22} = -1$$

$$\boxed{\{(3, -1)\}}$$

17. $\quad x + 2y = 3$
$\qquad 5x + 10y = 15$

$$D = \begin{vmatrix} 1 & 2 \\ 5 & 10 \end{vmatrix} = 10 - 10 = 0$$

$$Dy = \begin{vmatrix} 3 & 2 \\ 15 & 10 \end{vmatrix} = 30 - 30 = 0$$

$$Dy = \begin{vmatrix} 1 & 3 \\ 5 & 15 \end{vmatrix} = 15 - 15 = 0$$

The system is $\boxed{\text{dependent}}$.

$$\boxed{\{x, y \mid x + 2y = 3\}}$$

19. $\quad 3x - 4y = 4$
$\qquad 2x + 2y = 12$

$$D = \begin{vmatrix} 3 & -4 \\ 2 & 2 \end{vmatrix} = 6 - (-8) = 14$$

$$Dx = \begin{vmatrix} 4 & -4 \\ 12 & 2 \end{vmatrix} = 8 - (-48) = 56$$

$$Dy = \begin{vmatrix} 3 & 4 \\ 2 & 12 \end{vmatrix} = 36 - 8 = 28$$

$$x = \frac{56}{14} = 4$$

$$y = \frac{28}{14} = 2$$

$$\boxed{\{(4, 2)\}}$$

21. $\quad 2x = 3y + 2 \qquad \rightarrow \qquad 2x - 3y = 2$
$\qquad 5x = 51 - y \qquad\qquad\qquad 5x + 4y = 51$

$$D = \begin{vmatrix} 2 & -3 \\ 5 & 4 \end{vmatrix} = 8 - (-15) = 23$$

$$Dx = \begin{vmatrix} 2 & -3 \\ 51 & 4 \end{vmatrix} = 8 - (-153) = 161$$

$$Dy = \begin{vmatrix} 2 & 2 \\ 5 & 51 \end{vmatrix} = 102 - 10 = 92$$

$$x = \frac{161}{23} = 7$$

$$y = \frac{92}{23} = 4$$

$$\boxed{\{(7, 4)\}}$$

23. $3x = 2 - 3y$ $\rightarrow$ $3x + 3y = 2$
         $2y = 3 - 2x$ $2x + 2y = 3$

$$D = \begin{vmatrix} 3 & 3 \\ 2 & 2 \end{vmatrix} = 6 - 6 = 0$$

$$Dx = \begin{vmatrix} 2 & 3 \\ 3 & 2 \end{vmatrix} = 4 - 4 = -5$$

$$Dy = \begin{vmatrix} 3 & 2 \\ 2 & 3 \end{vmatrix} = 9 - 4 = 5$$

The system is $\boxed{\text{inconsistent}}$ since $D =$ and $Dx \neq 0$ and $Dy \neq 0$.

$\boxed{\varnothing}$

25. $4y = 16 - 3x$ $\rightarrow$ $3x + 4y = 16$
         $5x = 12 - 3y$ $5x + 3y = 12$

$$D = \begin{vmatrix} 3 & 4 \\ 5 & 3 \end{vmatrix} = 9 - 20 = -11$$

$$Dx = \begin{vmatrix} 16 & 4 \\ 12 & 3 \end{vmatrix} = 48 - 48 = 0$$

$$Dy = \begin{vmatrix} 3 & 16 \\ 5 & 12 \end{vmatrix} = 36 - 80 = 44$$

$$x = \frac{0}{-11} = 0$$

$$y = \frac{44}{-11} = -4$$

$\boxed{\{(0, -4)\}}$

27. $3x + 2y = 4$

         $x = 5$

$$D = \begin{vmatrix} 3 & 2 \\ 1 & 0 \end{vmatrix} = 0 - 2 = -2$$

$$Dx = \begin{vmatrix} 4 & 2 \\ 5 & 0 \end{vmatrix} = 0 - 10 = -10$$

$$Dy = \begin{vmatrix} 3 & 4 \\ 1 & 5 \end{vmatrix} = 15 - 4 = 11$$

$$x = \frac{-10}{-2} = 5$$

$$y = \frac{11}{-2} = -\frac{11}{2}$$

$$\boxed{\left\{ \left(5, -\frac{11}{2} \right) \right\}}$$

29. $\begin{vmatrix} 3 & 2 \\ y & 4 \end{vmatrix} = 0$

         $12 - 2y = 0$
         $-2y = -12$

         $\boxed{y = 6}$

31. $\begin{vmatrix} 3x - 1 & x + 2 \\ 4 & -4 \end{vmatrix} = 2x + 5$

$-4(3x - 1) - 2(x + 2) = 2x + 5$

$-16x = 5$

$\boxed{x = -\dfrac{5}{16}}$

33. $\begin{vmatrix} a_1 & a_1 \\ a_2 & a_2 \end{vmatrix} = a_1 a_2 - a_1 a_2 = 0$

35. $\begin{vmatrix} a_1 & kb_2 \\ a_2 & Kk_2 \end{vmatrix} = a_1(kb_2) - a_2(kb_1)$

$= k(a_1 b_2) - k(a_2 b_1)$

$= k(a_1 b_2 - a_2 b_1)$

$= k \begin{vmatrix} a_1 & b_1 \\ a_2 & b_2 \end{vmatrix}$ since $\begin{vmatrix} a_1 & b_1 \\ a_2 & b_2 \end{vmatrix} = a_1 b_2 - a_2 b_1$

Thus $\begin{vmatrix} a_1 & kb_1 \\ a_2 & kb_2 \end{vmatrix} = k \begin{vmatrix} a_1 & b_1 \\ a_2 & b_2 \end{vmatrix}$

37. $\begin{vmatrix} a_1 & b + ka_1 \\ a_2 & b_2 + ka_2 \end{vmatrix} = a_1(b_2 + ka_2) - a_2(b_1 + ka_1)$

$= a_1 b_2 + a_1 ka_2 - a_2 b_1 - a_2 ka_1$

$= a_1 b_2 - a_2 b_1 + (a_1 ka_2 - a_2 ka_1)$

$= a_1 b_2 - a_2 b_1 + 0$

$= \begin{vmatrix} a_1 & b_1 \\ a_2 & b_2 \end{vmatrix}$

Thus,

$\begin{vmatrix} a_1 & b_1 \\ a_2 & b_2 \end{vmatrix} = \begin{vmatrix} a_1 & b_1 + ka_1 \\ a_2 & b_2 + ka_2 \end{vmatrix}$

39. $\begin{vmatrix} a_1 & b_1 \\ a_2 & b_2 \end{vmatrix} = a_1 b_2 - a_2 b_1$

$\begin{vmatrix} b_1 & a_1 \\ b_2 & a_2 \end{vmatrix} = b_1 a_2 - b_2 a_1 = -(a_1 b_2 - a_2 b_1) = -\begin{vmatrix} a_1 & b_1 \\ a_2 & b_2 \end{vmatrix}$

$\boxed{\text{The sign of the value changes.}}$

41. original system:

$$a_1 x + b_1 y = c_1$$
$$a_2 x + b_2 y = c_2$$

$$D = \begin{vmatrix} a_1 & b_1 \\ a_2 & b_2 \end{vmatrix} = a_1 b_2 - a_2 b_1$$

$$Dx = \begin{vmatrix} c_1 & b_1 \\ c_2 & b_2 \end{vmatrix} = c_1 b_2 - c_2 b_1$$

$$x = \frac{c_1 b_2 - c_2 b_1}{a_1 b_2 - a_2 b_1}$$

$$Dy = \begin{vmatrix} a_1 & c_1 \\ a_1 & c_2 \end{vmatrix} = a_1 c_2 - a_2 c_1$$

$$y = \frac{a_1 c_2 - a_2 c_1}{a_1 b_2 - a_2 b_1}$$

new system: multiply first equation by K

$$Ka_1 x + Kb_1 y = Kc_1$$
$$a_2 x + b_2 y = c_2$$

$$D = \begin{vmatrix} Ka_1 & Kb_1 \\ a_2 & b_2 \end{vmatrix} = Ka_1 b_2 - Ka_2 b_1 = K(a_1 b_2 - a_2 b_1)$$

$$Dx = \begin{vmatrix} Kc_1 & Kb_1 \\ c_2 & b_2 \end{vmatrix} = Kc_1 b_2 - Kb_1 v_2 = K(c_1 b_2 - b_1 c_2)$$

$$x = \frac{K(c_1 b_2 - b_2 c_2)}{K(a_1 b_2 - a_2 b_1)} = \frac{c_1 b_2 - b_1 c_2}{a_1 b_2 - a_2 b_1}$$

$$Dy = \begin{vmatrix} Ka_1 & Kc_1 \\ a_2 & c_2 \end{vmatrix} = KA_1 c_2 - Ka_2 c_1 = K(a_1 c_2 - a_2 c_1)$$

$$y = \frac{K(a_1 c_2 - a_2 c_1)}{K(a_1 b_2 - a_2 b_1)} = \frac{a_1 c_2 - a_2 c_1}{a_1 b_2 - a_2 b_1}$$

The values of x and y remain the same.

43. $y = m_1 x + d_1$
$y = m_1 x + d_2$

a. The lines are parallel because they have equal slopes ($m_1 = m_2$) and different y-intercepts ($d_1 \neq d_2$).

b. $-m_1 x + y = d_1$
$-m_1 x + y = d_2$

$$\begin{vmatrix} a_1 & b_1 \\ a_2 & b_2 \end{vmatrix} = \begin{vmatrix} -m_1 & 1 \\ -m_1 & 1 \end{vmatrix} = -m_1 - (-m_1) = 0$$

45. $\boxed{D}$ is true; $\begin{vmatrix} \dfrac{1}{2} & \dfrac{1}{4} \\ \dfrac{1}{3} & -\dfrac{1}{2} \end{vmatrix} = \dfrac{1}{2}\left(-\dfrac{1}{2}\right) - \dfrac{1}{3}\left(\dfrac{1}{4}\right) = -\dfrac{1}{4} - \dfrac{1}{12} = -\dfrac{4}{12} = -\dfrac{1}{3}$

47. $\begin{vmatrix} -17.3412 & -3.1283 \\ -9.6005 & -94.3026 \end{vmatrix} = -17.3412(-94.3026) - (-9.6005)(-3.1283)$

$$= \boxed{1605.29}$$

49.
$$0.073x + 0.092y = 0.153$$
$$-0.439x - 0.235y = -0.649$$

$$D = \begin{vmatrix} 0073 & 0.092 \\ -0.439 & -0.235 \end{vmatrix} = 0.073(-0.235) - (0.092)(-0.439) = 0.023233$$

$$Dx = \begin{vmatrix} 0.153 & 0.092 \\ -0.649 & -0.235 \end{vmatrix} = 0.153(-0.235) - (0.092)(-0.649) = 0.023753$$

$$Dy = \begin{vmatrix} 0.073 & 0.153 \\ -0.439 & -0.649 \end{vmatrix} = (0.073)(-0.649) - (0.153)(-0.439) = 0.01979$$

$$x = \frac{Dx}{D} = 1.022$$

$$y = \frac{Dy}{D} = 0.852$$

$$\boxed{\{(1.022, 0.852)\}}$$

Review Problems

54. Let x = number of miles.
$$1.40 + 1.60x = 35$$
$$1.60x = 33.60$$
$$x = 21$$

$$\boxed{21 \text{ miles}}$$

55. Let
$$R = \text{resistance}$$
$$L = \text{length}$$
$$D = \text{diameter}$$
$$R = \frac{kL}{D^2}$$

Given $L = 34$, $D = 3$, $R = 4$
$$4 = \frac{k\,30}{3^2}$$
$$k = \frac{6}{5}$$
$$R = \frac{6L}{5D^2}$$

At $L = 10$, $D = 2(1) = 2$, $R = ?$
$$R = \frac{6(10)}{5(2^2)} = 3$$

$$\boxed{3 \text{ ohms}}$$

56.
$$\frac{|5 - 2x|}{3} \geq 3$$
$$5 - 2x \geq 9$$
$$-2x \geq 4$$
$$x \leq -2$$

$$\boxed{\{x \mid x \leq -2\}}$$

$$\boxed{(-\infty, -2]}$$

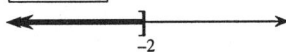

Section 4.5 Solving Linear Systems of Equations in Three Variables Using Determinants and Cramer's Rule

Problem Set 4.5, pp. 303-305

1.
$$\begin{vmatrix} 1 & -1 & 2 \\ 2 & 1 & 3 \\ 0 & -2 & 1 \end{vmatrix} = 1 \begin{vmatrix} 1 & 3 \\ -2 & 1 \end{vmatrix} - 2 \begin{vmatrix} -1 & 2 \\ -2 & 1 \end{vmatrix} + 0 \begin{vmatrix} -1 & 2 \\ 1 & 3 \end{vmatrix} \quad \text{(expanding about column 1)}$$

$$= (1 + 6) - 2(-1 + 4) + 0$$
$$= 7 - 2(3) = 7 - 6 = \boxed{1}$$

3.
$$\begin{vmatrix} 3 & 4 & -2 \\ -1 & 2 & 1 \\ -1 & -1 & 5 \end{vmatrix} = 3 \begin{vmatrix} 2 & 1 \\ -1 & 5 \end{vmatrix} - (-1) \begin{vmatrix} 4 & -2 \\ -1 & 5 \end{vmatrix} + (-1) \begin{vmatrix} 4 & -2 \\ 2 & 1 \end{vmatrix} \quad \text{(expand column 1)}$$

$$= 3(10 + 1) + 1(20 - 2) - 1(4 + 4)$$
$$= 3(11) + (18) - (8)$$
$$= 33 + 18 - 8$$
$$= \boxed{43}$$

5.
$$\begin{vmatrix} 4 & 1 & 0 \\ 1 & -1 & -1 \\ -2 & -1 & 0 \end{vmatrix} = 0 \begin{vmatrix} 1 & -1 \\ -2 & -1 \end{vmatrix} - (-1) \begin{vmatrix} 4 & 1 \\ -2 & -1 \end{vmatrix} + 0 \begin{vmatrix} 4 & 1 \\ 1 & -1 \end{vmatrix} \quad \text{(expand column 3)}$$

$$= 0 + 1(-4 + 2) + 0$$
$$= \boxed{-2}$$

7.
$$\begin{vmatrix} 2 & 4 & 2 \\ -1 & 5 & -1 \\ 3 & 2 & 3 \end{vmatrix} = 2 \begin{vmatrix} 5 & -1 \\ 2 & 3 \end{vmatrix} - (-1) \begin{vmatrix} 4 & 2 \\ 2 & 3 \end{vmatrix} + 3 \begin{vmatrix} 4 & 2 \\ 5 & -1 \end{vmatrix} \quad \text{(exapnd column 1)}$$

$$= 2(15 + 2) + 1(12 - 4) + 3(-4 - 10)$$
$$= 2(17) + 1(8) + 3(-14)$$
$$= 34 + 8 - 42$$
$$= \boxed{0}$$

9.
$$\begin{vmatrix} -4 & 0 & 3 \\ 6 & -2 & -1 \\ 8 & 0 & 4 \end{vmatrix} = -0 \begin{vmatrix} 6 & -1 \\ 8 & 4 \end{vmatrix} - 2 \begin{vmatrix} -4 & 3 \\ 8 & 4 \end{vmatrix} - 0 \begin{vmatrix} -4 & 3 \\ 6 & -1 \end{vmatrix} \quad \text{(expand column 2)}$$

$$= 0 - 2(-16 - 24) - 0$$
$$= -2(-40)$$
$$= \boxed{80}$$

11.
$$\begin{vmatrix} 2 & 0 & 0 \\ 3 & -1 & -1 \\ 2 & -2 & 4 \end{vmatrix} = 2 \begin{vmatrix} -1 & -1 \\ -2 & 4 \end{vmatrix} - 0 \begin{vmatrix} 3 & -1 \\ 2 & 4 \end{vmatrix} + 0 \begin{vmatrix} 3 & -1 \\ 2 & -2 \end{vmatrix} \quad \text{(expand row 1)}$$

$$= 2(-4 - 2) - 0 + 0$$
$$= 2(-6)$$
$$= \boxed{-12}$$

13.

$$3x + 3y - z = 10$$
$$x + 9y + 2z = 16$$
$$x - y + 6z = 14$$

$$D = \begin{vmatrix} 3 & 3 & -1 \\ 1 & 9 & 2 \\ 1 & -1 & 6 \end{vmatrix} = (3)\begin{vmatrix} 9 & 2 \\ -1 & 6 \end{vmatrix} - (1)\begin{vmatrix} 3 & -1 \\ -1 & 6 \end{vmatrix} + (1)\begin{vmatrix} 3 & -1 \\ 9 & 2 \end{vmatrix} \quad \text{(column 1)}$$

$$= 3(54 + 2) - (18 - 1) + (6 + 9)$$
$$= 3(56) - (17) + (15)$$
$$= 166$$

$$Dx = \begin{vmatrix} 10 & 3 & -1 \\ 16 & 9 & 2 \\ 14 & -1 & 6 \end{vmatrix} = 10\begin{vmatrix} 9 & 2 \\ -1 & 6 \end{vmatrix} - 16\begin{vmatrix} 3 & -1 \\ -1 & 6 \end{vmatrix} - 16\begin{vmatrix} 3 & -1 \\ -1 & 6 \end{vmatrix} + 14\begin{vmatrix} 3 & -1 \\ 9 & 2 \end{vmatrix} \quad \text{(column 1)}$$

$$= 10(56) - 16(17) + 14(15)$$
$$= 560 - 272 + 210$$
$$= 498$$

$$Dy = \begin{vmatrix} 3 & 10 & -1 \\ 1 & 16 & 2 \\ 1 & 14 & 6 \end{vmatrix} = 3\begin{vmatrix} 16 & 2 \\ 14 & 6 \end{vmatrix} - 1\begin{vmatrix} 10 & -1 \\ 14 & 6 \end{vmatrix} + 1\begin{vmatrix} 10 & -1 \\ 16 & 2 \end{vmatrix} \quad \text{(column 1)}$$

$$= 3(96 - 28) - (60 + 14) + (20 + 16)$$
$$= 3(68) - 74 + 36)$$
$$= 166$$

$$Dz = \begin{vmatrix} 3 & 3 & 10 \\ 1 & 9 & 16 \\ 1 & -1 & 14 \end{vmatrix} = 3\begin{vmatrix} 9 & 16 \\ -1 & 14 \end{vmatrix} - 1\begin{vmatrix} 3 & 10 \\ -1 & 14 \end{vmatrix} + 1\begin{vmatrix} 3 & 10 \\ 9 & 16 \end{vmatrix} \quad \text{(column 1)}$$

$$= 3(126 + 16) - (42 + 10) + (48 - 90)$$
$$= 3(142) - 52 - 42$$
$$= 426 - 52 - 42$$
$$= 332$$

$$x = \frac{Dx}{D} = \frac{498}{166} = 3$$
$$y = \frac{Dy}{D} = \frac{166}{166} = 1$$
$$z = \frac{Dz}{D} = \frac{332}{166} = 2$$

$$\boxed{\{(3, 1, 2)\}}$$

15.

$$x - y + 2z = 4$$
$$3x + 2y - 4z = 2$$
$$x + y + z = 3$$

$$D = \begin{vmatrix} 1 & -1 & 2 \\ 3 & 2 & -4 \\ 1 & 1 & 1 \end{vmatrix} = 1\begin{vmatrix} -1 & 2 \\ 2 & -4 \end{vmatrix} - 1\begin{vmatrix} 1 & 2 \\ 3 & -4 \end{vmatrix} + 1\begin{vmatrix} 1 & -1 \\ 3 & 2 \end{vmatrix} \quad \text{(row 3)}$$

$$= (4 - 4) - (-4 - 6) + (2 + 3)$$
$$= 0 + 10 + 5 = 15$$

$$Dx = \begin{vmatrix} 4 & -1 & 2 \\ 2 & 2 & -4 \\ 3 & 1 & 1 \end{vmatrix} = 3\begin{vmatrix} -1 & 2 \\ 2 & -4 \end{vmatrix} - 1\begin{vmatrix} 4 & 2 \\ 2 & -4 \end{vmatrix} + 1\begin{vmatrix} 4 & -1 \\ 2 & 2 \end{vmatrix} \quad \text{(row 3)}$$

$$= 3(4-4) - 1(-16-4) + 1(8+2)$$
$$= 0 + 20 + 10 = 30$$

$$Dy = \begin{vmatrix} 1 & 4 & 2 \\ 3 & 2 & -4 \\ 1 & 3 & 1 \end{vmatrix} = 1\begin{vmatrix} 4 & 2 \\ 2 & -4 \end{vmatrix} - 3\begin{vmatrix} 1 & 2 \\ 3 & -4 \end{vmatrix} + 1\begin{vmatrix} 1 & 4 \\ 3 & 2 \end{vmatrix} \quad \text{(row 3)}$$

$$= (-16-4) - 3(-4-6) + (2-12)$$
$$= -20 + 30 - 10 = 0$$

$$Dz = \begin{vmatrix} 1 & -1 & 4 \\ 3 & 2 & 2 \\ 1 & 1 & 3 \end{vmatrix} = 1\begin{vmatrix} -1 & 4 \\ 2 & 2 \end{vmatrix} - 1\begin{vmatrix} 1 & 4 \\ 3 & 2 \end{vmatrix} - 3\begin{vmatrix} 1 & -1 \\ 3 & 2 \end{vmatrix} \quad \text{(row 3)}$$

$$= (-2-8) - (2-12) - 3(2+3)$$
$$= -10 + 10 - 15 = -15$$

$$x = \frac{Dx}{D} = \frac{30}{15} = 2$$
$$y = \frac{Dy}{D} = \frac{0}{15} = 0$$
$$z = \frac{Dz}{D} = \frac{15}{15} = 1$$

$$\boxed{\{(2, 0, 1)\}}$$

17. $x - y = 8$
    $x + y - z = 1$
    $x - 2z = 0$

$$D = \begin{vmatrix} 1 & -1 & 0 \\ 1 & 1 & -1 \\ 1 & 0 & -2 \end{vmatrix} = -(-1)\begin{vmatrix} 1 & -1 \\ 1 & -2 \end{vmatrix} + 1\begin{vmatrix} 1 & 0 \\ 1 & -2 \end{vmatrix} - 0\begin{vmatrix} 1 & 0 \\ 1 & -1 \end{vmatrix} \quad \text{(column 2)}$$

$$= (-2+1) + (-2-0) - 0$$
$$= -1 - 2 = -3$$

$$Dx = \begin{vmatrix} 8 & -1 & 0 \\ 1 & 1 & -1 \\ 0 & 0 & -2 \end{vmatrix} = 0\begin{vmatrix} -1 & 0 \\ 1 & -1 \end{vmatrix} - 0\begin{vmatrix} 8 & 0 \\ 1 & -1 \end{vmatrix} - 2\begin{vmatrix} 8 & -1 \\ 1 & 1 \end{vmatrix} \quad \text{(row 3)}$$

$$= 0 - 0 - 2(8+1)$$
$$= -18$$

$$Dy = \begin{vmatrix} 1 & 8 & 0 \\ 1 & 1 & -1 \\ 1 & 0 & -2 \end{vmatrix} = -8\begin{vmatrix} 1 & -1 \\ 1 & -2 \end{vmatrix} + 1\begin{vmatrix} 1 & 0 \\ 1 & -2 \end{vmatrix} - 0\begin{vmatrix} 1 & 0 \\ 1 & -1 \end{vmatrix} \quad \text{(row 1)}$$

$$= -8(-2+1) + 1(-2-0) - 0$$
$$= 8 - 2 = 6$$

$$Dz = \begin{vmatrix} 1 & -1 & 8 \\ 1 & 1 & 1 \\ 1 & 0 & 0 \end{vmatrix} = 1\begin{vmatrix} -1 & 8 \\ 1 & 1 \end{vmatrix} - 0\begin{vmatrix} 1 & 8 \\ 1 & 1 \end{vmatrix} + 0\begin{vmatrix} 1 & -1 \\ 1 & 1 \end{vmatrix} \quad \text{(row 3)}$$

$$= (-1 - 8) - 0 + 0$$
$$= -9$$

$$x = \frac{Dx}{D} = \frac{-18}{-3} = 6$$

$$y = \frac{Dy}{D} = \frac{6}{-3} = -2$$

$$z = \frac{Dz}{D} = \frac{-9}{-3} = 3$$

$$\boxed{\{(6, -2, 3)\}}$$

19. $3x + y = -5$
$2x + z = -1$
$x - 2y = -4$

$$D = \begin{vmatrix} 3 & 1 & 0 \\ 2 & 0 & 1 \\ 1 & -2 & 0 \end{vmatrix} = 0\begin{vmatrix} 2 & 0 \\ 1 & -2 \end{vmatrix} - 1\begin{vmatrix} 3 & 1 \\ 1 & -2 \end{vmatrix} + 0\begin{vmatrix} 3 & 1 \\ 2 & 0 \end{vmatrix} \quad \text{(column 3)}$$

$$= 0 - 1(-6 - 1) + 0$$
$$= 7$$

$$Dx = \begin{vmatrix} -5 & 1 & 0 \\ -1 & 0 & 1 \\ -4 & -2 & 0 \end{vmatrix} = 0\begin{vmatrix} -1 & 0 \\ -4 & -2 \end{vmatrix} - 1\begin{vmatrix} -5 & 1 \\ -4 & -2 \end{vmatrix} + 0\begin{vmatrix} -5 & 1 \\ -1 & 0 \end{vmatrix} \quad \text{(column 3)}$$

$$= 0 - (10 + 4) + 0$$
$$= -14$$

$$Dy = \begin{vmatrix} 3 & -5 & 0 \\ 2 & -1 & 1 \\ 1 & -4 & 0 \end{vmatrix} = 0\begin{vmatrix} 2 & -1 \\ 1 & -4 \end{vmatrix} - 1\begin{vmatrix} 3 & -5 \\ 1 & -4 \end{vmatrix} + 0\begin{vmatrix} 3 & -5 \\ 2 & -1 \end{vmatrix} \quad \text{(column 3)}$$

$$= 0 - (-12 + 5) + 0$$
$$= 7$$

$$Dz = \begin{vmatrix} 3 & 1 & -5 \\ 2 & 0 & -1 \\ 1 & -2 & -4 \end{vmatrix} = 3\begin{vmatrix} 0 & -1 \\ -2 & -4 \end{vmatrix} - 2\begin{vmatrix} 1 & -5 \\ -2 & -4 \end{vmatrix} + 1\begin{vmatrix} 1 & -5 \\ 0 & -1 \end{vmatrix} \quad \text{(column 1)}$$

$$= 3(0 - 2) - 2(-4 - 10) + (-1 - 0)$$
$$= -6 + 28 - 1$$
$$= 21$$

$$x = \frac{Dx}{D} = \frac{-14}{7} = -2$$

$$y = \frac{Dy}{D} = \frac{7}{7} = 1$$

$$z = \frac{Dz}{D} = \frac{21}{7} = 3$$

$$\boxed{\{(-2, 1, 3)\}}$$

21. $x - z = -1$
  $x + y = 0$
  $y + z = 3$

$$D = \begin{vmatrix} 1 & 0 & -1 \\ 1 & 1 & 0 \\ 0 & 1 & 1 \end{vmatrix} = 1 \begin{vmatrix} 1 & 0 \\ 1 & 1 \end{vmatrix} - 1 \begin{vmatrix} 0 & -1 \\ 1 & 1 \end{vmatrix} + 0 \begin{vmatrix} 0 & -1 \\ 1 & 0 \end{vmatrix} \quad \text{(column 1)}$$

$$= (1 - 0) - (0 + 1) + 0 = 1 - 1 = 0$$

$$D_x = \begin{vmatrix} -1 & 0 & -1 \\ 0 & 1 & 0 \\ 3 & 1 & 1 \end{vmatrix} = -0 \begin{vmatrix} 0 & 0 \\ 3 & 1 \end{vmatrix} + 1 \begin{vmatrix} -1 & -1 \\ 3 & 1 \end{vmatrix} - 1 \begin{vmatrix} -1 & -1 \\ 0 & 0 \end{vmatrix} \quad \text{(column 2)}$$

$$= 0 + (-1 + 3) - 1(0 - 0) = 2$$

$$D_y = \begin{vmatrix} 1 & -1 & -1 \\ 1 & 0 & 0 \\ 0 & 3 & 1 \end{vmatrix} = -1 \begin{vmatrix} -1 & -1 \\ 3 & 1 \end{vmatrix} + 0 \begin{vmatrix} 1 & -1 \\ 0 & 1 \end{vmatrix} - 0 \begin{vmatrix} 1 & -1 \\ 0 & 3 \end{vmatrix} \quad \text{(row 2)}$$

$$= -(-1 + 3) = -2$$

$$D_z = \begin{vmatrix} 1 & 0 & -1 \\ 1 & 1 & 0 \\ 0 & 1 & 3 \end{vmatrix} = 1 \begin{vmatrix} 1 & 0 \\ 1 & 3 \end{vmatrix} - 1 \begin{vmatrix} 0 & -1 \\ 1 & 3 \end{vmatrix} + 0 \begin{vmatrix} 0 & -1 \\ 1 & 0 \end{vmatrix} \quad \text{(column 1)}$$

$$= (3 - 0) - 1(0 + 1) + 0 = 3 - 1 = 2$$

$D = 0$, $D_x = 2 \neq 0$, $D_y = -2 \neq 0$, $D_z = 2 \neq 0$

Thus, the system is $\boxed{\text{inconsistent}}$.

There is no solution; $\boxed{\varnothing}$.

23. $3x + y - 3z = 3$
  $2x + 3y - z = 2$

  $6x + 9y - 3z = 6$ (*Note*: $(\div 3) \to 2x + 3y - z = 2$; equations 2 and $\frac{3}{2}$ are the same)

$$D = \begin{vmatrix} 3 & 1 & -3 \\ 2 & 3 & -1 \\ 6 & 9 & -3 \end{vmatrix} = 3 \begin{vmatrix} 3 & -1 \\ 9 & -3 \end{vmatrix} - 1 \begin{vmatrix} 2 & -1 \\ 6 & -3 \end{vmatrix} + (-3) \begin{vmatrix} 2 & 3 \\ 6 & 9 \end{vmatrix} \quad \text{(row 1)}$$

$$= 3(-9 + 9) - (-6 + 6) - 3(18 - 18)$$
$$= 3(0) - 0 - 3(0) = 0$$

$$D_x = \begin{vmatrix} 3 & 1 & -3 \\ 2 & 3 & -1 \\ 6 & 9 & -3 \end{vmatrix} = D = 0$$

$$D_y = \begin{vmatrix} 3 & 3 & -3 \\ 2 & 2 & -1 \\ 6 & 6 & -3 \end{vmatrix} = 3 \begin{vmatrix} 2 & -1 \\ 6 & -3 \end{vmatrix} - 3 \begin{vmatrix} 2 & -1 \\ 6 & -3 \end{vmatrix} + (-3) \begin{vmatrix} 2 & 2 \\ 6 & 6 \end{vmatrix}$$

$$= 3(-6 + 6) - 3(-6 + 6) - 3(12 - 12)$$
$$= 3(0) - 3(0) - 3(0) = 0$$

(*Note*: When 2 columns are identical, the value of the determinant is zero.)

$$D_z = \begin{vmatrix} 3 & 1 & 3 \\ 2 & 3 & 2 \\ 6 & 9 & 6 \end{vmatrix} = 0 \qquad \text{(columns 1 and 3 are identical)}$$

Thus, since $D = 0$, $D_x = 0$, D_y, $= 0$ and $D_z = 0$ the system is $\boxed{\text{dependent}}$.

$\boxed{\{x, y, z \mid 2x + 3y - z = 2\}}$

25. $y + z = -3$
   $x + z = -5$
   $x + y = 4$

$$D = \begin{vmatrix} 0 & 1 & 1 \\ 1 & 0 & 1 \\ 1 & 1 & 0 \end{vmatrix} = 0\begin{vmatrix} 0 & 1 \\ 1 & 0 \end{vmatrix} - 1\begin{vmatrix} 1 & 1 \\ 1 & 0 \end{vmatrix} + 1\begin{vmatrix} 1 & 1 \\ 0 & 1 \end{vmatrix}$$

$$= 0 - (0 - 1) + (1 - 0) = 1 + 1 = 2$$

$$D_x = \begin{vmatrix} -3 & 1 & 1 \\ -5 & 0 & 1 \\ 4 & 1 & 0 \end{vmatrix} = -1\begin{vmatrix} -5 & 1 \\ 4 & 0 \end{vmatrix} + 0\begin{vmatrix} -3 & 1 \\ 4 & 0 \end{vmatrix} - 1\begin{vmatrix} -3 & 1 \\ -5 & 1 \end{vmatrix}$$

$$= -(0 - 4) + 0 - (-3 + 5)$$
$$= 4 + 0 - 2 = 2$$

$$D_y = \begin{vmatrix} 0 & -3 & 1 \\ 1 & -5 & 1 \\ 1 & 4 & 0 \end{vmatrix} = 0\begin{vmatrix} -5 & 1 \\ 4 & 0 \end{vmatrix} - 1\begin{vmatrix} -3 & 1 \\ 4 & 0 \end{vmatrix} + 1\begin{vmatrix} -3 & 1 \\ -5 & 1 \end{vmatrix}$$

$$= 0 - (0 - 4) + (-3 + 5)$$
$$= 0 + 4 + 2 = 6$$

$$D_z = \begin{vmatrix} 0 & 1 & -3 \\ 1 & 0 & -5 \\ 1 & 1 & 4 \end{vmatrix} = 0\begin{vmatrix} 0 & -5 \\ 1 & 4 \end{vmatrix} - 1\begin{vmatrix} 1 & -3 \\ 1 & 4 \end{vmatrix} + 1\begin{vmatrix} 1 & -3 \\ 0 & -5 \end{vmatrix}$$

$$= 0 - (4 + 3) + (-5 - 0)$$
$$= 0 - 7 - 5 = -12$$

$x = \dfrac{D_x}{D} = \dfrac{2}{2} = 1$

$y = \dfrac{D_y}{D} = \dfrac{6}{2} = 3$

$z = \dfrac{D_z}{D} = \dfrac{-12}{2} = -6$

$\boxed{\{(1, 3, -6)\}}$

27. $\boxed{C}$ is true

29. You must show

$$\begin{vmatrix} a_1 & b_1 & c_1 \\ a_1 & b_1 & c_1 \\ a_3 & b_3 & c_3 \end{vmatrix} = 0$$

$$\begin{vmatrix} a_1 & b_1 & c_1 \\ a_1 & b_1 & c_1 \\ a_3 & b_3 & c_3 \end{vmatrix} = a_1 \begin{vmatrix} b_1 & c_1 \\ b_3 & c_3 \end{vmatrix} - a_1 \begin{vmatrix} b_1 & c_1 \\ b_3 & c_3 \end{vmatrix} + a_3 \begin{vmatrix} b_1 & c_1 \\ b_1 & c_1 \end{vmatrix} \quad \text{(expand about column 1)}$$

(rows 1 and 2 are identical)

$$\begin{aligned} &= a_1(b_1 c_3 - b_3 c_1) - a(b_1 c_3 - b_3 c_1) + a_3(b_1 c_1 - b_1 c_1) \\ &= a_1 b_1 c_3 - a_1 b_3 c_1 - a_1 b_1 c_3 + a_1 b_2 c_1 + a_3(0) \\ &= (a_1 b_1 c_3 - a_1 b_1 c_3) - (a_1 b_3 c_1 - a_1 b_3 c_1) + 0 \\ &= 0 - 0 + 0 = 0 \end{aligned}$$

31. You must show

$$\begin{vmatrix} Ka_1 & Kb_1 & Kc_1 \\ a_1 & b_2 & c_2 \\ a_3 & b_3 & c_3 \end{vmatrix} = K \begin{vmatrix} a_1 & b_1 & c_1 \\ a_2 & b_2 & c_2 \\ a_3 & b_3 & c_3 \end{vmatrix}$$

(multiply each element of row by K)

$$\begin{vmatrix} Ka_1 & Kb_1 & Kc_1 \\ a_2 & b_2 & c_2 \\ a_3 & b_3 & c_3 \end{vmatrix} = Ka_1 \begin{vmatrix} b_2 & c_2 \\ b_3 & c_3 \end{vmatrix} - Kb_1 \begin{vmatrix} a_2 & c_2 \\ a_3 & c3 \end{vmatrix} + Kc_1 \begin{vmatrix} a_2 & b_2 \\ a_3 & c_3 \end{vmatrix}$$

$$= K \left[a_1 \begin{vmatrix} b_2 c_2 & b_3 \\ & c_3 \end{vmatrix} - b_1 \begin{vmatrix} a_2 & c_2 \\ a_3 & c_3 \end{vmatrix} + c_1 \begin{vmatrix} a_2 & b_2 \\ a_3 & c_3 \end{vmatrix} \right]$$

$$= K \begin{vmatrix} a_1 & b_1 & c_1 \\ a_2 & b_2 & c_2 \\ a_3 & b_3 & c_3 \end{vmatrix}$$

33. You must show

$$\begin{vmatrix} a_1 & b_1 & c_1 \\ a_2 + ka_1 & b_2 + kb_1 & c_2 + kc_1 \\ a_3 & b_3 & c_3 \end{vmatrix} = \begin{vmatrix} a_1 & b_1 & c_1 \\ a_2 & b_2 & c_2 \\ a_3 & b_3 & c_3 \end{vmatrix}$$

$$\begin{vmatrix} a_1 & b_1 & c_1 \\ a_2 + ka_1 & b_2 + kb_1 & c_2 + kc_1 \\ a_3 & b_3 & c_3 \end{vmatrix} = -(a_2 + ka_1) \begin{vmatrix} b_1 & c_1 \\ b_3 & c_3 \end{vmatrix} + (b_2 + kb_1) \begin{vmatrix} a_1 & c_1 \\ a_3 & c_3 \end{vmatrix} - (c_2 + kc_1) \begin{vmatrix} a_1 & b_1 \\ a_3 & b_3 \end{vmatrix}$$

$$= -a_2 \begin{vmatrix} b_1 & c_1 \\ b_3 & c_3 \end{vmatrix} - ka_1 \begin{vmatrix} b_1 & c_1 \\ b_3 & c_3 \end{vmatrix} + b_2 \begin{vmatrix} a_1 & c_1 \\ a_3 & c_3 \end{vmatrix} + kb_1 \begin{vmatrix} a_1 & c_1 \\ b_3 & c_3 \end{vmatrix} - c_2 \begin{vmatrix} a_1 & b_1 \\ a_3 & b_3 \end{vmatrix} - kc_1 \begin{vmatrix} a_1 & b_1 \\ a_3 & b_3 \end{vmatrix}$$

$$= -a_2 \begin{vmatrix} b_1 & c_1 \\ b_3 & c_3 \end{vmatrix} + b_2 \begin{vmatrix} a_1 & c_1 \\ a_3 & c_3 \end{vmatrix} - c_2 \begin{vmatrix} a & b_1 \\ a_3 & b_3 \end{vmatrix} - ka_1 \begin{vmatrix} b_1 & c_1 \\ b_3 & c_3 \end{vmatrix} + kb1 \begin{vmatrix} a_1 & c_1 \\ a_3 & c_3 \end{vmatrix} - kc_1 \begin{vmatrix} a_1 & b_1 \\ a_3 & b_3 \end{vmatrix}$$

$$= \begin{vmatrix} a_1 & b_1 & c_1 \\ a_2 & b_2 & c_2 \\ a_3 & b_3 & c_3 \end{vmatrix} - k \left[a_1 \begin{vmatrix} b_1 & c_1 \\ b_3 & c_3 \end{vmatrix} - b_1 \begin{vmatrix} a_1 & c_1 \\ a_3 & c_3 \end{vmatrix} + c_1 \begin{vmatrix} a_1 & b_1 \\ a_3 & b_3 \end{vmatrix} \right]$$

$$= \begin{vmatrix} a_1 & b_1 & c_1 \\ a_2 & b_2 & c_2 \\ a_3 & b_3 & c_3 \end{vmatrix} - k \left[a_1 b_1 c_3 - a_1 b_3 c_1 - b_1 a_1 c_3 + b_1 a_3 c_1 + c_1 a_1 b_3 - c_1 b_1 a_3 \right]$$

$$= \begin{vmatrix} a_1 & b_1 & c_1 \\ a_2 & b_2 & c_2 \\ a_3 & b_3 & c_3 \end{vmatrix} - k(0)$$

$$= \begin{vmatrix} a_1 & b_1 & c_1 \\ a_2 & b_2 & c_2 \\ a_3 & b_3 & c_3 \end{vmatrix}$$

35. $\begin{vmatrix} 3 & -2 & 4 \\ 1.296 & 3.481 & -1.726 \\ 8.079 & 6.924 & 7.253 \end{vmatrix}$

$$= 3 \begin{vmatrix} 3.841 & -1.726 \\ 6.924 & 7.253 \end{vmatrix} - (-2) \begin{vmatrix} 1.296 & -1.726 \\ 8.079 & 7.253 \end{vmatrix} + 4 \begin{vmatrix} 1.296 & 3.481 \\ 8.079 & 6.924 \end{vmatrix}$$

$$= 3[(3.481)(7.253) - (6.924)(-1.726)] + 2[(1.296)(7.253) - (8.079)(-1.726)]$$
$$+ [(1.296)(6.924) - (8.079)(3.481)]$$

$$\approx \boxed{81.6861}$$

37. Let

$$x = \text{initial amount of grass in each field}$$
$$y = \text{daily growth in each field}$$
$$z = \text{amount of grass consumed each day by each cow}$$

Using the given conditions:

$$Bx + CBy - CAz = 0$$
$$Ex + FEy - FDz = 0$$
$$Hx + IHy - IGz = 0$$

Given that x, y, and z are not all zero, then

$$\begin{vmatrix} B & CB & -CA \\ E & FE & -FD \\ H & IH & -IG \end{vmatrix} = 0$$

Equiavlently,

$$\begin{vmatrix} B & BC & CA \\ E & EF & FD \\ H & HI & IG \end{vmatrix} = 0$$

Review Problems

41. $-6 \le \dfrac{2}{3}x + 4 \;<\; 8$

$-10 \le \dfrac{2}{3}x \;<\; 4$

$-15 \le x \;<\; 6$

$\boxed{\{x \mid -15 \le x < 6\}}$

$\boxed{[-15, 6)}$

42. Let

$r \;=\;$ rate of automobile on outgoing trip

$r - 14 \;=\;$ rate of automobile on return trip

	rate, R $\times$	time, t $=$	Distance, D
outging	r	3	$3r$
return	$r - 14$	4	$4(r - 14)$

$3r \;=\; 4(r - 14)$

$3r \;=\; 4r - 56$

$-r \;=\; -56$

$r \;=\; 56 \quad$ (outgoing)

$\boxed{56 \text{ mph}}$

43. $\dfrac{y + 1}{14} - \dfrac{3y + 2}{7} \;=\; 1$

$y + 1 - 2(3y + 2) \;=\; 14$

$y + 1 - 6y - 4 \;=\; 14$

$-5y \;=\; 17$

$y \;=\; -\dfrac{17}{5}$

$\boxed{\left\{ -\dfrac{17}{5} \right\}}$

Section 4.6 Matrix Solutions to Linear Systems

Problem Set 4.6, pp. 313-314

1. $x + y = 6$
$x - y = 2$

$\begin{bmatrix} 1 & 1 & | & 6 \\ 1 & -1 & | & 2 \end{bmatrix} \quad R_1 - R_2 \rightarrow \quad \begin{bmatrix} 1 & 1 & | & 6 \\ 0 & 2 & | & 4 \end{bmatrix} \quad \tfrac{1}{2}R_2 \rightarrow \quad \begin{bmatrix} 1 & 1 & | & 6 \\ 0 & 1 & | & 2 \end{bmatrix}$

$\begin{aligned} x + y &= 6 \\ y &= 2 \end{aligned} \qquad \begin{aligned} x + 2 &= 6 \quad \text{(substitute for } y) \\ x &= 4 \end{aligned}$

$\boxed{\{(4, 2)\}}$

3. $2x + y = 3$
$x - 3y = 12$

$$\begin{bmatrix} 2 & 1 & | & 3 \\ 1 & -3 & | & 12 \end{bmatrix} \xrightarrow{R_1 - 2R_2 \rightarrow} \begin{bmatrix} 2 & 1 & | & 3 \\ 0 & 7 & | & -21 \end{bmatrix} \xrightarrow{\frac{1}{7}R_2 \rightarrow} \begin{bmatrix} 2 & 1 & | & 3 \\ 0 & 1 & | & -3 \end{bmatrix}$$

$$\begin{aligned} 2x + y &= 3 \\ y &= -3 \end{aligned} \qquad \begin{aligned} 2x - 3 &= 3 \quad \text{(substitute for } y) \\ 2x &= 6 \\ x &= 3 \end{aligned}$$

$\boxed{\{(3, -3)\}}$

5. $5x + 7y = -25$
$11x + 6y = -8$

$$\begin{bmatrix} 5 & 7 & | & -25 \\ 11 & 6 & | & -8 \end{bmatrix} \xrightarrow{11R_1 - 5R_2 \rightarrow} \begin{bmatrix} 5 & 7 & | & -25 \\ 0 & 47 & | & -235 \end{bmatrix} \xrightarrow[\frac{1}{47}R_2 \rightarrow]{\frac{1}{5}R_1 \rightarrow} \begin{bmatrix} 1 & 7/5 & | & 5 \\ 0 & 1 & | & -5 \end{bmatrix}$$

$$\begin{aligned} x + \frac{7}{5}y &= -5 \\ y &= -5 \end{aligned} \qquad \begin{aligned} x + \frac{7}{5} &= -5 \quad \text{(substitute for } y) \\ x &= 2 \end{aligned}$$

$\boxed{\{(2, -5)\}}$

7. $4x - 2y = 5$
$-2x + y = 6$

$$\begin{bmatrix} 4 & -2 & | & 5 \\ -2 & 1 & | & 6 \end{bmatrix} \qquad \begin{array}{c} \frac{1}{4}R_1 \rightarrow \\[4pt] 11R_1 - 5R_2 \rightarrow \end{array} \qquad \begin{bmatrix} 1 & -1/2 & | & 5/4 \\ 0 & 0 & | & 17 \end{bmatrix}$$

$$\begin{aligned} x - \frac{1}{2}y &= \frac{5}{4} \\ 0 &= 17 \quad \text{(False)} \quad \text{No solution} \end{aligned}$$

The system is $\boxed{\text{inconsistent}}$.

$\boxed{\varnothing}$

9. $x - 2y = 1$
$-2x + 4y = -2$

$$\begin{bmatrix} 1 & -2 & | & 1 \\ -2 & 4 & | & -2 \end{bmatrix} \xrightarrow{2R_1 + 2R_2 \rightarrow} \begin{bmatrix} 1 & -2 & | & 1 \\ 0 & 0 & | & 0 \end{bmatrix}$$

$$\begin{aligned} x - 2y &= 1 \\ 0 &= 0 \quad \text{(True)} \quad \text{Infinetly many solutions} \end{aligned}$$

The system is $\boxed{\text{dependent}}$.

$\boxed{\{(x, y) \mid x - 2y = 1\}}$

11. $x + y - z = -2$
$2x - y + z = 5$
$-x + 2y + 2z = 1$

$$\begin{bmatrix} 1 & 1 & -1 & | & -2 \\ 2 & -1 & 1 & | & 5 \\ -1 & 2 & 2 & | & 1 \end{bmatrix} \quad \begin{matrix} R_2 + 2R_3 \to \\ \\ R_1 + R_3 \to \end{matrix} \quad \begin{bmatrix} 1 & 1 & -1 & | & -2 \\ 0 & 3 & 5 & | & 7 \\ 0 & 3 & 1 & | & -1 \end{bmatrix}$$

$$R_2 - R_3 \to \begin{bmatrix} 1 & 1 & -1 & | & -2 \\ 0 & 3 & 5 & | & 7 \\ 0 & 0 & 4 & | & 8 \end{bmatrix} \quad \begin{matrix} \frac{1}{3}R_2 \to \\ \\ \frac{1}{4}R_3 \to \end{matrix} \quad \begin{bmatrix} 1 & 1 & -1 & | & -2 \\ 0 & 1 & 5/3 & | & 7/3 \\ 0 & 0 & 1 & | & 2 \end{bmatrix}$$

$$\begin{aligned} x + y - z &= -2 \\ y + \tfrac{5}{3}z &= \tfrac{7}{3} \\ z &= 2 \qquad y + \tfrac{5}{3}(2) = \tfrac{7}{3} \\ & \qquad\qquad\quad y = -\tfrac{3}{3} = -1 \qquad x - 1 - 2 = -2 \\ & \qquad\qquad\qquad\qquad\qquad\qquad\qquad x = 1 \end{aligned}$$

$\boxed{\{(1, -1, 2)\}}$

13. $x + 3y = 0$
$x + y + z = 1$
$3x - y - z = 11$

$$\begin{bmatrix} 1 & 3 & 0 & | & 0 \\ 1 & 1 & 1 & | & 1 \\ 3 & -1 & -1 & | & 11 \end{bmatrix} \quad \begin{matrix} R_1 - R_2 \to \\ \\ 3R_1 - R_3 \to \end{matrix} \quad \begin{bmatrix} 1 & 3 & 0 & | & 0 \\ 0 & 2 & -1 & | & -1 \\ 0 & 10 & 1 & | & -11 \end{bmatrix}$$

$$5R_2 - R_3 \to \begin{bmatrix} 1 & 3 & 0 & | & 0 \\ 0 & 2 & -1 & | & -1 \\ 0 & 0 & 6 & | & 6 \end{bmatrix} \quad \begin{matrix} \frac{1}{2}R_2 \to \\ \\ -\frac{1}{6}R_3 \to \end{matrix} \quad \begin{bmatrix} 1 & 3 & 0 & | & 0 \\ 0 & 1 & -1/2 & | & -1/2 \\ 0 & 0 & 1 & | & -1 \end{bmatrix}$$

$$\begin{aligned} x + 3y &= 0 \\ y - \tfrac{1}{2}z &= -\tfrac{1}{2} \\ z &= -1 \qquad y - \tfrac{1}{2}(-1) = -\tfrac{1}{2} \\ & \qquad\qquad\qquad\quad y = -1 \qquad x + 3(-1) = 0 \\ & \qquad\qquad\qquad\qquad\qquad\qquad\qquad x = 3 \end{aligned}$$

$\boxed{\{(3, -1, -1)\}}$

15. $2x + 2y + 7z = -1$
$2x + y + 2z = 2$
$4x + 6y + z = 15$

$$\begin{bmatrix} 2 & 2 & 7 & | & -1 \\ 2 & 1 & 2 & | & 2 \\ 4 & 6 & 1 & | & 15 \end{bmatrix} \quad \begin{matrix} R_1 - R_2 \to \\ \\ 2R_1 - R_3 \to \end{matrix} \quad \begin{bmatrix} 2 & 2 & 7 & | & -1 \\ 0 & 1 & 5 & | & -3 \\ 0 & -2 & 13 & | & -17 \end{bmatrix}$$

$$2R_2 + R_3 \to \quad \begin{bmatrix} 2 & 2 & 7 & | & -1 \\ 0 & 1 & 5 & | & -3 \\ 0 & 0 & 23 & | & -23 \end{bmatrix} \quad \begin{matrix} \frac{1}{2}R_1 \to \\ \\ -\frac{1}{23}R_3 \to \end{matrix} \quad \begin{bmatrix} 1 & 1 & 7/2 & | & -1/2 \\ 0 & 1 & 5 & | & -3 \\ 0 & 0 & 1 & | & -1 \end{bmatrix}$$

$$\begin{aligned} x + y + \frac{7}{2}z &= -\frac{1}{2} \\ y + 5z &= -3 \\ z &= -1 \end{aligned}$$

$$\begin{aligned} y + 5(-1) &= -3 \\ y &= 2 \end{aligned} \qquad \begin{aligned} x + 2 + \frac{7}{2}(-1) &= -\frac{1}{2} \\ x &= 1 \end{aligned}$$

$\boxed{\{(1, 2, -1)\}}$

17. $x + y + z = 6$
$x - z = -2$
$y + 3z = 11$

$$\begin{bmatrix} 1 & 1 & 1 & | & 6 \\ 1 & 0 & -1 & | & -2 \\ 1 & 1 & 3 & | & 11 \end{bmatrix} \quad R_1 - R_2 \to \quad \begin{bmatrix} 1 & 1 & 1 & | & 6 \\ 0 & 1 & 2 & | & 8 \\ 0 & 1 & 3 & | & 11 \end{bmatrix}$$

$$R_3 - R_2 \to \quad \begin{bmatrix} 1 & 1 & 1 & | & 6 \\ 0 & 1 & 2 & | & 8 \\ 0 & 0 & 1 & | & 3 \end{bmatrix}$$

$$\begin{aligned} x + y + z &= 6 \\ y + 2z &= 8 \\ z &= 3 \end{aligned}$$

$$\begin{aligned} y + 2(3) &= 8 \\ y &= 2 \end{aligned} \qquad \begin{aligned} x + 2 + 3 &= 6 \\ x &= 1 \end{aligned}$$

$\boxed{\{(1, 2, 3)\}}$

19. $x - y + 3z = 4$
$x + 2y - 2z = 10$
$3x - y + 5z = 14$

$$\begin{bmatrix} 1 & -1 & 3 & | & 4 \\ 1 & 2 & -2 & | & 10 \\ 3 & -1 & 5 & | & 14 \end{bmatrix} \quad \begin{array}{c} R_2 - R_1 \to \\ \\ R_3 - 3R_1 \to \end{array} \quad \begin{bmatrix} 1 & -1 & 3 & | & 4 \\ 0 & 3 & -5 & | & 6 \\ 0 & 2 & -4 & | & 2 \end{bmatrix}$$

$$\begin{array}{c} \frac{1}{3}R_2 \to \\ \\ \frac{1}{2}R_3 \to \end{array} \quad \begin{bmatrix} 1 & -1 & 3 & | & 4 \\ 0 & 1 & -5/3 & | & 2 \\ 0 & 1 & -2 & | & 1 \end{bmatrix} \quad R_2 - R_3 \to \quad \begin{bmatrix} 1 & -1 & 3 & | & 4 \\ 0 & 1 & -5/3 & | & 2 \\ 0 & 0 & 1/3 & | & 1 \end{bmatrix}$$

$$3R_3 \to \quad \begin{bmatrix} 1 & -1 & 3 & | & 4 \\ 0 & 1 & -5/3 & | & 2 \\ 0 & 0 & 1 & | & 3 \end{bmatrix}$$

$$\begin{aligned} x - y + 3z &= 4 \\ y - \frac{5}{3}z &= 2 \\ z &= 3 \end{aligned} \qquad \begin{aligned} y - \frac{5}{3}(3) &= 2 \\ y &= 7 \qquad \begin{aligned} x - 7 + 3(3) &= 4 \\ x &= 2 \end{aligned} \end{aligned}$$

$\boxed{\{(2, 7, 3)\}}$

21. $x - 2y + z = 4$
$5x - 10y + 5z = 20$
$-2x + 4y - 2z = -8$

$$\begin{bmatrix} 1 & -2 & 1 & | & 4 \\ 5 & -10 & 5 & | & 20 \\ -2 & 4 & -2 & | & -8 \end{bmatrix} \quad \begin{array}{c} 5R_1 - R_2 \to \\ \\ 2R_1 + R_3 \to \end{array} \quad \begin{bmatrix} 1 & -2 & 1 & | & 4 \\ 0 & 0 & 0 & | & 0 \\ 0 & 0 & 0 & | & 0 \end{bmatrix}$$

$$\begin{aligned} x - 2y + z &= 4 \\ 0 &= 0 \\ 0 &= 0 \quad \text{(True)} \quad \text{Infinitely many solutions.} \end{aligned}$$

The system is $\boxed{\text{dependent}}$.

$\boxed{\{(x, y, z) \mid x - 2y + z = 4\}}$

23. $x + y = 1$
$y + 2z = -2$
$2x - z = 0$

$$\begin{bmatrix} 1 & 1 & 0 & | & 1 \\ 0 & 1 & 2 & | & -2 \\ 2 & 0 & -1 & | & 0 \end{bmatrix} \quad 2R_1 - R_3 \rightarrow \qquad \begin{bmatrix} 1 & 1 & 0 & | & 1 \\ 0 & 1 & 2 & | & -2 \\ 0 & 2 & 1 & | & 2 \end{bmatrix}$$

$$2R_2 - R_3 \rightarrow \begin{bmatrix} 1 & 1 & 0 & | & 1 \\ 0 & 1 & 2 & | & -2 \\ 0 & 0 & 3 & | & -6 \end{bmatrix} \quad \frac{1}{3}R_3 \rightarrow \begin{bmatrix} 1 & 1 & 0 & | & 1 \\ 0 & 1 & 2 & | & -2 \\ 0 & 0 & 1 & | & -2 \end{bmatrix}$$

$$\begin{aligned} x + y &= 1 \\ y + 2z &= -2 \\ z &= -2 \end{aligned} \qquad \begin{aligned} y + 2(-2) &= -2 \\ y &= 2 \end{aligned} \qquad \begin{aligned} x + 2 &= 1 \\ x &= -1 \end{aligned}$$

$\boxed{\{(-1, 2, -2)\}}$

25. $\boxed{D}$ is true; $\quad x - \dfrac{3}{2}y = 5$
$\qquad\qquad\qquad\qquad\qquad 0 = 6 \quad$ False
The system is inconsistent. $\varnothing$

27. $\begin{aligned} x + y + z + w &= 5 \\ x + 2y - z - 2w &= -1 \\ x - 3y - 3z - w &= -1 \\ 2x - y + 2z - w &= -2 \end{aligned}$

$$\begin{bmatrix} 1 & 1 & 1 & 1 & | & 5 \\ 1 & 2 & -2 & -3 & | & -1 \\ 1 & -3 & 1 & -14 & | & -1 \\ 2 & -1 & 0 & -12 & | & -2 \end{bmatrix} \quad \begin{array}{l} R_2 - R_1 \rightarrow \\ R_3 - R_1 \rightarrow \\ R_4 - 2R_2 \rightarrow \end{array} \quad \begin{bmatrix} 1 & 1 & 1 & 1 & | & 5 \\ 0 & 1 & -2 & -3 & | & -6 \\ 0 & -4 & -4 & -2 & | & -6 \\ 0 & -3 & 0 & -3 & | & -12 \end{bmatrix}$$

$$\begin{array}{l} 4R_2 + R_3 \rightarrow \\ \\ 3R_3 + R_4 \rightarrow \end{array} \begin{bmatrix} 1 & 1 & 1 & 1 & | & 5 \\ 0 & 1 & -2 & -3 & | & -6 \\ 0 & 0 & -12 & -14 & | & -30 \\ 0 & 0 & -6 & -12 & | & -30 \end{bmatrix} \quad R_3 - 2R_4 \rightarrow \begin{bmatrix} 1 & 1 & 1 & 1 & | & 5 \\ 0 & 1 & -2 & -3 & | & -6 \\ 0 & 0 & -12 & -14 & | & -30 \\ 0 & 0 & 0 & 10 & | & 30 \end{bmatrix}$$

$$-\frac{1}{12}R_3 \rightarrow \quad \frac{1}{10}R_4 \rightarrow \quad \begin{bmatrix} 1 & 1 & 1 & 1 & | & 5 \\ 0 & 1 & -2 & -3 & | & -6 \\ 0 & 0 & 1 & 7/6 & | & 5/2 \\ 0 & 0 & 0 & 1 & | & 3 \end{bmatrix}$$

$$\begin{aligned} x + y + z + w &= 5 \\ y - 2z - 3w &= -6 \\ z + \frac{7}{6}w &= \frac{5}{2} \\ w &= 3 \end{aligned}$$

$$\begin{aligned} z + \frac{7}{6}(3) &= \frac{5}{6} \\ z &= -1 \end{aligned} \qquad \begin{aligned} y - 2(-1) - 3(3) &= -6 \\ y &= 1 \end{aligned}$$

$$\begin{aligned} x + 1 + (-1) + 3 &= 5 \\ x &= 2 \end{aligned}$$

$$\boxed{\{(2, 1, -1, 3)\}}$$

29.

$$\det \begin{bmatrix} 1 & 4 & -2 \\ 3 & 2 & 0 \\ -1 & 4 & 3 \end{bmatrix} = \begin{vmatrix} 1 & 4 & -2 \\ 3 & 2 & 0 \\ -1 & 4 & 3 \end{vmatrix} = \boxed{-58}$$

Review Problems

33. Let

$$\begin{aligned} x &= \text{number of liters of 30\% acid solution} \\ 20 - x &= \text{number of liters of 18\% acid solution} \end{aligned}$$

$$\begin{aligned} 0.30x + 0.18(20 - x) &= 0.22(20) \\ 0.30x + 3.6 - 0.18x &= 4.4 \\ 0.12x &= 0.8 \\ x &= 6\frac{2}{3} \quad (30\%) \end{aligned}$$

$$20 - x = 20 - 6\frac{2}{3} = 13\frac{1}{3} \quad (18\%)$$

$$\boxed{6\frac{2}{3} \text{ liters at 30\%; } 13\frac{1}{3} \text{ liters at 18\%}}$$

34.

$$\begin{aligned} -3x &\geq 6 \\ x &\leq -2 \end{aligned} \qquad \text{or} \qquad \begin{aligned} 3x - 1 &< -10 \\ 3x &< -9 \\ x &< -3 \end{aligned}$$

$$\boxed{\{x \mid x \leq -2\}}$$

$$\boxed{(-\infty, -2]}$$

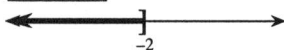

35. $y \le 3x + 2$ and $2x + 3y > 12$

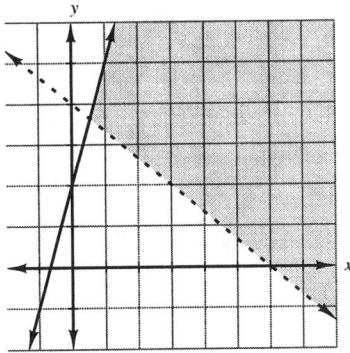

Chapter 4   Review Problems

Review Problems, pp. 316-318

1.

$$
\begin{aligned}
2x - y &= 2 \\
x + 2y &= 11
\end{aligned}
\qquad (\times 2) \to
$$

$$
\begin{aligned}
4x - 2y &= 4 \\
\underline{x + 2y} &= \underline{11} \\
5x &= 15 \\
x &= 3
\end{aligned}
\qquad
\begin{aligned}
x + 2y &= 11 \\
3 + 2y &= 11 \\
2y &= 8 \\
y &= 4
\end{aligned}
$$

$$\boxed{\{(3, 4)\}}$$

2.

$$
\begin{aligned}
y &= 4 - x \\
3x + 3y &= 12
\end{aligned}
\qquad (\div -3) \to
$$

$$
\begin{aligned}
x + y &= 4 \\
\underline{-x - y} &= \underline{-4} \\
0 &= 0
\end{aligned}
\quad \text{(True) Infinitely many solutions.}
$$

The system is $\boxed{\text{dependent}}$.

$$\boxed{\{(x, y) \mid y = 4 - x\}}$$

3.

$$
\begin{aligned}
5x + 3y &= -3 \\
2x + 7y &= -7
\end{aligned}
\qquad
\begin{aligned}
(\times 7) &\to \\
(\times -3) &\to
\end{aligned}
$$

$$
\begin{aligned}
35x + 21y &= -21 \\
\underline{-6x - 21y} &= \underline{21} \\
29x &= 0 \\
x &= 0
\end{aligned}
$$

$$
\begin{aligned}
2x + 7y &= -7 \\
2(0) + 7y &= -7 \\
y &= -1
\end{aligned}
$$

$$\boxed{\{(0, -1)\}}$$

4.

$$
\begin{aligned}
\frac{2x + y}{3} - \frac{x + 2y}{2} &= \frac{23}{6} \\
x &= 2 + \frac{3x - 4y}{5}
\end{aligned}
$$

(Simplify):

$$
\begin{aligned}
2(2x + y) - 3(x + 2y) &= 23 \\
5x &= 10 + 3x - 4y
\end{aligned}
$$

(Simplify again!):
$$4x + 2y - 3x - 6y = 23$$
$$2x + 4y = 10$$

(and again!):
$$x - 4y = 23$$
$$\underline{2x + 4y = 10}$$
$$3x = 33 \quad \text{(Add)}$$
$$x = 11$$

$$x - 4y = 23$$
$$11 - 4y = 23 \quad \text{(substitute 11 for } x)$$
$$-4y = 12$$
$$y = -3$$

$$\boxed{\{(11, -3)\}}$$

5.
$$x - 2y + 3 = 0 \qquad (\times -2) \rightarrow$$
$$2x - 4y + 7 = 0$$

$$-2x + 4y = 6$$
$$\underline{2x - 4y = -7}$$
$$0 = -1 \qquad \text{False (Contradiction)}$$

The system is $\boxed{\text{inconsistent}}$. $\boxed{\varnothing}$

6.
$$\frac{1}{8}x + \frac{3}{4}y = \frac{19}{8} \qquad (\times 16) \rightarrow \qquad 2x + 12y = 38$$
$$-\frac{1}{2}x + \frac{3}{4}y = \frac{1}{2} \qquad (\times 4) \qquad \underline{-2x + 3y = 2}$$
$$15y = 40$$
$$y = \frac{8}{3}$$

$$(\text{Equation 1})(\div 2) \qquad x + 6y = 19$$
$$x + 6\left(\frac{8}{3}\right) = 19$$
$$x = 3$$

$$\boxed{\left\{\left(3, \frac{8}{3}\right)\right\}}$$

7.
$$\frac{1}{5}x - \frac{1}{4}y = 0 \qquad (\times 20) \rightarrow \qquad 4x - 5y = 0$$
$$\frac{1}{3}x - \frac{1}{6}y = 3 \qquad (\times -12) \qquad \underline{-4x + 2y = -36}$$
$$-3y = -36$$
$$y = 12$$

$$4x - 5(12) = 0$$
$$4x = 60$$
$$x = 15$$

$$\boxed{\{(15, 12)\}}$$

8.

$$\begin{array}{rcl} 0.4x + 0.5y &=& 0.2 \\ 0.2x + 0.3y &=& 0.14 \end{array} \qquad (\times -2) \rightarrow \begin{array}{rcl} 0.4x + 0.5y &=& 0.2 \\ \underline{-0.4x - 0.6y} &=& \underline{-0.28} \\ -0.1y &=& -0.08 \\ y &=& 0.8 \end{array}$$

$$\begin{array}{rcl} 0.4x + 0.5y &=& 0.2 \\ 0.4x + 0.5(0.8) &=& 0.2 \\ 0.4x &=& -0.2 \\ x &=& -0.5 \end{array}$$

$$\boxed{\{(-0.5, 0.8)\}}$$

9.

$$\begin{array}{rcl} 3x - y + 4z &=& 4 \\ 4x + 4y - 3z &=& 3 \\ 2x + 3y + 2z &=& -4 \end{array}$$

(Equations 1 and 2):
$$\begin{array}{rcl} (\times 4) \quad 12x - 4y + 16z &=& 16 \\ \underline{4x + 4y - 3z} &=& \underline{3} \\ 16x + 13z &=& 19 \quad \text{(Eq 4)} \end{array}$$

(Equations 1 and 3):
$$\begin{array}{rcl} (\times 3) \quad 9x - 3y + 12z &=& 12 \\ \underline{2x + 3y + 2z} &=& \underline{-4} \\ 11x + 14z &=& 8 \quad \text{(Eq 5)} \end{array}$$

(Equations 4 and 5):
$$\begin{array}{rcl} (\times 14) \quad 224x + 182z &=& 266 \\ (\times -13) \quad -143x - 182z &=& -104 \\ 81x &=& 162 \\ x &=& 2 \end{array}$$

$$\begin{array}{rcl} 11x + 14z &=& 8 \\ 11(2) + 14z &=& 8 \\ 14z &=& -14 \\ z &=& -1 \end{array}$$

$$\begin{array}{rcl} 3x - y + 4z &=& 4 \\ 3(2) - y + 4(-1) &=& 4 \\ -y &=& 2 \\ y &=& -2 \end{array}$$

$$\boxed{\{(2, -2, -1)\}}$$

10.

$$\begin{array}{rcl} x - z + 2 &=& 0 \\ y + 3z &=& 11 \\ x + y + z &=& 6 \end{array}$$

(Equations 1 and 3):
$$\begin{array}{rcl} (\times -1) \quad -x + z &=& 2 \\ \underline{x + y + z} &=& \underline{6} \\ y + 2z &=& 8 \quad \text{(Eq 4)} \end{array}$$

(Equations 2 and 4):
$$\begin{array}{rcl} y + 3z &=& 11 \\ (\times -1) \quad -y - 2z &=& -8 \\ z &=& 3 \end{array}$$

(Equation 4):
$$\begin{array}{rcl} y + 2z &=& 8 \\ y + 2(3) &=& 8 \\ y &=& 2 \end{array}$$

(Equation 1):
$$\begin{array}{rcl} x - z &=& -2 \\ x - 3 &=& -2 \\ x &=& 1 \end{array}$$

$$\boxed{\{(1, 2, 3)\}}$$

11. Let

$$
\begin{aligned}
t &= \text{tens' digit} \\
u &= \text{units' digit} \\
10t + u &= \text{original number} \\
10t + t &= \text{number with digits reversed}
\end{aligned}
$$

$$
\begin{aligned}
t + u &= 7 \\
10u + t &= (10t + u) - 9
\end{aligned}
$$

(Simplify):

$$
\begin{array}{ll}
t + u = 7 & \qquad t + u = 7 \\
-9t + 9u = -9 \quad (\div 9) \rightarrow & \qquad \underline{-t + u = -1} \\
& \qquad \quad 2u = 6 \\
& \qquad \quad\ u = 3
\end{array}
$$

$$
\begin{aligned}
t + u &= 7 \\
t + 3 &= 7 \\
t &= 4
\end{aligned}
$$

$t = 4,\ u = 3$

number: $\boxed{43}$

12. Let

$$
\begin{aligned}
x &= \text{cost of 1 pen} \\
y &= \text{cost of 1 pad}
\end{aligned}
$$

$$
\begin{array}{lll}
8x + 6y = 16.10 & \rightarrow & 8x + 6y = 16.10 \\
3x + 2y = 5.85 & (\times -3) \rightarrow & \underline{-9x - 6y = -17.55} \\
& & -x = -1.45 \\
& & \ \ x = 1.45 \quad \text{(pen)}
\end{array}
$$

$$
\begin{aligned}
3x + 2y &= 5.85 \\
3(1.45) + 2y &= 5.85 \\
2y &= 1.50 \\
y &= 0.75 \quad \text{(pad)}
\end{aligned}
$$

$\boxed{\text{cost of one pen: \$1.45; cost of one pad: 75¢}}$

13. Let

x = speed of the airplane in still air
y = speed of the wind

	Rate, r $\times$	time, t (hr) $=$	Distance, D (miles)
against wind	$x - y$	2 hr 15 min $= 2\frac{1}{4}$ hr	360
with wind	$x + y$	1 hr 20 min $= 1\frac{1}{3}$ hr	360

$$2\frac{1}{4}(x-y) = 360$$

$$1\frac{1}{3}(x+y) = 360$$

(Simplify):

$$\frac{9}{4}(x-y) = 360 \qquad \left(\times \frac{4}{9}\right) \rightarrow \quad x - y = 160$$

$$\frac{4}{3}(x+y) = 360 \qquad \left(\times \frac{3}{4}\right) \rightarrow \quad \underline{x + y = 270}$$

$$2x = 430$$
$$x = 215 \quad \text{(airplane)}$$

$$215 + y = 270$$
$$y = 55 \quad \text{(wind)}$$

> Plane: 215 mph; Wind: 55 mph

14. $y = mx + b$

$(1, 6)$: $\qquad 6 = m + b$
$(-1, -12)$: $\quad \underline{-12 = -m + b}$
$\qquad\qquad\quad -6 = 2b$
$\qquad\qquad\quad -3 = b$

$$m + b = 6$$
$$m - 3 = 6$$
$$m = 9$$

> $m = 9, b = -3$

15. Let

x = age of newer home $x - 20$ = age of newer home 20 years ago
y = age of older home $y - 30$ = age of older home 20 years ago
$\qquad\qquad\qquad\qquad\qquad\qquad$ $x + 60$ = age of newer home 60 years from now
$\qquad\qquad\qquad\qquad\qquad\qquad$ $y + 60$ = age of older home 60 years from now

$$y - 20 = 10(x - 20)$$
$$y + 60 = 2(x + 60)$$

(Simplify):

$$y - 10x = -180$$
$$(\times -1) \quad \underline{-y + 2x = -60}$$
$$\qquad -8x = -240$$
$$\qquad x = 30 \qquad y = 2x + 60$$
$$\qquad\qquad\qquad\qquad y = 2(30) + 60 = 120$$

> Older home: 120 years old
> Newer home: 30 years old

16. Let

x = number of people upstairs
y = number of people downstairs

	upstairs	downstairs	Equation
start	x	y	
"one moves upstairs"	$x+1$	$y-1$	$x+1=10(y-1)$
"eight move downstairs"	$x-8$	$y+8$	$x-8=y+8$

$$
\begin{aligned}
x+1 &= 10(y-1) && \rightarrow & x-10y &= -11 \\
x-8 &= y+8 && \rightarrow & x-y &= 16
\end{aligned}
$$

$(\times -1) \rightarrow$

$$
\begin{aligned}
-x+10y &= 11 \\
\underline{x-y} &= \underline{16} \\
9y &= 27 \\
y &= 3
\end{aligned}
$$

$$
\begin{aligned}
x-y &= 16 \\
x-3 &= 16 \\
x &= 19
\end{aligned}
$$

Number of people: | upstairs: 19; downstairs: 3 |

17. Let

h = hundreds' digit
t = tens' digit
u = units' digit

$$
\begin{aligned}
h+t+u &= 12 && \rightarrow & h+t+u &= 12 \\
t &= 3u+1 && \rightarrow & t-3u &= 1 \\
2h+u &= 8 && \rightarrow & 2h+u &= 8
\end{aligned}
$$

(Equations 1 and 2):

$$
\begin{aligned}
h+t+u &= 12 \\
(\times -1) \quad \underline{-t+3u} &= \underline{-1} \\
h+4u &= 11 \quad \text{(Eq 4)}
\end{aligned}
$$

(Equations 3 and 4):

$$
\begin{aligned}
2h+u &= 8 \\
(\times -2) \quad \underline{-2h-8u} &= \underline{-22} \\
-7u &= -14 \\
u &= 2
\end{aligned}
$$

(Eq 2):
$$
\begin{aligned}
t-3u &= 1 \\
t-6 &= 1 \\
t &= 7
\end{aligned}
$$

(Eq 3):
$$
\begin{aligned}
2h+u &= 8 \\
2h+2 &= 8 \\
h &= 3
\end{aligned}
$$

$h=3,\ t=7,\ u=2$
number: | 372 |

18. $y = ax^2 + bx + c$
$(1,2)$: $2 = a(1^2) + b(1) + c$
$(-1, 6)$: $6 = a(-1)^2 + b(-1) + c$
$(2, 3)$: $3 = a(2^2) + b(2) + c$

$$
\begin{aligned}
a+b+c &= 2 \\
a-b+c &= 6 \\
4a+2b+c &= 3
\end{aligned}
$$

(Equations 1 and 2):

$$(\times -1) \quad \begin{array}{rcl} -a-b-c &=& -2 \\ a-b+c &=& 6 \\ \hline -2b &=& 4 \\ b &=& -2 \end{array}$$

(Equations 1 and 3):

$$(\times -1) \quad \begin{array}{rcl} -a-b-c &=& -2 \\ 4a+2b+c &=& 3 \\ \hline 3a+b &=& 1 \\ 3a+(-2) &=& 1 \\ 3a &=& 3 \\ a &=& 1 \end{array}$$

(Equation 1):

$$\begin{array}{rcl} a+b+c &=& 2 \\ 1+(-2)+c &=& 2 \\ c &=& 3 \end{array}$$

$$\boxed{a=1,\, b=-2,\, c=3}$$

$$y = x^2 - 2x + 3$$

19. Let

$$\begin{array}{rcl} x &=& \text{number of nickels} \\ y &=& \text{numer of dimes} \\ z &=& \text{number of quarters} \end{array}$$

$$\begin{array}{rcl} x+y+z &=& 85 \\ 0.05x+0.10y+0.25z &=& 6.25 \\ x &=& 3y \end{array}$$

(Substitute $x=3y$ into equations 1 and 2 and simplify):

$$\begin{array}{rclcrcl} 4y+z &=& 85 & \rightarrow & 4y+z &=& 85 \\ 0.25y+0.25z &=& 6.25 & (\div -0.25) \rightarrow & -y-z &=& -25 \\ & & & & \hline 3y &=& 60 \\ & & & & y &=& 20 \quad \text{(dimes)} \end{array}$$

$$\begin{array}{rcl} y+z &=& 25 \\ 20+z &=& 25 \\ z &=& 5 \quad \text{(quarters)} \end{array}$$

$$x = 3y = 3(20) = 60 \quad \text{(nickels)}$$

$$\boxed{60 \text{ nickels}, 20 \text{ dimes}, 5 \text{ quarters}}$$

20. Let

$$\begin{array}{rcl} x &=& \text{length of } BD \\ x &=& \text{length of } BC \\ x &=& \text{length of } AD \quad . \text{ (opposite sides of parallelogram are equal)} \\ \tfrac{1}{2}x &=& \text{length of } ED \quad \text{(the diagonals of parallelogram bisect each other)} \\ y &=& \text{length of } CD \\ z &=& \text{lenght of } AC \\ \tfrac{1}{2}z &=& \text{length of } CE \end{array}$$

perimeter of CDE:

$$\begin{array}{rcl} CE+ED+DC &=& 12 \\ \tfrac{1}{2}z+\tfrac{1}{2}x+y &=& 12 \quad (\times 2) \rightarrow \quad x+2y+z = 24 \end{array}$$

perimeter of CDA:

$$\begin{array}{rcl} AC+AD+DC &=& 20 \\ z+x+y &=& 20 \quad \rightarrow \quad x+y+z = 20 \end{array}$$

perimeter of CDB:

$$CB + BD + DC = 18$$
$$x + x + y = 18 \qquad \rightarrow \qquad 2x + y = 18$$

System of equations:

$$x + 2y + z = 24$$
$$x + y + z = 20$$
$$2x + y = 18$$

(Equations 1 and 2):

$$\begin{array}{rcl} x + 2y + z & = & 24 \\ (\times -1) \quad \underline{-x - y - z} & - & \underline{-20} \\ y & = & 4 \quad (CD) \end{array}$$

(Equation 3):

$$\begin{array}{rcl} 2x + y & = & 18 \\ 2x + 4 & = & 18 \\ 2x & = & 14 \\ x & = & 7 \quad (CB) \end{array}$$

(Equation 2):

$$\begin{array}{rcl} x + y + z & = & 20 \\ 7 + 4 + z & = & 20 \\ z & = & 9 \quad (CA) \end{array}$$

$\boxed{CB: \ 7 \text{ ft}; \ CD: \ 4 \text{ ft}; \ CA: \ 9 \text{ ft}}$

21. $7x + 14 = 4y + 4$ (Base angles of an isosceles triangle are equal.)
$(4x + 3y + 5) + (4y + 4) + (7x + 14) = 180$

(Simplify):

$$\begin{array}{rclcll} 7x - 4y & = & 10 & (\times 7) \rightarrow & 49x - 28y & = & -70 \\ 11x + 7y & = & 157 & (\times 4) \rightarrow & \underline{44x + 28y} & = & \underline{628} \\ & & & & 93x & = & 558 \\ & & & & x & = & 6 \end{array}$$

$$\begin{array}{rcl} 7x - 4y & = & 10 \\ 7(6) - 4y & = & 10 \\ y & = & 13 \end{array}$$

$m \angle A = (4x + 3y + 5)^\circ = [4(6) + 3(13) + 5]^\circ = \boxed{68^\circ}$

$m \angle B = (4y + 4)^\circ = [4(13) + 4]^\circ = \boxed{56^\circ}$

$m \angle C = (7x + 14)^\circ = [7(6) + 14]^\circ = \boxed{56^\circ}$

22. Let

$$\begin{array}{rcl} x & = & \text{number of rabbits} \\ y & = & \text{number of pheasants} \end{array}$$

$$\begin{array}{rclcll} x + y & = & 35 & (\times -1) \rightarrow & -x - y & = & -35 \\ 4x + 2y & = & 94 & (\div 2) \rightarrow & \underline{2x + y} & = & \underline{47} \\ & & & & x & = & 12 \quad \text{(rabbits)} \end{array}$$

$$\begin{array}{rcl} x + y & = & 35 \\ 12 + y & = & 35 \\ y & = & 23 \quad \text{(pheasants)} \end{array}$$

$\boxed{12 \text{ rabbits}, 23 \text{ pheasants}}$

23. $\begin{vmatrix} 3 & 2 \\ -1 & 5 \end{vmatrix} = 15 - (-2) = \boxed{17}$ **24.** $\begin{vmatrix} -2 & -3 \\ -4 & -8 \end{vmatrix} = 16 - 12 = \boxed{4}$

25. $\begin{vmatrix} 2 & 4 & -3 \\ -1 & 7 & -4 \\ 1 & -6 & -2 \end{vmatrix} = 2\begin{vmatrix} 7 & -4 \\ -6 & -2 \end{vmatrix} - (-1)\begin{vmatrix} 4 & -3 \\ -6 & -2 \end{vmatrix} + 1\begin{vmatrix} 4 & -3 \\ 7 & -4 \end{vmatrix}$

$$= 2(-14 - 24) + 1(-8 - 18) + (-16 + 21)$$
$$= 2(-38) + (-26) + (5)$$
$$= \boxed{-97}$$

26. $\begin{vmatrix} 4 & 7 & 0 \\ -5 & 6 & 0 \\ 3 & 2 & -4 \end{vmatrix} = 0\begin{vmatrix} -5 & 6 \\ 3 & 2 \end{vmatrix} - 0\begin{vmatrix} 4 & 7 \\ 3 & 2 \end{vmatrix} + (-4)\begin{vmatrix} 4 & 7 \\ -5 & 6 \end{vmatrix}$

$$= 0 - 0 - 4(24 - (-35)) = -4(59) = \boxed{-236}$$

27. $2x - y = 2$
$x + 2y = 11$

$$D = \begin{vmatrix} 2 & -1 \\ 1 & 2 \end{vmatrix} = 4 + 1 = 5$$

$$Dx = \begin{vmatrix} 2 & -1 \\ 11 & 2 \end{vmatrix} = 4 + 11 = 15$$

$$Dy = \begin{vmatrix} 2 & 2 \\ 1 & 11 \end{vmatrix} = 22 - 2 = 20$$

$$x = \frac{Dx}{D} = \frac{15}{5} = 3$$

$$y = \frac{Dy}{D} = \frac{20}{5} = 4$$

$$\boxed{\{(3, 4)\}}$$

28. $\quad\quad 4x = 12 - 3y \quad\quad \rightarrow \quad\quad 4x + 3y = 12$
$\quad\quad 2x - 6 = 5y \quad\quad\quad\quad\quad\quad 2\text{i}x - 5y = 6$

$$D = \begin{vmatrix} 4 & 3 \\ 2 & -5 \end{vmatrix} = -20 - 6 = -26$$

$$Dx = \begin{vmatrix} 12 & 3 \\ 6 & -5 \end{vmatrix} = -60 - 18 = -78$$

$$Dy = \begin{vmatrix} 4 & 12 \\ 2 & 6 \end{vmatrix} = 24 - 24 = 0$$

$$x = \frac{Dx}{D} = \frac{-78}{-26} = 3$$

$$y = \frac{Dy}{D} = \frac{0}{-26} = 0$$

$$\boxed{\{(3, 0)\}}$$

29.
$$4x + 2y + 3z = 9$$
$$3x + 5y + 4z = 19$$
$$9x + 3y + 2z = 3$$

$$D = \begin{vmatrix} 4 & 2 & 3 \\ 3 & 5 & 4 \\ 9 & 3 & 2 \end{vmatrix} = 4\begin{vmatrix} 5 & 4 \\ 3 & 2 \end{vmatrix} - 3\begin{vmatrix} 2 & 3 \\ 3 & 2 \end{vmatrix} + 9\begin{vmatrix} 2 & 3 \\ 5 & 4 \end{vmatrix}$$

$$= 4(10 - 12) - 3(4 - 9) + 9(8 - 15)$$
$$= 4(-2) - 3(-5) + 9(-7)$$
$$= -8 + 15 - 63 = -56$$

$$Dx = \begin{vmatrix} 9 & 2 & 3 \\ 19 & 5 & 4 \\ 3 & 3 & 2 \end{vmatrix} = 9\begin{vmatrix} 5 & 4 \\ 3 & 2 \end{vmatrix} - 19\begin{vmatrix} 2 & 3 \\ 3 & 2 \end{vmatrix} + 3\begin{vmatrix} 2 & 3 \\ 5 & 4 \end{vmatrix}$$

$$= 9(10 - 12) - 19(4 - 9) + 3(8 - 15)$$
$$= 9(-2) - 19(-5) + 3(-7)$$
$$= -18 + 95 - 21 = 56$$

$$Dy = \begin{vmatrix} 4 & 9 & 3 \\ 3 & 19 & 4 \\ 9 & 3 & 2 \end{vmatrix} = 4\begin{vmatrix} 19 & 4 \\ 3 & 2 \end{vmatrix} - 3\begin{vmatrix} 9 & 3 \\ 3 & 2 \end{vmatrix} + 9\begin{vmatrix} 9 & 3 \\ 19 & 4 \end{vmatrix}$$

$$= 4(38 - 12) - 3(18 - 9) + 9(36 - 57)$$
$$= 4(26) - 3(9) + 9(-21)$$
$$= 104 - 27 - 189 = -112$$

$$Dz = \begin{vmatrix} 4 & 2 & 9 \\ 3 & 5 & 19 \\ 9 & 3 & 3 \end{vmatrix} = 4\begin{vmatrix} 5 & 19 \\ 3 & 3 \end{vmatrix} - 3\begin{vmatrix} 2 & 9 \\ 3 & 3 \end{vmatrix} + 9\begin{vmatrix} 2 & 9 \\ 5 & 19 \end{vmatrix}$$

$$= 4(15 - 57) - 3(6 - 27) + 9(38 - 45)$$
$$= 4(-42) - 3(-21) + 9(-7)$$
$$= -168 + 63 - 63 = -168$$

$$x = \frac{Dx}{D} = \frac{56}{-56} = -1$$
$$y = \frac{Dy}{D} = \frac{-112}{-56} = 2$$
$$z = \frac{Dz}{D} = \frac{-168}{-56} = 3$$
$$\boxed{\{(-1, 2, 3)\}}$$

30.
$$2x + z = -4$$
$$-2y + z = -4$$
$$2x - 4y + z = -20$$

$$D = \begin{vmatrix} 2 & 0 & 1 \\ 0 & -2 & 1 \\ 2 & -4 & 1 \end{vmatrix} = 2\begin{vmatrix} -2 & 1 \\ -4 & 1 \end{vmatrix} - 0\begin{vmatrix} 0 & 1 \\ -4 & 1 \end{vmatrix} + 2\begin{vmatrix} 0 & 1 \\ -2 & 1 \end{vmatrix}$$

$$= 2(-2 + 4) - 0 + 2(0 + 2) = 4 + 4 = 8$$

$$Dx = \begin{vmatrix} -4 & 0 & 1 \\ -4 & -2 & 1 \\ -20 & -4 & 1 \end{vmatrix} = -0\begin{vmatrix} -4 & 1 \\ -20 & 1 \end{vmatrix} + (-2)\begin{vmatrix} -4 & 1 \\ -20 & 1 \end{vmatrix} - (-4)\begin{vmatrix} -4 & 1 \\ -4 & 1 \end{vmatrix}$$

$$= -0 - 2(-4 + 20) + 4(-4 + 4) = -32$$

$$D_y = \begin{vmatrix} 2 & -4 & 1 \\ 0 & -4 & 1 \\ 2 & -20 & 1 \end{vmatrix} = 2\begin{vmatrix} -4 & 1 \\ -20 & 1 \end{vmatrix} - 0\begin{vmatrix} -4 & 1 \\ -20 & 1 \end{vmatrix} + 2\begin{vmatrix} -4 & 1 \\ -4 & 1 \end{vmatrix}$$

$$= 2(-4 + 20) - 0 + 2(-4 + 4) = 32$$

$$D_z = \begin{vmatrix} 2 & 0 & -4 \\ 0 & -2 & -4 \\ 2 & -4 & -20 \end{vmatrix} = 2\begin{vmatrix} -2 & -4 \\ -4 & -20 \end{vmatrix} - 0\begin{vmatrix} 0 & -4 \\ -4 & -20 \end{vmatrix} + 2\begin{vmatrix} 0 & -4 \\ -2 & -4 \end{vmatrix}$$

$$= 2(40 - 16) - 0 + 2(0 - 8) = 48 - 16 = 32$$

$$x = \frac{D_x}{D} = \frac{-32}{8} = -4$$

$$y = \frac{D_y}{D} = \frac{32}{8} = 4$$

$$z = \frac{D_z}{D} = \frac{32}{8} = 4$$

$$\boxed{\{(-4, 4, 4)\}}$$

31. $3x - 2y - 16$
$12x - 8y = -5$

$$D = \begin{vmatrix} 3 & -2 \\ 12 & -8 \end{vmatrix} = -24 + 24 = 0$$

$$D_x = \begin{vmatrix} 16 & -2 \\ -5 & -8 \end{vmatrix} = -128 - 10 = -138$$

$$D_y = \begin{vmatrix} 3 & 16 \\ 12 & -5 \end{vmatrix} = -15 - 192 = -207$$

$D = 0, D_x = -138 \neq 0, D_y = -207 \neq 0$

Thus, the system is $\boxed{\text{inconsistent}}$.

There is no solution. $\boxed{\varnothing}$

32. $x - y - 2z - 1$
$2x - 2y - 4z = 2$
$-x + y + 2z = -1$

$$D = \begin{vmatrix} 1 & -1 & -2 \\ 2 & -2 & -4 \\ -1 & 1 & 2 \end{vmatrix} = 1\begin{vmatrix} -2 & -4 \\ 1 & 2 \end{vmatrix} - 2\begin{vmatrix} -1 & -2 \\ 1 & 2 \end{vmatrix} + (-1)\begin{vmatrix} -1 & -2 \\ -2 & -4 \end{vmatrix}$$

$$= (-4 + 4) - 2(-2 + 2) - 1(4 - 4) = 0$$

$$D_x = \begin{vmatrix} 1 & -1 & -2 \\ 2 & -2 & -4 \\ -1 & 1 & 2 \end{vmatrix} = D - 0$$

$$D_y = \begin{vmatrix} 1 & 1 & -2 \\ 2 & 2 & -4 \\ -1 & -1 & 2 \end{vmatrix} = 0 \quad \text{(columns 1 and 2 are identical)}$$

$$D_z = \begin{vmatrix} 1 & -1 & 1 \\ 2 & -2 & 2 \\ -1 & 1 & -1 \end{vmatrix} = 0 \quad \text{(columns 1 and 3 are identical)}$$

$D = 0,\ D_x = 0,\ D_y = 0,\ D_z = 0$

Thus, the system is $\boxed{\text{dependent}}$.

$\boxed{\{(x, y, z) \mid x - y - 2z = 1\}}$

33. $x + 4y = 7$
$3x + 5y = 0$

$$\begin{bmatrix} 1 & 4 & | & 7 \\ 3 & 5 & | & 0 \end{bmatrix} \xrightarrow{3R_1 - R_2} \begin{bmatrix} 1 & 4 & | & 7 \\ 0 & 7 & | & 21 \end{bmatrix} \xrightarrow{\frac{1}{7}R_2} \begin{bmatrix} 1 & 4 & | & 7 \\ 0 & 1 & | & 3 \end{bmatrix}$$

$$\begin{aligned} x + 4y &= 7 \\ y &= 3 \end{aligned} \qquad \begin{aligned} x + 4(3) &= 7 \\ x &= -5 \end{aligned}$$

$\boxed{\{(-5, 3)\}}$

34. $6x - 3y = 1$
$5x + 6y = 15$

$$\begin{bmatrix} 6 & -3 & | & 1 \\ 5 & 6 & | & 15 \end{bmatrix} \xrightarrow{6R_2 - 5R_1} \begin{bmatrix} 6 & -3 & | & 1 \\ 0 & 51 & | & 85 \end{bmatrix} \begin{array}{c} \xrightarrow{\frac{1}{6}R_1} \\ \xrightarrow{\frac{1}{51}R_2} \end{array} \begin{bmatrix} 1 & -1/2 & | & 1/6 \\ 0 & 1 & | & 5/3 \end{bmatrix}$$

$$\begin{aligned} x - \tfrac{1}{2}y &= \tfrac{1}{6} \\ y &= \tfrac{5}{3} \end{aligned} \qquad \begin{aligned} x - \tfrac{1}{2}\left(\tfrac{5}{3}\right) &= \tfrac{1}{6} \\ x &= 1 \end{aligned}$$

$\boxed{\left\{\left(1, \tfrac{5}{3}\right)\right\}}$

35. $x + y + 2x = 0$
    $2x - y - z = 1$
    $x + 2y + 3z = 1$

$$\begin{bmatrix} 1 & 1 & 2 & | & 0 \\ 2 & -1 & -1 & | & 1 \\ 1 & 2 & 3 & | & 1 \end{bmatrix} \quad \begin{matrix} R_2 - 2R_1 \to \\ R_3 - R_1 \to \end{matrix} \quad \begin{bmatrix} 1 & 1 & 2 & | & 0 \\ 0 & -3 & -5 & | & 1 \\ 0 & 1 & 1 & | & 1 \end{bmatrix}$$

$$R_2 + 3R_3 \to \begin{bmatrix} 1 & 1 & 2 & | & 0 \\ 0 & -3 & -5 & | & 1 \\ 0 & 0 & -2 & | & 4 \end{bmatrix} \quad \begin{matrix} -\frac{1}{30}R_2 \to \\ -\frac{1}{2}R_3 \to \end{matrix} \quad \begin{bmatrix} 1 & 1 & 2 & | & 0 \\ 0 & 1 & 5/3 & | & -1/3 \\ 0 & 0 & 1 & | & -2 \end{bmatrix}$$

$$\begin{aligned} x + y + 2z &= 0 \\ y + \frac{5}{3}z &= -\frac{1}{3} \\ z &= -1 \end{aligned}$$

$$\begin{aligned} y + \frac{5}{3}(-2) &= -\frac{1}{3} \\ y &= (3) \end{aligned} \qquad \begin{aligned} x + 3 + 2(-2) &= 0 \\ x &= 1 \end{aligned}$$

$$\boxed{\{(1, 3, -2)\}}$$

36. $2x - 3y + 2z = -7$
    $x + 4y - z = 10$
    $3x + 2y + z = 4$

$$\begin{bmatrix} 2 & -3 & 2 & | & -7 \\ 1 & 4 & -1 & | & 10 \\ 3 & 2 & 1 & | & 4 \end{bmatrix} \quad \begin{matrix} 2R_2 - R_1 \to \\ R_3 - 3R_1 \to \end{matrix} \quad \begin{bmatrix} 2 & -3 & 2 & | & -7 \\ 0 & 11 & -4 & | & 27 \\ 0 & -10 & 4 & | & -26 \end{bmatrix}$$

$$10R_2 + 11R_3 \to \begin{bmatrix} 2 & -3 & 2 & | & -7 \\ 0 & 11 & -4 & | & 27 \\ 0 & 0 & 4 & | & -16 \end{bmatrix} \quad \begin{matrix} \frac{1}{2}R_1 \to \\ \frac{1}{11}R_2 \to \\ \frac{1}{4}R_3 \to \end{matrix} \quad \begin{bmatrix} 1 & -3/2 & 1 & | & -7/2 \\ 0 & 1 & -4/11 & | & 27/11 \\ 0 & 0 & 1 & | & -4 \end{bmatrix}$$

$$\begin{aligned} x - \frac{3}{2}y + z &= -\frac{7}{2} \\ y - \frac{4}{11}z &= \frac{27}{11} \\ z &= -4 \end{aligned}$$

$$\begin{aligned} y - \frac{4(-4)}{11} &= \frac{27}{11} \\ y &= 1 \end{aligned} \qquad \begin{aligned} x - \frac{3}{2}(1) + (-4) &= -\frac{7}{2} \\ x &= 2 \end{aligned}$$

$$\boxed{\{(2, 1, -4)\}}$$

37. $3x - y + 2z = 2$
$ x + 4z = -1$
$ 3x - 2y = -1$

$$\begin{bmatrix} 3 & -1 & 2 & | & 2 \\ 1 & 0 & 4 & | & -1 \\ 3 & -2 & 0 & | & -1 \end{bmatrix} \quad \begin{matrix} 3R_2 - R_1 \to \\ R_3 - R_1 \to \end{matrix} \quad \begin{bmatrix} 3 & -1 & 2 & | & 2 \\ 0 & 1 & 10 & | & -5 \\ 0 & -1 & -2 & | & -3 \end{bmatrix}$$

$$\begin{matrix} \\ \\ R_3 + R_2 \to \end{matrix} \begin{bmatrix} 3 & -1 & 2 & | & 2 \\ 0 & 1 & 10 & | & -5 \\ 0 & 0 & 8 & | & -8 \end{bmatrix} \quad \begin{matrix} \frac{1}{3}R_1 \to \\ \\ \frac{1}{8}R_3 \to \end{matrix} \begin{bmatrix} 1 & -1/3 & 2/3 & | & 2/3 \\ 0 & 1 & 10 & | & -5 \\ 0 & 0 & 1 & | & -1 \end{bmatrix}$$

$x - \dfrac{1}{3}y + \dfrac{2}{3}z = \dfrac{2}{3}$
$\phantom{x - \dfrac{1}{3}y +} y + 10z = -5$
$\phantom{x - \dfrac{1}{3}y + 10} z = -1 \qquad\qquad y + 10(-1) = -5$

$ y = 5 \qquad x - \dfrac{1}{3}(5) + \dfrac{2}{3}(-1) = \dfrac{2}{3}$

$ x = 3$

$\boxed{\{(3, 5, -1)\}}$

38. $2x - y - 3z = 1$
$6x - 3y - 9z = 3$
$4x - 2y - 6z = 2$

$$\begin{bmatrix} 2 & -1 & -3 & | & 1 \\ 6 & -3 & -9 & | & 3 \\ 4 & -2 & -6 & | & 2 \end{bmatrix} \quad \begin{matrix} \frac{1}{2}R_1 \to \\ R_2 - 3R_1 \to \\ R_3 - 2R_1 \to \end{matrix} \begin{bmatrix} 1 & -1/2 & -3/2 & | & -1/2 \\ 0 & 0 & 0 & | & 0 \\ 0 & 0 & 0 & | & 0 \end{bmatrix}$$

$x - \dfrac{1}{2}y - \dfrac{3}{2}z = \dfrac{1}{2}$
$\phantom{x - \dfrac{1}{2}y - \dfrac{3}{2}z} 0 = 0 \quad$ (True) Infinitely many solutions

The system is $\boxed{\text{dependent}}$.
$\boxed{\{(x, y, z) \mid 2x - y - 3z = 1\}}$

39. $3x + 2y + z = 7$
$x + y - z = 2$
$6x + 4y + 2z = 10$

$$\begin{bmatrix} 3 & 2 & 1 & | & 7 \\ 1 & 1 & -1 & | & 2 \\ 6 & 4 & 2 & | & 10 \end{bmatrix} \quad \begin{matrix} 3R_2 - R_1 \to \\ R_3 - 2R_1 \to \end{matrix} \begin{bmatrix} 3 & 2 & 1 & | & 7 \\ 0 & 1 & -4 & | & -1 \\ 0 & 0 & 0 & | & -4 \end{bmatrix}$$

$3x + 2y + z = 7$
$ y - 4z = -1$
$ 0 = -4 \quad$ (False) Contradiction

The system is $\boxed{\text{inconsistent}}$.
There is no solution; $\boxed{\varnothing}$.

Cumulative Review Problems (Chapters 1-4)

Cumulative Review, pp. 318-319

1. $\{x \mid |x| > 3\}$

$\{x \mid x < -3 \text{ or } x > 3\}$

$\boxed{(-\infty, -3) \text{ or } (3, \infty)}$

2. $\dfrac{(6000)(800{,}000)}{0.000\,04}$

$= \dfrac{(6 \times 10^3)(8 \times 10^5)}{4 \times 10^{-5}}$

$= 12 \times 10^{13}$

$= \boxed{1.2 \times 10^{14}}$

3.
$$\begin{aligned}
3x - 2[5x - 7(x - 1)] &= 5[x - 4(1 - 3x)] - 4x \\
3x - 2(5x - 7x + 7) &= 5(x - 4 + 12x) - 4x \\
3x - 2(-2x + 7) &= 5(13x - 4) - 4x \\
3x + 4x - 14 &= 65x - 20 - 4x \\
7x - 14 &= 61x - 20 \\
-54x &= -6 \\
x &= \frac{1}{9}
\end{aligned}$$

$\boxed{\left\{\dfrac{1}{9}\right\}}$

4.
$$\begin{array}{lcll}
\{x \mid 2x + 5 & < & 1\} & \cup \\
2x & < & -4 \\
x & < & -2
\end{array}$$
$$\begin{array}{lcl}
\{x \mid 7 - 2x & \le & 1\} \\
-2x & \le & -6 \\
x & \ge & 3
\end{array}$$

$\boxed{\{x \mid x < -2 \text{ or } x \ge 3\}}$

$\boxed{(-\infty, -2) \cup [3, \infty)}$

5. Let
$$\begin{aligned}
x &= \text{first number} \\
x + 4 &= \text{second number} \\
2(x + 4) &= \text{third number}
\end{aligned}$$

$$\begin{aligned}
2[2(x + 4)] &= 3[x + (x + 4)] \\
4x + 16 &= 6x + 12 \\
-2x &= -4 \\
x &= 2 \quad \text{(first)} \\
x + 4 &= 2 + 4 = 6 \quad \text{(second)} \\
2(x + 4) &= 2(6) = 12 \quad \text{(third)}
\end{aligned}$$

The numbers are $\boxed{2, 6, \text{ and } 12}$.

6. Let
$$\begin{aligned}
x &= \text{amount inveseted at 12\% for 2 years} \\
10{,}000 - x &= \text{amount invested at 7\% for 3 years}
\end{aligned}$$

$$\begin{aligned}
0.12x(2) + (10000 - x)(0.07)(3) &= 2340 \\
0.24x + 2100 - 0.21x &= 2340 \\
0.03x &= 240 \\
x &= 8000 \quad \text{(at 12\%)} \\
10000 - x &= 2000 \quad \text{(at 7\%)}
\end{aligned}$$

$\boxed{\$8000 \text{ at } 12\% \text{ for 2 years; } \$2000 \text{ at } 7\% \text{ for 3 years}}$

7. $f(t) = -t^2 + 26t + 106$

$$
\begin{aligned}
f(10) - f(5) &= [-10^2 + 26(10) + 106] - [-5^2 + 26(5) + 106] \\
&= (-100 + 260 + 160) - (-25 + 130 + 106) \\
&= 266 - 211 \\
&= \boxed{55}
\end{aligned}
$$

$\boxed{\text{The number of people who caught the flu from day 5 through day 10 is 55.}}$

8. Let

x	$=$	Fred's age now	$x + 7$ $=$ Fred's age in 7 years
y	$=$	Carlos' age now	$y - 5$ $=$ Carlos' age 5 years ago

$$
\begin{aligned}
x + 7 &= 3(y - 5) & \rightarrow & & x - 3y &= -22 \\
x + y &= 38 & \rightarrow & & \underline{-x - y} &= \underline{-38} \\
& & & & -4y &= -60 \\
& & & & y &= 15 \quad \text{(Carlos)}
\end{aligned}
$$

$$
\begin{aligned}
x + y &= 38 \\
x + 15 &= 38 \\
x &= 23 \quad \text{(Fred)}
\end{aligned}
$$

$\boxed{\text{Fred: 23 years old; Carlos: 15 years old}}$

9. Let

$$
\begin{aligned}
x &= \text{first consecutive integer} \\
x + 1 &= \text{second consecutive integer} \\
x + 2 &= \text{third consecutive integer}
\end{aligned}
$$

$$
\begin{aligned}
\text{perimeter} &= 4x - 4 \\
x + (x + 1) + (x + 2) &= 4x - 4 \\
3x + 3 &= 4x - 4 \\
-x &= -7 \\
x &= 7 \\
x + 1 &= 8 \\
x + 2 &= 9
\end{aligned}
$$

The lengths of the sides of the triangles are $\boxed{7 \text{ m}, 8 \text{ m}, \text{ and } 9 \text{ m}}$.

10.

terms	Number of dots	Pattern
1	8	$\dfrac{1^2 + 7(1) + 8}{2} = 8$
2	13	$\dfrac{2^2 + 7(2) + 8}{2} = 13$
3	19	$\dfrac{3^2 + 7(3) + 8}{2} = 19$
4	26	$\dfrac{4^2 + 7(4) + 8}{2} = 26$
etc		
10	$\boxed{89 \text{ dots}}$	$\dfrac{10^2 + 7(10) + 8}{2} = 89$
n	$\boxed{\dfrac{n^2 + 7n + 8}{2} \text{ dots}}$	

11. $2x < -12$ $5x > -80$
                $x < -6$ $x > -16$

  a. $x = -6$ $\boxed{\text{I}}$mpossible

  b. $x = -5$ $\boxed{\text{I}}$mpossible

  c. $x = -10$ $\boxed{\text{P}}$ossible

  d. $x = -15.99$ $\boxed{\text{P}}$ossible

  e. $x = -16$ $\boxed{\text{I}}$mpossible

12. $f(x) = x^2 - 1$

x	-2	-1	0	1	2
$f(x)$	3	0	-1	0	3

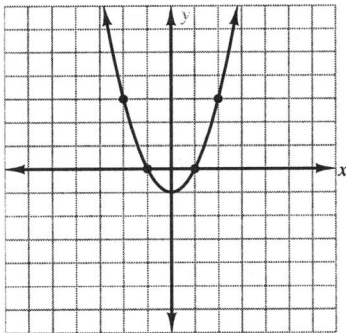

$\boxed{\text{The graph defines } y \text{ as a function of } x, \text{ since for each value of } x \text{ there is exactly one value of } y.}$

13. a. $f(-2) = 4, f(-1) = 2, f(3) = 0, f(4) = -2$

  $f(-2) - f(1) - \left| f(3) - f(4) \right|$

  $= 4 - 2 - \left| 0 - (-2) \right|$

  $= 4 - 2 - 2$

  $= \boxed{0}$

  b. $\boxed{\text{Domain} = [-2, 4]}$

  $\boxed{\text{Range} = [-2, 4]}$

14. $y \le -\dfrac{2}{3}x + 1$

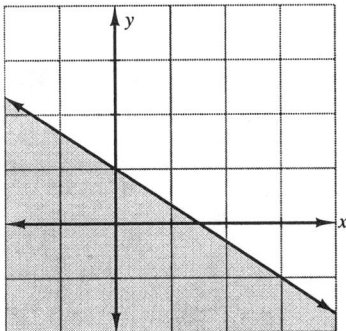

15. line tangent to $x^2 + y^2 = 25$ at $(3, -4)$ is the line perpendicular to the radius at $(3, -4)$

slope of radius: $(0, 0)$, $(3, -4)$
$$m = \frac{-4 - 0}{3 - 0} = -\frac{4}{3}$$

slope of line perpendicular:
$$m = -\frac{1}{-\frac{4}{3}} = \frac{3}{4}$$

$$
\begin{aligned}
y + 4 &= \frac{3}{4}(x - 3) \\
y + 4 &= \frac{3}{4}x - \frac{9}{4} \\
\boxed{y = \frac{3}{4}x - \frac{25}{4}}
\end{aligned}
$$

16. Let
$$
\begin{aligned}
D &= \text{distance object falls} \\
t &= \text{time in which it falls} \\
D &= kt^2
\end{aligned}
$$

Given: $D = 490$ meters, $t = 10$ seconds
$$
\begin{aligned}
490 &= k(10^2) \\
4.9 &= k \\
D &= 4.9 t^2
\end{aligned}
$$

If $t = 5$ seconds, $D = ?$
$$D = 4.9(5^2) = 122.5$$
$$\boxed{122.5 \text{ meters}}$$

17. $3x - 6y - 8 = 0$ $\rightarrow$ $y = \frac{1}{2}x - \frac{4}{3}$

$2y = x + 1$ $\rightarrow$ $y = \frac{1}{2}x + \frac{1}{2}$

$\boxed{\text{Yes}}$ the lines are parallel since the slopes are equal and the y-intercepts are different.

18. $|x| < 2$ or $|y| < 4$
           $-2 < x < 2$ or $-4 < y < 4$

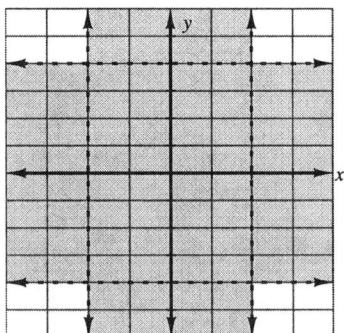

19.
$$\begin{aligned} x - y &= 7 \\ 2x + y &= 2 \end{aligned}$$

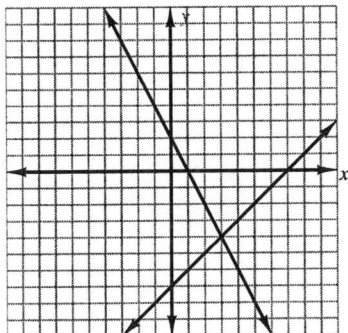

$$\{(3, -4)\}$$

20.
$$\frac{x+3}{6} + \frac{y-1}{3} = \frac{1}{3} \qquad (\times 6) \rightarrow \qquad (x+3) + 2(y-1) = 2$$

$$\frac{x-1}{2} - \frac{y-2}{5} = -\frac{1}{5} \qquad (\times 10) \rightarrow \qquad 5(x-1) - 2(y-2) = -2$$

(Simplify):
$$\begin{aligned} x + 2y &= 1 \\ 5x - 2y &= -1 \end{aligned} \qquad \rightarrow \qquad \begin{aligned} x + 2y &= 1 \\ \underline{5x - 2y} &= \underline{-1} \\ 6x &= 0 \\ x &= 0 \end{aligned}$$

$$\begin{aligned} x + 2y &= 1 \\ 0 + 2y &= 1 \\ y &= \frac{1}{2} \end{aligned}$$

$$\left\{ \left(0, \frac{1}{2} \right) \right\}$$

21.
$$\begin{aligned} 2x - 4y + 3z &= 0 \\ 5x + 3y - 2z &= 19 \\ x - 2y - 5z &= 13 \end{aligned}$$

(Equations 1 and 3):
$$\begin{aligned} 2x - 4y + 3z &= 0 \\ (\times -2) \quad \underline{-2x + 4y + 10z} &= \underline{-26} \\ 13z &= -26 \\ z &= -2 \end{aligned}$$

(Equations 2 and 3):
$$\begin{aligned} 5x + 3y - 2z &= 19 \\ (\times -5) \quad \underline{-5x + 10y + 25z} &= \underline{-65} \\ 13y + 23z &= -46 \\ 13y + 23(-2) &= -46 \\ 13y &= 0 \\ y &= 0 \end{aligned}$$

(Equation 1):
$$\begin{aligned} x - 2y - 5z &= 13 \\ x - 2(0) - 5(-2) &= 13 \\ x &= 3 \end{aligned}$$

$$\{(3, 0, -2)\}$$

22. $6x + 7y = -9$
$4x + 6y = 0$

$$D = \begin{vmatrix} 6 & 7 \\ 4 & 6 \end{vmatrix} = 36 - 28 = 8$$

$$Dx = \begin{vmatrix} -9 & 7 \\ 4 & 0 \end{vmatrix} = -54 - 0 = -54$$

$$Dy = \begin{vmatrix} 6 & -9 \\ 4 & 0 \end{vmatrix} = 0 + 36 = 36$$

$$x = \frac{Dx}{D} = \frac{-54}{8} = -\frac{27}{4}$$

$$y = \frac{Dy}{D} = \frac{36}{8} = \frac{9}{2}$$

$$\boxed{\left\{ \left(-\frac{27}{4}, \frac{9}{2} \right) \right\}}$$

23. $2x - 7y - z = 35$
$x + y = 1$
$2y + z = -5$

$$D = \begin{vmatrix} 2 & -7 & -1 \\ 1 & 1 & 0 \\ 0 & 2 & 1 \end{vmatrix} = 2 \begin{vmatrix} 1 & 0 \\ 2 & 1 \end{vmatrix} - 1 \begin{vmatrix} -7 & -1 \\ 2 & 1 \end{vmatrix} + 0 \begin{vmatrix} -7 & -1 \\ 1 & 0 \end{vmatrix}$$

$$= 2(1 - 0) - (-7 + 2) + 0$$
$$= 2 + 5 = 7$$

$$Dx = \begin{vmatrix} 35 & -7 & -1 \\ 1 & 1 & 0 \\ -5 & 2 & 1 \end{vmatrix} = -1 \begin{vmatrix} 1 & 1 \\ -5 & 2 \end{vmatrix} - 0 \begin{vmatrix} 35 & -7 \\ -5 & 2 \end{vmatrix} + 1 \begin{vmatrix} 35 & -7 \\ 1 & 1 \end{vmatrix}$$

$$= -(2 + 5) - 0 + (35 + 7)$$
$$= -7 + 42 = 35$$

$$Dy = \begin{vmatrix} 2 & 35 & -1 \\ 1 & 1 & 0 \\ 0 & -5 & 1 \end{vmatrix} = 2 \begin{vmatrix} 1 & 0 \\ -5 & 1 \end{vmatrix} - 1 \begin{vmatrix} 35 & -1 \\ -5 & 1 \end{vmatrix} + 0 \begin{vmatrix} 35 & -1 \\ 1 & 0 \end{vmatrix}$$

$$= 2(1 + 0) - (35 - 5) + 0$$
$$= 2 - 30 = -28$$

$$Dz = \begin{vmatrix} 2 & -7 & 35 \\ 1 & 1 & 1 \\ 0 & 2 & -5 \end{vmatrix} = 2 \begin{vmatrix} 1 & 1 \\ 2 & -5 \end{vmatrix} - 1 \begin{vmatrix} -7 & 35 \\ 2 & -5 \end{vmatrix} + 0 \begin{vmatrix} -7 & 35 \\ 1 & 1 \end{vmatrix}$$

$$= 2(-5 - 2) - (35 - 70) + 0$$
$$= -14 + 35 = 21$$

$$x = \frac{Dx}{D} = \frac{35}{7} = 5$$

$$y = \frac{Dy}{D} = -\frac{28}{7} = -4$$

$$z = \frac{Dz}{D} = \frac{21}{7} = 3$$

$$\boxed{\{(5, -4, 3)\}}$$

24. $2x + 5y + 3z = 0$
$3x + y + z = 1$
$5x + 2y - 4z = 6$

$$\begin{bmatrix} 2 & 5 & 3 & | & 0 \\ 3 & 1 & 1 & | & 1 \\ 5 & 2 & -4 & | & 6 \end{bmatrix} \quad \begin{array}{l} 3R_1 - 2R_2 \to \\ 5R_1 - 2R_3 \to \end{array} \quad \begin{bmatrix} 2 & 5 & 3 & | & 0 \\ 0 & 13 & 7 & | & -2 \\ 0 & 21 & 23 & | & -12 \end{bmatrix}$$

$$21R_2 - 13R_3 \to \begin{bmatrix} 2 & 5 & -3 & | & 0 \\ 0 & 13 & 7 & | & -2 \\ 0 & 0 & -152 & | & 114 \end{bmatrix} \quad \begin{array}{l} \frac{1}{2}R_1 \to \\ \frac{1}{13}R_2 \to \\ -\frac{1}{152}R_3 \to \end{array} \quad \begin{bmatrix} 1 & 5/2 & 3/2 & | & 0 \\ 0 & 1 & 7/13 & | & -2/13 \\ 0 & 0 & 1 & | & -3/4 \end{bmatrix}$$

$$\begin{aligned} x + \frac{5}{2}y + \frac{3}{2}z &= 0 \\ y + \frac{7}{13}z &= -\frac{2}{13} \\ z &= -\frac{3}{4} \end{aligned} \qquad \begin{aligned} y + \frac{7}{13}\left(-\frac{3}{4}\right) &= -\frac{2}{13} \\ y &= \frac{1}{4} \end{aligned}$$

$$\begin{aligned} x + \frac{5}{2}\left(\frac{1}{4}\right) + \frac{3}{2}\left(-\frac{3}{4}\right) &= 0 \\ x &= \frac{1}{2} \end{aligned}$$

$$\boxed{\left\{ \left(\frac{1}{2}, \frac{1}{4}, -\frac{3}{4} \right) \right\}}$$

25. Let

$$\begin{aligned} t &= \text{tens' digit} \\ u &= \text{units' digit} \\ 10t + u &= \text{original number} \\ 10u + t &= \text{number with digits reversed} \end{aligned}$$

$$\begin{aligned} t + u &= 9 \\ 10u + t &= (10t + u) + 27 \end{aligned}$$

(Simplify):

$$\begin{array}{lll} t + u = 9 & & t + u = 9 \\ -9t + 9u = 27 & \to & \underline{-t + u = 3} \\ & & \quad\quad 2u = 12 \\ & & \quad\quad u = 6 \\ & & \\ & & t + u = 9 \\ & & \quad\quad t = 3 \end{array}$$

$t = 3, u = 6$

number: $\boxed{36}$

26. Let

$$x = \text{numerator}$$
$$y = \text{denominator}$$
$$\frac{x}{y} = \text{original fraction}$$

$$\frac{x+3}{y+3} = \frac{3}{4} \qquad \rightarrow \qquad 4(x+3) = 3(y+3)$$
$$\frac{x-3}{y-3} = \frac{2}{3} \qquad \rightarrow \qquad 3(x-3) = 2(y-3)$$

(Simplfy):

$$4x - 3y = -3 \qquad (\times 2) \rightarrow \qquad 8x - 6y = -6$$
$$3x - 2y = 3 \qquad (\times -3) \rightarrow \qquad \underline{-9x + 6y = -9}$$
$$-x = -15$$
$$x = 15$$

$$3x - 2y = 3$$
$$3(15) - 2y = 3$$
$$-2y = -42$$
$$y = 21$$
$$\frac{x}{y} = \boxed{\frac{15}{21}}$$

27. Let

$$x = \text{first number}$$
$$y = \text{second number}$$

$$\frac{1}{3}(x+y) = 14 \qquad \rightarrow \qquad x + y = 42$$
$$\frac{1}{2}(x-y) = 3 \qquad \rightarrow \qquad \underline{x - y = 6}$$
$$2x = 48$$
$$x = 24$$

$$x + y = 42$$
$$24 + y = 42$$
$$y = 18$$

The numbers are $\boxed{18 \text{ and } 24}$.

28. Let

$$x = \text{unit cost of a small grapefruit}$$
$$y = \text{unit cost of a large grapefruit}$$

$$12x + 30y = 18.60 \qquad \rightarrow \qquad 12x + 30y = 18.60$$
$$8x + 6y = 5.40 \qquad (\times -5) \rightarrow \qquad \underline{-40x - 30y = -27}$$
$$-28x = -8.40$$
$$x = 0.30 \quad \text{(small)}$$

$$8x + 6y = 5.40$$
$$8(0.30) + 6y = 5.40$$
$$6y = 3.00$$
$$y = 0.50 \quad \text{(large)}$$

$\boxed{\text{cost of small grapfruit: } \$0.30;}$

$\boxed{\text{cost of large grapefruit: } \$0.50}$

29. Let

x = speed of boat in still water
y = speed of current

	Rate, r	$\times$	time, t (hours)	$=$	Distance, D (miles)
with current	$x+y$		2		14
against current	$x-y$		2		2

$$
\begin{aligned}
2(x+y) &= 14 \\
2(x-y) &= 2
\end{aligned}
\qquad
\begin{aligned}
&\rightarrow \\
&\rightarrow
\end{aligned}
\qquad
\begin{aligned}
x+y &= 7 \\
\underline{x-y} &= \underline{1} \\
2x &= 8 \\
x &= 4 \quad \text{(boat in still water)}
\end{aligned}
$$

$$
\begin{aligned}
x+y &= 7 \\
4+y &= 7 \\
y &= 3 \quad \text{(current)}
\end{aligned}
$$

Speed of boat in still water, 4 mph;

Speed of current, 3 mph

30. Let

x = length of smallest side of triangle
y = lenght of medium side of triangle
z = length of largest side of triangle

$$
\begin{aligned}
x+y+z &= 197 \\
x+y &= z+49 \\
z-x &= 22
\end{aligned}
\qquad
\begin{aligned}
&\rightarrow \\
&\rightarrow
\end{aligned}
\qquad
\begin{aligned}
x+y+z &= 197 \\
x+y-z &= 49 \\
-x+z &= 22
\end{aligned}
$$

(Equations 1 and 2):

$$
\begin{aligned}
x+y+z &= 197 \\
(\times -1) \quad \underline{-x-y+z} &= \underline{-49} \\
2z &= 148 \\
z &= 74 \quad \text{(longest)}
\end{aligned}
$$

$$
\begin{aligned}
-x+z &= 22 \\
-x+74 &= 22 \\
-x &= -52 \\
x &= 52 \quad \text{(smallest)}
\end{aligned}
$$

$$
\begin{aligned}
x+y+z &= 197 \\
52+y+74 &= 197 \\
y &= 71 \quad \text{(medium)}
\end{aligned}
$$

The lengths of the sides of the triangle are 52 m, 71 m and 74 m .

Algebra for College Students

Chapter 5 Polynomials, Polynomial Functions, and Factoring

Section 5.1 Polynomials: Sums and Differences

Problem Set 5.1, pp. 327-329

1. $y = 0.071x^2 + 0.73x$

$\boxed{\text{Binomial, degree 2}}$

3. $w^3 + 2p^2 - 6w + 6p - 6wp$

$\boxed{\text{Five terms, degree 3}}$

5. $15x + 6x^2 - x^3$

$\boxed{\text{Trinomial, degree 3}}$

7. $(7x^2 - 3x) + (4x^2 - 7x - 8)$
$= 7x^2 + 4x^2 - 3x - 7x - 8$
$= \boxed{11x^2 - 10x - 8}$

9. $(-7r^3 + 3r - 2 + 8n^2) + (-3r^2 + 7r + 4)$
$= -7r^3 + 8r^2 - 3r^2 + 3r + 7r - 2 + 4$
$= \boxed{-7r^3 + 5r^2 + 10r + 2}$

11. $(-3x^2y + 2xy) + (-5x^3y + 4x^2y - 7) + (7xy + 7)$
$= -5x^3y - 3x^2y + 4x^2y + 2xy + 7xy - 7 + 7$
$= \boxed{-5x^3y + x^2y + 9xy}$

13. $(0.15x^4 + 0.01x^3 - 0.8x^2) + (22.1x^3 + 0.01x^2 + x) + (1.33x^4 - 0.27x^3 + 0.99)$
$= 0.15x^4 + 1.33x^4 + 0.01x^3 + 22.1x^3 - 0.27x^3 - 0.8x^2 + 0.01x^2 + x + 0.99$
$= \boxed{1.48x^4 + 21.84x^3 - 0.79x^2 + x + 0.99}$

15. $\left(\dfrac{1}{3}x^9y^2 - \dfrac{1}{5}x^5y + \dfrac{1}{2}x^2y^3 + 7\right) + \left(-\dfrac{1}{5}x^9y^2 + \dfrac{1}{4}x^4y + \dfrac{3}{5}x^5y - \dfrac{3}{4}x^2y^3 - \dfrac{1}{2}\right)$
$= \left(\dfrac{1}{3} - \dfrac{1}{5}\right)x^9y^2 + \left(-\dfrac{1}{5} + \dfrac{3}{5}\right)x^5y + \left(\dfrac{1}{4}\right)x^4y + \left(\dfrac{1}{2} - \dfrac{3}{4}\right)x^2y^3 + \left(7 - \dfrac{1}{2}\right)$
$= \left(\dfrac{5}{15} - \dfrac{3}{15}\right)x^9y^2 + \left(\dfrac{2}{5}\right)x^5y + \left(\dfrac{1}{4}\right)x^4y + \left(\dfrac{2}{4} - \dfrac{3}{4}\right)x^2y^3 + \left(\dfrac{14}{2} - \dfrac{1}{2}\right)$
$= \boxed{\dfrac{2}{15}x^9y^2 + \dfrac{2}{5}x^5y + \dfrac{1}{4}x^4y - \dfrac{1}{4}x^2y^3 + \dfrac{13}{2}}$

17. $(7x^4y - 8x^3y^2 + 5x^2y^3 + 8x) + (11x^3y^2 - x^2y^3 + 8) + (-7x^4y - 3x^2y^3 - 8x)$
$= 7x^4y - 7x^4y - 8x^3y^2 + 11x^3y^2 + 5x^2y^3 - x^2y^3 - 3x^2y^3 + 8x - 8x + 8$
$= 0.x^4y + 3x^3y^2 + x^2y^3 + 0x + 8$
$= \boxed{3x^3y^2 + x^2y^3 + 8}$

19. $(0.07x^4 + 0.11x^3 - 0.2x^2) + (-0.01x^3 + x^2 + 5x) + (1.8x^4 + 0.001x^2 + 0.17)$
      $+ (0.37x^3 + 0.85) + (-0.11x^4 + 10x^4 + 10x^2 - 0.03)$
$= (0.07 + 1.8 - 0.11)x^4 + (0.11 - 0.01 + 0.37)x^3 + (-0.2 + 1 + 0.001 + 10)x^2$
      $+ (5)x + (0.17 + 0.85 - 0.03)$
$= \boxed{1.76x^4 + 0.47x^3 + 10.801x^2 + 5x + 0.99}$

21. $(-3r^2 - 5r^4 + 2r) + (19r^2 + 3 - 2r) + (-8r + 5r^4 + 2r^2) + (-6 + 4r)$
$= (-5 + 5)r^4 + (-3 + 19 + 2)r^2 + (2 - 2 - 8 + 4)r + (3 - 6)$
$= \boxed{18r^2 - 4r - 3}$

23. $(y^{2n} + 7y^n - 3) + (-5y^{2n} + 3y^n + 8)$
$= (1 - 5)y^{2n} + (7 + 3)y^n + (-3 + 8)$
$= \boxed{-4y^{2n} + 10y^n + 5}$

25. $(17x^3 - 5x^2 + 4x - 3) - (5x^3 - 9x^2 - 8x + 11)$
$= (17 - 5)x^3 + (-5 + 9)x^2 + (4 - 8)x + (-3 - 11)$
$= \boxed{12x^3 + 4x^2 + 12x - 14}$

27. $(13r^5 + 9r^4 - 5r^2 + 3r + 6) - (-9r^5 - 7r^3 + 8r^2 + 11)$
$= (13 + 9)r^5 + 9r^4 + 7r^3 + (-5 - 8)r^2 + 3r + (6 - 11)$
$= \boxed{22r^5 + 9r^4 + 7r^3 - 13r^2 + 3r - 5}$

29. $(-6x^3y^2 - 8x^2y + 11xy - 3) - (7x^3y^2 - 5x^2y + 9xy - 3)$
$= (-6 - 7)x^3y^2 + (-8 + 5)x^2y + (11 - 9)xy + (-3 + 3)$
$= \boxed{-13x^3y^2 - 3x^2y + 2xy}$

31. $(-5x^3y + 13xy + 6) - (8x^2y - 9xy - 11)$
$= -5x^3y - 8x^2y + (13 + 9)xy + (6 + 11)$
$= \boxed{-5x^3y - 8x^2y + 22xy + 17}$

33. $(-x^4y^3 - x^3y^2 + xy - 1) - (x^4y^3 - x^3y^2 - xy + 1)$
$= (-1 - 1)x^4y^3 + (-1 + 1)x^3y^2 + (1 + 1)xy + (-1 - 1)$
$= \boxed{-2x^4y^3 + 2xy - 2}$

35. $(3x^{2n} + 7x^n - 4) - (-2x^{2n} + 5x^n - 4)$
$= (3 + 2)x^{2n} + (7 - 5)x^n + (-4 + 4)$
$= \boxed{5x^{2n} + 2x^n}$

37. $(y^{2n} - y^n - 3) - (2y^{2n} - 3y^n + 5)$
$= (1 - 2)y^{2n} + (-1 + 3)y^n + (-3 - 5)$
$= \boxed{-y^{2n} + 2y^n - 8}$

39. $(5x^2 - 7x - 8) + (2x^2 - 3x + 7) - (x^2 - 4x - 3)$
$= (5 + 2 - 1)x^2 + (-7 - 3 + 4)x + (-8 + 7 + 3)$
$= \boxed{6x^2 - 6x + 2}$

41. $(6y^4 - 5y^3 + 2y) - (4y^3 + 3y^2 - 1) + (y^4 - 2y^2 + 7y - 3)$
$= (6 + 1)y^4 + (-5 - 4)y^3 + (-3 - 2)y^2 + (2 + 7)y + (1 - 3)$
$= \boxed{7y^4 - 9y^3 - 5y^2 + 9y - 2}$

43. $(2x^3 - 2x^2y + 3xy^2 + 4y^3) - (4x^2y - 3xy^2 - 5y^3) + (3x^3 - x^2y + 5xy^2 - y^3)$
$= (2 + 3)x^3 + (-2 - 4 - 1)x^2y + (3 + 3 + 5)xy^2 + (4 + 5 - 1)y^3$
$= \boxed{5x^3 - 7x^2y + 11xy^2 + 8y^3}$

45. $(y^{3n} - 7y^{2n} + 3) - (-3y^{3n} - 2y^{2n} - 1) + (6y^{3n} - y^{2n} + 1)$
$= (1 + 3 + 6)y^{3n} + (-7 + 2 - 1)y^{2n} + (3 + 1 + 1)$
$= \boxed{10y^{3n} - 6y^{2n} + 5}$

47. $(12a^2b^2 + 15a^2 + 3) + (19a^2b^2 - 7a^2 + 1) - (-11a^2b^2 - 5a^2 - 2)$
$= (12 + 19 + 11)a^2b^2 + (15 - 7 + 5)a^2 + (3 + 1 + 2)$
$= \boxed{42a^2b^2 + 13a^2 + 6}$

49. $(9 - 2a^3b^2 - a^2b^3 - ab^4) + (4 - 3a^2b^2 - 7a^2b^3 + 5ab^4) - (13 - 8a^2b^3 - 5a^3b^2 - 4ab^4)$
$= (9 + 4 - 13) + (-2 - 3 + 5)a^3b^2 + (-1 - 7 + 8)a^2b^3 + (-1 + 5 + 4)ab^4$
$= \boxed{8ab^4}$

51. $(8r^2s + 2r^3s^2 - 3rs) + (2rs - 11r^2s - 3r^3s^2) - (rs - 4r^2s + r^3s^2)$
$= (2 - 3 - 1)r^3s^2 + (8 - 11 + 4)r^2s + (-3 + 2 - 1)rs$
$= \boxed{-2r^3s^2 + r^2s - 2rs}$

53.   Profit $= (25x^2 - 30x) - (2x^2 - 10x - 5)$
$= (25 - 2)x^2 + (-30 + 10)x + 5$
$= \boxed{23x^2 - 20x + 5}$

55.   Difference $= h_{\text{moon}} - h_{\text{earth}}$
$= (-2.7t^2 + 48t + 6) - (-16t^2 + 48t + 6)$
$= (-2.7 + 16)t^2 + (48 - 48)t + (6 - 6)$
$= \boxed{13.3t^2}$

57. $[5x^2 - (8xy + 11y^2)] - [6x^2 - (-8xy - 9y^2)]$
$= [5x^2 - 8xy - 11y^2] - [6x^2 + 8xy + 9y^2]$
$= (5 - 6)x^2 + (8 - 8)xy + (-11 - 9)y^2$
$= \boxed{-x^2 - 16xy - 20y^2}$

59. $[(4xy - 8y^2) - (-3y^2 - 5xy)] - [(2y^3 - xy) - (-y^2 - xy)]$
$= [(-8 + 3)y^2 + (4 + 5)xy] - [(-2 + 1)y^2 + (-3 + 1)xy]$
$= (-5y^2 + 9xy) - (-y^2 - 2xy)$
$= (-5 + 1)y^2 + (9 + 2)xy$
$= \boxed{-4y^2 + 11xy}$

61. $(\text{polynomial}) - (4x^2 + 2x - 3) = (5x^2 - 5x + 8)$
$(\text{polynomial}) = (5x^2 - 5x + 8) + (4x^2 + 2x - 3)$
$= (5 + 4)x^2 + (-5 + 2)x + (8 - 3)$
$= \boxed{9x^2 - 3x + 5}$

63. $(y^3 + 8y - 3) - (y^3 - 6y + 5) = -1$
$(1 - 1)y^3 + (8 + 6)y + (-3 - 5) = -1$
$14y - 8 = -1$
$14y = 7$
$y = \dfrac{1}{2}$

$\boxed{\left\{\dfrac{1}{2}\right\}}$

65. B is true since the coefficients of the cubed terms could cancel leaving the resultant polynomial of a lower degree.
$\boxed{B}$ is true.

67. $(34.6x^n + 21.2x^{n-2} + 15.0473) + (19.68x^n - 174.63x^{n-2} - 19.48)$
$= (34.6 + 19.68)x^n + (21.2 - 174.63)x^{n-2} + (15.0473 - 19.48)$
$= \boxed{54.28x^n - 153.43x^{n-2} - 4.4327}$

69. $(43.19x^n - 34.6x^{n-2} - 7.36) - (-6.25x^n - 43.197x^{n-2} - 54.026)$
$= (43.19 + 6.25)x^n + (-34.6 + 43.197)x^{n-1} + (-7.36 + 54.026)$
$= \boxed{49.44x^n + 8.597x^{n-1} + 46.666}$

71. 1, 8, 27, 64, ___, ___, ___
$1^3, 2^3, 3^3, 4^3, 5^3 = 125, 6^3 = 216, 7^3 = 343$
$\boxed{125, 216, 343; x^3}$

73. $-1, 6, 25, 62, \underline{\quad}, \underline{\quad}, \underline{\quad}$
$1 - 2, 8 - 2, 27 - 2, 64 - 2, \underline{\quad}, \underline{\quad}, \underline{\quad}$
$1^3 - 2, 2^3 - 2, 3^3 - 2, 4^3 - 2, 5^3 - 2 = 123, 6^3 - 2 = 214, 7^3 - 2 = 341$
$\boxed{123, 214, 314, x^3 - 2}$

75. $2, 10, 30, 68, \underline{\quad}, \underline{\quad}, \underline{\quad}$
$1 + 1, 8 + 2, 27 + 3, 64 + 4, \underline{\quad}, \underline{\quad}, \underline{\quad}$
$1^3 + 1, 2^3 + 2, 3^3 + 3, 4^3 + 4, 5^3 + 5 = 130; 6^3 + 6 = 222, 7^3 + 7 = 350$
$\boxed{130, 222, 350; x^3 + x}$

Review Problems

78.
$$\frac{1}{2} - \frac{7y}{4} \geq \frac{3y}{4} - \frac{5}{6}$$
$$\frac{3y}{4} + \frac{7y}{4} \leq \frac{1}{2} + \frac{5}{6}$$
$$\frac{10y}{4} \leq \frac{8}{6}$$
$$\frac{30y}{12} \leq \frac{16}{12}$$
$$30y \leq 16$$
$$y \leq \frac{16}{30}$$
$$y \leq \frac{8}{15}$$
$$\boxed{\left\{ y \mid y \leq \frac{8}{15} \right\} \text{ or } \left(-\infty, \frac{8}{15} \right]}$$

79. Let
$$x = \text{amount invested at } 7\%$$
$$11{,}200 - x = \text{amount invested at } 9\%$$

$$.07x + .09(11200 - x) = 924$$
$$.07x + 1008 - .09x = 924$$
$$-.02x = -84$$
$$x = \frac{84}{.02}$$
$$x = 4200 \quad (7\%)$$
$$11{,}200 - x = 7000 \quad (9\%)$$
$$\boxed{\$4200 \text{ at } 7\%; \$7000 \text{ at } 9\%}$$

80. Let
$$x = \text{number of liters of } 45\% \text{ salt solution}$$
$$40 - x = \text{number of liters at } 60\% \text{ salt solution}$$

$$0.45x + 0.60(40 - x = 0.48(40)$$
$$0.45x + 24 - 0.60x = 19.2$$
$$-0.15x = -4.8$$
$$x = \frac{4.8}{.15}$$
$$x = 32 \quad (45\%)$$
$$\boxed{32 \text{ liters of } 45\% \text{ salt solutions}}$$

Section 5.2 Multiplication of Polynomials

Problem Set 5.2, pp. 337-340

1. $y^7 \cdot y^5 = y^{7+5} = \boxed{y^{12}}$

3. $(3x^8)(5x^6) = (3)(5)x^{8+6} = \boxed{15x^{14}}$

5. $(3x^2y^4)(5xy^7) = 15x^{2+1}y^{4+7} = \boxed{15x^3y^{11}}$

7. $(-3xy^2z^5)(2xy^7z^4) = -6x^{1+1}y^{2+7}z^{5+4} = \boxed{-6x^2y^9z^9}$

9. $4x^2(3x - 2) = 4x^2(3x) + 4x^2(-2) = \boxed{12x^3 - 8x^2}$

11. $2xy(4x^2y + 7x - 2y - 3)$
$= 2xy(4x^2y) + 2xy(7x) + 2xy(-2y) + 2xy(-3)$
$= \boxed{8x^3y^2 + 14x^2y - 4xy^2 - 6xy}$

13. $\dfrac{1}{3}x^3y^7\left(\dfrac{1}{2}xy^6 + \dfrac{2}{5}x^4y^2 + 6\right)$
$= \dfrac{1}{3}x^3y^7\left(\dfrac{1}{2}xy^6\right) + \dfrac{1}{3}x^3y^7\left(\dfrac{2}{5}x^4y^2\right) + \dfrac{1}{3}x^3y^7(6)$
$= \boxed{\dfrac{1}{6}x^4y^{13} + \dfrac{2}{15}x^7y^9 + 2x^3y^7}$

15. $16x^4y^5z^3\left(-\dfrac{1}{8}xz + \dfrac{1}{16}x^4yz^2 - \dfrac{1}{32}x^6y^2z\right)$
$= 16x^4y^5z^3\left(-\dfrac{1}{8}xz\right) + 16x^4y^5z^3\left(\dfrac{1}{16}x^4yz^2\right) + 16x^4y^5z^3\left(-\dfrac{1}{32}x^6y^2z\right)$
$= \boxed{-2x^5y^5z^4 + x^8y^6z^5 - \dfrac{1}{2}x^{10}y^7z^4}$

17. $(6uv^3w - 8uv + w^4)(-5u^5v^3w)$
$= (6uv^3w)(-5u^5v^3w) - 8uv(-5u^5v^3w) + w^4(-5u^5v^3w)$
$= \boxed{-30u^6v^6w^2 + 40u^6v^4w - 5u^5v^3w^5}$

19. $(a^2c^3 + b^2c^3 - a^2b^2)(-3a^2c^3)$
$= (a^2c^3)(-3a^2c^3) + (b^2c^3)(-3a^2c^3) - a^2b^2(-3a^2c^3)$
$= \boxed{-3a^4c^6 - 3a^2b^2c^6 + 3a^4b^2c^3}$

21. $y^{n-3}(y^{2n+7} - 3y^4 - 1)$
$= y^{n-3}(y^{2n+7}) + (y^{n-3})(-y^4) + y^{n-3}(-1)$
$= \boxed{y^{3n+4} - 3y^{n+1} - y^{n-3}}$

23. $2x^{n-4}y^{5n+3}(-6x^{3n+4}y^{-5n-3} - x^4y^5 + 2)$
$= -12x^{n-4+3n+4}y^{-5n+3-5n-3} - 2x^{n-4+4}y^{5n+3+5} + 4x^{n-4}y^{5n+3}$
$= \boxed{-12x^{4n} - 2x^ny^{5n+8} + 4x^{n-4}y^{5n+3}}$

25. $(x + y)(x^2 - xy + y^2)$
$= x(x^2 - xy + y^2) + y(x^2 - xy + y^2)$
$= x^3 - x^2y + xy^2 + x^2y - xy^2 + y^3$
$= \boxed{x^3 + y^3}$

27. $(x + 5)(x^2 - 5x + 25)$
$= x(x^2 - 5x + 25) + 5(x^2 - 5x + 25)$
$= x^3 - 5x^2 + 25x + 5x^2 - 25x + 125$
$= \boxed{x^3 + 125}$

29. $(a^2b^4 + 3)(a^4b^8 - 3a^2b^4 + 9)$
$= a^2b^4(a^4b^8 - 3a^2b^4 + 9) + 3(a^4b^8 - 3a^2b^4 + 9)$
$= a^6b^{12} - 3a^4b^8 + 9a^2b^4 + 3a^4b^8 - 9a^2b^4 + 27$
$= \boxed{a^6b^{12} + 27}$

31. $(9r^4 - r^3 - r^2 + r - 1)(3r^2 + 5r - 1)$
$= (9r^4 - r^3 - r^2 + r - 1)(3r^2) + (9r^4 - r^3 - r^2 + r - 1)(5r) - (9r^4 - r^3 - r^2 + r - 1)$
$= 27r^6 - 3r^5 - 3r^4 + 3r^3 - 3r^2 + 45r^5 - 5r^4 - 5r^3 + 5r^2 - 5r - 9r^4 + r^3 + r^2 - r + 1$
$= \boxed{27r^6 + 42r^5 - 17r^4 - r^3 + 3r^2 - 6r + 1}$

33. $(2y + 3x^2y - 4x)(5x^2y - 3y + 4x + x^3y - xy^2)$
$= 2y(5x^2y - 3y + 4x + x^3y - xy^2) + 3x^2y(5x^2y - 3y + 4x + x^3y - xy^2) - 4x(5x^2y - 3y + 4x + x^3y - xy^2)$
$= 10x^2y^2 - 6y^2 + 8xy + 2x^3y^2 - 2xy^3 + 15x^4y^2 - 9x^2y^2 + 12x^3y + 3x^5y^2 - 3x^3y^3 + 20x^3y + 12xy$
$\quad - 16x^2 - 4x^4y + 4x^2y^2$
$= \boxed{3x^5y^2 + 15x^4y^2 - 4x^4y - 3x^3y^3 + 2x^3y^3 + 2x^3y^2 - 8x^3y + 5x^2y^2 - 16x^2 + 20xy - 6y^2 - 2xy^3}$

35. $(9x^2 - 4)(3x^2 + 5)$ $=$ $(9x^2)(3x^2) + (9x^2)(5) - 4(3x^2) - 4(5)$
$\qquad\qquad\qquad\qquad\qquad$ **F** $\qquad$ **O** $\qquad$ **I** $\qquad\qquad$ **L**
$\qquad\qquad\qquad\qquad = \ 27x^4 + 45x^2 - 12x^2 - 20$
$\qquad\qquad\qquad\qquad = \ \boxed{27x^4 + 33x^2 - 20}$

37. $(5z^2 + 1)(3z - 4)$ $= \ (5z^2)(3z) + (5z^2)(-4) + 1(3z) - 1(4)$
$\qquad\qquad\qquad\qquad = \ \boxed{15z^3 - 20z^2 + 3z - 4}$

39. $(3x^3 + 2y^2)(5x + 4y) = \boxed{15x^4 + 12x^3y + 10xy^2 + 8y^5}$

41. $(2x + 7)(2x - 7) = 4x^2 - 14x + 14x - 49 = \boxed{4x^2 - 49}$

43. $(a + c)(b + d) = \boxed{ab + ad + bc + cd}$

45. $(x - y)^2 = (x - y)(x - y) = x^2 - xy - xy + y^2 = \boxed{x^2 - 2xy + y^2}$

47. $(3a - 2b)^2 = (3a)^2 - 2(3a)(2b) + (2b)^2 = \boxed{9a^2 - 12ab + 4b^2}$

49. $(5x^3 - 1)^2 = (5x^3)^2 - 2(5x^3)(1) + 1^2 = \boxed{25x^6 - 10x^3 + 1}$

51. $(4r^2s - s)^2 = (4r^2s)^2 - 2(4r^2s)(s) + s^2 = \boxed{16r^4s^2 - 8r^2s^2 + s^2}$

53. $(5xy^2 - 2x)^2 = (5xy^2)^2 - 2(5xy^2)(2x) + (2x)^2 = \boxed{25x^2y^4 - 20x^2y^2 + 4x^2}$

55. $(9y^n - 2)(y^n + 4) = (9y^n)(y^n) + (9y^n)(4) - 2y^n - 2(4) = \boxed{9y^{2n} + 34y^n - 8}$

57. $(4x^{2a} - y^{5a})(3x^{5a} - y^{2a}) = \boxed{12x^{7a} - 4x^{2a}y^{2a} - 3x^{5a}y^{5a} + y^{7a}}$

59. $(4x^2y + 5x)(4x^2y - 5x) = (4x^2y)^2 - (5x)^2 = \boxed{16x^4y^2 - 25x^2}$

61. $(-3a^4b^2 + 5c)(3a^4b^2 + 5c)$
$= (5c - 3a^4b^2)(5c + 3a^4b^2)$
$= 5(c)^2 - (3a^4b^2)^2$
$= \boxed{25c^2 - 9a^8b^4}$

63. $(4a^2b^3c - 7ab)(4a^2b^3c + 7ab)$
$= (4a^2b^3c)^2 - (7ab)^2$
$= \boxed{16a^4b^6c^2 - 49a^2b^2}$

65. $(y^{3n} + 1)(y^{3n} - 1)$
$= (y^{3n})^2 - 1$
$= \boxed{y^{6n} - 1}$

67. $(7x^{n-1} + y^{3n})(7x^{n-1} - y^{3n})$
$= (7x^{n-1})^2 - (y^{3n})^2$
$= \boxed{49x^{2n-2} - y^{6n}}$

69. $(x + 6)^2 = x^2 + 2x(6) + 36 = \boxed{x^2 + 12x + 36}$

71. $(x^n + 4)^2 = \boxed{x^{2n} + 8x^n + 16}$

73. $(x^{2n} - 6y^n)^2 = \boxed{x^{4n} - 12x^{2n}y^n + 36y^{2n}}$

75. $(3x^2y + 2xy)^2 = 9x^4y^2 + 2(3x^2y)(2xy) + 4x^2y^2$
$$= \boxed{9x^4y^2 + 12x^3y^2 + 4x^2y^2}$$

77. $(5m^2n^2 - 3mn^2)^2 = 25m^4n^6 - 2(5m^2n^3)(3mn^2) + 9m^2n^4$
$$= \boxed{25m^4n^6 - 30m^3n^5 + 9m^2n^4}$$

79. $(3x + 7 + 5y)(3x + 7 - 5y)$
$= (3x + 7)^2 - (5y)^2$
$= \boxed{9x^2 + 42x + 49 - 25y^2}$

81. $[5y - (2x + 3)][5y + (2x + 3)]$
$= (5y)^2 - (2x + 3)^2$
$= 25y^2 - (4x^2 + 12x + 9)$
$= \boxed{25y^2 - 4x^2 - 12x - 9}$

83. $(2x + y + 1)^2$
$= [(2x + y) + 1]^2$
$= (2x + y)^2 + 2(2x + y) + 1$
$= \boxed{4x^2 + 4xy + y^2 + 4x + 2y + 1}$

85. $[(3x - 1) + y]^2$
$= (3x - 1)^2 + 2(3x - 1)y + y^2$
$= \boxed{9x^2 - 6x + 1 + 6xy - 2y + y^2}$

87. $[(3x - 1) + y][(3x - 1) - y]$
$= (3x - 1)^2 - y^2$
$= \boxed{9x^2 - 6x + 1 - y^2}$

89. $(x + y - 3)(x - y + 3)$
$= [x + (y - 3)][x - (y - 3)]$
$= x^2 - (y - 3)^2$
$= x^2 - (y^2 - 6y + 9)$
$= \boxed{x^2 - y^2 + 6y - 9}$

91. $(2a + b)^3$
$= (2a + b)^2(2a + b)$
$= (4a^2 + 4ab + b^2)(2a + b)$
$= (8a^3 + 8a^2b + 2ab^2) + (4a^2b + 4ab^2 + b^3)$
$= \boxed{8a^3 + 12a^2b + 6ab^2 + b^3}$

93. $(3x - y)^3$
$= (3x - y)^2(3x - y)$
$= (9x^2 - 6xy + y^2)(3x - y)$
$= (27x^3 - 18x^2y + 3xy^2) + (-9x^2y + 6xy^2 - y^3)$
$= \boxed{27x^3 - 27x^2y + 9xy^2 - y^3}$

95. $(x^4 - 2x^5)^3$
$= (x^4 - 2x^5)^2(x^4 - 2x^5)$
$= (x^8 - 4x^9 + 4x^{10})(x^4 - 2x^5)$
$= (x^{12} - 4x^{13} + 4x^{14}) + (-2x^{13} + 8x^{14} - 8x^{15})$
$= \boxed{-8x^{15} + 12x^4 - 6x^{13} + x^{12}}$

97. area of shaded region
= area of large rectangle − area of small rectangle
$= 2x(5x + 1) - x(4x - 1)$
$= 10x^2 + 2x - 4x^2 + x$
$= \boxed{6x^2 + 3x}$

99. area of shaded region
= area of triangle − area of square
$= \dfrac{1}{2}8x(6x + 4) - 2x(2x)$
$= 4x(6x + 4) - 4x^2$
$= 24x^2 + 16x - 4x^2$
$= \boxed{20x^2 + 16x}$

101.

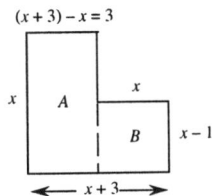

$(x + 3) - x = 3$

Area of figure
$= $ area $A + $ area B
$= 3x + x(x - 1)$
$= 3x + x^2 - x$
$= \boxed{x^2 + 2x}$

103. area of figure
$= $ area of rectangle $ + $ area of small triangle $ + $ area of large triangle

$= (x + 2)[(x - 5) + (x + 3)] + \dfrac{1}{2}\, x(x - 5) + \dfrac{1}{2}\, x(x + 3)$

$= (x + 2)(2x - 2) + \dfrac{1}{2}\, x^2 - \dfrac{5}{2}\, x + \dfrac{1}{2}\, x^2 + \dfrac{3}{2}\, x$

$= 2x^2 - 2x + 4x - 4 + x^2 - x$

$= \boxed{3x^2 + x - 4}$

105. volume of figure
$= $ volume of large rectangular solid $ + $ volume of small rectangular solid
$= x(x + 1)(x + 3) + x[(x + 3) - x)][2x - 1 - (x + 1)]$
$= x(x^2 + 4x + 3) + x(3)(x - 2)$
$= x^3 + 4x^2 + 3x + 3x^2 - 6x$
$= \boxed{x^3 + 7x^2 - 3x}$

107. $\boxed{B}$ is true; $(3x + 7)(3x - 2)$
$\qquad\qquad\qquad = 9x^2 - 6x + 21x - 14$
$\qquad\qquad\qquad = 9x^2 + 15x - 14$
A is *not* true since $(x - 5)^2 = x^2 - 10x + 25$
C is *not* true since $(x^m + 2)(x^m + 4) = x^{2m} + 6x^m + 8$
D is *not* true since $7x^2 - 4x^3$ is the product of two monomials

109. $2y^{2n}(3y^{3n} + 4y^n - 1) - 5y^{2n}(y^{2n} - 3)$
$= 6y^{5n} + 8y^{3n} - 2y^{2n} - 5y^{4n} + 15y^{2n}$
$= \boxed{6y^{5n} - 5y^{4n} + 8y^{3n} + 13y^{2n}}$

111. $\dfrac{\text{(polynomial)}}{x + 1} = x^2 - x + 1$

$\qquad$ (polynomial) $= (x + 1)(x^2 - x + 1)$
$\qquad\qquad\qquad = x(x^2 - x + 1) + 1(x^2 - x + 1)$
$\qquad\qquad\qquad = x^3 - x^2 + x + x^2 - x + 1$
$\qquad\qquad\qquad = \boxed{x^3 + 1}$

113. $\begin{vmatrix} x+5 & x-4 \\ x-3 & x-2 \end{vmatrix}$

$= (x+5)(x-2) - (x-3)(x-4)$

$= (x^2 - 2x + 5x - 10) - (x^2 - 4x - 3x + 12)$

$= x^2 + 3x - 10 - x^2 + 7x - 12$

$= \boxed{10x - 22}$

115. $\begin{vmatrix} 3 & 5 & -2 \\ 1 & 2 & -1 \\ x & -3 & 4x \end{vmatrix} = -12$

$3\begin{vmatrix} 2 & -1 \\ -3 & 4x \end{vmatrix} - 1\begin{vmatrix} 5 & -2 \\ -3 & 4x \end{vmatrix} + x\begin{vmatrix} 5 & -2 \\ 2 & -1 \end{vmatrix} = -12$

$$
\begin{aligned}
3(8x - 3) - (20x - 6) + x(-5 + 4) &= -12 \\
24x - 9 - 20x + 6 - x &= -12 \\
3x - 3 &= -12 \\
3x &= -9 \\
\boxed{x = -3}
\end{aligned}
$$

117. $\begin{vmatrix} x-1 & x & -1 \\ x & x+1 & 1 \\ x+2 & x-1 & 2 \end{vmatrix} = 15$

$-1\begin{vmatrix} x & x+1 \\ x+2 & x-1 \end{vmatrix} - 1\begin{vmatrix} x-1 & x \\ x+2 & x-1 \end{vmatrix} + 2\begin{vmatrix} x-1 & x \\ x & x+1 \end{vmatrix} = 15$

$$
\begin{aligned}
-[(x^2 - x) - (x^2 + 3x + 2)] - [(x^2 - 2x + 1) - (x^2 + 2x)] + 2[(x^2 - 1) - x^2] &= 15 \\
-(x^2 - x - x^2 - 3x - 2) - (x^2 - 2x + 1 - x^2 - 2x) + 2(x^2 - 1 - x^2) &= 15 \\
-(-4x - 2) - (-4x + 1) + 2(-1) &= 15 \\
4x + 2 + 4x - 1 - 2 &= 15 \\
8x &= 16 \\
\boxed{x = 2}
\end{aligned}
$$

119. $N = \left(\dfrac{x-4}{4}\right)^2 y$

$= \left(\dfrac{x^2 - 8x + 16}{16}\right) y$

$= \boxed{\dfrac{x^2 y - 8xy + 16y}{16} \text{ or } \dfrac{x^2 y}{16} - \dfrac{xy}{2} + y}$

121. $C = 300 + 10x + 13y + 0.01(x + y)^2$

Revenue $= 25x + 20y$

Profit

$=$ Revenue $-$ Cost

$= (25x + 20y) - (300 + 10x + 13y + 0.01x^2 + 0.02xy + 0.01y^2)$

$= 25x + 20y - 300 - 10x - 13y - 0.01x^2 - 0.02xy - 0.01y^2$

$= \boxed{-0.01x^2 + 15x + 7y - 0.02xy - 0.01y^2 - 300}$

123. $(7.4x - 13.02)(9.682x + 14.3)$
$= (7.4x)(9.682x) + [(7.4)(14.3) - (13.02)(9.682)]x - (13.02)(14.3)$
$= \boxed{71.6468x^2 - 20.23964x - 186.186}$

125. $(17.06x - 19.3)^2$
$= (17.06)^2 x^2 - 2(17.06)(19.3)x + (19.3)^2$
$= \boxed{291.043x^2 - 658.516x + 372.49}$

127. $(13.06x - 17.83y)(13.06x + 17.83y)$
$= (13.06)^2 x^2 - (17.83)^2 y^2$
$= \boxed{170.563x^2 - 317.9089y^2}$

129. **a.** $(y - 1)(y + 1) = \boxed{y^2 - 1}$

 b. $(y - 1)(y^2 + y + 1) = y^3 + y^2 + y - y^2 - y - 1 = \boxed{y^3 - 1}$

 c. $(y - 1)(y^3 + y^2 + y + 1) = y^4 + y^3 + y^2 + y - y^3 - y^2 - y - 1 = \boxed{y^4 - 1}$

 d. $(y - 1)(y^4 + y^3 + y^2 + y + 1) = \boxed{y^5 - 1}$

131. $(5 - 2)^2 = 5(5) - 5(4) + 2(2)$
$(7 - 3)^2 = 7(7) - 7(6) + 3(3)$
$(11 - 5)^2 = 11(11) - 11(10) + 5(5) = \boxed{121 - 110 + 25}$

Review Problems

135.

$$\begin{aligned}
|4 - 3x| &\geq 10 \\
4 - 3x &\geq 10 \qquad \text{or} \qquad & 4 - 3x &\leq -10 \\
-3x &\geq 6 & -3x &\leq -14 \\
x &\leq -2 & x &\geq \frac{14}{3}
\end{aligned}$$

$$\boxed{\left\{\, x \mid x \leq -2 \text{ or } x \geq \frac{14}{3} \right\}}$$

$$\boxed{(-\infty, -2] \cup \left[\frac{14}{3},\ \infty\right)}$$

136. $[(2y^2 - 3y - 1) + (3y^2 + y - 1)] - (y^2 + 3y - 1)$
$= 5y^2 - 2y - 2 - y^2 - 3y + 1$
$= \boxed{4y^2 - 5y - 1}$

137. Let
$$\begin{aligned}
m\angle ABD &= 12m\angle DBC - 2 = 12x - 2 \\
m\angle ABD + m\angle DBC &= 180 \\
12x - 2 + x &= 180 \\
13x &= 182 \\
x &= 14 \quad (m\angle DBC) \\
12x - 2 &= 166 \quad (m\angle ABD)
\end{aligned}$$

$$\boxed{\angle ABD, 160°; \angle DBC, 14°}$$

Section 5.3 Problem Solving Involving Polynomial Multiplication

Problem Set 5.3, pp. 345-346

1. Let

$$x = \text{first whole number}$$
$$x + 2 = \text{next consecutive whole integer}$$

$$
\begin{aligned}
(x + 2)^2 - x^2 &= 44 \\
x^2 + 4x + 4 - x^2 &= 44 \\
4x + 4 &= 44 \\
4x &= 40 \\
x &= 10 \quad \text{(first)} \\
x + 2 &= 12 \quad \text{(next)}
\end{aligned}
$$

The numbers are $\boxed{10 \text{ and } 12}$.

3. Let

$$x = \text{first odd integer}$$
$$x + 2 = \text{next consecutive odd integer}$$

$$
\begin{aligned}
x(x + 2) &= (x + 2)^2 - 30 \\
x^2 + 2x &= x^2 + 4x + 4 - 30 \\
2x &= 4x - 26 \\
-2x &= -26 \\
x &= 13 \quad \text{(first)} \\
x + 2 &= 15 \quad \text{(next)}
\end{aligned}
$$

The integers are $\boxed{13 \text{ and } 15}$.

5. Let x, $x + 1$ and $x + 2$ equal three consecutive integers.

$$
\begin{aligned}
x(x + 1)(x + 2) &= (x + 1)^3 - 18 \\
x(x^2 + 3x + 2) &= (x + 1)(x + 1)^2 - 18 \\
x^3 + 3x^2 + 2x &= (x + 1)(x^2 + 2x + 1) - 18 \\
x^3 + 3x^2 + 2x &= x^3 + 3x^2 + 3x + 1 - 18 \\
2x &= 3x - 17 \\
-x &= -17 \\
x &= 17 \\
x + 1 &= 18 \\
x + 2 &= 19
\end{aligned}
$$

The integers are $\boxed{17, 18, \text{ and } 19}$.

7. Let

$$x = \text{side of second square}$$
$$x + 4 = \text{sides of first square}$$
$$
\begin{aligned}
x^2 &= (x + 4)^2 - 56 \\
x^2 &= x^2 + 8x + 16 - 56 \\
-8x &= -40 \\
x &= 5 \quad \text{(second square)}
\end{aligned}
$$

The length is $\boxed{5 \text{ meters}}$.

9. Let
$$x = \text{width of rectangular box}$$
$$x + 2 = \text{length of rectangular box}$$
$$x - 5 = \text{height of rectangular box}$$

area of top + area of bottom = sum of areas of two sides + 112

$$
\begin{aligned}
x(x+2) + x(x+2) &= x(x-5) + x(x-5) + 112 \\
x^2 + 2x + x^2 + 2x &= x^2 - 5x + x^2 - 5x + 112 \\
2x^2 + 4x &= 2x^2 - 10x + 112 \\
14x &= 112 \\
x &= 8 \quad \text{(width)} \\
x + 2 &= 8 + 2 = 10 \quad \text{(length)} \\
x - 5 &= 8 - 5 = 3 \quad \text{(height)}
\end{aligned}
$$

width: 8 cm; length: 10 cm; height: 3 cm

11.

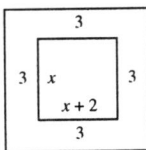

Let
$$x = \text{width of picture}$$
$$x + 2 = \text{length of picture}$$

area of (picture + frame) = area of picture + 108

$$
\begin{aligned}
(x+6)(x+8) &= x(x+2) + 108 \\
x^2 + 14x + 48 &= x^2 + 2x + 108 \\
12x &= 60 \\
x &= 5 \quad \text{(width)} \\
x + 2 &= 5 + 2 = 7 \quad \text{(length)}
\end{aligned}
$$

width: 5 cm; length: 7 cm

13. Let
$$x = \text{edge of middle cube}$$
$$x + 2 = \text{edge of largest cube}$$
$$x - 3 = \text{edge of smallest cube}$$

surface area of largest cube = surface area of smallest cube + 210

$$
\begin{aligned}
6(x+2)^2 &= 6(x-3)^2 + 210 \\
6(x^2 + 4x + 4) &= 6(x^2 - 6x + 9) + 210 \\
6x^2 + 24x + 24 &= 6x^2 - 36x + 54 + 210 \\
24x + 24 &= -36x + 264 \\
60x &= 240 \\
x &= 4 \quad \text{(middle cube)} \\
x + 2 &= 4 + 2 = 6 \quad \text{(largest cube)} \\
x - 3 &= 4 - 3 = 1 \quad \text{(smallest cube)}
\end{aligned}
$$

edge of smallest cube: 1 cm;
edge of middle cube: 4 cm;
edge of largest cube: 6 cm

15. Let

$$3x = \text{first multiple of 3}$$
$$3(x + 1) = \text{next consecutive multiple of 3}$$

$$
\begin{aligned}
[3(x + 1)]^2 - (3x)^2 &= 81 \\
9(x^2 + 2x + 1) - 9x^2 &= 81 \\
18x + 9 &= 81 \\
18x &= 72 \\
x &= 4 \\
x + 1 &= 5
\end{aligned}
$$

$$3x = 3(4) = 12$$
$$3(x + 1) = 3(5) = 15$$

The numbers are $\boxed{12 \text{ and } 15}$.

17. Let

$$x = \text{age of glass plate} \qquad x - 3 = \text{age of glass plate 3 years ago}$$
$$x + 4 = \text{age of china plate} \qquad x + 4 - 3 = x + 1 = \text{age of china plate 3 years ago}$$

product of ages three years ago = product of present ages − 75

$$
\begin{aligned}
(x - 3)(x + 1) &= x(x + 4) - 75 \\
x^2 - 2x - 3 &= x^2 + 4x - 75 \\
-6x &= -72 \\
x &= 12 \quad \text{(glass)} \\
x + 4 &= 12 + 4 = 16 \quad \text{(china)}
\end{aligned}
$$

$\boxed{\text{Glass plate: 12 years old; China plate: 16 years old}}$

19.

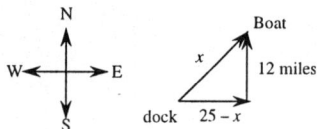

$$
\begin{aligned}
x^2 &= 12^2 + (25 - x)^2 \\
x^2 &= 144 + 625 - 50x + x^2 \\
50x &= 769 \\
x &= 15.38
\end{aligned}
$$

The boat is $\boxed{15.38 \text{ miles}}$ from the dock.

Review Problems

21.

$$\{y \mid y - 1 \le 7y - 1\} \qquad \cap \qquad \{y \mid 4y - 7 < 3 - y\}$$

$$
\begin{aligned}
-6y &\le 0 \\
y &\ge 0
\end{aligned}
\qquad\qquad
\begin{aligned}
5y &< 10 \\
y &< 2
\end{aligned}
$$

$\boxed{\{y \mid 0 \le y < 2\}}$

$\boxed{[0, 2)}$

22.

$$\{x \mid 2 - x > 4\} \qquad \cup \qquad \{x \mid 6 - 3x \le 2\}$$

$$
\begin{aligned}
-x &> 2 \\
x &< -2
\end{aligned}
\qquad\qquad
\begin{aligned}
-3x &\le -4 \\
x &\ge \frac{4}{3}
\end{aligned}
$$

$\boxed{\left\{ x \mid x < -2 \text{ or } x \ge \frac{4}{3} \right\}}$ $\boxed{(-\infty, -2) \cup \left[\frac{4}{3}, \infty\right)}$

23.
$$\begin{aligned} x - y + z &= 4 \\ 3x - y + z &= 6 \\ 2x + 2y - 3z &= -6 \end{aligned}$$

(Equations 1 and 2):

$(\times -1)$

$$\begin{aligned} -x + y - z &= -4 \\ \underline{3x - y + z} &= \underline{6} \\ 2x &= 2 \\ x &= 1 \end{aligned}$$

(Equations 1 and 3):

$(\times 3)$

$$\begin{aligned} 3x - 3y + 3z &= 12 \\ \underline{2x + 2y - 3z} &= \underline{-6} \\ 5x - y &= 6 \\ 5(1) - y &= 6 \\ -y &= 1 \\ y &= -1 \end{aligned}$$

(Equation 1):

$$\begin{aligned} x - y + z &= 4 \\ \underline{1 - (-1) + z} &= \underline{4} \\ z &= 2 \end{aligned}$$

$$\boxed{\{(1, -1, 2)\}}$$

Section 5.4 Division of Polynomials

Problem Set 5.4, pp. 355-357

1. $y^7 \div y^5 = \dfrac{y^7}{y^5} = y^{7-5} = \boxed{y^2}$

3. $\dfrac{15x^8}{5x^6} = 3x^{8-6} = \boxed{3x^2}$

5. $\dfrac{25x^3y^7}{-5xy^5} = -5x^{3-1}y^{7-5} = \boxed{-5x^2y^2}$

7. $(-54x^7y^4z^5) \div (3x^3yz^2) = \dfrac{-54x^7y^4z^5}{3x^3yz^2} = \boxed{-18x^4y^3z^3}$

9. $\dfrac{24x^7 - 15x^4 + 18x^3}{3x}$

$= \dfrac{24x^7}{3x} - \dfrac{15x^4}{3x} + \dfrac{18x^3}{3x}$

$= \boxed{8x^6 - 5x^3 + 6x^2}$

11. $\dfrac{125x^5 - 10x^2 + 15x - 20}{-5x^2}$

$= \dfrac{125x^5}{-5x^2} - \dfrac{10x^2}{-5x^2} + \dfrac{15x}{-5x^2} - \dfrac{20}{-5x^2}$

$= \boxed{-25x^3 + 2 - \dfrac{3}{x} - \dfrac{4}{x^2}}$

13. $\dfrac{64x^7 - 20x^3 + 24x - 80}{-4x^3}$

$= \dfrac{64x^7}{-4x^3} - \dfrac{-20x^3}{-4x^3} + \dfrac{24x}{-4x^3} - \dfrac{80}{-4x^3}$

$= \boxed{-16x^4 + 5 - \dfrac{6}{x^2} + \dfrac{20}{x^3}}$

15. $\dfrac{6x^7 - 3x^4 + x^2 - 5x + 2}{3x^3}$

$= \dfrac{6x^7}{3x^3} - \dfrac{3x^4}{3x^3} + \dfrac{x^2}{3x^3} - \dfrac{5x}{3x^3} + \dfrac{2}{3x^3}$

$= \boxed{2x^4 - x + \dfrac{1}{3x} - \dfrac{5}{3x^2} + \dfrac{2}{3x^3}}$

17. $\dfrac{16x^3y^2 - 28x^2y^3 - 20xy^5}{4x^2y}$

$= \dfrac{16x^3y^2}{4x^2y} - \dfrac{28x^2y^3}{4x^2y} - \dfrac{20xy^5}{4x^2y}$

$= \boxed{4xy - 7y^2 - \dfrac{5y^4}{x}}$

19. $\dfrac{36x^4y^3 - 18x^5z - 12xz^2}{6x^2yz}$

$= \dfrac{36x^4y^3}{6x^2yz} - \dfrac{18x^5z}{6x^2yz} - \dfrac{12xz^2}{6x^2yz}$

$= \boxed{\dfrac{6x^2y^2}{z} - \dfrac{3x^3}{y} - \dfrac{2z}{xy}}$

21. $\dfrac{16t^4 + 8t^3}{2t^2} = \dfrac{16t^4}{2t^2} + \dfrac{8t^3}{2t^2} = \boxed{8t^2 + 4t}$

$t = 4$: $8(4^2) + 4(4) = 128 + 16 = 144$

$\boxed{144 \text{ gallons}}$

23. $(a^2 + 8a + 15) \div (a + 5) = \boxed{a + 3}$

$$
\begin{array}{r}
a + 3 \\
a + 5 \overline{\smash{)}\, a^2 + 8a + 15} \\
\underline{a^2 + 5a } \\
3a + 15 \\
\underline{3a + 15} \\
0
\end{array}
$$

25. $(b^2 - 4b - 12) \div (b - 6) = \boxed{b + 2}$

$$
\begin{array}{r}
b + 2 \\
b - 6 \overline{\smash{)}\, b^2 - 4b - 12} \\
\underline{b^2 - 6b } \\
2b - 12 \\
\underline{2b - 12} \\
0
\end{array}
$$

27. $(24 - 10c - c^2) \div (2 - c) = \boxed{c + 12}$

$$
\begin{array}{r}
c + 12 \\
-c + 2 \overline{\smash{)}\, -c^2 - 10c + 24} \\
\underline{-c^2 + 2c } \\
-12c + 24 \\
\underline{-12c + 24} \\
0
\end{array}
$$

29. $(b^3 - 2b^2 - 5b + 6) \div (b - 3) = \boxed{b^2 + b - 2}$

$$
\begin{array}{r}
b^2 + b - 2 \\
b - 3 \overline{\smash{)}\, b^3 - 2b^2 - 5b + 6} \\
\underline{b^3 - 3b^2 } \\
b^2 - 5b \\
\underline{b^2 - 3b } \\
-2b + 6 \\
\underline{-2b + 6} \\
0
\end{array}
$$

31. $(6b^3 + 17b^2 + 27b + 20) \div (3b + 4) = \boxed{2b^2 + 3b + 5}$

$$
\begin{array}{r}
2b^2 + 3b + 5 \\
3b + 4 \overline{\smash{)}\, 6b^3 + 17b^2 + 27b + 20} \\
\underline{6b^3 + 8b^2 } \\
9b^2 + 27b \\
\underline{9b^2 + 12b} \\
15b + 20 \\
\underline{15b + 20} \\
0
\end{array}
$$

33. $(a^3 + 3a^2b + 2ab^2) \div (a + b) = \boxed{a^2 + 2ab}$

$$
\begin{array}{r}
a + b \overline{\smash{)}\, a^3 + 3a^2b + 2ab^2} \\
\underline{a^3 + a^2b } \\
2a^2b + 2ab^2 \\
\underline{2a^2b + 2ab^2} \\
0
\end{array}
$$

35. $(12x^2 + x - 4) \div (3x - 2) = \boxed{4x + 3 + \dfrac{2}{3x - 2}}$

$$
\begin{array}{r}
4x + 3 \\
3x - 2 \overline{\smash{)}\, 12x^2 + x - 4} \\
\underline{12x^2 - 8x } \\
9x - 4 \\
\underline{9x - 6} \\
2
\end{array}
$$

37. $\dfrac{2y^3 + 7y^2 + 9y - 20}{y + 3} = \boxed{2y^2 + y + 6 - \dfrac{38}{y+3}}$

$$
\begin{array}{r}
2y^2 + y + 6 \\
y + 3 \overline{\smash{\big)}\ 2y^3 + 7y^2 + 9y - 20} \\
\underline{2y^3 + 6y^2} \\
y^2 + 9y \\
\underline{y^2 + 3y} \\
6y - 20 \\
\underline{6y + 18} \\
-38
\end{array}
$$

39. $\dfrac{4x^4 - 4x^2 + 6x}{x - 4} = \boxed{4x^3 + 16x^2 + 60x + 246 + \dfrac{984}{x-4}}$

$$
\begin{array}{r}
4x^3 + 16x^2 + 60x + 246 \\
x - 4 \overline{\smash{\big)}\ 4x^4 + 0x^3 - 4x^2 + 6x + 0} \\
\underline{4x^4 - 16x^3} \\
16x^3 - 4x^2 \\
\underline{16x^3 - 64x^2} \\
60x^2 + 6x \\
\underline{60x^2 - 240x} \\
246x + 0 \\
\underline{246x - 984} \\
984
\end{array}
$$

41. $\dfrac{x^3 - 1}{x - 1} = \boxed{x^2 + x + 1}$

$$
\begin{array}{r}
x^2 + x + 1 \\
x - 1 \overline{\smash{\big)}\ x^3 + 0x^2 + 0x - 1} \\
\underline{x^3 - x^2} \\
x^2 + 0x \\
\underline{x^2 - x} \\
x - 1 \\
\underline{x - 1} \\
0
\end{array}
$$

43. $\dfrac{6a^3 + 13a^2 - 11a - 15}{3a^2 - a - 3} = \boxed{2a + 5}$

$$
\begin{array}{r}
2a + 5 \\
3a^2 - a - 3 \overline{\smash{\big)}\ 6a^3 + 13a^2 - 11a - 15} \\
\underline{6a^3 - 2a^2 - 6a} \\
15a^2 - 5a - 15 \\
\underline{15a^2 - 5a - 15} \\
0
\end{array}
$$

45. $\dfrac{y^4 + y^3 - 3y^2 - y + 2}{y^2 + 3y + 2} = \boxed{y^2 - 2y + 1}$

$$
\begin{array}{r}
y^2 - 2y + 1 \\
y^2 + 3y + 2 \overline{\smash{\big)}\ y^4 + y^3 - 3y^2 - y + 2} \\
\underline{y^4 + 3y^3 + 2y^2} \\
-2y^3 - 5y^2 - y \\
\underline{-2y^3 - 6y^2 - 4y} \\
y^2 + 3y + 2 \\
\underline{y^2 + 3y + 2} \\
0
\end{array}
$$

47. $\dfrac{18y^4 + 9y^3 + 3y^2}{3y^2 + 1} = \boxed{6y^2 + 3y - 1 + \dfrac{-3y + 1}{3y^2 + 1}}$

$$
\begin{array}{r}
6y^2 + 3y - 1 \\
3y^2 + 1 \overline{\smash{\big)}\ 18y^4 + 9y^3 + 3y^2 + 0y + 0} \\
\underline{18y^4 \quad\quad + 6y^2} \\
9y^3 - 3y^2 + 0y \\
\underline{9y \quad\quad + 3y} \\
-3y^2 - 3y + 0 \\
\underline{-3y^2 \quad\quad - 1} \\
-3y + 1
\end{array}
$$

49. $\dfrac{x^2 - y^2}{x - y} = \boxed{x + y}$

$$
\begin{array}{r}
x + y \\
x - y \overline{\smash{\big)}\ x^2 + 0xy - y^2} \\
\underline{x^2 - xy} \\
xy - y^2 \\
\underline{xy - y^2} \\
0
\end{array}
$$

51. $\dfrac{a^4 - b^4}{a - b} = \boxed{a^3 + a^2 b + ab^2 + b^3}$

$$
\begin{array}{r}
a^3 + a^2 b + ab^2 + b^3 \\
a - b \overline{\smash{\big)}\ a^4 + 0a^3 b + 0a^2 b^2 + 0ab^3 - b^4} \\
\underline{a^4 - a^3 b} \\
a^3 b + 0a^2 b^2 \\
\underline{a^2 b - a^2 b^2} \\
a^2 b^2 + 0ab^3 \\
\underline{a^2 b^2 - ab^3} \\
ab^3 - b^4 \\
\underline{ab^3 - b^4} \\
0
\end{array}
$$

53. $\dfrac{x^5 - 1}{x^2 - x + 2} = \boxed{x^3 + x^2 - x - 3 + \dfrac{-x + 5}{x^2 - x + 2}}$

$$
\begin{array}{r}
x^3 + x^2 - x - 3 \\
x^2 - x + 2 \overline{\smash{\big)}\ x^5 + 0x^5 + 0x^3 + 0x^2 + 0x - 1} \\
\underline{x^5 - x^4 + 2x^3} \\
x^4 - 2x^3 + 0x^2 \\
\underline{x^4 - x^3 + 2x^2} \\
-x^3 - 2x^2 + 0x \\
\underline{-x^3 + x^2 - 2x} \\
-3x^2 + 2x - 1 \\
\underline{-3x^2 + 3x - 6} \\
-x + 5
\end{array}
$$

55. $L = 2y - 1$
$W = y + 3$
$V = 11y^2 + 6y^3 + 6 - 19y$

$V = LWH$
$11y^2 + 6y^3 + 6 - 19y = (2y - 1)(y + 3)H$
$6y^3 + 11y^2 - 19y + 6 = (2y^2 + 5y - 3)H$
$\dfrac{6y^3 + 11y^2 - 19y + 6}{2y^2 + 5y - 3} = H$

$$
\begin{array}{r}
3y - 2 \\
2y^2 + 5y - 3 \overline{\smash{\big)}\ 6y^3 + 11y^2 - 19y + 6} \\
\underline{6y^3 + 15y^2 - 9y} \\
-4y^2 - 10y + 6 \\
\underline{-4y^2 - 10y + 6} \\
0
\end{array}
$$

$H = \boxed{3y - 2}$

57. $V = \frac{1}{3} Bh$

$V = 8y^3 - 12y^2 - 18y - 5$

$B = 12y^2 + 12y + 3$

$8y^3 - 12y^2 - 18y - 5 = \frac{1}{3}(12y^2 + 12y + 3)h$

$8y^3 - 12y^2 - 18y - 5 = (4y^2 + 4y + 1)h$

$\dfrac{8y^3 - 12y^2 - 18y - 5}{4y^2 + 4y + 1} = h$

$$
\begin{array}{r}
2y - 5 \\
4y^2 + 4y + 1 \overline{\smash{\big)}\,8y^3 - 12y^2 - 18y - 5} \\
\underline{8y^3 + 8y^2 + 2y} \\
-20y^2 - 20y - 5 \\
\underline{-20y^2 - 20y - 5} \\
0
\end{array}
$$

$h = \boxed{2y - 5}$

59. $(2x^2 + x - 10) \div (x - 2) = \boxed{2x + 5}$

$$
\begin{array}{r|rrr}
2 & 2 & 1 & -10 \\
 &   & 4 & +10 \\
\hline
 & 2 & 5 & 0
\end{array}
$$

61. $(3x^2 + 7x - 20) \div (x + 5) = \boxed{3x - 8 + \dfrac{20}{x + 5}}$

$$
\begin{array}{r|rrr}
-5 & 3 & 7 & -20 \\
 &   & -15 & 40 \\
\hline
 & 3 & -8 & 20
\end{array}
$$

63. $(4x^3 - 3x^2 + 3x - 1) \div (x - 1) = \boxed{4x^2 + x + 4 + \dfrac{3}{x - 1}}$

$$
\begin{array}{r|rrrr}
1 & 4 & -3 & 3 & -1 \\
 &   & 4 & 1 & 4 \\
\hline
 & 4 & 1 & 1 & 3
\end{array}
$$

65. $(6y^2 - 2y^3 + 4y^2 - 3y + 1) \div (y - 2) = \boxed{6y^4 + 12y^3 + 22y^2 + 48y + 93 + \dfrac{187}{y - 2}}$

$$
\begin{array}{r|rrrrrr}
2 & 6 & 0 & -2 & 4 & -3 & 1 \\
 &   & 12 & 24 & 44 & 96 & 186 \\
\hline
 & 6 & 12 & 22 & 48 & 93 & 187
\end{array}
$$

67. $(x^2 - 5x - 5x^3 + x^4) \div (5 + x) = (x^4 - 5x^3 + x^2 - 5x) \div (x + 5) = \boxed{x^3 - 10x^2 + 51x - 260 + \dfrac{1300}{x + 5}}$

$$
\begin{array}{r|rrrrr}
-5 & 1 & -5 & 1 & -5 & 0 \\
 &   & -5 & 50 & -255 & 1300 \\
\hline
 & 1 & -10 & 51 & -260 & 1300
\end{array}
$$

69. $\dfrac{z^5 + z^3 - 2}{z - 1} = \boxed{z^4 + z^3 + 2z^2 + 2z + 2}$

$$
\begin{array}{r|rrrrrr}
1 & 1 & 0 & 1 & 0 & 0 & -2 \\
 &   & 1 & 1 & 2 & 2 & 2 \\
\hline
 & 1 & 1 & 2 & 2 & 2 & 0
\end{array}
$$

71. $\dfrac{y^4 - 256}{y - 4} = \boxed{y^3 + 4y^2 + 16y + 64}$

$$
\begin{array}{r|rrrrr}
4 & 1 & 0 & 0 & 0 & -256 \\
 &   & 4 & 16 & 64 & 256 \\
\hline
 & 1 & 4 & 16 & 64 & 0
\end{array}
$$

73. $\dfrac{2y^5 - 3y^4 + y^3 - y^2 + 2y - 1}{y + 2} = \boxed{2y^4 - 7y^3 + 15y^2 - 31y + 64 - \dfrac{129}{y + 2}}$

$$
\begin{array}{r|rrrrrr}
-2 & 2 & -3 & 1 & -1 & 2 & -1 \\
   &   & -4 & 14 & -30 & 62 & -128 \\
\hline
   & 2 & -7 & 15 & -31 & 64 & -129
\end{array}
$$

75. $\boxed{C}$ is true

77. $(1.4w^3 - 2.6w^2 + 56.3) \div (w + 3.5) = \boxed{1.4w^2 - 7.5w + 26.25 - \dfrac{35.575}{w + 3.5}}$

$$
\begin{array}{r|rrrr}
-3.5 & 1.4 & -2.6 & 0 & 56.3 \\
     &   & -4.9 & 26.25 & -91.875 \\
\hline
     & 1.4 & -7.5 & 26.25 & -35.575
\end{array}
$$

79. $\dfrac{2x^2 - 7x + 9}{(\text{polynomial})} = 2x - 3 + \dfrac{3}{(\text{polynomial})}$

$2x^2 - 7x + 9 = (2x - 3)(\text{polynomial}) + 3$

$2x^2 - 7x + 6 = (2x - 3)(\text{polynomial})$

$\dfrac{2x^2 - 7x + 6}{2x + 6} = (\text{polynomial})$

$$
\begin{array}{r|rrr}
\frac{3}{2} & 2 & -7 & 6 \\
            &   & 3 & -6 \\
\hline
            & 2 & -4 & 0 \\
(\div 2)   & 1 & -2 & 0
\end{array}
\qquad \text{or} \qquad
\begin{array}{l}
x - 2 \\
2x^2 - 7x + 6 \overline{} \\
\underline{2x^2 - 3x} \\
-4x + 6 \\
\underline{-4x + 6} \\
 0
\end{array}
$$

$\text{polynomial} = \boxed{x - 2}$

81. $\dfrac{27y^{3n} - 1}{3y^n - 1} = \boxed{9y^{2n} + 3y^n + 1}$

$$
\begin{array}{l}
9y^{2n} + 3y^n + 1 \\
3y^n - 1 \overline{)27y^{3n} + 0y^{2n} + 0y^n - 1} \\
\underline{27y^{3n} - 9y^{2n}} \\
\phantom{3y^n - 1)27y^{3n}}9y^{2n} + 0y^n \\
\phantom{3y^n - 1)27y^{3n}}\underline{9y^{2n} - 3y^n} \\
\phantom{3y^n - 1)27y^{3n} + 9y^{2n}}3y^n - 1 \\
\phantom{3y^n - 1)27y^{3n} + 9y^{2n}}\underline{3y^n - 1} \\
\phantom{3y^n - 1)27y^{3n} + 9y^{2n} + 3y^n}0
\end{array}
$$

83. $(8x^2 + 27x + k) \div (6x + 5) = 3x + 2 + \dfrac{k - 10}{6x + 5}$

$$
\begin{array}{l}
3x + 2 \\
6x + 5 \overline{)18x^2 + 27x + k} \qquad\qquad k - 10 = 0 \quad \text{for a zero remainder} \\
\underline{18x^2 + 15x} \qquad\qquad\qquad \boxed{k = 10} \\
12x + k \\
\underline{12x + 10} \\
k - 10
\end{array}
$$

85. $(-2x^3 + 3x^3 + x + k) \div (x + 1) = -2x^2 + 5x - 4 + \dfrac{k+4}{x+1}$

$$
\begin{array}{r|rrrr}
-1 & -2 & 3 & 1 & k \\
 & & 2 & -5 & 4 \\
\hline
 & -2 & 5 & -4 & k+4
\end{array}
$$

The remainder is 3 when $k + 4 = 3$

$\boxed{k = -1}$

Review Problems

90.　　$\begin{aligned} 3x + 5y &= 0 \\ x + 4y &= 7 \end{aligned}$

$D = \begin{vmatrix} 3 & 5 \\ 1 & 4 \end{vmatrix} = 12 - 5 = 7$

$Dx = \begin{vmatrix} 0 & 5 \\ 7 & 4 \end{vmatrix} = 0 - 35 = -35$

$Dy = \begin{vmatrix} 3 & 0 \\ 1 & 7 \end{vmatrix} = 21 - 0 = 21$

$x = \dfrac{Dx}{D} = -\dfrac{35}{7} = -5$

$y = \dfrac{Dy}{D} = \dfrac{21}{7} = 3$

$\boxed{\{(-5,\ 3)\}}$

91. Let

$\begin{aligned} F &= \text{wind force} \\ A &= \text{area of the surface} \\ V &= \text{wind velocity} \end{aligned}$

$F = \dfrac{kA}{V^2}$

Given: $A = 1$ square foot, $F = 1.8$ pounds, $V = 20$ mph

$\begin{aligned} 1.8 &= \dfrac{k(1)}{(20)^2} \\ 720 &= k \\ F &= \dfrac{720A}{V^2} \end{aligned}$

When $A = 2(1$ square foot$)$, $V = 3(20$ mph$)$, $F = ?$

$\begin{aligned} F &= \dfrac{720(2)(1)}{3^2(20)^2} \\ F &= 0.4 \end{aligned}$

The force is $\boxed{0.4 \text{ pounds}}$.

92. $3x - y + 3z = 4$
$x + 2y + z = -1$
$2x - 3y + z = 1$

(Equations 1 and 2):

$3x - y + 3z = 4$
$(\times -3)\quad \underline{-3x - 6y - 3z = 3}$
$-7y = 7$
$y = -1$

(Equations 2 and 3):

$(\times -1)\quad -x - 2y - z = 1$
$\underline{2x - 3y + z = 1}$
$x - 5y = 2$
$x - 5(-1) = 2$
$x = -3$

(Equation 2):

$x + 2y + z = -1$
$-3 + 2(-1) + z = -1$
$z = 4$

$\boxed{\{(-3, -1, 4)\}}$

Section 5.5 Polynomials Functions

Problem Set 5.5, pp. 367-372

1. $f(x) = -5x^2 + 80x$
$f(3) = -5(3^2) + 80(3)$
$\boxed{f(3) = 195}$

$\boxed{\text{At 3 amperes the voltage power is 195 volts.}}$

3. $s(t) = -16t^2 + 64t + 80$
$s(0) = -0 + 0 + 80$
$\boxed{s(0) = 80}$; $\boxed{\text{The initial height (at } t = 0) \text{ of the ball is 80 feet.}}$

$s(1) = -16(1^2) + 64(1) + 80$
$\boxed{s(1) = 128}$; $\boxed{\text{After 1 second, the ball's height is 128 feet.}}$

$s(2) = -16(2^2) + 64(2) + 80$
$\boxed{s(2) = 144}$; $\boxed{\text{After 2 seconds, the ball's height is 144 feet.}}$

$s(3) = -16(3^2) + 64(3) + 80$
$\boxed{s(3) = 128}$; $\boxed{\text{After 3 seconds, the ball's height is 128 feet.}}$

$s(4) = -16(4^2) + 64(4) + 80$

$\boxed{s(4) = 80}$; $\boxed{\text{After 4 seconds, the ball's height is 80 feet.}}$

$s(5) = -16(5^2) + 64(5) + 80$

$\boxed{s(5) = 0}$; $\boxed{\text{After 5 seconds, the ball's height is 0 feet.}}$

$\boxed{\text{The ball is on the ground after 5 seconds.}}$

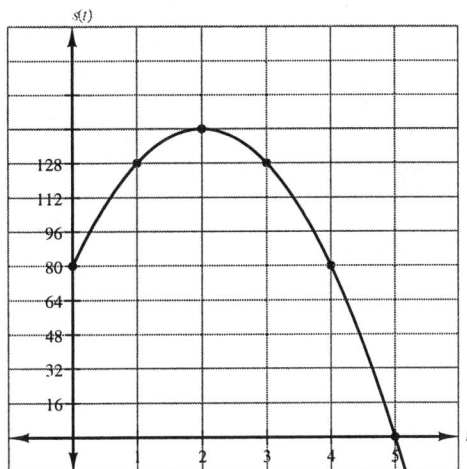

5. $f(x) = 5x^2 - 7x + 10$

$\dfrac{f(a + h) - f(a)}{h}$

$= \dfrac{[5(a+h)^2 - 7(a+h) + 10] - (5a^2 - 7a + 10)}{h}$

$= \dfrac{5a^2 + 10ah + 5h^2 - 7a - 7h + 10 - 5a^2 + 7a - 10}{h}$

$= \dfrac{10ah + 5h^2 - 7h}{h}$

$= \boxed{10a + 5h - 7}$

7. $f(x) = -4a^2 + 3x - 7$

$\dfrac{f(a + h) - f(a)}{h}$

$= \dfrac{[-4(a+h)^2 + 3(a+h) - 7] - (-4a^2 + 3a - 7)}{h}$

$= \dfrac{-4a^2 - 8ah - 4h^2 + 3a + 3h - 7 + 4a^2 - 3a + 7}{h}$

$= \dfrac{-8ah - 4h^2 + 3h}{h}$

$= \boxed{-8a - 4h + 3}$

9. $f(x) = x^3 - 2x + 4$

$\dfrac{f(a+h) - f(a)}{h}$

$= \dfrac{[(a+h)^3 - 2(a+h) + 4] - (a^3 - 2a + 4)}{h}$

$= \dfrac{a^3 + 3a^2 + 3ah^2 + h^3 - 2a - 2h + 4 - a^3 + 2a - 4}{h}$

$= \dfrac{3a^2h + 3ah^2 + h^3 - 2h}{h}$

$= \boxed{3a2 + 3ah + h^2 - 2}$

11. $f(x) = 5x - 3$

$\dfrac{f(a+h) - f(a)}{h}$

$= \dfrac{[5(a+h) - 3] - (5a - 3)}{h}$

$= \dfrac{5a + 5h - 3 - 5a + 3}{h}$

$= \dfrac{5h}{h}$

$= \boxed{5}$

13. $f(x) = -2x + 19$

$\dfrac{f(a+h) - f(a)}{h}$

$= \dfrac{[-2(a+h) + 19] - (-2a + 19)}{h}$

$= \dfrac{-2a - 2h + 19 + 2a - 19}{h}$

$= \dfrac{-2h}{h}$

$= \boxed{-2}$

15. $f(x) = 7$

$\dfrac{f(a+h) - f(a)}{h} = \dfrac{7 - 7}{h} = \dfrac{0}{h} = \boxed{0}$

17. $f(x) = 4x - 3$ and $g(x) = 5x^2 - x - 6$

$\begin{aligned}
(f+g)(x) &= (4x-3) + (5x^2 - x - 6) = \boxed{5x^2 + 3x - 9} \\
(g-g)(x) &= (4x-3) - (5x^2 - x - 6) \\
&= 4x - 3 - 5x^2 + x + 6 \\
&= \boxed{-5x^2 + 5x + 3} \\
(fg)(x) &= (4x-3)(5x^2 - x - 6) \\
&= 20x^3 - 4x^2 - 24x - 15x^2 + 3x + 18 \\
&= \boxed{20x^3 - 19x^2 - 21x + 18}
\end{aligned}$

19. $f(x) = -3x^2 - 9x$ and $g(x) = -8x^2 - x$

$\begin{aligned}
(f+g)(x) &= (-3x^2 - 9x) + (-8x^2 - x) = \boxed{-11x^2 - 10x} \\
(f-g)(x) &= (-3x^2 - 9x) - (-8x^2 - x) \\
&= -3x^2 - 9x + 8x^2 + x \\
&= \boxed{5x^2 - 8x} \\
(fg)(x) &= (-3x^2 - 9x)(-8x^2 - x) \\
&= 24x^4 + 3x^3 + 72x^3 + 9x^2 \\
&= \boxed{24x^4 + 75x^3 + 9x^2}
\end{aligned}$

21. $f(x) = 2x^2 - x - 10$ and $g(x) = x + 2$

$$(f/g)(x) = \frac{2x^2 - x - 10}{x + 2} = \boxed{2x - 5 \quad (x \neq -2)}$$

$$
\begin{array}{r|rrr}
-2 & 2 & -1 & -10 \\
   &   & -4 & 10 \\
\hline
   & 2 & -5 & 0
\end{array}
\quad \rightarrow \quad 2x - 5
$$

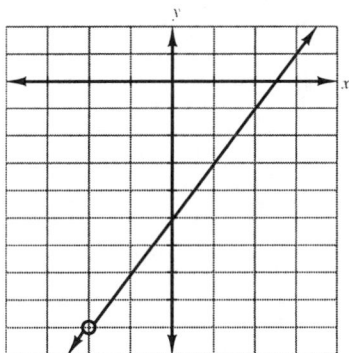

23. $f(x) = 2x^2 - 3x - 9$ and $g(x) = 2x + 3$

$$(f/g)(x) = \frac{2x^2 - 3x - 9}{2x + 3} = \boxed{x - 3 \ \left(x \neq -\frac{3}{2} \right)}$$

$$
\begin{array}{r|rrr}
-\dfrac{3}{2} & 2 & -3 & -9 \\
   &   & -3 & 9 \\
\hline
   & 2 & -6 & 0 \\
(\div 2) & 1 & -3 & 0
\end{array}
\quad \rightarrow \quad x - 3
$$

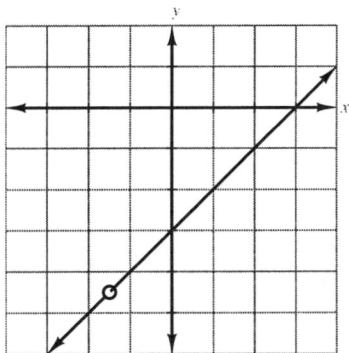

25. $f(x) = x^2 + 3x - 2,\ g(x) = 2x - 1$

$$
\begin{aligned}
f[g(x)] &= (2x - 1)^2 + 3(2x - 1) - 2 \\
&= 4x^2 - 4x + 1 + 6x - 3 - 2 \\
&= \boxed{4x^2 + 2x - 4}
\end{aligned}
$$

$$
\begin{aligned}
g[f(x)] &= 2(x^2 + 3x - 2) - 1 \\
&= 2x^2 + 6x - 4 - 1 \\
&= \boxed{2x^2 + 6x - 5}
\end{aligned}
$$

27. $f(x) = -3x^2 - 8x + 11,\ g(x) = 3x - 4$

$$
\begin{aligned}
f[g(x)] &= -3(x - 4)^2 - 8(3x - 4) + 11 \\
&= -3(9x^2 - 24x + 16) - 24x + 32 + 11 \\
&= -27x^2 + 72x - 48 - 24x + 32 + 11 \\
&= \boxed{-27x^2 + 48x - 5}
\end{aligned}
$$

$$
\begin{aligned}
g[f(x)] &= 3(-3x^2 - 8x + 11) - 4 \\
&= \boxed{-9x^2 - 24x + 29}
\end{aligned}
$$

29. $f(x) = 2x^3 - 7x^2 + 11x - 6, g(x) = 3x$

$$
\begin{aligned}
f[g(x)] &= 2(3x)^3 - 7(3x)^2 + 11(3x) - 6 \\
&= \boxed{54x^3 - 63x^2 + 33x - 6} \\
g[f(x)] &= 3(2x^3 - 7x^2 + 11x - 6) \\
&= \boxed{6x^3 - 21x^2 + 33x - 18}
\end{aligned}
$$

31. $f(x) = x^3 + 2x, g(x) = 2x - 1$

$$
\begin{aligned}
f[g(x)] &= (2x - 1)^3 + 2(2x - 1) \\
&= 8x^3 - 12x^2 + 6x - 1 + 4x - 2 \\
&= \boxed{8x^3 - 12x^2 + 10x - 3} \\
g[f(x)] &= 2(x^3 + 2x) - 1 \\
&= \boxed{2x^3 + 4x - 1}
\end{aligned}
$$

33. $f(x) = x^3 - 5x$ and $g(x) = x^2 - 6$

$$
\begin{aligned}
(g \circ f)(x) &= (x^3 - 5x)^2 - 6 = x^6 - 10x^4 + 25x^2 - 6 \\
(g \circ f)(2) &= 2^6 - 10(2^4) + 25(2^2) - 6 = 64 - 160 + 100 - 6 = \boxed{-2}
\end{aligned}
$$

35. $f(x) = x^2 + 2x - 1$ and $g(x) = 2x + 3$

$$
\begin{aligned}
(f \circ g)(x) &= (2x + 3)^2 + 2(2x + 3) - 1 \\
&= 4x^2 + 12x + 9 + 4x + 6 - 1 \\
&= 4x^2 + 16x + 14 \\
(f \circ g)(0) &= 0 + 0 + 14 = \boxed{14}
\end{aligned}
$$

For exercises 37-63, $f(x) = 2x + 3$, $g(x) = x - 1$, $h(x) = 2x^2 + x - 3$.

37. $(f + g)(x) = (2x + 3) + (x - 1) = \boxed{3x + 2}$

39. $(f + g)(-5) = 3(-5) + 2 = -15 + 2 = \boxed{-13}$

41. $(g - f)(x) = (x - 1) - (2x + 3) = x - 1 - 2x - 3 = \boxed{-x - 4}$

43. $(g - f)(-8) = -(-8) - 4 = 8 - 4 = \boxed{4}$

45. $(fg)(x) = (2x + 3)(x - 1) = 2x^2 + 3x - 2x - 3 = \boxed{2x^2 + x - 3}$

47. $(fg)(-3) = 2(-3)^2 + (-3) - 3 = 18 - 3 - 3 = \boxed{12}$

49. $\left(\dfrac{h}{f}\right)(x) = \dfrac{2x^2 + x - 3}{2x + 3} = \boxed{x - 1 \left(x \neq -\dfrac{3}{2}\right)}$

$$
\begin{array}{r|rrr}
-\dfrac{3}{2} & 2 & 1 & -3 \\
& & -3 & 3 \\
\hline
& 2 & -2 & 0
\end{array}
$$

$(\div 2) \quad 1 \quad -1 \quad \rightarrow x - 1$

51. $\left(\dfrac{h}{f}\right)(2) = 2 - 1 = \boxed{1}$

53.
$$
\begin{aligned}
(h \circ g)(x) &= 2(x - 1)^2 + (x - 1) - 3 \\
&= 2x^2 - 4x + 2 + x - (-3) \\
&= \boxed{2x^2 - 3x - 2}
\end{aligned}
$$

55. $(h \circ g)(-2) = 2(-2)^2 - 3(-2) - 2 = 8 + 6 - 2 = \boxed{12}$

57. $f(a + b) + g(a + b)$

$$
\begin{aligned}
&= [2(a + b) - 1] + [(a + b) - 1] \\
&= 2a + 2b + 3 + a + b - 1 \\
&= \boxed{3a + 3b + 2}
\end{aligned}
$$

59. $(f \circ f)(x) = 2(2x + 3) + 3 = \boxed{4x + 9}$

61. $(f + g + h)(x)$ $=$ $(2x + 3) + (x - 1) + (2x^2 + x - 3)$
 $=$ $\boxed{2x^2 + 4x - 1}$

63. $(h + fg)(x)$
$= (2x^2 + x - 3) + (2x + 3)(x - 1)$
$= 2x^2 + x - 3 + 2x^2 + x - 3$
$= \boxed{4x^2 + 2x - 6}$

65. $\boxed{B}$ is true; $f(x) = mx + b$

$$\frac{f(a + h) - f(a)}{h} = \frac{m(a + h) + b - (ma + b)}{h}$$

$$= \frac{ma + mb + b - ma - b}{h}$$

$$= \frac{mh}{h}$$

$= m$ which represents the slope of the linear functions

67. $f(t) = 10{,}000 + 100t^2$, $g(P) = 0.5P + 2$

   a. $\boxed{\text{Air pollution is a function of time.}}$

   b. $f(6) = 10{,}000 + 100(6^2) = 10{,}000 + 3600 = 13{,}600$
$g[f(6)]$ $=$ $g(13{,}600) = 0.5(13{,}600) + 2$
$=$ $6800 + 2$
$=$ 6802

average daily level of carbon monoxide 6 years from now will be $\boxed{6802 \text{ ppm}}$

   c. $f(10) = 10{,}000 + 100(10^2) = 10{,}000 + 10{,}000 = 20{,}000$
$g[f(10)]$ $=$ $g(20{,}000) = 0.5(20{,}000) + 2 = 10{,}000 + 2$
$=$ $10{,}002$

$\boxed{\text{In 10 years the carbon monoxide level will be } 10{,}002 \text{ ppm.}}$

   d. $g[f(t)]$ $=$ $0.5(10{,}000 + 100t^2) + 2$
$=$ $5000 + 50t^2 + 2$
$=$ $\boxed{50t^2 + 500^2}$

69. $f(u) = u^2 + 5u + 200$
$g(t) = 20t$

   a. $\boxed{\text{Manufactoring cost is a function of time.}}$

   b. $g(5) = 20(5) = 100$
$f[g(5)]$ $=$ $f(100) = 100^2 + 5(100) + 20$
$=$ $10{,}000 + 500 + 200 = 10{,}700$

$\boxed{\$10{,}700}$

   c. $g(7) = 20(7) = 140$
$f[g(7)]$ $=$ $f(140) = (140)^2 + 5(140) + 200 = 19600 + 700 + 200$
$=$ $\boxed{20{,}500}$

$\boxed{\text{By the end of the 7}^{\text{th}} \text{ hour, manufacturing cost is } \$20{,}500.}$

   d. $f[g(t)]$ $=$ $(20t)^2 + 5(20t) + 200$
$=$ $\boxed{400t^2 + 100t + 200}$

71. $f(x) = 2x^3 - 8x^2 - 7x - 15$

$$
\begin{array}{r|rrrr}
5 & 2 & -8 & -7 & -15 \\
  &   & 10 & 10 & 15 \\
\hline
  & 2 & 2 & 3 & 0
\end{array}
\quad \rightarrow \quad \text{Remainder} = 0
$$

$\boxed{\text{Yes}}$, 5 is a solution to $2x^3 - 8x^2 - 7x - 15$.

73. $f(x) = x^4 + 2x^3 - 3x^2 - 8x - 4 = 0$

$$
\begin{array}{r|rrrrr}
-2 & 1 & 2 & -3 & -8 & -4 \\
   &   & -2 & 0 & 6 & 4 \\
\hline
   & 1 & 0 & -3 & -2 & 0
\end{array}
\quad \rightarrow \quad \text{Remainder} = 0
$$

$\boxed{\text{Yes}}$, -2 is a solution to $x^4 + 2x^3 - 3x^2 - 8x - 4$

75. $f(x) = 3x^4 - x^3 + 9x^2 - x$

$$
\begin{array}{r|rrrrr}
\dfrac{1}{3} & 3 & -1 & 9 & -1 & 0 \\
             &   & 1 & 0 & 3 & \dfrac{2}{3} \\
\hline
             & 3 & 0 & 9 & 2 & \dfrac{2}{3}
\end{array}
\quad \rightarrow \quad \text{Remainder} \neq 0
$$

$\boxed{\text{No}}$, $\dfrac{1}{3}$ is *not* a solution to $3x^4 - x^3 + 9x^2 - x$

77. $P(x) = 5x^3 + 2x^2 - 4$

$$
\begin{array}{r|rrrr}
3 & 5 & 2 & 0 & -4 \\
  &   & 15 & 51 & 153 \\
\hline
  & 5 & 17 & 51 & 149
\end{array}
\quad \rightarrow \quad \text{Remainder} = 149
$$

$P(3) = \boxed{149}$

79. $P(x) = x^3 - 2x^2 + 5x + 6$

$$
\begin{array}{r|rrrr}
-3 & 1 & -2 & 5 & 6 \\
   &   & -3 & 15 & -60 \\
\hline
   & 1 & -5 & 20 & -54
\end{array}
\quad \rightarrow \quad \text{Remainder} = -54
$$

$P(-3) = \boxed{-54}$

81. $C(x) = -0.01x^2 + 6x + 400$

$$
\begin{aligned}
C(200) &= -0.01(200^2) + 6(200) + 400 \\
&= -400 + 1200 + 400 \\
&= 1200
\end{aligned}
$$

$\boxed{\$1200}$

83. $f(x) = x^2 - 3x - 4$, $g(x) = x^2 + 3x - 4$
$g(1) = 0$ (from the graph)
$f[g(1)] = f(0) = \boxed{-4}$ (from the graph)

85. $f(4) = 0$ (from the graph)
$g(-1) = -6$ (from the graph)
$f(4) > g(-1)$
$\boxed{f(4)}$

87. $\boxed{\text{D}}$ is true; $f(3) = -4$ (from the graph)

$$|f(3)| = |-4| = 4$$

89. $f(x) = 14x^3 - 17x^2 - 16x + 34$ for $1.5 \le x \le 3.5$

2.53	14	−17	−16	34
		35.42	46.6026	77.424578
	14	18.42	30.6026	111.42458

$\boxed{f(2.53) \approx 111.42}$

$\boxed{\text{A female moth of abdominal width 2.53 mm has approximately 111 eggs.}}$

91. $f(x) = x^2 - 5x + 3$ and $g(x) = 17x - 6$

$g(2.5) = 17(2.5) - 6 = 42.5 - 6 = 36.5$

$f[g(2.5)] = f(36.5) = (36.5)^2 - 5(36.5) + 3 = \boxed{1152.75}$

93. $f(x) = -x^3 + 450x^2 + 52{,}500x$

maximum revenue occurs at $x = 350$

$$\begin{aligned} f(350) &= -(350)^3 + 450(350)^2 + 52{,}500(350) \\ &= -42{,}875{,}000 + 55{,}125{,}000 + 18{,}375{,}000 \\ &= 30{,}625{,}000 \end{aligned}$$

The company's maximum revenue is $\boxed{\$30{,}625{,}000}$.

95. $f(x) = x - 1, f[g(x)] = x^2$

$$\begin{aligned} g(x) - 1 &= x^2 \\ \boxed{g(x) = x^2 + 1} \end{aligned}$$

97. $f(x) = 3x - 2, g(x) = x^5$

$g(x) - f(x) = x^5 - (3x - 2) = x^5 - 3x + 2$

$f[g(x) - f(x)] = f(x^5 - 3x + 2) = 3(x^5 - 3x + 2) - 2 = \boxed{3x^5 - 9x + 4}$

$f[g(x)] - f[f(x)]$

$= [3(x^5) - 2] - [3(3x - 2)]$

$= 3x^5 - 2 - 9x + 6$

$= \boxed{3x^5 - 9x + 4}$

99. $f(x) = 2x - 5$ and $g(x) = 3x + b$

$$\begin{aligned} f[g(x)] &= g[f(x)] \\ 2(3x + b) - 5 &= 3(2x - 5) + b \\ 6x + 2b - 5 &= 6x - 15 + b \\ \boxed{b = -10} \end{aligned}$$

101. $f(x) = x, f\left(\frac{1}{2}\right) = \frac{1}{2}$

$g(x) = x^2, g\left(\frac{1}{2}\right) = \frac{1}{4}$

$h(x) = x^3, h\left(\frac{1}{2}\right) = \frac{1}{8}$

$F(x) = x^4, F\left(\frac{1}{2}\right) = \frac{1}{16}$

$G(x) = x^5, G\left(\frac{1}{2}\right) = \frac{1}{32}$

$H(x) = x^6, H\left(\frac{1}{2}\right) = \frac{1}{64}$

$f\left(\frac{1}{2}\right) > g\left(\frac{1}{2}\right) > h\left(\frac{1}{2}\right) > F\left(\frac{1}{2}\right) > G\left(\frac{1}{2}\right) > H\left(\frac{1}{2}\right)$

From the left to right: $f(x) = x, g(x) = x^2, h(x) = x^3, F(x) = x^4, G(x) = x^5, H(x) = x^6$

Review Problems

106.   Domain = $\{ x \mid -3 \le x \le 4 \}$
or $[-3, 4]$

Range = $\{ y \mid -2 \le y \le 2 \}$
or $[-2, 2]$

107. $6x + y = 19$
$y = -6x + 19$
$f(x) = -6x + 19$
$f(-3) = -6(-3) + 19 = 18 + 9$
$f(-3) = 37$
$f(a) = -6a + 19$

108.   number < 100: 1, 2, 3, . . . , 99
odd: 1, 3, 5, . . . , 99
a multiple of 5: 5, 15, 25, 45, 55, 65, 75, 85, 95
divisible by 3: 15, 45, 75
sum of digits: $1 + 5 = 6, 4 + 5 = 9, 7 + 5 = 12$
odd sum of digits: $4 + 5 = 9$ $\rightarrow$ 45

Section 5.6 Greatest Common Factors and Factoring by Grouping

Problem Set 5.6, pp. 377-378

1. $4x - 20 = \boxed{4(x - 5)}$ GCF = 4   **3.** $18a + 27 = \boxed{9(2a + 3)}$ GCF = 9

5. $12x^2 + 4x = \boxed{4x(3x + 1)}$ GCF = $4x$ **7.** $9x - 18y = \boxed{9x - 2y}$ GCF = 9

9. $12x^2y - 8xy^2 = \boxed{4xy(3x - 2y)}$ GCF = $4xy$ **11.** $4xy - 7xy^2 = \boxed{xy(4 - 7y)}$ GCF = xy

13. $18x^4 + 9x^3 - 27x^2 = \boxed{9x^2(2x^2 + x - 3)}$ GCF = $9x^2$

15. $10y^7 - 16y^4 + 8y^3 = \boxed{2y^3(5y^4 - 8y + 4)}$ GCF $= 2y^3$

17. $15x^5y^3 - 25x^3y^4 = \boxed{5x^2y^3(3x^2 - 5y)}$ GCF $= 5x^2y^3$

19. $2a^2b - 5ab^2 + 7a^2b^2 = \boxed{ab(2a - 5b + 7ab)}$ GCF $= ab$

21. $24xy^3 - 36x^3y^2 + 12x^2y^4 = \boxed{12xy^2(2y - 3x^2 + xy^2)}$ GCF $= 12xy^2)$

23. $26x^5y^3 + 52x^7y^2 - 39x^8y^5 = \boxed{13x^5y^2(2y + 4x^2 - 3x^3y^3)}$ GCF $= 13x^5y^2$

25. $30x^4y + 50x^2y^2 + 20x^3z^3 = \boxed{10x^2(3x^2y + 5y^2 + 2xz^3)}$ GCF $= 10x^2$

27. $55x^2y^2z^4 - 77x^3y^2z = \boxed{11x^2y^2z(5z^3 - 7x)}$ GCF $= 11x$

29. $7x^3y^2 + 14x^2y - 42x^5y^3 + 21xy^4 = \boxed{7xy(x^2y + 2x - 6x^4y^2 + 3y^3)}$ GCF $= 7xy$

31. $70x^3y + 42x^2y - 28x^2y^2 - 84xy^3 = \boxed{14xy(5x^2 + 3x - 2xy - 6y^2)}$ GCF $= 14xy$

33. $7(x - 3) + y(x - 3) = \boxed{(x - 3)(7 + y)}$ GCF $= x - 3$

35. $-2a(x + 7y) + 4c(x + 7y)$
$= (x + 7y)(-2a + 4c)$ GCF $= 2(x + 7y)$
$= \boxed{-2(x + 7y)(a - 2c) \text{ or } 2(x + 7y)(2c - a)}$

37. $4x(a + b - c) - 2y(a + b - c)$
$= (a + b - c)(4x - 2y)$ GCF $= 2(a + b - c)$
$= \boxed{2(a + b - c)(2x - y)}$

39. $(x - 4y)z^2 + (x - 4y)z + (x - 4y)$
$= \boxed{(x - 4y)(z^2 + z + 1)}$ GCF $= x - 4y$

41. $(x - 4y)z^2 + (x - 4y)z + (4y - x)$
$= (x - 4y)z^2 + (x - 4y)z + (x - 4y)(-1)$ GCF $= x - 4y$
$= \boxed{(x - 4y)(z^2 + z - 1)}$

43. $5x^3(2a - 7b) + 15x^2(2a - 7b)$
$= \boxed{5x^2(2a - 7b)(x + 3)}$ GCF $= 5x^2(2a - 7b)$

A second method involves first factoring out $(2a - 7b)$:
$= (2a - 7b)(5x^3 + 15x^2)$
$= (2a - 7b)(5x^2)(x + 3)$
$= 5x^2(2a - 7b)(x + 3)$

45. $77x^3y^2(7a - 5b) + 11x^2y(7a - 5b)$
$= \boxed{11x^2y(7a - 5b)(7xy + 1)}$ GCF $= 11x^2y(7a - 5b)$

or by first factoring out $(7a - 5b)$:
$77x^3y^2(7a - 5b) + 11x^2y(7a - 5b)$:
$= (7a - 5b)(77x^3y^2 - 11x^2y)$
$= (7a - 5b)(11x^2y)(7xy + 1)$
$= 11x^2y(7a - 5b)(7xy + 1)$

47. $3a(17x - y) + 7b(17x - y) - 8c(17x - y)$
$= \boxed{(17x - y)(3a + 7b - 8c)}$

49. $(7x - 4y)a - b(4y - 7x)$
$= (7x - 4y)a - b(-1)(7x - 4y)$
$= (7x - 4y)a + b(7x - 4y)$
$= \boxed{(7x - 4y)(a + b)}$

51. $11a(7x - 4y) + (4y - 7x)$
$= 11a(7x - 4y) + (-1)(7x - 4y)$
$= 11a(7x - 4y) - (7x - 4y)$
$= \boxed{(7x - 4y)(11a - 1)}$

53. $a^2c + 5ac + 2a + 10$
$= ac(a + 5) + 2(a + 5)$
$= \boxed{(a + 5)(ac + 2)}$

55. $3Y^2 + 4YZ + 24Y + 32Z$
$= Y(3Y + 4Z) + 8(3Y + 4Z)$
$= \boxed{(3Y + 4Z)(Y + 8)}$

57. $2a - 6b + ac - 3bc$
$= 2(a - 3b) + c(a - 3b)$
$= \boxed{(a - 3b)(2 + c)}$

59. $c + d + 2c^2 + 2cd$
$= (c + d) + 2c(c + d)$
$= \boxed{(c + d)(1 + 2c)}$

61. $4x^2y + 16xy - x - 4$
$= 4xy(x + 4) - (x + 4)$
$= \boxed{(x + 4)(4xy - 1)}$

63. $3ax + 3ay - 2x - 2y$
$= 3a(x + y) - 2(x + y)$
$= \boxed{(x + y)(3a - 2)}$

65. $4x^2 + 4xy - 3xy^2 - 3y^2$
$= 4x(x + y) - 3y^2(x + y)$
$= \boxed{(x + y)(4x - 3y^2)}$

67. $4x - 3y - 12ax + 9ay$
$= (4x - 3y) - 3a(4x - 3y)$
$= \boxed{(4x - 3y)(1 - 3a)}$

69. $ab - c - ac + b$
$= ab + b - ac - c$
$= b(a + 1) - c(a + 1)$
$= \boxed{(a + 1)(b - c)}$

71. $9a^3 + 10b^2 - 6a^2 - 15ab$
$= 9a^3 - 6a^2b + 10b^2 - 15ab$
$= 3a^2(3a - 2b) + 5b(2b - 3a)$
$= 3a^2(3a - 2b) = 5b(-1)(3a - 2b)$
$= 3a^2(3a - 2b) - 5b(3a - 2b)$
$= \boxed{(3a - 2b)(3a^2 - 5b)}$

73. $ax + ay + az - bx - by - bz + cx + cy + cz$
$= a(x + y + z) - b(x + y + z) + c(x + y + z)$
$= \boxed{(x + y + z)(a - b + c)}$

75. $4x^{n+3} - 5x^{n+2}$
$= 4x(x^{n+2}) - 5(x^{n+2})$
$= \boxed{x^{n+2}(4x - 5)}$ GCF $= x^{n+2}$

77. $3x^{5n} - 9x^{4n} + 6x^{3n}$
$= x^{2n}(3x^{3n}) - 3x^n(3x^{3n}) + 2(3x^{3n})$
$= \boxed{3x^{3n}(x^{2n} - 3x^n + 2)}$ GCF $= 3x^{3n}$

79. $7x^{2n}y^m + 3x^ny^{m+1}$
$= x^ny^m(7x^n) + x^ny^m(3y)$
$= \boxed{x^ny^m(7x^n + 3y)}$ GCF $= x^ny^m$

81. $x^{n+5} + x^{n+4} + x^{n+2}$
$= x^{n+2}x^3 + x^{n+2}x^2 + x^{n+2}$
$= \boxed{x^{n+2}(x^3 + x + 1)}$ GCF $= x^{n+2}$

83. $9x^{n+3}y^nz - 18x^{n+2}y^{n+1}z^2 - 6x^ny^{n+3}z^4$
$= (3x^ny^nz)(3x^3) - (3x^ny^nz)(6x^2yz) - (3x^ny^nz)(2y^3z^3)$
$= \boxed{3x^ny^nz(3x^3 - 6x^2yz - 2y^3z^3)}$ GCF $= 3x^ny^nz$

85. $A = P + Pr + (P + Pr)r$
$= P(1 + r) + P(1 + r)r$
$= P(1 + r)(1 + r)$
$= \boxed{P(1 + r)^2}$

87. $S.A. = 2\pi rh + 2\pi r^2$
$= 2\pi r(h) + 2\pi r(r)$
$= \boxed{2\pi r(h + r)}$

89. $\boxed{B}$ is true; $4a^2b - 8ab^2$
$= 4ab(a) - 4ab(2b)$
$= 4ab(a - 2b)$ or $-4ab(2b - a)$

Review Problems

92. $\dfrac{15y^3 - 5y^2}{5y^2} - \dfrac{12y^4 - 6y^2}{6y^2}$

$= \dfrac{15y^3}{5y^2} - \dfrac{5y^2}{5y^2} - \dfrac{12y^4}{6y^2} + \dfrac{6y^2}{6y^2}$

$= 3y - 1 - 2y^2 + 1$

$= \boxed{3y - 2y^2}$

93. Let

$\qquad\qquad x$ = amount (in millimeters) of water (0% solution)

$\qquad x + 500$ = amount of 25% solution

$\qquad\qquad 0x + 0.40(500)$ = $0.25(x + 500)$

$\qquad\qquad\qquad\qquad 200$ = $0.25x + 125$

$\qquad\qquad\qquad\qquad\ \ 75$ = $0.25x$

$\qquad\qquad\qquad\qquad 300$ = x

$\boxed{300 \text{ millimeters}}$ of water must be added

94. $\left| \dfrac{3x - 8}{2} \right| \geq 4$

$\dfrac{3x - 8}{2} \geq 4$ $\qquad$ or $\qquad$ $\dfrac{3x - 8}{2} \leq -4$

$3x - 8 \geq 8$ $\qquad\qquad\qquad\quad$ $3x - 8 \leq -8$

$3x \geq 16$ $\qquad\qquad\qquad\qquad\ \ $ $3x \leq 0$

$x \geq \dfrac{16}{3}$ $\qquad\qquad\qquad\qquad\ $ $x \leq 0$

$\boxed{\left\{ x \mid x \leq 0 \text{ or } x \geq \dfrac{16}{3} \right\}}$

$\boxed{(-\infty, 0] \text{ or } \left[\dfrac{16}{3}, \ \infty \right)}$

Section 5.7 Factoring Trinomials

Problem Set 5.7, pp. 386-387

Not all possible factorizations are given. Only the pair that satisfies the conditions whose product is *c* and whose sum is *b* are given here.

1. $x^2 + 5x + 6 = \boxed{(x + 3)(x + 2)}$

product: $3(3) = 6$

sum: $3 + 2 = 5$

Only the pair 3 and 2 satisfy the conditions.

3. $a^2 + 8a + 15 = \boxed{(a + 5)(a + 3)}$

product: $5(3) = 15$

sum: $5 + 3 = 8$

5. $x^2 + 9x + 20 = \boxed{(x + 5)(x + 4)}$

product $= 5(4) = 20$

sum: $5 + 4 = 9$

7. $d^2 + 10d + 16 = \boxed{(d+8)(d+2)}$
product: $8(2) = 16$
sum: $8 + 2 = 10$

9. $t^2 + 6t + 9 = (t+3)(t+3) = \boxed{(t+3)^2}$
product: $3(3) = 9$
sum: $3 + 3 = 6$

11. $Y^2 - 14Y + 49 = (Y-7)(Y-7) = \boxed{(Y-7)^2}$
product: $-7(-7) = 49$
sum: $-7 + (-7) = -14$

13. $x^2 - 8x + 15 = \boxed{(x-5)(x-3)}$
product: $(-5)(-3) = 15$
sum: $-5 + (-3) = -8$

15. $y^2 - 12y + 20 = \boxed{(y-10)(y-2)}$
product: $(-10)(-2) = 20$
sum: $-10 + (-2) = -12$

17. $y^2 + 5y - 14 = \boxed{(y+7)(y-2)}$
product: $7(-2) = -14$
sum: $7 + (-2) = 5$

19. $x^2 + x - 30 = \boxed{(x+6)(x-5)}$
product: $(6 + (-5) = 1$
sum: $6 + (-5) = 1$

21. $d^2 - 3d - 28 = \boxed{(d-7)(d+4)}$
product: $-7(4) = -28$
sum: $-7 + 4 = -3$

23. $R^2 - 5R - 36 = \boxed{(R-9)(R+4)}$
product: $-9(4) = -5$
sum: $-9 + 4 = -5$

25. $-12x + 35 + x^2 = x^2 - 12x + 35 = \boxed{(x-7)(x-5)}$
product: $-7(-5) = 35$
sum: $-7 + (-5) = -12$

27. $x^2 - x + 7$
not factorable
$\boxed{\text{prime}}$

29. $X^2 + 12XY + 20Y^2 = \boxed{(x+10Y)(X+2Y)}$
product: $(10Y)(2Y) = 20Y^2$
sum: $10Y + 2Y = 12Y$

31. $W^2 - 10WZ - 11Z^2 = \boxed{(W-11Z)(W+Z)}$
product: $(-11Z)(Z) = -11Z^2$
sum: $-11Z + Z = -10Z$

33. $a^2 - ab + b^2$
not factorable
$\boxed{\text{prime}}$

35. $2x^3 + 6x^2 + 4x$
$= 2x(x^2 + 3x + 2)$
$= \boxed{2x(x+2)(x+1)}$

37. $2M^2 + 9M + 7 = \boxed{(2M+7)(M+1)}$
first term: $2M(M) = 2M^2$
middle term: $7M + 2M = 9M$
last term: $7(1) = 7$

39. $4x^2 + 9x + 2 = \boxed{(4x-1)(x+2)}$
first term: $4x(x) = 4x^2$
middle term: $8x + x = 9x$
last term: $1(2) = 2$

41. $5T^2 + 17T + 6 = \boxed{(5T+2)(T+3)}$
first term: $5T(T) = 5T^2$
middle term: $3(5T) + 2T = 17T$
last term: $2(3) = 6$

43. $6b^2 + 19b + 15 = \boxed{(3b+5)(2b+3)}$
first term: $(3b)(2b) = 6b^2$
middle term: $(3b)(3) + 5(2b) = 19b$
last term: $5(3) = 51$

45. $6x^2 + 17x + 12 = \boxed{(3x+4)(2x+3)}$
first term: $3x(2x) = 6x^2$
middle term: $9x + 8x = 17x$
last term: $4(3) = 12$

47. $9y^2 - 30y + 25 = (3y-5)(3y-5)) = \boxed{(3y-5)^2}$
first term: $3y(3y) = 9y^2$
middle term: $-15y - 15y = -30y$
last term: $5(5) = 25$

49. $4a^2 - 27a + 18 = \boxed{(4a-3)(a-6)}$
first term: $4a(a) = 4a^2$
middle term: $-24a - 3a = -27a$
last term: $-3(-6) = 18$

51. $16S^2 - 6S - 27 = \boxed{(2S - 3)(8S + 9)}$
first term: $2S(8S) = 16S^2$
middle term: $18S - 24S = -6S$
last term: $-3(9) = -27$

53. $4y^2 - y - 18 = \boxed{(4y - 9)(y + 2)}$
first term: $4y(y) = 4y^2$
middle term: $8y - 9y = -y$
last term: $-9(2) = -18$

55. $9M^2 - 3MN - 2N^2 = \boxed{(3M + N)(3M - 2N)}$
first term: $3M(3M) = 9M^2$
middle term: $-6MN + 3MN = -3MN$
last term: $N(-2N) = -2N^2$

57. $10x^2 + 29xy - 21y^2 = \boxed{(5x - 3y)(2x + 7y)}$
first term: $5x(2x) = 10x^2$
middle term: $35xy - 6xy = +29xy$
last term: $-3y(7y) = -21y^2$

59. $6a^2 + 14a + 3$
not factorable
$\boxed{\text{prime}}$

61. $15w^3 - 25w^{-2} + 10w$
$= 5w(3w^2 - 5w + 2)$
$= \boxed{5w(3w + 1)(w - 2)}$

63. $-x^2 + xy + 6y^2$
$= -(x^2 - xy - 6y^2)$
$= \boxed{-(x - 3y)(x + 2y)}$

65. $4a^4b - 24a^3b - 64a^2b$
$= 4a^2b(a^2 - 6ab - 16)$
$= \boxed{4a^2b(a + 2)(a - 8)}$

67. $36x^3y - 6x^2y - 20xy$
$= 2xy(18x^2 - 3x - 10)$
$= \boxed{2xy(6x - 5)(3x + 2)}$

69. $4x^2 - 12xy + 9y^2$
$= (2x - 3y)(2x - 3y)$
$= \boxed{(2x - 3y)^2}$

71. $35a^2 - 41ab - 24b^2$
$= \boxed{(7a + 3b)(5a - 8b)}$
first term: $7a(5a) = 35a^2$
middle term: $-56ab + 15ab = -41ab$
last term: $3b(-8b) = -24b^2$

73. $8x^2 - 14xy - 39y^2$
$= \boxed{(4x - 13y)(2x + 3y)}$
first term: $4x(2x) = 8x^2$
middle term: $12xy - 26xy = -14xy$
last term: $(-13y)(3y) = -39y^2$

75. $6x^2y^2 + 13xy + 6$
$= \boxed{(2xy + 3)(3xy + 2)}$
first term: $2xy(3xy) = 6x^2y^2$
middle term: $4xy + 9xy = 13xy$
last term: $3(2) = 6$

77. $6a^3b^3 + 12a^2b^2 - 90ab$
$= 6ab(a^2b^2 + 2ab - 15)$
$= \boxed{6ab(ab + 5)(ab - 3)}$

79. $13x^3y^3 + 39x^3y^2 - 52x^3y$
$= 13x^3y(y^2 + 3y - 4)$
$= \boxed{13x^3y(y + 4)(y - 1)}$

81. $y^4 + 5y^2 + 6$
$= x^2 + 5x + 6$ (Let $x = y^2$)
$= (x + 3)(x + 2)$
$= \boxed{(y^2 + 3)(y^2 + 2)}$ (substitute y^2 for x)

83. $5m^4 + m^2 - 6$
$= 5x^2 + x - 6$ (Let $x = m^2$)
$= (5x + 6)(x - 1)$
$= (5m^2 + 6)(m^2 - 1)$ (substitute m^2 for x)
$= \boxed{(5m^2 + 6)(m + 1)(m - 1)}$

85. $2n^4 + mn^2 - 6m^2$
$= 2x^2 + mx - 6m^2$ (Let $x = n^2$)
$= (2x - 3m)(x + 2m)$
$= \boxed{(2n^2 - 3m)(n^2 + 2m)}$ (substitute n^2 for x)

87. $y^6 - 9y^3 - 36$
$= x^2 - 9x - 36$ (Let $x = y^3$)
$= (x - 12)(x + 3)$
$= \boxed{(y^3 - 12)(y + 3)}$ (substitute y^3 for x)

89. $y^8 + 10y^4 - 39$
$= x^2 + 10y - 39$ (Let $x = y^4$)
$= (x + 13)(x - 3)$
$= \boxed{(y^4 + 13)(y^4 - 3)}$ (substitute y^4 for x)

91. $(a - 3b)^2 - 5(a - 3b) - 36$
$= x^2 - 5x - 36$ (Let $x = a - 3b$)
$= (x + 4)(x - 9)$
$= \boxed{(a - 3b + 4)(a - 3b - 9)}$ (substitute $a - 3b$ for x)

93. $5(a + b)^2 + 12(a + b) + 7$
$= 5x^2 + 12x + 7$ (Let $x = a + b$)
$= (5x + 7)(x + 1)$
$= [5(a + b + 7](a + b + 1)$ (substitute $a + b$ for x)
$= \boxed{(5a + 5b + 7)(a + b + 1)}$

95. $18(x + y)^2 - 3(x + y)b - 28b^2$
$= 18a^2 - 3ab - 28b^2$ (Let $a = x + y$)
$= (6a + 7b)(3a - 4b)$
$= [6(x + y) + 7b][3(x + y) - 4b]$ (substitute $x + y$ for a)
$= \boxed{(6x + 6y + 7b)(3x + 3y - 4b)}$

97. $6a^2(a - b)^2 - 13ab(a - b) + 6b^2$
$= 6a^2x^2 - 13axb + 6b^2$ (Let $x = a - b$)
$= (3ax - 2b)(2ax - 3b)$
$= [3a(a - b) - 2b][2a(a - b) - 3b]$ (substitute $a - b$ for x)
$= \boxed{(3a^2 - 3ab - 2b)(2a^2 - 2ab - 3b)}$

99. $x^{2n} + 6x^n + 8$
$= a^2 + 6a + 8$ (Let $a = x^x$)
$= (a + 4)(a + 2)$
$= \boxed{(x^n + 4)(x^n + 2)}$ (substitute x^x for a)

101. $9x^{2n} + x^n - 8$
$= 9a^2 + a - 8$ (Let $a = x^x$)
$= (9a - 8)(a + 1)$
$= \boxed{(9x^n - 8)(x^n + 2)}$ (substitute x^x for a)

103. $3y^{2n} - 8y^n + 5$
$= 9a^2 - 8a + 5$ (Let $a = y^n$)
$= (3a - 5)(a - 1)$
$= \boxed{(3y^n - 5)(y^n - 1)}$ (substitute y^n for a)

105. $y^{3n} + 10y^{2n} + 16y^n$
$= y^n(y^{2n} + 10y^n + 16)$
$= \boxed{y^n(y^n + 8)(y^n + 2)}$

107. $-16t^2 + 16t + 32$
$= -161(t^2 - t - 2)$
$= \boxed{-16(t - 2)(t + 1)}$

109. $x^2 + Kx - 6$

possible factors of –6	sum of digits	possible value of K
–6(1)	–6 + 1 = –5	–5
6(–1)	6 + (–1) = 5	5
–3(2)	–3 + 2 = –1	–1
3(–2)	3 + (–2) = 1	1

The integers are $\boxed{-5, 5, -1, \text{ and } 1}$.

111. $4x^2 + Kx - 1$

possible factors of $4x^2 + Kx - 1$	sum	possible values of 2
$(4x - 1)(x + 1)$	$-x + 4x = 3x$	3
$(4x + 1)(x - 1)$	$x - 4x = -3x$	-3
$(2x - 1)(2x + 1)$	$-2x + 2x = 0$	0

The integers are $\boxed{-3, 3, \text{ and } 0}$.

113. $ax + b$ is a factor of $4x^2 + 4x + 1$ and $2x^2 - 5x + 6$
$4x^2 + 4x + 1 = (2x + 1)(2x + 1)$

Since $ax + b$ is a factor of $4x^2 + 4x + 1$,

$ax + b = 2x + 1$. Thus, $\boxed{a = 2 \text{ and } b = 1}$.

Also, $ax + b = 2x + 1$ is a factor of $2x^2 - 5x + c$.
$2x^2 - 5x + c = (2x + 1)(\text{factor})$

Only a second factor of $(x - 3)$ will result in $2x^2$ when the first terms are multiplied and $-5x$ when the sum of the outside and inside terms is taken.

Thus, $2x^2 - 5x + c = (2x + 1)(x - 3) = 2x^2 - 5x - 3$ which means that $\boxed{c = -3}$.

Review Problems

115. $I = Prt + P$
$I = P(rt + 1)$
$$\frac{I}{rt + 1} = P$$
$$\boxed{P = \frac{I}{1 + rt}}$$

116.
$$
\begin{aligned}
-2x &\leq 6 \\
x &\geq -3
\end{aligned}
\qquad \text{and} \qquad
\begin{aligned}
-2x + 3 &< -7 \\
-2x &< -10 \\
x &> 5
\end{aligned}
$$
$\boxed{x \mid x > 5\}}$
$\boxed{(5, \infty)}$

117. the line through $(3, -2)$ and perpendicular to the line whose equation is $x + 2y = 3$:
slope: $2y = -x + 3$
$$y = -\frac{1}{2}x + 3,\ m = -\frac{1}{2}$$
slope of line perpendicular: $m = \dfrac{-1}{-\frac{1}{2}} = 2$

$$
\begin{aligned}
y + 2 &= 2(x - 3) \\
y + 2 &= 2(x - 6) \\
y &= 2x - 8
\end{aligned}
$$
$\boxed{2x - y = 8}$

Section 5.8 Factoring Special Forms

Problem Set 5.8, pp. 393-396

1. $B^2 - 1 = B^2 - 1^2 = \boxed{(B+1)(B-1)}$ **3.** $25 - a^2 = 5^2 - a^2 = \boxed{(5+a)(5-a)}$

5. $36x^2 - 49 = (6x)^2 - 7^2 = \boxed{(6x+7)(6x-7)}$ **7.** $36x^2 - 49y^2 = (6x)^2 - (7y)^2 = \boxed{(6x+7)(6x-7)}$

9. $x^2y^2 - a^2b^2 = (xy)^2 - (ab)^2 = \boxed{(xy+ab)(xy-ab)}$

11. $x^2y^6 - a^4b^2 = (xy^3)^2 - (a^2b)^2 = \boxed{(xy^3+a^2b)(xy^3-a^2b)}$

13. $4x^2y^6 - 25a^4b^2 = (2xy^3)^2 - (5a^2b)^2 = \boxed{(2xy^3+5a^2b)(2xy^3-5a^2b)}$

15. $81a^2b^4c^6 - 49x^8y^2 = (9ab^2c^3)^2 - (7x^4y)^2 = \boxed{(9ab^2c^3+7x^4y)(9ab^2c^3-7x^4y)}$

17. $(x+3)^2 - y^2 = \boxed{(x+3+y)(x+3-y)}$

19. $(x+y)^2 - 36 = (x+y)^2 - 6^2 = \boxed{(x+y+6)(x+y-6)}$

21. $x^2 + 4$ is $\boxed{\text{prime}}$, not factorable

23. $16y^2 - (3x-1)^2$
$= (4y)^2 - (3x-1)^2$
$= (4y + 3x - 1)[4y - (3x-1)]$
$= \boxed{(4y+3x-1)(4y-3x+1)}$

25. $(x+1)^2 - (x+3)^2$
$= [(x+1)+(x+3)][(x+1)-(x+3)]$
$= (2x+4)(x+1-x-3)$
$= 2(x+2)(-2$
$= \boxed{-4(x+2)}$

27. $(2x-1)^2 - (3x+2)^2$
$= [(2x-1)+(3x+2)][(2x-1)-(3x+2)]$
$= (5x+1)(2x-1-3x-2)$
$= (5x+1)(-x-3)$
$= (5x+1)(-1)(x+3)$
$= \boxed{-(5x+1)(x+3)}$

29. $a^{14} - 9 = (a^7)^2 - 3^2 = \boxed{(a^7+3)(a^7-3)}$

31. $25x^2 + 36y^2$ is $\boxed{\text{prime}}$, not factorable   **33.** $-16 + x^2 = x^2 - 16 = x^2 - 4^2 = \boxed{(x+4)(x-4)}$

35. $-25A^2 + x^2y^2 = x^2y^2 - 25A^2 = x^2y^2 - (5A)^2 = \boxed{(xy+5A)(xy-5A)}$

37. $y^4 - 1 = (y^2+1)(y^2-1) = \boxed{(y^2+1)(y+1)(y-1)}$

39. $1 - 81b^2 = 1 - (9b)^2 = \boxed{(1+9b)(1-9b)}$ **41.** $x^2y^3 - 16y$
$= y(x^2y^2 - 16)$
$= y[(xy)^2 - 4^2]$
$= \boxed{y[(xy+4)(xy-4)]}$

43. $3x^3 - 3x = 3x(x^2-1) = \boxed{3x(x+1)(x-1)}$ **45.** $3x^3y - 12xy = 3xy(x^2-4) = \boxed{3x(x+2)(x-2)}$

47. $x^{2n} - 144 = (x^n)^2 - 12^2 = \boxed{(x^n + 12)(x^n - 12)}$

49. $x^{4n} - y^{4n}$
$= (x^{2n})^2 - (y^{2n})^2$
$= (x^{2n} + y^{2n})(x^{2n} - y^{2n})$
$= (x^{2n} + y^{2n})[(x^n)^2 - (y^n)^2]$
$= \boxed{(x^{2n} + y^{2n})(x^n + y^n)(x^n - y^n)}$

51. $p^3 + q^3 = \boxed{(p + q)(p^2 - pq + q^2)}$

53. $27x^3 - 64y^3$
$= (3x)^3 - (4y)^3$
$= (3x - 4y)[(3x)^2 + (3x)(4y) + (4y)^2]$
$= \boxed{(3x - 4y)(9x^2 + 12xy + 16y^2)}$

55. $27R^3 - 1$
$= (3R)^3 - 1$
$= (3R - 1)[(3R)^2 + 3R + 1]$
$= \boxed{(3R - 1)(9R^2 + 3R + 1)}$

57. $125b^3 + 64$
$= (5b)^3 + 4^3$
$= (5b + 4)[(5b)^2 - (5b)(4) + 4^2]$
$= \boxed{(5b + 4)(25b^2 - 20b + 16)}$

59. $125 + 64d^3$
$= 5^3 + (4d)^3$
$= (5 + 4d)[5^2 - 5(4d) + (4d)^2]$
$= \boxed{(5 + 4d)(25 - 20d + 16d^2)}$

61. $8a^3b^3 - 27$
$= (2ab)^3 - 3^3$
$= (2ab - 3)[(2ab)^2 + (2ab)(3) = 3^2]$
$= \boxed{(2ab - 3)(4a^2b^2 + 6ab + 9)}$

63. $64 - 27Y^3Z^3$
$= 4^3 - (3YZ)^3$
$= (4 - 3YZ)[4^2 + 4(3YZ) + (3YZ)^2]$
$= \boxed{(4 - 3YZ)(16 + 12YZ + 9Y^2Z^2)}$

65. $8x^3 + 27y^{12}$
$= (2x)^3 + (3y^4)^3$
$= (2x + 3y^4)[(2x)^2 - (2x)(3y^4) + (3y^4)^2]$
$= \boxed{(2x + 3y^4)(4x^2 - 6xy^4 + 9y^8)}$

67. $a^3b^6 - c^6d^{12}$
$= (ab^2)^3 - (c^2d^4)^3$
$= (ab^2 - c^2d^4)[(ab^2)^2 + (ab^2)(c^2d^4) + (c^2d^4)^2]$
$= \boxed{(ab^2 - c^2d^4)(a^2b^4 + ab^2c^2d^4 + c^4d^8)}$

69. $125x^2 + 27y^3$
$= 5(5x)^2 + (3y)^3$ is $\boxed{\text{prime}}$, not factorable

71. $x^6 + 1$
$= (x^2)^3 + 1^3$
$= (x^2 + 1)[(x^2)^2 - x^2 + 1]$
$= \boxed{(x^2 + 1)(x^4 - x^2 + 1)}$

73. $(x + y)^3 + (x - y)^3$
$[(x + y + (x - y)][(x + y)^2 - (x + y)(x - y) + (x - y)^2]$
$= (x + y + x - y)(x^2 + 2xy + y^2 - x^2 + y^2 + x^2 - 2xy + y^2)$
$= \boxed{2x(x^2 + 3y^2)}$

75. $(2x - y)^3 - (2x + y)^3$
$= [(2x - y) - (2x + y)][(2x - y)^2 + (2x - y)(2x + y) + (2x + y)^2]$
$= (2x - y - 2x - y)(4x^2 - 4xy + y^2 + 4x^2 - y^2 + 4x^2 + 4xy + y^2)$
$= \boxed{-2y(12x^2 + y^2)}$

77. $x^{3n} - 8$
$= (x^n)^3 - 2^3$
$= (x^n - 2)[(x^n)^2 + x^n(2) + 2^2]$
$= \boxed{(x^n - 2)(x^{2n} + 2x^n + 4)}$

79. $125x^{3n} + 27y^{12m}$
$= (5x^n)^3 + (3y^{4m})^3$
$= (5x^n + 3y^{4m})[(5x^n)^2 - 5x^n(3y^{4m}) + (3y^{4m})^2]$
$= \boxed{(5x^n + 3y^{4m})(25x^{2n} - 15x^ny^{4m} + 9y^{8m})}$

81. $1000x^3y^{15} - 64w^6z^9$
$= 8(125x^3y^{15} - 8w^6z^9)$
$= 8[(5xy^5)^3 - (2w^2z^3)^3]$
$= 8(5xy^5 - 2w^2z^3)[(5xy^5)^2 + (5xy^5)(2w^2z^3) + (2w^2z^3)^2]$
$= \boxed{8(5xy^5 - 2w^2z^3)(25x^2y^{10} + 10xy^5w^2z^3 + 4w^4z^6)}$

83. $y^2 + 4y + 4 = y^2 + 2 \cdot y \cdot 2 + 2^2 = \boxed{(y + 2)^2}$

85. $z^2 - 10z + 25 = z^2 - 2 \cdot z \cdot 5 + 5^2 = \boxed{(z - 5)^2}$

87. $9x^2 - 12xy + 4y^2 = (3x)^2 - 2 \cdot 3\mathrm{ix} \cdot 2y + (2y)^2 = \boxed{(3x - 2y)^2}$

89. $25a^2b^2 - 20ab + 4$
$= (5ab)^2 - 2 \cdot 5ab \cdot 2 + 2^2$
$= \boxed{(5ab - 2)^2}$

91. $x^4 - 4x^2 + 4 = (x^2)^2 - 2 \cdot x^2 \cdot 2 + 2^2 = \boxed{(x^2 + 2)^2}$

93. $4x^4 - 20x^2y^2 + 25y^4$
$= (2x^2)^2 - 2 \cdot 2x^2 \cdot 5y^2 + (5y^2)$
$= \boxed{(2x^2 - 5y^2)^2}$

95. $(x + y)^2 + 2(x + y) + 1 = [(x + y) + 1]^2 = \boxed{(x + y + 1)^2}$

97. $(v - w)^2 + 4(v - w) + 4$
$= (v - w)^2 + 2 \cdot (v - w) \cdot 2 + 2^2$
$= [(v - w) + 2]^2$
$= \boxed{(v - w + 2)^2}$

99. $x^2 - 6x + 9 - y^2$
$= (x - 3)^2 - y^2$
$= \boxed{(x - 3 + y)(x - 3 - y)}$

101. $9x^2 - 30x + 25 - 36y^2$
$= (3x - 5)^2 - (6y)^2$
$= \boxed{(3x - 5 + 6y)(3x - 5 - 6y)}$

103. $r^2 - (16s^2 - 24s + 9($
$= r^2 - (4s - 3)^2$
$= (r + 4s - 3)[r - (4s - 3)]$
$= \boxed{(r + 4s - 3)(r - 4s + 3)}$

105. $y^2 - x^2 - 4x - 4$
$= y^2 - (x^2 + 4x + 4)$
$= y^2 - (x + 2)^2$
$= (y + x + 2)[y - (x + 2)]$
$= \boxed{(y + x + 2)(y - x - 2)}$

107. $z^2 - x^2 + 4xy - 4y^2$
$= z^2 - (x^2 - 4xy + 4y^2)$
$= z^2 - (x - 2y)^2$
$= (z + x - 2y)[z - (x - 2y)]$
$= \boxed{(z + x - 2y)(z - x + 2y)}$

109. $x^3 - y^3 - x + y$
$= (x^3 - y^3) - (x - y)$
$= (x - y)(x^2 + xy + y^2) - (x - y)$
$= \boxed{(x - y)(x^2 + xy + y^2 - 1)}$

111. $x^3 + y^3 - 3x - 3y$
$= (x^3 + y^3) - 3(x + y)$
$= (x + y)(x^2 - xy + y^2) - 3(x + y)$
$= \boxed{(x + y)(x^2 - xy + y^2 - 3)}$

113. $4x^{2n} + 20x^ny^m + 25y^{2m}$
$= (2x^n)^2 + 2 \cdot 2x^n \cdot 5y^m + (5y^m)^2$
$= \boxed{(2x^n + 5y^m)^2}$

115. $x^2 - 12x + K = x^2 - 2 \cdot x \cdot 6 + K$
In order to obtain the middle term of $-12x = -2 \cdot x \cdot 6$, K must be 6^2 or 36.

Thus, $x^2 - 12x + K = (x - 6)^2 = x^2 - 12x + 36$, and $\boxed{K = 36}$.

117. $Kx^2 + 8xy + y^2 = (\square x)^2 + 2 \cdot (\square x) \cdot y + y^2$
In order to obtain the middle term of $8xy = 2 \cdot 4x \cdot y$ the $\square x$ must be $4x$ or $\square = 4$ and $\square^2 = 4^2 = 16$.

Thus, $Kx^2 + 8xy + y^2 = (4x + y)^2 = 16x^2 + 8xy + y^2$, and $\boxed{K = 16}$.

119. Area of shaded region
= area of large square – area of small square
$$= x^2 - y^2$$
$$= \boxed{(x+y)(x-y)}$$

121. area of shaded region
= area of large square – 4(area of each small square)
$$= x^2 - 4y^2$$
$$= \boxed{(x+2y)(x-2y)}$$

123. $10^2 - 9^2 = (10+9)(10-9) = (19)(1) = 19$ True

125.

$x^6 - y^6$
$= (x^3)^2 - (y^3)^2$
$= (x^3 + y^3)(x^3 - y^3)$
$= (x+y)(x^2 - xy + y^2)(x-y)(x^2 + xy + y^2)$
$= (x+y)(x-y)(x^2 - xy + y^2)(x^2 - 1xy + y^2)$

$x^6 - y^6$
$= (x^2)^3 - (y^2)^3$
$= (x^2 - y^2)(x^4 + x^2 y^2 + y^4)$
$= (x+y)(x-y)(x^4 + x^2 y^2 + y^4)$

Since both factorizations are equal and involve $(x+y)(x-y)$, we can conclude that
$$\boxed{(x^2 - xy + y^2)(x^2 + xy + y^2) = x^4 + x^2 y^2 + y^4}$$

127. Using the formula $(c+d)^3 = c^3 + 3c^2 d + 3cd^2 + d^3$:
$27a^3 + 27a^2 b + 9ab^2 + b^3$
$= (3a)^3 + 3(3a)^2 b + 3(3a)b^2 + b^3$
$= \boxed{(3a+b)^3}$ $(c = 3a, d = b)$

129. Let

$$
\begin{aligned}
A_{\text{UL}} &= \text{area of upper left square} \\
A_{\text{UR}} &= \text{area of upper right rectangle} \\
A_{\text{LL}} &= \text{area of lower left rectangle} \\
A_{\text{LR}} &= \text{area of lower right square} \\
A_{\text{S}} &= \text{area of large square}
\end{aligned}
$$

$$
\begin{aligned}
A_{\text{UL}} + A_{\text{UR}} + A_{\text{LL}} + A_{\text{LR}} &= A_{\text{S}} \\
a_2 + ab + ab + b_2 &= (a+b)^2
\end{aligned}
$$
$$\boxed{a^2 + 2ab + b^2 = (a+b)^2}$$

131. $\boxed{D}$ is true; A is *not* true since $9x^2 + 15x + 25$ is prime and $(3x+5)^2 = 9x^2 + 30x + 25$
B is *not* true since $x^3 - 27 = (x-3)(x^2 + 3x + 9) \neq (x-3)(x^2 + 6x + 9)$
C is *not* true since $x^3 - 64 = (x-4)(x^2 + 4x + 16) \neq (x-4)^3$

Review Problems

134. $y > x - 2$, $x + y \geq -2$ and $y \leq 0$

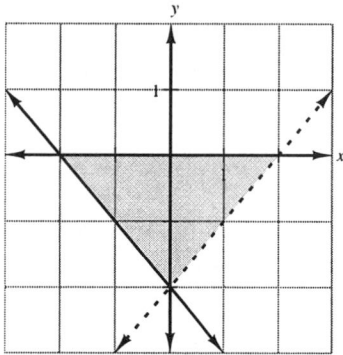

135. $f(x) = 7x^2 - 5x + 4$

$$\dfrac{f(a + h) - f(a)}{h}$$

$$= \dfrac{[7(a + h)^2 - 5(a + h) + 4] - (7a^2 - 5a + 4)}{h}$$

$$= \dfrac{7a^2 + 14ah + 7h^2 - 5a - 5h + 4 - 7a^2 + 5a - 4}{h}$$

$$= \dfrac{15ah + 7h^2 - 5h}{h}$$

$$= \boxed{14a + 7h - 5}$$

136.   Let

$$t = \text{time (in hours) when case will be 400 miles apart}$$

$$60t + 40t = 400$$
$$100t = 400$$
$$t = 4$$

$$\boxed{4 \text{ hours}}$$

Section 5.9   A General Factoring Strategy

Problem Set 5.9, pp. 399-401

1. $c^3 - 16c = c(c^2 - 16) = \boxed{c(c + 4)(c - 4)}$ **3.** $3x^2 + 18x + 27 = 3(x^2 + 6x + 9) = \boxed{3(x + 3)^2}$

5. $81x^3 - 3 = 3(27x^3 - 1) = \boxed{3(3x - 1)(9x^2 + 3x + 1)}$

7. $B^2C - 16C + 32 - 2B^2$ **9.** $-x^2 + 12x - 27 = -(x^2 - 12x + 27) = \boxed{-(x - 9)(x - 3)}$
$= C(B^2 - 16) - 2(B^2 - 16)$
$= (B^2 - 16)(C - 2)$
$= \boxed{(B + 4)(B - 4)(C - 2)}$

11. $4a^2b - 2ab - 30b$ **13.** $a(y^2 - 4) - 4(y^2 - 4)$
$= 2b(2a^2 - a - 15)$ $= (y^2 - 4)(a - 4)$
$= \boxed{2b(2a + 5)(a - 3)}$ $= \boxed{(y + 2)(y - 2)(a - 4)}$

15. $11x^5 - 11xy^2 = 11x(x^4 - y^2) = \boxed{11x(x^2 + y)(x^2 - y)}$

17. $3x^2 + 3x + 3y - 3y^2$
$= 3x^2 - 3y^2 + 3x + 3y$
$= 3(x^2 - y^2) + 3(x + y)$
$= 3[(x - y)(x + y) + (x + y)]$
$= \boxed{3(x + y)(x - y + 1)}$

19. $25x^2 - xy + 36y^2$ is $\boxed{\text{prime}}$, not factorable

21. $ax^3 + 8a = a(x^3 + 8) = \boxed{a(x + 2)(x^2 - 2x + 4)}$

23. $s^2 - 12s + 36 - 49t^2$
$= (s - 6)^2 - (7t)^2$
$= \boxed{(s - 6 + 7t)(s - 6 - 7t)}$

25. $4m^{10} + 12m^5n^3 + 9n^6$
$= (2m^5)^2 + 2(2m)^5(3n)^3 + (3n^3)^2$
$= \boxed{(2m^5 + 3n^3)}$

27. $9s^2t^2 - 36t^2$
$= 9t^2(s^2 - 4)$
$= \boxed{9t^2(s + 2)(s - 2)}$

29. $ax + ay + bx + by$
$= a(x + y) + b(x + y)$
$= \boxed{(x + y)(a + b)}$

31. $5x^2yz - 5y^3z$
$= 5yz(x^2 - y^2)$
$= \boxed{5yz(x + y)(x - y)}$

33. $20a^3b - 245ab^3$
$= 5ab(a^2 - 49b^2)$
$= \boxed{5ab(a + 7b)(a - 7b)}$

35. $63y^2 + 30y - 72$
$= 3(21y^2 + 10y - 24)$
$= \boxed{3(7y - 6)(3y + 4)}$

37. $r^6 + 4r^3s + 4s^2$
$= (r^3)^2 + 2 \cdot r^3 \cdot 2s + (2s)^2$
$= \boxed{(r^3 + 2s)^2}$

39. $4ax^3 - 32a$
$= 4a(x^3 - 8)$
$= \boxed{4a(x - 2)(x^2 + 2x + 4)}$

41. $100x^4 + 120x^3y + 36x^2y^2$
$= 4x^2(25x^2 + 30xy + 9y^2)$
$= \boxed{4x^2(5x + 3y)^2}$

43. $49x^2 + 126xy + 81y^2$
$= (7x)^2 + 2 \cdot (7x)(9y) + (9y)^2$
$= \boxed{(7x + 9y)^2}$

45. $71bx^4 - 71b$
$= 71b(x^4 - 1)$
$= 71b(x^2 + 1)(x^2 - 1)$
$= \boxed{71b(x^2 + 1)(x + 1)(x - 1)}$

47. $x^2 + 25$ is $\boxed{\text{prime}}$, not factorable

49. $r^3 - s^3 + r - s$
$= (r - s)(r^2 + rs + s^2) + (r - s)$
$= \boxed{(r - s)(r^2 + rs + s^2 + 1)}$

51. $x^{2n} - y^{2m}$
$= \boxed{(x^n + y^m)(x^n - y^m)}$

53. $a^2 + 4a + 4 - 16b^2$
$= (a + 2)^2 - (4b)^2$
$= \boxed{(a + 2 + 4b)(a + 2 - 4b)}$

55. $27r^3s + 72r^2s^2 + 48rs^2$
$= 3rs(9r^2 + 24rs + 16s^2)$
$= \boxed{3rs(3r + 4s)^2}$

57. $-6by^3 + 24by$
$= -6by(y^2 - 4)$
$= \boxed{-6by(y + 2)(y - 2)}$

59. $16x^2 + 49y^2$ is $\boxed{\text{prime}}$, not factorable

61. $(3x - y)^2 - 100a^2$
$= (3x - y)^2 - (10a)^2$
$= \boxed{(3x - y + 10a)(3x - y - 10a)}$

63. $(5x + 3y)^2 - 6(5x + 3y) + 9$
$= a^2 - 6a + 9$ (Let $a = 5x + 3y$)
$= (a - 3)^2$
$= \boxed{(5x + 3y - 3)^2}$ (substitute $5x + 3y$ for a)

65. $6y^4 - 11y^2 - 10$
$= \boxed{(2y^2 - 5)(3y^2 + 2)}$

67. $48y^4 - 243$
$= 3(16y^4 - 81)$
$= 3(4y^2 + 9)(4y^2 - 9)$
$= \boxed{3(4y^2 + 9)(2y + 3)(2y - 3)}$

69. $20bx^4 + 220bx^2y + 605by^2$
$= 5b(4x^4 + 44x^2y + 121y^2)$
$= \boxed{5b(2x^2 + 11y)^2}$

71. $18x^3 + 63x^2 - 36x$
$= 9x(2x^2 + 7x - 4)$
$= \boxed{9x(2x - 1)(x + 4)}$

73. $4x^7 + 32xy^3$
$= 4x(x^6 + 8y^3)$
$= 4x[(x^2)^3 + (2y)^3]$
$= \boxed{4x(x^2 + 2y)(x^4 - 2x^2y + 4y^2)}$

75. $36(c - d)y^2 - 6(c - d)yx - 20(c - d)x^2$
$= 2(c - d)(18y^2 - 3yx - 10x^2)$
$= \boxed{2(c - d)(6y - 5x)(3y + 2x)}$

77. $x^8 - y^{12}$
$= (x^4)^2 - (y^6)^2$
$= (x^4 + y^6)(x^4 - y^6)$
$= \boxed{(x^4 + y^6)(x^2 + y^3)(x^2 - y^3)}$

79. $4x^2y + 5 - 20x^2 - y$
$= 4x^2y - y - 20x^2 + 5$
$= y(4x^2 - 1) - 5(4x^2 - 1)$
$= (4x^2 - 1) - (y - 5)$
$= \boxed{(2x + 1)(2x - 1)(y - 5)}$

81. $x^4 - 25x^2 - 144$ is $\boxed{\text{prime}}$, not factorable

83. $4r^2s^2 - 4r^2 - 9s^2 + 9$
$= 4r^2(s^2 - 1) - 9(s^2 - 1)$
$= (s^2 - 1)(4r^2 - 9)$
$= \boxed{(s + 1)(s - 1)(2r + 3)(2r - 3)}$

85. $y^{2n+1} + 2y^{n+1} + y$
$= y(y^{2n} + 2y^n + 1)$
$= \boxed{y(y^n + 1)^2}$

87. $rs^2 - 2a - r + 2as^2$
$= rs^2 - r + 2as^2 - 2a$
$= r(s^2 - 1) + 2a(s^2 - 1)$
$= (s^2 - 1)(r + 2a)$
$= \boxed{(s + 1)(s - 1)(r + 2a)}$

89. $7y^2 - 28$
$= 7(y^2 - 4)$
$= \boxed{7(y + 2)(y - 2)}$

91. $r^3s^3 - r^3$
$= r^3(s^3 - 1)$
$= \boxed{r^3(s - 1)(s^2 + s + 1)}$

93. $(4x^2 - 12xy + 9y^2) + (72by - 48bx) - 25b^2$
$= (2x - 3y)^2 - 24b(2x - 3y) - 25b^2$
$= a^2 - 24ab - 25b^2$ (Let $a = 2x - 3y$)
$= (a - 25b)(a + b)$
$= \boxed{(2x - 3y - 25b)(2x - 3y + b)}$ (substitute $2x - 3y$ for a)

95. $16a^3b + 4a^2b^2 - 42ab^3$
$= 2ab(8ab^2 + 2ab - 12b^2)$
$= \boxed{2ab(4a + 7b)(2a - 3b)}$

97. $2x^{n+2} - 7x^{n+1} + 3x^n$
$= x^n(2x^2 - 7x + 3)$
$= \boxed{x^n(2x - 1)(x - 3)}$

99. $a^6b^6 - a^3b^3$
$= a^3b^3(a^3b^3)$
$= \boxed{a^3b^3(ab - 1)(a^2b^2 + ab + 1)}$

101. $10x^3 - 6x^2 - 21x$

$= \boxed{x(10x^2 - 6x - 21)}$

103. $\dfrac{4}{3}\pi r^3 - \dfrac{4}{3}\pi s^3$

$= \dfrac{4}{3}\pi(r^3 - s^3)$

$= \boxed{\dfrac{4}{3}\pi(r - s)(r^2 + rs + s^2)}$

105. $\dfrac{2x^3 + 9x^2 - 26x + 12}{2x - 3} = x^2 + 6x - 4$

$$
\begin{array}{r}
x^2 + 6x - 4 \\
2x - 3\overline{\smash{\big)}\ 2x^3 + 9x^2 - 26x + 12} \\
\underline{2x^3 - 3x^2} \\
12x^2 - 26x \\
\underline{12x^2 - 18x} \\
-8x + 12 \\
\underline{-8x + 12} \\
0
\end{array}
$$

$2x^3 + 9x^2 - 26x + 12 = \boxed{(2x - 3)(x^2 + 6x - 4)}$

107. $\dfrac{y^3 - 5y^2 + 10y - 8}{y - 2} = y^2 - 3y + 4$

$$
\begin{array}{r|rrrr}
2 & 1 & -5 & 10 & -8 \\
  &   & 2 & -6 & 8 \\
\hline
  & 1 & -3 & 4 & 0
\end{array}
$$

$y^3 - 5y^2 + 10y - 8 = \boxed{(y - 2)(y^2 - 3y + 4)}$

109. $\dfrac{x^4 + 4x^3 + 3x^2 - 2x - 1}{x^2 + 3x + 1} = x^2 + x - 1$

$$
\begin{array}{r}
x^2 + x - 1 \\
x^2 + 3x + 1\overline{\smash{\big)}\ x^4 + 4x^3 + 3x^2 - 2x - 1} \\
\underline{x^4 + 3x^3 + x^2} \\
x^3 + 2x^2 - 2x \\
\underline{x^3 + 3x^2 + x} \\
-x^2 - 3x - 1 \\
\underline{-x^2 - 3x - 1} \\
0
\end{array}
$$

$x^4 + 4x^3 + 3x^2 - 2x - 1 = \boxed{(x^2 + 3x + 1)(x^2 + x - 1)}$

111. $\boxed{B}$ is true

113. $x^4 + 2x^2y^2 + 9y^4$

$= (x^4 + 6x^2 + 9y^4) - 4x^2y^2$

$= (x^2 + 3y^2)^2 - (2xy)^2$

$= \boxed{(x^2 + 3y^2 + 2xy)(x^2 + 3y^2 - 2xy)}$

115. a. Using the formula
$$a^5 + b^5 = (a + b)(a^4 - a^3b + a^2b^2 - ab^3 + b^4):$$
$$32x^5 + 1$$
$$= (2x)^5 + 1$$
$$= (2x + 1)[(2x)^4 - (2x)^3(1) + (2x)^2(1^2) - (2x)(1^3) + 1^4]$$
$$= \boxed{(2x + 1)(16x^4 - 8x^3 + 4x^2 - 2x + 1)}$$

b. $\dfrac{a^5 + b^5}{a + b} = a^4 - a^3b + a^2b^2 - ab^3 + b^4$

$$
\begin{array}{r}
a^4 - a^3b + a^2b^2 - ab^3 + b^4 \\
a + b \,\overline{\big)\, a^5 + 0a^4b + 0a^3b^2 + 0a^2b^3 + 0ab^4 + b^5} \\
\underline{a^5 + a^4b} \qquad\qquad\qquad\qquad\qquad\qquad \\
-a^4b + 0a^3b^2 \qquad\qquad\qquad\qquad\quad \\
\underline{-a^4b - a^3b^2} \qquad\qquad\qquad\qquad\quad \\
a^3b^2 + 0a^2b^3 \qquad\qquad\qquad \\
\underline{a^3b^2 + a^2b^3} \qquad\qquad\qquad \\
-a^2b^3 + 0ab^4 \qquad\quad \\
\underline{-a^2b^3 - ab^4} \qquad\quad \\
ab^4 + b^5 \\
\underline{ab^4 + b^5} \\
0
\end{array}
$$

117. a. Using the formula
$$a^7 + b^7 = (a + b)(a^6 - a^5b + a^4b^2 - a^3b^3 + a^2b^4 - ab^5 + b^6)$$
$$x^7y^{14} + 128z^{21}$$
$$= (xy^2)^7 + (2z^3)^7$$
$$= (xy^2 + 2z^3)[(xy^2)^6 - (xy^2)^5(2z^3) + (xy^2)^4(2z^3)^2 - (xy^2)^3(2z^3)^3 + (xy^2)^2(2z^3)^4 - (xy^2)(2z^3)^5 + (2z^3)^6]$$
$$= \boxed{(xy^2 + 2z^3)(x^6y^{12} - 2x^5y^{10}z^3 + 4x^4y^8z^6 - 8x^3y^6z^9 + 16x^2y^4z^{12} - 32xy^2z^{15} + 64z^{18})}$$

b. $(a + b)(a^6 - a^5b + a^4b^2 - a^3b^3 + a^2b^4 - ab^5 + b^6)$
$$= a^7 - a^6b + a^5b^2 - a^4b^3 + a^3b^4 - a^2b^5 + ab^6 + a^6b - a^5b^2 + a^4b^3 - a^3b^4 + a^2b^5 - ab^6 + b^7)$$
$$= a^7 + b^7$$

119. $a^n + b^n = (a + b)(a^{n-1} - a^{n-2}b + a^{n-3}b^2 - a^{n-4}b^3 + \ldots + b^{n-1})$
   a. $n = 3$:
   $$a^3 + b^3 = (a + b)(a^2 - ab + b^2)$$
   b. $n = 5$:
   $$a^5 + b^5 = (a + b)(a^4 - a^3b + a^2b^2 - ab^3 + b^4)$$
   c. $n = 7$:
   $$a^7 + b^7 = (a + b)(a^6 - a^5b + a^4b^2 - a^3b^3 + a^2b^4 -\neq ab^5 + b^6)$$
   d. $n = 9$:
   $$\boxed{a^9 + b^9 = (a + b)(a^8 - a^7b + a^6b^2 - a^5b^3 + a^4b^4 - a^3b^5 + a^2b^6 - ab^7 + b^8)}$$
   e. $a = x^2,\ b = y^3,\ n = 9$:
   $$x^{18} + y^{27} = (x^2)^9 + (y^3)^9$$
   $$\boxed{x^{18} + y^{27} = (x^2 + y^3)(x^{16} - x^{14}y^3 + x^{12}y^6 - x^{10}y^9 + x^8y^{12} - x^6y^{15} + x^4y^{18} - x^2y^{21} + y^{24})}$$

121. $x^4 - y^4 - 2x^3y + 2xy^3$
$$= (x^2 + y^2)(x^2 - y^2) - 2xy(x^2 - y^2)$$
$$= (x^2 - y^2)(x^2 + y^2 - 2xy)$$
$$= (x^2 - y^2)(x^2 - 2xy + y^2)$$
$$= (x^2 - y^2)(x - y)^2$$
$$= (x + y)(x - y)(x - y)^2$$
$$= \boxed{(x + y)(x - y)^3}$$

Review Problems

127. $\left[\dfrac{7+(-16)}{17-101}\right]\left[\dfrac{12+(-2)}{3+(-2)^3}\right] = \left(\dfrac{-9}{|-3|}\right)\left(\dfrac{10}{3-8}\right) = \left(\dfrac{-9}{3}\right)\left(\dfrac{10}{-5}\right) = (-3)(-2) = \boxed{6}$

128. Let
$$W \;=\; \text{weight of an object}$$
$$d \;=\; \text{distance from Earth's center}$$
$$W \;=\; \dfrac{K}{d^2}$$

Given: $d = 12{,}000$ meters, $W = 40$ pounds
$$40 \;=\; \dfrac{K}{(12{,}000)^2}$$
$$5.76 \times 10^9 \;=\; K$$
$$W \;=\; \dfrac{5.76 \times 10^9}{d^2}$$

At $d = 8000$ meters, find W.
$$W \;=\; \dfrac{5.76 \times 10^9}{(8000)^2} = 90$$

The object's weight is $\boxed{90\ \text{lb}}$.

129. $f(x) = 4x - 5$ and $g(x) = -x^2 + 7x$
$g(-2) = -(-2)^2 + 7(-2) = -4 - 14 = -18$
$f[g(-2)] = f(-18) = 4(-18) - 5 = -72 - 5 = \boxed{-77}$

Section 5.10   Solving Quadratic Equations by Factoring

Problem Set 5.10, pp. 410-413

For problems 1-45, the check is left to the reader.

1. $(x - 7)(x + 3) = 0$

$x - 7 = 0$	$x + 3 = 0$
$x = 7$	$x = -3$

            or

A check of both values in the origianl equation confirmes that the solution set is $\boxed{\{-3, 7\}}$.

3. $(2x + 3)(5x - 1) = 0$

$2x + 3 = 0$	$5x - 1 = 0$
$2x = -3$	$5x = 1$
$x = -\dfrac{3}{2}$	$x = \dfrac{1}{5}$

     or

$$\boxed{\left\{ -\dfrac{3}{2}, \dfrac{1}{5} \right\}}$$

5. $3x^2 + 10x - 8 = 0$
$(x + 4)(3x - 2) = 0$

$x + 4 = 0$ or $3x - 2 = 0$
$x = -4$ $3x = 2$
$x = \dfrac{2}{3}$

$$\boxed{\left\{-4, \dfrac{2}{3}\right\}}$$

7. $5x^2 - 8x + 3 = 0$
$(5x - 3)(x - 1) = 0$

$5x - 3 = 0$ or $x - 1 = 0$
$5x = 3$ $x = 1$
$x = \dfrac{3}{5}$

$$\boxed{\left\{\dfrac{3}{5}, 1\right\}}$$

9. $6x^2 - x - 35 = 0$
$(3x + 7)(2x - 5) = 0$

$3x + 7 = 0$ or $2x - 5 = 0$
$3x = -7$ $2x = 5$
$x = -\dfrac{7}{3}$ $x = \dfrac{5}{2}$

$$\boxed{\left\{-\dfrac{7}{3}, \dfrac{5}{2}\right\}}$$

11. $5x^2 + 26x + 5 = 0$
$(5x + 1)(x + 5) = 0$

$5x + 1 = 0$ or $x + 5 = 0$
$5x = -1$ $x = -5$
$x = -\dfrac{1}{5}$

$$\boxed{\left\{-\dfrac{1}{5}, -5\right\}}$$

13. $5x^2 - 3x - 2 = 0$
$(5x + 2)(x - 1) = 0$

$5x + 2 = 0$ or $x - 1 = 0$
$5x = -2$ $x = 1$
$x = -\dfrac{2}{5}$

$$\boxed{\left\{-\dfrac{2}{5}, 1\right\}}$$

15. $3x^2 + 5x - 2 = 0$
$(x + 2)(3x - 1) = 0$

$x + 2 = 0$ or $3x - 1 = 0$
$x = -2$ $3x = 1$
$x = \dfrac{1}{3}$

$$\boxed{\left\{-2, \dfrac{1}{3}\right\}}$$

17.
$$5x^2 - 8x - 21 = 0$$
$$(5x + 7)(x - 3) = 0$$

$$5x + 7 = 0 \qquad \text{or} \qquad x - 3 = 0$$
$$5x = -7 \qquad\qquad\qquad x = 3$$
$$x = -\frac{7}{5}$$

$$\boxed{\left\{-\frac{7}{5}, 3\right\}}$$

19.
$$x^2 - x = 2$$
$$x^2 - x - 2 = 0$$
$$(x - 2)(x + 1) = 0$$

$$x - 2 = 0 \qquad \text{or} \qquad x + 1 = 0$$
$$x = 2 \qquad\qquad\qquad x = -1$$

$$\boxed{\{-1, 2\}}$$

21.
$$3x^2 - 17x = -10$$
$$3x^2 - 17x + 10 = 0$$
$$(3x - 2)(x - 5) = 0$$

$$3x - 2 = 0 \qquad \text{or} \qquad x - 5 = 0$$
$$3x = 2 \qquad\qquad\qquad x = 5$$
$$x = \frac{2}{3}$$

$$\boxed{\left\{\frac{2}{3}, 5\right\}}$$

23.
$$x(x - 3) = 54$$
$$x^2 - 3x - 54 = 0$$
$$(x - 9)(x + 6) = 0$$

$$x - 9 = 0 \qquad \text{or} \qquad x + 6 = 0$$
$$x = 9 \qquad\qquad\qquad x = -6$$

$$\boxed{\{-6, 9\}}$$

25.
$$x(2x + 1) = 3$$
$$2x^2 + x - 3 = 0$$
$$(2x + 3)(x - 1) = 0$$

$$2x + 3 = 0 \qquad \text{or} \qquad x - 1 = 0$$
$$2x = -3 \qquad\qquad\qquad x = 1$$
$$x = -\frac{3}{2}$$

$$\boxed{\left\{-\frac{3}{2}, 1\right\}}$$

27.
$$x^2 = \frac{5}{6}x + \frac{2}{3}$$
$$6x^2 = 5x + 4$$
$$6x^2 - 5x - 4 = 0$$
$$(2x + 1)(3x - 4) = 0$$

$$2x + 1 = 0 \qquad \text{or} \qquad 3x - 4 = 0$$
$$2x = -1 \qquad\qquad\qquad 3x = 4$$
$$x = -\frac{1}{2} \qquad\qquad\qquad x = \frac{4}{3}$$

$$\boxed{\left\{-\frac{1}{2}, \frac{4}{3}\right\}}$$

29. $(x+1)^2 - 5(x+2) = 3x + 7$
$x^2 + 2x + 1 - 5x - 10 = 3x + 7$
$x^2 - 6x - 16 \ = \ 0$
$(x-8)(x+2) \ = \ 0$

$x - 8 \ = \ 0$	$x + 2 \ = \ 0$
$x \ = \ 8$	$x \ = \ -2$

$\boxed{\{-2, 8\}}$

31. $\frac{1}{6}x^2 + x - \frac{1}{2} \ = \ -2$
$x^2 + 6x - 2 \ = \ -12$
$x^2 + 6x + 9 \ = \ 0$
$(x+3)^2 \ = \ 0$
$x + 3 \ = \ 0$
$x \ = \ -3$

$\boxed{\{-3\}}$

33. $x + (x+2)^2 \ = \ 130$
$x + x^2 + 4x + 4 \ = \ 130$
$x^2 + 5x - 126 \ = \ 0$
$(x+14)(x-9) \ = \ 0$

$x + 14 \ = \ 0$	$x - 9 \ = \ 0$
$x \ = \ -14$	$x \ = \ 9$

$\boxed{\{-14, 9\}}$

35. $3(x^2 - 4x - 1) \ = \ 2(x+1)$
$3x^2 - 12x - 3 \ = \ 2x + 2$
$3x^2 - 14x - 5 \ = \ 0$
$(3x+1)(x-5) \ = \ 0$

$3x + 1 \ = \ 0$	$x - 5 \ = \ 0$
$3x \ = \ -1$	$x \ = \ 5$
$x \ = \ -\frac{1}{3}$	

$\boxed{\left\{-\frac{1}{3}, 5\right\}}$

37. $9x^2 + 6x \ = \ -1$
$9x^2 + 6x + 1 \ = \ 0$
$(3x+1)^2 \ = \ 0$
$3x + 1 \ = \ 0$
$3x \ = \ -1$
$x \ = \ -\frac{1}{3}$

$\boxed{\left\{-\frac{1}{3}\right\}}$

39.
$$
\begin{aligned}
25 &= 30x - 9x^2 \\
9x^2 - 30x + 25 &= 0 \\
(3x - 5)^2 &= 0 \\
3x - 5 &= 0 \\
3x &= 5 \\
x &= \frac{5}{3}
\end{aligned}
$$

$$\boxed{\left\{\frac{5}{3}\right\}}$$

41.
$$
\begin{aligned}
(x + 2)(x - 5) &= 0 \\
x^2 - 3x - 10 &= 8 \\
x^2 - 3x - 18 &= 0 \\
(x + 3)(x - 6) &= 0 \\
x + 3 &= 0 \qquad \text{or} \qquad x - 6 = 0 \\
x &= -3 \qquad\qquad\qquad\quad x = 6
\end{aligned}
$$

$$\boxed{\{-3, 6\}}$$

43.
$$
\begin{aligned}
2x - [(x + 2)(x - 3) + 8] &= 0 \\
2x - (x^2 - x - 6 + 8) &= 0 \\
2x - (x^2 - x + 2) &= 0 \\
2x - x^2 + x - 2 &= 0 \\
-x^2 + 3x - 2 &= 0 \\
x^2 - 3x + 2 &= 0 \\
(x - 1)(x - 2) &= 0 \\
x - 1 &= 0 \qquad \text{or} \qquad x - 2 = 0 \\
x &= 1 \qquad\qquad\qquad\quad x = 2
\end{aligned}
$$

$$\boxed{\{1, 2\}}$$

45.
$$
\begin{aligned}
3[(x + 2)^2 - 4x] &= 15 \\
3(x^2 + 4x + 4 - 4x) &= 15 \\
3(x^2 + 4) &= 15 \\
3x^2 + 12 &= 15 \\
3x^2 - 3 &= 0 \\
3(x^2 - 1) &= 0 \\
3(x + 1)(x - 1) &= 0 \\
x + 1 &= 0 \qquad \text{or} \qquad x - 1 = 0 \\
x &= -1 \qquad\qquad\qquad\quad x = 1
\end{aligned}
$$

$$\boxed{\{-1, 1\}}$$

47. $s(t) = -16t^2 + 64t + 80$

    a. The ball reaches the ground at $s(t) = 0$.
$$
\begin{aligned}
0 &= -16t^2 + 64t + 80 \\
16t^2 - 64t - 80 &= 0 \\
16(t^2 - 4t - 5) &= 0 \\
16(t - 5)(t + 1) &= 0 \\
t - 5 &= 0 \qquad \text{or} \qquad x - 1 = 0 \\
t &= 5 \qquad\qquad\qquad\quad t = -1 \qquad \text{(Reject } t = -1 \text{ since it is physically impossible)}
\end{aligned}
$$

    It takes the ball $\boxed{5 \text{ seconds}}$ to reach the ground.

b. $s(t) = 80$

$$80 = -16t^2 + 64t + 80$$
$$16t^2 - 64t = 0$$
$$16t(t - 4) = 0$$
$$t - 4 = 0 \quad\quad \text{or} \quad\quad t = 0 \quad \text{(Reject } t = 0 \text{ since } t = 0 \text{ represents the initial time}$$
$$t = 4 \quad\quad\quad\quad\quad\quad\quad\quad\quad\quad\quad\quad \text{when the ball was thrown upward)}$$

The ball passes the edge of the top of the building after $\boxed{4 \text{ seconds}}$.

49. $D = \dfrac{7}{10}x^2 + \dfrac{3}{4}x$

($D = 294$ feet):

$$\frac{7}{10}x^2 + \frac{3}{4}x = 295$$
$$14x^2 + 15x = 5900$$
$$14x^2 + 15x - 5900 = 0$$
$$(x - 20)(14x + 295) = 0$$
$$x - 20 = 0 \quad\quad \text{or} \quad\quad 14x + 295 = 0$$
$$x = 20 \quad\quad\quad\quad\quad\quad\quad 14x = -295$$
$$x = -\frac{295}{14} \quad \text{(Reject since not possible)}$$

Murray should travel at $\boxed{20 \text{ mph}}$.

51. $P = 3500 + 475t - 10t^2$ where $= \leq t \leq 20$

($P = 7250$):

$$7250 = 3500 + 475t - 10t^2$$
$$10t^2 - 475t + 3750 = 0$$
$$5(2t^2 - 95t + 750) = 0$$
$$5(t - 10)(2t - 75) = 0$$
$$t - 10 = 0 \quad\quad \text{or} \quad\quad 2t - 75 = 0$$
$$t = 10 \quad\quad\quad\quad\quad\quad 2t = 75$$
$$t = 37.5 \quad \text{(Reject since } 0 \leq t \leq 20\text{)}$$

$\boxed{10 \text{ years}}$

53. $P = -\dfrac{1}{50}A^2 + 2A + 22$

($P = 72$):

$$72 = -\frac{1}{5}A^2 + 2A + 22$$
$$\frac{1}{50}A^2 - 2A + 50 = 0$$
$$A^2 - 100A + 2500 = 0$$
$$(A - 50)^2 = 0$$
$$A - 50 = 0$$
$$A = 50$$

$\boxed{\text{Arousal level should be 50.}}$

55. Let x and $x + 2$ equal two consecutive odd integers.

$$(x + 2)^2 - 3x = 34$$
$$x^2 + 4x - 4 - 3x = 34$$
$$x^2 + x - 30 = 0$$
$$(x + 6)(x - 5) = 0$$
$$x + 6 = 0 \quad\quad \text{or} \quad\quad x - 5 = 0$$
$$x = -6 \quad\quad\quad\quad\quad\quad x = 5$$
$$\text{(Reject since } -6 \text{ is even } not \text{ odd)} \quad\quad x + 2 = 5 + 2 = 7$$

The integers are $\boxed{5 \text{ and } 7}$.

57. Let $x, x + 1$, and $x + 2$ equal three consecutive positive integers.

$$
\begin{aligned}
(x + 1)^2 - 5(x + 2) &= 3x + 7 \\
x^2 + 2x + 1 - 5x - 10 &= 3x + 7 \\
x^2 - 6x - 16 &= 0 \\
(x - 8)(x + 2) &= 0
\end{aligned}
$$

$$
\begin{array}{lll}
x - 8 = 0 & \text{or} & x + 2 = 0 \\
x = 8 & & x = -2 \quad \text{(Reject since –2 is } \textit{not } \text{positive)} \\
x + 1 = 9 & & \\
x + 2 = 10 & &
\end{array}
$$

The integers are $\boxed{8, 9, \text{ and } 10}$.

59. Let $x, x + 2$, and $x + 4$ equal three consecutive even integers.

$$
\begin{aligned}
x^2 + (x + 2)^2 &= (x + 4)^2 \\
x^2 + x^2 + 4x + 4 &= x^2 + 8x + 16 \\
x^2 - 4x - 12 &= 0 \\
(x - 6)(x + 2) &= 0
\end{aligned}
$$

$$
\begin{array}{lll}
x - 6 = 0 & \text{or} & x + 2 = 0 \\
x = 6 & & x = -2 \\
x + 2 = 6 + 2 = 8 & & x + 2 = -2 + 2 = 0 \\
x + 4 = 6 + 4 = 10 & & x + 4 = -2 + 4 = 2
\end{array}
$$

The integers are $\boxed{6, 8, \text{ and } 10}$ or $\boxed{-2, 0 \text{ and } 2}$.

61. Let

$$
\begin{aligned}
x &= \text{first number} \\
6x - 11 &= \text{second number}
\end{aligned}
$$

$$
\begin{aligned}
x(6x - 11) &= 7 \\
6x^2 - 11x &= 7 \\
6x^2 - 11x - 7 &= 0 \\
(3x - 7)(2x + 1) &= 0
\end{aligned}
$$

$$
\begin{array}{lll}
3x - 7 = 0 & \text{or} & 2x + 1 = 0 \\
x = \dfrac{7}{3} & & x = -\dfrac{1}{2} \\[2mm]
6x - 11 = 6\left(\dfrac{7}{3}\right) - 11 = 14 - 11 = 3 & & 6x - 11 = 6\left(-\dfrac{1}{2}\right) - 11 = -3 - 11 = -14
\end{array}
$$

The numbers are $\boxed{\dfrac{7}{3} \text{ and } 3}$ or $\boxed{-\dfrac{1}{2} \text{ and } -14}$.

63. total areas of pool and path = 600 square meters

$$
\begin{aligned}
(10 + 2x)(20 + 2x) &= 600 \\
200 + 20x + 40x + 4x^2 &= 600 \\
4x^2 + 60x - 400 &= 0 \\
4(x^2 + 15x - 100) &= 0 \\
(x - 5)(x + 20) &= 0
\end{aligned}
$$

$$
\begin{array}{lll}
x - 5 = 0 & \text{or} & x + 20 = 0 \\
x = 5 & & x = -20 \quad \text{(Reject since not possible)}
\end{array}
$$

The width of the path is $\boxed{5 \text{ meters}}$.

65. area of new garden = 48 square yards

$$
\begin{aligned}
(4 + x)(6 + x) &= 48 \\
24 + 10x + x^2 &= 48 \\
x^2 + 10x - 24 &= 0 \\
(x - 2)(x + 12) &= 0
\end{aligned}
$$

$$
\begin{array}{lll}
x - 2 = 0 & \text{or} & x + 12 = 0 \\
x = 2 & & x = -12 \quad \text{(reject)}
\end{array}
$$

The length and width should each be increased by $\boxed{2 \text{ yards}}$.

67. Volume = 128 cubic inches

$$
\text{length} \times \text{width} \times \text{height} = 128
$$

$$
\begin{aligned}
(x - 4)(x - 4)(2) &= 128 \\
(\div 2) \qquad (x - 4)^2 &= 64 \\
x^2 - 8x + 16 - 64 &= 0 \\
x^2 - 8x - 48 &= 0 \\
(x - 12)(x + 4) &= 0
\end{aligned}
$$

$$
\begin{array}{lll}
x - 12 = 0 & \text{or} & x + 4 = 0 \\
x = 12 & & x = -4 \quad \text{(reject)}
\end{array}
$$

The dimensions of the piece of square cardboard are $\boxed{12 \text{ inches} \times 12 \text{ inches}}$.

69. By the Pythagorean theorem

$$
\begin{aligned}
x^2 + (2x + 2)^2 &= 13^2 \\
x^2 + 4x^2 + 8x + 4 &= 169 \\
5x^2 + 8x - 165 &= 0 \\
(x - 5)(5x + 33) &= 0
\end{aligned}
$$

$$
\begin{array}{lll}
x - 5 = 0 & \text{or} & 5x + 33 = 0 \\
x = 5 \quad \text{(width)} & & x = -\dfrac{33}{5} \quad \text{(Reject)}
\end{array}
$$

$$
2x + 2 = 2(5) + 2 = 12 \qquad \text{(length)}
$$

$\boxed{\text{width: } 5 \text{ feet;}}$

$\boxed{\text{length: } 12 \text{ feet}}$

71. Let

$$
\begin{aligned}
x &= \text{the length of the shorter leg} \\
2x - 1 &= \text{the lngth of the longer leg} \\
2x + 1 &= \text{the length of the hypotenuse}
\end{aligned}
$$

$$
\begin{aligned}
x^2 + (2x - 1)^2 &= (2x + 1)^2 \\
x^2 + 4x^2 - 4x + 1 &= 4x^2 + 4x + 1 \\
x^2 - 8x &= 0 \\
x(x - 8) &= 0
\end{aligned}
$$

$$
\begin{array}{lll}
x = 0 \quad \text{(Reject)} & \text{or} & x - 8 = 0 \\
& & x = 8 \\
& & 2x - 1 = 2(8) - 1 = 15 \\
& & 2x + 1 = 2(8) + 1 = 17
\end{array}
$$

The lengths of the sides of the triangular are $\boxed{8 \text{ inches, } 15 \text{ inches and } 17 \text{ inches}}$.

73. Let

$$
\begin{aligned}
x &= \text{width of rectangular solid} \\
4x + 1 &= \text{height of rectangular solid} \\
\text{length} &= 3 \text{ yards, volume} = 54 \text{ cubic yards}
\end{aligned}
$$

$$
\begin{aligned}
3x(4x + 1) &= 54 \\
(\div 3) \quad x(4x + 1) &= 18 \\
4x^2 + x - 18 &= 0 \\
(x - 2)(4x + 9) &= 0
\end{aligned}
$$

$$
\begin{array}{lll}
x - 2 = 0 & \text{or} & 4x + 9 = 0 \\
x = 2 \quad \text{(width)} & & x = -\dfrac{9}{4} \quad \text{(Reject)}
\end{array}
$$

$$
4x + 1 = 4(2) + 1 = 9 \quad \text{(height)}
$$

width: 2 yards;

height: 9 yards

75. Let

$$
\begin{aligned}
x &= \text{height of rectangular solid} \\
2x &= \text{length of rectangular solid} \\
4x - 5 &= \text{width of rectangular solid} \\
\text{surface area} &= 162 \text{ square yards}
\end{aligned}
$$

$$
\begin{aligned}
2(\text{height} \times \text{length}) + 2(\text{height} \times \text{width}) + 2(\text{length} \times \text{width}) &= 162 \\
2x(2x) + 2x(4x - 5) + 2(2x)(4x - 5) &= 162 \\
4x^2 + 8x^2 - 10x + 16x^2 - 20x &= 162 \\
28x^2 - 30x - 162 &= 0 \\
2(14x^2 - 15x - 81) &= 0 \\
2(x - 3)(14x + 27) &= 0
\end{aligned}
$$

$$
\begin{array}{lll}
x - 3 = 0 & \text{or} & 14x + 27 = 0 \\
x = 3 \quad \text{(height)} & & x = -\dfrac{27}{14} \quad \text{(Reject)}
\end{array}
$$

$$
\begin{aligned}
2x &= 2(3) = 6 \quad \text{(length)} \\
4x - 5 &= 4(3) - 5 = 7 \quad \text{(width)}
\end{aligned}
$$

height: 3 yards;

length: 6 yards;

width: 7 yards

77. Let

$$
\begin{aligned}
x &= \text{width of rectangle} \\
x + 3 &= \text{length of rectangle}
\end{aligned}
$$

$$
\begin{aligned}
(\text{area}) &= (\text{perimeter}) \\
x(x + 3) &= 2x + 2(x + 3) \\
x^2 + 3x &= 2x + 2x + 6 \\
x^2 - x - 6 &= 0 \\
(x - 3)(x + 2) &= 0
\end{aligned}
$$

$$
\begin{array}{lll}
x - 3 = 0 & \text{or} & x + 2 = 0 \\
x = 3 \quad \text{(width)} & & x = -2 \quad \text{(reject)} \\
x + 3 = 6 \quad \text{(length)}
\end{array}
$$

The dimensions of the rectangle are 3 yards $\times$ 6 yards .

79. Let

$$
\begin{aligned}
x &= \text{age of chair} \\
x + 12 &= \text{age of chair in 12 years} \\
x + 24 &= \text{age of chair in 24 years}
\end{aligned}
$$

$$
\begin{aligned}
(x + 12)^2 - 24(x + 24) &= 9 \\
x^2 + 24x + 144 - 24x - 576 - 9 &= 0 \\
x^2 - 441 &= 0 \\
(x - 21)(x + 21) &= 0
\end{aligned}
$$

$$
\begin{array}{ccccc}
x - 21 = 0 & & \text{or} & & x + 21 = 0 \\
x = 21 & & & & x = -21 \quad \text{(Reject)}
\end{array}
$$

$\boxed{\text{No}}$, the chair is 21 years old *not* 50 years old.

81. $\boxed{\text{C}}$ is true; area of large square – area of picture = area of matting

$$(x + 6)^2 - x^2 = 60 \quad \text{True}$$

83.

$$\left| x^2 + 6x + 1 \right| = 8$$

$$
\begin{array}{ccc}
x^2 + 6x + 1 = 8 & \text{or} & x^2 + 6x + 1 = -8 \\
x^2 + 6x - 7 = 0 & & x^2 + 6x + 9 = 0 \\
(x + 7)(x - 1) = 0 & & (x + 3)^2 = 0 \\
x + 7 = 0 \ \text{or} \ x - 1 = 0 & & x + 3 = 0 \\
x = -7 \qquad x = 1 & & x = -3
\end{array}
$$

$\boxed{\{-7, -3, 1\}}$

85.

$$
\begin{vmatrix}
2x & x \\
-1 & x
\end{vmatrix} = 1
$$

$$
\begin{aligned}
2x(x) - (-1)(x) &= 1 \\
2x^2 + x &= 1 \\
2x^2 + x - 1 &= 0 \\
(2x - 1)(x + 1) &= 0
\end{aligned}
$$

$$
\begin{array}{ccccc}
2x - 1 = 0 & & \text{or} & & x + 1 = 0 \\
x = \dfrac{1}{2} & & & & x = -1
\end{array}
$$

$\boxed{\left\{ -1, \dfrac{1}{2} \right\}}$

87. Let

$$
\begin{aligned}
x &= \text{width of path} \\
x + 30x + x &= 30 + 2x = \text{length of total area} \\
x + 20 + x &= 20 + 2x = \text{width of total area}
\end{aligned}
$$

$$
\begin{aligned}
(20 + 2x)(30 + 2x) &= 20(30) + 336 \\
600 + 40x + 60x + 4x^2 &= 600 + 336 \\
4x^2 + 100x + 600 &= 336 + 600 \\
4x^2 + 100x - 336 &= 0 \\
4(x^2 + 25x - 84) &= 0 \\
4(x - 3)(x + 28) &= 0
\end{aligned}
$$

$$
\begin{array}{ccccc}
x - 3 = 0 & & \text{or} & & x + 28 = 0 \\
x = 3 \quad \text{(width)} & & & & x = -28 \quad \text{(Reject)}
\end{array}
$$

The width of the path is $\boxed{3 \text{ yards}}$.

89. Let

$$x = \text{height of rectangle}$$
$$4x = \text{length of rectangle (also, base of triangle)}$$
$$x+1 = \text{altitude of triangle's cross section}$$

area of entire cross section = 60 square yards

area of rectangle + area of triangle = area of entire cross section

$$
\begin{aligned}
x(4x) + \frac{1}{2}(4x)(x+1) &= 60 \\
4x^2 + 2x^2 + 2x &= 60 \\
6x^2 + 2x - 60 &= 0 \\
2(3x^2 + x - 30) &= 0 \\
2(x-3)(3x+10) &= 0 \\
x - 3 = 0 \qquad\text{or}\qquad 3x + 10 &= 0 \\
x = 3 \qquad\qquad x &= -\frac{10}{3} \quad \text{(Reject)}
\end{aligned}
$$

The height of the rectangle is $\boxed{3 \text{ yards}}$.

91. Let

$$
\begin{aligned}
x &= \text{width of border} \\
x + 4 + x &= 4 + 2x = \text{width of entire area} \\
x + 12 + x &= 12 + 2x = \text{length of entire area}
\end{aligned}
$$

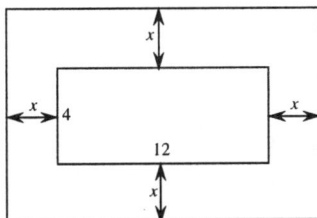

$\boxed{\text{area of border}}$ is $\boxed{4 \text{ times}}$ $\boxed{\text{the area of the lake}}$

$$
\begin{aligned}
(12 + 2x)(4 + 2x) - 4(12) &= 4 \cdot (4)(12) \\
48 + 32x + 4x^2 - 48 &= 192 \\
4x^2 + 32x - 192 &= 0 \\
4(x^2 + 8x - 48) &= 0 \\
4(x - 4)(x + 12) &= 0 \\
x - 4 = 0 \qquad\text{or}\qquad x + 12 &= 0 \\
x = 4 \qquad\qquad x &= -12 \quad \text{(Reject)}
\end{aligned}
$$

The width of the border is $\boxed{4 \text{ kilometers}}$.

93. $\dfrac{x^3 - x^2 - x + 1}{x^3 - x^2 + x - 1} = 0$

The only way that the algebraic fraction can equal 0 is if its numerator equals 0.

$$
\begin{aligned}
x^3 - x^2 - x + 1 &= 0 \\
x^2(x - 1) - (x - 1) &= 0 \\
(x - 1)(x^2 - 1) &= 0 \\
(x - 1)(x + 1)(x - 1) &= 0 \\
(x - 1)(x - 1)^2 &= 0 \\
x + 1 = 0 \qquad\text{or}\qquad (x - 1)^2 &= 0 \\
x = -1 \qquad\qquad x - 1 &= 0 \\
x &= 1
\end{aligned}
$$

However, $x = 1$ causes the denominator $(x^3 - x^2 + x - 1)$ to equal 0.

Thus, $x = -1$

$\boxed{\{-1\}}$

Review Problems

96. $4x^2 - 25y^2 - 4x + 10y$
$= 4x^2 - 4x - 25y^2 + 10y$
$= 4x^2 - 4x + 1 - 1 - 25y^2 + 10y$ (add $1 - 1 = 0$)
$= (4x^2 - 4x + 1) - (25y^2 - 10y + 1)$
$= (2x - 2)^2 - (5y - 1)^2$
$= (2x - 1 + 5y - 1)[(2x - 1) - (5y - 1)]$
$= (2x + 5y - 2)(2x - 5y)$
$= \boxed{(2x - 5y)(2x + 5y - 2)}$

97. $\begin{vmatrix} 5 & 2 & 34 \\ -1 & 3 & 22 \\ 0 & 0 & 4 \end{vmatrix}$

$= 0 \begin{vmatrix} 2 & 34 \\ 3 & 22 \end{vmatrix} - 0 \begin{vmatrix} 5 & 34 \\ -1 & 22 \end{vmatrix} + 4 \begin{vmatrix} 5 & 2 \\ -1 & 3 \end{vmatrix}$

$= 0 - 0 + 4(15 + 2) = 4(17) = \boxed{68}$

98. $2x - 3y + z = 0$
$3x + y + 2z = -2$
$x - 2y + z = -2$

(Equations 1 and 2):
$\quad 2x - 3y + z = 0$
$(\times 3)\quad \underline{9x + 3y + 6z = 0}$
$\quad 11x + 7z = -6$ (Equation 4)

(Equations 2 and 3):
$(\times 2)\quad 6x + 2y + 4z = -4$
$\quad \underline{x - 2y + z = -2}$
$\quad 7x + 5z = -6$ (Equation 5)

(Equations 4 and 5):
$(\times -5)\quad -55x - 35z = 30$
$(\times 7)\quad \underline{49x + 35z = -42}$
$\quad -6x = -12$
$\quad x = 2$

$7x + 5z = -6$ (Equation 5)
$7(2) + 5z = -6$
$5z = -20$
$z = -4$

$x - 2y + z = -2$ (Equation 3)
$2 - 2y + (-4) = -2$
$-2y = 0$
$y = 0$

$\boxed{\{(2, 0, -4)\}}$

Chapter 5   Review Problems

Review Problems, pp. 416-418

1. **a.** $4y^2 - 8y^3 + 9y$

$\boxed{\text{Trinomial; degree 3}}$

b. $12x^4y^3z \;\rightarrow\; 4 + 3 + 1 = 8$

$\boxed{\text{Monomial; degree 8}}$

c. $8y^4y^2 - 7xy^6 \rightarrow 1 + 6 = 7$

$\boxed{\text{Binomial; degree 7}}$

d. $7x^5 + 3x^3 - 2x^2 + x - 3$

$\boxed{\text{5 terms; degres 5}}$

2. **a.** $(4x^2 - 5xy + 3y^2) + (7xy - 10y^2 + 13x^2)$

$= \boxed{17x^2 + 2xy - 7y^2}$

b. $(5y^2 - 8xy + 7x^2) - (13xy - 12y^2 + 11x^2)$

$= 5y^2 - 8xy + 7x^2 - 13xy + 12y^2 - 11x^2$

$= \boxed{-4x^2 - 21xy + 17y^2}$

c. $(4x^{2n} - 5x^n + 3) + (7x^{2n} - 6(x^n - 8)$

$\qquad\qquad\qquad\qquad - (2x^{2n} - 4x^n - 9)$

$= 4x^{2n} - 5x^n + 3 + 7x^{2n} - 6x^n - 8 - 2x^{2n} + 4x^n + 9$

$= \boxed{9x^{2n} - 7x^n + 4}$

3. $[(6 + 3a^4b^3 + 4ab^3) + (-5 - 2a^4b^3 - 3ab^3)] - (10 - 6a^4b^3 - 2ab^3)$

$= 1 + a^4b^3 + ab^3 - 10 + 6a^4b^3 + 2ab^3$

$= \boxed{7a^4b^3 + 3ab^3 - 9}$

4. $(4x^2yz^5)(-3x^4yz^2) = \boxed{-12x^6y^2z^7}$

5. $-4a^3b^2c(3a^2b^3c^2 - \dfrac{1}{2}abc^5 - 2a^4b^2c)$

$= -4a^3b^2c(3a^2b^3c^2) - 4a^3b^3c\left(-\dfrac{1}{2}abc^5\right) - 4a^3b^2c(-2a^4b^2c)$

$= \boxed{-12a^5b^5c^3 + 2a^4b^3c^6 + 8a^7b^4c^2}$

6. $(4x - 2)(3x - 5) = 12x^2 - 20x - 6x + 10 = \boxed{12x^2 - 26x + 10}$

7. $(2a^2b + c)^2$

$= (2a^2b)^2 + 2(2a^2b)(c) + c^2$

$= \boxed{4a^4b^2 + 4a^2bc + c^2}$

8. $(3x^2y + 2y)(2x^2y - 3y)$

$= 6x^4y^2 - 9x^2y^2 + 4x^2y^2 - 6y^2$

$= \boxed{6x^4y^2 - 5x^2y^2 - 6y^2}$

9. $(4x^3y^2 + 1)(4x^3y^2 - 1)$

$= (4x^3y^2)^2 - 1^2$

$= \boxed{16x^6y^4 - 1}$

10. $(2xy + 2)(x^2y - 3y + 4)$

$= 2xy(x^2y - 3y + 4) + 2(x^2y - 3y + 4)$

$= 2x^3y^2 - 6xy^2 + 8xy + 2x^2y - 6y + 8$

$= \boxed{2x^3y^2 + 2x^2y - 6xy^2 + 8xy - 6y + 8}$

11. $(3x^{2n} - y^{5n})(4x^{2n} + 2y^{5n})$

$= 12x^{4n} + 6x^{2n}y^{5n} - 4x^{2n}y^{5n} - 2y^{10n}$

$= \boxed{12x^{4n} + 2x^{2n}y^{5n} - 2y^{10n}}$

12. $[5y - (2x + 7)][5y + (2x + 7)]$

$= (5y)^2 - (2x + 7)^2$

$= 25y^2 - (4x^2 + 28x + 49)$

$= \boxed{25y^2 - 4x^2 - 28x - 49}$

13. $(3x + y + 1)^2$

$= (3x + y + 1)(3x + y + 1)$

$= 3x(3x + y + 1) + y(3x + y + 1) + (3x + y + 1)$

$= 9x^2 + 3xy + 3x + 3xy + y^2 + y + 3x + y + 1$

$= \boxed{9x^2 + 6xy + 6x + 2y + y^2 + 1}$

14. Let x, $x + 1$, and $x + 2$ equal three consecutive integers.

$$
\begin{aligned}
\text{(product of the three integers)} &= \text{(cube of middle integer)} \quad \text{less } 12 \\
x(x + 1)(x + 2) &= (x + 1)^3 - 12 \\
x(x^2 + 3x + 2) &= x^3 + 3x^2 + 3x + 1 - 12 \\
x^3 + 3x^2 + 2x &= x^3 + 3x^2 + 3x - 11 \\
2x &= 3x - 11 \\
-x &= -11 \\
x &= 11 \\
x + 1 &= 11 + 1 = 12 \\
x + 2 &= 11 + 2 = 13
\end{aligned}
$$

The integers are $\boxed{11,\ 12,\ \text{and } 13}$.

15. Let

$$
\begin{aligned}
x &= \text{length of edge of cube} \\
x + 4 &= \text{length of rectangular box} \\
x - 3 &= \text{width of rectangular box} \\
x &= \text{height of rectangular box}
\end{aligned}
$$

$$
\begin{aligned}
\text{(surface area of cube)} &= \text{(surface area of rectangular box)} \\
6x^2 &= 2(x + 4)(x - 3) + 2(x - 3)x + 2(x + 4)x \\
6x^2 &= 2x^2 + 2x - 24 + 2x^2 - 6x + 2x^2 + 8x \\
0 &= 2x - 24 - 6x + 8x \\
-4x &= -24 \\
x &= 6 \quad \text{(height of box)} \\
x + 4 &= 6 + 4 = 10 \quad \text{(length of box)} \\
x - 3 &= 6 - 3 = 3 \quad \text{(width of box)}
\end{aligned}
$$

$\boxed{\text{length of rectangular box: } 10 \text{ feet;}}$

$\boxed{\text{width of rectangular box: } 3 \text{ feet}}$

16. $\dfrac{16x^4y^3z^7}{-8x^2yz^5} = -2x^{4-2}y^{3-1}z^{7-5} = \boxed{-2x^2y^2z^2}$

17. $\dfrac{16a^4b^2 - 8a^3b^2 + 20ab}{4ab}$

$= \dfrac{16a^4b^2}{4ab} - \dfrac{8a^3n^2}{4ab} + \dfrac{20ab}{4ab}$

$= \boxed{4a^3b - 2a^2b + 5}$

18. $\dfrac{15y^3 + 29y^2 - 4y - 30}{5y - 2} = \boxed{3y^2 + 7y + 2 - \dfrac{26}{5y - 2}}$

$$
\require{enclose}
\begin{array}{r}
3y^2 + 7y + 2 \\
5y - 2 \enclose{longdiv}{15y^3 + 29y^2 - 4y - 30} \\
\underline{15y^3 - 6y^2} \\
35y^2 - 4y \\
\underline{35y^2 - 14y} \\
10y - 30 \\
\underline{10y - \ 4} \\
-26
\end{array}
$$

19. $\dfrac{x^4 - y^4}{x - y} = \boxed{x^3 + x^2y + xy^2 + y^3}$

$$
\require{enclose}
\begin{array}{r}
x^3 + x^2y + xy^2 + y^3 \\
x - y \enclose{longdiv}{x^4 + 0x^3y + 0x^2y^2 + 0xy^3 - y^4} \\
\end{array}
$$

$$
\begin{array}{r}
\underline{x^4 - x^3y} \\
x^3y + 0x^2y^2 \\
\underline{x^3y - x^2y^2} \\
x^2y^2 + 0xy^3 \\
\underline{x^2y^2 - xy^3} \\
xy^3 - y^4 \\
\underline{xy^3 - y^4} \\
0
\end{array}
$$

20. $\dfrac{25x^2y^3z^4 - 45xy^4z^3}{5xyz^2}$

$= \dfrac{25x^2y^3z^4}{5xyz^2} - \dfrac{45xy^4z^3}{5xyz^2}$

$= \boxed{5xy^2z^2 - 9y^3z}$

21. $\dfrac{4x^4 + 4x^3 + 14x^2 + 6x + 12}{2x^2 + 3} = \boxed{2x^2 + 2x + 4}$

$$
\begin{array}{r}
2x^2 + 2x + 4 \\
2x^2 + 3 \enclose{longdiv}{4x^4 + 4x^3 + 14x^2 + 6x + 12} \\
\end{array}
$$

$$
\begin{array}{r}
\underline{4x^4 + 6x^2} \\
4x^3 + 8x^2 + 6x \\
\underline{4x^3 + 6x} \\
8x^2 + 12 \\
\underline{8x^2 + 12} \\
0
\end{array}
$$

22. $\dfrac{4x^3 - 3x^2 - 2x + 1}{x + 1} = \boxed{4x^2 - 7x + 5 - \dfrac{4}{x + 1}}$

$$
\begin{array}{r|rrrr}
-1 & 4 & -3 & -2 & 1 \\
   &   & -4 & 7 & -5 \\
\hline
   & 4 & -7 & 5 & -4
\end{array}
$$

23. $(3y^4 - 2y^2 - 10y) \div (y - 2) = \boxed{3y^3 + 6y^2 + 10y + 10 + \dfrac{20}{y - 2}}$

$$
\begin{array}{r|rrrrr}
2 & 3 & 0 & -2 & -10 & 0 \\
  &   & 6 & 12 & 20 & 20 \\
\hline
  & 3 & 6 & 10 & 10 & 20
\end{array}
$$

24. a.   area of shaded figure
= area of rectangle − 4(area of small square)
$= (3x + 2)(5x + 3) - 4(x)(x)$
$= 15x^2 + 9x + 10x + 6 - 4x^2$
$= \boxed{11x^2 + 19x + 6}$

b.   Let

$$\begin{aligned}
x &= \text{height of box} \\
3x + 2 - 2x &= x + 2 = \text{width of box} \\
5x + 3 - 2x &= 3x + 3 = \text{length of box}
\end{aligned}$$

$$\begin{aligned}
\text{Volume} &= \text{length} \times \text{width} \times \text{height} \\
&= (3x + 3)(x + 2)x \\
&= x(3x^2 + 9x + 6) \\
&= \boxed{3x^3 + 9x^2 + 6x}
\end{aligned}$$

25. Let

$$\begin{aligned}
18 - 2x &= \text{width of shaded rectangle} \\
26 - 2x &= \text{length of shaded rectangle}
\end{aligned}$$

area of shaded rectangle
$= (26 - 2x)(18 - 2x)$
$= 468 - 52x - 36x + 4x^2$
$= 468 - 88x + 4x^2$
$= \boxed{4x^2 - 88x + 468}$

26. $f(x) = 63 + 20\left(\dfrac{x-120}{40}\right) - 0.4\left(\dfrac{x-120}{40}\right)^2 - 1.2\left(\dfrac{x-120}{40}\right)^3$ where $0 \le x \le 240$

$f(140) = 63 + 20\,\dfrac{(140-120)}{40} - 0.4\left(\dfrac{140-120}{40}\right)^2 - 1.2\left(\dfrac{140-120}{40}\right)^2$

$f(140) = 63 + 20\left(\dfrac{20}{40}\right) - 0.4\left(\dfrac{20}{40}\right)^2 - 1.2\left(\dfrac{20}{40}\right)^2$

$f(140) = 63 + 20\left(\dfrac{1}{2}\right) - 0.4\left(\dfrac{1}{4}\right) - 1.2\left(\dfrac{1}{8}\right)$

$f(140) = 63 + 10 - 0.1 - 0.15$

$f(140) = \boxed{72.75}$

$\boxed{\text{The height of a 140-year old tree is 72.75 feet.}}$

27. $s(t) = -16t^2 + 16t + 32$

t	0	0.5	1	1.5	2
$s(t)$	32	36	32	20	0

$s(0) = -0 + 0 + 32 = \boxed{32}$

   The initial height of the diver above the water (at $t = 0$) is 32 feet.

$s(0.5) = -16(0.5)^2 + 16(0.5) + 32 = \boxed{36}$

   After 0.5 seconds, the diver's height is 36 feet.

$s(1) = -16(1^2) + 16(1) + 32 = \boxed{32}$

   After 1 second, the diver's height is 32 feet.

$s(1.5) = -16(1.5)^2 + 16(1.5) + 32 = \boxed{20}$

   After 1.5 seconds, the diver's height is 20 feet.

$s(2) = -16(2^2) + 16(2) + 32 = 0$

   After 2 seconds, the diver's height is 0 feet.

$\boxed{\text{The diver hits the water at 2 seconds.}}$

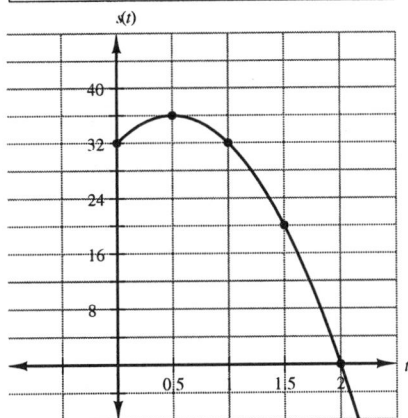

28. $f(x) = 5x^2 - 8x - 6$

$\dfrac{f(a + h) - f(a)}{h}$

$= \dfrac{[5(a + h)^2 - 8(a + h) - 6] - (5a^2 - 8a - 6)}{h}$

$= \dfrac{5a^2 + 10ah + 5h^2 - 8a - 8h - 6 - 5a^2 + 8a + 6}{h}$

$= \dfrac{10ah + 5h^2 - 8h}{h}$

$= \boxed{10a + 5h - 8}$

29. $f(x) = -4x^3 + 7x^2 - 10x + 11$

$\dfrac{f(a + h) - f(a)}{h}$

$= \dfrac{[-4(a + h)^3 + 7(a + h)^2 - 10(a + h) + 11] - (-4a^3 + 7a^2 - 10a + 11)}{h}$

$= \dfrac{-4a^3 - 12a^2h - 12ah^2 - 4h^3 + 7a^2 + 14ah + 7h^2 - 10a - 10h + 11 + 4a^3 - 7a^2 + 10a - 11}{h}$

$= \dfrac{-12a^2h - 12ah^2 - 4h^3 + 14ah + 7h^2 - 10h}{h}$

$= \boxed{-12a^2 - 12ah - 4h^2 + 14a + 7h - 10}$

30. $f(x) = 5x - 3$ and $g(x) = 8x^2 - 7x + 9$

$(f + g)(x) = (5x - 3) + (8x^2 - 7x + 9) = \boxed{8x^2 - 2x + 6}$

$$\begin{aligned}
(f - g)(x) &= (5x - 3) - (8x^2 + 7x - 9) \\
&= 5x - 3 - 8x^2 + 7x - 9 \\
&= \boxed{-8x^2 + 12x - 12}
\end{aligned}$$

$$\begin{aligned}
(fg)(x) &= (5x - 3)(8x^2 - 7x + 9) \\
&= 5x(8x^2 - 7x + 9) - 3(8x^2 - 7x + 9) \\
&= 40x^3 - 35x^2 + 45x - 24x^2 + 21x - 27 \\
&= \boxed{40x^3 - 59x^2 + 66x - 27}
\end{aligned}$$

31. $f(x) = 3x^2 + 11x - 4$ and $g(x) = x + 4$

$$\left(\frac{f}{g}\right)(x) = \frac{3x^2 + 11x - 4}{x + 4} = \boxed{3x - 1 \quad (x \neq -4)}$$

$$\begin{array}{r|rrr}
-4 & 3 & 11 & -4 \\
   &   & -12 & 4 \\
\hline
   & 3 & -1 & 0
\end{array}$$

32. $f(x) = 5x^2 - 3x + 11$ and $g(x) = 7x - 2$

$$\begin{aligned}
f[g(x)] &= 5(7x - 2)^2 - 3(7x - 2) + 11 \\
&= 5(49x^2 - 28x + 4) - 21x + 6 + 11 \\
&= 245x^2 - 140x + 20 - 21x + 17 \\
&= \boxed{245x^2 - 161x + 37} \\
g[f(x)] &= 7(5x^2 - 3x + 11) - 2 \\
&= 35x^2 - 21x + 77 - 2 \\
&= \boxed{35x^2 - 21x + 75}
\end{aligned}$$

For problems 33-38, $f(x) = 3x + 7$, $g(x) = 5x - 4$, $h(x) = 8x^2 + 7x - 5$.

33. $(f + h)(x) = (3x + 7) + (8x^2 + 7x - 5) = 8x^2 + 10x + 2$

$(f + h)(-2) = 8(-2)^2 + 10(-2) + 2 = 32 - 20 + 2 = \boxed{14}$

34. $(fg)(x) = (3x + 7)(5x - 4) = 15x^2 + 23x - 28$

$(fg)(4) = (3 \cdot 4 + 7)(5 \cdot 4 - 4) = (12 + 7)(20 - 4) = (19)(16) = \boxed{304}$

$\qquad$ or $15(4^2) + 23(4) - 28 = 304$

35. $g(-3) = 5(-3) - 4 = -19$

$$\begin{aligned}
(h \circ g)(-3) &= h[g(-3)] = h(-19) = 8(-19)^2 + 7(-19) - 5 = \boxed{2750} \\
\text{or } (h \circ g)(x) &= 8(5x - 4)^2 + 7(5x - 4) - 5 \\
&= 8(25x^2 - 40x + 16) + 35x - 28 - 5 \\
&= 200x^2 - 320x + 128 - 135x - 33 \\
&= 200x^2 - 285x + 95 \\
(h \circ g)(-3) &= 200(-3)^2 - 285(-3) + 95 = 2750
\end{aligned}$$

36. $f(-5) = 3(-5) + 7 = -8$

$$\begin{aligned}
(h \circ f)(x) &= 8(3x + 7)^2 + 7(3x + 7) - 5 = \boxed{451} \\
&= 8(9x^2 + 42x + 49) + 21x + 49 - 5 \\
&= 72x^2 + 336x + 392 + 21x + 44 \\
&= 72x^2 + 357x + 436 \\
(h \circ f)(-5) &= 72(-5)^2 + 357(-5) + 436 = 451
\end{aligned}$$

37. $h(a+b) - g(a+b)$

$= [8(a+b)^2 + 7(a+b) - 5] - [5(a+b) - 4]$

$= 8a^2 + 16ab + 8b^2 + 7a + 7b - 5 - 5a - 5b + 4$

$= \boxed{8a^2 + 16ab + 8b^2 + 2a + 2b - 1}$

38. $\left(\dfrac{f}{g}\right)(x) = \dfrac{3x+7}{4x-4}$

$\left(\dfrac{f}{g}\right)(0) = \dfrac{0+7}{0-4} = \boxed{-\dfrac{7}{4}}$

39. $f(t) = 6000 + 200t^2$ and $g(P) = 0.5P + 1$

a. $f(5) = 6000 + 2000(5^2) = 6000 + 200(25) = 11,000$

$g[f(5)] = g(11,000) = 0.5(11,000) + 1 = \boxed{5501}$

$\boxed{\text{In 5 years, the carbon monoxide level will be 5501 ppm.}}$

40.

$$\begin{array}{r|rrrrr}
-5 & -3 & 2 & 5 & -9 & 10 \\
& & 15 & -85 & 4000 & -1955 \\
\hline
& -3 & 17 & -80 & 391 & -1945 \quad \to \quad R \ne 0
\end{array}$$

The remainder is not zero.

Thus, -5 is $\boxed{not \text{ a solution}}$.

41. $P(x) = 3x^3 + 4x^2 - 3x + 2$

$$\begin{array}{r|rrrr}
-4 & 3 & 4 & -3 & 2 \\
& & -12 & 32 & -116 \\
\hline
& 3 & -8 & 29 & -114
\end{array}$$

$\boxed{P(-4) = -114}$

42. $5a^4b^2c^3 - 20a^3b^3c^4 + 15a^2b^2c^3$

$= \boxed{5a^2b^2c^3(a^2 - 4abc + 3)}$

43. $21x^{3n+1} - 35x^{3n}$

$= \boxed{7x^{3n}(3x - 5)}$

44. $(x+5y)z^2 + (x+5y)z - 42(x+5y)$

$= (x+5y)(z^2 + z - 42)$

$= \boxed{(x+5y)(z+7)(z-6)}$

45. $bc - d - bd + c$

$= bc + c - bd - d$

$= c(b+1) - d(b+1)$

$= \boxed{(b+1)(c-d)}$

46. $(a-4b)x^2 + (a-4b)x + (4b-a)$

$= (a-4b)x^2 + (a-4b)x + (a-4b)(-1)$

$= \boxed{(a-4b)(x^2 + x - 1)}$

47. $3x^2 + 15x - 2xy - 10y$

$= 3x(x+5) - 2y(x+5)$

$= \boxed{(x+5)(3x-2y)}$

48. $a^2 + 37a + 36$

$= \boxed{(a+36)(a+1)}$

49. $-2x^3 + 36x^2 - 64x$

$= -2x(x^2 - 18x + 32)$

$= \boxed{-2x(x-2)(x-16)}$

50. $8y^4 - 14y^2 - 15$

$= \boxed{(4y^2 + 3)(2y^2 - 5)}$

51. $x^2(b^2 - 9) - 25(b^2 - 9)$

$= (b^2 - 9)(x^2 - 25)$

$= \boxed{(b+3)(b-3)(x+5)(x-5)}$

52. $4(a+b)^2 - 27(a+b) + 18$

$= 4x^2 - 27x + 18$ (Let $x = a + b$)

$= (4x - 3)(x - 6)$

$= [4(a+b) - 3][(a+b) - 6]$ (substitute $a + b$ for x)

$= \boxed{(4a + 4b - 3)(a + b - 6)}$

53. $x^{2n} - 5x^n - 36$

$= \boxed{(x^n + 4)(x^n - 9)}$

54. $10x^2 - 160$

$= 10(x^2 - 16)$

$= \boxed{10(x + 4)(x - 4)}$

55. $(x+4)^2 - (3x-1)^2$

$= [(x+4) + (3x-1)][(x+4) - (3x-1)]$

$= \boxed{(4x + 3)(-2x + 5)}$

56. $9x^2y^6 - 25a^4b^2$

$= (3xy^3)^2 - (5a^2b)^2$

$= \boxed{(3xy^3 + 5a^2b)(3xy^3 - 5a^2b)}$

57. $81x^4 - 100y^4$

$= \boxed{(9x^2 + 10y^2)(9x^2 - 10y^2)}$

58. $1 - 64y^3$

$= 1^3 - (4y)^3$

$= (1 - 4y)[1 + 1(4y) + (4y)^2]$

$= \boxed{(1 - 4y)(1 + 4y + 16y^2)}$

59. $9x^2 - 21xy + 10y^2$

$= \boxed{(3x - 2y)(3x - 5y)}$

60. $4a^3 + 32$

$= 4(a^3 + 8)$

$= 4(a^3 + 2^3)$

$= \boxed{4(a + 2)(a^2 - 2a + 4)}$

61. $x^2 + 6x + 9 - 4a^2$

$= (x+3)^2 - (2a)^2$

$= \boxed{(x + 3 + 2a)(x + 3 - 2a)}$

62. $x^{3n} + y^{3m}$

$= (x^n)^3 + (y^m)^3$

$= \boxed{(x^n + y^m)(x^{2n} - x^n y^m + y^{2m})}$

63. $(x+y)^3 - (2x+y)^3$

$= [(x+y) - (2x+y)][(x+y)^2 + (x+y)(2x+y)^2]$

$= (x + y - 2x - y)(x^2 + 2xy + y^2 + 2x^2 + 3xy + y^2 + 4x^2 + 4xy + y^2)$

$= \boxed{-x(7x^2 + 9xy + 3y^2)}$

64. $9x^2 + 30xy + 25y^2$

$= (3x)^2 + 2 \cdot 3x \cdot 5y + (5y)^2$

$= \boxed{(3x + 5y)^2}$

65. $x^3 + y + y^3 + x$

$= x^3 + y^3 + x + y$

$= (x + y)(x^2 - xy + y^2) + (x + y)$

$= \boxed{(x + y)(x^2 - xy + y^2 + 1)}$

66. $x^2 + 49$ is $\boxed{\text{prime}}$, not factorable

67. $2xy - 2x^{10}y$

$= 2xy(1 - x^9)$

$= 2xy[1^3 - (x^3)^3]$

$= 2xy(1 - x^3)(1 + x^3 + x^6)$

$= \boxed{2xy(1 - x)(1 + x + x^2)(1 + x^3 + x^6)}$

68. $x^{2n} - 1$

$= (x^n)^2 - 1^2$

$= \boxed{(x^n + 1)(x^n - 1)}$

69. $x^4 - 6x^2 + 9$

$= \boxed{(x^2 - 3)^2}$

70. $-x^2 + 4x + 21$
$= -(x^2 - 4x - 21)$
$= \boxed{-(x-7)(x+3)}$

71. $6x^3y^2 - 6xy^4 - 6x^2y^2 + 6xy^3$
$= 6xy^2(x^2 - y^2 - x + y)$
$= 6xy^2[(x^2 - y^2) - (x - y)]$
$= 6xy[(x+y)(x-y) - (x-y)]$
$= \boxed{6xy(x-y)(x+y-1)}$

72. $x^4 - x^3 - x + 1$
$= x^3(x-1) - (x-1)$
$= (x-1)(x^3 - 1)$
$= (x-1)(x-1)(x^2 + x + 1)$
$= \boxed{(x-1)^2(x^2+x+1)}$

73. $27x^{3n} - 8$
$= (3x^n)^3 - 2^3$
$= (3x^n - 2)[(3x^n)^2 + (3x^n)(2) + 2^2]$
$= \boxed{(3x^n - 2)(9x^{2n} + 6x^n + 4)}$

74. $9x^{2n} + 24x^ny^m + 16y^{2m}$
$= (3x^n)^2 + 2(3x^n)(4y^m) + (4y^m)^2$
$= \boxed{(3x^n + 4y^m)^2}$

75. $a^{4m} - b^{4m}$
$= (a^{2m})^2 - (b^{2m})^2$
$= (a^{2m} + b^{2m})(a^{2m} - b^{2m})$
$= \boxed{(a^{2m} + b^{2m})(a^m + b^m)(a^m - b^m)}$

76. $r^2 - 4rs + 4s^2 - 13r + 26s + 12$
$= (r^2 - 4rs + 4s^2) + (-13r + 26s) + 12$
$= (r - 2s)^2 - 13(r - 2s) + 12$
$= x^2 - 13x + 12 \quad \text{(Let } x = r - 2s)$
$= (x - 12)(x - 1)$
$= \boxed{(r - 2s - 12)(r - 2s - 1)}$

77. $27b^3 - 125c^3$
$= (3b)^3 - (5c)^3$
$= (3b - 5c)[(3b)^2 + (3b)(5c) + (5c)^2]$
$= \boxed{(3b - 5c)(9b^2 + 15bc + 25c^2)}$

78. $x + y + 3x^2 + 3xy$
$= (x + y) + 3x(x + y)$
$= \boxed{(x + y)(1 + 3x)}$

79. $a^4b + 9 - 9a^4 - b$
$= a^4b - 9a^4 + 9 - b$
$= a^4(b - 9) - (b - 9)$
$= (b - 9)(a^4 - 1)$
$= (b - 9)(a^2 + 1)(a^2 - 1)$
$= \boxed{(b - 9)(a^2 + 1)(a + 1)(a - 1)}$

80. $\dfrac{4x^3 - 16x^2 - 9x + 36}{x - 4} = 4x^2 - 9$

$$
\begin{array}{r|rrrr}
4 & 4 & -16 & -9 & 36 \\
  &   & 16 & 0 & -36 \\
\hline
  & 4 & 0   & -9 & 0
\end{array}
$$

$4x^3 - 16x^2 - 9x + 36 = (x - 4)(4x^2 - 9) = \boxed{(x - 4)(2x + 3)(2x - 3)}$

81. $\dfrac{4x^3 + 28x^2 + 9x - 90}{2x + 5} = 2x^2 + 9x - 18$

$$
\begin{array}{r}
2x^2 + 9x - 18 \\
2x + 5 \overline{\smash{\big)}\, 4x^3 + 28x^2 + 9x - 90} \\
\underline{4x^3 + 10x^2} \\
18x^2 + 9x \\
\underline{18x^2 + 45x} \\
-36x - 90 \\
\underline{-36x - 90} \\
0
\end{array}
$$

$4x^3 + 28x^2 + 9x - 90 = (2x + 5)(2x^2 + 9x - 18) = \boxed{(2x + 5)(2x - 3)(x + 6)}$

82.
$$x(12x + 31) = 15$$
$$12x^2 + 31x - 15 = 0$$
$$(x + 3)(12x - 5) = 0$$

$x + 3 = 0$ or $12x - 5 = 0$

$x = -3$ $x = \dfrac{5}{12}$

$$\boxed{\left\{-3, \dfrac{5}{12}\right\}}$$

83.
$$(2x + 6)(x + 6) - 2x^2 = x^2 - 4$$
$$2x^2 + 18x + 36 - 2x^2 = x^2 - 4$$
$$-x^2 + 18x + 40 = 0$$
$$x^2 - 18x - 40 = 0$$
$$(x + 2)(x - 20) = 0$$

$x + 2 = 0$ or $x - 20 = 0$

$x = 2$ $x = 20$

$$\boxed{\{-2, 20\}}$$

84.
$$x^2 + \dfrac{3}{2}x + \dfrac{1}{2} = 0$$
$$2x^2 + 3x + 1 = 0$$
$$(2x + 1)(x + 1) = 0$$

$2x + 1 = 0$ or $x + 1 = 0$

$x = -\dfrac{1}{2}$ $x = -1$

$$\boxed{\left\{-1, -\dfrac{1}{2}\right\}}$$

85.
$$2x^2 - 3x = 0$$
$$x(2x - 3) = 0$$

$x = 0$ or $2x - 3 = 0$

$x = \dfrac{3}{2}$

$$\boxed{\left\{0, \dfrac{3}{2}\right\}}$$

86.
$$s(t) = -16t^2 + 80t$$
$$s(t) = 0$$
$$0 = -16t^2 + 80t$$
$$16t^2 - 80t = 0$$
$$16t(t - 5) = 0$$

$t = 0$ (Reject)   or $t - 5 = 0$

$t = 5$

after $\boxed{5 \text{ seconds}}$ the object will strike the ground

87. $P = 100 + 25x - 5x^2$
($P = 120$ millimeter):
$$120 = 100 + 25x - 5x^2$$
$$5x^2 - 25x + 20 = 0$$
$$5(x^2 - 5x + 4) = 0$$
$$5(x - 4)(x - 1) = 0$$

$x - 4 = 0$ or $x - 1 = 0$

$x = 4$ $x = 1$ (reject since $4 > 1$)

$\boxed{4 \text{ milligrams}}$

88. Let x and $x + 2$ equal two consecutive even integers

(square of larger) $- 7$(smaller) $= 44$

$$
\begin{aligned}
(x + 2)^2 - 7(x) &= 44 \\
x^2 + 4x + 4 - 7x - 44 &= 0 \\
x^2 - 3x - 40 &= 0 \\
(x - 8)(x + 5) &= 0
\end{aligned}
$$

$x - 8 = 0$	or	$x + 5 = 0$
$x = 8$		$x = -5$ (reject, *not* even)
$x + 2 = 10$		

The integers are $\boxed{8 \text{ and } 10}$.

89. Let $x =$ an even integer.

$$
\begin{aligned}
x(3x - 16) &= 12 \\
3x^2 - 16x - 12 &= 0 \\
(x - 6)(3x + 2) &= 0
\end{aligned}
$$

$x - 6 = 0$	or	$3x + 2 = 0$
$x = 6$		$x = -\dfrac{2}{3}$
		(reject, *not* an integer)

The number is $\boxed{6}$.

90. Let

$$
\begin{aligned}
x &= \text{length of the shorter leg} \\
x + 7 &= \text{length of the longer leg} \\
\text{hypotenuse} &= 13 \text{ meters}
\end{aligned}
$$

$$
\begin{aligned}
x^2 + (x + 7)^2 &= 13^2 \\
x^2 + x^2 + 14x + 49 &= 169 \\
2x^2 + 14x - 120 &= 0 \\
2(x^2 + 7x - 60) &= 0 \\
2(x + 12)(x - 5) &= 0
\end{aligned}
$$

$x + 12 = 0$	or	$x - 5 = 0$
$x = -12$ (Reject)		$x = 5$ (shorter legt)
		$x + 7 = 12$ (longer leg)

The length of the longer leg is $\boxed{12 \text{ meters}}$.

91. Let

$$
\begin{aligned}
x &= \text{width of first pool} \\
x + 9 &= \text{length of first pool} \\
x + 6 &= \text{width of second pool} \\
2(x + 9) &= \text{length of second pool}
\end{aligned}
$$

(area of first pool) $+$ (area of second pool) $=$ total area

$$
\begin{aligned}
x(x + 9) + (x + 6)(2x + 18) &= 528 \\
x^2 + 9x + 2x^2 + 18x + 12x + 108 &= 528 \\
3x^2 + 39x - 420 &= 0 \\
3(x^2 + 13x - 140) &= 0 \\
3(x + 20)(x - 7) &= 0
\end{aligned}
$$

$x + 20 = 0$	or	$x - 7 = 0$
$x = -20$ (Reject)		$x = 7$ (width of first pool)
		$x + 9 = 16$ (length of first pool)

Dimensions of first pool:

$\boxed{7 \text{ yards} \times 16 \text{ yards}}$

92. Let
$$x = \text{face value of first dice}$$
$$x + 3 = \text{face value of second dice}$$

$$(\text{product}) = 2(\text{sum})$$
$$x(x + 3) = 2(x + x + 3)$$
$$x^2 + 3x = 4x + 6$$
$$x^2 - x - 6 = 0$$
$$(x - 3)(x + 2) = 0$$

$$x - 3 = 0 \qquad \text{or} \qquad x + 2 = 0$$
$$x = 3 \quad \text{(first dice)} \qquad\qquad x = -2 \quad \text{(Reject)}$$
$$x + 3 = 6 \quad \text{(second dice)}$$

The two numbers thrown are $\boxed{3 \text{ and } 6}$.

93. Let
$$x = \text{width of frame}$$
$$x + 10 + x = 10 + 2x = \text{width of picture plus frame}$$
$$x + 16 + x = 16 + 2x = \text{length of picture plus frame}$$

$$(10 + 2x)(16 + 2x) = 280$$
$$160 + 20x + 32x + 4x^2 = 280$$
$$4x^2 + 52x - 120 = 0$$
$$4(x^2 + 13x - 30) = 0$$
$$4(x + 15)(x - 2) = 0$$

$$x + 15 = 0 \qquad \text{or} \qquad x - 2 = 0$$
$$x = -15 \quad \text{(Reject)} \qquad\qquad x = 2$$

The width of the frame is $\boxed{2 \text{ centimeters}}$.

94. Let
$$x = \text{width of frame}$$
$$4 + 2x = \text{width of frame plus picture}$$
$$7 + 2x = \text{length of frame plus picture}$$
$$\text{area of frame} = 26 \text{ square centimeters}$$

$$\text{total area} = \text{area of picture} + \text{area of frame}$$
$$(4 + 2x)(7 + 2x) = 4(7) + 26$$
$$28 + 22x + 4x^2 = 28 + 26$$
$$4x^2 + 22x - 26 = 0$$
$$2(2x^2 + 11x - 13) = 0$$
$$2(x - 1)(2x + 13) = 0$$

$$x - 1 = 0 \qquad \text{or} \qquad 2x + 13 = 0$$
$$x = 1 \qquad\qquad\qquad x = -\frac{13}{2}$$

The width of the frame is $\boxed{1 \text{ centimeter}}$.

95. Let

$$
\begin{aligned}
x &= \text{length of side of small square} \\
\text{area of resulting figure} &= \text{area of square} - 4(\text{area of small square}) \\
55 &= 8(8) - 4x^2 \\
55 &= 64 - 4x^2 \\
4x^2 - 9 &= 0 \\
(2x - 3)(2x + 3) &= 0 \\
2x - 3 = 0 \quad &\text{or} \quad 2x + 3 = 0 \\
x = \frac{3}{2} = 1.5 \qquad\qquad & \qquad\qquad x = -\frac{3}{2} = -1.5 \quad \text{(Reject)}
\end{aligned}
$$

The size of the small square is $\boxed{1.5 \text{ meters} \times 1.5 \text{ meters}}$.

96. Let

$$
\begin{aligned}
x &= \text{depth of trough} \\
26 - 2x &= \text{width of trough} \\
(26 - 2x)(x) &= 84 \\
26x - 2x^2 - 84 &= 0 \\
-2(x^2 - 13x + 42) &= 9 \\
-2(x - 7)(x - 6) &= 0 \\
x - 7 = 0 \quad &\text{or} \quad x - 6 = 0 \\
x = 7 \quad \text{(depth)} \qquad & \qquad x = 6 \quad \text{(width)}
\end{aligned}
$$

$\boxed{7 \text{ decimeters} \times 12 \text{ decimeters, or, } 6 \text{ decimeters} \times 14 \text{ decimeters}}$

97. Let $x, x + 2$ and $x + 4$ equal the three consecutive even integers (the length of three sides of a right triangle)

$$
\begin{aligned}
x^2 + (x + 2)^2 &= (x + 4)^2 \\
x^2 + x^2 + 4x + 4 &= x^2 + 8x + 16 \\
x^2 - 4x - 12 &= 0 \\
(x - 6)(x + 2) &= 0 \\
x - 6 = 0 \quad &\text{or} \quad x + 2 = 0 \\
x = 6 \qquad\qquad & \qquad\qquad x = -2 \quad \text{(Reject)} \\
x + 2 &= 8 \\
x + 4 &= 10
\end{aligned}
$$

The lengths of the sides are $\boxed{6, 8 \text{ and } 10}$.

98.

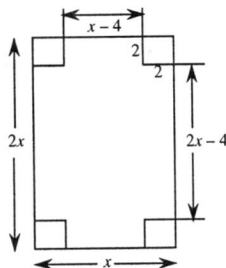

Let

$$
\begin{aligned}
x &= \text{width of rectangular piece} \\
2x &= \text{length of rectangular piece} \\
x - 4 &= \text{width of box} \\
2x - 4 &= \text{length of box} \\
2 &= \text{height of box} \\
\text{volume} &= 480 \text{ cubic inches}
\end{aligned}
$$

$$
\begin{aligned}
(x-4)(2x-4)(2) &= 480 \\
4(x-4)(x-2) &= 480 \\
(\div 4) \quad (x-4)(x-2) &= 120 \\
x^2 - 6x + 8 &= 120 \\
(x+8)(x-14) &= 0
\end{aligned}
$$

$$
\begin{array}{lll}
x+8 = 0 & \text{or} & x-14 = 0 \\
\quad x = -8 \;\;\text{(Reject)} & & \quad\quad x = 14 \quad\text{(width)} \\
& & \quad 2x = 28 \quad\text{(length)}
\end{array}
$$

dimensions of piece of tin: $\boxed{14 \text{ inches} \times 28 \text{ inches}}$

99. Let

$$
\begin{aligned}
x+3 &= \text{length of one leg} \\
(x+3)+3 &= x+6 = \text{length of hypotenuse} \\
x+6-6 &= x = \text{length of other leg}
\end{aligned}
$$

$$
\begin{aligned}
(x+3)2 + x^2 &= (x+6)^2 \\
x^2 + 6x + 9 + x^2 &= x^2 + 12x + 36 \\
x^2 - 6x - 27 &= 0 \\
(x-9)(x+3) &= 0
\end{aligned}
$$

$$
\begin{array}{lll}
x-9 = 0 & \text{or} & x+3 = 0 \\
\quad x = 9 \;\;\text{(leg)} & & \quad x = -3 \;\;\text{(Reject)} \\
x+3 = 12 \;\;\text{(leg)} & & \\
x+6 = 15 \;\;\text{(hypotenuse)} & &
\end{array}
$$

$$
\begin{aligned}
\text{Area} &= \frac{1}{2}(\text{leg}) \times (\text{leg}) \\
&= \frac{1}{2}(9)(12) \\
&= 54
\end{aligned}
$$

$\boxed{54 \text{ m}^2}$

100.

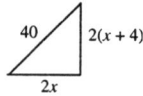

Let

$$
\begin{array}{ll}
x &= \text{rate of second person } (2x = \text{distance traveled in 2 hours}) \\
x+4 &= \text{rate of first person } (2(x+4) = \text{distance traveled in 2 hours})
\end{array}
$$

$$
\begin{aligned}
(2x)^2 + [2(x+4)]^2 &= 40^2 \\
4x^2 + 4(x^2 + 8x + 16) &= 1600 \\
4x^2 + 4x^2 + 32x + 64 &= 1600 \\
8x^2 + 32x - 1536 &= 0 \\
8(x^2 + 4x - 192) &= 0 \\
8(x-12)(x+16) &= 0
\end{aligned}
$$

$$
\begin{array}{lll}
x-12 = 0 & \text{or} & x+16 = 0 \\
\quad x = 12 & & \quad x = -16 \;\;\text{(Reject)}
\end{array}
$$

The rate of the slower person is $\boxed{12 \text{ mph}}$.

Cumulative Review Problems (Chapters 1-5)

Cumulative Problems, pp. 418-420

1. $(x + 1) + y = x + (1 + y)$

   $\boxed{\text{Associative property of addition}}$

2. $\dfrac{(7500)(4000)}{(0.08)(0.15)} = \dfrac{(7.5 \times 10^3)(4 \times 10^3)}{(8 \times 10^{-2})(1.5 \times 10^{-1})} = \left(\dfrac{7.5}{1.5}\right)\left(\dfrac{4}{8}\right) \times \dfrac{10^6}{10^{-3}} = 5\left(\dfrac{1}{2}\right) \times 10^9$

   $= \boxed{2.5 \times 10^9}$

3. $\dfrac{6 - 2^2 - (-2)^3}{8 - 3(2) - (-3)}$

   $= \dfrac{6 - 4 + 8}{8 - 6 + 3}$

   $= \dfrac{10}{5}$

   $= \boxed{2}$

4. $\left|2x - 8\right| = \left|6x - 7\right|$

   $\begin{aligned} 2x - 8 &= 6x - 7 \\ -4x &= 1 \\ x &= -\dfrac{1}{4} \end{aligned}$

   or

   $\begin{aligned} 2x - 8 &= -(6x - 7) \\ 2x - 8 &= -6x + 7 \\ 8x &= 15 \\ x &= \dfrac{15}{8} \end{aligned}$

   $\boxed{\left\{-\dfrac{1}{4}, \dfrac{15}{8}\right\}}$

5. $\begin{aligned} 8(y + 2) - 3(2 - y) &= 4(2y + 6) - 2 \\ 8y + 16 - 6 + 3y &= 8y + 24 - 2 \\ 3y + 10 &= 22 \\ 3y &= 12 \\ y &= 4 \end{aligned}$

   $\boxed{\{4\}}$

6. $\begin{aligned} -2 < \dfrac{2}{7}x - 8 &\le 2 \\ 6 < \dfrac{2}{7}x &\le 10 \\ 42 < 2x &\le 70 \\ 21 < x &\le 35 \end{aligned}$

   $\boxed{\{x \mid 21 < x \le 35\}}$ or $\boxed{(21, 35]}$

7. $\begin{aligned} x &= \dfrac{ax + b}{c} \\ cx &= ax + b \\ cx - ax &= b \\ x(c - a) &= b \end{aligned}$

   $\boxed{x = \dfrac{b}{c - a}, \, c \ne a}$

8.   Let

$$x \;=\; \text{speed of slower car}$$
$$x+4 \;=\; \text{speed of faster car}$$

(distance of slower car) + (distance of faster car) = total distance

$$
\begin{aligned}
3x + 3(x + 4) &= 252 \\
6x + 12 &= 252 \\
6x &= 240 \\
x &= 40 \quad \text{(slower)} \\
x + 4 &= 44 \quad \text{(faster)}
\end{aligned}
$$

$\boxed{\text{slower: 40 mph}}$

$\boxed{\text{faster: 44 mph}}$

9.   Let

$$H \;=\; \text{number of marbles Haraold had to start}$$
$$M \;=\; \text{number of marbles Maude had to start}$$

Round	Harald	Maude	Equation
Start	H	M	$H = M$
1	$H + 20$	$M - 20$	
2	$\frac{1}{3}(H + 20)$	$(M - 20) + \frac{2}{3}(H + 20)$	Maude = 4 (Harold)

$$
\begin{aligned}
\text{(Maude)} &= 4\,\text{(Harold)} \\
(M - 20) + \frac{2}{3}(H + 20) &= 4\left[\frac{1}{3}(H + 20)\right] \\
3(M - 20) + 2(H + 20) &= 4(H + 20) \\
3M - 60 + 2H + 40 &= 4H + 80 \\[6pt]
3M - 2H &= 100 \\
H &= M \\[6pt]
3H - 2H &= 100 \\
H &= 100 \\
M &= 100
\end{aligned}
$$

$\boxed{\text{100 marbles each}}$ to start

10. line passing through $(-2, -3)$ and $(2, 5)$:

$$
\begin{aligned}
m &= \frac{5 + 3}{2 + 2} = \frac{8}{4} = 2 \\
y + 3 &= 2(x + 2) \\
y + 3 &= 2x + 4 \\
\boxed{y = 2x + 1}
\end{aligned}
$$

11. a. From the graph,

$f(5) = 3, f(2) = -3, f(-3) = 3, f(-2) = -2$

$f(5) - f(2) - \left| f(-3) - f(-2) \right|$

$= 3 - (-3) - \left| 3 - (-2) \right|$

$= 3 + 3 - 5$

$= \boxed{1}$

b. $f(x_1) = -3,\ x_1 = 2$ since $f(2) = -3$

$f(x_2) = 5,\ x_2 = -6$ since $f(-6) = 5$

$x_1 - x_2 = 2 - (-6) = \boxed{8}$

c. $\boxed{\text{Domain} = [-6, 5]}$

$\boxed{\text{Range} = [-3, 5]}$

12. $x = 2(y - 5)$

$4x + 40 = y - 7$ (substitute for x) $\rightarrow$

$$\begin{aligned} 4[2(y - 5)] + 40 &= y - 7 \\ 8y - 40 + 40 &= y - 7 \\ 7y &= -7 \\ y &= -1 \end{aligned}$$

$x = 2(y - 5)$

$x = 2(-1 - 5)$

$x = -12$

$\boxed{\{(-12, -1)\}}$

13. $\begin{aligned} 6x + 4y + 4z &= 2 \\ 7x + 5y + z &= 14 \\ 5x + 4y + 2z &= 4 \end{aligned}$

(Equations 1 and 2):

$(\div 2)$ $3x + 2y + 2z = 1$

$\times -2)$ $\underline{-14x - 10y - 2z = -28}$

$-11x - 8y = -27$ (Equation 4)

(Equations 2 and 3):

$(\times -3)$ $-21x - 15y - 3z = -42$

$\underline{5x + 4y + 3z = 4}$

$-16x - 11y = -38$ (Equation 5)

(Equations 4 and 5):

$(\times 11)$ $-121x - 88y = -297$

$(\times -8)$ $\underline{128x + 88y = 304}$

$7x = 7$

$x = 1$

Equation 4 $(\times -1)$: $\begin{aligned} 11x + 8y &= 27 \\ 11 + 8y &= 27 \\ 8y &= 16 \\ y &= 2 \end{aligned}$

(Equation 2): $\begin{aligned} 7x + 5y + z &= 14 \\ 7(1) + 5(2) + z &= 14 \\ z &= -13 \end{aligned}$

$\boxed{\{(1, 2, -3)\}}$

14.

$$
\begin{aligned}
4x - 5y &= 2 \\
6x + 2y + 1 &= 0
\end{aligned}
\qquad
\begin{aligned}
\rightarrow \\
\rightarrow
\end{aligned}
\qquad
\begin{aligned}
4x - 5y &= 2 \\
6x + 2y &= -1
\end{aligned}
$$

$$
D = \begin{vmatrix} 4 & -5 \\ 6 & 2 \end{vmatrix} = 8 + 30 = 38
$$

$$
D_x = \begin{vmatrix} 2 & -5 \\ -1 & 2 \end{vmatrix} = 4 - 5 = -1
$$

$$
D_y = \begin{vmatrix} 4 & 2 \\ 6 & -1 \end{vmatrix} = -4 - 12 = -16
$$

$$
x = \frac{D_x}{D} = \frac{-1}{38}
$$

$$
y = \frac{D_y}{D} = -\frac{16}{38} = -\frac{8}{19}
$$

$$
\boxed{\left\{ \left(-\frac{1}{38}, -\frac{8}{19} \right) \right\}}
$$

15.
$$
\begin{aligned}
6x - y - 3z &= 2 \\
-3x + y - 3z &= 1 \\
-2x + 3y + z &= -6
\end{aligned}
$$

$$
\begin{bmatrix}
6 & -1 & -3 & | & 2 \\
-3 & 1 & -3 & | & 1 \\
-2 & 3 & 1 & | & -6
\end{bmatrix}
\begin{array}{l} \\ 2R_2 + R_1 \rightarrow \\ 3R_3 + R_1 \rightarrow \end{array}
\begin{bmatrix}
6 & -1 & -3 & | & 2 \\
0 & 1 & -9 & | & 4 \\
0 & 8 & 0 & | & -16
\end{bmatrix}
$$

$$
\begin{array}{l} \\ \\ R_3 - 8R_2 \rightarrow \end{array}
\begin{bmatrix}
6 & -1 & -3 & | & 2 \\
0 & 1 & -9 & | & 4 \\
0 & 0 & 72 & | & -48
\end{bmatrix}
\begin{array}{l} \frac{1}{6}R_1 \rightarrow \\ \\ \frac{1}{72}R_3 \rightarrow \end{array}
\begin{bmatrix}
1 & -1/6 & -1/2 & | & 1/3 \\
0 & 1 & -9 & | & 4 \\
0 & 0 & 1 & | & -2/3
\end{bmatrix}
$$

$$
\begin{aligned}
x - \tfrac{1}{6}y - \tfrac{1}{2}z &= \tfrac{1}{3} \\
y - 9z &= 4 \\
z &= -\tfrac{2}{3}
\end{aligned}
\qquad
\begin{aligned}
y - 9\left(-\tfrac{2}{3} \right) &= 4 \\
y + 6 &= 4 \\
y &= -2
\end{aligned}
$$

$$
\begin{aligned}
x - \tfrac{1}{6}(-2) - \tfrac{1}{2}\left(-\tfrac{2}{3} \right) &= \tfrac{1}{3} \\
x + \tfrac{1}{3} + \tfrac{1}{3} &= \tfrac{1}{3} \\
x &= -\tfrac{1}{3}
\end{aligned}
$$

$$
\boxed{\left\{ \left(-\frac{1}{3}, -2, -\frac{2}{3} \right) \right\}}
$$

16. Let

V = volume
T = Temperature
P = pressure
$V = \dfrac{KT}{P}$

Given: T = 400 Kelvin, P = 12 pounds/square inch,
V = 80 cubic inches

$80 = \dfrac{K\,400}{12}$

$\dfrac{12}{5} = K$

$V = \dfrac{12T}{5P}$

At V = 54 cubic inches, T = 350 Kelvin, P = ?

$54 = \dfrac{12(350)}{5P}$

$P = 15\dfrac{5}{9}$

$\boxed{15\dfrac{5}{9}\text{ psi}}$

17. Let

x = cost of can of paint
y = cost of each paintbrush

$\begin{aligned} 2x + 3y &= 53 \\ 3x + y &= 55 \end{aligned}$ $\rightarrow$ $(\times -3) \rightarrow$ $\begin{aligned} 2x + 3y &= 53 \\ \underline{-9x - 3y} &= \underline{-165} \\ -7x &= -112 \\ x &= 16 \end{aligned}$

$\begin{aligned} 3x + y &= 55 \\ 3(16) + y &= 55 \\ y &= 7 \end{aligned}$

$\boxed{\text{can of paint: } \$16; \text{ each paint brush: } \$7}$

18. $y = ax^2 + bx + c$

$(1, 5)$: $5 = a(1)^2 + b(1) + c$ $\rightarrow a + b + c = 5$
$(-2, 14)$: $14 = a(-2)^2 + b(-2) + c$ $\rightarrow 4a - 2b + c = 14$
$(-1, 7)$: $7 = a(-1)^2 + b(-1) + c$ $\rightarrow a - b + c = 7$

(Equations 1 and 3):

$\begin{aligned} a + b + c &= 5 \\ (\times -1) \quad \underline{-a + b - c} &= \underline{-7} \\ 2b &= -2 \\ b &= -1 \end{aligned}$

(Equations 1 and 2):

$\begin{aligned} (\times -1) \quad -a - b - c &= -5 \\ \underline{4a - 2b + c} &= \underline{14} \\ 3a - 3b &= 9 \\ (\div 3) \quad a - b &= 3 \\ a - (-1) &= 3 \quad \text{(substitute } -1 \text{ for } b) \\ a &= 2 \end{aligned}$

(Equations 1):

$\begin{aligned} a + b + c &= 5 \\ 2 + (-1) + c &= 5 \\ c &= 4 \end{aligned}$

$\boxed{a = 2, b = -1, c = 4}$

$y = 2x^2 - x + 4$

19. Let

$$
\begin{aligned}
x &= \text{number of votes for loser} \\
x + 160 &= \text{number of votes for winner}
\end{aligned}
$$

$$
\begin{aligned}
x + (x + 160) &= 2800 \\
2x &= 2640 \\
x &= 1320 \quad \text{(loser)} \\
x + 160 &= 1480 \quad \text{(winner)}
\end{aligned}
$$

$\boxed{\text{loser: } 1320 \text{ votes}}$

$\boxed{\text{winner: } 1480 \text{ votes}}$

20. t = year after 1965
$C = 1.44t + 318.1$
double the preindustrial level of concentration
= 2(280 parts per million)
= 560 parts per million

$$
\begin{aligned}
560 &= 1.44t + 318.1 \\
144t &= 241.9 \\
t &\approx 167.986 \approx 168 \\
\text{year} &\approx 1965 + 168 = \boxed{2133}
\end{aligned}
$$

21. $(6x + 1)(2x^2 + 2x - 7)$
$= 6x(2x^2 + 2x - 7) + (2x^2 + 2x - 7)$
$= 12x^3 + 12x^2 - 42x + 2x^2 + 2x - 7$
$= \boxed{12x^3 + 12x^2 - 40x - 7}$

22. $(x^4 - 2x^3 + 2x - 4) \div (x^2 - 1) = \boxed{x^2 - 2x + 1 + \dfrac{2x - 3}{x^2 - 1}}$

$$
\begin{array}{r}
x^2 - 2x + 1 \\
x^2 - 1 \overline{\smash{\big)}\, x^4 - 2x^3 + 0x^2 + 4x - 4} \\
\underline{x^4 \qquad\;\; - x^2} \\
-2x^3 + x^2 + 4x \\
\underline{-2x^3 \qquad + 2x} \\
x^2 + 2x - 4 \\
\underline{x^2 \qquad - 1} \\
2x - 3
\end{array}
$$

23. $(x^4 - 7x^2 + 3x + 22) \div (x + 2) = \boxed{x^3 - 2x^2 - 3x + 9 + \dfrac{4}{x + 2}}$

$$
\begin{array}{r|rrrrr}
-2 & 1 & 0 & -7 & 3 & 22 \\
   &   & -2 & 4 & 6 & -18 \\
\hline
   & 1 & -2 & -3 & 9 & 4
\end{array}
$$

24. $f(x) = -6x^2 - 5x + 3$

$$\frac{f(a+h) - f(a)}{h}$$

$$= \frac{[-6(a+h)^2 - 5(a+h) + 3] - (-6a^2 - 5a + 3)}{h}$$

$$= \frac{-6a^2 - 12ah - 6h^2 - 5a - 5h + 3 + 6a^2 + 5a - 3}{h}$$

$$= \frac{-12ah - 6h^2 - 5h}{h}$$

$$= \boxed{-12a - 6h - 5}$$

25. $f(x) = -2x^2 + 3x - 7$ and $g(x) = -4x + 3$

$$\begin{aligned}
f[g(x)] &= -2(-4x + 3)^2 + 3(-4x + 3) - 7 \\
&= -2(16x^2 - 24x + 9) - 12x + 9 - 7 \\
&= -32x^2 + 48x - 18 - 12x + 2 \\
&= \boxed{-32x^2 + 36x - 16}
\end{aligned}$$

26. $x^4 - x^2 - 12 = (x^2 - 4)(x^2 + 3) = \boxed{(x + 2)(x - 2)(x^2 + 3)}$

27. $x^3 - 3x^2 - 9x + 27$

$$\begin{aligned}
&= x^2(x - 3) - 9(x - 3) \\
&= (x - 3)(x^2 - 9) \\
&= (x - 3)(x + 3)(x - 3) \\
&= \boxed{(x + 3)(x - 3)^2}
\end{aligned}$$

28. $\begin{vmatrix} 2x & x+1 \\ 3 & x-2 \end{vmatrix} = 1$

$2x(x - 2) - 3(x + 1) = 1$

$2x^2 - 4x - 3x - 3 = 1$

$2x^2 - 7x - 4 = 0$

$(2x + 1)(x - 4) = 0$

$$\begin{array}{ccl}
2x + 1 = 0 & \text{or} & x - 4 = 0 \\
x = -\dfrac{1}{2} & & x = 4
\end{array}$$

$$\boxed{\left\{ -\frac{1}{2}, 4 \right\}}$$

29. Let x and $x + 2$ equal two consecutive odd integers.

$$
\begin{aligned}
[x + (x + 2)]^2 &= x^2 + (x + 2)^2 + 30 \\
(2x + 2)^2 &= x^2 + x^2 + 4x + 4 + 30 \\
4x^2 + 8x + 4 &= 2x^2 + 4x + 34 \\
2x^2 + 4x - 30 &= 0 \\
2(x^2 + 2x - 15) &= 0 \\
2(x + 5)(x - 3) &= 0
\end{aligned}
$$

$$
\begin{array}{lclcrcl}
x + 5 &=& 0 & \text{or} & x - 3 &=& 0 \\
x &=& -5 & & x &=& 3 \\
x + 2 &=& -3 & & x + 2 &=& 5
\end{array}
$$

The integers are $\boxed{-5 \text{ and } -3}$ or $\boxed{3 \text{ and } 5}$.

30.

Let

$$
\begin{aligned}
x &= \text{length of the ladder} \\
x - 1 &= \text{height of the top of the ladder}
\end{aligned}
$$

$$
\begin{aligned}
x^2 &= 7^2 + (x - 1)^2 \\
x^2 &= 49 + x^2 - 2x + 1 \\
2x &= 50 \\
x &= 25 \quad \text{(length)}
\end{aligned}
$$

The ladder's length is $\boxed{25 \text{ meters}}$.

Algebra for College Students

Chapter 6   Rational Expressions and Rational Functions

Section 6.1 Rational Expressions, Rational Functions, and Their Simplification

Problem Set 6.1, pp. 433-436

1. $f(x) = \dfrac{x}{x - 7}$

$f(x)$ is undefined when the denominator equals zero.

$$x - 7 = 0$$
$$x = 7$$

Domain of $f = \{ x \mid x \neq 7 \}$ or $(\infty, 7) \cup (7, \infty)$

3. $f(x) = \dfrac{3x + 8}{3x - 15}$

$f(x)$ is undefined when the denominator equals zero.

$$3x - 15 = 0$$
$$3x = 15$$
$$x = 5$$

Domain of $f = \{ x \mid x \neq 5 \}$ or $(-\infty, 5) \cup (5, \infty)$

5. $f(x) = \dfrac{x + 5}{x}$

$f(x)$ is undefined when the denominator equals zero.

$$x = 0$$

Domain of $f = \{ x \mid x \neq 0 \}$ or $(-\infty, 0) \cup (0, \infty)$

7. $f(x) = \dfrac{2x^2 + 5x - 3}{6}$

No real numbers will cause the denominator of 6 to equal 0. No real values of x need to be excluded. The domain of $f(x)$ is the set of all real numbers.

Domain of $f = \{ x \mid x \in R \}$ or $(-\infty, \infty)$

9. $f(x) = \dfrac{x + 5}{(x - 3)(x + 1)}$

$f(x)$ is undefined when the denominator equals zero.

$$(x - 3)(x + 1) = 0$$

$$x - 3 = 0 \qquad \text{or} \qquad x + 1 = 0$$
$$x = 3 \qquad\qquad\qquad x = -1$$

Domian of $f = \{ x \mid x \neq -1, 3 \}$ or $(-\infty, -1) \cup (-1, 3) \cup (3, \infty)$

11. $f(x) = \dfrac{x+17}{(x-1)(2x+6)}$

$f(x)$ is undefined when the denominator equals zero.

$(x-1)(2x+6) = 0$

$\begin{array}{ccc} x-1 & = & 0 \\ x & = & 1 \end{array}$ or $\begin{array}{ccc} 2x+6 & = & 0 \\ 2x & = & -6 \\ x & = & -3 \end{array}$

Domain of $f = \{\, x \mid x \neq -3, 1 \,\}$ or $(-\infty, -3) \cup (-3, 1) \cup (1, \infty)$

13. $f(x) = \dfrac{x+4}{x^2 - 25}$

$f(x)$ is undefined when the denominator equals zero.

$x^2 - 25 = 0$

$(x-5)(x+5) = 0$

$\begin{array}{ccc} x-5 & = & 0 \\ x & = & 5 \end{array}$ or $\begin{array}{ccc} x+5 & = & 0 \\ x & = & -5 \end{array}$

Domain of $f = \{\, x \mid x \neq -5, 5 \,\}$ or $(-\infty, 5) \cup (-5, 5) \cup (5, \infty)$

15. $f(x) = \dfrac{x+4}{x^2 + 25}$

No real number substituted into the denominator will cause $x^2 + 25$ to equal zero. No values of x need to be excluded. The domain of f is the set of all real numbers.

Domain of $f = \{\, x \mid x \in R \,\}$ or $(-\infty, \infty)$

17. $f(x) = \dfrac{x-3}{(x-3)(x+4)}$

$f(x)$ is undefined when

$(x-3)(x+4) = 0$

$\begin{array}{ccc} x-3 & = & 0 \\ x & = & 3 \end{array}$ or $\begin{array}{ccc} x+4 & = & 0 \\ x & = & -4 \end{array}$

Domain of $f = \{\, x \mid x \neq -4, 3 \,\}$ or $(-\infty, -4) \cup (-4, 3) \cup (3, \infty)$

19. $f(x) = \dfrac{x+8}{x^2 - x - 12}$

$f(x)$ is undefined when

$x^2 - x - 12 = 0$

$(x-4)(x+3) = 0$

$\begin{array}{ccc} x-4 & = & 0 \\ x & = & 4 \end{array}$ or $\begin{array}{ccc} x+3 & = & 0 \\ x & = & -3 \end{array}$

Domain of $f = \{\, x \mid x \neq -3, 4 \,\}$ or $(-\infty, -3) \cup (-3, 4) \cup (4, \infty)$

21. $f(x) = \dfrac{x^2 + 3}{2x^2 + 3x - 2}$

$f(x)$ is undefined when

$2x^2 + 3x - 2 = 0$

$(x+2)(2x-1) = 0$

$\begin{array}{ccc} x+2 & = & 0 \\ x & = & -2 \end{array}$ or $\begin{array}{ccc} 2x-1 & = & 0 \\ x & = & \dfrac{1}{2} \end{array}$

Domain of $f = \left\{\, x \mid x \neq -2, \dfrac{1}{2} \,\right\}$ or $(-\infty, -2) \cup \left(-2, \dfrac{1}{2}\right) \cup \left(\dfrac{1}{2}, \infty\right)$

23. $f(x) = \dfrac{2}{x-2}$

a. $f(x)$ is undefined when

$$x - 2 = 0$$
$$x = 0$$

$\boxed{D_f = \{x \mid x \neq 2\} \text{ or } (-\infty, 2) \cup (2, \infty)}$

b.

x	1	1.5	1.9	1.99	1.999
$f(x) = \dfrac{2}{x-2}$	−2	−4	−20	−200	−2000

x	2.001	2.01	2.1	2.5	3
$f(x) = \dfrac{2}{x-2}$	2000	200	20	4	2

c. $x = 2$

$$f(x) = \dfrac{2}{x-2}$$

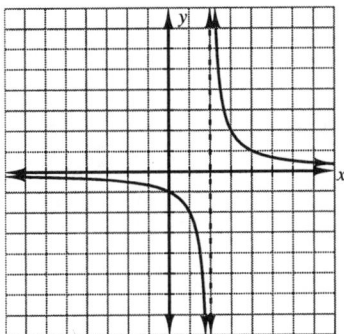

25. a. $\boxed{h(x) = \dfrac{3}{x^2 - 1}}$

$h(0) = -3$

$h(\pm 2) = \dfrac{3}{4-1} = 1$

b. $\boxed{g(x) = \dfrac{3}{x^2 + 1}}$

$g(0) = 3$

$g(\pm 1) = 1.5$

c. $\boxed{f(x) = \dfrac{2}{(x-1)^3}}$

$f(0) = -2$

$f(2) = 2$

27. $\qquad f(x) = \dfrac{3}{x^2 - 1}$

$f(x)$ is undefined when

$$x^2 - 1 = 0$$
$$x = \pm 1$$

$\boxed{D_f = \{x \mid x \neq \pm 1\} \ \text{ or } \ (\infty, -1) \cup (-, 1) \cup (1, \infty)}$

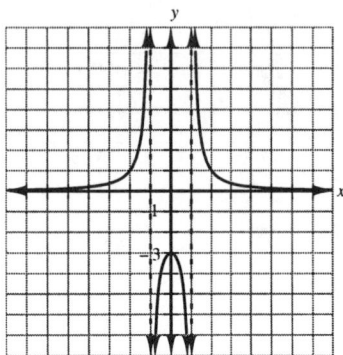

29. $C = \dfrac{4x}{100 - x}$, C(in thousands of dollars)

    a. $x\% = 80\%$

$$C - \frac{4(80)}{100 - 80} = \frac{320}{20} = 16$$

         cost: $\boxed{\$16{,}000}$

    b. $x\% = 95\%$

$$C = \frac{4(95)}{100 - 95} = \frac{380}{5} = 76$$

         cost: $\boxed{\$76{,}000}$

    c. $100 - x = 0$

$$x = \boxed{100} \ \text{ is not permissible since } C(x) \text{ is undefined when } x = 100.$$

    d.   As $x\%$ approaches 100%, the cost becomes increasingly high;

         $\boxed{\text{The cost becomes prohibitive.}}$

31. $\dfrac{5bc}{7bc} = \dfrac{5}{7}\left(\dfrac{b}{b}\right)\left(\dfrac{c}{c}\right) = \boxed{\dfrac{5}{7}}$ **33.** $\dfrac{3a + 9}{a + 3} = \dfrac{3(a + 3)}{(x + 3)} = \boxed{3}$

35. $\dfrac{a - c}{a^2 - ac} = \dfrac{(a - c)}{a(a - c)} = \boxed{\dfrac{1}{a}}$ **37.** $\dfrac{6x + 3}{18} - \dfrac{3(2x + 1)}{3(6)} = \boxed{\dfrac{2x + 1}{6}}$

39. $\dfrac{3b}{3b + 3c} = \dfrac{3(b)}{3(b + c)} = \boxed{\dfrac{b}{b + c}}$ **41.** $\dfrac{12ab^2}{6ab^3 - 6ab^4} = \dfrac{12b^2}{6ab^3(1 - b)} = \boxed{\dfrac{2}{b(1 - b)}}$

43. $\dfrac{3x + 3y}{4x + 4y} = \dfrac{3(x + y)}{4(x + y)} = \boxed{\dfrac{3}{4}}$ **45.** $\dfrac{5a2b + 5a^2c}{5ab + 15ac} = \dfrac{5a^2(b + c)}{15a(b + c)} = \boxed{\dfrac{a}{3}}$

47. $\dfrac{x^2 - 4}{2x - 4} = \dfrac{(x + 2)(x - 2)}{2(x - 2)} = \boxed{\dfrac{x + 2}{2}}$ **49.** $\dfrac{4b - 8}{b^2 - 4b + 4} = \dfrac{4(b - 2)}{(b - 2)^2} = \boxed{\dfrac{4}{b - 2}}$

51. $\dfrac{y^2 - 8y + 16}{3y - 12} = \dfrac{(y-4)^2}{3(y-4)} = \boxed{\dfrac{y-4}{3}}$

53. $\dfrac{x^2 - 2xy + y^2}{x^2 - y^2} = \dfrac{(x-y)^2}{(x+y)(x-y)} = \boxed{\dfrac{x-y}{x+y}}$

55. $\dfrac{a^4 - b^4}{a^2 - 2ab + b^2}$

$= \dfrac{(a^2 + b^2)(a^2 - b^2)}{(a-b)^2}$

$= \dfrac{(a^2 + b^2)(a+b)(a-b)}{(a-b)^2}$

$= \boxed{\dfrac{(a^2 + b^2)(a+b)}{a-b}}$

57. $\dfrac{y^2 - 4y - 5}{y^2 + 5y + 4} = \dfrac{(y-5)(y+1)}{(y+4)(y+1)} = \boxed{\dfrac{y-5}{y+4}}$

59. $\dfrac{6b^2 - b - 2}{3b^2 + 4b - 4} = \dfrac{(3b-2)(2b+1)}{(3b-2)(b+2)} = \boxed{\dfrac{2b+1}{b+2}}$

61. $\dfrac{x^2 - 9}{x^2 + x - 6} = \dfrac{(x+3)(x-3)}{(x+3)(x-2)} = \boxed{\dfrac{x-3}{x-2}}$

63. $\dfrac{a^3 + 64}{a^2 - 16} = \dfrac{(a+4)(a^2 - 4a + 16)}{(a+4)(a-4)} = \boxed{\dfrac{a^2 - 4a + 16}{a-4}}$

65. $\dfrac{x^3 - 8}{x^2 + 2x - 8} = \dfrac{(x-2)(x^2 + 2x + 4)}{(x-2)(x+4)} = \boxed{\dfrac{x^2 + 2x + 4}{x+4}}$

67. $\dfrac{x^3 + x^2 - 20x}{x^2 + 2x^2 - 15x} = \dfrac{x(x^2 + x - 20)}{x(x^2 + 2x - 15)} = \dfrac{x^2 + x - 20}{x^2 + 2x - 15} = \dfrac{(x+5)(x-4)}{(x+5)(x-3)} = \boxed{\dfrac{x-4}{x-3}}$

69. $\dfrac{2x+3}{2x-5}$ $\boxed{\text{cannot be reduced}}$

71. $\dfrac{x}{x+y}$ $\boxed{\text{cannot be reduced}}$

73. $\dfrac{x^2 - 5x + 6}{x^2 - 7x - 18} = \dfrac{(x-2)(x-3)}{(x-9)(x+2)}$

Although both numerator and denominator can be factored, there are no identical factors in the numerator and the denominator. Thus, the algebraic fraction $\boxed{\text{cannot be simplified or reduced}}$.

75. $\dfrac{a^2 - 16}{a^2 - 4a + 3ab - 12b} = \dfrac{(a+4)(a-4)}{a(a-4) + 3b(a-4)} = \dfrac{(a+4)(a-4)}{(a+3b)(a-4)} = \boxed{\dfrac{a+4}{a+3b}}$

77. $\dfrac{3m^2 + 9m°mx - 3x}{m^3 + 27} = \dfrac{3m(m+3) - x(m+3)}{(m+3)(m^2 - 3m + 9)} = \dfrac{(m+3)(3m-x)}{(m+3)(m^2 - 3m + 9)} = \boxed{\dfrac{3m-x}{m^2 - 3m + 9}}$

79. $\dfrac{3-y}{y-3} = \dfrac{-(y-3)}{y-3} = \boxed{-1}$

You can immediately obtain -1 using the property that the quotient of two polynomials that have exactly opposite signs and are additive inverse is -1.

81. $\dfrac{a^2 - 4}{2-a} = \dfrac{(a+2)(a-2)}{-(a-2)} = -(a+2) = \boxed{-a-2}$

83. $\dfrac{3-x}{x^2 - 7x + 12} = \dfrac{-(x-3)}{(x-4)(x-3)} = \boxed{-\dfrac{1}{x-4} \text{ or } \dfrac{1}{4-x}}$

The answer can be equivalently expressed by attaching the negative sign to the denominator rather than the numerator.

85. $\dfrac{x^{2n} - 2x^n - 3}{x^{2n} + x^n - 12} = \dfrac{(x^n - 3)(x^n + 1)}{(x^n - 3)(x^n + 4)} = \boxed{\dfrac{x^n + 1}{x^{in} + 4}}$

87. $\dfrac{x^{2n} - y^{2n}}{x^{2n} + 2x^n y^n + y^{2n}} = \dfrac{(x^n + y^n)(x^n - y^n)}{(x^n + y^n)^2} = \boxed{\dfrac{x^n - y^n}{x^n + y^n}}$

89. $\dfrac{a^{2n} - b^{2n}}{a^{2n} + a^n b^n} = \dfrac{(a^n + b^n)(a^n - b^n)}{a^n(a^n + b^n)} = \boxed{\dfrac{a^n - b^n}{a^n}}$

91. $\dfrac{(x^3 - y^3)(x^3 - xy^2)}{x^3 + x^2 y + xy^2}$

$\quad = \dfrac{(x - y)(x^2 + xy + y^2)(x)(x^2 - y^2)}{x(x^2 + xy + y^2)}$

$\quad = (x - y)(x^2 - y^2)$

$\quad = (x - y)(x + y)(x - y)$

$\quad = \boxed{(x + y)(x - y)^2}$

93. $\dfrac{(x^2 + 9)(54 - 2x^3)}{2x^4 - 162}$

$\quad = \dfrac{(x^2 + 9)(-2)(x^3 - 27)}{2(x^4 - 81)}$

$\quad = -\dfrac{(x^2 + 9)(x^3 - 27)}{(x^4 - 81)}$

$\quad = -\dfrac{(x^2 + 9)(x - 3)(x^2 + 3x + 9)}{(x^2 + 9)(x^2 - 9)}$

$\quad = -\dfrac{(x - 3)(x^2 + 3x + 9)}{(x - 3)(x + 3)} = \boxed{\dfrac{-(x^2 + 3x + 9)}{x + 3}}$

95. $f(x) = \dfrac{x^2 - x - 2}{x + 1} = \dfrac{(x - 2)(x + 1)}{x + 1} = \boxed{x - 2 \quad (x \neq -1)}$

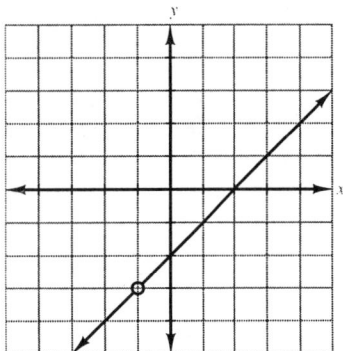

97. $\boxed{D}$ is true; A is not true: $\dfrac{x^2 - 25}{x - 5} = \dfrac{(x - 5)(x + 5)}{x - 5} = x + 5 \quad not\ x - 5$

$\qquad\qquad\qquad\quad\ B$ is not true: $\dfrac{x^2 + 7}{7} = \dfrac{x^2}{7} + 1 \quad not\ x^2 + 1$

$\qquad\qquad\qquad\quad\ C$ is *not* true: *Not all* rational functions have vertical asymptotes.
$\qquad\qquad\qquad\quad\ D$ is true.

99. $f(x) = \dfrac{600x}{300x - 300} \ = \ \dfrac{600x}{300(x - 1)} = \boxed{\dfrac{2x}{x - 1}}$

$\qquad\quad x = \dfrac{4}{5} = 0.8 \ = \ 80\%$

$\qquad\quad f(80) = \dfrac{2(80)}{80 - 1} \ = \ \dfrac{160}{79} \approx 2.0253 \approx 2$

To distribute telephone books to $\dfrac{4}{5}$ of the population would take $\boxed{\text{approximately 2 hours}}$.

101.
$$dx - 2x = d^2 - 4d + 4$$
$$x(d - 2) = d^2 - 4d + 4$$
$$x = \frac{(d - 2)^2}{d - 2}$$
$$\boxed{x = d - 2, d \neq 2}$$

103.
$$d^2(x - 1) = 5d + 10 + 4(x - 1)$$
$$d^2 x - d^2 = 5d + 10 + 4x - 4$$
$$d^2 x - 4x = d^3 + 5d + 6$$
$$x(d^2 - 4) = (d + 3)(d + 2)$$
$$x = \frac{(d + 3)(d + 2)}{(d - 2)(d + 2)}$$
$$\boxed{x = \frac{d + 3}{d - 2}, d \neq -2, 2}$$

105. a.
$$\frac{ax^2 - (x - 4)^2}{(x - 1)(x + 2)}$$
$$= \frac{[3x + (x - 4)][3x - (x - 4)]}{(x - 1)(x + 2)}$$
$$= \frac{(4x - 4)(3x - x + 4)}{(x - 1)(x + 2)}$$
$$= \frac{4(x - 1)(2x + 4)}{(x - 1)(x + 2)}$$
$$= \frac{4(2)(x + 2)}{x + 2}$$
$$= \boxed{8}$$

b.   Let
$$x = 3351$$
$$x - 4 = 3351 - 4 = 3347$$
$$x - 1 = 3351 - 1 = 3350$$
$$x + 2 = 3351 + 2 = 3353$$

$$\frac{9(3351)^2 - (3347)^2}{3350 \cdot 3353} = \frac{9x^2 - (x - 4)^2}{(x - 1)(x + 2)} = \boxed{8} \quad \text{(from part } \mathbf{a}\text{)}$$

107.   slope of line passing through $(5, 25)$ and $(5 + w, (5 + w)^2)$:
$$m = \frac{(5 + w)^2 - 25}{(5 + w) - 5} = \frac{[(5 + w) + 5][(5 + w) - 5]}{w} = \frac{(w + 10)(w)}{w} = \boxed{w + 10, w \neq 0}$$

109. $f(x) = 0$ $\qquad$ and $\qquad$ $g(x) = 0$
$\qquad$ $\{2, 3, 5, 7, 9\}$ $\qquad\qquad\qquad$ $\{-1, 3, 5, 7, 8\}$

For $\dfrac{f(x)}{g(x)} = 0$ the solution set is all x such that $f(x) = 0$ and $g(x) \neq 0$.

Eliminating 3, 5, and 7 from $\{2, 3, 5, 7, 9\}$ (since 3, 5, and 7 are three of the values that cause $g(x)$ to equal 0), the solution set for $\dfrac{f(x)}{g(x)} = 0$ is $\boxed{\{2, 9\}}$.

Review Problems

116. method 1:
$$x^7 - x$$
$$= x(x^6 - 1)$$
$$= x[(x^3)^2 - 1^2]$$
$$= x(x^3 + 1)(x^3 - 1)$$
$$= \boxed{x(x + 1)(x^2 - x + 1)(x - 1)(x^2 + x + 1)}$$

or method 2:
$$x^7 - x$$
$$= x(x^6 - 1)$$
$$= x[(x^2)^3 - 1^3]$$
$$= x(x^2 - 1)(x^4 + x^2 + 1)$$
$$= \boxed{x(x + 1)(x - 1)(x^4 + x^2 + 1)}$$
The two answers are identical, and
$$x^4 + x^2 + 1 = (x^2 - x + 1)(x^2 + x + 1)$$

117. Let
$$x = \text{length of side of smaller tirangle}$$
$$x + 10 = \text{length of side of large triangle}$$

$$\begin{aligned}
(\text{sum of perimeter}) &= 186 \\
3x + 3(x + 10) &= 186 \\
6x + 30 &= 186 \\
6x &= 156 \\
x &= 26 \quad \text{(smaller triangle)} \\
x + 10 &= 36 \quad \text{(larger triangle)}
\end{aligned}$$
length of each side of the larger triangle: $\boxed{36 \text{ m}}$

118. $(3x^n - 5y^n)(6x^n + y^n)$
$$= 18x^{2n} + 3x^n y^n - 30x^n y^n - 5y^{2n}$$
$$= \boxed{18x^{2n} - 27x^n y^n - 5y^{2n}}$$

Section 6.2 Multiplying and Dividing Rational Expressions

Problem Set 6.2, pp. 442-444

1. $\dfrac{x}{y} \cdot \dfrac{4y}{x-y} = \dfrac{4xy}{y(x-y)} = \boxed{\dfrac{4x}{x-y}}$

3. $\dfrac{x+y}{2} \cdot 10 = \dfrac{10(x+y)}{2} = \boxed{5(x+y)}$

5. $3xy^2 \cdot \dfrac{x-y}{3xy} = \dfrac{3xy^2(x-y)}{3xy} = \boxed{y(x-y)}$

7. $\dfrac{3y^3}{2x(a+b)} \cdot \dfrac{4x^3(a+b)}{a-b} = \dfrac{12x^3 y^3(a+b)}{2x(a+b)(a-b)} = \boxed{\dfrac{6x^2 y^3}{a-b}}$

9. $\dfrac{(a-1)(a+5)}{(a-3)(a+5)} \cdot \dfrac{(a-3)(a-2)}{(a+1)(a-3)} = \dfrac{a-1}{a-3} \cdot \dfrac{a-2}{a+1} = \boxed{\dfrac{(a-1)(a-2)}{(a-3)(a+1)}}$

11. $\dfrac{y^2-4}{y-2}\cdot\dfrac{y-2}{y^2+y-6}$

$=\dfrac{(y+2)(y-2)}{y-2}\cdot\dfrac{y-2}{(y+3)(y-2)}$

$=(y+2)\cdot\dfrac{1}{(y+3)}$

$=\boxed{\dfrac{y+2}{y+3}}$

13. $\dfrac{5x+5y}{x-y}\cdot\dfrac{3x-3y}{10}$

$=\dfrac{5(x+y)}{x-y}\cdot\dfrac{3(x-y)}{10}$

$=\dfrac{15(x+y)(x-y)}{10(x-y)}$

$=\boxed{\dfrac{3(x+y)}{2}}$

15. $\dfrac{b^2+b}{b^2-4}\cdot\dfrac{b^2+5b+6}{b^2-1}$

$=\dfrac{b(b+1)(b+3)(b+2)}{(b+2)(b-2)(b+1)(b-1)}$

$=\boxed{\dfrac{b(b+3)}{(b-2)(b-1)}}$

17. $\dfrac{m^2-4}{m^2+4m+4}\cdot\dfrac{2m+4}{m^2+m-6}$

$=\dfrac{(m+2)(m-2)(2)(m+2)}{(m+2)^2(m+3)(m-2)}$

$=\boxed{\dfrac{2}{m+3}}$

19. $\dfrac{b+2}{b^2+7b+6}\cdot(b+1)=\dfrac{(b+2)(b+1)}{(b+6)(b+1)}=\boxed{\dfrac{b+2}{b+6}}$

21. $\dfrac{x^3-6}{x^2-4}\cdot\dfrac{x+2}{3x}=\dfrac{(x-2)(x^2+2x+4)(x+2)}{(x+2)(x-2)(3x)}=\boxed{\dfrac{x^2+2x+4}{3x}}$

23. $\dfrac{2a^2-13a-7}{a^2-6a-7}\cdot\dfrac{a^2-a-2}{2a^2-5a-3}$

$=\dfrac{(2a+1)(a-7)(a-2)(a+1)}{(a-7)(a+1)(2a+1)(a-3)}$

$=\boxed{\dfrac{a-2}{a-3}}$

25. $\dfrac{2y^2+9y-35}{6y^2-13y-5}\cdot\dfrac{3y^2+10y+3}{y^2+10y+21}$

$=\dfrac{(2y-5)(y+7)(3y+1)(y+3)}{(2y-5)(3y+1)(y+7)(y+3)}$

$=\boxed{1}$

27. $\dfrac{y^2+10y+25}{y-4}\cdot\dfrac{y^2-y-12}{y+5}\cdot\dfrac{1}{y+3}$

$=\dfrac{(y+5)^2(y-4)(y+3)}{(y-4)(y+5)(y+3)}$

$=\boxed{y+5}$

29. $\dfrac{m^2+m-12}{m^2+m-30}\cdot\dfrac{m^2+5m+6}{m^2-2m-3}\cdot\dfrac{m^2+7m+6}{m+3}$

$=\dfrac{(m+4)(m-3)(m+3)(m+2)(m+6)(m+1)}{(m+6)(m-5)(m-3)(m+1)(m+3)}$

$=\boxed{\dfrac{(m+4)(m+2)}{m-5}}$

31. $\dfrac{20a^2b^3c^4}{x^2-16}\cdot\dfrac{4-x}{5abc}=\dfrac{20a^2b^3c^4(-1)(x-4)}{5abc(\ +4)(x-4)}=\boxed{\dfrac{-4ab^2c^3}{x+4}}$

33. $\dfrac{x^2+5x+4}{x^2+x-12}\cdot\dfrac{3-x}{x+1}$

$=\dfrac{(x+4)(x+1)(-1)(x-3)}{(x+4)(x-3)(x+1)}$

$=\boxed{-1}$

35. $\dfrac{x^3-27}{x^3-3x}\cdot\dfrac{3-x^2}{x-3}$

$=\dfrac{(x-3)(x^2+3x+9)(-1)(x^2-3)}{x(x^2-3)(x-3)}$

$=\boxed{\dfrac{-(x^2+3x+9)}{x}}$

37. $(y^2-25)\cdot\dfrac{3}{y-5}$

$=\dfrac{(y+5)(y-5)(3)}{y-5}=\boxed{3(y+5)}$

39. $(y^2-3y+2)\cdot\dfrac{1}{y-2}$

$=\dfrac{(y-2)(y-1)}{y-2}$

$=\boxed{y-1}$

41. $\dfrac{pr - ps + qr - qs}{pr + ps + qr + qs} \cdot \dfrac{mr + ms - nr - ns}{mr - ms + nr - nr}$

$= \dfrac{p(r-s) + q(r-s)}{p(r+s) + q(r+s)} \cdot \dfrac{m(r+s) - n(r+s)}{m(r-s) + n(r-s)}$

$= \dfrac{(r-s)(p+q)(r+s)(m-n)}{(r+s)(p+q)(r-s)(m+n)}$

$= \boxed{\dfrac{m-n}{m+n}}$

43. $\dfrac{x^3 - 4x^2 + x - 4}{2x^3 - 8x^2 + x - 4} \cdot \dfrac{2x^3 + 2x^2 + x + 1}{x^4 - x^3 + x^2 - x}$

$= \dfrac{x^2(x-4) + (x-4)}{2x^2(x-4) + (x-4)} \cdot \dfrac{2x^2(x+1) + (x+1)}{x^3(x-1) + x(x-1)}$

$= \dfrac{(x-4)(x^2+1)(x+1)(2x^2+1)}{(x-4)(2x^2+1)(x-1)(x^3+x)}$

$= \dfrac{(x^2+1)(x+1)}{(x-1)(x)(x^2+1)}$

$= \boxed{\dfrac{x+1}{x(x-1)}}$

45. $\dfrac{y^{n+1} - 3y^n}{y^{2n} + 2y^n} \cdot \dfrac{y^{n+1} + 2y}{y^2 - 3y}$

$= \dfrac{y^n(y-3)(y)(y^n+2)}{y^n(y^n+2)y(y-3)}$

$= \boxed{1}$

47. $\dfrac{y^{2n} - 1}{y^{2n} + 3y^n + 2} \cdot \dfrac{y^{2n} - y^n - 6}{y^{2n} + y^n - 12}$

$= \dfrac{(y^n+1)(y^n-1)(y^n-3)(y^n+2)}{(y^n+2)(y^n+1)(y^n+4)(y^n-3)}$

$= \boxed{\dfrac{y^n - 1}{y^n + 4}}$

49. $\dfrac{ax - ay + 3x - 3y}{x^3 + y^3} \cdot \dfrac{xy - x^2 - y^2}{ab + 3b + ac + 3c}$

$= \dfrac{a(x-y) + 3(x-y)}{x^3 + y^3} \cdot \dfrac{(-1)(x^2 - xy + y^2)}{b(a+3) + c(a+3)}$

$= \dfrac{(x-y)(a+3)(-1)(x^2 - xy + y^2)}{(x+y)(x^2 - xy + y^2)(a+3)(b+c)}$

$= \boxed{\dfrac{-(x-y)}{(x+y)(b+c)}}$

51. $\dfrac{2a + 4b}{3ab} \div \dfrac{6a + 12b}{6a^2 b}$

$= \dfrac{2(a+2b)}{3ab} \cdot \dfrac{6a^2 b}{6(a+2b)}$

$= \dfrac{2(6)a^2 b(a+2b)}{3(6)ab(a+2b)}$

$= \boxed{\dfrac{2a}{3}}$

53. $\dfrac{x+y}{x^2 - xy} \div \dfrac{3x + 3y}{x - y}$

$= \dfrac{x+y}{x(x-y)} \cdot \dfrac{x-y}{3(x+y)}$

$= \dfrac{(x+y)(x-y)}{3x(x-y)(x+y)}$

$= \boxed{\dfrac{1}{3x}}$

55. $\dfrac{x^2 - y^2}{(x+y)^2} \div \dfrac{x-y}{4x + 4y}$

$= \dfrac{x^2 - y^2}{(x+y)^2} \cdot \dfrac{4x + 4y}{x - y}$

$= \dfrac{(x+y)(x-y)(4)(x+y)}{(x+y)^2(x-y)}$

$= \boxed{4}$

57. $\dfrac{x^3 - 27}{a^3 + 8} \div \dfrac{x - 3}{a + 2}$

$= \dfrac{x^3 - 27}{a^3 + 8} \cdot \dfrac{a + 2}{x - 3}$

$= \dfrac{(x-3)(x^2 + 3x + 9)(a+2)}{(a+2)(a^2 - 2a\ 4)(x-3)}$

$= \boxed{\dfrac{x^2 + 3x + 9}{a^2 - 2a + 4}}$

59. $\dfrac{4b^2 + 20ab + 25}{5b} \div \dfrac{4b^2 - 25}{4b^2}$

$= \dfrac{(2b+5)^2}{5b} \cdot \dfrac{4b^2}{4b^2 - 25}$

$= \dfrac{(2b+5)^2(4b^2)}{5b(2b+5)(2b-5)}$

$= \boxed{\dfrac{4b(2b+5)}{5(2b-5)}}$

61. $\dfrac{a^2 - 4a - 21}{a^2 - 10a + 25} \div \dfrac{a^2 + 2a - 3}{a^2 - 6a + 5}$

$= \dfrac{a^2 - 4a - 21}{a^2 - 10a + 25} \cdot \dfrac{a^2 - 6a + 5}{a^2 + 2a - 3}$

$= \dfrac{(a - 7)(a + 3)(a - 5)(a - 1)}{(a - 5)^2(a + 3)(a - 1)}$

$= \boxed{\dfrac{a - 7}{a - 5}}$

63. $\dfrac{9x^2 - 12x + 4}{2x^2 + 3x - 5} \div \dfrac{3x^2 - 8x + 4}{2x^2 + 7x + 5}$

$= \dfrac{9x^2 - 12x + 4}{2x^2 + 3x - 5} \cdot \dfrac{2x^2 + 7x + 5}{3x^2 - 8x + 4}$

$= \dfrac{(3x - 2)^2(2x + 5)(x + 1)}{(2x + 5)(x - 1)(3x - 2)(x - 2)}$

$= \boxed{\dfrac{(3x - 2)(x + 1)}{(x - 1)(x - 2)}}$

65. $(bx + by + 3x + 3y) \div \dfrac{x^2 - y^2}{b + 3}$

$= b(x + y) + 3(x + y) \cdot \dfrac{b + 3}{x^2 - y^2}$

$= \dfrac{(x + y)(b + 3)(b + 3)}{(x + y)(x - y)}$

$= \boxed{\dfrac{(b + 3)^2}{x - y}}$

67. $\dfrac{x - 2}{8x^3 - 27} \div \dfrac{ax - 2a + bx - 2b}{3 - 2x}$

$= \dfrac{x - 2}{8x^3 - 27} \cdot \dfrac{3 - 2x}{ax + bx - 2a - 2b}$

$= \dfrac{x - 2}{(2x)^3 - 3^3} \cdot \dfrac{(-1)(2x - 3)}{x(a + b) - 2(a + b)}$

$= \dfrac{(x - 2)(-1)(2x - 3)}{(2x - 3)(4x^2 + 6x + 9)(a + b)(x - 2)}$

$= \boxed{\dfrac{-1}{(4x^2 + 6x + 9)(a + b)}}$

69. $\dfrac{2y^2 + 13y + 20}{6y^2 - 13y - 5} \div \dfrac{8 - 10y - 3y^2}{9y^2 - 3y - 2}$

$= \dfrac{2y^2 + 13y + 20}{6y^2 - 13y - 5} \cdot \dfrac{9y^2 - 3y - 2}{(7)(3y^2 + 10y - 8)}$

$= \dfrac{(2y + 5)(y + 4)(3y + 1)(3y - 2)}{(3y + 1)(2y - 5)(-1)(3y - 2)(y + 4)}$

$= \boxed{\dfrac{-(2y + 5)}{2y - 5}}$

71. $\dfrac{4y^3 - 12y^2}{4y^n} \div \dfrac{y^{n+1} - 2y}{y^{2n} - 4}$

$= \dfrac{4y^3 - 12y^2}{4y^n + 8} \cdot \dfrac{y^{2n} - 4}{y^{n+1} - 2y}$

$= \dfrac{4y^2(y - 3)(y^n + 2)(y^n - 2)}{4(y^n + 2)(y)(y^n - 2)}$

$= \boxed{y(y - 3)}$

73. $\dfrac{y^{2n} - 1}{2y^{2n} + y^n - 3} \div \dfrac{y^{2n} - y^n - 2}{2y^{2n} - y^n - 6}$

$= \dfrac{y^{2n} - 1}{2y^{2n} + y^n - 3} \cdot \dfrac{2y^{2n} - y^n - 6}{y^{2n} - y^n - 2}$

$= \dfrac{(y^n + 1)(y^n - 1)(2y^n + 3)(y^n - 2)}{(2y^n + 3)(y^n - 1)(y^n - 2)(y^n + 1)}$

$= \boxed{1}$

75. $\dfrac{y - 1}{y^2 - 3y} \div \dfrac{1}{y^3 - 9y}$

$= \dfrac{y - 1}{y^2 - 3y} \cdot \dfrac{y(y^2 - 9)}{1}$

$= \dfrac{(y - 1)(y)(y + 3)(y - 3)}{y(y - 3)}$

$= \boxed{(y - 1)(y + 3)}$

77. $\dfrac{x^3(x - y)^3}{x^3 - y^3} \div \dfrac{x^3 - 2x^2y + xy^2}{x^2 + xy + y^2}$

$= \dfrac{x^3(x - y)^3}{x^3 - y^3} \cdot \dfrac{x^2 + xy + y^2}{x(x^2 - 2xy + y^2)}$

$= \dfrac{x^3(x - y)^3(x^2 + xy + y^2)}{(x - y)(x^2 + xy + y^2)(x)(x - y)^2}$

$= \boxed{x^2}$

79. $\left(\dfrac{a - b}{4c} \div \dfrac{b - a}{c}\right) \div \dfrac{a - b}{c^2}$

$= \left(\dfrac{a - b}{4c} \cdot \dfrac{c}{(-1)(a - b)}\right) \div \dfrac{a - b}{c^2}$

$= \left(-\dfrac{1}{4}\right) \cdot \dfrac{c^2}{a - b}$

$= \dfrac{-c^2}{4(a - b)}$

$= \boxed{\dfrac{-c^2}{4(a - b)}}$

81. $\dfrac{a^2 - 8a + 15}{2a^3 - 10a^2} \cdot \dfrac{2a^2 + 3a}{3a^3 - 27a} \div \dfrac{14a + 21}{a^2 - 6a - 27}$

$= \dfrac{a^2 - 8a + 15}{2a^3 - 10a^2} \cdot \dfrac{2a^2 + 3a}{3a(a^2 - 9)} \cdot \dfrac{a^2 - 6a - 27}{14a + 21}$

$= \dfrac{(a - 5)(a - 3)(a)(2a + 3)(a - 9)(a + 3)}{2a^2(a - 5)(3a)(a + 3)(a - 3)(7)(2a + 3)}$

$= \boxed{\dfrac{a - 9}{42a^2}}$

83. area of rectangle $\div$ area of triangle

$= \left(\dfrac{1}{x^2 - 9} \cdot x^2 + 6x + 9 \right) \div \left(\dfrac{1}{2} \cdot \dfrac{1}{x^2 + 6x + 9} \cdot 2(x + 3) \right)$

$= \dfrac{(x + 3)^2}{(x + 3)(x - 3)} \div \dfrac{2(x + 3)}{2(x + 3)^2}$

$= \dfrac{x + 3}{x - 3} \cdot (x + 3)$

$= \boxed{\dfrac{(x + 3)^2}{x - 3}}$

85. $f(x) = \dfrac{12 - 4x}{x^2 - 9}$ and $g(x) = \dfrac{x^2 - 2x + 1}{x^2 - 1}$

$(fg)(x) = \dfrac{12 - 4x}{x^2 - 9} \cdot \dfrac{x^2 - 2x + 1}{x^2 - 1}$

$= \dfrac{-4(x - 3)(x - 1)^2}{(x + 3)(x - 3)(x + 1)(x - 1)}$

$= \boxed{\dfrac{-4(x - 1)}{(x + 3)(x + 1)}} \, ; \; \{x \mid x \neq \pm 3 \text{ and } x \neq \pm 1\}$

87. $\boxed{D}$ is true; A is *not* true: All rational expressions can be multiplied.

B is *not* true: $x \div y = \dfrac{x}{y} \neq \dfrac{1}{x} \cdot y$

$\qquad\qquad\qquad (y \neq 0) \quad not \; (x \neq 0)$

C is *not* true: $\dfrac{x}{y} \div \dfrac{y}{x} = \dfrac{x}{y} \cdot \dfrac{x}{y} = \dfrac{x^2}{y^2} \neq 1$

D is true.

Review Problems

90. $f(x) = 4x^2 - 3x + 5$ and $g(x) = 2x - 1$

$f[g(x)] = 4(2x - 1)^2 - 3(2x - 1) + 5$

$f[g(x)] = 4(4x^2 - 4x + 1) - 6x + 3 + 5$

$\boxed{f[g(x)] = 16x^2 - 22x + 12}$

$g[f(x)] = 2(4x^2 - 3x + 5) - 1$

$\boxed{g[f(x)] = 8x^2 - 6x + 9}$

91. $f(x) = 2x^2 + 9$

$\dfrac{f(a + h) - f(a)}{h}$

$= \dfrac{[2(a + h)^2 - 8(a + h) + 9] - (2a^2 - 8a + 9)}{h}$

$= \dfrac{2a^2 + 4ah + 2h^2 - 8a - 8h + 9 - 2a^2 + 8a - 9}{h}$

$= \dfrac{4ah + 2h^2 - 8h}{h}$

$= \boxed{4a + 2h - 8}$

92. $(x - 14)(2x + 1) = (x - 4)(5x + 2)$

$\qquad 2x^2 - 27x - 14 = 5x^2 - 18x - 8$

$\qquad\quad -3x^2 - 9x - 6 = 0$

$\qquad\quad -3(x^2 + 3x + 2) = 0$

$\qquad\quad -3(x + 2)(x + 1) = 0$

$\qquad\qquad\qquad x + 2 = 0 \qquad \text{or} \qquad x + 1 = 0$

$\qquad\qquad\qquad\qquad x = -2 \qquad\qquad\qquad x = -1$

$\boxed{\{-2, -1\}}$

Section 6.3 Adding and Subtracting Rational Expressions

Problem Set 6.3, pp. 453-456

1. $\dfrac{3a^2}{x+3} + \dfrac{4b^2}{x+3} = \boxed{\dfrac{3a^2 + 4b^2}{x+3}}$

3. $\dfrac{4x}{x+2y} + \dfrac{8y}{x+2y} = \dfrac{4x+8y}{x+2y} = \dfrac{4(x+2y)}{x+2y} = \boxed{4}$

5. $\dfrac{2x}{(2x-y)(2x+y)} + \dfrac{y}{(2x-y)(2x+y)} = \dfrac{2x+y}{(2x-y)(2x+y)} = \boxed{\dfrac{1}{2x-y}}$

7. $\dfrac{a^2+7a+3}{a^2+9a+9} + \dfrac{2a+6}{a^2+9a+9} = \dfrac{a^2+7a+3+2a+6}{a^2+9a+9} = \dfrac{a^2+9a+9}{a^2+9a+9} = \boxed{1}$

9. $\dfrac{x^2}{x+y} - \dfrac{y^2}{x+y} = \dfrac{x^2-y^2}{x+y} = \dfrac{(x+y)(x-y)}{x+y} = \boxed{x-y}$

11. $\dfrac{5b}{b-c} - \dfrac{5c}{b-c} = \dfrac{5b-5c}{b-c} = \dfrac{5(b-c)}{b-c} = \boxed{5}$

13. $\dfrac{x^2}{x^2+2x-15} - \dfrac{25}{x^2+2x-15} = \dfrac{x^2-25}{x^2+2x-15} = \dfrac{(x+5)(x-5)}{(x+5)(x-3)} = \boxed{\dfrac{x-5}{x-3}}$

15. $\dfrac{5x+5y}{2x} - \dfrac{3x+y}{2x} = \dfrac{5x+5y-(3x+y)}{2x} = \dfrac{5x+5y-3x-y}{2x} = \dfrac{2x+4y}{2x} = \dfrac{2(x+2y)}{2x} = \boxed{\dfrac{x+2y}{x}}$

17. $\dfrac{y}{x+y} - \dfrac{x+y}{x+y} = \dfrac{y-(x+y)}{x+y} = \dfrac{y-x-y}{x+y} = \boxed{-\dfrac{x}{x+y}}$

19. $\dfrac{a^2-4a}{a^2-a-6} - \dfrac{a-6}{a^2-a-6} = \dfrac{a^2-4a-(a-6)}{a^2-a-6} = \dfrac{a^2-4a-a+6}{a^2-a-6} = \dfrac{a^2-5a+6}{a^2-a-6} = \dfrac{(a-3)(a-2)}{(a-3)(a+)} = \boxed{\dfrac{a-2}{a+2}}$

21. $\dfrac{9x}{10} - \dfrac{7x}{10} + \dfrac{3x}{10} = \dfrac{9x-7x+3x}{10} = \dfrac{5x}{10} = \boxed{\dfrac{x}{2}}$

23. $\dfrac{3a+b}{a+b} + \dfrac{2a+3b}{a+b} - \dfrac{4a+3b}{a+b}$

$= \dfrac{(3a+b)+(2a+3b)-(4a+3b)}{a+b}$

$= \dfrac{3a+b+2a+3b-4a-3b}{a+b}$

$= \dfrac{a+b}{a+b}$

$= \boxed{1}$

25. $\dfrac{3a^3+4b^3}{a^2-b^2} - \dfrac{5b^3+2a^3}{a^2-b^2}$

$= \dfrac{(3a^3+4b^3)-(5b^3+2a^3)}{a^2-b^2}$

$= \dfrac{3a^3+4b^3-5b^3-2a^3}{a^2-b^2}$

$= \dfrac{a^3-b^3}{a^2-b^2}$

$= \dfrac{(a-b)(a^2+ab+b^2)}{(a-b)(a+b)}$

$= \boxed{\dfrac{a^2+ab+b^2}{a+b}}$

27. $\dfrac{b}{ac+ad-bc-bd} - \dfrac{a}{ac+ad-bc-bd}$

$= \dfrac{b-a}{a(c+d)-b(c+d)}$

$= \dfrac{-(a-b)}{(c+d)(a-b)}$

$= \boxed{-\dfrac{1}{c+d}}$

29. $\dfrac{2y}{y-5} - \left(\dfrac{2}{y-5} + \dfrac{y-2}{y-5}\right)$

$= \dfrac{2y}{y-5} - \left(\dfrac{2+y-2}{y-5}\right)$

$= \dfrac{2y}{y-5} - \dfrac{y}{y-5}$

$= \dfrac{2y-y}{y-5}$

$= \boxed{\dfrac{y}{y-5}}$

31. $\dfrac{3}{6x^3} - \dfrac{2}{9x^2}$ (LCD is $18x^3$)

$= \dfrac{3}{6x^3} \cdot \dfrac{3}{3} - \dfrac{2}{9x^2} \cdot \dfrac{2x}{2x}$

$= \dfrac{9}{18x^3} - \dfrac{4x}{18x^3}$

$= \boxed{\dfrac{9-4x}{18x^3}}$

33. $\dfrac{2a}{3c^2} - \dfrac{3b}{4cd}$ (LCD is $12c^2 d$)

$= \dfrac{2a}{3c^2} \cdot \dfrac{4d}{4d} - \dfrac{3b}{4cd} \cdot \dfrac{3c}{3c}$

$= \dfrac{8ad}{12c^2 d} - \dfrac{9bc}{12c^2 d}$

$= \boxed{\dfrac{8ad - 9bc}{12c^2 d}}$

35. $\dfrac{3a}{2c^2} - \dfrac{2a}{3cd} + \dfrac{a}{6d^{\,2}}$ (LCD is $6c^2 d^2$)

$= \dfrac{3a}{2c^2} \cdot \dfrac{3d^{\,2}}{3d^2} - \dfrac{2a}{3cd} \cdot \dfrac{2cd}{2cd} + \dfrac{a}{6d^{\,2}} \cdot \dfrac{c^2}{c^2}$

$= \dfrac{9ad^{\,2}}{6c^2 d^{\,2}} - \dfrac{4acd}{6c^2 d^{\,2}} + \dfrac{ac^2}{6c^2 d^{\,2}}$

$= \boxed{\dfrac{9ad^{\,2} - 4acd + ac^{\,2}}{6c^2 d^{\,2}}}$

37. $\dfrac{2b-2c}{b^2 c} + \dfrac{b-c}{bc^2}$ (LCD is $b^2 c^{\,2}$)

$= \dfrac{2b-2c}{b^2 c} \cdot \dfrac{c}{c} + \dfrac{b-c}{bc^2} \cdot \dfrac{b}{b}$

$= \dfrac{(2b-2c)c}{b^2 c^2} + \dfrac{(b-c)b}{b^2 c^2}$

$= \dfrac{2bc - 2c^2 + b^2 - bc}{b^2 c^2}$

$= \boxed{\dfrac{b^2 + bc - 2c^{\,2}}{b^2 c^2}}$

39. $\dfrac{b-2y}{4b^2 y} + \dfrac{2b+y}{6by^2}$ (LCD is $12b^2 y^2$)

$= \dfrac{b-2y}{4b^2 y} \cdot \dfrac{3y}{3y} + \dfrac{2b+y}{6by^2} \cdot \dfrac{2b}{2b})$

$= \dfrac{3y(b-2y)}{12b^2 y^2} + \dfrac{2b(2b+y)}{12b^2 y^2}$

$= \dfrac{3by - 6y^2 + 4b^2 + 2by}{12b^2 y^2}$

$= \boxed{\dfrac{5by - 6y^2 + 4b^2}{12b^2 y^2}}$

41. $\dfrac{4x-3y}{6xy} - \dfrac{x-4z}{8xz} - \dfrac{3y-z}{4yz}$ (LCD is $24xyz$)

$= \dfrac{4x-3y}{6xy} \cdot \dfrac{4z}{4z} - \dfrac{x-4z}{8xz} \cdot \dfrac{3y}{3y} - \dfrac{3y-z}{4yz} \cdot \dfrac{6x}{6x}$

$= \dfrac{4z(4x-3y) - 3y(x-4z) - 6x(3y-z)}{24xyz}$

$= \dfrac{16xz - 12yz - 3ixy + 12yz - 18xy + 6xz}{24xy}$

$= \dfrac{22xz - 21xy}{24xyz}$

$= \dfrac{x(22z - 21y)}{24xyz}$

$= \boxed{\dfrac{22z - 21y}{24yz}}$

43. $\dfrac{10}{x+4} - \dfrac{2}{x-6}$ (LCD is $(x+4)(x-6)$)

$= \dfrac{10}{x+4} \cdot \dfrac{x-6}{x-6} - \dfrac{2}{x-6} \cdot \dfrac{x+4}{x+4}$

$= \dfrac{10(x-6) - 2(x+4)}{(x+4)(x-6)}$

$= \dfrac{10x - 60 - 2x - 8}{(x+4)(x-6)}$

$= \boxed{\dfrac{8x - 68}{(x+4)(x-6)}}$

45. $\dfrac{b}{b-c} - \dfrac{c}{b+c}$ (LCD is $(b-c)(b+c)$)

$= \dfrac{b}{b-c} \cdot \dfrac{b+c}{b+c} - \dfrac{c}{b+c} \cdot \dfrac{b-c}{b-c}$

$= \dfrac{b(b+c) - c(b-c)}{(b-c)(b+c)}$

$= \dfrac{b^2 + bc - bc + c^2}{(b-c)(b+c)}$

$= \boxed{\dfrac{b^2 + c^2}{(b-c)(b+c)}}$

47. $\dfrac{3}{a+1} - \dfrac{3}{a}$ (LCD is $a(a+1)$)

$= \dfrac{3}{a+1} \cdot \dfrac{a}{a} - \dfrac{3}{a} \cdot \dfrac{a+1}{a+1}$

$= \dfrac{3a - 3(a+1)}{a(a+1)}$

$= \dfrac{3a - 3a - 3}{a(a+1)}$

$= \boxed{\dfrac{-3}{a(a+1)}}$

49. $\dfrac{5x}{x-2} - \dfrac{x-1}{x+2}$ (LCD is $(x-2)(x+2)$)

$= \dfrac{5x}{x-2} \cdot \dfrac{x+2}{x+2} - \dfrac{x-1}{x+2} \cdot \dfrac{x-2}{x-2}$

$= \dfrac{5x(x+2) - (x-1)(x-2)}{(x-2)(x+2)}$

$= \dfrac{5x^2 + 10x - x^2 + 3x - 2}{(x-2)(x+2)}$

$= \boxed{\dfrac{4x^2 + 13x - 2}{(x-2)(x+2)}}$

51. $\dfrac{3x}{x-3} - \dfrac{x+4}{x+2}$ (LCD is $(x-3)(x+2)$)

$= \dfrac{3x}{x-3} \cdot \dfrac{x+2}{x+2} - \dfrac{x+4}{x+2} \cdot \dfrac{x-3}{x-3}$

$= \dfrac{3x(x+2) - (x+4)(x-3)}{(x-3)(x+2)}$

$= \dfrac{3x^2 + 6x - x^2 - x + 12}{(x-3)(x+2)}$

$= \boxed{\dfrac{2x^2 + 5x + 12}{(x-3)(x+2)}}$

53. $\dfrac{a-b}{a+b} - \dfrac{a+b}{a-b}$ (LCD is $(a+b)(a-b)$)

$= \dfrac{a-b}{a+b} \cdot \dfrac{a-b}{a-b} - \dfrac{a+b}{a-b} \cdot \dfrac{a+b}{a+b}$

$= \dfrac{(a-b)(a-b) - (a+b)(a+b)}{(a+b)(a-b)}$

$= \dfrac{a^2 - 2ab + b^2 - a^2 - 2ab - b^2}{(a+b)(a-b)}$

$= \boxed{\dfrac{-4ab}{(a+b)(a-b)}}$

55. $\dfrac{4}{x+2} - \dfrac{3}{x+1} + \dfrac{2}{x}$ (LCD is $x(x+1)(x+2)$)

$= \dfrac{4}{x+2} \cdot \dfrac{x}{x} \cdot \dfrac{x+1}{x+1} - \dfrac{3}{x+1} \cdot \dfrac{x}{x} \cdot \dfrac{x+2}{x+2}$

$\qquad + \dfrac{2}{x} \cdot \dfrac{x+1}{x+1} \cdot \dfrac{x+2}{x+2}$

$= \dfrac{4x(x+1) - 3x(x+2) + 2(x+1)(x+2)}{x(x+1)(x+2)}$

$= \dfrac{4x^2 + 4x - 3x^2 - 6x + 2x^2 + 6x + 4}{x(x+1)(x+2)}$

$= \boxed{\dfrac{3x^2 + 4x + 4}{x(x+1)(x+2)}}$

57. $\dfrac{5}{2b-8} + \dfrac{3}{4b-2}$

$= \dfrac{5}{2(b-4)} + \dfrac{3}{2(2b-1)}$ (LCD is $2(b-4)(2b-1)$)

$= \dfrac{5}{2(b-4)} \cdot \dfrac{2b-1}{2b-1} + \dfrac{3}{2(2b-1)} \cdot \dfrac{b-4}{b-4}$

$= \dfrac{5(2b-1) + 3(b-4)}{2(b-4)(2b-1)}$

$= \dfrac{10b - 5 + 3b - 12}{2(b-4)(2b-1)}$

$= \boxed{\dfrac{13b - 17}{2(b-4)(2b-1)}}$

59. $\dfrac{4}{x^2 + 6x + 9} + \dfrac{4}{x + 3} = \dfrac{4}{(x + 3)^2} + \dfrac{4}{x + 3}$ (LCD is $(x + 3)^2$)

$= \dfrac{4}{(x + 3)^2} + \dfrac{4}{x + 3} \cdot \dfrac{x + 3}{x + 3}$

$= \dfrac{4 + 4(x + 3)}{(x + 3)^2}$

$= \boxed{\dfrac{4x + 16}{(x + 3)^2}}$

61. $\dfrac{c}{c^2 - 10c + 25} - \dfrac{c - 4}{2c - 10}$

$= \dfrac{c}{(c - 5)^2} - \dfrac{c - 4}{2(c - 5)}$ (LCD is $2(c - 5)^2$)

$= \dfrac{c}{(c - 5)^2} \cdot \dfrac{2}{2} - \dfrac{c - 4}{2(c - 5)} \cdot \dfrac{c - 5}{c - 5}$

$= \dfrac{2c - (c - 4)(c - 5)}{2(c - 5)^2}$

$= \dfrac{2c - c^2 + 9c - 20}{2(c - 5)^2}$

$= \boxed{\dfrac{-c^2 + 11c - 20}{2(c - 5)^2}}$

63. $\dfrac{a - b}{3a + 3b} + \dfrac{a + b}{2a - 2b}$

$= \dfrac{a - b}{3(a + b)} + \dfrac{a + b}{2(a - b)}$ (LCD is $6(a + b)(a - b)$)

$= \dfrac{a - b}{3(a + b)} \cdot \dfrac{2(a - b)}{2(a - b)} + \dfrac{a + b}{2(a - b)} \cdot \dfrac{3(a + b)}{3(a + b)}$

$= \dfrac{2(a - b)(a - b) + 3(a + b)(a + b)}{6(a + b)(a - b)}$

$= \dfrac{2a^2 - 4ab + 2b^2 + 3a^2 + 6ab + 3b^2}{6(a + b)(a - b)}$

$= \boxed{\dfrac{5a^2 + 2ab + 5b^2}{6(a + b)(a - b)}}$

65. $\dfrac{b + 2}{b^2 + b - 2} + \dfrac{2}{b^2 - 1}$

$= \dfrac{b + 2}{(b + 2)(b - 1)} + \dfrac{2}{(b + 1)(b - 1)}$

$= \dfrac{1}{b - 1} + \dfrac{2}{(b + 1)(b - 1)}$ (LCD is $(b + 1)(b - 1)$)

$= \dfrac{1}{b - 1} \cdot \dfrac{b + 1}{b + 1} + \dfrac{2}{(b + 1)(b - 1)}$

$= \boxed{\dfrac{b + 3}{(b + 1)(b - 1)}}$

67. $\dfrac{y + 3}{y^2 - y - 2} - \dfrac{y - 1}{y^2 + 2y + 1}$

$= \dfrac{y + 3}{(y + 1)(y - 2)} - \dfrac{y - 1}{(y + 1)^2}$ (LCD is $(y + 1)^2(y - 2)$)

$= \dfrac{y + 3}{(y + 1)(y - 2)} \cdot \dfrac{y + 1}{y + 1} - \dfrac{y - 1}{(y + 1)^2} \cdot \dfrac{y - 2}{y - 2}$

$= \dfrac{(y + 3)(y + 1)}{(y + 1)^2(y - 2)} - \dfrac{(y - 1)(y - 2)}{(y + 1)^2(y - 2)}$

$= \dfrac{y^2 + 4y + 3 - (y^2 - 3y + 2)}{(y + 1)^2(y - 2)}$

$= \boxed{\dfrac{7y + 1}{(y + 1)^2(y - 2)}}$

69. $\dfrac{x^2 + x + 2}{x^3 - 1} - \dfrac{1}{x - 1}$

$= \dfrac{x^2 + x + 2}{(x - 1)(x^2 + x + 1)} - \dfrac{1}{x - 1}$ (LCD is $(x - 1)(x^2 + x + 1)$)

$= \dfrac{x^2 + x + 2}{(x - 1)(x^2 + x + 1)} - \dfrac{1}{x - 1} \cdot \dfrac{x^2 + x + 1}{x^2 + x + 1}$

$= \dfrac{x^2 + x + 2 - (x^2 + x + 1)}{(x - 1)(x^2 + x + 1)}$

$= \boxed{\dfrac{1}{(x - 1)(x^2 + x + 1)}}$

71. $\dfrac{1}{y - x} + \dfrac{1}{x - y} = \dfrac{1}{y - x} + \dfrac{1}{x - y} \cdot \dfrac{-1}{-1}$

$= \dfrac{1}{y - x} + \dfrac{-1}{-x + y}$

$= \dfrac{1}{y - x} - \dfrac{1}{x - y}$

$= \dfrac{1 - 1}{y - x}$

$= \dfrac{0}{y - x}$

$= \boxed{0}$

73. $\dfrac{y^2}{y - 7} + \dfrac{6y + 7}{7 - y}$

$= \dfrac{y^2}{y - 7} + \dfrac{6y + 7}{7 - y} \cdot \dfrac{-1}{-1}$

$= \dfrac{y^2}{y - 7} + \dfrac{(-1)(6y + 7)}{y - 7}$

$= \dfrac{y^2 - 6y - 7}{y - 7}$

$= \dfrac{(y - 7)(y + 1)}{y - 7}$

$= \boxed{y + 1}$

75. $\dfrac{x}{1} - \dfrac{3}{x - 2}$

$= \dfrac{x}{1} \cdot \dfrac{x - 2}{x - 2} - \dfrac{3}{x - 2}$

$= \dfrac{x(x - 2) - 3}{x - 2}$

$= \boxed{\dfrac{x^2 - 2x - 3}{x - 2}}$

77. $\dfrac{x + 3y}{x^2 - 7xy + 12y^2} - \dfrac{x - 3y}{x^2 - xy - 12y^2}$

$= \dfrac{x + 3y}{(x - 4y)(x - 3y)} - \dfrac{x - 3y}{(x - 4y)(x + 3y)}$ (LCD is $(x - 4y)(x - 3y)(x + 3y)$)

$= \dfrac{x + 3y}{(x - 4y)(x - 3y)} \cdot \dfrac{(x + 3y)}{(x + 3y)} - \dfrac{x - 3y}{(x - 4y)(x + 3y)} \cdot \dfrac{(x - 3y)}{(x - 3y)}$

$= \dfrac{(x + 3y)(x + 3y) - (x - 3y)(x - 3y)}{(x - 4y)(x - 3y)(x + 3y)}$

$= \dfrac{x^2 + 6xy + 9y^2 - (x^2 - 6xy + 9y^2)}{(x - 4y)(x - 3y)(x + 3y)}$

$= \boxed{\dfrac{12xy}{(x - 4y)(x - 3y)(x + 3y)}}$

79. $\dfrac{y^n}{y^{2n} - 1} + \dfrac{2}{y^n - 1}$

$= \dfrac{y^n}{(y^n + 1)(y^n - 1)} + \dfrac{2}{y^n - 1}$ (LCD is $(y^n + 1)(y^n - 1)$)

$= \dfrac{y^n}{(y^n + 1)(y^n - 1)} + \dfrac{2}{y^n - 1} \cdot \dfrac{y^n + 1}{y^n + 1}$

$= \dfrac{y^n + 2(y^n + 1)}{(y^n + 1)(y^n - 1)}$

$= \boxed{\dfrac{3y^n + 2}{(y^n + 1)(y^n - 1)}}$

81. $\dfrac{2}{y^n + 1} - \dfrac{y^n}{y^{2n} - 1}$

$= \dfrac{2}{(y^n + 1)} - \dfrac{y^n}{(y^n + 1)(y^n - 1)}$ (LCD is $(y^n + 1)(y^n - 1)$)

$= \dfrac{2}{y^n + 1} \cdot \dfrac{y^n - 1}{y^n - 1} - \dfrac{y^n}{(y^n + 1)(y^n - 1)}$

$= \dfrac{2y^n - 2 - y^n}{(y^n + 1)(y^n - 1)}$

$= \boxed{\dfrac{y^n - 2}{(y^n + 1)(y^n - 1)}}$

83. $\dfrac{y}{y-4} + \dfrac{y}{y+4} - \dfrac{16}{(y+4)(y-4)}$ (LCD is $(y+4)(y-4)$)

$= \dfrac{y}{y-4} \cdot \dfrac{y+4}{y+4} + \dfrac{y}{y+4} \cdot \dfrac{y-4}{y-4} - \dfrac{16}{(y+4)(y-4)}$

$= \dfrac{y(y+4) + y(y-4) - 16}{(y+4)(y-4)}$

$= \boxed{\dfrac{2y^2 - 16}{(y+4)(y-4)}}$

85. $\dfrac{a+2}{a} - \dfrac{a+1}{a+3} + 1$ (LCD is $a(a+3)$)

$= \dfrac{a+2}{a} \cdot \dfrac{a+3}{a+3} - \dfrac{a+1}{a+3} \cdot \dfrac{a}{a} + 1 \cdot \dfrac{a(a+3)}{a(a+3)}$

$= \dfrac{(a+2)(a+3) - a(a+1) + a(a+3)}{a(a+3)}$

$= \dfrac{a^2 + 5a + 6 - a^2 - a + a^2 + 3a}{a(a+3)}$

$= \dfrac{a^2 + 7a + 6}{a(a+3)}$

$= \boxed{\dfrac{(a+6)(a+1)}{a(a+3)}}$

87. $\dfrac{y-3}{y-2} + \dfrac{7-4y}{(2y-5)(y-2)} - \dfrac{y+1}{2y-5}$ (LCD is $(2y-5)(y-2)$)

$= \dfrac{y-3}{y-2} \cdot \dfrac{2y-5}{2y-5} + \dfrac{7-4y}{(2y-5)(y-2)} - \dfrac{y+1}{2y-5} \cdot \dfrac{y-2}{y-2}$

$= \dfrac{(y-3)(2y-5) + 7 - 4y - (y+1)(y-2)}{(2y-5)(y-2)}$

$= \dfrac{2y^2 - 11y + 15 + 7 - 4y - y^2 + y + 2}{(2y-5)(y-2)}$

$= \dfrac{y^2 - 14y + 24}{(2y-5)(y-2)}$

$= \dfrac{(y-12)(y-2)}{(2y-5)(y-2)}$

$= \boxed{\dfrac{y-12}{2y-5}}$

89. $\dfrac{1}{c^2 - ac - cb + ab} - \dfrac{1}{b^2 - ab - bc + ac} + \dfrac{1}{a^2 - ab - ac + bc}$

$= \dfrac{1}{c(c-a) - b(c-a)} - \dfrac{1}{b(b-a) - c(b-a)} + \dfrac{1}{a(a-b) - c(a-b)}$

$= \dfrac{1}{(c-a)(c-b)} - \dfrac{1}{(b-a)(b-c)} + \dfrac{1}{(a-b)(a-c)}$ (LCD is $(a-b)(a-c)(b-c)$)

$= \dfrac{1}{(c-a)(c-b)} \cdot \dfrac{-1}{-1} \cdot \dfrac{-1}{-1} \cdot \dfrac{a-b}{a-b} - \dfrac{1}{(b-a)(b-c)} \cdot \dfrac{-1}{-1} \cdot \dfrac{a-c}{a-c} + \dfrac{1}{(a-b)(a-c)} \cdot \dfrac{b-c}{b-c}$

$= \dfrac{a-b-(-1)(a-c) + b-c}{(a-b)(a-c)(b-c)}$

$= \dfrac{a-b + a - c + b - c}{(a-b)(a-c)(b-\text{ic})}$

$= \dfrac{2a - 2c}{(a-b)(a-c)(b-c)}$

$= \dfrac{2(a-c)}{(a-b)(a-c)(b-c)}$

$= \boxed{\dfrac{2}{(a-b)(b-c)}}$

91. $\dfrac{3x}{2y} + \dfrac{4y}{y^2} \cdot \dfrac{3y}{4} = \dfrac{3x}{2y} + \dfrac{12y^2}{4y^2} = \dfrac{3x}{2y} + \dfrac{3}{1}$ (LCD is $2y$)

$= \dfrac{3x}{2y} + \dfrac{3}{1} \cdot \dfrac{2y}{2y}$

$= \dfrac{3x + 6y}{2y}$

$= \boxed{\dfrac{3(x+2y)}{2y}}$

93. $\dfrac{6}{5a} \cdot \dfrac{2}{10a^2} + \dfrac{4}{a} \div \dfrac{8}{3a^2}$

$= \dfrac{6 \cdot 2}{5a \cdot 10a^2} + \dfrac{4}{a} \cdot \dfrac{3a^2}{8}$

$= \dfrac{6}{25a^3} + \dfrac{3a}{2}$ (LCD is $50a^3$)

$= \dfrac{6}{25a^3} \cdot \dfrac{2}{2} + \dfrac{3a}{2} \cdot \dfrac{25a^3}{25a^3}$

$= \boxed{\dfrac{12 + 75a^4}{50a^3}}$

95. $\dfrac{6}{(x+4)(x-4)} \cdot \dfrac{x+4}{12} - \dfrac{1}{2}$

$= \dfrac{1}{2(x-4)} - \dfrac{1}{2}$ (LCD is $2(x-4)$)

$= \dfrac{1}{2(x-4)} - \dfrac{1}{2} \cdot \dfrac{x-4}{x-4}$

$= \dfrac{1-(x-4)}{2(x-4)}$

$= \boxed{\dfrac{-x+5}{2(x-4)}}$

97. $\left(2 - \dfrac{6}{y+1}\right)\left(1 + \dfrac{3}{y-2}\right)$

$= \left[2\dfrac{(y+1)}{(y+1)} - \dfrac{6}{y+1}\right]\left[1\dfrac{(y-2)}{(y-2)} + \dfrac{3}{y-2}\right]$

$= \left(\dfrac{2y+2-6}{y+1}\right)\left(\dfrac{y-2+3}{y-2}\right)$

$= \left(\dfrac{2y-4}{y+1}\right)\left(\dfrac{y+1}{y-2}\right)$

$= \dfrac{2(y-2)(y+1)}{(y+1)(y-2)}$

$= \boxed{2}$

99. $\dfrac{3y^2}{y^2+5y+6} \cdot \dfrac{y+3}{y} + \dfrac{y^2-9}{6y} \div \dfrac{y^2-6y+9}{3y^2}$

$= \dfrac{3y^2}{(y+3)(y+2)} \cdot \dfrac{y+3}{y} + \dfrac{(y+3)(y-3)}{6y} \div \dfrac{(y-3)^2}{3y^2}$

$= \dfrac{3y^2(y+3)}{y(y+3)(y+2)} + \dfrac{(y+3)(y-3)3y^2}{6y(y-3)^2}$

$= \dfrac{3y}{y+2} + \dfrac{y(y+3)}{2(y-3)}$ (LCD is $2(y-3)(y+2)$)

$= \dfrac{3y}{y+2} \cdot \dfrac{2(y-3)}{2(y-3)} + \dfrac{y(y+3)}{2(y-3)} \cdot \dfrac{y+2}{y+2}$

$= \dfrac{6y(y-3) + y(y+3)(y+2)}{2(y-3)(y+2)}$

$= \dfrac{6y^2 - 18y + y^3 + 5y^2 + 6y}{2(y-3)(y+2)}$

$= \boxed{\dfrac{y^3 + 11y^2 - 12y}{2(y-3)(y+2)}}$

101. $\left(\dfrac{1}{b+2} + \dfrac{1}{b+4}\right) \cdot (b^2 + 6b + 8)$

$= \left(\dfrac{1}{b+2} \cdot \dfrac{b+4}{b+4} + \dfrac{1}{b+4} \cdot \dfrac{b+2}{b+2}\right) \cdot (b^2 + 6b + 8)$

$= \dfrac{2b+6}{(b+2)(b+4)} \cdot \dfrac{(b+2)(b+4)}{1}$

$= \boxed{2b+6}$

103. $\left(\dfrac{1}{a^3-b^3} \cdot \dfrac{ac+ad-bc-bd}{1}\right) - \dfrac{c-d}{a^2+ab+b^2}$

$= \left[\dfrac{1}{(a-b)(a^2+ab+b^2)} \cdot \dfrac{(c+d)(a-b)}{1}\right]$
$\qquad - \dfrac{c-d}{a^2+ab+b^2}$

$= \dfrac{c+d}{a^2+ab+b^2} - \dfrac{c-d}{a^2+ab+b^2}$

$= \dfrac{(c+d)-(c-d)}{a^2+ab+b^2}$

$= \boxed{\dfrac{2d}{a^2+ab+b^2}}$

105. $\left(\dfrac{1}{x^2} - \dfrac{1}{y^2}\right)\left(x - y + \dfrac{2y^2}{x+y}\right) \div \left(x^3 - \dfrac{x^3y + y^4}{x+y}\right)$

$= \left(\dfrac{1}{x^2} \cdot \dfrac{y^2}{y^2} - \dfrac{1}{y^2} \cdot \dfrac{x^2}{x^2}\right) \cdot \left(x \cdot \dfrac{x+y}{x+y} - y \cdot \dfrac{x+y}{x+y} + \dfrac{2y^2}{x+y}\right) \div \left(x^3 \cdot \dfrac{x+y}{x+y} - \dfrac{x^3y + y^4}{x+y}\right)$

$= \left(\dfrac{y^2 - x^2}{x^2y^2}\right)\left(\dfrac{x^2+y^2}{x+y}\right) \div \left(\dfrac{x^4-y^4}{x+y}\right)$

$= \dfrac{(y-x)(y+x)(x^2+y^2)(x+y)}{(x^2y^2)(x+y)(x^4-y^4)}$

$= \dfrac{(-1)(x-y)(x^2+y^2)(x+y)}{x^2y^2(x^2+y^2)(x+y)(x-y)}$

$= \boxed{-\dfrac{1}{x^2y^2}}$

107. $\left(\dfrac{1}{x+h}-\dfrac{1}{x}\right)\div h$

$=\left(\dfrac{1}{x+h}\cdot\dfrac{x}{x}-\dfrac{1}{x}\cdot\dfrac{x+h}{x+h}\right)\div h$

$=\left[\dfrac{x-x-h}{x(x+h)}\right]\div\dfrac{h}{1}$

$=\dfrac{-h}{x(x+h)}\cdot\dfrac{1}{h}$

$=\boxed{\dfrac{-1}{x(x+h)}}$

For 109-111, $f(x)=\dfrac{x+1}{x+2}$, $g(x)=\dfrac{x}{x^2-1}$, $h(x)=\dfrac{x-2}{1-x}$.

109. $(g+h)(x)\ =\ \dfrac{x}{x^2-1}+\dfrac{x-2}{1-x}$

$=\dfrac{x}{(x+1)(x-1)}-\dfrac{(x-2)}{x-1}\cdot\dfrac{x+1}{x+1}$

$=\dfrac{x-(x-2)(x+1)}{(x+1)(x-1)}$

$=\boxed{\dfrac{-x^2+2x+2}{(x-1)(x+1)}\ ,\ (x\neq-1,1)}$

$\boxed{D_{g+h}=\{x\ |\ x\neq-1,1\}\ \text{ or }\ (-\infty,-1)\ \cup(-1,1)\ \cup(1,\infty)}$

111. $(g-h+f)(x)\ =\ \dfrac{x}{x^2-1}-\dfrac{x-2}{1-x}+\dfrac{x+1}{x+2}$

$=\dfrac{x}{(x+1)(x-1)}\cdot\dfrac{x+2}{x+2}+\dfrac{x-2}{(x-1)}\cdot\dfrac{(x+1)(x+2)}{(x+1)(x+2)}+\dfrac{x+1}{x+2}\cdot\dfrac{(x+1)(x-1)}{(x+1)(x-1)}$

$=\dfrac{x(x+2)+(x-2)(x+1)(x+2)+(x+2)^2(x-1)}{(x+1)(x-1)(x+2)}$

$=\dfrac{x^2+2x+x^3+x^2-4x-4+x^3+x^2-x-1}{(x+1)(x-1)(x+2)}$

$=\boxed{\dfrac{2x^3+3x^2-3x-5}{(x-1)(x+1)(x+2)}\ ,\ (x\neq-2,-1,1)}$

$\boxed{D_{g-h+f}=\{x\ |\ x\neq-2,-1,1\}\ \text{ or }\ (-\infty,-2)\ \cup(-2,-1)\ \cup(-1,1)\ \cup(1,\infty)}$

113. $\dfrac{1}{a}+\dfrac{1}{b}=\dfrac{1}{a}\cdot\dfrac{b}{b}+\dfrac{1}{b}\cdot\dfrac{a}{a}=\boxed{\dfrac{b+a}{ab}}\neq\dfrac{1}{a+b}$

115. $\dfrac{1}{x}+7=\dfrac{1}{x}+7\cdot\dfrac{x}{x}=\boxed{\dfrac{1+7x}{x}}\neq\dfrac{1}{x+7}$

117. $\dfrac{a+bx}{a}=\dfrac{a}{a}+\dfrac{bx}{a}=\boxed{1+\dfrac{bx}{a}}\neq1+bx$

119. $\dfrac{a}{x}+\dfrac{a}{b}=\dfrac{a}{x}\cdot\dfrac{b}{b}+\dfrac{a}{b}\cdot\dfrac{x}{x}=\boxed{\dfrac{ab+ax}{bx}}\neq\dfrac{a}{x+b}$

121.
$$R = (x^2 + x)\left[3 - \frac{x}{20(x+1)}\right]$$
$$= x(x+1)\left[3 \cdot \frac{20(x+1)}{20(x+1)} - \frac{x}{20(x+1)}\right]$$
$$= x(x+1)\left(\frac{60x + 60 - x}{20(x+1)}\right)$$
$$= \frac{x(x+1)(59x+60)}{20(x+1)}$$
$$= \boxed{\frac{x(59x+60)}{20}}$$

123. a. $x = 1, y = 2, z = 3$

$$\frac{y-z}{1+yz} = \frac{2-3}{1+6} = -\frac{1}{7}$$
$$\frac{z-x}{1+xz} = \frac{3-1}{1+3} = \frac{2}{4} = \frac{1}{2}$$
$$\frac{x-y}{1+xy} = \frac{1-2}{1+2} = -\frac{1}{3}$$

125.
$$\frac{1}{x^n - 1} - \frac{1}{x^n + 1} - \frac{1}{x^{2n} - 1}$$
$$= \frac{1}{x^n - 1} - \frac{1}{x^n + 1} - \frac{1}{(x^n + 1)(x^n - 1)}$$
$$= \frac{1}{x^n - 1} \cdot \frac{x^n + 1}{x^n + 1} - \frac{1}{x^n + 1} \cdot \frac{x^n - 1}{x^n - 1} - \frac{1}{(x^n + 1)(x^n - 1)}$$
$$= \frac{x^n + 1 - x^n + 1 - 1}{(x^n + 1)(x^n - 1)}$$
$$= \frac{1}{(x^n + 1)(x^n - 1)}$$
$$= \boxed{\frac{1}{x^{2n} - 1}}$$

127.   sum of areas = area of trapezoid + area of right triangle
$$= \frac{1}{2} \cdot \frac{1}{y}\left[\left(\frac{1}{y} + 2 - \frac{1}{y}\right) + \left(2 - \frac{1}{y}\right)\right] + \frac{1}{2}[5y - (y+1)] \cdot \frac{1}{y}$$
$$= \frac{1}{2y}\left(2 + 2 - \frac{1}{y}\right) + \frac{1}{2y}(4y - 1)$$
$$= \frac{4 - \frac{1}{y} + 4y - 1}{2y}$$
$$= \frac{4y + 3 - \frac{1}{y}}{2y} \cdot \frac{y}{y}$$
$$= \boxed{\frac{4y^2 + 3y - 1}{2y^2}}$$

129. a.
$$\left(1 - \frac{1}{x}\right)\left(1 - \frac{1}{x+1}\right)\left(1 - \frac{1}{x+2}\right)\left(1 - \frac{1}{x+3}\right)$$
$$= \left(\frac{x-1}{x}\right)\left(\frac{x+1-1}{x+1}\right)\left(\frac{x+2-1}{x+2}\right)\left(\frac{x+3-1}{x+3}\right)$$
$$= \left(\frac{x-1}{x}\right)\left(\frac{x}{x+1}\right)\left(\frac{x+1}{x+2}\right)\left(\frac{x+2}{x+3}\right)$$
$$= \boxed{\frac{x-1}{x+3}}$$

b. $\left(1-\dfrac{1}{x}\right)\left(1-\dfrac{1}{x+1}\right)\left(1-\dfrac{1}{x+2}\right)\left(1-\dfrac{1}{x+3}\right)\left(1-\dfrac{1}{x+4}\right)$

$\quad = \left(\dfrac{x-1}{x+3}\right)\left(\dfrac{x+4-1}{x+4}\right)$ (from part a)

$\quad = \left(\dfrac{x-1}{x+3}\right)\left(\dfrac{x+3}{x+4}\right)$

$\quad = \boxed{\dfrac{x-1}{x+4}}$

c. $\left(1-\dfrac{1}{x}\right)\left(1-\dfrac{1}{x+1}\right)\left(1-\dfrac{1}{x+2}\right)\cdots\left(1-\dfrac{1}{x+99}\right)\left(1-\dfrac{1}{x+100}\right)$

$\quad = \left(\dfrac{x-1}{x}\right)\left(\dfrac{x}{x+1}\right)\left(\dfrac{x+1}{x+2}\right)\cdots\left(\dfrac{x+98}{x+99}\right)\left(\dfrac{x+99}{x+100}\right)$

(*Note* the pattern from parts **a** and **b**: the answer is the numerator from the first factor and the denominator from the last factor. The remaining factors cancel.)

$\quad = \boxed{\dfrac{x-1}{x+100}}$

Review Problems

133. $x^5 - x^3 - 12x = x(x^4 - x^2 - 12) = x(x^2 - 4)(x^2 + 3) = \boxed{x(x+2)(x-2)(x^2+3)}$

134. line passing through $(-2, 4)$ and $(1, 6)$

$$\begin{aligned}
m &= \frac{6-4}{1-(-2)} = \frac{2}{3} \\
y - 4 &= \frac{2}{3}(x+2) \\
3y - 12 &= 2x + 4 \\
0 &= 2x - 3y + 16 \\
\boxed{2x - 3y} &\boxed{= -16}
\end{aligned}$$

135. Let

$$\begin{aligned}
x &= \text{width of the photograph} \\
2x + 1 &= \text{length of the photograph} \\
x + 2 &= \text{width of photograph plus frame} \\
2x + 1 &= 2x + 3 = \text{length of photograph plus frame}
\end{aligned}$$

$$\begin{aligned}
\text{combined area} &= 45 \text{ square centimeters} \\
(x+2)(2x+3) &= 45 \\
2x^2 + 7x + 6 &= 45 \\
2x^2 + 7x - 39 &= 0 \\
(2x+13)(x-3) &= 0 \\
2x + 13 = 0 \qquad &\text{or} \qquad x - 3 = 0
\end{aligned}$$

$\quad\quad\quad\quad x = -\dfrac{13}{2}$ (Reject) $\qquad\qquad x = 3$ (width of photograph)

$\qquad\qquad\qquad\qquad\qquad\qquad\qquad\qquad\qquad 2x + 1 = 7$ (length of photograph)

Dimensions of photograph: $\boxed{\text{width: } 3 \text{ cm; length: } 7 \text{ cm}}$

Section 6.4 Complex Fractions

Problem Set 6.4, pp. 464-466

1. $\dfrac{\dfrac{3}{y}}{y-\dfrac{1}{y}} = \dfrac{\dfrac{3}{y}}{y-\dfrac{1}{y}} \cdot \dfrac{y}{y} = \dfrac{\dfrac{3}{y}\cdot y}{y\cdot y - \dfrac{1}{y}\cdot y} = \boxed{\dfrac{3}{y^2-1}}$

3. $\dfrac{3-\dfrac{2}{b}}{\dfrac{2}{b}+\dfrac{3}{b}} = \dfrac{3-\dfrac{2}{b}}{\dfrac{2}{b}+\dfrac{3}{b}} \cdot \dfrac{b}{b} = \dfrac{3b-2}{2+3} = \boxed{\dfrac{3b-2}{5}}$

5. $\dfrac{\dfrac{1}{y}+\dfrac{1}{y^2}}{1+\dfrac{1}{y}} \cdot \dfrac{y^2}{y^2}$

$= \dfrac{\dfrac{1}{y}\cdot y^2 + \dfrac{1}{y^2}\cdot y^2}{1\cdot y^2 + \dfrac{1}{y}\cdot y^2}$

$= \dfrac{y+1}{y^2+y}$

$= \dfrac{y+1}{y(y+1)}$

$= \boxed{\dfrac{1}{y}}$

7. $\dfrac{\dfrac{x}{y}+\dfrac{y}{x}}{\dfrac{1}{y}+\dfrac{1}{x}} \cdot \dfrac{xy}{xy}$

$= \dfrac{\dfrac{x}{y}\cdot xy + \dfrac{y}{x}\cdot xy}{\dfrac{1}{y}\cdot xy + \dfrac{1}{x}\cdot xy}$

$= \boxed{\dfrac{x^2+y^2}{x+y}}$

9. $\dfrac{\dfrac{b^2-c^2}{b}}{\dfrac{b-c}{b^2}} = \dfrac{\dfrac{b^2-c^2}{b}}{\dfrac{b-c}{b^2}} \cdot \dfrac{b^2}{b^2}$

$= \dfrac{b(b^2-c^2)}{b-c} = \dfrac{b(b+c)(b-c)}{b-c}$

$= \boxed{b(b+c)}$

11. $\dfrac{\dfrac{x}{y}-\dfrac{y}{x}}{\dfrac{x^2}{y}-y} = \dfrac{\dfrac{x}{y}-\dfrac{y}{x}}{\dfrac{x^2}{y}-y} \cdot \dfrac{xy}{xy}$

$= \dfrac{x^2-y^2}{x^3-xy^2}$

$= \dfrac{x^2-y^2}{x(x^2-y^2)} = \boxed{\dfrac{1}{x}}$

13. $\dfrac{y+5+\dfrac{6}{y}}{y-\dfrac{9}{y}} = \dfrac{y+5+\dfrac{6}{y}}{y-\dfrac{9}{y}} \cdot \dfrac{y}{y} = \dfrac{y^2+5y+6}{y^2-9}$

$= \dfrac{y^2}{y-7} + \dfrac{6y+7}{7-y} \cdot \dfrac{-1}{-1}$

$= \dfrac{(y+3)(y+2)}{(y+3)(y-3)} = \boxed{\dfrac{y+2}{y-3}}$

15. $\dfrac{1-\dfrac{1}{a}}{\dfrac{a+1}{a}} = \dfrac{1-\dfrac{1}{a}}{\dfrac{a+1}{a}} \cdot \dfrac{a}{a} = \boxed{\dfrac{a-1}{a+1}}$

17. $\dfrac{\dfrac{1}{x}+\dfrac{1}{y}}{x+y} = \dfrac{\dfrac{1}{x}+\dfrac{1}{y}}{x+y} \cdot \dfrac{xy}{xy} = \dfrac{y+x}{(x+y)xy} = \boxed{\dfrac{1}{xy}}$

19. $\dfrac{\dfrac{b^2-c^2}{1}}{\dfrac{1}{b}+\dfrac{1}{c}} = \dfrac{\dfrac{b^2-c^2}{1}}{\dfrac{1}{b}+\dfrac{1}{c}} \cdot \dfrac{bc}{bc} = \boxed{\dfrac{bc(b^2-c^2)}{c+b}}$

21. $\dfrac{3-\dfrac{1}{c}}{3c-1} = \dfrac{3-\dfrac{1}{c}}{3c-1} \cdot \dfrac{c}{c} = \dfrac{3c-1}{(3c-1)c} = \boxed{\dfrac{1}{c}}$

23. $\dfrac{b-3}{b-\dfrac{3}{b-2}} = \dfrac{b-3}{b-\dfrac{3}{b-2}} \cdot \dfrac{b-2}{b-2}$

$= \dfrac{(b-3)(b-2)}{b(b-2)-3}$

$= \dfrac{(b-3)(b-2)}{b^2-2b-3} = \dfrac{(b-3)(b-2)}{(b+1)(b-3)}$

$= \boxed{\dfrac{b-2}{b+1}}$

25. $\dfrac{y+\dfrac{12}{y-7}}{y-3} = \dfrac{y+\dfrac{12}{y-7}}{y-3} \cdot \dfrac{y-7}{y-7}$

$= \dfrac{y(y-7)+12}{(y-3)(y-7)}$

$= \dfrac{y^2-7y+12}{(y-3)(y-7)}$

$= \dfrac{(y-4)(y-3)}{(y-3)(y-7)}$

$= \boxed{\dfrac{y-4}{y-7}}$

27. $\dfrac{\dfrac{3}{y-2}-\dfrac{4}{y+2}}{\dfrac{7}{y^2-4}} = \dfrac{\dfrac{3}{y-2}-\dfrac{4}{y+2}}{\dfrac{7}{(y+2)(y-2)}} \cdot \dfrac{(y+2)(y-2)}{(y+2)(y-2)}$

$= \dfrac{3(y+2)-4(y-2)}{7} = \boxed{\dfrac{-y+14}{7}}$

29. $\dfrac{\dfrac{4}{b-2}+1}{\dfrac{3}{b^2-4}+1} = \dfrac{\dfrac{4}{b-2}+1}{\dfrac{3}{(b+2)(b-2)}+1} \cdot \dfrac{(b+2)(b-2)}{(b+2)(b-2)}$

$= \dfrac{4(b+2)+(b+2)(b-2)}{3+(b+2)(b-2)}$

$= \dfrac{(b+2)[4+(b-2)]}{3+b^2-4} = \dfrac{(b+2)(b+2)}{b^2-1}$

$= \boxed{\dfrac{(b+2)^2}{b^2-1}}$

31. $\dfrac{\dfrac{6}{x^2+2x-15}-\dfrac{1}{x-3}}{\dfrac{1}{x+5}+1}$

$= \dfrac{\dfrac{6}{(x+5)(x-3)}-\dfrac{1}{x-3}}{\dfrac{1}{x+5}+1} \cdot \dfrac{(x+5)(x-3)}{(x+5)(x-3)}$

$= \dfrac{6-(x+5)}{x-3+(x+5)(x-3)}$

$= \dfrac{6-x-5}{(x-3)[1+(x+5)]}$

$= \boxed{\dfrac{-x+1}{(x-3)(x+6)}}$

33. $\dfrac{\dfrac{3}{x+2y} - \dfrac{2y}{x^2+2xy}}{\dfrac{3y}{x^2+2xy} + \dfrac{5}{x}}$

$= \dfrac{\dfrac{3}{x+2y} - \dfrac{2y}{x(x+2y)}}{\dfrac{3y}{x(x+2y)} + \dfrac{5}{x}} \cdot \dfrac{x(x+2y)}{x(x+2y)}$

$= \dfrac{3x - 2y}{3y + 5(x+2y)}$

$= \dfrac{3x - 2y}{3y + 5x + 10y}$

$= \boxed{\dfrac{3x - 2y}{5x + 13y}}$

35. $\dfrac{\dfrac{1}{x^3 - 125}}{\dfrac{1}{x^2 - 25} - \dfrac{1}{x^2 + 5x + 25}}$

$= \dfrac{\dfrac{1}{(x-5)(x^2+5x+25)}}{\dfrac{1}{(x+5)(x-5)} - \dfrac{1}{x^2+5x+25}} \cdot \left[\dfrac{(x+5)(x-5)(x^2+5x+25)}{(x+5)(x-5)(x^2+5x+25)} \right]$

$= \dfrac{x+5}{x^2 + 5x + 25 - (x+5)(x-5)}$

$= \dfrac{x+5}{x^2 + 5x + 25 - x^2 + 25}$

$= \dfrac{x+5}{5x + 50}$

$= \boxed{\dfrac{x+5}{5(x+10)}}$

37. $2 + \dfrac{1}{1 + \dfrac{2}{x + \dfrac{1}{x}}}$

$= 2 + \dfrac{1}{1 + \dfrac{2}{x + \dfrac{1}{x}} \cdot \dfrac{x}{x}}$

$= 2 + \dfrac{1}{1 + \dfrac{2x}{x^2 + 1}} \cdot \dfrac{x^2 + 1}{x^2 + 1}$

$= 2 + \dfrac{x^2 + 1}{x^2 + 1 + 2x} = 2 + \dfrac{x^2 + 1}{x^2 + 2x + 1}$

$= 2 \cdot \dfrac{x^2 + 2x + 1}{x^2 + 2x + 1} + \dfrac{x^2 + 1}{x^2 + 2x + 1}$

$= \dfrac{2x^2 + 4x + 2 + x^2 + 1}{x^2 + 2x + 1}$

$= \boxed{\dfrac{3x^2 + 4x + 3}{x^2 + 2x + 1}}$

39. $\dfrac{\dfrac{1}{a^2 + b^2}}{1 - \dfrac{a}{a + \dfrac{b^2}{a - b}}}$

$= \dfrac{\dfrac{1}{a^2 + b^2}}{1 - \dfrac{a}{a + \dfrac{b^2}{a - b}} \cdot \dfrac{a - b}{a - b}}$

$= \dfrac{\dfrac{1}{a^2 + b^2}}{1 - \dfrac{a(a - b)}{a(a - b) + b^2}}$

$= \dfrac{\dfrac{1}{a^2 + b^2}}{1 - \dfrac{a^2 - ab}{a^2 - ab + b^2}}$

$= \dfrac{\dfrac{1}{a^2 + b^2}}{1 \cdot \dfrac{a^2 - ab + b^2}{a^2 - ab + b^2} - \dfrac{a^2 - ab}{a^2 - ab + b^2}}$

$= \dfrac{\dfrac{1}{a^2 + b^2}}{\dfrac{b^2}{a^2 - ab + b^2}} = \dfrac{1}{a^2 + b^2} \cdot \dfrac{a^2 - ab + b^2}{b^2}$

$= \boxed{\dfrac{a^2 - ab + b^2}{b^2(a^2 + b^2)}}$

41. $[x(x - 1)^{-1} + 1][3(x^2 - 1)^{-1} + 4]^{-1}$

$= \left[\dfrac{x}{(x - 1)^1} + 1 \right] \left[\dfrac{3}{(x^2 - 1)^1} + 4 \right]^{-1}$

$= \left[\dfrac{x}{x - 1} + 1 \right] \cdot \dfrac{1}{\left[\dfrac{3}{x^2 - 1} + 4 \right]^1}$

$= \left(\dfrac{x}{x - 1} + 1 \right) \cdot \dfrac{1}{\dfrac{3}{x^2 - 1} + 4} \cdot \dfrac{x^2 - 1}{x^2 - 1}$

$= \left(\dfrac{x}{x - 1} + 1 \right) \cdot \left[\dfrac{x^2 - 1}{3 + 4(x^2 - 1)} \right]$

$= \left(\dfrac{x}{x - 1} + \dfrac{x - 1}{x - 1} \right) \left(\dfrac{x^2 - 1}{3 + 4x^2 - 4} \right)$

$= \left(\dfrac{2x - 1}{x - 1} \right) \left(\dfrac{x^2 - 1}{4x^2 - 1} \right)$

$= \dfrac{(2x - 1)(x + 1)(x - 1)}{(x - 1)(2x + 1)(2x - 1)}$

$= \boxed{\dfrac{x + 1}{2x + 1}}$

43. $y + \dfrac{y}{y + \dfrac{1}{y}} = y + \dfrac{y}{y + \dfrac{1}{y}} \cdot \dfrac{y}{y} = y + \dfrac{y^2}{y^2 + 1}$

$= y \cdot \dfrac{y^2 + 1}{y^2 + 1} + \dfrac{y^2}{y^2 + 1}$

$= \dfrac{y(y^2 + 1) + y^2}{y^2 + 1} = \boxed{\dfrac{y^3 + y^2 + y}{y^2 + 1}}$

45. $1 - \dfrac{1}{1 - \dfrac{1}{x-2}} = 1 - \dfrac{1}{1 - \dfrac{1}{x-2}} \cdot \dfrac{x-2}{x-2} = 1 - \dfrac{x-2}{x-2-1}$

$= 1 - \dfrac{x-2}{x-3} = \dfrac{x-3}{x-3} - \dfrac{x-2}{x-3} = \dfrac{x-3-(x-2)}{x-3}$

$= \boxed{-\dfrac{1}{x-3}}$

47. $\dfrac{\dfrac{2y}{y-y-1} - y^{-1}}{2y + \dfrac{2y}{1-y-1}} = \dfrac{\dfrac{2y}{y-\dfrac{1}{y}} - \dfrac{1}{y}}{2y + \dfrac{2y}{1-\dfrac{1}{y}}} = \dfrac{\dfrac{2y}{y-\dfrac{1}{y}} \cdot \dfrac{y}{y} - \dfrac{1}{y}}{2y + \dfrac{2y}{1-\dfrac{1}{y}} \cdot \dfrac{y}{y}}$

$= \dfrac{\dfrac{2y^2}{y^2-1} - \dfrac{1}{y}}{2y + \dfrac{2y^2}{y-1}}$

$= \dfrac{\dfrac{2y^2}{(y+1)(y-1)} - \dfrac{1}{y}}{2y + \dfrac{2y^2}{y-1}} \cdot \dfrac{y(y+1)(y-1)}{y(y+1)(y-1)}$

$= \dfrac{2y^3 - (y+1)(y-1)}{2y^2(y+1)(y-1) + 2y^3(y+1)}$

$= \dfrac{2y^3 - (y^2-1)}{2y^2(y+1)[(y-1)+y]}$

$= \boxed{\dfrac{2y^3 - y^2 + 1}{2y^2(y+1)(2y-1)}}$

49. $\dfrac{2d}{\dfrac{d}{r_1} + \dfrac{d}{r_2}} = \dfrac{2d}{\dfrac{d}{r_1} + \dfrac{d}{r_2}} \cdot \dfrac{r_1 r_2}{r_1 r_1} = \dfrac{2dr_1 r_2}{dr_2 + dr_1}$

$= \dfrac{2dr_1 r_2}{d(r_2 + r_1)} = \boxed{\dfrac{2r_1 r_2}{r_2 + r_1}}$

$r_1 = 30 \text{ mph}$

$r_2 = 20 \text{ mph}$

$\dfrac{2r_1 r_2}{r_2 + r_1} = \dfrac{2(30)(20)}{30 + 20} = \dfrac{1200}{50} = \boxed{24 \text{ mph}}$

51. $f(x) = \dfrac{2}{x}$

$\dfrac{f(a+h) - f(a)}{h}$

$= \dfrac{\dfrac{2}{a+h} - \dfrac{2}{a}}{h}$

$= \left[\dfrac{2}{a+h} \cdot \dfrac{a}{a} - \dfrac{2}{a} \cdot \dfrac{a+h}{a+h} \right] \div h$

$= \dfrac{2a - 2a - 2h}{a(a+h)} \cdot \dfrac{1}{h}$

$= \dfrac{-2h}{a(a+h)(h)}$

$= \boxed{\dfrac{-2}{a(a+h)}}$

53. $f(x) = \dfrac{1}{x-1}$

$\dfrac{f(a+h) - f(a)}{h}$

$= \dfrac{\dfrac{1}{a+h-1} - \dfrac{1}{a-1}}{h}$

$= \left[\dfrac{1}{a+h-1} \cdot \dfrac{a-1}{a-1} - \dfrac{1}{a-1} \cdot \dfrac{a+h-1}{a+h-1} \right] \div h$

$= \dfrac{a - 1 - a - h + 1}{(a-1)(a+h-1)} \cdot \dfrac{1}{h}$

$= \dfrac{-h}{(a-1)(a+h-1)(h)}$

$= \boxed{\dfrac{-1}{(a-1)(a+h-1)}}$

55. $f(x) = \dfrac{2x}{x-3}$

$\dfrac{f(a+h) - f(a)}{h}$

$= \dfrac{\dfrac{2(a+h)}{a+h-3} - \dfrac{2a}{a-3}}{h}$

$= \left[\dfrac{2a+2h}{a+h-3} \cdot \dfrac{a-3}{a-3} - \dfrac{2a}{a-3} \cdot \dfrac{a+h-3}{a+h-3} \right] \div h$

$= \dfrac{2a^2 - 6a + 2ah - 6h - 2a^2 - 2ah + 6a}{(a-3)(a+h-3)} \cdot \dfrac{1}{h}$

$= \dfrac{-6h}{(a-3)(a+h-3)(h)}$

$= \boxed{\dfrac{-6}{(a-3)(a+h-3)}}$

57. $f(x) = \dfrac{x-2}{x+3}$

$\dfrac{f(a+h) - f(a)}{h}$

$= \dfrac{\dfrac{a+h-2}{a+h+3} - \dfrac{a-2}{a+3}}{h}$

$= \left[\dfrac{a+h-2}{a+h+3} \cdot \dfrac{a+3}{a+3} - \dfrac{a-2}{a+3} \cdot \dfrac{a+h+3}{a+h+3} \right] \div h$

$= \dfrac{a^2 + ah - 2a + 3a + 3h - 6 - a^2 - ah - 3a + 2a + 2h + 6}{(a+3)(a+h+3)} \cdot \dfrac{1}{h}$

$= \dfrac{5h}{(a+3)(a+h+3)(h)}$

$= \boxed{\dfrac{5}{(a+3)(a+h+3)}}$

59. $f(x) = \dfrac{1}{x^2}$

$\dfrac{f(a+h) - f(a)}{h}$

$= \dfrac{\dfrac{1}{(a+h)^2} - \dfrac{1}{a^2}}{h}$

$= \left[\dfrac{1}{(a+h)^2} \cdot \dfrac{a^2}{a^2} - \dfrac{1}{a^2} \cdot \dfrac{(a+h)^2}{(a+h)^2} \right] \div h$

$= \dfrac{a^2 - a^2 - 2ah - h^2}{a^2(a+h)^2} \cdot \dfrac{1}{h}$

$= \dfrac{-2ah - h^2}{a^2(a+h)^2 h}$

$= \boxed{\dfrac{-2a - h}{a^2(a+h)^2}}$

61. $f(x) = \dfrac{1 + \dfrac{1}{x}}{1 - \dfrac{1}{x}}$ and $g(x) = x + 3$

$\begin{aligned}
(f \circ g)(x) &= f[g(x)] \\
&= \dfrac{1 + \dfrac{1}{x+3}}{1 - \dfrac{1}{x+3}} = \dfrac{\dfrac{x+3}{x+3} + \dfrac{1}{x+3}}{\dfrac{x+3}{x+3} - \dfrac{1}{x+3}} \\
&= \left(\dfrac{x+3+1}{x+3}\right) \div \left(\dfrac{x+3-1}{x+3}\right) \\
&= \dfrac{x+4}{x+3} \cdot \dfrac{x+3}{x+2} \\
&= \boxed{\dfrac{x+4}{x+2},\ (x \neq -2, -3)}
\end{aligned}$

$\boxed{D_{f \circ g} = \{x \mid x \neq -2, -3\}\ \text{ or }\ (-\infty, -3) \cup (-3, -2) \cup (-2, \infty)}$

63. $f(x) = 1 - \dfrac{1}{1 - \dfrac{1}{x}}$ and $g(x) = x - 1$

$\begin{aligned}
(f \circ g)(x) &= f[g(x)] = 1 - \dfrac{1}{1 - \dfrac{1}{x-1}} \\
&= 1 - \dfrac{1}{\dfrac{x-1}{x-1} - \dfrac{1}{x-1}} \\
&= 1 - \dfrac{1}{\dfrac{x-1-1}{x-1}} \\
&= 1 - 1 \div \dfrac{x-2}{x-1} \\
&= 1 - \dfrac{x-1}{x-2} \\
&= \dfrac{x-2}{x-2} - \dfrac{x-1}{x-2} \\
&= \dfrac{x-2-x+1}{x-2} \\
&= \boxed{-\dfrac{1}{x-2}\ (x \neq 1, 2)}
\end{aligned}$

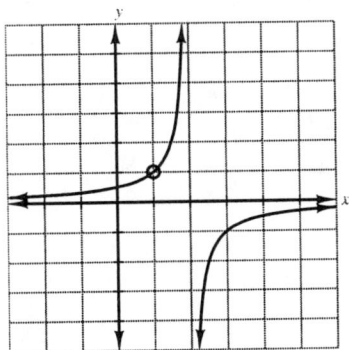

65. $f(x) = \dfrac{x}{1 - \dfrac{1}{x}}$ and $g(x) = 1 + \dfrac{1}{x}$

$$(f \circ g)(x) \;=\; f[g(x)] = \dfrac{1 + \dfrac{1}{x}}{1 - \dfrac{1}{1 + \dfrac{1}{x}}}$$

$$= \dfrac{\dfrac{x}{x} + \dfrac{1}{x}}{1 - \dfrac{1}{\dfrac{x}{x} + \dfrac{1}{x}}}$$

$$= \left(\dfrac{x+1}{x}\right) \div \left(1 - \dfrac{1}{\dfrac{x+1}{x}}\right)$$

$$= \left(\dfrac{x+1}{x}\right) \div \left(\dfrac{x+1-x}{x+1}\right)$$

$$= \left(\dfrac{x+1}{x}\right) \div \left(\dfrac{1}{x+1}\right)$$

$$= \dfrac{x+1}{x} \cdot \dfrac{x+1}{1}$$

$$= \boxed{\dfrac{(x+1)2}{x} \;\; (x \neq 0, -1)}$$

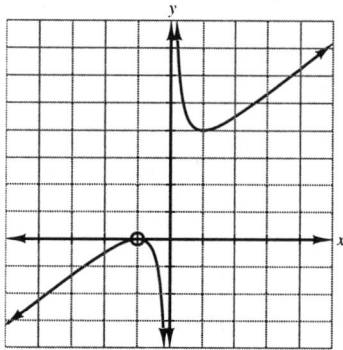

67. $\boxed{\text{C}}$ is true; $\dfrac{1 - x^{-1}}{1 - x^{-2}} = \dfrac{1 - \dfrac{1}{x}}{1 - \dfrac{1}{x^2}} = \dfrac{\dfrac{x}{x} - \dfrac{1}{x}}{\dfrac{x^2}{x^2} - \dfrac{1}{x^2}}$

$$= \left(\dfrac{x-1}{x}\right) \div \left(\dfrac{x^2-1}{x^2}\right)$$

$$= \left(\dfrac{x-1}{x}\right) \cdot \dfrac{x^2}{(x-1)(x+1)}$$

$$= \dfrac{x}{x+1}, \text{ true}$$

69. a. Harmonic mean $= \dfrac{1}{\dfrac{\dfrac{1}{x} + \dfrac{1}{y}}{2}}$

$ = \dfrac{2}{\dfrac{1}{x} + \dfrac{1}{y}} \cdot \dfrac{xy}{xy}$

$ = \dfrac{2xy}{y + x}$

$ = \dfrac{2xy}{x + y}$

b.

1260		
$a = 504$		360
315	280	$b = 252$

$a = \dfrac{2(1260)(315)}{1260 + 315} = \dfrac{793{,}800}{1575} = 504$

b: Since 280 is the harmonic mean of 315 and b:

$\dfrac{2(315)b}{315 + b} = 280$

$630b = 280(315 + b)$

$630b = 280b + 88{,}200$

$350b = 88{,}200$

$b = 252$

Using similar methods, the table can be completed giving this square:

1260	840	630
504	420	360
315	280	252

Review Problems

72. $f(x) = \dfrac{6x}{x^2 - 4}$ and $g(x) = \dfrac{3}{2 - x}$

$$
\begin{aligned}
(f + g)(x) &= \frac{6x}{x^2 - 4} + \frac{3}{2 - x} \\
&= \frac{6x}{(x + 2)(x - 2)} - \frac{3}{(x - 2)} \cdot \frac{x + 2}{x + 2} \\
&= \frac{6x - 3 - 6}{(x + 2)(x - 2)} \\
&= \frac{3x - 6}{(x + 2)(x - 2)} = \frac{3(x - 2)}{(x + 2)(x - 2)} = \boxed{\frac{3}{x + 2}} \quad (x \neq -2, 2)
\end{aligned}
$$

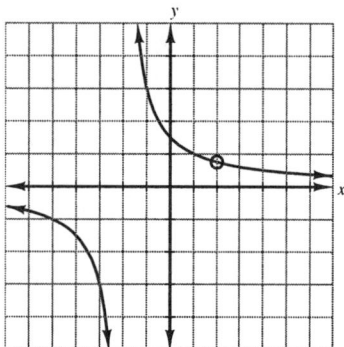

73. Let

$$
\begin{aligned}
x &= \text{length of one leg of right triangle} \\
x + 7 &= \text{length of other leg of right triangle} \\
\text{hypotenuse} &= 13 \text{ meters}
\end{aligned}
$$

$$
\begin{aligned}
x^2 + (x + 7)^2 &= 13^2 \\
x^2 + x^2 + 14x + 49 &= 169 \\
2x^2 + 14x - 120 &= 0 \\
2(x^2 + 7x - 60) &= 0 \\
2(x + 12)(x - 5) &= 0 \\
x + 12 &= 0 \qquad \text{or} \qquad x - 5 = 0 \\
x &= -12 \quad \text{(Reject)} \qquad\qquad x = 5 \\
&\qquad\qquad\qquad\qquad\qquad\qquad x + 7 = 12
\end{aligned}
$$

lengths of the legs of the triangle: $\boxed{5 \text{ meters and } 12 \text{ meters}}$

74. Let

$$
\begin{aligned}
x &= \text{amount at } 11\% \text{ for 2 years} \\
20{,}000 - x &= \text{amount at } 9\% \text{ for 3 years}
\end{aligned}
$$

$$
\begin{aligned}
0.11(x)(2) + (0.09)(20{,}000 - x)(3) &= 5025 \\
0.22x + 5400 - 0.27x &= 5025 \\
-0.05x &= -375 \\
x &= 7500 \quad (11\%) \\
20{,}000 - x &= 12{,}500 \quad (9\%)
\end{aligned}
$$

$\boxed{\$7500 \text{ at } 11\% \text{ for 2 years; } \$12{,}500 \text{ at } 9\% \text{ for 3 years}}$

Section 6.5 Equations Containing Rational Expressions

Problem Set 6.5, pp. 471-474

1.
$$\frac{2}{3x}+\frac{1}{4} = \frac{11}{6x}-\frac{1}{3}$$
$$12x\left(\frac{2}{3x}+\frac{1}{4}\right) = 12x\left(\frac{11}{6x}-\frac{1}{3}\right)$$
$$12x\cdot\frac{2}{3x}+12x\cdot\frac{1}{4} = 12x\cdot\frac{11}{6x}-12x\cdot\frac{1}{3}$$
$$8+3x = 22-4x$$
$$7x = 14$$
$$x = 2$$
$\boxed{\{2\}}$

3.
$$\frac{4}{5}+\frac{7}{2a} = \frac{13}{2a}-\frac{4}{20}$$
$$20a\left(\frac{4}{5}+\frac{7}{2a}\right) = 20a\left(\frac{13}{2a}-\frac{4}{20}\right)$$
$$16a+70 = 130-4ia$$
$$20a = 60$$
$$a = 3$$
$\boxed{\{3\}}$

5.
$$\frac{2}{y-3}+\frac{3y+1}{y+3} = 3$$
$$(y-3)(y+3)\left[\frac{2}{y-3}+\frac{3y+1}{y+3}\right]=(y-3)(y+3)\cdot 3$$
$$2(y+3)+(3y+1)(y-3) = 3(y-3)(y+3)$$
$$2y+6+3y^2-8y-3 = 3y^2-27$$
$$-6+3 = -27$$
$$-6 = -30$$
$$y = 5$$
$\boxed{\{5\}}$

7.
$$\frac{2x-4}{x-4}-2 = \frac{20}{x+4}$$
$$(x-4)(x+4)\left[\frac{2x-4}{x-4}-2\right] = (x-4)(x+4)\frac{20}{x+4}$$
$$(x+4)(2x-4)-2(x-4)(x+4) = 20(x-4)$$
$$2x^2+4x-16-2x^2+32 = 20x-80$$
$$4x+16 = 20x-80$$
$$-16x = -96$$
$$x = 6$$
$\boxed{\{6\}}$

9.
$$\frac{3c}{c+1} = \frac{5c}{c-1}-2$$
$$(c+1)(c-1)\frac{3c}{c+1} = (c+1)(c-1)\left[\frac{5c}{c-1}-1\right]$$
$$3c(c-1) = 5c(c+1)-2(c+1)(c-1)$$
$$3c^2-3c = 5c^2+5c-2c^2+2$$
$$3c^2-3c = 3c^2+5c+2$$
$$-3c = 5c+2$$
$$-8c = 2$$
$$c = \frac{2}{-8}=-\frac{1}{4}$$
$\boxed{\left\{-\frac{1}{4}\right\}}$

11.

$$\frac{x+5}{x^2-4}-\frac{3}{2x-4}=\frac{1}{2x+4}$$

$$\frac{x+5}{(x+2)(x-2)}-\frac{3}{2(x-2)}=\frac{1}{2(x+2)}$$

$$2(x+2)(x-2)\left[\frac{x+5}{(x+2)(x-2)}-\frac{3}{2(x-2)}\right]=2(x+2)(x-2)\cdot\frac{1}{2(x+2)}$$

$$2(x-5)+3(x+2)=x-2$$
$$2x+10-3x-6=x-2$$
$$-x+4=x-2$$
$$6=2x$$
$$3=x$$

$\boxed{\{3\}}$

13.

$$\frac{4}{a+2}+\frac{2}{a-4}=\frac{30}{a^2-2a-8}$$

$$(a+2)(a-4)\left[\frac{4}{a+2}+\frac{2}{a-4}\right]=(a+2)(a-4)\frac{30}{(a+2)(a-4)}$$

$$4(a-4)+2(a+2)=30$$
$$6a-12=30$$
$$6a=42$$
$$a=7$$

$\boxed{\{7\}}$

15.

$$\frac{y+5}{y+1}-\frac{y}{y+2}=\frac{4y+1}{y^2+3y+2}$$

$$\frac{y+5}{y+1}-\frac{y}{y+2}=\frac{4y+1}{(y+1)(y+2)}$$

$$(y+1)(y+2)\left[\frac{y+5}{y+1}-\frac{y}{y+2}\right]=(y+1)(y+2)\frac{4y+1}{(y+1)(y+2)}$$

$$(y+2)(y+5)-y(y+1)=4y+1$$
$$y^2+7y+10-y^2-y=4y+1$$
$$6y+10=4y+1$$
$$2y=-9$$
$$y=\frac{-9}{2}$$

$\boxed{\left\{-\frac{9}{2}\right\}}$

17.

$$\frac{c}{c+4}=\frac{2}{5}-\frac{4}{c+4}$$

$$5(c+4)\frac{c}{c+4}=5(c+4)\left[\frac{2}{5}-\frac{4}{c+4}\right]$$

$$5c=2(c+4)-20$$
$$5c=2c+8-20$$
$$5c=2c-12$$
$$3c=-12$$
$$c=-4$$

The value −4 makes two denominators in the original equation equal to zero. Thus, −4 is not a solution. The equation has no solution.

$\boxed{\varnothing}$

19.

$$\frac{3y}{y-3} - \frac{5y}{y-3} = -2$$

$$(y-3)\left[\frac{3y}{y-3} - \frac{5y}{y-3}\right] = (y-3)(-2)$$

$$3y - 5y = -2y + 6$$

$$-2y = -2y + 6$$

$$0 = 6$$

This contradiction indicates that the equation has no solution.

$$\boxed{\varnothing}$$

21.

$$\frac{6}{x} - \frac{x}{3} = 1$$

$$3x\left(\frac{6}{x} - \frac{x}{3}\right) = 3x(1)$$

$$18 - x^2 = 3x$$

$$0 = x^2 + 3x - 18$$

$$0 = (x+6)(x-3)$$

$$x + 6 = 0 \qquad \text{or} \qquad x - 3 = 0$$

$$x = -6 \qquad\qquad\qquad x = 3$$

$$\boxed{\{-6, 3\}}$$

23.

$$\frac{2}{x} + \frac{9}{x+4} = 1$$

$$x(x+4)\left(\frac{2}{x} + \frac{9}{x+4}\right) = x(x+4) \cdot 1$$

$$2(x+4) + 9x = x(x+4)$$

$$2x + 8 + 9x = x^2 + 4x$$

$$11x + 8 = x^2 + 4x$$

$$0 = x^2 - 7x + 8$$

$$0 = (x+1)(x-8)$$

$$x + 1 = 0 \qquad \text{or} \qquad x - 8 = 0$$

$$x = -1 \qquad\qquad\qquad x = 8$$

$$\boxed{\{-1, 8\}}$$

25.

$$\frac{1}{x-1} + \frac{1}{x-4} = \frac{5}{4}$$

$$4(x-1)(x-4)\left[\frac{1}{x-1} + \frac{1}{x-4}\right] = 4(x-1)(x-4) \cdot \frac{5}{4}$$

$$4(x-4) + 4(x-1) = 5(x-1)(x-4)$$

$$4x - 16 + 4x - 4 = 5x^2 - 25x + 20$$

$$8x - 20 = 5x^2 - 25x + 20$$

$$0 = 5x^2 - 33x + 40$$

$$0 = (5x-8)(x-5)$$

$$5x - 8 = 0 \qquad \text{or} \qquad x - 5 = 0$$

$$x = \frac{8}{5} \qquad\qquad\qquad x = 5$$

$$\boxed{\left\{\frac{8}{5}, 5\right\}}$$

27.

$$\frac{x^2 + 10}{x - 5} = \frac{7x}{x - 5}$$

$$(x - 5)\frac{x^2 + 10}{x - 5} = (x - 5)\frac{7x}{x - 5}$$

$$x^2 + 10 = 7x$$

$$x^2 - 7x + 10 = 0$$

$$(x - 5)(x - 2) = 0$$

$$x - 5 = 0 \quad\quad \text{or} \quad\quad\quad\quad x - 2 = 0$$

$$x = 5 \;\text{(Reject } x = 5 \text{ (causes division by 0))}\quad x = 2$$

$$\boxed{\{2\}}$$

29.

$$\frac{7}{y + 5} - \frac{3}{y - 1} = \frac{8}{y - 6}$$

$$(y + 5)(y - 1)(y - 6)\left[\frac{7}{y + 5} - \frac{3}{y - 1}\right] = (y + 5)(y - 1)(y - 6)\frac{8}{y - 6}$$

$$7(y - 1)(y - 6) - 3(y + 5)(y - 6) = 8(y + 5)(y - 1)$$

$$7y^2 - 49y + 42 - 3y^2 + 3y + 90 = 8y^2 + 32y - 40$$

$$4y^2 - 46y + 132 = 8y^2 + 32y - 40$$

$$0 = 4y^2 + 78y - 172$$

$$\frac{1}{2} \cdot 0 = \frac{1}{2}(4y^2 + 78y - 172)$$

$$0 = 2y^2 + 39y - 86$$

$$0 = (2y + 43)(y - 2)$$

$$2y + 43 = 0 \quad\quad \text{or} \quad\quad y - 2 = 0$$

$$y = -\frac{43}{2} \quad\quad\quad\quad y = 2$$

$$\boxed{\left\{-\frac{43}{2}, 2\right\}}$$

31.

$$\frac{24}{10 + y} + \frac{24}{10 - y} = 5$$

$$(10 + y)(10 - y)\left[\frac{24}{10 + y} + \frac{24}{10 - y}\right] = (10 + y)(10 - y) \cdot 5$$

$$24(10 - y) + 24(10 + y) = 5(100 - y^2)$$

$$240 - 24y + 240 + 24y = 500 - 5y^2$$

$$480 = 500 - 5y^2$$

$$5y^2 - 20 = 0$$

$$y^2 - 4 = 0$$

$$(y + 2)(y - 2) = 0$$

$$y + 2 = 0 \quad\quad \text{or} \quad\quad y - 2 = 0$$

$$y = -2 \quad\quad\quad\quad y = 2$$

$$\boxed{\{-2, 2\}}$$

33.

$$\frac{x}{x - 5} + \frac{17}{25 - x^2} = \frac{1}{x + 5}$$

$$(x - 5)(x + 5)\left[\frac{x}{x - 5} + \frac{17}{25 - x^2}\right] = (x - 5)(x + 5)\left[\frac{1}{x + 5}\right]$$

$$x(x + 5) + (-1)17 = x - 5$$

$$x^2 + 5x - 17 = x - 5$$

$$x^2 + 4x - 12 = 0$$

$$(x + 6)(x - 2) = 0$$

$$x + 6 = 0 \quad\quad \text{or} \quad\quad x - 2 = 0$$

$$x = -6 \quad\quad\quad\quad x = 2$$

$$\boxed{\{-6, 2\}}$$

35.

$$\frac{5}{y-3} = \frac{30}{y^2-9} + 1$$

$$(y+3)(y-3)\left[\frac{5}{y-3}\right] = (y+3)(y-3)\left[\frac{30}{y^2-9} + 1\right]$$

$$5(y+3) = 30 + (y+3)(y-3)$$

$$5y+15 = 30 + y^2 - 9$$

$$0 = y^2 - 5y + 6$$

$$0 = (y-3)(y-2)$$

$$y-3 = 0 \qquad\qquad \text{or} \qquad\qquad y-2 = 0$$

$$y = 3 \quad \text{Reject 3 (it causes division} \qquad y = 2$$
$$\text{by zero in the original equation).}$$

$$\boxed{\{2\}}$$

37.

$$\frac{8}{x+1} + \frac{2}{1-x^2} = \frac{x}{x-1}$$

$$(x+1)(x-1)\left[\frac{8}{x+1} + \frac{2}{(1-x)(1+x)}\right] = \left(\frac{x}{x-1}\right)(x+1)(x-1)$$

$$8(x-1) + (-1)2 = x(x+1)$$

$$8x - 8 - 1 = x^2 + x$$

$$0 = (x-5)(x-2)$$

$$x-5 = 0 \qquad\qquad \text{or} \qquad\qquad x-2 = 0$$

$$x = 5 \qquad\qquad\qquad\qquad x = 2$$

$$\boxed{\{2, 5\}}$$

39.

$$\frac{x}{x+5} + \frac{x}{5-x} = \frac{15+5x}{x^2-25}$$

$$(x+5)(x-5)\left[\frac{x}{x+5} + \frac{x}{5-x}\right] = \left[\frac{15+5x}{(x+5)(x-5)}\right](x+5)(x-5)$$

$$x(x-5) + (-1)x(x+5) = 15 + 5x$$

$$x^2 - 5x - x^2 - 5x = 15 + 5x$$

$$-10x = 15 + 5x$$

$$-15x = 15$$

$$x = -1$$

$$\boxed{\{-1\}}$$

41.

$$\frac{x+2}{x^2-x} - \frac{6}{x^2-1} = 0$$

$$x(x-1)(x+1)\left[\frac{x+2}{x(x-1)} - \frac{6}{(x-1)(x+1)}\right] = x(x-1)(x+1)\cdot 0$$

$$(x+1)(x+2) - 6x = 0$$

$$x^2 + 3x + 2 - 6x = 0$$

$$x^2 - 3x + 2 = 0$$

$$(x-1)(x-2) = 0$$

$$x-1 = 0 \qquad\qquad \text{or} \qquad\qquad x-2 = 0$$

$$x = 1 \text{ (reject; it causes division by 0)} \qquad x = 2$$

$$\boxed{\{2\}}$$

43. Even though this equation appears "complicated," it is not quadratic.

$$\frac{1}{x^3-8}-\frac{2}{x^2+2x+4}=\frac{3}{(2-x)(x^2+2x+4)}$$

$$(x-2)(x^2+2x+4)\cdot\left[\frac{1}{(x-2)(x^2+2x+4)}-\frac{2}{x^2+2x+4}\right]=(x-2)(x^2+2x+4)\cdot\left[\frac{3}{(2-x)(x^2+2x+4)}\right]$$

$$1-2(x-2)=(-1)3$$
$$1-2x+4=-3$$
$$-2x=-8$$
$$x=4$$

$$\boxed{\{4\}}$$

45.

$$5y^{-2}+1=6y^{-1}$$
$$\frac{5}{y^2}+1=\frac{6}{y}$$
$$y^2\left(\frac{5}{y^2}+1\right)=y^2\left(\frac{6}{y}\right)$$
$$5+y^2=6y$$
$$y^2-6y+5=0$$
$$(y-5)(y-1)=0$$
$$y-5=0 \qquad\text{or}\qquad y-1=0$$
$$y=5 \qquad\qquad\qquad y=1$$

$$\boxed{\{1,5\}}$$

47. $P=20-\dfrac{4}{t+1}$

$P=19$, since P is given in thousands.

$$19=20-\frac{4}{t+1}$$
$$(t+1)19=(t+1)\left[20-\frac{4}{t+1}\right]$$
$$19t+19=20(t+1)-4$$
$$19t+19=20t+16$$
$$3=t$$

The population will be 19,000 in 3 years after 1990, or in $\boxed{1993}$.

49. $y=\dfrac{100}{x}$

($y=4$ milligrams):

$$4=\frac{100}{x}$$
$$4x=100$$
$$x=25$$

$$\boxed{25\text{ species per thousand individuals}}$$

51.

$$\frac{D}{d}=q+\frac{R}{d}$$
$$D=dq+R$$
$$D-R=dq$$
$$\frac{D-R}{q}=d$$

$$\boxed{d=\frac{D-R}{q}}$$

53.
$$W = \frac{10x}{150 - x}$$

$$(150 - x)W = (150 - x)\frac{10x}{150 - x}$$

$$150W - Wx = 10x$$

$$150W = Wx + 10x$$

$$150W = x(W + 10)$$

$$\frac{150W}{W + 10} = x$$

$$\boxed{x = \frac{150W}{W + 10}}$$

If $W = 5$,

$$x = \frac{150(5)}{5 + 10} = \frac{750}{15} = 50$$

$\boxed{50\%}$ can be raised in 5 weeks.

55.
$$\frac{1}{R} = \frac{1}{R_1} + \frac{1}{R_2}$$

$$(RR_1R_2)\left(\frac{1}{R}\right) = (RR_1R_2)\left(\frac{1}{R_1} + \frac{1}{R_2}\right)$$

$$R_1R_2 = RR_2 + RR_1$$

$$R_1R_2 - RR_2 = RR_1$$

$$R_2(R_1 - R) = RR_1$$

$$\boxed{R_2 = \frac{RR_1}{R_1 - R}}$$

57.
$$S = \frac{a}{1 - r}$$

$$(1 - r)s = (1 - r)\frac{a}{1 - r}$$

$$S - Sr = a$$

$$S - a = Sr$$

$$\frac{S - a}{S} = r$$

$$\boxed{r = \frac{S - a}{S}}$$

59. $A = \dfrac{r - s}{r + s}$

$$(r + s)A = (r + s)\left(\frac{r - s}{r + s}\right)$$

$$Ar + As = r - s$$

$$As + s = r - Ar$$

$$s(A + 1) = r(1 - A)$$

$$\boxed{s = \frac{r(1 - A)}{A + 1}}$$

61.
$$S = \frac{rL - a}{r - 1}$$

$$(r - 1)S = (r - 1)\left(\frac{rL - a}{r - 1}\right)$$

$$Sr - S = rL - a$$

$$Sr - rL = S - a$$

$$r(S - L) = S - a$$

$$\boxed{r = \frac{S - a}{S - L}}$$

63.
$$\frac{x}{2} = \frac{4b}{1 + 2b}$$

$$2\left(\frac{x}{2}\right) = 2\left(\frac{4b}{1 + 2b}\right)$$

$$\boxed{x = \frac{8b}{1 + 2b}}$$

65.
$$C = \frac{130x}{200 - x} \quad (C, \text{ in millions of dollars})$$

cost: \$32.5 million

$$32.5 = \frac{130}{200 - x}$$

$$(200 - x)32.5 = (200 - x)\frac{130}{200 - x}$$

$$6500 - 32.5x = 130x$$

$$6500 = 162.5x$$

$$40 = x$$

$\boxed{40\%}$

67.
$$\frac{13w - 6z}{w - 3z} = 3$$

$$(w + 3z)\left(\frac{13w - 6z}{w + 3z}\right) = (w + 3z)3$$

$$13w - 6z = 3w + 9z$$

$$10w = 5z$$

$$w = \frac{3}{2}z \quad \text{(solve for } w \text{ in terms of } z\text{)}$$

$$\frac{w + z}{w - z} = \frac{\frac{3}{2}z + z}{\frac{3}{2}z - z} = \frac{3z + 2z}{3z - 2z} = \frac{5z}{z} = \boxed{5} \quad \text{(substitute for } w\text{)}$$

69. $\left|\dfrac{y + 1}{y + 8}\right| = \dfrac{2}{3}$

$$\frac{y + 1}{y + 8} = \frac{2}{3} \qquad \text{or} \qquad \frac{y + 1}{y + 8} = -\frac{2}{3}$$

$$3(y + 8) \cdot \frac{y + 1}{y + 8} = 3(y + 8) \cdot \frac{2}{8} \qquad\qquad 3(y + 1) = -2(y + 8)$$

$$3(y + 1) = 2(y + 8) \qquad\qquad\qquad 3y + 3 = -2y - 16$$

$$3y + 3 = 2y + 16 \qquad\qquad\qquad\qquad 5y = -19$$

$$y = 13 \qquad\qquad\qquad\qquad\qquad y = -\frac{19}{5}$$

$$\boxed{\left\{-\frac{19}{5}, 13\right\}}$$

71.
$$x = \frac{1 - a}{1 + a} \qquad \text{and} \qquad a = \frac{1 + y}{1 - y}$$

$$a(1 - y) = 1 + y$$

$$a - ay = 1 + y$$

$$a - 1 = ay + y$$

$$a - 1 = y(a + 1)$$

$$\frac{a - 1}{a - 1} = y$$

$$x + y = \frac{1 - a}{1 + a} + \frac{a - 1}{a + 1} = \frac{1 - a + a - 1}{a + 1} = \frac{0}{a + 1} = 0$$

73.
$$\frac{4x - b}{x - 5} = 3$$

The solution set for $\dfrac{4x - b}{x - 5} = 3$ is $\emptyset$ if $x = 5$.

$$4x - b = 3(x - 5)$$

$$4(5) - b = 3(5 - 5)$$

$$20 - b = 3(0)$$

$$20 - b = 0$$

$$20 = b$$

$$\boxed{b = 20}$$

Review Problems

76. $2\left|1 - 2y\right| - 3 \;<\; 1$

$\qquad\quad 2\left|1 - 2y\right| \;<\; 4$

$\qquad\qquad (1 - 2y) \;<\; 2$

$\qquad -2 < 1 - 2y \;<\; 2$

$\qquad\quad -3 < -2y \;<\; 1$

$\qquad\qquad \dfrac{3}{2} > y \;>\; -\dfrac{1}{2}$

$\qquad\quad -\dfrac{1}{2} < y \;<\; \dfrac{3}{2}$

$\boxed{\left\{ y \mid -\dfrac{1}{2} < y < \dfrac{3}{2} \right\}}$

77. $(x - y)^{-2}(x + y)$

$= [1 - (-1)]^{-2}[2(1) + (-1)]$ $(x = 1 \text{ and } y = -1)$

$= (2)^{-2}\,(1)$

$= \dfrac{1}{2^2} = \boxed{\dfrac{1}{4}}$

78. $(x^2 - x - 1)(x^3 - x^2 + 1)$

$= x^2(x^3 - x^2 + 1) - x(x^3 - x^2 + 1) - 1(x^3 - x^2 + 1)$

$= x^5 - x^4 + x^2 - x^4 + x^3 - x - x^3 + x^2 - 1$

$= \boxed{x^5 - 2x^4 + 2x^2 - x - 1}$

Section 6.6 Problem Solving

Problem Set 6.6, pp. 483-485

1. Let $x =$ number

$\qquad \dfrac{3 + x}{8 + x} \;=\; \dfrac{2}{3}$

$\qquad 3(3 + x) \;=\; 2(8 + x)$

$\qquad 9 + 3x \;=\; 16 + 2x$

$\qquad\qquad x \;=\; 7$

The number is $\boxed{7}$.

3. Let x and $3x$ equal the numbers.

$\qquad \dfrac{1}{x} + \dfrac{1}{3x} \;=\; \dfrac{1}{3}$

$\qquad 3x\left(\dfrac{1}{x} + \dfrac{1}{3x}\right) \;=\; 3x\left(\dfrac{1}{3}\right)$

$\qquad\qquad 3 + 1 \;=\; x$

$\qquad\qquad\quad 4 \;=\; x$

$\qquad\qquad\quad x \;=\; 4$

$\qquad\qquad 3x \;=\; 3(4) = 12$

The numbers are $\boxed{4 \text{ and } 12}$.

5. Let x and $x + 1$ equal two consecutive integers.

$\qquad\qquad \dfrac{1}{x} + \dfrac{1}{x + 1} \;=\; 15 \cdot \dfrac{1}{x(x + 1)}$

$\qquad x(x + 1)\left[\dfrac{1}{x} + \dfrac{1}{x + 1}\right] \;=\; x(x + 1) \cdot \dfrac{15}{x(x + 1)}$

$\qquad\qquad x + 1 + x \;=\; 15$

$\qquad\qquad\qquad 2x \;=\; 14$

$\qquad\qquad\qquad\; x \;=\; 7$

$\qquad\qquad\quad x + 1 \;=\; 7 + 1 = 8$

The integers are $\boxed{7 \text{ and } 8}$.

7.   Let x and $3x$ equal the numbers.

$$\frac{135}{x} = \frac{135}{3x} + 10$$

$$3x\left(\frac{135}{x}\right) = 3x\left(\frac{135}{3x} + 10\right)$$

$$405 = 135 + 30x$$
$$270 = 30x$$
$$9 = x$$
$$x = 9$$
$$3x = 3(9) = 27$$

The numbers are $\boxed{9 \text{ and } 27}$.

9.   Let x = number.

$$1 - \frac{1}{x} = \frac{3}{x}$$

$$x\left(1 - \frac{1}{x}\right) = x \cdot \frac{3}{x}$$

$$x - 1 = 3$$
$$x = 4$$

The number is $\boxed{4}$.

11.   Let

$$x = \text{lesser number}$$
$$3x + 1 = \text{greater number}$$

$$\frac{3x + 1}{x} = 3 + \frac{1}{x}$$

$$x\left(\frac{3x + 1}{x}\right) = x\left(3 + \frac{1}{x}\right)$$

$$3x + 1 = 3x + 1$$
$$1 = 1 \quad \{x \mid x \in R\}$$

$\boxed{\text{Any pair of numbers such that the second number is 1 more than 3 times the first number will}}$
$\boxed{\text{satisfy the conditions}}$ of this problem.

13.   Let

$$x = \text{lesser number}$$

$$34 - x = \text{greater number}$$

$$\frac{34 - x}{x} = 4 + \frac{4}{x}$$

$$x\left(\frac{34 - x}{x}\right) = x\left(4 + \frac{4}{x}\right)$$

$$34 - x = 4x + 4$$
$$30 = 5x$$
$$6 = x$$

$$x = 6$$

$$34 - 6 = 28$$
$$34 - x = 28$$

The numbers are $\boxed{6 \text{ and } 28}$.

15.   Let x = the number.

$$x + \frac{1}{x} = 2\frac{1}{30}$$

$$\frac{1}{x} + x = \frac{61}{30}$$

$$30x\left(\frac{1}{x} + x\right) = 30x\left(\frac{61}{30}\right)$$

$$30 + 30x^2 = 61x$$

$$30x^2 - 61x + 30 = 0$$

$$(5x - 6)(6x - 5) = 0$$

$$5x - 6 = 0 \qquad \text{or} \qquad 6x - 5 = 0$$

$$x = \frac{6}{5} \qquad\qquad\qquad x = \frac{5}{6}$$

The numbers are $\boxed{\dfrac{6}{5} \text{ and } \dfrac{5}{6}}$.

17. Let x and $x + 1$ equal two consecutive positive integers.

$$\frac{1}{x} + \frac{1}{x+1} = \frac{5}{6}$$

$$6x(x+1)\left[\frac{1}{x} + \frac{1}{(x+1)}\right] = 6x(x+1) \cdot \frac{5}{6}$$

$$6x + 6 + 6x = 5x^2 + 5x$$

$$0 = 5x^2 - 7x - 6$$

$$0 = (5x+3)(x-2)$$

$$5x + 3 = 0 \qquad \text{or} \qquad x - 2 = 0$$

$$x = -\frac{3}{5} \quad \text{(reject; not positive)} \qquad x = 2$$

$$x + 1 = 2 + 1 = 3$$

The numbers are $\boxed{2 \text{ and } 3}$.

19. Let x = the number.

$$\frac{x}{3} - 1 = \frac{6}{x}$$

$$3x\left(\frac{x}{3} - 1\right) = 3x\left(\frac{6}{x}\right)$$

$$x^2 - 3x - 18 = 0$$

$$(x - 6)(x + 3) = 0$$

$$x - 6 = 0 \qquad \text{or} \qquad x + 3 = 0$$

$$x = 6 \qquad\qquad x = -3$$

The numbers are $\boxed{-3 \text{ and } 6}$.

21.

$$\frac{1}{R} = \frac{1}{R_1} + \frac{1}{R_2}$$

$$R_1 = 2R_2$$

($R = 60$ ohms):

$$\frac{1}{60} = \frac{1}{2R_2} + \frac{1}{R_2}$$

$$60R_2\left(\frac{1}{60}\right) = 60R_2\left(\frac{1}{2R_2} + \frac{1}{R_2}\right)$$

$$R_2 = 30 + 60$$

$$R_2 = 90$$

$$R_1 = 2(90) = 180$$

$$\boxed{R_1 = 180 \text{ ohms}, R_2 = 90 \text{ ohms}}$$

23. Let x = speed of wind

$$240 + x \ = \ \text{speed with the wind}$$
$$240 - x \ = \ \text{speed against the wind}$$

	D	R	$T = \dfrac{D}{R}$
with wind	640	$240 + x$	$\dfrac{640}{240 + x}$
against wind	560	$240 - x$	$\dfrac{560}{240 - x}$

Time traveling 560 miles against the wind = Time traveling 640 miles with the wind

$$\frac{560}{240 - x} \ = \ \frac{640}{240 + x}$$
$$560(240 + x) \ = \ 640(240 - x)$$
$$134400 + 560x \ = \ 153600 - 640x$$
$$1200x \ = \ 19200$$
$$x \ = \ 16$$

The speed of wind is $\boxed{16 \text{ mph}}$.

25. Let

$$x \ = \ \text{car's rate}$$
$$x + 1 \ = \ \text{train's rate}$$

	D (km)	R	$T = \dfrac{D}{R}$
car	480	x	$\dfrac{480}{x}$
train	540	$x + 10$	$\dfrac{540}{x + 10}$

$$\text{(train's time)} \ = \ \text{(car's time)}$$
$$\frac{540}{x + 10} \ = \ \frac{480}{x}$$
$$540x \ = \ 480x + 4800$$
$$60x \ = \ 4800$$
$$x \ = \ 80$$

The car was traveling $\boxed{80 \text{ kph}}$.

27. Let

$$x = \text{rate of the helicopter}$$
$$3x = \text{rate of jet}$$

	D (miles)	R	$T = \dfrac{D}{R}$
helicopter	600	x	$\dfrac{600}{x}$
jet	900	$3x$	$\dfrac{900}{3x}$

$$\text{Time in helicopter} + \text{time in jet} = 6$$

$$\frac{600}{x} + \frac{900}{3x} = 6$$

$$3x \cdot \frac{600}{x} + 3x \cdot \frac{900}{3x} = 3x \cdot 6$$

$$1800 + 900 = 18x$$

$$2700 = 18x$$

$$150 = x$$

$$x = 150 \quad \text{(helicopter)}$$

$$3x = 3(150) = 450 \quad \text{(jet)}$$

rate of the jet: $\boxed{450 \text{ mph}}$

29. Let x = distance traveled by Bunkers in one direction (outgoing or return)

	D (km)	R	$T = \dfrac{D}{R}$
outgoing	x	75	$\dfrac{x}{75}$
return	x	50	$\dfrac{x}{50}$

$$\text{Time spent on outgoing trip} + \text{time spent on return trip} = 10$$

$$\frac{x}{75} + \frac{x}{50} = 10$$

$$150 \cdot \frac{x}{75} + 150 \cdot \frac{x}{50} = 150 \cdot 10$$

$$2x + 3x = 1500$$

$$5x = 1500$$

$$x = 300$$

The Bunkers ventured $\boxed{300 \text{ km}}$ from Queens.

31. Let

$$x = \text{car's rate}$$
$$2x = \text{train's rate}$$

	D (miles)	R	$T = \dfrac{D}{R}$	
car	300	x	$\dfrac{300}{x}$	1 P.M. to 7 P.M. = 6 hr
train	300	2x	$\dfrac{300}{2x}$	1 P.M. to 4 P.M. = 3 hr

$$\text{time for car} - \text{time for train} = 3$$
$$\frac{300}{x} - \frac{300}{2x} = 3$$
$$2x \cdot \frac{300}{x} - 2x \cdot \frac{300}{2x} = 2x \cdot 3$$
$$600 - 300 = 6x$$
$$-6x = -300$$
$$x = 50 \quad \text{(car)}$$
$$2x = 2(50) = 100 \quad \text{(train)}$$

rate of the train: $\boxed{100 \text{ mph}}$

33. Let

$$x = \text{car's usual speed}$$
$$x + 10 = \text{car's increased speed}$$

	D (miles)	R	$T = \dfrac{D}{R}$
usual	60	x	$\dfrac{60}{x}$
increased	60	x + 10	$\dfrac{60}{x + 10}$

Time at increased speed = Time at usual speed $- \dfrac{1}{2}$

$$\frac{60}{x + 10} = \frac{60}{x} - \frac{1}{2}$$
$$2x(x + 10)\frac{60}{x + 10} = 2x(x + 10)\left[\frac{60}{x} - \frac{1}{2}\right]$$
$$120x = 120x + 1200 - x^2 - 10x$$
$$x^2 + 10x - 1200 = 0$$
$$(x + 40)(x - 30) = 0$$
$$x + 40 = 0 \qquad \text{or} \qquad x - 30 = 0$$
$$x = -40 \quad \text{(reject)} \qquad\qquad x = 30 \quad \text{(usual)}$$

Car's usual speed: $\boxed{30 \text{ mph}}$

35. Let
$$x = \text{Bennie's outgoing rate}$$
$$x - 1 = \text{Bennie's return rate}$$
total time: 9 A.M. to 5 P.M. $- 1$ hour rest $= 8 - 1 = 7$ hours

	D (miles)	R	$T = \dfrac{D}{R}$
outgoing	12	x	$\dfrac{12}{x}$
return	12	$x-1$	$\dfrac{12}{x-1}$

Time (outgoing) + time (return) = 7

$$\frac{12}{x} + \frac{12}{x-1} = 7$$

$$x(x-1) \cdot \frac{12}{x} + x(x-1) \cdot \frac{12}{x-1} = x(x-1) \cdot 7$$

$$12(x-1) + 12x = 7(x^2 - x)$$

$$12x - 12 + 12x = 7x^2 - 7x$$

$$-7x^2 + 31x - 12 = 0$$

$$7x^2 - 31x + 12 = 0$$

$$(x - 4)(7x - 3) = 0$$

$$x - 4 = 0 \qquad \text{or} \qquad 7x - 3 = 0$$

$$x = 4 \quad \text{(outgoing)} \qquad\qquad x = \frac{3}{7} \quad \text{(Reject since this will lead to a negative return rate.)}$$

$$x - 1 = 4 - 1 = 3 \quad \text{(return)} \qquad\qquad x - 1 = \frac{3}{7} - 1 = -\frac{4}{7}$$

rate on the return trip: $\boxed{3 \text{ mph}}$

37. Let
$$x = \text{time for Mrs. Lovett working alone}$$
$$x + 4 = \text{time for Mr. Todd working alone}$$

	Fractional part of job completed in 1 hour	time spent on job (days)	Fractional part of job completed
Mrs. Lovett	$\dfrac{1}{x}$	2	$\dfrac{2}{x}$
Mr. Todd	$\dfrac{1}{x+4}$	$2 + 7 = 9$	$\dfrac{9}{x+4}$

(Part of job done by Mrs. Lovett in 2 days) + (Part of job done by Mr. Todd in 9 days) = One complete job

$$\frac{2}{x} + \frac{9}{x+4} = 1$$

$$x(x+4) \cdot \frac{2}{x} + x(x+4) \cdot \frac{9}{x+4} = x(x+4) \cdot 1$$

$$2(x+4) + 9x = x^2 + 4x$$

$$0 = x^2 - 7x - 8$$

$$0 = (x - 8)(x + 1)$$

$$x - 8 = 0 \qquad \text{or} \qquad x + 1 = 0$$

$$x = 8 \qquad\qquad x = -1 \quad \text{(Reject)}$$

Time for Mrs. Lovett working alone: $\boxed{8 \text{ days}}$.

39. Let = number of minutes needed to fill the sink.

	Fractional part of job completed in 1 hour	Time spent working together	Fractional part of job completed in x minutes.
Fill (5 minutes)	$\dfrac{1}{5}$	x	$\dfrac{x}{5}$
Drain ($2 \cdot 5 = 10$ minutes)	$\dfrac{1}{10}$	x	$\dfrac{x}{10}$

(Fraction of sink filled by faucet) – (fraction of sink emptied by drain) = 1

$$\frac{x}{5} - \frac{x}{10} = 1$$
$$10 \cdot \frac{x}{5} - 10 \cdot \frac{x}{10} = 10 \cdot 1$$
$$2x - x = 10$$
$$x = 10$$

$\boxed{10 \text{ minutes}}$

41. $\dfrac{3}{4} \cdot$ (perimeter of first square) = (reciprocal of second square's area) $+ \dfrac{1}{2} \cdot$ (perimeter of third square)

$$\frac{3}{4} \cdot 4 \cdot \frac{1}{4x - 8} = \frac{1}{6 \cdot 6} + \frac{1}{2} \cdot 4 \cdot \frac{1}{3x - 6}$$
$$\frac{3}{4(x - 1)} = \frac{1}{36} + \frac{2}{3(x - 2)}$$
$$36(x - 2) \cdot \frac{3}{4(x - 2)} = 36(x - 2) \cdot \frac{1}{36} + 36(x - 2) \cdot \frac{2}{3(x - 2)}$$
$$27 = x - 2 + 24$$
$$-x = -5$$
$$\boxed{x = 5}$$

43. Let x = number of seagulls in the flock.

$$x + x + \frac{1}{2}x + \frac{1}{4}x + 1 = 100$$
$$4\left(2x + \frac{1}{2}x + \frac{1}{4}x + 1\right) = 4(100)$$
$$8x + 2x + x + 4 = 400$$
$$11x = 396$$
$$x = 36$$

$\boxed{36 \text{ seagulls}}$

45. Let x = cost of the computer.
(cost per share for 5 students) – (cost per share for 8 students) = (Reduction in cost for each of the original 55 students)

$$\frac{x}{5} - \frac{x}{8} = 120$$
$$(\times 40) \qquad 8x - 5x = 4800$$
$$3x = 4800$$
$$x = 1600$$

cost of computer: $\boxed{\$1600}$

47. Let x = time to wash one car.

	Fractional part of job completed in 1 minute	Time spent working together	Fractional part of job completed in x minutes
first person (28 minutes)	$\dfrac{1}{28}$	x	$\dfrac{x}{28}$
second person (23 minutes)	$\dfrac{1}{23}$	x	$\dfrac{x}{23}$

(first person) + (second person) = (1 complete job)

$$\dfrac{x}{28} + \dfrac{x}{23} = 1$$

$(\times 23 \cdot 28) \quad 23x + 28x = 644$

$$51x = 644$$
$$x \approx 12.63 \quad \text{(one car)}$$
$$2x \approx 25.25 \quad \text{(two cars)}$$

Time for both people working together to wash 2 cars: $\boxed{\text{approximately 25 minutes}}$.

49. Let

$$x = \text{first number}$$
$$y = \text{second number}$$

(sum of numbers) = (sum of reciprocals)

$$x + y = \dfrac{1}{x} + \dfrac{1}{y}$$

$$x + y = \dfrac{1}{x} \cdot \dfrac{y}{y} + \dfrac{1}{y} \cdot \dfrac{x}{x} \quad \text{(LCD is } xy \text{ for right side)}$$

$$x + y = \dfrac{y + x}{xy}$$

$$xy(x + y) = x + y$$

$$xy = \dfrac{x + y}{x + y} = 1 \quad \text{(divide both sides by } x + y)$$

$$\text{product} = \boxed{1}$$

51. Let

$$x = \text{train's speed on old schedule}$$
$$x + 2 = \text{train's speed on new schedule}$$

	D (miles)	R	$T = \dfrac{D}{R}$
train on old schedule	351	x	$\dfrac{351}{x}$
train on new schedule	351	$x + 2$	$\dfrac{351}{x + 2}$

Time on new schedule = time on old schedule $- \dfrac{1}{4}$

$$\dfrac{351}{x + 2} = \dfrac{351}{x} - \dfrac{1}{4}$$

$$4x(x + 2)\left[\dfrac{351}{x + 2}\right] = 4x(x + 2)\left[\dfrac{351}{x} - \dfrac{1}{4}\right]$$

$$1404x = 10404x + 2808 - x^2 - 2x$$
$$x^2 + 2x - 2808 = 0$$
$$(x + 54)(x - 52) = 0$$

$$x + 54 = 0 \qquad \text{or} \qquad x - 52 = 0$$
$$x = -54 \quad \text{(Reject)} \qquad\qquad x = 52 \quad \text{(old)}$$
$$\qquad\qquad\qquad\qquad\qquad\qquad x + 2 = 52 + 2 = 54 \quad \text{(new)}$$

Train's speed on new schedule: $\boxed{54 \text{ mph}}$

Review Problems

55. $|6 - 4x| - 4 = 3$

$\qquad\quad |6 - 4x| = 7$

$6 - 4x = 7$	or	$6 - 4x = -7$
$-4x = 1$		$-4x = -13$
$x = -\dfrac{1}{4}$		$x = \dfrac{13}{4}$

$$\boxed{\left\{ -\frac{1}{4}, \frac{13}{4} \right\}}$$

56. $\dfrac{4y + 1}{8y^3 - 1} + \dfrac{2y}{4y^2 + 2y + 1}$

$= \dfrac{4y + 1}{(2y - 1)(4y^2 + 2y + 1)} + \dfrac{2y}{(4y^2 + 2y + 1)} \cdot \dfrac{2y - 1}{2y - 1}$ (LCD is $(2y - 1)(4y^2 + 2y + 1)$)

$= \dfrac{4y + 1 + 4y^2 - 2y}{(2y - 1)(4y^2 + 2y + 1)}$

$= \dfrac{4y^2 + 2y + 1}{(2y - 1)(4y^2 + 2y + 1)}$

$= \boxed{\dfrac{1}{2y - 1}}$

57. $s = 16t^2 + 80t$

($s = 224$ feet):

$16t^2 + 80t = 244$
$16t^2 + 80t - 244 = 0$
$16(t^2 + 5t - 14) = 0$
$16(t + 7)(t - 2) = 0$

$t + 7 = 0$	or	$t - 2 = 0$
$t = -7$ (reject)		$t = 2$

$\boxed{2 \text{ seconds}}$

Chapter 6   Review Problems

Review Problems, pp. 488-490

1. $f(x) = \dfrac{7x}{9x - 18}$

$f(x)$ is undefined when

$\qquad 9x - 18 = 0$

$\qquad\qquad x = 2$

$\boxed{\text{Domain of } f = \{ x \mid x \neq 2 \} \text{ or } (-\infty, 2) \cup (2, \infty)}$

2. $f(x) = \dfrac{x + 3}{(x - 1)(x + 5)}$

$f(x)$ is undefined when

$(x - 1)(x + 5) = 0$

$x - 1 = 0$	or	$x + 5 = 0$
$x = 1$		$x = -5$

$\boxed{\text{Domain of } f = \{ x \mid x \neq -5, 1 \} \text{ or } (-\infty, -5) \cup (-5, 1) \cup (1, \infty)}$

3. $f(x) = \dfrac{7x + 14}{2x^2 + 5x - 3}$

$f(x)$ is undefined when

$$2x^2 - 5x - 3 = 0$$
$$(2x - 1)(x + 3) = 0$$

$2x - 1 = 0$ or $x + 3 = 0$

$x = \dfrac{1}{2}$ $x = -3$

Domain of $f = \left\{ x \mid x \neq -3, \dfrac{1}{2} \right\}$ or $(-\infty, -3) \cup \left(-3, \dfrac{1}{2}\right) \cup \left(\dfrac{1}{2}, \infty\right)$

4. $f(x) = \dfrac{x^2 - 25}{x^2 + 4}$

$f(x)$ is undefined when

$$x^2 + 4 = 0$$

No real number substituted into the denominator will cause $x^2 + 4$ to equal zero. No values of x need to be excluded. The domain of f is the set of all real numbers.

Domain of $f = \{ x \mid x \in R \}$ or $(-\infty, \infty)$

5. $f(x) = \dfrac{1}{x + 2}$

a. $f(x)$ is undefined when

$$x + 2 = 0$$
$$x = -2$$

$D_f = \{ x \mid x \neq -2 \}$ or $(-\infty, -2) \cup (-2, \infty)$

b.

x	-3	-2.5	-2.1	-2.01	-2.001
$f(x) = \dfrac{1}{x + 2}$	-1	-2	-10	-100	-1000

x	-1.999	-1.99	-1.9	-1.5	1
$f(x) = \dfrac{1}{x + 2}$	1000	100	10	2	1

c. $x = -2$ (dotted)

$f(x) = \dfrac{1}{x + 2}$

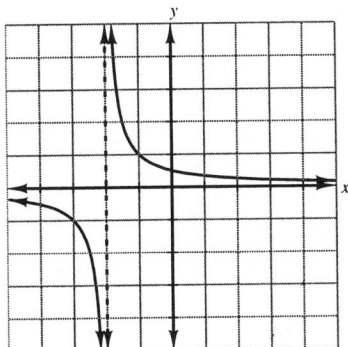

6. $f(x) = \dfrac{1}{x^2 - 4}$

$f(x)$ is undefined when

$$x^2 - 4 = 0$$
$$x = \pm 2$$

$\boxed{D_f = \{x \mid x \neq -2, 2\} \text{ or } (-\infty, -2) \cup (-2, 2) \cup (2, \infty)}$

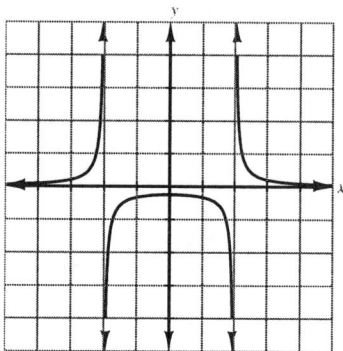

7. $f(x) = \dfrac{-2}{x - 1}$

$f(0) = \dfrac{-2}{-1} = 2 \qquad f(2) = \dfrac{-2}{2-1} = -2$

8. $\dfrac{2x - 3xy}{9y^2 - 4} = \dfrac{x(2 - 3y)}{(3y + 2)(3y - 2)} = \boxed{\dfrac{-x}{3y + 2}}$

9. $\dfrac{x^2 + 6x - 7}{x^2 - 49} = \dfrac{(x + 7)(x - 1)}{(x + 7)(x - 7)} = \boxed{\dfrac{x - 1}{x - 7}}$

10. $\dfrac{6m^2 + 7m + 2}{2m^2 - 9m - 5} = \dfrac{(3m + 2)(2m + 1)}{(2m + 1)(m - 5)} = \boxed{\dfrac{3m + 2}{m - 5}}$

11. $\dfrac{x^{2n} - x^n - 2}{x^{2n} + 3x^n + 2} = \dfrac{(x^n - 2)(x^n + 1)}{(x^n + 1)(x^n - 5)} = \boxed{\dfrac{x^n - 2}{x^n + 2}}$

12. $\dfrac{y^3 - 8}{y^2 - 4} = \dfrac{(y - 2)(y^2 + 2y + 4)}{(y + 2)(y - 2)} = \boxed{\dfrac{y^2 + 2y + 4}{y + 2}}$

13. $f(x) = \dfrac{x^2 - 7x + 12}{x - 4} = \dfrac{(x - 4)(x - 3)}{x - 4} = \boxed{x - 3 \ (x \neq 4)}$

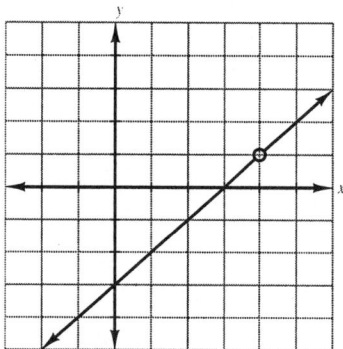

14. $\dfrac{x^2 - 9x + 14}{x^3 + 2x^2} \cdot \dfrac{x^2 - 4}{(x - 2)^2} = \dfrac{(x - 2)(x - 7)}{x^2(x + 2)} \cdot \dfrac{(x + 2)(x - 2)}{(x - 2)^2} = \boxed{\dfrac{x - 7}{x^2}}$

15. $\dfrac{5xy - 10y^2}{x^2 - 3xy + 2y^2} \cdot \dfrac{x + 2y}{x^2} \cdot \dfrac{3x^2 - 3xy}{xy + 2y^2} = \dfrac{5y(x - 2y)}{(x - 2y)(x - y)} \cdot \dfrac{x + 2y}{x^2} \cdot \dfrac{3x(x - y)}{y(x + 2y)} = \boxed{\dfrac{15}{x}}$

16. $\dfrac{y^4 - 81}{y^2 + 9} \cdot \dfrac{4y - 20}{y^2 - 8y + 15}) = \dfrac{(y^2 + 9)(y + 3)(y - 3)}{y^2 + 9} \cdot \dfrac{4(y - 5)}{(y - 5)(y - 3)} = \boxed{4(y + 3)}$

17. $\dfrac{y^{2n} + y^n - 12}{y^{2n} - 1} \cdot \dfrac{y^{2n} + 3y^n + 2}{y^{2n} - y^n - 6} = \dfrac{(y^n + 4)(y^n - 3)}{(y^n + 1)(y^n - 1)} \cdot \dfrac{(y^n + 1)(y^n + 2)}{(y^n - 3)(y^n + 2)} = \boxed{\dfrac{y^n + 4}{y^n - 1}}$

18. $\dfrac{x^2 + 16x + 64}{2x^2 - 128} \div \dfrac{3x^2 + 30x + 48}{x^2 - 6x - 16}$

$= \dfrac{x^2 + 16x + 64}{2x^2 - 128} \cdot \dfrac{x^2 - 6x - 16}{3x^2 + 30x + 48}$

$= \dfrac{(x + 8)^2}{2(x + 8)(x - 8)} \cdot \dfrac{(x - 8)(x + 8)}{3(x + 8)(x + 2)} = \boxed{\dfrac{1}{6}}$

19. $\dfrac{a^3 - 27}{a^2 + 3a + 9} \div (ab + ac - 3b - 3c)$

$= \dfrac{a^3 - 27}{a^2 + 3a + 9} \cdot \dfrac{1}{ab + ac - 3b - 3c}$

$= \dfrac{(a - 3)(a^2 + 3a + 9)}{a^2 + 3a + 9} \cdot \dfrac{1}{(b + c)(a - 3)} = \boxed{\dfrac{1}{b + c}}$

20. $\dfrac{y^3 - 8}{y^4 - 16} \div \dfrac{y^3 + 2y + 4}{y^2 + 4} = \dfrac{y^3 - 8}{y^4 - 16} \cdot \dfrac{y^2 + 4}{y^2 + 2y + 4}$

$= \dfrac{(y - 2)(y^2 + 2y + 4)}{(y^2 + 4)(y + 2)(y - 2)} \cdot \dfrac{y^2 + 4}{y^2 + 2y + 4}$

$= \boxed{\dfrac{1}{y + 2}}$

21. $\dfrac{2x - 7}{x^2 - 9} - \dfrac{x - 4}{x^2 - 9} = \dfrac{2x - 7 - (x - 4)}{x^2 - 9}$

$= \dfrac{x - 3}{x^2 - 9} = \dfrac{x - 3}{(x - 3)(x + 3)} = \boxed{\dfrac{1}{x + 3}}$

22. $\dfrac{1}{x} + \dfrac{2}{x - 5}$ (LCD is $x(x - 5)$)

$= \dfrac{1}{x} \cdot \dfrac{x - 5}{x - 5} + \dfrac{2}{x - 5} \cdot \dfrac{x}{x}$

$= \dfrac{x - 5 + 2x}{x(x - 5)} = \boxed{\dfrac{3x - 5}{x(x - 5)}}$

23. $\dfrac{3a^2}{9a^2 - 16b^2} - \dfrac{a}{3a + 4b}$

$= \dfrac{3a^2}{(3a + 4b)(3a - 4b)} - \dfrac{a}{3a + 4b}$ (LCD is $(3a + 4b)(3a - 4b)$)

$= \dfrac{3a^2}{(3a + 4b)(3a - 4b)} - \dfrac{a}{3a + 4b} \cdot \dfrac{3a - 4b}{3a - 4b}$

$= \dfrac{3a^2 - a(3a - 4b)}{(3a + 4b)(3a - 4b)}$

$= \dfrac{3a^2 - 3a^2 + 4ab}{(3a + 4b)(3a - 4b)}$

$= \boxed{\dfrac{4ab}{(3a + 4b)(3a - 4b)}}$

25. $\dfrac{x}{x+3} + \dfrac{x}{x-3} - \dfrac{9}{x^2-9}$

$= \dfrac{x}{x+3} + \dfrac{x}{x-3} - \dfrac{9}{(x+3)(x-3)}$ (LCD is $(x+3)(x-3)$)

$= \dfrac{x}{x+3} \cdot \dfrac{x-3}{x-3} + \dfrac{x}{x-3} \cdot \dfrac{x+3}{x+3} - \dfrac{9}{(x+3)(x-3)}$

$= \dfrac{x(x-3) + x(x+3) - 9}{(x+3)(x-3)}$

$= \dfrac{x^2 - 3x + x^2 + 3x - 9}{(x+3)(x-3)}$

$= \boxed{\dfrac{2x^2 - 9}{(x+3)(x-3)}}$

26. $= \dfrac{4}{a^2 + a - 2} - \dfrac{2}{a^2 - 4} + \dfrac{3}{a^2 - 4a + 4}$

$= \dfrac{4}{(a+2)(a-1)} - \dfrac{2}{(a+2)(a-2)} + \dfrac{3}{(a-2)^2}$ (LCD is $(a-2)^2(a+2)(a-1)$)

$= \dfrac{4}{(a+2)(a-1)} \cdot \dfrac{(a-2)^2}{(a-2)^2} - \dfrac{2}{(a+2)(a-2)} \cdot \dfrac{(a-2)(a-1)}{(a-2)(a-1)} + \dfrac{3}{(a-2)^2} \cdot \dfrac{(a+2)(a-1)}{(a+2)(a-1)}$

$= \dfrac{4(a-2)^2 - 2(a-2)(a-1) + 3(a+2)(a-1)}{(a-2)^2(a+2)(a-1)}$

$= \dfrac{4a^2 - 16a + 16 - 2a^2 + 6a - 4 + 3a^2 + 3a - 6}{(a-2)^2(a+2)(a-1)}$

$= \boxed{\dfrac{5a^2 - 7a + 6}{(a-2)^2(a+2)(a-1)}}$

27. $\dfrac{3b^2}{b^2 + 5b + 6} \cdot \dfrac{b+3}{b} + \dfrac{b-9}{6b}$

$= \dfrac{3b^2}{(b+3)(b+2)} \cdot \dfrac{b+3}{b} + \dfrac{b-9}{6b}$

$= \dfrac{3b}{b+2} + \dfrac{b-9}{6b}$ (LCD is $6b(b+2)$)

$= \dfrac{3b}{b+2} \cdot \dfrac{6b}{6b} + \dfrac{b-9}{6b} \cdot \dfrac{b+2}{b+2}$

$= \dfrac{(3b)(6b) + (b-9)(b+2)}{6b(b+2)}$

$= \dfrac{18b^2 + b^2 - 7b - 18}{6b(b+2)} = \boxed{\dfrac{19b^2 - 7b - 18}{6b(b+2)}}$

28. $\left(\dfrac{1}{x+2} + \dfrac{1}{x+4} \right) \cdot (x^2 + 6x + 8)$

$= \left(\dfrac{1}{x+2} \cdot \dfrac{x+4}{x+4} + \dfrac{1}{x+4} \cdot \dfrac{x+2}{x+2} \right) \cdot (x^2 + 6x + 8)$

$= \dfrac{2x+6}{(x+2)(x+4)} \cdot (x+2)(x+4) = \boxed{2x+6}$

29. $\dfrac{\dfrac{b}{x} - \dfrac{b}{y}}{\dfrac{b}{x} + \dfrac{b}{y}} = \dfrac{\dfrac{b}{x} - \dfrac{b}{y}}{\dfrac{b}{x} + \dfrac{b}{y}} \cdot \dfrac{xy}{xy}$

$= \dfrac{by - bx}{by + bx} = \dfrac{b(y-x)}{b(y+x)}$

$= \boxed{\dfrac{y-x}{y+x}}$

30. $\dfrac{3-\dfrac{1}{x+3}}{3+\dfrac{1}{x+3}}=\dfrac{3-\dfrac{1}{x+3}}{3+\dfrac{1}{x+3}}\cdot\dfrac{x+3}{x+3}$

$=\dfrac{3(x+3)-1}{3(x+3)+1}=\dfrac{3x+9-1}{3x+9+1}$

$=\boxed{\dfrac{3x+8}{3x+10}}$

31. $\dfrac{\dfrac{1}{y+5}+1}{\dfrac{6}{y^2+2y-15}-\dfrac{1}{y-3}}$

$=\dfrac{\dfrac{1}{y+5}+1}{\dfrac{6}{(y+5)(y-3)}-\dfrac{1}{y-3}}\cdot\dfrac{(y+5)(y-3)}{(y+5)(y-3)}$

$=\dfrac{y-3+(y+5)(y-3)}{6-(y+5)}$

$=\dfrac{y-3+y^2+2y-15}{6-y-5}$

$=\dfrac{y^2+3y-18}{1-y}=\boxed{\dfrac{(y+6)(y-3)}{1-y}}$

32. $\dfrac{4-\dfrac{1}{y^2}}{4+\dfrac{4}{y}+\dfrac{1}{y^2}}=\dfrac{4-\dfrac{1}{y^2}}{4+\dfrac{4}{y}+\dfrac{1}{y^2}}\cdot\dfrac{y^2}{y^2}=\dfrac{4y^2-1}{4y^2+4y+1}$

$=\dfrac{(2y+1)(2y-1)}{(2y+1)^2}=\boxed{\dfrac{2y-1}{2y+1}}$

33. $1+\dfrac{y}{1+\dfrac{1}{y}}=1+\dfrac{y}{1+\dfrac{1}{y}}\cdot\dfrac{y}{y}=1+\dfrac{y^2}{y+1}$

$=1\cdot\dfrac{y+1}{y+1}+\dfrac{y^2}{y+1}=\boxed{\dfrac{y^2+y+1}{y+1}}$

34. $\dfrac{\dfrac{4}{16x^2-1}+\dfrac{3}{4x+1}}{\dfrac{x}{4x-1}+\dfrac{5}{4x+1}}$

$=\dfrac{(4x+1)(4x-1)}{(4x+1)(4x-1)}\cdot\dfrac{\dfrac{4}{(4x+1)(4x-1)}+\dfrac{3}{4x+1}}{\dfrac{x}{4x-1}+\dfrac{5}{4x+1}}$

$=\dfrac{4+3(4x-1)}{x(4x+1)+5(4x-1)}$

$=\dfrac{12x+1}{4x^2+x+20x-5}$

$=\boxed{\dfrac{12x+1}{4x^2+21x-5}}$

For 35-39, $f(x) = \dfrac{4}{x^2-9}$, $g(x) = \dfrac{1}{x-2}$, $h(x) = \dfrac{x}{x-3}$.

35.
$$
\begin{aligned}
(g+f)(x) &= \frac{1}{x-2} + \frac{4}{x^2-9} \\
&= \frac{1}{x-2} \cdot \frac{x^2-9}{x^2-9} + \frac{4}{x^2-9} \cdot \frac{x-2}{x-2} \quad \text{(LCD is } (x-2)(x^2-9)) \\
&= \frac{x^2-9+4x-9}{(x-2)(x^2-9)} \\
&= \boxed{\frac{x^2+4x-17}{(x-2)(x^2-9)}}
\end{aligned}
$$

$(g+f)(x)$ is undefined when
$(x-2)(x+3)(x-3) = 0$

$x-2 = 0$ or $x+3 = 0$ or $x-3 = 0$
$x = 2$ $x = -3$ $x = 3$

$\boxed{D_{g+f} = \{x \mid x \neq -3, 2, 3\} \text{ or } (-\infty, -3) \cup (-3, 2) \cup (2, 3) \cup (3, \infty)}$

36.
$$
\begin{aligned}
(h-f)(x) &= \frac{x}{x-3} - \frac{4}{x^2-9} \\
&= \frac{x}{x-3} \cdot \frac{x+3}{x+3} - \frac{4}{(x-3)(x+3)} \quad \text{(LCD is } (x-3)(x+3)) \\
&= \frac{x^2+3x-3}{(x-3)(x+3)} \\
&= \boxed{\frac{(x+4)(x-1)}{(x-3)(x+3)}}
\end{aligned}
$$

$(h-f)(x)$ is undefined when
$(x-3)(x+3) = 0$
$x = 3, -3$

$\boxed{D_{h-f} = \{x \mid x \neq -3, 3\} \text{ or } (-\infty, -3) \cup (-3, 3) \cup (3, \infty)}$

37.
$$
\begin{aligned}
(h-g)(x) &= \frac{x}{x-3} - \frac{1}{x-2} \\
&= \frac{x}{x-3} \cdot \frac{x-3}{x-2} - \frac{4}{x-2} \cdot \frac{x-3}{x-3} \quad \text{(LCD is } (x-2)(x-3)) \\
&= \frac{x^2-2x-x+3}{(x-2)(x-3)} \\
&= \boxed{\frac{x^2-3x+3}{(x-2)(x-3)}}
\end{aligned}
$$

$(h-g)(x)$ is undefined when
$(x-2)(x-3) = 0$
$x = 2, 3$

$\boxed{D_{h-g} = \{x \mid x \neq 2, 3\} \text{ or } (-\infty, 2) \cup (2, 3) \cup (3, \infty)}$

38.
$$
\begin{aligned}
(gh)(x) &= \frac{1}{x-2} \cdot \frac{x}{x-3} \\
&= \boxed{\frac{x}{(x-2)(x-3)}}
\end{aligned}
$$

$(gh)(x)$ is undefined when
$(x-2)(x-3) = 0$
$x = 2, 3$

$\boxed{D_{gh} = \{x \mid x \neq 2, 3\} \text{ or } (-\infty, 2) \cup (2, 3) \cup (3, \infty)}$

39.

$$\left(\frac{h}{f}\right)(x) = \left(\frac{x}{x-3}\right) \div \left(\frac{4}{x^2-9}\right)$$

$$= \frac{x}{x-3} \cdot \frac{(x-3)(x+3)}{4}$$

$$= \boxed{\frac{x(x+3)}{4}}$$

$\left(\dfrac{h}{f}\right)(x)$ is undefined when

$$\begin{array}{ccccccc} x-3 & = & 0 & \quad \text{and} \quad & x^2-9 & = & 0 \\ x & = & 3 & & x & = & -3, 3 \end{array}$$

$$\boxed{D_{h/f} = \{x \mid x \neq -3, 3\} \ \text{ or } \ (-\infty, -3) \cup (-3, 3) \cup (3, \infty)}$$

40. $f(x) = -\dfrac{5}{x}$

$$= \frac{-\dfrac{5}{a+h} - \left(-\dfrac{5}{a}\right)}{h}$$

$$= \left[-\frac{5}{a+h} \cdot \frac{a}{a} + \frac{5}{a} \cdot \frac{a+h}{a+h}\right] \div h$$

$$= \frac{-5a + 5a + 5h}{a(a+h)} \cdot \frac{1}{h}$$

$$= \frac{5h}{a(a+h)h}$$

$$= \boxed{\frac{5}{a(a+h)}}$$

41. $f(x) = \dfrac{2 + \dfrac{1}{x}}{2 - \dfrac{1}{x}}$ and $g(x) = x + 4$

$$f \circ g = f[g(x)] = \frac{2 + \dfrac{1}{x+4}}{2 - \dfrac{1}{x+4}}$$

$$= \left[2 \cdot \frac{x+4}{x+4} + \frac{1}{x+4}\right] \div \left[2 \cdot \frac{x+4}{x+4} - \frac{1}{x+4}\right]$$

$$= \frac{2x+8+1}{x+4} \div \frac{2x+8-1}{x+4}$$

$$= \frac{2x+9}{x+4} \cdot \frac{x+4}{2x+7}$$

$$= \boxed{\frac{2x+9}{2x+7}}$$

$f \circ g$ is undefined when

$$\begin{array}{ccccccc} x+4 & = & 0 & \quad \text{or} \quad & 2x+7 & = & 0 \\ x & = & -4 & & x & = & -\dfrac{7}{2} \end{array}$$

$$\boxed{D_{f \circ g} = \left\{x \mid x \neq -4, -\frac{7}{2}\right\} \ \text{ or } (-\infty, -4) \cup \left(-4, -\frac{7}{2}\right) \cup \left(-\frac{7}{2}, \infty\right)}$$

42. $f(x) = 1 - \dfrac{1}{1 - \dfrac{1}{x}}$ and $g(x) = x - 3$

$$(f \circ g)(x) = f[g(x)] \quad = \quad 1 - \dfrac{1}{1 - \dfrac{1}{x-3}}$$

$$= \quad 1 - \dfrac{1}{\dfrac{x-3}{x-3} - \dfrac{1}{x-3}}$$

$$= \quad 1 - \dfrac{1}{\dfrac{x-4}{x-3}}$$

$$= \quad 1 - \dfrac{x-3}{x-4}$$

$$= \quad \dfrac{x-4}{x-4} - \dfrac{x-3}{x-4}$$

$$= \quad \dfrac{x-4-x-3}{x-4} = \boxed{\dfrac{-1}{x-4} \quad x \neq 3}$$

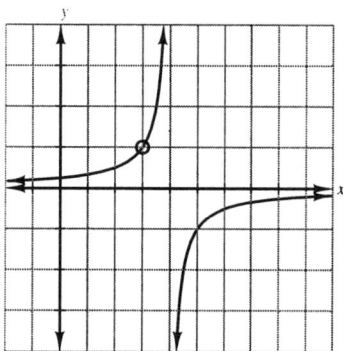

43. $f(x) = \dfrac{x}{x^2 - 2x - 3}$ and $g(x) = \dfrac{3}{x^2 - 2x - 3}$

$$(f - g)(x) \quad = \quad \dfrac{x}{x^2 - 2x - 3} - \dfrac{3}{x^2 - 2x - 3}$$

$$= \quad \dfrac{x-3}{(x-3)(x+1)}$$

$$= \quad \boxed{\dfrac{1}{x+1} \quad (x \neq 3)}$$

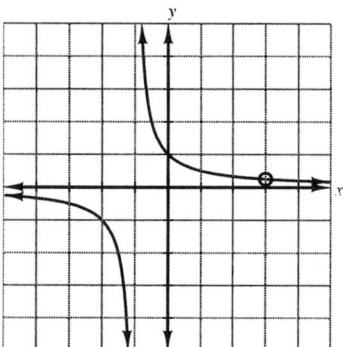

44.
$$\frac{2y}{y-2} - 3 = \frac{4}{y-2}$$
$$(y-2)\left[\frac{2y}{y-2}\right] = (y-2)\frac{4}{y-2}$$
$$2y - 3(y-2) = 4$$
$$2y - 3y + 6 = 4$$
$$-y + 6 = 4$$
$$2 = y$$

Since 2 causes a denominator to be 0 in the original equation, 2 is not a solution. The equation has no solution. $\boxed{\varnothing}$

45.
$$\frac{1}{y-5} - \frac{3}{y+5} = \frac{6}{y^2 - 25}$$
$$\frac{1}{y-5} - \frac{3}{y+5} = \frac{6}{(y+5)(y-5)}$$
$$(y+5)(y-5)\left[\frac{1}{y-5} - \frac{3}{y+5}\right] = (y+5)(y-5) \cdot \frac{6}{(y+5)(y-5)}$$
$$y + 5 - 3(y-5) = 6$$
$$y + 5 - 3y + 15 = 6$$
$$-2y + 20 = 6$$
$$-2y = -14$$
$$y = 7$$

$\boxed{\{7\}}$

46.
$$\frac{x+5}{x+4} - \frac{x}{x+2} = \frac{4x+1}{x^2 + 3x + 2}$$
$$\frac{x+5}{x+1} - \frac{x}{x+2} = \frac{4x+1}{(x+1)(x+2)}$$
$$(x+1)(x+2)\left[\frac{x+5}{x+1} - \frac{x}{x+2}\right] = (x+1)(x+2) \cdot \frac{4x+1}{(x+1)(x+2)}$$
$$(x+5)(x+2) - x(x+1) = 4x+1$$
$$x^2 + 7x + 10 - x^2 - x = 4x+1$$
$$6x + 10 = 4x+1$$
$$2x = -9$$
$$x = -\frac{9}{2}$$

$\boxed{\left\{-\frac{9}{2}\right\}}$

47.
$$\frac{2}{3} - \frac{5}{3y} = \frac{1}{y^2}$$
$$3y^2 \cdot \frac{2}{3} - 3y^2 \cdot \frac{5}{3y} = 3y^2 \cdot \frac{1}{y^2}$$
$$2y^2 - 5y = 3$$
$$2y^2 - 5y - 3 = 0$$
$$(2y+1)(y-3) = 0$$
$$2y + 1 = 0 \qquad\text{or}\qquad y - 3 = 0$$
$$y = -\frac{1}{2} \qquad\qquad\qquad y = 3$$

$\boxed{\left\{-\frac{1}{2}, 3\right\}}$

48.
$$\frac{2}{y-1} = \frac{1}{4} + \frac{7}{y+2}$$

$$4(y-1)(y+2) \cdot \frac{2}{y-1} = 4(y-1)(y+2)\left[\frac{1}{4} + \frac{7}{y+2}\right]$$

$$8y+16 = y^2 + y - 2 + 28y - 28$$

$$0 = y^2 + 21y - 46$$

$$0 = (y+23)(y-2)$$

$$y + 23 = 0 \qquad \text{or} \qquad y - 2 = 0$$

$$y = -23 \qquad\qquad\qquad y = 2$$

$$\boxed{\{-23, 2\}}$$

49.
$$\frac{2y+7}{y+5} - \frac{y-8}{y-4} = \frac{y+18}{y^2+y-20}$$

$$(y+5)(y-4)\left[\frac{2y+7}{y+5} - \frac{y-8}{y-4}\right] = (y+5)(y-4) \cdot \frac{y+18}{(y+5)(y-4)}$$

$$(y-4)(2y+7) - (y+5)(y-8) = y + 18$$

$$2y^2 - y - 28 - y^2 + 3y + 40 = y + 18$$

$$y^2 + y - 6 = 0$$

$$(y+3)(y-2) = 0$$

$$y+3 = 0 \qquad \text{or} \qquad y - 2 = 0$$

$$y = -3 \qquad\qquad\qquad y = 2$$

$$\boxed{\{-3, 2\}}$$

50.
$$P = 30 - \frac{9}{t+1}$$

$P = 29$ (since P is in thousands):

$$29 = 30 - \frac{9}{t+1}$$

$$(t+1)29 = (t+1)\left[30 - \frac{9}{t+1}\right]$$

$$29t + 29 = 30t + 30 - 9$$

$$29t + 29 = 30t + 21$$

$$1985 + 9 = 1993$$

The population of 29,000 will occur 8 years from 1985, in $\boxed{1993}$.

51.
$$P = \frac{R-C}{n}$$

$$nP = n\left(\frac{R-C}{n}\right)$$

$$Pn = R - C$$

$$\boxed{C = R - Pn}$$

52.
$$T = \frac{A-p}{pr}$$

$$prT = pr\left(\frac{A-P}{pr}\right)$$

$$prT = A - p$$

$$prT + p = A$$

$$p(rT+1) = A$$

$$\boxed{p = \frac{A}{rT+1}}$$

53.
$$A = \frac{rs}{r+s}$$
$$(r+s)A = (r+s)\frac{rs}{r+s}$$
$$Ar + As = rs$$
$$As = rs - Ar$$
$$As = r(s-A)$$
$$\frac{As}{s-A} = r$$
$$\boxed{r = \frac{As}{s-A}}$$

54. Let
$$x = \text{lesser number}$$
$$6x - 4 = \text{greater number}$$

$$\frac{6x-4}{x} = 5 + \frac{1}{x}$$
$$x\left(\frac{6x-4}{x}\right) = x\left(5+\frac{1}{x}\right)$$
$$6x - 4 = 5x + 1$$
$$x = 5 \quad \text{(lesser)}$$
$$6x - 4 = 6(5) - 4 = 26 \quad \text{(greater)}$$
The numbers are $\boxed{5 \text{ and } 26}$.

55. Let x and $x + 1$ equal two consecutive integers.
$$\frac{1}{x} + \frac{1}{x+1} = 11 \cdot \frac{1}{x(x+1)}$$
$$x(x+1) \cdot \left[\frac{1}{x}+\frac{1}{x+1}\right] = x(x+1) \cdot \frac{11}{x(x+1)}$$
$$x + 1 + x = 11$$
$$2x = 10$$
$$x = 5$$
$$x + 1 = 5 + 1 = 6$$
The integers are $\boxed{5 \text{ and } 6}$.

56. Let $x = $ number.
$$\frac{1}{x} - 2 = 5$$
$$x\left(\frac{1}{x}-2\right) = x \cdot 5$$
$$1 - 2x = 5x$$
$$1 = 7x$$
$$\frac{1}{7} = x$$
The number is $\boxed{\dfrac{1}{7}}$.

57. Let

$$x = \text{wind speed}$$
$$320 + x = \text{speed of plane with wind}$$
$$320 - x = \text{speed of plane against the wind}$$

	D (miles)	R	$T = \dfrac{D}{R}$
with wind	1400	$320 + x$	$\dfrac{1400}{320 + x}$
against wind	1160	$320 - x$	$\dfrac{1160}{320 - 6}$

Time plane travels 1400 miles with wind = Time plane travels 1160 against the wind

$$\frac{1400}{320 + x} = \frac{1160}{320 - x}$$
$$(320 + x)(320 - x) \cdot \frac{1400}{320 + x} = (320 + x)(320 - x) \cdot \frac{1160}{320 - x}$$
$$1400(320 - x) = 1160(320 + x)$$
$$448{,}000 - 1400x = 371{,}200 + 1160x$$
$$76{,}800 = 2560x$$
$$30 = x$$

The wind speed is $\boxed{30 \text{ mph}}$.

58. Let

$$x = \text{walking rate}$$
$$3x = \text{cycling rate}$$
$$\text{total time} = 7 \text{ hours}$$

	D (miles)	R	$T = \dfrac{D}{R}$
cycling	8	x	$\dfrac{8}{x}$
walking	60	$3x$	$\dfrac{60}{3x}$

Time spent cycling + time spent walking = 7 hours

$$\frac{60}{3x} + \frac{8}{x} = 7$$
$$\frac{20}{x} + \frac{8}{x} = 7$$
$$x\left(\frac{20}{x} + \frac{8}{x}\right) = x \cdot 7$$
$$20 + 8 = 7x$$
$$28 = 7x$$
$$4 = x$$
$$x = 4 \quad \text{(walking)}$$
$$3x = 3(4) = 12 \quad \text{(cycling)}$$

Cycling rate: $\boxed{12 \text{ mph}}$

59.
$$\frac{1}{R_T} = \frac{1}{R_1} + \frac{1}{R_2} + \frac{1}{R_3}$$
$$R_2 = 5R_1$$
$$R_3 = R_1 + 3$$
$$R_T = 2 \text{ ohms}$$
$$\frac{1}{2} = \frac{1}{R_1} + \frac{1}{5R_2} + \frac{1}{R_1 + 1}$$
$$10R_1(R_1 + 1) \cdot \frac{1}{2} = 10R_1(R_1 + 1)\left[\frac{1}{R_1} + \frac{1}{5R_1} + \frac{1}{R_1 + 1}\right]$$
$$5R_1{}^2 + 5R_1 = 10R_1 + 10 + 2R_1 + 2 + 10R_1$$
$$5R_1 - 17R_1 - 12 = 0$$

$$5R_1 + 3 = 0 \qquad\qquad \text{or} \qquad\qquad R_1 - 4 = 0$$
$$R_1 = -\frac{3}{4} \quad \text{(Reject)} \qquad\qquad\qquad R_1 = 4$$
$$\qquad\qquad R_2 = 5R_1 = 5(4) = 20$$
$$\qquad\qquad R_3 = R_1 + 3 = 4 + 3 = 7$$

$$\boxed{R_1 = 4 \text{ ohms}, R_2 = 20 \text{ ohms}, R_3 = 5 \text{ ohms}}$$

60. Let x = the number.
$$6x\left[x + \frac{1}{x}\right] = 6x \cdot \frac{13}{6}$$
$$6x^2 - 6 = 13x$$
$$6x^2 - 13x + 6 = 0$$
$$(2x - 3)(3x - 2) = 0$$
$$2x - 3 = 0 \qquad\qquad \text{or} \qquad\qquad 3x - 2 = 0$$
$$x = \frac{3}{2} \qquad\qquad\qquad\qquad x = \frac{2}{3}$$

The number is $\boxed{\dfrac{3}{2} \text{ or } \dfrac{2}{3}}$.

61. Let
$$x = \text{rate of boat in still water}$$
$$x - 3 = \text{rate of boat against current}$$
$$x + 3 = \text{rate of boat with current}$$
rate of stream: 3 mph
total time: 3 hours

	D (miles)	R	$T = \dfrac{D}{R}$
against current	12	$x - 3$	$\dfrac{12}{x - 3}$
with current	12	$x + 3$	$\dfrac{12}{x + 3}$

(time upstream) + (time downstream) = 3
$$\frac{12}{x - 3} + \frac{12}{x + 3} = 3$$
$$(x - 3)(x + 3)\left[\frac{12}{x - 3} + \frac{12}{x + 3}\right] = (x - 3)(x + 3) \cdot 3$$
$$12x + 36 + 12x - 36 = 3x^2 - 27$$
$$-3x^2 + 24x + 27 = 0$$
$$-3(x^2 - 8x - 9) = 0$$
$$-3(x - 9)(x + 1) = 0$$
$$x - 9 = 0 \qquad\qquad \text{or} \qquad\qquad x + 1 = 0$$
$$x = 9 \qquad\qquad\qquad\qquad x = -1 \quad \text{(reject)}$$

boat's rate in still water: $\boxed{9 \text{ mph}}$

62. Let

$$x = \text{normal speed of train}$$
$$x - 10 = \text{decreased speed of train}$$

	D (miles)	R	$T = \dfrac{D}{R}$
normal	300	x	$\dfrac{300}{x}$
decreased	300	$x - 10$	$\dfrac{300}{x - 10}$

(time at decreased speed) = (time at normal speed + 1 hour)

$$\frac{300}{x - 10} = \frac{300}{x} + 1$$
$$x(x - 10) \cdot \frac{300}{x - 10} = x(x - 10)\left[\frac{300}{x} + 1\right]$$
$$300x = 300x - 3000 + x^2 - 10x$$
$$0 = x^2 - 10x - 3000$$
$$0 = (x - 60)(x + 50)$$
$$x - 60 = 0 \qquad \text{or} \qquad x + 50 = 0$$
$$x = 60 \qquad\qquad\qquad x = -50 \quad \text{(Reject)}$$

normal rate of travel: $\boxed{60 \text{ mph}}$

63. moving sidewalk: 1 meter/second
Let

$$x = \text{rate of second person on stationary walkway}$$
$$x + 1 = \text{rate of first person on moving walkway}$$

	D (miles)	R	$T = \dfrac{D}{R}$
stationary	300	x	$\dfrac{300}{x}$
moving	300	$x + 1$	$\dfrac{300}{x + 1}$

(time of first person on stationary walkway) – (time of second person on moving walkway) = 50 seconds

$$\frac{300}{x} - \frac{300}{x + 1} = 50$$
$$\frac{x(x + 1)}{50}\left[\frac{300}{x} - \frac{300}{x + 1}\right] = \frac{x(x + 1)}{50} \cdot 50$$
$$6x + 6 - 6x = x^2 + x$$
$$0 = x^2 + x - 6$$
$$x + 3 = 0 \qquad \text{or} \qquad x - 2 = 0$$
$$x = -3 \quad \text{(reject)} \qquad\qquad x = 2 \quad \text{(stationary)}$$
$$x + 1 = 2 + 1 = 3 \quad \text{(moving)}$$

walking speeds: $\boxed{2 \text{ meters/second and 3 meters/second}}$

64. Let

$$x = \text{number of days for Norman's mother to do the job alone}$$
$$x - 9 = \text{number of days for Norman to do the job alone}$$

total time to do the job together: 20 days

	Fractional part of job completed in one day	Time spent working together	Fractional part of job completed
stationary	300	x	$\dfrac{300}{x}$
moving	300	$x + 1$	$\dfrac{300}{x + 1}$

$$(\text{Norman}) + (\text{mother}) = (1 \text{ complete job})$$

$$\frac{20}{x-9} + \frac{20}{x} = 1$$

$$x(x-9)\left[\frac{20}{x-9} + \frac{20}{x}\right] = x(x-9) \cdot 1$$

$$20x + 20x - 180 = x^2 - 9x$$

$$0 = x^2 - 49x + 180$$

$$0 = (x - 45)(x - 4)$$

$$\begin{array}{lll} x - 45 = 0 & \text{or} & x - 4 = 0 \\ \quad x = 45 \quad \text{(mother)} & & \quad x = 4 \quad \text{(reject since } x - 9 \text{ is negative)} \\ x - 9 = 45 - 9 = 36 \quad \text{(Norman)} & & \end{array}$$

mother: 45 days; Norman: 36 days

65. Let $x =$ the time to fill the pool.

	Fractional part of job completed in 1 hour	Time spent workng together	Fractional part of job completed in x hours
Pipe A (8 hours)	$\dfrac{1}{8}$	x	$\dfrac{x}{8}$
Pipe B (12 hours)	$\dfrac{1}{12}$	x	$\dfrac{x}{12}$

$$(\text{pipe A-fill}) - (\text{pipe B-drain}) = \left(\text{pool } \frac{1}{2} \text{ full}\right)$$

$$\frac{x}{8} - \frac{x}{12} = \frac{1}{2}$$

$$24\left[\frac{x}{8} - \frac{x}{12}\right] = 24 \cdot \frac{1}{2}$$

$$3x - 2x = 12$$

$$x = 12$$

time elasped: 12 hours

66. Let

$$x \; = \; \text{price per mask (original price), (first month)}$$
$$x - 1 \; = \; \text{reduced price (second month)}$$
$$\text{number of masks sold} \; = \; \frac{\text{amount of money taken in (sales)}}{\text{price per mask}}$$

	sales ($)	price per mask	number masks sold
first month	2000	x	$\dfrac{2000}{x}$
second month	2700	$x - 1$	$\dfrac{2700}{x - 1}$

$$(\text{number at reduced price}) \; = \; (\text{number at original price}) + (100 \text{ masks})$$

$$\frac{2700}{x - 1} = \frac{2000}{x} + 100$$

$$\frac{x(x - 1)}{100} \cdot \frac{2700}{x - 1} = \frac{x(x - 1)}{100} \left[\frac{2000}{x} + 100 \right]$$

$$27x = 20x - 20 + x^2 - x$$

$$0 = x^2 - 8x - 20$$

$$0 = (x - 10)(x + 2)$$

$$x - 10 = 0 \qquad \text{or} \qquad x + 2 = 0$$
$$x = 10 \qquad\qquad\qquad\qquad x = -2 \quad \text{(Reject)}$$

original price: $\boxed{\$10}$

Cumulative Review Problems (Chapters 1-6)

Cumulative Review, pp. 490-491

1. $\boxed{\text{D}}$ is true

2.
$$\frac{8(-6) - (-7)(4) - (-2)}{6 - (3)(4)}$$
$$= \frac{-48 + 28 + 2}{6 - 12}$$
$$= \frac{-18}{-6}$$
$$= \boxed{3}$$

3. $\{x \mid 2x + 5 \le 11\} \qquad \cap \qquad \{x \mid -3x > 18\}$
$$\phantom{\{x \mid} 2x \le 6 \qquad\qquad\qquad\qquad x < -6$$
$$\phantom{\{x \mid} x \le 6$$
$$\boxed{\{x \mid x < -6\} \text{ or } (-\infty, -6)}$$

4. $\dfrac{7(y-1)}{4} - (2y+3) = \dfrac{y+1}{2} - \dfrac{3}{4}(4-y)$

(×4) $7y - 7 - 8y - 12 = 2y + 2 - 12 + 3y$

$-6y = 9$

$y = -\dfrac{3}{2}$

$\boxed{\left\{ -\dfrac{3}{2} \right\}}$

5. Let

x = pounds of meat with 30% fat content
$100 - x$ = pounds of meat with 10% fat content

$0.30x + 0.10(100 - x) = 0.25(100)$

$0.30x + 10 - 0.10x = 25$

$0.20x = 15$

$x = 75$

$\boxed{75 \text{ pounds}}$

6. Let

A = Ana's weight
$230 - A$ = Juan's weight
$274 - A$ = Mike's weight

(Ana's weight) < (Juan's weight)

$A < 230 - A$

$2A < 230$

$A < 115$ (Ana)

$-A > -115$

$230 - A > 115$ (Juan)

$274 - A > 159$ (Mike)

(Ana's weight) < (Juan's weight) < (Mikes weight)

Thus, $\boxed{\text{Mike weighs the most, which is necessarily the case, since Ana's weight < Juan's weight < Mike's weight}}$.

7. $\boxed{\begin{array}{l} \text{Domain} = \{ x \mid -4 \le x \le 4 \} \\ \quad \text{or } [-4, 4] \\ \text{Range} = \{ y \mid -2 \le y < 2 \} \\ \quad \text{or } [-2, 2) \end{array}}$

The graph defines y as a function of x, since for each x-value there is exactly one value of y.

8. $\boxed{D}$ is true; $f(a) + f(b) + f(c) + f(d) = 5$

$3 + 0 + 0 + 2 = 5$

$5 = 5$ True

9. line passing through the point $(2, -3)$ and perpendicular to $3x - y = -12$:

slope of line $3x - y = -12$

$y = 3x + 12$

$m = 3$

slope of line perpendicular: $m = -\dfrac{1}{3}$

$y + 3 = -\dfrac{1}{3}(x - 2)$

$3y + 9 = -x + 2$

$\boxed{x + 3y = -7}$

10.
$$4x + 3y + 3z = 4$$
$$3x + 2z = 2$$
$$2x - 5y = {}^\circ 4$$

(Equations 1 and 2):
$$(\times 2) \quad 8x + 6y + 6z = 8$$
$$(\times -3) \quad \underline{-9x - 6z = -6}$$
$$-x = 2$$
$$x = -2$$

(Equation 3):
$$2x - 5y = -4$$
$$2(-2) - 5y = -4$$
$$-5y = 0$$
$$y = 0$$

(Equation 2):
$$3x + 2z = 2$$
$$3(-2) + 2z = 2$$
$$2z = 8$$
$$z = 4$$

$$\boxed{\{(-2, 0, 4)\}}$$

11. Let
$$x = \text{rate of plane in still air}$$
$$y = \text{rate of wind}$$

	R	T (hours)	$D = RT$	(1950 km)
against wind	$x - y$	3	$3(x - y)$	
with wind	$x + y$	2	$2(x + y)$	

$$3(x - y) = 1950 \quad \rightarrow \quad x - y = 650$$
$$2(x - y) = 1950 \quad \rightarrow \quad \underline{x + y = 975}$$
$$2x = 1625$$
$$x = 812.5 \quad \text{(plane)}$$
$$x + y = 975$$
$$812.5 + y = 975$$
$$y = 162.5 \quad \text{(wind)}$$

$$\boxed{\text{speed of plane in still air: } 812.5 \text{ kph; speed of wind: } 162.5 \text{ kph}}$$

12.
$$\begin{vmatrix} 5 & 2 & 1 \\ 3 & 0 & -2 \\ -4 & -1 & 2 \end{vmatrix} = -2\begin{vmatrix} 3 & -2 \\ -4 & 2 \end{vmatrix} + 0\begin{vmatrix} 5 & 1 \\ -4 & 2 \end{vmatrix} - (-1)\begin{vmatrix} 5 & 1 \\ 3 & -2 \end{vmatrix}$$
$$= -2(6 - 8) + 0 + 1(-10 - 3)$$
$$= 4 - 13$$
$$= \boxed{-9}$$

13.
$$\begin{aligned} x - y &= 1 \\ 2y + 3z &= 8 \\ 2x + z &= 6 \end{aligned}$$

$$\left[\begin{array}{ccc|c} 1 & -1 & 0 & 1 \\ 0 & 2 & 3 & 8 \\ 0 & 0 & 2 & 6 \end{array}\right] \quad R_3 - 2R_1 \to \quad \left[\begin{array}{ccc|c} 1 & -1 & 0 & 1 \\ 0 & 2 & 3 & 8 \\ 0 & 2 & 1 & 4 \end{array}\right]$$

$$R_2 - R_3 \to \left[\begin{array}{ccc|c} 1 & -1 & 0 & 1 \\ 0 & 2 & 3 & 8 \\ 0 & 0 & 3 & 4 \end{array}\right] \quad \begin{array}{c} \frac{1}{2}R_2 \to \\ \frac{1}{2}R_3 \to \end{array} \left[\begin{array}{ccc|c} 1 & -1 & 0 & 1 \\ 1 & 1 & 3/2 & 4 \\ 0 & 0 & 1 & 2 \end{array}\right]$$

$$\begin{aligned} x - y &= 1 \\ y + \frac{3}{2}z &= 4 \\ z &= 2 \qquad y + \frac{3}{2}(2) = 4 \\ & \qquad\qquad\quad y = 1 \qquad x - 1 = 1 \\ & \qquad\qquad\qquad\qquad\qquad\quad x = 2 \end{aligned}$$

$$\boxed{\{(2, 1, 2)\}}$$

14. Let
$$\begin{aligned} I &= \text{illumination} \\ d &= \text{distance from source} \\ I &= \frac{K}{d^2} \end{aligned}$$
Given: $d = 4$ feet, $I = 75$ foot-candles
$$\begin{aligned} 75 &= \frac{K}{16} \\ 1200 &= K \\ I &= \frac{1200}{d^2} \end{aligned}$$
At $d = 9$ feet, $I = ?$
$$I = \frac{1200}{81} = 14\frac{22}{27}$$

$$\boxed{14\tfrac{22}{27} \text{ foot-candles, or approximately } 14.815 \text{ foot-candles}}$$

15. $(2x^5 + 6x^4 - x^3 + 3x^2 - x) \div (2x^2 + 1) = \boxed{x^3 + 3x^2 - x}$

$$
\begin{array}{r}
x^3 + 3x^2 - x \\
2x^2 + 1 \,\overline{\big)\, 2x^5 + 6x^4 - x^3 + 3x^2 - x} \\
\underline{2x^5 \qquad\quad + x^3} \\
6x^4 - 2x^3 + 3x^2 \\
\underline{6x^4 \qquad\quad + 3x^2} \\
-2x^3 \qquad\quad - x \\
\underline{-2x^3 \qquad\quad - x} \\
0
\end{array}
$$

16. $f(x) = 2x^3 - 5x + 1$

$$\frac{f(a+h) - f(a)}{h}$$

$$= \frac{[(2a+h)^3 - 5(a+h) + 1] - (2a^3 - 5a + 1)}{h}$$

$$= \frac{2a^3 + 6a^2h + 6ah^2 + h^3 - 5a - 5h + 1 - 2a^3 + 5a - 1}{h}$$

$$= \frac{6a^2h + 6ah^2 + h^3 - 5h}{h}$$

$$= \boxed{6a^2 + 6ah + h^2 - 5}$$

17. $f(x) = 3x^3 - 7x + 1$ and $g(x) = 2x - 5$

$f[g(x)] - g[f(x)]$

$= [3(2x-5)^2 - 7(2x-5) + 1] - [2(3x^2 - 7x + 1) - 5]$

$= 12x^2 - 60x + 75 - 14x + 35 + 1 - 6x^2 + 14x + 3$

$= \boxed{6x^2 - 60x + 114}$

18. $2x^3 + 13x^2 + 17x - 12 = 0$

$$
\begin{array}{r|rrrr}
-4 & 2 & 13 & 17 & -12 \\
   &   & -8 & -20 & 12 \\
\hline
   & 2 & 5 & -3 & 0 \\
\end{array}
\quad \rightarrow \text{Remainder} = 0
$$

$\boxed{\text{Yes; } -4 \text{ is a solution}}$

19. $6x^4 - 96 = 6(x^4 - 16) = 6(x^2 + 4)(x^2 - 4) = \boxed{6(x^2 + 4)(x + 2)(x - 2)}$

20. $4x^{2n} + 14x^n - 30 = 2(x^{2n} + 7x^n - 15) = \boxed{2(2x^n - 3)(x^n + 5)}$

21. Let

$$
\begin{array}{rcl}
x & = & \text{width of rectangle} \\
2x + 3 & = & \text{length of rectangle} \\
\text{area} & = & 65 \text{ square yards}
\end{array}
$$

$$
\begin{array}{rcl}
x(2x + 3) & = & 65 \\
2x^2 + 3x - 65 & = & 0 \\
(2x + 13)(x - 5) & = & 0
\end{array}
$$

$2x + 13 = 0$ or $x - 5 = 0$

$x = -\dfrac{13}{2}$ (Reject) $x = 5$ (width)

$2x + 3 = 2(5) + 3 = 13$ (length)

Rectangle's dimensions: $\boxed{\text{width: } 5 \text{ yards; length: } 13 \text{ yards}}$

22. $I = 18t - 12t^2$

$(I = 6 \text{ amperes}):$

$$
\begin{array}{rcl}
6 & = & 18t - 12t^2 \\
12t^2 - 18t + 6 & = & 0 \\
6(2t^2 - 3t + 1) & = & 0 \\
6(2t - 1)(t - 1) & = & 0
\end{array}
$$

$2t - 1 = 0$ or $t - 1 = 0$

$t = \dfrac{1}{2} = 0.5$ $t = 1$

$\boxed{0.5 \text{ seconds or } 1 \text{ second}}$

23. $\dfrac{x^3 + 65y^3}{16y^2 - x^2} \div \dfrac{x^2 + 3xy - 4y^2}{x + 4y}$

$= \dfrac{x^3 + 64y^3}{(-1)(x^2 - 16y^2)} \cdot \dfrac{x + 4y}{x^2 - 13xy - 4y^2}$

$= \dfrac{(x + 4y)(x^2 - 4xy + 16y^2)(x + 4y)}{(-1)(x + 4y)(x - 4y)(x + 4y)(x - y)}$

$= \dfrac{x^3 - 4xy + 16y^2}{(-1)(x - 4y)(x - y)}$

$= \boxed{\dfrac{x^2 - 4xy + 16y^2}{(x - 4y)(y - x)}}$

24. $\dfrac{2x + 5}{x - 2} - \dfrac{4x + 12}{x^2 + x - 6}$

$= \dfrac{2x + 5}{x - 2} - \dfrac{4(x + 3)}{(x + 3)(x - 2)}$

$= \dfrac{2x + 5}{x - 2} - \dfrac{4}{x - 2}$

$= \boxed{\dfrac{2x + 1}{x - 2}}$

25. $\dfrac{1 + \dfrac{1}{x - 2}}{\dfrac{6}{x^2 + 3x - 10} - \dfrac{1}{x - 2}}$

$= \dfrac{(x - 2)(x + 5)}{(x - 2)(x + 5)} \cdot \dfrac{1 + \dfrac{1}{x - 2}}{\dfrac{6}{(x - 3)(x + 5)} - \dfrac{1}{x - 2}}$

$= \dfrac{(x - 2)(x + 5) + (x + 5)}{6 - (x + 5)}$

$= \dfrac{x^2 + 3x - 10 + x + 5}{-x + 1}$

$= \dfrac{x^2 + 4x - 5}{-(x - 1)}$

$= \dfrac{(x + 5)(x - 1)}{-(x - 1)}$

$= -(x + 5)$

$= \boxed{-x - 5}$

26. $\dfrac{5}{2x + 1} + \dfrac{1}{x + 2} = -2$

$(2x + 1)(x + 2)\left[\dfrac{5}{2x + 1} + \dfrac{1}{x + 2}\right] = (2x + 1)(x + 2)(-2)$

$5x + 10 + 2x + 1 = -4x^2 - 10x - 4$

$4x^2 + 17x + 15 = 0$

$(4x + 5)(x + 3) = 0$

$4x + 5 = 0 \qquad \text{or} \qquad x + 3 = 0$

$x = -\dfrac{5}{4} \qquad\qquad\qquad x = -3$

$\boxed{\left\{-3, -\dfrac{5}{4}\right\}}$

27. $f(x) = \dfrac{7}{x}$

$\dfrac{f(a + h) - f(a)}{h}$

$= \dfrac{\dfrac{7}{a+h} - \dfrac{7}{a}}{h}$

$= \dfrac{a(a+h)}{a(a+h)} \cdot \dfrac{\dfrac{7}{a+h} - \dfrac{7}{a}}{h}$

$= \dfrac{7a - 7a - 7h}{a(a+h)h}$

$= \dfrac{-7h}{a(a+h)h}$

$= \boxed{\dfrac{-7}{a(a+h)}}$

28. $f(x) = \dfrac{2x^2 + x - 3}{x - 1} = \dfrac{(2x+3)(x-1)}{x-1} = \boxed{2x + 3 \ (x \neq 1)}$

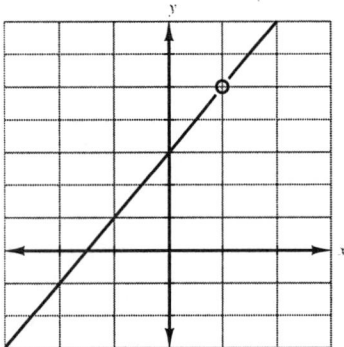

29. $f(x) = \dfrac{1}{x^2 - 1}$

$f(x)$ is undefined when

$\begin{aligned} x^2 - 1 &= 0 \\ (x+1)(x-1) &= 0 \\ x &= -1, 1 \end{aligned}$

$\boxed{D_f = \{x \mid x \neq -1, 1\}}$

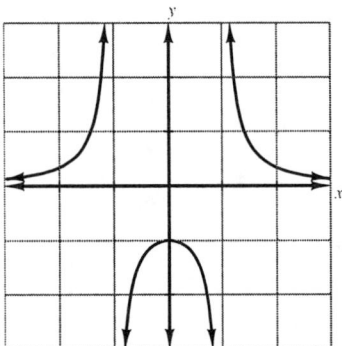

30. Let

$$x \;=\; \text{rate of first part of cycling}$$
$$x+5 \;=\; \text{rate of last portion of cycling}$$

total cycling time: 8 hours

	D (miles)	R	$T = \dfrac{D}{R}$
first part	75	x	$\dfrac{75}{x}$
last portion	$135 - 75 = 60$	$x + 5$	$\dfrac{60}{x+5}$

(first part) + (last portion) = 8 hours

$$\frac{75}{x} + \frac{60}{x+5} \;=\; 8$$

$$x(x+5)\left[\frac{75}{x} + \frac{60}{x+5}\right] \;=\; x(x+5)(8)$$

$$75x + 375 + 60x \;=\; 8x^2 + 40x$$

$$0 \;=\; 8x^2 - 95x - 375$$

$$0 \;=\; (x-15)(8x+25) = 0$$

$$x - 15 \;=\; 0 \qquad \text{or} \qquad 8x + 25 \;=\; 0$$

$$x \;=\; 15 \quad \text{(first part)} \qquad\qquad x \;=\; -\frac{25}{8} \quad \text{(Reject)}$$

$$x + 5 \;=\; 15 + 5 = 20 \quad \text{(last portion)}$$

The rate for the first portion is $\boxed{15 \text{ mph}}$ and for the last portion is $\boxed{20 \text{ mph}}$.

Algebra for College Students

Chapter 7   Exponents and Radicals

Section 7.1 Integral Exponents

Problem Set 7.1, pp. 499-501

1. $(5x^3y^4)^2(-3x^7y^{11})$
$= 5^2(x^3)^2(y^4)^2(-3x^7y^{11})$
$= 25x^6y^8(-3x^7y^{11})$

$= \boxed{-75x^{13}y^{19}}$

3. $(4ab^3)^3(-3a^{-5}b^8)$
$= 64a^3b^9(-3a^{-5}b^8)$
$= -192a^{-2}b^{17}$

$= \boxed{\dfrac{-192b^{17}}{a^2}}$

5. $(54r^3s^9)(-3\,r^2s^{-4})^{-3}$
$= (54r^3s^9)(-3)^{-3}(r^2)^{-3}(s^{-4})^{-3}$
$= (54r^3s^9)[(-3)^{-3}r^{-6}s^{12}]$

$= 54(-3)^{-3}r^{-3}s^{21}$

$= \dfrac{54s^{21}}{(-3)^3r^3}$

$= \dfrac{54s^{21}}{-27r^3}$

$= \boxed{\dfrac{-2s^{21}}{r^3}}$

7. $(-a^{-2}b^3c)(ab^{-1}c^{-4})^{-3}$
$= (-a^{-2}b^3c)(a^{-3}b^3\,c^{12})$
$= (-a^{-5}b^6c^{13})$

$= \boxed{\dfrac{-b^6c^{13}}{a^5}}$

9. $(3x^{-3}y^{-4}z)^3(3xy^{-5}z)^2(-3x^{-4}z^{12})$

$= (27x^{-9}y^{-12}z^3)(9\,x^2y^{-10}z^2)(-3\,x^4z^{12})$

$= (27)(9)(-3)x^{-11}y^{-22}z^{17}$

$= \boxed{\dfrac{-729z^{17}}{x^{11}y^{22}}}$

11. $\dfrac{-27x^{-8}y^4z}{15x^4y^4z^{-4}}$

$= -\dfrac{9}{5}\,x^{-8-4}y^{4-4}z^{1-(-4)}$

$= -\dfrac{9}{5}\,x^{-12}y^0z^5$

$= \boxed{\dfrac{-9z^5}{5x^{12}}}$

13. $\dfrac{(-4x^4yz^3)^3}{-4x^2y^{-3}z^9}$

$= \dfrac{-64x^{12}y^3z^9}{-4x^2y^{-3}z^9}$

$= 16x^{12-2}y^{3+3}z^{9-9}$

$= \boxed{16x^{10}y^6}$

15. $\dfrac{(-4xy^{-3}z^2)^{-3}}{(2xy^{-3}z^2)^{-2}}$

$= \dfrac{(-4)^{-3}x^{-3}y^9z^{-6}}{2^{-2}x^{-2}y^6z^{-4}}$

$= \dfrac{(-4)^{-3}}{2^{-2}}\,x^{-3-(-2)}y^{9-6}z^{-6-(-4)}$

$= \dfrac{2^2}{(-4)^3}x^{-1}y^3z^{-2}$

$= \dfrac{4}{-64}\,x^{-1}y^3z^{-2}$

$= \boxed{\dfrac{-y^3}{16xz^2}}$

17. $\dfrac{-4a^4b^{-1}c^2}{(-ab^{-2}c^4)^{-3}}$

$= \dfrac{4^{-2}a^{-6}b^2c^{-4}}{-a^{-3}b^6c^{-12}}$

$= -\dfrac{1}{4^2}\,a^{-3}b^{-4}c^8$

$= \boxed{-\dfrac{c^3}{16a^3b^4}}$

19. $\left(\dfrac{10a^8}{7b^5}\right)^2$

$= \dfrac{10^2(a^8)^2}{7^2(b^5)^2}$

$= \boxed{\dfrac{100a^{16}}{49b^{10}}}$

21. $\left(\dfrac{-25a^3b^4c}{-75ab^6c^{-3}}\right)^{-2}$

$= \left(\dfrac{1}{3}\,a^2b^{-2}c^4\right)^{-2}$

$= \left(\dfrac{1}{3}\right)^{-2}a^{-4}b^4c^{-8}$

$= \dfrac{1}{\left(\dfrac{1}{3}\right)^2}\,a^{-4}b^4c^{-8}$

$= \boxed{\dfrac{9b^4}{a^4c^8}}$

23. $\left(\dfrac{-75a^3b^4c}{25a^{-2}b^4c^5}\right)$

$= (-3a^5c^{-4})^3$

$= (-3)^3a^{15}c^{-12}$

$= \boxed{\dfrac{-27a^5}{c^{12}}}$

25. $\left(\dfrac{-30x^3y^2}{6xy^7}\right)\cdot\left(\dfrac{48xy^{-5}}{16xy^{-8}}\right)$

$= (-5x^2y^{-5})(3x^0y^3)$

$= -15x^2y^{-2}$

$= \boxed{\dfrac{-15x^2}{y^2}}$

27. $\left(\dfrac{54a^4b^3c^5}{-3a^4b^{-6}c}\right)\cdot\left(\dfrac{5a^{-5}c}{-45b^3c^{-7}}\right)$

$= (-18a^0b^9c^4)\left(-\dfrac{1}{9}\,a^{-5}b^{-3}c^8\right)$

$= 2a^{-5}b^6c^{12}$

$= \boxed{\dfrac{2b^6c^{12}}{a^5}}$

29. $y^{4n-1}\cdot y^{8n+5}$

$= y^{4n-1+8n+5}$

$= \boxed{y^{12n+4}}$

31. $(y^{-5n}\cdot y^{3n})^{-4}$

$= (y^{-5n+3n})^{-4}$

$= (y^{-2n})^{-4}$

$= y^{(-2n)(-4)}$

$= \boxed{y^{8n}}$

33. $(4x^{3n+1}y^{2n})^3(-2x^{2n}y^{4n-3})^{-2}$

$= 64x^{9n+3}y^{6n}(-2)^{-2}x^{-4n}y^{-8n+6}$

$= \dfrac{64}{(-2)^2}x^{9n+3-4n}y^{6n-8n+6}$

$= \dfrac{64}{4}\,x^{5n+3}y^{6-2n}$

$= \boxed{16x^{5n+3}y^{6-2n}}$

35. $(x^{-2}y^5)^{-3n}$

$= x^{(-2)(-3n)}y^{(5)(-3n)}$

$= x^{6n}y^{-15n}$

$= \boxed{\dfrac{x^{6n}}{y^{15n}}}$

37. $(y^{2n-5}\cdot y^{4n+1})2n$

$= (y^{6n-4})2n$

$= y^{(6n-4)(2\,in)}$

$= \boxed{y^{12n^2-8n}}$

39. $\left(\dfrac{y^{1-n}}{y^{4-n}}\right)^{-2}$

$= (y^{1-n-4(-n)})^{-2}$

$= (y^{-3})^{-2}$

$= \boxed{y^6}$

41. $\left(\dfrac{a^{3n}}{a^{2n-2}}\right)^{-3}$

$= (a^{3n-(2n-2)})^{-3}$

$= (a^{n+2})^{-3}$

$= \boxed{a^{-3n-6}}$

Equivalently: $(a^{n+2})^{-3} = \dfrac{1}{(a^{n+2})^3}$

$= \dfrac{1}{a^{3n+6}}$

43. $\dfrac{a^{2n-1}b^n}{a^{2n}b^{n+3}}$

$= a^{2n-1-2n}b^{n-(n+3)}$

$= a^{-1}b^{-3}$

$= \boxed{\dfrac{1}{ab^3}}$

45. $\left(\dfrac{x^n y^{3n+2}}{x^{n-1}y^{3n-1}}\right)^{-2}$

$= (x^{n-(n-1)}y^{3n+2-(3n-1)})^{-2}$

$= (xy^3)^{-2}$

$= x^{-2}y^{-6}$

$= \boxed{\dfrac{1}{x^2y^6}}$

47. $T = \left(\dfrac{h}{12.3}\right)^3$

($h = 5$ feet 10 inches $= 70$ inches):

$T = \left(\dfrac{70}{12.3}\right)^3$

$\approx (5.691)^3$

≈ 184

Threshold weight: $\boxed{184\ \text{pounds}}$

49. $r = 3 \times 10^5$ kilometers 1 second

$d = 4.58 \times 10^9$ kilometers

$d = rt$

$t = \dfrac{d}{r} = \dfrac{4.58 \times 10^9}{3 \times 10^5}$

$= 1.52667 \times 10^4 = \boxed{15{,}266.7\ \text{seconds}}$

$15266.7 \text{ seconds} \times \dfrac{1 \text{ minute}}{60 \text{ seconds}} \times \dfrac{1 \text{ hour}}{60 \text{ minutes}}$

$\approx \boxed{4.24\ \text{hours}}$

51. a. $d = \dfrac{3(2^{n-2}) + 4}{10}$

Venus: $n = 2$

$d = \dfrac{3(2^{2-2}) + 4}{10} = \dfrac{3(2^0) + 4}{10} = \dfrac{7}{10} = 0.7$

distance from sun to Venus: 0.7 astronomical units

Earth: $n = 3$

$d = \dfrac{3(2^{3-2}) + 4}{10} = \dfrac{3(2^1) + 4}{10} = \dfrac{10}{10} = 1$

distance from sun to Earth: 1 astronomical units

difference in distance: $1 - 0.7 = 0.3$

$\boxed{0.3\ \text{astronomical units}}$

b. Pluto: $n = 9$

$d = \dfrac{3(2^{9-2}) + 4}{10} = \dfrac{3(2^7) + 4}{10} = 38.8$ astronomical units

distance from sun to Pluto $\approx 4.6 \times 10^9$ miles

Equating both distances:

1 astronomical unit $\approx \dfrac{4.6 \times 10^9}{38.8}$ miles $\approx 1.185567 \times 10^7$ miles

$\boxed{1\ \text{astronomical unit is approximately } 1.185567 \times 10^7 \text{ miles}}$

53. $\dfrac{(a+b)^3(a+b)^{-2}}{(a+b)^4}$

$= \dfrac{(a+b)^{3-2}}{(a+b)^4}$

$= \dfrac{(a+b)^1}{(a+b)^4}$

$= (a+b)^{1-4}$

$= (a+b)^{-3}$

$= \boxed{\dfrac{1}{(a+b)^3}}$

55. $A = \pi r^2$

$(r = 3x^4 \text{ centimeters})$:

$A = \pi(3x^4)^2$

$= \pi(9x^8)$

$= 9\pi x^8$

The area is $\boxed{9\pi x^8 \text{ cm}^2}$.

57. $A = s^2$

$(s = 6a^{-1}b^{-2} \text{ inches})$:

$A = (6a^{-1}b^{-2})^2$

$= 36a^{-2}b^{-4}$

$= \dfrac{36}{a^2 b^4}$

The area is $\boxed{\dfrac{36}{a^2 b^4} \text{ in.}^2}$.

59. $V = \pi r^2$

diameter $= 4a^2$ centimeters,

radius $= \dfrac{1}{2}(4a^2) = 2a^2$ centimeters,

height $= 10$ meters $= 10 \text{ m} \left(\dfrac{100 \text{ cm}}{1 \text{ m}} \right) = 1000$ centimeters

$V = \pi(2a^2)^2(1000) = \pi(4a^4)(1000) = 4000\pi a^4$

The volume is $\boxed{4000\pi a^4 \text{ cm}^3}$.

61. $\boxed{\text{D}}$ is true; A is *not* true: $2^2 \cdot 2^2 = 2^6 \text{ } not \text{ } 2^8$

B is *not* true: $5^6 \cdot 5^2 = 5^8 \text{ } not \text{ } 25^8$

C is *not* true: $2^{-3} \cdot 3^2 = 8 \cdot 9 = 72 \neq 6^5$

D is true

63. Let

$\begin{aligned} F &= \text{force} \\ d &= \text{distance between them} \\ m_1 &= \text{mass of object 1} \\ m_2 &= \text{mass of object 2} \\ F &= \dfrac{km_1 m_2}{d^2} \end{aligned}$

Given: $\begin{aligned} m_1 &= m_2 = 30 \text{ kilograms} \\ d &= 1 \text{ meter} \\ F &= 6 \times 10^{-8} \text{ newton} \end{aligned}$

$6 \times 10^{-8} = \dfrac{k(30)(30)}{1^2}$

$6 \times 10^{-8} = 900k$

$\dfrac{6}{900} \times 10^{-8} = k$

$k = \dfrac{6 \times 10^{-8}}{9 \times 10^2} = \dfrac{2}{3} \times 10^{-10}$

$F = \dfrac{\frac{2}{3} \times 10^{-10} m_1 m_2}{d^2}$

If m_1 = 1 kilograms
            m_2 = mass of earth
            d = 6×10^6 meters
            F = 9.8 newton
Find m_2.

$$9.8 = \frac{\frac{2}{3} \times 10^{-10}(1)m_2}{(6 \times 10^6)^2}$$

$$9.8 = \frac{\frac{2}{3} \times 10^{-10}m_2}{36 \times 10^{12}}$$

$$9.8 = \frac{1}{54} \times 10^{-22}m_2$$

$$m_2 = \frac{9.8}{\frac{2}{3} \times 10^{-22}} \approx 529.2 \times 10^{22}$$

$$= 5.292 \times 10^2 \times 10^{22}$$

$$= 5.292 \times 10^{24}$$

Mass of Earth is approximately $\boxed{5.292 \times 10^{24} \text{ kg}}$

65. $x = (1 - 9 + 8 + 9)^{1989}$ and $y = (1 - 9 + 8 - 9)^{1989}$
            $x = 9^{1989}$ $y = (-9)^{1989}$
                                                                   $= (-1)^{1989}9^{1989}$
                                                                   $= -9^{1989}$

$$x + y = 9^{1989} + (-9^{1989}) = 0$$
Thus, $(1 - 9 + 8 + 9)^{x+y} = (1 - 9 + 8 + 9)^0 = \boxed{1}$

For 67-71, $a^3 = 3$, $b^2 = 2$, $c^2 = 5$.

67. $$\frac{1}{a^6} < \frac{1}{b^6}$$

$$\frac{1}{(a^3)^2} < \frac{1}{(b^2)^3}$$

$$\frac{1}{3^2} < \frac{1}{23} \quad (a^3 = 3, \ b2 = 2)$$

$$\frac{1}{9} < \frac{1}{8} \quad \boxed{\text{True}}$$

69. $$\left(\frac{ab}{c^2}\right)^6 < 1$$

$$\frac{a^6 b^6}{c^{12}}$$

$$\frac{9(8)}{5} < 1 \quad (a^6 = 9, \ b^6 = 8, \ c^{12} = (c^2)^6 = 5^6 = 15{,}625)$$

$$\frac{72}{15{,}625} < 1 \quad \boxed{\text{True}}$$

71. $\dfrac{2}{a^3 b^2} > c^{-4} + 2^{-4}$

$\dfrac{2}{a^3 b^2} > \dfrac{1}{c^4} + \dfrac{1}{2^4}$

$\dfrac{2}{(3)(2)} > \dfrac{1}{25} + \dfrac{1}{16}$ $(a^3 = 3,\ b^2 = 2,\ c^4 = (c^2)^2 = 5^2 = 25)$

$\dfrac{1}{3} > \dfrac{41}{400}$

$0.33.. > 0.1025$ $\boxed{\text{True}}$

Review Problems

75. $\dfrac{2y}{y^2 - y - 6} - \dfrac{6(y-1)}{2y^2 - 9y + 9} \div \dfrac{y^2 + y - 2}{2y - 3}$

$= \dfrac{2y}{(y+2)(y-3)} - \dfrac{6(y-1)}{(2y-3)(y-3)} \div \dfrac{(y+2)(y-1)}{2y-3}$

$= \dfrac{2y}{(y+2)(y-3)} - \dfrac{6(y-1)}{(2y-3)(y-3)} \cdot \dfrac{2y-3}{(y+2)(y-1)}$

$= \dfrac{2y}{(y+2)(y-3)} - \dfrac{6}{(y-3)(y+2)}$

$= \dfrac{2y - 6}{(y-3)(y+2)}$

$= \dfrac{2(y-3)}{(y-3)(y+2)}$

$= \boxed{\dfrac{2}{y+2}}$

76. $1 + \dfrac{x}{2 + \dfrac{1}{x}}$

$= 1 + \dfrac{x}{2 + \dfrac{1}{x}} \cdot \dfrac{x}{x}$

$= 1 + \dfrac{x^2}{2x + 1}$

$= 1 \cdot \dfrac{2x+1}{2x+1} + \dfrac{x^2}{2x+1}$

$= \dfrac{x^2 + 2x + 1}{2x + 1}$

$= \boxed{\dfrac{(x+1)^2}{2x+1}}$

77. $125 - 8x^3 y^3$

$= 5^3 - (2xy)^3$

$= (5 - 2xy)\,[5^2 + 5(2xy) + (2xy)^2]$

$= \boxed{(5 - 2xy)(25 + 10xy + 4x^2 y^2)}$

Section 7.2 Radicals and Radical Functions

Problem Set 7.2, pp. 510-511

1. **a.** $\sqrt{49} = \boxed{7}$ because $7^2 = 49$

   b. $-\sqrt{49} = -(\sqrt{49}) = \boxed{-7}$

   c. $\sqrt{49}$ is $\boxed{\text{not a real number}}$ because there is no real number whose square is -49.

d. $\sqrt[4]{16} = \boxed{2}$ because $2^4 = 16$

e. $\sqrt[5]{-1} = \boxed{-1}$ becuase $(-1)^5 = -1$

f. $\sqrt{\dfrac{1}{9}} = \boxed{\dfrac{1}{3}}$ because $\left(\dfrac{1}{3}\right)^2 = \dfrac{1}{9}$

g. $\sqrt[3]{-\dfrac{1}{64}} = \boxed{-\dfrac{1}{4}}$ because $\left(-\dfrac{1}{4}\right)^3 = -\dfrac{1}{64}$

h. $\sqrt[5]{\dfrac{1}{32}} = \boxed{\dfrac{1}{2}}$ because $\left(\dfrac{1}{2}\right)^5 = \dfrac{1}{32}$

i. $\sqrt[4]{-16}$ is $\boxed{\text{not a real number}}$

j. $\left(\sqrt[3]{2}\right)^3 = \boxed{2}$

k. $\left(\sqrt[5]{-3}\right)^5 = \boxed{-3}$

3. a. $\sqrt{36x^2} = \sqrt{(6x)^2} = \boxed{6x}$

   b. $\sqrt{100y^8} = \sqrt{(10y^4)^2} = \boxed{10y^4}$

   c. $\sqrt[3]{27a^{12}} = \sqrt[3]{(3a^4)^3} = \boxed{3a^4}$

   d. $\sqrt[5]{32x^{10}} = \sqrt[5]{(2x^2)^5} = \boxed{2x^2}$

   e. $\sqrt[4]{81a^4b^{12}} = \sqrt[4]{(3ab^3)^4} = \boxed{3ab^3}$

5. $N = 2\sqrt{Q} - 9$

   $(Q = 121)$:

   $N = 2\sqrt{121} - 9 = 2(11) - 9 = 22 - 9 = 13$

   $\boxed{13 \text{ syllables}}$

7. $C = 10x + 200\sqrt{x}$ (C, in dollars; x, in tons)

$x = 180{,}000{,}000 \text{ pounds} = 180{,}000{,}000 \text{ pound} \times \dfrac{1 \text{ ton}}{2000 \text{ pounds}} = 90{,}000 \text{ tons} = 9 \times 10^4 \text{ tons}$

$C = 10(9 \times 10^4) + 200\sqrt{9 \times 10^4}$
$= 9 \times 10^5 + 200(3 \times 10^2)$
$= 900{,}000 + 60{,}000$
$= 960{,}000$

The cost is $\boxed{\$960{,}000}$.

9. $f(x) = 4\left(\sqrt{x}\right)^5 + 17{,}300$ (f, in dollars; x, in years)

$f(16) = 4(\sqrt{16})^5 + 17{,}300 = 4(4^5) + 17{,}300 = 4096 + 17{,}300 = \boxed{\$21{,}396}$

$\boxed{\text{The yearly income for a person with 16 years of education is \$21,396.}}$

11. $H = (10.45 + \sqrt{1000} - w)(33 - t)$ *(H, in kilometers/square meter/hour, t in degrees, w in meters/second)*

when H = 2000 kilocalories/squre meter/hour flesh freezes in 1 minute
If w = 4 meters/second, $t = 0°C$, is $H \geq 200$?

$$H = (10.45 + \sqrt{100(4)} - 4)(33 - 0)$$
$$= (10.45 + \sqrt{400} - 4)(33)$$
$$= (10.45 + 20 - 4)(33)$$
$$= (26.45)(33) = 872.85$$

Since $872.45 < 2000$, exposed flesh will not freeze under these conditions.

13. $I = \sqrt{x} + \sqrt{y} + 2xy$ *(x and y in milligrams)*
$(x = 9$ milligrams, $y = 25$ milligrams$)$:
$I = \sqrt{9} + \sqrt{25} + 2(9)(25) = 3 + 5 + 450 + 458$
The production index is $\boxed{458}$.

15. $2\sqrt{x} - \sqrt[3]{y} + 4\sqrt[5]{z}$
$$= 2\sqrt{36} - \sqrt[3]{-8} + 4\sqrt[5]{1} \quad (x = 36, y = -8, z = 1)$$
$$= 2(6) - (-2) + 4(1)$$
$$= 12 + 2 + 4$$
$$= \boxed{18}$$

17.
$$A = \sqrt[3]{-\frac{b}{2} + \sqrt{\frac{b^2}{4} + \frac{a^3}{27}}}$$
$$= \sqrt[3]{\frac{-2}{2} + \sqrt{\frac{2^2}{4} + \frac{(-3)^3}{27}}} \quad (b = 2, a = -3)$$
$$= \sqrt[3]{-1 + \sqrt{1 - 1}}$$
$$= \sqrt[3]{-1 + \sqrt{0}}$$
$$= \sqrt[3]{-1}$$
$$= \boxed{-1}$$

19. $\dfrac{-b \pm \sqrt{b^2 - 4ac}}{2a}$
$$= \frac{-3 \pm \sqrt{(3)^2 - 4(5)(-8)}}{2(5)} \quad (a = 5, b = 3, c = -8)$$
$$= \frac{-3 \pm \sqrt{9 - (-160)}}{10}$$
$$= \frac{-3 \pm \sqrt{169}}{10}$$
$$= \frac{-3 \pm 13}{10}$$
$$\frac{-3 + 13}{10} = \frac{10}{10} = \boxed{1} \quad \text{or} \quad \frac{-3 - 13}{10} = \frac{-16}{10} = \boxed{-\frac{8}{5}}$$

21. $f(x) = \sqrt{x+1}$

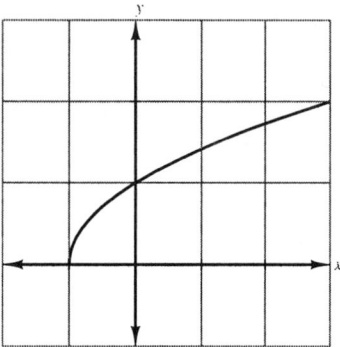

To determine the domain of f, the radicand must be greater than or equal to 0.

Thus, $x+1 \geq 0$
$$x \geq -1$$

When $x \geq -1$, the range of f consists of all real numbers greater than or equal to 0.

$D_f = \text{Domain} = \{x \mid x \geq -1\} = [-1, \infty)$

$R_f = \text{Range} = \{y \mid y \geq 0\} = [0, \infty)$

23. $f(x) = \sqrt{1-x}$

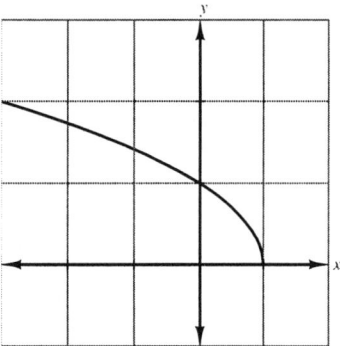

For the domain of f,
$$1 - x \geq 0$$
$$-x \geq -1$$
$$x \leq 1$$
When $x \leq 1, f(x) \geq 0$ $(x = 1, f(x) = 0)$
Thus,

$D_f = \text{Domain} = \{x \mid x \leq 1\} = (-\infty, 1]$

$R_f = \text{Range} = \{y \mid y \geq 0\} = [0, \infty)$

25. $f(x) = \sqrt{4x - 2}$

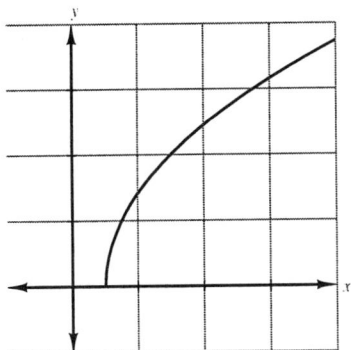

For the domain of f,

$$4x - 2 \geq 0$$
$$4x \geq 2$$
$$x \geq \frac{1}{2}$$

when $x \geq \frac{1}{2}$, $f(x) \geq 0$ $\left(x = \frac{1}{2}, f(x) = 0\right)$

$$D_f = \text{Domain} = \left\{ x \mid x \geq \frac{1}{2} \right\} = \left[\frac{1}{2}, \ \infty \right)$$

$$R_f = \text{Range} = \{ y \mid y \geq 0 \} = [0, \infty)$$

27. $f(x) = -\sqrt{x + 1}$

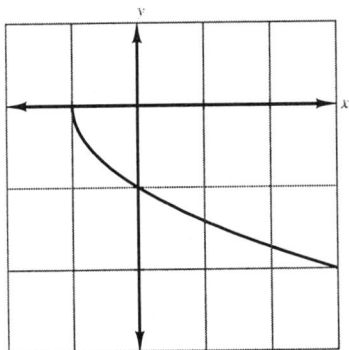

For the domain of f,

$$x + 1 \geq 0$$
$$x \geq -1$$

when $x \geq -1$, $f(x) \leq 0$ $(x = -1, f(x) = 0)$

$$D_f = \text{Domain} = \{ x \mid x \geq -1 \} = [-1, \infty)$$

$$R_f = \text{Range} = \{ y \mid y \leq 0 \} = (-\infty, 0]$$

29. $f(x) = -2\sqrt{x+1}$

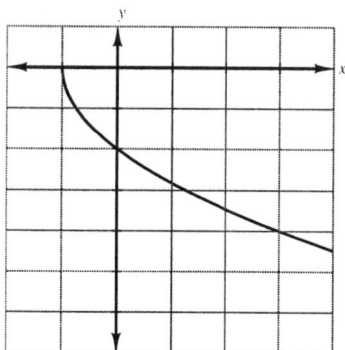

For the domain of f,

$$x + 1 \;\;\geq\;\; 0$$
$$x \;\;\geq\;\; -1$$

when $x \geq -1, f(x) \leq 0$ $(x = -1, f(x) = 0)$

$\boxed{D_f = \text{Domain} = \{x \mid x \geq -1\} = [-1, \infty)}$

$\boxed{R_f = \text{Range} = \{y \mid y \leq 0\} = (-\infty, 0]}$

31. $f(x) = \sqrt{x^2} = |x|$

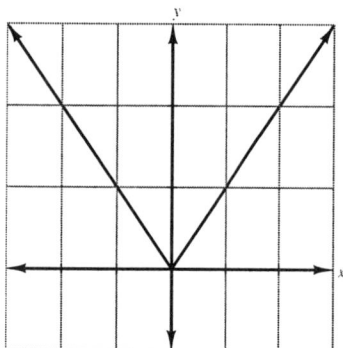

For the domain f, $|x| \geq 0$, which is true for all real values of x

when $|x| \geq 0, f(x) = \geq 0$

$\boxed{D_f = \text{Domain} = \{x \mid x \in R\} = (-\infty, \infty)}$

$\boxed{R_f = \text{Range} = \{y \mid y \geq 0\} = [0, \infty)}$

33. $f(x) = \sqrt[3]{x}$

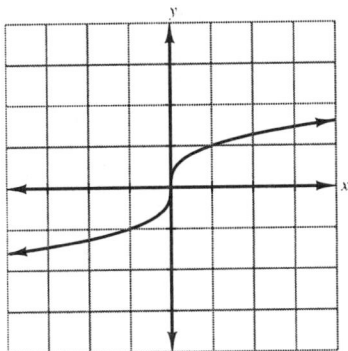

$f(x)$ is defined for all real values of x

$D_f = \text{Domain} = \{x \mid x \in R\} = (-\infty, \infty)$

$R_f = \text{Range} = \{y \mid y \in R\} = (-\infty, \infty)$

35. $f(x) = -\sqrt[3]{x}$

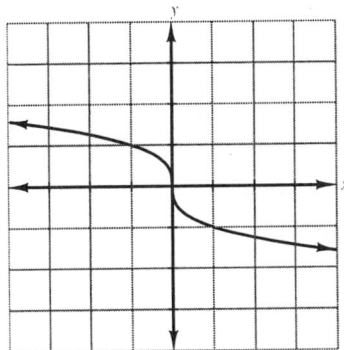

$f(x)$ is defined for all real values of x

$D_f = \text{Domain} = \{x \mid x \in R\} = (-\infty, \infty)$

$R_f = \text{Range} = \{y \mid y \in R\} = (-\infty, \infty)$

37. $f(x) = \sqrt{x} - 1$

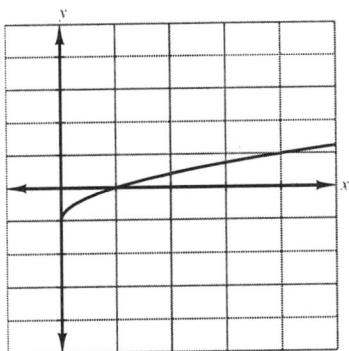

$f(x)$ is defined when $x \geq 0$

where $x \geq 0$, $f(x) \geq -1$ $(x = 0, f(x) = -1)$

$D_f = \text{Domain} = \{x \mid x \geq 0\} = [0, \infty)$

$R_f = \text{Range} = \{y \mid y \geq -1\} = [-1, \infty)$

39. $f(x) = \sqrt[3]{x} - 1$

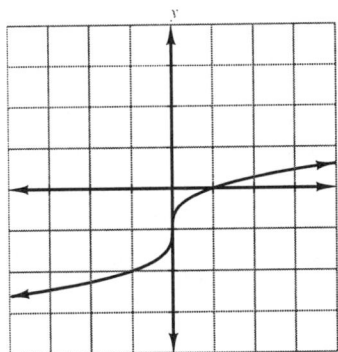

$f(x)$ is defined for all real values of x

$D_f = \text{Domain} = \{x \mid x \in R\} = (-\infty, \infty)$

$R_f = \text{Range} = \{y \mid y \in R\} = (-\infty, \infty)$

41. $f(x) = \begin{cases} \sqrt{x+1} & \text{if } x \geq 1 \\ 1-x & \text{if } x < 1 \end{cases}$

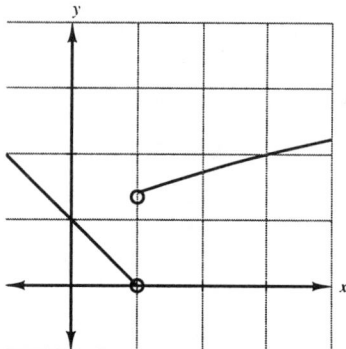

The completed graph indicates that the domain of f is the set of all real numbers. The range of f is the set of all numbers greater than zero but not equal to zero.

$D_f = \text{Domain} = \{x \mid x \in R\} = (-\infty, \infty)$

$R_f = \text{Range} = \{y \mid y > 0\} = (0, \infty)$

43. $f(x) = \begin{cases} \sqrt{x}-1 & \text{if } x \geq 0 \\ -1 & \text{if } x < 0 \end{cases}$

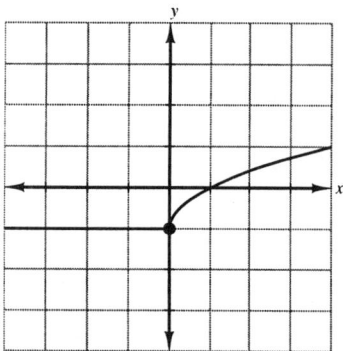

The completed graph indicates that the domian of f is the set of all real numbers. The range of f is the set of al numbers greater than or equal to -1.

$D_f = \text{Domain} = \{x \mid x \in R\} = (-\infty, \infty)$

$R_f = \text{Range} = \{y \mid y \geq -1\} = [-1, \infty)$

45. $f(x) = \begin{cases} 2 & \text{if } x < 4 \\ \sqrt{x} & \text{if } x \geq 4 \end{cases}$

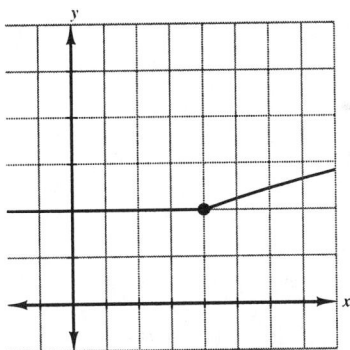

The completed graph indicates that the domain of f is the set of all real numbers. The range of f is the set of all numbers greater than or equal to 2.

$D_f = \text{Domain} = \{x \mid x \in R\} = (-\infty, \infty)$

$R_f = \text{Range} = \{y \mid y \geq 2\} = [2, \infty)$

47. $f(x) = \dfrac{1}{x-3}$ and $g(x) = \sqrt{x}$

$(f \circ g)x = f[g(x)] = \dfrac{1}{\sqrt{x}-3}$

The domain of $f \circ g$ consists of x such that $x \geq 0$ because $\sqrt{x}$ is a real number only when $x \geq 0$.

Also $\sqrt{x} - 3 \neq 0$
$\sqrt{x} \neq 3$
$x \neq 9$

Thus, $D_{f \circ g} = \{x \mid x \geq 0 \text{ and } x \neq 9\} = [0, 9) \cup (9, \infty)$

$(g \circ f)x = g[f(x)] = \dfrac{1}{\sqrt{x-3}}$

The domain of $g \circ f$ consists of x such that $x - 3 > 0$ or $x > 3$. $x \neq 3$ since the denominator cannot equal 0.

$D_{g \circ f} = \{x \mid x > 3\} = (3, \infty)$

49. $f(x) = \sqrt{x}$ and $g(x) = 8 - 4x$

$(f \circ g)(x) = \sqrt{8 - 4x} = 2\sqrt{2 - x}$

The domain of $f \circ g$ consists of x such that
$2 - x \geq 0$
$-x \geq -2$
$x \leq 2$

$D_{f \circ g} = \{x \mid x \leq 2\} = (-\infty, 2]$

$(g \circ f)(x) = 8 - 4\sqrt{x} = 4(2 - \sqrt{x})$

The domain of $g \circ f$ consists of the set of all real numbers such that $x \geq 0$.

$D_{f \circ g} = \{x \mid x \geq 0\} = [0, \infty)$

51. $f(x) = \sqrt[3]{x}$ and $g(x) = 8 - 4x$

$$\boxed{(f \circ g)(x) = \sqrt[3]{8 - 4x}}$$

The domain of the set of all real numbers.

$$\boxed{D_{f \circ g} = \{x \mid x \in R\} = (-\infty, \infty)}$$

$$\boxed{(g \circ f)(x) = 8 - 4\sqrt[3]{x}}$$

The domain of the set of all real numbers.

$$\boxed{D_{f \circ g} = \{x \mid x \in R\} = (-\infty, \infty)}$$

53. $\sqrt[3]{\sqrt[4]{16} + \sqrt{625})}$

$$= \sqrt[3]{2 + 25}$$

$$= \sqrt[3]{27}$$

$$= \boxed{3}$$

55. $\boxed{D}$ is true;   *A* is *not* true; $f(x) = \sqrt{x} - 1$ and $g(x) = \sqrt{x} - 1$ have different graphs.

B is *not* true; The domain of $f(x) = \sqrt[3]{x} + 5$ is the set of all real numbers.

C is *not* true; The index of $\sqrt{x}$ is 2 *not* 1.

D is true

57.

Keystroke	$2\boxed{\sqrt{}}$	$\boxed{\sqrt{}}$	$\boxed{\sqrt{}}$	$\boxed{\sqrt{}}$
Display	1.4142136	1.1892071	1.0905077	1.0442738

$\boxed{\sqrt{}}$	$\boxed{\sqrt{}}$	$\boxed{\sqrt{}}$	$\boxed{\sqrt{}}$	$\boxed{\sqrt{}}$
1.0218971	1.0108893	1.0054299	1.0027113	etc

Pattern: $\boxed{\text{The numbers approach 1. The decimal part is halfed.}}$

Review Problems

59. $\dfrac{x^3 + 64}{x^2 - 16} = \dfrac{(x + 4)(x^2 - 4x + 16)}{(x + 4)(x - 4)} = \boxed{\dfrac{x^2 - 4x + 16}{x - 4}}$

60. $f(x) = \dfrac{x+10}{x-5}$, $g(x) = \dfrac{x^2-26}{x-5}$, and $h(x) = \dfrac{2x+4}{x-5}$

$$
\begin{aligned}
f(x) + g(x) - h(x) &= \frac{x+10}{x-5} + \frac{x^2-26}{x-5} - \frac{2x+4}{x-5} \\
&= \frac{x+10+x^2-26-2x-4}{x-5} \\
&= \frac{x^2-x-20}{x-5} \\
&= \frac{(x-5)(x+4)}{x-5} \\
&= \boxed{x+4 \quad (x \neq 5)}
\end{aligned}
$$

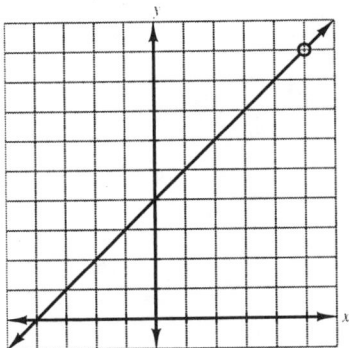

61. $x + 3y - z = 5$
$2x - 5y - z = -8$
$-x + 2y + 3z = 13$

(Equations 1 and 2):

$$
\begin{array}{rrcr}
& x + 3y - z &=& 5 \\
(\times -1) & -2x + 5y + z &=& 8 \\
\hline
& -x + 8y &=& 13 \quad \text{(Equation 4)}
\end{array}
$$

(Equations 1 and 3):

$$
\begin{array}{rrcr}
(\times 3) & 3x + 9y - 3z &=& 15 \\
& -x + 2y + 3z &=& 13 \\
\hline
& 2x + 11y &=& 28 \quad \text{(Equation 5)}
\end{array}
$$

(Equations 4 and 5):

$$
\begin{array}{rrcr}
(\times 2) & -2x + 16y &=& 26 \\
& 2x + 11y &=& 28 \\
\hline
& 27y &=& 54 \\
& y &=& 2
\end{array}
$$

(Equation 4):

$$
\begin{aligned}
-x + 8y &= 13 \\
-x + 8(2) &= 13 \\
-x &= -3 \\
x &= 3
\end{aligned}
$$

(Equation 1):

$$
\begin{aligned}
x + 3y - z &= 5 \\
3 + 3(2) - z &= 5 \\
-z &= -4 \\
z &= 4
\end{aligned}
$$

$\boxed{\{(3, 2, 4)\}}$

Section 7.3 Inverse Properties of nth Powers and nth Roots; Rational Exponents

Problem Set 7.3, pp. 517-519

1. $\sqrt[3]{4^3} = \boxed{4}$ ($\sqrt[n]{x^n} = n$ when n is odd)   **3.** $\sqrt[3]{(-2)^3} = \boxed{-2}$

5. $\sqrt[4]{2^4} = |2| = \boxed{2}$ ($\sqrt[n]{x^n} = |x|$ when n is even)   **7.** $\sqrt[4]{(-2)^4} = |-2| = \boxed{2}$

9. $\sqrt[5]{(-2)^5} = \boxed{-2}$ **11.** $\sqrt[17]{(-9)^{17}} = \boxed{-9}$

13. $\sqrt[8]{(-9)^{18}} = |-9| = \boxed{9}$ **15.** $\sqrt[n]{(-8)^n}$ (n is odd) $= \boxed{-8}$

17. $\sqrt[n]{(-8)^n}$ (n is even) $= |-8| = \boxed{8}$

19. $f(x) = \sqrt{x^2}$ and $g(x) = x$

    a. From the graph
        $\sqrt{x^2} = x$ when $f(x) = g(x)$ or when $\boxed{x \ge 0}$

    b. $\sqrt{x^2} \ne -x$ when $f(x) \ne g(x)$ or when $\boxed{x < 0}$

21. $36^{1/2} = \sqrt{36} = \boxed{6}$ **23.** $8^{1/3} = \sqrt[3]{8} = \boxed{2}$

25. $(-27)^{1/3} = \sqrt[3]{-27} = \sqrt[3]{(-3)^3} = \boxed{-3}$ **27.** $64^{-1/3} = \dfrac{1}{64^{1/2}} = \dfrac{1}{\sqrt{64}} = \boxed{\dfrac{1}{8}}$

29. $\left(\dfrac{1}{64}\right)^{-1/3} = \dfrac{1}{\left(\frac{1}{64}\right)^{1/3}} = \dfrac{1}{\frac{1}{4}} = \boxed{4}$ **31.** $(-64)^{-1/3} = \dfrac{1}{(-64)^{-1/3}} = \dfrac{1}{\sqrt[3]{-64}} = \dfrac{1}{-4} = \boxed{-\dfrac{1}{4}}$

33. $\left(-\dfrac{27}{64}\right)^{-1/3} = \dfrac{1}{\left(\frac{-27}{64}\right)^{1/3}} = \dfrac{1}{\sqrt[3]{-\frac{27}{64}}} = \dfrac{1}{-\frac{3}{4}} = \boxed{-\dfrac{4}{3}}$

35. $16^{5/2} = (\sqrt{16})^5 = 4^5 = \boxed{1024}$ **37.** $8^{2/3} = (\sqrt[3]{8})^2 = 2^2 = \boxed{4}$

39. $(-1)^{17/3} = (\sqrt[3]{-1})^{17} = (-1)^{17} = \boxed{-1}$ **41.** $-25^{3/2} = -(\sqrt{25})^3 = -(5^3) = \boxed{-125}$

43. $\left(\dfrac{27}{64}\right)^{2/3} = \left(\sqrt[3]{\dfrac{27}{64}}\right)^2 = \left(\dfrac{3}{4}\right)^2 = \boxed{\dfrac{9}{16}}$ **45.** $\left(\dfrac{1}{32}\right)^{-3/5} = \dfrac{1}{\left(\frac{1}{32}\right)^{3/5}} = \dfrac{1}{\sqrt[5]{\frac{1}{32}^3}} = \dfrac{1}{\left(\frac{1}{2}\right)^3} = \dfrac{1}{\frac{1}{8}} = \boxed{8}$

47. $16^{-5/2} = \dfrac{1}{16^{5/2}} = \dfrac{1}{(\sqrt{16})^5} = \dfrac{1}{4^5} = \boxed{\dfrac{1}{1024}}$ **49.** $(-1)^{17/125} = \left(\dfrac{125}{\sqrt{-1}}\right)^{17} = (-1)^{17} = \boxed{-1}$

51. $[3 + (27^{2/3} + 32^{2/5})]^{3/2} - 9^{1/2}$

$= [3 + (\sqrt[3]{27})^2 + (\sqrt[5]{32})^2]^{3/2} - 9^{1/2}$

$= [3 + 3^2 + 2^2]^{3/2} - 9^{1/2}$

$= 16^{3/2} - 9^{1/2}$

$= (\sqrt{16})^3 - \sqrt{9}$

$= 4^3 - 3$

$= 64 - 3$

$= \boxed{61}$

53. $N = 30t^{3/4} - 10$ (t, in weeks)

$(t = 16)$:
$N = 30(16^7) - 10$

$= 30(\sqrt[4]{16})^3 - 10$

$= 30(2)^3 - 10$

$= 30(8) - 10$

$= 230$

Since $230 < 250$, Pierre (in spite of his name) will not be ready for his trip to Paris.
$\boxed{\text{No}}$

55. $A = 2W^{2/5}H^{3/4}$ (W, in kilograms; H, in meters)

$(W = 243$ kilograms, $H = 2$ meters$)$:

$A = 2(243)^{2/5}(2)^{3/4}$

$= 2(\sqrt[5]{243})^2(\sqrt[4]{2^3})$

$= 2(3)^2(\sqrt[4]{8})$

$= \boxed{18(\sqrt[4]{8})\text{ m}^2}$

57. $T = 10(60 + 120x^{-1})Q^{-1/2}$ (T, in minutes)

$(Q = 100; x = 6)$:

$T = 10(60 + 120 \cdot 6^{-1})100^{-1/2}$

$= 10\left(60 + 120 \cdot \frac{1}{6}\right) \cdot \frac{1}{100^{1/2}}$

$= 10(60 + 20) \cdot \frac{1}{\sqrt{100}}$

$= 10(80) \cdot \frac{1}{10}$

$= 80$

$\boxed{80 \text{ minutes}}$

59. $v = \left(\dfrac{5r}{2}\right)^{1/2}$ (v, in mph; r, in feet)

$(r = 250$ feet$)$:

$v = \left(\dfrac{5 \cdot 250}{2}\right)^{1/2} = \sqrt{625} = 25$

$\boxed{25 \text{ mph}}$

61. $O = (0.26 + 0.2t)(36 + 16t - t^2)^{-1/2}$

$t = 16$ since 7:00 P.M. is 16 hours after 3:00 A.M.

$O = [0.26 + 0.02(16)] \cdot [36 + 16(16) - 16^2]^{-1/2}$

$= (0.58)(36)^{-1/2}$

$= (0.58) \cdot \dfrac{1}{36^{1/2}}$

$= (0.58) \dfrac{1}{\sqrt{36}}$

$= \dfrac{0.58}{6}$

$= 0.09\overline{6}$

The ozone level is approximately 0.097 parts per million.

63. $3x^{1/2} - (2x)^0 + 16x^{-2}$ when $x = 4$

$3(4)^{1/2} - (2 \cdot 4)^0 + 16(4^{-2})$

$= 3(2) - 1 + 16 \cdot \dfrac{1}{16}$

$= 6 - 1 + 1$

$= \boxed{6}$

65. $\dfrac{2x^{-1/3}}{3x^{5/3}}$ when $x = 4$

$= \dfrac{2(4^{-1/3})}{3(4^{5/3})}$

$= \dfrac{2}{3} \cdot \dfrac{1}{4^{1/3} \cdot 4^{5/3}}$

$= \dfrac{2}{3} \cdot \dfrac{1}{4^{6/3}}$

$= \dfrac{2}{3} \cdot \dfrac{1}{4^2}$

$= \dfrac{2}{3(16)}$

$= \boxed{\dfrac{1}{24}}$

67. $\sqrt[3]{73} \approx \boxed{4.1793}$

69. $\sqrt[3]{-19} \approx \boxed{-2.6684}$

71. $\sqrt[4]{973} \approx \boxed{5.5851}$

73. $\sqrt[4]{-19}$ is $\boxed{\text{not possible; for } n \text{ even } (n = 4),\, b(b = -19) \text{ must be} \geq 0}$

75. $31^{-3/4} = \dfrac{1}{31^{3/4}} = \dfrac{1}{\sqrt[4]{31^3}} \approx \boxed{0.0761}$

77. $\dfrac{7 - \sqrt{23}}{2} \approx \boxed{1.1021}$

79. $r = \left(\dfrac{A}{P}\right)^{1/t} - 1$

$(P = \$80{,}000,\ A = \$120{,}000,\ t = 4 \text{ years})$:

$r = \left(\dfrac{120{,}000}{80{,}000}\right)^{1/4} - 1 \approx 0.107 = \boxed{10.7\%}$

81. $\boxed{A}$ is true; $(-b)^{1/n} = (-1)^{1/4}(b^{1/n}) = -b^{1/n}$ if n is odd

83. $x \uparrow y = x^y$ and $x \downarrow y = y^x$

$$\left[\left(64 \uparrow \frac{1}{2}\right) \uparrow \frac{1}{3}\right] \downarrow 5$$

$$= \left[(64^{1/2}) \uparrow \frac{1}{3}\right] \downarrow 5$$

$$= \left(8 \uparrow \frac{1}{3}\right) \downarrow 5$$

$$= (8^{1/3}) \downarrow 5$$

$$= 2 \downarrow 5$$

$$= 5^2$$

$$= \boxed{25}$$

85.

$$\left(\frac{1}{81}\right)^{1/4} \boxed{?} \left(\frac{1}{81}\right)^{1/2}$$

$$\frac{1}{81^{1/4}} \boxed{?} \frac{1}{81^{1/2}}$$

$$\frac{1}{3} > \frac{1}{9}$$

$\boxed{\left(\dfrac{1}{81}\right)^{1/4} \text{ is greater}}$ $\left(Note:\ 4 > 2 \text{ and } 0 < \dfrac{1}{81} < 1\right)$

In general if $n > m$ and $0 < b < 1$

$\boxed{b^{1/n} \text{ is greater}}$

Review Problems

90. $x^2 y - xy^2 - xy + y^2$

$$= xy(x - y) - y(x - y)$$

$$= (x - y)(xy - y)$$

$$= (x - y)y(x - 1)$$

$$= \boxed{y(x - y)(x - 1)}$$

91.

$$\frac{5}{y + 2} - \frac{3}{y + 5} = \frac{9}{y^2 + 7y + 10}$$

$$\frac{5(y + 5)}{(y + 2)(y + 5)} - \frac{3(y + 2)}{(y + 5)(y + 2)} = \frac{9}{(y + 2)(y + 5)}$$

$$(y + 2)(y + 5)\left[\frac{5(y + 5)}{(y + 2)(y + 5)} - \frac{3(y + 2)}{(y + 5)(y + 2)}\right] = (y + 2)(y + 5)\left[\frac{9}{(y + 2)(y + 5)}\right]$$

$$5(y + 5) - 3(y + 2) = 9$$

$$5y + 25 - 3y - 6 = 9$$

$$2y + 19 = 9$$

$$2y = -10$$

$$y = -5$$

Since -5 causes denominator to be zero in the original equation, -5 is not a solution. The equation has no solution. $\boxed{\varnothing}$.

92. $\dfrac{4y + 1}{8y^3 - 1} + \dfrac{2y}{4y^2 + 2y + 1}$

$$= \frac{4y + 1}{(2y - 1)(4y^2 + 2y + 1)} + \frac{2y}{4y^2 + 2y + 1} \cdot \frac{2y - 1}{2y - 1}$$

$$= \frac{4y + 1 + 2y(2y - 1)}{(2y - 1)(4y^2 + 2y + 1)}$$

$$= \frac{4y + 1 + 4y^2 - 2y}{(2y - 1)(4y^2 + 2y + 1)}$$

$$= \frac{4y^2 + 2y + 1}{(2y - 1)(4y^2 + 2y + 1)}$$

$$= \boxed{\frac{1}{2y - 1}}$$

Section 7.4   More About Rational Exponents and Radicals

Problem Set 7.4, pp. 525-527

1. $(7y^{1/3})(2y^{1/4})$
$= 14y^{1/3 + 1/4}$
$= 14y^{4/12 + 3/12}$
$= \boxed{14y^{-7/12}}$

3. $(3x^{3/4})(-5x^{-1/2})$
$= -15x^{3/4 - 1/2}$
$= -15x^{3/4 - 2/4}$
$= \boxed{-15x^{1/4}}$

5. $\dfrac{20x^{1/2}}{5x^{1/4}}$
$= 4x^{1/2 - 1'4}$
$= 4x^{2/4 - 1/4}$
$= \boxed{4x^{1/4}}$

7. $\dfrac{80y^{1.6}}{10y^{1/4}}$
$= 8y^{1/6 - 1/4}$
$= 8y^{2/12 - 3/12}$
$= 8y^{-1/12}$

$= \boxed{\dfrac{8}{y^{1/12}}}$

9. $(2x^{1/5}y^2z^{2/5})^5$
$= 2^5(x^{1/5})^5(y^2)^5(z^{2/5})^5$
$= \boxed{32xy^{10}z^2}$

11. $(25x^4y^6)^{1/2}$
$= 25^{1/2}(x^4)^{1/2}(y^6)^{1/2}$
$= 25^{1/2}x^2y^3$
$= \boxed{5x^2y^3}$

13. $(16xy^{1/4}z^{2/3})^{1/4}$

$= 16^{1/4}x^{1/4}y^{1/16}z^{1/6}$

$= \boxed{2x^{1/4}y^{1/16}z^{1/6}}$

15. $\left(\dfrac{2x^{1/4}}{5y^{1/3}}\right)^3$
$= \dfrac{(2x^{1/4})^3}{(5y^{1/3})^3}$
$= \boxed{\dfrac{8x^{3/4}}{125y}}$

17. $\left(\dfrac{x^3}{y^5}\right)^{-1/2}$
$= \dfrac{x^{-3/2}}{y^{-5/2}}$
$= \boxed{\dfrac{y^{5/2}}{x^{3/2}}}$

19. $\left(\dfrac{27a^{-3}}{64b^{-3}}\right)^{-1/3}$
$= \dfrac{27^{-1/3}a}{64^{-1/3}b}$
$= \dfrac{64^{1/3}a}{27^{1/3}b}$
$= \boxed{\dfrac{4a}{3b}}$

21. $\dfrac{8a^2 b^{-2/3}}{6(ab)^{1/2}}$

$= \dfrac{8a^2 b^{-2/3}}{6a^{1/2} b^{1/2}}$

$= \dfrac{4}{3} a^{2-1/2} b^{-2/3-1/2}$

$= \dfrac{4}{3} a^{3/2} b^{-7/6}$

$= \boxed{\dfrac{4a^{3/2}}{3b^{7/6}}}$

23. $\left(\dfrac{27a^3 b^{-3/2}}{8a^{-1} b^3}\right)^{-2/3}$

$= \left(\dfrac{27}{8} a^4 b^{-9/2}\right)^{-2/3}$

$= \left(\dfrac{27}{8}\right)^{-2/3} a^{-8/3} b^3$

$= \dfrac{b^3}{\left(\dfrac{27}{8}\right)^{2/3} a^{8/3}}$

$= \dfrac{b^3}{\left(\dfrac{3}{2}\right)^2 a^{8/3}}$

$= \dfrac{b^3}{\dfrac{9}{4} a^{8/3}}$

$= \boxed{\dfrac{4b^3}{9a^{8/3}}}$

25. $(x^{1/a})^{a^2}$

$= x^{1/a \cdot a^2}$

$= \boxed{x^a}$

27. $(x^{3b} y^b)^{1/b}$

$= (x^{3b})^{1/b} (y^b)^{1/b}$

$= \boxed{x^3 y}$

29. $\dfrac{(r^n s^{1/n})^n}{r^{n^2} n}$

$= \dfrac{r^{n^2} s}{r^{n^2} s^n}$

$= \boxed{s^{1-n}}$

or $\left(\dfrac{1}{s^{n-1}}\right)$

31. $\dfrac{x^{1/a} y^{5/a}}{x^{2/a}}$

$= x^{1/a - 2/a} y^{5/a}$

$= x^{-1/a} y^{5/a}$

$= \boxed{\dfrac{y^{5/a}}{x^{1/a}}}$

33. $\left(\dfrac{x^{4a} y^{6a}}{z^{8a} w^{10a}}\right)^{-3/2a}$

$= \dfrac{(x^{4a} y^{6a})^{-3/2a}}{(z^{8a} w^{10a})^{-3/2a}}$

$= \dfrac{x^{4a \cdot (-3/2a)} y^{6a \cdot (-3/2a)}}{(z^{8a})^{-3/2a} (w^{10a})^{-3/2a}}$

$= \dfrac{x^{-6} y^{-9}}{z^{-12} w^{-15}}$

$= \boxed{\dfrac{z^{12} w^{15}}{x^6 y^9}}$

35. $\left(\dfrac{r^{-3/4a}}{s^{2/3 a}}\right)^{-12 a} (r^{-6a^2} s^{-8a^2})^{-5/2a^2}$

$= \dfrac{r^9}{s^{-8}} (r^{-15} s^{20})$

$= \dfrac{r^{-6} s^{20}}{s^{-8}}$

$= r^{-6} s^{28}$

$= \boxed{\dfrac{s^{28}}{r^6}}$

37. $\sqrt[3]{3} \cdot \sqrt{3}$

$= 3^{1/3} \cdot 3^{1/2}$

$= 3^{5/6}$

$= \boxed{\sqrt[6]{3^5}}$ or $\sqrt[6]{243}$

39. $\sqrt[4]{2} \cdot \sqrt{2}$

$= 2^{1/4} \cdot 2^{1/2}$

$= 2^{3/4}$

$= \sqrt[4]{2^3}$

$= \boxed{\sqrt[4]{8}}$

41. $\dfrac{\sqrt{3}}{\sqrt[3]{3}}$

$= \dfrac{3^{1/2}}{3^{1/3}}$

$= 3^{1/2-1/3}$

$= 3^{1/6}$

$= \boxed{\sqrt[6]{3}}$

43. $\dfrac{\sqrt{7}}{\sqrt[3]{7}}$

$= \dfrac{7^{1/2}}{7^{1/3}}$

$= 7^{1/6}$

$= \boxed{\sqrt[6]{7}}$

45. $\dfrac{\sqrt{8}}{\sqrt[3]{2}}$

$= \dfrac{8^{1/2}}{2^{1/3}}$

$= \dfrac{(2^3)^{1/2}}{2^{1/3}}$

$= \dfrac{2^{3/2}}{2^{1/3}}$

$= 2^{3/2-1/3}$

$= 2^{7/6}$

$= \sqrt[6]{2^7}$

$= \boxed{\sqrt[6]{128}}$

47. $\dfrac{\sqrt[3]{8}}{\sqrt[6]{4}}$

$= \dfrac{2}{4^{1/6}}$

$= \dfrac{2}{(2^2)^{1/6}}$

$= \dfrac{2}{2^{1/3}}$

$= 2^{1-1/3}$

$= 2^{2/3}$

$= \sqrt[3]{2^2}$

$= \boxed{\sqrt[3]{4}}$

49. $\sqrt[9]{a^3}$

$= (a^3)^{1/9}$

$= a^{1/3}$

$= \boxed{\sqrt[3]{a}}$

51. $\sqrt[9]{27}$

$= \sqrt[9]{3^3}$

$= (3^3)^{1/9} = 3^{1/3}$

$= \boxed{\sqrt[3]{3}}$

53. $\sqrt[12]{16}$

$= \sqrt[12]{2^4}$

$= (2^4)^{1/12}$

$= 2^{1/3}$

$= \boxed{\sqrt[3]{2}}$

55. $\sqrt[4]{x^2y^2}$

$= (x^2y^2)^{1/4}$

$= x^{1/2}y^{1/2}$

$= (xy)^{1/2}$

$= \boxed{\sqrt{xy}}$

57. $\sqrt[9]{2^3x^3y^6}$

$= (2^3x^3y^6)^{1/9}$

$= 2^{1/3}x^{1/3}y^{2/3}$

$= (2xy^2)^{1/3}$

$= \boxed{\sqrt[3]{2xy^2}}$

59. $\sqrt[9]{27x^3y^6}$

$= \sqrt[9]{3^3x^3y^6}$

$= (3^3x^3y^6)^{1/9}$

$= 3^{1/3}x^{1/3}y^{2/3} = (3xy^2)^{1/3}$

$= \boxed{\sqrt[3]{3xy^2}}$

61. $\sqrt[6]{x^2 - 2xy + y^2}$

$= \sqrt[6]{(x-y)^2}$

$= [(x-y)^2]^{1/6}$

$= (x-y)^{1/3}$

$= \boxed{\sqrt[3]{x-y}}$

63. $(2^{5/2} \cdot 2^{3/4} \div 2^{1/4})$

$= \dfrac{2^{5/2} \cdot 2^{3/4}}{2^{1/4}}$

$= \dfrac{2^{13/4}}{2^{1/4}}$

$= 2^{12/4} = 2^3$

$= \boxed{8}$

The son is $\boxed{8 \text{ years old}}$.

65. $x^{1/3}(x^{2/3} + x^{4/3})$

$= x^{1/3} x^{2/3} + x^{1/x \cdot 4/3}$

$= \boxed{x + x^{5/3}}$

67. $3y^{1/2}(5y^{3.2} + 4y^{1/2} - 2y^{5/2})$

$= 3y^{1/2}(5y^{3/2}) + 3y^{1/2}(4y^{1/2}) - 3y^{1/2}(2y^{5/2})$

$= \boxed{15y^2 + 12y - 6y^3}$

69. $2x^{1/3}y^{4.5}(7x^{5.3}y^{-4/5} - 3x^{-1/3}y^{6/5})$

$= 2x^{1/3}y^{4/5}(7x^{5/3}y^{-4/5}) + 2x^{1/3}y^{4/5}(-3x^{-1/3}y^{6/5})$

$= 14x^2 y^0 - 6x^0 y^2$

$= \boxed{14x^2 - 6y^2}$

71. $(x^{1/3} - 5)(x^{1/3} + 2)$

$= (x^{1/3})^2 + 2x^{1/3} - 5x^{1/3} - 10$

$= \boxed{x^{2/3} - 3x^{1/3} - 10}$

73. $(5x^{2/3} - 3)(2x^{2/3} + 1)$

$= 10(x^{2/3})^2 + 5x^{2/3} - 6x^{2/3} - 3$

$= \boxed{10x^{4/3} - x^{2/3} - 3}$

75. $(3x^{1/2} + 4y^{2/3})(5x^{1/2} - 2y^{2/3})$

$= (3x^{1/2})(5x^{1/2}) + (3x^{1/2})(-2y^{2/3})$

$\qquad + (4y^{2/3})(5x^{1/2}) + (4y^{2/3})(-2y^{2/3})$

$= 15x - 6x^{1/2}y^{2/3} + 20x^{1/2}y^{2/3} - 8y^{4/3}$

$= \boxed{15x + 14x^{1/2}y^{2/3} - 8y^{4/3}}$

77. $(x^{2/3} + 3)^2$

$= (x^{2/3})^2 + 2x^{2/3} \cdot 3 + 3^2$

$= \boxed{x^{4/3} + 6x^{2/3} + 9}$

79. $(x^{1/2} - y^{1/2})^2$

$= (x^{1/2})^2 - 2x^{1/2}y^{1/2} + (y^{1/2})^2$

$= \boxed{x - 2x^{1/2}y^{1/2} + y}$

81. $(3x^{1/2} + 4y^{1/2})^2$

$= (3x^{1/2})^2 + 2(3x^{1/2})(4y^{1/2}) + (4y^{1/2})^2$

$= \boxed{9x + 24x^{1/2}y^{1/2} + 16y}$

83. $(x^{1/2} + 5^{1/2})(x^{1/2} - 5^{1/2})$

$= (x^{1/2})^2 - (5^{1/2})^2$

$= \boxed{x - 5}$

85. $(x^{2/3} + y^{2.3})(x^{2/3} - y^{2/3})$

$= (x^{2/3})^2 - (y^{2/3})^2$

$= \boxed{x^{4/3} - y^{4/3}}$

87. $(3x^{3/2} + 4^{1/2})(3x^{3/2} - 4^{1/2})$

$= (3x^{3/2})^2 - (4^{1/2})^2$

$= \boxed{9x^3 - 4}$

89. $(x^{1/3} - y^{1/3})(x^{2/3} + x^{1/3}y^{1/3} + y^{2/3})$

$= x + x^{2/3}y^{1/3} + x^{1/3}y^{2/3} - x^{2/3}y^{1/3} - x^{1/3}y^{2/3} - y$

$= \boxed{x - y}$

91. $(x^{1/2} + 1)[(x^{1/4} + 1)(x^{1/4} - 1)]$

$= (x^{1/2} + 1)(x^{1/2} - 1)$

$= (x^{1/2})^2 - 1^2$

$= \boxed{x - 1}$

93. $\dfrac{12x^{1/3}y^{4/3} - 24x^{4/3}y^{1/3}}{4x^{1/3}y^{1/3}}$

$= \dfrac{12x^{1/3}y^{3/4}}{4x^{1/3}y^{1/3}} - \dfrac{24x^{4/3}y^{1/3}}{4x^{1/3}y^{1/3}}$

$= 3x^0 y^1 - 6x^1 y^0$

$= \boxed{3y - 6x}$

95. $\dfrac{27x^{8/5}y^{4/5} - 36x^{3/5}y^{9/5}}{9x^{3/5}y^{4/5}}$

$= \dfrac{27x^{8/5}y^{4/5}}{9x^{3/5}y^{4/5}} - \dfrac{36x^{3/5}y^{9/5}}{9x^{3/5}y^{4/5}}$

$= 3x^1 y^0 - 4x^0 y^1$

$= \boxed{3x - 4y}$

97. $12(x-5)^{3/2} - 18(x-5)^{1/2}$
$= 6(x-5)^{1/2}[2(x-5)-3]$
$= \boxed{6(x-5)^{1/2}(2x-13)}$

99. $7(x-1)^{12/5} - 14(x-1)^{7/5}$
$= 7(x-1)^{7/5}[(x-1)^{5/5}-2]$
$= 7(x-1)^{7/5}(x-1-2)$
$= \boxed{7(x-1)^{7/5}(x-3)}$

101. $15y(y+3)^{3/2} + 20(y+3)^{1/2}$
$= 5(y+3)^{1/2}[3y(y+3)+4]$
$= \boxed{5(y+3)^{1/2}(3y^2+9y+4)}$

103. $x^{1/2} + x^{1/4} - 6$
$= (x^{1/4})^2 + x^{1/4} - 6$
$= \boxed{(x^{1/4}+3)(x^{1/4}-2)}$

105. $8x^{2/5} + 10x^{1/5} - 3$
$= 8(x^{1/5})^2 + 10x^{1/5} - 3$
$= \boxed{(4x^5-1)(2x^{1/5}+3)}$

107. $9x^{2/5} + 12x^{1/3} + 4$
$= 9(x^{1/5})^2 + 12x^{1/5} + 4$
$= (3x^{1/5}+2)(3x^{1/5}+2)$
$= \boxed{(3x^{1/5}+2)^2}$

109. $9x^{2/3} - 16$
$= (3x^{1/3})^2 - 4^2$
$= \boxed{(3x^{1/3}+4)(3x^{1/3}-4)}$

111. $\dfrac{3}{y^{1/2}} - y^{1/2}$ [LCD is $y^{1/2}$]
$= \dfrac{3}{y^{1/2}} - y^{1/2} \cdot \dfrac{y^{1/2}}{y^{1/2}}$
$= \boxed{\dfrac{3-y}{y^{1/2}}}$

113. $y^{3/4} - \dfrac{4}{y^{1/4}}$ [LCD is $y^{1/4}$]
$= y^{3/4} \cdot \dfrac{y^{1/4}}{y^{1/4}} - \dfrac{4}{y^{1/4}}$
$= \boxed{\dfrac{y-4}{y^{1/4}}}$

115. $\dfrac{4y^3}{(y^2-1)^{1/2}} + 3y(y^2-1)^{1/2}$ [LCD is $(y^2-1)^{1/2}$]
$= \dfrac{4y^2}{(y^2-1)^{1/2}} + 3y(y^2-1)^{1/2} \cdot \dfrac{(y^2-1)^{1/2}}{(y^2-1)^{1/2}}$
$= \dfrac{4y^3 + 3y(y^2-1)}{(y^2-1)^{1/2}}$
$= \dfrac{4y^3 + 3y^3 - 3y}{(y^2-1)^{1/2}}$
$= \boxed{\dfrac{7y^3 - 3y}{(y^2-1)^{1/2}}}$

117. $\dfrac{y^5}{(y^2+1)^{1/2}} + 3y^3(y^2+1)^{1/2}$ [LCD is $(y^2+1)^{1/2}$]
$= \dfrac{y^5}{(y^2+1)^{1/2}} + 3y^3(y^2+1)^{1/2} \cdot \dfrac{(y^2+1)^{1/2}}{(y^2+1)^{1.2}}$
$= \dfrac{y^5 + 3y^3(y^2+1)}{(y^2+1)^{1/2}}$
$= \dfrac{y^5 + 3y^5 + 3y^3}{(y^2+1)^{1/2}}$
$= \boxed{\dfrac{4y^5 + 3y^3}{(y^2+1)^{1/2}}}$

119. Let x = number
$\dfrac{x^{-2/3}}{\sqrt[3]{x}} = \dfrac{1}{4}$
$\dfrac{x^{-2/3}}{x^{1/3}} = \dfrac{1}{4}$
$x^{-1} = \dfrac{1}{4}$
$\dfrac{1}{x} = \dfrac{1}{4}$
$x = 4$
The number is $\boxed{4}$.

121. $\boxed{\sqrt[3]{\sqrt{b}} = \sqrt[3]{b^{1/2}} = (b^{1/2})^{1/3} = b^{1/6} = \sqrt[6]{b}}$ $(b \ge 0)$

Review Problems

125.
$$x + 4y - 2z = -3$$
$$2x + y + z = 3$$
$$-5x - 2y + 3z = -14$$

$$\begin{bmatrix} 1 & 4 & -2 & | & -3 \\ 2 & 1 & 1 & | & 3 \\ -5 & -2 & 3 & | & -14 \end{bmatrix} \begin{matrix} R_2 - 2R_1 \to \\ \\ 5R_1 + R_3 \to \end{matrix} \begin{bmatrix} 1 & 4 & -2 & | & -3 \\ 0 & -7 & 5 & | & 9 \\ 0 & 18 & -7 & | & -29 \end{bmatrix}$$

$$\begin{matrix} \\ \\ 18R_2 + 7R_3 \to \end{matrix} \begin{bmatrix} 1 & 4 & -2 & | & -3 \\ 0 & -7 & 5 & | & 9 \\ 0 & 0 & 41 & | & -41 \end{bmatrix} \begin{matrix} -\frac{1}{7}R_2 \to \\ \\ \frac{1}{41}R_3 \to \end{matrix} \begin{bmatrix} 1 & 4 & -2 & | & -3 \\ 0 & 1 & -5/7 & | & -9/7 \\ 0 & 0 & 1 & | & -1 \end{bmatrix}$$

$$x + 4y - 2z = -3$$
$$y - \frac{5}{7}z = -\frac{9}{7}$$
$$z = -1 \qquad\qquad y - \frac{5}{7}(-1) = -\frac{9}{7}$$
$$y = -\frac{14}{7} = -2$$

$$x + 4(-2) - 2(-1) = -3$$
$$x - 6 = -3$$
$$x = 3$$

$$\boxed{\{(3, -2, -1)\}}$$

126. $f(x) = -7x^2 + 3x - 5$

$$\frac{f(a + h) - f(a)}{h}$$

$$= \frac{[-7(a + h)^2 + 3(a + h) - 5] - (-7a^2 + 3a - 5)}{h}$$

$$= \frac{-7a^2 - 14ah - 7h^2 + 3a + 3h - 5 + 7a^2 - 3a + 5}{h}$$

$$= \frac{-14ah - 7h^2 + 3h}{h}$$

$$= \boxed{-14a - 7h + 3}$$

127. $\begin{vmatrix} -3 & 0 & 4 \\ 5 & 2 & -3 \\ 7 & 0 & 6 \end{vmatrix} = -0 \begin{vmatrix} 5 & -3 \\ 7 & 6 \end{vmatrix} + 2 \begin{vmatrix} -3 & 4 \\ 7 & 6 \end{vmatrix} - 0 \begin{vmatrix} -3 & 4 \\ 5 & -3 \end{vmatrix}$

$$= -0 + 2(-18 - 28) - 0$$
$$= 2(-46)$$
$$= \boxed{-92}$$

Section 7.5 Simplifying Radical Expressions

Problem Set 7.5, pp. 534-535

1. $\sqrt{20} = \sqrt{4 \cdot 5} = \sqrt{2^2 \cdot 5} = \boxed{2\sqrt{5}}$

3. $\sqrt{80} = \sqrt{16 \cdot 5} = \sqrt{4^2 \cdot 5} = \boxed{4\sqrt{5}}$

5. $\sqrt{250} = \sqrt{25 \cdot 10} = \sqrt{5^2 \cdot 10} = \boxed{5\sqrt{10}}$

7. $7\sqrt{28} = 7\sqrt{4 \cdot 7} = 7\sqrt{2^2 \cdot 7} = 7(2\sqrt{7}) = \boxed{14\sqrt{7}}$

9. $2\sqrt{98} = 2\sqrt{49 \cdot 2} = \boxed{14\sqrt{2}}$

11. $\sqrt[3]{54} = \sqrt[3]{27 \cdot 2} = \sqrt[3]{3^3 \cdot 2} = \boxed{3\sqrt[3]{2}}$

13. $\sqrt[5]{64} = \sqrt[5]{32 \cdot 2} = \sqrt[5]{2^5 \cdot 2} = \boxed{2\sqrt[5]{2}}$

15. $6\sqrt[3]{16} = 6\sqrt[3]{8 \cdot 2} = 6\sqrt[3]{2^3 \cdot 2} = 6(2)(\sqrt[3]{2}) = \boxed{12\sqrt[3]{2}}$

17. $-5\sqrt[3]{128} = -5(\sqrt[3]{64 \cdot 2}) = \boxed{-20\sqrt[3]{2}}$

19. $\sqrt{12y} = \sqrt{4 \cdot 3y} = \sqrt{2^2 \cdot 3y} = \boxed{2\sqrt{3y}}$

21. $\sqrt{48x^2} = \sqrt{16 \cdot 3x^2} = \boxed{4x\sqrt{3}}$

23. $\sqrt{48x^3} = \sqrt{16 \cdot 3x^3} = \sqrt{4^2 \cdot 3x^2 \cdot x} = \boxed{4x\sqrt{3x}}$

25. $\sqrt{20xy^3} = \sqrt{4 \cdot 5xy^3} = \sqrt{2^2 \cdot 5xy^2 \cdot y} = \boxed{2y\sqrt{5xy}}$

27. $\sqrt{75xy^2z^5} = \sqrt{25 \cdot 3xy^2 \cdot z^5} = \sqrt{5^2 \cdot 3xy^2(z^2)^2z} = \boxed{5yz^2\sqrt{3xz}}$

29. $4\sqrt{50x^7} = 4\sqrt{25 \cdot 2(x^3)^2 \cdot x} = 4 \cdot 5x^3\sqrt{2x} = \boxed{20x^3\sqrt{2x}}$

31. $\sqrt[3]{32x^3} = \sqrt[3]{8 \cdot 4x^3} = \sqrt[3]{2^3 \cdot 4x \cdot x^3} = \boxed{2x\sqrt[3]{4}}$

33. $\sqrt[3]{-32xy^5z^6} = \sqrt[3]{-8 \cdot 4xy^5z^6} = \sqrt[3]{(-2)^3 \cdot 4xy^3y^2(z^2)^3} = \boxed{-2yz^2\sqrt[3]{4xy^2}}$

35. $-2x^2y(\sqrt[3]{54x^3y^7z^2}) = -2x^2y(\sqrt[3]{27 \cdot 2x^3(y^2)^3yz^2}) = -2x^2y(3xy^2)\sqrt[3]{2yz^2} = \boxed{-6x^3y^3\sqrt[3]{2yz^2}}$

37. $-5\sqrt[4]{48y^7} = -5\sqrt[4]{16 \cdot 3y^7} = -5\sqrt[4]{2^4 \cdot 3y^4y^3} = -5(2)y\sqrt[4]{3y^3} = \boxed{-10y\sqrt[4]{3y^3}}$

39. $-3y(\sqrt[5]{64x^3y^6}) = -3y(\sqrt[5]{2^5 \cdot 2x^3y^5y}) = -3y(2y)\sqrt[5]{2x^3y} = \boxed{-6y^2\sqrt[5]{2x^3y}}$

41. $\sqrt{\dfrac{1}{4}} = \dfrac{\sqrt{1}}{\sqrt{4}} = \boxed{\dfrac{1}{2}}$

43. $\sqrt[3]{\dfrac{8}{125}} = \dfrac{\sqrt[3]{8}}{\sqrt[3]{125}} = \boxed{\dfrac{2}{5}}$

45. $\sqrt[4]{\dfrac{1}{16}} = \dfrac{\sqrt[4]{1}}{\sqrt[4]{16}} = \boxed{\dfrac{1}{2}}$

47. $-2\sqrt{\dfrac{16}{100}} = -2\dfrac{\sqrt{16}}{\sqrt{100}} = -2 \cdot \dfrac{4}{10} = \boxed{-\dfrac{4}{5}}$

49. $-\sqrt[5]{-\dfrac{1}{32}} = -\dfrac{\sqrt[5]{-1}}{\sqrt[5]{32}} = -\dfrac{(-1)}{2} = \boxed{\dfrac{1}{2}}$

51. $\sqrt{\dfrac{80}{25}} = \dfrac{\sqrt{80}}{\sqrt{25}} = \dfrac{\sqrt{16 \cdot 5}}{5} = \boxed{\dfrac{4\sqrt{5}}{5}}$

53. $\sqrt{\dfrac{20}{81}} = \dfrac{\sqrt{4 \cdot 5}}{\sqrt{81}} = \boxed{\dfrac{2\sqrt{5}}{9}}$

55. $-5\sqrt{\dfrac{28}{121}} = -5\dfrac{\sqrt{28}}{\sqrt{121}} = -5\dfrac{\sqrt{4 \cdot 7}}{11} = -5\dfrac{(2)\sqrt{7}}{11} = \boxed{\dfrac{-10\sqrt{7}}{11}}$

57. $-3\sqrt[3]{\dfrac{16}{27}} = -3\dfrac{\sqrt[3]{8 \cdot 2}}{\sqrt[3]{27}} = \dfrac{-6\sqrt[3]{2}}{3} = \boxed{-2\sqrt[3]{2}}$

59. $\sqrt{\dfrac{1}{3}} = \dfrac{1}{\sqrt{3}} = \dfrac{1}{\sqrt{3}} \cdot \dfrac{\sqrt{3}}{\sqrt{3}} = \dfrac{\sqrt{3}}{\sqrt{3^2}} = \boxed{\dfrac{\sqrt{3}}{3}}$

61. $\sqrt{\dfrac{2}{5}} = \dfrac{\sqrt{2}}{\sqrt{5}} = \dfrac{\sqrt{2}}{\sqrt{5}} \cdot \dfrac{\sqrt{5}}{\sqrt{5}} = \dfrac{\sqrt{10}}{\sqrt{5^2}} = \boxed{\dfrac{\sqrt{10}}{5}}$

63. $\sqrt{\dfrac{3}{8}} = \dfrac{\sqrt{3}}{\sqrt{8}} = \dfrac{\sqrt{3}}{\sqrt{8}} \cdot \dfrac{\sqrt{2}}{\sqrt{2}} = \dfrac{\sqrt{6}}{\sqrt{16}} = \boxed{\dfrac{\sqrt{6}}{4}}$

65. $\sqrt{\dfrac{5}{18}} = \dfrac{\sqrt{5}}{\sqrt{18}} = \dfrac{\sqrt{5}}{\sqrt{18}} \cdot \dfrac{\sqrt{2}}{\sqrt{2}} = \dfrac{\sqrt{10}}{\sqrt{36}} = \boxed{\dfrac{\sqrt{10}}{6}}$

67. $\sqrt{\dfrac{35}{4} + \dfrac{7}{9}} = \sqrt{\dfrac{315}{36} + \dfrac{28}{36}} = \sqrt{\dfrac{343}{36}} = \dfrac{\sqrt{343}}{\sqrt{36}} = \boxed{\dfrac{\sqrt{343}}{6}}$

69. $-6\sqrt{\dfrac{5}{18}} = -6\dfrac{\sqrt{5}}{\sqrt{18}} = \dfrac{-6\sqrt{5}}{\sqrt{18}} \cdot \dfrac{\sqrt{2}}{\sqrt{2}} = \dfrac{-6\sqrt{10}}{\sqrt{36}} = \dfrac{-6\sqrt{10}}{6} = \boxed{-\sqrt{10}}$

71. $\dfrac{1}{5}\sqrt{\dfrac{75}{63}} = \dfrac{1}{5}\dfrac{\sqrt{75}}{\sqrt{63}} = \dfrac{\sqrt{25 \cdot 3}}{5\sqrt{9 \cdot 7}} = \dfrac{5\sqrt{3}}{5 \cdot 3\sqrt{7}} = \dfrac{\sqrt{3}}{3\sqrt{7}} = \dfrac{\sqrt{3}}{3\sqrt{7}} \cdot \dfrac{\sqrt{7}}{\sqrt{7}} = \dfrac{\sqrt{21}}{3\sqrt{49}} = \dfrac{\sqrt{21}}{3 \cdot 7} = \boxed{\dfrac{\sqrt{21}}{21}}$

73. $\sqrt[3]{\dfrac{1}{2}} = \dfrac{\sqrt[3]{1}}{\sqrt[3]{2}} = \dfrac{1}{\sqrt[3]{2}} \cdot \dfrac{\sqrt[3]{2^2}}{\sqrt[3]{2^2}} = \dfrac{\sqrt[3]{2^2}}{\sqrt[3]{2^3}} = \boxed{\dfrac{\sqrt[3]{4}}{2}}$

75. $\sqrt[3]{\dfrac{2}{3}} = \dfrac{\sqrt[3]{2}}{\sqrt[3]{3}} = \dfrac{\sqrt[3]{2}}{\sqrt[3]{3}} \cdot \dfrac{\sqrt[3]{3^2}}{\sqrt[3]{3^2}} = \dfrac{\sqrt[3]{2 \cdot 3^2}}{\sqrt[3]{3^3}} = \boxed{\dfrac{\sqrt[3]{18}}{3}}$

77. $-6\sqrt[3]{\dfrac{3}{2}} = -6\dfrac{\sqrt[3]{3}}{\sqrt[3]{2}} = \dfrac{-6\sqrt[3]{3}}{\sqrt[3]{2}} \cdot \dfrac{\sqrt[3]{2^2}}{\sqrt[3]{2^2}} = \dfrac{-6\sqrt[3]{3 \cdot 2^2}}{\sqrt[3]{2^3}} = \dfrac{-6\sqrt[3]{12}}{2} = \boxed{-3\sqrt[3]{12}}$

79. $\sqrt{2 - \dfrac{1}{3}} = \sqrt{\dfrac{6}{3} - \dfrac{1}{3}} = \sqrt{\dfrac{5}{3}} = \dfrac{\sqrt{5}}{\sqrt{3}} = \dfrac{\sqrt{5}}{\sqrt{3}} \cdot \dfrac{\sqrt{3}}{\sqrt{3}} = \boxed{\dfrac{\sqrt{15}}{3}}$

81. $\sqrt{4 + \dfrac{3}{5}} = \sqrt{\dfrac{20}{5} + \dfrac{3}{5}} = \sqrt{\dfrac{23}{5}} = \dfrac{\sqrt{23}}{\sqrt{5}} = \dfrac{\sqrt{23}}{\sqrt{5}} \cdot \dfrac{\sqrt{5}}{\sqrt{5}} = \boxed{\dfrac{\sqrt{115}}{5}}$

83. $\left(\sqrt{\dfrac{1}{2}}\right)^3 = \left(\sqrt{\dfrac{1}{2}}\right)^2\left(\sqrt{\dfrac{1}{2}}\right) = \dfrac{1}{2}\dfrac{\sqrt{1}}{\sqrt{2}} = \dfrac{1}{2\sqrt{2}} = \dfrac{1}{2\sqrt{2}} \cdot \dfrac{\sqrt{2}}{\sqrt{2}} = \dfrac{\sqrt{2}}{2\sqrt{4}} = \dfrac{\sqrt{2}}{2(2)} = \boxed{\dfrac{\sqrt{2}}{4}}$

85. $\sqrt{\dfrac{3x}{7y}} = \dfrac{\sqrt{3x}}{\sqrt{7y}} \cdot \dfrac{\sqrt{7y}}{\sqrt{7y}} = \dfrac{\sqrt{21xy}}{\sqrt{(7y)^2}} = \boxed{\dfrac{\sqrt{21xy}}{7y}}$

87. $\sqrt{\dfrac{3}{32y^2}} = \dfrac{\sqrt{3}}{\sqrt{16 \cdot 2y^2}} = \dfrac{\sqrt{3}}{4y\sqrt{2}} = \dfrac{\sqrt{3}}{4y\sqrt{2}} \cdot \dfrac{\sqrt{2}}{\sqrt{2}} = \dfrac{\sqrt{6}}{4y(2)} = \boxed{\dfrac{\sqrt{6}}{8y}}$

89. $\dfrac{3}{\sqrt{32y^2}} = \dfrac{3}{\sqrt{16 \cdot 2y^2}} = \dfrac{3}{4y\sqrt{2}} \cdot \dfrac{\sqrt{2}}{\sqrt{2}} = \boxed{\dfrac{3\sqrt{2}}{8y}}$

91. $\sqrt{\dfrac{11}{20y^3}} = \dfrac{\sqrt{11}}{\sqrt{4 \cdot 5y^2 y}} = \dfrac{\sqrt{11}}{2y\sqrt{5y}} = \dfrac{\sqrt{11}}{2y\sqrt{5y}} \cdot \dfrac{\sqrt{5y}}{\sqrt{5y}} = \dfrac{\sqrt{55y}}{2y(5y)} = \boxed{\dfrac{\sqrt{55y}}{10y^2}}$

93. $\dfrac{5}{\sqrt{12y}} = \dfrac{5}{\sqrt{12y}} \cdot \dfrac{\sqrt{3y}}{\sqrt{3y}} = \boxed{\dfrac{5\sqrt{3y}}{6y}}$

95. $\dfrac{\sqrt{13}}{\sqrt{32y^3}} = \sqrt{\dfrac{13}{16 \cdot 2y^2 y}} = \dfrac{\sqrt{13y}}{4y\sqrt{2y}} \cdot \dfrac{\sqrt{2y}}{\sqrt{2y}} = \boxed{\dfrac{\sqrt{26y}}{8y^2}}$

97. $\sqrt{\dfrac{3x}{32y^3}} = \dfrac{\sqrt{3x}}{\sqrt{16 \cdot 2y^2 y}} = \dfrac{\sqrt{3x}}{4y\sqrt{2y}} = \dfrac{\sqrt{3x}}{4y\sqrt{2y}} \cdot \dfrac{\sqrt{2y}}{\sqrt{2y}} = \dfrac{\sqrt{6xy}}{4y\sqrt{(2y)^2}} = \dfrac{\sqrt{6xy}}{(4y)(2y)} = \boxed{\dfrac{\sqrt{6xy}}{8y^2}}$

99. $\dfrac{3x}{\sqrt{8x^3}} = \dfrac{3x}{\sqrt{4 \cdot 2x^2 x}} = \dfrac{3x}{2x\sqrt{2x}} = \dfrac{3}{2\sqrt{2x}} \cdot \dfrac{\sqrt{2x}}{\sqrt{2x}} = \dfrac{3\sqrt{2x}}{2(2x)} = \boxed{\dfrac{3\sqrt{2x}}{4x}}$

101. $\sqrt{\dfrac{27xy^3}{32z^5}} = \dfrac{\sqrt{9 \cdot 3xy^2 y}}{\sqrt{16 \cdot 2(z^2)^2 z}} = \dfrac{3y\sqrt{3xy}}{4z^2\sqrt{2z}} = \dfrac{3y\sqrt{3xy}}{4z^2\sqrt{2z}} \cdot \dfrac{\sqrt{2z}}{\sqrt{2z}} = \dfrac{3y\sqrt{6xyz}}{4z^2(2z)} = \boxed{\dfrac{3y\sqrt{6xyz}}{8z^3}}$

103. $\dfrac{27xy^3}{\sqrt{32x^5}} = \dfrac{27xy^3}{\sqrt{16 \cdot 2(x^2)^2 x}} = \dfrac{27xy^3}{4x^2\sqrt{2x}} = \dfrac{27y^3}{4x\sqrt{2x}} = \dfrac{27y^3}{4x\sqrt{2x}} \cdot \dfrac{\sqrt{2x}}{\sqrt{2x}} = \dfrac{27y^3\sqrt{2x}}{(4x)(2x)} = \boxed{\dfrac{27y^3\sqrt{2x}}{8x^2}}$

105. $\dfrac{\sqrt{7y}}{\sqrt{8x^3}} = \dfrac{\sqrt{7y}}{\sqrt{4 \cdot 2x^2 x}} = \dfrac{\sqrt{7y}}{2x\sqrt{2x}} \cdot \dfrac{\sqrt{2x}}{\sqrt{2x}} = \boxed{\dfrac{\sqrt{14xy}}{4x^2}}$

107. $\dfrac{\sqrt{20x^2 y}}{\sqrt{5x^3 y^3}} = \dfrac{\sqrt{4 \cdot 5x^2 y}}{\sqrt{5x^2 xy^2 y}} = \dfrac{2x\sqrt{5y}}{xy\sqrt{5xy}} = \dfrac{2\sqrt{5y}}{y\sqrt{5xy}} \cdot \dfrac{\sqrt{5xy}}{\sqrt{5xy}} = \dfrac{2\sqrt{25xy^2}}{y(5xy)} = \dfrac{2(5y)\sqrt{x}}{5xy^2} = \boxed{\dfrac{2\sqrt{x}}{xy}}$

109. $\sqrt[3]{\dfrac{7}{2x}} = \dfrac{\sqrt[3]{7}}{\sqrt[3]{2x}} \cdot \dfrac{\sqrt[3]{(2x)^2}}{\sqrt[3]{(2x)^2}} = \dfrac{\sqrt[3]{7(2x)^2}}{\sqrt[3]{(2x^3)}} = \boxed{\dfrac{\sqrt[3]{28x^2}}{2x}}$

111. $\dfrac{7}{\sqrt[3]{2x^2}} \cdot \dfrac{\sqrt[3]{2^2 x}}{\sqrt[3]{2^2 x}} = \dfrac{7\sqrt[3]{4x}}{\sqrt[3]{2^3 x^3}} = \boxed{\dfrac{7\sqrt[3]{4x}}{2x}}$

113. $\dfrac{7x^2}{\sqrt[3]{2x^5}} = \dfrac{7x^2}{\sqrt[3]{2x^3 x^2}} = \dfrac{7x^2}{x\sqrt[3]{2x^2}} = \dfrac{7x}{\sqrt[3]{2x^2}} \cdot \dfrac{\sqrt[3]{2^2 x}}{\sqrt[3]{2^2 x}} = \dfrac{7x\sqrt[3]{4x}}{2x} = \boxed{\dfrac{7\sqrt[3]{4x}}{2}}$

115. $\dfrac{2}{\sqrt[3]{16x^2y}} = \dfrac{2}{\sqrt[3]{8\cdot 2x^2y}} = \dfrac{2}{2\sqrt[3]{2x^2y}} = \dfrac{1}{\sqrt[3]{2x^2y}}\cdot\dfrac{\sqrt[3]{2^2xy^2}}{\sqrt[3]{2^2xy^2}} = \dfrac{\sqrt[3]{2^2xy^2}}{\sqrt[3]{2^3x^3y^3}} = \boxed{\dfrac{\sqrt[3]{4xy^2}}{2xy}}$

117. $\sqrt[5]{\dfrac{3}{8x^3}} = \sqrt[5]{\dfrac{3}{2^3x^3}} = \dfrac{\sqrt[5]{3}}{\sqrt[5]{2^3x^3}}\cdot\dfrac{\sqrt[5]{2^2x^2}}{\sqrt[5]{2^2x^2}} = \dfrac{\sqrt[5]{3\cdot2^2x^2}}{\sqrt[5]{2^5x^5}} = \boxed{\dfrac{\sqrt[5]{12x^2}}{2x}}$

119. $\dfrac{7}{\sqrt[5]{32x^7y^4}} = \dfrac{7}{\sqrt[5]{2^5x^5x^2y^4}} = \dfrac{7}{2x\sqrt[5]{x^2y^4}}\cdot\dfrac{\sqrt[5]{x^3y}}{\sqrt[5]{x^3y}} = \dfrac{7\sqrt[5]{x^3y}}{2x\sqrt[5]{x^5y^5}} = \dfrac{7\sqrt[5]{x^3y}}{2x(xy)} = \boxed{\dfrac{7\sqrt[5]{x^3y}}{2x^2y}}$

121. $\sqrt{1+\dfrac{1}{y}} = \sqrt{\dfrac{y}{y}+\dfrac{1}{y}} = \sqrt{\dfrac{y+1}{y}} = \dfrac{\sqrt{y+1}}{\sqrt{y}}\cdot\dfrac{\sqrt{y}}{\sqrt{y}} = \boxed{\dfrac{\sqrt{y^2+y}}{y}}$

123. $\sqrt{y^2+\dfrac{1}{y^3}} = \sqrt{y^2\cdot\dfrac{y^3}{y^3}+\dfrac{1}{y^3}} = \sqrt{\dfrac{y^5+1}{y^3}} = \dfrac{\sqrt{y^5+1}}{\sqrt{y^2\cdot y}} = \dfrac{\sqrt{y^5+1}}{y\sqrt{y}}\cdot\dfrac{\sqrt{y}}{\sqrt{y}} = \boxed{\dfrac{\sqrt{y^6+y}}{y^2}}$

125. $\sqrt{16x^2+16y^2} = \sqrt{16(x^2+y^2)} = \boxed{4\sqrt{x^2+y^2}}$ **127.** $\sqrt{9x-27y} = \sqrt{9(x-3y)} = \boxed{3\sqrt{x-3y}}$

129. $\sqrt{x^2-18x+81} = \sqrt{(x-9)^2} = \boxed{x-9}$ **131.** $\sqrt{4x^2y^2-4xyz+z^2} = \sqrt{(2xy-z)^2} = \boxed{2xy-z}$

133. $\boxed{A}$ is true; $\sqrt[3]{5\sqrt{5}} = (5\cdot 5^{1/2})^{1/3} = (5^{3/2})^{1/3} = 5^{1/2} = \sqrt{5}$

135. Let

$$L = \text{length of the rectangle}$$

$$4^2 + L^2 = 10^2$$
$$L^2 = 100 - 16 = 84$$
$$L = \sqrt{84} = \sqrt{4\cdot 21} = 2\sqrt{21}$$

$\boxed{2\sqrt{21}\ \text{ft}}$

137. $S = \dfrac{2xyz}{\sqrt[5]{2x^4y^2}}\cdot\dfrac{\sqrt[5]{2^4xy^3}}{\sqrt[5]{2^4xy^3}} = \dfrac{2xyz\sqrt[5]{16xy^3}}{\sqrt[5]{2^5x^5y^5}} = \dfrac{2xyz\sqrt[5]{16xy^3}}{2xy} = \boxed{z\sqrt[5]{16xy^3}}$

139.

number	double the number
x	$2x$
square root	square root
$\sqrt{x}$	$\sqrt{2x} = \sqrt{2}\sqrt{x}$

The square root of the doubled number is $\boxed{\text{the square root of the number multiplied by }\sqrt{2}}$.

141.

number	cube root
x	$\sqrt[3]{x}$
$8x$	$\sqrt[3]{8x} = 2\sqrt[3]{x}$

$\boxed{\text{Multiply the number by 8.}}$

Review Problems

145. $\dfrac{8x^3y^2}{-24x^7y^{-5}} = -\dfrac{1}{3}\,x^{-4}y^7 = \boxed{\dfrac{-y^7}{3x^4}}$

146. $(4y^3 - 22y^2 + 44y - 35) \div (2y - 5) = \boxed{2y^2 - 6y + 7}$

$$
\begin{array}{r}
2y^2 - 6y + 7 \\
2y - 5 \overline{\smash{)}\,4y^3 - 22y^2 + 44y - 35} \\
\underline{4y^3 - 10y^2} \\
-12y^2 + 44y \\
\underline{-12y^2 + 30y} \\
14y - 35 \\
\underline{14y - 35} \\
0
\end{array}
$$

147. $\dfrac{2y-1}{y^2-4} = \dfrac{y+6}{y^2+3y+2} + \dfrac{y+3}{y^2-y-2}$

$\dfrac{2y-1}{(y+2)(y-2)} = \dfrac{y+6}{(y+2)(y+1)} + \dfrac{y+3}{(y+1)(y-2)}$

$(y+2)(y-2)(y+1) \cdot \dfrac{2y-1}{(y+2)(y-2)} = (y+2)(y-2)(y+1) \cdot \left[\dfrac{y+6}{(y+2)(y+1)} + \dfrac{y+3}{(y+1)(y-2)} \right]$

$\begin{aligned}
(y+1)(2y-1) &= (y-2)(y+6) + (y+2)(y+3) \\
2y^2 + y - 1 &= y^2 + 4y - 12 + y^2 + 5y + 6 \\
2y^2 + y - 1 &= 2y^2 + 9y - 6 \\
y - 1 &= 9y - 6 \\
-8y &= -5 \\
y &= \dfrac{-5}{-8} = \dfrac{5}{8}
\end{aligned}$

$\boxed{\left\{ \dfrac{5}{8} \right\}}$

Section 7.6   Operations with Radicals

Problem Set 7.6, pp. 541-542

1. $\sqrt{3}\,\sqrt{6} = \sqrt{18} = \sqrt{9 \cdot 2} = \boxed{3\sqrt{2}}$

3. $\sqrt{28}\,\sqrt{\dfrac{3}{7}} = \sqrt{(28)\left(\dfrac{3}{7}\right)} = \sqrt{12} = \sqrt{4 \cdot 3} = \boxed{2\sqrt{3}}$

5. $(2\sqrt{5})(3\sqrt{20}) = 6\sqrt{100} = 6(10) = \boxed{60}$

7. $(2\sqrt{10})(5\sqrt{6}) = 10\sqrt{60} = 10\sqrt{4 \cdot 15} = \boxed{20\sqrt{15}}$

9. $(-8\sqrt{12})\left(\dfrac{3}{2}\sqrt{2}\right) = (-8)\left(\dfrac{3}{2}\right)\sqrt{12 \cdot 2} = -12\sqrt{24} = -12\sqrt{4 \cdot 6} = -12(2)\sqrt{6} = \boxed{-24\sqrt{6}}$

11. $\sqrt[3]{9} \cdot \sqrt[3]{6} = \sqrt[3]{54} = \sqrt[3]{27 \cdot 2} = \boxed{3\sqrt[3]{2}}$

13. $(4\sqrt[3]{5})(6\sqrt[3]{50}) = 24\sqrt[3]{250} = 24\sqrt[3]{125 \cdot 2} = 24(5)\sqrt[3]{2} = \boxed{120\sqrt[3]{2}}$

15. $\dfrac{\sqrt{20}}{\sqrt{5}} = \sqrt{4} = \boxed{2}$

17. $\dfrac{\sqrt{54x^3}}{\sqrt{6x}} = \sqrt{\dfrac{54x^3}{6x}} = \sqrt{9x^2} = \boxed{3x}$

19. $\dfrac{-10\sqrt[3]{15}}{-2\sqrt[3]{5}} = \boxed{5\sqrt[3]{3}}$

21. $\dfrac{-12\sqrt[4]{8}}{6\sqrt[4]{2}} = \boxed{-2\sqrt[4]{4}}$

23. $\dfrac{18\sqrt{20}+16\sqrt{28}}{2\sqrt{2}} = \dfrac{18\sqrt{20}}{2\sqrt{2}} + \dfrac{16\sqrt{28}}{2\sqrt{2}} = \boxed{9\sqrt{10}+8\sqrt{14}}$

25. $\sqrt{9x}\,\sqrt{2x} = \sqrt{18x^2} = \sqrt{9\cdot 2x^2} = \boxed{3x\sqrt{2}}$

27. $\sqrt{12x^2y}\,\sqrt{5xy} = \sqrt{60x^3y^2} = \sqrt{4\cdot 15x^2xy^2} = \boxed{2xy\sqrt{15x}}$

29. $\sqrt[3]{4x^3y}\,\sqrt[3]{2xy} = \sqrt[3]{8x^3y^2} = \boxed{2x\sqrt[3]{y^2}}$

31. $(-2xy^{-2}\sqrt{3x})(xy\sqrt{6x}) = -2x^2y^3\sqrt{9\cdot 2x^2} = -2x^2y^3(3x)\sqrt{2} = \boxed{-6x^3y^3\sqrt{2}}$

33. $(2x^2y\sqrt[4]{8xy}\,)-3xy^2\sqrt[4]{2x^2y^3}) = -6x^3y^3\sqrt[4]{16x^3y^4} = -6x^3y^3(2y)\sqrt[4]{x^3} = \boxed{-12x^3y^4\sqrt[4]{x^3}}$

35. $\dfrac{\sqrt{200x^3}}{\sqrt{10x^{-1}}} = \sqrt{\dfrac{200x^3}{10x^{-1}}} = \sqrt{20x^4} = \sqrt{4\cdot 5x^4} = \boxed{2x^2\sqrt{5}}$

37. $\dfrac{\sqrt[3]{108x^4y^5}}{\sqrt[3]{2xy^3}} = \sqrt[3]{\dfrac{108x^4y^5}{2xy^3}} = \sqrt[3]{54x^3y^2} = \sqrt[3]{27\cdot 2x^3y^2} = \boxed{3x\sqrt[3]{2y^2}}$

39. $\dfrac{\sqrt{98x^2y}}{\sqrt{2x^{-3}y}} = \sqrt{\dfrac{98x^2y}{2x^{-3}y}} = \sqrt{49x^5} = \sqrt{49(x^2)^2x} = \boxed{7x^2\sqrt{x}}$

41. $\dfrac{\sqrt{x^2-y^2}}{\sqrt{x-y}} = \sqrt{\dfrac{x^2-y^2}{x-y}} = \sqrt{\dfrac{(x+y)(x-y)}{x-y}} = \boxed{\sqrt{x+y}}$

43. $\dfrac{\sqrt{a^2-4b^2}}{\sqrt{a+2b}} = \sqrt{\dfrac{(a+2b)(a-2b)}{a+2b}} = \boxed{\sqrt{a-2b}}$

45. $7\sqrt{3}+8\sqrt{3} = (7+8)\sqrt{3} = \boxed{15\sqrt{3}}$

47. $8\sqrt{7}-7\sqrt{7}-\sqrt{7} = (8-7-1)\sqrt{7} = \boxed{0}$

49. $3\sqrt{13}-2\sqrt{5}-2\sqrt{5}-2\sqrt{13}+4\sqrt{5} = 3\sqrt{13}-2\sqrt{13}-2\sqrt{5}+4\sqrt{5} = \boxed{\sqrt{13}+2\sqrt{5}}$

51. $\sqrt{2} - \sqrt{11} + 6\sqrt{2} + 4\sqrt{11}$
$= \sqrt{2} + 6\sqrt{2} - \sqrt{11} + 4\sqrt{11}$
$= (1+6)\sqrt{2} + (-1+4)\sqrt{11}$
$= \boxed{7\sqrt{2} + 3\sqrt{11}}$

53. $3\sqrt{15} - 2\sqrt{7} - 2\sqrt{15} + 2\sqrt{7}$
$= 3\sqrt{15} - 2\sqrt{15} - 2\sqrt{7} + 2\sqrt{7}$
$= 13\sqrt{15} + 0\sqrt{7}$
$= \boxed{13\sqrt{15}}$

55. $\sqrt{50} + \sqrt{18}$
$= \sqrt{25 \cdot 2} + \sqrt{9 \cdot 2}$
$= 5\sqrt{2} + 3\sqrt{2}$
$= \boxed{8\sqrt{2}}$

57. $3\sqrt{18} - 5\sqrt{50}$
$= 3\sqrt{9 \cdot 2} - 5\sqrt{25 \cdot 2}$
$= 9\sqrt{2} - 25\sqrt{2}$
$= \boxed{-16\sqrt{2}}$

59. $3\sqrt{8} - \sqrt{32} + 3\sqrt{72} - \sqrt{75}$
$= 3\sqrt{4 \cdot 2} - \sqrt{16 \cdot 2} + 3\sqrt{36 \cdot 2} - \sqrt{25 \cdot 3}$
$= 3(2)\sqrt{2} - 4\sqrt{2} + 3(6)\sqrt{2} - 5\sqrt{3}$
$= 6\sqrt{2} - 4\sqrt{2} + 18\sqrt{2} - 5\sqrt{3}$
$= \boxed{20\sqrt{2} - 5\sqrt{3}}$

61. $8\sqrt{\dfrac{1}{2}} - \dfrac{1}{2}\sqrt{8}$
$= 8 \cdot \dfrac{1}{\sqrt{2}} \cdot \dfrac{\sqrt{2}}{\sqrt{2}} - \dfrac{1}{2}\sqrt{4 \cdot 2}$
$= \dfrac{8\sqrt{2}}{2} - \dfrac{1}{2}(2)\sqrt{2}$
$= 4\sqrt{2} - \sqrt{2}$
$= \boxed{3\sqrt{2}}$

63. $\dfrac{\sqrt{63}}{3} + 7\sqrt{3}$
$= \dfrac{\sqrt{9 \cdot 7}}{3} + 7\sqrt{3}$
$= \dfrac{3\sqrt{7}}{3} + 7\sqrt{3}$
$= \boxed{\sqrt{7} + 7\sqrt{3}}$

65. $\sqrt{25x} + \sqrt{16x}$
$= 5\sqrt{x} + 4\sqrt{x}$
$= \boxed{9\sqrt{x}}$

67. $\sqrt[4]{32} + 3\sqrt[4]{1250}$
$= \sqrt[4]{16 \cdot 2} + 3\sqrt[4]{625 \cdot 2}$
$= 2\sqrt[4]{2} + 3 \cdot 5\sqrt[4]{2}$
$= 2\sqrt[4]{2} + 15\sqrt[4]{2}$
$= \boxed{17\sqrt[4]{2}}$

69. $4\sqrt[3]{40} - 3\sqrt[3]{320} + 2\sqrt[3]{625}$
$= 4\sqrt[3]{8 \cdot 5} - 3\sqrt[3]{64 \cdot 5} + 2\sqrt[3]{125 \cdot 5}$
$= 8\sqrt[3]{5} - 12\sqrt[3]{5} + 10\sqrt[3]{5}$
$= \boxed{6\sqrt[3]{5}}$

71. $\dfrac{1}{4}\sqrt{2}+\dfrac{2}{3}\sqrt{8}$

$=\dfrac{\sqrt{2}}{4}+\dfrac{2}{3}\sqrt{4\cdot 2}$

$=\dfrac{\sqrt{2}}{4}+\dfrac{2\cdot 2\sqrt{2}}{3}$

$=\dfrac{\sqrt{2}}{4}+\dfrac{4\sqrt{2}}{3}$

$=\dfrac{3\sqrt{2}}{12}+\dfrac{16\sqrt{2}}{12}$

$=\boxed{\dfrac{19\sqrt{2}}{12}}$

73. $\dfrac{\sqrt{45}}{4}-\sqrt{80}+\dfrac{\sqrt{20}}{3}$

$=\dfrac{\sqrt{9\cdot 5}}{4}-\sqrt{16\cdot 5}+\dfrac{\sqrt{4\cdot 5}}{3}$

$=\dfrac{3\sqrt{5}}{4}-\dfrac{4\sqrt{5}}{1}+\dfrac{2\sqrt{5}}{3}$

$=\dfrac{9\sqrt{5}}{12}-\dfrac{48\sqrt{5}}{12}+\dfrac{8\sqrt{5}}{12}$

$=\boxed{\dfrac{-31\sqrt{5}}{12}}$

75. $7\sqrt[3]{2}+8\sqrt[3]{16}-2\sqrt[3]{54}$

$=7\sqrt[3]{2}+8\sqrt[3]{8\cdot 2}-2\sqrt[3]{27\cdot 2}$

$=7\sqrt[3]{2}+8(2)\sqrt[3]{2}-2(3)\sqrt[3]{2}$

$=7\sqrt[3]{2}+16\sqrt[3]{2}-6\sqrt[3]{2}$

$=\boxed{17\sqrt[3]{2}}$

77. $2\sqrt[3]{3}+2\sqrt[3]{81}-\sqrt[3]{54}$

$=2\sqrt[3]{2}-\sqrt[3]{8\cdot 2}-\sqrt[3]{27\cdot 2})$

$=2\sqrt[3]{2}-2\sqrt[3]{2}-3\sqrt[3]{2}$

$=\boxed{-3\sqrt[3]{2}}$

79. $5\sqrt{12x}-2\sqrt{3x}$

$=5\sqrt{4\cdot 3x}-2\sqrt{3x}$

$=10\sqrt{3x}-2\sqrt{3x}$

$=\boxed{8\sqrt{3x}}$

81. $\sqrt[3]{54x^5}+2x\sqrt[3]{16x^2}-7\sqrt[3]{2x^5}$

$=\sqrt[3]{27\cdot 2x^3x^2}+2x\sqrt[3]{8\cdot 2x^2}-7\sqrt[3]{2x^3x^2}$

$=3x\sqrt[3]{2x^2}+4x\sqrt[3]{2x^2}-7x\sqrt[3]{2x^2}$

$=\boxed{0}$

83. $16\sqrt{\dfrac{5}{8}}+6\sqrt{\dfrac{5}{2}}$

$=\dfrac{16\sqrt{5}}{\sqrt{8}}\cdot\dfrac{\sqrt{2}}{\sqrt{2}}+\dfrac{6\sqrt{5}}{\sqrt{2}}\cdot\dfrac{\sqrt{2}}{\sqrt{2}}$

$=\dfrac{16\sqrt{10}}{4}+\dfrac{6\sqrt{10}}{2}$

$=4\sqrt{10}+3\sqrt{10}$

$=\boxed{7\sqrt{10}}$

85. $12\sqrt{\dfrac{2}{3}}+24\sqrt{\dfrac{1}{6}}$

$=\dfrac{12\sqrt{2}}{\sqrt{3}}+\dfrac{24}{\sqrt{6}}$

$=\dfrac{12\sqrt{2}}{\sqrt{3}}\cdot\dfrac{\sqrt{3}}{\sqrt{3}}+\dfrac{24}{\sqrt{6}}\cdot\dfrac{\sqrt{6}}{\sqrt{6}}$

$=\dfrac{12\sqrt{6}}{3}+\dfrac{24\sqrt{6}}{6}$

$=4\sqrt{6}+4\sqrt{6}$

$=\boxed{8\sqrt{6}}$

87. $2x\sqrt{\dfrac{y^3}{x}} - 2y^2\sqrt{\dfrac{x}{y}}$

$= \dfrac{2x\sqrt{y^2 \cdot y}}{\sqrt{x}} - 2y^2\sqrt{\dfrac{x}{y}}$

$= 2xy\dfrac{\sqrt{y}}{\sqrt{x}} \cdot \dfrac{\sqrt{x}}{\sqrt{x}} - 2y^2\dfrac{\sqrt{x}}{\sqrt{y}} \cdot \dfrac{\sqrt{y}}{\sqrt{y}}$

$= \dfrac{2xy\sqrt{xy}}{x} - \dfrac{2y^2\sqrt{xy}}{y}$

$= 2y\sqrt{xy} - 2y\sqrt{xy} = \boxed{0}$

89. $\sqrt{\dfrac{x}{y}} - \dfrac{\sqrt{xy}}{y}$

$= \dfrac{\sqrt{x}}{\sqrt{y}} \cdot \dfrac{\sqrt{y}}{\sqrt{y}} - \dfrac{\sqrt{xy}}{y}$

$= \dfrac{\sqrt{xy}}{y} - \dfrac{\sqrt{xy}}{y}$

$= \boxed{0}$

91. $\sqrt{\dfrac{5}{2}} - \sqrt{\dfrac{9}{10}} + 15\sqrt{\dfrac{2}{5}}$

$= \dfrac{\sqrt{5}}{\sqrt{2}} \cdot \dfrac{\sqrt{2}}{\sqrt{2}} - \dfrac{3}{\sqrt{10}} \cdot \dfrac{\sqrt{10}}{\sqrt{10}} + \dfrac{15\sqrt{2}}{\sqrt{5}} \cdot \dfrac{\sqrt{5}}{\sqrt{5}}$

$= \dfrac{\sqrt{10}}{2} - \dfrac{3\sqrt{10}}{10} + \dfrac{15\sqrt{10}}{5}$

$= \dfrac{5\sqrt{10}}{10} - \dfrac{3\sqrt{10}}{10} + \dfrac{30\sqrt{10}}{10}$

$= \dfrac{32\sqrt{10}}{10}$

$= \boxed{\dfrac{16\sqrt{10}}{5}}$

93. $7\sqrt{3} + \sqrt[4]{12^2}$

$= 7\sqrt{3} + (12^2)^{1/4}$

$= 7\sqrt{3} + 12^{1/2}$

$= 7\sqrt{3} + \sqrt{12}$

$= 7\sqrt{3} + \sqrt{4 \cdot 3}$

$= 7\sqrt{3} + 2\sqrt{3}$

$= \boxed{9\sqrt{3}}$

95. $8m\sqrt{m} - \sqrt[10]{m^{15}}$

$= 8m\sqrt{m} - (m^{15})^{1/10}$

$= 8m\sqrt{m} - m^{3/2}$

$= 8m\sqrt{m} - \sqrt{m^3}$

$= 8m\sqrt{m} - \sqrt{m^2 \cdot m}$

$= 8m\sqrt{m} - m\sqrt{m}$

$= \boxed{7m\sqrt{m}}$

97. $\sqrt{45} + \sqrt[4]{25}$

$= \sqrt{9 \cdot 5} + (5^2)^{1/4}$

$= 3\sqrt{5} + 5^{1/2}$

$= 3\sqrt{5} + \sqrt{5}$

$= \boxed{4\sqrt{5}}$

99. $P = 2L + 2W = 2 \cdot \dfrac{12}{\sqrt{6}} + 2 \cdot 7\sqrt{\dfrac{3}{2}}$

$= \dfrac{24}{\sqrt{6}} \cdot \dfrac{\sqrt{6}}{\sqrt{6}} + \dfrac{14\sqrt{3}}{\sqrt{2}} \cdot \dfrac{\sqrt{2}}{\sqrt{2}}$

$= \dfrac{24\sqrt{6}}{6} + \dfrac{14\sqrt{6}}{2}$

$= 4\sqrt{6} + 7\sqrt{6}$

$= \boxed{11\sqrt{6}}$

101. Let $x =$ the irrational number

$$x - (\sqrt{18} - \sqrt{50}) = \sqrt{2}$$
$$x - 2\sqrt{18} + \sqrt{50} = \sqrt{2}$$
$$x - 2\sqrt{9 \cdot 2} + \sqrt{25 \cdot 2} = \sqrt{2}$$
$$x - 6\sqrt{2} + 5\sqrt{2} = \sqrt{2}$$
$$x - \sqrt{2} = \sqrt{2}$$
$$x = \sqrt{2} + \sqrt{2}$$
$$x = 2\sqrt{2}$$

The number is $\boxed{2\sqrt{2}}$.

103. $\left| \dfrac{3x + \sqrt{32}}{2} \right| = \sqrt{50}$

$\left| \dfrac{3x + 4\sqrt{2}}{2} \right| = 5\sqrt{2}$

$$\dfrac{3x + 4\sqrt{2}}{2} = 5\sqrt{2} \qquad \text{or} \qquad \dfrac{3x + 4\sqrt{2}}{2} = -5\sqrt{2}$$
$$3x + 4\sqrt{2} = 10\sqrt{2} \qquad\qquad 3x + 4\sqrt{2} = -10\sqrt{2}$$
$$3x = 6\sqrt{2} \qquad\qquad\qquad 3x = -14\sqrt{2}$$
$$x = 2\sqrt{2} \qquad\qquad\qquad x = \dfrac{-14}{3}\sqrt{2}$$

$$\boxed{\left\{ 2\sqrt{2},\; \dfrac{-14}{3}\sqrt{2} \right\}}$$

Review Problems

108.

$$H = \dfrac{62.4\, Ns}{33,000}$$
$$33,000\,H = 62.4\, Ns$$
$$\dfrac{33,000\,H}{62.4\,N} = s$$
$$\boxed{s = \dfrac{33,000\,H}{62.4\,N}}$$
$$s = \dfrac{33,000(468)}{62.4(1500)} = \dfrac{15,444,000}{93,600} = 165$$

The dam is $\boxed{165 \text{ feet}}$ high.

109. $\dfrac{(x^a y^{1/a})^a}{(x^{1/a} y^a)^{a^2}} = \dfrac{x^{a^2} y}{x^a y^{a^3}} = \boxed{x^{a^2 - a} y^{1 - a^3}}$

110. $\dfrac{\dfrac{1}{y^2} - \dfrac{1}{100}}{\dfrac{1}{y} - \dfrac{1}{10}} \cdot \dfrac{100 y^2}{100 y^2} = \dfrac{100 - y^2}{100 y - 10 y^2} = \dfrac{(10 + y)(10 - y)}{10y(10 - y)} = \boxed{\dfrac{10 + y}{10y}}$

Section 7.7 Further Operations with Radicals

Problem Set 7.7, pp. 547-549

1. $\sqrt{2}(\sqrt{3} + \sqrt{7}) = \sqrt{2}\sqrt{3} + \sqrt{2}\sqrt{7} = \boxed{\sqrt{6} + \sqrt{14}}$

3. $4\sqrt{3}(2\sqrt{5} + 3\sqrt{7}) = 4\sqrt{3}(2\sqrt{5}) + 4\sqrt{3}(3\sqrt{7}) = \boxed{8\sqrt{15} + 12\sqrt{21}}$

5. $5\sqrt{6}(7\sqrt{8} - 2\sqrt{12}$
$= 35\sqrt{48} - 10\sqrt{72}$
$= 35\sqrt{16 \cdot 3} - 10\sqrt{36 \cdot 2}$
$= 35(4)\sqrt{3} - 10(6)\sqrt{2}$
$= \boxed{140\sqrt{3} - 60\sqrt{2}}$

7. $4\sqrt{x}(7\sqrt{2} - 3\sqrt{y})$
$= 4\sqrt{x}(7\sqrt{2}) - 4\sqrt{x}(3\sqrt{y})$
$= \boxed{28\sqrt{28} - 12\sqrt{xy}}$

9. $\sqrt{2x}(\sqrt{6x} - 3\sqrt{x})$
$= \sqrt{12x^2} - 3\sqrt{2x^2}$
$= \sqrt{4 \cdot 3x^2} - 3\sqrt{2x^2}$
$= \boxed{2x\sqrt{3} - 3x\sqrt{2}}$

11. $\sqrt{3x}(2y\sqrt{12xy} + 2\sqrt{12xy^3})$
$= 2y\sqrt{36x^2y} + 2\sqrt{36x^2y^3}$
$= 12xy\sqrt{y} + 12xy\sqrt{y}$
$= \boxed{24xy\sqrt{y}}$

13. $\sqrt{3}(1 - \sqrt{5x}) + \sqrt{5}(\sqrt{3x} - 1)$
$= \sqrt{3} - \sqrt{15x} + \sqrt{15x} - \sqrt{5}$
$= \boxed{\sqrt{3} - \sqrt{5}}$

15. $\sqrt[3]{2}(\sqrt[3]{6} + 4\sqrt[3]{5})$
$= \sqrt[3]{2}\sqrt[3]{6} + \sqrt[3]{2}(4\sqrt[3]{5})$
$= \boxed{\sqrt[3]{12} + 4\sqrt[3]{10}}$

17. $4\sqrt[3]{3}(9\sqrt[3]{9} + 7\sqrt[3]{7})$
$= 36\sqrt[3]{27} + 28\sqrt[3]{21}$
$= 36(3) + 28\sqrt[3]{21}$
$= \boxed{108 + 28\sqrt[3]{21}}$

19. $(\sqrt{2} + \sqrt{7})(\sqrt{3} + \sqrt{5})$
$= \sqrt{2}\sqrt{3} + \sqrt{2}\sqrt{5} + \sqrt{7}\sqrt{3} + \sqrt{7}\sqrt{5}$
$= \boxed{\sqrt{6} + \sqrt{10} + \sqrt{21} + \sqrt{35}}$

21. $(\sqrt{2} - \sqrt{7})(\sqrt{3} - \sqrt{5})$
$= \sqrt{2}\sqrt{3} - \sqrt{2}\sqrt{5} - \sqrt{7}\sqrt{3} + \sqrt{7}\sqrt{5}$
$= \boxed{\sqrt{6} - \sqrt{10} - \sqrt{21} + \sqrt{35}}$

23. $(4\sqrt{2} + 5\sqrt{7})(2\sqrt{3} + 3\sqrt{5})$
$= 4\sqrt{2}(2\sqrt{3}) + 4\sqrt{2}(3\sqrt{5}) + 5\sqrt{7}(2\sqrt{3} + 5\sqrt{7}(3\sqrt{5})$
$= \boxed{8\sqrt{6} + 12\sqrt{10} + 10\sqrt{21} + 15\sqrt{35}}$

25. $(3\sqrt{2} - 2\sqrt{8})(2\sqrt{3} - 4\sqrt{5})$
$= 6\sqrt{6} - 12\sqrt{10} - 4\sqrt{24} + 8\sqrt{40}$
$= 6\sqrt{6} - 12\sqrt{10} - 4\sqrt{4 \cdot 6} + 8\sqrt{4 \cdot 10}$
$= 6\sqrt{6} - 12\sqrt{10} - 4(2)\sqrt{6} + 8(2)\sqrt{10}$
$= 6\sqrt{6} - 12\sqrt{10} - 8\sqrt{6} + 16\sqrt{10}$
$= \boxed{4\sqrt{10} - 2\sqrt{6}}$

27. $(\sqrt{5} + 7)(\sqrt{5} - \sqrt{7}$
$= \sqrt{25} - 49$
$= 5 - 49$
$= \boxed{-44}$

29. $(2 + 5\sqrt{3})(2 - 5\sqrt{3})$
$= 4 - 25\sqrt{9}$
$= 4 - 75$
$= \boxed{-71}$

31. $(\sqrt{7} + \sqrt{5})(\sqrt{7} - \sqrt{5})$
$= \sqrt{49} - \sqrt{25}$
$= 7 - 5$
$= \boxed{2}$

33. $(\sqrt{x} + \sqrt{y})(\sqrt{x}) - \sqrt{y})$
$= (\sqrt{x})^2 - (\sqrt{y})^2$
$= x - y$

35. $(3\sqrt{x} + 7\sqrt{y})(3\sqrt{x} - 7\sqrt{y})$
$= (3\sqrt{x})^2 - (7\sqrt{y})^2$
$= 9x - 49y$

37. $(\sqrt{2x} + \sqrt{50})^2$
$= (\sqrt{2x} + \sqrt{25 \cdot 2})^2$
$= (\sqrt{2x} + 5\sqrt{2})^2$
$= (\sqrt{2x} + 5\sqrt{2})(\sqrt{2x} + 5\sqrt{2})$
$= (\sqrt{2x^2} + 5\sqrt{4x} + 5\sqrt{4x} + 25(\sqrt{2})^2$
$= 2x + 10\sqrt{4x} + 25 \cdot 2$
$= 2x + 10 \cdot 2\sqrt{x} + 50$
$= \boxed{2x + 20\sqrt{x} + 50}$

39. $(\sqrt{2a + b} + \sqrt{c})(\sqrt{2a + b} - \sqrt{c})$
$= (\sqrt{2a + b})^2 - (\sqrt{c})^2$
$= \boxed{2a + b - c}$

41. $(\sqrt{x - 4} + \sqrt{x + 4})^2$
$= (\sqrt{x - 4} + \sqrt{x + 4})(\sqrt{x - 4} + \sqrt{x + 4})$
$= x - 4 + \sqrt{x - 4}\sqrt{x + 4} + \sqrt{x - 4}\sqrt{x + 4} + x + 4$
$= 2x + 2\sqrt{(x - 4)(x + 4)}$
$= \boxed{2x + 2\sqrt{x^2 - 16}}$

43. $(\sqrt{y + 1} + \sqrt{y - 7})^2$
$= \{\sqrt{y + 1} + \sqrt{y - 7})(\sqrt{y + 1} + \sqrt{y - 7})$
$= y + 1 + 2\sqrt{(y + 1)(y - 7)} + y - 7$
$= \boxed{2y - 6 + 2\sqrt{y^2 - 6y - 7}}$

45. $(y - \sqrt{2y - 5})^2$
$= (y - \sqrt{2y - 5})(y - \sqrt{2y - 5})$
$= y^2 - y\sqrt{2y - 5} - y\sqrt{2y - 5} + 2y - 5$
$= \boxed{y^2 + 2y - 5 - 2y\sqrt{2y - 5}}$

47. $(x + \sqrt{1 - x^2})^2$
$= (x + \sqrt{1 - x^2})(x + \sqrt{1 - x^2})$
$= x^2 + x\sqrt{1 - x^2} + x\sqrt{1 - x^2} + 1 - x^2$
$= \boxed{2x\sqrt{1 - x^2} + 1}$

49. $\dfrac{15}{\sqrt{6}-1}$

$= \dfrac{15}{\sqrt{6}-1} \cdot \dfrac{\sqrt{6}+1}{\sqrt{6}+1}$

$= \dfrac{15(\sqrt{6}+1)}{(\sqrt{6})^2 - 1^2}$

$= \dfrac{15(\sqrt{6}+1)}{6-1}$

$= \dfrac{15(\sqrt{6}+1)}{5}$

$= \boxed{3(\sqrt{6}+1)}$

51. $\dfrac{3}{\sqrt{7}+3}$

$= \dfrac{3}{\sqrt{7}+3} \cdot \dfrac{\sqrt{7}-3}{\sqrt{7}-3}$

$= \dfrac{3(\sqrt{7}-3)}{(\sqrt{7})^2 - 3^2}$

$= \dfrac{3(\sqrt{7}-3)}{7-9}$

$= \dfrac{3(\sqrt{7}-3)}{-2}$

$= \dfrac{-3(\sqrt{7}-3)}{2}$

$= \boxed{\dfrac{-3\sqrt{7}+9}{2}}$

53. $\dfrac{12}{\sqrt{7}+\sqrt{3}}$

$= \dfrac{12}{\sqrt{7}+\sqrt{3}} \cdot \dfrac{\sqrt{7}-\sqrt{3}}{\sqrt{7}-\sqrt{3}}$

$= \dfrac{12(\sqrt{7}-\sqrt{3})}{(\sqrt{7})^2 - (\sqrt{3})^2}$

$= \dfrac{12(\sqrt{7}-\sqrt{3})}{4}$

$= 3(\sqrt{7}-\sqrt{3})$

$= \boxed{3\sqrt{7}-3\sqrt{3}}$

55. $\dfrac{13}{\sqrt{5}-\sqrt{3}}$

$= \dfrac{13}{\sqrt{5}-\sqrt{3}} \cdot \dfrac{\sqrt{5}+\sqrt{3}}{\sqrt{5}+\sqrt{3}}$

$= \dfrac{13(\sqrt{5}+\sqrt{3})}{(\sqrt{5})^2 - (\sqrt{3})^2}$

$= \dfrac{13(\sqrt{5}+\sqrt{3})}{5-3}$

$= \boxed{\dfrac{13\sqrt{5}+13\sqrt{3}}{2}}$

57. $\dfrac{\sqrt{8}}{\sqrt{5}+\sqrt{3}}$

$= \dfrac{\sqrt{8}}{\sqrt{5}+\sqrt{3}} \cdot \dfrac{\sqrt{5}-\sqrt{3}}{\sqrt{5}-\sqrt{3}}$

$= \dfrac{\sqrt{8}(\sqrt{5}) - \sqrt{8}(\sqrt{3})}{(\sqrt{5})^2 - (\sqrt{3})^2}$

$= \dfrac{\sqrt{40} - \sqrt{24}}{5-3}$

$= \dfrac{2\sqrt{10} - 2\sqrt{6}}{2}$

$= \boxed{\sqrt{10} - \sqrt{6}}$

59. $\dfrac{\sqrt{5}}{\sqrt{11}-\sqrt{5}}$

$= \dfrac{\sqrt{5}}{\sqrt{11}-\sqrt{5}} \cdot \dfrac{\sqrt{11}+\sqrt{5}}{\sqrt{11}+\sqrt{5}}$

$= \dfrac{\sqrt{55} + (\sqrt{5})^2}{(\sqrt{11})^2 - (\sqrt{5})^2}$

$= \dfrac{\sqrt{55} + 5}{11-5}$

$= \boxed{\dfrac{\sqrt{55}+5}{6}}$

61. $\dfrac{12}{3 - 2\sqrt{5}}$

$= \dfrac{12}{3 - 2\sqrt{5}} \cdot \dfrac{3 + 2\sqrt{5}}{3 + 2\sqrt{5}}$

$= \dfrac{12(3 + 2\sqrt{5})}{3^2 - 4 \cdot 5}$

$= \dfrac{12(3 + 2\sqrt{5})}{9 - 4 \cdot 5}$

$= \dfrac{12(3 + 2\sqrt{5})}{-11}$

$= \dfrac{36 + 24\sqrt{5}}{-11}$

$= \boxed{\dfrac{-36 - 24\sqrt{5}}{11}}$

63. $\dfrac{16}{2\sqrt{5} - 4\sqrt{3}}$

$= \dfrac{16}{2\sqrt{5} - 4\sqrt{3}} \cdot \dfrac{2\sqrt{5} + 4\sqrt{3}}{2\sqrt{5} + 4\sqrt{3}}$

$= \dfrac{16(2\sqrt{5} + 4\sqrt{3})}{(2\sqrt{5})^2 - (4\sqrt{3})^2}$

$= \dfrac{16(2\sqrt{5} + 4\sqrt{3})}{4 \cdot 5 - 16 \cdot 3}$

$= \dfrac{16(2\sqrt{5} + 4\sqrt{3})}{-28}$

$= \dfrac{-4(2\sqrt{5} + 4\sqrt{3})}{7}$

$= \boxed{\dfrac{-8\sqrt{5} - 16\sqrt{3}}{7}}$

65. $\dfrac{\sqrt{5}}{7\sqrt{3} + 4\sqrt{5}}$

$= \dfrac{\sqrt{5}}{7\sqrt{3} + 4\sqrt{5}} \cdot \dfrac{7\sqrt{3} - 4\sqrt{5}}{7\sqrt{3} - 4\sqrt{5}}$

$= \dfrac{7\sqrt{15} - 4(\sqrt{5})^2}{(7\sqrt{3})^2 - (4\sqrt{5})^2}$

$= \dfrac{7\sqrt{15} - 20}{49 \cdot 3 - 16 \cdot 5}$

$= \boxed{\dfrac{7\sqrt{15} - 20}{67}}$

67. $\dfrac{11}{\sqrt{x} + 7}$

$= \dfrac{11}{\sqrt{x} + 7} \cdot \dfrac{\sqrt{x} - 7}{\sqrt{x} - 7}$

$= \dfrac{11\sqrt{x} - 77}{(\sqrt{x})^2 - 7^2}$

$= \boxed{\dfrac{11\sqrt{x} - 77}{x - 49}}$

69. $\dfrac{\sqrt{y}}{\sqrt{y} - 2}$

$= \dfrac{\sqrt{y}}{\sqrt{y} - 2} \cdot \dfrac{\sqrt{y} + 2}{\sqrt{y} + 2}$

$= \dfrac{(\sqrt{y})^2 + 2\sqrt{y}}{(\sqrt{y})^2 - 2^2}$

$= \boxed{\dfrac{y + 2\sqrt{y}}{y - 4}}$

71. $\dfrac{2\sqrt{5} + 3}{2\sqrt{5} - 5}$

$= \dfrac{2\sqrt{5} + 3}{2\sqrt{5} - 5} \cdot \dfrac{2\sqrt{5} + 5}{2\sqrt{5} + 5}$

$= \dfrac{(2\sqrt{5})^2 + 10\sqrt{5} + 6\sqrt{5} + 15}{(2\sqrt{5})^2 - 5^2}$

$= \dfrac{4 \cdot 5 + 10\sqrt{5} + 6\sqrt{5} + 15}{4 \cdot 5 - 25}$

$= \dfrac{35 + 16\sqrt{5}}{-5}$

$= \boxed{\dfrac{-35 - 16\sqrt{5}}{5}}$

73. $\dfrac{\sqrt{7}-\sqrt{2}}{\sqrt{7}+\sqrt{2}}$

$= \dfrac{\sqrt{7}-\sqrt{2}}{\sqrt{7}+\sqrt{2}} \cdot \dfrac{\sqrt{7}-\sqrt{2}}{\sqrt{7}-\sqrt{2}}$

$= \dfrac{(\sqrt{7}-\sqrt{2})^2}{(\sqrt{7})^2-(\sqrt{2})^2}$

$= \dfrac{7-\sqrt{14}-\sqrt{14}+2}{7-2}$

$= \boxed{\dfrac{9-2\sqrt{14}}{5}}$

75. $\dfrac{\sqrt{y}-2}{\sqrt{y}-5}$

$= \dfrac{\sqrt{y}-2}{\sqrt{y}-5} \cdot \dfrac{\sqrt{y}+5}{\sqrt{y}+5}$

$= \dfrac{(\sqrt{y}-2)(\sqrt{y}+5)}{(\sqrt{y}-5)(\sqrt{y}+5)}$

$= \boxed{\dfrac{y+3\sqrt{y}-10}{y-25}}$

77. $\dfrac{3\sqrt{x}}{\sqrt{x}-\sqrt{y}}$

$= \dfrac{3\sqrt{x}}{\sqrt{x}-\sqrt{y}} \cdot \dfrac{\sqrt{x}+\sqrt{y}}{\sqrt{x}+\sqrt{y}}$

$= \dfrac{3\sqrt{x}\sqrt{x}+3\sqrt{x}\sqrt{y}}{(\sqrt{x}-\sqrt{y})(\sqrt{x}+\sqrt{y})}$

$= \boxed{\dfrac{3x+3\sqrt{xy}}{x-y}}$

79. $\dfrac{3\sqrt{7}+2\sqrt{3}}{2\sqrt{7}-\sqrt{3}}$

$= \dfrac{3\sqrt{7}+2\sqrt{3}}{2\sqrt{7}-\sqrt{3}} \cdot \dfrac{2\sqrt{7}+\sqrt{3}}{2\sqrt{7}+\sqrt{3}}$

$= \dfrac{6\cdot7+3\sqrt{21}+4\sqrt{21}+2\cdot3}{4\cdot7-3}$

$= \boxed{\dfrac{48+7\sqrt{21}}{25}}$

81. $\dfrac{5\sqrt{x}-2\sqrt{y}}{\sqrt{x}-3\sqrt{y}}$

$= \dfrac{5\sqrt{x}-2\sqrt{y}}{\sqrt{x}-3\sqrt{y}} \cdot \dfrac{\sqrt{x}+3\sqrt{y}}{\sqrt{x}+3\sqrt{y}}$

$= \dfrac{(5\sqrt{x}-2\sqrt{y})(\sqrt{x}+3\sqrt{y})}{(\sqrt{x}-3\sqrt{y})(\sqrt{x}+3\sqrt{y})}$

$= \boxed{\dfrac{5x+13\sqrt{xy}-6y}{x-9y}}$

83. $\dfrac{x^3+y^3}{\sqrt{x+y}}$

$= \dfrac{x^3+y^3}{\sqrt{x+y}} \cdot \dfrac{\sqrt{x+y}}{\sqrt{x+y}}$

$= \dfrac{(x+y)(x^2-xy+y^2)\sqrt{x+y}}{x+y}$

$= \boxed{\sqrt{x+y}(x^2-xy+y^2)}$

85. $\dfrac{6x-6y}{\sqrt{x}-\sqrt{y}}$

$= \dfrac{6x-6y}{\sqrt{x}-\sqrt{y}} \cdot \dfrac{\sqrt{x}+\sqrt{y}}{\sqrt{x}+\sqrt{y}}$

$= \dfrac{6(x-y)(\sqrt{x}+\sqrt{y})}{x-y}$

$= \dfrac{6(\sqrt{x}+\sqrt{y})}{(\sqrt{x}-\sqrt{y})(\sqrt{x}+\sqrt{y})}$

$= \boxed{6\sqrt{x}+6\sqrt{y}}$

87. $\dfrac{\sqrt{a+b}+c}{\sqrt{a+b}-c}$

$= \dfrac{\sqrt{a+b}+c}{\sqrt{a+b}-c} \cdot \dfrac{\sqrt{a+b}+c}{\sqrt{a+b}+c}$

$= \dfrac{(\sqrt{a+b})^2+2(\sqrt{a+b})c+c^2}{(\sqrt{a+b})^2-c^2}$

$= \boxed{\dfrac{a+b+2c\sqrt{a+b}+c^2}{a+b-c^2}}$

89. $f(x) = \sqrt{2x}$

$$\frac{f(a + h) - f(a)}{h}$$

$$= \frac{\sqrt{2(a + h)} - \sqrt{2a}}{h}$$

$$= \frac{\sqrt{2a + 2h} - \sqrt{2a}}{h} \cdot \frac{\sqrt{2a + 2h} + \sqrt{2a}}{\sqrt{2a + 2h} + \sqrt{2a}}$$

$$= \frac{(\sqrt{2a + 2h})^2 - (\sqrt{2a})^2}{h(\sqrt{2a + 2h} + \sqrt{2a})}$$

$$= \frac{2a + 2h - 2a}{h(\sqrt{2a + 2h} + \sqrt{2a})}$$

$$= \frac{2h}{h(\sqrt{2a + 2h} + \sqrt{2a})}$$

$$= \boxed{\frac{2}{\sqrt{2a + 2h} + \sqrt{2a}}}$$

91. via.

4.	$\overline{p.}$	R.	6.	$\rightarrow$	$(4 + \sqrt{6})$
4.	$\overline{m.}$	R.	6.	$\rightarrow$	$(4 - \sqrt{6})$
16.	$\overline{m.}$	6.		$\rightarrow$	$16 - 6$

Production 10 $\rightarrow$ 10

or, $\boxed{(4 + \sqrt{6})(4 - \sqrt{6}) = 16 - 6 = 10}$

93.
$$3x - 2(x - 1) = (\sqrt{14} - 1)(\sqrt{14} + 1)$$
$$3x - 2x + 2 = 14 - 1$$
$$x + 2 = 13$$
$$x = 11$$
$$\boxed{\{11\}}$$

95. Let

$$x = \text{the number}$$
$$0.7x + x = (\sqrt{69} + 1)(\sqrt{69} - 1)$$
$$1.7x = 69 - 1$$
$$1.7x = 68$$
$$x = \frac{68}{7.1} = 40$$

The number is $\boxed{40}$.

97. Let

$$x = \text{side of square} = 3 + 2\sqrt{7}$$
$$\boxed{\text{Perimeter}} = 4s = 4(3 + 2\sqrt{7}) = \boxed{12 + 8\sqrt{7}}$$
$$\boxed{\text{Area}} = s^2$$
$$= (3 + 2\sqrt{7})^2 = (3 + 2\sqrt{7})(3 + 2\sqrt{7})$$
$$= 9 + 6\sqrt{7} + 6\sqrt{7} + 4 \cdot 7$$
$$= \boxed{37 + 12\sqrt{7}}$$

99. $3y^2 + 2 \;=\; 6y$

Show $y = \dfrac{3 + \sqrt{3}}{3}$ is a solution.

$$3\left(\frac{3 + \sqrt{3}}{3}\right)^2 + 2 \;=\; 6\left(\frac{3 + \sqrt{3}}{3}\right)$$

$$3\,\frac{(9 + 6\sqrt{3} + 3)}{9} + 2 \;=\; 2(3 + \sqrt{3})$$

$$\frac{12 + 6\sqrt{3}}{3} + 2 \;=\; 6 + 2\sqrt{3}$$

$$4 + 2\sqrt{3} + 2 \;=\; 6 + 2\sqrt{3}$$

$$6 + 2\sqrt{3} \;=\; 6 + 2\sqrt{3} \quad \text{True}$$

$$\boxed{\dfrac{3 + \sqrt{3}}{3} \text{ is a solution}}$$

101.
$$\sqrt{2\sqrt{15} + 8} \;=\; \sqrt{3} + \sqrt{5}$$
$$\left(\sqrt{2\sqrt{15} + 8}\right)^2 \;=\; \left(\sqrt{3} + \sqrt{5}\right)^2$$
$$2\sqrt{15} + 8 \;=\; (\sqrt{3} + \sqrt{5})(\sqrt{3} + \sqrt{5})$$
$$2\sqrt{15} + 8 \;=\; 3 + \sqrt{15} + \sqrt{15} + 5$$
$$2\sqrt{15} + 8 \;=\; 2\sqrt{15} + 8 \quad \text{True}$$

103. $\dfrac{1}{\sqrt{2} + \sqrt{3} + \sqrt{4}}$

$$= \frac{1}{(\sqrt{2} + \sqrt{3}) + 2} \cdot \frac{(\sqrt{2} + \sqrt{3}) - 2}{(\sqrt{2} + \sqrt{3}) - 2}$$

$$= \frac{\sqrt{2} + \sqrt{3} - 2}{(\sqrt{2} + \sqrt{3}^2) - 4}$$

$$= \frac{\sqrt{2} + \sqrt{3} - 2}{2 + 2\sqrt{6} + 3 - 4}$$

$$= \frac{\sqrt{2} + \sqrt{3} - 2}{2\sqrt{6} + 1}$$

$$= \frac{\sqrt{2} + \sqrt{3} - 2}{2\sqrt{6} + 1} \cdot \frac{2\sqrt{6} - 1}{2\sqrt{6} - 1}$$

$$= \frac{\sqrt{2}(2\sqrt{6} - 1) + \sqrt{3}(2\sqrt{6} - 1) - 2(2\sqrt{6} - 1)}{(2\sqrt{6})^2 - 1^2}$$

$$= \frac{2\sqrt{12} - \sqrt{2} + 2\sqrt{18} - \sqrt{3} - 4\sqrt{6} + 2}{4(6) - 1}$$

$$= \frac{2(2\sqrt{3}) - \sqrt{2} + 2(3\sqrt{2}) - \sqrt{3} - 4\sqrt{6} + 2}{24 - 1}$$

$$= \frac{4\sqrt{3} - \sqrt{3} - \sqrt{2} + 6\sqrt{2} - 4\sqrt{6} + 2}{23}$$

$$= \boxed{\dfrac{3\sqrt{3} + 5\sqrt{2} - 4\sqrt{6} + 2}{23}}$$

105. Given: area of $BEGC = 2\,m^2$
area of $ADFB = 3\,m^2$

Since the area of ADFB = 3, DF = $\sqrt{3}$ and AD = $\sqrt{3}$.
Since the area of BEGC = 2, EG = $\sqrt{2}$ and CG = $\sqrt{2}$.

Area of shaded region
$= (DH)(HG)$
$= (DF + FH)(CH - CG)$
$= (DF + EG)(AD - CG)$
$= (\sqrt{3} + \sqrt{2})(\sqrt{3} - \sqrt{2}) = 3 - 2 = 1$
The area of shaded region is $\boxed{1\,m^2}$.

Review Problems

114. $\dfrac{\dfrac{1}{y-2} - \dfrac{1}{y+2}}{1 + \dfrac{1}{y^2-4}}$

$= \dfrac{\dfrac{1}{y-2} - \dfrac{1}{y+2}}{1 + \dfrac{1}{(y+2)(y-2)}} \cdot \dfrac{(y+2)(y-2)}{(y+2)(y-2)}$

$= \dfrac{y+2 - (y-2)}{(y+2)(y-2)+1}$

$= \dfrac{4}{y^2-4+1}$

$= \boxed{\dfrac{4}{y^2-3}}$

115. $\dfrac{x^3+y^3}{x^2-xy+y^2} \div \dfrac{x^2-y^2}{x^2-2xy+y^2}$

$= \dfrac{x^3+y^3}{x^2-xy+y^2} \cdot \dfrac{x^2-2xy+y^2}{x^2-y^2}$

$= \dfrac{(x+y)(x^2-xy+y^2)}{x^2-xy+y^2} \cdot \dfrac{(x-y)^2}{(x+y)(x-y)}$

$= \boxed{x-y}$

116. Let

$$x = \text{time it takes both pipes working together}$$
$$1 \text{ complete job} = \text{pool is completely emptied}$$

	Fractional part of the job completed in 1 minute	Time spent working together	Fractional part of job completed in x minutes
pipe to fill (30 minutes)	$\dfrac{1}{30}$	x	$\dfrac{x}{30}$
pipe to drain (20 minutes)	$\dfrac{1}{20}$	x	$\dfrac{x}{20}$

Fraction of the pool emptied by the second pipe in x minutes – Fraction of the pool filled by the first pipe in x minutes = 1.

$$\frac{x}{20} - \frac{x}{30} = 1$$
$$60\left(\frac{x}{20} - \frac{x}{30}\right) = 60(1)$$
$$3x - 2x = 60$$
$$x = 60$$

It will take $\boxed{60 \text{ minutes (or 1 hour)}}$ before the pool has no water.

Section 7.8 Radical Equations

Problem Set 7.8, pp. 557-559

1.
$$\sqrt{3x - 1} = 4$$
$$3x - 1 = 16$$
$$3x = 17$$
$$x = \frac{17}{3}$$

$$\boxed{\left\{\frac{17}{3}\right\}}$$

check: $\sqrt{3\left(\frac{17}{3}\right) - 1} = 4$
$$\sqrt{17 - 1} = 4$$
$$\sqrt{16} = 4$$
$$4 = 4 \quad \text{The solution checks.}$$

3.
$$\sqrt{2x + 4} - 6 = 0$$
$$\sqrt{2x + 4} = 6$$
$$(\sqrt{2x + 4})^2 = 6^2$$
$$2x + 4 = 36$$
$$2x = 32$$
$$x = 16$$

check: $\sqrt{2(16) + 4} - 6 = 0$
$$\sqrt{32 + 4} - 6 = 0$$
$$\sqrt{36} - 6 = 0$$
$$6 - 6 = 0 \quad \text{The solution checks.}$$
$$0 = 0 \ \sqrt{}$$

$$\boxed{\{16\}}$$

5.
$$\sqrt{3x - 1} = -4$$
$$(\sqrt{3x - 1})^2 = (-4)^2$$
$$3x - 1 = 16$$
$$3x = 17$$
$$x = \frac{17}{3}$$

check: $\sqrt{3\left(\frac{17}{3}\right) - 1} = -4$
$$\sqrt{16} = -4$$
$$4 \neq -4$$
$\dfrac{17}{3}$ is an extraneous solution.

$$\boxed{\varnothing}$$

7.
$$\sqrt{2x+4}+6 = 0$$
$$\sqrt{2x+4} = -6$$
$$2x+4 = 36$$
$$x = 16$$

check: $\sqrt{2(16)+4}+6 = 0$
$$12 \ne 0$$
16 is an extraneous solution.
$$\boxed{\varnothing}$$

9.
$$\sqrt{y-7} = 7-\sqrt{y}$$
$$(\sqrt{y-7})^2 = (7-\sqrt{y})^2$$
$$y-7 = (7-\sqrt{y})(7-\sqrt{y})$$
$$y-7 = 49-14\sqrt{y}+y$$
$$-56 = -14\sqrt{y}$$
$$4 = \sqrt{y}$$
$$(4)^2 = (\sqrt{y})^2$$
$$16 = y$$

check: $\sqrt{16-7} = 7-\sqrt{16}$
$$\sqrt{9} = 7-\sqrt{16}$$
$$3 = 7-4$$
$$3 = 3 \quad \text{The solution checks.}$$
$$\boxed{\{16\}}$$

11.
$$\sqrt{r+3} = \sqrt{r}-3$$
$$(\sqrt{r+3})^2 = (\sqrt{r}-3)^2$$
$$r+3 = r-6\sqrt{r}+9$$
$$-6 = -6\sqrt{r}$$
$$1 = \sqrt{r}$$
$$1 = r$$

check: $\sqrt{1+3} = \sqrt{1}-3$
$$2 \ne -2$$
1 is an extraneous solution.
$$\boxed{\varnothing}$$

13.
$$\sqrt{y+8} = \sqrt{y-4}+2$$
$$(\sqrt{y+8})^2 = (\sqrt{y-4}+2)^2$$
$$y+8 = (\sqrt{y-4}+2)(\sqrt{y-4}+2)$$
$$y+8 = y-4+4\sqrt{y-4}+4$$
$$y+8 = y+4\sqrt{y-4}$$
$$8 = 4\sqrt{y-4}$$
$$2 = \sqrt{y-4}$$
$$2^2 = (\sqrt{y-4})^2$$
$$4 = y-4$$
$$8 = y$$

check: $\sqrt{8+8} = \sqrt{8-4}+2$
$$\sqrt{16} = \sqrt{4}+2$$
$$4 = 4 \quad \text{The solution checks.}$$
$$\boxed{\{8\}}$$

15. $\sqrt{p-5} - \sqrt{p-8} = 3$

$\qquad\qquad \sqrt{p-5} = 3 + \sqrt{p-8}$

$\qquad\quad (\sqrt{p-5})^2 = (3 + \sqrt{p-8})^2$

$\qquad\qquad\quad p-5 = 9 + 6\sqrt{p-8} + p - 8$

$\qquad\qquad\quad p-5 = 1 + p + 6\sqrt{p-8}$

$\qquad\qquad\quad -6 = 6\sqrt{p-8}$

$\qquad\qquad\quad -1 = \sqrt{p-8}$

$\qquad\qquad\quad\ 9 = p$

check: $\sqrt{9-5} - \sqrt{9-8} = 3$

$\qquad\qquad\quad \sqrt{4} - \sqrt{1} = 3$

$\qquad\qquad\qquad\ 2 - 1 = 3$

$\qquad\qquad\qquad\quad\ 1 \neq 3$

9 is an extraneous solution.

$\boxed{\varnothing}$

17. $\qquad\qquad \sqrt{x+4} + 2 = \sqrt{x+20}$

$x + 4 + 4\sqrt{x+4} + 4 = x + 20$

$\quad x + 8 + 4\sqrt{x+4} = x + 20$

$\qquad\qquad 4\sqrt{x+4} = 12$

$\qquad\qquad\ \sqrt{x+4} = 3$

$\qquad\qquad\quad x+4 = 9$

$\qquad\qquad\qquad\ x = 5 \qquad$ checks

$\boxed{\{5\}}$

19. $\qquad\quad \sqrt{x+6} = 2 - \sqrt{x-2}$

$\qquad (\sqrt{x+6})^2 = (2 - \sqrt{x-2})^2$

$\qquad\qquad x+6 = (2 - \sqrt{x-2})(2 - \sqrt{x-2})$

$\qquad\qquad x+6 = 4 - 4\sqrt{x-2} + x - 2$

$\qquad\qquad x+6 = x + 2 - 4\sqrt{x-2}$

$\qquad\qquad\quad 4 = -4\sqrt{x-2}$

$\qquad\qquad\ -1 = \sqrt{x-2}$

$\qquad\quad (-1)^2 = (\sqrt{x-2})^2$

$\qquad\qquad\quad 1 = x - 2$

$\qquad\qquad\quad 3 = x$

check: $\sqrt{3+6} = 2 - \sqrt{3-2}$

$\qquad\qquad\quad 3 = 2 - 1$

$\qquad\qquad\quad 3 \neq 1$

3 is an extraneous solution.

$\boxed{\varnothing}$

21.
$$\sqrt{x-4} + \sqrt{x+4} = 4$$
$$(\sqrt{x-4} + \sqrt{x+4})^2 = 4^2$$
$$(\sqrt{x-4})^2 + 2\sqrt{x-4}\sqrt{x+4} + (\sqrt{x+4})^2 = 16$$
$$x-4 + 2\sqrt{x^2-16} + x + 4 = 16$$
$$2x + 2\sqrt{x^2-16} = 16$$
$$x + \sqrt{x^2-16} = 8$$
$$\sqrt{x^2-16} = 8-x$$
$$x^2-16 = 64 - 16 + x^2$$
$$16x = 80$$
$$x = 5 \quad \text{checks}$$

$\boxed{\{5\}}$

23.
$$\sqrt{10y} = y$$
$$(\sqrt{10y})^2 = (y)^2$$
$$10y = y^2$$
$$0 = y^2 - 10y$$
$$0 = y(y-10)$$
$$y = 0 \quad \text{or} \quad y = 10$$

check 0: $\sqrt{10(0)} = 0$
$$0 = 0 \quad \text{checks}$$
check 10: $\sqrt{10(10)} = 10$
$$10 = 10 \quad \text{checks}$$

$\boxed{\{0, 10\}}$

25.
$$5 = x - \sqrt{x-3}$$
$$\sqrt{x-3} = x-3$$
$$(\sqrt{x-3})^2 = (x-5)^2$$
$$x-3 = x^2 - 10x + 25$$
$$0 = x^2 - 11x + 28$$
$$0 = (x-7)(x-4)$$
$$x = 7 \quad \text{or} \quad x = 4$$

check 7: $5 = 7 - \sqrt{7-3}$
$$5 = 7 - \sqrt{4}$$
$$5 = 7 - 2$$
$$5 = 5 \quad \text{checks}$$
check 4: $5 = 4 - \sqrt{4-3}$
$$5 = 4 - \sqrt{1}$$
$$5 \neq 3$$
4 is extraneous.

$\boxed{\{7\}}$

27.
$$\sqrt{2y-5}+y-2 = 0$$
$$(\sqrt{2y-5})^2 = (2-y)^2$$
$$2y-5 = 4-4y+y^2$$
$$0 = y^2-6y+9$$
$$0 = (y-3)^2$$
$$y = 3 \quad \text{(extraneous)}$$

$\boxed{\varnothing}$

29.
$$\sqrt{2x+6}-3x = -11$$
$$\sqrt{2x+6} = 3x-11$$
$$(\sqrt{2x+6})^2 = (3x-11)^2$$
$$2x+6 = 9x^2-66x+121$$
$$0 = 9x^2-68x+115$$
$$0 = (x-5)(9x-23)$$
$$x = 5 \quad \text{or} \quad x = \frac{23}{9}$$

check 5:
$$\sqrt{2(5)+6}-3(5) = -11$$
$$4-15 = -11$$
$$-11 = -11 \quad \text{checks}$$

check $\frac{23}{9}$:
$$\sqrt{2\left(\frac{23}{9}\right)+6}-3 = -11$$
$$\sqrt{\frac{46}{9}+\frac{54}{9}}-\frac{23}{3} = -11$$
$$\sqrt{\frac{100}{9}}-\frac{23}{3} = -11$$
$$\frac{10}{3}-\frac{23}{3} \neq -11$$

$\frac{23}{9}$ is extraneous.

$\boxed{\{5\}}$

31.
$$2(z-4) = \sqrt{3z-2}$$
$$[2(z-4)]^2 = (\sqrt{3z-2})^2$$
$$4(z^2-8z+16) = 3z-2$$
$$4z^2-35z+66 = 0$$
$$(z-6)(4z-11) = 0$$
$$z = 6 \quad \text{or} \quad z = \frac{11}{4} \quad \text{(extraneous)}$$

$\boxed{\{6\}}$

33.
$$\sqrt{y} + 1 = \sqrt{5y - 1}$$
$$(\sqrt{y} + 1)^2 = (\sqrt{5y - 1})^2$$
$$(\sqrt{y} + 1)(\sqrt{y} + 1) = 5y - 1$$
$$y + 2\sqrt{y} + 1 = 5y - 1$$
$$2\sqrt{y} = 4y - 2$$
$$\frac{1}{2}(2\sqrt{y}) = \frac{1}{2}(4y - 2)$$
$$\sqrt{y} = 2y - 1$$
$$(\sqrt{y})^2 = (2y - 1)^2$$
$$y = 4y^2 - 4y + 1$$
$$0 = 4y^2 - 5y + 1$$
$$0 = (4y - 1)(y - 1)$$
$$y = \frac{1}{4} \quad \text{or} \quad y = 1$$

check $\frac{1}{4}$:
$$\sqrt{\frac{1}{4} + 1} = \sqrt{5\left(\frac{1}{4}\right) - 1}$$
$$\frac{1}{2} + 1 = \sqrt{\frac{5}{4} - 1}$$
$$\frac{1}{2} + 1 = \sqrt{\frac{1}{4}}$$
$$1\frac{1}{2} \neq \frac{1}{2}$$

$\frac{1}{4}$ is extraneous.

check 1:
$$\sqrt{1} + 1 = \sqrt{5(1) - 1}$$
$$1 + 1 = \sqrt{4}$$
$$2 = 2 \quad \text{checks}$$

$$\boxed{\{1\}}$$

35.
$$\sqrt{4z - 3} - \sqrt{8z + 1} + 2 = 0$$
$$\sqrt{4z - 3} = \sqrt{8z + 1} - 2$$
$$(\sqrt{4z - 3})^2 = (\sqrt{8z + 1} - 2)^2$$
$$4z - 3 = (\sqrt{8z + 1} - 2)(\sqrt{8z + 1} - 2)$$
$$4z - 3 = 8z + 1 - 4\sqrt{8z + 1} + 4$$
$$4z - 3 = 8z + 5 - 4\sqrt{8z + 1}$$
$$-4z - 8 = -4\sqrt{8z + 1}$$
$$\frac{1}{4}(-4z - 8) = \frac{1}{4}(-4\sqrt{8z + 1})$$
$$z + 2 = \sqrt{8z + 1}$$
$$(z + 2)^2 = (\sqrt{8z + 1})^2$$
$$z^2 + 4z + 4 = 8z + 1$$
$$z^2 - 4z + 3 = 0$$
$$(z - 1)(z - 3) = 0$$
$$z = 1 \quad \text{or} \quad z = 3 \quad \text{(Both check.)}$$

$$\boxed{\{1, 3\}}$$

37.

$$2\sqrt{3x-2} + \sqrt{2x-3} = 5$$
$$(2\sqrt{3x-2})^2 = (5 - \sqrt{2x-3})^2$$
$$4(3x-2) = 25 - 10\sqrt{2x-3} + 2x - 3$$
$$12x - 8 = 2x + 22 - 10\sqrt{2x-3}$$
$$10x - 30 = -10\sqrt{2x-3}$$
$$x - 3 = -\sqrt{2x-3}$$
$$(x-3)^2 = (-\sqrt{2x-3})^2$$
$$x^2 - 6x + 9 = 2x - 3$$
$$x^2 - 8x + 12 = 0$$
$$(x-6)(x-2) = 0$$
$$x = 6 \quad \text{or} \quad x = 2$$

check 6: $2\sqrt{3(6)-2} + \sqrt{2(6)-3} = 5$
$$2(4) + 3 = 5$$
$$11 \neq 5$$

6 is extraneous.

check 2: $2\sqrt{3(2)-2} + \sqrt{2(2)-3} = 5$
$$2(2) + 1 = 5$$
$$5 = 5 \quad \text{checks}$$

$\boxed{\{2\}}$

39.

$$\sqrt{2y+3} + \sqrt{y+2} = 2$$
$$(\sqrt{2y+3})^2 = (2 - \sqrt{y+2})^2$$
$$2y + 3 = 4 - 4\sqrt{y+2} + y + 2$$
$$y - 3 = -4\sqrt{y+2}$$
$$(y-3)^2 = (-4\sqrt{y+2})^2$$
$$y^2 - 6y + 9 = 16(y+2)$$
$$y^2 - 22y - 23 = 0$$
$$(y+1)(y-23) = 0$$
$$y = -1 \quad \text{or} \quad y = 23 \quad \text{(extraneous)}$$

$\boxed{\{-1\}}$

41.

$$\sqrt{3-y} + \sqrt{2+y} = 1$$
$$\sqrt{3-y} = 1 - \sqrt{2+y}$$
$$3 - y = 1 - 2\sqrt{2+y} + 2 + y$$
$$-2y = -2\sqrt{2+y}$$
$$y = \sqrt{2+y}$$
$$y^2 = 2 + y$$
$$y^2 - y - 2 = 0$$
$$(y+1)(y-2) = 0$$
$$y = -1 \quad \text{or} \quad y = 2 \quad \text{(Both are extraneous.)}$$

$\boxed{\varnothing}$

43.

$$\sqrt{6y-2} = \sqrt{2y+3} - \sqrt{4y-1}$$

$$(\sqrt{6y-2})^2 = (\sqrt{2y+3} - \sqrt{4y-1})^2$$

$$6y-2 = 2y+3 - 2\sqrt{2y+3}\sqrt{4y-1} + 4y-1$$

$$6y-2 = 6y+2 - 2\sqrt{2y+3}\sqrt{4y-1}$$

$$-4 = -2\sqrt{2y+3}\sqrt{4y-1}$$

$$2 = \sqrt{2y+3}\sqrt{4y-1}$$

$$2^2 = (\sqrt{2y+3}\sqrt{4y-1})^2$$

$$4 = (2y+3)(4y-1)$$

$$4 = 8y^2 + 10y - 3$$

$$0 = 8y^2 + 10y - 7$$

$$0 = (4y+7)(2y-1)$$

$$4y+7 = 0 \quad \text{or} \quad 2y-1 = 0$$

$$y = -\frac{7}{4} \quad \text{or} \quad y = \frac{1}{2}$$

check $-\frac{7}{4}$: $\sqrt{6\left(-\frac{7}{4}\right)-2} = \sqrt{2\left(-\frac{7}{4}\right)+3} - \sqrt{4\left(-\frac{7}{4}\right)-1}$

$$\sqrt{\frac{-21}{2}-\frac{4}{2}} = \sqrt{-\frac{7}{2}+\frac{6}{2}} - \sqrt{-7-1}$$

$$\sqrt{-\frac{25}{2}} = \sqrt{\frac{-1}{2}} - \sqrt{-8}$$

$$\frac{5i}{\sqrt{2}} = \frac{i}{\sqrt{2}} - 2\sqrt{2i} \cdot \frac{\sqrt{2}}{\sqrt{2}} \quad \text{(\textit{Note}: } i = \sqrt{-1} \text{ from Section 7.9)}$$

$$\frac{5i}{\sqrt{2}} = \frac{i}{\sqrt{2}} - \frac{4i}{\sqrt{2}}$$

$$\frac{5i}{\sqrt{2}} \neq -\frac{3i}{\sqrt{2}}$$

$-\frac{7}{4}$ is extraneous.

check $\frac{1}{2}$: $\sqrt{6\left(\frac{1}{2}\right)-2} = \sqrt{2\left(\frac{1}{2}\right)+3} - \sqrt{4\left(\frac{1}{2}\right)-1}$

$$\sqrt{1} = \sqrt{4} - \sqrt{1}$$

$$1 = 1$$

$$\boxed{\left\{\frac{1}{2}\right\}}$$

45.

$$\sqrt{5z+1} = \sqrt{3z+4} + \sqrt{z-6}$$

$$5z+1 = 3z+4 + 2\sqrt{3z+4}\sqrt{iz-6} + z-6$$

$$z+3 = 2\sqrt{3z+4}\sqrt{z-6}$$

$$z^2 + 6z + 9 = 4(3z+4)(z-4)$$

$$z^2 + 6z + 9 = 12z^2 - 56z - 96$$

$$0 = 11z^2 - 62z - 105$$

$$0 = (z-7)(11z+15)$$

$$z = 7 \quad \text{(checks)} \quad z = -\frac{15}{11}$$

check for $-\dfrac{15}{11}$:

$$\sqrt{5\left(-\dfrac{15}{11}\right)+1} = \sqrt{3\left(-\dfrac{15}{11}\right)+4} + \sqrt{-\dfrac{15}{11}-6}$$

$$\sqrt{-\dfrac{64}{11}} = \sqrt{-\dfrac{1}{11}} + \sqrt{-\dfrac{81}{11}}$$

$$\dfrac{8i}{\sqrt{11}} = \dfrac{i}{\sqrt{11}} + \dfrac{9i}{\sqrt{11}} \qquad (\textit{Note: } i = \sqrt{-1} \text{ from Section 7.9})$$

$$\dfrac{8i}{\sqrt{11}} \neq \dfrac{10i}{\sqrt{11}}$$

$-\dfrac{15}{11}$ is extraneous.

$\boxed{\{7\}}$

47.
$$\left(\sqrt{y+1+\sqrt{7y+4}}\right)^2 = 3^2$$
$$y+1+\sqrt{7y+4} = 9$$
$$\sqrt{7y+4} = 8-y$$
$$\left(\sqrt{7y+4}\right)^2 = (8-y)^2$$
$$7y+4 = 64-16y+y^2$$
$$0 = y^2-23y+60$$
$$0 = (y-20)(y-3)$$
$$y-20 = 0 \quad \text{or} \quad y-3=0$$
$$y = 20 \quad \text{or} \quad y=3$$

check 20:
$$\sqrt{20+1+\sqrt{7(20)+4}}-3 = 0$$
$$\sqrt{21+\sqrt{144}}-3 = 0$$
$$\sqrt{21+12}-3 = 0$$
$$\sqrt{33}-3 \neq 0$$

20 is extraneous.

check 3:
$$\sqrt{3+1\sqrt{7(3)+4}}-3 = 0$$
$$\sqrt{4+\sqrt{25}}-3 = 0$$
$$\sqrt{4+5}-3 = 0$$
$$\sqrt{9}-3 = 0 \qquad \text{(checks)}$$
$$0 = 0$$

$\boxed{\{3\}}$

49.
$$\sqrt[3]{z^2-1} = 2$$
$$\left(\sqrt[3]{z^2-1}\right)^3 = 2^3$$
$$z^2-1 = 8$$
$$z^2-9 = 0$$
$$(z+3)(z-3) = 0$$
$$z = -3 \quad \text{or} \quad z=3$$

$\boxed{\{-3,3\}}$ (both check)

51.
$$\sqrt[3]{x^2 + 6x} = -2$$
$$x^2 + 6x = -8$$
$$x^2 + 6x + 8 = 0$$
$$(x + 4)(x + 2) = 0$$
$$x = -4 \quad \text{or} \quad x = -2 \quad \text{(both check)}$$

$$\boxed{\{-4, -2\}}$$

53. $P = 8\sqrt[3]{x} + 12$
($x = 84$ tons):

$$184 = 8\sqrt[3]{x} + 12$$
$$72 = 8\sqrt[3]{x}$$
$$9 = \sqrt[3]{x}$$
$$9^3 = x$$
$$729 = x$$
$$1935 + 729 = 2664$$

$$\boxed{\text{in the year 2664}}$$

55. $N = \sqrt{100 - x^2} - 6$ where $0 \le x \le 10$
If there is no demand, no calculators sell, so $N = 0$.

$$\sqrt{100 - x^2} - 6 = 0$$
$$\sqrt{100 - x^2} = 6$$
$$100 - x^2 = 36$$
$$0 = x^2 - 64$$
$$0 = (x - 8)(x + 8)$$
$$x = 8 \quad \text{or} \quad x = -8 \quad \text{(meaningless)}$$

The smallest price is $\boxed{\$8}$.

57. Let $x =$ the number.
$$\sqrt{2x - 3} = \sqrt{x + 2} + 1$$
$$(\sqrt{2x - 3})^2 = (\sqrt{x + 2} + 1)^2$$
$$2x - 3 = x + 2 + 2\sqrt{x + 2} + 1$$
$$x - 6 = 2\sqrt{x + 2}$$
$$(x - 6)^2 = (2\sqrt{x + 2})^2$$
$$x^2 - 12x + 36 = 4(x + 2)$$
$$x^2 - 12x + 36 = 4x + 8$$
$$x^2 - 16x + 28 = 0$$
$$(x - 14)(x - 2) = 0$$
$$x = 14 \quad \text{or} \quad x = 2 \quad \text{(extraneous)}$$

The number is $\boxed{14}$.

59.

$$M = \frac{M_0}{\sqrt{1 - \frac{r^2}{c^2}}}$$

Find v when $M = 4M_0$.

$$4M_0 = \frac{M_0}{\sqrt{1 - \frac{v^2}{c^2}}}$$

$$4 = \frac{1}{\sqrt{1 - \frac{v^2}{c^2}}}$$

$$4^2 = \frac{1}{\left(\sqrt{1 - \frac{v^2}{c^2}}\right)^2}$$

$$16 = \frac{1}{1 - \frac{v^2}{c^2}}$$

$$16 = \frac{1}{1 - \frac{v^2}{c^2}} \cdot \frac{c^2}{c^2}$$

$$16 = \frac{c^2}{c^2 - v^2}$$

$$16(c^2 - v^2) = c^2$$

$$16c^2 - 16v^2 = c^2$$

$$-16v^2 = -15c^2$$

$$v^2 = \frac{15}{6}c^2$$

$$\boxed{v = \frac{\sqrt{15}}{4}c \approx 0.968\,c \approx 180{,}094 \text{ miles/sec}}\ \text{since } c \approx 186{,}000 \text{ miles/sec}$$

61.

$$L = L_0 \sqrt{1 - \frac{v^2}{c^2}}$$

$$L_0 = 6 \text{ feet}$$

$$\frac{v}{c} = 0.8$$

$$L = 6\sqrt{1 - (0.8)^2} = 6\sqrt{1 - 0.64} = 6\sqrt{0.36} = 6(0.6) = 3.6 \text{ feet}$$

$$(0.6 \text{ feet} = 0.6(12) \text{ inches} = 7.2 \text{ inches})$$

The person's relativistic length is $\boxed{3 \text{ feet } 7.2 \text{ inches}}$.

63. $\boxed{D}$ is true; $-\sqrt{x} = 9$

$$\sqrt{x} = -9$$

$$(\sqrt{x})^2 = (-9)^2$$

$$x = 81$$

$$-\sqrt{81} = 9$$

$$-9 \neq 9 \quad \text{(81 is extraneous)}$$

Thus $-\sqrt{x} = 9$ has no solution.

65. $5 - \dfrac{2}{x} = \sqrt{5 - \dfrac{2}{x}}$

Since only $0 = \sqrt{0}$ and $1 = \sqrt{1}$ (that is, $x = \sqrt{x}$ is true only for $x = 0$ and $x = 1$), then

$$5 - \frac{2}{x} = 0 \qquad \text{or} \qquad 5 - \frac{2}{x} = 1$$

$$5x - 2 = 0 \qquad\qquad\qquad 5x - 2 = x$$

$$x = \frac{2}{5} \qquad\qquad\qquad\qquad 4x = 2$$

$$\qquad\qquad\qquad\qquad\qquad\qquad x = \frac{1}{2}$$

The solution set is $\boxed{\left\{ \dfrac{2}{5}, \dfrac{1}{2} \right\}}$.

67. $y = \sqrt{x - 2} + 2$ if $x \neq y$

Try substituting prime numbers for x.

If $x = 2$: $y = \sqrt{x-2} + 2 = \sqrt{2-2} + 2 = 2$

We are given that $x \neq y$, so try another prime number for x.

If $x = 3$: $y = \sqrt{x-2} + 2 = \sqrt{3-2} + 2 = 3$ (Since $x = y$, eliminate this selection.)

If $x = 5$: $y = \sqrt{x-2} + 2 = \sqrt{5-2} + 2 = \sqrt{3} + 2$, not prime
Continuing in this manner, using trial and error, you should eliminate $x = 7$. Test the next prime.

If $x = 11$: $y = \sqrt{x-2} + 2 = \sqrt{11-2} + 2 = 5$
Since 5 is prime and not equal to 11, the smallest prime numbers are $x = 11$ and $y = 5$
$\boxed{\{(11, 5)\}}$

69. Let $x =$ the number.
   1. $x + 1$
   2. $\sqrt{x + 1}$
   3. $\sqrt{x + 1} + 1$
   4. $\sqrt{\sqrt{x + 1} + 1}$
   5. $\sqrt{\sqrt{x + 1} + 1} + 1$
   6. $\sqrt{\sqrt{\sqrt{x + 1} + 1} + 1}$

7.

$$\sqrt{\sqrt{\sqrt{\sqrt{x+1}+1}+1}+1} = 2$$

$$\left(\sqrt{\sqrt{\sqrt{\sqrt{x+1}+1}+1}+1}\right)^2 = 2^2$$

$$\sqrt{\sqrt{\sqrt{x+1}+1}+1}+1 = 4$$

$$\sqrt{\sqrt{\sqrt{x+1}+1}+1} = 3$$

$$\left(\sqrt{\sqrt{\sqrt{x+1}+1}+1}\right)^2 = 3^2$$

$$\sqrt{\sqrt{x+1}+1}+1 = 9$$

$$\sqrt{\sqrt{x+1}+1} = 8$$

$$(\sqrt{x+1})^2 = 8^2$$

$$x+1 = 64$$

$$x = 63$$

The number is $\boxed{63}$.

Review Problems

74. $(4x^{-4}y^{1/2})^{-3/2}$

$= 4^{-3/2}(x^{-4})^{-3/2}(y^{1/2})^{-3/2}$

$= \dfrac{1}{4^{3/2}}\,x^6 y^{-3/4}$

$= \dfrac{x^6}{(\sqrt{4})^3 y^{3/4}}$

$= \boxed{\dfrac{x^6}{8y^{3/4}}}$

75. $\dfrac{\sqrt[3]{5}}{\sqrt[4]{5}}$

$= \dfrac{5^{1/3}}{5^{1/4}}$

$= 5^{1/3\,-\,1/4}$

$= 5^{4/12\,-\,3/12} = 5^{1/2}$

$= \boxed{\sqrt[12]{5}}$

76. $\dfrac{2}{y-3} - \dfrac{y}{y^2-y-6} \div \dfrac{y^2+y}{y^2-2y-3}$

$= \dfrac{2}{y-3} - \dfrac{y}{y^2-y-6} \cdot \dfrac{y^2-2y-3}{y^2+1}$

$= \dfrac{2}{y-3} - \dfrac{y}{(y-3)(y-2)} \cdot \dfrac{(y-3)(y+1)}{y(y+1)}$

$= \dfrac{2}{y-3} - \dfrac{1}{y+2}$

$= \dfrac{2}{y-3} \cdot \dfrac{y+2}{y+2} - \dfrac{1}{y+2} \cdot \dfrac{y-3}{y-3}$

$= \dfrac{2(y+2)-(y-3)}{(y-3)(y+2)}$

$= \dfrac{2y+4-y+3}{(y-3)(y+2)}$

$= \boxed{\dfrac{y+7}{(y-3)(y+2)}}$

Section 7.9 Imaginary and Complex Numbers

Problem Set 7.9, pp. 567-569

1. $\sqrt{4} = \sqrt{4}\, i = \boxed{2i}$

3. $\sqrt{-17} = \boxed{\sqrt{17}\, i}$

5. $\sqrt{-28} = \sqrt{28}\, i = \sqrt{4 \cdot 7}\, i = \boxed{2\sqrt{7}i}$

7. $\sqrt{-45} = \sqrt{45}\, i = \sqrt{9 \cdot 5}\, i = \boxed{3\sqrt{5}i}$

9. $\sqrt{-\dfrac{4}{9}} = \sqrt{\dfrac{4}{9}}\, i = \boxed{\dfrac{2}{3}i}$

11. $5\sqrt{-12} = 5\sqrt{4 \cdot 3}\, i = 5(2)\sqrt{3}\, i = \boxed{10\sqrt{3}i}$

13. $\sqrt{-49} + \sqrt{-100} = \sqrt{49}\, i + \sqrt{100}\, i = 7i + 10i = \boxed{17i}$

15. $\sqrt{-64} - \sqrt{-25} = 8i - 5i = \boxed{3i}$

17. $3\sqrt{-49} + 5\sqrt{-100} = 3(7i) + 5(10i) = 21i + 50i = \boxed{71i}$

19. $\sqrt{-72} + \sqrt{-50} = \sqrt{36 \cdot 2}\, i + \sqrt{25 \cdot 2}\, i = 6\sqrt{2}\, i + 5\sqrt{2}\, i = \boxed{11\sqrt{2}\, i}$

21. $5\sqrt{-8} - 3\sqrt{-18} = 5\sqrt{4 \cdot 2}\, i - 3\sqrt{9 \cdot 2}\, i = 10\sqrt{2}\, i - 9\sqrt{2}\, i = \boxed{\sqrt{2}\, i}$

23. $\dfrac{3}{5}\sqrt{-50} + \dfrac{1}{2}\sqrt{-32} = \dfrac{3}{5}\sqrt{25 \cdot 2}\, i + \dfrac{1}{2}\sqrt{16 \cdot 2}\, i = \dfrac{3}{5}(5)\sqrt{2}\, i + \dfrac{1}{2}(4)\sqrt{2}\, i = 3\sqrt{2}\, i + 2\sqrt{2}\, i = \boxed{5\sqrt{2}\, i}$

25. $\sqrt{-7}\sqrt{-2} = (\sqrt{7}i)(\sqrt{2}i) = \sqrt{14}ii^2 = \sqrt{14}(-1) = \boxed{-\sqrt{14}}$

27. $\sqrt{-9}\sqrt{-4} = (3i)(2i) = 6i^2 = 6(-1) = \boxed{-6}$

29. $\sqrt{-7}\sqrt{-25} = (\sqrt{7}i)(5i) = 5\sqrt{7}\, i^2 = 5\sqrt{7}(-1) = \boxed{-5\sqrt{7}}$

31. $\sqrt{-8}\sqrt{-3} = (\sqrt{4 \cdot 2}\, i)(\sqrt{3}\, i) = 2\sqrt{6}\, i^2 = 2\sqrt{6}(-1) = \boxed{-2\sqrt{6}}$

33. $(2\sqrt{-8})(3\sqrt{-6}) = (2\sqrt{8}\, i)(3\sqrt{6}\, i) = 6\sqrt{48}\, i^2 = 6\sqrt{16 \cdot 3}(-1) = 6(4)\sqrt{3}(-1) = \boxed{-24\sqrt{3}}$

35. $(3\sqrt{-5})(-4\sqrt{-12}) = (3\sqrt{5}\, i)(-4\sqrt{12}\, i) = -12\sqrt{60}\, i^2 = -12\sqrt{4 \cdot 15}(-1) = -12(2)\sqrt{15}(-1) = \boxed{24\sqrt{5}}$

37. $\dfrac{\sqrt{-125}}{\sqrt{5}} = \sqrt{\dfrac{125}{5}}\, i = \sqrt{25}\, i = \boxed{5i}$

39. $\dfrac{\sqrt{-24}}{\sqrt{-6}} = \sqrt{\dfrac{24}{6}} \cdot \dfrac{i}{i} = \sqrt{4} = \boxed{2}$

41. $\dfrac{\sqrt{-200}}{\sqrt{5}} = \sqrt{\dfrac{200}{5}}\, i = \sqrt{40}\, i = \sqrt{4 \cdot 10} = \boxed{2\sqrt{10}i}$

43. $\sqrt{\dfrac{-1}{2}} + \sqrt{\dfrac{-9}{2}} = \sqrt{\dfrac{1}{2}}\, i + \sqrt{\dfrac{9}{2}}\, i = \dfrac{1}{\sqrt{2}}i + \dfrac{3}{\sqrt{2}}i = \dfrac{1}{\sqrt{2}} \cdot \dfrac{\sqrt{2}}{\sqrt{2}}i + \dfrac{3}{\sqrt{2}} \cdot \dfrac{\sqrt{2}}{\sqrt{2}}i = \dfrac{\sqrt{2}}{2}i + \dfrac{3\sqrt{2}}{2}i = \dfrac{4\sqrt{2}}{2}i = \boxed{2\sqrt{2}i}$

45. a. $i^{14} = (i^2)^7 = (-1)^7 = \boxed{-1}$

   b. $i^{31} = i^{30}i = (i^2)^{15}i = (-1)^{15}i = \boxed{-i}$

   c. $i^{22} = (i^2)^{11} = (-1)^{11} = \boxed{-1}$

   d. $i^{37} = i^{36} \cdot i = (i^2)^{18} \cdot i = (-1)^{18}i = 1i = \boxed{i}$

47. $8x + 3yi = 40 - 6i$

   $\begin{array}{lll} 8x = 40 & \text{and} & 3y = -6 \quad \text{(Equate real and imaginary parts.)} \\ \boxed{x = 5} & \text{and} & \boxed{y = -2} \end{array}$

49. $3x + 8i = -15 - 4yi$

   $\begin{array}{lll} 3x = -15 & \text{and} & 8 = -4y \quad \text{(Equate real and imaginary parts.)} \\ \boxed{x = -5} & \text{and} & \boxed{-2 = y} \end{array}$

51. $(3x - 5) - 4i = 10 - 20yi$

   $\begin{array}{lll} 3x - 5 = 10 & \text{and} & -4 = -20y \\ \boxed{x = 5} & \text{and} & \boxed{y = \dfrac{1}{5}} \end{array}$

53. $(3 + 2i) + (5 + i) = (3 + 5) + (2 + 2)i = \boxed{8 + 3i}$

55. $(7 + 2i) + (1 - 4i) = (7 + 1) + (2 - 4)i = \boxed{8 - 2i}$

57. $(3 + 2i) - (5 + i) = (3 - 5) + (2 - 1)i = \boxed{-2 + i}$

59. $(7 + 2i) - (1 - 4i) = (7 - 1) + (2 + 4)i = \boxed{6 + 6i}$

61. $(2 + \sqrt{3}\,i) + (7 + 4\sqrt{3}\,i) = (2 + 7) + (\sqrt{3} + 4\sqrt{3})i = \boxed{9 + 5\sqrt{3}\,i}$

63. $(5 + 2\sqrt{32}\,i) + (11 - 5\sqrt{8}\,i)$

   $= (5 + 11) + (2\sqrt{32} - 5\sqrt{8})i$

   $= 16 + (2\sqrt{16 \cdot 2} - 5\sqrt{4 \cdot 2})i$

   $= 16 + (8\sqrt{2} - 10\sqrt{2})i$

   $= \boxed{16 - 2\sqrt{2}\,i}$

65. $(5 + 2\sqrt{32}\,i) - (11 - 5\sqrt{8}\,i)$

   $= (5 + 2\sqrt{32}\,i) + (-11 + 5\sqrt{8}\,i)$

   $= (5 - 11) + (2\sqrt{32} + 5\sqrt{8})i$

   $= -6 + (2\sqrt{16 \cdot 2} + 5\sqrt{4 \cdot 2})i$

   $= -6 + (8\sqrt{2} + 10\sqrt{2})i$

   $= \boxed{-6 + 18\sqrt{2}\,i}$

67. $(5\sqrt{20} + 3\sqrt{63}\,i) - (\sqrt{80} - 2\sqrt{28}\,i)$

   $= (5\sqrt{20} + 3\sqrt{63}\,i) + (-\sqrt{80} + 2\sqrt{28}\,i)$

   $= (5\sqrt{20} - \sqrt{80}) + (3\sqrt{63} + 2\sqrt{28})i$

   $= (5\sqrt{4 \cdot 5} - \sqrt{16 \cdot 5}) + (3\sqrt{9 \cdot 7} + 2\sqrt{4 \cdot 7})i$

   $= (10\sqrt{5} - 4\sqrt{5}) + (9\sqrt{7} + 4\sqrt{7})i$

   $= \boxed{6\sqrt{5} + 13\sqrt{7}\,i}$

69. $\left(\dfrac{1}{3} - \dfrac{2}{5}i\right) - \left(\dfrac{1}{2} - \dfrac{3}{4}i\right)$

   $= \left(\dfrac{1}{3} - \dfrac{1}{2}\right) + \left(-\dfrac{2}{5} + \dfrac{3}{4}\right)i$

   $= \left(\dfrac{2}{6} - \dfrac{3}{6}\right) + \left(\dfrac{-8}{20} + \dfrac{15}{20}\right)i$

   $= \boxed{-\dfrac{1}{6} + \dfrac{7}{20}i}$

71. $(7 + 3i)(5 + 2i)$
$= 35 + 14i + 15i + 6i^2$
$= 35 + 14i + 15i + (-6)$
$= \boxed{29 + 29i}$

73. $(3 + 4i)(4 - 7i)$
$= 12 - 21i + 16i - 28i^2$
$= 12 - 5i - 28(-1)$
$= \boxed{40 - 5i}$

75. $(-5 - 4i)(3 + 7i)$
$= -15 - 47i - 28i^2$
$= -15 - 47 - (-28)$
$= \boxed{13 - 47i}$

77. $(7 - 5i)(-2 - 3i)$
$= -14 - 21i + 10i + 15i^2$
$= -14 - 21i + 10i + (-15)$
$= \boxed{-29 - 11i}$

79. $(3 + 5i)(3 - 5i)$
$= 9 - 15i + 15i - 25i^2$
$= 9 - 25(-1)$
$= 9 + 25$
$= \boxed{34}$

81. $(-5 + 3i)(-5 - 3i)$
$= (-5)^2 - (3i)^2$
$= 25 - 9i^2$
$= 25 - 9(-1)$
$= \boxed{34}$

83. $(2 + 3i)^2$
$= (2 + 3i)(2 + 3i)$
$= 4 + 12i + 9i^2$
$= 4 + 12i + 9(-1)$
$= \boxed{-5 + 12i}$

85. $(1 + i)^3$
$= (1 + i)[(1 + i)(1 + i)]$
$= (1 + i)(1 + 2i + i^2)$
$= (1 + i)(1 + 2i - 1)$
$= (1 + i)2i$
$= 2i + 2i^2$
$= 2i + 2(-1)$
$= \boxed{-2 + 2i}$

87. $(3 + 4i)(5 - 2i) + 8 + 6i$
$= (15 + 14i - 8i^2) + 8 + 6i$
$= 23 + 14i + 8 + 6i$
$= \boxed{31 + 20i}$

89. $(8 + 9i)(2 - i) - (1 - i)(1 + i)$
$= (16 + 10i - 9i^2) - (1 - i^2)$
$= (25 + 10i) - (2)$
$= \boxed{23 + 10i}$

91. $(2 + i)^2 - (3 - i)^2$

$= (2 + i)(2 + i) - (3 - i)(3 - i)$

$= (4 + 4i + i^2) - (9 - 6i + i^2)$

$= (3 + 4i) - (8 - 6i)$

$= (3 + 4i) + (-8 + 6i)$

$= \boxed{-5 + 10i}$

93. $\dfrac{2i}{3 - i}$

$= \dfrac{2i}{3 - i} \cdot \dfrac{3 + i}{3 + i}$

$= \dfrac{2i(3 + i)}{9 - i^2}$

$= \dfrac{2i(3 + i)}{10}$

$= \dfrac{i(3 + i)}{5}$

$= \dfrac{3i + i^2}{5}$

$= \dfrac{3i - 1}{5}$

$= \boxed{-\dfrac{1}{5} + \dfrac{3}{5}i}$

95. $\dfrac{3-i}{2i}$

$= \dfrac{3-i}{2i} \cdot \dfrac{i}{i}$

$= \dfrac{3i - i^2}{2i^2}$

$= \dfrac{3i - (-1)}{2(-1)}$

$= \dfrac{3i + 1}{-2}$

$= \boxed{-\dfrac{1}{2} - \dfrac{3}{2}i}$

97. $\dfrac{1+i}{1-i}$

$= \dfrac{1+i}{1-i} \cdot \dfrac{1+i}{1+i}$

$= \dfrac{1 + 2i + i^2}{1 - i^2}$

$= \dfrac{1 + 2i + (-1)}{1 - (-1)}$

$= \dfrac{2i}{2}$

$= \boxed{i}$

99. $\dfrac{2+3i}{3-i}$

$= \dfrac{2+3i}{3-i} \cdot \dfrac{3+i}{3+i}$

$= \dfrac{6 + 11i + 3i^2}{9 - i^2}$

$= \dfrac{3 + 11i}{10}$

$= \boxed{\dfrac{3}{10} + \dfrac{11}{10}i}$

101. $\dfrac{3+2i}{2+i}$

$= \dfrac{3+2i}{2+i} \cdot \dfrac{2-i}{2-i}$

$= \dfrac{6 + i - 2i^2}{4 - i^2}$

$= \dfrac{8 + i}{5}$

$= \boxed{\dfrac{8}{5} + \dfrac{1}{5}i}$

103. $\dfrac{-4+7i}{-2-5i}$

$= \dfrac{-4+7i}{-2-5i} \cdot \dfrac{-2+5i}{-2+5i}$

$= \dfrac{8 - 34i + 35i^2}{4 - 25i^2}$

$= \dfrac{-27 - 34i}{29}$

$= \boxed{-\dfrac{27}{29} - \dfrac{34}{29}i}$

105. $\dfrac{-4+7i}{-5i}$

$= \dfrac{-4+7i}{-5i} \cdot \dfrac{i}{i}$

$= \dfrac{-4i + 7i^2}{-5i^2}$

$= \dfrac{-4i - 7}{5}$

$= \boxed{-\dfrac{7}{5} - \dfrac{4}{5}i}$

107. $\boxed{D}$ is true;   *A* is *not* true; The conjugate of $7 + 3i$ is $7 - 3i$ *not* $-7 - 3i$.

  B is *not* true; $(3 + 7i)(3 - 7i) = 9 - 49i^2 = 9 + 49 = 58$
    58 is *not* an imaginary number
  C is *not* true; $\dfrac{7+3i}{5+3i} = \dfrac{7+3i}{5+3i} \cdot \dfrac{5-3i}{5-3i} = \dfrac{35 - 6i - 9i^2}{25 - 9i^2} = \dfrac{44 - 6i}{34}$ *not* $\dfrac{7}{5}$
  D is true; $x^2 + y^2 = (x + yi)(x - yi) = x^2 - y^2(i^2) = x^2 + y^2$

109. $\sqrt[3]{1} = -\frac{1}{2} - \frac{\sqrt{3}}{2}i$

$(\sqrt[3]{1})^2 = \left(-\frac{1}{2} - \frac{\sqrt{3}}{2}i\right)^3$

$\left(-\frac{1}{2} - \frac{\sqrt{3}}{2}i\right)^3$

$= \left(-\frac{1}{2} - \frac{\sqrt{3}}{2}i\right)\left[\left(-\frac{1}{2} - \frac{\sqrt{3}}{2}i\right)\left(-\frac{1}{2} - \frac{\sqrt{3}}{2}i\right)\right]$

$= \left(-\frac{1}{2} - \frac{\sqrt{23}}{2}i\right)\left(\frac{1}{4} + \frac{\sqrt{3}}{4}i + \frac{\sqrt{3}}{4}i + \frac{3}{4}i^2\right)$

$= \left(-\frac{1}{2} - \frac{\sqrt{3}}{2}i\right)\left(-\frac{1}{2} + \frac{\sqrt{3}}{2}i\right)$

$= \frac{1}{4} - \frac{3}{4}i^2$

$= \frac{1}{4} - \frac{3}{4}(-1)$

$= \frac{1}{4} + \frac{3}{4} = 1$

Thus, $\sqrt[3]{1} = -\frac{1}{2} - \frac{\sqrt{3}}{2}i$.

111. $\left(-\frac{\sqrt{6}}{2} + \frac{\sqrt{2}}{2}i\right)^2$

$= \left(-\frac{\sqrt{6}}{2} + \frac{\sqrt{2}}{2}i\right)\left(-\frac{\sqrt{6}}{2} + \frac{\sqrt{2}}{2}i\right)$

$= \frac{6}{4} - \frac{\sqrt{12}}{4}i - \frac{\sqrt{12}}{4}i + \frac{2}{4}i^2$

$= \frac{3}{2} - \frac{2\sqrt{12}}{4}i + \frac{1}{2}(-1)$

$= \frac{2}{2} - \frac{\sqrt{12}}{2}i$

$= 1 - \frac{\sqrt{4 \cdot 3}}{2}i$

$= 1 - \frac{2\sqrt{3}}{2}i$

$= 1 - \sqrt{3}i$

$$\left(\frac{\sqrt{6}}{2} - \frac{\sqrt{2}}{2}i\right)^2$$

$$= \left(\frac{\sqrt{6}}{2} - \frac{\sqrt{2}}{2}i\right)\left(\frac{\sqrt{6}}{2} - \frac{\sqrt{2}}{2}i\right)$$

$$= \frac{6}{4} - \frac{\sqrt{12}}{4}i - \frac{\sqrt{12}}{4}i + \frac{2}{4}i^2$$

$$= \frac{3}{2} - \frac{2\sqrt{12}}{4}i + \frac{1}{2}(-1)$$

$$= \frac{2}{2} - \frac{\sqrt{12}}{2}i$$

$$= 1 - \frac{\sqrt{4 \cdot 3}}{2}i$$

$$= 1 - \frac{2\sqrt{3}}{2}i$$

$$= 1 - \sqrt{3}i$$

The two square roots of $1 - \sqrt{3}i$ are $-\frac{\sqrt{6}}{2} + \frac{\sqrt{2}}{2}i$ and $\frac{\sqrt{6}}{2} - \frac{\sqrt{2}}{2}i$.

113. $x - 3 \geq 0$ when $x \geq 3$ and $x - 3 < 0$ when $x < 3$.

Thus, $\boxed{\sqrt{x - 3} \text{ is real when } x \geq 3 \text{ and imaginary when } x < 3}$.

115. $x^2 - 2x + 5 = 0;\ x = 1 - 2i$

$(1 - 2i)^2 - 2(1 - 2i) + 5$

$= (1 - 2i)(1 - 2i) - 2(1 - 2i) + 5$

$= 1 - 4i + 4i^2 - 2 + 4i + 5$

$= 1 - 4i - 4 - 2 + 4i + 5$

$= 0$ True

117. Show that the product of $2 + 3i$ and $\frac{2}{13} - \frac{3}{13}i$ is 1.

$(2 + 3i)\left(\frac{2}{13} - \frac{3}{13}i\right)$

$= \frac{4}{13} - \frac{6}{13}i + \frac{6}{13}i - \frac{9}{13}i^2$

$= \frac{4}{13} - \frac{9}{13}(-1)$

$= \frac{4}{13} + \frac{9}{13}$

$= \frac{13}{13}$

$= 1$ True

119. $\dfrac{(1 + i)(-1 + 2i) + (2 - i)}{2 - 3i}$

$= \dfrac{-1 + i + 2i^2 + 2 - i}{2 - 3i}$

$= \dfrac{-1}{2 - 3i} \cdot \dfrac{2 + 3i}{2 + 3i}$

$= \dfrac{-2 - 3i}{4 - 9i^2}$

$= \dfrac{-2 - 3i}{13}$

$= \boxed{-\dfrac{2}{13} - \dfrac{3}{13}i}$

121. $-\sqrt{-9}\sqrt{-4} = -(3i)(2i)$
$= -6i^2 = -6(-1) = 6$

Let $x =$ the number.
$$
\begin{aligned}
15x - 3(x + 6) &= 6 \\
15x - 3x - 18 &= 6 \\
12x &= 24 \\
x &= 2
\end{aligned}
$$
The number is $\boxed{2}$.

123. i

opposite of $i = -i$

reciprocal of $i = \dfrac{1}{i} = \dfrac{1}{i} \cdot \dfrac{i}{i} = \dfrac{i}{i^2} = \dfrac{i}{-1} = -i$

Thus, the opposite and reciprocal of i are the same number,

125. $a + bi$ and $a - bi$ are a number and its conjugate.
$a + bi = a - bi$ when $b = -b$ or $2b = 0$ or $b = 0$
Thus a complex number that is its own conjugate is $\boxed{a + bi \text{ when } b = 0}$

127.   The flaw is in $\boxed{\text{Line 4, since } \sqrt{\dfrac{a}{b}} = \dfrac{\sqrt{a}}{\sqrt{b}} \text{ only if } a \geq 0 \text{ and } b > 0}$.

Review Problems

135. $f(x) = 3x + 5$ and $g(x) = 4x^2 - 7x + 6$
$$
\begin{aligned}
(g \circ f)(x) &= g[f(x)] = 4(3x + 5)^2 - 7(3x + 5) + 6 \\
&= 4(9x^2 + 30x + 25) - 21x - 35 + 6 \\
&= 36x^2 + 120x + 100 - 21x - 29 \\
&= \boxed{36x^2 + 99x + 71}
\end{aligned}
$$

136. $V = kr^3$
Given: $V = 36\pi$ cubic meters, $r = 3$ meters
$$
\begin{aligned}
3\pi &= k(3^3) \\
\frac{4}{3}\pi &= k \\
V &= \frac{4}{3}\pi r^3
\end{aligned}
$$
Find v when $r = 5$ meters
$$
\begin{aligned}
V &= \frac{4}{3}\pi(5^3) \\
V &= \frac{4}{3}\pi(125) \\
V &= \frac{500}{3}\pi
\end{aligned}
$$
The volume is $\boxed{\dfrac{500}{3}\pi\,\text{m}^3}$.

137. $f(x) = x^2 - x - 12$ and $g(x) = \dfrac{x-4}{3}$

$$\left(\dfrac{f}{g}\right)(x) = \dfrac{x^2 - x - 12}{\dfrac{x-4}{3}} = (x-3)(x+3) \cdot \left(\dfrac{3}{x-4}\right) = \boxed{3x + 9 \quad (x \neq 4)}$$

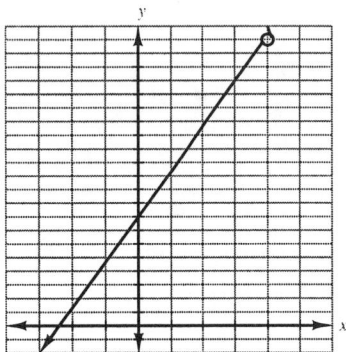

Chapter 7   Review Problems

Review Problems, pp. 572-574

1. $7^{-3} = \dfrac{1}{7^3} = \boxed{\dfrac{1}{343}}$

2. $\left(\dfrac{3}{4}\right)^{-3} = \dfrac{3^{-3}}{4^{-3}} = \dfrac{4^3}{3^3} = \boxed{\dfrac{64}{27}}$

3. $(4^5 \cdot 4^{-7})^{-2} = (4^{-2})^{-2} = 4^4 = \boxed{256}$

4. $\sqrt[3]{-27} = \sqrt[3]{(-3)^3} = \boxed{-3}$

5. $\sqrt[3]{\dfrac{27}{64}} = \sqrt[3]{\left(\dfrac{3}{4}\right)^3} = \boxed{\dfrac{3}{4}}$

6. $16^{3/2} = (\sqrt{16})^3 = 4^3 = \boxed{64}$

7. $8^{-2/3} = \dfrac{1}{8^{2/3}} = \dfrac{1}{(\sqrt[3]{8})^2} = \dfrac{1}{2^2} = \boxed{\dfrac{1}{4}}$

8. $\left(\dfrac{1}{32}\right)^{4/5} = \left(\sqrt[3]{\dfrac{1}{32}}\right)^4 = \left(\dfrac{1}{2}\right)^4 = \boxed{\dfrac{1}{16}}$

9. $\dfrac{3^3}{3^{-2}} = 3^{3-(-2)} = 3^5 = \boxed{243}$

10. $(5^{-3} \cdot 5^3)^{-4} = (5^0)^{-4} = 1^{-4} = \boxed{1}$

11. $\left(\dfrac{2^{-1}}{2^{-3}}\right)^{-4} = (2^{-1-(-3)})^{-4} = (2^2)^{-4} = 2^{-8} = \dfrac{1}{2^8} = \boxed{\dfrac{1}{256}}$

12. $A = 2000x^{-1} - 70x^{3/4} + 3000$
($x = 16$):
$$\begin{aligned}
A &= (2000)(16)^{-1} - 70(16)^{3/4} + 3000 \\
&= \dfrac{2000}{16} - 70(\sqrt[4]{16})^3 + 3000 \\
&= 125 - 70(2)^3 + 3000 \\
&= 125 - 560 + 3000 = 2565
\end{aligned}$$
$\boxed{2{,}565 \text{ clocks}}$ can be sold.

13. $(54x^3y^7)(-3x^2y^{-5})^{-3} = (54x^3y^7)(-3)^{-3}x^{-6}y^{15} = (54)(-3)^{-3}x^{-3}y^{22} = \dfrac{54y^{22}}{(-3)^3x^3} = \dfrac{54y^{22}}{-27x^3} = \boxed{\dfrac{-2y^{22}}{x^3}}$

14. $\dfrac{15x^4y^2}{-45x^7y^{-3}} = -\dfrac{1}{3}x^{4-7}y^{2-(-3)} = -\dfrac{1}{3}x^{-3}y^5 = \boxed{\dfrac{-y^5}{3x^3}}$

15. $\left(\dfrac{2x^{-2}}{3y^3}\right)^{-4} = \dfrac{(2x^{-2})^{-4}}{(3y^3)^{-4}} = \dfrac{2^{-4}x^8}{3^{-4}y^{-12}} = \dfrac{3^4x^8y^{12}}{2^4} = \boxed{\dfrac{81x^8y^{12}}{16}}$

16. $\left(\dfrac{-1a^2bc^2}{-30ab^{-2}c^6}\right)^{-2} = \left(\dfrac{1}{3}ab^3c^{-4}\right)^{-2} = \dfrac{a^{-2}b^{-6}c^8}{3^{-2}} = \dfrac{3^2c^8}{a^2b^6} = \boxed{\dfrac{9c^8}{a^2b^6}}$

17. $(7x^{1/3})(4x^{1/4}) = 28x^{1/3+1/4} = 28x^{4/12+3/12} = \boxed{28x^{7/12}}$

18. $\dfrac{80y^{3/4}}{-20y^{1/5}} = -4y^{3/4-1/5} = -4y^{15/20-4/20} = \boxed{-4y^{11/20}}$

19. $\left(\dfrac{32x^5}{y^{-4}}\right)^{-1/5} = \dfrac{(32)^{-1/5}(x^5)^{-1/5}}{(y^{-4})^{-1/5}} = \dfrac{(32)^{-1/5}x^{-1}}{y^{4/5}} = \dfrac{1}{(32)^{1/5}xy^{4/5}} = \dfrac{1}{\sqrt[4]{32}xy^{4/5}} = \boxed{\dfrac{1}{2xy^{4/5}}}$

20. $(9x^{-2}y^{1/2})^{-3/2} = (9)^{-3/2}(x^{-2})^{-3/2}(y^{1/2})^{-3/2} = 9^{-3/2}x^3y^{-3/4} = \dfrac{x^3}{9^{3/2}y^{3/4}} = \dfrac{x^3}{(\sqrt{9})^3y^{3/4}} = \boxed{\dfrac{x^3}{27y^{3/4}}}$

21. $(x^{2n+5} \cdot x^{3n-2})^{4n} = (x^{2n+5+3n-2})^{4n} = (x^{5n+3})^{4n} = x^{4n(5n+3)} = \boxed{x^{20n^2+12n}}$

22. $\left(\dfrac{x^{1-n}}{x^{7-n}}\right)^3 = (x^{1-n-(7-n)})^3 = (x^{-6})^3 = x^{-18} = \boxed{\dfrac{1}{x^{18}}}$

23. $\left(\dfrac{x^{3n+2}y^{2n}}{x^{3n}}\right)^4 = (x^{3n+2-3n}y^{2n})^4 = (x^2y^{2n})^4 = (x^2)^4(y^{2n})^4 = \boxed{x^8y^{8n}}$

24. $(x^{b^2}y^b)^{1/b} = (x^{b^2})^{1/b}(y^b)^{1/b} = \boxed{x^by}$

25. $\dfrac{(x^ay^{1/a})^a}{x^{a^2}y^a} = \dfrac{(x^a)^a(y^{1/a})^a}{x^{a^2}y^a} = \dfrac{x^{a^2}y}{x^{a^2}y^a} = x^{a^2-a^2}y^{1-a} = \boxed{y^{1-a}\ \left(\text{or}\ \dfrac{1}{y^{a-1}}\right)}$

26. $x^{3/4}(x^{1/4}-x^{1/2}) = x^{3/4+1/4} - x^{3/4+1/2} = \boxed{x - x^{5/4}}$

27. $(3x^{2/3}-2y^{1/2})(5x^{2/3}+3y^{1/2}) = 15x^{4/3} + 9x^{2/3}y^{1/2} - 10x^{2/3}y^{1/2} - 6y = \boxed{15x^{4/3} - x^{2/3}y^{1/2} - 6y}$

28. $(x^{2/3}-5)^2 = (x^{2/3}-5)(x^{2/3}-5) = x^{4/3} - 5x^{2/3} - 5x^{2/3} + 25 = \boxed{x^{4/3} - 10x^{2/3} + 25}$

29. $\dfrac{30x^{4/3}y^{1/3} - 20x^{1/3}y^{4/3}}{10x^{1/3}y^{1/3}} = \dfrac{30x^{4/3}y^{1/3}}{10x^{1/3}y^{1/3}} - \dfrac{20x^{1/3}y^{4/3}}{10x^{1/3}y^{1/3}} = 3x^{4/3-1/3}y^{1/3-1/3} - 2x^{1/3-1/3}y^{4/3-1/3} = \boxed{3x - 2y}$

30. $3x^{2/3} - 5x^{1/3} + 2 = \boxed{(3x^{1/3}-2)(x^{1/3}-1)}$ **31.** $4x^{2/3} - 25 = (2x^{1/3})^2 - 5^2 = \boxed{(2x^{1/3}+5)(2x^{1/3}-5)}$

32. $30(x+4)^{4/3} - 20(x+4)^{1/3} = 10(x+4)^{1/3}[3(x+4)-2] = \boxed{10(x+4)^{1/3}(3x+10)}$

33. $\dfrac{5}{y^{1/2}} + y^{1/2} = \dfrac{5}{y^{1/2}} + y^{1/2} \cdot \dfrac{y^{1/2}}{y^{1/2}} = \boxed{\dfrac{5+y}{y^{1/2}}}$

34. $\dfrac{y^2}{(y^2+4)^{1/2}} - (y^2+4)^{1/2} = \dfrac{y^2}{(y^2+4)^{1/2}} - (y^2+4)^{1/2} \cdot \dfrac{(y^2+4)^{1/2}}{(y^2+4)^{1/2}} = \dfrac{y^2 - (y^2+4)}{(y^2+4)^{1/2}} = \boxed{\dfrac{-4}{(y^2+4)^{1/2}}}$

35. $f(x) = \sqrt{x+2}$

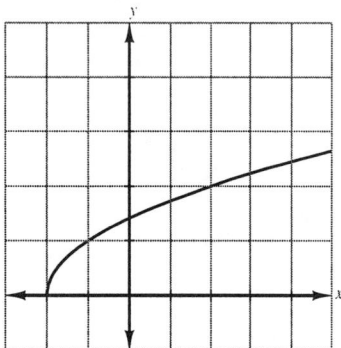

36. $f(x) = \begin{cases} \sqrt{x-4} & \text{if } x \ge 4 \\ 4-x & \text{if } x < 4 \end{cases}$

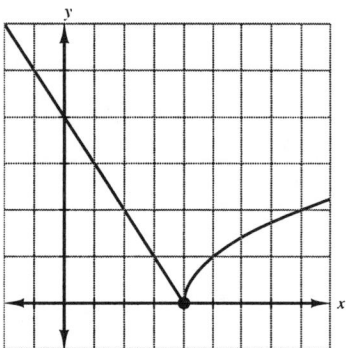

37. $f(x) = \sqrt{18 - 3x}$

$18 - 3x \ge 0$

$-3x \ge -18$

$x \le 6$

$\boxed{D_f = \text{Domain} = \{x \mid x \ge 6\} = (-\infty, 6]}$

38. $f(x) = \dfrac{1}{x-4}$ and $g(x) = \sqrt{x}$

$$\boxed{(f \circ g)(x) = \dfrac{1}{\sqrt{x}-4}}$$

$\quad \sqrt{x} \geq 0 \qquad\qquad \text{and} \qquad\qquad \sqrt{x}-4 \neq 0$

$\quad\quad x \geq 0 \qquad\qquad\qquad\qquad\qquad \sqrt{x} \neq 4$

$\qquad\qquad\qquad\qquad\qquad\qquad\qquad\qquad x \neq 16$

$$\boxed{D_{f \circ g} = \{x \mid x \geq 0 \text{ and } x \neq 16\} = [0, 16) \cup (16, \infty)}$$

$$\boxed{(g \circ f)(x)} = \sqrt{\dfrac{1}{x-4}} = \boxed{\dfrac{1}{\sqrt{x-4}}}$$

$\quad x-4 \geq 0 \qquad\qquad \text{and} \qquad\qquad \sqrt{x-4} \neq 0$

$\quad\quad x \geq 4 \qquad\qquad\qquad\qquad\qquad\quad x \neq 4$

Thus $x > 4$

$$\boxed{D_{g \circ f} = \{x \mid x > 4\} = (4, \infty)}$$

39. $f(x) = \sqrt[3]{x} - 1$

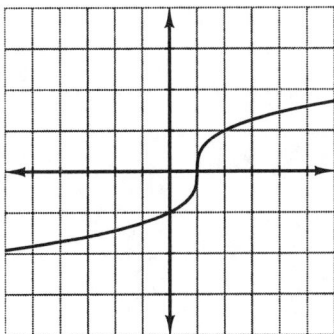

40. $f(x) = \begin{cases} \sqrt[3]{x} - 1 & \text{if } x \geq 2 \\ 1 & \text{if } x < 2 \end{cases}$

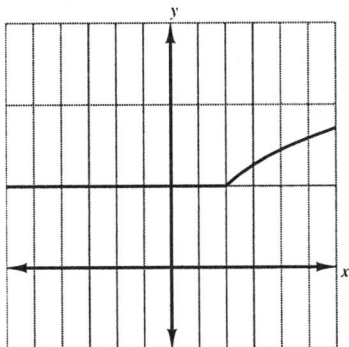

41. $f(A) = 28.6\sqrt[3]{A}$ (A, in square miles)

$\quad f(16) = 28.6\sqrt[3]{16} \approx \boxed{72.07}$

$\boxed{\text{The number of plant species on a 16 mi}^2 \text{ island in the Galápagos chain is about 72.}}$

42. $f(x) = 500(2 + 0.4x^{2/3} + 0.2x)$

$f(8) = 500[2 + 0.4(8^{2/3}) + 0.2(8)] = \boxed{2600}$

$\boxed{\text{After 8 years, a bacteria colony that started out with 1000 bacteria has a population of 2600 bacteria.}}$

43. $f(x) = \sqrt{x^2}$, $g(x) = |x|$ and $h(x) = x$

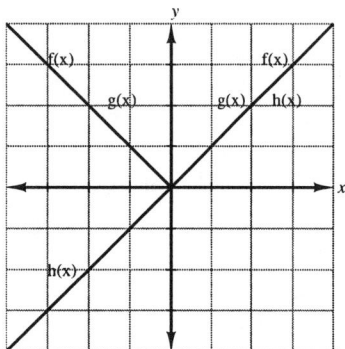

$\boxed{f(x) = g(x)}$

$\boxed{h(x) = f(x) = g(x) \text{ for } x \geq 0}$

44. $\sqrt{(-5)^2} - \sqrt[3]{(-5)^3} = |-5| - (-5) = 5 + 5 = \boxed{10}$ **45.** $\sqrt{(-2)^4} - \sqrt[5]{(-1)^5} = |-2| - (-1) = 2 + 1 = \boxed{3}$

46. $\left(\dfrac{-8}{27}\right)^{-2/3} - 32^{1/5} = \left[\left(-\dfrac{2}{3}\right)^3\right]^{-2/3} - (2^5)^{1/5} = \left(-\dfrac{2}{3}\right)^{-2} - 2^1 = \left(-\dfrac{2}{3}\right)^2 - 2 = \dfrac{9}{4} - 2 = \boxed{\dfrac{1}{4}}$

47. $\sqrt{20xy^3} = \sqrt{4 \cdot 5xy^2y} = \boxed{2y\sqrt{5xy}}$ **48.** $\sqrt{\dfrac{3}{7}} = \dfrac{\sqrt{3}}{\sqrt{7}} \cdot \dfrac{\sqrt{7}}{\sqrt{7}} = \dfrac{\sqrt{21}}{\sqrt{49}} = \boxed{\dfrac{\sqrt{21}}{7}}$

49. $\sqrt{\dfrac{3x}{7y}} = \dfrac{\sqrt{3x}}{\sqrt{7y}} \cdot \dfrac{\sqrt{7y}}{\sqrt{7y}} = \dfrac{\sqrt{21xy}}{\sqrt{49y^2}} = \boxed{\dfrac{\sqrt{21xy}}{7y}}$

50. $\sqrt{\dfrac{5}{8x^3}} = \dfrac{\sqrt{5}}{\sqrt{8x^3}} = \dfrac{\sqrt{5}}{\sqrt{4 \cdot 2x^2x}} = \dfrac{\sqrt{5}}{2x\sqrt{2x}} = \dfrac{\sqrt{5}}{2x\sqrt{2x}} \cdot \dfrac{\sqrt{2x}}{\sqrt{2x}} = \dfrac{\sqrt{10x}}{2x\sqrt{4x^2}} = \dfrac{\sqrt{10x}}{2x(2x)} = \boxed{\dfrac{\sqrt{10x}}{4x^2}}$

51. $\sqrt[5]{64x^3y^{12}z^6} = \sqrt[5]{32 \cdot 2x^3(y^2)^5y^2z^5z} = \boxed{2y^2z\sqrt[5]{2x^3y^2z}}$

52. $\dfrac{3y}{\sqrt{8y^3}} = \dfrac{3y}{\sqrt{4 \cdot 2y^2y}} = \dfrac{3y}{2y\sqrt{2y}} = \dfrac{3}{2\sqrt{2y}} \cdot \dfrac{\sqrt{2y}}{\sqrt{2y}} = \dfrac{3\sqrt{2y}}{2(2y)} = \boxed{\dfrac{3\sqrt{2y}}{4y}}$

53. $\dfrac{5}{\sqrt[5]{32x^4y}} = \dfrac{5}{2\sqrt[5]{x^4y}} \cdot \dfrac{\sqrt[5]{xy^4}}{\sqrt[5]{xy^4}} = \dfrac{5\sqrt[5]{xy^4}}{2\sqrt[5]{x^5y^5}} = \boxed{\dfrac{5\sqrt[5]{xy^4}}{2xy}}$

54. $\sqrt[3]{\dfrac{4}{9}} = \dfrac{\sqrt[3]{2^2}}{\sqrt[3]{3^2}} \cdot \dfrac{\sqrt[3]{3}}{\sqrt[3]{3}} = \dfrac{\sqrt[3]{12}}{\sqrt[3]{3^3}} = \boxed{\dfrac{\sqrt[3]{12}}{3}}$ **55.** $\sqrt{\dfrac{6x^5}{y^3}} = \dfrac{\sqrt{x^4(6x)}}{\sqrt{y^2y}} = \dfrac{x^2\sqrt{6x}}{y\sqrt{y}} \cdot \dfrac{\sqrt{y}}{\sqrt{y}} = \boxed{\dfrac{x^2\sqrt{6xy}}{y^2}}$

56. $\sqrt[3]{24x^4y^{12}} = \sqrt[3]{8x^3y^{12}(3x)} = \boxed{2xy^4\sqrt[3]{3x}}$

57. $\dfrac{3x}{\sqrt[3]{y^2}} = \dfrac{3x}{\sqrt[3]{y^2}} \cdot \dfrac{\sqrt[3]{y}}{\sqrt[3]{y}} = \boxed{\dfrac{3x\sqrt[3]{y}}{y}}$

58. $\dfrac{3xy}{\sqrt[4]{y}} = \dfrac{3xy}{\sqrt[4]{y}} \cdot \dfrac{\sqrt[4]{y^3}}{\sqrt[4]{y^3}} = \dfrac{3xy}{y}\sqrt[4]{y^3} = \boxed{3x\sqrt[4]{y^3}}$

59. $\sqrt[3]{\dfrac{2x^8}{9y}} = \sqrt[3]{\dfrac{x^6 - 2x^2}{3^2y}} = x^2\dfrac{\sqrt[3]{2x^2}}{\sqrt[3]{3^2y}} \cdot \dfrac{\sqrt[3]{3y^2}}{\sqrt[3]{3y^2}} = \boxed{\dfrac{x^2\sqrt[3]{6x^2y^2}}{3y}}$

60. $5\sqrt{18} + 3\sqrt{8} - \sqrt{2}$

$= 5\sqrt{9 \cdot 2} + 3\sqrt{4 \cdot 2} - \sqrt{2}$

$= 5(3)\sqrt{2} + 3(2)\sqrt{2} - \sqrt{2}$

$= 15\sqrt{2} + 6\sqrt{2} - \sqrt{2}$

$= \boxed{20\sqrt{2}}$

61. $12\sqrt{\dfrac{1}{3}} + 4\sqrt{\dfrac{1}{12}}$

$= \dfrac{12\sqrt{1}}{\sqrt{3}} + \dfrac{4\sqrt{1}}{\sqrt{12}}$

$= \dfrac{12}{\sqrt{3}} + \dfrac{4}{\sqrt{12}}$

$= \dfrac{12}{\sqrt{3}} + \dfrac{4}{\sqrt{4 \cdot 3}}$

$= \dfrac{12}{\sqrt{3}} + \dfrac{4}{2\sqrt{3}}$

$= \dfrac{12}{\sqrt{3}} + \dfrac{2}{\sqrt{3}}$

$= \dfrac{12}{\sqrt{3}} \cdot \dfrac{\sqrt{3}}{\sqrt{3}} + \dfrac{2}{\sqrt{3}} \cdot \dfrac{\sqrt{3}}{\sqrt{3}}$

$= \dfrac{12\sqrt{3}}{3} + \dfrac{2\sqrt{3}}{3}$

$= \boxed{\dfrac{14\sqrt{3}}{3}}$

62. $3y^2\sqrt{\dfrac{x}{y}} + \dfrac{5y}{x}\sqrt{x^3y}$

$= 3y^2\dfrac{\sqrt{x}}{\sqrt{y}} + \dfrac{5y}{x}\sqrt{x^2xy}$

$= 3y^2\dfrac{\sqrt{x}}{\sqrt{y}} \cdot \dfrac{\sqrt{y}}{\sqrt{y}} + \dfrac{5}{x}x\sqrt{xy}$

$= \dfrac{3y^2\sqrt{xy}}{y} + 5y\sqrt{xy}$

$= 3y\sqrt{xy} + 5\sqrt{xy} = \boxed{8y\sqrt{xy}}$

63. $(7\sqrt{10})(-3\sqrt{5}) = -21\sqrt{50} = -21\sqrt{25 \cdot 2} = -21(5)\sqrt{2} = \boxed{-105\sqrt{2}}$

64. $(7\sqrt[3]{2})\left(-\dfrac{2}{7}\sqrt[3]{4}\right) = (7)\left(-\dfrac{2}{7}\right)\sqrt[3]{4 \cdot 2} = -2\sqrt[3]{8} = -2(2) = \boxed{-4}$

65. $\sqrt[3]{4xy^2}\sqrt[3]{8xy^5} = \sqrt[3]{32x^2y^7} = \sqrt[3]{8\cdot 4x^2(y^2)^3y} = \boxed{2y^2\sqrt[3]{4x^2y}}$

66. $\dfrac{\sqrt{98x^3}}{\sqrt{2x^{-2}}} = \sqrt{\dfrac{98}{2}\,x^{3-(-2)}} = \sqrt{49x^5} = \sqrt{49(x^2)^2x} = \boxed{7x^2\sqrt{x}}$

67. $5\sqrt{3}(2\sqrt{6}+4\sqrt{15}) = 10\sqrt{18}+20\sqrt{45} = 10\sqrt{9\cdot 2}+20\sqrt{9\cdot 5} = 10(3)\sqrt{2}+20(3)\sqrt{5} = \boxed{30\sqrt{2}+60\sqrt{5}}$

68. $(5\sqrt{2}-4\sqrt{3}(7\sqrt{2}+3\sqrt{3}) = 35(2)+15\sqrt{6}-28\sqrt{6}-12(3) = 70+15\sqrt{6}-28\sqrt{6}-36 = \boxed{34-13\sqrt{6}}$

69. $\sqrt{2x}(\sqrt{6x}+3\sqrt{3x}) = \sqrt{12x^2}+3\sqrt{2x^2} = \sqrt{4\cdot 3x^2}+3\sqrt{2x^2} = \boxed{2x\sqrt{3}+3x\sqrt{2}}$

70. $\dfrac{\sqrt[4]{64x^7}}{\sqrt[4]{2x^2}} = \sqrt[4]{\dfrac{64x^7}{2x^2}} = \sqrt[4]{32x^5} = \sqrt[4]{16\cdot 2x^4x} = \boxed{2x\sqrt[4]{2x}}$

71. $(\sqrt{7x}+\sqrt{3y})(\sqrt{7x}-\sqrt{3y}) = \boxed{7x-3y}$

72. $11\sqrt{3}-7\sqrt[6]{3^3} = 11\sqrt{3}-7(3^3)^{1/6} = 11\sqrt{3}-7\cdot 3^{1/2} = 11\sqrt{3}-7\sqrt{3} = \boxed{4\sqrt{3}}$

73. $2x\sqrt[4]{32y^5}-3y\sqrt[4]{162x^4y}+\sqrt[4]{2x^4y^4}$
$= 2x\sqrt[4]{2^4y^4(2y)}-3\sqrt[4]{3^4x^4(2y)}+\sqrt[4]{x^4y^4(2)}$
$= 2x(2y\sqrt[4]{2y})-3y(3x\sqrt[4]{2y})+xy\sqrt[4]{2}$
$= (4xy-9xy)\sqrt[4]{2y}+xy\sqrt[4]{2}$
$= \boxed{-5xy\sqrt[4]{2y}+xy\sqrt[4]{2}}$
$\boxed{\text{or } xy\sqrt[4]{2}\,(1-5\sqrt[4]{y})}$

74. $3x\sqrt[3]{24x^2y^{12}}+4y\sqrt[3]{81x^5y^9}$
$= 3x\sqrt[3]{8y^{12}(3x^2)}+4y\sqrt[3]{27x^3y^9(3x^2)}$
$= 3x(2y^4)\sqrt[3]{3x^2}+4y(3xy^3)\sqrt[3]{3x^2}$
$= (6xy^4+12xy^4)\sqrt[3]{3x^2}$
$= \boxed{18xy^4\sqrt[3]{3x^2}}$

75. $\sqrt[3]{\dfrac{y}{2}}-\sqrt[3]{108y} = \dfrac{\sqrt[3]{y}}{\sqrt[3]{2}}\cdot\dfrac{\sqrt[3]{4}}{\sqrt[3]{4}}-\sqrt[3]{27(4y)} = \dfrac{\sqrt[3]{4y}}{2}-3\sqrt[3]{4y} = \boxed{-\dfrac{5}{2}\sqrt[3]{4y}}$

76. $5\sqrt[3]{40}+3\sqrt[3]{72} = 5\sqrt[3]{8\cdot 5}+3\sqrt[3]{8\cdot 9} = 5(2\sqrt[3]{5})+3(2\sqrt[3]{9}) = \boxed{10\sqrt[3]{5}+6\sqrt[3]{9}}$

77. $8x\sqrt[4]{x} - 2\sqrt[8]{x^{10}}$

$= 8x\sqrt[4]{x} - 2\sqrt[8]{x^8(x^2)}$

$= 8x\sqrt[4]{x} - 2x\sqrt[4]{x}$ $(\sqrt[8]{x^2} = \sqrt[4]{x})$

$= \boxed{6x\sqrt[4]{x}}$

78. $\dfrac{20\sqrt{15} - 8\sqrt{27}}{4\sqrt{3}}$

$= \dfrac{20\sqrt{15}}{4\sqrt{3}} - \dfrac{8\sqrt{27}}{4\sqrt{3}}$

$= 5\sqrt{5} - 2\sqrt{9}$

$= 5\sqrt{5} - 2(3)$

$= \boxed{5\sqrt{5} - 6}$

79. $(\sqrt{7x-1} - \sqrt{3x})^2$

$= (\sqrt{7x-1})^2 - 2(\sqrt{7x-1})(\sqrt{3x}) + (\sqrt{3x})^2$

$= 7x - 1 - 2\sqrt{21x^2 - 3x} + 3x$

$= \boxed{10x - 1 - 2\sqrt{21x^2 - 3x}}$ or $10x - 1 - 2\sqrt{3x}\sqrt{7x-1}$

80. $(\sqrt{3x} - 4\sqrt{y})(5\sqrt{3x} + \sqrt{y})$

$= (\sqrt{3x})^2 + \sqrt{3x}\sqrt{y} - 20\sqrt{3x}\sqrt{y} - 4(\sqrt{y})^2$

$= \boxed{3x - 4y - 19\sqrt{3xy}}$

81. $(\sqrt[3]{5} + 1)(\sqrt[3]{25} - \sqrt[3]{5} + 1)$

$= \sqrt[3]{125} - \sqrt[3]{25} + \sqrt[3]{5} + \sqrt[3]{25} - \sqrt[3]{5} + 1$

$= \sqrt[3]{5^3} + 1$

$= 5 + 1$

$= \boxed{6}$

82. $\dfrac{6}{\sqrt{3}-1} = \dfrac{6}{\sqrt{3}-1} \cdot \dfrac{\sqrt{3}+1}{\sqrt{3}+1} = \dfrac{6(\sqrt{3}+1)}{3-1}) = \dfrac{6(\sqrt{3}+1)}{2} = 3(\sqrt{3}+1) = \boxed{3\sqrt{3}+3}$

83. $\dfrac{\sqrt{7}}{\sqrt{5}+\sqrt{3}} = \dfrac{\sqrt{7}}{\sqrt{5}+\sqrt{3}} \cdot \dfrac{\sqrt{5}-\sqrt{3}}{\sqrt{5}-\sqrt{3}} = \dfrac{\sqrt{35}-\sqrt{21}}{5-3} = \boxed{\dfrac{\sqrt{35}-\sqrt{21}}{2}}$

84. $\dfrac{7}{2\sqrt{5}-3\sqrt{7}}$

$= \dfrac{7}{2\sqrt{5}-3\sqrt{7}} \cdot \dfrac{2\sqrt{5}+3\sqrt{7}}{2\sqrt{5}+3\sqrt{7}}$

$= \dfrac{7(2\sqrt{5}+3\sqrt{7})}{4(5)-9(7)}$

$= \dfrac{7(2\sqrt{5}+3\sqrt{7})}{-43}$

$= \dfrac{14\sqrt{5}+21\sqrt{7}}{-43} - \dfrac{14\sqrt{5}+21\sqrt{7}}{43}$

$= \boxed{\dfrac{-14\sqrt{5}-21\sqrt{7}}{43}}$

85. $\dfrac{\sqrt{y}+5}{\sqrt{y}-3}$

$= \dfrac{\sqrt{y}+5}{\sqrt{y}-3} \cdot \dfrac{\sqrt{y}+3}{\sqrt{y}+3}$

$= \dfrac{y+3\sqrt{y}+5\sqrt{y}+15}{y-9}$

$= \boxed{\dfrac{y+8\sqrt{y}+15}{y-9}}$

86. $\dfrac{\sqrt{7}+\sqrt{3}}{\sqrt{7}-\sqrt{3}} = \dfrac{\sqrt{7}+\sqrt{3}}{\sqrt{7}-\sqrt{3}} \cdot \dfrac{\sqrt{7}+\sqrt{3}}{\sqrt{7}+\sqrt{3}}$

$= \dfrac{7+\sqrt{21}+\sqrt{21}+3}{7-3}$

$= \dfrac{10+2\sqrt{21}}{4}$

$= \dfrac{2(5+\sqrt{21})}{4}$

$= \boxed{\dfrac{5+\sqrt{21}}{2}}$

87. $\dfrac{2\sqrt{x}}{\sqrt{x}+\sqrt{y}}$

$= \dfrac{2\sqrt{x}}{\sqrt{x}+\sqrt{y}} \cdot \dfrac{\sqrt{x}-\sqrt{y}}{\sqrt{x}-\sqrt{y}}$

$= \dfrac{2x-2\sqrt{xy}}{x-y}$

88. $\dfrac{\sqrt{3a}+\sqrt{b}}{\sqrt{5a}+\sqrt{2b}} = \dfrac{\sqrt{3a}+\sqrt{b}}{\sqrt{5a}+\sqrt{2b}} \cdot \dfrac{\sqrt{5a}-\sqrt{2b}}{\sqrt{5a}-\sqrt{2b}}$

$= \dfrac{\sqrt{15a^2}-\sqrt{6ab}+\sqrt{5ab}-\sqrt{2b^2}}{5a-2b}$

$= \boxed{\dfrac{a\sqrt{15}+\sqrt{5ab}-\sqrt{6ab}-b\sqrt{2}}{5a-2b}}$

$\boxed{\text{or } \dfrac{a\sqrt{15}-b\sqrt{2}+(\sqrt{5}-\sqrt{6})\sqrt{ab}}{5a-2b}}$

89. $\dfrac{3\sqrt{7}+1}{\sqrt{3}-4\sqrt{5}} = \dfrac{3\sqrt{7}+1}{\sqrt{3}-4\sqrt{5}} \cdot \dfrac{\sqrt{3}+4\sqrt{5}}{\sqrt{3}+4\sqrt{5}}$

$= \dfrac{3\sqrt{21}+12\sqrt{35}+\sqrt{3}+4\sqrt{5}}{3-16(5)}$

$= \dfrac{3\sqrt{21}+12\sqrt{35}+\sqrt{3}+4\sqrt{5}}{-77}$

$= \boxed{\dfrac{-3\sqrt{21}-12\sqrt{35}-\sqrt{3}-4\sqrt{5}}{77}}$

90. $\dfrac{4\sqrt{3a}-5\sqrt{2b}}{3\sqrt{6a}-6\sqrt{8b}}=\dfrac{4\sqrt{3a}-5\sqrt{2b}}{3\sqrt{6a}-12\sqrt{2b}}$

$=\dfrac{4\sqrt{3a}-5\sqrt{2b}}{3(\sqrt{6a}-4\sqrt{2b})}\cdot\dfrac{\sqrt{6a}+4\sqrt{2b}}{\sqrt{6a}+4\sqrt{2b}}$

$=\dfrac{4\sqrt{18a^2}+16\sqrt{6ab}-5\sqrt{12ab}-20(2b)}{3[6a-16(2b)]}$

$=\dfrac{4(3a\sqrt{2})+16\sqrt{6ab}-5(2\sqrt{3ab})-40b}{3(6a-32b)}$

$=\dfrac{12a\sqrt{2}+16\sqrt{6ab}-10\sqrt{3ab}-40b}{6(3a-16b)}$

$=\boxed{\dfrac{6a\sqrt{2}+8\sqrt{6ab}-5\sqrt{3ab}-20b}{3(3a-16b)}}$

91. $f(x)=\sqrt{2x}$

$\dfrac{f(a+h)-f(a)}{h}$

$=\dfrac{\sqrt{2(a+h)}-\sqrt{2a}}{h}$

$=\dfrac{\sqrt{2a+2h}-\sqrt{2a}}{h}\cdot\dfrac{\sqrt{2a+2h}+\sqrt{2a}}{\sqrt{2a+2h}+\sqrt{2a}}$

$=\dfrac{(2a+2h)-(2a)}{h(\sqrt{2a+2h}+\sqrt{2a})}$

$=\dfrac{2h}{h(\sqrt{2a+2b}+\sqrt{2a})}$

$=\boxed{\dfrac{2}{\sqrt{2a+2h}+\sqrt{2a}}}$

92. $\sqrt[3]{5\sqrt{5}}=5^{1/3}\cdot5^{1/2}=5^{1/3+1/2}=5^{2/6+3/6}=5^{5/6}=\sqrt[6]{5^5}=\boxed{\sqrt[6]{3125}}$

93. $\dfrac{\sqrt[3]{3}}{\sqrt[4]{3}}=\dfrac{3^{1/3}}{3^{1/4}}=3^{1/3-1/4}=3^{4/12-3/12}=3^{1/12}=\boxed{\sqrt[12]{3}}$

94. $\sqrt[12]{x^4}=(x^4)^{1/12}=x^{1/3}=\boxed{\sqrt[3]{x}}$

95.

$\sqrt{2x+4}=6$

$(\sqrt{2x+4})^2=6^2$

$2x+4=36$

$2x=32$

$x=16$

check: $\sqrt{2(16)+4}=6$

$\sqrt{36}=6$

$6=6$

The solution checks.

$\boxed{\{16\}}$

96. $\sqrt[3]{2y-1} - 3 = 0$

$\sqrt[3]{2y-1} = 3$

$(\sqrt[3]{2y-1})^3 = 3^3$

$2y - 1 = 27$

$2y = 28$

$y = 14$

check: $(\sqrt[3]{2y-1}) - 3 = 0$

$\sqrt[3]{27} - 3 = 0$

$3 - 3 = 0$ checks

$\boxed{\{14\}}$

97. $\sqrt{5y-5} = \sqrt{4y-3}$

$(\sqrt{5-5})^2 = (\sqrt{4y-3})^2$

$5y - 5 = 4y - 3$

$y - 5 = -3$

$y = 2$

check: $\sqrt{5(2)-5} = \sqrt{4(2)-3}$

$\sqrt{5} = \sqrt{5}$ checks

$\boxed{\{2\}}$

98. $\sqrt{4x} + 4 = 0$

$\sqrt{4x} = -4$

$(\sqrt{4x})^2 = (-4)^2$

$4x = 16$

$x = 4$

check: $\sqrt{4(4)} + 4 = 0$

$\sqrt{16} + 4 = 0$

$4 + 4 = 0$

$8 \ne 0$

4 is an extraneous solution.

$\boxed{\varnothing}$

99.

$$\sqrt{y-3} = 1 - \sqrt{y+2}$$
$$(\sqrt{y-3})^2 = (1 - \sqrt{y+2})^2$$
$$y - 3 = (1 - \sqrt{y+2})(1 - \sqrt{y+2})$$
$$y - 3 = 1 - 2\sqrt{y+2} + y + 2$$
$$y - 3 = y + 3 - 2\sqrt{y+2}$$
$$-6 = -2\sqrt{y+2}$$
$$3 = \sqrt{y+2}$$
$$3^2 = (\sqrt{y+2})^2$$
$$9 = y + 2$$
$$7 = y$$

check: $\sqrt{7-3} = 1 - \sqrt{7+2}$
$$\sqrt{4} = 1 - \sqrt{9}$$
$$2 = 1 - 3$$
$$2 \neq -2$$

7 is an extraneous solution.

$$\boxed{\varnothing}$$

100.

$$\sqrt{2y+1} - y + 1 = 0$$
$$\sqrt{2y+1} = y - 1$$
$$(\sqrt{2y+1})^2 = (y-1)^2$$
$$2y + 1 = y^2 - 2y + 1$$
$$0 = y^2 - 4y$$
$$0 = y(y-4)$$
$$y = 0 \qquad \text{or} \qquad y - 4 = 0$$
$$y = 4$$

check 0: $\sqrt{2(0)+1} - 0 + 1 = 0$
$$1 - 0 + 1 = 0$$
$$2 \neq 0$$

0 is extraneous

check 4: $\sqrt{2(4)+1} - 4 + 1 = 0$
$$3 - 4 + 1 = 0$$
$$0 = 0 \quad \text{checks}$$

$$\boxed{\{4\}}$$

101.

$$2 - \sqrt{3y+1} + \sqrt{y-1} = 0$$
$$\sqrt{y-1} = \sqrt{3y+1} - 2$$
$$(\sqrt{y-1})^2 = (\sqrt{3y+1} - 2)^2$$
$$y-1 = 3y + 1 - 4\sqrt{3y+1} + 4$$
$$y-1 = 3y + 5 - 4\sqrt{3y+1}$$
$$-2y - 6 = -4\sqrt{3y+1}$$
$$y + 3 = 2\sqrt{3y+1}$$
$$(y+3)^2 = (2\sqrt{3y+1})^2$$
$$y^2 + 6y + 9 = 4(3y+1)$$
$$y^2 + 6y + 9 = 12y + 4$$
$$y^2 - 6y + 5 = 0$$
$$(y-1)(y-5) = 0$$

$$y - 1 = 0 \qquad \text{or} \qquad y - 5 = 0$$
$$y = 1 \qquad\qquad\qquad y = 5$$

check 1: $2 - \sqrt{3(1)+1} + \sqrt{1-1} = 0$
$$2 - 2 + 0 = 0$$
$$0 = 0 \quad \text{checks}$$

check 5: $2 - \sqrt{3(5)+1} + \sqrt{5-1} = 0$
$$2 - 4 + 2 = 0$$
$$0 = 0 \quad \text{checks}$$

$\boxed{\{1, 5\}}$

102.

$$\sqrt[3]{3(x+3)^2} - 3 = 0$$
$$\sqrt[3]{3(x+3)^2} = 3$$
$$\left[\sqrt[3]{3(x+3)^2}\right]^3 = 3^3$$
$$3(x+3)^2 = 3^3$$
$$x^2 + 6x + 9 = 9$$
$$x^2 + 6x = 0$$
$$x(x+6) = 0$$
$$x = 0 \qquad \text{or} \qquad x + 6 = 0$$
$$x = -6$$

The solutions check.

$\boxed{\{-6, 0\}}$

103.

$$C = 200\sqrt{x} + 10$$
$(C = 1010):$
$$1010 = 200\sqrt{x} + 10$$
$$1000 = 200\sqrt{x}$$
$$5 = \sqrt{x}$$
$$5^2 = (\sqrt{x})^2$$
$$25 = x$$

$\boxed{\text{25 tons}}$

104. $L = L_0 \sqrt{1 - \dfrac{v^2}{c^2}}$

$L_0 = 600$ meters

$\dfrac{v}{c} = 90\% = 0.9$

Find L.

$L = 600 \sqrt{1 - (0.9)^2}$

$ = 600\sqrt{1 - 0.81}$

$ = 600\sqrt{0.19}$

$ \approx 261.5$

The length of the ship is $\boxed{\text{approximately 261.5 meters}}$.

105. $M = \dfrac{M_0}{\sqrt{1 - \dfrac{v^2}{c^2}}}$

$M_0 = 200$ pounds, $M = 1.5M_0$ since M_0 increases by 50%

$1.5M_0 = \dfrac{M_0}{\sqrt{1 - \dfrac{v^2}{c^2}}}$

$1.5 = \dfrac{1}{\sqrt{1 - \dfrac{v^2}{c^2}}}$

$1.5^2 = \dfrac{1}{1 - \dfrac{v^2}{c^2}}$

$2.25 = \dfrac{1}{1 - \dfrac{v^2}{c^2}} \cdot \dfrac{c^2}{c^2}$

$2.25 = \dfrac{c^2}{c^2 - v^2}$

$2.25 c^2 - 2.25 v^2 = c^2$

$-2.25 v^2 = -1.25 c^2$

$v^2 = \dfrac{5}{9} c^2$

$\boxed{v = \dfrac{\sqrt{5}}{3} c \approx 0.7454 c \approx 138{,}636 \text{ mi/sec}}$

106. $T = \dfrac{T_0}{\sqrt{1 - \dfrac{v^2}{c^2}}}$

As $v \approx c$ the denominator approaches $1 - \dfrac{c^2}{c^2} = 1 - 1 = 0$

$\boxed{\text{Time slows down}}$ or Each instant of time seems to take literally forever.

107.
$$t = \sqrt{\frac{d}{16}}$$
($t = 4$ seconds):
$$4 = \sqrt{\frac{d}{16}}$$
$$16 = \frac{d}{16}$$
$$256 = d$$
The length of the bridge is $\boxed{\text{approximately 256 feet}}$.

108. $\sqrt{-20} = \sqrt{20}\, i = \sqrt{4 \cdot 5}\, i = \boxed{2\sqrt{5}\, i}$

109. $2\sqrt{-100} + 3\sqrt{-36} = 2\sqrt{100}\, i + 3\sqrt{36}\, i = 2(10)i + 3(6)i = 20i + 18i = \boxed{38i}$

110. $\sqrt{-5}\sqrt{-9} - \sqrt{5}\, i\sqrt{9}\, i = \sqrt{45}\, i^2 = \sqrt{9 \cdot 5}(-1) = \boxed{-3\sqrt{5}}$

111. $\dfrac{10\sqrt{-24}}{-2\sqrt{6}} = \dfrac{10\sqrt{6 \cdot 4}\, i}{-2\sqrt{6}} = \dfrac{20\sqrt{6}\, i}{-2\sqrt{6}} = \boxed{-10i}$

112. $(7x - 5) - 3i = 9 + (3y - 2)i$

$\qquad 7x - 5 = 9 \qquad$ and $\qquad -3 = 3y - 2$

$\qquad\quad 7x = 14 \qquad\qquad\qquad\quad -1 = 3y$

$\qquad\quad \boxed{x = 2} \qquad$ and $\qquad \boxed{-\dfrac{1}{3} = y}$

113. $(7 + 12i) + (5 - 10i) = \boxed{12 + 2i}$

114. $(7 - 12i) - (-3 - 7i) = (7 - 12i) + (3 + 7i) = \boxed{10 - 5i}$

115. $(7 - 5i)(2 + 3i) = 14 + 21i - 10i - 15i^2 = 14 + 21i - 10i - 15(-1) = \boxed{29 + 11i}$

116. $(2 + 5i)^2 = (2 + 5i)(2 + 5i) = 4 + 10i + 10i + 25i^2 = 4 + 20i + 25(-1) = \boxed{-21 + 20i}$

117. $\dfrac{3i}{5 + i} = \dfrac{3i}{5 + i} \cdot \dfrac{5 - i}{5 - i} = \dfrac{15i - 3i^2}{25 - i^2} = \dfrac{15i - 3(-1)}{25 - (-1)} = \dfrac{3 + 15i}{26} = \boxed{\dfrac{3}{26} + \dfrac{15i}{26}}$

118. $\dfrac{3 - 4i}{4 + 2i} = \dfrac{3 - 4i}{4 + 2i} \cdot \dfrac{4 - 2i}{4 - 2i} = \dfrac{12 - 6i - 16i + 8i^2}{16 - 4i^2} = \dfrac{12 - 22i + 8(-1)}{16 - 4(-)} = \dfrac{4 - 22i}{20} = \dfrac{4}{20} - \dfrac{22}{20}\, i = \boxed{\dfrac{1}{5} - \dfrac{11}{10}\, i}$

119. $\dfrac{5 + i}{3i} = \dfrac{5 + i}{3i} \cdot \dfrac{i}{i} = \dfrac{5i + i^2}{3i^2} = \dfrac{5i + (-1)}{3(-1)} = \dfrac{-1 + 5i}{-3} = \dfrac{-1}{-3} + \dfrac{5i}{-3} = \boxed{\dfrac{1}{3} - \dfrac{5i}{3}}$

120. $i^{23} = i^{22}i = (i^2)^{11}i = (-1)^{11}i = (-1)i = \boxed{-i}$

Cumulative Review Problems (Chapters 1-7)

Cumulative Review, pp. 574-575

1. $\dfrac{(250{,}000{,}000)(63{,}000)}{(0.0007)(0.09)(500)}$

$= \dfrac{(2.5 \times 10^8)(6.3 \times 10^4)}{(7 \times 10^{-4})(9 \times 10^{-2})(5 \times 10^2)}$

$= \dfrac{(2.5 \times 10^8)(6.3 \times 10^4)}{(5 \times 10^2)(6.3 \times 10^{-5})}$

$= (0.5 \times 10^6)(1 \times 10^9)$

$= \boxed{5.0 \times 10^{14}}$

2. $\dfrac{-60 \div (-2)(-3) + 5 - (-7) + 3}{81 \div (-9) + 3(-5) + 2 - (-2)}$

$= \dfrac{30(-3) + 5 + 7 + 3}{-9 - 15 + 2 + 2}$

$= \dfrac{-90 + 15}{-24 + 4}$

$= \dfrac{-75}{-20}$

$= \boxed{\dfrac{15}{4}}$

3. $\left| \dfrac{2}{5}(x-2) \right| > 4$

$\begin{array}{llll}
\dfrac{2}{5}(x-2) > 4 & \quad\text{or}\quad & \dfrac{2}{5}(x-2) < -4 \\[2mm]
x - 2 > 10 & & x - 2 < -10 \\[1mm]
x > 12 & & x < -8
\end{array}$

$\boxed{\{x \mid x < -8 \text{ or } x > 12\} = (-\infty, -8) \cup (12, \infty)}$

4. $\begin{aligned}
-7[3(2x-1) - 4(x+1)] &= 3(2-4x) - (4x-8) \\
-7(6x - 3 - 4x - 4) &= 6 - 12x - 4x + 8 \\
-7(2x - 7) &= -16x + 14 \\
-14x + 49 &= -16x + 14 \\
2x &= -35 \\
x &= -\dfrac{35}{2}
\end{aligned}$

$\boxed{\left\{ -\dfrac{35}{2} \right\}}$

5. Let x = total number of votes cast.
Candidate A: $0.40x + 40$
Candidate B: 0.50 (Candidate A) $- 40$
$= 0.50(0.40x + 40) - 40$
$= 0.20x + 20 - 40$
$= 0.20x - 20$
Candidate C: 0.50 (Candidate B) $+ 20$
$= 0.50(0.20x - 20) + 20$
$= 0.10x - 10 + 20$
$= 0.10x + 10$
Candidate D: 180

(Candidate A) + (Candidate B) + (Candidate C) + (Candidate D) = (Total votes)

$$(0.40x + 40) + (0.20x - 20) + (0.10x + 10) + 180 = x$$
$$0.70x + 210 = x$$
$$-0.30x = -210$$
$$x = 700$$

The total number of votes cast is $\boxed{700}$.
Candidate A: $0.40(700) + 40 = 320$
Candidate B: $0.20(700) - 20 = 120$
Candidate C: $0.10(700) + 10 = 80$
Candidate D: 180
$\boxed{\text{Candidate A won with 320 votes}}$

6. $\boxed{\text{Domain} = \{\, x \mid x \neq -2, 2\} = (-\infty, 2) \cup (-2, 2) \cup (2, \infty)}$
$\boxed{\text{Range} = \{\, y \mid y \leq -1 \text{ or } y > 2\} = (-\infty, -1] \cup (2, \infty)}$

7. line passing through points $(-4, -1)$ and $(3, 4)$
$$m = \frac{4 - (-1)}{3 - (-4)} = \frac{5}{7}$$
$$y + 1 = \frac{5}{7}(x + 4) \quad \text{or} \quad y - 4 = \frac{5}{7}(x - 3)$$
$$y + 1 = \frac{5}{7}x + \frac{20}{7}$$
$$y = \frac{5}{7}x + \frac{13}{7}$$
$$7y = 5x + 13$$
$$\boxed{5x - 7y = -13}$$

8.
$$2x - y + z = -5$$
$$x - 2y - 3z = 6$$
$$x + y - 2z = 1$$

(Equations 1 and 2):
$$(\times 3) \quad 6x - 3y + 3z = -15$$
$$\underline{x - 2y - 3z = 6}$$
$$7x - 5y = -9 \quad \text{(Equation 4)}$$

(Equations 1 and 3):
$$(\times 2) \quad 4x - 2y + 2z = -10$$
$$\underline{x + y - 2z = 1}$$
$$5x - y = -9 \quad \text{(Equation 5)}$$

(Equation 4 and 5):
$$7x - 5y = -9$$
$$(\times -5) \quad \underline{-25x + 5y = 45}$$
$$-18x = 36$$
$$x = -2$$

(Equation 5):
$$5x - y = -9$$
$$5(-2) - y = -9$$
$$-y = 1$$
$$y = -1$$

(Equation 1):
$$2x - y + z = -5$$
$$2(-2) + 1 + z = -5$$
$$z = -2$$

$$\boxed{\{(-2, -1, -2)\}}$$

9.
$$3x - 4y - 2z = 3$$
$$x + 2y + 2z = 0$$
$$2x + 2y + 5z = -2$$

$$\begin{bmatrix} 3 & -4 & -2 & | & 3 \\ 1 & 2 & 2 & | & 0 \\ 2 & 2 & 5 & | & -2 \end{bmatrix} \begin{matrix} \\ 3R_2 - R_1 \to \\ 3R_3 - 2R_1 \to \end{matrix} \begin{bmatrix} 3 & -4 & -2 & | & 3 \\ 0 & 10 & 8 & | & -3 \\ 0 & 14 & 19 & | & -12 \end{bmatrix}$$

$$\begin{matrix} \\ \\ 10R_3 - 14R_2 \to \end{matrix} \begin{bmatrix} 3 & -4 & -2 & | & 3 \\ 0 & 10 & 8 & | & -3 \\ 0 & 0 & 78 & | & -78 \end{bmatrix} \begin{matrix} \frac{1}{3}R_1 \to \\ \frac{1}{10}R_2 \to \\ \frac{1}{78}R_3 \to \end{matrix} \begin{bmatrix} 1 & -4/3 & -2/3 & | & 1 \\ 0 & 1 & 4/5 & | & -3/10 \\ 0 & 0 & 1 & | & -1 \end{bmatrix}$$

$$x - \frac{4}{3}y - \frac{2}{3}z = 1$$
$$y + \frac{4}{5}z = -\frac{3}{10}$$
$$z = -1$$

$$y + \frac{4}{5}(-1) = -\frac{3}{10}$$
$$y = -\frac{3}{10} + \frac{8}{10} = \frac{5}{10} = \frac{1}{2}$$

$$x - \frac{4}{3}\left(\frac{1}{2}\right) - \frac{2}{3}(-1) = 1$$
$$x - \frac{2}{3} + \frac{2}{3} = 1$$
$$x = 1$$

$$\boxed{\left\{ \left(1, \frac{1}{2}, -1\right) \right\}}$$

10. $x + 2y < 2$ or $2y - x \geq 4$

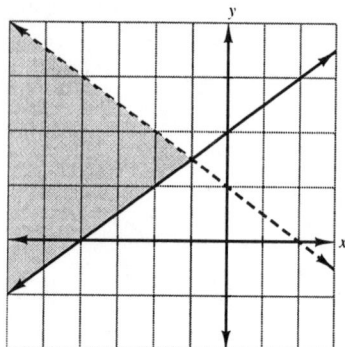

11. Let

$$
\begin{aligned}
u &= \text{the units' digit} \\
t &= \text{the tens' digit} \\
10t + u &= \text{original number} \\
10u + t &= \text{digits reversed number}
\end{aligned}
$$

$$
\begin{aligned}
u + t &= 11 \\
10u + t &= (10t + u) - 27 \quad \rightarrow \quad 9u - 9t = -27
\end{aligned}
$$

(Rearrange and simplify):

$$
\begin{aligned}
u + t &= 11 \\
\underline{u - t} &= \underline{-3} \\
2u &= 8 \\
u &= 4
\end{aligned}
$$

$$
\begin{aligned}
u + t &= 11 \\
4 + t &= 11 \\
t &= 7
\end{aligned}
$$

original number: $\boxed{74}$

12. $f(x) = x^3 - 1$ and $g(x) = x - 1$

$g[f(x)] - f[g(x)]$

$= [(x^2 - 1) - 1] - [(x - 1)^3 - 1]$

$= (x^3 - 2) - (x^3 - 3x^2 + 3x - 1 - 1)$

$= x^3 - 2 - x^3 + 3x^2 - 3x + 2$

$= \boxed{3x^2 - 3x}$

13. $2xy^3 - 14xy^2 + 24xy$

$= 2xy(y^2 - 7y + 12)$

$= \boxed{2xy(y - 4)(y - 3)}$

14. $x^4 - xy^3 + x^2 - xy$

$= x(x^3 - y^3) + x(x - y)$

$= x(x - y)(x^2 + xy + y^2) + x(x - y)$

$= \boxed{x(x - y)(x^2 + xy + y^2 + 1)}$

15.
$$\frac{1}{x} + \frac{1}{y} = \frac{1}{r}$$

$$\frac{1}{y} = \frac{1}{r} - \frac{1}{x}$$

$$\frac{1}{y} = \frac{1}{r} \cdot \frac{x}{x} - \frac{1}{x} \cdot \frac{r}{r}$$

$$\frac{1}{y} = \frac{x - r}{xr}$$

$$(xry) \cdot \frac{1}{y} = (xry) \cdot \frac{x - r}{xr}$$

$$xr = y(x - r)$$

$$\frac{xr}{x - r} = y$$

$$\boxed{y = \frac{xr}{x - r}}$$

16. Let
$$\begin{aligned}
x &= \text{width of rectangular pool} \\
2x &= \text{length of rectangular pool} \\
x + 5 + 5 &= x + 10 = \text{width of pool plus path} \\
2x + 5 + 5 &= 2x + 10 = \text{length of pool plus path} \\
\text{total area} &= 1000 \text{ square feet}
\end{aligned}$$

$$\begin{aligned}
(2x + 10)(x + 10) &= 1000 \\
2x^2 + 30x + 1000 &= 1000 \\
2x^2 + 30x - 900 &= 0 \\
2(x^2 + 15x - 450) &= 0 \\
2(x + 30)(x - 15) &= 0
\end{aligned}$$

$$\begin{array}{lll}
x + 30 = 0 & \text{or} & x - 15 = 0 \\
x = 30 \quad \text{(Reject)} & & x = 15 \\
& & 2x = 2(15) = 30
\end{array}$$

$$\boxed{\text{width: 15 feet; length: 30 feet}}$$

17. $\dfrac{x + 5}{x + 3} + \dfrac{6x}{x^3 - x^2 - 12x} \div \dfrac{3x}{3x - 12}$

$$= \frac{x + 5}{x + 3} + \frac{6x}{x(x^2 - x - 12)} \cdot \frac{3(x - 4)}{3x}$$

$$= \frac{x + 5}{x + 3} + \frac{6}{(x - 4)(x + 3)} \cdot \frac{(x - 4)}{x}$$

$$= \frac{x + 5}{x + 3} + \frac{6}{x(x + 3)}$$

$$= \frac{x(x + 5) + 6}{x(x + 3)}$$

$$= \frac{x^2 + 5x + 6}{x(x + 3)}$$

$$= \frac{(x + 3)(x + 2)}{x(x + 3)}$$

$$= \boxed{\frac{x + 2}{x}}$$

18. $\dfrac{x+5}{x^3-5x^2-x+5} - \dfrac{x+1}{x^3-x^2-25x+25}$

$= \dfrac{x+5}{x^2(x-5)-(x-5)} - \dfrac{x+1}{x^2(x-1)-25(x-1)}$

$= \dfrac{x+5}{(x-5)(x^2-1)} - \dfrac{x+1}{(x-1)(x^2-25)}$

$= \dfrac{x+5}{(x-5)(x+1)(x-1)} - \dfrac{x+1}{(x-1)(x+5)(x-5)}$

$= \dfrac{(x+5)}{(x-5)(x+1)(x-1)} \cdot \dfrac{x+5}{x+5} - \dfrac{x+1}{(x-1)(x+5)(x-5)} \cdot \dfrac{x+1}{x+1}$

$= \dfrac{(x^2+10x+25)-(x^2+2x+1)}{(x+5)(x-5)(x+1)(x-1)}$

$= \dfrac{8x+24}{(x+5)(x-5)(x+1)(x-1)}$

$= \boxed{\dfrac{8(x+3)}{(x+5)(x-5)(x+1)(x-1)} \quad \text{or} \quad \dfrac{8(x+3)}{(x^2-25)(x^2-1)}}$

19. $f(x) = \dfrac{3+\dfrac{1}{x}}{3-\dfrac{1}{x}}$ and $g(x) = x+5$

$(f \circ g)(x) = \dfrac{3+\dfrac{1}{x+5}}{3-\dfrac{1}{x+5}}$

$= \dfrac{3+\dfrac{1}{x+5}}{3-\dfrac{1}{x+5}} \cdot \dfrac{x+5}{x+5}$

$= \dfrac{3x+15+1}{3x+15-1}$

$= \boxed{\dfrac{3x+16}{3x+14}}$

$x+5 \ne 0$ and $3x+14 \ne 0$

$x \ne -5$ $x \ne -\dfrac{14}{3}$

$\boxed{D_{f \circ g} = \left\{ x \mid x \ne -5, -\dfrac{14}{3} \right\}}$

20.

$$\frac{x}{x-3} - \frac{3x}{x^2 - x - 6} = \frac{4x^2 - 4x - 18}{x^2 - x - 6}$$

$$\frac{x}{x-3} \cdot \frac{x+2}{x+2} - \frac{3x}{(x-3)(x+2)} = \frac{4x^2 - 4x - 18}{(x-3)(x+2)}$$

$$\frac{x^2 + 2x - 3x}{(x-3)(x+1)} = \frac{4x^2 - 4x - 18}{(x-3)(x+2)}$$

$$\frac{x^2 - x}{(x-3)(x+2)} - \frac{4x^2 - 4x - 18}{(x-3)(x+2)} = 0$$

$$\frac{-3x^2 + 3x + 18}{(x-3)(x+2)} = 0$$

$$\frac{-3(x^2 - x - 6)}{(x-3)(x+2)} = 0$$

$$\frac{-3(x-3)(x+2)}{(x-3)(x+2)} = 0$$

$$-3 = 0 \quad \text{False}$$

No solution

$$\boxed{\varnothing}$$

21. Let x = the number.

$$x + \frac{1}{x} = \frac{73}{24}$$

$$24x\left(x + \frac{1}{x}\right) = 24x\left(\frac{73}{24}\right)$$

$$24x^2 + 24 = 73x$$

$$24x^2 - 73x + 24 = 0$$

$$(8x - 3)(3x - 8) = 0$$

$$8x - 3 = 0 \qquad \text{or} \qquad 3x - 8 = 0$$

$$x = \frac{3}{8} \qquad\qquad\qquad x = \frac{8}{3}$$

The number is $\boxed{\dfrac{3}{8} \text{ or } \dfrac{8}{3}}$.

22. Let

$$x = \text{number of hours for Edgar to mow lawn alone}$$

$$x + 1\frac{2}{3} = \text{number of hours for Clive to mow lawn alone}$$

	Fractional part of job completed in 1 hour	Time spent working together	Fractional part of job completed in 2 hours
(takes x hours working alone)	$\dfrac{1}{x}$	2	$\dfrac{2}{x}$
(takes $x + 1\frac{2}{3}$ hours working alone)	$\dfrac{1}{x + 1\frac{2}{3}}$	2	$\dfrac{2}{x + 1\frac{2}{3}}$

$$(\text{Edgar}) + (\text{Clive}) = (\text{one complete job})$$

$$\frac{2}{x} + \frac{2}{x + \frac{5}{3}} = 1$$

$$\frac{2}{x} + \frac{2}{\frac{3x+5}{3}} = 1$$

$$x(3x+5)\left[\frac{2}{x} + \frac{6}{3x+5}\right] = x(3x+5)(1)$$

$$2(3x+5) + 6x = 3x^2 + 5x$$

$$-3x^2 + 7x + 10 = 0$$

$$3x^2 - 7x - 10 = 0$$

$$(x+1)(3x-10) = 0$$

$$(x+1)(3x-10) = 0$$

$$x + 1 = 0 \qquad \text{or} \qquad 3x - 10 = 0$$

$$x = -1 \quad (\text{Reject}) \qquad\qquad x = \frac{10}{3} = 3\frac{1}{3}$$

$$x + 1\frac{2}{3} = 3\frac{1}{3} + 1\frac{2}{3} = 5$$

It takes Edgar $\boxed{3\frac{1}{3} \text{ hours}}$ to mow the lawn working alone.

23. $\dfrac{(-8x^3 y^7)^{2/3}}{6x^4 y^{-1/3}}$

$= \dfrac{(-8)^{2/3}(x^3)^{2/3}(y^7)^{2/3}}{6x^4 y^{-1/3}}$

$= \dfrac{[(-2)^3]^{2/3} x^2 y^{14/3}}{6x^4 y^{-1/3}}$

$= \dfrac{4}{6} x^{2-4} y^{14/3 + 1/3}$

$= \dfrac{2}{3} x^{-2} y^{45/3}$

$= \boxed{\dfrac{2y^5}{3x^2}}$

24. $2(5x-1)^{1/2}(3x+5)^{-1/3} + 2(5x-1)^{-1/2}(3x+5)^{2/3}$

$= 2(5x-1)^{1-1/2}(3x+5)^{-1/3} + 2(5x-1)^{-1/2}(3x+5)^{1-1/3}$

$= 2(5x-1)(5x-1)^{-1/2}(3x+5)^{-1/3} + 2(5x-1)^{-1/2}(3x+5)(3x+5)^{-1/3}$

$= 2(5x-1)^{-1/2}(3x+5)^{-1/3}[(5x-1)+(3x+5)]$

$= 2(5x-1)^{-1/2}(3x+5)^{-1/3}(8x+4)$

$= \boxed{8(2x+1)(5x-1)^{-1/2}(3x+5)^{-1/3}}$

25. $f(x) = \begin{cases} \sqrt{x-2} & \text{if } x \geq 2 \\ 2x - x^2 & \text{if } x < 2 \end{cases}$

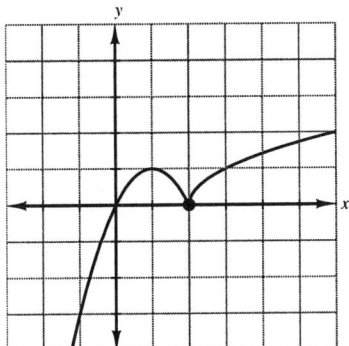

26. $\sqrt[4]{\dfrac{16a}{3b^3}} = \sqrt[4]{\dfrac{24a}{3b^3} \cdot \dfrac{3^3 b}{3^3 b}} = \sqrt[4]{\dfrac{2^4(27ab)}{3^4 b^4}} = \boxed{\dfrac{2^4}{3b}\sqrt[4]{27ab}}$

27. $3\sqrt{2y^3} - y\sqrt{200y} + \sqrt{32y^3}$

$= 3\sqrt{y^2(2y)} - y\sqrt{100(2y)} + \sqrt{16y^2(2y)}$

$= 3y\sqrt{2y} - 10y\sqrt{y}^2 + 4y\sqrt{2y}$

$= (3y - 10y + 4y)\sqrt{2y}$

$= \boxed{-3y\sqrt{2y}}$

28. $(\sqrt[3]{x} + \sqrt[3]{y})(\sqrt[3]{x^2} - \sqrt[3]{xy} + \sqrt[3]{y^2})$

$= \sqrt[3]{x^3} - \sqrt[3]{x^2 y} + \sqrt[3]{xy^2} + \sqrt[3]{x^2 y} - \sqrt[3]{xy^2} + \sqrt[3]{y^3}$

$= \sqrt[3]{x^3} + \sqrt[3]{y^3}$

$= \boxed{x + y}$

29.

$$\frac{2}{\sqrt{x+5}} = \sqrt{x+5} + 1$$

$$\sqrt{x+5} \cdot \frac{2}{\sqrt{x+5}} = \sqrt{x+5}\left(\sqrt{x+5} + 1\right)$$

$$2 = (x+5) + \sqrt{x+5}$$

$$-x - 3 = \sqrt{x+5}$$

$$[-(x+3)]^2 = (\sqrt{x+5})^2$$

$$x^2 + 6x + 9 = x + 5$$

$$x^2 + 5x + 4 = 0$$

$$(x+4)(x+1) = 0$$

$$x + 4 = 0 \qquad \text{or} \qquad x + 1 = 0$$

$$x = -4 \qquad\qquad\qquad x = -1$$

Check: $x = -4$: $\dfrac{2}{\sqrt{-1+5}} = \sqrt{-4+5} + 1$

$$2 = 2 \quad \text{checks}$$

−4 is a solution.

$x = -1$: $\dfrac{2}{\sqrt{-1+5}} = \sqrt{-1+5} + 1$

$$\frac{2}{2} = 2 + 1$$

$$1 = 3 \quad \text{False}$$

−1 is extraneous.

$$\boxed{\{-4\}}$$

30. $\dfrac{16 - 15i}{6 - i} = \dfrac{16 - 15i}{6 - i} \cdot \dfrac{6 + i}{6 + i} = \dfrac{96 + 16i - 90i - 15i^2}{36 - i^2}$

$$= \frac{96 - 74i + 15}{36 + 1} = \frac{111 - 74i}{37} = \boxed{3 - 2i}$$

Algebra for College Students

Chapter 8   Quadratic Equations and Functions

Section 8.1 Equations That Are Quadratic in Form

Problem Set 8.1, pp. 586-587

1. $(x^2 + 3x)^2 - 8(x^2 + 3x) - 20 = 0$

$t^2 - 8t - 20 = 0$ (Let $t = x^2 + 3x$)

$(t - 10)(t + 2) = 0$

$t - 10 = 0$	or	$t + 2 = 0$
$x^2 + 3x = 10$ (Replace t by $x^2 + 3x$)		$x^2 + 3x = -2$
$x^2 + 3x - 10 = 0$		$x^2 + 3x + 2 = 0$
$(x + 5)(x - 2) = 0$		$(x + 2)(x + 1) = 0$

$x + 5 = 0$ or $x - 2 = 0$ $x + 2 = 0$ or $x + 1 = 0$

$x = -5$ $x = 2$ $x = -2$ $x = -1$

$\boxed{\{-5, -2, -1, 2\}}$

3. $x^4 - 5x^2 + 4 = 0$

$(x^2)^2 - 5x^2 + 4 = 0$

$t^2 - 5t + 4 = 0$ (Let $t = x^2$)

$(t - 4)(t - 1) = 0$

$t - 4 = 0$ or $t - 1 = 0$

$t = 4$ $t = 1$

$x^2 = 4$ (Replace t by x^2) $x^2 = 1$

$x^2 - 4 = 0$ $x^2 - 1 = 0$

$(x - 2)(x + 2) = 0$ $(x - 1)(x + 1) = 0$

$x - 2 = 0$ or $x + 2 = 0$ $x - 1 = 0$ or $x + 1 = 0$

$x = 2$ $x = -2$ $x = 1$ $x = -1$

$\boxed{\{-2, -1, 1, 2\}}$

5. $9x^4 = 25x^2 - 16$

$9x^4 - 25x^2 + 16 = 0$

$9(x^2)^2 - 25x^2 + 16 = 0$

$9t^2 - 25t + 16 = 0$ (Let $t = x^2$)

$(9t - 16)(t - 1) = 0$

$9t - 16 = 0$ or $t - 1 = 0$

$t = \dfrac{16}{9}$ $t = 1$

$x^2 = \dfrac{16}{9}$ (Replace t with x^2) $x^2 = 1$

$\left(x - \dfrac{4}{3}\right)\left(x + \dfrac{4}{3}\right) = 0$ $(x - 1)(x + 1) = 0$

$x - \dfrac{4}{3} = 0$ or $x + \dfrac{4}{3} = 0$ $x - 1 = 0$ or $x + 1 = 0$

$x = \dfrac{4}{3}$ $x = -\dfrac{4}{3}$ $x = 1$ $x = -1$

$\boxed{\left\{-\dfrac{4}{3}, -1, 1, \dfrac{4}{3}\right\}}$

7. $y - 7\sqrt{y} + 10 = 0$

$$(\sqrt{y})^2 - 7\sqrt{y} + 10 = 0$$
$$t^2 - 7t + 10 = 0 \quad \text{(Let } t = \sqrt{y}\text{)}$$
$$(t - 5)(t - 2) = 0$$

$t - 5 = 0$	or	$t - 2 = 0$
$t = 5$		$t = 2$
$\sqrt{y} = 5$ (Replace t with $\sqrt{y}$)		$\sqrt{y} = 2$
$(\sqrt{y})^2 = 5^2$		$(\sqrt{y})^2 = 2^2$
$y = 25$		$y = 4$

$$\boxed{\{4, 25\}}$$

9. $x^{-2} - x^{-1} - 12 = 0$

$$(x^{-1})^2 - x^{-1} - 12 = 0$$
$$t^2 - t - 12 = 0 \quad \text{(Let } t = x^{-1}\text{)}$$
$$(t - 4)(t + 3) = 0$$

$t - 4 = 0$	or	$t + 3 = 0$
$t = 4$		$t = -3$
$x^{-1} = 4$ (Replace x^{-1} with t)		$x^{-1} = 3$
$\dfrac{1}{x} = 4$		$\dfrac{1}{x} = 3$
$1 = 4x$		$1 = 3x$
$\dfrac{1}{4} = x$		$\dfrac{1}{3} = x$

$$\boxed{\left\{\frac{1}{4}, \frac{1}{3}\right\}}$$

11. $6(w - 1)^{-2} + (w - 1)^{-1} - 2 = 0$

$$6t^2 + t - 2 = 0 \quad \text{(Let } t = (w - 1)^{-1}\text{)}$$
$$(3t + 2)(2t - 1) = 0$$

$3t + 2 = 0$	or	$2t - 1 = 0$
$t = -\dfrac{2}{3}$		$t = \dfrac{1}{2}$
$(w - 1)^{-1} = -\dfrac{2}{3}$ (Replace t with $(w-1)^{-1}$)		$(w - 1)^{-1} = \dfrac{1}{2}$
$\dfrac{1}{w - 1} = -\dfrac{2}{3}$		$\dfrac{1}{w - 1} = \dfrac{1}{2}$
$3(w - 1) \cdot \dfrac{1}{w - 1} = 3(w - 1) \cdot -\dfrac{2}{3}$		$2(w - 1) \cdot \dfrac{1}{w - 1} = 2(w - 1) \cdot \dfrac{1}{2} \quad (w \neq 1)$
$3 = -2w + 2$		$2 = w - 1$
$2w = -1$		$3 = w$
$w = -\dfrac{1}{2}$		

$$\boxed{\left\{-\frac{1}{2}, 3\right\}}$$

13. $\left(\dfrac{y^2-8}{y}\right)^2 + 5\left(\dfrac{y^2-8}{y}\right) - 14 = 0$

$\qquad t^2 + 5t - 14 = 0 \quad \left(\text{Let } t = \dfrac{y^2-8}{y}\right)$

$\qquad (t+7)(t-2) = 0$

$\qquad t+7 = 0 \qquad\qquad \text{or} \qquad\qquad\qquad\qquad t-2 = 0$

$\qquad\qquad t = -7 \qquad\qquad\qquad\qquad\qquad\qquad t = 2$

$\qquad \dfrac{y^2-8}{y} = -7 \quad \left(\text{Replace } t \text{ with } \dfrac{y^2-8}{y}\right) \qquad \dfrac{y^2-8}{y} = 2$

$\qquad y\cdot\dfrac{y^2-8}{y} = y\cdot(-7) \qquad\qquad\qquad y\cdot\dfrac{y^2-8}{y} = y\cdot 2 \quad (y\neq 0)$

$\qquad\qquad y^2-8 = -7y \qquad\qquad\qquad\qquad\qquad y^2-8 = 2y$

$\qquad y^2+7y-8 = 0 \qquad\qquad\qquad\qquad y^2-2y-8 = 0$

$\qquad (y+8)(y-1) = 0 \qquad\qquad\qquad (y-4)(y+2) = 0$

$\qquad y+8 = 0 \ \text{ or } \ y-1 = 0 \qquad y-4 = 0 \ \text{ or } \ y+2 = 0$

$\qquad\qquad y = -8 \qquad\quad y = 1 \qquad\quad y = 4 \qquad\qquad y = -2$

$\boxed{\{-8,-2,1,4\}}$

15. $y^{2/3} - y^{1/3} - 6 = 0$

$\qquad t^2 - t - 6 = 0 \quad (\text{Let } t = y^{1/3})$

$\qquad (t-3)(t+2) = 0$

$\qquad t-3 = 0 \qquad\qquad \text{or} \qquad\qquad t+2 = 0$

$\qquad\qquad t = 3 \qquad\qquad\qquad\qquad\qquad t = -2$

$\qquad y^{1/3} = 3 \quad (\text{Replace } t \text{ with } y^{1/3}) \qquad y^{1/3} = -2$

$\qquad (y^{1/3})^3 = 3^3 \qquad\qquad\qquad\qquad (y^{1/3})^3 = (-2)^3$

$\qquad\qquad y = 27 \qquad\qquad\qquad\qquad\qquad y = -8$

$\boxed{\{-8,27\}}$

17. $2y - 3y^{1/2} + 1 = 0$

$\qquad 2t^2 - 3t + 1 = 0 \quad (\text{Let } t = y^{1/2})$

$\qquad (2t-1)(t-1) = 0$

$\qquad 2t-1 = 0 \qquad\qquad \text{or} \qquad\qquad t-1 = 0$

$\qquad\qquad t = \dfrac{1}{2} \qquad\qquad\qquad\qquad\qquad t = 1$

$\qquad y^{1/2} = \dfrac{1}{2} \quad (\text{Replace } t \text{ with } y^{1/2}) \qquad y^{1/2} = 1$

$\qquad (y^{1/2})^2 = \left(\dfrac{1}{2}\right)^2 \qquad\qquad\qquad (y^{1/2})^2 = 1^2$

$\qquad\qquad y = \dfrac{1}{4} \qquad\qquad\qquad\qquad\qquad y = 1$

$\boxed{\left\{\dfrac{1}{4},1\right\}}$

19. $\qquad\qquad (y-1)^{1/2} = 2(y-1)^{1/4} + 15$

$\qquad (y-1)^{1/2} - 2(y-1)^{1/4} - 15 = 0$

$\qquad\qquad t^2 - 2t - 15 = 0 \quad (\text{Let } t = (y-1)^{1/4})$

$\qquad\qquad (t-5)(t+3) = 0$

$\qquad\qquad t-5 = 0 \qquad\qquad \text{or} \qquad\qquad t+3 = 0$

$\qquad\qquad\qquad t = 5 \qquad\qquad\qquad\qquad\qquad t = -3$

$\qquad\qquad (y-1)^{1/4} = 5 \quad (\text{Replace } t \text{ with } (y-1)^{1/4}) \qquad (y-1)^{1/4} = -3$

$\qquad\qquad [(y-1)^{1/4}]^4 = 5^4 \qquad\qquad\qquad\qquad \text{Not possible}$

$\qquad\qquad\qquad y-1 = 625$

$\qquad\qquad\qquad\qquad y = 626$

$\boxed{\{626\}}$

21.
$$
\begin{aligned}
5x^{2/3} + 11x^{1/3} &= -2 \\
5x^{2/3} + 11x^{1/3} + 2 &= 0 \\
5t^2 + 11t + 2 &= 0 \quad \text{(Let } t = x^{1/3}\text{)} \\
(5t + 1)(t + 2) &= 0
\end{aligned}
$$

$$
\begin{array}{ll}
5t + 1 = 0 \qquad \text{or} & t + 2 = 0 \\[4pt]
t = -\dfrac{1}{5} & t = -2 \\[8pt]
x^{1/3} = -\dfrac{1}{5} \quad \text{(Replace } t \text{ with } x^{1/3}\text{)} & x^{1/3} = -2 \\[8pt]
(x^{1/3})^3 = \left(-\dfrac{1}{5}\right)^3 & (x^{1/3})^3 = (-2)^3 \\[8pt]
x = -\dfrac{1}{125} & x = -8
\end{array}
$$

$$\boxed{\left\{-8, -\frac{1}{125}\right\}}$$

23.
$$
\begin{aligned}
w^{-4} &= 5w^{-2} - 4 \\
w^{-4} - 5w^{-2} + 4 &= 0 \\
t^2 - 5t + 4 &= 0 \quad \text{(Let } t = w^{-2}\text{)} \\
(t - 4)(t - 1) &= 0
\end{aligned}
$$

$$
\begin{array}{ll}
t - 4 = 0 \qquad \text{or} & t - 1 = 0 \\
t = 4 & t = 1 \\
w^{-2} = 4 \quad \left(\text{Replace } t \text{ with } w^{-2}\right) & w^{-2} = 1 \\[6pt]
\dfrac{1}{w^2} = 4 & \dfrac{1}{w^2} = 1 \\[6pt]
1 = 4w^2 & 1 = w^2 \\
0 = 4w^2 - 1 & 0 = w^2 - 1 \\
0 = (2w - 1)(2w + 1) & 0 = (w - 1)(w + 1)
\end{array}
$$

$$
\begin{array}{ll}
2w - 1 = 0 \ \text{or} \ 2w + 1 = 0 & w - 1 = 0 \ \text{or} \ w + 1 = 0 \\[4pt]
w = \dfrac{1}{2} \qquad w = -\dfrac{1}{2} & w = 1 \qquad w = -1
\end{array}
$$

$$\boxed{\left\{-1, -\frac{1}{2}, \frac{1}{2}, 1\right\}}$$

25.
$$
\begin{aligned}
6\left(\frac{2y}{y-3}\right)^2 &= 5\left(\frac{2y}{y-3}\right) + 6 \\
6t^2 &= 5t + 6 \quad \left(\text{Let } t = \frac{2y}{y-3}\right) \\
6t^2 - 5t - 6 &= 0 \\
(3t + 2)(2t - 3) &= 0
\end{aligned}
$$

$$
\begin{array}{ll}
3t + 2 = 0 \qquad \text{or} & 2t - 3 = 0 \\[4pt]
t = -\dfrac{2}{3} & t = \dfrac{3}{2} \\[8pt]
\dfrac{2y}{y-3} = -\dfrac{2}{3} \quad \left(\text{Replace } t \text{ with } \frac{2y}{y-3}\right) & \dfrac{2y}{y-3} = \dfrac{3}{2} \\[8pt]
3(y-3) \cdot \dfrac{2y}{y-3} = 3(y-3) \cdot \left(-\dfrac{2}{3}\right) & 2(y-3) \cdot \dfrac{2y}{y-3} = 2(y-3) \cdot \dfrac{3}{2} \\[8pt]
6y = -2y + 6 & 4y = 3y - 9 \\
8y = 6 & y = -9 \\[4pt]
y = \dfrac{3}{4} &
\end{array}
$$

$$\boxed{\left\{-9, \frac{3}{4}\right\}}$$

27.
$$2x^3 + 3x^2 - 8x - 12 = 0$$
$$x^2(2x + 3) - 4(2x + 3) = 0$$
$$(2x + 3)(x^2 - 4) = 0$$
$$(2x + 3)(x + 2)(x - 2) = 0$$

$2x + 3 = 0$ or $x + 2 = 0$ or $x - 2 = 0$

$x = -\dfrac{3}{2}$ $x = -2$ $x = 2$

$$\boxed{\left\{ -2, -\dfrac{3}{2}, 2 \right\}}$$

29.
$$y^3 - 3y^2 - 9y + 27 = 0$$
$$y^2(y - 3) - 9(y - 3) = 0$$
$$(y - 3)(y^2 - 9) = 0$$
$$(y - 3)^2(y + 3) = 0$$

$y - 3 = 0$ or $y + 3 = 0$

$y = 3$ $y = -3$

$$\boxed{\{-3, 3\}}$$

31.
$$8w^3 - 12w^2 = 2w - 3$$
$$8w^3 - 12w^2 - 2w + 3 = 0$$
$$4w^2(2w - 3) - (2w - 3) = 0$$
$$(2w - 3)(4w^2 - 1) = 0$$
$$(2w - 3)(2w - 1)(2w + 1) = 0$$

$2w - 3 = 0$ or $2w - 1 = 0$ or $2w + 1 = 0$

$w = \dfrac{3}{2}$ $w = \dfrac{1}{2}$ $w = -\dfrac{1}{2}$

$$\boxed{\left\{ -\dfrac{1}{2}, \dfrac{1}{2}, \dfrac{3}{2} \right\}}$$

33.
$$9z^3 + 8 = 4z + 18z^2$$
$$9z^3 - 18z^2 - 4z + 8 = 0$$
$$9z^2(z - 2) - 4(z - 2) = 0$$
$$(z - 2)(9z^2 - 4) = 0$$
$$(z - 2)(3z + 2)(3z - 2) = 0$$

$z - 2 = 0$ or $3z + 2 = 0$ or $3z - 2 = 0$

$z = 2$ $z = -\dfrac{2}{3}$ $z = \dfrac{2}{3}$

$$\boxed{\left\{ -\dfrac{2}{3}, \dfrac{2}{3}, 2 \right\}}$$

35.
$$x^3 - 4x^2 - 17x + 60 = 0 \qquad \text{(One solution is an integer between 1 and 3 inclusive)}$$
Try $x = 3$:

```
3│    1    -4    -17     60
            3    - 3    -60
    ─────────────────────────
      1    -1    -20      0   →  Remainder = 0
                             So  x = 3 is a solution
```

$$x^3 - 4x^2 - 17x + 60 = 0$$
$$(x - 3)(x^2 - x - 20) = 0$$
$$(x - 3)(x - 5)(x + 4) = 0$$

$x - 3 = 0$ or $x - 5 = 0$ or $x + 4 = 0$

$x = 3$ $x = 5$ $x = -4$

$$\boxed{\{-4, 3, 5\}}$$

37. $2x^3 - 9x^2 + 7x + 6 = 0$

The equation has a solution that is the solution of

$$3x - 4 = -2x + 6$$
$$5x = 10$$
$$x = 2$$

Using synthetic division with $x = 2$ as a solution:

$$
\begin{array}{r|rrrr}
2 & 2 & -9 & 7 & 6 \\
  &   & 4 & -10 & -6 \\
\hline
  & 2 & -5 & -3 & 0
\end{array}
$$

$$2x^3 - 9x^2 + 7x + 6 = 0$$
$$(x - 2)(2x^2 - 5x - 3) = 0$$
$$(x - 2)(2x + 1)(x - 3) = 0$$
$$x - 2 = 0 \quad \text{or} \quad 2x + 1 = 0 \quad \text{or} \quad x - 3 = 0$$
$$x = 2 \qquad\qquad x = -\frac{1}{2} \qquad\qquad x = 3$$

$$\boxed{\left\{ -\frac{1}{2}, 2, 3 \right\}}$$

39. $x^3 + 2x^2 - 5x - 6 = 0$

The equation has a solution that exceeds by 2 the solution of:

$$4(y - 1) + 5(y + 2) = 3(y - 8)$$
$$4y - 4 + 5y + 10 = 3y - 24$$
$$9y + 6 = 3y - 24$$
$$6y = -30$$
$$y = -5$$
$$y + 2 = -5 + 2 = -3$$

-3 is a solution

Using synthetic division:

$$
\begin{array}{r|rrrr}
-3 & 1 & 2 & -5 & -6 \\
   &   & -3 & 3 & 6 \\
\hline
   & 1 & -1 & -2 & 0
\end{array}
$$

$$x^3 + 2x^2 - 5x - 6 = 0$$
$$(x + 3)(x^2 - x - 2) = 0$$
$$(x + 3)(x - 2)(x + 1) = 0$$
$$x + 3 = 0 \quad \text{or} \quad x - 2 = 0 \quad \text{or} \quad x + 1 = 0$$
$$x = -3 \qquad\qquad x = 2 \qquad\qquad x = -1$$

$$\boxed{\{-3, -1, 2\}}$$

41. $\boxed{\text{D}}$ is true;   A is *not* true: $2x^{2/3} + 7x^{1/3} - 15 = 0$ is *not* a polynomial equation.

B is *not* true: $(x^2 + 3x)^4 - 8(x^2 + 3x)^3 - 20 = 0$ is *not* quadratic in form.

C is *not* true: $x^6 - 9x^3 + 8 = 0$

$$t^2 - 9t + 8 = 0 \quad \text{(Let } t = x^3)$$
$$(t - 8)(t - 1) = 0$$
$$t = 8 \quad \text{or} \quad t = 1$$
$$x^3 = 8 \qquad\qquad x^3 = 1 \quad \text{(Replace } t \text{ with } x^3)$$
$$x = 2 \qquad\qquad x = 1$$

$\{1, 2\}$ is the solution set, *not* $\{8, 1\}$.

D is true.

43. $\sqrt{\dfrac{x+4}{x-1}} + \sqrt{\dfrac{x-1}{x+4}} = \dfrac{5}{2}$

Using substitution on the given equation results in:

$$t + \dfrac{1}{t} = \dfrac{5}{2} \quad \left(\text{Let } t = \sqrt{\dfrac{x+4}{x-1}}\right)$$

$$2t\left(t + \dfrac{1}{t}\right) = 2t\left(\dfrac{5}{2}\right)$$

$$2t^2 + 2 = 5t$$

$$2t^2 - 5t + 2 = 0$$

$$(2t - 1)(t - 2) = 0$$

$$2t - 1 = 0 \qquad \text{or} \qquad t - 2 = 0$$

$$t = \dfrac{1}{2} \qquad\qquad\qquad t = 2$$

$$\sqrt{\dfrac{x+4}{x-1}} = \dfrac{1}{2} \quad \text{or} \quad \sqrt{\dfrac{x+4}{x-1}} = 2$$

$$\dfrac{x+4}{x-1} = \dfrac{1}{4} \qquad\qquad \dfrac{x+4}{x-1} = 4$$

$$4x + 16 = x - 1 \qquad\quad x + 4 = 4x - 4$$

$$3x = -17 \qquad\qquad\quad 8 = 3x$$

$$x = -\dfrac{17}{3} \qquad\qquad \dfrac{8}{3} = x$$

The solution set is $\boxed{\left\{-\dfrac{17}{3}, \dfrac{8}{3}\right\}}$

45. $x + y + \sqrt{x+y} = 12$

$x - y + \sqrt{x-y} = 6$

Using the given substitutions and the given equations, we have:

$$t^2 + t = 12 \quad \text{(Let } t = \sqrt{x+y})$$
$$v^2 + v = 6 \quad \text{(Let } v = \sqrt{x-y})$$

Solving the quadratic equations:

$$t^2 + t - 12 = 0 \qquad\qquad v^2 + v - 6 = 0$$
$$(t + 4)(t - 3) = 0 \qquad\qquad (v + 3)(v - 2) = 0$$
$$t + 4 = 0 \quad \text{or} \quad t - 3 = 0 \qquad v + 3 = 0 \quad \text{or} \quad v - 2 = 0$$
$$t = -4 \qquad\qquad t = 3 \qquad\qquad v = -3 \qquad\qquad v = 2$$

Since t and v both represent positive square roots,

$$t = 3 \qquad \text{and} \qquad v = 2$$
$$\sqrt{x+y} = 3 \qquad\qquad\qquad \sqrt{x-y} = 2$$
$$x + y = 9 \qquad\qquad\qquad x - y = 4$$

We now solve the resulting system.

$$x + y = 9$$
$$\underline{x - y = 4}$$
Add: $2x \quad\;\; = 13$

$$x = \dfrac{13}{2}$$

Substituting, we find that $y = \dfrac{5}{2}$.

The solution set is $\boxed{\left\{\left(\dfrac{13}{2}, \dfrac{5}{2}\right)\right\}}$.

Review Problems

50. $128a^3 b - 50ab$

$= 2ab(64a^2 - 25)$

$= \boxed{2ab(8a + 5)(8a - 5)}$

51. $\dfrac{3}{y - 1} - \dfrac{2}{y + 1} + \dfrac{2y}{y^2 - 1}$

$= \dfrac{3}{y - 1} \cdot \dfrac{y + 1}{y + 1} - \dfrac{2}{y + 1} \cdot \dfrac{y - 1}{y - 1} + \dfrac{2y}{(y + 1)(y - 1)}$

$= \dfrac{3y + 3 - 2y + 2 + 2y}{(y - 1)(y + 1)}$

$= \boxed{\dfrac{3y + 5}{y^2 - 1}}$

52. $(4x^5 y^{10})(-2x^3 y^2)^{-3}$

$= 4x^5 y^{10}(-2)^{-3}x^{-9}y^{-6}$

$= \dfrac{4}{-8} \, x^{-4}y^4$

$= \boxed{\dfrac{-y^4}{2x^4}}$

Section 8.2 Solving Quadratic Equations of the Form $ax^2 + c = 0$

Problem Set 8.2, pp. 593-594

1. $y^2 = 100$

$y = \pm\sqrt{100}$

$y = \pm 10$

$\boxed{\{-10, 10\}}$

3. $y^2 = 7$

$y = \pm\sqrt{7}$

$\boxed{\{-\sqrt{7}, \sqrt{7}\}}$

5. $x^2 = 75$

$x = \pm\sqrt{75}$

$= \pm\sqrt{25 \cdot 3}$

$= \pm 5\sqrt{3}$

$\boxed{\{-5\sqrt{3}, 5\sqrt{3}\}}$

7. $z^2 = -4$

$z = \pm\sqrt{-4}$

$= \pm\sqrt{4}\, i$

$= \pm 2i$

$\boxed{\{-2i, 2i\}}$

9. $4y^2 + 3 = 103$

$4y^2 = 100$

$y^2 = 25$

$y = \pm\sqrt{25}$

$= \pm 5$

$\boxed{\{-5, 5\}}$

11. $3x^2 = 25$

$x^2 = \dfrac{25}{3}$

$x = \pm\sqrt{\dfrac{25}{3}}$

$= \pm\dfrac{5}{\sqrt{3}} \cdot \dfrac{\sqrt{3}}{\sqrt{3}}$

$= \pm\dfrac{5\sqrt{3}}{3}$

$\boxed{\left\{-\dfrac{5\sqrt{3}}{3}, \dfrac{5\sqrt{3}}{3}\right\}}$

13.
$$7x^2 + 2 = 13$$
$$7x^2 = 11$$
$$x^2 = \frac{11}{7}$$
$$x = \pm\sqrt{\frac{11}{7}}$$
$$= \pm\frac{\sqrt{11}}{\sqrt{7}} \cdot \frac{\sqrt{7}}{\sqrt{7}}$$
$$= \pm\frac{\sqrt{77}}{7}$$

$$\boxed{\left\{ -\frac{\sqrt{77}}{7}, \frac{\sqrt{77}}{7} \right\}}$$

15.
$$4x^2 + 7 = 3(x^2 + 1)$$
$$4x^2 + 7 = 3x^2 + 3$$
$$x^2 = -4$$
$$x = \pm\sqrt{-4}$$
$$= \pm 2i$$

$$\boxed{\{ -2i, 2i \}}$$

17.
$$4(x^2 + 2x) + 7 = 3x^2 + 8x + 2$$
$$4x^2 + 8x + 7 = 3x^2 + 8x + 2$$
$$x^2 = -5$$
$$x = \pm\sqrt{-5}$$
$$= \pm\sqrt{5}\, i$$

$$\boxed{\{ -\sqrt{5}\, i, \sqrt{5}\, i \}}$$

19.
$$(x + 4)(x + 1) = 5x - 71$$
$$x^2 + 5x + 4 = 5x - 71$$
$$x^2 = -75$$
$$x = \pm\sqrt{-75}$$
$$= \pm\sqrt{(25)(3)(-1)}$$
$$= \pm 5\sqrt{3}\, i$$

$$\boxed{\{ -5\sqrt{3}\, i, 5\sqrt{3}\, i \}}$$

21.
$$3y^4 - 2y^2 - 5 = 0$$
$$3t^2 - 2t - 5 = 0 \quad \text{(Let } t = y^2\text{)}$$
$$(3t - 5)(t + 1) = 0$$

$3t - 5 = 0$ or	$t + 1 = 0$
$t = \dfrac{5}{3}$	$t = -1$
$y^2 = \dfrac{5}{3}$ (Replace t with y^2)	$y^2 = -1$
$y = \pm\dfrac{\sqrt{5}}{\sqrt{3}} \cdot \dfrac{\sqrt{3}}{\sqrt{3}}$	$y = \sqrt{-1}$
$= \pm\dfrac{\sqrt{15}}{3}$	$= \pm i$

$$\boxed{\left\{ -i, i, -\frac{\sqrt{15}}{3}, \frac{\sqrt{15}}{3} \right\}}$$

23.
$$\frac{Y^4}{8} + \frac{Y^2}{4} = 3$$
$$\frac{t^2}{8} + \frac{t}{4} = 3 \quad \text{(Let } t = y^2\text{)}$$
$$t^2 + 2t - 24 = 0$$
$$(t + 6)(t - 4) = 0$$

$t = -6$ or	$t = 4$
$y^2 = -6$ (Replace t with y^2)	$y^2 = 4$
$y = \pm\sqrt{-6}$	$y = \pm\sqrt{4}$
$= \pm\sqrt{6}\, i$	$= \pm 2$

$$\boxed{\{ -\sqrt{6}\, i, \sqrt{6}\, i, -2, 2 \}}$$

25. $3(x^2 - 1) + \sqrt{x^2 - 1} = 2$

   $3t^2 + t - 2 = 0$ (Let $t = \sqrt{x^2 - 1}$)

   $(3t - 2)(t + 1) = 0$

   $3t - 2 = 0$ or $t + 1 = 0$

   $t = \dfrac{2}{3}$ $t = -1$

   $\sqrt{x^2 - 1} = \dfrac{2}{3}$ $\sqrt{x^2 - 1} = -1$

   $(\sqrt{x^2 - 1})^2 = \left(\dfrac{2}{3}\right)^2$ $(\sqrt{x^2 - 1})^2 = (-1)^2$

   $x^2 - 1 = \dfrac{4}{9}$ $x^2 - 1 = 1$

   $x^2 = \dfrac{13}{9}$ $x^2 = 2$

   $x = \pm\sqrt{\dfrac{13}{9}}$ $x = \pm\sqrt{2}$

   $= \pm\dfrac{\sqrt{13}}{3}$

Since both sides were squared in the solution process, all potential answers must be checked.

check $\pm\sqrt{2}$: $(\pm\sqrt{2})^2 = 2$

   $3(2 - 1) + \sqrt{2 - 1} = 2$

   $3 + 1 \neq 2$

$\sqrt{2}$ and $-\sqrt{2}$ are extraneous.

check $\pm\dfrac{\sqrt{13}}{3}$: $\left(\pm\dfrac{\sqrt{13}}{3}\right)^2 = \dfrac{13}{9}$

   $3\left(\dfrac{13}{9} - 1\right) + \sqrt{\dfrac{13}{9} - 1} = 2$

   $3\left(\dfrac{4}{9}\right) + \sqrt{\dfrac{4}{9}} = 2$

   $\dfrac{4}{3} + \dfrac{2}{3} = 2$

   $2 = 2$

$\dfrac{\sqrt{13}}{3}$ and $-\dfrac{\sqrt{13}}{3}$ check

$$\boxed{\left\{-\dfrac{\sqrt{13}}{3}, \dfrac{\sqrt{13}}{3}\right\}}$$

27.
$$\sqrt{y^2 + 7} = 1 + \sqrt{y^2 + 2}$$
$$(\sqrt{y^2 + 7})^2 = (1 + \sqrt{y^2 + 2})^2$$
$$y^2 + 7 = 1 + 2\sqrt{y^2 + 2} + y^2 + 2$$
$$y^2 + 7 = y^2 + 3 + 2\sqrt{y^2 + 2}$$
$$4 = 2\sqrt{y^2 + 2}$$
$$2 = \sqrt{y^2 + 2}$$
$$4 = y^2 + 2$$
$$y^2 = 2$$
$$y = \pm\sqrt{2} \quad \text{(Both check)}$$
$$\boxed{\{-\sqrt{2}, \sqrt{2}\}}$$

29.
$$\sqrt[3]{x^2 - 1} + 2 = 0$$
$$\sqrt[3]{x^2 - 1} = -2$$
$$x^2 - 1 = -8$$
$$x^2 = -7$$
$$x = \pm\sqrt{-7}$$
$$= \pm\sqrt{7}\, i \quad \text{(Both check)}$$
$$\boxed{\{-\sqrt{7}\, i, \sqrt{7}\, i\}}$$

31.
$$\sqrt{x^4 - 2} = x$$
$$(\sqrt{x^4 - 2})^2 = x^2$$
$$x^4 - 2 = x^2$$
$$x^4 - x^2 - 2 = 0$$
$$t^2 - t - 2 = 0 \quad \text{(Let } t = x^2\text{)}$$
$$(t - 2)(t + 1) = 0$$

$t - 2 = 0$	or	$t + 1 = 0$
$t = 2$		$t = -1$
$x^2 = 2$		$x^2 = -1$
$x = \pm\sqrt{2}$		$x = \pm i$

check $-\sqrt{2}$:

check $\sqrt{2}$:
$$\sqrt{(\sqrt{2})^4 - 2} = \sqrt{2}$$
$$\sqrt{4 - 2} = \sqrt{2}$$
$$\sqrt{2} = \sqrt{2}$$

check $-\sqrt{2}$:
$$\sqrt{(-\sqrt{2})^4 - 2} = -\sqrt{2}$$
$$\sqrt{4 - 2} = -\sqrt{2}$$
$$\sqrt{2} \neq -\sqrt{2}$$

$\sqrt{2}$ checks

$-\sqrt{2}$ is extraneous

check i :

$$\sqrt{i^4 - 2} = i$$
$$\sqrt{1 - 2} = i \quad i^4 = (i^2)^2 = (-1)^2 = 1$$
$$\sqrt{-1} = i$$
$$i = i$$

check $-i$:

$$\sqrt{(-i)^4 - 2} = -i$$
$$\sqrt{1 - 2} = -i$$
$$i \neq -i$$

$-i$ is extraneous

i checks

$$\boxed{\{\sqrt{2}, i\}}$$

33. Let $\quad x \quad = \quad$ width of original rectangle

$\qquad 3x \quad = \quad$ length of original rectangle

$\qquad x - 1 \quad = \quad$ width of new rectangle

$\qquad 3x + 3 \quad = \quad$ length of new rectangle

area of new rectangle $\ = \ 72$

$$
\begin{aligned}
(x - 1)(3x + 3) &= 72 \\
3x^2 - 3 &= 72 \\
3x^2 &= 75 \\
x^2 &= 25 \\
x &= 5 \qquad \text{or} \qquad x = -5 \ \text{(reject)} \\
3x &= 3(5) = 15
\end{aligned}
$$

Width: 5 yd   Length: 15 yd

35.
$$
\begin{aligned}
A &= \pi r^2 \\
100\pi &= \pi r^2 \\
100 &= r^2 \\
r &= 10 \quad \text{or} \quad r = -10 \ \text{(reject)}
\end{aligned}
$$
Circumference: $\quad 2\pi r = 2\pi(10)$
$$
= \boxed{20\pi \text{ ft}}
$$

37.
$$
B = 10^5(1 + 2t^5)
$$
($B = 33 \times 10^5$ bacteria):
$$
\begin{aligned}
10^5(1 + 2t^2) &= 33 \times 10^5 \\
\frac{1}{10^5}\left[10^5(1 + 2t^2)\right] &= (33 \times 10^5)\left(\frac{1}{10^5}\right) \\
1 + 2t^2 &= 33 \\
2t^2 &= 32 \\
t^2 &= 16 \\
t = 4 \quad \text{or} \quad t &= -4 \ \text{(reject)}
\end{aligned}
$$
After 4 hours

39. Let $\qquad\qquad x \quad = \quad$ width of rectangular sheet

$\qquad\qquad\qquad x + 8 \quad = \quad$ length of rectangular sheet

$\quad x + 8 - 2(2) = x + 8 - 4 \quad = \quad$ length of open box

$\qquad x - 2(2) = x - 4 \quad = \quad$ width of open box

height of open box $\ = \ 2$ inches
$$
\begin{aligned}
V &= LWH \\
256 &= (x + 8 - 4)(x - 4)2 \\
256 &= (x + 4)(x - 4)2 \\
128 &= (x + 4)(x - 4) \\
128 &= x^2 - 16 \\
144 &= x^2 \\
\pm\sqrt{144} &= x \\
x = 12 \quad \text{or} \quad x &= -12 \ \text{(reject)} \\
x + 8 &= 12 + 8 \\
&= 20
\end{aligned}
$$
Width: 12 in.   Length: 20 in.

41. Let $x = $ measure of each leg
$$
\begin{aligned}
x^2 + x^2 &= 3^2 \\
2x^2 &= 9 \\
x^2 &= \frac{9}{2} \\
x &= \pm\frac{3}{\sqrt{2}} \cdot \frac{\sqrt{2}}{\sqrt{2}} = \pm\frac{3\sqrt{2}}{2} \text{ ft}
\end{aligned}
$$
Each leg measures $\dfrac{3\sqrt{2}}{2}$ ft

43.
$$
\begin{aligned}
A &= 4\pi r^2 \\
4\pi r^2 &= A \\
r^2 &= \frac{A}{4\pi} \\
r &= \sqrt{\frac{A}{4\pi}} \qquad \left(\text{Reject} - \sqrt{\frac{A}{4\pi}}\right) \\
r &= \frac{\sqrt{A}}{2\sqrt{\pi}} \cdot \frac{\sqrt{\pi}}{\sqrt{\pi}} = \boxed{\frac{\sqrt{A\pi}}{2\pi}}
\end{aligned}
$$

45. Distance traveled $= 16t^2$

Distance traveled $= 300 - 200 = 100$

$$16t^2 = 100$$

$$t^2 = \frac{100}{16}$$

$$t = \frac{10}{4}$$

$$= 2.5$$

The object will be 200 meters from the ground in

$\boxed{2.5 \text{ seconds}}$.

47.

$$y = \frac{b}{a}\sqrt{a^2 - x^2}$$

$$y^2 = \frac{b^2}{a^2}(a^2 - x^2)$$

$$a^2 y^2 = a^2 \left[\frac{b^2}{a^2}(a^2 - x^2)\right]$$

$$a^2 y^2 = b^2 a^2 - b^2 x^2$$

$$b^2 x^2 = b^2 a^2 - a^2 y^2$$

$$x^2 = \frac{b^2 a^2 - a^2 y^2}{b^2}$$

$$x = \pm\sqrt{\frac{b^2 a^2 - a^2 y^2}{b^2}}$$

$$= \pm\sqrt{\frac{a^2(b^2 - y^2)}{b^2}}$$

$$\boxed{x = \pm\frac{a}{b}\sqrt{b^2 - y^2}}$$

49.

$$m_v = \frac{m_0}{\sqrt{1 - \dfrac{v^2}{c^2}}}$$

$$m_v{}^2 = \frac{m_0}{1 - \dfrac{v^2}{c^2}} \cdot \frac{c^2}{c^2}$$

$$= \frac{m_0{}^2 c^2}{c^2 - v^2}$$

$$(c^2 - v^2)(m_v{}^2) = (c^2 - v^2)\frac{m_0{}^2 c^2}{c^2 - v^2}$$

$$m_v{}^2 c^2 - m_v{}^2 v^2 = m_0{}^2 c^2$$

$$m_v{}^2 c^2 - m_0{}^2 c^2 = m_v{}^2 v^2$$

$$\frac{m_v{}^2 c^2 - m_0{}^2 c^2}{m_v{}^2} = v^2$$

$$v = \sqrt{\frac{m_v{}^2 c^2 - m_0{}^2 c^2}{m_v{}^2}}$$

$$= \sqrt{\frac{c^2(m_v{}^2 - m_0{}^2)}{m_v{}^2}}$$

$$\boxed{v = \frac{c\sqrt{m_v{}^2 - m_0{}^2}}{m_v}}$$

51. $\boxed{C}$ is true.

If $x =$ the person's age, we must find

$$\sqrt{(x+1)(x-1) + 1} = \sqrt{x^2 - 1 + 1}$$

$$= \sqrt{x^2}$$

$$= x \quad \text{(Since } x > 0\text{)}$$

Review Problems

54. $\sqrt[5]{32x^6y^2} - \sqrt[5]{x^6y^2} = \sqrt[5]{2^5 x^5(xy^2)} - \sqrt[5]{x^5(xy^2)}$

$$= 2x\sqrt[5]{xy^2} - x\sqrt[5]{xy^2}$$

$$= (2x - x)\sqrt[5]{xy^2}$$

$$= \boxed{x\sqrt[5]{xy^2}}$$

55. All telephone area codes have the same sum of $2 + 5 + 2 = 9$. One of the area codes is 252.

A second area code begins with 6. The remaining 2 numbers must add to $9 - 6 = 3$. The possibilities are 0 and 3 or 1 and 2.

However this number cannot contain a 1 since the remaining area code ends in 1. Thus the second area code must be 603 or 630.

For the remaining area code, the possible remaining digits are 1, 4, 7, 9. The only combinations that sum to 9 and end in 1 are $\boxed{711 \text{ and } 171}$.

56.
$$\begin{aligned} x + 3z &= 15 \\ 2x - 3y &= -6 \\ 2y - 4z &= -16 \end{aligned}$$

(Equations 1 and 2)::

$$\begin{aligned} (\times -2) \quad -2x - 6z &= -30 \\ \underline{2x - 3y } &= \underline{-6} \\ -3y - 6z &= -36 \quad \text{(Equation 4)} \end{aligned}$$

(Equations 4 and 3):

$$\begin{aligned} (\div 3) \quad y + 2z &= 12 \\ \underline{y - 2z} &= \underline{-8} \\ 2y &= 4 \\ y &= 2 \end{aligned}$$

(Equation 4):
$$\begin{aligned} y + 2z &= 12 \\ 2 + 2z &= 10 \\ 2z &= 8 \\ z &= 4 \end{aligned}$$

(Equation 2):
$$\begin{aligned} 2x - 3y &= -6 \\ 2x - 3(2) &= -6 \\ 2x &= 0 \\ x &= 0 \end{aligned}$$

$$\boxed{\{(0, 2, 4)\}}$$

Section 8.3 Further Applications of the Square Root Method

Problem Set 8.3, pp. 602-604

1. $16x^2 + 9y^2 = 144$ or $\dfrac{x^2}{9} + \dfrac{y^2}{16} = 1$

x-intercepts:
$$\begin{aligned} 16x^2 + 0 &= 144 \\ 16x^2 &= 144 \\ x^2 &= 9 \\ x &= \pm 3 \end{aligned}$$

The points $(3, 0)$ and $(-3, 0)$ are on the graph.

y-intercepts:
$$0 + 9y^2 = 144$$
$$9y^2 = 144$$
$$y^2 = 16$$
$$y = \pm 4$$

The points $(0, 4)$ and $(0, -4)$ are on the graph.

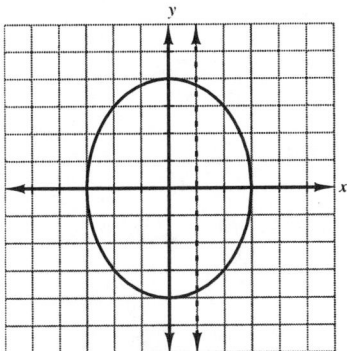

A vertical line drawn at $x = 1$ (any x-value between -3 and 3 could have been chosen) intersects the graph more than once. Thus, the relation does *not* define y as a function of x.

From the graph,

Domain $= \{x \mid -3 \le x \le 3\}$ or $[-3, 3]$ Range $= \{y \mid -4 \le y \le 4\}$ or $[-4, 4]$

3. $9x^2 + 4y^2 = 36$ or $\dfrac{x^2}{4} + \dfrac{y^2}{9} = 1$

x-intercepts:
$$9x^2 + 0 = 36 \quad (y = 0)$$
$$9y^2 = 36$$
$$y^2 = 4$$
$$y = \pm 2$$

The points $(2, 0)$ and $(-2, 0)$ are on the graph.

y-intercepts:
$$0 + 4y^2 = 36$$
$$4y^2 = 36$$
$$y^2 = 9$$
$$y = \pm 3$$

The points $(0, 3)$ and $(0, -3)$ are on the graph.

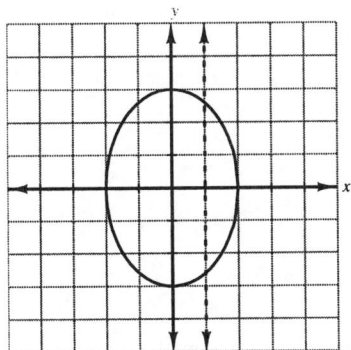

A vertical line drawn at $x = 1$ intersects the graph more than once. Thus, the relation does *not* define y as a function of x.

From the graph,

Domain $= \{x \mid -2 \le x \le 2\}$ or $[-2, 2]$ Range $= \{y \mid -3 \le y \le 3\}$ or $[-3, 3]$

5. $x^2 + y^2 = 16$ or $\dfrac{x^2}{16} + \dfrac{y^2}{16} = 1$

   x-intercepts: $x^2 + 0 = 16$ $(y = 0)$
                       $x^2 = 16$
                         $x = \pm 4$

   The points (4, 0) and (–4, 0) are on the graph.

   y-intercepts: $0 + y^2 = 16$
                       $y^2 = 16$
                         $y = \pm 4$

   The points (0, 4) and (0, –4) are on the graph.

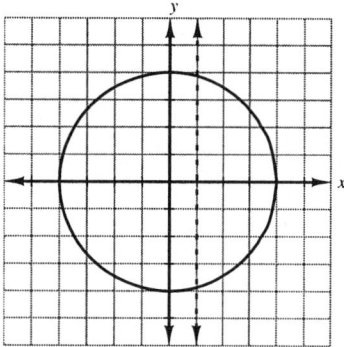

   A vertical line drawn at $x = 1$ intersects the graph more than once. Thus, the relation does *not* define *y* as a function of *x*.

   From the graph,

   Domain $= \{x \mid -4 \le x \le 4\}$ or $[-4, 4]$ Range $= \{y \mid -4 \le y \le 4\}$ or $[-4, 4]$

7. $x^2 + y^2 = 16$
   a. $y^2 = 16 - x^2$

   $y = \pm\sqrt{16 - x^2}$

   b. For certain values of *x*, there is more than one value of *y*.

   c. $y = \sqrt{16 - x^2}$

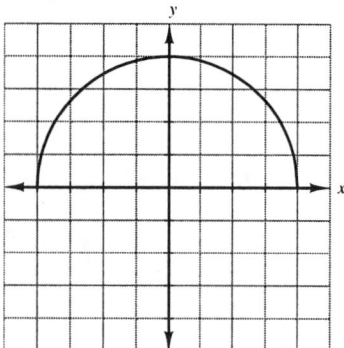

   Yes; *y* is now a function of *x* since for each value of *x* there is only one value of *y*.

   d. Domain $= \{x \mid -4 \le x \le 4\}$ or $[-4, 4]$ Range $= \{y \mid 0 \le y \le 4\}$ or $[0, 4]$

9. $\boxed{D}$ is true; A is *not* true: $y = -\sqrt{25 - x^2}$ does define y as a function of x.

B is *not* true;   If $x^2 + y^2 = 144$

then $y = \pm\sqrt{144 - x^2}$

C is *not* true; $9y^2 - 4x^2 = 36$

x-intercept: $0 - 4x^2 = 36$

$x^2 = -9$

$x = \pm i$

There are no x-intercepts.

11. $(4, -3)$ and $(-6, 2)$

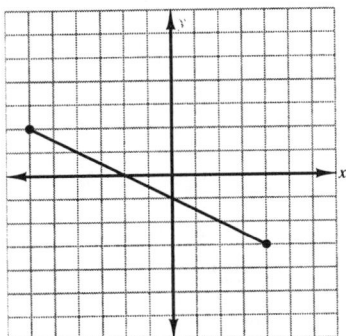

$$\begin{aligned} d &= \sqrt{(-6-4)^2 + (2+3)^2} \\ &= \sqrt{(-10)^2 + (5)^2} \\ &= \sqrt{100 + 25} \\ &= \sqrt{125} \\ &= \sqrt{25(5)} \\ &= \boxed{5\sqrt{5}} \end{aligned}$$

13. $(3, 2)$ and $(6, 7)$

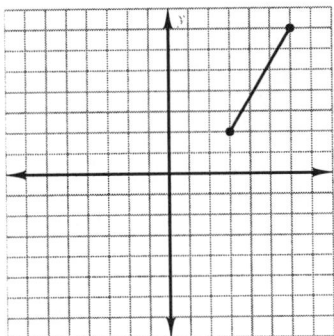

$$\begin{aligned} d &= \sqrt{(6-3)^2 + (7-2)^2} \\ &= \sqrt{9 + 25} \\ &= \boxed{\sqrt{34}} \end{aligned}$$

15. $(0, 0)$ and $(5, -12)$

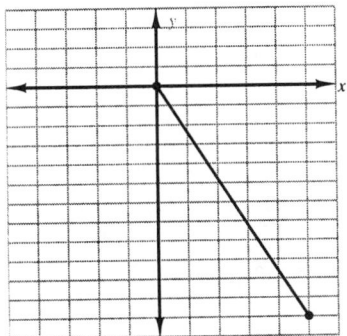

$$\begin{aligned} d &= \sqrt{(0-5)^2 + (0+12)^2} \\ &= \sqrt{25 + 144} \\ &= \sqrt{169} \\ &= \boxed{13} \end{aligned}$$

17. $(0, -3)$ and $(-3, 3)$

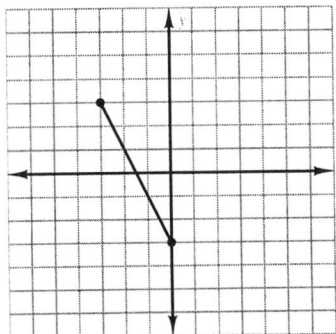

$$\begin{aligned} d &= \sqrt{(-3-0)^2 + (3+3)^2} \\ &= \sqrt{9 + 36} \\ &= \sqrt{45} \\ &= \boxed{3\sqrt{5}} \end{aligned}$$

19. (1, –2) and (–3, 6)

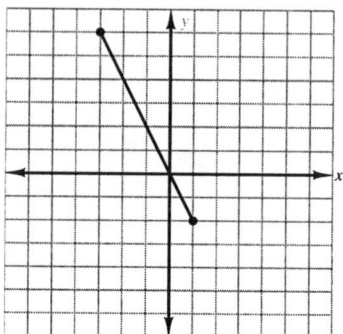

$$d = \sqrt{(-3-1)^2 + (6+2)^2}$$
$$= \sqrt{16 + 64}$$
$$= \sqrt{80}$$
$$= \boxed{4\sqrt{5}}$$

21. A(5, 7), B(1, 10), C(–3, –8)

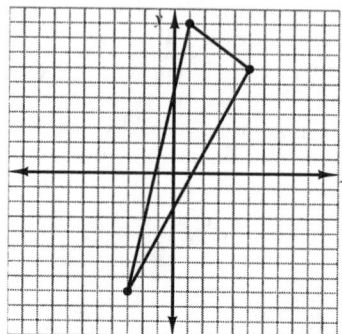

$$AB = \sqrt{(1-5)^2 + (10-7)^2}$$
$$= \sqrt{16 + 9}$$
$$= \sqrt{25}$$
$$= 5$$

$$CB = \sqrt{(1+3)^2 + (10+8)^2}$$
$$= \sqrt{16 + 324}$$
$$= \sqrt{340}$$
$$= \sqrt{4 \cdot 85}$$
$$= 2\sqrt{85}$$

$$CA = \sqrt{(5+3)^2 + (7+8)^2}$$
$$= \sqrt{64 + 225}$$
$$= \sqrt{289}$$
$$= 17$$

$$\text{Perimeter} = 5 + 2\sqrt{85} + 17$$
$$= \boxed{22 + 2\sqrt{85}}$$

23. A(2, 3), B(–1, –1), C(3, –4)

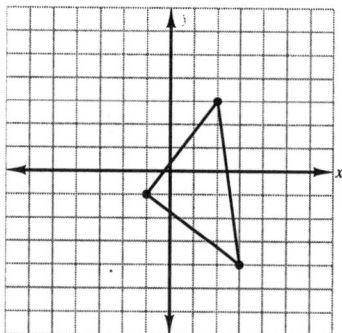

$$AB = \sqrt{(2+1)^2 + (3+1)^2}$$
$$= \sqrt{(9 + 16)}$$
$$= \sqrt{25}$$
$$= 5$$

$$AC = \sqrt{(2-3)^2 + (3+4)^2}$$
$$= \sqrt{1+49}$$
$$= \sqrt{50}$$
$$= 5\sqrt{2}$$
$$BC = \sqrt{(-1-3)^2 + (-1+4)^2}$$
$$= \sqrt{16+9}$$
$$= \sqrt{25}$$
$$= 5$$

Since the two sides have equal measure $(AB = BC)$, the triangle is isosceles.

25. $(5, y)$ and $(5, 1)$:

$$\sqrt{(5-5)^2 + (y_1-1)^2} = 8$$
$$\sqrt{(y_1-1)^2} = 8$$
$$(y_1-1)^2 = 64$$
$$y_1 - 1 = \pm\sqrt{64}$$
$$y_1 - 1 = 8 \quad \text{or} \quad y_1 - 1 = -8$$
$$\boxed{y_1 = 9 \quad \text{or} \quad y_1 = -7}$$

27. $(x_1, 3)$ and $(0, 0)$:

$$\sqrt{(x_1-0)^2 + (3-0)^2} = 4$$
$$\sqrt{x_1{}^2 + 9} = 4$$
$$x_1{}^2 + 9 = 16$$
$$x_1{}^2 = 7$$
$$\boxed{x_1 = \pm\sqrt{7}}$$

29.
$$AB = \sqrt{(1+4)^2 + (0+6)^2}$$
$$= \sqrt{25+36}$$
$$= \sqrt{61}$$
$$BC = \sqrt{(11-1)^2 + (12-0)^2}$$
$$= \sqrt{100+144}$$
$$= \sqrt{244}$$
$$= \sqrt{4\cdot61}$$
$$= 2\sqrt{61}$$
$$AC = \sqrt{(11+4)^2 + (12+6)^2}$$
$$= \sqrt{225+324}$$
$$= \sqrt{549}$$
$$= \sqrt{9\cdot61}$$
$$= 3\sqrt{61}$$
$$AB + BC = \sqrt{61} + 2\sqrt{61} = 3\sqrt{61} = AC$$

Thus A, B, and C are collinear.

31.
$$x^4 - 8x^2 - 9 = 0$$
$$(x^2-9)(x^2+1) = 0$$
$$x^2 - 9 = 0 \quad \text{or} \quad x^2 + 1 = 0$$
$$x^2 = 9 \qquad\qquad x^2 = -1$$
$$x = \pm3 \qquad\qquad x = \pm i$$

Check: $x = 3$ or $x = -3$, $x^2 = 9$
$$9^2 - 8(9) - 9 = 0$$
$$81 - 72 - 9 = 0$$
$$81 - 81 = 0$$
$$0 = 0 \quad \text{True}$$

Check: $x = i$ or $x = -i$, $x^2 = 1$
$$(-1)^2 - 8(-1) - 9 = 0$$
$$1 + 8 - 9 = 0$$
$$9 - 9 = 0$$
$$0 = 0 \quad \text{True}$$

The solutions check.
$$\boxed{\{-3, 3, -i, i\}}$$

For 33-37, the check is left to the student.

33.
$$x^4 + x^2 - 12 = 0$$
$$(x^2 + 4)(x^2 - 3) = 0$$
$$x^2 + 4 = 0 \quad \text{or} \quad x^2 - 3 = 0$$
$$x^2 = -4 \qquad\qquad x^2 = 3$$
$$x = \pm 2i \qquad\qquad x = \pm\sqrt{3}$$

$$\boxed{\{-\sqrt{3}, \sqrt{3}, -2i, 2i\}}$$

35.
$$2x^4 + x^2 - 6 = 0$$
$$(2x^2 - 3)(x^2 + 2) = 0$$
$$2x^2 - 3 = 0 \quad \text{or} \quad x^2 + 2 = 0$$
$$x^2 = \frac{3}{2} \qquad\qquad x^2 = -2$$
$$x = \pm\sqrt{\frac{3}{2}} \qquad\qquad x = \pm i\sqrt{2}$$
$$= \pm\frac{\sqrt{3}}{\sqrt{2}} \cdot \frac{\sqrt{2}}{\sqrt{2}}$$
$$= \pm\frac{\sqrt{6}}{2}$$

$$\boxed{\left\{-\frac{\sqrt{6}}{2}, \frac{\sqrt{6}}{2}, -i\sqrt{2}, i\sqrt{2}\right\}}$$

37.
$$x^4 - 4 = 0$$
$$(x^2 - 2)(x^2 + 2) = 0$$
$$x^2 - 2 = 0 \quad \text{or} \quad x^2 + 2 = 0$$
$$x^2 = 2 \qquad\qquad x^2 = -2$$
$$x = \pm\sqrt{2} \qquad\qquad x = \pm i\sqrt{2}$$

$$\boxed{\{\sqrt{2}, \sqrt{2}, -i\sqrt{2}, i\sqrt{2}\}}$$

39.
$$(x + 7)^2 = 9$$
$$x + 7 = \pm 3$$
$$x + 7 = 3 \quad \text{or} \quad x + 7 = -3$$
$$x = -4 \qquad\qquad x = -10$$
$$\boxed{\{-10, -4\}}$$

41.
$$(3y - 1)^2 = 16$$
$$3y - 1 = \pm 4$$
$$3y - 1 = 4 \quad \text{or} \quad 3y - 1 = -4$$
$$3y = 5 \qquad\qquad 3y = -3$$
$$y = \frac{5}{3} \qquad\qquad y = -1$$

$$\boxed{\left\{-1, \frac{5}{3}\right\}}$$

43.
$$(2x + 7)^2 = 5$$
$$2x + 7 = \pm\sqrt{5}$$
$$2x + 7 = \sqrt{5} \quad \text{or} \quad 2x + 7 = -\sqrt{5}$$
$$2x = -7 + \sqrt{5} \qquad\qquad 2x = -7 - \sqrt{5}$$
$$x = \frac{-7 + \sqrt{5}}{2} \qquad\qquad x = \frac{-7 - \sqrt{5}}{2}$$

$$\boxed{\left\{\frac{-7 - \sqrt{5}}{2}, \frac{-7 + \sqrt{5}}{2}\right\}}$$

45.
$$(5y - 4)^2 = 24$$
$$5y - 4 = \pm\sqrt{24}$$
$$5y - 4 = \sqrt{24} \quad \text{or} \quad 5y - 4 = -\sqrt{24}$$
$$5y = 4 + 2\sqrt{6} \qquad\qquad 5y = 4 - 2\sqrt{6}$$
$$y = \frac{4 + 2\sqrt{6}}{5} \qquad\qquad y = \frac{4 - 2\sqrt{6}}{5}$$

$$\boxed{\left\{\frac{4 - 2\sqrt{6}}{5}, \frac{4 + 2\sqrt{6}}{5}\right\}}$$

47.
$$(x - 3)^2 = -4$$
$$x - 3 = \pm\sqrt{-4}$$
$$x - 3 = \sqrt{-4} \quad \text{or} \quad x - 3 = -\sqrt{-4}$$
$$x = 3 + 2i \qquad\qquad x = 3 - 2i$$

$$\boxed{\{3 - 2i, 3 + 2i\}}$$

49. $(2y + 5)^2 = -5$

$$2y + 5 = \pm\sqrt{-5}$$

$2y + 5 = \sqrt{-5}$ or $2y + 5 = -\sqrt{-5}$

$\qquad\quad 2y = -5 + i\sqrt{5}$ $\qquad\qquad\qquad 2y = -5 - i\sqrt{5}$

$\qquad\quad y = \dfrac{-5 + i\sqrt{5}}{2}$ $\qquad\qquad\qquad y = \dfrac{-5 - i\sqrt{5}}{2}$

$$\boxed{\left\{\dfrac{-5 - i\sqrt{5}}{2}, \dfrac{-5 + i\sqrt{5}}{2}\right\}}$$

51. $(3z - 2)^2 = -50$

$$3z - 2 = \pm\sqrt{-50}$$

$3z - 2 = \sqrt{-50}$ or $3z - 2 = -\sqrt{-50}$

$\qquad\quad 3z = 2 + 5i\sqrt{2}$ $\qquad\qquad\qquad 3z = 2 - 5i\sqrt{2}$

$\qquad\quad z = \dfrac{2 + 5i\sqrt{2}}{3}$ $\qquad\qquad\qquad z = \dfrac{2 - 5i\sqrt{2}}{3}$

$$\boxed{\left\{\dfrac{2 - 5i\sqrt{2}}{3}, \dfrac{2 + 5i\sqrt{2}}{3}\right\}}$$

53. $3(5x - 4)^2 - 81 = 0$

$$3(5x - 4)^2 = 81$$

$$(5x - 4)^2 = 27$$

$$5x - 4 = \pm\sqrt{27}$$

$5x - 4 = \sqrt{27}$ or $5x - 4 = -\sqrt{27}$

$\qquad\quad 5x = 4 + 3\sqrt{3}$ $\qquad\qquad\qquad 5x = 4 - 3\sqrt{3}$

$\qquad\quad x = \dfrac{4 + 3\sqrt{3}}{5}$ $\qquad\qquad\qquad x = \dfrac{4 - 3\sqrt{3}}{5}$

$$\boxed{\left\{\dfrac{4 - 3\sqrt{3}}{5}, \dfrac{4 + 3\sqrt{3}}{5}\right\}}$$

55. $4(2x + 5)^2 + 100 = 0$

$$4(2x + 5)^2 = -100$$

$$(2x + 5)^2 = -25$$

$$2x + 5 = \pm\sqrt{-25}$$

$2x + 5 = \sqrt{-25}$ or $2x + 5 = -\sqrt{-25}$

$\qquad\quad 2x = -5 + 5i$ $\qquad\qquad\qquad 2x = -5 - 5i$

$\qquad\quad x = \dfrac{-5 + 5i}{2}$ $\qquad\qquad\qquad x = \dfrac{-5 - 5i}{2}$

$$\boxed{\left\{\dfrac{-5 - 5i}{2}, \dfrac{-5 + 5i}{2}\right\}}$$

57.
$$4(5x - 3)^2 - 1 = 0$$
$$4(5x - 3)^2 = 1$$
$$(5x - 3)^2 = \frac{1}{4}$$
$$5x - 3 = \pm\sqrt{\frac{1}{4}}$$
$$5x - 3 = \pm\frac{1}{2}$$

$$5x - 3 = \frac{1}{2} \qquad \text{or} \qquad 5x - 3 = -\frac{1}{2}$$
$$5x = 3 + \frac{1}{2} \qquad\qquad 5x = 3 - \frac{1}{2}$$
$$5x = \frac{7}{2} \qquad\qquad\qquad 5x = \frac{5}{2}$$
$$x = \frac{7}{10} \qquad\qquad\qquad x = \frac{1}{2}$$

$$\boxed{\left\{\frac{1}{2}, \frac{7}{10}\right\}}$$

59.
$$\left(x - \frac{1}{3}\right)^2 = \frac{1}{3}$$
$$x - \frac{1}{3} = \pm\sqrt{\frac{1}{3}}$$

$$x - \frac{1}{3} = \sqrt{\frac{1}{3}} \qquad \text{or} \qquad x - \frac{1}{3} = -\sqrt{\frac{1}{3}}$$
$$x = \frac{1}{3} + \frac{1}{\sqrt{3}} \cdot \frac{\sqrt{3}}{\sqrt{3}} \qquad\qquad x = \frac{1}{3} - \frac{1}{\sqrt{3}} \cdot \frac{\sqrt{3}}{\sqrt{3}}$$
$$x = \frac{1}{3} + \frac{\sqrt{3}}{3} \qquad\qquad\qquad x = \frac{1}{3} - \frac{\sqrt{3}}{3}$$
$$x = \frac{1 + \sqrt{3}}{3} \qquad\qquad\qquad x = \frac{1 - \sqrt{3}}{3}$$

$$\boxed{\left\{\frac{1 - \sqrt{3}}{3}, \frac{1 + \sqrt{3}}{3}\right\}}$$

61. $\boxed{D}$ is true; A is *not* true:
$$(x - 5)^2 = 12$$
$$x - 5 = \pm\sqrt{12}$$
$$x - 5 = \pm 2\sqrt{3}$$
B is *not* true: $(x + 5)^2 = 0$ has one solution, *not* two.
C is *not* true: The square root method *may* be applied as the first step.
D is true.

63. A(–3, 6), B(2, 3), C(11, 2), D(6, 11)

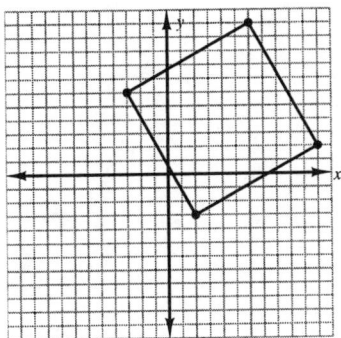

a. AD: $m = \dfrac{11 - 6}{6 + 3}$

$\qquad\qquad = \dfrac{5}{9}$

BC: $m = \dfrac{2 + 3}{11 - 2}$

$\qquad\qquad = \dfrac{5}{9}$

AD is parallel to BC.

AB: $m = \dfrac{-3 - 6}{2 + 3}$

$\qquad\qquad = -\dfrac{9}{5}$

DC: $m = \dfrac{11 - 2}{6 - 11}$

$\qquad\qquad = -\dfrac{9}{5}$

AB is parallel to DC.

b. AD: $d = \sqrt{(11 - 6)^2 + (6 + 3)^2}$

$\qquad\qquad = \sqrt{25 + 81}$

$\qquad\qquad = \sqrt{106}$

AB: $d = \sqrt{(2 + 3)^2 + (-3 - 6)^2}$

$\qquad\qquad = \sqrt{25 + 81}$

$\qquad\qquad = \sqrt{106}$

A parallelogram with two equal adjacent sides is a rhombus.

c. AB: $m = -\dfrac{9}{5}$

BC: $m = \dfrac{5}{9}$

Since the slopes are negative reciprocals, AB and BC are perpendicular.

65. $P(x_1, y_1)$; $(-1, 0)$ and $(-1, 5)$

$$\sqrt{(x_1 + 1)^2 + (y_1 - 0)^2} = \sqrt{(x_1 + 1)^2 + (y_1 - 5)^2}$$
$$(x_1 + 1)^2 + y_1^2 = (x_1 + 1)^2 + (y_1 - 5)^2$$
$$y_1^2 = (y_1 - 5)^2$$
$$y_1^2 = y_1^2 - 10y_1 + 25$$
$$0 = -10y_1 + 25$$
$$y_1 = \frac{25}{10}$$
$$y_1 = \boxed{\frac{5}{2}}$$

67. $A(0, 0)$, $B(a, 0)$, $C(b, d)$, $D(a + b, a)$

   a. AB: $m = \dfrac{0 - 0}{a - 0}$

             $= 0$

       CD: $m = \dfrac{d - b}{a + b - a}$

             $= \dfrac{0}{a}$

             $= 0$

   AB and CD are parallel.

       AC: $m = \dfrac{d - 0}{b - 0}$

             $= \dfrac{d}{b}$

       BD: $m = \dfrac{d - 0}{a + b - a}$

             $= \dfrac{d}{b}$

   AC and BD are parallel.

   b. 1. $AB = a$ $AC = \sqrt{b^2 + d^2}$

       2. $\dfrac{d}{a + b} \cdot \dfrac{d}{b - a} = -1$

            $\dfrac{d^2}{b^2 - a^2} = -1$

               $d^2 = a^2 - b^2$

       3. $AC = \sqrt{b^2 + d^2}$

            $= \sqrt{b^2 + a^2 - b^2}$

            $= \sqrt{a^2}$

            $= a$

   Thus $AC = AB$, and ABCD is a rhombus

69. $\left| x^2 + 6 \right| = 2$

$$x^2 + 6 = 2 \quad \text{or} \quad x^2 + 6 = -2$$
$$x^2 = -4 \qquad\qquad\quad x^2 = -8$$
$$x^2 = \pm\sqrt{-4} \qquad\qquad x^2 = \pm\sqrt{-8}$$
$$x^2 = \pm 2i \qquad\qquad\quad x^2 = \pm 2i\sqrt{2}$$

$$\boxed{\{-2i\sqrt{2},\, 2i\sqrt{2},\, -2i,\, 2i\,\}}$$

Review Problems

75. $\dfrac{y^2 - 2y + 1}{3y^2 + 7y - 20} \cdot \dfrac{3y^2 - 2y - 5}{y^2 + 3y - 4} \div \dfrac{y^2 - 4y + 3}{y + 4}$

$= \dfrac{(y - 1)(y - 1)}{(3y - 5)(y + 4)} \cdot \dfrac{(3y - 5)(y + 1)}{(y + 4)(y - 1)} \cdot \dfrac{(y + 4)}{(y - 3)(y - 1)}$

$= \boxed{\dfrac{y + 1}{(y + 4)(y - 3)}}$

76.

$$\frac{2}{\sqrt{3}+1} = \frac{2}{\sqrt{3}+1} \cdot \frac{\sqrt{3}-1}{\sqrt{3}-1}$$

$$= \frac{2(\sqrt{3}-1)}{3-1}$$

$$= \frac{2(\sqrt{3}-1)}{2}$$

$$= \boxed{\sqrt{3}-1}$$

77. Let

$$x = \text{width of rectangle}$$
$$x+4 = \text{length of rectangle}$$
$$\text{Area} = 96 \text{ square yards}$$

$$\begin{aligned}
x(x+4) &= 96 \\
x^2 + 4x - 96 &= 0 \\
(x+12)(x-8) &= 0 \\
x+12 &= 0 \qquad\qquad \text{or} \qquad\qquad x-8 = 0 \\
x &= -12 \text{ (reject)} \qquad\qquad\qquad x = 8 \\
x+4 &= 8+4 \\
x+4 &= 12
\end{aligned}$$

$$\boxed{\text{Width: 8 yd}\quad \text{Length: 12 yd}}$$

Section 8.4 Solving Quadratic Equations by Completing the Square

Problem Set 8.4, p. 611

1.
$$\begin{aligned}
x^2 - 4x &= 21 \\
x^2 - 4x + 4 &= 21 + 4 \\
(x-2)^2 &= 25 \\
x-2 &= \pm 5 \\
x &= 2 \pm 5 \\
x = 7 \quad &\text{or} \quad x = -3
\end{aligned}$$

$$\boxed{\{-3, 7\}}$$

3.
$$\begin{aligned}
x(x-6) &= 16 \\
x^2 - 6x + 9 &= 16 + 9 \\
(x-3)^2 &= 25 \\
x-3 &= \pm 5 \\
x &= 3 \pm 5 \\
x = 8 \quad &\text{or} \quad x = -2
\end{aligned}$$

$$\boxed{\{-2, 8\}}$$

5.

$$2y^2 - 5y = 3$$

$$y^2 - \frac{5}{2}y = \frac{3}{2}$$

$$y^2 - \frac{5}{2}y + \frac{25}{16} = \frac{3}{2} + \frac{25}{16} \quad \left[\text{Note: } \left(\frac{1}{2}\right)\left(-\frac{5}{2}\right) = -\frac{5}{2} \text{ and } \left(-\frac{5}{2}\right)^2 = \frac{25}{16} \right]$$

$$\left(y - \frac{5}{4}\right)^2 = \frac{24}{16} + \frac{25}{16}$$

$$\left(y - \frac{5}{4}\right)^2 = \frac{49}{16}$$

$$y - \frac{5}{4} = \pm\frac{7}{4}$$

$$y = \frac{5}{4} \pm \frac{7}{4}$$

$$y = \frac{12}{4} \quad \text{or} \quad y = -\frac{2}{4}$$

$$y = 3 \qquad\qquad y = -\frac{1}{2}$$

$$\boxed{\left\{ -\frac{1}{2}, 3 \right\}}$$

7.

$$9z^2 - 30z + 25 = 0$$

$$z^2 - \frac{30}{9}z + \frac{25}{9} = -\frac{25}{9} + \frac{25}{9} \quad \left[\frac{1}{2}\left(-\frac{10}{3}\right) = -\frac{5}{3} \text{ and } \left(-\frac{5}{3}\right)^2 = \frac{25}{9} \right]$$

$$\left(z - \frac{5}{3}\right)^2 = 0$$

$$z = \frac{5}{3}$$

$$\boxed{\left\{ \frac{5}{3} \right\}}$$

9.

$$y^2 - 6y + 2 = 0$$

$$y^2 - 6y = -2$$

$$y^2 - 6y + 9 = -2 + 9 \quad \left[\left(\frac{1}{2}\right)(-6) = -3 \text{ and } (-3)^2 = 9 \right]$$

$$(y - 3)^2 = 7$$

$$y - 3 = \pm\sqrt{7}$$

$$y = 3 \pm \sqrt{7}$$

$$\boxed{\left\{ 3 + \sqrt{7}, 3 - \sqrt{7} \right\}}$$

11.

$$x^2 + x - 1 = 0$$

$$x^2 + x + \frac{1}{4} = 1 + \frac{1}{4}$$

$$\left(x + \frac{1}{2}\right)^2 = \frac{5}{4} \quad \left[\tfrac{1}{2}(1) = \tfrac{1}{2} \text{ and } \left(\tfrac{1}{2}\right)^2 = \tfrac{1}{4}\right]$$

$$x + \frac{1}{2} = \pm\frac{\sqrt{5}}{2}$$

$$x = -\frac{1}{2} \pm \frac{\sqrt{5}}{2}$$

$$x = \frac{-1 \pm \sqrt{5}}{2}$$

$$\boxed{\left\{\frac{-1 + \sqrt{5}}{2}, \frac{-1 - \sqrt{5}}{2}\right\}}$$

13.

$$2z^2 + z = 5$$

$$z^2 + \frac{1}{2}z = \frac{5}{2}$$

$$z^2 + \frac{1}{2}z + \frac{1}{16} = \frac{5}{2} + \frac{1}{16} \quad \left[\text{Note: } \tfrac{1}{2}\left(\tfrac{1}{2}\right) = \tfrac{1}{4} \text{ and } \left(\tfrac{1}{4}\right)^2 = \tfrac{1}{16}\right]$$

$$\left(z + \frac{1}{4}\right)^2 = \frac{41}{16}$$

$$z + \frac{1}{4} = \pm\sqrt{\frac{41}{16}}$$

$$z + \frac{1}{4} = \pm\frac{\sqrt{41}}{4}$$

$$z = -\frac{1}{4} \pm \frac{\sqrt{41}}{4}$$

$$z = \frac{-1 \pm \sqrt{41}}{4}$$

$$\boxed{\left\{\frac{-1 + \sqrt{41}}{4}, \frac{-1 - \sqrt{41}}{4}\right\}}$$

15.

$$2x^2 - 2x = 3$$

$$x^2 - x + \frac{1}{4} = \frac{3}{2} + \frac{1}{4} \quad \left[\tfrac{1}{2}(-1) = -\tfrac{1}{2} \text{ and } \left(-\tfrac{1}{2}\right)^2 = \tfrac{1}{4}\right]$$

$$\left(x - \frac{1}{2}\right)^2 = \frac{7}{4}$$

$$x - \frac{1}{2} = \pm\frac{\sqrt{7}}{2}$$

$$x = \frac{1 \pm \sqrt{7}}{2}$$

$$\boxed{\left\{\frac{1 + \sqrt{7}}{2}, \frac{1 - \sqrt{7}}{2}\right\}}$$

17.
$$y^2 + 2y + 2 = 0$$
$$y^2 + 2y + 1 = -2 + 1$$
$$(y+1)^2 = -1$$
$$y + 1 = \pm\sqrt{-1}$$
$$y + 1 = -i$$
$$y = -1 \pm i$$

(For practice) check $-1 - i$:
$$(-1 - i)^2 + 2(-1 - i) + 2 = 0$$
$$1 + 2i + i^2 - 2 - 2i + 2 = 0$$
$$1 + i^2 = 0$$
$$1 + (-1) = 0$$
$$0 = 0$$

$$\boxed{\{-1 + i, -1 - i\}}$$

19.
$$x^2 - x + 1 = 0$$
$$x^2 - x + \frac{1}{4} = -1 + \frac{1}{4} \quad \left[\tfrac{1}{2}(-1) = -\tfrac{1}{2} \text{ and } \left(-\tfrac{1}{2}\right)^2 = \tfrac{1}{4}\right]$$
$$\left(x - \frac{1}{2}\right)^2 = -\frac{3}{4}$$
$$x - \frac{1}{2} = \pm\sqrt{-\frac{3}{4}}$$
$$x - \frac{1}{2} = \pm\frac{\sqrt{3}\,i}{2}$$
$$x = \frac{1 \pm \sqrt{3}\,i}{2}$$

$$\boxed{\left\{\frac{1 + \sqrt{3}\,i}{2}, \frac{1 - \sqrt{3}\,i}{2}\right\}}$$

21.
$$8z^2 - 4z = -1$$
$$z^2 - \frac{1}{2}z = -\frac{1}{8} \quad \left[\text{Note: } \tfrac{1}{2}\left(-\tfrac{1}{2}\right) = -\tfrac{1}{4}; \left(-\tfrac{1}{4}\right)^2 = \tfrac{1}{16}\right]$$
$$z^2 - \frac{1}{2}z + \frac{1}{16} = -\frac{1}{8} + \frac{1}{16}$$
$$\left(z - \frac{1}{4}\right)^2 = -\frac{1}{16}$$
$$z - \frac{1}{4} = \pm\sqrt{-\frac{1}{16}}$$
$$z - \frac{1}{4} = \pm\frac{i}{4}$$
$$z = \frac{1}{4} \pm \frac{i}{4}$$
$$= \frac{1 \pm i}{4}$$

$$\boxed{\left\{\frac{1 + i}{4}, \frac{1 - i}{4}\right\}}$$

23.
$$3y^2 + 2y + 4 = 0$$
$$y^2 + \frac{2}{3}y + \frac{4}{3} = 0$$
$$y^2 + \frac{2}{3}y = -\frac{4}{3}$$

$$\left[\text{Note: } \tfrac{1}{2}\left(\tfrac{2}{3}\right) = \tfrac{1}{3}; \left(\tfrac{1}{3}\right)^2 = \tfrac{1}{9}\right]$$

$$y^2 + \frac{2}{3}y + \frac{1}{9} = -\frac{4}{3} + \frac{1}{9}$$
$$\left(y + \frac{1}{3}\right)^2 = -\frac{12}{9} + \frac{1}{9}$$
$$\left(y + \frac{1}{3}\right)^2 = -\frac{11}{9}$$
$$y + \frac{1}{3} = \pm\sqrt{-\frac{11}{9}}$$
$$= \pm\frac{\sqrt{11}\,i}{3}$$
$$y = -\frac{1}{3} \pm \frac{\sqrt{11}\,i}{3}$$
$$= \frac{-1 \pm \sqrt{11}\,i}{3}$$

$$\boxed{\left\{\frac{-1 + \sqrt{11}\,i}{3}, \frac{-1 - \sqrt{11}\,i}{3}\right\}}$$

25. Let x = side of square

$$\text{Area} = \text{side} + 3\frac{3}{4}$$
$$x^2 = x + \frac{15}{4}$$
$$x^2 - x + \frac{1}{4} = \frac{15}{4} + \frac{1}{4}$$
$$\left(x - \frac{1}{2}\right)^2 = 4$$
$$x - \frac{1}{2} = \pm\frac{1}{2}$$
$$x = \frac{1}{2} \pm 2$$
$$x = \frac{5}{2} \quad\text{or}\quad x = -\frac{3}{2} \text{ (reject; the square's side cannot be negative)}$$

Side of square: $\boxed{\dfrac{5}{2}}$

Review Problems

29.
$$6\sqrt{\frac{1}{2}} + 24\sqrt{\frac{1}{8}} - 3\sqrt{2} = \frac{6}{\sqrt{2}} \cdot \frac{\sqrt{2}}{\sqrt{2}} + \frac{24}{2\sqrt{2}} \cdot \frac{\sqrt{2}}{\sqrt{2}} - 3\sqrt{2}$$
$$= \frac{6\sqrt{2}}{2} + \frac{24\sqrt{2}}{4} - 3\sqrt{2}$$
$$= 3\sqrt{2} + 6\sqrt{2} - 3\sqrt{2}$$
$$= \boxed{6\sqrt{2}}$$

30.
$$\frac{3y+2}{y+5} + \frac{32y-9}{2y^2+7y-15} + \frac{y-2}{3-2y} = \frac{3y+2}{y+5} + \frac{32y-9}{(y+5)(2y-3)} + \frac{y-2}{3-2y}$$

$$= \frac{3y+2}{y+5} \cdot \frac{2y-3}{2y-3} + \frac{32y-9}{(y+5)(2y-3)} + \frac{y-2}{3-2y} \cdot \frac{-1}{-1} \cdot \frac{y+5}{y+5}$$

$$= \frac{(3y+2)(2y-3) + 32y - 9 - (y-2)(y+5)}{(y+5)(2y-3)}$$

$$= \frac{6y^2 - 5y - 6 + 32y - 9 - y^2 - 3y + 10}{(y+5)(2y-3)}$$

$$= \frac{5y^2 + 24y - 5}{(y+5)(2y-3)}$$

$$= \frac{(5y-1)(y+5)}{(y+5)(2y-3)}$$

$$= \boxed{\frac{5y-1}{2y-3}}$$

31.
$$\frac{y-5}{6} - \frac{1}{3} < \frac{y+2}{5}$$

$$30\left(\frac{y-5}{6} - \frac{1}{3}\right) < 30\left(\frac{y+2}{5}\right)$$

$$5(y-5) - 10 < 6(y+2)$$

$$5y - 25 - 10 < 6y + 12$$

$$5y - 35 < 6y + 12$$

$$-y < 47$$

$$y > -47$$

$$\boxed{\{\, y \mid y > -47 \,\}}$$

Section 8.5 Solving Quadratic Equations by the Quadratic Formula

Problem Set 8.5, pp. 622-625

1.
$$y^2 - 3y = 10$$
$$y^2 - 3y - 10 = 0$$
$$a = 1, \ b = -3, \ c = -10$$

$$y = \frac{-b \pm \sqrt{b^2 - 4ac}}{2a}$$

$$= \frac{3 \pm \sqrt{9 - 4(1)(-10)}}{4}$$

$$= \frac{3 \pm \sqrt{49}}{2}$$

$$= \frac{3 \pm 7}{2}$$

$$x = \frac{10}{2} \qquad \text{or} \qquad x = -\frac{4}{2}$$
$$= 5 \qquad\qquad\qquad\qquad = -2$$

$$\boxed{\{-2, 5\}}$$

3.
$$3 + \frac{7}{x} = \frac{6}{x^2}$$
$$x^2\left(3 + \frac{7}{x}\right) = x^2\left(\frac{6}{x^2}\right)$$
$$3x^2 + 7x - 6 = 0$$

$a = 3,\ b = 7,\ c = -6$

$$x = \frac{-b \pm \sqrt{b^2 - 4ac}}{2a}$$
$$= \frac{-7 \pm \sqrt{49 - 4(3)(-6)}}{6}$$
$$= \frac{-7 \pm \sqrt{49 - (-72)}}{6}$$
$$= \frac{-7 \pm \sqrt{121}}{6}$$
$$= \frac{-7 \pm \sqrt{121}}{6}$$
$$= \frac{-7 \pm 11}{6}$$

$$x = \frac{-7 + 11}{6} \qquad \text{or} \qquad x = \frac{-7 - 11}{6}$$
$$= \frac{4}{6} \qquad\qquad\qquad\qquad = \frac{-18}{6}$$
$$= \frac{2}{3} \qquad\qquad\qquad\qquad = -3$$

$$\boxed{\left\{-3, \frac{2}{3}\right\}}$$

5.
$$2z^2 - 7z = 1$$
$$2z^2 - 7z - 1 = 0$$
$$z = \frac{7 \pm \sqrt{49 - 4(2)(-1)}}{4}$$
$$= \frac{7 \pm \sqrt{57}}{4}$$

$$\boxed{\left\{\frac{7 + \sqrt{57}}{4}, \frac{7 - \sqrt{57}}{4}\right\}}$$

7.
$$2y^2 + y = 5$$
$$2y^2 + y - 5 = 0$$
$$y = \frac{-1 \pm \sqrt{1 - 4(2)(-5)}}{4}$$
$$= \frac{-1 \pm \sqrt{41}}{4}$$

$$\boxed{\left\{\frac{-1 + \sqrt{41}}{4}, \frac{-1 - \sqrt{41}}{4}\right\}}$$

9.
$$\frac{5}{x^2} = 3 + \frac{8}{x}$$
$$x^2\left(\frac{5}{x^2}\right) = x^2\left(3 + \frac{8}{x}\right)$$
$$5 = 3x^2 + 8x$$
$$0 = 3x^2 + 8x - 5$$

$$a = 3, \quad b = 8, \quad c = -5$$
$$x = \frac{-b \pm \sqrt{b^2 - 4ac}}{2a}$$
$$= \frac{-8 \pm \sqrt{64 - 4(3)(-5)}}{6}$$
$$= \frac{-8 \pm \sqrt{64 - (-60)}}{6}$$
$$= \frac{-8 \pm \sqrt{124}}{6}$$
$$= \frac{-8 \pm \sqrt{4 \cdot 31}}{6}$$
$$= \frac{-8 \pm 2\sqrt{31}}{6}$$
$$= \frac{2(-4 \pm \sqrt{31})}{6}$$
$$= \frac{-4 \pm \sqrt{31}}{3}$$
$$\boxed{\left\{\frac{-4 + \sqrt{31}}{3}, \frac{-4 - \sqrt{31}}{3}\right\}}$$

11.
$$2z^2 = 2z - 1$$
$$2z^2 - 2z + 1 = 0$$
$$a = 2, \quad b = -2, \quad c = 1$$
$$z = \frac{-b \pm \sqrt{b^2 - 4ac}}{2a}$$
$$= \frac{-(-2) \pm \sqrt{4 - 4(2)(1)}}{4}$$
$$= \frac{2 \pm \sqrt{-4}}{4}$$
$$= \frac{2 \pm 2i}{4}$$
$$= \frac{2(1 \pm i)}{4}$$
$$= \frac{1 \pm i}{2}$$
$$\boxed{\left\{\frac{1 + i}{2}, \frac{1 - i}{2}\right\}}$$

13.
$$5y^2 = 2y - 3$$
$$5y^2 - 2y + 3 = 0$$
$$y = \frac{2 \pm \sqrt{4 - 4(5)(3)}}{10}$$
$$= \frac{2 \pm \sqrt{-56}}{10}$$
$$= \frac{2 \pm 2\sqrt{14}\, i}{10}$$
$$= \frac{1 \pm \sqrt{14}\, i}{5}$$
$$\boxed{\left\{\frac{1 + \sqrt{14}\, i}{5}, \frac{1 - \sqrt{14}\, i}{5}\right\}}$$

15.
$$4 + \frac{2}{y} = -\frac{5}{y^2}$$
$$y^2\left(4 + \frac{2}{y}\right) = y^2\left(-\frac{5}{y^2}\right)$$
$$4y^2 + 2y + 5 = 0$$
$$y = \frac{-2 \pm \sqrt{4 - 4(4)(5)}}{8}$$
$$= \frac{-2 \pm \sqrt{-76}}{8}$$
$$= \frac{-2 \pm 2\sqrt{19}\, i}{8}$$
$$= \frac{-1 \pm \sqrt{19}\, i}{4}$$
$$\boxed{\left\{\frac{-1 + \sqrt{19}\, i}{4}, \frac{-1 - \sqrt{19}\, i}{4}\right\}}$$

17.

$$\frac{3}{z} = \frac{2z}{z-1}$$

$$3(z-1) = z(2z)$$

$$3z - 3 = 2z^2$$

$$0 = 2z^2 - 3z + 3$$

$$a = 2, \ b = -3, \ c = 3$$

$$z = \frac{-b \pm \sqrt{b^2 - 4ac}}{2a}$$

$$= \frac{-(-3) \pm \sqrt{9 - 4(2)(3)}}{4}$$

$$= \frac{3 \pm \sqrt{-15}}{4}$$

$$= \frac{3 \pm \sqrt{15}\, i}{4}$$

$$\boxed{\left\{ \frac{3 + \sqrt{15}\, i}{4}, \frac{3 - \sqrt{15}\, i}{4} \right\}}$$

19.

$$8y^3 - 1 = 0$$

$$(2y)^3 - 1^3 = 0$$

$$(2y - 1)[(2y^2 + (2y)(1) + 1^2] = 0$$

$$(2y - 1)(4y^2 + 2y + 1) = 0$$

$$2y - 1 = 0 \qquad \text{or} \qquad 4y^2 + 2y + 1 = 0$$

$$y = \frac{1}{2} \qquad\qquad y = \frac{-2 \pm \sqrt{4 - 4(4)(1)}}{8}$$

$$= \frac{-2 \pm \sqrt{-12}}{8}$$

$$= \frac{-2 \pm \sqrt{4(3)(-1)}}{8}$$

$$= \frac{-2 \pm 2\sqrt{3}\, i}{8}$$

$$= \frac{-1 \pm \sqrt{3}\, i}{4}$$

$$\boxed{\left\{ \frac{-1 + \sqrt{3}\, i}{4}, \frac{-1 - \sqrt{3}\, i}{4}, \frac{1}{2} \right\}}$$

21.

$$4(y^2 + 4) = 9 + 12y$$
$$4y^2 + 16 = 9 + 12y$$
$$4y^2 - 12y + 7 = 0$$

$$y = \frac{12 \pm \sqrt{144 - 4(4)(7)}}{8}$$

$$= \frac{12 \pm \sqrt{32}}{8}$$

$$= \frac{12 \pm 4\sqrt{2}}{8}$$

$$= \frac{3 \pm \sqrt{2}}{2}$$

$$\boxed{\left\{\frac{3 + \sqrt{2}}{2}, \frac{3 - \sqrt{2}}{2}\right\}}$$

23.

$$x = \frac{25 - 5x}{3x + 3}$$
$$x(3x + 3) = 25 - 5x$$
$$3x^2 + 3x = 25 - 5x$$

$$3x^2 + 8x - 25 = 0$$

$$a = 3, \quad b = 8, \quad c = -25$$

$$x = \frac{-b \pm \sqrt{b^2 - 4ac}}{2a}$$

$$= \frac{-8 \pm \sqrt{64 - 4(3)(-25)}}{6}$$

$$= \frac{-8 \pm \sqrt{64 - (-300)}}{6}$$

$$= \frac{-8 \pm \sqrt{364}}{6}$$

$$= \frac{-8 \pm \sqrt{4 \cdot 91}}{6}$$

$$= \frac{-8 \pm 2\sqrt{91}}{6}$$

$$= \frac{-4 \pm \sqrt{91}}{3}$$

$$\boxed{\left\{\frac{-4 + \sqrt{91}}{3}, \frac{-4 - \sqrt{91}}{3}\right\}}$$

25.

$$\frac{1}{y^2 - 3y + 2} = \frac{1}{y + 2} + \frac{5}{y^2 - 4}$$

$$\frac{1}{(y - 2)(y - 1)} = \frac{1}{y + 2} + \frac{5}{(y + 2)(y - 2)}$$

$$(y + 2)(y - 2)(y - 1)\left[\frac{1}{(y - 2)(y - 1)}\right] = (y + 2)(y - 2)(y - 1)\left[\frac{1}{y + 2} + \frac{5}{(y + 2)(y - 2)}\right]$$

$$y + 2 = (y - 2)(y - 1) + 5(y - 1)$$
$$y + 2 = y^2 - 3y + 2 + 5y - 5$$
$$y + 2 = y^2 + 2y - 3$$
$$0 = y^2 + y - 5$$

$$a = 1, \quad b = 1, \quad c = -5$$

$$y = \frac{-b \pm \sqrt{b^2 - 4ac}}{2a}$$

$$= \frac{-1 \pm \sqrt{1 - 4(1)(-5)}}{2}$$

$$= \frac{-1 \pm \sqrt{21}}{2}$$

$$\boxed{\left\{\frac{-1 + \sqrt{21}}{2}, \frac{-1 - \sqrt{21}}{2}\right\}}$$

27. $\sqrt{2}x^2 + 3x - 2\sqrt{2} = 0$

$a = \sqrt{2}, \ b = 3, \ c = -2\sqrt{2}$

$$x = \frac{-3 \pm \sqrt{9 - 4(\sqrt{2})(-2\sqrt{2})}}{2\sqrt{2}}$$

$$= \frac{-3 \pm \sqrt{9 - 4(-2\sqrt{4})}}{2\sqrt{2}}$$

$$= \frac{-3 \pm \sqrt{9 - 4(-4)}}{2\sqrt{2}}$$

$$= \frac{-3 \pm \sqrt{25}}{2\sqrt{2}}$$

$$= \frac{-3 \pm 5}{2\sqrt{2}}$$

$x = \frac{2}{2\sqrt{2}}$ $\qquad$ or $\qquad$ $x = \frac{-8}{2\sqrt{2}}$

$x = \frac{1}{\sqrt{2}} \cdot \frac{\sqrt{2}}{\sqrt{2}}$ $\qquad\qquad$ $x = \frac{-4}{\sqrt{2}} \cdot \frac{\sqrt{2}}{\sqrt{2}}$

$x = \frac{\sqrt{2}}{2}$ $\qquad\qquad\qquad$ $x = -2\sqrt{2}$

$$\boxed{\left\{-2\sqrt{2}, \ \frac{\sqrt{2}}{2}\right\}}$$

29. $x^2 + \sqrt{7}x + 2 = 0$

$$x = \frac{-\sqrt{7} \pm \sqrt{7 - 4(1)(2)}}{2}$$

$$= \frac{-\sqrt{7} \pm \sqrt{-1}}{2}$$

$$= \frac{-\sqrt{7} \pm i}{2}$$

$$\boxed{\left\{\frac{-\sqrt{7} + i}{2}, \frac{-\sqrt{7} - i.}{2}\right\}}$$

31. $i x^2 - 5x + 2i = 0$

$a = i, \ b = -5, \ c = 2i$

$$x = \frac{-(-5) \pm \sqrt{25 - 4(i)(2i)}}{2i}$$

$$= \frac{5 \pm \sqrt{25 - 8i^2}}{2i}$$

$$= \frac{5 \pm \sqrt{25 - 8(-1)}}{2i}$$

$$= \frac{5 \pm \sqrt{33}}{2i}$$

$$= \frac{5 \pm \sqrt{33}}{2i} \cdot \frac{i}{i}$$

$$= \frac{5i \pm \sqrt{33} \, i}{2i^2}$$

$$= \frac{5i \pm \sqrt{33} \, i}{2(-1)}$$

$$= \frac{5i \pm \sqrt{33} \, i}{-2}$$

$$\boxed{\left\{\frac{5i + \sqrt{33} \, i}{-2}, \frac{5i - \sqrt{33} \, i}{-2}\right\}}$$

33. $\left| y^2 + 2y \right| = 3$

$$y^2 + 2y = 3$$
$$y^2 + 2y - 3 = 0$$
$$y = \frac{-2 \pm \sqrt{4 - 4(1)(-3)}}{2}$$
$$= \frac{-2 \pm \sqrt{16}}{2}$$
$$= \frac{-2 \pm 4}{2}$$
$$= 1, \ -3$$

or

$$y^2 + 2y = -3$$
$$y^2 + 2y + 3 = 0$$
$$y = \frac{-2 \pm \sqrt{4 - 4(1)(3)}}{2}$$
$$= \frac{-2 \pm \sqrt{-8}}{2}$$
$$= \frac{-2 \pm 2\sqrt{2}\,i}{2}$$
$$= -1 \pm \sqrt{2}\,i$$

$$\boxed{\{-3, 1, -1 + \sqrt{2}\,i, -1 - \sqrt{2}\,i\}}$$

35.
$$\frac{1}{y+1} - \frac{1}{y} = \frac{1}{2}$$
$$2y(y+1)\left[\frac{1}{y+1} - \frac{1}{y} \right] = 2y(y+1)\left(\frac{1}{2} \right)$$
$$2y - 2y - 2 = y^2 + y$$
$$0 = y^2 + y^2 + 2$$
$$y = \frac{-1 \pm \sqrt{1 - 4(1)(2)}}{2}$$
$$= \frac{-1 \pm \sqrt{-7}}{2}$$
$$= \frac{-1 \pm \sqrt{7}\,i}{2}$$

$$\boxed{\left\{ \frac{-1 + \sqrt{7}\,i}{2}, \frac{-1 - \sqrt{7}\,i}{2} \right\}}$$

37.
$$125z^2 - 1 = 0$$
$$(5z - 1)(25z^2 + 5z + 1) = 0$$
$$z = \frac{1}{5} \quad \text{or} \quad 25z^2 + 5z + 1 = 0$$
$$z = \frac{-5 \pm \sqrt{25 - 4(25)(1)}}{50}$$
$$= \frac{-5 \pm \sqrt{-75}}{50}$$
$$= \frac{-5 \pm 5\sqrt{3}\,i}{50}$$
$$= \frac{-1 \pm \sqrt{3}\,i}{10}$$

$$\boxed{\left\{ \frac{1}{5}, \frac{-1 + \sqrt{3}\,i}{10}, \frac{-1 - \sqrt{3}\,i}{10} \right\}}$$

39. $7x^2 = 2 - \sqrt{7}\,x$

$$7x^2 + \sqrt{7}\,x - 2 = 0$$
$$x = \frac{-\sqrt{7} \pm \sqrt{(\sqrt{7})^2 - 4(7)(-2)}}{2(7)}$$
$$= \frac{-\sqrt{7} \pm \sqrt{63}}{14}$$
$$= \frac{-\sqrt{7} \pm 3\sqrt{7}}{14}$$

$$x = \frac{-\sqrt{7} + 3\sqrt{7}}{14} \quad \text{or} \quad x = \frac{-\sqrt{7} - 3\sqrt{7}}{14}$$
$$= \frac{2\sqrt{7}}{14} \qquad\qquad\qquad = \frac{-4\sqrt{7}}{14}$$
$$= \frac{\sqrt{7}}{7} \qquad\qquad\qquad\quad = \frac{-2\sqrt{7}}{7}$$

$$\boxed{\left\{ \frac{-2\sqrt{7}}{7}, \frac{\sqrt{7}}{7} \right\}}$$

41. $x^2 - ix + 12 = 0$

$$x = \frac{-(-i) \pm \sqrt{(-i)^2 - 4(12)}}{2(1)}$$

$$= \frac{i \pm \sqrt{-49}}{2}$$

$$= \frac{i \pm 7i}{2}$$

$$x = \frac{i + 7i}{2} \quad \text{or} \quad x = \frac{i - 7i}{2}$$

$$= \frac{8i}{2} \qquad\qquad\qquad = \frac{-6i}{2}$$

$$= 4i \qquad\qquad\qquad = -3i$$

$$\boxed{\{-3i, 4i\}}$$

43.

$$ix^2 + 2 = (2 + 2i)x$$

$$ix^2 - (2 + 2i)x + 2 = 0$$

$$x = \frac{-[-(2 + 2i)] \pm \sqrt{[-(2 + 2i)]^2 - 4(i)(2)}}{2i}$$

$$= \frac{2 + 2i \pm \sqrt{4 + 8i + 4i^2 - 8i}}{2i}$$

$$= \frac{2 + 2i \pm \sqrt{4 + 4(-1)}}{2i}$$

$$= \frac{2 + 2i \pm 0}{2i}$$

$$= \frac{2 + 2i}{2i}$$

$$= \frac{1 + i}{i} \cdot \frac{i}{i}$$

$$= \frac{i + i^2}{i^2}$$

$$= \frac{i - 1}{-1}$$

$$= 1 - i$$

$$\boxed{\{1 - i\}}$$

45. $\boxed{\text{D}}$ is true

47. $(x^2 + 2x - 3)^2 + 6(x^2 + 2x - 3) + 8 = 0$

(Let $t = x^2 + 2x - 3$):

$$t^2 + 6t + 8 = 0$$

$$t = \frac{-6 \pm \sqrt{36 - 4(1)(8)}}{2}$$

$$= \frac{-6 \pm \sqrt{4}}{2}$$

$$= \frac{-6 \pm 2}{2}$$

$$t = -2 \qquad\qquad \text{or} \qquad\qquad t = -4$$
$$x^2 + 2x - 3 = -2 \qquad\qquad\qquad x^2 + 2x - 3 = -4$$
$$x^2 + 2x - 1 = 0 \qquad\qquad\qquad x^2 + 2x + 1 = 0$$

$$x = \frac{-2 \pm \sqrt{4 - 4(1)(-1)}}{2} \qquad\qquad x = \frac{-2 \pm \sqrt{4 - 4(1)(1)}}{2}$$

$$= \frac{-2 \pm \sqrt{8}}{2} \qquad\qquad\qquad = \frac{-2 \pm \sqrt{0}}{2}$$

$$= \frac{-2 \pm 2\sqrt{2}}{2} \qquad\qquad\qquad = -1$$

$$= -1 \pm \sqrt{2}$$

$$\boxed{\{-1, -1 + \sqrt{2}, -1 - \sqrt{2}\,\}}$$

49. $\left(y - \dfrac{3}{y}\right)^2 - \left(y - \dfrac{3}{y}\right) - 2 = 0$

$\left(\text{Let } t = y - \dfrac{3}{y}\right)$:

$$t^2 - t - 2 = 0$$

$$(t - 2)(t + 1) = 0$$

$$t = 2 \qquad\qquad \text{or} \qquad\qquad t = -1$$

$$y - \frac{3}{y} = 2 \qquad\qquad\qquad y - \frac{3}{y} = -1$$

$$y^2 - 3 = 2y \qquad\qquad\qquad y^2 - 3 = -y$$

$$y^2 - 2y - 3 = 0 \qquad\qquad\qquad y^2 + y - 3 = 0$$

$$(y - 3)(y + 1) = 0 \qquad\qquad\qquad y = \frac{-1 \pm \sqrt{1 - 4(1)(-3)}}{2}$$

$$y = 3 \quad \text{or} \quad y = -1 \qquad\qquad = \frac{-1 \pm \sqrt{13}}{2}$$

$$\boxed{\left\{-1, 3, \frac{-1 + \sqrt{13}}{2}, \frac{-1 - \sqrt{13}}{2}\right\}}$$

51 $y^{-2} - 4y^{-1} - 3 = 0$

(Let $t = y^{-1}$):

$$t^2 - 4t - 3 = 0$$

$$t = \frac{4 \pm \sqrt{16 - 4(1)(-3)}}{2}$$

$$= \frac{4 \pm \sqrt{28}}{2}$$

$$= \frac{4 \pm 2\sqrt{7}}{2}$$

$$= 2 \pm \sqrt{7}$$

$$y^{-1} = t$$

$$y^{-1} = 2 \pm \sqrt{7}$$

$$\frac{1}{y} = 2 \pm \sqrt{7}$$

$$1 = (2 \pm \sqrt{7})y$$

$$y = \frac{1}{2 \pm \sqrt{7}}$$

$$y = \frac{1}{2 + \sqrt{7}} \cdot \frac{2 - \sqrt{7}}{2 - \sqrt{7}} \qquad \text{or} \qquad y = \frac{1}{2 - \sqrt{7}} \cdot \frac{2 + \sqrt{7}}{2 + \sqrt{7}}$$

$$= \frac{2 - \sqrt{7}}{4 - 7} \qquad\qquad\qquad\qquad = \frac{2 + \sqrt{7}}{4 - 7}$$

$$= \frac{2 - \sqrt{7}}{-3} \qquad\qquad\qquad\qquad = \frac{2 + \sqrt{7}}{-3}$$

$$= \frac{-2 + \sqrt{7}}{3} \qquad\qquad\qquad\qquad = \frac{-2 - \sqrt{7}}{3}$$

$$\boxed{\left\{ \frac{-2 + \sqrt{7}}{3}, \frac{-2 - \sqrt{7}}{3} \right\}}$$

53. $$4x^4 - 16x^3 + 20x^2 = 0$$
$$4x^2(x^2 - 4x + 5) = 0$$

$$4x^2 = 0 \qquad \text{or} \qquad x^2 - 4x + 5 = 0$$

$$x = 0 \qquad \text{or} \qquad x = \frac{4 \pm \sqrt{16 - 4(1)(5)}}{2} = \frac{4 \pm \sqrt{-4}}{2}$$

$$= \frac{4 \pm 2i}{2} = 2 \pm i$$

$$\boxed{\{0, 2 + i, 2 - i\}}$$

55.
$$6y^5 = -4y^4 - 2y^3$$
$$6y^5 + 4y^4 + 2y^3 = 0$$
$$2y^3(3y^2 + 2y + 1) = 0$$
$$2y^3 = 0 \qquad \text{or} \qquad 3y^2 + 2y + 1 = 0$$
$$y = 0$$

$$y = \frac{-2 \pm \sqrt{4 - 4(3)(1)}}{6}$$
$$= \frac{-2 \pm \sqrt{-8}}{6}$$
$$= \frac{-2 \pm \sqrt{4 \cdot 2}\, i}{6}$$
$$= \frac{-2 \pm 2\sqrt{2}\, i}{6}$$
$$= \frac{-1 \pm \sqrt{2}\, i}{3}$$

$$\boxed{\left\{0, \frac{-1 + \sqrt{2}\, i}{3}, \frac{-1 - \sqrt{2}\, i}{3}\right\}}$$

57.
$$\sqrt{2y + 3} = y - 1$$
$$(\sqrt{2y + 3})^3 = (y - 1)^2$$
$$2y + 3 = y^2 - 2y + 1$$
$$0 = y^2 - 4y - 2$$
$$y = \frac{4 \pm \sqrt{16 - 4(1)(-2)}}{2}$$
$$= \frac{4 \pm \sqrt{24}}{2}$$
$$= \frac{4 \pm 2\sqrt{6}}{2} = 2 \pm \sqrt{6}$$

check $2 + \sqrt{6}$ (using a calculator)
$$\sqrt{2(2 + \sqrt{6}) + 3} = 2 + \sqrt{6} - 1$$
$$\sqrt{7 + 2\sqrt{6}} = 1 + \sqrt{6}$$
$$3.449 = 3.449 \quad \text{(checks)}$$

check $2 - \sqrt{6}$:
$$\sqrt{2(2 - \sqrt{6}) + 3} = 2 - \sqrt{6} - 1$$
$$\sqrt{7 - 2\sqrt{6}} = 1 - \sqrt{6}$$
$$1.449 \neq -1.449$$

$2 - \sqrt{6}$ is extraneous

$$\boxed{\{2 + \sqrt{6}\}}$$

59.
$$10\sqrt{x} = x + 23$$
$$(10\sqrt{x})^2 = (x + 23)^2$$
$$100x = x^2 + 46x + 529$$
$$0 = x^2 - 54x + 529$$
$$x = \frac{54 \pm \sqrt{2916 + 4(1)(592)}}{2}$$
$$= \frac{54 \pm \sqrt{800}}{2}$$
$$= \frac{54 \pm 20\sqrt{2}}{2} = 27 \pm 10\sqrt{2}$$

Use a calculator, both solutions check.

$$\boxed{\{27 + 10\sqrt{2}, \ 27 - 10\sqrt{2}\}}$$

61. $I = t^2 - 7t + 12$
($I = 2$ amperes):
$$t^2 - 7t + 12 = 2$$
$$t^2 - 7t + 10 = 0$$
$$t = \frac{7 \pm \sqrt{49 - 4(1)(10)}}{2}$$
$$= \frac{7 \pm \sqrt{9}}{2}$$
$$= \frac{7 \pm 3}{2}$$
$$t = 5 \quad \text{or} \quad t = 2$$

$$\boxed{2 \text{ seconds; } 5 \text{ seconds}}$$

63. $4x^2 - 2x + 3 = 0$;
$a = 4, \ b = -2, \ c = 3$
$b^2 - 4ac = 4 - 4(4)(3) = \boxed{-44}$

solutions $\boxed{\text{not real numbers}}$

65. $2y^2 + 11y = 6$
$$2y^2 + 11y - 6 = 0$$
$a = 2, \ b = 11, \ c = -6$
$b^2 - 4ac = 11^2 - 4(2)(-6) = \boxed{169}$

$\boxed{\text{rational numbers}}$

67.
$$3y^2 = 2y - 1$$
$$3y^2 - 2y + 1 = 0; \quad a = 3, \ b = -2, \ c = 1$$
$b^2 - 4ac = (-2)^2 - 4(3)(1) = 4 - 12 = \boxed{-8}$

solutions $\boxed{\text{not real numbers}}$

69. $3y^2 + 4y - 2 = 0$; $\quad a = 3, \ b = 4, \ c = -3$
$b^2 - 4ac = 16 - 4(3)(-2) = \boxed{40}$

$\boxed{\text{irrational numbers}}$

71. $hx^2 + 3x + 2 = 0$
Set $b^2 - 4ac = 0$
$$3^2 - 4(h)(2) = 0$$
$$9 - 8h = 0$$
$$\boxed{h = \frac{9}{8}}$$

73. $5x^2 + 3x + h = 0$
Set $b^2 - 4ac = 0$.
$$9 - 4 \cdot 5 \cdot h = 0$$
$$9 - 20h = 0$$
$$\boxed{h = \frac{9}{20}}$$

75. $3x^2 - 6x + 9 = 0$
Set $b^2 - 4ac > 0$.
$$36 - 4(3)a > 0$$
$$36 - 12a > 0$$
$$-12a > -36$$
$$\boxed{a < 3}$$

77. $y = -x^2 + 2x + 27$

($y = 27.5$ meters):
$$27.5 = -x^2 + 2x + 27$$
$$x^2 - 2x + 0.5 = 0$$

Compute the discriminant:
$$b^2 - 4ac = (-2)^2 - 4(1)(0.5) = 4 - 2 = 2 > 0$$

The discriminant is positive. There are two irrational solutions.
$\boxed{\text{Yes}}$ the diver will reach a height of 27.5 meters two times.
$$x = \frac{-(-2) \pm \sqrt{2}}{2} = \frac{2 \pm \sqrt{2}}{2} = 1 \pm \frac{\sqrt{2}}{2}$$

The diver will reach a height of 27.5 meters
at $1 + \dfrac{\sqrt{2}}{2}$ seconds (≈ 1.707 seconds) and
at $1 - \dfrac{\sqrt{2}}{2}$ seconds (≈ 0.293 seconds)

79. $3.87x^2 + 4.39x - 2.17 = 0$
$$x = \frac{-4.39 \pm \sqrt{(4.39)^2 - 4(3.87)(-2.17)}}{2(3.87)}$$
$$x \approx 0.372 \quad \text{or} \quad -1.507$$
$$\boxed{\{0.372, -1.507\}}$$

81. $1.2x^2 - 17.6x + 8.13 = 0$
$$x = \frac{-(-17.6) \pm \sqrt{(-17.6)^2 - 4(1.2)(8.13)}}{2(1.2)}$$
$$x \approx 14.189 \quad \text{or} \quad 0.477$$
$$\boxed{\{14.189, 0.477\}}$$

83. $x^2 + 12{,}357x + 6597 = 0$
$$x = \frac{-12{,}357 + \sqrt{(12{,}357)^2 - 4(1)(6597)}}{2(1)}$$
$$x \approx -0.534 \quad \text{or} \quad -12{,}356.466$$
$$\boxed{\{-0.534, -12{,}356.466\}}$$

85. $x^2 + (7.8 \times 10^{-3})x + 1.2 \times 10^{-6} = 0$
$$x = \frac{-7.8 \times 10^{-3} \pm \sqrt{(7.8 \times 10^{-3})^2 - 4(1)(1.2 \times 10^{-6})}}{2(1)}$$
$$x \approx -1.570 \times 10^{-4} \quad \text{or} \quad -7.643 \times 10^{-3}$$
$$\boxed{\{-1.570 \times 10^{-4}, -7.643 \times 10^{-3}\}}$$

For 87-95, x^2 + (sum of roots with sign changed)x + product of roots = 0.

87. 6, 2:

sum: $6 + 2 = 8$

product: $(6)(2) = 12$

$\boxed{x^2 - 8x + 12 = 0}$

89. $\frac{1}{2}$, 16:

sum: $\frac{1}{2} + 16 = \frac{33}{2}$

product: $\left(\frac{1}{2}\right)(16) = 8$

$\boxed{2x^2 - 33x + 16 = 0}$

91. $-\frac{3}{4}, -\frac{1}{2}$:

sum: $-\frac{3}{4} - \frac{1}{2} = -\frac{5}{4}$

product: $\left(-\frac{3}{4}\right)\left(-\frac{1}{2}\right) = \frac{3}{8}$

$\boxed{8x^2 + 10x + 3 = 0}$

93. $-4 - \sqrt{2}, -4 + \sqrt{2}$:

sum: $(-4 - \sqrt{2}) + (-4 + \sqrt{2}) = -8$

product: $(-4 - \sqrt{2})(-4 + \sqrt{2}) = 16 - 2 = 14$

$\boxed{x^2 + 8x + 14 = 0}$

95. $\frac{1}{2} + \frac{\sqrt{3}}{2}i, \frac{1}{2} - \frac{\sqrt{3}}{2}i$:

sum: $\left(\frac{1}{2} + \frac{\sqrt{3}}{2}i\right) + \left(\frac{1}{2} - \frac{\sqrt{3}}{2}i\right) = 1$

product: $\left(\frac{1}{2} + \frac{\sqrt{3}}{2}i\right)\left(\frac{1}{2} - \frac{\sqrt{3}}{2}i\right) = \frac{1}{4} - \frac{3}{4}i^2$

$= \frac{1}{4} - \frac{3}{4}(-1) = 1$

$\boxed{x^2 - x + 1 = 0}$

97.

$$10^{-4}x^2 + 2 \cdot 10^{-3}x + 10^{-2} = 0$$
$$10^{-2}(10^{-2}x^2 + 2 \cdot 10^{-1}x + 1) = 0$$
$$10^{-2}(t^2 + 2t + 1) = 0 \quad \text{(Let } t = 10^{-1}x)$$
$$10^{-2}(t + 1)^2 = 0$$
$$t + 1 = 0$$
$$t = -1$$
$$10^{-1}x = -1$$
$$10 \cdot 10^{-1}x = 10 \cdot (-1)$$
$$x = -10$$

$\boxed{\{-10\}}$

99.

$$4x^4 + 1 = 0$$
$$4x^4 + 4x^2 + 1 - 4x^2 = 0$$
$$(2x^2 + 1)^2 - (2x)^2 = 0$$
$$(2x^2 + 1 + 2x)(2x^2 + 1 - 2x) = 0$$
$$2x^2 + 2x + 1 = 0$$

$$x = \frac{-2 \pm \sqrt{2^2 - 4(2)(1)}}{2(2)}$$

$$x = \frac{-2 \pm 2i}{4}$$

$$x = \frac{-1 \pm i}{2}$$

or

$$2x^2 - 2x + 1 = 0$$

$$x = \frac{2 \pm \sqrt{(-2)^2 - 4(2)(1)}}{2(2)}$$

$$x = \frac{2 \pm 2i}{4}$$

$$x = \frac{1 \pm i}{2}$$

$\boxed{\left\{\frac{-1-i}{2}, \frac{-1+i}{2}, \frac{1-i}{2}, \frac{1+i}{2}\right\}}$

Review Problems

103. $\sqrt[4]{36x^2y^4} \ \sqrt[4]{12x^5y^3}$

$= \sqrt[4]{432x^7y^7}$

$= \sqrt[4]{16 \cdot 27 x^4 x^3 y^4 y^3}$

$= \boxed{2xy \sqrt[4]{27x^3y^3}}$

104. $(3{,}200{,}000{,}000{,}000)(0.000\ 006\ 5)$

$= (3.2 \times 10^{12})(6.5 \times 10^{-6})$

$= 20.8 \times 10^6$

$= 2.08 \times 10 \times 10^6$

$= \boxed{2.08 \times 10^7}$

105. $3 - \dfrac{3}{3 - \dfrac{3}{3-y}}$

$= 3 - \dfrac{3}{3 - \dfrac{3}{3-y}} \cdot \dfrac{3-y}{3-y}$

$= 3 - \dfrac{3(3-y)}{3(3-y) - 3}$

$= 3 - \dfrac{9 - 3y}{6 - 3y}$

$= 3 - \dfrac{3(3y)}{3(2-y)}$

$= 3 - \dfrac{3-y}{2-y}$

$= 3 \cdot \dfrac{2-y}{2-y} - \dfrac{3-y}{2-y}$

$= \dfrac{3(2-y) - (3-y)}{2-y} = \dfrac{6 - 3y - 3 + y}{2-y}$

$= \boxed{\dfrac{3 - 2y}{2-y}}$

Section 8.6 Applications and Problem Solving

Problem Set 8.6, pp. 634-636

1. Let x = positive number.

$$x^2 + 2x = 7$$
$$x^2 + 2x - 7 = 0$$

$$x = \frac{-2 \pm \sqrt{4 - 4(1)(-7)}}{2} = \frac{-2 \pm \sqrt{32}}{2}$$

$$= \frac{-2 \pm 4\sqrt{2}}{2} = -1 \pm 2\sqrt{2}$$

Positive number: $\boxed{-1 + 2\sqrt{2}}$

3. Let x = the number.

$$x^2 - 2x = -5$$
$$x^2 - 2x + 5 = 0$$
$$x = \frac{2 \pm \sqrt{4 - 4(1)(5)}}{2} = \frac{2 \pm \sqrt{-16}}{2}$$
$$= \frac{2 \pm 4i}{2} = 1 \pm 2i$$

$\boxed{x \text{ is not a real number}}$

5. Let x = the number.

$$x + \frac{1}{x} = 1$$
$$x^2 - x + 1 = 0$$
$$x = \frac{1 \pm \sqrt{-3}}{2} = \frac{1 \pm \sqrt{3}\, i}{2}$$

$\boxed{x \text{ is not a real number}}$

7. $s(t) = -16t^2 + 96t + 104$

The rocket hits the ground when $s(t) = 0$

$$0 = -16t^2 + 96t + 104$$
$$16t^2 - 96t - 104 = 0$$
$$(\div 8) \qquad 2t^2 - 12t - 13 = 0$$
$$t = \frac{-(-12) \pm \sqrt{(-12)^2 - 4(2)(-13)}}{2(2)}$$
$$t = \frac{12 \pm \sqrt{248}}{4}$$
$$t = \frac{12 \pm 2\sqrt{62}}{4} = \frac{6 \pm \sqrt{62}}{2}$$

$$t = \frac{6 + \sqrt{62}}{2} \qquad \text{or} \qquad t = \frac{6 - \sqrt{62}}{2}$$
$$t \approx 6.9 \qquad\qquad\qquad\quad t = -0.9 \quad \text{(reject)}$$

The rocket hits the ground $\boxed{\text{after 6.9 seconds}}$.

9. $C(x) = -0.4x^2 + 20$

$R(x) = 0.8x$

$$\text{(total cost)} = \text{(total revenue)}$$
$$C(x) = R(x)$$
$$-0.4x^2 + 20 = 0.8x$$
$$-0.4x^2 - 0.8x + 20 = 0$$
$$x = \frac{-(-0.8) \pm \sqrt{(-0.8)^2 - 4(-0.4)(20)}}{2(-0.4)}$$
$$x = \frac{0.8 \pm \sqrt{32.64}}{-0.8} = -1 \pm 10\sqrt{051}$$

$$x = -1 - 10\sqrt{0.51} \qquad \text{or} \qquad x = -1 + 10\sqrt{0.51}$$
$$x \approx -8 \quad \text{(reject)} \qquad\qquad\qquad x \approx 6$$

The break-even point occurs at approximately $\boxed{6 \text{ units}}$.

11. $d = 0.044v^2 + 1.1v$
($d = 165$ feet):

$$165 = 0.044v^2 + 1.1v$$
$$0 = 0.044v^2 + 1.1v - 165$$
$$v = \frac{-1.1 \pm \sqrt{1.1^2 - 4(0.044)(-165)}}{2(0.44)}$$
$$v \approx \frac{-1.1 \pm 5.5}{0.088}$$
$$v \approx 50 \quad \text{or} \quad v = -75 \quad \text{(reject)}$$

The speed of a car requiring 165 feet to stop is $\boxed{50 \text{ mph}}$.

13. Let

$$x = \text{width of the border}$$
$$5 + 2x = \text{width of pool plus border}$$
$$9 + 2x = \text{length of pool plus border}$$

(area of pool plus border) − (area of pool) = (area of border)

$$(5 + 2x)(9 + 2x) - (5)(9) = 40$$
$$45 + 28x + 4x^2 - 45 = 0$$
$$4x^2 + 28x - 40 = 0$$
$$x^2 + 7x - 10 = 0$$
$$x = \frac{-7 \pm \sqrt{7^2 - 4(1)(-10)}}{2(1)}$$
$$x = \frac{-7 \pm \sqrt{89}}{2}$$
$$x = \frac{-7 + \sqrt{89}}{2} \quad \text{or} \quad x = \frac{-7 - \sqrt{89}}{2} \quad \text{(reject)}$$
$$x \approx 1.2$$

The border should be $\boxed{\dfrac{-7 + \sqrt{89}}{2} \text{ m}}$ or $\boxed{\text{approximately } 1.2 \text{ m}}$ wide.

15. Let

$$x = \text{width}$$
$$2x + 3 = \text{length}$$
$$\text{Area} = 10 \text{ square inches}$$

$$x(2x + 3) = 10$$
$$2x^2 + 3x - 10 = 0$$
$$x = \frac{-3 \pm \sqrt{9 - 4(2)(-10)}}{4} = \frac{-3 \pm \sqrt{89}}{4} \quad \left(\text{reject } \frac{-3 - \sqrt{89}}{4}\right)$$

width: $\boxed{\dfrac{-3 + \sqrt{89}}{4} \text{ in.} \approx 1.6 \text{ in.}}$

17. Let

$$x = \text{width of frame}$$
$$24 + 2x = \text{width of picture plus frame}$$
$$30 + 2x = \text{length of picture plus frame}$$

$$\text{Area of frame} = \frac{1}{4}(\text{area of picture})$$

$$(30 + 2x)(24 + 2x) - 30(24) = \frac{1}{4}(30)(24)$$
$$720 + 108x + 4x^2 - 720 = 180$$
$$4x^2 + 108x - 180 = 0$$
$$x^2 + 27x - 45 = 0$$
$$x = \frac{-27 \pm \sqrt{(27)^2 - 4(1)(-45)}}{2}$$
$$= \frac{-27 \pm \sqrt{729 + 180}}{2} = \frac{-27 \pm \sqrt{909}}{2}$$
$$= \frac{-27 \pm \sqrt{9(101)}}{2} = \frac{-27 \pm 3\sqrt{101}}{2} \quad \left(\text{reject } \frac{-27 - 3\sqrt{101}}{2}\right)$$

$$\text{width of frame} = \boxed{\frac{-27 + 3\sqrt{101}}{2} \text{ in.} \approx 1.6 \text{ in.}}$$

19. Area of triangle − Area of rectangle = shaded region

$$\frac{1}{2}[(y + 5 + y + 1 + 3)(2y)] - y(y + 1) = 10$$
$$\frac{1}{2}(2y + 9)(2y) - y(y + 1) = 10$$
$$2y^2 + 9y - y^2 = 10$$
$$y^2 + 8y - 10 = 0$$
$$y = \frac{-8 \pm \sqrt{64 + 4(1)(-10)}}{2} = \frac{-8 \pm \sqrt{104}}{2}$$
$$= \frac{-8 \pm 2\sqrt{26}}{2} = -4 \pm \sqrt{26} \quad (\text{reject } -4 - \sqrt{26})$$

$$\boxed{y = -4 + \sqrt{26} \text{ yd} \approx 1.1 \text{ yd}}$$

21.

Let
$$\begin{aligned} x &= \text{length of the shortest leg of a right triangle} \\ x + 1 &= \text{length of other leg of right triangle} \\ (x + 1) + 7 = x + 8 &= \text{length of hypotenuse} \end{aligned}$$

$$x^2 + (x + 1)^2 = (x + 8)^2$$
$$x^2 + x^2 + 2x + 1 = x^2 + 16x + 64$$
$$x^2 - 14x - 63 = 0$$
$$x = \frac{-(-14) \pm \sqrt{(-14)^2 - 4(1)(-63)}}{2(1)}$$
$$x = \frac{14 \pm \sqrt{448}}{2}$$
$$x = 7 \pm 4\sqrt{7} \quad (\text{reject } 7 - 4\sqrt{7})$$

The length of the shorter leg is $\boxed{7 + 4\sqrt{7} \text{ in.} \approx 17.6 \text{ in.}}$.

23. Let

$$x = \text{length of side of the garden}$$
$$x + 3 = \text{length of diagonal}$$

By the Pythagorean theorem,

$$x^2 + x^2 = (x + 3)^2$$
$$2x^2 = x^2 + 6x + 9$$
$$x^2 - 6x - 9 = 0$$

$$x = \frac{-(-6) \pm \sqrt{(-6)^2 - 4(1)(-9)}}{2(1)}$$

$$x = \frac{6 \pm \sqrt{72}}{2}$$

$$x = \frac{6 \pm 6\sqrt{2}}{2} = 3 \pm 3\sqrt{2}$$

$$x = 3 + 3\sqrt{2} \quad \text{(Reject } 3 - 3\sqrt{2})$$

The length of the side is $\boxed{3 + 3\sqrt{2} \text{ yd} \approx 7.2 \text{ yd}}$.

25. Let

$$x = \text{width of the rectangular plot}$$
$$x + 30 = \text{length of the rectangular plot}$$
$$\text{diagonal} = 240 \text{ meters}$$

$$x^2 + (x + 30)^2 = 240^2$$
$$x^2 + x^2 + 60x + 900 = 57600$$
$$2x^2 + 60x - 56700 = 0$$
$$x^2 + 30x - 28350 = 0$$

$$x = \frac{-30 \pm \sqrt{30^2 - 4(1)(-28350)}}{2(1)}$$

$$x = \frac{-30 \pm \sqrt{114,300}}{2} = \frac{-30 \pm 30\sqrt{127}}{2} = -15 \pm 15\sqrt{127}$$

$$x = -15 + 15\sqrt{127} \quad \text{(width) (reject } -15 - 15\sqrt{127})$$

$$x + 30 = 15 + 15\sqrt{127} \quad \text{(length)}$$

$\boxed{\text{width: } -15 + 15\sqrt{127} \text{ m} \approx 154.0 \text{ m}}$

$\boxed{\text{length: } 15 + 15\sqrt{127} \text{ m} \approx 184.0 \text{ m}}$

27. Let

$$
\begin{aligned}
x &= \text{length of original price of tin} \\
x - 10 &= \text{width of open top box} \\
x - 10 &= \text{length of open top box}
\end{aligned}
$$

height: 5 inches
volume: 520 cubic inches

$$
\begin{aligned}
V &= LWH \\
520 &= (x-10)(x-10)5 \\
104 &= x^2 - 20x + 100 \\
0 &= x^2 - 20x - 4 \\
x &= \frac{20 \pm \sqrt{400 - 4(1)(-4)}}{2} \\
&= \frac{20 \pm \sqrt{416}}{2} \\
&= \frac{20 \pm 4\sqrt{26}}{2} \\
&= 10 \pm 2\sqrt{26} \quad \text{(reject } 10 - 2\sqrt{26})
\end{aligned}
$$

Length of tin: $\boxed{10 + 2\sqrt{26} \text{ inches} \approx 20.2 \text{ inches}}$

29. Let

$$
\begin{aligned}
PB &= x \\
AB &= AP + PB = 10 + x \\
\frac{AB}{AD} &= \frac{BC}{QC}
\end{aligned}
$$

$$
\begin{aligned}
\frac{10 + x}{10} &= \frac{10}{x} \\
10x + x^2 &= 100 \\
x^2 + 10x - 100 &= 0 \\
x &= \frac{-10 \pm \sqrt{100 - 4(1)(-100)}}{2} \\
&= \frac{-10 \pm \sqrt{500}}{2} = \frac{-10 \pm 10\sqrt{5}}{2} \\
&= -5 \pm 5\sqrt{5} \quad \text{(reject } -5 - 5\sqrt{5}) \\
AB &= 10 + x = 10 + (-5 + 5\sqrt{5}) = 5 + 5\sqrt{5}
\end{aligned}
$$

The length of side AB is $\boxed{5 + 5\sqrt{5} \text{ ft} \approx 16.2 \text{ ft}}$.

31. area of square $APQD = 100$ square feet
length of side of square

$$QPQD = \sqrt{100}\text{ ft} = 10\text{ ft}$$
$$AP = 10$$

Let

$$x = \text{speed of boat in still water}$$
$$x - 2 = \text{speed of boat against current (going upstream)}$$
$$x + 2 = \text{speed of boat with current (going downstream)}$$

	Distance (miles)	Rate	Time $= \dfrac{\text{Distance}}{\text{Rate}}$
upstream	7	$x - 2$	$\dfrac{7}{x-2}$
downstream	7	$x + 2$	$\dfrac{7}{x+2}$

(time upstream) + (time downstream) = (3 hours)

$$\frac{7}{x-2} + \frac{7}{x+2} = 3$$
$$(x-2)(x+2)\left(\frac{7}{x-2} + \frac{7}{x+2}\right) = (x-2)(x+2) = 3$$
$$7(x+2) + 7(x-2) = 3(x^2 - 4)$$
$$7x + 14 + 7x - 14 = 3x^2 - 12$$
$$0 = 3x^2 - 14x - 12$$
$$x = \frac{-(-14) \pm \sqrt{(-14)^2 - 4(3)(-12)}}{2(3)}$$
$$x = \frac{14 \pm \sqrt{340}}{6} = \frac{14 \pm 2\sqrt{85}}{6}$$
$$x = \frac{7 + \sqrt{85}}{3} \quad \left(\text{reject } \frac{7 - \sqrt{85}}{3}\right)$$

The speed of the boat in still water is $\boxed{\dfrac{7 + \sqrt{85}}{3} \text{ mph} \approx 5.4 \text{ mph}}$.

33. Let

$$x = \text{speed of hiker down the trail}$$
$$x - 1 = \text{speed of hiker up the trail}$$

	Distance (miles)	Rate	Time $= \dfrac{\text{Distance}}{\text{Rate}}$
up	4	$x - 1$	$\dfrac{4}{x-1}$
down	4	x	$\dfrac{4}{x}$

(time up) + (time down) = (5 hours)

$$\frac{4}{x-1} + \frac{4}{x} = 5$$
$$x(x-1)\left[\frac{4}{x-1} + \frac{4}{x}\right] = x(x-1) \cdot 5$$
$$4x + 4(x-1) = 5x(x-1)$$
$$8x - 4 = 5x^2 - 5x$$
$$0 = 5x^2 - 13x + 4$$

$$x = \frac{13 \pm \sqrt{(-13)^2 - 4(5)(4)}}{2(5)}$$

$$x = \frac{13 \pm \sqrt{89}}{10} \quad \left(\text{reject } \frac{13 - \sqrt{89}}{10} \text{ since } \frac{13 - \sqrt{89}}{10} - 1 < 0 \right)$$

$$x = \frac{13 + \sqrt{89}}{10} \quad \text{(down)}$$

$$x - 1 = \frac{13 + \sqrt{89}}{10} - \frac{10}{10} = \frac{3 + \sqrt{89}}{10} \quad \text{(up)}$$

$$\boxed{\text{Up: } \frac{3 + \sqrt{89}}{10} \text{ mph} \approx 1.2 \text{ mph; Down: } \frac{13 + \sqrt{89}}{10} \text{ mph} \approx 2.2 \text{ mph}}$$

35. Let = speed of each boat in still water.

	Distance (miles)	Rate	Time = $\dfrac{\text{Distance}}{\text{Rate}}$
downstream	75	$x + 5$	$\dfrac{75}{x + 5}$
upstream	44	$x - 4$	$\dfrac{44}{x - 5}$

Time spent by the boat going downstream (with current) + 1 hour = Time spent by the boat going upstream (against current).

$$\frac{75}{x + 5} + 1 = \frac{44}{x - 5}$$

$$(x + 5)(x - 5) \left[\frac{75}{x + 5} + 1 \right] = (x + 5)(x - 5) \left[\frac{44}{x - 5} \right]$$

$$75x - 375 + x^2 - 25 = 44x + 220$$

$$x^2 + 75x - 400 = 44x + 220$$

$$x^2 + 31x - 620 = 0$$

$$x = \frac{-31 \pm \sqrt{961 - 4(1)(-620)}}{2}$$

$$= \frac{-31 \pm \sqrt{3441}}{2} \quad \left(\text{reject } \frac{-31 - \sqrt{3441}}{2} \right)$$

Speed of each boat: $\boxed{\dfrac{-31 + \sqrt{3411}}{2} \text{ mph} \approx 13.8 \text{ mph}}$

37. Let

$$x = \text{number of hours for slower person to complete job alone}$$
$$x - 1 = \text{number of hours for faster person to complete job alone}$$

	Fractional part of job completed in 1 hours	Time spent working together	Fractional part of the job completed in 4 hours
slower person	$\dfrac{1}{x}$	4	$\dfrac{4}{x}$
faster person	$\dfrac{1}{x-1}$	4	$\dfrac{4}{x-1}$

(Fractional part by slower person in 4 hours) + (Fractional part by faster person in 4 hours) = (one whole job)

$$\frac{4}{x} + \frac{4}{x-1} = 1$$

$$x(x-1)\left[\frac{4}{x} + \frac{4}{x-1}\right] = x(x-1) \cdot 1$$

$$4(x-1) + 4x = x(x-1)$$

$$4x - 4 + 4x = x^2 - x$$

$$0 = x^2 - 9x + 4$$

$$x = \frac{-(-9) \pm \sqrt{(-9)^2 - 4(1)(4)}}{2(1)}$$

$$x = \frac{9 \pm \sqrt{65}}{2} \quad \left(\text{reject } \frac{9 - \sqrt{65}}{2} \text{ since } x - 1 = \frac{7 - \sqrt{65}}{2} < 0\right)$$

$$x = \frac{9 + \sqrt{65}}{2} \quad \text{(slower)}$$

$$x - 1 = \frac{7 + \sqrt{65}}{2} \quad \text{(faster)}$$

slower: $\boxed{\dfrac{9 + \sqrt{65}}{2} \text{ hr} \approx 8.5 \text{ hr}}$

faster: $\boxed{\dfrac{7 + \sqrt{65}}{2} \text{ hr} \approx 7.5 \text{ hr}}$

39. Let

$$x = \text{time to fill pool}$$
$$x + 2 = \text{time to empty pool}$$

	Fractional part of pool filled in one hour	Time spent (hours) working together	Fractional part of pool filled or emptied in 8 hours
fill pool	$\dfrac{1}{x}$	8	$\dfrac{8}{x}$
empty pool	$\dfrac{1}{x+2}$	8	$\dfrac{8}{x+2}$

(fractional part of pool filled in 8 hours) − (fractional part of pool emptied in 8 hours) = (1 full pool)

$$\frac{8}{x} - \frac{8}{x+2} = 1$$

$$x(x+2)\left[\frac{8}{x} - \frac{8}{x+2}\right] = x(x+2) \cdot 1$$

$$8(x+2) - 8x = x(x+2)$$

$$8x + 16 - 8x = x^2 + 2x$$

$$0 = x^2 + 2x - 16$$

$$x = \frac{-2 \pm \sqrt{2^2 - 4(1)(-16)}}{2(1)}$$

$$x = \frac{-2 \pm \sqrt{68}}{2} = \frac{-2 \pm 2\sqrt{17}}{2} = -1 \pm \sqrt{17}$$

$$x = -1 + \sqrt{17} \quad \text{(reject } -1 - \sqrt{17}\text{)}$$

$$x + 2 = -1 + \sqrt{17} + 2 = 1 + \sqrt{17}$$

Time for outlet pipe to empty pool: $\boxed{1 + \sqrt{17} \text{ hr} \approx 5.1 \text{ hr}}$

41. Let

$$x = \text{rate of the faster horse}$$
$$x - 1.5 = \text{rate of the slower horse}$$

	Distance (miles)	Rate	Time = $\dfrac{\text{Distance}}{\text{Rate}}$
faster horse	1.25	x	$\dfrac{1.25}{x}$
slower horse	1.25	$x - 1.5$	$\dfrac{1.25}{x-1.5}$

Time for the slower horse	−	Time for the faster horse	=	5 seconds
$\dfrac{1.25}{x-1.5}$	−	$\dfrac{1.25}{x}$	=	$\dfrac{5}{3600}$

$$3600x(x-1.5)\left(\frac{1.25}{x-1.5} - \frac{1.25}{x}\right) = 3600x(x-1.5)\frac{5}{3600}$$

$$4500x - 4500x + 6750 = 5x^2 - 7.5x$$

$$0 = 5x^2 - 7.5x - 6750$$

Using the quadratic formula with $a = 5$, $b = -7.5$, and $c = -6750$.

$$x = \frac{-(-7.5) \pm \sqrt{(-7.5)^2 - 4(5)(-6750)}}{2(5)}$$

$$x = \frac{150}{4} \quad \text{or} \quad x = -\frac{143}{4}$$

Rejecting the negative value, $x = \dfrac{150}{4} = 37.5$. The faster horse travels at 37.5 mph.

Its time is $\dfrac{1.25}{x} = \dfrac{1.25}{37.5} = \dfrac{1}{30}$ of an hour = 2 minutes.

The faster horse does not break the record of 1 minute, 56 seconds.
$\boxed{\text{No}}$

43. Let x = the depth of the water. The reed's length is shown in the diagram as $x + 1$.

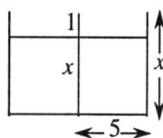

When the reed is drawn to the shore, we obtain a right triangle as shown in the diagram.

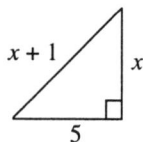

Using the Gougu (Pythagorean) theorem:
$$x^2 + 5^2 = (x + 1)^2$$
$$x^2 + 25 = x^2 + 2x + 1$$
$$24 = 2x$$
$$12 = x$$
The water is $\boxed{12 \text{ feet deep}}$.

Review Problems

44.
$$2x + 3y = -2$$
$$x - 4y = 6$$

$$D = \begin{vmatrix} 2 & 3 \\ 1 & -4 \end{vmatrix} = 2(-4) - (1)(3) = -8 - 3 = -11$$

$$Dx = \begin{vmatrix} -2 & 3 \\ 6 & -4 \end{vmatrix} = -2(-4) - (3)(6) = 8 - 18 = -10$$

$$Dy = \begin{vmatrix} 2 & -2 \\ 1 & 6 \end{vmatrix} = 2(6) - (1)(-2) = 12 + 2 = 14$$

$$x = \frac{Dx}{D} = \frac{-10}{-11} = \frac{10}{11}$$

$$y = \frac{Dy}{D} = \frac{14}{-11} = -\frac{14}{11}$$

$$\boxed{\left\{ \left(\frac{10}{11}, \frac{-14}{11} \right) \right\}}$$

45. $f(x) = 3x + 4$ and $g(x) = x^2 - 2x + 5$

$$\begin{aligned}
(g \circ f)(x) = g[f(x)] &= (3x + 4)^2 - 2(3x + 4) + 5 \\
&= 9x^2 + 24x - 6x - 8 + 5 \\
&= \boxed{9x^2 + 18x + 13}
\end{aligned}$$

46. $x + \dfrac{5}{2x + 2} + \dfrac{3}{x + 1}$

$$= x \cdot \frac{2(x + 1)}{2(x + 1)} + \frac{5}{2(x + 1)} + \frac{3}{x + 1} \cdot \frac{2}{2} \quad \text{(LCD is } 2(x + 1))$$

$$= \frac{2x^2 + 2x + 5 + 6}{2(x + 1)}$$

$$= \boxed{\frac{2x^2 + 2x + 11}{2x + 2}}$$

Section 8.7 Quadratic Functions

Problem Set 8.7, pp. 648-650

1. $y = x^2 + 6x + 5$
    x-intercepts (set $y = 0$):
    $$x^2 + 6x + 5 = 0$$
    $$(x + 5)(x + 1) = 0$$
    $$x = -5 \quad \text{or} \quad x = -1$$
    y-intercept (set $x = 0$):
    $$y = 5$$
    vertex:
    $$x = -\frac{b}{2a} = -\frac{6}{2} = -3$$
    $$y = (-3)^2 + 6(-3) + 5 = -4$$
    $$(-3, -4)$$

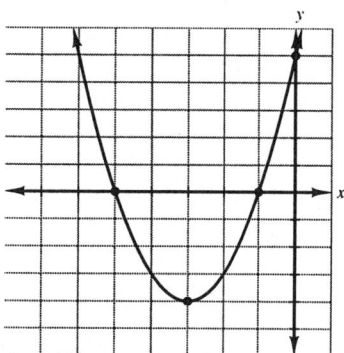

3. $y = x^2 + 4x + 3$
    x-intercepts:
    $$x^2 + 4x + 3 = 0$$
    $$(x + 3)(x + 1) = 0$$
    $$x = -3 \quad \text{or} \quad x = -1$$
    y-intercept:
    $$y = 3$$
    vertex:
    $$x = -\frac{b}{2a} = -\frac{4}{2} = -2$$
    $$y = (-2)^2 + 4(-2) + 3 = -1$$
    $$(-2, -1)$$

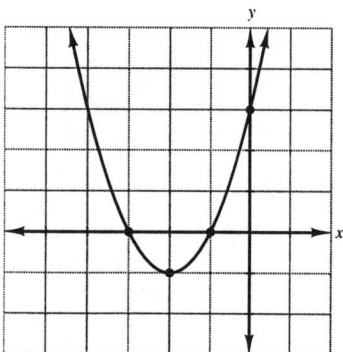

5. $y = -x^2 - 4x - 5$

x-intercepts:

$-x^2 - 4x - 5 = 0$

$x^2 + 4x + 5 = 0$

$b^2 - 4ac = 16 - 4(1)(5) = -4 < 0$, so the equation has no real solutions.

The graph has no x-intercepts.

y-intercept:

$y = -5$

vertex:

$x = -\dfrac{b}{2a} = \dfrac{(-4)}{2(-1)} = \dfrac{4}{-2} = -2$

$y = -(-2)^2 - 4(-2) - 5 = -4 + 8 - 5 = -1$

vertex:

$(-2, -1)$

Other points:

If $x = -3$, $y = -(-3)^2 - 4(-3)^2 - 4(-3) - 5 = -2$

If $x = -1$, $y = -(-1)^2 - 4(-1) - 5 = -2$

$(-3, -2)$ and $(-1, -2)$

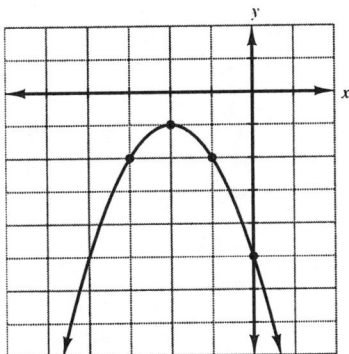

7. $y = -x^2 - 4x - 3$

x-intercepts:

$-x^2 - 4x - 3 = 0$

$x = -3$ or $x = -1$

y-intercept:

$y = -3$

vertex:

$x = -\dfrac{b}{2a} = \dfrac{-(-4)}{2(-1)} = -2$

$(-2, 1)$

$y = -(-2)^2 - 4(-2) - 3 = 1$

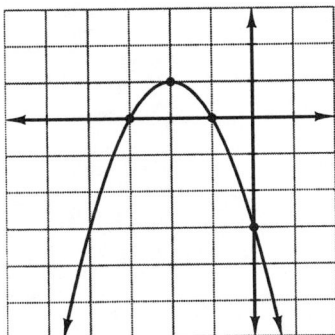

9. $y = 2x^2 - 3$

   x-intercepts:

   $$2x^2 - 3 = 0$$
   $$2x^2 = 3$$
   $$x^2 = \frac{3}{2}$$
   $$x = \pm \sqrt{\frac{3}{2}} \approx \pm 1.2$$

   y-intercept:
   $$y = -3$$

   vertex:
   $$x = -\frac{b}{2a} = -\frac{0}{2(2)} = 0$$
   $$y = 0 - 3 = -3$$
   $$(0, -3)$$

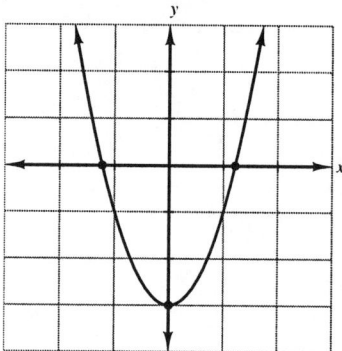

11. $y = x^2 - 4x + 4$

   x-intercept:

   $$x^2 - 4x + 4 = 0$$
   $$(x - 2)^2 = 0$$
   $$x = 2$$

   y-intercept:
   $$y = 4$$

   vertex:
   $$x = -\frac{b}{2a} = -\frac{(-4)}{2(1)} = 2$$
   $$(2, 0)$$
   $$y = 2^2 - 4(2) + 4 = 0$$

   Other points:
   If $x = 4$, $y = 4^2 - 4 \cdot 4 + 4 = 4$
   $$(4, 4)$$

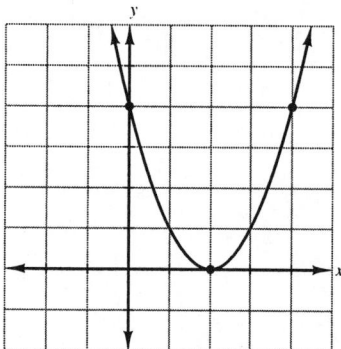

13. $y = -3x^2 + 6x - 1$

x-intercepts:

$-3x^2 + 6x - 1 = 0$

$a = -3,\ b = 6,\ c = -1$

$x = \dfrac{-b \pm \sqrt{b^2 - 4ac}}{2a}$

$\quad = \dfrac{-6 \pm \sqrt{36 - 4(-3)(-1)}}{2(-3)}$

$\quad = \dfrac{-6 \pm \sqrt{24}}{-6} = \dfrac{-6 \pm 2\sqrt{6}}{-6} = \dfrac{-3 \pm \sqrt{6}}{-3}$

$x \approx 0.18 \quad \text{or} \quad x \approx 1.82$

y-intercept:

$y = -1$

vertex:

$x = -\dfrac{b}{2a} = -\dfrac{6}{2(-3)} = 1$

$(1, 2)$

$y = -3(1) + 6(1) - 1 = 2$

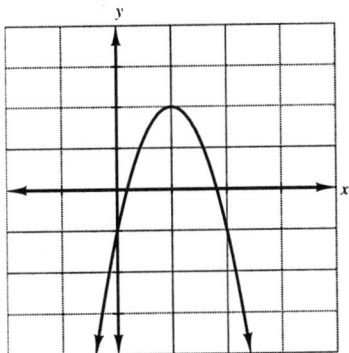

15. $y = 3x^2 + 5x - 2$

    x-intercepts:

$$3x^2 + 5x - 2 = 0$$

$$x = \frac{1}{3} \text{ or } x = -2$$

    y-intercept:

$$y = -2$$

    vertex: $x = -\dfrac{b}{2a} = -\dfrac{5}{6}$

$$\left(-\frac{5}{6}, -4\frac{1}{12}\right)$$

$$y = 3\left(-\frac{5}{6}\right)^2 + 5\left(-\frac{5}{6}\right) - 2 = -4\frac{1}{12}$$

$$y = 3\left(-\frac{5}{6}\right)^2 + 5\left(-\frac{5}{6}\right) - 2$$

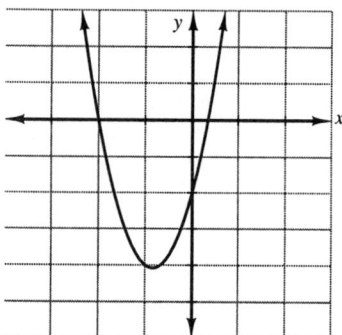

17. $y = x^2 - 3x - 2$

    x-intercepts:

$$x^2 - 3x - 2 = 0$$

$$x = \frac{3 \pm \sqrt{17}}{2}$$

$$x \approx 3.6 \text{ or } x \approx -0.6$$

    y-intercept:

$$y = -2$$

    vertex: $x = -\dfrac{b}{2a} = -\dfrac{(-3)}{2} = \dfrac{3}{2}$

$$y = \left(\frac{3}{2}\right)^2 - 3\left(\frac{3}{2}\right) - 2 = -4\frac{1}{4}$$

$$\left(\frac{3}{2}, -4\frac{1}{4}\right)$$

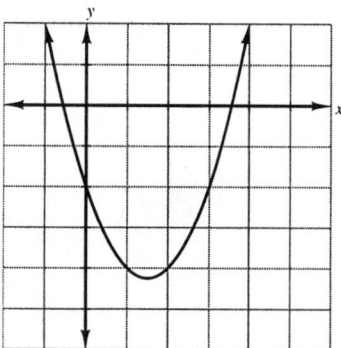

19. $y = x^2 - 2x - 7$

x-intercepts:
$$x^2 - 2x - 7 = 0$$
$$x = 1 \pm 2\sqrt{2}$$
$$x \approx 3.8 \text{ or } x \approx -1.8$$

y-intercept:
$$y = -7$$

vertex:
$$x = -\frac{b}{2a} = \frac{-(-2)}{2(1)} = 1$$
$$y = 1^2 - 2(1) - 7 = -8$$

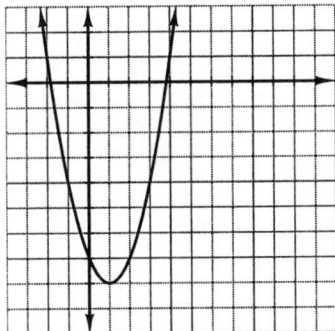

21. $y = 2x^2 + 4x + 5$

No x-intercepts: $(b^2 - 4ac = 16 - 4(2)(5) = -24 < 0)$

y-intercept:
$$y = 5$$

vertex:
$$x = -\frac{b}{a} = -\frac{4}{2(2)} = -1$$
$$y = 2(-1)^2 + 4(-1) + 5 = 3$$
$$(-1, 3)$$

another point:
$$\text{If } x = -2, y = 2(-2)^2 + 4(-2) + 5 = 5$$
$$(-2, 5)$$

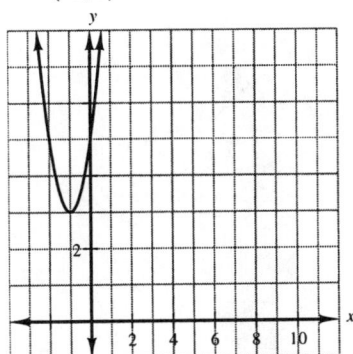

23. $y = 3x^2 - 6x + 4$

$3x^2 - 6x + 4 = 0$

$b^2 - 4ac = (-6)^2 - 4(3)(4) = -1240$

No x-intercepts

y-intercept:

$\quad y = 4$

vertex:

$\quad x = -\dfrac{b}{2a} = -\dfrac{(-6)}{2(3)} = 1$

$\quad y = 3(1) - 6(1) + 4 = 1$

$\quad (1, 1)$

another point:

$\quad$ If $x = 2$, $y = 3(2^2) - 6(2) + 4 = 4$

$\quad (2, 4)$

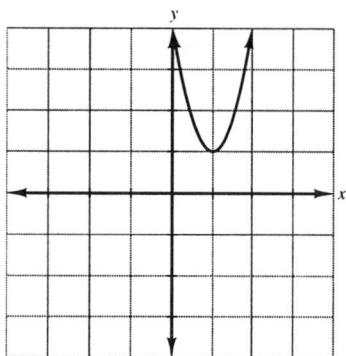

25. $v = -20,000x^2 + 1.28$, $0 \le x \le 0.008$

x	0	0.001	0.002	0.003	0.004
v	1.28	1.26	1.2	1.1	0.96

v	0.005	0.006	0.007	0.008
x	0.78	0.56	0.3	0

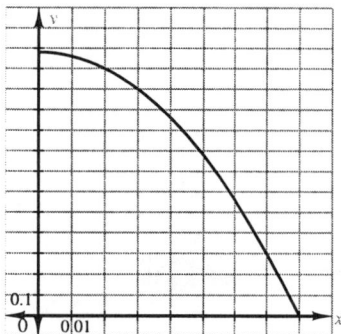

27. $f(x) = -x^2 + 180x - 4500$

The maximum value occurs at the vertex.

To maximize profit: $x = -\dfrac{b}{2a} = -\dfrac{180}{2(-1)} = 90.$

$\boxed{90 \text{ units}}$ will maximize profit.

maximum profit:

$$\begin{aligned} y &= -90^2 + 180(90) - 4500 \\ &= -8100 + 16200 - 4500 \\ &= 3600 \end{aligned}$$

maximum profit: $\boxed{\$3600}$

x-intercepts:

$$\begin{aligned} -x^2 + 180x - 4500 &= 0 \\ x^2 - 180x + 4500 &= 0 \\ (x - 150)(x - 30) &= 0 \\ x = 150 \quad \text{or} \quad x &= 30 \end{aligned}$$

y-intercept: $y = -4500$

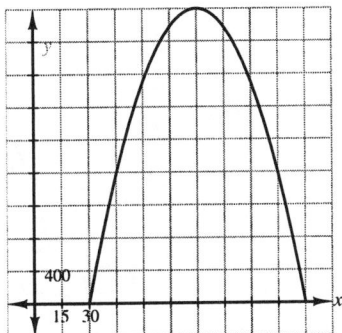

29. $P(I) = -5I^2 + 80I$

Maximum power occurs at the vertex.

$I = -\dfrac{b}{2a} = -\dfrac{80}{2(-5)} = -\dfrac{80}{-10} = 8$

$P(8) = -5(8)^2 + 80(8) = 320$

$\boxed{\text{A current of 8 amps produces a maximum power of 320 volts.}}$

$-5I^2 + 80I = 0$

$-5I(I - 16) = 0$

$I = 0 \quad \text{or} \quad I = 16$

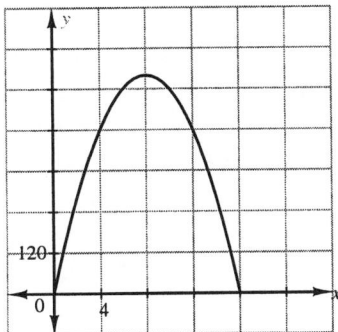

31. $C(t) = 20t^2 - 200t + 640$

The minimum value occurs at the vertex:

$$t = -\frac{b}{2a} = -\frac{(-200)}{2(20)} = 5$$

Concentration is at a maximum after $\boxed{5 \text{ days}}$.

33. Let

$$\begin{aligned} x &= \text{ one number} \\ 48 - x &= \text{ other number} \end{aligned}$$

$$y = x(48 - x) = -x^2 + 48x$$

product is a maximum at the vertex:

$$\begin{aligned} x &= -\frac{b}{2a} = -\frac{48}{2(-1)} = 24 \\ 48 - x &= 48 - 24 = 24 \end{aligned}$$

one number: 24

other number: 24

$\boxed{24 \text{ and } 24}$

35. $A = x(20 - x) = -x^2 + 20x$

The area is a maximum at the vertex.

$$\begin{aligned} x &= -\frac{b}{2a} = -\frac{20}{2(-1)} = 10; \quad \boxed{x = 10} \\ 20 - x &= 20 - 10 = 10 \end{aligned}$$

maximum area: $10(10) = \boxed{100 \text{ ft}^2}$

37.

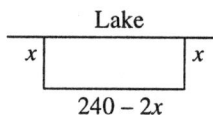

Let

$$\begin{aligned} x &= \text{ measure of one side} \\ 240 - 2x &= \text{ measure of other side} \end{aligned}$$

$$\begin{aligned} A &= x(240 - 2x) = -2x^2 + 240x \\ x &= -\frac{b}{2a} = -\frac{240}{2(-2)} = 60 \\ 240 - 2x &= 240 - 2(60) = 120 \end{aligned}$$

$\boxed{60 \text{ m} \times 120 \text{ m}}$

maximum area: $(60 \text{ m})(120 \text{ m}) = \boxed{7200 \text{ m}^2}$

39. $\boxed{C}$ is true;

Let

$$\begin{aligned} x &= \text{ one side of rectangle} \\ 10 - x &= \text{ adjacent side of rectangle} \\ A &= x(10 - x) = 10x - x^2 \end{aligned}$$

Maximum area occurs at

$$\begin{aligned} x &= \frac{-10}{2(-1)} = 5 \\ 10 - x &= 10 - 5 = 5 \end{aligned}$$

Thus, the maximum area occurs when the rectangle is a square.

41. Let

$$x \; = \; \text{one number}$$
$$20 - x \; = \; \text{other number}$$

$$P = x^2 + 12(20 - x) \quad \text{such that } P \text{ is a minimum}$$
$$P = x^2 - 12x + 240$$

minimum P occurs at the vertex:
$$x = \frac{-(-12)}{2(1)} = 6$$
$$20 - x = 20 - 6 = 14$$

The numbers are $\boxed{6 \text{ and } 14}$.

43. $C(x) = \dfrac{1}{4} x^2 - 10x + 800,\; 0 < x < 40$

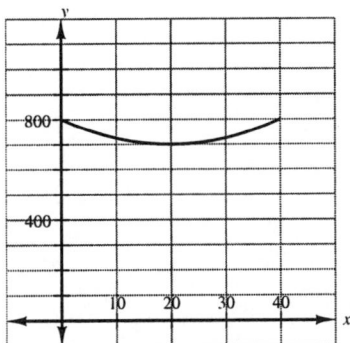

Using the $\boxed{\text{TRACE}}$ function, we can see that minimum occurs at $\boxed{(20, 700)}$.

45.

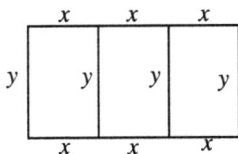

Let

$$x \; = \; \text{width of each equal part}$$
$$3x \; = \; \text{length of large rectangle}$$
$$y \; = \; \text{length of each part (this is also the width of the large rectangle)}$$

total length of fencing = 20 yards
$$6x + 4y \; = \; 200$$
$$y \; = \; -\frac{3}{2} x + 50$$

Area of the rectangle:
$$A \; = \; x\left(-\frac{3}{2} x + 50\right) = -\frac{3}{2} x^2 + 50x$$

The area is a maximum when:

$$x = -\frac{b}{2a} = -\frac{50}{2\left(-\frac{3}{2}\right)} = \frac{50}{3} = 16\frac{2}{3}$$

length of the rectangle: $3x = 3\left(\frac{50}{3}\right) = 50$ yards

width of large rectangle: $y = -\frac{3}{2}x + 50 = -\frac{3}{2}\left(\frac{50}{3}\right) + 50 = 25$ yards

maximum area enclosed:

$$A = (3x)y = (50 \text{ yd})(25 \text{ yd}) = \boxed{1250 \text{ yd}^2}$$

Review Problems

51. $x^{4a+1} - xy^{4a}$

$= x(x^{4a} - y^{4a})$

$= x[(x^{2a})^2 - (y^{2a})^2]$

$= x(x^{2a} + y^{2a})(x^{2a} - y^{2a})$

$= x(x^{2a} + y^{2a})[(x^a)^2 - (y^a)^2]$

$= \boxed{x(x^{2a} + y^{2a})(x^a + y^a)(x^a - y^a)}$

52. $\dfrac{3}{y} + \dfrac{2}{y-b} = \dfrac{5}{y+b}$

$$y(y+b)(y-b)\left[\frac{3}{y} + \frac{2}{y-b}\right] = y(y+b)(y-b)\left[\frac{5}{y+b}\right]$$

$$3(y+b)(y-b) + 2y(y+b) = 5y(y-b)$$

$$3y^2 - 3b^2 + 2y^2 + 2by = 5y^2 - 5by$$

$$5y^2 - 3b^2 + 2by = 5y^2 - 5by$$

$$-3b^2 + 2by = -5by$$

$$-3b^2 = -7by$$

$$\frac{-3b^2}{-7b} = y$$

$$\frac{3b}{7} = y$$

$$\boxed{y = \frac{3b}{7}}$$

53. line passing through $(-4, 5)$ and perpendicular to $4x - 5y = 7$:

$$-5y = -4x + 7$$

$$y = \frac{4}{5}x - \frac{7}{5} \quad \text{slope: } \frac{4}{5}$$

slope of the line perpendicular: $-\dfrac{1}{\dfrac{4}{5}} = -\dfrac{5}{4}$

$$y - 5 = -\frac{5}{4}(x + 4)$$

$$y - 5 = -\frac{5x}{4} - 5$$

$$y = -\frac{5x}{4}$$

$$4y = -5x$$

$$\boxed{5x + 4y = 0}$$

Section 8.8 Solving Quadratic and Rational Inequalities

Problem Set 8.8, pp. 660-662

1. $x^2 - 5x + 4 > 0$
$\quad\quad x^2 - 5x + 4 = 0$
$\quad (x - 4)(x - 1) = 0$
$\quad x = 4 \quad\quad x = 1$

$$
\begin{array}{c|c|c}
\text{T} & \text{F} & \text{T} \\
\hline
\quad 1 & \quad 4 &
\end{array}
$$

Test 0: $0^2 - 5 \cdot 0 + 4 > 0$
$\quad\quad\quad\quad\quad\quad\quad 4 > 0 \quad$ True
Test 2: $2^2 - 5 \cdot 2 + 4 > 0$
$\quad\quad\quad\quad\quad\quad\quad -2 > 0 \quad$ False
Test 5: $5^2 - 5 \cdot 5 + 4 > 0$
$\quad\quad\quad\quad\quad\quad\quad 4 > 0 \quad$ True
$\boxed{\{x \mid x < 1 \ \text{ or } \ x > 4\}} \quad \boxed{(-\infty, 1) \cup (4, \infty)}$

3. $x^2 + 5x + 4 > 0$
$\quad (x + 1)(x + 4) = 0$
$\quad x = -1 \quad x = -4$

$$
\begin{array}{c|c|c}
\text{T} & \text{F} & \text{T} \\
\hline
\quad -4 & \quad -1 &
\end{array}
$$

Test -5: $(-5)^2 + 5(-5) + 4 > 0$
$\quad\quad\quad\quad\quad\quad\quad\quad\quad 4 > 0 \quad$ False
Test -2: $(-2)^2 + 5(-2) + 4 > 0$
$\quad\quad\quad\quad\quad\quad\quad\quad\quad -2 > 0 \quad$ False
Test 0: $0 + 4 > 0$
$\quad\quad\quad\quad\quad\quad 4 > 0 \quad$ True
$\boxed{\{x \mid x < -4 \ \text{ or } \ x > -1\}} \quad \boxed{(-\infty, -4) \cup (-1, \infty)}$

5. $x^2 - 6x + 9 < 0$
$\quad\quad\quad (x - 3)^2 < 0$
$(x - 3)^2$ is not less than 0 for any value of x.

$$
\begin{array}{c|c}
\text{F} & \text{F} \\
\hline
\quad -3 &
\end{array}
$$

$\boxed{\varnothing}$

7. $x^2 - 6x + 8 \; < \; 0$
$\quad (x-2)(x-4) \; \leq \; 0$
$\quad x \; = \; 2 \qquad x \; = \; 4$

$$\frac{\quad \text{F} \quad | \quad \text{T} \quad | \quad \text{F} \quad}{\qquad 2 \qquad\quad 4}$$

Test 0: $0^2 - 6 \cdot 0 + 8 \; \leq \; 0$
$\qquad\qquad\qquad\qquad 8 \; \leq \; 0 \qquad$ False
Test 3: $3^2 - 6 \cdot 3 + 8 \; \leq \; 0$
$\qquad\qquad\qquad\qquad -1 \; \leq \; 0 \qquad$ True
Test 5: $5^2 - 6 \cdot 5 + 8 \; \leq \; 0$
$\qquad\qquad\qquad\qquad 3 \; \leq \; 0 \qquad$ False

$\boxed{\{x \mid 2 \leq x \leq 4\}} \quad \boxed{[2,\,4]}$

9. $3x^2 + 10x - 8 \; \leq \; 0$
$\quad (3x-2)(x+4) \; \leq \; 0$
$\quad x \; = \; \dfrac{2}{3} \qquad x \; = \; -4$

$$\frac{\quad \text{F} \quad | \quad \text{T} \quad | \quad \text{F} \quad}{\quad -4 \qquad 2/3}$$

Test -5: $3(-5)^2 + 10(-5) - 8 \; \leq \; 0$
$\qquad\qquad\qquad\qquad\quad 17 \; \leq \; 0 \qquad$ False
Test 0: $0 - 8 \; \leq \; 0$
$\qquad\qquad\quad -8 \; \leq \; 0 \qquad$ True
Test 1: $3(1) + 10(1) - 8 \; \leq \; 0$
$\qquad\qquad\qquad\qquad 5 \; \leq \; 0 \qquad$ False

$\boxed{\left\{ x \mid -4 \leq x \leq \dfrac{2}{3} \right\}} \quad \boxed{\left[-4,\,\dfrac{2}{3}\right]}$

11. $2x^2 + x \; < \; 15$
$\quad 2x^2 + x - 15 \; < \; 0$
$\quad (2x-5)(x+3) \; < \; 0$
$\quad x \; = \; \dfrac{5}{2} \qquad x \; = \; -3$

$$\frac{\quad \text{F} \quad | \quad \text{T} \quad | \quad \text{F} \quad}{\quad -3 \qquad 5/2}$$

Test -4: $2(16) + (-4) - 15 \; < \; 0$
$\qquad\qquad\qquad\qquad 13 \; < \; 0 \qquad$ False
Test 0: $0 - 15 \; < \; 0$
$\qquad\qquad\quad -15 \; < \; 0 \qquad$ True
Test 3: $2(9) + 3 - 15 \; < \; 0$
$\qquad\qquad\qquad\qquad 6 \; < \; 0 \qquad$ False

$\boxed{\left\{ x \mid -3 < x < \dfrac{5}{2} \right\}} \quad \boxed{\left(-3,\,\dfrac{5}{2}\right)}$

13.
$$4x^2 + 7x < -3$$
$$4x^2 + 7x + 3 < 0$$
$$(4x + 3)(x + 1) < 0$$
$$x = -\frac{3}{4} \qquad x = -1$$

$$\begin{array}{c|c|c} F & T & F \\ \hline & -1 & -3/4 \end{array}$$

Test -2: $4(4) + 7(-2) + 3 < 0$
$$5 < 0 \quad \text{False}$$

Test $-\frac{7}{8}$: $4\left(\frac{49}{64}\right) + 7\left(-\frac{7}{8}\right) + 3 < 0$
$$-0.0625 < 0 \quad \text{True}$$

Test 0: $0 + 3 < 0$
$$3 < 0 \quad \text{False}$$

$$\boxed{\left\{ x \mid -1 < x < -\frac{3}{4} \right\}} \qquad \boxed{\left(-1, -\frac{3}{4}\right)}$$

15.
$$5x \le 2 - 3x^2$$
$$3x^2 + 5x - 2 \le 0$$
$$(3x - 1)(x + 2) \le 0$$
$$x = \frac{1}{3} \qquad x = -2$$

$$\begin{array}{c|c|c} F & T & F \\ \hline & -2 & 1/3 \end{array}$$

Test 0 (middle interval):
$$5 \cdot 0 \le 2 - 3 \cdot 0^2$$
$$0 \le 2 \quad \text{True}$$

$$\boxed{\left\{ x \mid -2 \le x \le \frac{1}{3} \right\}} \qquad \boxed{\left[-2, \frac{1}{3}\right]}$$

17.
$$x^2 - 4x \ge 0$$
$$x(x - 4) \ge 0$$
$$x = 0 \qquad x = 4$$

$$\begin{array}{c|c|c} T & F & T \\ \hline & 0 & 4 \end{array}$$

Test 1 (middle interval):
$$1^2 - 4(1) \ge 0$$
$$-3 \ge 0 \quad \text{False}$$

$$\boxed{\{ x \mid x \le 0 \ \text{ or } \ x \ge 4 \}} \qquad \boxed{(-\infty, 0] \cup [4, \infty)}$$

19.
$$2x^2 + 3x > 0$$
$$x(2x + 3) > 0$$
$$x = 0 \qquad x = -\frac{3}{2}$$

T	F	T

−3/2   0

Test −1 (middle interval):
$$(1) + 3(-1) > 0$$
$$-1 > 0 \quad \text{False}$$

$$\left\{ x \mid x < -\frac{3}{2} \text{ or } x > 0 \right\} \qquad \left(-\infty, -\frac{3}{2}\right) \cup (0, \infty)$$

21.
$$-x^2 + x \geq 0$$
$$x^2 - x \leq 0$$
$$x(x - 1) \leq 0$$
$$x = 0 \qquad x = 1$$

F	T	F

0   1

Test $\frac{1}{2}$ (middle interval):
$$-\frac{1}{4} + \frac{1}{2} \geq 0$$
$$\frac{1}{4} \geq 0 \quad \text{True}$$

$$\{ x \mid 0 \leq x \leq 1 \} \qquad [0, 1]$$

23. $C = -100x^2 + 800x + 500$
$$-100x^2 + 800x + 500 < 1700$$
$$-100x^2 + 800x - 1200 < 0$$
$$x^2 - 8x + 12 > 0$$
$$(x - 2)(x - 6) > 0$$
$$x = 2 \qquad x = 6$$

T	F	T

2   6

Test 4 (middle interval):
$$(4 - 2)(4 - 6) > 0$$
$$-4 > 0 \quad \text{False}$$
$$\{ x \mid x < 2 \text{ or } x > 6 \}$$

They can manufacture fewer than 2 items or more than 6 items .

25. Let
$$x = \text{width of rectangle}$$
$$2x + 5 = \text{length of rectangle}$$

$$\begin{aligned}
\text{Area} &\geq 33 \\
x(2x + 5) &\geq 33 \\
2x^2 + 5x - 33 &\geq 0 \\
(2x + 11)(x - 3) &> 0
\end{aligned}$$
$$x = -\frac{11}{2} \qquad x = 3$$

$$\underline{\quad T \quad | \quad F \quad | \quad T \quad}$$
$$\qquad -11/2 \qquad 3$$

Test 0 (middle interval):
$$\begin{aligned}
0 - 33 &\geq 0 \\
-33 &\geq 0 \quad \text{False}
\end{aligned}$$
$$x \leq -\frac{11}{2} \text{ (reject)} \quad \text{or} \quad x \geq 3$$

The width must be greater than or equal to 3 meters.

27. $f(x) = \sqrt{x^2 - x - 12}$
Domain:
$$\begin{aligned}
x^2 - x - 12 &\geq 0 \\
(x - 4)(x + 3) &\geq 0
\end{aligned}$$
$$x = 4 \qquad x = -3$$

$$\underline{\quad T \quad | \quad F \quad | \quad T \quad}$$
$$\qquad -3 \qquad 4$$

Test 0:
$$\begin{aligned}
0 - 12 &\geq 0 \\
-12 &\geq 0 \quad \text{False}
\end{aligned}$$
$$x \leq -3 \quad \text{or} \quad x \geq 4$$
Domain of $f(x) = D_{f(x)} = \{x \mid x \leq 3 \text{ or } x \geq 4\} = (-\infty, -3] \cup [4, \infty)$

29. $f(x) = \dfrac{5}{\sqrt{2x^2 - x - 6}}$
Domain:
$$2x^2 - x - 6 > 0 \quad (2x^2 - x - 6 \neq 0 \text{ since for the denominator to equal zero, } f(x) \text{ is undefined})$$
$$(2x + 3)(x - 2) = 0$$
$$x = -\frac{3}{2} \quad \text{or} \quad x = 2$$

$$\underline{\quad T \quad | \quad F \quad | \quad F \quad}$$
$$\qquad -3/2 \qquad 2$$

Test 0:
$$\begin{aligned}
0 - 6 &> 0 \\
-6 &> 0 \quad \text{False}
\end{aligned}$$
$$x < -\frac{3}{2} \quad \text{or} \quad x > 2$$

Domain of $f(x) = D_{f(x)} = \left\{ x \mid x < -\frac{3}{2} \text{ or } x > 2 \right\} = \left(-\infty, -\frac{3}{2}\right) \cup (2, \infty)$

31. $\dfrac{x-4}{x+3}>0$

points where the quotient is zero or undefined:

$$x-4 = 0 \qquad \text{or} \qquad x+3 = 0$$
$$x = 4 \qquad\qquad\qquad x = -3$$

$$\begin{array}{c|c|c} \text{T} & \text{F} & \text{T} \\ \hline & -3 \quad\quad 4 & \end{array}$$

Test -4: $\dfrac{-4-4}{-4+3} > 0$

$\dfrac{-8}{-1} > 0$

$8 > 0$ True

Test 0: $\dfrac{0-4}{0+3} > 0$

$\dfrac{-4}{3} > 0$ False

Test 5: $\dfrac{5-4}{5+3} > 0$

$\dfrac{1}{8} > 0$ True

$\boxed{\{x \mid x<-3 \quad \text{or} \quad x>4\}}$ $\boxed{(-\infty, -3) \cup (4, \infty)}$

33. $\dfrac{x+3}{x+4}<4$

$$x+3 = 0 \qquad \text{or} \qquad x+4 = 0$$
$$x = -3 \qquad\qquad\qquad x = -4$$

$$\begin{array}{c|c|c} \text{F} & \text{T} & \text{F} \\ \hline & -4 \quad\quad -3 & \end{array}$$

Test -3.5 (middle interval):

$\dfrac{-0.5}{0.5} < 0$

$-1 < 0$ True

$\boxed{\{x \mid -4<x<-3\}}$ $\boxed{(-4, -3)}$

35. $\dfrac{-x+2}{x-4}\geq 0$

$$-x+2 = 0 \qquad \text{or} \qquad x-4 = 0$$
$$x = 2 \qquad\qquad\qquad x = 4$$

$$\begin{array}{c|c|c} \text{F} & \text{T} & \text{F} \\ \hline & 2 \quad\quad 4 & \end{array}$$

Test 0: $\dfrac{0+2}{0-4} \geq 0$

$-\dfrac{1}{2} \geq 0$ False

Test 3: $\dfrac{-3+2}{3-4} \geq 0$

$1 \geq 0$ True

Test 5: $\dfrac{-5+2}{5-4} \geq 0$

$-3 \geq 0$ False

$\boxed{\{x \mid 2\leq x<4\}}$ $\boxed{[2, 4)}$

37.
$$\frac{x+1}{x+3} < 2$$

$$\frac{x+1}{x+3} - 2 < 0$$

$$\frac{x+1}{x+3} - 2 \cdot \frac{x+3}{x+3} < 0$$

$$\frac{x+1-2x-6}{x+3} < 0$$

$$\frac{-x-5}{x+3} < 0$$

$$\begin{array}{lll} -x-5 = 0 & \text{or} & x+3 = 0 \\ x = -5 & \text{or} & x = -3 \end{array}$$

$$\begin{array}{c|c|c} \text{T} & \text{F} & \text{T} \\ \hline -5 & -3 & \end{array}$$

Test -6: $\quad \dfrac{-6+1}{-6+3} < 2$

$$\frac{-5}{-3} < 2$$

$$\frac{5}{3} < 2 \quad \text{True}$$

Test -4: $\quad \dfrac{-4+1}{-4+3} < 2$

$$\frac{-3}{-1} < 2$$

$$3 < 2 \quad \text{False}$$

Test 0: $\quad \dfrac{0+1}{0+3} < 2$

$$\frac{1}{3} < 2 \quad \text{True}$$

$$\boxed{\{x \mid x < -5 \ \text{ or } \ x > -3\}} \quad \boxed{(-\infty, -5) \cup (-3, \infty)}$$

39.
$$\frac{x+4}{2x-1} \le 3$$

$$\frac{x+4}{2x-1} - 3 \cdot \frac{(2x-1)}{(2x-1)} \le 0$$

$$\frac{-5x+7}{2x-1} \le 0$$

$$\begin{array}{lll} -5x+7 = 0 & \text{or} & 2x-1 = 0 \\ x = \dfrac{7}{5} & & x = \dfrac{1}{2} \end{array}$$

$$\begin{array}{c|c|c} \text{T} & \text{F} & \text{T} \\ \hline 1/2 & 7/5 & \end{array}$$

Test 1 (middle interval):
$$\frac{1+4}{2-1} \le 3$$

$$5 \le 3 \quad \text{False}$$

$$\boxed{\left\{ x \mid x < \frac{1}{2} \ \text{ or } \ x \ge \frac{7}{5} \right\}} \quad \boxed{\left(-\infty, \frac{1}{2}\right) \cup \left[\frac{7}{5}, \ \infty\right)}$$

41.
$$\frac{x-2}{x+2} \le 2$$

$$\frac{x-2}{x+2} - 2 \cdot \frac{x+2}{x+2} \le 0$$

$$\frac{x-2-2x-4}{x+2} \le 0$$

$$\frac{-x-6}{x+2} \le 0$$

$$-x - 6 = 0 \qquad \text{or} \qquad x + 2 = 0$$
$$x = -6 \qquad\qquad\qquad x = -2$$

$$\begin{array}{c|c|c} \text{T} & \text{F} & \text{T} \\ \hline & & \end{array}$$
$$\quad -6 \qquad -2$$

Test -4 (middle interval):
$$\frac{-4-2}{-4+2} \le 2$$

$$\frac{-6}{-2} \le 2$$

$$3 \le 2 \quad \text{False}$$

$$\boxed{\{x \mid x \le -6 \ \text{ or } \ x > -2\}} \qquad \boxed{(-\infty, -6] \cup (-2, \infty)}$$

43.
$$\frac{x}{x+5} \le 1$$

$$\frac{x}{x+5} - 1 \cdot \frac{x+5}{x+5} \le 0$$

$$\frac{x-x-5}{x+5} \le 9$$

$$\frac{-5}{x+5} \le 0$$

$$x = -5$$

$$\begin{array}{c|c} \text{F} & \text{T} \\ \hline & \end{array}$$
$$\quad -5$$

Test -6: $\quad \dfrac{-6}{-6+5} \le 1$

$$6 \le 1 \quad \text{False}$$

Test -4: $\quad \dfrac{-4}{-4+5} \le 1$

$$-4 \le 1 \quad \text{True}$$

$$\boxed{\{x \mid x > -5\}} \qquad \boxed{(-5, \infty)}$$

45. $\boxed{\text{C}}$ is true; $f(x) = \sqrt{x^2 - 6x + 10}$

$$\text{Domain of } f(x) = x^2 - 6x + 10 \ge 0$$
$$b^2 - 4ac = 36 - 4(1)(10) = -4 < 0$$

Thus $x^2 - 6x + 10$ is greater than zero for all real numbers.
Domain of $f(x) = D_{f(x)} = (-\infty, \infty)$

47.
$$
\begin{aligned}
ax^2 + bx + c &= 0 \\
b^2 - 4ac &< 0 \\
(2k)^2 - 4(1)(9) &< 0 \\
4k^2 - 36 &< 0 \\
k^2 - 9 &< 0 \\
(k + 3)(k - 3) &< 0
\end{aligned}
$$

F	T	F
-3		3

Test 0 (middle interval):
$$
\begin{aligned}
0 - 9 &< 0 \\
-9 &< 0 \quad \text{True}
\end{aligned}
$$

$x^2 + 2kx + 9 = 0$ has no real solutions for $\boxed{-3 < k < 3}$.

49.
$$\frac{x^2 - x - 2}{x^2 - 4x + 3} > 0$$

$$\frac{(x - 2)(x + 1)}{(x - 3)(x - 1)} > 0$$

$x = -1,\ x = 1,\ x = 2,\ x = 3$

T	F	T	F	T
-1	1	2	3	

Test -2:
$$
\begin{aligned}
\frac{(-2 - 2)(-2 + 1)}{(-2 - 3)(-2 - 1)} &> 0 \\
\frac{(-4)(-1)}{(-5)(-3)} &> 0 \\
\frac{4}{15} &> 0 \quad \text{True}
\end{aligned}
$$

Test 0:
$$
\begin{aligned}
\frac{0 - 2}{0 + 3} &> 0 \\
-\frac{2}{3} &> 0 \quad \text{False}
\end{aligned}
$$

Test $1\frac{1}{2}$:
$$
\begin{aligned}
\frac{\left(1\frac{1}{2} - 2\right)\left(1\frac{1}{2} + 1\right)}{\left(1\frac{1}{2} - 3\right)\left(1\frac{1}{2} - 1\right)} &> 0 \\
\frac{\left(-\frac{1}{2}\right)\left(\frac{5}{2}\right)}{\left(-\frac{3}{2}\right)\left(\frac{1}{2}\right)} &> 0 \\
\frac{5}{3} &> 0 \quad \text{True}
\end{aligned}
$$

Test $2\frac{1}{2}$: $\dfrac{\left(\frac{1}{2}\right)\left(\frac{7}{2}\right)}{\left(-\frac{1}{2}\right)\left(\frac{3}{2}\right)} > 0$

$-\dfrac{7}{3} > 0$ False

Test $3\frac{1}{2}$: $\dfrac{\left(\frac{3}{2}\right)\left(\frac{9}{2}\right)}{\left(\frac{1}{2}\right)\left(\frac{5}{2}\right)} > 0$

$\dfrac{27}{5} > 0$ True

$x < -1$ or $1 < x < 2$ or $x > 3$

$\boxed{\{x \mid x < -1 \ \text{or} \ 1 < x < 2 \ \text{or} \ x > 3\} = (-\infty, -1) \cup (1, 2) \cup (3, \infty)}$

51.

$$x^3 > x$$
$$x^3 - x > 0$$
$$x(x^2 - 1) > 0$$
$$x(x + 1)(x - 1) > 0$$
$$x = 0, \ x = -1, \ x = 1$$

$$\begin{array}{c|c|c|c}
\text{F} & \text{T} & \text{F} & \text{T} \\
\hline
-1 & 0 & 1 &
\end{array}$$

Test -2: $-2(-2 + 1)(-2 - 1) > 0$

$-2(-1)(-3) > 0$

$-6 > 0$ False

Test $-\dfrac{1}{2}$: $-\dfrac{1}{2}\left(\dfrac{1}{2}\right)\left(-\dfrac{3}{2}\right) > 0$

$\dfrac{3}{8} > 0$ True

Test $\dfrac{1}{2}$: $\dfrac{1}{2}\left(\dfrac{3}{2}\right)\left(-\dfrac{1}{2}\right) > 0$

$-\dfrac{3}{8} > 0$ False

Test 2: $2(3)(1) > 0$

$6 > 0$ True

$-1 < x < 0$ or $x > 1$

$\boxed{\{x \mid -1 < x < 0 \ \text{or} \ x > 1\} = (-1, 0) \cup (1, \infty)}$

53. $x^3 + 5x^2 - 4x - 20 \geq 0$

$x^2(x + 5) - 4(x + 5) \geq 0$

$(x + 5)(x^2 - 4) \geq 0$

$(x + 5)(x + 2)(x - 2) > 0$

$x = -5,\ x = -2,\ x = 2$

F	T	F	T

$-5 -2 2$

using $(x + 5)(x + 2)(x - 2) \geq 0$:

Test -6: $(-1)(-4)(-8) \geq 0$

$-32 \geq 0$ False

Test -3: $(2)(-1)(-5) \geq 0$

$10 \geq 0$ True

Test 0: $0 - 20 \geq 0$

$-20 \geq 0$ False

Test 3: $(8)(5)(1) \geq 0$

$40 \geq 0$ True

$-5 \leq x \leq -2$ or $x \geq 2$

$\boxed{\{x \mid -5 \leq x \leq -2 \ \text{ or } \ x \geq 2\} = [-5, -2] \cup [2, \infty)}$

Review Problems

59. $\dfrac{1}{x^2 - xz - xy + yz} + \dfrac{1}{yz - y^2 - xz + xy} - \dfrac{1}{xy - xz - yz + z^2}$

$= \dfrac{1}{x(x - z) - y(x - z)} + \dfrac{1}{y(z - y) - x(z - y)} - \dfrac{1}{x(y - z) - z(y - z)}$

$= \dfrac{1}{(x - z)(x - y)} + \dfrac{1}{(z - y)(y - z)} - \dfrac{1}{(y - z)(x - z)}$ [LCD is $(x - z)(x - y)(y - z)$]

$= \dfrac{1}{(x - z)(x - y)} \cdot \dfrac{(y - z)}{(y - z)} + \dfrac{1}{(z - y)(y - x)} \cdot \dfrac{-1}{-1} \cdot \dfrac{-1}{-1} \cdot \dfrac{(x - z)}{(x - z)} - \dfrac{1}{(y - z)(x - z)} \cdot \dfrac{(x - y)}{(x - y)}$

$= \dfrac{(y - z) + (x - z) - (x - y)}{(x - z)(x - y)(y - z)}$

$= \dfrac{y - z + x - z - x + y}{(x - z)(x - y)(y - z)} = \dfrac{2y - 2z}{(x - z)(x - y)(y - z)}$

$= \dfrac{2(y - z)}{(x - z)(x - y)(y - z)} = \boxed{\dfrac{2}{(x - z)(x - y)}}$

60. $8\sqrt[3]{81} + 3\sqrt[3]{24} - 2\sqrt[3]{\dfrac{3}{27}}$

$= 8\sqrt[3]{27 \cdot 3} + 3\sqrt[3]{8 \cdot 3} - 2\dfrac{\sqrt[3]{3}}{3}$

$= 24\sqrt[3]{3} + 6\sqrt[3]{3} - 2\dfrac{\sqrt[3]{3}}{3}$

$= 30\sqrt[3]{3} - 2\dfrac{\sqrt[3]{3}}{3}$

$= 30\sqrt[3]{3} \cdot \dfrac{3}{3} - 2\dfrac{\sqrt[3]{3}}{3}$

$= \dfrac{90\sqrt[3]{3}}{3} - 2\dfrac{\sqrt[3]{3}}{3} = \boxed{\dfrac{88\sqrt[3]{3}}{3}}$

61.

	3 points	2 points	1 point
1 way	$5 \cdot 3 = 15$	0	0
2 ways	$4 \cdot 3 = 12$	$1 \cdot 2 = 2$	1
		$0 \cdot 2 = 0$	3
4 ways	$3 \cdot 3 = 9$	$3 \cdot 2 = 6$	0
		$2 \cdot 2 = 4$	2
		$1 \cdot 2 = 2$	4
		$0 \cdot 2 = 0$	6
5 ways	$2 \cdot 3 = 6$	$4 \cdot 2 = 8$	1
		$3 \cdot 2 = 6$	3
		$2 \cdot 2 = 4$	5
		$1 \cdot 2 = 2$	7
		$0 \cdot 2 = 0$	9
7 ways	$1 \cdot 3 = 3$	$6 \cdot 2 = 12$	0
		$5 \cdot 2 = 10$	2
		$4 \cdot 2 = 8$	4
		$3 \cdot 2 = 6$	6
		$2 \cdot 2 = 4$	8
		$1 \cdot 2 = 2$	10
		$0 \cdot 2 = 0$	12
8 ways	$0 \cdot 3 = 0$	$7 \cdot 2 = 14$	1
		$6 \cdot 2 = 12$	3
		$5 \cdot 2 = 10$	5
		$4 \cdot 2 = 8$	7
		$3 \cdot 2 = 6$	9
		$2 \cdot 2 = 4$	11
		$1 \cdot 2 = 2$	13
		$0 \cdot 2 = 0$	15
total 27 ways			
$\boxed{27 \text{ ways}}$			

Chapter 8   Review Problems

Review Problems, pp. 663-665

1.

$$
\begin{aligned}
x^4 - 5x^2 + 4 &= 0 \\
t^2 - 5t + 4 &= 0 \quad \text{(Let } t = x^2) \\
(t - 4)(t - 1) &= 0
\end{aligned}
$$

$$
\begin{array}{lll}
t = 4 & \quad \text{or} \quad & t = 1 \\
x^2 = 4 & & x^2 = 1 \\
x = \pm\sqrt{4} & & x = \pm\sqrt{1} \\
x = \pm 2 & & x = \pm 1
\end{array}
$$

$$\boxed{\{-2, -1, 1, 2\}}$$

2.
$$(x^2 + 2x)^2 - 14(x^2 + 2x) = 15$$
$$(x^2 + 2x)^2 - 14(x^2 + 2x) - 15 = 0$$
$$t^2 - 14t - 15 = 0 \quad \text{(Let } t = x^2 + 2x)$$
$$(t - 15)(t + 1) = 0$$

$$t = 15 \qquad \text{or} \qquad t = -1$$
$$x^2 + 2x = 15 \qquad\qquad x^2 + 2x = -1$$
$$x^2 + 2x - 15 = 0 \qquad\qquad x^2 + 2x + 1 = 0$$
$$(x + 5)(x - 3) = 0 \qquad\qquad (x + 1)^2 = 0$$
$$x = -5 \quad x = 3 \qquad\qquad x = -1$$

$$\boxed{\{-5, -1, 3\}}$$

3.
$$x^{2/3} - x^{1/3} - 12 = 0$$
$$t^2 - t - 12 = 0 \quad \text{(Let } t = x^{1/3})$$
$$(t - 4)(t + 3) = 0$$

$$t = 4 \qquad \text{or} \qquad t = -3$$
$$x^{1/3} = 4 \qquad\qquad x^{1/3} = -3$$
$$(x^{1/3})^3 = 4^3 \qquad\qquad (x^{1/3})^3 = (-3)^3$$
$$x = 64 \qquad\qquad x = -27$$

$$\boxed{\{-27, 64\}}$$

4.
$$x + 7\sqrt{x} = 8$$
$$x + 7\sqrt{x} = 0$$
$$t^2 + 7t - 8 = 0 \quad \text{(Let } t = \sqrt{x})$$
$$(t + 8)(t - 1) = 0$$

$$t = -8 \qquad \text{or} \qquad t = 1$$
$$\sqrt{x} = -8 \qquad\qquad \sqrt{x} = 1$$
$$x = 64 \text{ (extraneous)} \quad x = 1 \text{ (checks)}$$

$$\boxed{\{1\}}$$

5.
$$x^{-2} + x^{-1} - 56 = 0$$
$$t^2 + t - 56 = 0 \quad \text{(Let } t = x^{-1})$$
$$(t + 8)(t - 7) = 0$$

$$t = -8 \qquad \text{or} \qquad t = 7$$
$$x^{-1} = -8 \qquad\qquad x^{-1} = 7$$
$$\frac{1}{x} = -8 \qquad\qquad \frac{1}{x} = 7$$
$$x = -\frac{1}{8} \qquad\qquad x = \frac{1}{7} \text{ (Both check)}$$

$$\boxed{\left\{-\frac{1}{8}, \frac{1}{7}\right\}}$$

6.
$$4x^3 + 8x^2 - x - 2 = 0$$
$$4x^2(x + 2) - (x + 2) = 0$$
$$(x + 2)(4x^2 - 1) = 0$$
$$(x + 2)(2x + 1)(2x - 1) = 0$$
$$x = -2 \quad \text{or} \quad x = -\frac{1}{2} \quad \text{or} \quad x = \frac{1}{2}$$

$$\boxed{\left\{-2, -\frac{1}{2}, \frac{1}{2}\right\}}$$

7.

$$
\begin{aligned}
3x^3 + 4x^2 - 27x - 36 &= 0 \\
x^2(3x + 4) - 9(3x + 4) &= 0 \\
(3x + 4)(x^2 - 9) &= 0 \\
(3x + 4)(x + 3)(x - 3) &= 0
\end{aligned}
$$

$x = -\dfrac{4}{3}$ or $x = -3$ or $x = 3$

$$\boxed{\left\{ -\dfrac{4}{3}, -3, 3 \right\}}$$

8. $3x^3 + 7x^2 - 22x - 8 = 0$

Try $x = 2$:

$$
\begin{array}{r|rrrr}
2 & 3 & 7 & -22 & -8 \\
  &   & 6 & 26 & 8 \\
\hline
  & 3 & 13 & 4 & 0
\end{array}
\quad \rightarrow \quad R = 0; \; x - 2 \text{ is a factor}
$$

$$
\begin{aligned}
3x^3 + 7x^2 - 22x - 8 &= 0 \\
(x - 2)(3x^2 + 13x + 4) &= 0 \\
(x - 2)(3x + 1)(x + 4) &= 0
\end{aligned}
$$

$x = 2, \; x = -\dfrac{1}{3}, \; x = -4$

$$\boxed{\left\{ -4, -\dfrac{1}{3}, 2 \right\}}$$

9.

$$
\begin{aligned}
\frac{2}{y + 2} &= \frac{y}{y + 1} + \frac{1}{y^2 + 3y + 2} \\
(y + 2)(y + 1)\left[\frac{2y}{y + 2} \right] &= (y + 2)(y + 1)\left[\frac{y}{y + 1} + \frac{1}{(y + 2)(y + 1)} \right] \\
2y(y + 1) &= y(y + 2) + 1 \\
2y^2 + 2y &= y^2 + 2y + 1 \\
y^2 &= 1 \\
y &= \pm\sqrt{1} \\
y &= \pm 1
\end{aligned}
$$

Reject -1 since it causes 0 in the original equation's denominator.

$$\boxed{\{1\}}$$

10.

$$
\begin{aligned}
2x^2 - 3 &= 0 \\
x^2 &= \frac{3}{2} \\
x &= \pm\sqrt{\frac{3}{2}} = \pm\sqrt{\frac{3}{2}} \cdot \frac{\sqrt{2}}{\sqrt{2}} = \pm\frac{\sqrt{6}}{2}
\end{aligned}
$$

$$\boxed{\left\{ -\frac{\sqrt{6}}{2}, \frac{\sqrt{6}}{2} \right\}}$$

11.

$$
\begin{aligned}
(x - 4)^2 &= 25 \\
x - 4 &= \pm 5
\end{aligned}
$$

$$
\begin{array}{lll}
x - 4 = 5 & \text{or} & x - 5 = -5 \\
x = 9 & & x = -1
\end{array}
$$

$$\boxed{\{-1, 9\}}$$

12.
$$\frac{2}{4y^2 + 1} = \frac{3}{5y^2 - 1}$$
$$2(5y^2 - 1) = 3(4y^2 + 1)$$
$$10y^2 - 2 = 12y^2 + 3$$
$$-5 = 2y^2$$
$$-\frac{5}{2} = y^2$$
$$y = \pm\sqrt{-\frac{5}{2}} = \pm\frac{\sqrt{5}\,i}{\sqrt{2}} \cdot \frac{\sqrt{2}}{\sqrt{2}}$$
$$y = \pm\frac{\sqrt{10}\,i}{2}$$
$$\boxed{\left\{-\frac{\sqrt{10}\,i}{2}, \frac{\sqrt{10}\,i}{2}\right\}}$$

13.
$$3x^2 + 2x = 4$$
$$3x^2 + 2x - 4 = 0$$
$$a = 3, \, b = 2, \, c = -4$$
$$x = \frac{-b \pm \sqrt{b^2 - 4ac}}{2a}$$
$$= \frac{-2 \pm \sqrt{4 - 4(3)(-4)}}{2(3)}$$
$$= \frac{-2 \pm \sqrt{52}}{6} = \frac{-2 \pm \sqrt{4 \cdot 13}}{6}$$
$$= \frac{-2 \pm 2\sqrt{13}}{6}$$
$$= \frac{-1 \pm \sqrt{13}}{3}$$
$$\boxed{\left\{\frac{-1 + \sqrt{13}}{3}, \frac{-1 - \sqrt{13}}{3}\right\}}$$

14.
$$\frac{5}{x + 1} + \frac{x - 1}{4} = 2$$
$$4(x + 1)\left[\frac{5}{x + 1} + \frac{x - 1}{4}\right] = 4(x + 1)(2)$$
$$20 + (x + 1)(x - 1) = 8(x + 1)$$
$$20 + x^2 - 1 = 8x + 8$$
$$x^2 - 8x + 11 = 0$$
$$a = 1, \, b = -8, \, c = 11$$
$$x = \frac{-b \pm \sqrt{b^2 - 4ac}}{2a} = \frac{8 \pm \sqrt{64 - 4(1)(11)}}{2}$$
$$= \frac{8 \pm \sqrt{20}}{2} = \frac{8 \pm \sqrt{4 \cdot 5}}{2} = \frac{8 \pm 2\sqrt{5}}{2}$$
$$= 4 \pm \sqrt{5}$$
$$\boxed{\{4 + \sqrt{5}, 4 - \sqrt{5}\}}$$

15.
$$x(x-2) = -5$$
$$x^2 - 2x + 5 = 0$$
$$a = 1, \ b = -2, \ c = 5$$
$$x = \frac{-b \pm \sqrt{b^2 - 4ac}}{2a} = \frac{2 \pm \sqrt{4 - 4(1)(5)}}{2}$$
$$= \frac{2 \pm \sqrt{-16}}{2}$$
$$= \frac{2 \pm 4i}{2} = 1 \pm 2i$$
$$\boxed{\{1 + 2i, \ 1 - 2i\}}$$

16.
$$(x-8)^2 = 80$$
$$x - 8 = \pm\sqrt{80}$$
$$x = 8 + \sqrt{80} \qquad \text{or} \qquad x = 8 - \sqrt{80}$$
$$x = 8 + 4\sqrt{5} \qquad\qquad\qquad x = 8 - 4\sqrt{5}$$
$$\boxed{\{8 - 4\sqrt{5}, \ 8 + 4\sqrt{5}\}}$$

17.
$$(2x+4)^2 = -16$$
$$2x + 4 = \pm\sqrt{-16}$$
$$2x + 4 = -4i \qquad \text{or} \qquad 2x + 4 = 4i$$
$$2x = -4 - 4i \qquad\qquad\qquad 2x = -4 + 4i$$
$$x = -2 - 2i \qquad\qquad\qquad x = -2 + 2i$$
$$\boxed{\{-2 - 2i, \ -2 + 2i\}}$$

18.
$$3(2x-6)^2 - 54 = 0$$
$$(2x-6)^2 - 18 = 0$$
$$(2x-6)^2 = 18$$
$$2x - 6 = \pm\sqrt{18}$$
$$2x - 6 = -3\sqrt{2} \qquad \text{or} \qquad 2x - 6 = 3\sqrt{2}$$
$$2x = 6 - 3\sqrt{2} \qquad\qquad\qquad 2x = 6 + 3\sqrt{2}$$
$$x = \frac{6 - 3\sqrt{2}}{2} \qquad\qquad\qquad x = \frac{6 + 3\sqrt{2}}{2}$$
$$\boxed{\left\{\frac{6 - 3\sqrt{2}}{2}, \ \frac{6 + 3\sqrt{2}}{2}\right\}}$$

19.
$$9(5y-2)^2 - 1 = 0$$
$$[3(5y-2)]^2 - 1 = 0$$
$$3(5y-2) + 1 = 0 \qquad \text{or} \qquad 3(5y-2) - 1 = 0$$
$$15y - 5 = 0 \qquad\qquad\qquad 15y - 7 = 0$$
$$y = \frac{1}{3} \qquad\qquad\qquad\qquad y = \frac{7}{15}$$
$$\boxed{\left\{\frac{7}{15}, \frac{1}{3}\right\}}$$

21. $6x^2 + \sqrt{3}x - 6 = 0$

$$x = \frac{-\sqrt{3} \pm \sqrt{(\sqrt{3})^2 - 4(6)(-6)}}{2(6)}$$

$$x = \frac{-\sqrt{3} \pm \sqrt{147}}{12} = \frac{-\sqrt{3} \pm 7\sqrt{3}}{12}$$

$$x = \frac{-\sqrt{3} - 7\sqrt{3}}{12} \quad \text{or} \quad x = \frac{-\sqrt{3} + 7\sqrt{3}}{12}$$

$$x = \frac{-8\sqrt{3}}{12} = \frac{-2\sqrt{3}}{3} \quad\quad\quad x = \frac{6\sqrt{3}}{12} = \frac{\sqrt{3}}{2}$$

$$\boxed{\left\{\frac{-2\sqrt{3}}{3}, \frac{\sqrt{3}}{2}\right\}}$$

22. $ix^2 - 7x + 8i = 0$

$$x = \frac{-(-7) \pm \sqrt{(-7)^2 - 4(i)(8i)}}{2i}$$

$$x = \frac{7 \pm \sqrt{49 + 32}}{2i}$$

$$x = \frac{7 \pm 9}{2i}$$

$$x = \frac{-2}{2i} \quad \text{or} \quad x = \frac{16}{2i}$$

$$x = -\frac{1}{i} = -\frac{1}{i} \cdot \frac{i}{i} = i \quad\quad x = \frac{8}{i} = \frac{8 \cdot i}{i \cdot i} = \frac{8i}{-1} = -i$$

$$\boxed{\{-8i, i\}}$$

23.
$$x^3 - 1 = 0$$
$$(x - 1)(x^2 + x + 1) = 0$$
$$x = 1 \quad \text{or} \quad x^2 + x + 1 = 0$$
$$x = \frac{-1 \pm \sqrt{1 - 4}}{2} = \frac{-1 \pm i\sqrt{3}}{2}$$

$$\boxed{\left\{\frac{-1 - i\sqrt{3}}{2}, \frac{-1 + i\sqrt{3}}{2}, 1\right\}}$$

24.
$$x^2 + 3ix + 4 = 0$$
$$x^2 + 3ix + 4(-i^2) = 0$$
$$(x + 4i)(x - i) = 0$$
$$x = -4i \quad \text{or} \quad x = i$$
$$\boxed{\{-4i, i\}}$$

25.
$$(x^2 - 1)^2 - 4(x^2 - 1) + 3 = 0$$
$$t^2 - 4t + 3 = 0 \quad \text{(Let } t = x^2 - 1\text{)}$$
$$(t - 3)(t - 1) = 0$$
$$t = 3 \quad \text{or} \quad t = 1$$
$$x^2 - 1 = 3 \quad\quad\quad x^2 - 1 = 1$$
$$x^2 = 4 \quad\quad\quad\quad x^2 = 2$$
$$x = \pm 2 \quad\quad\quad\quad x = \pm\sqrt{2}$$

$$\boxed{\{-\sqrt{2}, \sqrt{2}, -2, 2\}}$$

26.
$$(x^2 + 2x)^2 = 5(x^2 + 2x) - 6$$
$$(x^2 + 2x)^2 - 5(x^2 + 2x) + 6 = 0$$
$$t^2 - 5t + 6 = 0 \quad \text{(Let } t = x^2 + 2x\text{)}$$
$$(t - 3)(t - 2) = 0$$

$$t = 3 \qquad \text{or} \qquad t = 2$$
$$x^2 + 2x = 3 \qquad\qquad x^2 + 2x = 2$$
$$x^2 + 2x - 3 = 0 \qquad\qquad x^2 + 2x - 2 = 0$$

$$(x + 3)(x - 1) = 0 \qquad\qquad x = \frac{-2 \pm \sqrt{4 + 8}}{2}$$

$$x = -3 \text{ or } x = 1 \qquad\qquad x = \frac{-2 \pm 2\sqrt{3}}{2} = -1 \pm \sqrt{3}$$

$$\boxed{\{-1 - \sqrt{3}, -1 + \sqrt{3}, -3, 1\}}$$

27.
$$\frac{y}{y^2 + 6y + 5} + \frac{3}{y^2 + 3y - 10} = \frac{1}{y^2 - y - 2}$$
$$\frac{y}{(y + 5)(y + 1)} + \frac{3}{(y + 5)(y - 2)} = \frac{1}{(y - 2)(y + 1)}$$
$$(y + 5)(y + 1)(y - 2)\left[\frac{y}{(y + 5)(y + 1)} + \frac{3}{(y + 5)(y - 2)}\right] = (y + 5)(y + 1)(y - 2) \cdot \frac{1}{(y - 2)(y + 1)}$$
$$y(y - 2) + 3(y + 1) = y + 5$$
$$y^2 - 2y + 3y + 3 = y + 5$$
$$y^2 = 2$$
$$y = \pm\sqrt{2}$$

$$\boxed{\{-\sqrt{2}, \sqrt{2}\}}$$

28.
$$\frac{y}{3y^2 + 25y + 8} + \frac{6}{2y^2 + 13y - 24} = \frac{3}{6y^2 - 7y - 3}$$
$$\frac{y}{(3y + 1)(y + 8)} + \frac{6}{(2y - 3)(y + 8)} = \frac{3}{(3y + 1)(2y - 3)}$$
$$(3y + 1)(y + 8)(2y - 3)\left[\frac{y}{(3y + 1)(y + 8)} + \frac{6}{(2y - 3)(y + 8)}\right] = (3y + 1)(y + 8)(2y - 3) \cdot \frac{3}{(3y + 1)(2y - 3)}$$
$$y(2y - 3) + 6(3y + 1) = 3(y + 8)$$
$$2y^2 - 3y + 18y + 6 = 3y + 24$$
$$2y^2 + 12y - 18 = 0$$
$$y^2 + 6y - 9 = 0$$

$$y = \frac{-6 \pm \sqrt{36 - (-36)}}{2}$$

$$y = \frac{-6 \pm 6\sqrt{2}}{2} = -3 \pm 3\sqrt{2}$$

$$\boxed{\{-3 - \sqrt{2}, -3 + 3\sqrt{2}\}}$$

29.
$$2x^2 + 5x - 3 < 0$$
$$(2x - 1)(x + 3) < 0$$
$$x = \frac{1}{2} \quad \text{or} \quad x = -3$$

T	F	T

$$\underbrace{}_{-3} \quad \underbrace{}_{1/2}$$

Test -4: $2(-4)^2 + 5(-4) - 3 < 0$
$$9 < 0 \quad \text{False}$$
Test 0: $2(0)^2 + 5(0) - 3 < 0$
$$-3 < 0 \quad \text{True}$$
Test 1: $2(1)^2 + 5(1) - 3 < 0$
$$4 < 0 \quad \text{False}$$

$$\boxed{\left\{ x \mid -3 < x < \frac{1}{2} \right\}} \quad \boxed{\left(-3, \frac{1}{2} \right)}$$

30.
$$2x^2 + 9x + 4 \le 0$$
$$(2x + 1)(x + 4) = 0$$
$$x = -\frac{1}{2} \quad \text{or} \quad x = -4$$

T	F	T

$$\underbrace{}_{-4} \quad \underbrace{}_{-1/2}$$

Test -5: $2(-5)^2 + 9(-5) + 4 \ge 0$
$$9 \ge 0 \quad \text{True}$$
Test -3: $2(-3)^2 + 9(-3) + 4 \ge 0$
$$-5 \ge 0 \quad \text{False}$$
Test 0: $2(0)^2 + 9(0) + 4 \ge 0$
$$4 \ge 0 \quad \text{True}$$

$$\boxed{\left\{ x \mid x \le -4 \quad \text{or} \quad x \ge -\frac{1}{2} \right\}} \quad \boxed{(-\infty, -4] \cup \left[-\frac{1}{2}, \infty \right)}$$

31.
$$\frac{x + 7}{x - 3} > 0$$
$$x + 7 = 0 \quad \text{or} \quad x - 3 = 0$$
$$x = -7 \quad \text{or} \quad x = 3$$

T	F	T

$$\underbrace{}_{-7} \quad \underbrace{}_{3}$$

Test -8: $\dfrac{-8 + 7}{-8 - 3} > 0$

$$\frac{1}{11} > 0 \quad \text{True}$$

Test 0: $\dfrac{0 + 7}{0 - 3} > 0$

$$-\frac{7}{3} > 0 \quad \text{False}$$

Test 4: $\dfrac{4 + 7}{4 - 3} > 0$

$$11 > 0 \quad \text{True}$$

$$\boxed{\{ x \mid x < -7 \quad \text{or} \quad x > 3 \}} \quad \boxed{(-\infty, -7) \cup (3, \infty)}$$

32.

$$\frac{x}{x+3} \le 1$$

$$\frac{x}{x+3} - 1 \le 0$$

$$\frac{x}{x+3} - 1 \cdot \frac{x+3}{x+3} \le 0$$

$$\frac{-3}{x+3} \le 0$$

$$x+3 = 0$$

$$x = -3$$

F	T

-3

Test -4: $\dfrac{-4}{-4+3} \le 1$

$4 \le 1$ False

Test 0: $\dfrac{0}{0+3} \le 1$

$0 \le 1$ True

$\boxed{\{x \mid x > -3\}}$ $\boxed{(-3, \infty)}$

33. $4x^2 + 9y^2 = 36$

$$\frac{x^2}{9} + \frac{x^2}{4} = 1$$

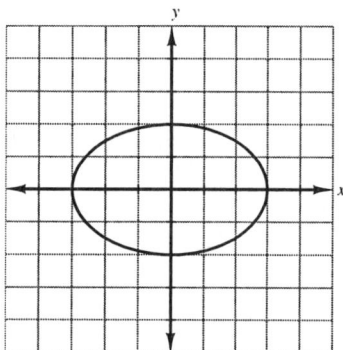

$\boxed{\text{Domain} = \{x \mid -3 \le x \le 3\} = [-3, 3]}$

$\boxed{\text{Range} = \{y \mid -2 \le y \le 2\} = [-2, 2]}$

34. $6x^2 + y^2 = 24$

$$y^2 = -6x^2 + 24$$

$\boxed{y = \pm\sqrt{24 - 6x^2}}$

$\boxed{\text{For certain values of } x \text{ there is more than one value of } y.}$

35. $\boxed{D}$ describes y as a function of x

$$y = \sqrt{4 - x^2}$$

36.

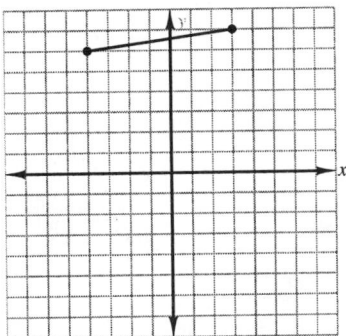

$$d = \sqrt{(x_2 - x_1)^2 + (y_2 - y_1)^2}$$
$$d = \sqrt{(-4 - 3)^2 + (6 - 7)^2}$$
$$= \sqrt{(-7)^2 + (-1)^2} = \sqrt{49 + 1} = \sqrt{50}$$
$$= \sqrt{25 \cdot 2} = \boxed{5\sqrt{2}}$$

37.
$$\sqrt{(-2 - 2)^2 + (y_1 + 1)^2} = 5$$
$$(-4)^2 + (y_1 + 1)^2 = 25$$
$$16 + y_1{}^2 + 2y_1 + 1 = 25$$
$$y_1{}^2 + 2y_1 - 8 = 0$$
$$(y_1 + 4)(y_1 - 2) = 0$$
$$y_1 + 4 = 0 \quad \text{or} \quad y_1 - 2 = 0$$
$$y_1 = -4 \qquad\qquad y_1 = 2$$
$$\boxed{y_1 = -4 \ \text{ or } \ 2}$$

38.
$$4x^2 + 8x = 5$$
$$4x^2 + 8x - 5 = 0$$
$$a = 4, \ b = 8, \ c = -5$$
$$b^2 - 4ac = 8^2 - 4(4)(-5) = 64 + 80 = \boxed{144}$$

solutions are $\boxed{\text{rational}}$

39.
$$y(y - 2) + 4 = 0$$
$$y^2 - 2y + 4 = 0$$
$$a = 1, \ b = -2, \ c = 4$$
$$b^2 - 4ac = (-2)^2 - 4(1)(4) = 4 - 16 = \boxed{-12}$$

solutions are $\boxed{\text{imaginary}}$

40.
$$\frac{z^2}{2} - \frac{4}{5}z = \frac{3}{10}$$
$$10\left(\frac{z^2}{2} - \frac{4}{5}z\right) = 10\left(\frac{3}{10}\right)$$
$$5z^2 - 8z = 3$$
$$5z^2 - 8z - 3 = 0$$
$$a = 5, \ b = -8, \ c = -3$$
$$b^2 - 4ac = 64 - 4(5)(-3) = 64 + 60 = \boxed{124}$$

solutions are $\boxed{\text{irrational}}$

41.
$$y = -x^2 + 2x + 24$$
$$-x^2 + 2x + 24 = 27$$
$$0 = x^2 - 2x + 3$$
$$b^2 - 4ac = (-2)^2 - 4(1)(3) = -8 < 0$$

$\boxed{\text{No}}$. The diver will not reach a height of 27 meters.

42.
$$A = \pi r^2$$
$$49\pi = \pi r^2$$
$$49 = r^2$$
$$r = \sqrt{49} = 7 \quad \text{(reject } -7\text{)}$$
$$C = 2\pi r = 2\pi(7) = 14\pi$$

Circumference: $\boxed{14\pi \, \text{cm}}$

43. Let
$$x = \text{width}$$
$$3x = \text{length}$$
$$(x-1)(3x+3) = 72$$
$$3x^2 - 3 = 72$$
$$3x^2 = 75$$
$$x^2 = 25$$
$$x = \pm\sqrt{25}$$
$$x = 5 \qquad \text{or} \qquad x = -5 \quad \text{(reject)}$$
$$3x = 3(5) = 15$$

$\boxed{\text{Width: 5 yd; Length: 15 yd}}$

44. Let x = a positive number.
$$x^2 - (4x + 7) = 0$$
$$x^2 - 4x - 7 = 0$$
$$x = \frac{-(-4) \pm \sqrt{16 - 4(-7)}}{2}$$
$$x = \frac{4 \pm \sqrt{44}}{2} = \frac{4 \pm 2\sqrt{11}}{2} = 2 \pm \sqrt{11}$$
$$x = 2 + \sqrt{11} \quad \text{(reject } 2 - \sqrt{11}. \text{ It is not a positive number.)}$$

The positive number is $\boxed{2 + \sqrt{11} \approx 5.3}$.

45.
$$s(t) = -16t^2 + 96t + 80$$
$$(s(t) = 128 \text{ feet):}$$
$$128 = -16t^2 + 96t + 80$$
$$16t^2 - 96t + 48 = 0$$
$$t^2 - 6t + 3 = 0$$
$$t = \frac{6 \pm \sqrt{36 - 4(3)}}{2} = \frac{6 \pm \sqrt{24}}{2} = \frac{6 \pm 2\sqrt{6}}{2} = 3 \pm \sqrt{6}$$

The ball will be 128 feet above ground after $\boxed{3 - \sqrt{6} \text{ sec} \approx 0.6 \text{ sec}}$ and $\boxed{3 + \sqrt{6} \text{ sec} \approx 5.4 \text{ sec}}$.

46. Let

$$x = \text{width of border}$$
$$12 + 2x = \text{width of pool and border}$$
$$20 + 2x = \text{length of pool and border}$$

$$\text{area of pool + border} = \text{area of pool and area of border}$$
$$(12 + 2x)(20 + 2x) = 12(20) + 160$$
$$240 + 64x + 4x^2 = 240 + 160$$
$$4x^2 + 64x - 240 = 0$$
$$x^2 + 16x - 60 = 0$$

$$x = \frac{-16 \pm \sqrt{16^2 - 4(-60)}}{2}$$

$$x = \frac{-16 \pm \sqrt{416}}{2}$$

$$x = \frac{-16 \pm 4\sqrt{26}}{2} = -8 \pm 2\sqrt{26}$$

$$x = -8 + 2\sqrt{26} \quad \text{(reject } -8 - 2\sqrt{26})$$

The width of the border is $\boxed{-8 + 2\sqrt{26} \text{ m} \approx 2.2 \text{ m}}$.

47.

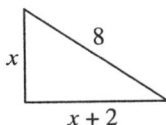

Let

$$x = \text{length of shorter leg of a right triangle}$$
$$x + 2 = \text{length of longer leg}$$

$$x^2 + (x + 2)^2 = 8^2$$
$$x^2 + x^2 + 4x + 4 = 64$$
$$2x^2 + 4x - 60 = 0$$
$$x^2 + 2x - 30 = 0$$

$$x = \frac{-2 \pm \sqrt{4 - 4(-30)}}{2} = \frac{-2 \pm \sqrt{124}}{2}$$

$$x = \frac{-2 \pm 2\sqrt{31}}{2} = -1 \pm \sqrt{31}$$

$$x = -1 + \sqrt{31} \quad \text{(reject } -1 - \sqrt{31})$$

The length of the shorter leg is $\boxed{-1 + \sqrt{31} \text{ cm} \approx 4.6 \text{ cm}}$.

48.

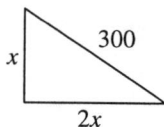

Let

$$x = \text{height of building}$$
$$2x = \text{length of shadow}$$

$$x^2 + (2x)^2 = 300^2$$
$$5x^2 = 90000$$
$$x^2 = 18,000$$
$$x = \pm 60\sqrt{5}$$
$$x = 60\sqrt{5} \quad \text{(reject } -60\sqrt{5}\text{)}$$

The building is $\boxed{60\sqrt{5} \text{ m} \approx 134.2 \text{ m}}$ high.

49. Let

$$x = \text{speed of boat in still water}$$
$$x - 5 = \text{speed of boat against current (upstream)}$$
$$x + 5 = \text{speed of boat with current (downstream)}$$

	Distance (miles)	Rate	Time = $\dfrac{\text{Distance}}{\text{Rate}}$
upstream	10	$x-5$	$\dfrac{10}{x-5}$
downstream	10	$x+5$	$\dfrac{10}{x+5}$

$$(\text{time upstream}) + (\text{time downstream}) = \left(2\tfrac{1}{2} \text{ hours}\right)$$

$$\frac{10}{x-5} + \frac{10}{x+5} = \frac{5}{2}$$

$$\frac{2}{5}(x-5)(x+5)\left[\frac{10}{x-5} + \frac{10}{x+5}\right] = \frac{2}{5}(x-5)(x+5) \cdot \frac{5}{2}$$

$$4(x+5) + 4(x-5) = (x-5)(x+5)$$
$$4x + 20 + 4x - 20 = x^2 - 25$$
$$8x = x^2 - 25$$
$$0 = x^2 - 8x - 25$$

$$x = \frac{-(-8) \pm \sqrt{(-8)^2 - 4(-25)}}{2}$$

$$x = \frac{8 \pm \sqrt{164}}{2} = \frac{8 \pm 2\sqrt{41}}{2} = 4 \pm \sqrt{41}$$

$$x = 4 + \sqrt{41} \quad \text{(reject } 4 - \sqrt{41}\text{)}$$

The speed of the boat in still water is $\boxed{4 + \sqrt{41} \text{ mph} \approx 10.4 \text{ mph}}$.

50. Let x = the number.

$$x + \frac{1}{x} = 6$$

$$x\left(x + \frac{1}{x}\right) = x \cdot 6$$

$$x^2 + 1 = 6x$$

$$x^2 - 6x + 1 = 0$$

$$x = \frac{-(-6) \pm \sqrt{(-6)^2 - 4(1)}}{2}$$

$$x = \frac{6 \pm \sqrt{32}}{2} = \frac{6 \pm 4\sqrt{2}}{2} = 3 \pm 2\sqrt{2}$$

The number is $\boxed{3 - 2\sqrt{2} \approx 0.2 \quad \text{or} \quad 3 + 2\sqrt{2} \approx 5.8}$.

51. Let

$$x = \text{number of hours for slower person to complete the job alone}$$
$$x - 1 = \text{number of hours for faster person to complete the job alone}$$

	Fractional part of job completed in 1 hour	Time spent working together	Fractional part of job completed in 2 hours
slower person	$\dfrac{1}{x}$	2	$\dfrac{2}{x}$
faster person	$\dfrac{1}{x-1}$	2	$\dfrac{2}{x-1}$

(Fractional part for slower person in 2 hours) + (Fractional part for faster person in 2 hours)
          = (one complete job)

$$\frac{2}{x} + \frac{2}{x-1} = 1$$

$$x(x-1)\left[\frac{2}{x} + \frac{2}{x-1}\right] = x(x-1) \cdot 1$$

$$2(x-1) + 2x = x(x-1)$$

$$2x - 2 + 2x = x^2 - x$$

$$0 = x^2 - 5x + 2$$

$$x = \frac{-(-5) \pm \sqrt{25 - 8}}{2} = \frac{5 \pm \sqrt{17}}{2}$$

$$x = \frac{5 + \sqrt{17}}{2} \quad \left(\text{reject } \frac{5 - \sqrt{17}}{2}\right)$$

$$x - 1 = \frac{5 + \sqrt{17}}{2} - 1 = \frac{3 + \sqrt{17}}{2}$$

$$\boxed{\text{slower person: } \frac{5 + \sqrt{17}}{2} \approx 4.6 \text{ hr}}$$

$$\boxed{\text{faster person: } \frac{3 + \sqrt{17}}{2} \approx 3.6 \text{ hr}}$$

52. $C(x) = 6.5x + 800$
$R(x) = -0.002x^2 + 10x$

$$
\begin{aligned}
\text{Profit} &= \text{Revenue} - \text{cost} \\
P(x) &= R(x) - C(x) \\
P(x) &= -0.002x^2 + 10x - (6.5x + 800) \\
&= -0.002x^2 + 3.5x - 800
\end{aligned}
$$

$(P(x) = \$700)$:
$$
\begin{aligned}
700 &= -0.002x^2 + 3.5x - 800 \\
0.002x^2 - 3.5x + 1500 &= 0
\end{aligned}
$$

$$
x = \frac{-(-3.5) \pm \sqrt{(-3.5)^2 - 4(0.002)(1500)}}{2(0.002)}
$$

$$
x = \frac{3.5 \pm \sqrt{0.25}}{0.004} = \frac{3.5 \pm 0.5}{0.004}
$$

$$
x = \frac{4.0}{0.004} = 1000 \qquad \text{or} \qquad x = \frac{3.0}{0.004} = 750
$$

A profit of $700 is achieved when $\boxed{750 \text{ units or } 1000 \text{ units}}$ are sold.

53. $y = x^2 + 5x + 4$
x-intercepts (set $y = 0$):
$$
\begin{aligned}
x^2 + 5x + 4 &= 0 \\
(x + 4)(x + 1) &= 0 \\
x + 4 &= 0 \qquad\qquad x + 1 = 0 \\
x &= -4 \quad \text{and} \quad x = -1
\end{aligned}
$$
y-intercepts (set $x = 0$):
$$
\begin{aligned}
y &= 0^2 + 5 \cdot 0 + 4 \\
y &= 4
\end{aligned}
$$
vertex:
$$
x = -\frac{b}{2a} = -\frac{5}{2(1)} = -\frac{5}{2}
$$
$$
\begin{aligned}
y &= \left(-\frac{5}{2}\right)^2 + 5\left(-\frac{5}{2}\right) + 4 \\
&= \frac{25}{4} - \frac{25}{2} + 4 = \frac{25}{4} - \frac{50}{4} + \frac{16}{4} = -\frac{9}{4}
\end{aligned}
$$
vertex:
$$
\left(-\frac{5}{2}, -\frac{9}{4}\right) = \left(-2\frac{1}{2}, -2\frac{1}{4}\right)
$$

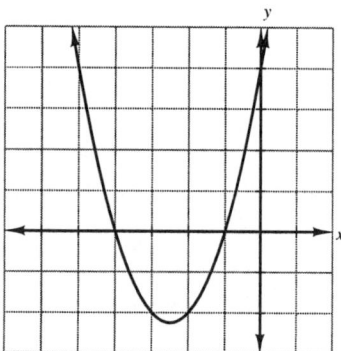

54. $y = -2x^2 + 12x - 10$

x-intercepts:

$$
\begin{aligned}
-2x^2 + 12x - 10 &= 0 \\
x^2 - 6x + 5 &= 0 \\
(x - 5)(x - 1) &= 0 \\
x = 5 \quad \text{and} \quad x &= 1
\end{aligned}
$$

y-intercepts: $y = -10$

vertex:

$$
\begin{aligned}
x &= -\frac{b}{2a} = -\frac{12}{2(-2)} = 3 \\
y &= -2 \cdot 3^2 + 12 \cdot 3 - 10 \\
&= -18 + 36 - 10 = -8
\end{aligned}
$$

vertex: $(3, 8)$

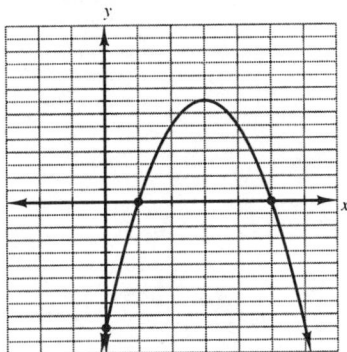

55. $C = 10t^2 - 100t + 320$

The minimum occurs when

$$
\begin{aligned}
t &= -\frac{b}{2a} = -\frac{(-100)}{2(10)} = \frac{100}{20} = 5 \\
C &= 10(5)^2 - 100(5) + 320 \\
&= 250 - 500 + 320 = 70
\end{aligned}
$$

$\boxed{\text{After 5 days}}$ the lowest concentration of $\boxed{70 \text{ bacteria per cm}^3}$ will occur.

56. $s(t) = -16t^2 + 256t$

Maximum height is reached when

$$
t = -\frac{b}{2a} = -\frac{256}{2(-16)} = \frac{-256}{-32} = 8
$$

Maximum height occurs after 8 seconds.

$$
\begin{aligned}
h &= -16(8)^2 + 256(8) = -1024 + 2048 \\
&= 1024
\end{aligned}
$$

Maximum height is $\boxed{1024 \text{ feet}}$.

57. Let
$$x = \text{one number}$$
$$20 - x = \text{other}$$
$$\text{Product} = x(20 - x) = 20x - x^2$$

The product is a maximum when:
$$x = -\frac{b}{2a} = -\frac{20}{2(-1)} = 10$$
$$20 - x = 10$$
The numbers are $\boxed{10 \text{ and } 10}$.

58. Let
$$x = \text{smaller number}$$
$$x + 10 = \text{larger number}$$

$P(x) = [2(\text{smaller number})][3(\text{larger number})]$
$P(x) = (2x)[3(x + 10)] = 6x^2 + 60x$
$P(x)$ is a minimum when
$$x = -\frac{b}{2a} = \frac{-60}{2(6)} = -5$$
$$x + 10 = -5 + 10 = 5$$

The numbers are $\boxed{-5 \text{ and } 5}$.

59.

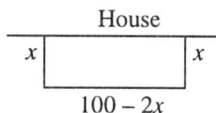

Let
$$x = \text{width of rectangular path}$$
$$100 - 2x = \text{length of rectangular garden}$$

$A(x) = x(100 - 2x) = 100x - 2x^2$
$A(x)$ is a maximum when
$$x = -\frac{b}{2a} = -\frac{100}{2(-2)} = 25$$
$$100 - 2x = 100 - 2(25) = 50$$

The dimensions of the garden are $\boxed{25 \text{ ft} \times 50 \text{ ft}}$.

Maximum area $= x(100 - 2x) = 25 \text{ ft}(50 \text{ ft}) = \boxed{1250 \text{ ft}^2}$.

60.

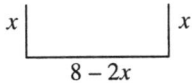

$$x \left| \underline{} \right| x$$
$$8 - 2x$$

Let
$$x = \text{height of rain gutter}$$
$$8 - 2x = \text{width of rain gutter}$$

gutter's cross-sectional area $= x(8 - 2x) = 8x - 2x^2$

Maximum x occurs when
$$x = -\frac{b}{2a} = -\frac{8}{2(-2)} = 2$$

$\boxed{x = 2 \text{ centimeters}}$

61. $C(x) = 300x + 90,000$
$R(x) = -0.02x^2 + 500x$

$$\begin{aligned} \text{Profit} &= \text{Revenue} - \text{Cost} \\ P(x) &= R(x) - C(x) \\ P(x) &= -0.02x^2 + 500x - (300x + 90,000) \\ &= -0.02x^2 + 200x - 90,000 \end{aligned}$$

Maximum profit occurs when
$$\begin{aligned} x &= -\frac{b}{2a} = -\frac{200}{2(-0.02)} = 5000 \\ P(x) &= -0.02(5000)^2 + 200(5000) - 90,000 \\ &= 410,000 \end{aligned}$$

$\boxed{5000 \text{ units}}$ must be manufactured to achieve a maximum profit of $\boxed{\$410,000}$

62. $f(x) = \sqrt{x - x^2}$
Domain:
$$\begin{aligned} x - x^2 &\geq 0 \\ x(1 - x) &\geq 0 \\ x = 0 \qquad x &= 1 \end{aligned}$$

F	T	F
0	1	

Test $\frac{1}{2}$ (middle interval):
$$\frac{1}{2}\left(\frac{1}{2}\right) \geq 0$$
$$\frac{1}{4} \geq 0 \quad \text{True}$$

$0 \leq x \leq 1$

$\boxed{\{x \mid 0 \leq x \leq 1\} = [0, 1]}$

63. $f(x) = \sqrt{x^2 + 6x + 9} = \sqrt{(x+3)^2} = |x+3|$

Domain:
$$x^2 + 6x + 9 \geq 0$$
$$(x+3)^2 \geq 0$$
$(x+3)^2$ is ≥ 0 for all real values of x.

Thus, all real values of x are in the domain of x.

$$\boxed{\{x \mid x \in R\} = (-\infty, \infty)}$$

64. $f(x) = \dfrac{5}{\sqrt{3(2x^2 - 5) - x}}$

Domain:
$$3(2x^2 - 5) - x > 0$$
$$6x^2 - x - 15 > 0$$
$$(3x - 5)(2x + 3) > 0$$
$$x = \frac{5}{3} \qquad x = -\frac{3}{2}$$

$$\underline{\quad T \quad | \quad F \quad | \quad T \quad}$$
$$\quad\;\; -3/2 \quad\;\; 5/3$$

Test 0 (middle interval):
$$0 - 15 > 0$$
$$-15 > 0 \quad \text{False}$$

$$\boxed{\left\{ x \mid x < -\frac{3}{2} \;\text{ or }\; x > \frac{5}{3} \right\} = \left(-\infty, -\frac{3}{2}\right) \cup \left(\frac{5}{3}, \infty\right)}$$

65. $f(x) = \dfrac{1}{\sqrt{x^2 - 4x + 4}}$

Domain:
$$x^2 - 4x + 4 > 0$$
$$(x-2)^2 > 0 \quad \text{and} \quad (x-2)^2 \neq 0$$
$$(x-2)^2 \text{ is } > 0 \qquad\qquad x \neq 2$$
for all real values of x
except $x = 2$

$$\boxed{\{x \mid x \neq 2\} = (-\infty, 2) \cup (2, \infty)}$$

66. $f(x) = \sqrt{10x - x^2 - 25} = \sqrt{-x^2 + 10x - 25} = \sqrt{-(x-5)^2}$
$$-(x-5)^2 \geq 0$$
$$(x-5)^2 \leq 0$$
$(x-5)^2$ is ≥ 0 for all real values of x except $x = 5$
Thus, the domain of x includes only [5]

Domain $= \boxed{\{x \mid x = 5\} = [5]}$

67. Let

$$\begin{aligned}
x &= \text{width of rectangle} \\
x + 2 &= \text{length of rectangle} \\
(\text{area}) &> (\text{perimeter}) \\
x(x + 2) &> 2x + 2(x + 2) \\
x^2 + 2x &> 4x + 4 \\
x^2 - 2x - 4 &> 0
\end{aligned}$$

$$x = \frac{-(-2) \pm \sqrt{4 + 16}}{2} = \frac{2 \pm 2\sqrt{5}}{2} = 1 \pm \sqrt{5}$$

$$\begin{array}{c|c|c}
T & F & T \\
\hline
\multicolumn{1}{c}{1 - \sqrt{5}} & 1 + \sqrt{5} &
\end{array}$$

Test 0 (middle interval):

$$\begin{aligned}
0 - 4 &> 0 \\
-4 &> 0 \quad \text{False}
\end{aligned}$$

(Also reject $1 - \sqrt{5}$ since $1 - \sqrt{5} < 0$.)

Thus, the $\boxed{\text{width} > 1 + \sqrt{5} \text{ m} \approx 3.24 \text{ m}}$.

68. Let x and $x + 1$ equal consecutive integers.

$$\begin{aligned}
(\text{sum of squares}) &\le 25 \\
x^2 + (x + 1)^2 &\le 25 \\
2x^2 + 2x + 1 &\le 25 \\
2x^2 + 2x - 24 &\le 0 \\
x^2 + x - 12 &\le 0 \\
(x + 4)(x - 3) &= 0 \\
x = -4 \qquad x &= 3
\end{aligned}$$

$$\begin{array}{c|c|c}
F & T & F \\
\hline
\multicolumn{1}{c}{-4} & 3 &
\end{array}$$

Test 0:

$$\begin{aligned}
0 - 12 &\le 0 \\
-12 &\le 0 \quad \text{True}
\end{aligned}$$

$$\begin{aligned}
-4 \le x &\le 3 \\
-3 \le x + 1 &\le 4
\end{aligned}$$

Pairs of consecutive integers are:

$$\boxed{-4, -3; -3, -2; -2, -1; -1, 0; 0, 1; 1, 2; 2, 3; 3, 4}$$

69. a. $s(t) = -16t^2 + 80t + 96$

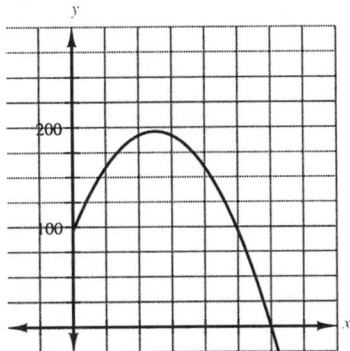

Cumulative Review Problems (Chapters 1-8)

Cumulative Review, pp. 665-666

1. $\left|1-x\right| > 3$

$$1 - x > 3 \qquad \text{or} \qquad 1 - x < -3$$
$$-x > 2 \qquad\qquad\qquad -x < -4$$
$$x < -2 \qquad\qquad\qquad x > 4$$

$\boxed{\{x \mid x < -2 \ \text{ or } \ x > 4\}}$ $\boxed{(-\infty, -2) \cup (4, \infty)}$

2. $x^2 = 2x^{1.3} + 35$

$$x^{2/3} - 2x^{1/3} - 35 = 0$$
$$t^2 - 2t - 35 = 0 \qquad (\text{Let } t = x^{1/3})$$
$$(t - 7)(t + 5) = 0$$

$$t = 7 \qquad\qquad \text{or} \qquad\qquad t = -5$$
$$x^{1/3} = 7 \qquad\qquad\qquad x^{1/3} = -5$$
$$x = 7^3 \qquad\qquad\qquad x = (-5)^3$$
$$x = 343 \qquad\qquad\qquad x = -125$$

$\boxed{\{-125, 343\}}$

3. $8(2x - 1) + 3(1 - x) + 2 = 5 - 2x(1 + x)$

$$16x - 8 + 3 - 3x + 2 = 5 - 2x - 2x^2$$
$$13 - 3 = 5 - 2x - 2ix^2$$
$$2x^2 + 15x - 8 = 0$$
$$(2x + 1)(x + 8) = 0$$

$$x = \frac{1}{2} \qquad\qquad \text{or} \qquad\qquad x = -8$$

$\boxed{\left\{-8, \dfrac{1}{2}\right\}}$

4. $\dfrac{1}{x} + \dfrac{1}{x - 4} = \dfrac{5}{6}$

$$6x(x - 4)\left[\frac{1}{x} + \frac{1}{x - 4}\right] = 6x(x - 4) \cdot \frac{5}{6}$$
$$6(x - 4) + 6x = 5x(x - 4)$$
$$6x - 24 + 6x = 5x^2 - 20x$$
$$0 = 5x^2 - 32x + 24$$

$$x = \frac{32 \pm \sqrt{32^2 - 4(5)(24)}}{10} = \frac{32 \pm \sqrt{544}}{10}$$

$$x = \frac{16 \pm 2\sqrt{34}}{5}$$

$\boxed{\left\{\dfrac{16 - 2\sqrt{34}}{5}, \dfrac{16 + 2\sqrt{34}}{5}\right\}}$

5. $\sqrt{2x+5} - \sqrt{x-1} - \sqrt{x+2} = 0$

$\begin{aligned}
\sqrt{2x+5} &= \sqrt{x-1} + \sqrt{x+2} \\
2x+5 &= x-1 + 2\sqrt{x-1}\sqrt{x+2} + x + 2 \\
2x+5 &= 2x-1 + 2\sqrt{(x-1)(x+2)} \\
4 &= 2\sqrt{(x-1)(x+2)} \\
2 &= \sqrt{(x-1)(x+2)} \\
4 &= x^2 + x - 2 \\
0 &= x^2 + x - 6 \\
0 &= (x+3)(-2) \\
x = -3 \quad &\text{or} \quad x = 2 \quad \text{(checks)}
\end{aligned}$

(−3 is an extraneous solution)

$\boxed{\{2\}}$

6. $\begin{aligned}
(x^2 - 2x)^2 + 2(x^2 - 2x) &= 3 \\
(x^2 - 2x)^2 + 2(x^2 - 2x) - 3 &= 0 \\
t^2 + 2t - 3 &= 0 \quad (\text{Let } t = x^2 - 2x) \\
(t+3)(t-1) &= 0
\end{aligned}$

$\begin{aligned}
t &= -3 & \text{or} & & t &= 1 \\
x^2 - 2x &= -3 & & & x^2 - 2x &= 1 \\
x^2 - 2x + 3 &= 0 & & & x^2 - 2x - 1 &= 0 \\
x &= \dfrac{2 \pm \sqrt{4-12}}{2} & & & x &= \dfrac{2 \pm \sqrt{4+4}}{2} \\
x &= \dfrac{2 \pm 2i\sqrt{2}}{2} & & & x &= \dfrac{2 \pm 2\sqrt{2}}{2} \\
x &= 1 \pm i\sqrt{2} & & & x &= 1 \pm \sqrt{2}
\end{aligned}$

$\boxed{\{1 - i\sqrt{2}, 1 + i\sqrt{2}, 1 - \sqrt{2}, 1 + \sqrt{2}\}}$

7. $\begin{aligned}
\left|5x+2\right| &= \left|4-3x\right|
\end{aligned}$

$\begin{aligned}
5x+2 &= 4-3x & \text{or} & & 5x+2 &= -(4-3x) \\
8x &= 2 & & & 5x+2 &= -4+3x \\
x &= \dfrac{1}{4} & & & 2x &= -6 \\
& & & & x &= -3
\end{aligned}$

$\boxed{\left\{-3, \dfrac{1}{4}\right\}}$

8. $\begin{aligned}
\{x \mid 3(2x+8) - 8x &\geq 32\} & \cap \\
\{x \mid 6x+24-8x &\geq 32\} & \cap \\
\{x \mid -2x &\geq 8\} & \cap \\
\{x \mid x &\leq -4\} & \cap
\end{aligned}$
$\qquad \begin{aligned}
\{x \mid 4(2x-3)-5 &< 2(3x-1)-15\} \\
\{x \mid 8x-12-5 &< 6x-2-15\} \\
\{x \mid 2x &< 0\} \\
\{x \mid x &< 0\}
\end{aligned}$

$\boxed{\{x \mid x \leq -4\} = (-\infty, -4]}$

9.
$$5\left|4-3x\right|-2 \le 23$$
$$5\left|4-3x\right| \le 25$$
$$\left|4-3x\right| \le 5$$
$$-5 \le 4-3x \le 5$$
$$-9 \le -3x \le 1$$
$$3 \ge x \ge -\frac{1}{3}$$
$$-\frac{1}{3} \le x \le 3$$

$$\boxed{\left\{ x \mid -\frac{1}{3} \le x \le 3 \right\} = \left[-\frac{1}{3}, 3\right]}$$

10.
$$8x^2 + 5x + 4 < 2x^2 - 5x - 8$$
$$6x^2 + 10x + 12 < 0$$
$$3x^2 + 5x + 6 < 0$$
$$b^2 - 4ac = 25 - 4(3)(6) < 0$$

Thus $3x^2 + 5x + 6$ is *greater than zero* for all real values of x.
No real values of x cause $3x^2 + 5x + 6$ to be less than or equal to zero.
$$\boxed{\varnothing}$$

11.
$$\frac{x-1}{x+2} < 2$$
$$\frac{x-1}{x+2} - 2 < 0$$
$$\frac{x-1}{x+2} - 2 \cdot \frac{x+2}{x+2} < 0$$
$$\frac{x-1-2x-4}{x+2} < 0$$
$$\frac{-x-5}{x+2} < 0$$
$$-\frac{(x+5)}{x+2} < 0$$
$$x = -5 \qquad x = -2$$

Test -3 (middle interval):
$$\frac{-3-1}{-3+2} < 2$$
$$\frac{-4}{-1} < 2$$
$$4 < 2 \quad \text{False}$$
$$\boxed{\{x \mid x < -5 \ \text{ or } \ x > -2\} = (-\infty, -5) \cup (-2, \infty)}$$

12. $x^2 - 2ix - 5 = 0$
$$x = \frac{2i \pm \sqrt{(-2i)^2 - 4(-5)}}{2}$$
$$x = \frac{2i \pm \sqrt{-4+20}}{2}$$
$$x = \frac{2i \pm 4}{2} = i \pm 2$$
$$x = 2 + i \quad \text{or} \quad x = -2 + i$$
$$\boxed{\{-2 + i, 2 + i\}}$$

13.
$$-3x + 2y + 4z = 6$$
$$7x - y + 3z = 23$$
$$2x + 3y + z = 7$$

(Equations 1 and 2):
$$-3x + 2y + 4z = 6$$
$$(\times 2) \quad \underline{14x - 2y + 6z = 46}$$
$$11x + 10z = 52 \quad \text{(Equation 4)}$$

(Equations 2 and 3):
$$(\times 3) \quad 21x - 3y + 9z = 69$$
$$\underline{2x + 3y + z = 7}$$
$$23x + 10z = 76 \quad \text{(Equation 5)}$$

(Equations 4 and 5):
$$(\times -1) \quad -11x - 10z = -52$$
$$\underline{23x + 10z = 76}$$
$$12x = 24$$
$$x = 2$$

(Equation 4):
$$11x + 10z = 52$$
$$22 + 10z = 52$$
$$10z = 30$$
$$z = 3$$

(Equation 1):
$$-3x + 2y + 4z = 6$$
$$-6 + 2y + 12 = 6$$
$$2y = 0$$
$$y = 0$$

$$\boxed{\{(2, 0, 3)\}}$$

14. $\left(\dfrac{-40x^{3n-2}y^{4-2n}}{20x^{3n-1}y^{-2-2n}}\right)^{-3}$

$= (-2x^{-1}y^6)^{-3}$

$= (-2)^{-3}x^3y^{-18}$

$= \boxed{\dfrac{-x^3}{8y^{18}}}$

15. $(5x - 2)(2x^2 + 3xy - y^2)$

$= 5x(2x^2 + 3xy - y^2) - 2(2x^2 + 3xy - y^2)$

$= \boxed{10x^3 + 15x^2y - 5xy^2 - 4x^2 - 6xy + 2y^2}$

16. $\dfrac{y+7}{3y^2 - 13y + 4} + \dfrac{5y+3}{3y^2 + 5y - 2} - \dfrac{2y-1}{y^2 - 2y - 8}$

$= \dfrac{y+7}{(3y-1)(y-4)} + \dfrac{5y+3}{(3y-1)(y+2)} - \dfrac{2y-1}{(y-4)(y+2)}$

$= \dfrac{y+7}{(3y-1)(y-4)} \cdot \dfrac{y+2}{y+2} + \dfrac{5y+3}{(3y-1)(y+2)} \cdot \dfrac{y-4}{y-4} - \dfrac{2y-1}{(y-4)(y+2)} \cdot \dfrac{3y-1}{3y-1}$

$= \dfrac{(y+7)(y+2) + (5y+3)(y-4) - (2y-1)(3y-1)}{(3y-1)(y-4)(y+2)}$

$= \dfrac{y^2 + 9y + 14 + 5y^2 - 17y - 12 - 6y^2 + 5y - 1}{(3y-1)(y-4)(y+2)}$

$= \dfrac{3y + 1}{(3y-1)(y-4)(y+2)}$

$= -\dfrac{(3y-1)}{(3y-1)(y-4)(y+2)}$

$= -\dfrac{(3y-1)}{(3y-1)(y-4)(y+2)}$

$= \boxed{-\dfrac{1}{(y-4)(y+2)} \quad \left(y \neq \dfrac{1}{3}\right)}$

17. $\dfrac{3x^3 - 5x^2 + x + 1}{x + \dfrac{1}{3}} = \boxed{3x^2 - 6x + 3}$

$$
\begin{array}{r|rrrr}
-\dfrac{1}{3} & 3 & -5 & 1 & 1 \\
& & -1 & 2 & -1 \\
\hline
& 3 & -6 & 3 & 0
\end{array}
$$

18. $\sqrt{\dfrac{4x^2}{5y^5}} = \sqrt{\dfrac{4x^2}{5y^5} \cdot \dfrac{5y}{5y}} = \boxed{\dfrac{2x\sqrt{5y}}{5y^3}}$

19. $\sqrt{\dfrac{2}{3}} + 5\sqrt{6} - \sqrt{\dfrac{3}{2}}$

$= \dfrac{\sqrt{2}}{\sqrt{3}} \cdot \dfrac{\sqrt{3}}{\sqrt{3}} + 5\sqrt{6} - \dfrac{\sqrt{3}}{\sqrt{2}} \cdot \dfrac{\sqrt{2}}{\sqrt{2}}$

$= \dfrac{\sqrt{6}}{3} + 5\sqrt{6} - \dfrac{\sqrt{6}}{2}$

$= 5\sqrt{6} - \dfrac{\sqrt{6}}{6} = \boxed{\dfrac{29}{6}\sqrt{6}}$

20. $\dfrac{5\sqrt{2} - 3\sqrt{3}}{2\sqrt{2} + 3\sqrt{3}}$

$= \dfrac{5\sqrt{2} - 3\sqrt{3}}{2\sqrt{2} + 3\sqrt{3}} \cdot \dfrac{2\sqrt{2} - 3\sqrt{3}}{2\sqrt{2} - 3\sqrt{3}}$

$= \dfrac{10(2) - 2\sqrt{6} + 9(3)}{4(2) - 9(3)}$

$= \dfrac{47 - 2\sqrt{6}}{-19}$

$= \boxed{\dfrac{-47 + 2\sqrt{6}}{19}}$

21. $x^2 - 12x + 36 - b^2$

$= (x - 6)^2 - b^2$

$= \boxed{(x - 6 - b)(x - 6 + b)}$

22. Let

$$\begin{aligned} x &= \text{the number} \\ 3x &= \text{other number} \end{aligned}$$

$$\begin{aligned} \dfrac{1}{x} + \dfrac{1}{3x} &= \dfrac{2}{9} \\ 9x\left(\dfrac{1}{x} + \dfrac{1}{3x}\right) &= 9x \cdot \dfrac{2}{9} \\ 9 + 3 &= 2x \\ 12 &= 2x \\ 6 &= x \\ 3x &= 3(6) = 18 \end{aligned}$$

The numbers are $\boxed{6 \text{ and } 18}$.

23. Let

$$
\begin{aligned}
x &= \text{width of strip} \\
12 - 2x &= \text{width of room less strip} \\
20 - 2x &= \text{length of room less strip}
\end{aligned}
$$

$$
\begin{aligned}
(\text{area of room less strip}) &= (\text{remaining area}) \\
(12 - 2x)(20 - 2x) &= 180 \\
240 - 64x + 4x^2 &= 180 \\
4x^2 - 64x + 60 &= 0 \\
x^2 - 16x + 15 &= 0 \\
(x - 15)(x - 1) &= 0
\end{aligned}
$$

$x = 15 \quad \text{or} \quad x = 1 \quad$ (reject since $12 - 2x < 0$ and $20 - 2x < 0$)

The width of the strip is $\boxed{1 \text{ ft}}$.

24. Let

$$
\begin{aligned}
x &= \text{numerator of a fraction} \\
x + 4 &= \text{denominator of a fraction} \\
\frac{x}{x + 4} &= \text{original fraction} \\
\frac{x + 2}{x + 4 + 2} &= \frac{x + 2}{x + 6} = \text{new fraction}
\end{aligned}
$$

$$
\begin{aligned}
\frac{x + 2}{x + 6} &= \frac{x}{x + 4} + \frac{1}{6} \\
6(x + 4)(x + 6) \cdot \frac{x + 2}{x + 6} &= 6(x + 4)(x + 6) \left[\frac{x}{x + 4} + \frac{1}{6} \right] \\
6(x + 4)(x + 2) &= 6(x + 6)x + (x + 4)(x + 6) \\
6x^2 + 36x + 48 &= 6x^2 + 36x + x^2 + 10x + 24 \\
0 &= x^2 + 10x - 24 \\
0 &= (x + 12)(x - 2) \\
x &= -12 \quad \text{or} \quad x = 2 \\
x + 4 &= -12 + 4 = -8 \qquad x + 4 = 2 + 4 = 6
\end{aligned}
$$

The original fraction is $\boxed{\dfrac{2}{6} \text{ or } \dfrac{-12}{-8}}$.

25. Let

$$
\begin{aligned}
x &= \text{speed of cyclist with no wind} \\
x + 4 &= \text{speed of cyclist with headwind} \\
x - 4 &= \text{speed of cyclist against headwind}
\end{aligned}
$$

	Rate	Time (hours)	Distance = Rate × Time
with headwind	$x + 4$	2	$2(x + 4)$
against headwind	$x - 4$	3	$3(x - 4)$

$$
\begin{aligned}
2(x + 4) &= 3(x - 4) \\
2x + 8 &= 3x - 12 \\
20 &= x
\end{aligned}
$$

average speed of cyclist with no wind: $\boxed{20 \text{ mph}}$

distance $= 2(x + 4) = 2(20 + 4) = \boxed{48 \text{ miles}}$

26. $f(x) = \sqrt{x}$

$\dfrac{f(a+h) - f(a)}{h}$

$= \dfrac{\sqrt{a+h} - \sqrt{a}}{h}$

$= \dfrac{\sqrt{a+h} - \sqrt{a}}{h} \cdot \dfrac{\sqrt{a+h} + \sqrt{a}}{\sqrt{a+h} + \sqrt{a}}$

$= \dfrac{a+h-a}{h(\sqrt{a+h} + \sqrt{a})}$

$= \boxed{\dfrac{1}{\sqrt{a+h} + \sqrt{a}}}$

27. $f(x) = \begin{cases} \sqrt{x-3} & \text{if } x \geq 3 \\ 3-x & \text{if } x < 3 \end{cases}$

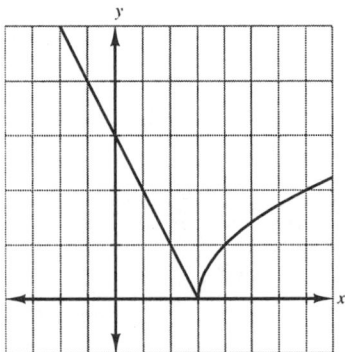

28. $f(x) = 1 - \dfrac{1}{1 - \dfrac{1}{x}}$ and $g(x) = x - 2$

$f(x) = 1 - \dfrac{1}{\dfrac{x-1}{x}} = 1 - \dfrac{x}{x-1} = \dfrac{x-1-x}{x-1} = \dfrac{-1}{x-1}$

$(f \circ g)(x) = \dfrac{-1}{(x-2)-1} = \boxed{\dfrac{-1}{x-3} \, (x \neq 2)}$

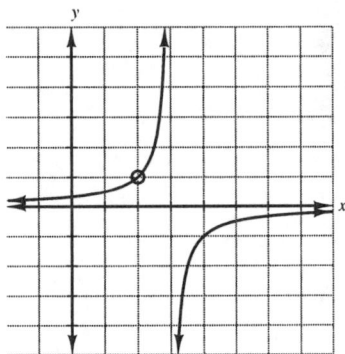

29. $f(x) = \dfrac{1}{x+1}$

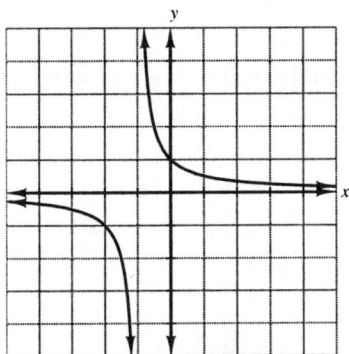

30. $f(x) = 2x^2 - 5x + 3$ and $g(x) = x^2 - x$

$g[f(x)] - f[g(x)]$

$= [(2x^2 - 5x + 3)^2 - (2x^2 - 5x + 3)] - [2(x^2 - x)^2 - 5(x^2 - x) + 3]$

$= [4x^2 - 10x^3 + 6x^2 - 10x^3 + 25x^2 - 15x + 6x^2 - 15x + 9] - [2(x^4 - 2x^3 + x^2) - 5x^2 + 5x + 3]$

$= 4x^4 - 20x^3 + 35x^2 - 25x + 6 - 2x^4 + 4x^3 - 2x^2 + 5x^2 - 5x - 3$

$= \boxed{2x^4 - 16x^3 + 38x^2 - 30x + 3}$

Algebra for College Students

Chapter 9   Conic Sections and Nonlinear Systems of Equations

Section 9.1 The Circle

Problem Set 9.1, pp. 674-676

1. center (3,2): $(h, k) = (3, 2)$, $r = 5$
$$(x-h)^2 + (y-k)^2 = r^2$$
$$(x-3)^2 + (y-2)^2 = 5^2$$
$$\boxed{(x-3)^2 + (y-2)^2 = 25}$$

3. center (−1, 4): $(h, k) = (-1, 4)$, $r = 2$
$$(x-h)^2 + (y-k)^2 = r^2$$
$$[x-(-1)]^2 + (y-4)^2 = 2^2$$
$$\boxed{(x+1)^2 + (y-4)^2 = 4}$$

5. center (−3, −1): $(h, k) = (-3, -1)$, $r = \sqrt{3}$
$$(x-h)^2 + (y-k)^2 = r^2$$
$$[x-(-3)]^2 + [y-(-1)]^2 = (\sqrt{3})^2$$
$$\boxed{(x+3)^2 + (y+1)^2 = 3}$$

7. center (−4, 0): $(h, k) = (-4, 0)$, $r = 2$
$$(x-h)^2 + (y-k)^2 = r^2$$
$$[x-(-4)]^2 + (y-0)^2 = 2^2$$
$$\boxed{(x+4)^2 + y^2 = 4}$$

9. center (0, 0): $(h, k) = (0, 0)$, $r = 7$
$$(x-h)^2 + (y-k)^2 = r^2$$
$$(x-0)^2 + (y-0)^2 = 7^2$$
$$\boxed{x^2 + y^2 = 49}$$

11.
$$x^2 + y^2 = 16$$
$$(x-0)^2 + y-0)^2 = 4^2$$
$$(x-h)^2 + (y-k)^2 = r^2$$
center: $(h, k) = \boxed{(0, 0)}$
radius: $\boxed{r = 4}$

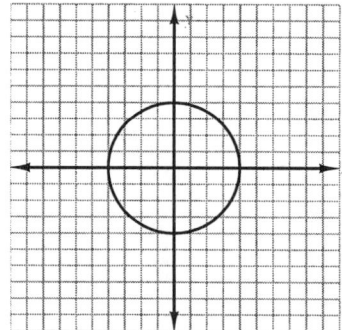

13.
$$(x-3)^2 + (y-1)^2 = 36$$
$$(x-3)^2 + (y-1)^2 = 6^2$$
$$(x-h)^2 + (y-k)^2 = r^2$$

center: $(h, k) = \boxed{(3, 1)}$

radius: $\boxed{r = 6}$

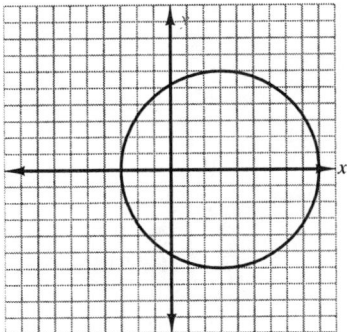

15.
$$(x+3)^2 + (y-2)^2 = 4$$
$$[x-(-3)]^2 + (y-2)^2 = 2^2$$
$$(x-h)^2 + (y-k)^2 = r^2$$

center: $(h, k) = \boxed{(-3, 2)}$

radius: $\boxed{r = 2}$

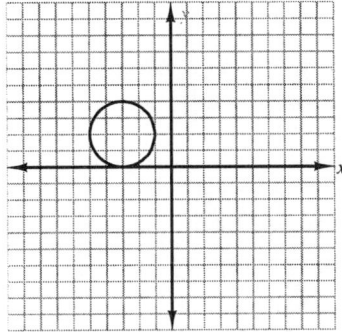

17.
$$(x+2)^2 + (y+2)^2 = 4$$
$$[x-(-2)]^2 + [y-(-2)]^2 = 2^2$$
$$(x-h)^2 + (y-k)^2 = r^2$$

center: $(h, k) = \boxed{(-2, -2)}$

radius: $\boxed{r = 2}$

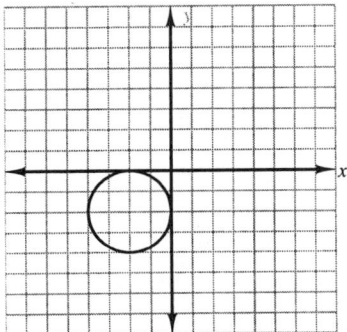

19.
$$x^2 + y^2 + 6x + 2y + 6 = 0$$
$$x^2 + 6x \underline{\quad} + y^2 + 2y \underline{\quad} = -6$$

$\qquad\qquad\uparrow\qquad\qquad\qquad\uparrow$

$$\left(\tfrac{1}{2}\right)(6) = 3 \qquad \left(\tfrac{1}{2}\right)(2) = 1$$
$$3^2 = 9 \qquad\qquad 1^2 = 1$$

$$x^2 + 6x + 9 + y^2 + 2y + 6 = -6 + 9 + 1$$
$$(x+3)^2 + (y+1)^2 = 4 = 2^2$$

center: $(h, k) = \boxed{(-3, -1)}$

radius: $\boxed{r = 2}$

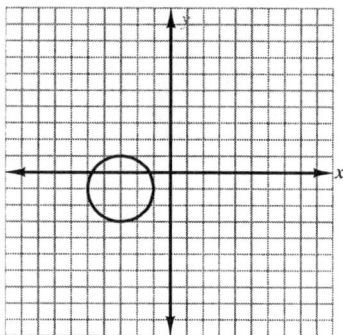

21.
$$x^2 + y^2 - 10x - 6y - 30 = 0$$
$$x^2 - 10x \underline{\quad} + y^2 - 6y \underline{\quad} = 30$$

$\qquad\qquad\uparrow\qquad\qquad\qquad\uparrow$

$$\left(\tfrac{1}{2}\right)(-10) = -5 \qquad \left(\tfrac{1}{2}\right)(-6) = -3$$
$$(-5)^2 = 25 \qquad\qquad (-3)^2 = 9$$

$$x^2 - 10x + 25 + y^2 - 6y + 9 = 30 + 25 + 9$$
$$(x-5)^2 + (y-3)^2 = 64 = 8^2$$

center: $(h, k) = \boxed{(5, 3)}$

radius: $\boxed{r = 8}$

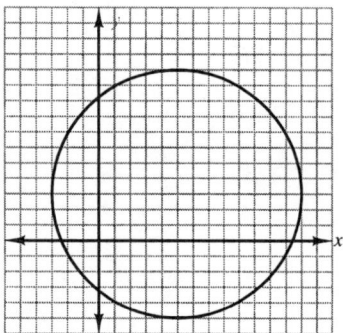

23.
$$x^2 + y^2 + 8x - 2y - 8 = 0$$
$$x^2 + 8x + 16 + y^2 - 2y + 1 = 8 + 16 + 1$$
$$(x + 4)^2 + (y - 1)^2 = 25 = 5^2$$

center: $(h, k) = \boxed{(-4, 1)}$

radius: $\boxed{r = 5}$

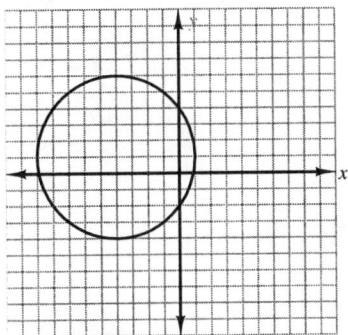

25.
$$x^2 - 2x + y^2 - 15 = 0$$
$$x^2 - 2x \underline{\quad} + y^2 = 15$$

$$\uparrow$$

$$\left(\frac{1}{2}\right)(-2) = -1$$
$$(-1)^2 = -1$$

$$x^2 - 2x + 1 + y^2 = 15 + 1$$
$$(x - 1)^2 + y^2 = 16 = 4^2$$

center: $(h, k) = \boxed{(1, 0)}$

radius: $\boxed{r = 4}$

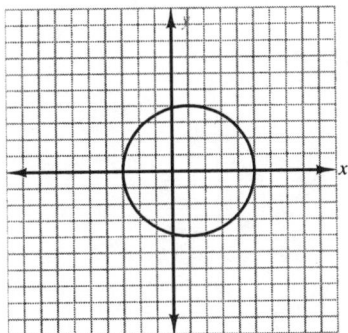

27. $x^2 + y^2 = 16$
$x - 4 = 4$

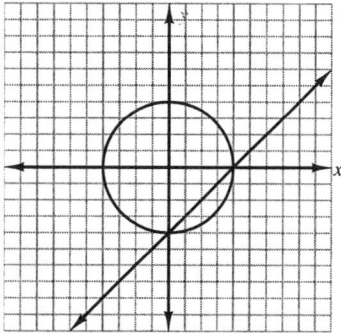

$$\{(0, -4), (4, 0)\}$$

29. $x^2 + y^2 = 25$
$x - y = 1$

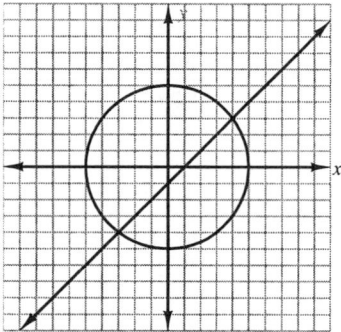

$$\{(-3, -4), (4, 3)\}$$

31. $x^2 + y^2 = 25$
$x - 2y = -5$

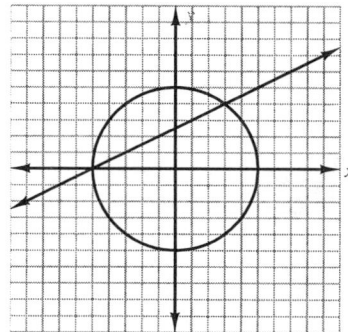

$$\{(-5, 0), (3, 4)\}$$

33. $(x - 3)^2 + (y + 1)^2 = 9$
$y = x - 1$

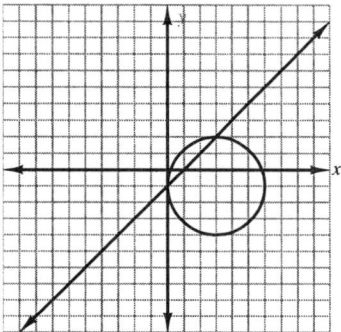

$$\{(0, -1), (3, 2)\}$$

35. $x^2 + y^2 \geq 9$

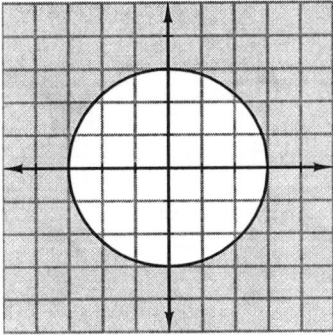

37. $x^2 + y^2 < 16$

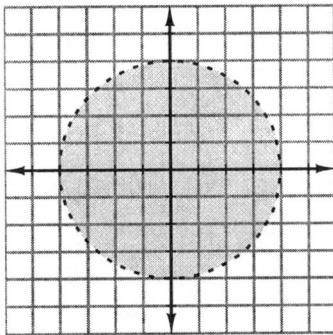

39. $y = \sqrt{9 - x^2}$

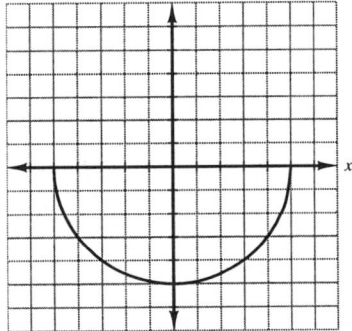

41. $y = -\sqrt{25 - x^2}$

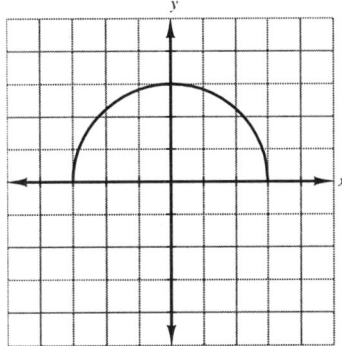

43. $y = \pm\sqrt{36 - x^2}$

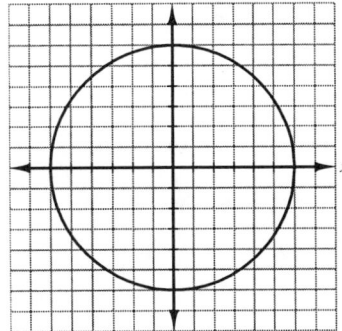

45. $2x + 3y \geq 6$
$\quad\ \ x^2 + y^2 \leq 16$

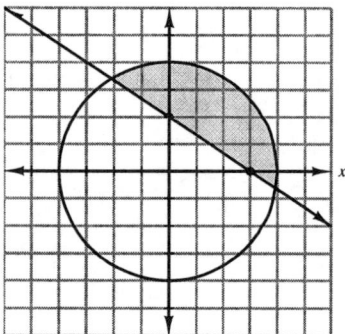

47. $x^2 + y^2 > 4$
$\quad\ \ x^2 + y^2 \leq 9$

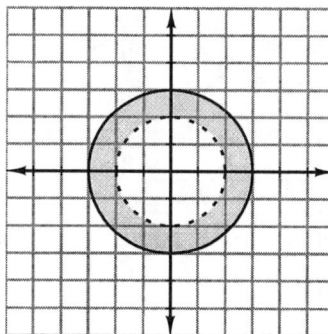

49. $x^2 + y^2 < 4$
$\qquad\ x \geq 0$

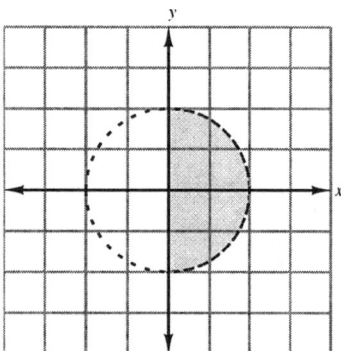

51. $x^2 + y^2 > 36$
$\qquad\ y \leq 0$

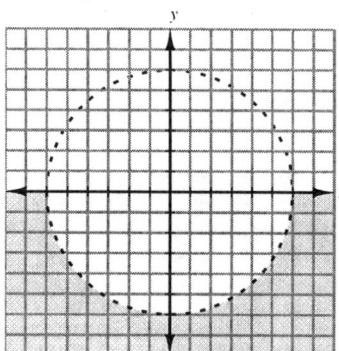

53. $(x - 2)^2 + y^2 \leq 4$
$\quad (x - 2)^2 + (y - 2)^2 \leq 4$

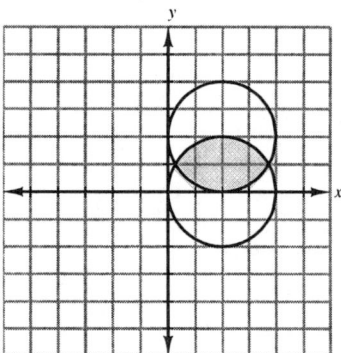

55. $\boxed{\text{D}}$ is true: A is *not* true: $x^2 + y^2 = 16 = 4^2$

$\qquad\qquad\qquad\qquad\qquad\quad$ radius is 4 *not* 16.

$\qquad\qquad\qquad\quad$ B is *not* true: $(x - 3)^2 + (y + 5)^2 = 36 = 6^2$
$\qquad\qquad\qquad\qquad\qquad\quad$ $h = 3,\ k\ -5,\ r = 6$
$\qquad\qquad\qquad\qquad\qquad\quad$ center is at $(3, -5)$ *not* $(3, 5)$

$\qquad\qquad\qquad\quad$ C is *not* true: $(x - 4)^2 + (y + 6)^2 = 25 = 5^2$
$\qquad\qquad\qquad\qquad\qquad\quad$ is a circle of radius 5., centered at $(4, -6)$,
$\qquad\qquad\qquad\qquad\qquad\quad$ *not* $(x - 4) + (y + 6) = 25$

$\qquad\qquad\qquad\quad$ D is true

57. $(x-3)^2 + (y+2)^2 \leq 9 = 3^2$
Area enclosed by a circle with center $(3, -2)$ and radius 3 is:
$$\pi r^2 = \pi 3^2 = 9\pi$$

59. Area of region bounded by the graph of $x^2 + y^2 = 25$ and $x^2 + y^2 = 36$
(i.e. between the two circles)
= difference between area of circle $x^2 + y^2 = 36 = 6^2$
  and area of circle $x^2 + y^2 = 25 = 5^2$
= $\pi 6^2 - \pi 5^2$
= $36\pi - 25\pi$
= $\boxed{11\pi}$

61. center: $(0, 0)$
radius: $r = \sqrt{(4-0)^2 + (3-0)^2}$
$= \sqrt{25}$
$= 5$
$(x-0)^2 + (y-0)^2 = 5^2$
$\boxed{x^2 + y^2 = 25}$

63. center: $(0, 0)$
radius: 4
$(x-0)^2 + (y-0)^2 = 4^2$
$\boxed{x^2 + y^2 = 16}$

65. center: $(3, 7)$
radius: $r = \sqrt{(6-3)^2 + (3-7)^2}$
$= \sqrt{9+16}$
$= \sqrt{25}$
$= 5$
$(x-3)^2 + (y-7)^2 = 5^2$
$\boxed{(x-3)^2 + (y-7)^2 = 25}$

67. $x^2 + y^2 = 25$
$y = \pm\sqrt{25-x^2}$

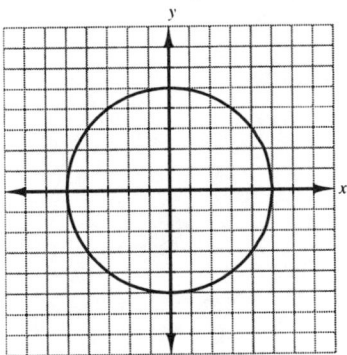

69. Let $2x =$ length of rectangle
$y =$ width of rectangle
$x^2 + y^2 = 25$
$y = \pm\sqrt{25-x^2}$
$y = \sqrt{25-x^2}$ (reject $y = -\sqrt{25-x^2}$)
Area $= (2x)(2y)$
$= 4xy$
$= \boxed{4x\sqrt{25-x^2}}$

Review Problems

74.
$$\frac{1-6x}{x} < 0$$
$$1 - 6x = 0 \quad \text{or} \quad x = 0$$
$$-6x = -1$$
$$x = \frac{1}{6}$$

$$\begin{array}{c|c|c} \text{T} & \text{F} & \text{T} \\ \hline 0 & 1/6 & \end{array}$$

Test -1:
$$\frac{1-6x}{x} < 0$$
$$\frac{1-6(-1)}{(-1)} < 0$$
$$-6 < 0 \quad \text{True}$$

Test $\frac{1}{10}$:
$$\frac{1-6\left(\frac{1}{10}\right)}{\left(\frac{1}{10}\right)} < 0$$

$$\frac{1-\frac{3}{5}}{\frac{1}{10}} < 0$$

$$\frac{\frac{2}{5}}{\frac{1}{10}} < 0 \quad \text{False}$$

Test 1:
$$\frac{1-6x}{x} < 0$$
$$-5 < 0 \quad \text{True}$$

$$\boxed{\left\{ x \mid x < 0 \text{ or } x > \frac{1}{6} \right\}}$$

75.
$$y = x^2 + 2x - 3$$

x-intercepts:
$$x^2 + 2x - 3 = 0$$
$$(x + 3)(x - 1) = 0$$
$$x = -3 \qquad x = 1$$

y-intercept:
$$y = 0^2 + 2 \cdot 0 - 3$$
$$y = -3$$

vertex:
$$x = -\frac{b}{2a} \qquad\qquad y = (-1)^2 + 2(-1) - 3$$
$$= -\frac{2}{2(1)} \qquad\qquad = -4$$
$$= -1$$

vertex: $(-1, -4)$

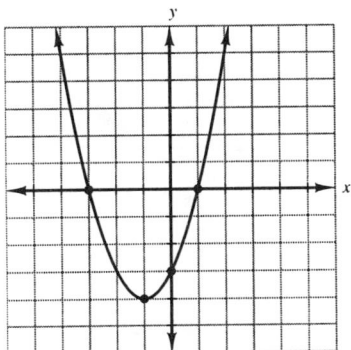

76. $x^6 - 5x^3 - 6 \;=\; (x^3 - 6)(x^3 + 1)$

$ \;=\; \boxed{(x^3 - 6)(x + 1)(x^2 - x + 1)}$

Section 9.2 The Ellipse

Problem Set 9.2, pp. 685-688

1. $\dfrac{x^2}{9} + \dfrac{y^2}{4} = 1$

y-intercepts (Let $x = 0$): $\dfrac{y^2}{4} = 1$

$\phantom{y\text{-intercepts (Let } x = 0):}\quad y^2 = 4$

$\phantom{y\text{-intercepts (Let } x = 0):}\quad\; y = \pm2$

x-intercepts (Let $y = 0$): $\dfrac{x^2}{9} = 1$

$\phantom{x\text{-intercepts (Let } y = 0):}\quad x^2 = 9$

$\phantom{x\text{-intercepts (Let } y = 0):}\quad\;\; x = \pm3$

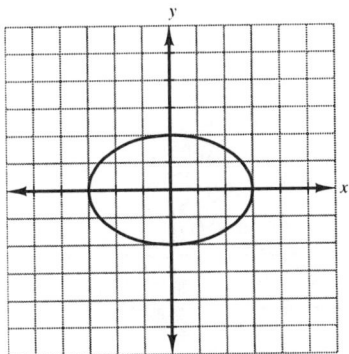

3. $\dfrac{x^2}{9} + \dfrac{y^2}{36} = 1$

y-intercepts (Let $x = 0$): $\dfrac{y^2}{36} = 1$

$\phantom{y\text{-intercepts (Let } x = 0):}\quad y^2 = 36$

$\phantom{y\text{-intercepts (Let } x = 0):}\quad\; y = \pm6$

x-intercepts (Let $y = 0$): $\dfrac{x^2}{9} = 1$

$\phantom{x\text{-intercepts (Let } y = 0):}\quad x^2 = 9$

$\phantom{x\text{-intercepts (Let } y = 0):}\quad\;\; x = \pm3$

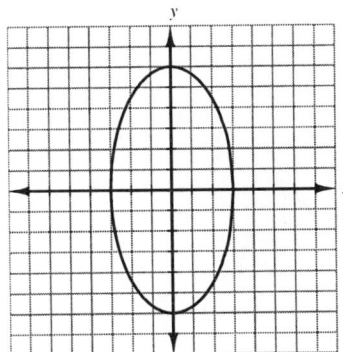

5.
$$\frac{x^2}{25} + \frac{y^2}{64} = 1$$

y-intercepts:
$$\frac{y^2}{64} = 1$$
$$y^2 = 64$$
$$y = \pm 8$$

x-intercepts:
$$\frac{x^2}{25} = 1$$
$$x^2 = 25$$
$$x = \pm 5$$

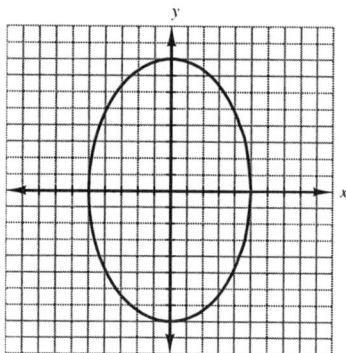

7.
$$\frac{x^2}{49} + \frac{y^2}{81} = 1$$

y-intercepts:
$$\frac{y^2}{81} = 1$$
$$y^2 = 81$$
$$y = \pm 9$$

x-intercepts:
$$\frac{x^2}{49} = 1$$
$$x^2 = 49$$
$$x = \pm 7$$

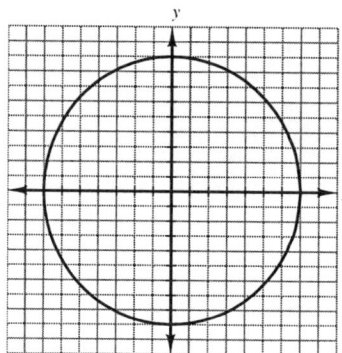

9.
$$25x^2 + 4y^2 = 100$$
$$\frac{25x^2}{100} + \frac{4y^2}{100} = 1$$
$$\frac{x^2}{4} + \frac{y^2}{25} = 1$$

y-intercepts:
$$\frac{y^2}{25} = 1$$
$$y^2 = 25$$
$$y = \pm 5$$

x-intercepts:
$$\frac{x^2}{4} = 1$$
$$x^2 = 4$$
$$x = \pm 2$$

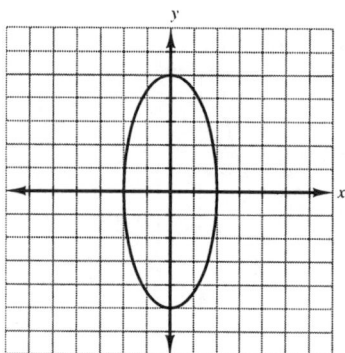

11.
$$4x^2 + 16y^2 = 64$$
$$\frac{x^2}{16} + \frac{y^2}{4} = 1$$

y-intercepts:
$$\frac{y^2}{4} = 1$$
$$y^2 = 4$$
$$y = \pm 2$$

x-intercepts:
$$\frac{x^2}{16} = 1$$
$$x^2 = 16$$
$$x = \pm 4$$

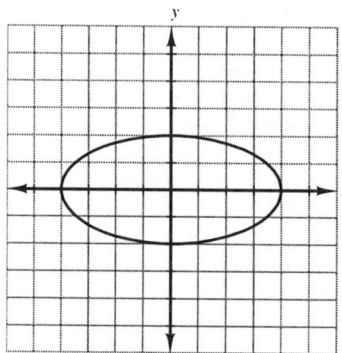

13.
$$25x^2 + 9y^2 = 225$$
$$\frac{x^2}{9} + \frac{y^2}{25} = 1$$

y-intercepts:
$$\frac{y^2}{25} = 1$$
$$y^2 = 25$$
$$y = \pm 5$$

x-intercepts:
$$\frac{x^2}{9} = 1$$
$$x^2 = 9$$
$$x = \pm 3$$

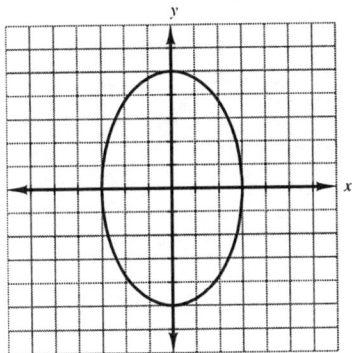

15.
$$x^2 + 2y^2 = 8$$
$$\frac{x^2}{8} + \frac{2y^2}{8} = 1$$
$$\frac{x^2}{8} + \frac{y^2}{4} = 1$$

y-intercepts:
$$\frac{y^2}{4} = 1$$
$$y^2 = 4$$
$$y = \pm 2$$

x-intercepts:
$$\frac{x^2}{8} = 1$$
$$x^2 = 8$$
$$x = \pm\sqrt{8}$$
$$= \pm 2\sqrt{2}$$
$$= \pm 2.8$$

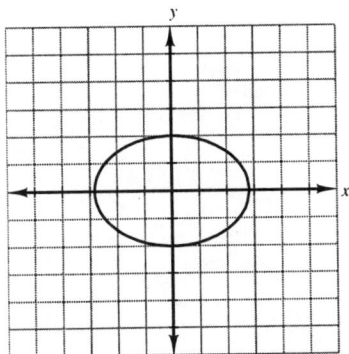

17.
$$y = \sqrt{16 - x^2}$$

y-intercept:
$$y = -\sqrt{16}$$
$$= -4$$

x-intercepts:
$$0 = -\sqrt{16 - 4x^2}$$
$$4x^2 = 16$$
$$x^2 = 4$$
$$x = \pm 2$$

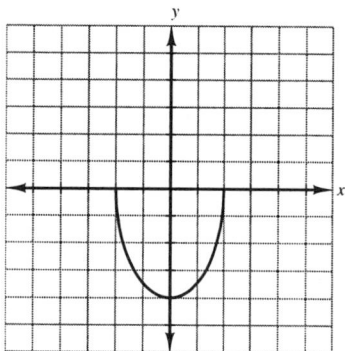

19.
$$\frac{(x-2)^2}{9} + \frac{(y-1)^2}{4} = 1$$

$$C(h, k) = C(2, 1)$$
$$a^2 = 9 \qquad b^2 = 4$$
$$a = 3 \qquad b = 2$$

vertices:
$$(2 + 3, 1) = (5, 1)$$
$$(2 - 3, 1) = (-1, 1)$$
$$(2, 1 + 2) = (2, 3)$$
$$(2, 1 - 2) = (2, -1)$$

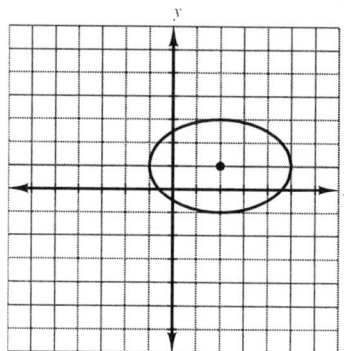

21.
$$\frac{(x+2)^2}{9} + \frac{(y-3)^2}{16} = 1$$

$$C(h, k) = C(-2, 3)$$
$$a^2 = 9 \qquad\qquad b^2 = 16$$
$$a = 3 \qquad\qquad b = 4$$

Vertices:
$$(-2 + 3, 4) = (1, 4)$$
$$(-2 - 3, 4) = (-5, 4)$$
$$(-2, 3 + 4) = (-2, 7)$$
$$(-2, 3 - 4) = (-2, -1)$$

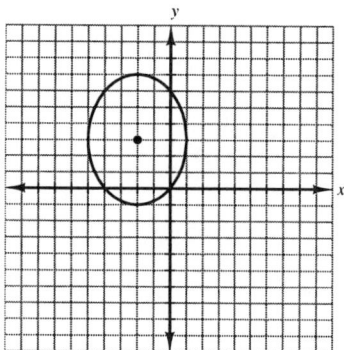

23. $\dfrac{x^2}{25} + \dfrac{(y-2)^2}{36} = 1$

$C(h, k) = C(0, 2)$

$a^2 = 25 \qquad\qquad b^2 = 36$

$a = 5 \qquad\qquad\ b = 6$

Vertices: $(0 + 5, 2) = (5, 2)$

$(0 - 5, 2) = (-5, 2)$

$(0, 2 + 6) = (0, 8)$

$(0, 2 - 6) = (0, -4)$

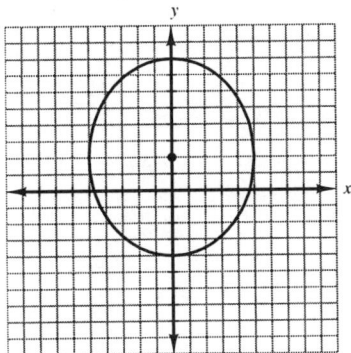

25. $9x^2 + 25y^2 - 36x + 50y - 164 = 0$

$9x^2 - 36x + 25y^2 + 50y = 164$

$9(x^2 - 4x) + 25(y^2 + 2y) = 164$

$9(x^2 - 4x + 4) + 25(y^2 + 2y + 1) = 164 + 36 + 25$

$9(x - 2)^2 + 25(y + 1)^2 = 225$

$\dfrac{9(x - 2)^2}{225} + \dfrac{25(y + 1)^2}{225} = \dfrac{225}{225}$

$\dfrac{(x - 2)^2}{25} + \dfrac{(y + 1)^2}{16} = 1$

$C(2, -1)$

$a = \sqrt{25} = 5, \qquad b = \sqrt{16} = 4$

Vertices: $(2 + 5, -1) = (7, -1)$

$(2 - 5, -1) = (-3, -1)$

$(2 - 5, -1) = (2, 3)$

$(2, -1 - 4) = (2, -5)$

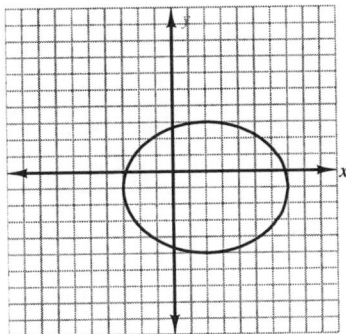

27.
$$9x^2 + 16y^2 - 18x + 64y - 71 = 0$$
$$9x^2 - 18x + 16y^2 + 64y = 71$$
$$9(x^2 - 2x) + 16(y^2 + 4y) = 71$$
$$9(x^2 - 2x + 1) + 16(y^2 + 4y + 4) = 71 + 9 + 64$$
$$9(x - 1)^2 + 16(y + 2)^2 = 144$$
$$\frac{(x - 1)^2}{16} + \frac{(y + 2)^2}{9} = 1$$

$C(1, -2)$

$$a = \sqrt{16} = 4, \qquad b = \sqrt{9} = 3$$

Vertices: $(1 + 4, -2) = (5, -2)$
$\qquad\quad (1 - 4, -2) = (-3, -2)$
$\qquad\quad (1, -2 + 3) = (1, 1)$
$\qquad\quad (1, -2 - 3) = (1, -5)$

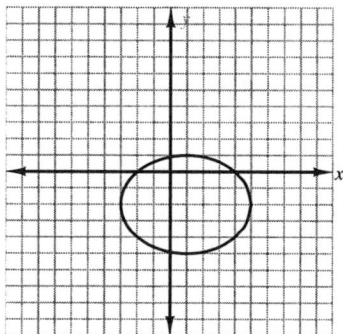

29.
$$9x^2 + 25y^2 - 100y - 125 = 0$$
$$9x^2 + 25(y^2 - 4y) = 125$$
$$9x^2 + 25(y^2 - 4y + 4) = 125 + 100$$
$$9x^2 + 25(y - 2)^2 = 225$$
$$\frac{x^2}{25} + \frac{(y - 2)^2}{9} = 1$$

$C(0, 2)$

$$a = \sqrt{25} = 5, \qquad b = \sqrt{9} = 3$$

Vertices: $(0 + 5, 2) = (5, 2)$
$\qquad\quad (0 - 5, 2) = (-5, 2)$
$\qquad\quad (0, 2 + 3) = (0, 5)$
$\qquad\quad (0, 2 - 3) = (0, -1)$

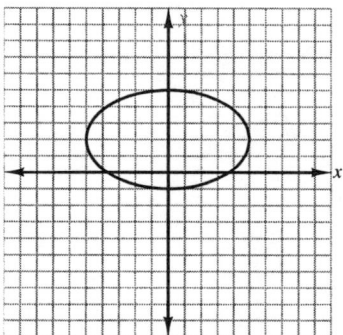

31.
$$4x^2 + y^2 + 16x - 6y - 39 = 0$$
$$4x^2 + 16x + y^2 - 6y = 39$$
$$4(x^2 + 4x) + (y^2 - 6y) = 39$$
$$4(x^2 + 4x + 4) + (y^2 - 6y + 9) = 39 + 16 + 9$$
$$4(x + 2)^2 + (y - 3)^2 = 64$$
$$\frac{(x + 2)^2}{16} + \frac{(y - 3)^2}{64} = 1$$

C(−2, 3)
$$a = \sqrt{16} = 4, \qquad b = \sqrt{64} = 8$$

Vertices: (−2 + 4, 3) = (2, 3)
(−2 − 4, 3) = (−6, 3)
(−2, 3 + 8) = (−2, 11)
(−2, 3 − 8) = (−2, −5)

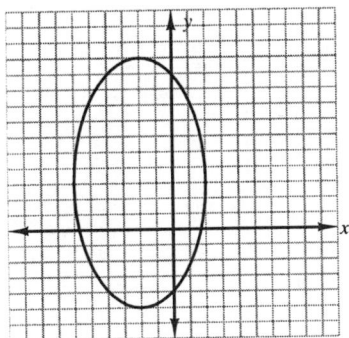

33. $x^2 + y^2 = 1$
$x^2 + 9y^2 = 9$

$$x^2 + y^2 = 1$$

C(0, 0), $r = \sqrt{1} = 1$

$$\frac{x^2}{9} + y^2 = 1$$

C(0, 0)
$a = \sqrt{9} = 3, \quad b = 1$
Vertices: (±3, 0)
(0, ±1)

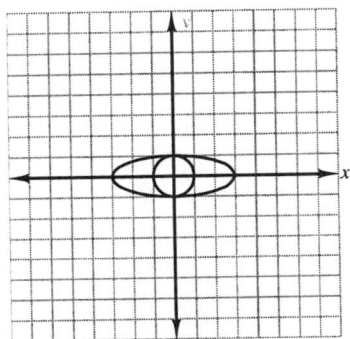

$$\boxed{\{(0, -1), (0, 1)\}}$$

Check: (0, −1):
$$x^2 + y^2 = 1 \qquad\qquad x^2 + 9y^2 = 9$$
$$0 + (-1)^2 = 1 \qquad\qquad 0 + 9(-1)^2 = 9$$
$$1 = 1 \text{ (checks)} \qquad\qquad 9 = 9 \text{ (checks)}$$

35. $\dfrac{x^2}{25} + \dfrac{y^2}{9} = 1$

$y = 3$

$\dfrac{x^2}{25} + \dfrac{y^2}{9} = 1$

$C(0, 0)$

$a = \sqrt{25} = 5,\ b = \sqrt{9} = 3$

Vertices: $(\pm 5, 0)$

$(0, \pm 3)$

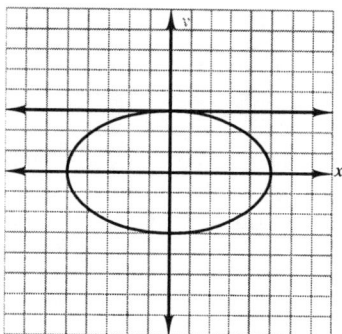

$\boxed{\{(0, 3)\}}$ (checks)

37. $4x^2 + y^2 = 4$

$\underline{2x - y = 2}$

$x^2 + \dfrac{y^2}{4} = 1$

$C(0, 0)$

$a = 1,\ b = 2$

Vertex: $(\pm 1, 0)$

$(0, \pm 2)$

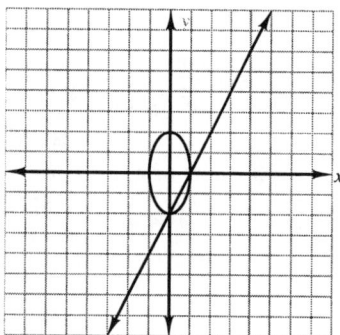

$\boxed{\{(0, -2), (1, 0)\}}$ (checks)

39. $\dfrac{x^2}{25} + \dfrac{y^2}{16} \leq 1$

Graph $\dfrac{x^2}{25} + \dfrac{y^2}{16} = 1$ with a solid line

$C(0, 0)$

x-intercepts: $(\pm 5, 0)$

y-intercepts: $(0, \pm 4)$

Using $(0, 0)$ as a test point:

$0 + 0 \ \leq \ 1$ True

(shade inside ellipse)

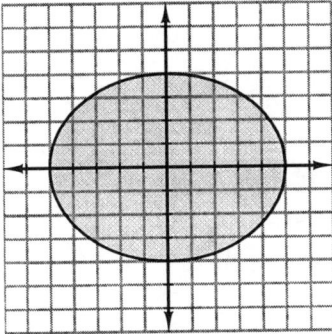

41. $\dfrac{x^2}{16} + \dfrac{y^2}{9} > 1$

Graph $\dfrac{x^2}{16} + \dfrac{y^2}{9} = 1$ with a dashed line

$C(0, 0)$

x-intercepts: $(\pm 4, 0)$

y-intercepts: $(0, \pm 3)$

Test point: $(0, 0)$

$0 + 0 \ > \ 1$ False

(shade outside ellipse)

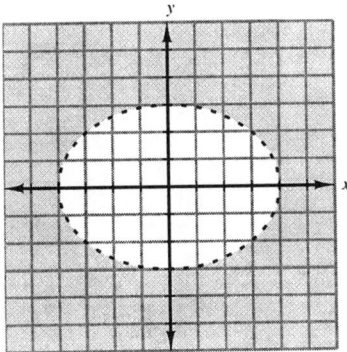

43. $4x^2 + 9y^2 < 36$

$\dfrac{x^2}{9} + \dfrac{y^2}{4} < 1$

Graph $\dfrac{x^2}{9} + \dfrac{y^2}{4} = 1$ with a dashed line

$C(0, 0)$

x-intercepts: $(\pm 3, 0)$

y-intercepts: $(0, \pm 2)$

Test point $(0, 0)$:

$0 + 0 < 1$ True

(shade inside ellipse)

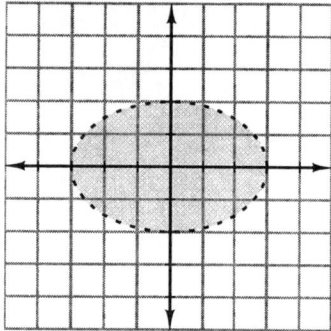

45. $\dfrac{(x-1)^2}{9} + \dfrac{(y+2)^2}{16} > 1$

Graph $\dfrac{(x-1)^2}{9} + \dfrac{(y+2)^2}{16} = 1$ with a dashed line

$C(1, -2)$

$a = \sqrt{9} = 3,\ b = \sqrt{16} = 4$

Vertices: $(1 \pm 3, -2) = (4, -2)$ and $(-2, -2)$

$(1, -2 \pm 4) = (1, 2)$ and $(1, -6)$

Test point: $(1, -2)$

$0 + 0 > 1$ False

(shade outside ellipse)

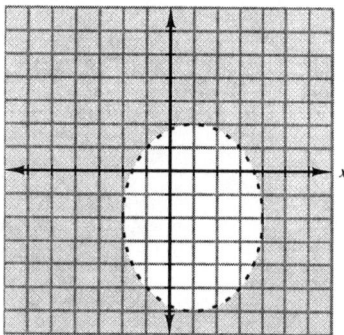

47. $\dfrac{(x+2)^2}{4} + \dfrac{(y-3)^2}{9} \le 1$

Graph $\dfrac{(x+2)^2}{4} + \dfrac{(y-3)^2}{9} = 1$ with a solid line

$C(-2, 3)$

$a = \sqrt{4} = 2, \ b = \sqrt{9} = 3$

Vertices: $(-2 \pm 2, 3) = (0, 3)$ and $(-4, 3)$

$(-2, 3 \pm 3) = (-2, 6)$ and $(-2, 0)$

Test point: $(-2, 3)$

$0 + 0 \le 1$ True

(shade inside ellipse)

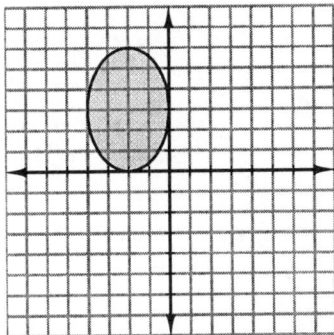

49. $x^2 + y^2 \le 9$

$\dfrac{x^2}{4} + \dfrac{y^2}{25} \ge 1$

Graph $x^2 + y^2 = 9$ Graph $\dfrac{x^2}{4} + \dfrac{y^2}{25} = 1$

with a solid line   with a solid line

Test point: $(0, 0)$ $C(0, 0)$

$0 + 0 \le 9$ True $a = \sqrt{4} = 2, \ b = \sqrt{25} = 5$

(inside circle) x-intercepts: $(\pm 2, 0)$

y-intercepts: $(0, \pm 5)$

Test point: $(0, 0)$

$0 + 0 \ge 1$ False

(outside ellipse)

Shade region inside circle and outside ellipse.

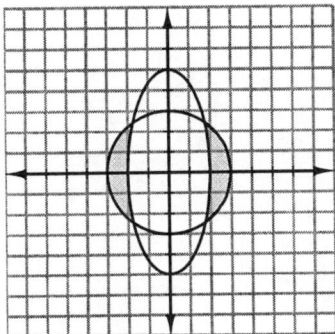

51. $\dfrac{x^2}{16} + \dfrac{y^2}{9} \le 1$

$\dfrac{x^2}{9} + \dfrac{y^2}{16} \le 1$

  Graph $\dfrac{x^2}{16} + \dfrac{y^2}{9} = 1$ Graph $\dfrac{x^2}{9} + \dfrac{y^2}{16} = 1$

    with a solid line   with a solid line

    $C(0,0)$ $C(0,0)$

    $a = \sqrt{16} = 4,\ b = \sqrt{9} = 3$ $a = \sqrt{9} = 3,\ b = \sqrt{16} = 4$

    x-intercepts: $(\pm 4, 0)$ x-intercepts: $(\pm 3, 0)$

    y-intercepts: $(0, \pm 3)$ y-intercepts: $(0, \pm 4)$

  Test point: $(0,0)$ Test point: $(0,0)$

    $0 + 0 \le 1$ True $0 + 0 \le 1$ True

    (inside ellipse)   (inside ellipse)

  Shade region common to both ellipses inside each ellipse.

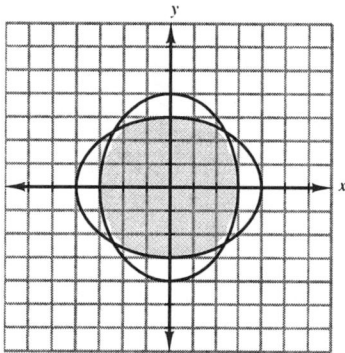

53. $\dfrac{x^2}{4} + y^2 < 1$

  $y \le 0$

  Graph $\dfrac{x^2}{4} + y^2 = 1$ with dashed line

    $C(0,0)$

    $a = \sqrt{4} = 2,\ b = 1$

    x-intercepts: $(\pm 2, 0)$

    y-intercepts: $(0, \pm 1)$

  Test point: $(0,0)$

    $0 + 0 < 1$ True

    (inside ellipse)

  Since $y \le 0$, shade region inside ellipse below the x-axis.

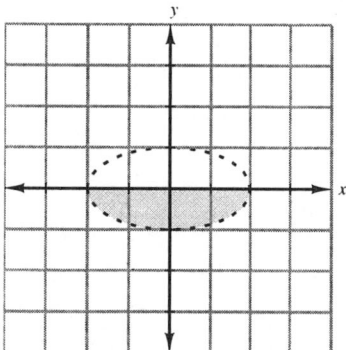

55.
$$9x^2 + 25y^2 \leq 225$$
$$x^2 + 4y^2 \geq 16$$

Graph $\dfrac{x^2}{25} + \dfrac{y^2}{9} = 1$ Graph $\dfrac{x^2}{16} + \dfrac{y^2}{4} = 1$

  with a solid curve   with a solid curve

  $C(0, 0)$ $C(0, 0)$

  $a = \sqrt{25} = 5,\ b = \sqrt{9} = 3$ $a = \sqrt{16} = 4,\ b = \sqrt{4} = 2$

  x-intercepts: $(\pm 5, 0)$ x-intercepts: $(\pm 4, 0)$

  y-intercepts: $(0, \pm 3)$ y-intercepts: $(0, \pm 2)$

Test point: $(0, 0)$ Test point: $(0, 0)$

  $0 + 0 \leq 1$ True $0 + 0 \geq 1$ False

  (inside ellipse)   (outside ellipse)

Shade the region between the ellipses.

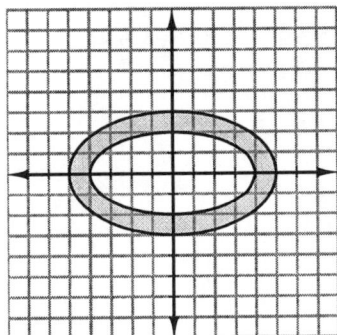

57. $\boxed{\text{C}}$ is true; $\dfrac{x^2}{a^2} + \dfrac{y^2}{b^2} = 1$

$$\dfrac{y^2}{b^2} = 1 - \dfrac{x^2}{a^2}$$

$$y^2 = b^2 \left(1 - \dfrac{x^2}{a^2} \right)$$

$$y = \pm b \sqrt{1 - \dfrac{x^2}{a^2}}$$

y is not a function of x.

59. $4x^2 + y^2 = 18$

$$y^2 = 18 - 4x^2$$

$$y = \pm\sqrt{18 - 4x^2}$$

Using a graphing calculator, graph

$$y = \sqrt{18 - 4x^2} \quad \text{and} \quad y = -\sqrt{18 - 4x^2}$$

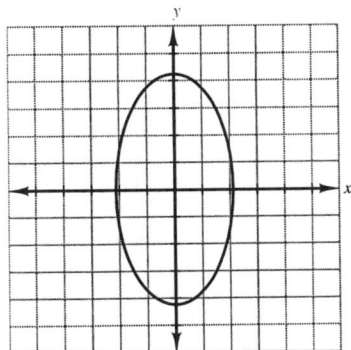

61. arch with height of 8 feet and width of 24 feet:

$b = 8$, $a = \frac{1}{2}(24) = 12$

$$\frac{x^2}{a^2} + \frac{y^2}{b^2} = 1$$

$$\frac{x^2}{12^2} + \frac{y^2}{8^2} = 1$$

$$\frac{x^2}{144} + \frac{y^2}{64} = 1$$

$$\frac{y^2}{64} = 1 - \frac{x^2}{144}$$

$$y^2 = 64\left(1 - \frac{x^2}{144}\right)$$

$$\boxed{y = 8\sqrt{1 - \frac{x^2}{144}}} \quad \left(\text{reject} -8\sqrt{1 - \frac{x^2}{144}} \text{ since only the positive } y \text{ expresses the semicircle of an arch}\right)$$

If 4 feet from end, then $x = 12 - 4 = 8$.

$$y = 8\sqrt{1 - \frac{8^2}{144}} = 8\sqrt{1 - \frac{64}{144}} = 8\sqrt{1 - \frac{4}{9}} = 8\sqrt{\frac{5}{9}} = \frac{8\sqrt{5}}{3}$$

The arch is $\boxed{\frac{8\sqrt{5}}{3} \text{ feet} \approx 5.96 \text{ feet}}$.

63. a. distance from (x, y) to $(-c, 0)$: $\sqrt{(x+c)^2 + (y-0)^2} = \sqrt{(x+c)^2 + y^2}$

distance from (x, y) to $(c, 0)$: $\sqrt{(x-c)^2 + (y-0)^2} = \sqrt{(x-c)^2 + y^2}$

sum of distances from (x, y) to foci $= 2a$

$$\sqrt{(x+c)^2 + y^2} + \sqrt{(x-c)^2 + y^2} = 2a$$

$$\sqrt{(x+c)^2 + y^2} = 2a - \sqrt{(x-c)^2 + y^2}$$

$$(x+c)^2 + y^2 = 4a^2 - 4a\sqrt{(x-c)^2 + y^2} + (x-c)^2 + y^2$$

$$x^2 + 2xc + c^2 + y^2 = 4a^2 - 4a\sqrt{(x-c)^2 + y^2} + x^2 - 2xc + c^2 + y^2$$

$$4a\sqrt{(x-c)^2 + y^2} = 4a^2 - 4xc$$

$$a\sqrt{(x-c)^2 + y^2} = a^2 - xc$$

$$a^2(x-c)^2 + a^2y^2 = a^4 - 2a^2xc + x^2c^2$$

$$a^2x^2 - 2a^2xc + a^2c^2 + a^2y^2 = a^4 - 2a^2xc + x^2c^2$$

$$a^2x^2 + a^2c^2 + a^2y^2 = a^4 + x^2c^2$$

$$a^2x^2 - x^2c^2 + a^2y^2 = a^4 - a^2c^2$$

$$x^2(a^2 - c^2) + a^2y^2 = a^2(a^2 - c^2)$$

$$\frac{x^2(a^2 - c^2)}{a^2(a^2 - c^2)} + \frac{a^2y^2}{a^2(a^2 - c^2)} = \frac{a^2(a^2 - c^3)}{a^2(a^2 - c^2)}$$

$$\boxed{\frac{x^2}{a^2} + \frac{y^2}{a^2 - c^2} = 1}$$

b. Let $b^2 = a^2 - c^2$

$$\boxed{\frac{x^2}{a^2} + \frac{y^2}{b^2} = 1}$$

Review Problems

66. $y - k = (x - h)^2$
$\qquad\quad y = x^2 - 2hx + h^2 + k$

vertex:

$$x = -\frac{b}{2a} = -\frac{(-2h)}{2(1)} = h$$

If $x = h$:

$$\begin{aligned} y &= h^2 - 2h(h) + h^2 + k \\ &= h^2 - 2h^2 + h^2 + k = k \end{aligned}$$

vertex: $\boxed{(h, k)}$

67. $s(t) = -16t^2 + 32t + 62$

a. maximum height occurs when $t = -\dfrac{b}{2a}$.

$$t = -\frac{32}{2(-16)} = \frac{-32}{-32} = 1$$

$\boxed{\text{After 1 second}}$

b. maximum height:

$$s = -16 \cdot 1^2 + 32 \cdot 1 + 62 = 78$$

$\boxed{\text{Maximum height of 78 ft occurs 1 second after the ball is thrown upward.}}$

68. $\boxed{4}$ $\qquad$ 13 $\qquad$ $\boxed{9}$ $\qquad$ $4 + 9 = 13$

$\qquad\qquad$ 16 $\quad$ 21 $\qquad\qquad$ $4 + 12 = 16$

$\qquad\qquad\quad$ $\boxed{12}$ $\qquad\qquad\quad$ $9 + 12 = 21$

Section 9.3 The Hyperbola

Problem Set 9.3, pp. 697-699

1. $\dfrac{x^2}{9} - \dfrac{y^2}{25} = 1$

$\dfrac{x^2}{a^2} - \dfrac{y^2}{b^2} = 1$

$\quad a^2 = 9 \qquad\qquad b^2 = 25$

$\quad a = \sqrt{9} = 3 \qquad b = \sqrt{25} = 5$

The sides of the rectangle used to draw the asymptotes pass through 3 and −3 on the x-axis and 5 and −5 on the y-axis.

x-intercepts (set $y = 0$): $\dfrac{x^2}{9} = 1$

$\qquad\qquad\qquad\qquad\qquad x^2 = 9$

$\qquad\qquad\qquad\qquad\qquad x = \pm 3$

y-intercepts (set $x = 0$): $\quad -\dfrac{y^2}{25} = 1$

$$y^2 = -25$$
$$y = \pm\sqrt{-25}$$

y is not a real number. No y-intercepts.

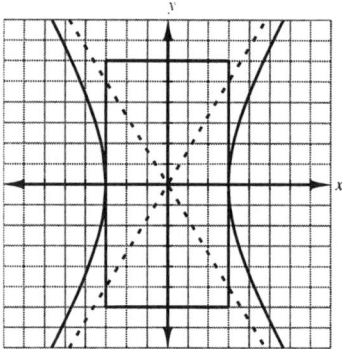

3. $\dfrac{x^2}{100} - \dfrac{y^2}{64} = 1$

$\qquad a^2 = 100 \qquad a = 10$
$\qquad b^2 = 64 \qquad\ \ b = 8$

The sides of the rectangle asymptotes pass through 10 and -10 on the x-axis and 8 and -8 on the y-axis.
x-intercepts: ± 10
No y-intercepts.

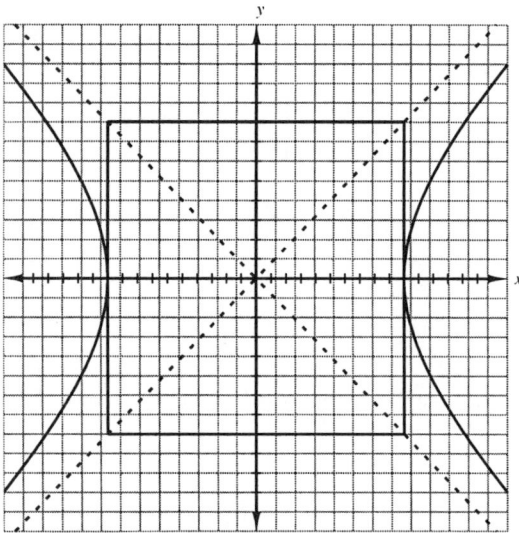

5. $\dfrac{y^2}{16} - \dfrac{x^2}{36} = 1$

$\dfrac{y^2}{a^2} - \dfrac{x^2}{b^2} = 1$

$a^2 = 16 \qquad\qquad b^2 = 36$

$a = 4 \qquad\qquad\ b = 6$

The sides of the rectangle used to draw the asymptotes pass through 4 and –4 on the y-axis and 6 and –6 on the x-axis.

x-intercepts (set $y = 0$): $-\dfrac{x^2}{36} = 1$

$x^2 = -36$

$x = \pm\sqrt{-36}$

x is not a real number. No x-intercepts.

y-intercepts (set $x = 0$): $\dfrac{y^2}{16} = 1$

$y^2 = 16$

$y = \pm 4$

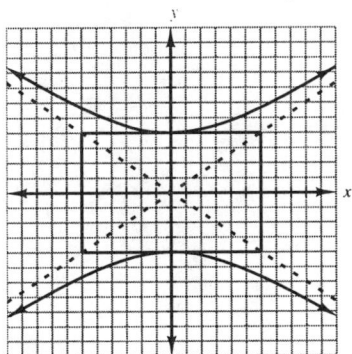

7. $\dfrac{y^2}{36} - \dfrac{x^2}{25} = 1$

$a^2 = 36 \qquad\qquad b^2 = 25$

$a = 6 \qquad\qquad\ b = 5$

The sides of the rectangle used to draw the asymptotes pass through 5 and –5 on the x-axis and 6 and –6 on the y-axis.

y-intercepts: ± 6

no x-intercepts

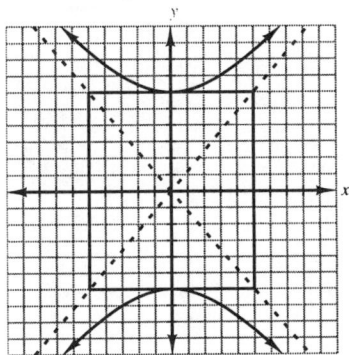

9. $9x^2 - 4y^2 = 36$

$$\frac{9x^2}{36} - \frac{4y^2}{36} = 1$$

$$\frac{x^2}{4} - \frac{y^2}{9} = 1$$

$a^2 = 4 \qquad b^2 = 9$

$a = 2 \qquad b = 3$

The sides of the rectangle and to draw the asymptotes pass through 2 and −2 on the x-axis and 3 and −3 on the y-axis.

x-intercepts: ±2

no y-intercepts

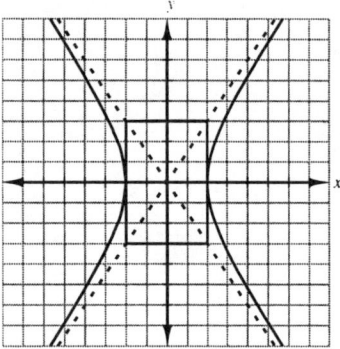

11. $9y^2 - 25x^2 = 225$

$$\frac{y^2}{25} - \frac{x^2}{9} = 1$$

$a^2 = 25 \qquad b^2 = 9$

$a = 5 \qquad b = 3$

The sides of the rectangle used to draw the asymptotes pass through 3 and −3 on the x-axis and 5 and −5 on the y-axis.

y-intercepts: ±5

no x-intercepts

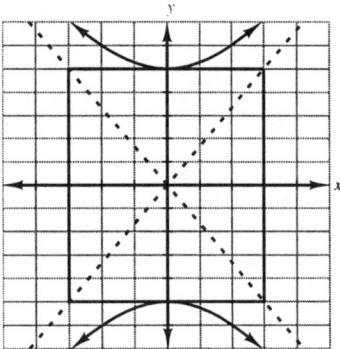

13.
$$4x^2 = 4 + y^2$$
$$4x^2 - y^2 = 4$$
$$\frac{4x^2}{4} - \frac{y^2}{4} = 1$$
$$\frac{x^2}{1} - \frac{y^2}{4} = 1$$
$$a^2 = 1 \qquad b^2 = 4$$
$$a = 1 \qquad b = 2$$

The sides of the rectangle used to draw the asymptotes pass through 1 and -1 on the x-axis and 2 and -2 on the y-axis.

x-intercepts: ± 1

no y-intercepts

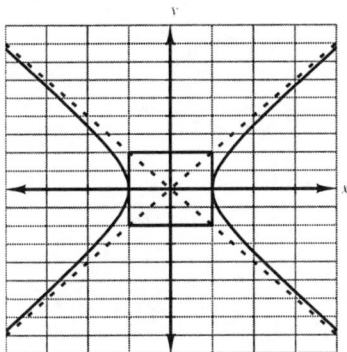

15. $\dfrac{(x-2)^2}{4} - \dfrac{(y-3)^2}{9} = 1$

$$C(2, 3)$$
$$a^2 = 4 \qquad a = 2$$
$$b^2 = 9 \qquad b = 3$$

Vertices: $(2 \pm 2, 3) = (4, 3)$ and $(0, 3)$

No y-intercepts

The rectangle used to draw the asymptotes also passes thorugh $(2, 3 \pm 3) = (2, 6)$ and $(2, 0)$

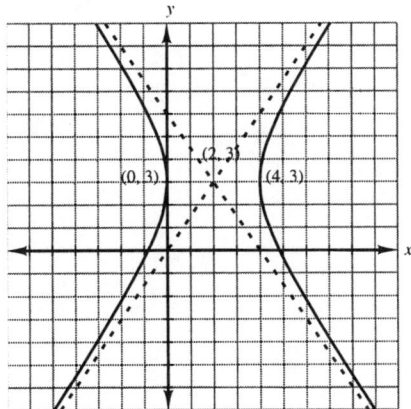

17. $\dfrac{(x-1)^2}{16} - \dfrac{(y+2)^2}{9} = 1$

$C(1, -2)$

$\begin{array}{llll} a^2 & = & 16 & a = 4 \\ b^2 & = & 9 & b = 3 \end{array}$

Vertices: $(1 \pm 4, -2) = (5, -2)$ and $(-3, -2)$

No y-intercepts

The rectangle used to draw the asymptotes also passes through $(1, -2 \pm 3) = (1, 1)$ and $(1, -5)$

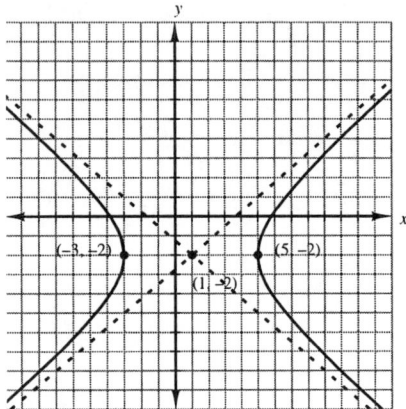

19. $\dfrac{(y+2)^2}{25} - \dfrac{(x+1)^2}{4} = 1$

$C(-1, 2)$

$\begin{array}{llll} a^2 & = & 25 & a = 5 \\ b^2 & = & 4 & b = 2 \end{array}$

Vertices: $(-1, 2 \pm 5) = (-1, 7)$ and $(-1, 3)$

no x-intercepts

The rectangle used to draw the asymptotes also passes through $(-1 \pm 2, 2) = (1, 2)$ and $(-3, 2)$.

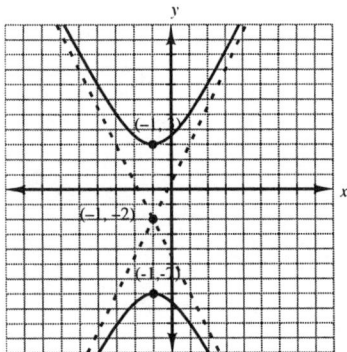

21. $\dfrac{(x-3)^2}{16} - \dfrac{y^2}{4} = 1$

$C(3, 0)$

$a^2 = 16 \qquad a = 4$

$b^2 = 4 \qquad b = 2$

Vertices: $(3 \pm 4, 0) = (7, 0)$ and $(-1, 0)$

no y-intercepts

The rectangle used to draw the asymptotes also passes through $(3, 0 \pm 2) = (3, 2)$ and $(3, -2)$

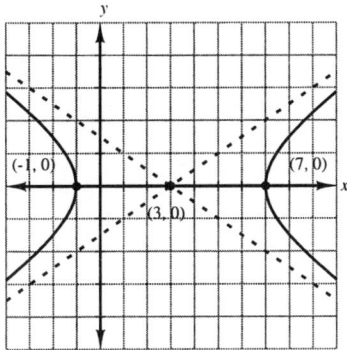

23. $\quad x^2 - 8y^2 + 6x + 32y - 39 = 0$

$\qquad\qquad x^2 + 6x - 8y^2 + 32y = 39$

$(x^2 + 6x + 9) - 8(y^2 - 4y + 4) = 39 + 9 - 32$

$\qquad (x+3)^2 - 8(y-2)^2 = 16$

$\qquad\qquad \dfrac{(x+3)^2}{16} - \dfrac{(y-2)^2}{2} = 1$

$C(-3, 2)$

$a^2 = 16 \qquad a = 4$

$b^2 = 2 \qquad b = \sqrt{2}$

Vertices: $(-3 \pm 4, 2) = (1, 2)$ and $(-7, 2)$

no y-intercepts

The rectangle used to draw the asymptotes also passes through $(-3, 2 \pm \sqrt{2})$

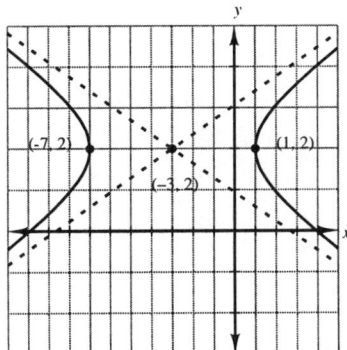

25.
$$9x^2 - 4y^2 - 18x - 24y - 63 = 0$$
$$9x^2 - 18x - 4y^2 - 24y = 63$$
$$9(x^2 - 2x + 1) - 4(y^2 + 6y + 9) = 63 + 9 - 36$$
$$9(x - 1)^2 - 4(y + 3)^2 = 36$$
$$\frac{(x - 1)^2}{4} - \frac{(y + 3)^2}{9} = 1$$

$C(1, -3)$
$a^2 = 4, a = 2$
$b^2 = 9, b = 3$
Vertices: $(1 \pm 2, -3) = (3, -3)$ and $(-1, -3)$
no y-intercepts

The rectangle used to draw the asymptotes also passes through $(1, -3 \pm 3) = (1, 0)$ and $(1, -6)$

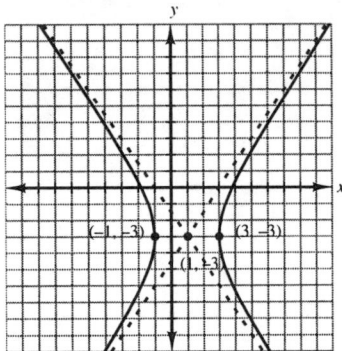

27.
$$9x^2 - 4y^2 - 72x + 8y + 176 = 0$$
$$9x^2 - 72x - 4y^2 + 8y = -176$$
$$9(x^2 - 8x + 16) - 4(y^2 - 2y + 1) = -176 + 144 - 4$$
$$9(x - 4)^2 - 4(y - 1)^2 = -36$$
$$4(y - 1)^2 - 9(x - 4)^2 = 36$$
$$\frac{(y - 1)^2}{9} - \frac{(x - 4)^2}{4} = 1$$

$C(4, 1)$
$a^2 = 9, a = 3$
$b^2 = 4, b = 2$
Vertices: $(4, 1 \pm 3) = (4, 4)$ and $(4, -2)$
no x-intercepts

The rectangle used to draw the asymptotes also passes through $(4 \pm 2, 1) = (6, 1)$ and $(2, 1)$

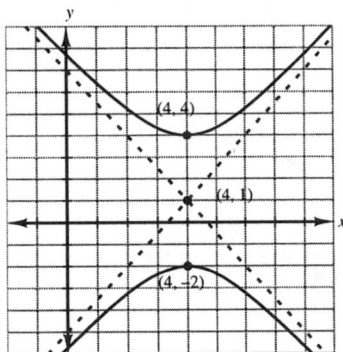

29. $\frac{x^2}{9} - \frac{y^2}{16} \leq 1$

Graph $\frac{x^2}{9} - \frac{y^2}{16} = 1$ with a solid line

$C(0, 0)$
$a^2 = 9,\ a = 3$
$b^2 = 16,\ b = 4$

Vertices: $(\pm 3, 0)$
no y-intercepts (use $(0, \pm 4)$ for the rectangle)
Test point: $(0, 0)$
$\quad 0 - 0 \leq 1$ True
shade middle region

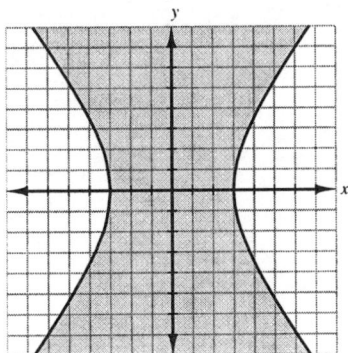

31. $\frac{x^2}{4} - \frac{y^2}{25} > 1$

Graph $\frac{x^2}{4} - \frac{y^2}{25} = 1$ with a dashed line

$C(0, 0)$
$a^2 = 4,\ a = 2$
$b^2 = 25,\ b = 5$

Vertices: $(\pm 2, 0)$
no y-intercepts (use $(0, -5)$ for the rectangle)
Test point: $(0, 0)$
$\quad 0 - 0 > 1$ False
$\quad$ shade outside region

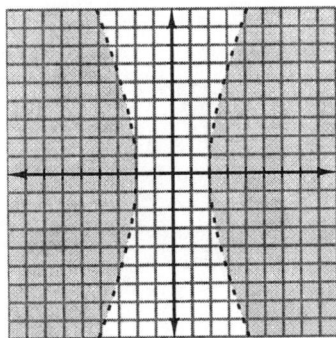

33.
$$x^2 > 4 + 4y^2$$
$$x^2 - 4y^2 > 4$$
$$\frac{x^2}{4} - y^2 > 1$$

Graph $\frac{x^2}{4} - y^2 = 1$ with a dashed line

$C(0, 0)$
$a^2 = 4$, $a = 2$
$b^2 = 1$, $b = 1$

Vertices: $(\pm 2, 0)$

no y-intercepts (use $(0, \pm 1)$ for the rectangle)

Test point: $(0, 0)$

$0 - 0 > 1$ False

shade outside region

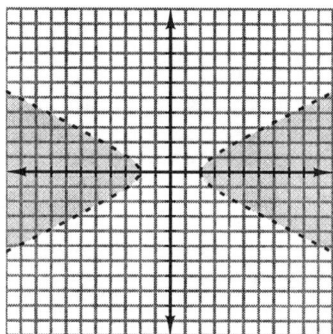

35.
$$16y^2 \leq 144 + 9x^2$$
$$16y^2 - 9x^2 \leq 144$$
$$\frac{y^2}{9} - \frac{x^2}{16} \leq 1$$

Graph $\frac{y^2}{9} - \frac{x^2}{16} = 1$ with a solid line

$C(0, 0)$
$a^2 = 9$, $a = 3$
$b^2 = 16$, $b = 4$

Vertices: $(0, \pm 3)$

no x-intercepts (use $(\pm 4, 0)$ for the rectangle)

Test point: $(0, 0)$

$0 - 0 \leq 1$ True

shade middle region

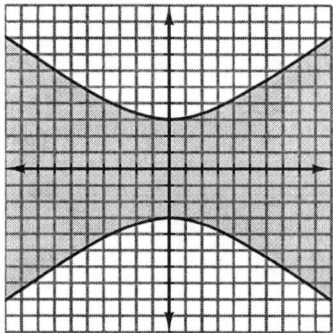

37. $\dfrac{(x-4)^2}{4} - \dfrac{(y-2)^2}{9} < 1$

Graph $\dfrac{(x-4)^2}{4} - \dfrac{(y-2)^2}{9} = 1$ with a dashed line

$C(4, 2)$

Vertices: $(4 \pm 2, 2) = (6, 2)$ and $(2, 2)$

Test point: $(4, 2)$

$0 - 0 < 1$ True

shade middle region

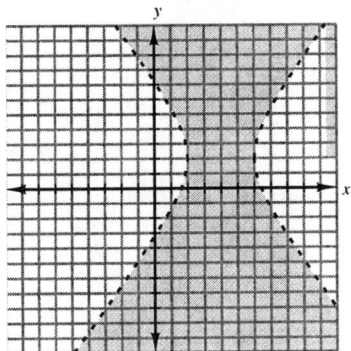

39. $\dfrac{x^2}{4} - \dfrac{y^2}{9} \geq 1$

$\dfrac{x^2}{9} + \dfrac{y^2}{25} < 1$

Begin by graphing the solutions of $\dfrac{x^2}{4} - \dfrac{y^2}{9} \geq 1$.

The boundary is the graph of

$\dfrac{x^2}{4} - \dfrac{y^2}{9} = 1$ (solid line) with

$C(0, 0)$ and Vertices: $(\pm 2, 0)$

Test point: $(0, 0)$

$0 - 0 \geq 1$ False

The middle region is not included.

On the same axes, graph

$$\frac{x^2}{9} + \frac{y^2}{25} < 1$$

The boundary is $\frac{x^2}{9} + \frac{y^2}{25} = 1$ (dashed line)

with $C(0, 0)$
and vertices $(\pm 3, 0)$ and $(0, \pm 5)$
Test point: $(0, 0)$
$0 + 0 < 1$ True
The region inside the ellipse is included.

The graph of the solution set is the intersection of the graphs of the two inequalities.

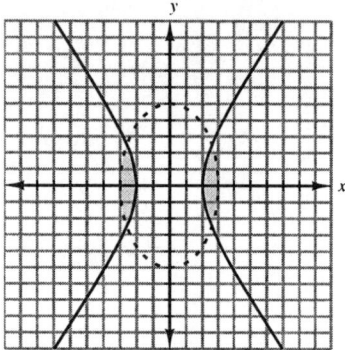

41. $4y^2 - 9x^2 \ < \ 36$
$3x + y \ \leq \ 1$

$4y^2 - 9x^2 \ < \ 36$
$\frac{y^2}{9} - \frac{x^2}{4} \ < \ 1$

Begin by graphing the solution of $\frac{y^2}{9} - \frac{x^2}{4} < 1$.

The boundary is the graph of $\frac{y^2}{9} - \frac{x^2}{4} = 1$ (dashed line)

with $C(0, 0)$
and vertices $(0, \pm 3)$

Testing $(0, 0)$ we find that the middle region is included.

On the same axes graph $3x + y \leq 1$ (with a solid line).

Again use $(0, 0)$ as a test point. We find that the half-plane including $(0, 0)$ is included (the region to the left of the line).

The graph of the solution set is the intersection of the graphs of the two inequalities.

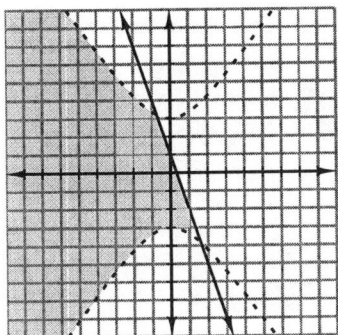

43. $x^2 - y^2 \leq 1$
       $x^2 + y^2 < 4$

Begin by graphing the solution of $x^2 - y^2 \leq 1$. Use a solid line for the boundary (of the hyperbola) $C(0, 0)$ and the vertices are $(\pm 1, 0)$.
   Test point: $(0, 0)$
       $0 - 0 \leq 1$ True
The middle region including $(0, 0)$ is included.

On the same axes graph the solution of $x^2 + y^2 < 4$ using a dashed line for the boundary (of the circle).
   $C(0, 0)$ and $r = \sqrt{4} = 2$
   Test point: $(0, 0)$
       $0 + 0 < 4$ True
The region inside the circle is included.

The graph of the solution set is the intersection of the graphs of the two inequalities

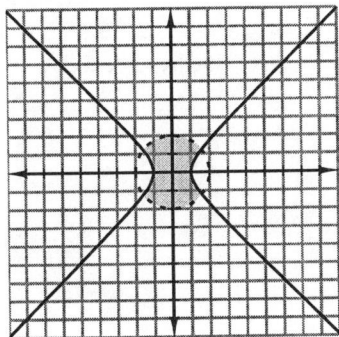

45. $-x^2 + \dfrac{y^2}{4} \geq 1$ $\qquad \rightarrow \qquad$ $\dfrac{y^2}{4} - x^2 \geq 1$

$\qquad |y| < 4$

Begin by graphing solution of $\dfrac{y^2}{4} - x^2 \geq 1$.

The boundary is the graph of the hyperbola $\dfrac{y^2}{4} - x^2 = 1$ (solid line) with $C(0, 0)$ and Vertices $(0, \pm 2)$

Test point: $(0, 0)$
$\qquad 0 - 0 \geq 1$ $\quad$ False
The middle region is not included.

On the same axis graph
$\qquad |y| < 4$
$\qquad -4 < y < 4$
The boundaries are $y = 4$ (dashed line) and $y = -4$ (dashed line)
The region included is between $y = -4$ and $y = 4$ (not including the boundary).

The graph of the solution set is the intersection of the graph of the two inequalities.

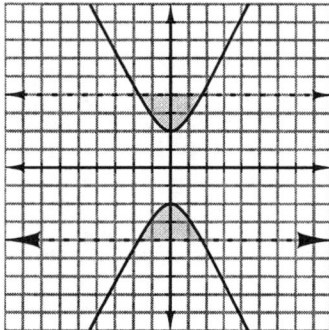

47. $4x^2 - y^2 \leq 4$ $\qquad \rightarrow \qquad$ $x^2 - \dfrac{y^2}{4} \leq 1$

$\qquad x^2 + 4y^2 \geq 4$ $\qquad \rightarrow \qquad$ $\dfrac{x^2}{4} + y^2 \geq 1$

Begin by graphing the solution of $x^2 - \dfrac{y^2}{4} \leq 1$. The boundary is the graph of (the hyperbola) $x^2 - \dfrac{y^2}{4} = 1$

(solid line) with $C(0, 0)$ and vertices $(\pm 1, 0)$.

Test point: $(0, 0)$
$\qquad 0 - 0 \leq 1$ $\quad$ True
The middle region including $(0, 0)$ is included.

On the same axes graph the solution of $\frac{x^2}{4} + y^2 \geq 1$. The boundary is the graph (of the ellipse)

$\frac{x^2}{4} + y^2 = 1$ (solid line) with $C(0, 0)$ and vertices $(\pm 2, 0)$ and $(0, \pm 1)$.

Test point: $(0, 0)$

$0 + 0 \; \geq \; 1$ False

The region inside the ellipse is not included.

The graph of the solution set is the intersection of the graphs of the two inequalities, the region outside the ellipse but in the middle region of the hyperbola.

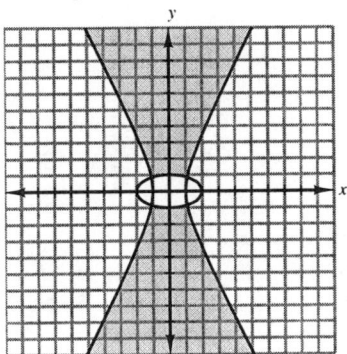

49. $\boxed{D}$ is true; $\frac{x^2}{9} - \frac{y^2}{25} = 1$

$a^2 = 9, \; a = 3$
$b^2 = 5, \; b = 5$

The asymptotes are the lines drawn through the opposite corners of the rectangle whose sides pass through $(3, 0)$ and $(-3, 0)$ and $(0, 5)$ and $(0, -5)$.

The opposite corners are $(3, 5)$ and $(-3, -5)$ and $(-3, 5)$ and $(3, -5)$.

Equation for the lines drawn through the opposite corners are:

$$y - 5 \; = \; \left(\frac{-5 - 5}{-3 - 3}\right)(x - 3)$$

$$y - 5 \; = \; \frac{5}{3}(x - 3)$$

$$y - 5 \; = \; \frac{5}{3}x - 5$$

$$y \; = \; \frac{5}{3}x$$

and $$y - 5 \; = \; \left(\frac{-5 - 5}{3 + 3}\right)(x + 3$$

$$y - 5 \; = \; -\frac{5}{3}(x + 3)$$

$$y - 5 \; = \; -\frac{5}{3}x - 5$$

$$y \; = \; -\frac{5}{3}x$$

The equations for the asymptotes are $y = \frac{5}{3}x$ and $y = -\frac{5}{3}x$

(Note: $a = 3, b = 5$

$$y = \pm \frac{b}{a}x$$

$$y = \pm \frac{5}{3}x$$ are the equations for the asymptotes)

51. $\dfrac{y^2}{16} - \dfrac{x^2}{9} = 1$

$\dfrac{y^2}{16} = 1 + \dfrac{x^2}{9}$

$y^2 = 16\left(1 + \dfrac{x^2}{9}\right)$

$\boxed{y = \pm 4\sqrt{1 + \dfrac{x^2}{9}}}$

53. $2x^2 - 5y^2 + 10 = 0$

$2x^2 - 5y^2 = -10$

$5y^2 - 2x^2 = 10$

$\dfrac{y^2}{2} - \dfrac{x^2}{5} = 1$

$\dfrac{y^2}{2} = 1 + \dfrac{x^2}{5}$

$y^2 = 2\left(1 + \dfrac{x^2}{5}\right)$

$y = \pm\sqrt{2}\sqrt{1 + \dfrac{x^2}{5}}$

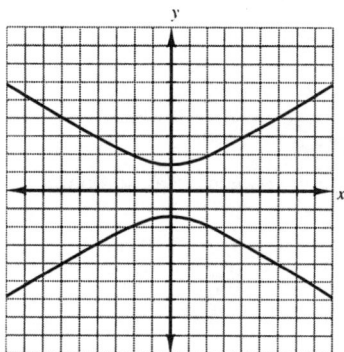

55. $625y^2 - 400x^2 = 250{,}000$

y-intercepts (set $x = 0$):

$625y^2 = 250{,}000$

$y^2 = 400$

$y = \pm 20$

vertex of parabola's upper branch: (0, 20)
vertex of parabola's lower branch: (0, –20)

$$d = \sqrt{(0-0)^2 + (20+20)^2} = \sqrt{40^2} = 40$$

At their closest point, the homes are $\boxed{\text{40 yards apart}}$.

57. letter F_1

$y = 2 \ (0 \le x \le 1)$
$y = 1 \ (0 \le x \le 0.75)$
$x = 0 \ (0 \le y \le 2)$

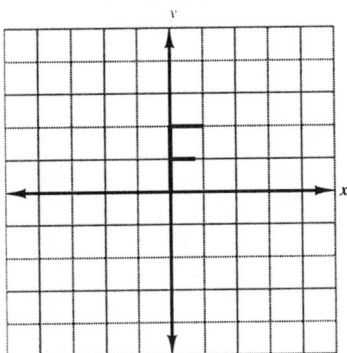

59. letter K:

$x = 0 \ (-1 \le y \le 1)$

$y^2 = x^3 \rightarrow y = \pm x\sqrt{x} \ (0 \le x \le 1)$

x	0	0.5	1
y	0	± 0.35	± 1

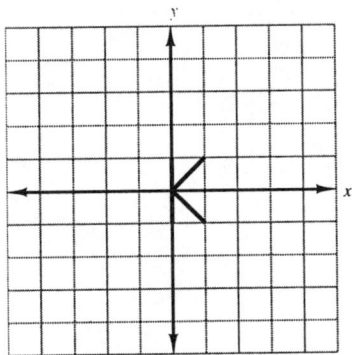

61. letter Q:

$x^2 + y^2 = 1$ is a circle with $C(0, 0)$ and $r = 1$

$y = \pm\sqrt{1 - x^2}$

$y = -x \ \left(\dfrac{1}{2} \le x \le 1\right)$

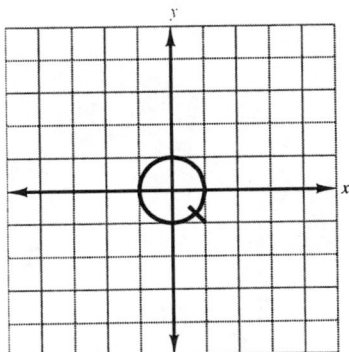

63. letter Y:

$$y = -x + 3 \quad (1 \le x \le 2)$$
$$y = x - 1 \quad (2 \le x \le 3)$$
$$x = 2 \quad (0 \le y \le 1)$$

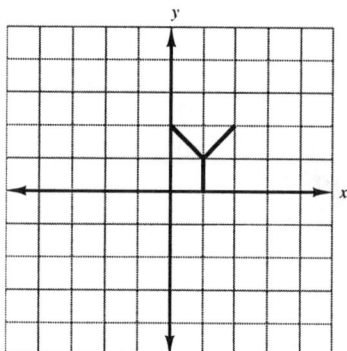

Review Problems

68. $x^2 + 7x - 18 \;\le\; 0$
$(x + 9)(x - 2) \;\le\; 0$
$x = -9 \;$ or $\; x = 2$

F		T		F
	−9		2	

Test −10: $(-10)^2 + 7(-10) - 18 \;\le\; 0$
$\qquad\qquad\qquad\qquad\qquad 12 \;\le\; 0 \quad$ False
Test 0: $(0)^2 + 7(0) - 18 \;\le\; 0$
$\qquad\qquad\qquad\qquad -18 \;\le\; 0 \quad$ True
Test 3: $(3)^2 + 7(3) - 18 \;\le\; 0$
$\qquad\qquad\qquad\qquad\quad 12 \;\le\; 0 \quad$ False

$\boxed{\{x \mid -9 \le x \le 2\}} \quad \boxed{[-9, 2]}$

69. $\qquad\qquad 3y^2 \;=\; -1 - y$
$\qquad 3y^2 + y + 1 \;=\; 0$
$a = 3, \, b = 1, \, c = 1$

$$y \;=\; \frac{-b \pm \sqrt{b^2 - 4ac}}{2a} = \frac{-1 \pm \sqrt{1^2 - 4(3)(1)}}{2(3)}$$

$$= \; \frac{-1 \pm \sqrt{-11}}{6}$$

$$= \; \frac{-1 \pm \sqrt{11}\, i}{6}$$

$$\boxed{\left\{ \frac{-1 + \sqrt{11}\, i}{6}, \frac{-1 - \sqrt{11}\, i}{6} \right\}}$$

70.
$$\sqrt{y} + \sqrt{y+1} - 2 = 0$$
$$\sqrt{y+1} = 2 - \sqrt{y}$$
$$(\sqrt{y+1})^2 = (2 - \sqrt{y})^2$$
$$y+1 = (2 - \sqrt{y})(2 - \sqrt{y})$$
$$y+1 = 4 - 4\sqrt{y} + y$$
$$-3 = -4\sqrt{y}$$
$$(-3)^2 = (-4\sqrt{y})^2$$
$$9 = 16y$$
$$\frac{9}{16} = y$$

check:
$$\sqrt{\frac{9}{16}} + \sqrt{\frac{9}{16} + 1} - 2 = 0$$
$$\sqrt{\frac{9}{16}} + \sqrt{\frac{25}{16}} - 2 = 0$$
$$\frac{3}{4} + \frac{5}{4} - 2 = 0$$
$$0 = 0$$

$$\boxed{\left\{ \frac{9}{16} \right\}}$$

Section 9.4 More About Parabolas: Quadratic Functions and Relations

Problem Set 9.4, pp. 710-712

1.
$$y = (x-2)^2 - 4$$

y-intercept (let $x = 0$):
$$y = (0-2)^2 - 4$$
$$= 4 - 4$$
$$= 0$$

x-intercept (let $y = 0$):
$$0 = (x-2)^2 - 4$$
$$4 = (x-2)^2$$
$$\pm 2 = x - 2$$
$$x = 2 \pm 2$$
$$x = 4 \quad \text{or} \quad x = 0$$

Vertex:

The form of this equation is $y = a(x-h)^2 + k$ where $a = 1$, $h = 2$, $k = -4$.

Thus, the vertex is $(2, -4)$.

Since a is positive, the parabola will open upward.

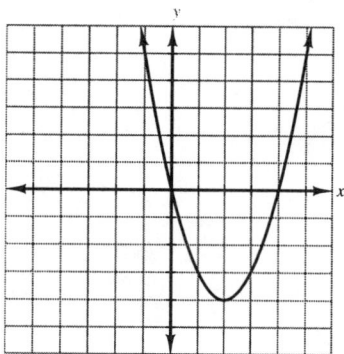

3. $x = (y - 2)^2 - 4$

y-intercept: (let $x = 0$):

$$0 = (y - 2)^2 - 4$$
$$4 = (y - 2)^2$$
$$\pm 2 = y - 2$$
$$y = 2 \pm 2$$
$$y = 4 \text{ or } y = 0$$

x-intercept: (let $y = 0$)

$$x = (-2)^2 - 4 = 4 - 4 = 0$$

Vertex:

The form of this equation is $x = a(y - k)^2 + h$
where $a = 1$, $h = -4$, $k = 2$.
Thus, the vertex is $(-4, 2)$.

Since $a = 1 > 0$, the parabola will open to the right.

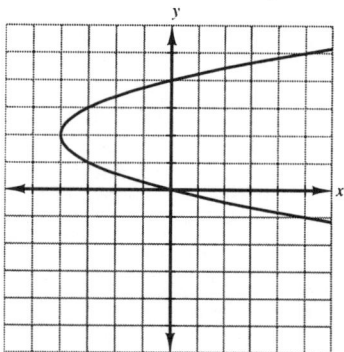

5. $y = x^2 - 4$

y-intercept: (let $x = 0$):

$$y = 0 - 4 = -4$$

x-intercept (let $y = 0$):

$$0 = x^2 - 4$$
$$4 = x^2$$
$$\pm 2 = x$$
$$x = 2 \text{ or } x = -2$$

Vertex:

    The form of this equation is $y = a(x-h)^2 + k$

    where $a = 1,$ $h = 0,$ $k = 0.$

    Thus, the vertex is $(0, 0).$

Since $a = 1 > 0,$ the parabola opens upward.

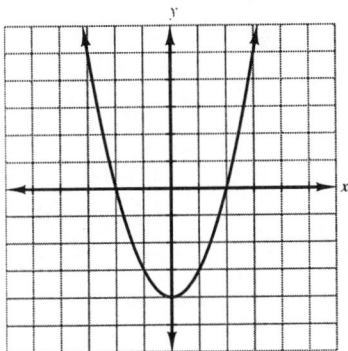

7. $x = y^2 + 4$

y-intercept: (let $x = 0$):

    $0 = y^2 + 4$

    $-4 = y^2$

    no y-intercept.

x-intercept ($y = 0$):

    $x = 0 + 4 = 4$

Vertex:

    The form of this equation is $x = a(y-k)^2 + h$

    where $a = 1,$ $h = 4,$ $k = 0.$

    Thus, the vertex is $(4, 0).$

Since $a = 1 > 0,$ the parabola will open to the right.

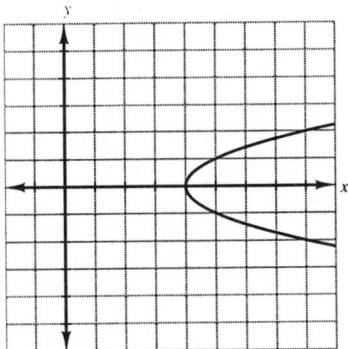

9. $y = -2(x-1)^2 - 1$

y-intercept ($x = 0$):

    $y = -2(0-1)^2 - 1 = -2(1) - 1 = -3$

x-intercept ($y = 0$):

    $0 = -2(x-1)^2 - 1$

    $-\dfrac{1}{2} = (x-1)^2$

    no x-intercept.

Vertex:

> The form of this equation is $y = a(x - h)^2 + k$
> where $a = -2$, $h = 1$, $k = -1$.
> Thus, the vertex is $(1, -1)$.

Since $a = -2 < 0$, the parabola opens downward.

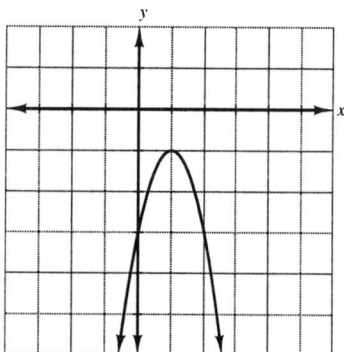

11. $x = -2(y - 1)^2 - 1$

y-intercept $(x = 0)$:
$$0 = -2(y - 1)^2 - 1$$
$$-\frac{1}{2} = (y - 1)^2$$

no y-intercept.

x-intercept $(y = 0)$:
$$x = -2(0 - 1)^2 - 1 = -2(1) = -3$$

Vertex:

> The form of this equation is $x = a(y - k)^2 + h$
> where $a = -2$, $h = -1$, $k = 1$.
> Thus, the vertex is $(-1, 1)$.

Since $a = -2 < 0$, the parabola will open to the left.

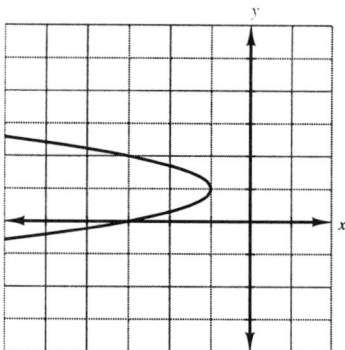

13. $x = y^2 + 2y - 3$

y-intercept $(x = 0)$:
$$0 = y^2 + 2y - 3$$
$$0 = (y + 3)(y - 1)$$
$$y = -3 \quad \text{or} \quad y = 1$$

x-intercept $(y = 0)$:
$$x = 0 + 0 - 3 = -3$$

vertex:

$$y = -\frac{b}{2a} = -\frac{2}{2(1)} = -1$$
$$x = (-1)^2 + 2(-1) - 3 = 1 - 2 - 3 = -4$$

Thus, the vertex is $(-4, -1)$.

Since $a = 1 > 0$, the parabola opens to the right.

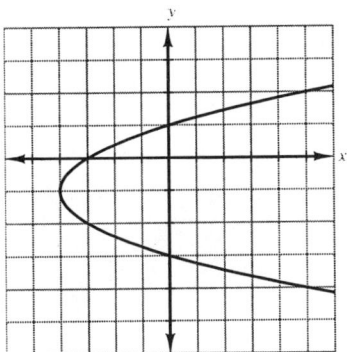

15. $x = y^2$

y-intercept $(x = 0)$:
$$0 = y^2$$
$$y = 0$$

x-intercept $(y = 0)$:
$$x = 0$$

vertex: $(0, 0)$

Since $a = 1 > 0$, the parabola opens to the right.

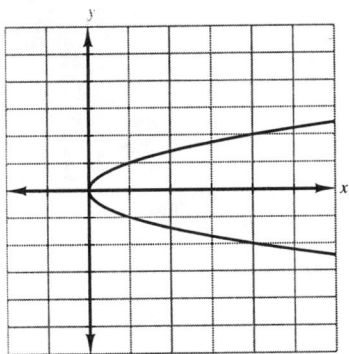

17. $x = -y^2 - 2y + 3$

y-intercept $(x = 0)$:
$$0 = -y^2 - 2y + 3$$
$$y^2 + 2y - 3 = 0$$
$$(y + 3)(y - 1) = 0$$
$$y = -3 \quad \text{or} \quad y = 1$$

x-intercept $(y = 0)$:
$$x = -0 - 0 + 3 = 3$$

vertex:

$$y = -\frac{b}{2a} = \frac{-(-2)}{2(-1)} = -1$$
$$x = -(-1)^2 - 2(-1) + 3 = -1 + 2 + 3 = 4$$

Thus, the vertex is $(4, -1)$.

Since $a = -1 < 0$, the parabola will open to the left.

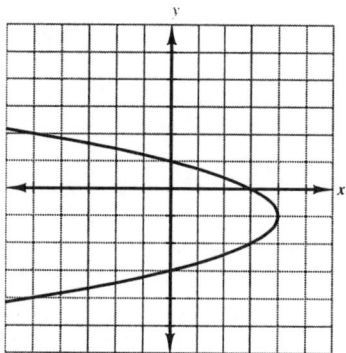

19. $x \geq -y^2 + 2y + 3$

The boundary parabola is the graph of
$$x = -y^2 + 2y + 3 \quad \text{(Use a \textit{solid} curve)}$$
y-intercepts $(x = 0)$:
$$0 = -y^2 + 2y + 3$$
$$y^2 - 2y - 3 = 0$$
$$(y - 3)(y + 1) = 0$$
$$y = 3 \quad \text{or} \quad y = -1$$
x-intercept $(y = 0)$:
$$x = -0 + 0 + 3 = 3$$
vertex:
$$y = -\frac{b}{2a} = \frac{-2}{2(-1)} = 1$$
$$x = -1^2 + 2(1) + 3 = 4$$
Thus, the vertex is $(4, 1)$.
Since $a = -1 < 0$, the parabola opens to the left.

Test point: $(0, 0)$
$$0 \geq 0 + 0 + 3 \quad \text{False}$$
The region *not* including $(0, 0)$ is shaded. (The region shaded is *outside* and including the boundary.)

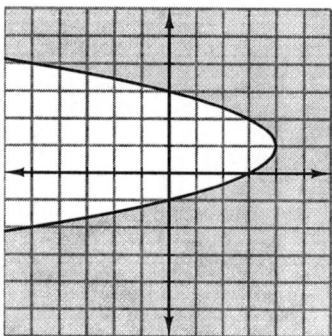

21. $x < 3(y-1)^2 + 2$
The boundary parabola is the graph of
$x = 3(y-1)^2 + 2$ (Use a *dashed* curve)
y-intercepts $(x = 0)$:
$0 = 3(y-1)^2 + 2$
$-\dfrac{2}{3} = (y-1)^2$
no y-intercept
x-intercept $(y = 0)$:
$x = 3(0-1)^2 + 2 = 3(1) + 2 = 5$
vertex: $h = 2$, $k = 1$ $\rightarrow$ $(2, 1)$
Since $a = 3 > 0$, the parabola opens to the right.

Test point: $(0, 0)$
$0 \geq 3(0-1)^2 + 2 = 5$ True
The region including $(0, 0)$ is shaded. (The region shaded is *outside* the boundary.)

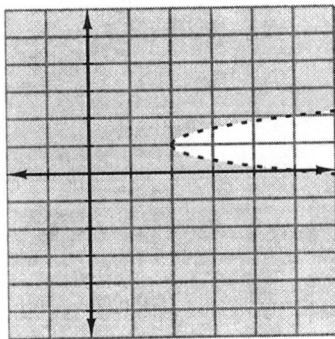

23. $x > (y-2)^2 + 3$

The boundary parabola is the graph of
$x = (y-2)^2 + 3$ (Use a *dashed* curve)
y-intercepts $(x = 0)$:
$0 = (y-2)^2 + 3$
$-3 = (y-2)^2$
no y-intercept
x-intercept $(y = 0)$:
$x = (0-2)^2 + 3 = 4 + 3 = 7$
vertex: $h = 3$, $k = 2$ $\rightarrow$ $(3, 2)$
Since $a = 1 > 0$, the parabola opens to the right.

Test point: $(0, 0)$
$0 > (0-2)^2 + 3 = 7$ False.
The region *not* including $(0, 0)$ is shaded. (The region shaded is *inside* the boundary.)

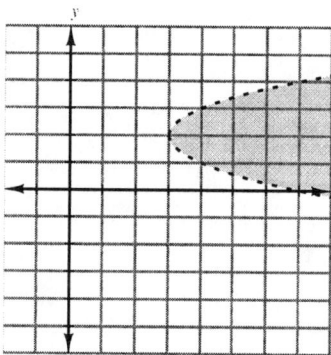

25. $y < (x-2)^2 - 4$

The boundary parabola is the graph of

$$y = (x-2)^2 - 4$$ (Use a *dashed* curve)

y-intercepts $(x = 0)$:

$$y = (0-2)^2 - 4 = 4 - 4 = 0$$

x-intercept $(y = 0)$:

$$0 = (x-2)^2 - 4$$
$$4 = (x-2)^2$$
$$\pm 2 = x - 2$$
$$x = 2 \pm 2$$
$$x = 4 \quad \text{or} \quad x = 0$$

vertex: $h = 2, \quad k = -4 \quad \rightarrow \quad (2, -4)$

Since $a = 1 > 0$, the parabola opens upward.

Test point: $(1, 1)$

$$1 < (1-2)^2 - 4 = 1 - 4 = -3 \quad \text{False.}$$

The region *not* including $(1, 1)$ is shaded. (The region shaded is *outside* the boundary.)

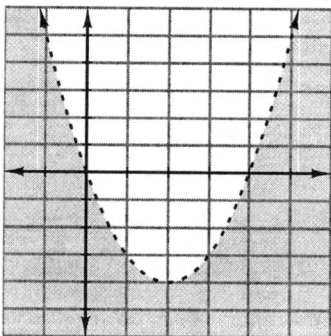

27. $x \leq y^2 - 6y + 1$

The boundary parabola is the graph of
$x = y^2 - 6y + 1$ (Use a *solid* curve)
y-intercepts $(x = 0)$:
$$0 = y^2 - 6y + 1$$
$$y = \frac{-(-6) \pm \sqrt{(6)^2 - 4(1)}}{2} = \frac{6 \pm \sqrt{32}}{2} = 3 \pm 2\sqrt{2}$$
$$y \approx 5.83 \quad \text{or} \quad y \approx 0.17$$
x-intercept $(y = 0)$:
$$x = 0 - 0 + 1 = 1$$
vertex:
$$y = -\frac{b}{2a} = \frac{-(-6)}{2(1)} = 3$$
$$x = (3)^2 - 6(3) + 1 = -8$$
Thus, the vertex is $(-8, 3)$.
Since $a = 1 > 0$, the parabola opens to the right.

Test point: $(1, 1)$
$$1 \leq 1 - 6 + 1 = -4 \qquad \text{False}$$
The region *not* including $(1, 1)$ is shaded. (The region
shaded is *outside* and including the boundary.)

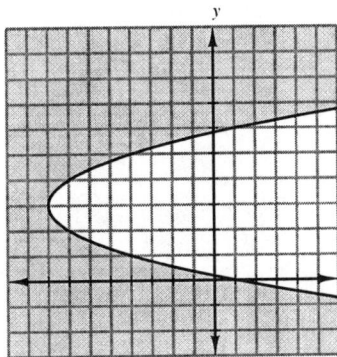

29. $x \geq y^2 - 6y + 5$
        $x^2 + y^2 \leq 4$

Begin by graphing the solution of $x \geq y^2 - 6y + 5$.
The boundary parabola is the graph of
$x = y^2 - 6y + 5$ (Use a *solid* curve)
y-intercepts $(x = 0)$:
$$0 = y^2 - 6y + 5$$
$$0 = (y - 5)(y - 1)$$
$$y = 5 \quad \text{or} \quad y = 1$$
x-intercept $(y = 0)$:
$$x = 0 - 0 + 5 = 5$$
vertex:
$$y = -\frac{b}{2a} = \frac{-(-6)}{2(1)} = 3$$
$$x = (3)^2 - 6(3) + 5 = -4$$
Thus, the vertex is $(-4, 3)$.
$a = 1 > 0$: the parabola opens to the right.

Test point: $(0, 0)$
$$0 \geq 0 - 0 + 5 = 5 \quad \text{False.}$$
The region including $(0, 0)$ is *not* included. (i.e., inside and including the boundary)

On the same axes, graph $x^2 + y^2 \leq 4$.
The boundary is the graph of the circle $x^2 + y^2 = 4$, with center $(0, 0)$ and $r = 2$. $\qquad$ (Use a *solid* curve)
Test point: $(0, 0)$
$$0 + 0 \leq 4 \quad \text{True}$$
The region including $(0, 0)$ is included (i.e. inside region of the circle including the boundary)

The graph of the solution set is the intersection of the graphs of the two inequalities (the region inside and including the boundaries of the parabola and the circle.)

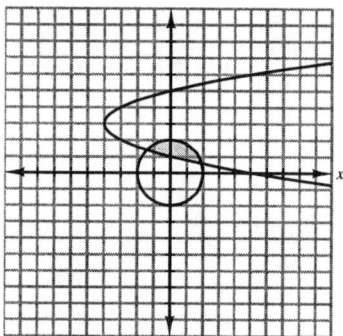

31.
$$y > -(x - 3)^2 + 4$$
$$\frac{x^2}{9} + \frac{y^2}{4} \leq 1$$

Begin by graphing the solution of
$$y > -(x - 3)^2 + 4$$

The boundary parabola is the graph of
$$y = -(x - 3)^2 + 4$$
y-intercepts $(x = 0)$:
$$y = -(0 - 3)^2 + 4 = -9 + 4 = -5$$
x-intercepts $(y = 0)$:
$$0 = -(x - 3)^2 + 4$$
$$(x - 3)^2 = 4$$
$$x = 3 \pm 2$$
$$x = 5 \quad \text{or} \quad x = 1$$

Vertex: $h = 3$, $k = 4 \to (3, 4)$
$\qquad a = -1 < 0$: parabola opens downward

Test point: $(0, 0)$
$$0 > -(0 - 3)^2 + 4 = -5 \quad \text{False}$$

The region including $(0, 0)$ is *not* included. (i.e. outside the boundary)

One the same axes, graph $\dfrac{x^2}{9} + \dfrac{y^2}{4} \leq 1$

The boundary is the ellipse $\frac{x^2}{9} + \frac{y^2}{4} = 1$ with center $(0, 0)$

Vertices $(\pm 3, 0)$ and $(0, \pm 2)$
(Use a *solid* curve)

Test point: $(0, 0)$
$$0 + 0 \ \leq \ 1 \quad \text{True}$$

The region including $(0, 0)$ is included (i.e. inside region of the ellipse including the boundary)
The graph of the solution set is the intersection of the graph of the two inequalities, the region inside and including the boundary of the ellipse but outside and not including the boundary of the parabola.

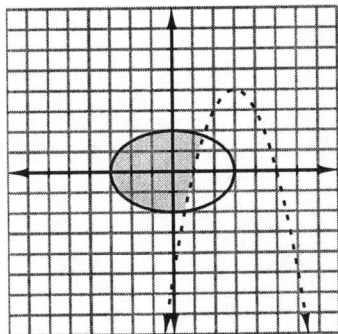

33. $x \leq (y + 2)^2 - 1$
$(x - 2)^2 + (y + 2)^2 \geq 1$

Begin by graphing the solutions of
$$x \ \leq \ (y + 2)^2 - 1$$

The boundary parabola is the graph of $x = (y + 2)^2 - 1$ \quad (Use a *solid* curve.)
y-intercept $(x = 0)$:
$$\begin{aligned}
0 &= (y + 2)^2 - 1 \\
1 &= (y + 2)^2 \\
\pm 1 &= y + 2 \\
y &= -2 \pm 1 \\
y &= -1 \quad \text{or} \quad y = -3
\end{aligned}$$
x-intercept $(y = 0)$:
$$x = (0 + 2)^2 - 1 = 4 - 1 = 3$$

Vertex: $h = -1, \ k = -2 \rightarrow (-1, -2)$
$a = 1 > 0$: parabola opens to the right

Test point: $(0, 0)$
$$0 \ \leq \ (0 + 2)^2 - 1 = 3 \quad \text{True}$$

The region including $(0, 0)$ is included (i.e. outside and including the boundary)

On the same axes, graph $(x-2)^2 + (y+2)^2 \geq 1$

The boundary is the circle
$$(x-2)^2 + (y+2)^2 = 1 \quad \text{with center } (2,-2) \text{ and } r = 1 \quad (\text{Use a } solid \text{ curve})$$

Test point: $(2,-2)$
$$0 + 0 \geq 1 \quad \text{False}$$
The region *not* including $(2,-2)$ is included.
(i.e. outside and including the boundary)

The graph of the solution set is the intersection of the graph of the two inequalities. (Only the region outside and including the boundary of the parabola is common to both solutions.)

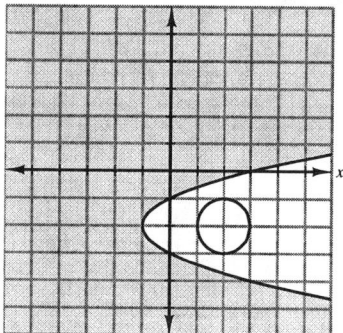

35. $\boxed{\text{B}}$ is true; $x = ay^2 + by + c$

$$x = a\left(y^2 + \frac{by}{a} + \frac{b^2}{4a^2}\right) + \left(c - \frac{b^2}{4a}\right)$$

$$x = a\left(y + \frac{b}{2a}\right)^2 + \left(c - \frac{b^2}{4a}\right)$$

If the vertex is $(3, 2)$,

then $\qquad k = -\dfrac{b}{2a} = 2$

and $\qquad h = c - \dfrac{b^2}{4a} = 3$

Thus, $\qquad x = a(y-2)^2 + 3$

y-intercept $(x = 0)$:
$$0 = a(y-2)^2 + 3$$
$$-\frac{3}{a} = (y-2)^2$$

If $a > 0$, then
$$(y-2)^2 < 0$$
and the equation has no y-intercepts.

37. $\quad y = -x^2 + 5$
$\qquad y = x^2 - 3$

$y = -x^2 + 5$: Vertex $(0, 0)$
$\qquad y$-intercept $(x = 0)$: $\ y = 5$
$\qquad x$-intercept $(y = 0)$: $\ 0 = -x^2 + 5$
$\qquad\qquad\qquad\qquad\qquad\quad x^2 = 5$
$\qquad\qquad\qquad\qquad\qquad\quad x = \pm\sqrt{5}$

$y = x^2 - 3$:   Vertex $(0, 0)$

                    y-intercept $(x = 0)$: $y = -3$

                    x-intercept $(y = 0)$: $0 = x^2 - 3$

$$3 = x^2$$

$$x = \pm\sqrt{3}$$

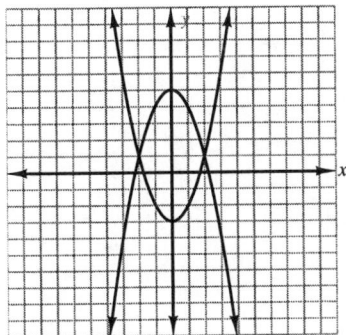

$$\{(-2, 1), (2, 1)\}$$

39. $x = y^2 + 3y$

       $y = x + 1$

$x = y^2 + 3y$: Vertex: $y = -\dfrac{b}{2a} = \dfrac{-3}{2(1)} = -\dfrac{3}{2}$

$$x = \left(-\frac{3}{2}\right)^2 + 3\left(-\frac{3}{2}\right) = \frac{9}{4} - \frac{9}{2} = -\frac{9}{4}$$

$$\left(-\frac{9}{4}, -\frac{3}{2}\right)$$

        y-intercept $(x = 0)$:

$$0 = y^2 + 3y$$

$$0 = y(y + 3)$$

$$y = 0 \quad \text{or} \quad y = -3$$

        x-intercept $(y = 0)$:

$$x = 0$$

$y = x + 1$

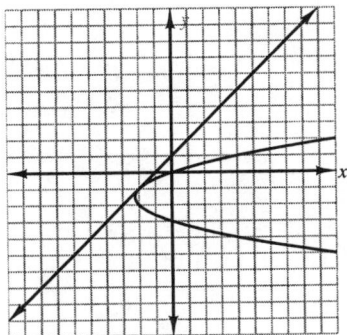

$$\{(-2, -1)\}$$

41. $y = 0.4x^2 - 0.2x - 3.1$

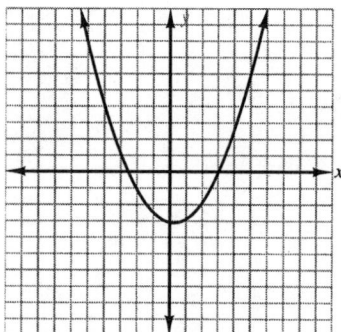

x-intercepts: $\boxed{\{(3.045, 0), (-2.545, 0)\}}$

43. $Ax^2 + Cy^2 + Dx + Ey + F = 0$

$\boxed{\text{circle: } A = C}$

$$Ax^2 + Ay^2 + Dx + Ey + F = 0$$

$$A\left(x^2 + \frac{D}{A}x\right) + A\left(y^2 + \frac{Ey}{A}\right) = -F$$

$$A\left(x^2 + \frac{D}{A}x + \frac{D^2}{4A^2}\right) + A\left(y^2 + \frac{E}{A}y + \frac{E^2}{4A^2}\right) = -F + \frac{D^2}{4A} + \frac{E^2}{4A}$$

$$A\left(x + \frac{D}{2A}\right)^2 + A\left(y + \frac{E}{2A}\right)^2 = -F + \frac{1}{4A}(D^2 + E^2)$$

$$\left(x + \frac{D}{2A}\right)^2 + \left(y + \frac{E}{2A}\right)^2 = -\frac{F}{A} + \frac{D^2 + E^2}{4A^2}$$

$\boxed{\text{ellipse: } A \neq C, \ A \text{ and } C \text{ have same signs}}$

$\boxed{\text{hyperbola: } A \text{ and } C \text{ have opposite signs}}$

$\boxed{\text{parabola: } A = 0 \text{ or } C = 0}$

For example, if $A = 0$

$$Cy^2 + Dx + Ey + F = 0$$
$$Dx = -Cy^2 - Ey - F$$

45. Let $c = $ the *y*-intercept.

y-intercept $(x = 0)$: $(0, c)$

If $(h, k) = (0, c)$

then the equation of the parabola is

$$y = a(x - 0)^2 + c \qquad \text{or} \qquad x = a(y - c)^2 + 0$$

$$\boxed{y = ax^2 + c} \qquad\qquad\qquad \boxed{x = a(y - c)^2}$$

47. parabola with focus $(0, 4)$
and directrix, $y = -4$

$$p = -(-4) = 4$$
$$x^2 = 4py$$
$$x^2 = 4(4)y$$
$$x^2 = 16y$$

$$\boxed{y = \frac{1}{16}x^2}$$

Review Problems

50. $\begin{aligned} 3x - 2y &= 1 \\ 5y + 3z &= -7 \\ 2x + 5y &= 5 \end{aligned}$

(Equations 1 and 3):

$\begin{aligned} (\times 3) \quad 15x - 10y &= 5 \\ (\times 2) \quad \underline{4x + 10y \;\; = \;\; 90} \\ 19x &= 95 \\ x &= 5 \end{aligned}$

(Equation 1):

$\begin{aligned} 3x - 2y &= 1 \\ 15 - 2y &= 1 \\ -2y &= -14 \\ y &= 7 \end{aligned}$

(Equation 2):

$\begin{aligned} 5y + 3z &= -7 \\ 5(7) + 3z &= -7 \\ 3z &= -42 \\ z &= -14 \end{aligned}$

$\boxed{\{(5, 7, -14)\}}$

51. $\begin{aligned} \sqrt{2x+3} - \sqrt{x+1} &= 1 \\ \sqrt{2x+3} &= \sqrt{x+1} + 1 \\ 2x + 3 &= x + 1 + 2\sqrt{x+1} + 1 \\ 2x + 3 &= x + 2 + 2\sqrt{x+1} \\ x + 1 &= 2\sqrt{x+1} \\ x^2 + 2x + 1 &= 4x + 4 \\ x^2 - 2x - 3 &= 0 \\ (x - 3)(x + 1) &= 0 \\ x = 3 \quad \text{or} \quad x &= -1 \quad \text{(both check)} \end{aligned}$

$\boxed{\{-1, 3\}}$

52. $\begin{aligned} (x^2 - 2x)^2 - 14(x^2 - 2x) &= 15 \\ (x^2 - 2x)^2 - 14(x^2 - 2x) - 15 &= 0 \\ t^2 - 14t - 15 &= 0 \quad \text{(Let } t = x^2 - 2x) \\ (t - 15)(t + 1) &= 0 \end{aligned}$

$\begin{aligned} t &= 15 \qquad\qquad\text{or} \\ x^2 - 2x &= 15 \\ x^2 - 2x - 15 &= 0 \\ (x - 5)(x + 3) &= 0 \\ x = 5 \quad \text{or} \quad x &= 3 \end{aligned}$
$\qquad\qquad\begin{aligned} t &= -1 \\ x^2 - 2x &= -1 \\ x^2 - 2x + 1 &= 0 \\ (x - 1)^2 &= 0 \\ x &= 1 \end{aligned}$

$\boxed{\{-3, 1, 5\}}$

Section 9.5 Solving Nonlinear Systems of Equations by the Substitution Method

Problem Set 9.5, pp. 720-722

1. $x^2 + 4y^2 = 29$
    $\underline{x - 3y = 2}$ (solve for x) → $x = 2 + 3y$
                       (substitute for x) → $(2 + 3y)^2 + 4y^2 = 29$
                                                           $4 + 12y + 9y^2 + 4y^2 = 29$
                                                           $13y^2 + 12y - 25 = 0$
                                                           $(13y + 25)(y - 1) = 0$

                                                $13y + 25 = 0$ or $y - 1 = 0$
                                                $y = -\dfrac{25}{13}$ or $y = 1$

    If $y = -\dfrac{25}{13}$: $x = 2 + 3y$ If $y = 1$: $x = 2 + 3y$

    $\qquad\qquad\qquad = 2 + 3\left(-\dfrac{25}{13}\right)$ $\qquad\qquad = 2 + 3(1)$

    $\qquad\qquad\qquad = \dfrac{26}{13} - \dfrac{75}{13}$ $\qquad\qquad = 5$

    $\qquad\qquad\qquad = -\dfrac{49}{13}$ $(5, 1)$

    $\left(-\dfrac{49}{13}, \dfrac{-25}{13}\right)$

    $\boxed{\left\{\ \left(-\dfrac{49}{13}, \dfrac{-25}{13}\right), (5, 1)\ \right\}}$

3. $x^2 + y^2 = 40$
    $\underline{3x - y + 20 = 0}$ (solve for y) → $y = 3x + 20$
                        (substitute for y) → $x^2 + (3x + 20)^2 = 40$
                                                  $x^2 + 9x^2 + 120x + 400 = 40$
                                                  $10x^2 + 120x + 360 = 0$
                                                  $x^2 - 12x + 36 = 0$
                                                  $(x + 6)^2 = 0$
                                                  $x = -6$
                                                  $y = 3(-6) + 20$
                                                  $= 2$

    $\boxed{\{(-6, 2)\}}$

5. $2x + y = 12$ (solve for y) → $y = 12 - 2x$
    $\underline{x^2 + 6y = 39}$ $x^2 + 6(12 - 2x) = 39$
                                                $x^2 + 72 - 12x = 39$
                                                $x^2 - 12x + 33 = 0$

    $x = \dfrac{-b \pm \sqrt{b^2 - 4ac}}{2a} = \dfrac{12 \pm \sqrt{144 - 4(1)(33)}}{2}$

    $\qquad = \dfrac{12 \pm \sqrt{12}}{2} = \dfrac{12 \pm 2\sqrt{3}}{2}$

    $\qquad = 6 \pm \sqrt{3}$

If $x = 6 + \sqrt{3}$: $y = 12 - 2x = 12 - 2(6 + \sqrt{3})$

$\quad\quad = 12 - 12 - 2\sqrt{3} = -2\sqrt{3}$

$(6 + \sqrt{3}, -2\sqrt{3})$

If $x = 6 - \sqrt{3}$: $y = 12 - 2x = 12 - 2(6 - \sqrt{3})$

$\quad\quad = 2\sqrt{3}$

$(6 - \sqrt{3}, 2\sqrt{3})$

$\boxed{\{(6 + \sqrt{3}, -2\sqrt{3}), (6 - \sqrt{3}, 2\sqrt{3})\}}$

7. $\quad x^2 + 4y + 2x - 11 = 0$

$\quad\quad\quad\quad\quad \underline{y = x + 5}$ $\quad$ (substitute for y) $\rightarrow$ $\quad\quad x^2 + 4(x + 5) + 2x - 11 = 0$

$\quad\quad\quad\quad\quad\quad\quad\quad\quad\quad\quad\quad\quad\quad\quad\quad\quad\quad\quad\quad x^2 + 6x + 9 = 0$

$\quad\quad\quad\quad\quad\quad\quad\quad\quad\quad\quad\quad\quad\quad\quad\quad\quad\quad\quad\quad\quad (x + 3)^2 = 0$

$\quad\quad\quad\quad\quad\quad\quad\quad\quad\quad\quad\quad\quad\quad\quad\quad\quad\quad\quad\quad\quad\quad\quad x = -3$

$\quad\quad\quad\quad\quad\quad\quad\quad\quad\quad\quad\quad\quad\quad\quad\quad\quad\quad\quad\quad\quad\quad\quad y = -3 + 5 = 2$

$\boxed{\{(-3, 2)\}}$

9. $\quad x^2 + y^2 = 52$

$\quad\quad \underline{3x - 2y = 0}$ $\quad\quad$ (solve for x) $\rightarrow$ $\quad\quad\quad\quad\quad\quad 3x = 2y$

$\quad\quad\quad\quad\quad\quad\quad\quad\quad\quad\quad\quad\quad\quad\quad\quad\quad\quad\quad\quad x = \dfrac{2y}{3}$

$\quad\quad\quad\quad\quad\quad\quad$ (substitute for x) $\rightarrow$ $\quad\quad \left(\dfrac{2y}{3}\right)^2 + y^2 = 52$

$\quad\quad\quad\quad\quad\quad\quad\quad\quad\quad\quad\quad\quad\quad\quad\quad\quad\quad \dfrac{4y^2}{9} + y^2 = 52$

$\quad\quad\quad\quad\quad\quad\quad\quad\quad\quad\quad\quad\quad\quad\quad 9\left(\dfrac{4y^2}{9} + y^2\right) = 9(52)$

$\quad\quad\quad\quad\quad\quad\quad\quad\quad\quad\quad\quad\quad\quad\quad\quad 4y^2 + 9y^2 = 468$

$\quad\quad\quad\quad\quad\quad\quad\quad\quad\quad\quad\quad\quad\quad\quad\quad\quad\quad 13y^2 = 468$

$\quad\quad\quad\quad\quad\quad\quad\quad\quad\quad\quad\quad\quad\quad\quad\quad\quad\quad\quad y^2 = 36$

$\quad\quad\quad\quad\quad\quad\quad\quad\quad\quad\quad\quad\quad\quad\quad\quad\quad\quad\quad y = \pm 6$

If $y = 6$: $x = \dfrac{2y}{3} = \dfrac{2(6)}{3} = 4$

If $y = -6$: $x = \dfrac{2y}{3} = \dfrac{2(-6)}{3} = -4$

$\boxed{\{(4, 6), (-4, -6)\}}$

11. $\quad y = x^2 - 1$

$\quad\quad \underline{y = -4x - 5}$

$\quad\quad\quad\quad\quad\quad\quad$ (substitute for y) $\rightarrow$ $\quad\quad\quad\quad x^2 - 1 = -4x - 5$

$\quad\quad\quad\quad\quad\quad\quad\quad\quad\quad\quad\quad\quad\quad\quad x^2 + 4x + 4 = 0$

$\quad\quad\quad\quad\quad\quad\quad\quad\quad\quad\quad\quad\quad\quad\quad (x + 2)^2 = 0$

$\quad\quad\quad\quad\quad\quad\quad\quad\quad\quad\quad\quad\quad\quad\quad\quad\quad x = -2$

$\quad\quad\quad\quad\quad\quad\quad\quad\quad\quad\quad\quad\quad\quad\quad\quad\quad y = -4(-2) - 5 = 3$

$\boxed{\{(-2, 3)\}}$

13. $x^2 + 2y = 19$

$\underline{2x - y = 1}$ (solve for y) $\rightarrow$ $y = 2x - 1$

(substitute for y) $\rightarrow$ $x^2 + 2(2x - 1) = 19$

$x^2 + 4x - 21 = 0$

$(x + 7)(x - 3) = 0$

$x = -7$ or $x = 3$

If $x = -7$: $y = 2(-7) - 1 = -15$

If $x = 3$: $y = 2(3) - 1 = 5$

$\boxed{\{(-7, -15), (3, 5)\}}$

15. $x - y = 2$ (solve for x) $\rightarrow$ $x = y + 2$

$\underline{x^2 - 3y^2 = 8}$ (substitute for x) $\rightarrow$ $(y + 2)^2 - 3y^2 = 8$

$y^2 + 4y + 4 - 3y^2 = 8$

$-2y^2 + 4y - 4 = 0$

$y^2 - 2y + 2 = 0$

$$y = \frac{-b \pm \sqrt{b^2 - 4ac}}{2a} = \frac{2 \pm \sqrt{4(1)(2)}}{2(1)}$$

$$= \frac{2 \pm \sqrt{-4}}{2} = \frac{2 \pm 2i}{2} = 1 \pm i$$

If $y = 1 + i$: $x = y + 2 = (1 + i) + 2 = 3 + i$

If $y = 1 - i$: $x = y + 2 = (1 - i) + 2 = 3 - i$

$\boxed{\{(3 + i, 1 + i), (3 - i, 1 - i)\}}$

17. $y - 4x = 0$

$\underline{\quad y = x^2 + 5}$

(substitute for y) $\rightarrow$ $(x^2 + 5) - 4x = 0$ $x^2 - 4x + 5 = 0$

$$x = \frac{4 \pm \sqrt{16 - 4(1)(5)}}{2} = \frac{4 \pm \sqrt{-4}}{2} = \frac{4 \pm 2i}{2} = 2 \pm i$$

If $x = 2 + i$: $y = 4x = 4(2 + i) = 8 + 4i$

If $x = 2 - i$: $y = 4x = 4(2 - i) = 8 - 4i$

$\boxed{\{(2 + i, 8 + 4i), (2 - i, 8 - 4i)\}}$

19. $y = -x^2 + 2x$

$\underline{y = -x\quad\quad}$

(substitute for y) $\rightarrow$ $-x = -x^2 + 2x$

$x^2 - 3x = 0$

$x(x - 3) = 0$

$x = 0$ $x = 3$

If $x = 0$: $y = 0$

If $x = 3$: $y = -3$

$\boxed{\{(0, 0), (3, -3)\}}$

21.
$$2y^2 = x + 4$$
$$x = y^2$$

(substitute for x) $\rightarrow$
$$2y^2 = y^2 + 4$$
$$y^2 = 4$$
$$y = \pm 2$$

If $y = 2$: $x = 2^2 = 4$
If $y = -2$: $x = (-2)^2 = 4$
$$\boxed{\{(4, 2), (4, -2)\}}$$

23.
$$x^2 + y^2 = 100$$
$$x + y = 10\sqrt{2}$$

(solve for y) $\rightarrow$ $y = 10\sqrt{2} - x$

(substitute for y) $\rightarrow$
$$x^2 + (10\sqrt{2} - x)^2 = 100$$
$$x^2 + 200 - 20\sqrt{2}x + x^2 = 100$$
$$2x^2 - 20\sqrt{2}x + 100 = 0$$
$$x^2 - 10\sqrt{2}x + 50 = 0$$

$a = 1$, $b = -10\sqrt{2}$, $c = 50$
$$x = \frac{-b \pm \sqrt{b^2 - 4ac}}{2a}$$
$$= \frac{10\sqrt{2} \pm \sqrt{200 - 4 \cdot 1 \cdot 50}}{2}$$
$$= \frac{10\sqrt{2} \pm \sqrt{0}}{2}$$
$$= 5\sqrt{2}$$
$$y = 10\sqrt{2} - x = 10\sqrt{2} - 5\sqrt{2} = 5\sqrt{2}$$
$$\boxed{\{(5\sqrt{2}, 5\sqrt{2})\}}$$

25.
$$x^2 + y^2 = 25$$
$$y = 6$$

(substitute for y) $\rightarrow$
$$x^2 + 36 = 25$$
$$x^2 = -11$$
$$x = \pm\sqrt{11}i$$
$$\boxed{\{(\sqrt{11}\,i, 6), (-\sqrt{11}\,i, 6)\}}$$

27.
$$2x^2 - y + 12 = 0$$
$$x^2 + 2y - 4 = 0$$

(solve for y) $\rightarrow$
(substitute for y) $\rightarrow$
$$y = 2x^2 + 12$$
$$x^2 + 2(2x^2 + 12) - 4 = 0$$
$$5x^2 + 20 = 0$$
$$5x^2 = -20$$
$$x^2 = -4$$
$$x = \pm\sqrt{-4} = \pm 2i$$

If $x = 2i$: $y = 2x^2 + 12 = 2(2i)^2 + 12$
$= 2(4i^2) + 12 = 8(-1) + 12 = 4$
If $x = -2i$: $y = 2x^2 + 12 = 2(-2i)^2 + 12$
$= 2(4i^2) + 12 = 4$
$$\boxed{\{(2i, 4), (-2i, 4)\}}$$

29.

$$\begin{aligned} x - 3y &= -1 \end{aligned}$$ (solve for x) $\rightarrow$ $\qquad$ $x = 3y - 1$

$$7 - 3x^2 + 2xy + y^2 = 0$$ (substitute for x) $\rightarrow$

$$\begin{aligned} 7 - 3(3y-1)^2 + 2(3y-1)y + y^2 &= 0 \\ 7 - 27y^2 + 18y - 3 + 6y^2 - 2y + y^2 &= 0 \\ -20y^2 + 16y + 4 &= 0 \\ 5y^2 - 4y - 1 &= 0 \\ (5y + 1)(y - 1) &= 0 \\ y = -\frac{1}{5} \text{ or } y &= 1 \end{aligned}$$

If $y = -\dfrac{1}{5}$: $x = 3y - 1 = 3\left(-\dfrac{1}{5}\right) - 1 = -\dfrac{3}{5} - \dfrac{5}{5} = -\dfrac{8}{5}$

If $y = 1$: $x = 3y - 1 = 3(1) - 1 = 2$

$$\boxed{\left\{ \left(-\frac{8}{5}, -\frac{1}{5}\right), (2, 1) \right\}}$$

31.

$$\begin{aligned} x^2 + y^2 &= 13 \\ y - 3x &= -11 \end{aligned}$$ (solve for y) $\rightarrow$

(substitute for y) $\rightarrow$

$$\begin{aligned} y &= 3x - 11 \\ x^2 + (3x - 11)^2 &= 13 \\ x^2 + 9x^2 - 66x + 121 &= 13 \\ 10x^2 - 66x + 108 &= 0 \\ 5x^2 - 33x + 54 &= 0 \\ (5x - 18)(x - 3) &= 0 \\ x = \frac{18}{5} \quad x &= 3 \end{aligned}$$

If $x = \dfrac{18}{5}$: $y = 3\left(\dfrac{18}{5}\right) - 11 = -\dfrac{1}{5}$

If $x = 3$: $y = 3(3) - 11 = -2$

$$\boxed{\left\{ \left(\frac{18}{5}, -\frac{1}{5}\right), (3, -2) \right\}}$$

33.

$$\begin{aligned} y^2 - 8x &= 0 \\ 2x - y - 6 &= 0 \end{aligned}$$ (solve for y) $\rightarrow$

(substitute for y) $\rightarrow$

$$\begin{aligned} y &= 2x - 6 \\ (2x - 6)^2 - 8x &= 0 \\ 4x^2 - 24x + 36 - 8x &= 0 \\ 4x^2 - 32x + 36 &= 0 \\ x^2 - 8x + 9 &= 0 \end{aligned}$$

$$\begin{aligned} x &= \frac{-b \pm \sqrt{b^2 - 4ac}}{2a} \\ &= \frac{8 \pm \sqrt{64 - 4(1)(9)}}{2} \\ &= \frac{8 \pm \sqrt{28}}{2} \\ &= \frac{8 \pm 2\sqrt{7}}{2} = 4 \pm \sqrt{7} \end{aligned}$$

If $x = 4 + \sqrt{7}$: $y = 2x - 6 = 2(4 + \sqrt{7}) - 6$
$\qquad = 8 + 2\sqrt{7} - 6 = 2 + 2\sqrt{7}$

If $x = 4 - \sqrt{7}$: $y = 2x - 6 = 2(4 - \sqrt{7}) - 6$
$\qquad = 8 - 2\sqrt{7} - 6 = 2 - 2\sqrt{7}$

$$\boxed{\{(4 + \sqrt{7}, 2 + 2\sqrt{7}), (4 - \sqrt{7}, 2 - 2\sqrt{7})\}}$$

35.

$$2y = -1 - x$$
$$\sqrt{x^2 + y^2} + y = 1$$

(solve for x) $\rightarrow$

$$x = -1 - 2y$$

(substitute for x) $\rightarrow$

$$\sqrt{(-1 - 2y)^2 + y^2} + y = 1$$
$$\sqrt{1 + 4y + 4y^2 + y^2} + y = 1$$
$$\sqrt{5y^2 + 4y + 1} = 1 - y$$
$$(\sqrt{5y^2 + 4y + 1})^2 = (1 - y)^2$$
$$5y^2 + 4y + 1 = 1 - 2y + y^2$$
$$4y^2 + 6y = 0$$
$$2y(2y + 3) = 0$$
$$2y = 0 \quad \text{or} \quad 2y + 3 = 0$$
$$y = 0 \qquad\qquad y = -\frac{3}{2}$$

If $y = 0$: $x = -1 - 2y = -1 - 2(0) = -1$

If $y = -\frac{3}{2}$: $x = -1 - 2\left(-\frac{3}{2}\right) = 2$

We must check for extraneous solutions.

Check: $(-1, 0)$

$$2y = -1 - x \qquad \sqrt{x^2 + y^2} + y = 1$$
$$2(0) = -1 - (-1) \qquad \sqrt{(-1)^2 + 0^2} + 0 = 1$$
$$0 = 0 \qquad\qquad \sqrt{1} = 1$$

Check: $\left(2, -\frac{3}{2}\right)$

$$2y = -1 - x \qquad \sqrt{x^2 + y^2} + y = 1$$
$$2\left(-\frac{3}{2}\right) = -1 - 2 \qquad \sqrt{2^2 + \left(-\frac{3}{2}\right)^2} + \left(-\frac{3}{2}\right) = 1$$
$$-3 = -3 \qquad\qquad \sqrt{\frac{25}{4}} - \frac{3}{2} = 1$$
$$\qquad\qquad\qquad \frac{5}{2} - \frac{3}{2} = 1$$

$$\boxed{\left\{ (-1, 0), \left(2, -\frac{3}{2}\right) \right\}}$$

37.

$$xy = 4$$

(solve for x) $\rightarrow$

$$x = \frac{4}{y}$$

$$x^2 + y^2 = 10$$

(substitute for x) $\rightarrow$

$$\left(\frac{4}{y}\right)^2 + y^2 = 10$$
$$\frac{16}{y^2} + y^2 = 10$$
$$y^2 \left(\frac{16}{y^2} + y^2\right) = y^2(10)$$
$$16 + y^4 = 10y^2$$
$$y^4 - 10y^2 + 16 = 0$$
$$\text{Let } t = y^2$$
$$t^2 - 10t + 16 = 0$$
$$(t - 8)(t - 2) = 0$$
$$t = 8 \quad \text{or} \quad t = 2$$
$$y^2 = 8 \quad \text{or} \quad y^2 = 2$$
$$y = \pm\sqrt{8} \quad y = \pm\sqrt{2}$$
$$y = \pm2\sqrt{2}$$
$$x = \frac{4}{y}$$

If $y = 2\sqrt{2}$: $x = \dfrac{4}{2\sqrt{2}} = \dfrac{2}{\sqrt{2}} = \dfrac{2}{\sqrt{2}} \cdot \dfrac{\sqrt{2}}{\sqrt{2}}$

$\quad\quad = \dfrac{2\sqrt{2}}{2} = \sqrt{2}$

If $y = -2\sqrt{2}$: $x = \dfrac{4}{-2\sqrt{2}} = -\sqrt{2}$

If $y = \sqrt{2}$: $x = \dfrac{4}{\sqrt{2}} = \dfrac{4}{\sqrt{2}} \cdot \dfrac{\sqrt{2}}{2} = \dfrac{4\sqrt{2}}{2}$

$\quad\quad = 2\sqrt{2}$

If $y = -\sqrt{2}$: $x = \dfrac{4}{-\sqrt{2}} = -2\sqrt{2}$

$\boxed{\{(\sqrt{2}, 2\sqrt{2}), (-\sqrt{2}, -2\sqrt{2}), (2\sqrt{2}, \sqrt{2}), (-2\sqrt{2}, -\sqrt{2})\}}$

39. $x = 40 - 3y$

$\quad\quad x = \dfrac{y^2}{10}$

(substitute for x) $\rightarrow$ $\dfrac{y^2}{10} = 40 - 3y$

$\quad\quad\quad\quad y^2 = 400 - 30y$

$y^2 + 30y - 400 = 0$

$(y - 10)(y + 40) = 0$

$\quad\quad y = 10 \quad \text{or} \quad y = -40 \quad \text{(reject)}$

At $\boxed{\$10}$ supply equals demand.

$\quad\quad x = 40 - 3(10) = 10$

$\boxed{\text{Ten items}}$ will be sold at the equilibrium price.

41. $x - y - 1 = 0 \quad \rightarrow \quad y = x - 1$

$\quad\quad\quad y = 2\sqrt{x + 3}$

$\boxed{\{(7.47, 6.47)\}}$

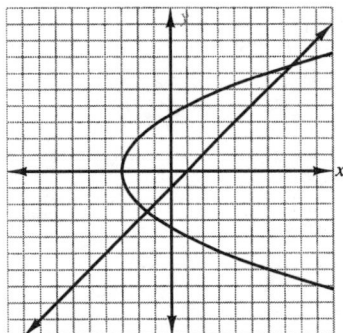

43. $x^3 - y^2 = 61$ (Equation 1)
   $x - y = 1$ (Equation 2)

a. $\dfrac{x^3 - y^3}{x - y} = \dfrac{61}{1}$

   $\dfrac{(x - y)(x^2 + xy + y^2)}{(x - y)} = 61$

   $x^2 + xy + y^2 = 61$ (Equation 3)

b. $x = 1 + y$ (solve for x in Equation 2)

   $(1 + y)^2 + (1 + y)y + y^2 = 61$ (substitute for x in Equation 3)

$$
\begin{aligned}
3y^2 + 3y - 60 &= 0 \\
y^2 + y - 20 &= 0 \\
(y + 5)(y - 4) &= 0 \\
y = -5 \ \ \text{or} \ \ y &= 4
\end{aligned}
$$

   If $y = -5$: $x = 1 + (-5) = -4$
   If $y = 4$: $x = 1 + 4 = 5$
   $\boxed{\{(-4, -5), (5, 4)\}}$

45. $x + 2y + 5 = 0$ (Equation 1)
   $x^2 y^2 + xy - 6 = 0$ (Equation 2)

a. $(xy + 3)(xy - 2) = 0$
   $xy = -3$ $xy = 2$

b. $x = -2y - 5$ (solve for x in Equation 1)

$$
\begin{array}{ll}
(-2y - 5)y = -3 & \qquad (-2y - 5)y = 2 \\
-2y^2 - 5y + 3 = 0 & \qquad -2y^2 - 5y - 2 = 0 \\
2y^2 + 5y - 3 = 0 & \qquad 2y^2 + 5y + 2 = 0 \\
(2y - 1)(y + 3) = 0 & \qquad (2y + 1)(y + 2) = 0 \\
y = \dfrac{1}{2} \ \text{or} \ y = -3 & \qquad y = -\dfrac{1}{2} \ \text{or} \ y = -2
\end{array}
$$

   If $y = \dfrac{1}{2}$: $x = -2\left(\dfrac{1}{2}\right) - 5 = -6$
   If $y = -3$: $x = -2(-3) - 5 = 1$
   If $y = -\dfrac{1}{2}$: $x = -2\left(-\dfrac{1}{2}\right) - 5 = -4$
   If $y = -2$: $x = -2(-2) - 5 = -1$

   $\boxed{\left\{ \left(-6, \dfrac{1}{2}\right), (1, -3), \left(-4, -\dfrac{1}{2}\right), (-1, -2) \right\}}$

47.

$$
\begin{aligned}
xy &= -ab \\
2x - y &= a + 2b
\end{aligned}
$$

   (solve for y) $\rightarrow$ $y = 2x - a - 2b$
   (substitute for y) $\rightarrow$ $x(2x - a - 2b) = -ab$

$$
\begin{aligned}
2x^2 - ax - 2bx + ab &= 0 \\
x(2x - a) - b(2x - a) &= 0 \\
(2x - a)(x - b) &= 0 \\
2x - a = 0 \ \ \text{or} \ \ x - b &= 0 \\
x = \dfrac{a}{2} \qquad\qquad x &= b
\end{aligned}
$$

   If $x = \dfrac{a}{2}$: $y = 2\left(\dfrac{a}{2}\right) - a - 2b = -2b$
   If $x = b$: $y = 2b - a - 2b = -a$
   $\boxed{\left\{ \left(\dfrac{a}{2}, -2b\right), (b, -a) \right\}}$

49.

$$3\sqrt{x-2y}+\frac{3}{\sqrt{x-2y}} = 10$$

$$3t+\frac{3}{t} = 10 \quad \text{(Let } t = \sqrt{x-2y}.\text{)}$$

$$3t^2+3 = 10t \quad \text{(Multiply by } t.\text{)}$$

$$3t^2-10t+3 = 0$$

$$(3t-1)(t-3) = 0$$

$$t = \frac{1}{3} \quad \text{or} \quad t = 3$$

$$\sqrt{x-2y} = \frac{1}{9} \qquad\qquad \sqrt{x-2y} = 3$$

$$x-2y = \frac{1}{9} \qquad\qquad x-2y = 0$$

$$x = 2y+\frac{1}{9} \qquad\qquad x = 2y+9$$

Since we want (a, b) such that $x = ay + b$, the equation on the left yields $a = 2$ and $b = \frac{1}{9}$. The equation on the right yields $a = 2$ and $b = 9$.

The solutions are $\boxed{\left(2, \frac{1}{9}\right) \text{ or } (2, 9)}$.

51.

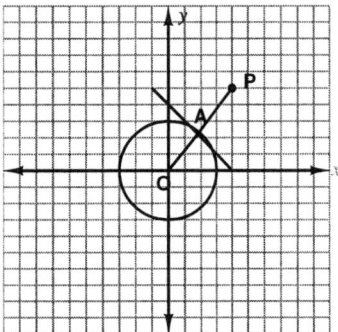

As shown in the diagram, AP is the minimum distance. Using the distance formula

$$OP = \sqrt{(4-0)^2+(5-0)^2} = \sqrt{16+25} = \sqrt{41}$$

Also, the circle has radius 3, so $OA = 3$.

The minimum distance

$$AP = OP - OA$$

$$= \sqrt{41}-3$$

Since $a = \sqrt{a^2}$ for $a > 0$, this distance can be expressed as

$$\sqrt{(\sqrt{41}-3)^2} = \sqrt{41-6\sqrt{41}+9}$$

$$= \boxed{\sqrt{50-6\sqrt{41}} \approx 3.403}$$

Review Problems

54. $\dfrac{x^{x+1}y^{2n-1}}{(x^n y^{1-n})^3}$

$= \dfrac{x^{n+1}y^{2n-1}}{x^{3n}y^{3-3n}}$

$= x^{n+1-3n}y^{2n-1-(3-3n)}$

$= \boxed{x^{1-2n}y^{5n-4}}$ or $\boxed{\dfrac{y^{5n-4}}{x^{2n-1}}}$

55. line through $(1,4)$ and parallel to $2x - 5y + 7 = 0$:

$(x_1, y_1) = (1, 4)$

Slope of $2x - 5y + 7 = 0$:

$\begin{aligned} -5y &= -2x - 7 \\ y &= \frac{2}{5}x + \frac{7}{5}, \; m = \frac{2}{5} \end{aligned}$

Slope of line parallel: $\dfrac{2}{5}$

$y - y_1 = m(x - x_1)$

$\boxed{y - 4 = \dfrac{2}{5}(x - 1)}$ Point Slope Equation

56.

$\begin{aligned} \sqrt{3 - 3y} &= 3 + \sqrt{3y + 2} \\ (\sqrt{3 - 3y})^2 &= (3 + \sqrt{3y + 2})^2 \\ 3 - 3y &= 9 + 6\sqrt{3y + 2} + 3y + 2 \\ 3 - 3y &= 3y + 11 + 6\sqrt{3y + 2} \\ -8 - 6y &= 6\sqrt{3y + 2} \\ 4 + 3y &= -3\sqrt{3y + 2} \\ (4 + 3y)^2 &= (-3\sqrt{3y + 2})^2 \\ 16 + 24y + 9y^2 &= 9(3y + 2) \\ 16 + 24y + 9y^2 &= 27y + 18 \\ 9y^2 - 3y - 2 &= 0 \\ (3y - 2)(3y + 1) &= 0 \\ y = \frac{2}{3} \quad \text{or} \quad y &= -\frac{1}{3} \end{aligned}$

Check $\dfrac{2}{3}$:

$$\sqrt{3 - 3\left(\frac{2}{3}\right)} = 3 + \sqrt{3\left(-\frac{2}{3}\right) + 2}$$

$$\sqrt{3 - 2} = 3 + \sqrt{-2 + 2}$$

$$1 = 3 \text{ False}$$

$-\dfrac{2}{3}$ is extraneous

Check $-\dfrac{1}{3}$:

$$\sqrt{3-3\left(-\dfrac{1}{3}\right)} = 3 + \sqrt{3\left(-\dfrac{1}{3}\right)+2}$$

$$\sqrt{3+1} = 3 + \sqrt{-1+2}$$

$$2 = 4 \quad \text{False}$$

$-\dfrac{1}{3}$ is extraneous

$$\boxed{\varnothing}$$

Section 9.6 Solving Nonlinear Systems by the Addition Method

Problem Set 9.6, pp. 727-728

1.
$$\begin{aligned}
x^2 + y^2 &= 13 \\
x^2 - y^2 &= 5 \\
\hline
2x^2 &= 18 \quad \text{(add)} \\
x^2 &= 9 \\
x &= \pm 3
\end{aligned}$$

If $x = 3$:
$$\begin{aligned}
x^2 + y^2 &= 13 \\
9 + y^2 &= 13 \\
y^2 &= 4 \\
y &= \pm 2
\end{aligned}$$

$(3, 2), (3, -2)$

If $x = -3$:
$$\begin{aligned}
(-3)^2 + y^2 &= 13 \\
y &= \pm 2
\end{aligned}$$

$(-3, 2), (-3, -2)$

$$\boxed{\{(3, 2), (3, -2), (-3, 2), (-3, -2)\}}$$

3.
$$\begin{aligned}
x^2 - y^2 &= 11 \quad (\times -2) \rightarrow \\
2x^2 - 5y^2 &= 7 \qquad\qquad \rightarrow
\end{aligned}$$
$$\begin{aligned}
-2x^2 + 2y^2 &= -22 \\
2x^2 - 5y^2 &= 7 \\
\hline
-3y^2 &= -15 \quad \text{(add)} \\
y^2 &= 5 \\
y &= \pm\sqrt{5}
\end{aligned}$$

If $y = \pm\sqrt{5}$:
$$\begin{aligned}
x^2 - 5 &= 11 \\
x^2 &= 16 \\
x &= \pm 4
\end{aligned}$$

Similarly, if $y = -\sqrt{5}$: $\quad x = \pm 4$

$$\boxed{\{(4, \sqrt{5}), (-4, \sqrt{5}), (4, -\sqrt{5}), (-4, -\sqrt{5})\}}$$

5.
$$\begin{aligned}
4x^2 - y^2 &= 4 \\
4x^2 + y^2 &= 4 \\
\hline
8x^2 &= 8 \quad \text{(add)} \\
x^2 &= 1 \\
x &= \pm 1
\end{aligned}$$

If $x = \pm 1$:
$$\begin{aligned}
4 - y^2 &= 4 \\
y^2 &= 0 \\
y &= 0
\end{aligned}$$

$$\boxed{\{(1, 0), (-1, 0)\}}$$

7.
$$\begin{aligned} y &= x^2 + 4 \\ x^2 + y^2 &= 7 \end{aligned} \qquad \text{(rearrange)} \rightarrow$$

$$\begin{aligned} -x^2 + y &= 4 \\ \underline{x^2 + y^2 = 16} \\ y + y^2 &= 20 \quad \text{(add)} \\ y^2 + y - 20 &= 0 \\ (y + 5)(y - 4) &= 0 \\ y &= -5 \quad \text{or} \quad y = 4 \end{aligned}$$

If $y = -5$:
$$\begin{aligned} y &= x^2 + 4 \\ x^2 + 4 &= -5 \\ x^2 &= -9 \\ x &= \pm\sqrt{-9} \\ &= \pm 3i \end{aligned}$$

If $y = 4$:
$$\begin{aligned} y &= x^2 + 4 \\ 4 &= x^2 + 4 \\ x^2 &= 0 \\ x &= 0 \end{aligned}$$

$$\boxed{\{(3i, -5), (-3i, -5), (0, 4)\}}$$

9.
$$\begin{aligned} y &= x^2 - 4 \\ x^2 + y^2 &= 10 \end{aligned} \qquad \text{(rearrange)} \rightarrow$$

$$\begin{aligned} -x^2 + y &= -4 \\ \underline{x^2 + y^2 = 10} \\ y + y^2 &= 6 \quad \text{(add)} \\ y^2 + y - 6 &= 0 \\ (y + 3)(y - 2) &= 0 \\ y &= -3 \quad \text{or} \quad y = 2 \end{aligned}$$

If $y = -3$:
$$\begin{aligned} -3 &= x^2 - 4 \\ x^2 &= 1 \\ x &= \pm 1 \end{aligned}$$

If $y = 2$:
$$\begin{aligned} 2 &= x^2 - 4 \\ x^2 &= 6 \\ x &= \pm\sqrt{6} \end{aligned}$$

$$\boxed{\{(1, -3), (-1, -3), (\sqrt{6}, 2), (-\sqrt{6}, 2)\}}$$

11.
$$\begin{aligned} x^2 - 16y^2 &= 8 \qquad (\times 3) \rightarrow \\ y^2 - 3x^2 &= 23 \end{aligned}$$

$$\begin{aligned} 3x^2 - 48y^2 &= 24 \\ \underline{-3x^2 + y^2 = 23} \\ -47y^2 &= 47 \quad \text{(add)} \\ y^2 &= -1 \\ y &= \pm\sqrt{-1} \\ &= \pm i \end{aligned}$$

If $y = i$:
$$\begin{aligned} x^2 - 16y^2 &= 8 \\ x^2 - 16i^2 &= 8 \\ x^2 - 16(-1) &= 8 \\ x^2 &= -8 \\ x &= \pm\sqrt{-8} \\ x &= \pm 2\sqrt{2}\, i \end{aligned}$$

If $y = -i$:
$$\begin{aligned} x^2 - 16(-i)^2 &= 8 \\ x^2 - 16i^2 &= 8 \\ x &= \pm 2\sqrt{2}\, i \end{aligned}$$

$$\boxed{\{(2\sqrt{2}\, i, i), (-2\sqrt{2}\, i, i), (2\sqrt{2}\, i, -i), (-2\sqrt{2}\, i, -i)\}}$$

13.

$$5x^2 = 8 - 6y^2 \quad \text{(rearrange)} \rightarrow \quad 5x^2 + 6y^2 = 8$$
$$y^2 = 4 - x^2 \quad \text{(rearrange)} \rightarrow \quad x^2 + y^2 = 4$$

$$\rightarrow \quad 5x^2 + 6y^2 = 8$$
$$(\times -5) \rightarrow \quad \underline{-5x^2 - 5y^2 = -20}$$
$$y^2 = -12 \quad \text{(add)}$$
$$y = \pm\sqrt{-12}$$
$$y = \pm 2\sqrt{3}$$
$$y^2 = 4 - x^2$$
$$-12 = 4 - x^2$$
$$x^2 = 16$$
$$x = \pm 4$$

$$\boxed{\{(4, 2\sqrt{3}\,i), (4, -2\sqrt{3}\,i), (-4, 2\sqrt{3}\,i), (-4, -2\sqrt{3}\,i)\}}$$

15.

$$4x^2 - y = 3 \quad (\times -2) \rightarrow \quad -8x^2 + 2y = -6$$
$$8x^2 - y^2 = -9 \quad \rightarrow \quad \underline{8x^2 - y^2 = -9}$$
$$2y - y^2 = -15 \quad \text{(add)}$$
$$-y^2 + 2y + 15 = 0$$
$$y^2 - 2y - 15 = 0$$
$$(y - 5)(y + 3) = 0$$
$$y = 5 \quad \text{or} \quad y = -3$$

If $y = 5$:
$$4x^2 - y = 3$$
$$4x^2 = 8$$
$$x^2 = 2$$
$$x = \pm\sqrt{2}$$

If $y = -3$:
$$4x^2 - y = 3$$
$$4x^2 - (-3) = 3$$
$$4x^2 = 0$$
$$x = 0$$

$$\boxed{\{(\sqrt{2}, 5), (-\sqrt{2}, 5), (0, -3)\}}$$

17.

$$y^2 + 3xy = 1 \quad (\times 4) \rightarrow \quad 4y^2 + 12xy = 4$$
$$y^2 + 4xy = 2 \quad (\times -3) \rightarrow \quad \underline{-3y^2 - 12xy = -6}$$
$$y^2 = -2 \quad \text{(add)}$$
$$y = \pm\sqrt{-2} = \pm\sqrt{2}\,i$$

If $y = \sqrt{2}\,i$:
$$y^2 + 3xy = 1$$
$$(\sqrt{2}\,i)^2 + 3x(\sqrt{2}\,i) = 1$$
$$2i^2 + 3\sqrt{2}\,ix = 1$$
$$-2 + 3\sqrt{2}\,ix = 1$$
$$3\sqrt{2}\,ix = 3$$
$$x = \frac{3}{3\sqrt{2}\,i}$$
$$= \frac{1}{\sqrt{2}\,i} \cdot \frac{\sqrt{2}\,i}{\sqrt{2}\,i}$$
$$= \frac{\sqrt{2}\,i}{2i^2}$$
$$= \frac{\sqrt{2}\,i}{2(-1)}$$
$$= \frac{-\sqrt{2}\,i}{2}$$

$$\text{If } y = -\sqrt{2}\, i: \quad y^2 + 3xy = 1$$
$$(-\sqrt{2}\, i)^2 + 3x(-\sqrt{2}\, i) = 1$$
$$2i^2 - 3\sqrt{2}\, ix = 1$$
$$-2 - 3\sqrt{2}\, ix = 1$$
$$-3\sqrt{2}\, ix = 3$$
$$x = \frac{-1}{\sqrt{2}\, i} \cdot \frac{\sqrt{2}\, i}{\sqrt{2}\, i}$$
$$= \frac{-\sqrt{2}\, i}{2i^2}$$
$$= \frac{\sqrt{2}\, i}{2}$$

$$\boxed{\left\{ \left(-\frac{\sqrt{2}\, i}{2}, \sqrt{2}\, i \right), \left(\frac{\sqrt{2}\, i}{2}, -\sqrt{2}\, i \right) \right\}}$$

19.
$$
\begin{array}{ll}
2xy + 3y = 3 & \rightarrow \\
-xy + 2y = -1 & (\times 2) \rightarrow
\end{array}
$$

$$
\begin{array}{rl}
2xy + 3y &= 3 \\
-2xy + 4y &= -2 \\
\hline
y &= 1 \quad \text{(add)} \\
2x(1) - 3(1) &= 3 \\
2x &= 6 \\
x &= 3
\end{array}
$$

$$\boxed{\{(3, 1)\}}$$

21.
$$
\begin{array}{ll}
x^2 + 2xy - y^2 = 7 & \rightarrow \\
x^2 - y^2 = 3 & (\times -1) \rightarrow
\end{array}
$$

$$
\begin{array}{rl}
x^2 + 2xy - y^2 &= 7 \\
-x^2 + y^2 &= -3 \\
\hline
2xy &= 4 \quad \text{(add)} \\
y &= \frac{2}{x} \quad \text{(solve for } y)
\end{array}
$$

(Substitute for y in Equation 2):
$$x^2 - \frac{4}{x^2} = 3$$
$$x^2 \left(\frac{x^2 - 4}{x^2} \right) = 3x^2$$
$$x^4 - 4 = 3x^2$$
$$x^4 - 3x^2 - 4 = 0$$
$$(x^2 - 4)(x^2 + 1) = 0$$

$$
\begin{array}{ccc}
x^2 - 4 = 0 & \quad\text{or}\quad & x^2 + 1 = 0 \\
x = \pm 2 & & x = \pm i
\end{array}
$$

$$
\begin{array}{rl}
\text{If } x \pm 2: \quad (\pm 2)^2 + y^2 &= 3 \\
4 - y^2 &= 3 \\
-y^2 &= -1 \\
y^2 &= 1 \\
y &= \pm 1
\end{array}
$$
$$(\text{if } x = +2, y = 1) \ [(2, -1) \text{ is extraneous}]$$
$$(\text{if } x = -2, y = -1) \ [(-2, 1) \text{ is extraneous}]$$

If $x = \pm i$: $(\pm i)^2 - y^2 = 3$

$$-1 - y^2 = 3$$
$$-y^2 = 4$$
$$y^2 = -4$$
$$y = \pm 2i$$

(if $x = i$, $y = -2i$) [$(i, 2i)$ is extraneous]
(if $x = -i$, $y = 2i$) [$(-i, -2i)$ extraneous]

$$\boxed{\{(2, 1), (-2, -1), (i, -2i), (-i, 2i)\}}$$

23. $5x^2 - xy + 5y^2 = 89$ $\rightarrow$ $5x^2 - xy + 5y^2 = 89$

$x^2 + y^2 = 17$ $(\times -5) \rightarrow$ $\underline{-5x^2 - 5y^2 = -85}$

$$-xy = 4$$
$$y = -\frac{4}{x}$$

$$x^2 + \frac{16}{x^2} = 17$$
$$x^4 - 17x^2 + 16 = 0$$
$$(x^2 - 16)(x^2 - 1) = 0$$

$x^2 - 16 = 0$ or $x^2 - 1 = 0$

$x = \pm 4$ $x = \pm 1$

If $x = \pm 4$: $(\pm 4)^2 + y^2 = 17$

$$16 + y^2 = 17$$
$$y^2 = 1$$
$$y = \pm 1$$

(if $x = 4$, $y = -1$) [$(4, 1)$ is extraneous]
(if $x = -4$, $y = 1$) [$(-4, -1)$ is extraneous]

If $x = \pm 1$: $(\pm 1)^2 + y^2 = 17$

$$1 + y^2 = 17$$
$$y^2 = 16$$
$$y = \pm 4$$

(if $x = 1$, $y = -4$) [$(1, 4)$ is extraneous]
(if $x = -1$, $y = 4$) [$(-1, -4)$ is extraneous]

$$\boxed{\{(4, -1), (-4, 1), (1, -4), (-1, 4)\}}$$

25. $x^2 - 2xy - y^2 = -8$ $(\times -1) \rightarrow$ $-x^2 + 2xy + y^2 = 8$

$x^2 + 5xy - y^2 = 20$ $\rightarrow$ $\underline{x^2 + 5xy - y^2 = 20}$

$$7xy = 28$$
$$y = \frac{4}{x}$$

$$x^2 - 2x\left(\frac{4}{x}\right) - \left(\frac{4}{x}\right)^2 = -8$$

$$x^2 - 8 - \frac{16}{x^2} = -8$$

$$x^2 - \frac{16}{x^2} = 0$$

$$x^4 - 16 = 0$$

$$(x^2 - 4)(x^2 + 4) = 0$$

$x = \pm 2$ or $x = \pm 2i$

If $x = 2$: $4 - 4y - y^2 = -8$

$$0 = y^2 + 4y - 12$$
$$0 = (y + 6)(y - 2)$$
$$y = -6 \quad \text{or} \quad y = 2$$

$(2, 2)$ checks

$(2, -6)$ is extraneous

If $x = -2$: $4 + 4y - y^2 = -8$
$$0 = y^2 - 4y - 12$$
$$0 = (y - 6)(y + 2)$$
$$y = 6 \quad \text{or} \quad y = -2$$

$(-2, 6)$ is extraneous
$(-2, -2)$ checks

If $x = 2i$: $-4 - 4iy - y^2 = -8$
$$0 = y^2 + 4iy - 4$$
$$0 = (y + 2i)^2$$
$$y = -2i$$

$(2i, -2i)$

If $x = -2i$: $-4 + 4iy - y^2 = -8$
$$0 = y^2 - 4iy - 4$$
$$0 = (y - 2i)^2$$
$$y = 2i$$

$(-2i, 2i)$

$$\boxed{\{(2, 2), (-2, -2), (2i, -2i), (-2i, 2i)\}}$$

27. $\begin{aligned} 2x^2 + 5.793y^2 &= 39.748 \\ 3x^2 + 4.251y^2 &= 34.078 \end{aligned}$ $\begin{aligned} (\times 3) &\rightarrow \\ (\times -2) &\rightarrow \end{aligned}$ $\begin{aligned} 6x^2 + 17.379y^2 &= 119.244 \\ \underline{-6x^2 - 8.502y^2} &= \underline{-68.156} \\ 8.877y^2 &= 51.088 \\ y^2 &= \dfrac{51.088}{8.877} \end{aligned}$

$$y = \pm\sqrt{\frac{51.088}{8.877}} \approx \pm 2.40$$
$$2x^2 + 5793y^2 = 39.748$$
$$x^2 = \frac{39.748 - 5.793y^2}{2}$$
$$x = \pm\sqrt{\frac{39.748 - 5.793y^2}{2}}$$
$$x \approx \pm 1.79$$

$$\boxed{\{(-1.79, -2.40), (-1.79, 2.40), (1.79, -2.40), (1.79, 2.40)\}}$$

29. $\begin{aligned} 2x^2 - 2xy + y^2 &= 2 \\ \underline{3x^2 + 2xy - y^2} &= \underline{3} \end{aligned}$

Add: $\begin{aligned} 5x^2 &= 5 \\ x^2 &= 1 \\ x &= \pm 1 \end{aligned}$

$2x^2 - 2xy + y^2 = 2$ (Given equation)
$2 - 2y + y^2 = 2$ (Let $x = 1$.)
$y^2 - 2y = 0$
$y(y - 2) = 0$
$y = 0 \quad \text{or} \quad y = 2$ (1, 0) and (1, 2) are solutions.

$2x^2 - 2xy + y^2 = 2$ (Given equation)
$2 + 2y + y^2 = 2$ (Let $x = -1$.)
$y^2 + 2y = 0$
$y(y + 2) = 0$
$y = 0 \quad \text{or} \quad y = -2$ (-1, 0) and (-1, -2) are solutions.

The paths intersect at $\boxed{(1, 0), (-1, 0), (1, 2), \text{nd } (-1, -2)}$.

31. $\dfrac{2}{x^2} - \dfrac{3}{y^2} = -6$

$\dfrac{3}{x^2} + \dfrac{4}{y^2} = 59$

$\left(\text{Let } t = \dfrac{1}{x^2} \text{ and } z = \dfrac{1}{y^2}\right):$

$2t - 3z = -6$	$(\times 4) \rightarrow$	$8t - 12z = -24$
$3t + 4z = 59$	$(\times 3) \rightarrow$	$\underline{9t + 12z = 177}$

$$17t = 153$$
$$t = 9$$
$$2t - 3z = -6$$
$$18 - 3z = -6$$
$$-3z = -24$$
$$z = 8$$

$t = 9:\quad \dfrac{1}{x^2} = 9 \qquad\qquad z = 8:\quad \dfrac{1}{y^2} = 8$

$\qquad\qquad x^2 = \dfrac{1}{9} \qquad\qquad\qquad\qquad y^2 = \dfrac{1}{8}$

$\qquad\qquad x = \pm\dfrac{1}{3} \qquad\qquad\qquad\qquad y = \pm\dfrac{\sqrt{2}}{4}$

$$\left\{\left(\dfrac{1}{3}, \dfrac{\sqrt{2}}{4}\right), \left(\dfrac{1}{3}, -\dfrac{\sqrt{2}}{4}\right), \left(-\dfrac{1}{3}, \dfrac{\sqrt{2}}{4}\right), \left(-\dfrac{1}{3}, -\dfrac{\sqrt{2}}{4}\right)\right\}$$

Review Problems

34. Let

$$x = \text{width of the frame}$$
$$4 + 2x = \text{width of photo + frame}$$
$$7 + 2x = \text{length of photo + frame}$$

$$\text{area of frame} = 26 \text{ square inches}$$
$$\text{area of photo + frame} = \text{area of photo + area of frame}$$
$$(4 + 2x)(7 + 2x) = (4)(7) + 26$$
$$28 + 22x + 4x^2 = 28 + 26$$
$$4x^2 + 22x - 26 = 0$$
$$2x^2 + 11x - 13 = 0$$
$$(2x + 13)(x - 1) = 0$$
$$x = -\dfrac{13}{2} \quad \text{(reject)} \quad \text{or} \quad x = 1$$

The width of the frame is $\boxed{1 \text{ inch}}$.

35. Let

$$x = \text{amount invested at 6\%}$$
$$x = \text{amount invested at 8\%}$$
$$10{,}000 - 2x = \text{amount invested at 7\%}$$

$$0.06x(3) + 0.095x(2) + (10{,}000 - 2x)(0.07)(2) = 1670$$
$$0.18x + 0.19x + 1400 - 0.28x = 1670$$
$$0.09x = 270$$
$$x = 3000 \quad \text{(at 6\%)}$$
$$x = 3000 \quad \left(\text{at } 9\tfrac{1}{2}\%\right)$$
$$10000 - 2x = 4000 \quad \text{(at 7\%)}$$

$\boxed{\$3000 \text{ at } 6\% \text{ for 3 years, } \$3000 \text{ at } 9\tfrac{1}{2}\% \text{ for 2 years, } \$4000 \text{ at } 7\% \text{ for 2 years}}$

36. $x^5 - 9x^2 - 8x^2 + 72$
$= x^3(x^2 - 9) - 8(x^2 - 9)$
$= (x^2 - 9)(x^3 - 8)$
$= \boxed{(x + 3)(x - 3)(x - 2)(x^2 + 2x + 4)}$

Section 9.7 Problem Solving

Problem Set 9.7, pp. 732-733

1. Let x and y equal the numbers.
$$\begin{array}{rl} x^2 + y^2 &= 4 \\ \underline{x^2 - y^2 \ \ } &= \underline{4 \ \ } \\ 2x^2 &= 8 \quad \text{(add)} \\ x^2 &= 4 \end{array}$$

$$\begin{array}{rl} x &= \pm\sqrt{4} \\ x &= \pm 2 \\ x^2 + y^2 &= 4 \\ 4 + y^2 &= 0 \\ y &= 0 \end{array}$$

Numbers: $\boxed{2 \text{ and } 0 \text{ or } -2 \text{ and } 0}$

3. Let x and y equal the numbers.
$$x^2 + y^2 = 45$$
$$\underline{xy = 18} \qquad \rightarrow \qquad x = \frac{18}{y}$$
$$\left(\frac{18}{y}\right)^2 + y^2 = 45$$
$$y^4 - 45y^2 + 324 = 0$$

$$\begin{array}{rl} t^2 - 45t + 324 &= 0 \quad \text{(Let } t = y^2) \\ (t - 9)(t - 36) &= 0 \end{array}$$

$$\begin{array}{rlcrl} t &= 9 & \text{or} & t &= 36 \\ y^2 &= 9 & & y^2 &= 36 \\ y &= \pm 3 & & y &= \pm 6 \\ x &= \dfrac{18}{y} & & & \end{array}$$

If $y = 3, x = 6$.   If $y = -3, x = -6$.
If $y = 6, x = 3$.   If $y = -6, x = -3$.
Numbers: $\boxed{3 \text{ and } 6 \text{ or } -3 \text{ and } -6}$

5. Let
$$\begin{array}{rl} t &= \text{tens' digit} \\ u &= \text{units' digit} \end{array}$$

$$\begin{array}{rl} (10t + u)(t + u) &= 115 \\ 2u + 4t &= 14 \end{array}$$

$$\begin{array}{rlcrl} 10t^2 + 11tu + u^2 &= 115 & & & \\ u + 2t &= 7 & & u &= 7 - 2t \end{array}$$

$$10t^2 + 11t(7 - 2t) + (7 - 2t)^2 = 115$$
$$10t^2 + 77t - 22t^2 + 49 - 28t + 4t^2 = 115$$
$$-8t^2 + 49t - 66 = 0$$
$$8t^2 - 49t + 66 = 0$$

$$t = \frac{33}{8} \text{ (reject)} \quad \text{or} \quad t = 2$$

$$u = 7 - 2t = 7 - 2(2) = 3$$

Number: $\boxed{23}$

7. Let x and y equal the numbers.

$$x + y = 5 \qquad \rightarrow \qquad x = 5 - y$$
$$\frac{1}{x} + \frac{1}{y} = \frac{5}{6} \qquad\qquad\qquad \frac{1}{5-y} + \frac{1}{y} = \frac{5}{6}$$

$$6y(5-y)\left[\frac{1}{5-y} + \frac{1}{y}\right] = 6y(5-y)\left(\frac{5}{6}\right)$$
$$6y + 30 - 6y = 25y - 5y^2$$
$$5y^2 - 25y + 30 = 0$$
$$y^2 - 5y + 6 = 0$$
$$(y-2)(y-3) = 0$$
$$y = 2 \quad \text{or} \quad y = 3$$
$$x = 5 - y$$

If $y = 2$, $x = 3$.
If $y = 3$, $x = 2$.
Numbers: $\boxed{2 \text{ and } 3}$

9. Let

$$x = \text{length}$$
$$y = \text{width}$$

Perimeter is 20 feet.

$$2x + 2y = 20$$

Area is 21 ft^2.

$$xy = 21$$
$$(\div 2) \quad x + y = 10 \qquad y = 10 - x$$
$$xy = 21$$
$$x(10 - x) = 21$$
$$10x - x^2 = 21$$
$$0 = x^2 - 10x + 21$$
$$0 = (x-7)(x-3)$$
$$x = 7 \quad \text{or} \quad x = 3$$

If $x = 7$: $y = 10 - x = 10 - 7 = 3$
If $x = 3$: $y = 10 - x = 10 - 3 = 7$
Dimensions: $\boxed{7 \text{ ft} \times 3 \text{ ft}}$

11. Let x and y equal the length of the legs of the right triangle.

hypotenuse: 5 yards

$$\begin{aligned} x + y + 5 &= 12 \\ x^2 + y^2 &= 25 \end{aligned}$$

$$\begin{aligned} x + y &= 7 \\ x^2 + y^2 &= 25 \end{aligned}$$

$$\begin{aligned} (7 - y)^2 + y^2 &= 25 \\ y = 4 \text{ or } y &= 3 \end{aligned}$$

Since $x = 7 - y$, legs measure $\boxed{4 \text{ yards and } 3 \text{ yards}}$.

13. Let

$$\begin{aligned} x &= \text{length} \\ y &= \text{width} \\ xy &= 480 \end{aligned}$$

$$\begin{aligned} \text{area} &= 480 \text{ squre feet} \\ (x + 10)(y - 2) &= 480 + 20 = 500 \\ xy &= 480 \quad \rightarrow \quad x = \frac{480}{y} \\ xy - 2x + 10y &= 520 \end{aligned}$$

$$\begin{aligned} \left(\frac{480}{y}\right) y - 2\left(\frac{480}{y}\right) + 10y &= 520 \\ 480y - 960 + 10y^2 &= 520y \\ y^2 - 4y - 96 &= 0 \\ (y - 12)(y + 8) &= 0 \\ y = 12 \text{ or } y &= -8 \quad \text{(reject)} \end{aligned}$$

If $y = 12$: $x = \dfrac{480}{y} = \dfrac{480}{12} = 40$.

Length: 40 feet

Width: 12 feet

Dimensions: $\boxed{40 \text{ ft} \times 12 \text{ ft}}$

15. Let x and y equal the length of the legs of the right triangle.

$$\begin{aligned} \text{hypotenuse} &= 10 \text{ inches} \\ x^2 + y^2 &= 100 \\ \frac{1}{2} xy &= 24 \\ y &= \frac{48}{x} \\ x^2 + \frac{48}{x^2} &= 100 \\ x^4 - 100x^2 + 2304 &= 0 \\ (x^2 - 64)(x^2 - 36) &= 0 \end{aligned}$$

$$\begin{aligned} x &= \pm 8 \quad \text{(reject } -8) & \text{or} && x &= \pm 6 \quad \text{(reject } -6) \\ x &= 8 & && x &= 6 \end{aligned}$$

If $x = 8$, $y = \dfrac{48}{8} = 6$

If $x = 6$, $y = \dfrac{48}{6} = 9$

$\{(6, 8), (8, 6)\})$

Legs: $\boxed{6 \text{ inches and } 8 \text{ inches}}$

17. Let x and y equal the length and width of the rectangle.
diagonal = 13 feet
perimeter = 34 feet

$$x^2 + y^2 = 169$$
$$2x + 2y = 34$$
$$x + y = 17 \quad \rightarrow$$

$$y = 17 - x$$
$$x^2 + (17 - x)^2 = 169$$
$$x^2 + 289 - 34x + x^2 = 169$$
$$2x^2 - 34x + 120 = 0$$
$$x^2 - 17x + 60 = 0$$
$$(x - 12)(x - 5) = 0$$
$$x = 12 \quad \text{or} \quad x = 5$$

If $x = 12$, $y = 17 - 12 = 5$.
If $x = 5$, $y = 17 - 5 = 12$.
$\{(12, 5), (5, 12)\}$
Dimensions: $\boxed{12 \text{ feet by } 5 \text{ feet}}$

19. Area is 21.
$$x^2 - y^2 = 21$$
Perimeter is 24.
$$3x + 3y + (x - y) = 24$$

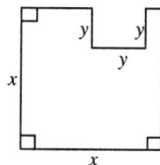

$$x^2 - y^2 = 21$$
$$4x + 2y = 24$$

$$x^2 - y^2 = 21$$
$$2x + y = 12$$
$$y = 12 - 2x$$

$$x^2 - (12 - 2x)^2 = 21$$
$$x^2 - 144 + 48x - 4x^2 = 21$$
$$-3x^2 + 48x - 165 = 0$$
$$x^2 - 16x + 55 = 0$$
$$(x - 11)(x - 5) = 0$$
$$x = 11 \quad \text{or} \quad x = 5$$

If $x = 5$, $y = 12 - 10 = 2$
If $x = 1$, $y = 12 - 22 = -10$
$\{(5, 2), (11, -10)\}$
Thus, $x = 5$ and $y = 2$.
$\boxed{x = 5 \text{ feet}; \ y = 2 \text{ feet}}$

21.

$$
\begin{array}{ll}
a^2 + b^2 = 10^2 & \quad (\times -1) \ \to \\
a^2 + (b+9)^2 = 17^2 & \quad \to
\end{array}
$$

$$
\begin{array}{rcl}
-a^2 - b^2 &=& -100 \\
\underline{a^2 + b^2 + 18b + 81} &=& \underline{289} \\
18b + 81 &=& 189 \\
18b &=& 108 \\
b &=& 6
\end{array}
$$

$$
\begin{array}{rcll}
a^2 + 36 &=& 100 & \\
a^2 &=& 64 & \\
a &=& \pm 8 & \text{(reject } -8) \\
a &=& 8 &
\end{array}
$$

$$\boxed{a = 8,\, b = 6}$$

23. $CD = 8$ inches

$$
\begin{array}{rcl}
x + y &=& 8 \\
y^2 &=& 4^2 + x^2
\end{array}
\qquad \to \qquad y = 8 - x
$$

$$
\begin{array}{rcl}
(8 - x)^2 &=& 16 + x^2 \\
64 - 16x + x^2 &=& 16 + x^2 \\
-16x &=& -48 \\
x &=& 3 \\
3 + y &=& 8 \\
y &=& 5
\end{array}
$$

Area $= \dfrac{1}{2}x(4) = 2x = 2(3) = \boxed{6\ \text{in.}^2}$

25. $(x - h)^2 + y^2 = r^2$

$$
\begin{array}{lrcl}
(0, -2): & (0 - h)^2 + (-2)^2 &=& r^2 \\
& h^2 + 4 &=& r^2 \\
(6, 0): & (6 - h)^2 + 0 &=& r^2 \\
& (6 - h)^2 &=& r^2
\end{array}
$$

$$
\begin{array}{rcl}
h^2 + 4 &=& (6 - h)^2 \\
h^2 + 4 &=& 36 - 12h + h^2 \\
12h &=& 32 \\
h &=& \dfrac{8}{3} \\[2mm]
r^2 &=& h^2 + 4 = \dfrac{64}{9} + 4 = \dfrac{100}{9} \\[2mm]
r &=& \dfrac{10}{3} \qquad \left(\text{reject} - \dfrac{10}{3}\right) \\[2mm]
(x - h)^2 + y^2 &=& r^2
\end{array}
$$

$$\boxed{\left(x - \dfrac{8}{3}\right)^2 + y^2 = \dfrac{100}{9}}$$

Review Problems

28. Let $x =$ number of years after 1990

$$
\begin{array}{rcl}
4000 + 600x &=& 8000 - 400x \\
1000x &=& 4000 \\
x &=& 4
\end{array}
$$

Same population in $1990 + 4 = \boxed{1994}$

29. $(3y^3 - 2y^2 + y + 1) \div (y^2 + 5)$

$$= \boxed{3y - 2 + \dfrac{11 - 4y}{y^2 + 5}}$$

$$
\begin{array}{r}
3y - 2 \\
y^2 + 5 \overline{\smash{)}\, 3y^3 - 2y^2 + y + 1} \\
\underline{3y^3 \qquad + 15y} \\
-2y^2 - 14y + 1 \\
\underline{-2y^2 \qquad -10} \\
-14y + 11
\end{array}
$$

30. $\left(x - 2 - \dfrac{4}{x+1}\right) \div \left(x - 1 - \dfrac{3}{x+1}\right)$

$$= \left[\dfrac{x(x+1)}{x+1} - \dfrac{2(x+1)}{(x+1)} - \dfrac{4}{x+1}\right] \div \left[\dfrac{x(x+1)}{(x+1)} - \dfrac{1(x+1)}{(x+1)} - \dfrac{3}{x+1}\right]$$

$$= \left(\dfrac{x^2 + x - 2x - 2 - 4}{x+1}\right) \div \left(\dfrac{x^2 + x - x - 1 - 3}{x+1}\right)$$

$$= \dfrac{x^2 - x - 6}{x+1} \div \dfrac{x^2 - 4}{x+1}$$

$$= \dfrac{(x-3)(x+2)}{x+1} \div \dfrac{(x+2)(x-2)}{x+1}$$

$$= \dfrac{(x-3)(x+2)}{x+1} \cdot \dfrac{x+1}{(x+2)(x-2)}$$

$$= \boxed{\dfrac{x-3}{x-2}}$$

Chapter 9   Review Problems

Review Problems, pp. 735-737

1.   Center $(-2, 4)$, radius of 6:
$$
\begin{aligned}
(x - h)^2 + (y - k)^2 &= r^2 \\
h = -2, \ k = 4, \ r &= 6
\end{aligned}
$$
$$\boxed{(x + 2)^2 + (y - 4)^2 = 36}$$

2.   Center at the origin, radius of 3:
$$
\begin{aligned}
(x - h)^2 + (y - k)^2 &= r^2 \\
h = 0, \ k = 0, \ r &= 3
\end{aligned}
$$
$$\boxed{x^2 + y^2 = 9}$$

3.
$$
\begin{aligned}
x^2 + y^2 - 4x + 2y - 4 &= 0 \\
x^2 - 4x + y^2 + 2y &= 4 \\
x^2 - 4x + 4 + y^2 + 2y + 1 &= 4 + 4 + 1 \\
(x - 2)^2 + (y + 1)^2 &= 9 = 3^2 \\
h = 2, \ k = -1, \ r &= 3
\end{aligned}
$$
$$\boxed{C(2, -1), \ r = 3}$$

4. $x^2 + y^2 = 16$

Circle: $h = 0$, $k = 0$, $r = \sqrt{16} = 4$
Center at $(0, 0)$ and radius of 4

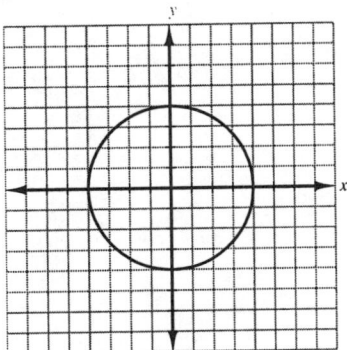

5.
$$\begin{aligned} y &= -\sqrt{16 - x^2} \\ y^2 &= 16 - x^2 \\ x^2 + y^2 &= 16 \quad \text{is a circle with center at the origin and radius of 4.} \end{aligned}$$

$y = -\sqrt{16 - x^2}$ represents the semi-circle below the x-axes with center at the origin and radius of 4

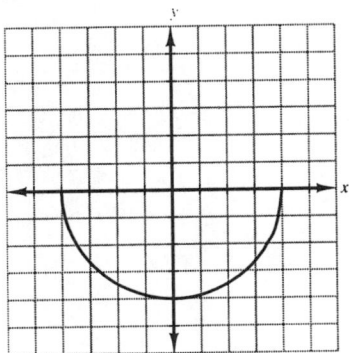

6. $(x + 2)^2 + (y - 3)^2 = 9$

$h = -2$, $k = 3$ and $r = \sqrt{9} = 3$
Circle: Center at $(-2, 3)$ and radius 3

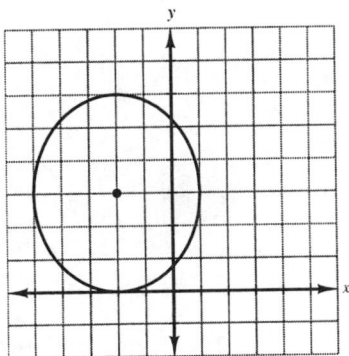

7.
$$\begin{aligned}
x^2 + y^2 + 2x + 6y - 15 &= 0 \\
x^2 + 2x + y^2 + 6y &= 15 \\
x^2 + 2x + 1 + y^2 + 6y + 9 &= 15 + 1 + 9 \\
(x + 1)^2 + (y + 3)^2 &= 25
\end{aligned}$$

$h = -1$, $k = -3$, $r = \sqrt{25} = 5$

Circle: center at $(-1, -3)$ and radius 5

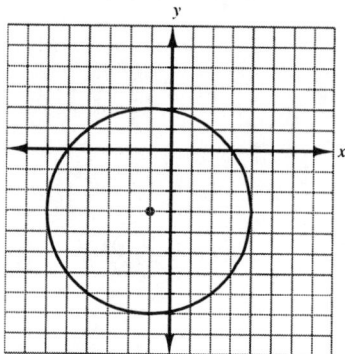

8. $\dfrac{x^2}{9} + \dfrac{y^2}{25} = 1$

$a^2 = 9$, $a = 3$

$b^2 = 25$, $b = 5$

ellipse: center at $(0, 0)$ and

           vertices at $(\pm 3, 0)$ and $(0, \pm 5)$

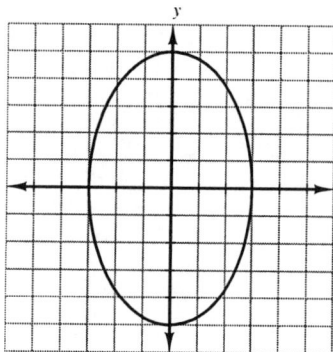

9. $\dfrac{(x-1)^2}{16} + \dfrac{(y+2)^2}{9} = 1$

$a^2 = 16,\ a = 4$

$b^2 = 9,\ b = 3$

ellipse: center at $(1, -2)$ and
vertices at
$(1 \pm 4, -2) = (5, -2)$ and $(-3, -2)$
$(1, -2 \pm 3) = (1, 1)$ and $(1, -5)$

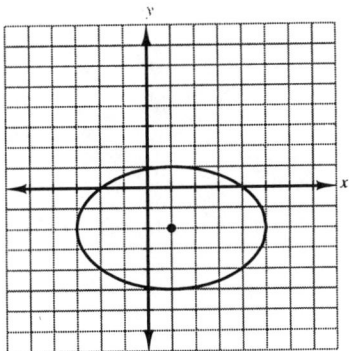

10. $\begin{aligned} 4x^2 + 24x + 9y^2 + 36 &= 0 \\ 4(x^2 + 6x) + 9(y^2 - 4y) &= -36 \\ 4(x^2 + 6x + 9) + 9(y^2 - 4y + 4) &= -36 + 36 + 36 \\ 4(x+3)^2 + 9(y-2)^2 &= 36 \\ \dfrac{(x+3)^2}{9} + \dfrac{(y-2)^2}{4} &= 1 \end{aligned}$

$a^2 = 9,\ a = 3$

$b^2 = 4,\ b = 2$

ellipse: center at $(-3, 2)$ and
vertices at
$(-3 \pm 3, 2) = (0, 2)$ and $(-6, 2)$
$(-3, 2 \pm 2) = (-3, 4)$ and $(-3, 0)$

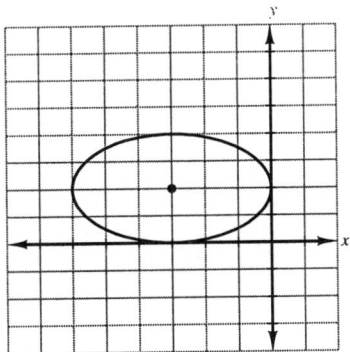

11. $\dfrac{x^2}{16} - \dfrac{y^2}{9} = 1$

$a^2 = 16,\ a = 4$

hyperbola: center at $(0, 0)$ and
vertices at $(\pm 4, 0)$
no y-intercepts

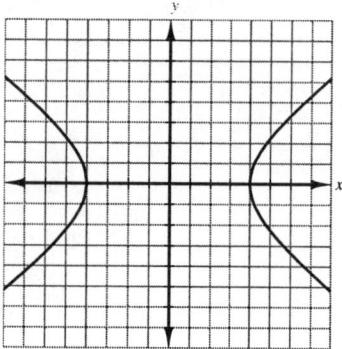

12. $\dfrac{y^2}{9} - \dfrac{x^2}{4} = 1$

$a^2 = 9,\ a = 3$

hyperbola: center $(0, 0)$ and
vertices at $(0, \pm 3)$
no x-intercepts

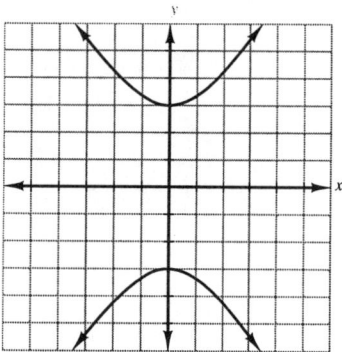

13. $x^2 - y^2 = 4$

$a^2 = 1,\ a = 1$

hyperbola: Center at $(0, 0)$ and
vertices at $(\pm 1, 0)$
no y-intercept

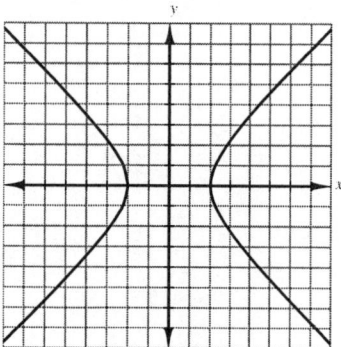

14. $\dfrac{(x-2)^2}{25} - \dfrac{(y+3)^2}{16} = 1$

$h = 2,\ k = -3$

$a^2 = 25,\ a = 5$

hyperbola: center at $(2, -3)$ and
                  vertices at $(2 \pm 5, -3) = (7, -3)$ and $(-3, -3)$
                  no y-intercepts

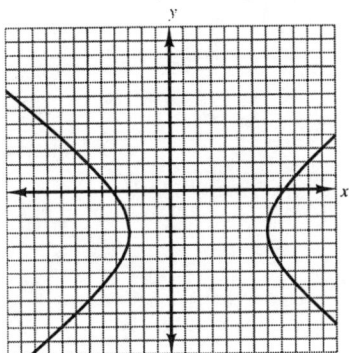

15.
$$f(x) = \sqrt{x^2 + 4}$$
$$y^2 = x^2 + 4$$
$$y^2 - x^2 = 4$$
$$\dfrac{y^2}{4} - \dfrac{x^2}{4} = 1 \quad \text{represents a hyperbola with center at } (0, 0), \text{ and vertices at } (0, \pm 2) \text{ and no } x\text{-}$$

intercepts.

Since $f(x) = +\sqrt{x^2 + 4}$, $f(x)$ represents only the portion of the hyperbola above the x-axis.

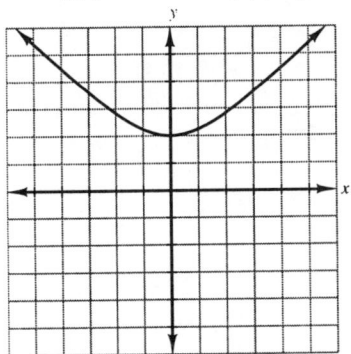

16.
$$4x^2 - 8x - y^2 - 4y - 16 = 0$$
$$4(x^2 - 2x) - (y^2 + 4y) = 16$$
$$4(x^2 - 2x + 1) - (y^2 + 4y + 4) = 16 + 4 - 4$$
$$4(x - 1)^2 - (y + 2)^2 = 16$$
$$\frac{(x - 1)^2}{4} - \frac{(y + 2)^2}{16} = 1$$

$h = 1, \ k = -2$

$a^2 = 4, \ a = 2$

hyperbola: center at $(1, -2)$ and
vertices at $(1 \pm 2, -2) = (3, -2)$ and $(-1, -2)$
no y-intercept

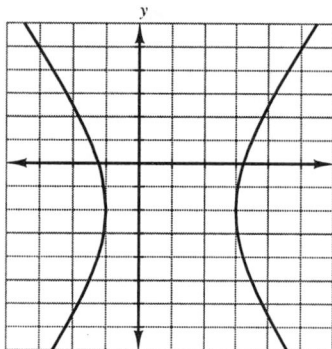

17. $y = -(x + 1)^2 + 4$
$h = -1, \ k = 4, \ a = -1$
parabola: vertex at $(-1, 4)$
$a = -1 < 1$; opens downward

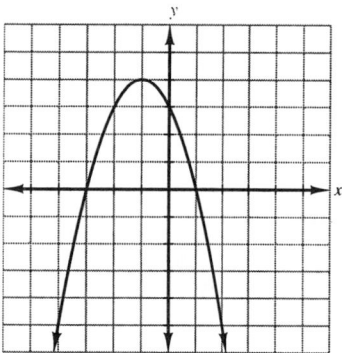

18. $x = y^2 - 8y + 12$
$x = y^2 - 8y + 16 + 12 - 16$
$x = (y - 4)^2 - 4$
$h = -4$, $k = 4$, $a = 1$
parabola: vertex at $(-4, 4)$
            $a = 1 > 0$; opens to the right

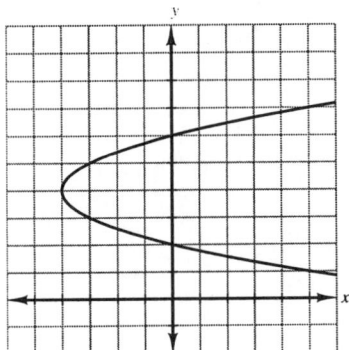

19. $x^2 + y^2 \geq 25$
boundary: $x^2 + y^2 = 25$ (*solid* curve)
            circle with center at $(0, 0)$, radius of 5
Test point: $(0, 0)$
            $0 + 0 \geq 0$ False
Shade the region outside the circle (including the boundary).

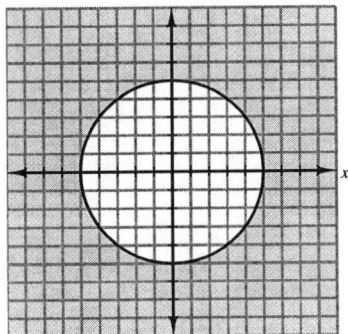

20. $\frac{x^2}{25} + \frac{y^2}{9} < 1$

boundary: $\frac{x^2}{25} + \frac{y^2}{9} = 1$ (*dashed* curve)

ellipse with center at $(0, 0)$ and
vertices at $(\pm 5, 0)$ and $(0, \pm 3)$

Test point: $(0, 0)$
$0 + 0 < 1$ True

Shade the region inside the ellipse (*not* including the boundary).

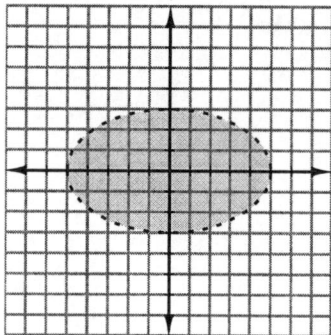

21. $4y^2 \geq 100 - 25x^2$
         $25x^2 + 4y^2 \geq 100$
         $\frac{x^2}{4} + \frac{y^2}{25} \geq 1$

boundary: $\frac{x^2}{4} + \frac{y^2}{25} = 1$ (*solid* curve)

ellipse with center at $(0, 0)$ and
vertices $(\pm 2, 0)$ and $(0, \pm 5)$

Test point: $(0, 0)$
$0 + 0 \geq 1$ False

Shade the region outside the ellipse (including the boundary).

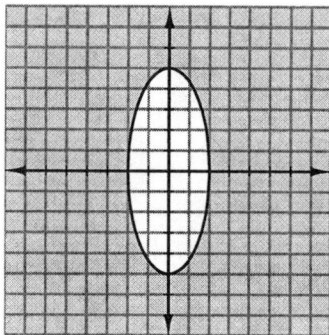

22. $\dfrac{x^2}{25} - \dfrac{y^2}{9} \geq 1$

boundary: $\dfrac{x^2}{25} - \dfrac{y^2}{9} = 1$ (*solid* curve)

hyperbola with center at $(0, 0)$ and
vertices $(\pm 5, 0)$
no y-intercepts

Test point: $(0, 0)$
$0 - 0 \geq 1$ False
(The middle region is *not* shaded.)
Shade the region *not* including $(0, 0)$.
(including the boundary)

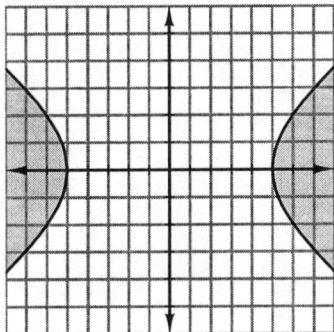

23.

$$4x^2 \;>\; 100 + 25y^2$$
$$4x^2 - 25y^2 \;>\; 100$$
$$\dfrac{x^2}{25} - \dfrac{y^2}{4} \;>\; 1$$

boundary: $\dfrac{x^2}{25} - \dfrac{y^2}{4} = 1$ (*dashed* curve)

hyperbola with center at $(0, 0)$ and
vertices $(\pm 5, 0)$ and
no y-intercepts

Test point: $(0, 0)$
$0 - 0 > 1$ False
(The middle region is *not* shaded.)
Shade the region not including $(0, 0)$.
(also *not* including the boundary)

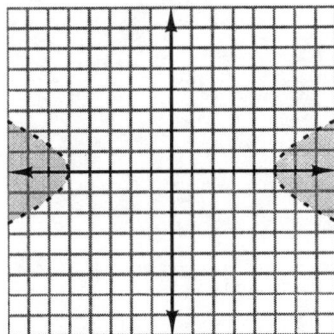

24. $x \le (y+2)^2 - 1$
　　boundary:　$x = (y+2)^2 - 1$　(*solid* curve)
　　　　　　　$h = -1,\ k = -2,\ a = 1$
　　　　　　　parabola with vertex at $(-1, -2)$ and opens to the right
　　Test point:　$(0, 0)$
　　　　　　　$0 \le (0+2)^2 - 1 = 3$　　True
　　Shade region including $(0, 0)$ (i.e. shade region outside and including the boundary)

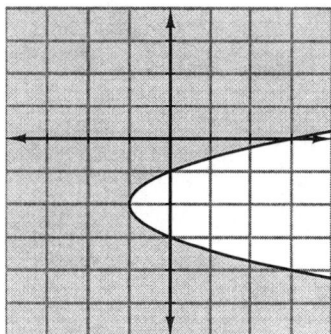

25. $x > y^2 + 4y + 6$
　　$x > y^2 + 4y + 4 + 6 - 4$
　　$x > (y+2)^2 + 2$
　　boundary:　$x = (y+2)^2 + 2$　(*dashed* curve)
　　　　　　　$h = 2,\ k = -2,\ a = 1$
　　　　　　　parabola with vertex at $(2, -2)$ and opens to the right
　　Test point:　$(0, 0)$
　　　　　　　$0 > (0+2)^2 + 2 = 6$　　False
　　Shade region *not* including $(0, 0)$ (i.e. shade region inside and not including boundary).

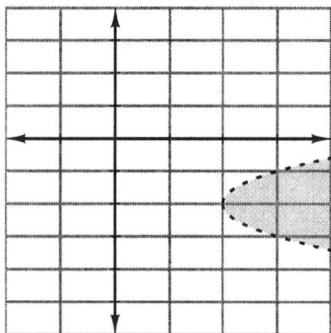

26. $x^2 + y^2 < 9$
　　$2x + 3y > 6$

　　Begin by graphing the solution of $x^2 + y^2 < 9$.
　　　　Boundary:　$x^2 + y^2 = 9$　(*dashed* curve)
　　　　Circle with center at $(0, 0)$ and radius of 3
　　　　Test point, $(0, 0)$:　$0 < 9$　　True
　　　　　　Shaded region:　interior region of the circle not including the boundary

　　On the same axes graph the solution of
　　　　$2x + 3y > 6$
　　　　Test point, $(0, 0)$:　$0 > 6$　　False
　　　　　　Shade region:　half-plane not including $(0, 0)$ (above and not including the line)

The graph of the solution set is the intersection of the graph of the two inequalities (the region above $2x + 3y = 6$ and inside the circle but not including the boundaries).

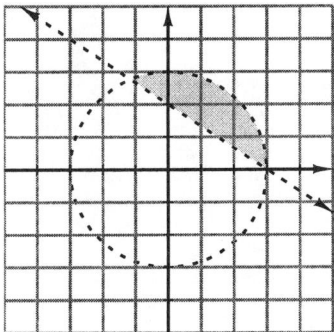

27. $4x^2 + 9y^2 \le 36$ $\rightarrow$ $\dfrac{x^2}{9} + \dfrac{y^2}{4} \le 1$

$-2 \le x \le 2$

Begin by graphing the solution of

$\dfrac{x^2}{9} + \dfrac{y^2}{4} \le 1$

Boundary: $\dfrac{x^2}{9} + \dfrac{y^2}{4} = 1$ (*solid* curve)

ellipse with center at $(0, 0)$ and
vertices $(\pm 3, 0)$ and $(0, \pm 2)$
Test point, $(0, 0)$: $0 \le 1$ True
   Shaded region: interior region of the ellipse including the boundary

On the same axes graph the solution of $-2 \le x \le 2$ which is the region between and including $x = 2$ and $x = -2$.

The graph of the solution set is the intersection of the graph of the two inequalities (the region is the interior of the ellipse which is between the vertical lines $x = 2$ and $x = -2$).

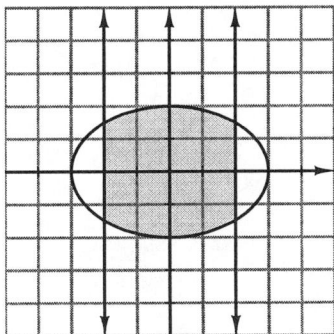

28. $\dfrac{x^2}{9} + y^2 < 1$

$x^2 - \dfrac{y^2}{4} \geq 1$

Begin by graphing the solution of $\dfrac{x^2}{9} + y^2 < 1$.

Boundary: $\dfrac{x^2}{9} + y^2 = 1$ (*dashed* curve)

ellipse with center at $(0, 0)$ and
vertices $(\pm 3, 0)$ and $(0, \pm 1)$
Test point, $(0, 0)$: $0 + 0 < 1$ True
Shaded region: interior region of the ellipse not including the bouandary.

On the same axes graph the solution of $x^2 - \dfrac{y^2}{4} \geq 1$

Boundary: $x^2 - \dfrac{y^2}{4} = 1$ (*solid* curve)

hyperbola with center $(0, 0)$,
vertices $(\pm 1, 0)$ and no y-intercepts
Test point, $(0, 0)$: $0 - 0 \geq 1$ False
Shaded region: outside region of the hyperbola, does *not* include $(0, 0)$ (i.e. is *not* the middle region)

The graph of the solution set (the intersection of the graph of the two inequalities) is the region in the interior of the ellipse but not in the middle region of the hyperbola.

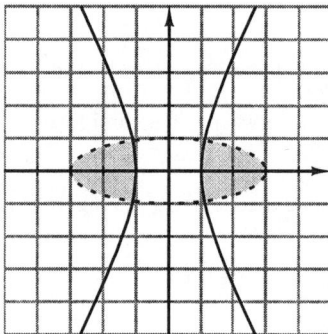

29. $x^2 + y^2 \leq 9$

$25x^2 + 4y^2 \geq 100$ $\rightarrow$ $\dfrac{x^2}{4} + \dfrac{y^2}{25} \geq 1$

Begin by graphing the solutions of $x^2 + y^2 \leq 9$.
Boundary: $x^2 + y^2 = 9$ (*solid* curve)
circle with center at $(0, 0)$ and radius of 3
Test point, $(0, 0)$: $0 \leq 9$ True
Shaded region: interior region of the circle including the boundary.

On the same axes graph the solutions of $\dfrac{x^2}{4} + \dfrac{y^2}{25} \geq 1$.

Boundary: $\dfrac{x^2}{4} + \dfrac{y^2}{25} = 1$

ellipse with center $(0, 0)$ and
vertices $(\pm 2, 0)$ and $(0, \pm 5)$
Test point, $(0, 0)$: $0 \geq 1$ False
   Shaded region: region outside of the ellipse (not including $(0, 0)$) including the boundary

The graph of the solution set (the intersection of the graphs of the two inequalities) is the region outside of the ellipse but inside the circle including both boundaries.

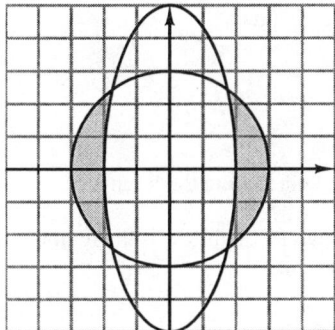

30. $x^2 + y^2 \leq 4$
   $x^2 - y^2 < 1$

Begin by graphing the solution of $x^2 + y^2 \leq 4$
   Boundary: $x^2 + y^2 = 4$ (*solid* curve)
   circle with center at $(0, 0)$ and radius 2.
   Test point, $(0, 0)$: $0 \leq 4$ True
      Shaded region: interior region of the circle including the boundary

On the same axes graph the solution of $x^2 - y^2 < 1$.
   Boundary: $x^2 - y^2 = 1$ (*dashed* curve)
   hyperbola with center $(0, 0)$
   vertices $(\pm 1, 0)$ and no y-intercepts.
   Test point, $(0, 0)$: $0 < 1$ True
      Shaded region: middle region of the hyperbola including $(0, 0)$ but not including the boundaries.

The graph of the solution set (the intersection of the graphs of the two inequalities) is the region inside the circle (including the boundary) and also in the middle region of the hyperbola (not including the boundary).

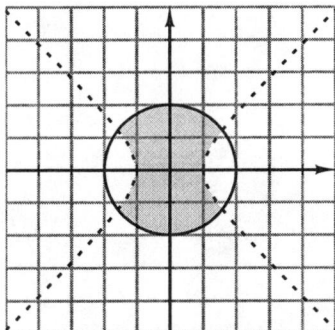

31.

$$9x^2 + y^2 \le 9 \qquad \rightarrow \qquad x^2 + \frac{y^2}{9} \le 1$$

$$x \le y^2 - 4y \qquad \rightarrow \qquad x \le y^2 - 4y + 4 - 4$$

$$x \le (y-2)^2 - 4$$

Begin by graphing the solution of $x^2 + \frac{y^2}{9} \le 1$.

 Boundary: $x^2 + \frac{y^2}{9} = 1$ (*solid* curve)

ellipse with center $(0, 0)$ and
vertices $(\pm 1, 0)$ and $(0, \pm 3)$
Test point, $(0, 0)$: $0 \le 1$ True
  Shaded region: interior region of the ellipse including the boundary

On the same axes graph the solution of $x \le (y-2)^2 - 4$
 Boundary: $x = (y-2)^2 - 4$ (*solid* cuve)
 parabola with vertex $(-4, 2)$ and opening to the right ($a = 1$)
 Test point, $(0, 1)$: $0 \le 1 - 4 = -3$ False
  Shaded region: region outside parabola (not including 0, 1) including the boundary

The graph of the solution set (the intersection of the graphs of the two inequalities) is the region outside of the parabola but inside the ellipse including both boundaries.

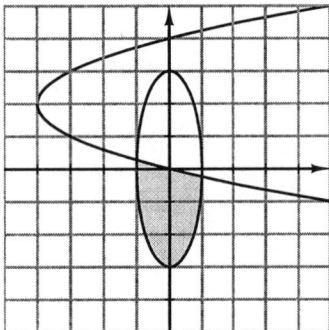

32. $x^2 + (y-1)^2 = 1$
$(x-1)^2 + y^2 = 1$

Begin by graphing, $x^2 + (y-1)^2 = 1$.
    Circle with center at $(0, 1)$ and radius 1

On the same axes graph $(x-1)^2 + y^2 = 1$
    Circle with center at $(1, 0)$ and radius 1.

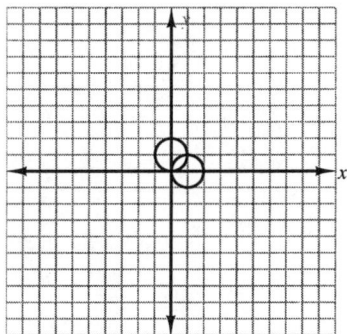

From the graph, the points of intersection are
    $x = 0$, $y = 0$ and $x = 1$, $y = 1$
    $\boxed{\{(0, 0), (1, 1)\}}$
Both solutions check.

33. $x^2 + y^2 = 17$
$x + y = 5$

Begin by graphing $x^2 + y^2 = 17$
    Circle with center at $(0, 0)$ and radius $\sqrt{17}$

On the same axes graph line $y = 5 - x$.

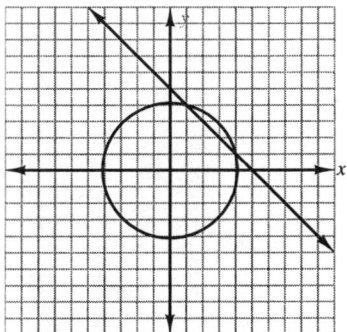

From the graph, the points of intersection are
    $x = 4$, $y = 1$ and $x = 1$, $y = 4$
    $\boxed{\{(4, 1), (1, 4)\}}$
Both solutions check.

34.
$$x^2 + y^2 = 9$$
$$x^2 + 9y^2 = 9 \qquad \rightarrow \qquad \frac{x^2}{9} + y^2 = 1$$

Begin by graphing $x^2 + y^2 = 9$
  Circle with center $(0, 0)$ and radius 3

On the same axes graph $\frac{x^2}{9} + y^2 = 1$
  Ellipse with center $(0, 0)$ and vertices $(\pm 3, 0)$ and $(0, \pm 1)$

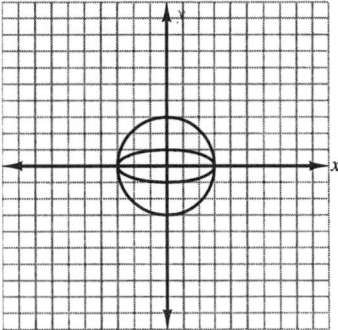

From the graph the points of intersection are
  $x = -3$, $y = 0$ and $x = 3$, $y = 0$
  $\boxed{\{(-3, 0), (3, 0)\}}$
Both solutions check.

35.
$$4x^2 + y^2 = 4 \qquad \rightarrow \qquad x^2 + \frac{y^2}{4} = 1$$
$$x + y = 3 \qquad \rightarrow \qquad y = -x + 3$$

Begin by graphing $x^2 + \frac{y^2}{4} = 1$
  ellipse with center $(0, 0)$ and
  vertices $(\pm 1, 0)$ and $(0, \pm 2)$

On the same axes graph $y = -x + 3$

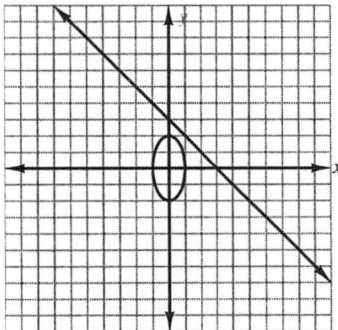

From the graph we find there are *no* points of intersection.
  $\boxed{\varnothing}$

36. $y = x^2$
$x = y^2$

Begin by graphing $y = x^2$ which is a parabola with vertex at $(0, 0)$ opening upward.

On the same axes graph $x = y^2$ which is also a parabola with vertex at $(0, 0)$ but opening to the right.

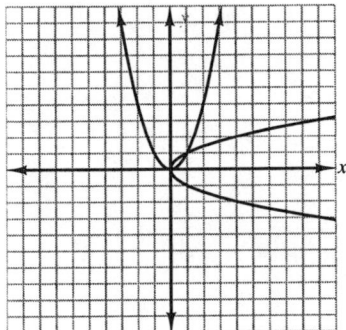

From the graph the points of intersection are
$x = 0, y = 0$ and $x = 1, y = 1$
$\boxed{\{(0, 0), (1, 1)\}}$
Both solutions check.

37.
$$
\begin{aligned}
y &= x^2 + 2x + 1 & \to & & y &= (x+1)^2 \\
x + y - 1 &= 0 & \to & & y &= -x + 1
\end{aligned}
$$

Begin by graphing $y = (x + 1)^2$ which is a parabola with vertex at $(-1, 0)$ opening upward.

On the same axes graph the line $y = -x + 1$.

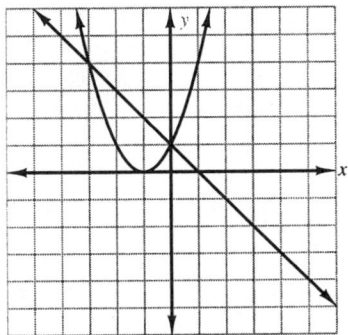

From the graph the points of intersection are
$x = 0, y = 1$ and $x = -3, y = 4$
$\boxed{\{(0, 1), (-3, 4)\}}$

38.

$$5y = x^2 - 1$$
$$x - y = 1 \qquad \rightarrow$$

$$y = x - 1$$
$$5(x - 1) = x^2 - 1$$
$$5x - 5 = x^2 - 1$$
$$0 = x^2 - 5x + 4$$
$$0 = (x - 4)(x - 1)$$
$$x = 4 \text{ or } x = 1$$
$$y = 4 - 1 = 3 \quad y = 1 - 1 = 0$$

$$\boxed{\{(1, 0), (4, 3)\}}$$

39.

$$y = x^2 - 2x - 1$$
$$y - x = 3 \qquad \rightarrow$$

$$y = x + 3$$
$$x + 3 = x^2 - 2x - 1$$
$$0 = x^2 - 3x - 4$$
$$0 = (x - 4)(x + 1)$$
$$x = 4 \text{ or } x = -1$$
$$y = 4 + 3 = 7 \quad y = -1 + 3 = 2$$

$$\boxed{\{(-1, 2), (4, 7)\}}$$

40.

$$x^2 + y^2 = 2$$
$$x + y = 0 \qquad \rightarrow$$

$$y = -x$$
$$x^2 + (-x)^2 = 2$$
$$2x^2 = 2$$
$$x^2 = 1$$
$$x = \pm 1$$
$$x = -1 \qquad \text{or} \quad x = 1$$
$$y = -(-1) = 1 \qquad y = -1$$

$$\boxed{\{(-1, 1), (1, -1)\}}$$

41.

$$x^2 + y^2 = 4$$
$$2x - y = 0 \qquad \rightarrow$$

$$y = 2x$$
$$x^2 + (2x)^2 = 4$$
$$5x^2 = 4$$
$$x^2 = \frac{4}{5}$$
$$x = \pm \frac{2\sqrt{5}}{5}$$
$$x = -\frac{2\sqrt{5}}{5} \qquad \text{or} \quad x = \frac{2\sqrt{5}}{5}$$
$$y = -\frac{4\sqrt{5}}{5} \qquad y = \frac{4\sqrt{5}}{5}$$

$$\boxed{\left\{ \left(-\frac{2\sqrt{5}}{5}, -\frac{4\sqrt{5}}{5}\right), \left(\frac{2\sqrt{5}}{5}, \frac{4\sqrt{5}}{5}\right) \right\}}$$

42.

$$x^2 + 2y^2 = 4$$
$$x - y - 1 = 0$$

$$\rightarrow$$

$$y = x - 1$$
$$x^2 + 2(x-1)^2 = 4$$
$$x^2 + 2x^2 - 4x + 2 = 4$$
$$3x^2 - 4x - 2 = 0$$

$$x = \frac{4 \pm \sqrt{16 + 24}}{6}$$

$$x = \frac{2 \pm \sqrt{10}}{3}$$

$$x = \frac{2 - \sqrt{10}}{3} \qquad \text{or} \qquad x = \frac{2 + \sqrt{10}}{3}$$

$$y = \frac{2 - \sqrt{10}}{3} - 1 = \frac{-1 - \sqrt{10}}{3} \qquad\qquad y = \frac{2 + \sqrt{10}}{3} - 1 = \frac{-1 + \sqrt{10}}{3}$$

$$\boxed{\left\{ \left(\frac{2 - \sqrt{10}}{3}, \frac{-1 - \sqrt{10}}{3} \right), \left(\frac{2 + \sqrt{10}}{3}, \frac{-1 + \sqrt{10}}{3} \right) \right\}}$$

43.

$$2x^2 + y^2 = 24 \qquad \rightarrow \qquad 2x^2 + y^2 = 24$$
$$x^2 + y^2 = 15 \qquad \rightarrow \qquad -x^2 - y^2 = -15$$

$$x^2 = 9$$
$$x = \pm 3$$
$$(\pm 3)^2 + y^2 = 15$$
$$9 + y^2 = 15$$
$$y^2 = 6$$
$$y = \pm\sqrt{6}$$

$$\text{if } x = -3, \qquad y = \pm\sqrt{6}$$
$$\text{if } x = 3, \qquad y = \pm\sqrt{6}$$

$$\boxed{\{(-3, -\sqrt{6}), (-3, \sqrt{6}), (3, -\sqrt{6}), (3, \sqrt{6})\}}$$

44.

$$xy - 4 = 0$$
$$y - x = 0$$

$$\rightarrow$$

$$y = x$$
$$x(x) - 4 = 0$$
$$x^2 = 4$$
$$x = \pm 2$$

$$x = -2 \qquad \text{or} \quad x = 2$$
$$y = -2 \qquad\qquad y = 2$$

$$\boxed{\{(-2, -2), (2, 2)\}}$$

45.

$$x^2 + y^2 = 2$$
$$y = x^2$$

$$\rightarrow$$

$$x^2 = y$$
$$y + y^2 = 2$$
$$y^2 + y - 2 = 0$$
$$(y + 2)(y - 1) = 0$$

$$y = -2 \qquad \text{or} \qquad y = 1$$
$$x^2 = -2 \qquad\qquad x^2 = 1$$
$$x = \pm\sqrt{2}i \qquad\qquad x = \pm 1$$

$$\boxed{\{(1, 1), (-1, 1), (-\sqrt{2}i, -2), (\sqrt{2}i, -2)\}}$$

46.

$$y^2 = 4x$$
$$x - 2y + 3 = 0 \quad\longrightarrow$$

$$x = 2y - 3$$
$$y^2 = 4(2y - 3)$$
$$y^2 = 8y - 12$$
$$y^2 - 8y + 12 = 0$$
$$(y - 6)(y - 2) = 0$$

$$y = 6 \qquad\qquad\text{or}\qquad\qquad y = 2$$
$$x = 2(6) - 3 = 9 \qquad\qquad x = 2(2) - 3 = 1$$

$$\boxed{\{(1, 2), (9, 6)\}}$$

47.

$$\frac{x^2}{4} + \frac{y^2}{9} = 1$$
$$2x - y = 0 \quad\longrightarrow$$

$$y = 2x$$
$$\frac{x^2}{4} + \frac{4x^2}{9} = 1$$
$$\frac{25x^2}{36} = 1$$
$$x^2 = \frac{36}{25}$$
$$x = \pm\frac{6}{5}$$

$$x = -\frac{6}{5} \qquad\qquad\text{or}\qquad\qquad x = \frac{6}{5}$$
$$y = 2\left(-\frac{6}{5}\right) = -\frac{12}{5} \qquad\qquad y = 2\left(\frac{6}{5}\right) = \frac{12}{5}$$

$$\boxed{\left\{\left(-\frac{6}{5}, -\frac{12}{5}\right), \left(\frac{6}{5}, \frac{12}{5}\right)\right\}}$$

48.

$$4x^2 + y^2 = 1 \quad\longrightarrow$$
$$x^2 + 4y^2 = 1$$

$$-16x^2 - 4y^2 = -4$$
$$x^2 + 4y^2 = 1$$
$$-15x^2 = -3$$
$$x^2 = \frac{1}{5}$$
$$x = \pm\frac{\sqrt{5}}{5}$$
$$4\left(\frac{1}{5}\right) + y^2 = 1$$
$$y^2 = \frac{1}{5}$$
$$y = \pm\frac{\sqrt{5}}{5}$$

$$\text{if } x = -\frac{\sqrt{5}}{5}, \qquad y = \pm\frac{\sqrt{5}}{5}$$

$$\boxed{\left\{\left(-\frac{\sqrt{5}}{5}, -\frac{\sqrt{5}}{5}\right), \left(-\frac{\sqrt{5}}{5}, \frac{\sqrt{5}}{5}\right), \left(\frac{\sqrt{5}}{5}, -\frac{\sqrt{5}}{5}\right), \left(\frac{\sqrt{5}}{5}, \frac{\sqrt{5}}{5}\right)\right\}}$$

49.

$$\begin{aligned} x^2 + y^2 &= 9 \\ (x-2)^2 + y^2 &= 21 \end{aligned} \quad \rightarrow \quad \begin{aligned} -x^2 - y^2 &= -9 \\ (x-2)^2 + y^2 &= 21 \\ (x-2)^2 - x^2 &= 12 \\ x^2 - 4x + 4 - x^2 &= 12 \\ -4x &= 8 \\ x &= -2 \\ (-2)^2 + y^2 &= 9 \\ y^2 &= 5 \\ y &= \pm\sqrt{5} \end{aligned}$$

$$\boxed{\{(-2, -\sqrt{5}), (-2, \sqrt{5})\}}$$

50.

$$\begin{aligned} x^2 + 2xy - y^2 &= 14 \\ x^2 - y^2 &= -16 \end{aligned} \quad \rightarrow \quad \begin{aligned} x^2 + 2xy - y^2 &= 14 \\ -x^2 \quad\quad\; + y^2 &= 16 \\ 2xy &= 30 \\ y &= \frac{15}{x} \end{aligned}$$

$$x^2 - \left(\frac{15}{x}\right)^2 = -16$$

$$x^2 - \frac{225}{x^2} = -16$$

$$x^4 + 16x^2 - 225 = 0$$

$$(x^2 + 25)(x^2 - 9) = 0$$

$x^2 + 25 = 0$	$x^2 - 9 = 0$
$x = \pm 5i$	$x = \pm 3$
$x = -5i, \quad y = \dfrac{15}{-5i} = 3i$	$x = -3, \quad y = \dfrac{15}{-3} = -5$
$x = 5i, \quad y = \dfrac{15}{5i} = -3i$	$x = 3, \quad y = \dfrac{15}{3} = 5$

$$\boxed{\{(-5i, 3i), (5i, -3i), (-3, -5), (3, 5)\}}$$

51.

$$\begin{aligned} x^2 + xy - y^2 &= 5 \\ x^2 - 3xy - y^2 &= -3 \end{aligned} \quad \rightarrow \quad \begin{aligned} -x^2 - xy + y^2 &= -5 \\ x^2 - 3xy - y^2 &= -3 \\ -4xy &= -8 \\ y &= \frac{2}{x} \end{aligned}$$

$$x^2 + x\left(\frac{2}{x}\right) - \left(\frac{2}{x}\right)^2 = 5$$

$$x^2 + 2 - \frac{4}{x^2} = 5$$

$$x^2 - \frac{4}{x^2} = 3$$

$$x^4 - 3x^2 - 4 = 0$$

$$(x^2 - 4)(x^2 + 1) = 0$$

$x^2 - 4 = 0$	$x^2 + 1 = 0$
$x = \pm 2$	$x = \pm i$
$x = -2, \quad y = \dfrac{2}{-2} = -1$	$x = -i, \quad y = \dfrac{2}{-i} = 2i$
$x = 2, \quad y = \dfrac{2}{2} = 1$	$x = i, \quad y = \dfrac{2}{i} = -2i$

$$\boxed{\{(-2, -1), (2, 1), (i, -2i), (-i, 2i)\}}$$

52. Let x and y equal the numbers.

$$x^2 + y^2 = 74$$
$$2x = 9 + y \qquad \rightarrow \qquad y = 2x - 9$$
$$x^2 + (2x - 9)^2 = 74$$
$$x^2 + 4x^2 - 36x + 81 = 74$$
$$5x^2 - 36x + 7 = 0$$
$$(5x - 1)(x - 7) = 0$$

$$x = \frac{1}{5} \qquad\qquad \text{or} \qquad x = 7$$

$$y = 2\left(\frac{1}{5}\right) - 9 \qquad\qquad y = 2(7) - 9$$

$$y = -\frac{43}{5} \qquad\qquad \text{or} \qquad y = 5$$

The numbers are $\boxed{\dfrac{1}{5} \text{ and } -\dfrac{43}{5} \text{ or } 7 \text{ and } 5}$

53. Let x and y equal the dimensions of a rectangle

$$2x + 2y = 26 \qquad \rightarrow \qquad y = 13 - x$$
$$xy = 40 \qquad \rightarrow \qquad x(13 - x) = 40$$
$$13x - x^2 = 40$$
$$0 = x^2 - 13x + 40$$
$$0 = (x - 8)(x - 5)$$

$$x = 8 \qquad\qquad \text{or} \qquad x = 5$$
$$y = 13 - 8 = 5 \qquad\qquad y = 13 - 5 = 8$$

The dimensions are $\boxed{5 \text{ meters} \times 8 \text{ meters}}$

54. Let x and y equal the length of the legs of the right triangle.

$$\text{hypotenuse} = 10 \text{ meters}$$
$$\text{perimeter} = 24 \text{ meters}$$
$$x^2 + y^2 = 10^2$$
$$x + y + 10 = 24 \qquad \text{(solve for } y\text{)} \rightarrow \qquad y = 14 - x$$
$$\text{(substitute for } y\text{)} \rightarrow \qquad x^2 + (14 - x)^2 = 100$$
$$x^2 + 196 - 28x + x^2 = 100$$
$$2x^2 - 28x + 96 = 0$$
$$x^2 - 14x + 48 = 0$$
$$(x - 8)(x - 6) = 0$$

$$x = 8 \qquad\qquad \text{or} \qquad x = 6$$
$$y = 14 - 8 = 6 \qquad\qquad y = 14 - 6 = 8$$

The lengths of the legs of the right triangle are $\boxed{6 \text{ meters and } 8 \text{ meters}}$.

55. Let x and y equal the lengths of the sides of two squares.

The perimeters are $4x$ and $4y$ (sum of perimeter = 32 cm)
The areas are x^2 and y^2 (sum of areas = 34 cm²)

$$x^2 + y^2 = 34$$
$$4x + 4y = 32$$

(solve for y) $\rightarrow$ $y = 8 - x$
(substitute for y) $\rightarrow$
$$x^2 + (8-x)^2 = 34$$
$$x^2 + 64 - 16x + x^2 = 34$$
$$2x^2 - 16x + 30 = 0$$
$$x^2 - 8x + 15 = 0$$
$$(x-5)(x-3) = 0$$

$x = 5$ or $x = 3$
$y = 8 - 5 = 3$ $y = 8 - 3 = 5$

The lengths of the sides of the square are $\boxed{3 \text{ cm and } 5 \text{ cm}}$.

56. $a^2 + b^2 = 13^2$ $(\times -1) \rightarrow$
$a^2 + (b+11)^2 = 20^2$

$$-a^2 - b^2 = -169$$
$$\underline{a^2 + b^2 + 22b + 121 = 400}$$
$$22b + 121 = 231$$
$$22b = 110$$
$$b = 5$$

$$a^2 + 25 = 169$$
$$a^2 = 144$$
$$a = \pm 12 \quad \text{(reject } -12\text{)}$$
$$a = 12$$

$\boxed{a = 12,\, b = 5}$

57. Let

$r = $ average speed
$t = $ time
$rt = 2000$
$r = t - 10$

(substitute for r) $\rightarrow$
$$(t-10)t = 2000$$
$$t^2 - 10t - 2000 = 0$$
$$(t-50)(t+40) = 0$$

$t = 50$ or $t = -40$ (reject)
$r = 50 - 10 = 10$

$\boxed{\text{rate: } 40 \text{ kph; time: } 50 \text{ hr}}$

58. $xy = 6$
$2x + y = 8$ $\rightarrow$

$$y = 8 - 2x$$
$$x(8 - 2x) = 6$$
$$8x - 2x^2 = 6$$
$$-2x^2 + 8x - 6 = 0$$
$$x^2 - 4x + 3 = 0$$
$$(x-3)(x-1) = 0$$

$x = 3$ or $x = 1$
$y = 8 - 2(3) = 2$ $y = 8 - 2(1) = 6$

$\boxed{(1,\,6) \text{ and } (3,\,2)}$

59.

$$2a = 40 \text{ feet} \quad \rightarrow \quad a = 20$$
$$b = 15 \text{ feet} \quad \rightarrow \quad b = 15$$
$$\frac{x^2}{a^2} + \frac{y^2}{b^2} = 1$$
$$\frac{x^2}{400} + \frac{y^2}{225} = 1$$
$$\frac{y^2}{225} = 1 - \frac{x^2}{400}$$
$$y^2 = 225\left(1 - \frac{x^2}{400}\right)$$

$$\boxed{y = 15\sqrt{1 - \frac{x^2}{400}}} \quad \left(\text{reject} - 15\sqrt{1 - \frac{x^2}{400}} \text{ since it represents the part of the ellipse below the } x\text{-axis}\right)$$

10 feet from the center: $x = 10$

$$y = 15\sqrt{1 - \frac{100}{400}} = 15\sqrt{1 - \frac{1}{4}} = 15\sqrt{\frac{3}{4}} = \boxed{\frac{15}{2}\sqrt{3}} \text{ feet} \approx 12.99 \text{ feet}$$

Cumulative Review Problems (Chapters 1-9)

Cumulative Review, pp. 737-738

1.

$$5y^4 + 3y^3 - 40y - 24 = 0$$
$$y^3(5y + 3) - 8(5y + 3) = 0$$
$$(5y + 3)(y^3 - 8) = 0$$
$$(5y + 3)(y - 2)(y^2 + 2y + 4) = 0$$

$$5y + 3 = 0 \quad \text{or} \quad y - 2 = 0 \quad \text{or} \quad y^2 + 2y + 4 = 0$$

$$y = -\frac{3}{5} \qquad\qquad y = 2 \qquad\qquad y = \frac{-2 \pm \sqrt{4 - 16}}{2}$$

$$y = -1 \pm i\sqrt{3}$$

$$\boxed{\left\{-\frac{3}{5},\, 2,\, -1 - i\sqrt{3},\, -1 + i\sqrt{3}\right\}}$$

2.

$$5x^{2/3} + 2x^{1/3} - 7 = 0$$
$$(5x^{1/3} + 7)(x^{1/3} - 1) = 0$$

$$x^{1/3} = -\frac{7}{5} \qquad\qquad \text{or} \qquad x^{1/3} = 1$$

$$x = \left(-\frac{7}{5}\right)^3 = \frac{-343}{125} \qquad\qquad x = 1^3 = 1$$

$$\boxed{\left\{-\frac{343}{125},\, 1\right\}}$$

3.

$$(5x - 4)^2 + 6 = 8$$
$$(5x - 4)^2 = 2$$
$$5x - 4 = \pm\sqrt{2}$$
$$5x = 4 \pm \sqrt{2}$$
$$x = \frac{4 \pm \sqrt{2}}{5}$$

$$\boxed{\left\{\frac{4 - \sqrt{2}}{5},\, \frac{4 + \sqrt{2}}{5}\right\}}$$

4. $x^2 - 6x + 13 = 0$

$$x = \frac{6 \pm \sqrt{36 - 52}}{2}$$

$$x = \frac{6 \pm \sqrt{-16}}{2}$$

$$x = \frac{6 \pm 4i}{2} = 3 \pm 2i$$

$\boxed{\{3 - 2i, \, 3 + 2i\}}$

5. $2x^2 - 5x - 3 \;\; > \;\; 0$

$(2x + 1)(x - 3) \;\; > \;\; 0$

$$\begin{array}{c|c|c} \text{T} & \text{F} & \text{T} \\ \hline & -1/2 \qquad 3 & \end{array}$$

Test -1: $(-2 + 1)(-1 - 3) \;\; > \;\; 0$

$(-1)(-4) \;\; > \;\; 0$

$4 \;\; > \;\; 0$ True

Test 0: $0 - 3 \;\; > \;\; 0$

$-3 \;\; > \;\; 0$ False

Test 4: $(8 + 1)(4 - 3) \;\; > \;\; 0$

$9(1) \;\; > \;\; 0$

$9 \;\; > \;\; 0$ True

$x < -\dfrac{1}{2}$ or $x > 3$

$$\boxed{\left\{ x \mid x < -\frac{1}{2} \text{ or } x > 3 \right\} = \left(-\infty, -\frac{1}{2}\right) \cup (3, \, \infty)}$$

6. $x^2 + y^2 \;\; = \;\; 25$

$x - 2y \;\; = \;\; -5 \qquad \rightarrow$

$$\begin{aligned} x &= 2y - 5 \\ (2y - 5)^2 + y^2 &= 25 \\ 4y^2 - 20y + 25 + y^2 &= 25 \\ 5y^2 - 20y &= 0 \\ 5y(y - 4) &= 0 \end{aligned}$$

$$\begin{array}{ccccc} y &=& 0 & \qquad \text{or} \qquad & y = 4 \\ x &=& 0 - 5 = -5 & & x = 8 - 5 = 3 \end{array}$$

$\boxed{\{(-5, 0), (3, 4)\}}$

7. $x^2 + y^2 \;\; = \;\; 13 \qquad\qquad \rightarrow \qquad -x^2 - y^2 \;\; = \;\; -13$

$x^2 + 4xy + y^2 \;\; = \;\; 37 \qquad\qquad \rightarrow \qquad x^2 + 4xy + y^2 \;\; = \;\; 37$

$$4xy \;\; = \;\; 24$$

$$y \;\; = \;\; \frac{6}{x}$$

$$x^2 + \left(\frac{6}{x}\right)^2 \;\; = \;\; 13$$

$$x^2 + \frac{36}{x^2} \;\; = \;\; 13$$

$$x^4 - 13x^2 + 36 \;\; = \;\; 0$$

$$(x^2 - 4)(x^2 - 9) \;\; = \;\; 0$$

$$\begin{array}{lll} x^2 - 4 &=& 0 \\ x &=& \pm 2 \end{array} \quad \text{or} \quad \begin{array}{lll} x^2 - 9 &=& 0 \\ x &=& \pm 3 \end{array}$$

$$\text{if } x = -2, \quad y = \frac{6}{-2} = -3 \qquad\qquad \text{if } x = -3, \quad y = \frac{6}{-3} = -2$$

$$\text{if } x = 2, \quad y = \frac{6}{2} = 3 \qquad\qquad\quad \text{if } x = 3, \quad y = \frac{6}{3} = 2$$

$$\boxed{\{(-2, -3), (2, 3), (-3, -2), (3, 2)\}}$$

8.
$$\begin{array}{rcl} x - 2y + 3z &=& 7 \\ 2x + y + z &=& 19 \\ -3x + 2y + 3z &=& 19 \end{array}$$

(Equations 1 and 2):
$$\begin{array}{rcl} x - 2y + 3z &=& 7 \\ 4x + 2y + 2z &=& 8 \\ \hline 5x \quad\quad + 5z &=& 15 \\ x + z &=& 3 \quad \text{(Eq. 4)} \end{array}$$

(Equations 1 and 3):
$$\begin{array}{rcl} x - 2y + 3z &=& 7 \\ -3x + 2y + 3z &=& 19 \\ \hline -2x \quad\quad + 6z &=& 26 \\ -x + 3z &=& 13 \quad \text{(Eq. 5)} \end{array}$$

(Equations 4 and 5):
$$\begin{array}{rcl} x + z &=& 3 \\ -x + 3z &=& 13 \\ \hline 4z &=& 16 \\ z &=& 4 \end{array}$$

(Equation 4):
$$\begin{array}{rcl} x + 4 &=& 3 \\ x &=& -1 \end{array}$$

(Equation 2):
$$\begin{array}{rcl} 2(-1) + y + 4 &=& 4 \\ y &=& 2 \end{array}$$

$$\boxed{\{(-1, 2, 4)\}}$$

9.
$$\begin{array}{rcl} \sqrt{3x + 4} + \sqrt{x + 5} - \sqrt{7 - 2x} &=& 0 \\ \sqrt{3x + 4} + \sqrt{x + 5} &=& \sqrt{7 - 2x} \\ 3x + 4 + 2\sqrt{3x + 4}\sqrt{x + 5} + x + 5 &=& 7 - 2x \\ 2\sqrt{3x + 4}\sqrt{x + 5} &=& -2 - 6x \\ \sqrt{3x + 4}\sqrt{x + 5} &=& -1 - 3x \\ (3x + 4)(x + 5) &=& (-1 - 3x)^2 \\ 3x^2 + 19x + 20 &=& 1 + 6x + 9x^2 \\ 0 &=& 6x^2 - 13x - 19 \\ 0 &=& (6x - 19)(x + 1) \end{array}$$

$$x = \frac{19}{6} \quad \text{or} \quad x = -1 \quad \text{(check)}$$

$$\frac{19}{6} \text{ is an extraneous solution.}$$

$$\boxed{\{-1\}}$$

10.

$$\frac{x-4}{x^2-5x} = \frac{2}{x^2-25}$$

$$\frac{x-4}{x(x-5)} = \frac{2}{(x+5)(x-5)}$$

$$x(x+5)(x-5) \cdot \frac{x-4}{x(x-5)} = x(x+5)(x-5) \cdot \frac{2}{(x+5)(x-5)}$$

$$(x+5)(x-4) = 2x$$

$$x^2 + x - 20 = 2x$$

$$x^2 - x - 20 = 0$$

$$(x-5)(x+4) = 0$$

$x = 5$ (reject; causes division by 0) or $x = -4$

$$\boxed{\{-4\}}$$

11.

$$(-3x^{-2}y)^2(-2x^2y^{-3})^{-2}$$

$$= (-3)^2 x^{-2(2)} y^2 (-2)^{-2} x^{2(-2)} y^{-3(-2)}$$

$$= \frac{9}{4} x^{-4-4} y^{2+6}$$

$$= \boxed{\frac{9y^8}{4x^8}}$$

12.

$$x\sqrt{48x} - 3\sqrt{12x^3} - \sqrt{300x^3}$$

$$= x\left(4\sqrt{3x}\right) - 3\left(2x\sqrt{3x}\right) - 10x\sqrt{3x}$$

$$= 4x\sqrt{3x} - 6x\sqrt{3x} - 10x\sqrt{3x}$$

$$= \boxed{-12x\sqrt{3x}}$$

13. $\dfrac{2i}{1+5i} = \dfrac{2i}{1+5i} \cdot \dfrac{1-5i}{1-5i} = \dfrac{2i-10i^2}{1-25i^2} = \dfrac{2i-10(-1)}{1-25(-1)} = \dfrac{10+2i}{26} = \boxed{\dfrac{5+i}{13}}$

14. $\dfrac{\sqrt{12x^3}}{\sqrt{5}} = 2x \cdot \dfrac{\sqrt{3x}}{\sqrt{5}} \cdot \dfrac{\sqrt{5}}{\sqrt{5}} = \boxed{\dfrac{2x}{5}\sqrt{15x}}$

15.

$$\frac{x^2+1}{x^3+1} - \frac{1}{x+1} + \frac{x}{x^2-x+1}$$

$$= \frac{x^2+1}{(x+1)(x^2-x+1)} - \frac{1}{x+1} \cdot \frac{x^2-x+1}{x^2-x+1} + \frac{x}{x^2-x+1} \cdot \frac{x+1}{x+1}$$

$$= \frac{x^2+1-x^2+x-1+x^2+x}{(x+1)(x^2-x+1)}$$

$$= \frac{x^2+2x}{(x+1)(x^2-x+1)}$$

$$= \boxed{\frac{x^2+2x}{x^3+1}}$$

16. $f(x) = 1 - \cfrac{1}{1 - \cfrac{1}{1 - \cfrac{1}{x}}} = 1 - \cfrac{1}{1 - \cfrac{1}{\frac{x-1}{x}}} = 1 - \cfrac{1}{1 - \frac{x}{x-1}} = 1 - \cfrac{1}{\frac{x-1-x}{x-1}} = 1 - \frac{x-1}{-1} = 1 + x - 1 = x$

$f(x) = x, g(x) = 1 - x$

$f[g(x)] = g(x) = \boxed{1 - x \quad (x \neq 0, x \neq 1)}$

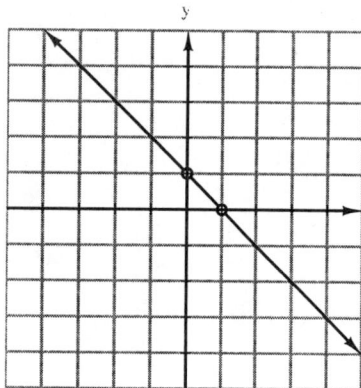

17. $12x^{3n}y^n + 14x^{2n}y^{2n} - 6x^ny^{3n} = 2x^ny^n(6x^{2n} + 7x^ny^n - 3y^{2n}) = \boxed{2x^ny^n(2x^n + 3y^n)(3x^n - y^n)}$

18. $2x - 3y \geq 6$

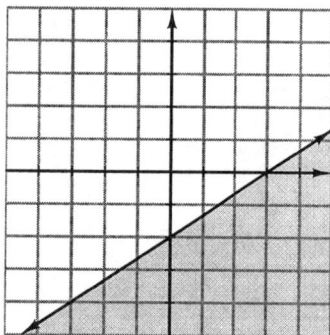

19. $f(x) = \begin{cases} 3 - x & \text{if } x \leq 1 \\ 2 & \text{if } x > 1 \end{cases}$

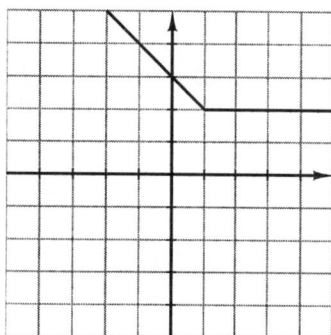

20. $9x^2 + 4y^2 = 36$

$\dfrac{x^2}{4} + \dfrac{y^2}{9} = 1$

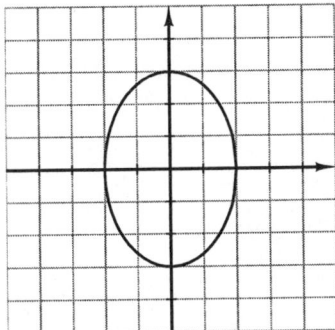

21. $\dfrac{(x-3)^2}{16} + \dfrac{(y-2)^2}{9} \le 1$

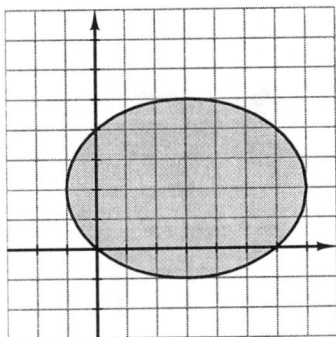

22. $x = y + 1$

$y = -(x+1)^2 + 4$

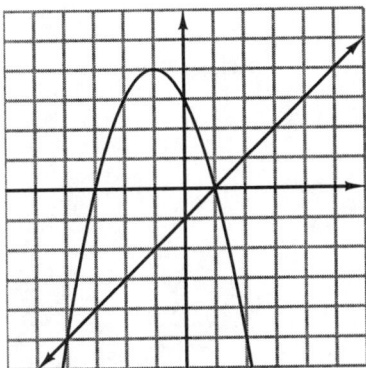

From the graph, the points of intersection are $x = -4$, $y = -5$ and $x = 1$, $y = 0$.

$\boxed{\{(-4,-5),\,(1,0)\}}$

Both solutions check.

23. $s(t) = -16t^2 + 16t + 32$

    a. maximum height occurs when $t = -\dfrac{b}{2a} = \dfrac{-16}{2(-16)} = \dfrac{1}{2}$

        maximum height: $s\left(\dfrac{1}{2}\right) = -16\left(\dfrac{1}{4}\right) + 16\left(\dfrac{1}{2}\right) + 32 = -4 + 8 + 32 = 36$

        maximum height of $\boxed{\text{36 feet at 0.5 sec}}$

    b. The diver reaches the water when $s(t) = 0$.

$$
\begin{aligned}
0 &= -16t^2 + 16t + 32 \\
16t^2 - 16t - 32 &= 0 \\
t^2 - t - 2 &= 0 \\
(t - 2)(t + 1) &= 0 \\
t &= 2 \text{ or } t = -1 \text{ (reject)}
\end{aligned}
$$

    $\boxed{\text{2 seconds}}$

    c. $s(t) = -16t^2 + 16t + 32$

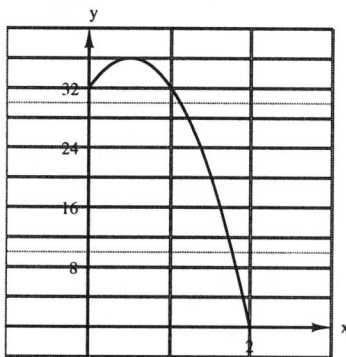

24. Let x = width of the border
$9 + 2x$ = width of pool plus border
$12 + 2x$ = length of pool plus border

$$
\begin{aligned}
\text{area of pool plus border} &= \text{area of pool + area of border} \\
(9 + 2x)(12 + 2x) &= 9(12) + 162 \\
108 + 42x + 4x^2 &= 108 + 162 \\
4x^2 + 42x - 162 &= 0 \\
2x^2 + 21x - 81 &= 0 \\
(2x + 27)(x - 3) &= 0 \\
x = -\dfrac{27}{2} \text{ (reject)} \quad &\text{or} \quad x = 3
\end{aligned}
$$

width of border: $\boxed{\text{3 meters}}$

25. Let x = amount invested at 7% for 2 years.
$10000 - x$ = amount invested at 5% for 3 years.

$$
\begin{aligned}
(0.07x)(2) + (0.05)(10000 - x)(3) &= 1440 \\
0.14x + 1500 - 0.15x &= 1440 \\
-0.01x &= -60 \\
x &= 6000 \ (7\%)
\end{aligned}
$$

$\boxed{\$6000 \text{ at } 7\% \text{ for 2 years; } \$4000 \text{ at } 5\% \text{ for 3 years}}$

26.
$$x = \frac{y-2}{2y}$$
$$2xy = y - 2$$
$$2xy - y = -2$$
$$y(2x - 1) = -2$$
$$y = \frac{-2}{2x - 1}$$
$$\boxed{y = \frac{2}{1 - 2x}}$$

27. From the graph, $g(3) = 4$, $g(-2) = -1$, $g(-4) = -2$.

a. $g(3) - g(-2) + \left| g(-4) \right| = 4 - (-1) + \left| -2 \right| = 4 + 1 + 2 = \boxed{7}$

b.
$$\boxed{\begin{array}{l} D_g = \{x \mid -4 \le x \le 5\} = [-4, 5] \\ R_y = \{y \mid -2 \le y \le 3 \text{ or } 4 \le y \le 5\} = [-2, 3 \cup [4, 5] \end{array}}$$

28.
$$f(x) = \sqrt{2x^2 - x - 6}$$
$$2x^2 - x - 6 \ge 0$$
$$(2x + 3)(x - 2) \ge 0$$
$$x = -\frac{3}{2} \quad x = 2$$

$$\begin{array}{c|c|c} \text{T} & \text{F} & \text{T} \\ \hline -3/2 & & 2 \end{array}$$

Test -2:
$$\begin{aligned} (-4 + 3)(-2 - 2) &\ge 0 \\ (-1)(-4) &\ge 0 \\ 4 &\ge 0 \quad \text{True} \end{aligned}$$

Test 3:
$$\begin{aligned} (6 + 3)(3 - 2) &\ge 0 \\ (9)(1) &\ge 0 \\ 9 &\ge 0 \quad \text{True} \end{aligned}$$

$$\boxed{\text{Domain} = \left[-\infty, -\frac{3}{2} \right] \cup [2, \infty)}$$

29. line containing $(1, 3)$ and parallel to $3x - 4y + 16 = 0$.
slope of
$$3x - 4y + 16 = 0$$
$$4y = 3x + 16$$
$$y = \frac{3}{4}x + 4$$
$$m = \frac{3}{4}$$

slope of line parallel: $\frac{3}{4}$

$$y - 3 = \frac{3}{4}(x - 1)$$
$$y - 3 = \frac{3}{4}x - \frac{3}{4}$$
$$y = \frac{3}{4}x + \frac{9}{4}$$
$$4y = 3x + 9$$
$$\boxed{3x - 4y = -9}$$

30. $cx + dy = c^2$
$dx + cy = d^2$

$$D = \begin{vmatrix} c & d \\ d & c \end{vmatrix} = c^2 - d^2$$

$$Dx = \begin{vmatrix} c^2 & d \\ d^2 & c \end{vmatrix} = c^3 - d^3$$

$$Dy = \begin{vmatrix} c & c^2 \\ d & d^2 \end{vmatrix} = cd^2 - c^2d$$

$$x = \frac{Dx}{D} = \frac{c^3 - d^3}{c^2 - d^2} = \frac{(c-d)(c^2 + cd + d^2)}{(c-d)(c+d)}$$

$$\boxed{x = \frac{c^2 + cd + d^2}{c+d}}$$

$$y = \frac{Dy}{D} = \frac{cd^2 - c^2d}{c^2 - d^2} = \frac{cd(d-c)}{(c-d)(c+d)} = \frac{-cd(c-d)}{(c-d)(c+d)}$$

$$\boxed{y = \frac{-cd}{c+d}}$$

Algebra for College Students

Chapter 10   Polynomial and Rational Functions

Section 10.1   Polynomial Equations Having Rational Solutions

Problem Set 10.1, pp. 748-750

1. a. All possible rational solutions: if c/d is a rational solution, then c must be a factor of -4 (the constant term) and d must be a factor of 1 (the leading coefficient). Therefore, the possible values for c and d are as follows:

For c (factors of -4): $\pm 1, \pm 2, \pm 4$
For d (factors of 1): ± 1

We obtain the possible values for c/d by dividing c by each value of d. Thus, the possible values for c/d are:

$$\boxed{\pm 1, \pm 2, \pm 4}$$

b. Possible number of positive and negative real solutions:

$$x^3 + x^2 - 4x - 4 = 0$$

Because there is one variation of sign, the polynomial equation has $\boxed{\text{one positive real solution}}$.

To find possibilities for negative roots, replace x by $-x$ in the given polynomial:

$$\begin{aligned} P(-x) &= (-x)^3 + (-x)^2 - 4(-x) - 4 \\ &= -x^3 + x^2 + 4x - 4 \end{aligned}$$

There are two variations of sign. Thus, there are $\boxed{\text{2 or 0 negative real solutions}}$.

c.
$$\begin{aligned} x^3 + x^2 - 4x - 4 &= 0 \\ x^2(x-1) - 4(x-1) &= 0 \\ (x-1)(x^2-4) &= 0 \\ (x-1)(x-2)(x+2) &= 0 \end{aligned}$$
$x = 1$ or $x = 2$ or $x = -2$
Solution set: $\boxed{\{-1, -2, 2\}}$

3. $2x^3 - 3x^2 - 11x + 6 = 0$

a. c (factors of 6): $\pm 1, \pm 2, \pm 3, \pm 6$
d (factors of 2): $\pm 1, \pm 2$
c/d: $c/\pm 1 = \pm 1, \pm 2, \pm 3, \pm 6$

$$c/\pm 2 = \pm \frac{1}{2}, \pm 1, \pm \frac{3}{2}, \pm 3$$

Possible rational solutions: $\boxed{\pm \frac{1}{2}, \pm 1, \pm \frac{3}{2}, \pm 2, \pm 3, \pm 6}$

b. $P(x) = 2x^3 - 3x^2 - 11x + 6$
2 variations of sign
$\boxed{\text{positive real solutions: 2 or 0}}$
$$\begin{aligned} P(-x) &= 2(-x)^3 - 3(-x)^2 - 11(-x) + 6 \\ &= -2x^3 - 3x^2 + 11x + 6 \end{aligned}$$
1 variation of sign
$\boxed{\text{Negative real solutions: 1}}$

c. $2x^3 - 3x^2 - 11x + 6 = 0$

$$
\begin{array}{r|rrrr}
-2 & 2 & -3 & -11 & 6 \\
   &   & -4 & 14 & -6 \\
\hline
   & 2 & -7 & 3 & 0
\end{array}
$$

$(x + 2)(2x^2 - 7x + 3) = 0$

$(x + 2)(2x - 1)(x - 3) = 0$

$x = -2$ or $x = \dfrac{1}{2}$ or $x = 3$

$$\boxed{\left\{-2, \frac{1}{2}, 3\right\}}$$

5. $3x^3 + 7x^2 - 22x - 8 = 0$

   a. c(factors of -8): $\pm1, \pm2, \pm4, \pm8$

      d(factors of 3): $\pm1, \pm3$

      c/d: $c/\pm1 = \pm1, \pm2, \pm4, \pm8$

           $c/\pm3 = \pm\dfrac{1}{3}, \pm\dfrac{2}{3}, \pm\dfrac{4}{3}, \pm\dfrac{8}{3}$

      Possible rational solutions: $\boxed{\pm1, \pm2, \pm4, \pm8, \pm\dfrac{1}{3}, \pm\dfrac{2}{3}, \pm\dfrac{4}{3}, \pm\dfrac{8}{3}}$

   b. $P(x) = 3x^3 + 7x^2 - 22x - 8$

              1 variation of sign

      $\boxed{\text{Positive real solutions: } 1}$

          $P(-x) = 3(-x)^3 + 7(-x)^2 - 22(-x) - 8$

                $= -3x^3 + 7x^2 + 22x - 8$

              2 varations of sign.

      $\boxed{\text{Negative real solution: } 2 \text{ or zero}}$

   c. $3x^3 + 7x^2 - 22x - 8 = 0$

$$
\begin{array}{r|rrrr}
2 & 3 & 7 & -22 & -8 \\
  &   & 6 & 26 & 8 \\
\hline
  & 3 & 13 & 4 & 0
\end{array}
$$

      $(x - 2)(3x^2 + 13x + 4) = 0$

      $(x - 2)(3x + 1)(x + 4) = 0$

      $x = 2$ or $x = -\dfrac{1}{3}$ or $x = -4$

$$\boxed{\left\{2, -\frac{1}{3}, -4\right\}}$$

7. $2x^3 - x^2 - 9x - 4 = 0$

   a. c(factors of -4): $\pm1, \pm2, \pm4$

      d(factors of 2): $\pm1, \pm2$

      c/d: $c/\pm1 = \pm1, \pm2, \pm4$

           $c/\pm2 = \pm\dfrac{1}{2}, \pm1, \pm2$

      Possible rational solutions: $\boxed{\pm1, \pm2, \pm4, \pm\dfrac{1}{2}}$

   b. $P(x) = 2x^3 - x^2 - 9x - 4$

              1 variation of sign

      $\boxed{\text{Positive real solutions: } 1}$

          $P(-x) = 2(-x)^3 - (-x)^2 - 9(x) - 4$

                $= -2x^3 - x^2 + 9x - 4$

              2 variations of sign

      $\boxed{\text{Negative real solutions: } 2 \text{ or } 0}$

c. $2x^3 - x^2 - 9x - 4 = 0$

$$-\frac{1}{2}\bigg|\ \ \begin{array}{cccc} 2 & -1 & -9 & -4 \\ & -1 & 1 & 4 \\ \hline 2 & -2 & -8 & 0 \end{array}$$

$$(2x + 1)(2x^2 - 2x - 8) = 0$$

$2x + 1 = 0$ or $2x^2 - 2x - 8 = 0$

$\qquad x = -\dfrac{1}{2}$ $\qquad x^2 - x - 4 = 0$

$$x = \frac{1 \pm \sqrt{1 + 16}}{2}$$

$$= \frac{1 \pm \sqrt{17}}{2}$$

$$\boxed{\left\{ -\frac{1}{2}, \frac{1 - \sqrt{17}}{2}, \frac{1 + \sqrt{17}}{2} \right\}}$$

9. $x^4 - 3x^3 - 20x^2 - 24x - 8 = 0$

a. c(factors of -8): $\pm 1, \pm 2, \pm 4, \pm 8$

d(factors of 1): ± 1

c/d: $\pm 1, \pm 2, \pm 4, \pm 8$

Possible rational solutions: $\boxed{\pm 1, \pm 2, \pm 4, \pm 8}$

b. $P(x) = x^4 - 3x^3 - 20x^2 - 24x - 8$

1 variation of sign.

$\boxed{\text{Positive real solutions: } 1}$

$P(-x) = (-x)^4 - 3(-x)^3 - 20(-x)^2 - 24(-x) - 8$

$= x^4 + 3x^3 - 20x^2 + 24x - 8$

3 variations of sign.

$\boxed{\text{Negative real solutions: } 3 \text{ or } 1}$

c. $x^4 - 3x^3 - 20x^2 - 24x - 8 = 0$

$$-1\bigg|\ \ \begin{array}{ccccc} 1 & -3 & -20 & -24 & -8 \\ & -1 & 4 & 16 & 8 \\ \hline 1 & -4 & -16 & -8 & 0 \end{array}$$

$$(x + 1)(x^3 - 4x^2 - 16x - 8) = 0$$

$$-2\bigg|\ \ \begin{array}{cccc} 1 & -4 & -16 & -8 \\ & -2 & 12 & 8 \\ \hline 1 & -6 & -4 & 0 \end{array}$$

$$(x + 1)(x + 2)(x^2 - 6x - 4) = 0$$

$x = -1$ or $x = -2$ or $x^2 - 6x - 4 = 0$

$$x = \frac{6 \pm \sqrt{36 + 16}}{2}$$

$$= \frac{6 \pm \sqrt{52}}{2}$$

$$= 3 \pm \sqrt{13}$$

$$\boxed{-1, -2, 3 + \sqrt{13}, 3 - \sqrt{13}}$$

11. $3x^4 - 11x^3 - x^2 + 19x + 6 = 0$

   a. c(factors of 6): $\pm1, \pm2 \pm3, \pm6$

        d(factors of 3): $\pm1, \pm3$

        c/d: $c/\pm1 = \pm1, \pm2, \pm3, \pm6$

$$c/\pm3 = \pm\frac{1}{3}, \pm\frac{2}{3}, \pm1, \pm2$$

      Possible rational solutions: $\boxed{\pm1, \pm2, \pm3, \pm6, \pm\frac{1}{3}, \pm\frac{2}{3}}$

   b. $P(x) = 3x^4 - 11x^3 - x^2 + 19x + 6$

            2 variations of sign

      $\boxed{\text{Positive real solutions: 2 or 0}}$

$$P(-x) = 3(-x)^4 - 11(-x)^3 - (-x)^2 + 19(-x) + 6$$
$$= 3x^4 - 11x^3 - x^2 - 19x + 6$$

            2 variations of sign

      $\boxed{\text{Negative real solutions: 2 or 0}}$

   c. $3x^4 - 11x^3 - x^2 + 19x + 6 = 0$

-1	3	-11	-1	19	6
		-3	14	-13	-6
	3	-14	13	6	0

$$(x+1)(3x^3 - 14x^2 + 13x + 6) = 0$$

2	3	-14	13	6
		6	-16	-6
	3	-8	-3	0

$$(x+1)(x-2)(3x^2 - 8x - 3) = 0$$
$$(x+1)(x-2)(x-3)(3x+1) = 0$$

$$x = -1 \quad \text{or} \quad x = 2 \quad \text{or} \quad x = 3 \quad \text{or} \quad x = -\frac{1}{3}$$

$$\boxed{\left\{ -1, 2, 3, -\frac{1}{3} \right\}}$$

13. $4x^4 - x^3 + 5x^2 - 2x - 6 = 0$

   a. c(factors of -6): $\pm1, \pm2, \pm3, \pm6$

        d(factors of 4): $\pm1, \pm2, \pm4$

        c/d: $c/\pm1 = \pm1, \pm2, \pm3, \pm6$

$$c/\pm2 = \pm\frac{1}{2}, \pm1, \pm\frac{3}{2}, \pm3$$

$$c/\pm4 = \pm\frac{1}{4}, \pm\frac{1}{2}, \pm\frac{3}{4}, \pm\frac{3}{2}$$

      Possible rational solutions: $\boxed{\pm1, \pm2, \pm3, \pm6, \pm\frac{1}{2}, \pm\frac{3}{2}, \pm\frac{1}{4}, \pm\frac{3}{4}}$

   b. $P(x) = 4x^4 - x^3 + 5x^2 - 2x - 6$

            3 variations of sign

      $\boxed{\text{Positive real solutions: 3 or 1}}$

$$P(-x) = 4(-x)^4 - (-x)^3 + 5(-x)^2 - 2(-x) - 6$$
$$= 4x^4 + x^3 + 5x^2 + 2x - 6$$

            1 variation of sign

      $\boxed{\text{Negative real solutions: 1}}$

c. $4x^4 - x^3 + 5x^2 - 2x - 6 = 0$

$$
\begin{array}{r|rrrrr}
1 & 4 & -1 & 5 & -2 & -6 \\
  &   & 4 & 3 & 8 & 6 \\
\hline
  & 4 & 3 & 8 & 6 & 0
\end{array}
$$

$(x-1)(4x^3 + 3x^2 + 8x + 6) = 0$

$$
\begin{array}{r|rrrr}
-\dfrac{3}{4} & 4 & 3 & 8 & 6 \\
              &   & -3 & 0 & -6 \\
\hline
              & 4 & 0 & 8 & 0
\end{array}
$$

$(x-1)(4x+3)(4x^2 + 8) = 0$

$x = 1$ or $x = -\dfrac{3}{4}$ or $4x^2 + 8 = 0$

$x = \pm i\sqrt{2}$

$$\left\{ -\frac{3}{4},\, 1,\, -\sqrt{2}\,i,\, \sqrt{2}\,i \right\}$$

15. $2x^5 + 7x^4 - 18x^2 - 8x + 8 = 0$

a. c(factors of 8): $\pm 1, \pm 2, \pm 4, \pm 8$

d(factors of 2): $\pm 1, \pm 2$

c/d: $c/\pm 1 = \pm 1, \pm 2, \pm 4, \pm 8$

$c/\pm 2 = \pm \dfrac{1}{2}, \pm 1, \pm 2, \pm 4$

Possible rational solutions: $\boxed{\pm 1, \pm 2, \pm 4, \pm 8, \pm \dfrac{1}{2}}$

b. $P(x) = 2x^5 + 7x^4 - 18x^2 - 8x + 8$

2 variations of sign

$\boxed{\text{Positive real solutions: 2 or 0}}$

$P(-x) = 2(-x)^5 + 7(-x)^4 - 18(-x)^2 - 8(-x) + 8$

$= -2x^5 + 7x^4 - 18x^2 + 8x + 8$

3 variations in sign

$\boxed{\text{Negative real solutions: 3 or 1}}$

c. $2x^5 + 7x^4 - 18x^2 - 8x + 8 = 0$

$$
\begin{array}{r|rrrrrr}
-2 & 2 & 7 & 0 & -18 & -8 & 8 \\
   &   & -4 & -6 & 12 & 12 & -8 \\
\hline
   & 2 & 3 & -6 & -6 & 4 & 0
\end{array}
$$

$(x+2)(2x^4 + 3x^3 - 6x^2 - 6x + 4) = 0$

$$
\begin{array}{r|rrrrr}
\dfrac{1}{2} & 2 & 3 & -6 & -6 & 4 \\
             &   & 1 & 2 & -2 & -4 \\
\hline
             & 2 & 4 & -4 & -8 & 0
\end{array}
$$

$(x+2)(2x-1)(2x^3 + 4x^2 - 4x - 8) = 0$
$(x+2)(2x-1)(x^3 + 2x^2 - 2x - 4) = 0$
$(x+2)(2x-1)[x^2(x+2) - 2(x+2)] = 0$
$(x+2)(2x-1)(x+2)(x^2 - 2) = 0$
$(x+2)^2(2x-1)(x^2 - 2) = 0$

$x = -2$ or $x = \dfrac{1}{2}$ or $x^2 - 2 = 0$

$x = \pm\sqrt{2}$

$$\left\{ -2,\, \frac{1}{2},\, -\sqrt{2},\, \sqrt{2} \right\}$$

17. $x^4 + x^3 - x - 1 = 0$

   a. c (factors of -1): ± 1

   d (factors of 1): ± 1

   c/d: ± 1

   Possible rational solutions: $\boxed{\pm 1}$

   b. $P(x) = x^4 + x^3 - x - 1$

   1 variation of sign

   $\boxed{\text{Positive real solutions: } 1}$

   $P(-x) = (-x)^4 + (-x)^3 - (-x) - 1$

   $= x^4 - x^3 + x - 1$

   3 variations of sign

   $\boxed{\text{Negative real solutions: } 3 \text{ or } 1}$

   c.
   $$x^4 + x^3 - x - 1 = 0$$
   $$x^3(x+1) - (x+1) = 0$$
   $$(x+1)(x^3 - 1) = 0$$
   $$(x+1)(x-1)(x^2 + x + 1) = 0$$

   $x = -1$ or $x = 1$ or $x = \dfrac{-1 \pm \sqrt{1-4}}{2}$

   $= \dfrac{-1 \pm i\sqrt{3}}{2}$

   $$\boxed{\left\{ 1, -1, \frac{-1 - \sqrt{3}\,i}{2}, \frac{-1 + \sqrt{3}\,i}{2} \right\}}$$

19. $\boxed{D}$ is true;

   A is *not* true: $x^3 + 5x^2 + 6x + 1 = 0$

   $P(x) = x^3 + 5x^2 + 6x + 1$

   no variations of sign

   Positive real solutions: 0 *not* 1

   B is *not* true: Descartes's rule gives the possible number of positive and negative real solutions.

   C is *not* true: Every polynomial equation of degree 3 *may have* at least one rational soluton.

   D is true.

21. $\boxed{B}$ is true;

   $3 - 5i$ and $3 + 5i$ are solutions.

   $4 + 3i$ and $4 - 3i$ are solutions.

   The degree of $P(x)$ is at least 4.

23. $24231_5 = 2(5^4) + 4(5^3) + 2(5^2) + 3(5) + 1$

   $P(x) = 2x^4 + 4x^3 + 2x^2 + 3x + 1$

   $P(5) = 24231_5$

   $= 2(625) + 4(125) + 2(25) + 3(5) + 1$

   $= 1250 + 500 + 50 + 15 + 1$

   $= \boxed{1816}$

25. $x^3 + 3x^2 - x - 3 = 0$
$y = x^3 + 3x^2 - x - 3$

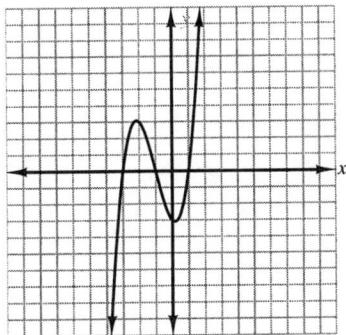

From the graph, $y = 0$ when $x = -3$, $x = -1$, and $x = 1$.

$$\boxed{\{-3, -1, 1\}}$$

27. $x^4 - 8x^3 + 7x^2 + 72x - 144 = 0$
$y = x^4 - 8x^3 + 7x^2 + 72x - 144$

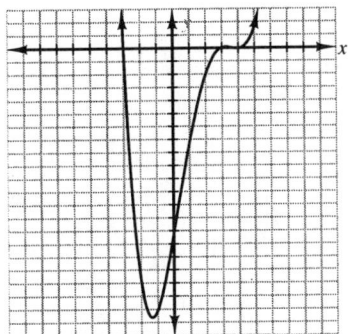

From the graph, $y = 0$ when $x = -3$, $x = 3$, and $x = 4$.

$$\boxed{\{-3, 3, 4\} \quad \text{(4 is a double root)}}$$

29. a. $\dfrac{P(x)}{x - c} = Q(x) + \dfrac{R}{x - c}$

$$\boxed{P(x) = (x - x)Q(x) + R} \qquad [\times (x - c)]$$

b. $P(c) = (c - c)Q(x) + R$
$P(c) = 0 + R$
$\quad\quad = R$

$$\boxed{R = P(c)}$$

Review Problems

39. $\left(\dfrac{x^{-3}y^{-5}}{xy^{-7}}\right)^{-6} = (x^{-4}y^2)^{-6}$

$\quad\quad\quad\quad\quad\quad = x^{24}y^{-12}$

$\quad\quad\quad\quad\quad\quad = \boxed{\dfrac{x^{24}}{y^{12}}}$

40. $\sqrt{8x^3} - 2x\sqrt{50x} \; = \; 2x\sqrt{2x} - 10x\sqrt{2x}$ where $x > 0$

$$= \; \boxed{-8x\sqrt{2x}}$$

41. Let x = amount invested at 4% for 3 years.
$20{,}000 - x$ = amount invested at 5% for 4 years.

$$
\begin{aligned}
0.04x(3) + 0.05(20{,}000 - x)(4) &= 2840 \\
0.12x + 4000 - 0.20x &= 2840 \\
-0.08x &= -1160 \\
x &= 14{,}500 \quad (4\%) \\
20{,}000 - x &= 20{,}000 - 14{,}500 = 5500 \quad (5\%)
\end{aligned}
$$

$\boxed{\$14{,}500 \text{ at } 4\%; \; \$5500 \text{ at } 5\%}$

Section 10.2   Graphing Polynomial Functions

Problem Set 10.2, pp. 756-758

1.
$$
\begin{aligned}
f(x) &= x^4 - x^2 \\
f(-x) &= (-x)^4 - (-x)^2 \\
&= x^4 - x^2
\end{aligned}
$$
Because $f(-x) = f(x)$, the graph has y–axis symmetry, i.e. f is $\boxed{\text{symmetric with respect to the } y\text{–axis}}$.

3.
$$
\begin{aligned}
f(x) &= 7x^3 + 5x \\
f(-x) &= 7(-x)^3 + 5(-x) \\
&= -7x^3 - 5x \\
&= -(7x^3 + 5x) \\
f(-x) &= -f(x)
\end{aligned}
$$
$\boxed{\text{symmetric with respect to the origin}}$

5.
$$
\begin{aligned}
f(x) &= x^4 - x \\
f(-x) &= (-x)^4 - (-x) \\
&= x^4 + x
\end{aligned}
$$
$f(-x) \neq f(x)$ and $f(-x) \neq -f(x)$: $\boxed{\text{not symmetric about the } y\text{–axis or the origin}}$.

7.
$$
\begin{aligned}
f(x) &= (x^2 - 4)^3 \\
f(-x) &= [(-x)^2 - 4]^3 \\
&= (x^2 - 4)^3
\end{aligned}
$$
$f(-x) = f(x)$: $\boxed{\text{symmetric with respect to the } y\text{–axis}}$.

9. $f(x) = x^3 - 2x^2$
   1.   Check for symmetry:
$$
\begin{aligned}
f(-x) &= (-x)^3 - 2(-x)^2 \\
&= -x^3 - 2x^2
\end{aligned}
$$
     $f(-x) \neq f(x), f(-x) \neq -f(x)$: not symmetric about the y–axis or the origin.
   2.   Find x–intercepts:
$$
\begin{aligned}
x^3 - 2x^2 &= 0 \\
x^2(x - 2) &= 0
\end{aligned}
$$
     $x = 0$ or $x = 2$: $(0, 0)$ (double root), $(2, 0)$

3.

$$\frac{x<0 \quad | \quad 0<x<2 \quad | \quad x>2}{0 \qquad\qquad 2}$$

Interval	Test Value	Sign of $f(x)$	Location of Graph (above or below x-axis)
$x<0$	-1: $f(-1)=-3$	$-$	below
$0<x<2$	1: $f(1)=-1$	$-$	below
$x>2$	3: $f(3)=9$	$+$	above

4. y-intercept at $(0,0)$

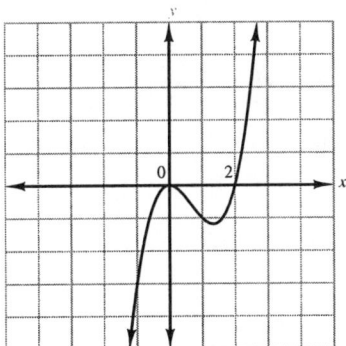

11. $f(x)=x^3+x^2-2x$

1.   symmetry:

$$\begin{aligned} f(-x) &= (-x)^3+(-x)^2-2(-x) \\ &= -x^3+x^2+2x \end{aligned}$$

$f(-x)\neq f(x), f(-x)\neq -f(x)$: not symmetric about the y-axis or the origin.

2. x–intercepts:

$$\begin{aligned} x^3+x^2-2x &= 0 \\ x(x^2+x-2) &= 0 \\ x(x+2)(x-1) &= 0 \\ x=0,-2,1 \;&\rightarrow\; (0,0),(-2,0),(1,0) \end{aligned}$$

3.   intervals

$$\frac{x<-2 \quad | \quad -2<x<0 \quad | \quad 0<x<1 \quad | \quad x>1}{-2 \qquad\qquad 0 \qquad\qquad 1}$$

Interval	Test Value	Sign of $f(x)$	Location of Graph (above or below x-axis)
$x<-2$	-3: $f(-3)=-12$	$-$	below
$-2<x<0$	-1: $f(-1)=2$	$+$	above
$0<x<1$	$\frac{1}{2}$: $f\left(\frac{1}{2}\right)=-\frac{5}{8}$	$-$	below
$x>1$	2: $f(2)=8$	$+$	above

4. *y*–intercept at $(0, 0)$

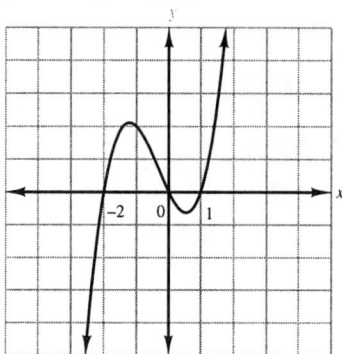

13. $f(x) = x^3 + 2x^2 - x - 2$

1. symmetry:
$$\begin{aligned} f(-x) &= (-x)^3 + 2(-x)^2 - (-x) - 1 \\ &= -x^3 + 2x^2 + x - 2 \end{aligned}$$

$f(-x) \neq f(x)$; $f(-x) \neq -f(x)$: not symmetric about the *y*–axis or the origin.

2. *x*–intercepts:
$$\begin{aligned} x^3 + 2x^2 - x - 2 &= 0 \\ x^2(x + 2) - (x + 2) &= 0 \\ (x + 2)(x^2 - 1) &= 0 \\ (x + 2)(x - 1)(x + 1) &= 0 \\ x = -2, 1, -1 &\rightarrow (-2, 0), (-1, 0), (1, 0) \end{aligned}$$

3. intervals

$$\underset{}{\underbrace{x < -2}} \underset{-2}{\bigg|} \underset{}{\underbrace{-2 < x < -1}} \underset{-1}{\bigg|} \underset{}{\underbrace{0 < x < 1}} \underset{1}{\bigg|} \underset{}{\underbrace{x > 1}}$$

Interval	Test Value	Sign of $f(x)$	Location of Graph (above or below *x*-axis)
$x < -2$	-3: $f(-3) = -8$	$-$	below
$-2 < x < -1$	-1.5: $f(-1.5) = \dfrac{5}{8}$	$+$	above
$-1 < x < 1$	0: $f(0) = -2$	$-$	below
$x > 1$	2: $f(2) = 12$	$+$	above

4. *y*–intercept: $f(0) = -2 \rightarrow (0, -2)$

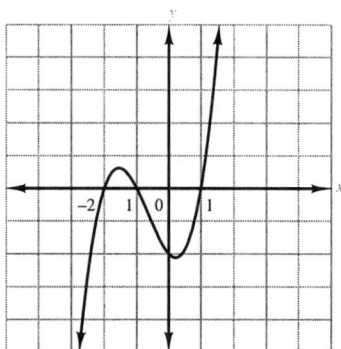

15. $f(x) = x^3 + 3x^2 - x - 3$

1. symmetry:

$$\begin{aligned} f(-x) &= (-x)^3 + 3(-x)^2 - (-x) - 3 \\ &= -x^3 + 3x^2 + x - 3 \end{aligned}$$

$f(-x) \neq f(x); f(-x) \neq -f(x)$: not symmetric about the y–axis or the origin.

2. x–intercepts:

$$\begin{aligned} x^3 + 3x^2 - x - 3 &= 0 \\ x^2(x+3) - (x+3) &= 0 \\ (x+3)(x^2-1) &= 0 \\ (x+3)(x+1)(x-1) &= 0 \end{aligned}$$

$x = -3, -1, 1 \;\rightarrow\; (-3, 0), (-1, 0), (1, 0)$

3. intervals

$x < -3$	$-3 < x < -1$	$-1 < x < 1$	$x > 1$
-3	-1	1	

Interval	Test Value	Sign of $f(x)$	Location of Graph (above or below x-axis)
$x < -3$	-4: $f(-4) = -15$	$-$	below
$-3 < x < -1$	-2: $f(-2) = 3$	$+$	above
$-1 < x < 1$	0: $f(0) = -3$	$-$	below
$x > 1$	2: $f(2) = 15$	$+$	above

4. y–intercept at $(0, -3)$

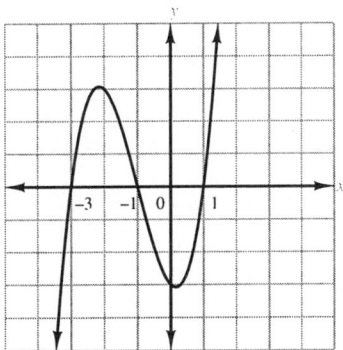

17. $f(x) = x^4 - 4x^3 + 3x^2$

1. symmetry:

$$\begin{aligned} f(-x) &= (-x)^4 - 4(-x)^3 + 3(-x)^2 \\ &= x^4 + 4x^3 + 3x^2 \end{aligned}$$

$f(-x) \neq f(x), f(-x) \neq -f(x)$: not symmetric about the y–axis or the origin.

2. x–intercepts:

$$\begin{aligned} x^4 - 4x^3 + 3x^2 &= 0 \\ x^2(x^2 - 4x + 3) &= 0 \\ x^2(x-3)(x-1) &= 0 \end{aligned}$$

$x = 0$ (double root), $x = 3$, $x = 1 \;\rightarrow\; (0, 0), (3, 0), (1, 0)$

3. intervals

$x < 0$	$0 < x < 1$	$1 < x < 3$	$x > 3$
0	1	3	

Interval	Test Value	Sign of $f(x)$	Location of Graph (above or below x-axis)
$x < 0$	-1: $f(-1) = -8$	$+$	above
$0 < x < 1$	$\frac{1}{2}$: $f\left(\frac{1}{2}\right) = \frac{5}{16}$	$+$	above
$1 < x < 3$	2: $f(2) = -4$	$-$	below
$x > 3$	4: $f(4) = 48$	$+$	above

4. y–intercept at $(0, 0)$

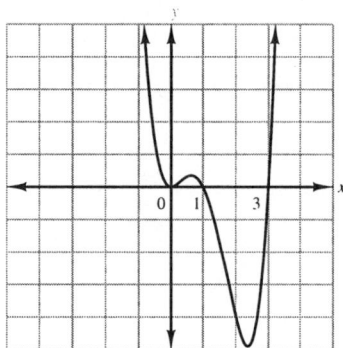

19. $f(x) = 4x^4 - 4x^3 - 25x^2 + x + 6$

   1.   symmetry:
$$\begin{aligned} f(-x) &= 4(-x)^4 - 4(-x)^3 - 25(-x)^2 + (-x) + 6 \\ &= 4x^4 - 4x^3 - 25x^2 - x + 6 \end{aligned}$$

       $f(-x) \neq f(x), f(-x) \neq -f(x)$: not symmetric about the y–axis or the origin.

   2. x–intercepts:
$$4x^4 - 4x^3 - 25x^2 + x + 6 = 0$$

$$\begin{array}{r|rrrrr} -2 & 4 & -4 & -25 & 1 & 6 \\ & & -8 & 24 & 2 & -6 \\ \hline & 4 & -12 & -1 & 3 & 0 \end{array}$$

$$(x + 2)(4x^3 - 12x^2 - x + 3) = 0$$

$$\begin{array}{r|rrrr} 3 & 4 & -12 & -1 & 3 \\ & & 12 & 0 & -3 \\ \hline & 4 & 0 & -1 & 0 \end{array}$$

$$(x + 2)(x - 3)(4x^2 - 1) = 0$$
$$(x + 2)(x - 3)(2x - 1)(2x + 1) = 0$$
$$x = -2, 3, \frac{1}{2}, -\frac{1}{2} \ \rightarrow \ (-2, 0), (3, 0), \left(\frac{1}{2}, 0\right), \left(-\frac{1}{2}, 0\right)$$

   3.   intervals:

$x < -2$	$-2 < x < -\dfrac{1}{2}$	$-\dfrac{1}{2} < x < \dfrac{1}{2}$	$\dfrac{1}{2} < x < 3$	$x > 3$

$$\quad -2 \qquad\qquad -\frac{1}{2} \qquad\qquad \frac{1}{2} \qquad\qquad 3$$

Interval	Test Value	Sign of $f(x)$	Location of Graph (above or below x-axis)
$x < -2$	-3: $f(-3) = 210$	$+$	above
$-2 < x < -\dfrac{1}{2}$	-1: $f(-1) = -12$	$-$	below
$-\dfrac{1}{2} < x < \dfrac{1}{2}$	0: $f(0) = 6$	$+$	above
$\dfrac{1}{2} < x < 3$	1: $f(1) = -18$	$-$	below
$x > 3$	4: $f(4) = 378$	$+$	above

4. y–intercept at $(0, 6)$.

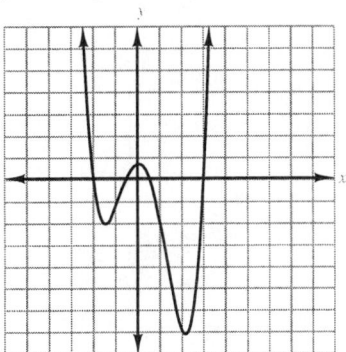

21. $f(x) = (x + 4)(x - 1)(x + 5)$

1.   symmetry:
$$f(-x) = (-x + 4)(-x - 1)(-x + 5)$$
$$= -(x - 4)(x + 1)(x - 5)$$
$f(-x) \neq f(x), f(-x) \neq -f(x)$: not symmetric about the y–axis or the origin.

2. x–intercepts:
$$(x + 4)(x - 1)(x + 5) = 0$$
$$x = -4, 1, -5 \rightarrow (-4, 0), (1, 0), (-5, 0)$$

3.   intervals:

$x < -5$	$-5 < x < -4$	$-4 < x < 1$	$x > 1$
-5	-4	1	

Interval	Test Value	Sign of $f(x)$	Location of Graph (above or below x-axis)
$x < -5$	-6: $f(-6) = -14$	$-$	below
$-5 < x < -4$	$-4\frac{1}{2}$: $f\left(-4\frac{1}{2}\right) = \frac{11}{8}$	$+$	above
$-4 < x < 1$	0: $f(0) = -20$	$-$	below
$x > 1$	2: $f(2) = 42$	$+$	above

4. y–intercept at $(0, -20)$

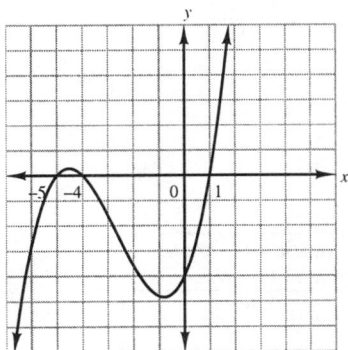

23. $f(x) = (x + 2)^2(x - 3)^2$

    1.   symmetry:

$$f(-x) = (-x + 2)^2(-x - 3)^2$$
$$= (x - 2)^2(x + 3)^2$$

$f(-x) \neq f(x), f(-x) \neq -f(x)$: not symmetric about y–axis or the origin.

    2. x–intercepts:

$$(x + 2)^2(x - 3)^2 = 0$$

$x = -2, 3$ (both double roots) $\rightarrow$ $(-2, 0), (3, 0)$

    3.   intervals

$x < -2$	$-2 < x < 3$	$x > 3$
-2	3	

Interval	Test Value	Sign of $f(x)$	Location of Graph (above or below x-axis)
$x < -2$	-3: $f(-3) = 36$	$+$	above
$-2 < x < 3$	0: $f(0) = 36$	$+$	above
$x > 3$	4: $f(4) = 36$	$+$	above

    4. y–intercept at $(0, 36)$

25. $f(x) = x^2(x + 2)(x - 1)^2(x - 2)$

    1.   symmetry:

$$f(-x) = (-x)^2(-x + 2)(-x - 1)^2(-x - 2)$$
$$= x^2(x - 2)(x + 1)^2(x + 2)$$

$f(-x) \neq f(x), f(-x) \neq -f(x)$: not symmetric about the y–axis or the origin.

    2. x–intercepts:

$$x^2(x + 2)(x - 1)^2(x - 2) = 0$$

$x = 0$ (double root), $-2, 1$ (double root), 2 $\rightarrow$ $(0, 0), (-2, 0), (1, 0), (2, 0)$

    3.   intervals

$x < -2$	$-2 < x < 0$	$-0 < x < 1$	$1 < x < 2$	$x > 2$
-2	0	1	2	

Interval	Test Value	Sign of $f(x)$	Location of Graph (above or below x-axis)
$x < -2$	-3: $f(-3) = 720$	$+$	above
$-2 < x < 0$	-1: $f(-1) = -12$	$-$	below
$0 < x < 1$	$\frac{1}{2}$: $f\left(\frac{1}{2}\right) = -\frac{15}{64}$	$-$	below
$1 < x < 2$	$1\frac{1}{2}$: $f\left(1\frac{1}{2}\right) = -\frac{63}{64}$	$-$	below
$x > 2$	3: $f(3) = 32$	$+$	above

4. y–intercept at $\left(0, -\dfrac{15}{64}\right)$

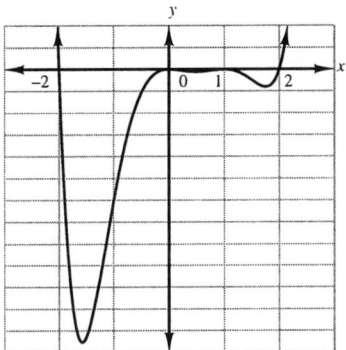

27. ⟦C⟧ is true;

$f(x) = (x - 1)^2 (x + 3)^2$ only touches the x–axis at the x–intercepts since both the $(x - 1)$ and the $(x + 3)$ terms are squared, and consequently the function is always positive.

29. $f(-x) = f(x)$: symmetric about the y–axis.

x–intercepts are $(-2, 0)$, $(-1, 0)$, $(1, 0)$, and $(2, 0)$

$$\begin{aligned} y &= a(x + 2)(x + 1)(x - 1)(x - 2) \\ &= a(x^2 - 1)(x^2 - 4) \end{aligned}$$

y–intercept at $(0, 4)$

$$\begin{aligned} 4 &= a(0 - 1)(0 - 4) \\ 4 &= 4a \end{aligned}$$

Thus,

$$\begin{aligned} y &= 1(x^2 - 1)(x^2 - 4) \\ \boxed{f(x) &= x^4 - 5x^2 + 4} \end{aligned}$$

31. $f(x) = -x^4 + 3x^3 - 2x^2$

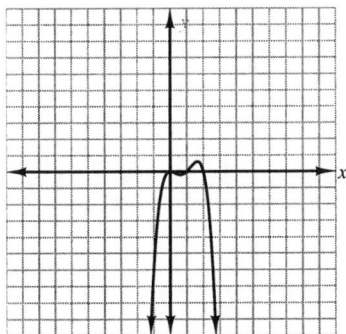

From the graph the ⟦x–intercepts are $0, 1, 2$⟧

$$\boxed{\begin{aligned} &f(x) < 0 \text{ on } (-\infty, 0) \cup (0, 1) \cup (2, \infty) \\ &f(x) > 0 \text{ on } (1, 2) \end{aligned}}$$

33.

$$x^4 - 4x^3 + 3x \geq 14(1-x)$$
$$x^4 - 4x^3 + 3x \geq 14 - 14x$$
$$x^4 - 4x^3 + 17x - 14 \geq 0$$

$$x^4 - 4x^3 + 17x - 14 = 0$$

$$\begin{array}{r|rrrrr}
1 & 1 & -4 & 0 & 17 & -14 \\
  &   & 1 & -3 & -3 & 14 \\
\hline
  & 1 & -3 & -3 & 14 & 0
\end{array}$$

$$(x-1)(x^3 - 3x^2 - 3x + 14) = 0$$

$$\begin{array}{r|rrrr}
-2 & 1 & -3 & -3 & 14 \\
   &   & -2 & 10 & -14 \\
\hline
   & 1 & -5 & 7 & 0
\end{array}$$

$$(x-1)(x+2)(x^2 - 5x + 7) = 0$$
$$x = 1 \quad \text{or} \quad x = -2 \quad \text{or} \quad x^2 - 5x + 7 = 0$$

$$x = \frac{5 \pm \sqrt{25-28}}{2}$$

$$x = \frac{5 \pm i\sqrt{3}}{2}$$

x–intercepts are $(1, 0)$ and $(-2, 0)$

$$\begin{array}{c|c|c}
T & F & T \\
\hline
-2 & 1 &
\end{array}$$

Test Value, -3: $(-4)(-1)(9+15+7) = 4(31) = 124$
$124 \geq 0$ True
Test Value, 0: $-14 \geq 0$ False
Test Value, 2: $(1)(4)(4-10+7) = 4(1) = 4$
$4 \geq 0$ True

$$\boxed{(-\infty, -2] \cup [1, \infty)}$$

Review Problems

39.

$$\begin{aligned}
ab - 2c + bc - 2a &= ab + bc - 2a - 2a \\
&= b(a+c) - 2(a+c) \\
&= \boxed{(a+c)(b-2)}
\end{aligned}$$

40. Let x = number of liters of 6% acid solution.
$3 - x$ = number of liters of 15% acid solution.

$$\begin{aligned}
0.06x + 0.15(3-x) &= 0.12(3) \\
0.06x + 0.45 - 0.15x &= 0.36 \\
-0.09x &= -0.09 \\
x &= 1 \quad \text{(6%)} \\
3 - x &= 3 - 1 = 2 \quad \text{(15%)}
\end{aligned}$$

$$\boxed{\text{1 liter of 6\% acid; mix with 2 liters of 15\% acid}}$$

41.
$$x^2 + y^2 \le 9$$
$$25x^2 + 4y^2 \ge 100 \quad \rightarrow \quad \frac{x^2}{4} + \frac{y^2}{25} \ge 1$$

1. $x^2 + y^2 \le 9$

    boundary *solid*; circle with center (0, 0) and radius 3; shade interior of circle.

2. $\frac{x^2}{4} + \frac{y^2}{25} \ge 1$

    boundary *solid*; ellipse with center (0, 0) and vertices (±2, 0) and (0, ±5); shade outside ellipse.

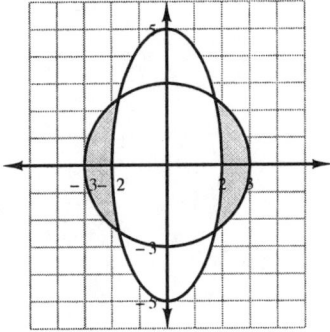

Section 10.3   Graphing Rational Functions

Problem Set 10.3, pp. 767-768

1. $f(x) = \dfrac{4x}{x-2}$

    1. Vertical asymptotes:
$$x - 2 = 0$$
$$= 2$$

    2. Values close to the asymptotes.

x	0	1	3	$\frac{3}{2}$	$\frac{5}{2}$
$f(x) = \dfrac{4x}{x-2}$	0	−4	12	−12	20

    3. horizontal asymptotes:
$$f(x) = \frac{4x}{x-2} = \frac{\frac{4x}{x}}{\frac{x-2}{x}} = \frac{4}{1 - \frac{2}{x}}$$

As $|x|$ becomes larger, $\dfrac{2}{x}$ approaches 0.

$f(x)$ approaches $\dfrac{4}{1-0} = 4$

$y = 4$ is a horizontal aymptote.

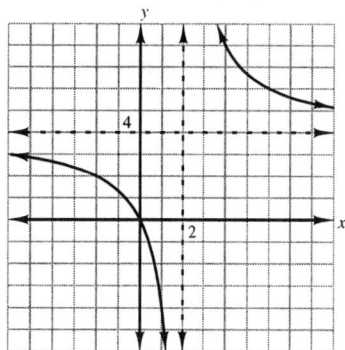

3. $f(x) = \dfrac{2x}{x^2 - 4}$

1.   vertical asymptotes:
$$x^2 - 4 = 0$$
$$x = \pm 2$$

2.   values close to asymptotes:

x	-3	-1	0	1	3	$-\dfrac{5}{2}$	$-\dfrac{3}{2}$	$\dfrac{3}{2}$	$\dfrac{5}{2}$
$f(x) = \dfrac{2x}{x^2 - 4}$	$-\dfrac{6}{5}$	$\dfrac{2}{3}$	0	$-\dfrac{2}{3}$	$\dfrac{6}{5}$	$-\dfrac{20}{9}$	$\dfrac{24}{7}$	$-\dfrac{24}{7}$	$\dfrac{20}{9}$

3.   horizontal asymptotes:

$$f(x) = \dfrac{2x}{x^2-4} = \dfrac{\dfrac{2x}{x^2}}{\dfrac{x^2-4}{x^2}} = \dfrac{\dfrac{2}{x}}{1 - \dfrac{4}{x^2}}$$

As $|x|$ becomes larger, $\dfrac{2}{x}$ approaches 0 and $\dfrac{4}{x^2}$ approaches 0 and $f(x)$ approaches $\dfrac{0}{1-0} = 0$.

$y = 0$ is a horizontal asymptote.

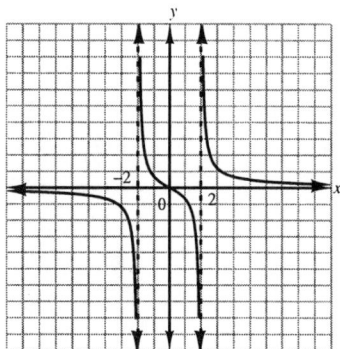

5. $f(x) = \dfrac{2x^2}{x^2 - 1}$

1. vertical asymptotes:
$$x^2 - 1 = 0$$
$$x = \pm 1$$

2. values close to asymptote:

x	± 2	0	$\pm \dfrac{3}{2}$	$\pm \dfrac{1}{2}$
$f(x) = \dfrac{2x^2}{x^2 - 1}$	$\dfrac{8}{3}$	0	$\dfrac{18}{5}$	$-\dfrac{2}{3}$

3. horizontal aymptote:

$$f(x) = \frac{2x^2}{x^2 - 1} = \frac{2x^2}{x^2 - 1} \cdot \frac{\dfrac{1}{x^2}}{\dfrac{1}{x^2}} = \frac{2}{1 - \dfrac{1}{x^2}}$$

As $|x|$ becomes larger, $\dfrac{1}{x^2}$ approaches 0. $f(x)$ approaches $\dfrac{2}{1 - 0} = 2$.

$y = 2$ is a horizontal asymptote.

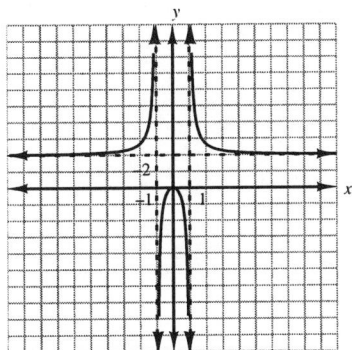

7. $f(x) = \dfrac{-x}{x + 1}$

1. vertical asymptote:
$$x + 1 = 0$$
$$x = -1$$

2. values close to asymptote:

x	-2	0	1	$-\dfrac{3}{2}$	$-\dfrac{1}{2}$
$f(x) = \dfrac{-x}{x + 1}$	-2	0	$-\dfrac{1}{2}$	-3	1

3. horizontal asymptote:

$$f(x) = \frac{-x}{x + 1} = \frac{-x}{x + 1} \cdot \frac{\dfrac{1}{x}}{\dfrac{1}{x}} = \frac{-1}{1 + \dfrac{1}{x}}$$

As $\left|x\right|$ becomes larger, $\dfrac{1}{x}$ approaches 0.

$f(x)$ approaches $\dfrac{-1}{1+0} = -1$

$y = -1$ is a horizontal asymptote.

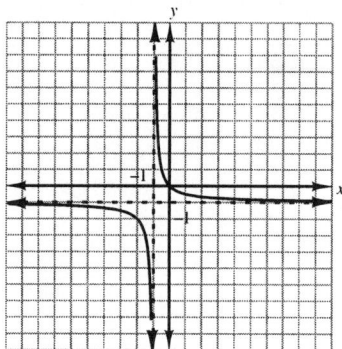

9. $f(x) = -\dfrac{1}{x^2 - 4}$

   1.   vertical asymptotes:

$$x^2 - 4 = 0$$
$$x = \pm 2$$

   2.   values close to asymptote:

x	± 3	± 1	0	$\pm\dfrac{5}{2}$	$\pm\dfrac{3}{2}$
$f(x) = -\dfrac{1}{x^2 - 4}$	$-\dfrac{1}{5}$	$\dfrac{1}{3}$	$\dfrac{1}{4}$	$-\dfrac{4}{9}$	$\dfrac{4}{7}$

   3.   horizontal asymptote:

$$f(x) = -\frac{1}{x^2 - 4} = -\frac{1}{x^2 - 4} \cdot \frac{\dfrac{1}{x^2}}{\dfrac{1}{x^2}} = \frac{-\dfrac{1}{x^2}}{1 - \dfrac{4}{x^2}}$$

As $\left|x\right|$ becomes larger, $\dfrac{1}{x^2}$ and $\dfrac{4}{x^2}$ approach 0.

$f(x)$ approaches $\dfrac{0}{1-0} = 0$

$y = 0$ is a horizontal asymptote.

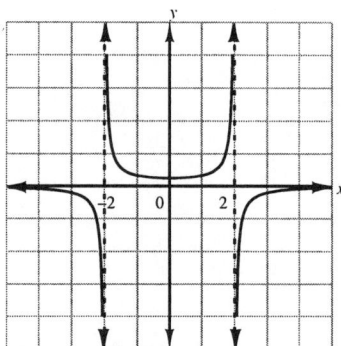

11. $f(x) = \dfrac{2}{x^2 + x - 2}$

    1. vertical asymptotes:

$$\begin{aligned} x^2 + x - 2 &= 0 \\ (x + 2)(x - 1) &= 0 \\ x &= -2, 1 \end{aligned}$$

    2. values close to aymptote:

x	-3	-1	0	2	$-\dfrac{5}{2}$	$-\dfrac{3}{2}$	$\dfrac{1}{2}$	$\dfrac{3}{2}$
$f(x) = \dfrac{2}{x^2 + x - 2}$	$\dfrac{1}{2}$	-1	-1	$\dfrac{1}{2}$	$\dfrac{8}{7}$	$-\dfrac{8}{5}$	$-\dfrac{8}{5}$	$\dfrac{8}{7}$

    3. horizontal asymptote:

$$f(x) \;=\; \frac{2}{x^2 + x - 2} = \frac{2}{x^2 + x - 2} \cdot \frac{\dfrac{1}{x^2}}{\dfrac{1}{x^2}} = \frac{\dfrac{2}{x^2}}{1 + \dfrac{1}{x} - \dfrac{2}{x^2}}$$

As $|x|$ becomes larger, $\dfrac{2}{x^2}$ and $\dfrac{1}{x}$ approach 0.

$f(x)$ approaches $\dfrac{0}{1 + 0 - 0} = 0$

$y = 0$ is a horizontal asymptote.

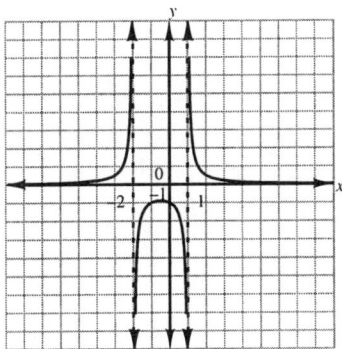

13. $f(x) = \dfrac{2x^2}{x^2 + 4}$

    1. vertical asymptote: $x^2 + 4$ cannot equal 0 for any real value of x. No vertical asymptote.

    2. horizontal asymptote:

$$f(x) \;=\; \frac{2x^2}{x^2 + 4} = \frac{2x^2}{x^2 + 4} \cdot \frac{\dfrac{1}{x^2}}{\dfrac{1}{x^2}} = \frac{2}{1 + \dfrac{4}{x^2}}$$

As $|x|$ becomes larger, $\dfrac{4}{x^2}$ approaches 0.

$f(x)$ approaches $\dfrac{2}{1 + 0} = 2$

$y = 2$ is a horizontal asymptote.

3.   values close to asymptote:

x	± 2	± 1	0
$f(x) = \dfrac{2x^2}{x^2 + 4}$	1	$\dfrac{2}{5}$	0

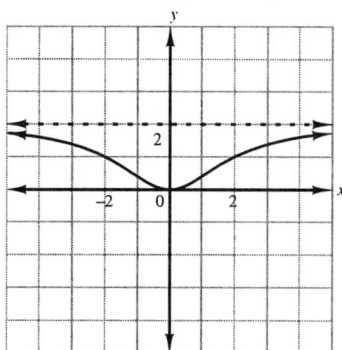

15. $f(x) = \dfrac{2x - 4}{x^2 + x - 6} = \dfrac{2(x - 2)}{(x + 3)(x - 2)} = \dfrac{2}{x + 3}$ $(x \neq 2)$

1.   vertical asymptote:

$$x + 3 = 0$$
$$x = -3 \quad (x \neq 2)$$

2.   values close to asymptote:

x	-4	$-\dfrac{7}{2}$	$-\dfrac{5}{2}$	-2
$f(x) = \dfrac{2}{x + 3}$ $(x \neq 2)$	-2	-4	4	2

3.   horizontal asymptote:

$$f(x) = \dfrac{2}{x + 3} = \dfrac{2}{x + 3} \cdot \dfrac{\dfrac{1}{x}}{\dfrac{1}{x}} = \dfrac{\dfrac{2}{x}}{1 + \dfrac{3}{x}} \quad (x \neq 2)$$

As $|x|$ becomes larger, $\dfrac{2}{x}$ and $\dfrac{3}{x}$ approach 0.

$f(x)$ approaches $\dfrac{0}{1 + 0} = 0$

$y = 0$ is a horizontal asymptote.

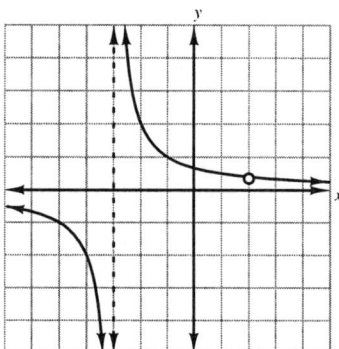

17. $f(x) = \dfrac{x^2}{x+1}$

1. vertical asymptote:

   $x + 1 \;=\; 0$

   $x \;=\; -1$

2. values close to asymptote:

x	-2	0	$-\dfrac{3}{2}$	$-\dfrac{1}{2}$
$f(x) = \dfrac{x^2}{x+1}$	-4	0	$-\dfrac{9}{2}$	$\dfrac{1}{2}$

3. horizontal asymptote: Because the degree of the numerator is greater than the degree of the denominator, the graph has no horizontal asymptote.

4. oblique asymptote:

   $$f(x) \;=\; x - 1 + \frac{1}{x+1}$$

   If $|x|$ is large, $\dfrac{1}{x+1}$ approaches 0.

   $f(x)$ approaches $x - 1 + 0 = x - 1$

   $y = x - 1$ is an oblique asymptote.

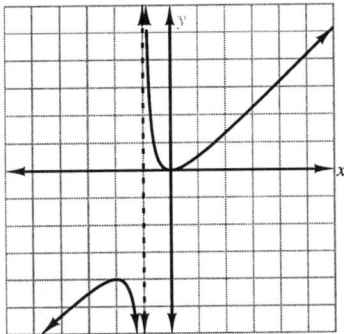

19. $f(x) = \dfrac{x^4}{x^2+2}$

1. vertical asymptote: $x^2 + 2$ cannot equal 0 for any real value of x. No vertical asymptote.

2. horizontal asymptote: Because the degree of the numerator is greater than the degree of the denominator, the graph has no horizontal asymptote.

3. values:

x	0	± 1	± 2
$f(x) = \dfrac{x^4}{x^2+2}$	0	$\dfrac{1}{3}$	$\dfrac{8}{3}$

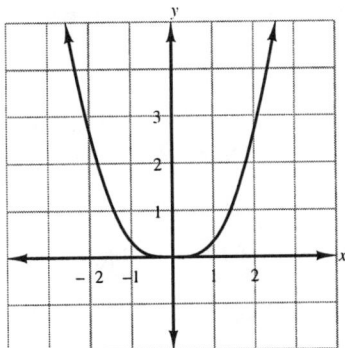

21. $f(x) = \dfrac{x^2 - x - 6}{x - 2}$

$\qquad\quad = \dfrac{(x - 3)(x + 2)}{x - 2}$

1. vertical asymptote:

$\qquad x - 2 = 0$

$\qquad\quad\ x = 2$

2. horizontal asymptote: Because the degree of the numerator is greater than the degree of the denominator, the graph has no horizontal asymptote:

3. values close to asymptote:

x	1	$\dfrac{3}{2}$	$\dfrac{5}{2}$	3
$f(x) = \dfrac{x^2 - x - 6}{x - 2}$	6	$\dfrac{21}{2}$	$-\dfrac{9}{2}$	0

4. oblique asymptote:

$\qquad f(x) = x + 1 - \dfrac{4}{x - 2}$

If $|x|$ is large, $\dfrac{-4}{x - 2}$ approaches 0.

$f(x)$ approaches $x + 1 + 0 = x + 1$.

$y = x + 1$ is an oblique asymptote.

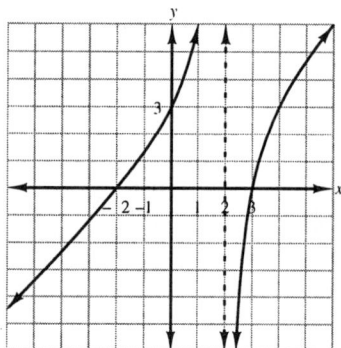

23. $\boxed{\text{D}}$ is true.

25. vertical asymptote: $x = 3$ $\rightarrow$ $y = \dfrac{a}{x - 3}$

horizontal asymptote: $y = 1$ $\rightarrow$ $y = \dfrac{\dfrac{a}{x}}{1 - \dfrac{3}{x}}$ $\rightarrow$ $\dfrac{a}{x} = 1 + \dfrac{b}{x}$

$\qquad\qquad\qquad\qquad\qquad\qquad\qquad\qquad\qquad\qquad\qquad\qquad a = x + b$

$\qquad\qquad\qquad\qquad\qquad\qquad\qquad\qquad y = \dfrac{x + b}{x - 3}$

x–intercept: $(-2, 0)$ $\rightarrow$ $0 = \dfrac{-2 + b}{-2 - 3}$

$\qquad\qquad\qquad\qquad\qquad\qquad\qquad\qquad 0 = -2 + b$

$\qquad\qquad\qquad\qquad\qquad\qquad\qquad\qquad 2 = b$

$\boxed{f(x) = \dfrac{x + 2}{x - 3}}$

Review Problems

29. $x^2 + xy - y^2 = 1$

$x - y = 3$ (solve for y) $\rightarrow$

$y = x - 3$

(substitute for y) $\rightarrow$ $x^2 + x(x-3) - (x-3)^2 = 1$

$x^2 + x^2 - 3x - x^2 + 6x - 9 = 1$

$x^2 + 3x - 10 = 0$

$(x+5)(x-2) = 0$

$x = -5$ or $x = 2$

$y = -5 - 3 = -8$ $y = 2 - 3 = -1$

$$\boxed{\{(-5, -8), (2, -1)\}}$$

30. $\dfrac{(x-2)^2}{9} + \dfrac{(y+1)^2}{25} = 1$

ellipse with center at $(2, -1)$ and vertices

$(2 \pm 3, -1) = (5, -1)$ and $(-1, -1)$ and

$(2, -1 \pm 5) = (2, 4)$ and $(2, -6)$

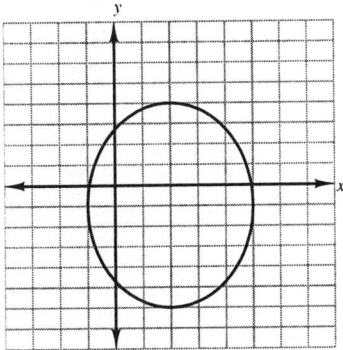

31. $\begin{vmatrix} 1 & y & y^2 \\ y^2 & 1 & y \\ y & y^2 & 1 \end{vmatrix} = 1 \begin{vmatrix} 1 & y \\ y^2 & 1 \end{vmatrix} - y \begin{vmatrix} y^2 & y \\ y & 1 \end{vmatrix} + y^2 \begin{vmatrix} y^2 & 1 \\ y & y^2 \end{vmatrix}$

$= (1 - y^3) - y(y^2 - y^2) + y^2(y^4 - y)$

$= -1(y^3 - 1) - 0 + y^3(y^3 - 1)$

$= (y^3 - 1)(1 - y^3)$

$= (y^3 - 1)^2$

$= \boxed{(y-1)^2(y^2 + y + 1)^2}$

Chapter 10   Review Problems

Review Problems, pp. 770

1. $3x^4 - 9x^3 + 14x^2 - 8x - 4 = 0$

c(factors of -4): $\pm 1, \pm 2, \pm 4$

d(factors of 3): $\pm 1, \pm 3$

c/d: $c/\pm 1 = \pm 1, \pm 2, \pm 4$

$c/\pm 3 = \pm \dfrac{1}{3}, \pm \dfrac{2}{3}, \pm \dfrac{4}{3}$

Possible rational roots: $\boxed{\pm 1, \pm 2, \pm 4, \pm \dfrac{1}{3}, \pm \dfrac{2}{3}, \pm \dfrac{4}{3}}$

2. $3x^4 - 2x^3 - 8x + 5 = 0$

$$P(x) = 3x^4 - 2x^3 - 8x + 5$$
$$\text{2 variations of sign}$$

$\boxed{\text{Positive real roots: 2 or 0}}$

$$P(-x) = 3(-x)^4 - 2(-x)^3 - 8(-x) + 5$$
$$= 3x^4 + 2x^3 + 8x + 5$$
$$\text{no variations of sign}$$

$\boxed{\text{Negative real roots: 0}}$

3. $2x^5 - 3x^3 - 5x^2 + 3x - 1 = 0$

$$P(x) = 2x^5 - 3x^3 - 5x^2 + 3x - 1$$
$$\text{3 variations of sign}$$

$\boxed{\text{Positive real roots: 3 or 1}}$

$$P(-x) = 2(-x)^5 - 3(-x)^3 - 5(-x)^2 + 3(-x) - 1$$
$$= -2x^5 + 3x^3 - 5x^2 - 3x - 1$$
$$\text{2 variations of sign}$$

$\boxed{\text{Negative real roots: 2 or 0}}$

4. $x^3 + 3x^2 - 4 = 0$

a. c (factors of -4): $\pm 1, \pm 2, \pm 4$
d (factors of 1): ± 1
c/d: $\pm 1, \pm 2, \pm 4$

Positive rational solutions: $\boxed{\pm 1, \pm 2, \pm 4}$

b. $$P(x) = x^3 + 3x^2 - 4$$
$$\text{1 variation of sign}$$

$\boxed{\text{Positive real solutions: 1}}$

$$P(-x) = (-x)^3 + 3(-x)^2 - 4$$
$$= -x^3 + 3x^2 - 4$$
$$\text{2 variations of sign}$$

$\boxed{\text{Negative real solutions: 2 or 0}}$

c. $$x^3 + 3x^2 - 4 = 0$$

$$\begin{array}{r|rrrr}
1 & 1 & 3 & 0 & -4 \\
  &   & 1 & 4 & 4 \\
\hline
  & 1 & 4 & 4 & 0
\end{array}$$

$$(x - 1)(x^2 + 4x + 4) = 0$$
$$(x - 1)(x + 2)^2 = 0$$
$$x = 1 \quad \text{or} \quad x = -2$$

$\boxed{\{1, -2\}}$

5. $3x^3 - 4x^2 - 17x + 6 = 0$

    a. c(factors of 6): $\pm 1, \pm 2, \pm 3, \pm 6$

        d(factors of 3): $\pm 1, \pm 3$

        c/d: $c/\pm 1 = \pm 1, \pm 2, \pm 3, \pm 6$

                  $c/\pm 3 = \pm \dfrac{1}{3}, \pm \dfrac{2}{3}, \pm 1, \pm 2$

        Positive rational solutions: $\boxed{\pm 1, \pm 2, \pm 3, \pm 6, \pm \dfrac{1}{3}, \pm \dfrac{2}{3}}$

    b.
$$\begin{aligned} P(x) &= 3x^3 - 4x^2 - 17x + 6 \\ &\quad \text{2 variations of sign} \end{aligned}$$

        $\boxed{\text{Positive real solutions: 2 or 0}}$

$$\begin{aligned} P(-x) &= 3(-x)^3 - 4(-x)^2 - 17(-x) + 6 \\ &= -3x^3 - 4x^2 + 17x + 6 \\ &\quad \text{1 variation of sign} \end{aligned}$$

        $\boxed{\text{Negative real solutions: 1}}$

    c. $3x^3 - 4x^2 - 17x + 6 = 0$

$$\begin{array}{r|rrrr} -2 & 3 & -4 & -17 & 6 \\ & & -6 & 20 & -6 \\ \hline & 3 & -10 & 3 & 0 \end{array}$$

        $(x + 2)(3x^2 - 10x + 3) = 0$

        $(x + 2)(3x - 1)(x - 3) = 0$

                $x = -2$ or $x = \dfrac{1}{3}$ or $x = 3$

        $\boxed{\left\{ -2, \dfrac{1}{3}, 3 \right\}}$

6. $2x^3 - 3x^2 + 6x + 4 = 0$

    a. c (factors of 4) $= \pm 1, \pm 2, \pm 4$

        d (factors of 2) $= \pm 1, \pm 2$

        c/d: $c/\pm 1 = \pm 1, \pm 2, \pm 4$

                  $c/\pm 2 = \pm \dfrac{1}{2}, \pm 1, \pm 2$

        Positive rational solutions: $\boxed{\pm 1, \pm 2, \pm 4, \pm \dfrac{1}{2}}$

    b.
$$\begin{aligned} P(x) &= 2x^2 - 3x^2 + 6x + 4 \\ &\quad \text{2 variations of sign} \end{aligned}$$

        $\boxed{\text{Positive real solutions: 2 or 0}}$

$$\begin{aligned} P(-x) &= 2(-x)^3 - 3(-x)^2 + 6(-x) + 4 \\ &= -2x^3 - 3x^2 - 6x + 4 \\ &\quad \text{1 variations of sign} \end{aligned}$$

        $\boxed{\text{Negative real solutions: 1}}$

c. $2x^3 - 3x^2 + 6x + 4 = 0$

$$-\frac{1}{2}\Big|\quad 2 \quad -3 \quad 6 \quad 4$$
$$\underline{\qquad\qquad -1 \quad 2 \quad -4}$$
$$\qquad\quad 2 \quad -4 \quad 8 \quad 0$$

$(2x + 1)(2x^2 - 4x + 8) = 0$

$$x = -\frac{1}{2} \qquad \text{or} \qquad 2x^2 - 4x + 8 = 0$$

$$x^2 - 2x + 4 = 0$$

$$x = \frac{2 \pm \sqrt{4 - 16}}{2}$$

$$= \frac{2 \pm 2i\sqrt{3}}{2}$$

$$= 1 \pm i\sqrt{3}$$

$$\boxed{\left\{ -\frac{1}{2}, 1 + i\sqrt{3}, 1 - i\sqrt{3} \right\}}$$

7. $x^4 - 5x^3 + 8x - 40 = 0$

a. c (factors of -40) $= \pm 1, \pm 2, \pm 4, \pm 5, \pm 8, \pm 10, \pm 20, \pm 40$

d (factors of 1) $= \pm 1$

c/d: $\pm 1, \pm 2, \pm 4, \pm 5, \pm 8, \pm 10, \pm 20, \pm 40$

Positive rational solutions: $\boxed{\pm 1, \pm 2, \pm 4, \pm 5, \pm 8, \pm 10, \pm 20, \pm 40}$

b. $P(x) = x^4 - 5x^3 + 8x - 40$

   3 variations of sign

   $\boxed{\text{Positive real solutions: 3 or 1}}$

   $P(-x) = (-x)^4 - 5(-x)^3 + 8(-x) - 4$

   $= x^4 + 5x^3 - 8x - 4$

   1 variation of sign

   $\boxed{\text{Negative real solutions: 1}}$

c.
$$x^4 - 5x^3 + 8x - 40 = 0$$
$$x^3(x - 5) + 8(x - 5) = 0$$
$$(x - 5)(x^3 + 8) = 0$$
$$(x - 5)(x + 2)(x^2 - 2x + 4) = 0$$
$$x = 5 \qquad \text{or} \qquad x = -2 \qquad \text{or} \qquad x^2 - 2x + 4 = 0$$

$$x = \frac{2 \pm \sqrt{4 - 16}}{2}$$

$$= \frac{2 \pm 2i\sqrt{3}}{2}$$

$$= 1 \pm i\sqrt{3}$$

$$\boxed{\left\{ -2, 5, 1 - i\sqrt{3}, 1 + i\sqrt{3} \right\}}$$

8. $3x^3 + 4x^2 - 7x + 2 = 0$

a. c (factors of 2) $= \pm 1, \pm 2$
d (factors of 3) $= \pm 1, \pm 3$
c/d: $c/\pm 1 = \pm 1, \pm 2$

$$c/\pm 3 = \pm \frac{1}{3}, \pm \frac{2}{3}$$

Positive rational solutions: $\boxed{\pm 1, \pm 2, \pm \frac{1}{3}, \pm \frac{2}{3}}$

b.
$$P(x) = 3x^3 + 4x^2 - 7x + 2$$
2 variations of sign

$\boxed{\text{Positive real solutions: 2 or 0}}$

$$P(-x) = 3(-x)^3 + 4(-x)^2 - 7(-x) + 2$$
$$= -3x^3 + 4x^2 + 7x + 2$$
1 variation of sign

$\boxed{\text{Negative real solutions: 1}}$

c. $3x^3 + 4x^2 - 7x + 2 = 0$

$$\begin{array}{r|rrrr} \frac{2}{3} & 3 & 4 & -7 & 2 \\ & & 2 & 4 & -2 \\ \hline & 3 & 6 & -3 & 0 \\ & 1 & 2 & -1 & 0 \end{array}$$

$$(3x - 2)(3x^2 + 6x - 3) = 0$$
$$x = \frac{2}{3} \quad \text{or} \quad 3x^2 + 6x - 3 = 0$$
$$x^2 + 2x - 1 = 0$$
$$x = \frac{-2 \pm \sqrt{4 + 4}}{2}$$
$$= \frac{-2 \pm 2\sqrt{2}}{2}$$
$$= -1 \pm \sqrt{2}$$

$$\boxed{\left\{ \frac{2}{3}, -1 + \sqrt{2}, -1 - \sqrt{2} \right\}}$$

9. $4x^4 + 7x^2 - 2 = 0$

a. c (factors of -2): $\pm 1, \pm 2$
d (factors of 4): $\pm 1, \pm 2, \pm 4$
c/d: $c/\pm 1 = \pm 1, \pm 2$

$$c/\pm 2 = \pm \frac{1}{2}, \pm 1$$

$$c/\pm 4 = \pm \frac{1}{4}, \pm \frac{1}{2}, \pm 1$$

Possible rational solutions: $\boxed{\pm 1, \pm 2, \pm \frac{1}{4}, \pm \frac{1}{2}}$

b.
$$P(x) = 4x^4 + 7x^2 - 2$$
1 variation of sign

$\boxed{\text{Positive real solutions: 1}}$

$$P(-x) = 4(-x)^4 + 7(-x)^2 - 2$$
$$= 4x^4 + 7x^2 - 2$$
1 variation of sign

$\boxed{\text{Negative real solutions: 1}}$

c.
$$4x^4 + 7x^2 - 2 = 0$$
$$(4x^2 - 1)(x^2 + 1) = 0$$
$$4x^2 - 1 = 0 \qquad \text{or} \qquad x^2 + 2 = 2$$
$$x = \pm\frac{1}{2} \qquad\qquad\qquad x = \pm i\sqrt{2}$$

$$\boxed{\left\{-\frac{1}{2}, \frac{1}{2}, -\sqrt{2}i, \sqrt{2}i \right\}}$$

10. $2x^4 + x^3 - 9x^2 - 4x + 4 = 0$

a. c (factors of 4): $\pm 1, \pm 2, \pm 4$
d (factors of 2): $\pm 1, \pm 2$
c/d: $c/\pm 1 = \pm 1, \pm 2, \pm 4$

$c/\pm 2 = \pm\frac{1}{2}, \pm 1, \pm 2$

Possible rational solutions: $\boxed{\pm 1, \pm 2, \pm 4, \pm\frac{1}{2}}$

b. $P(x) = 2x^4 + x^3 - 9x^2 - 4x + 4$
2 variations of sign
$\boxed{\text{Positive real solutions: 2 or 0}}$
$$P(-x) = 2(-x)^4 + (-x)^3 - 9(-x)^2 - 4(-x) + 4$$
$$= 2x^4 - x^3 - 9x^2 + 4x + 4$$
2 variations of sign
$\boxed{\text{Negative real solutions: 2 or 0}}$

c. $2x^4 + x^3 - 9x^2 - 4x + 4 = 0$

-1	2	1	-9	-4	4
		-2	1	8	-4
	2	-1	-8	4	0

$$(x + 1)(2x^3 - x^2 - 8x + 4) = 0$$

$\frac{1}{2}$	2	-1	-8	4
		1	0	-4
	2	0	-8	0

$$(x + 1)\left(x - \frac{1}{2}\right)(2x^2 - 8) = 0$$

$$x = -1 \quad \text{or} \quad x = \frac{1}{2} \quad \text{or} \quad 2x^2 - 8 = 0$$
$$x^2 = 4$$
$$x = \pm 2$$

$$\boxed{\left\{-1, \frac{1}{2}, -2, 2\right\}}$$

11. $3x^5 - 2x^4 - 15x^3 + 10x^2 + 12x - 8 = 0$

    a. c (factors of -8): $\pm 1, \pm 2, \pm 4, \pm 8$

        d (factors of 3): $\pm 1, \pm 3$

        c/d: $c/\pm 1 = \pm 1, \pm 2, \pm 4, \pm 8$

               $c/\pm 3 = \pm \dfrac{1}{3}, \pm \dfrac{2}{3}, \pm \dfrac{4}{3}, \pm \dfrac{8}{3}$

        Possible rational solutions: $\boxed{\pm 1, \pm 2, \pm 4, \pm 8, \pm \dfrac{1}{3}, \pm \dfrac{2}{3}, \pm \dfrac{4}{3}, \pm \dfrac{8}{3}}$

    b. $P(x) = 3x^5 - 2x^4 - 15x^3 + 10x^2 + 12x - 8$

              3 variations of sign

      $\boxed{\text{Positive real solutions: 3 or 1}}$

        $P(-x) = 3(-x)^5 - 2(-x)^4 - 15(-x)^3 + 10(-x)^2 + 12(-x) - 8$

                $= -3x^5 - 2x^4 + 15x^3 + 10x^2 - 12x - 8$

              2 variations of sign

      $\boxed{\text{Negative real solutions: 2 or 0}}$

    c. $3x^5 - 2x^4 - 15x^3 + 10x^2 + 12x - 8 = 0$

-2	3	-2	-15	10	12	-8
		-6	16	-2	-16	8
	3	-8	1	8	-4	0

        $(x + 2)(3x^4 - 8x^3 + x^2 + 8x - 4) = 0$

-1	3	-8	1	8	-4
		-3	11	-12	4
	3	-11	12	-4	0

        $(x + 2)(x + 1)(3x^3 - 11x^2 + 12x - 4) = 0$

1	3	-11	12	-4	0
		3	-8	4	
	3	-8	4	0	

        $(x + 2)(x + 1)(x - 1)(3x^2 - 8x + 4) = 0$

        $(x + 2)(x + 1)(x - 1)(3x - 2)(x - 2) = 0$

        $x = -2$ or $x = -1$ or $x = 1$ or $x = \dfrac{2}{3}$ or $x = 2$

        $\boxed{\left\{ -2, -1, 1, 2, \dfrac{2}{3} \right\}}$

12. $f(x) = x^6 - 4x^2 + 3$

      $f(-x) = (-x)^6 - 4(-x)^2 + 3$

             $= x^6 - 4x^2 + 3$

      $f(-x) = f(x)$: $\boxed{\text{symmetric with respect to the } y\text{-axis}}$

13. $f(x) = x^3 - 9x$

      $f(-x) = (-x)^3 - 9(-x)$

             $= -x^3 + 9x$

             $= -(x^3 - 9x)$

      $f(-x) = -f(x)$: $\boxed{\text{symmetric with respect to the origin}}$

14. $f(x) = x^4 - 4x^2 - 2x - 1$

      $f(-x) = (-x)^4 - 4(-x)^2 - 2(-x) - 1$

             $= x^4 - 4x^2 + 2x - 1$

      $f(-x) \neq f(x)$ and

      $f(-x) \neq -f(x)$

    $\boxed{\text{not symmetric about the } y\text{-axis or the origin}}$

15.
$$f(x) = (x^3 - 3)^2$$
$$f(-x) = [(-x)^3 - 3]^2$$
$$= (-x^3 - 3)^2$$
$$= (x^3 + 3)^2$$
$$f(-x) \neq f(x) \quad \text{and}$$
$$f(-x) \neq -f(x)$$

$\boxed{\text{not symmetric about the } y\text{–axis or the origin}}$

16. $f(x) = x^3 - x^2 - 9x + 9$

1. symmetry:
$$f(-x) = (-x)^3 - (-x)^2 - 9(-x) + 9$$
$$= -x^3 - x^2 + 9x + 9$$
$$f(-x) \neq f(x)$$
$$f(-x) \neq -f(x): \text{ not symmetric about the } y\text{–axis or the origin}$$

2. x–intercepts:
$$x^3 - x^2 - 9x + 9 = 0$$
$$x^2(x - 1) - 9(x - 1) = 0$$
$$(x - 1)(x^2 - 9) = 0$$
$$(x - 1)(x + 3)(x - 3) = 0$$
$$x = 1, -3, 3$$

3. intervals

$x < -3$	$-3 < x < 1$	$1 < x < 3$	$x > 3$
-3	1	3	

Interval	Test Value	Sign of $f(x)$	Location of Graph (above or below x-axis)
$x < -3$	-4: $f(-4) = -35$	$-$	below
$-3 < x < 1$	0: $f(0) = 9$	$+$	above
$1 < x < 3$	2: $f(2) = -5$	$-$	below
$x > 3$	4: $f(4) = 21$	$+$	above

4. y–intercept at $(0, 9)$

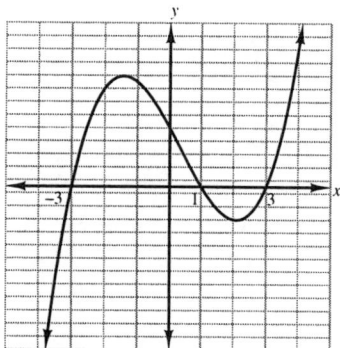

17. $f(x) = 4x - x^3$

1. symmetry:
$$f(-x) = 4(-x) - (-x)^3$$
$$= -4x + x^3$$
$$= -(4x - x^3)$$
$$f(-x) = -f(x): \quad \text{symmtric with respect to the origin}$$

2. x–intercepts:
$$4x - x^3 = 0$$
$$x(4 - x^2) = 0$$
$$x = 0 \quad \text{or} \quad 4 - x^2 = 0$$
$$x = \pm 2$$

3.   intervals

$x < -2$	$-2 < x < 0$	$0 < x < 2$	$x > 2$

$-2 \qquad\qquad 0 \qquad\qquad 2$

Interval	Test Value	Sign of $f(x)$	Location of Graph (above or below x-axis)
$x < -2$	-3: $f(-3) = 15$	$+$	above
$-2 < x < 0$	-1: $f(-1) = -3$	$-$	below
$0 < x < 2$	1: $f(1) = 3$	$+$	above
$x > 2$	3: $f(3) = -15$	$-$	below

4. y–intercept at $(0, 0)$

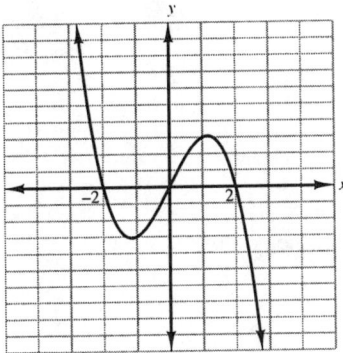

18. $f(x) = x^3 - 3x + 2$

1.   symmetry:

$$f(-x) = (-x)^3 - 3(-x) + 2$$
$$= -x^3 + 3x + 2$$

$$f(-x) \neq f(x) \quad \text{and}$$
$$f(-x) \neq -f(x): \quad \text{not symmetric about the } y\text{–axis or the origin}$$

2. x–intercepts:

$$x^3 - 3x + 2 = 0$$
$$(x + 2)(x^2 - 2x + 1) = 0$$

$$
\begin{array}{r|rrrr}
-2 & 1 & 0 & -3 & 2 \\
   &   & -2 & 4 & -2 \\
\hline
   & 1 & -2 & 1 & 0
\end{array}
$$

$$(x + 2)(x - 1)^2 = 0$$
$$x = -2, 1 \ (\text{double root})$$

3.   intervals

$x < -2$	$-2 < x < 1$	$x > 1$

$-2 \qquad\qquad 1$

Interval	Test Value	Sign of $f(x)$	Location of Graph (above or below x-axis)
$x < -2$	-3: $f(-3) = -16$	$-$	below
$-2 < x < 1$	0: $f(0) = 2$	$+$	above
$x > 1$	2: $f(2) = 4$	$+$	above

4.   *y*–intercept at (0, 2)

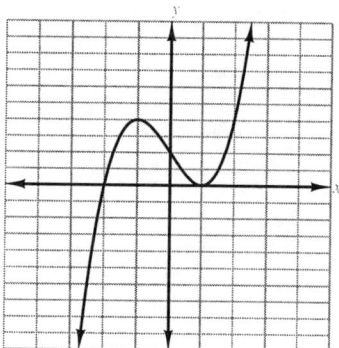

19. $f(x) = x^3 - 2x^2 - 3x$

  1.   symmetry:

$$f(-x) = (-x)^3 - 2(-x)^2 - 3(-x)$$
$$= -x^3 - 2x^2 + 3x$$
$$f(-x) \neq f(x) \quad \text{and}$$
$$f(-x) \neq -f(x): \quad \text{not symmetric about the } y\text{–axis or the origin}$$

  2.   *x*–intercepts:

$$x^3 - 2x^2 - 3x = 0$$
$$x(x^2 - 2x - 3) = 0$$
$$x(x - 3)(x + 1) = 0$$
$$x = 0, 3, -1$$

  3.   intervals

$x < -1$	$-1 < x < 0$	$0 < x < 3$	$x < 3$
-1	0	3	

Interval	Test Value	Sign of $f(x)$	Location of Graph (above or below x-axis)
$x < -1$	-2: $f(-2) = -10$	$-$	below
$-1 < x < 0$	$-\dfrac{1}{2}$: $f\left(-\dfrac{1}{2}\right) = \dfrac{7}{8}$	$+$	above
$0 < x < 3$	1: $f(1) = -4$	$-$	below
$x > 3$	4: $f(4) = 20$	$+$	above

  4.   *y*–intercept at (0, 0)

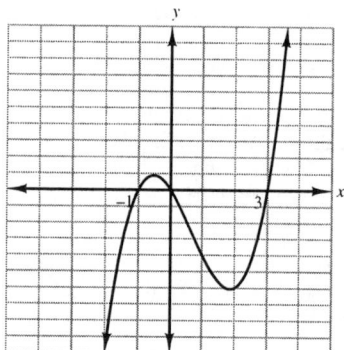

20. $f(x) = x^3 - 4x^2 + x + 6$

1. symmetry:

$$\begin{aligned} f(-x) &= (-x)^3 - 4(-x)^2 + (-x) + 6 \\ &= -x^3 - 4x^2 - x + 6 \end{aligned}$$

$f(-x) \ne f(x)$ and
$f(-x) = -f(x)$:   not symmetric about the y–axis or the origin

2. x–intercepts:

$$x^3 - 4x^2 + x + 6 = 0$$

$$\begin{array}{r|rrrr} 2 & 1 & -4 & 1 & 6 \\ & & 2 & -4 & -6 \\ \hline & 1 & -2 & -3 & 0 \end{array}$$

$$\begin{aligned} (x - 2)(x^2 - 2x - 3) &= 0 \\ (x - 2)(x - 3)(x + 1) &= 0 \\ x &= 2, 3, -1 \end{aligned}$$

3. intervals

$x < -1$		$-1 < x < 2$		$2 < x < 3$		$x > 3$
	-1		2		3	

Interval	Test Value	Sign of $f(x)$	Location of Graph (above or below x-axis)
$x < -1$	-2: $f(-2) = -20$	$-$	below
$-1 < x < 2$	0: $f(0) = 6$	$+$	above
$2 < x < 3$	$2\frac{1}{2}$: $f\left(2\frac{1}{2}\right) = -\frac{7}{8}$	$-$	below
$x > 3$	4: $f(4) = 10$	$+$	above

4. y–intercept at $(0, 6)$.

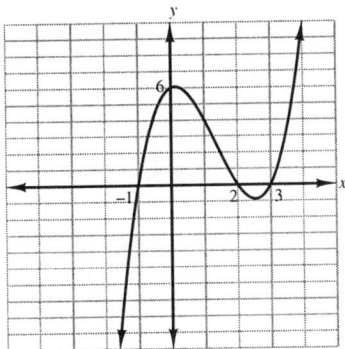

21. $f(x) = x^4 - 2x^3 - 7x^2 + 8x + 12$

1. symmetry:

$$\begin{aligned} f(-x) &= (-x)^4 - 2(-x)^3 - 7(-x)^2 + 8(-x) + 12 \\ &= x^4 + 2x^3 - 7x^2 - 8x + 12 \end{aligned}$$

$f(-x) \ne f(x)$ and
$f(-x) \ne -f(x)$:   not symmetric about the y–axis or the origin

2. x–intercepts:
$$x^4 - 2x^3 - 7x^2 + 8x + 12 = 0$$

$$
\begin{array}{r|rrrrr}
-2 & 1 & -2 & -7 & 8 & 12 \\
   &   & -2 & 8 & -2 & -12 \\
\hline
   & 1 & -4 & 1 & 6 & 0
\end{array}
$$

$$(x + 2)(x^3 - 4x^2 + x + 6) = 0$$

$$
\begin{array}{r|rrrr}
-1 & 1 & -4 & 1 & 6 \\
   &   & -1 & 5 & -6 \\
\hline
   & 1 & -5 & 6 & 0
\end{array}
$$

$$(x + 2)(x + 1)(x^2 - 5x + 6) = 0$$
$$(x + 2)(x + 1)(x - 3)(x - 2) = 0$$
$$x = -2, -1, 3, 2$$

3. intervals

| $x < -2$ | $-2 < x < -1$ | $-1 < x < 2$ | $2 < x < 3$ | $x > 3$ |

$-2 \qquad -1 \qquad 2 \qquad 3$

Interval	Test Value	Sign of $f(x)$	Location of Graph (above or below x-axis)
$x < -2$	-3: $f(-3) = 60$	$+$	above
$-2 < x < -1$	$-1\frac{1}{2}$: $f\left(-1\frac{1}{2}\right) = -\frac{63}{16}$	$-$	below
$-1 < x < 2$	0: $f(0) = 12$	$+$	above
$2 < x < 3$	$2\frac{1}{2}$: $f\left(2\frac{1}{2}\right) = -\frac{63}{16}$	$-$	below
$x > 3$	4: $f(4) = 60$	$+$	above

4. y–intercept at $(0, 12)$

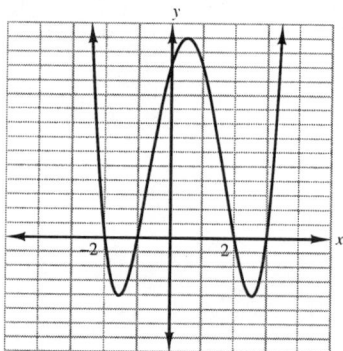

22. $f(x) = 4x^3 - 8x^2 - x + 2$

1. symmetry:
$$
\begin{aligned}
f(-x) &= 4(-x)^3 - 8(-x)^2 - (-x) + 2 \\
&= -4x^3 - 8x^2 + x + 2
\end{aligned}
$$
$$f(-x) \neq f(x) \quad \text{and}$$
$$f(-x) \neq -f(x): \text{ not symmetric about the } y\text{–axis or the origin}$$

2. x–intercepts:
$$
\begin{aligned}
4x^3 - 8x^2 - x + 2 &= 0 \\
4x^2(x - 2) - (x - 2) &= 0 \\
(x - 2)(4x^2 - 1) &= 0 \\
(x - 2)(2x - 1)(2x + 1) &= 0 \\
x &= 2, \frac{1}{2}, -\frac{1}{2}
\end{aligned}
$$

3. intervals

| $x < -\dfrac{1}{2}$ | $-\dfrac{1}{2} < x < \dfrac{1}{2}$ | $\dfrac{1}{2} < x < 2$ | $x > 2$ |

$$-\dfrac{1}{2} \qquad\qquad \dfrac{1}{2} \qquad\qquad 2$$

Interval	Test Value	Sign of $f(x)$	Location of Graph (above or below x-axis)
$x < -\dfrac{1}{2}$	-1: $f(-1) = -9$	$-$	below
$-\dfrac{1}{2} < x < \dfrac{1}{2}$	0: $f(0) = 2$	$+$	above
$\dfrac{1}{2} < x < 2$	1: $f(1) = -3$	$-$	below
$x > 2$	3: $f(3) = 35$	$+$	above

4. y–intercept at $(0, 2)$

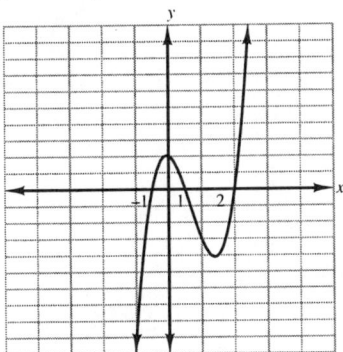

23. $f(x) = 4x^4 - 12x^3 + 9x^2$

1. symmetry:
$$\begin{aligned}
f(-x) &= 4(-x)^4 - 12(-x)^3 + 9(-x)^2 \\
&= 4x^4 + 12x^3 + 9x^2
\end{aligned}$$

$f(-x) \ne f(x)$ and
$f(-x) \ne -f(x)$:   not symmetric about the y–axis or the origin

2. x–intercepts:
$$\begin{aligned}
4x^4 - 12x^3 + 9x^2 &= 0 \\
x^2(4x^2 - 12x + 9) &= 0 \\
x^2(2x - 3)^2 &= 0 \\
x &= 0, \dfrac{3}{2} \quad \text{(both are double roots)}
\end{aligned}$$

3. intervals

| $x < 0$ | $0 < x < \dfrac{3}{2}$ | $x > \dfrac{3}{2}$ |

$$0 \qquad\qquad \dfrac{3}{2}$$

Interval	Test Value	Sign of $f(x)$	Location of Graph (above or below x-axis)
$x < 0$	-1: $f(-1) = -15$	$+$	above
$0 < x < \dfrac{3}{2}$	1: $f(1) = 1$	$+$	above
$x > \dfrac{3}{2}$	2: $f(2) = 4$	$+$	above

4.   *y*–intercept at (0, 0)

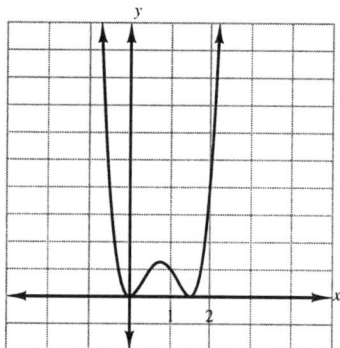

24. $\boxed{\text{C}}$ is true;

graph not symmetric about the *y*–axis or the origin
x–intercepts: $x = -6$, approximately -3.5, -2
y–intercept: approximately $(0, 1.5)$

$$
\begin{aligned}
f(-2) &= 0 \\
f(0) &\approx 1.5 \\
f[f(-2)] &= f(0) \\
&\approx 1.5
\end{aligned}
$$

$0 < f[f(-2)] < 2$

25. $f(x) = \dfrac{x^4}{x^2 - 3}$

$f(-x) = \dfrac{(-x)^4}{(-x)^2 - 3}$

$ = \dfrac{x^4}{x^2 - 3}$

$f(-x) = f(x)$: $\boxed{\text{symmetric with respect to the } y\text{–axis}}$

26. $f(x) = \dfrac{x^2 + 2}{x - 5}$

$f(-x) = \dfrac{(-x)^2 + 2}{(-x) - 5}$

$ = \dfrac{x^2 + 2}{-x - 5}$

$ = -\left(\dfrac{x^2 + 2}{x + 5}\right)$

$f(-x) \neq f(x)$ and

$f(-x) \neq -f(x)$: $\boxed{\text{not symmetric about the } y\text{–axis or the origin}}$

27. $f(x) = \dfrac{x - 1}{x + 3}$

$f(-x) = \dfrac{-x - 1}{-x + 3}$

$ = \dfrac{x + 1}{x - 3}$

$f(-x) \neq f(x)$ and

$f(-x) \neq -f(x)$: $\boxed{\text{not symmetric about the } y\text{–axis or the origin}}$

28. $f(x) = \dfrac{2x+1}{x-1}$

1. vertical asymptote:

$$x - 1 = 0$$
$$x = 1$$

2. values close to asymptote:

x	0	2	$\dfrac{1}{2}$	$\dfrac{3}{2}$
$f(x) = \dfrac{2x+1}{x-1}$	-1	5	-4	8

3. horizontal asymptotes:

$$f(x) = \frac{2x+1}{x-1} = \frac{2x+1}{x-1} \cdot \frac{\dfrac{1}{x}}{\dfrac{1}{x}} = \frac{2+\dfrac{1}{x}}{1-\dfrac{1}{x}}$$

As $|x|$ becomes larger, $\dfrac{1}{x}$ approaches 0.

$f(x)$ approaches $\dfrac{2+0}{1-0} = 2$

$y = 2$ is a horizontal asymptote.

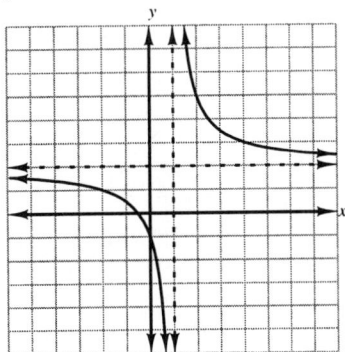

29. $f(x) = \dfrac{2x-3}{x^2-9}$

1. vertical asymptotes:

$$x^2 - 9 = 0$$
$$x = \pm 3$$

2. values close to asymptote:

x	-4	-2	0	2	4
$f(x) = \dfrac{2x-3}{x^2-9}$	$-\dfrac{11}{7}$	$\dfrac{7}{5}$	$\dfrac{1}{3}$	$-\dfrac{1}{5}$	$\dfrac{5}{7}$

3. horizontal asymptote:

$$f(x) = \frac{2x-3}{x^2-9} = \frac{2x-3}{x^2-9} \cdot \frac{\frac{1}{x^2}}{\frac{1}{x^2}} = \frac{\frac{2}{x}-\frac{3}{x^2}}{1-\frac{9}{x^2}}$$

As $|x|$ becomes larger, $\frac{2}{x}$, $\frac{3}{x^2}$, and $\frac{9}{x^2}$ approach 0.

$f(x)$ approaches $\frac{0-0}{1-0} = 0$

$y = 0$ is a horizontal asymptote.

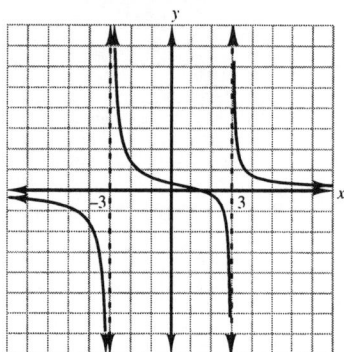

30. $f(x) = \dfrac{4x^2}{x^2-1}$

1. vertical asymptotes:
    $$x^2 - 1 = 0$$
    $$x = \pm 1$$

2. values close to asymptote:

x	± 2	0	$\pm\dfrac{3}{2}$	$\pm\dfrac{1}{2}$
$f(x) = \dfrac{4x^2}{x^2-1}$	$\dfrac{16}{3}$	0	$\dfrac{36}{5}$	$-\dfrac{4}{3}$

3. horizontal asymptote:

$$f(x) = \frac{4x^2}{x^2-1} = \frac{4x^2}{x^2-1} \cdot \frac{\frac{1}{x^2}}{\frac{1}{x^2}} = \frac{4}{1-\frac{1}{x^2}}$$

As $|x|$ becomes larger, $\frac{1}{x^2}$ approaches 0.

$f(x)$ approaches $\frac{4}{1-0} = 4$

$y = 4$ is a horizontal asymptote.

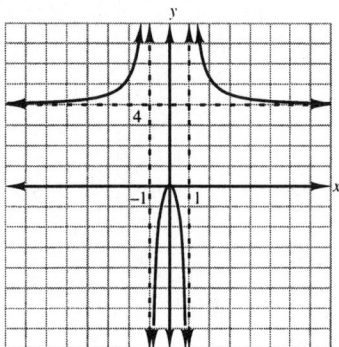

31. $f(x) = \dfrac{x^2}{x^2 + 1}$

1. vertical asymptotes: $x^2 + 1$ cannot equal 0 for any real value of x.
   No vertical asymptote.

2. values:

x	± 2	± 1	0
$f(x) = \dfrac{x^2}{x^2+1}$	$\dfrac{4}{5}$	$\dfrac{1}{2}$	0

3. horizontal asymptote:

$$f(x) \;=\; \frac{x^2}{x^2+1} = \frac{x^2}{x^2+1}\cdot\frac{\frac{1}{x^2}}{\frac{1}{x^2}} = \frac{1}{1+\frac{1}{x^2}}$$

As $|x|$ becomes larger, $\dfrac{1}{x^2}$ approaches 0.

$f(x)$ approaches $\dfrac{1}{1+0} = 1$.

$y = 1$ is a horizontal asymptote.

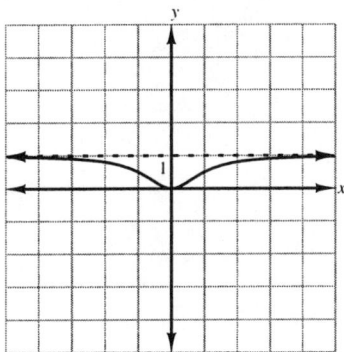

32. $f(x) = \dfrac{x-2}{x^2 - 2x + 3}$

1. vertical asymptote:
$$x^2 - 2x + 3 = 0$$
$$x = \frac{2 \pm \sqrt{4-12}}{2}$$
$$= \frac{2 \pm 2i\sqrt{2}}{2}$$
$$= 1 \pm i\sqrt{2}$$

$x^2 - 2x + 3$ cannot equal 0 for any real value of x.
No vertical asymptote.

2. values

x	0	1	2	3	4
$f(x) = \dfrac{x-2}{x^2-2x+3}$	$-\dfrac{2}{3}$	$-\dfrac{1}{2}$	0	$\dfrac{1}{6}$	$\dfrac{2}{11}$

3. horizontal asymptote:

$$f(x) = \frac{x-2}{x^2 - 2x + 3}$$

$$= \frac{x-2}{x^2 - 2x + 3} \cdot \frac{\frac{1}{x^2}}{\frac{1}{x^2}}$$

$$= \frac{\frac{1}{x} - \frac{2}{x^2}}{1 - \frac{2}{x} + \frac{3}{x^2}}$$

As $|x|$ becomes larger, $\frac{1}{x}$, $\frac{2}{x^2}$, $\frac{2}{x}$, and $\frac{3}{x^2}$ approach 0.

$f(x)$ approaches $\frac{0-0}{1-0+0} = 0$

$y = 0$ is a horizontal asymptote.

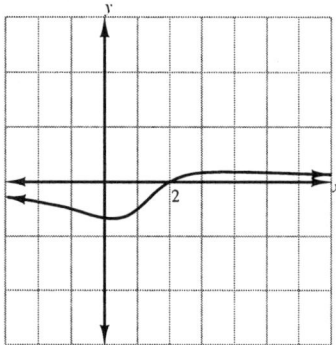

33. $f(x) = \frac{x^4}{x^2 + 2}$

1. vertical asymptote: $x^2 + 2$ cannot equal 0 for any real value of x.
   No vertical asymptote.
2. values

x	± 2	± 1	0
$f(x) = \frac{x^4}{x^2 + 2}$	$\frac{8}{3}$	$\frac{1}{3}$	0

3. horizontal asymptote: Because the degree of the numerator is greater than the degree of the denominator, the graph has no horizontal asymptote.

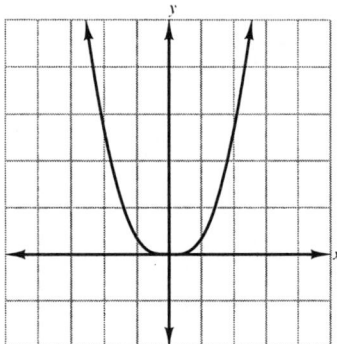

34. $f(x) = \dfrac{x^2 + x}{x - 1}$

1. vertical asymptote:
$$x - 1 = 0$$
$$x = 1$$

2. values:

x	0	$\dfrac{1}{2}$	$\dfrac{3}{2}$	2
$f(x) = \dfrac{x^2 + x}{x - 1}$	0	$-\dfrac{3}{2}$	$\dfrac{15}{2}$	6

3. horizontal asymptote: Because the degree of the numerator is greater than the degree of the denominator, the graph has no horizontal asymptote.

4. oblique asymptote:
$$f(x) = x + 2 + \frac{2}{x - 1}$$

If $|x|$ is large, $\dfrac{2}{x - 1}$ approaches 0.

$f(x)$ approaches $x + 2 + 0 = x + 2$
$y = x + 2$ is an oblique asymptote.

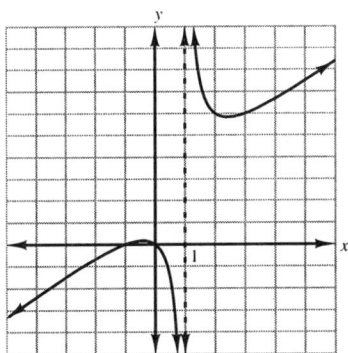

35. $f(x) = \dfrac{x^3 - 1}{x^2 - 4}$

1. vertical asymptotes:
$$x^2 - 4 = 0$$
$$x = \pm 2$$

2. values

x	-3	-1	0	1	3
$f(x) = \dfrac{x^3 - 1}{x^2 - 4}$	$-\dfrac{28}{5}$	$\dfrac{2}{3}$	$\dfrac{1}{4}$	0	$\dfrac{26}{5}$

3. horizontal asymptote: Because the degree of the numerator is greater than the degree of the denominator, the graph has no horizontal asymptote.

4.   oblique asymptote:

$$f(x) = x + \frac{4x - 1}{x^2 - 4}$$

$$= x + \frac{\dfrac{4}{x} - \dfrac{1}{x^2}}{1 - \dfrac{4}{x^2}}$$

If $\left| x \right|$ is large, $\dfrac{4}{x}$, $\dfrac{1}{x^2}$, $\dfrac{4}{x^2}$, and $\dfrac{4x - 1}{x^2 - 4}$ approaches 0.

$f(x)$ approaches $x + 0 = x$

$y = x$ is an oblique asymptote.

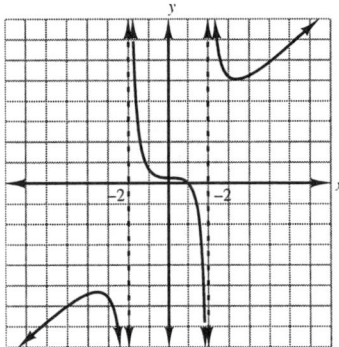

Cumulative Review Problems (Chapters 1-10)

Cumulative Review, p. 771

1. $x^3 - 4x^2 - 10x + 4 = 0$

$$\begin{array}{r|rrrr} -2 & 1 & -4 & -10 & 4 \\ & & -2 & 12 & -4 \\ \hline & 1 & -6 & 2 & 0 \end{array}$$

$(x + 2)(x^2 - 6x + 2) = 0$

$x = -2$ or $x^2 - 6x + 2 = 0$

$$x = \frac{6 \pm \sqrt{36 - 8}}{2}$$

$$= \frac{6 \pm 2\sqrt{7}}{2}$$

$$= 3 \pm \sqrt{7}$$

$$\boxed{\left\{ -2, 3 - \sqrt{7}, 3 + \sqrt{7} \right\}}$$

2.
$$\frac{x-4}{x+6} \geq 0$$
$$x = 4 \quad \text{or} \quad x = -6$$

T	F	T

$$\;-6\qquad 4$$

Test value -7:
$$\frac{-7-4}{-7+6} = 0$$
$$\frac{-11}{-1} \geq 0$$
$$11 \geq 0 \quad \text{True}$$

Test value 0:
$$\frac{0-4}{0+6} \geq 0$$
$$-\frac{2}{3} \geq 0 \quad \text{False}$$

Test value 5:
$$\frac{5-4}{5+6} \geq 0$$
$$\frac{1}{11} \geq 0 \quad \text{True}$$

$x < -6 \ (x \neq -6)$ or $x \geq 4$

$$\boxed{\{x \mid x < -6 \text{ or } x \geq 4\} = (-\infty, -6) \cup [4, \infty)}$$

3.
$$\begin{aligned} x + y - z &= -2 \quad (1) \\ -3x + 2y + 5z &= -7 \quad (2) \\ 2x - 3y + 4z &= 17 \quad (3) \end{aligned}$$

(Equations 1 and 2):
$$\begin{aligned} (\times 5) \quad 5x + 5y - 5z &= -10 \\ \underline{-3x + 2y + 5z} &= \underline{-7} \\ 2x + 7y &= -17 \quad \text{(Equation 4)} \end{aligned}$$

(Equations 1 and 3):
$$\begin{aligned} (\times 4) \quad 4x + 4y - 4z &= -8 \\ \underline{2x - 3y + 4z} &= \underline{17} \\ 6x + y &= 9 \quad \text{(Equation 5)} \end{aligned}$$

(Equations 4 and 5):
$$\begin{aligned} 2x + 7y &= -17 \\ (\times -7) \quad \underline{-42x - 7y} &= \underline{-63} \\ -40x &= -80 \\ x &= 2 \end{aligned}$$

(Equation 5):
$$\begin{aligned} 6(2) + y &= 9 \\ y &= -3 \end{aligned}$$

(Equation 1):
$$\begin{aligned} 2 - 3 - z &= -2 \\ -z &= -1 \\ z &= 1 \end{aligned}$$

$$\boxed{\{(2, -3, 1)\}}$$

4. $2x + y = 3$ (1)
           $y = 2x^2 - 1$ (2)

(Substitute for y into Equation 1):

$$2x + 2x^2 - 1 = 3$$
$$2x^2 + 2x - 4 = 0$$
$$x^2 + x - 2 = 0$$
$$(x + 2)(x - 1) = 0$$

$x = -2$ or $x = 1$
$y = 3 - 2(-2)$ $y = 3 - 2(1) = 1$
$= 7$

$$\boxed{\{(-2, 7), (1, 1)\}}$$

5.

$$2 + \frac{4}{x - 2} = \frac{8}{x^2 - 2x}$$

$$x(x - 2)\left[2 + \frac{4}{x - 2}\right] = x(x - 2) \cdot \frac{8}{x(x - 2)}$$

$$2x^2 - 4x + 4x = 8$$
$$2x^2 - 8 = 0 \quad (x \neq 0, 2)$$
$$x^2 - 4 = 0$$

$x = -2$ or $x = 2$ (extraneous); causes division by 0 in given equation

$$\boxed{\{-2\}}$$

6. $\sqrt{x} + 6 = x$
           $(\sqrt{x})^2 = (x - 6)^2$
           $x = x^2 - 12x + 36$
           $0 = x^2 - 13x + 36$
           $0 = (x - 9)(x - 4)$
           $x = 9$ $x = 4$ extraneous

$$\boxed{\{9\}}$$

7. $x^{2/3} + 3x^{1/3} - 18 = 0$
           $t^2 + 3t - 18 = 0$ (Let $t = x^{1/3}$)
           $(t + 6)(t - 3) = 0$

$t = -6$ or $t = 3$
$x^{1/3} = -6$ or $x^{1/3} = 3$
$x = -216$ $x = 27$

$$\boxed{\{-216, 27\}}$$

8. $3x^2 + 2y^2 = 7$ $(\times 7) \rightarrow$ $21x^2 + 14y^2 = 49$
           $5x^2 - 7y^2 = -9$ $(\times 2) \rightarrow$ $\underline{10x^2 - 14y^2 = -18}$
                                                            $31x^2 = 31$
                                                            $x^2 = 1$
                                                            $x = \pm 1$

$$3(\pm 1)^2 + 2y^2 = 7$$
$$2y^2 = 4$$
$$y^2 = 2$$
$$y = \pm\sqrt{2}$$

If $x = 1, y = \pm\sqrt{2}$
If $x = -1, y = \pm\sqrt{2}$

$$\boxed{\left\{\left(1, \sqrt{2}\right), \left(1, -\sqrt{2}\right), \left(-1, \sqrt{2}\right), \left(-1, -\sqrt{2}\right)\right\}}$$

9.

$$\dfrac{1-\dfrac{8}{x^2-1}}{\dfrac{2}{x+1}-\dfrac{1}{x-1}} = \dfrac{1-\dfrac{8}{(x-1)(x+1)}}{\dfrac{2}{x+1}-\dfrac{1}{x-1}}\cdot\dfrac{(x-1)(x+1)}{(x-1)(x+1)}$$

$$= \dfrac{x^2-1-8}{2(x-1)-(x+1)}$$

$$= \dfrac{x^2-9}{2x-2-x-1}$$

$$= \dfrac{(x-3)(x+3)}{x-3}$$

$$= \boxed{x+3 \quad (x\neq 3)}$$

10.

$$\dfrac{x-5}{x^2-x-2}-\dfrac{2}{x+1}+\dfrac{5}{x-2} = \dfrac{x-5}{(x-2)(x+1)}-\dfrac{2}{x+1}\cdot\dfrac{x-2}{x-2}+\dfrac{5}{x-2}\cdot\dfrac{x+1}{x+1}$$

$$= \dfrac{x-5-2x+4+5x+5}{(x-2)(x+1)}$$

$$= \dfrac{4x+4}{(x-2)(x+1)}$$

$$= \dfrac{4(x+1)}{(x-2)(x+1)}$$

$$= \boxed{\dfrac{4}{x-2} \quad (x\neq -1, 2)}$$

11.

$$\dfrac{5i}{3+4i} = \dfrac{5i}{3+4i}\cdot\dfrac{3-4i}{3-4i}$$

$$= \dfrac{15i-20i^2}{9-16i^2}$$

$$= \dfrac{20+15i}{25}$$

$$= \dfrac{4+3i}{5}$$

$$= \boxed{\dfrac{4}{5}+\dfrac{3}{5}i}$$

12.

$$\dfrac{9x}{\sqrt[3]{3x^2}} = \dfrac{9x}{\sqrt[3]{3x^2}}\cdot\dfrac{\sqrt[3]{9x}}{\sqrt[3]{9x}}$$

$$= \dfrac{9x\sqrt[3]{9x}}{3x}$$

$$= \boxed{3\sqrt[3]{9x}}$$

13.

$$\left(\dfrac{-4x^5y^7}{2xy^9}\right)^{-3} = (-2x^4y^{-2})^{-3}$$

$$= (-2)^{-3}x^{-12}y^6$$

$$= \boxed{\dfrac{-y^6}{8x^{12}}}$$

14.

$$f(x) = \dfrac{-3}{x}$$

$$\dfrac{f(a+h)-f(a)}{h} = \dfrac{\dfrac{-3}{a+h}+\dfrac{3}{a}}{h}$$

$$= \dfrac{\dfrac{-3}{a+h}+\dfrac{3}{a}}{h}\cdot\dfrac{a(a+h)}{a(a+h)}$$

$$= \dfrac{-3a+3(a+h)}{ha(a+h)}$$

$$= \dfrac{-3a+3a+3h}{ha(a+h)}$$

$$= \dfrac{3h}{ha(a+h)}$$

$$= \boxed{\dfrac{3}{a(a+h)}}$$

15. $x^6 + 7x^3 - 8 = (x^3 + 8)(x^3 - 1)$
$$= \boxed{(x+2)(x^2 - 2x + 4)(x-1)(x^2 + x + 1)}$$

16. $f(x) = \begin{cases} 4 - x & \text{if } x \le 1 \\ 3 & \text{if } x > 1 \end{cases}$

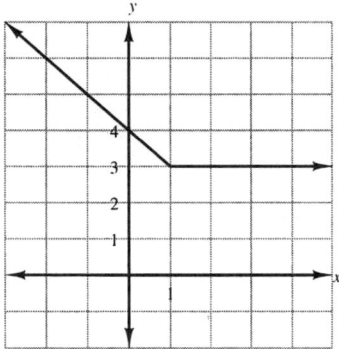

17. $\dfrac{x^2}{16} - \dfrac{y^2}{9} = 1$

hyperbola with center (0, 0) and vertices (±4, 0), no *y*–intercept

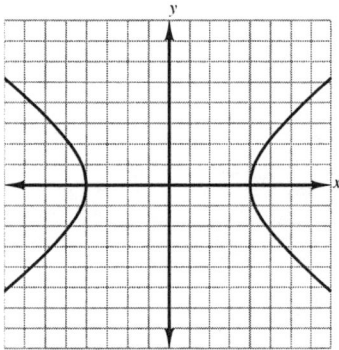

18. $\dfrac{(x-5)^2}{9} + \dfrac{(y-1)^2}{4} = 1$

ellipse with center (5, 1) and vertices
$$(5 \pm 3, 1) = (8, 1) \text{ and } (2, 1)$$
$$\text{and}\quad (5, 1 \pm 2) = (5, 3) \text{ and } (5, -1)$$

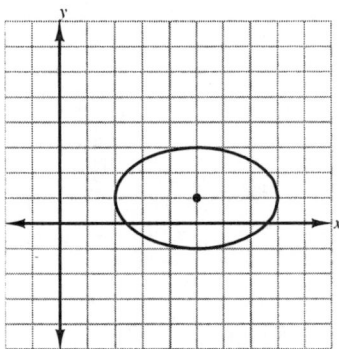

19. $f(x) = x^3 - 8x^2 + 19x - 12$
  x–intercepts:
  $$x^3 - 8x^2 + 19x - 12 = 0$$

$$
\begin{array}{r|rrrr}
1 & 1 & -8 & 19 & -12 \\
  &   & 1 & -7 & 12 \\
\hline
  & 1 & -7 & 12 & 0
\end{array}
$$

  $$(x - 1)(x^2 - 7x + 12) = 0$$
  $$(x - 1)(x - 3)(x - 4) = 0$$
  $$x = 1, 3, 4$$

intervals:

$x < 1$	$1 < x < 3$	$3 < x < 4$	$x > 4$
1	3	4	

Interval	Test Value
$x < 1$	0: $f(0) = -12$
$1 < x < 3$	2: $f(2) = 2$
$3 < x < 4$	$3\frac{1}{2}$: $f\left(3\frac{1}{2}\right) = -\frac{5}{8}$
$x > 4$	5: $f(5) = 8$

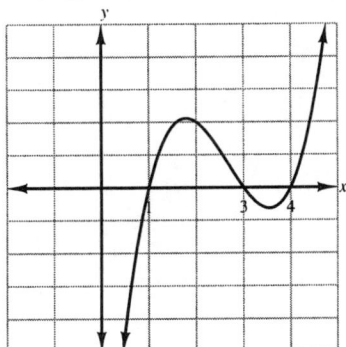

20. $f(x) = \dfrac{2x}{x+1}$

vertical asymptote:

$$x + 1 = 0$$
$$x = -1$$

horizontal asymptote:

$$f(x) = \dfrac{2x}{x+1} = \dfrac{2}{1 + \dfrac{1}{x}} \qquad \rightarrow \qquad \dfrac{2}{1 + 0} = 2$$

$y = 2$ is a horizontal asymptote

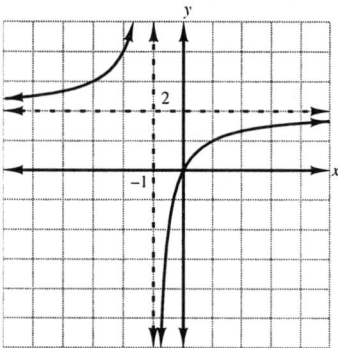

21. Let x = number of nickels

$18 - x$ = number of dimes

$$
\begin{aligned}
(\text{value of nickels}) &= (\text{value of dimes}) \\
0.05x &= 0.10(18 - x) \\
0.05x &= 1.8 - 0.10x \\
0.15x &= 1.8 \\
x &= 12 \quad \text{(nickels)} \\
18 - x &= 18 - 12 \\
&= 6 \quad \text{(dimes)}
\end{aligned}
$$

$\boxed{12 \text{ nickels, } 6 \text{ dimes}}$

22. Let x = side of square

$x + 6$ = length of rectangle

$x - 4$ = width of rectangle

$$
\begin{aligned}
\text{area of square} &= \text{area of rectangle} \\
x^2 &= (x + 6)(x - 4) \\
x^2 &= x^2 + 2x - 24 \\
0 &= 2x - 24 \\
12 &= x \quad \text{(square)} \\
x + 6 &= 18 \quad \text{(length)} \\
x - 4 &= 8 \quad \text{(width)}
\end{aligned}
$$

Dimensions:

Square: $12 \text{ cm} \times 12 \text{ cm}$
Rectangle: $18 \text{ cm} \times 8 \text{ cm}$

23. Let x = number of chickens (with 2 feet)
$52 - x$ = number of horses (with 4 feet)

$$
\begin{aligned}
2x + 4(52 - x) &= 132 \\
2x + 208 - 4x &= 132 \\
-2x &= -76 \\
x &= 38 \quad \text{(chickens)} \\
52 - 38 &= 14 \quad \text{(horses)}
\end{aligned}
$$

$\boxed{38 \text{ chickens, } 14 \text{ horses}}$

24. x = speed of current
$10 - x$ = speed of boat against current (upstream)
$10 + x$ = speed of boat with current (downstream)

	Rate	Distance (miles)	Time $= \dfrac{\text{Distance}}{\text{Rate}}$
upstream	$10 - x$	8	$\dfrac{8}{10-x}$
downstream	$10 + x$	8	$\dfrac{8}{10+x}$

$$
\begin{aligned}
\text{time upstream + time downstream} &= 2 \text{ hours} \\
\frac{8}{10-x} + \frac{8}{10+x} &= 2 \\
(10-x)(10+x)\left(\frac{8}{10-x} + \frac{8}{10+x}\right) &= (10-x)(10+x) \cdot 2 \\
8(10 + x) + 8(10 - x) &= 2(100 - x^2) \\
80 + 8x + 80 - 8x &= 200 - 2x^2 \\
2x^2 &= 40 \\
x^2 &= 20 \\
x &= 2\sqrt{5} \quad \left(\text{reject } -2\sqrt{5}\right)
\end{aligned}
$$

speed of current: $\boxed{2\sqrt{5} \text{ mph}}$

25. Let x = length of one leg of right triangle
$2x + 2$ = length of one other leg of right triangle

$$
\begin{aligned}
x^2 + (2x + 2)^2 &= 13^2 \\
x^2 + 4x^2 + 8x + 4 &= 169 \\
5x^2 + 8x - 165 &= 0 \\
(x - 5)(5x + 33) &= 0 \\
x = 5 \quad \text{or} \quad x &= -\frac{33}{5} \quad \text{(reject)} \\
2x + 2 &= 10 + 2 \\
&= 12
\end{aligned}
$$

$\boxed{5 \text{ in. and } 12 \text{ in.}}$

26. Let x = cost of one bath towel
    y = cost of one hand towel

$$
\begin{array}{llll}
3x + 2y = 31 & (\times -5) \rightarrow & -15x - 10y = -155 \\
2x + 5y = 39 & (\times 2) \rightarrow & \underline{4x + 10y = 78} \\
& & -11x = -77 \\
& & x = 7 \quad \text{(bath towel)}
\end{array}
$$

$$
\begin{array}{rl}
3(7) + 2y &= 37 \\
21 + 2y &= 31 \\
2y &= 10 \\
y &= 5 \quad \text{(hand towel)}
\end{array}
$$

Hand towel: \$5; Bath towel: \$7

27. Line passing through $(-2, -3)$ and perpendicular to line whose equation is $3x + 4y = 1$.
    slope of $3x + 4y = 1$

$$
y = -\frac{3}{4}x + \frac{1}{4}
$$

$$
\text{slope} = -\frac{3}{4}
$$

slope of line perpendicular: $\dfrac{-1}{-\dfrac{3}{4}} = \dfrac{4}{3}$

Equation of line:

$$
\begin{array}{rl}
y + 3 &= \frac{4}{3}(x + 2) \\[2mm]
y + 3 &= \frac{4}{3}x + \frac{8}{3} \\[2mm]
y &= \frac{4}{3}x - \frac{1}{3} \\[2mm]
3y &= 4x - 1
\end{array}
$$

$4x - 3y - 1 = 0$

28.

$$
\begin{array}{rl}
f(x) &= 3x - 2 \quad \text{and} \\
g(x) &= x^2 - 3x + 1 \\
f(-2) &= -6 - 2 = -8 \\
g(-1) &= (-1)^2 - 3(-1) + 1 = 1 + 3 + 1 = 5 \\
[f(-2) - g(-1)]^2 &= (-8 - 5)^2 \\
&= (-13)^2 \\
&= \boxed{169}
\end{array}
$$

29.
$$f(x) = 3x - 7$$
$$y = 3x - 7$$
$$x = 3y - 7 \quad \text{(Exchange } x \text{ and } y\text{)}$$
$$x + 7 = 3y$$
$$\frac{x + 7}{3} = y$$

$$\boxed{f^{-1}(x) = \frac{x + 7}{3}}$$

30. Let

x = number of hours person sleeps

c = amount of coffee

$$x = \frac{k}{c^2}$$

Given: $c = 2, x = 8$

$$8 = \frac{k}{4}$$
$$32 = k$$

$$x = \frac{32}{c^2}$$

If $c = 2(2) = 4$ find x

$$x = \frac{32}{4^2} = \frac{32}{16} = 2$$

$\boxed{2 \text{ hours of sleep}}$

Algebra for College Students

Chapter 11   Exponential and Logarithmic Functions

Section 11.1   Inverse Functions

Problem Set 11.1, pp. 749-752

1. $f(x) = 3x + 5$

    a.
$$y = 3x + 5$$
$$x = 3y + 5 \quad \text{(Exchange } x \text{ and } y.)$$
$$y = \frac{x-5}{3} \quad \text{(Solve for } y.)$$
$$\boxed{f^{-1}(x) - \frac{x-5}{3}} \quad \text{(Let } f^{-1}(x) = y.)$$

    b. $f[f^{-1}(x)] = 3\left(\dfrac{x-5}{3}\right) + 5 = x - 5 + 5 = x$

        $f^{-1}[f(x)] = \dfrac{(3x+5)-5}{3} = \dfrac{3x}{3} = x$

3. **a.** $f(x) = 7x - 2 = y$
$$y = 7x - 2$$
$$7y - 2 = x \quad \text{(Exchange } x \text{ and } y)$$
$$y = \frac{x+2}{7}$$
$$\boxed{f^{-1}(x) = \frac{x+2}{7}}$$

    b. $f[f^{-1}(x)] = \dfrac{7(x+2)}{7} - 2 = x$

        $f^{-1}[f(x)] = \dfrac{(7x-2)+2}{7} = x$

5. **a.** $f(x) = \dfrac{1}{3}x = y$
$$y = \frac{1}{3}x$$
$$\frac{1}{3}y = x \quad \text{(Exchange } x \text{ and } y)$$
$$y = 3x$$
$$\boxed{f^{-1}(x) = 3x}$$

    b. $f[f^{-1}(x)] = \dfrac{1}{3}(3x) = x$

        $f^{-1}[f(x)] = 3\left(\dfrac{1}{3}x\right) = x$

7. **a.** $f(x) = \dfrac{1}{2}x + 3$
$$y = \frac{1}{2}x + 3$$
$$\frac{1}{2}y + 3 = x \quad \text{(Exchange } x \text{ and } y)$$
$$y = 2x - 6$$
$$\boxed{f^{-1}(x) = 2x - 6}$$

    b. $f[f^{-1}(x)] = \dfrac{1}{2}[2(x-3)] + 3 = x$

        $f^{-1}[f(x)] = 2\left(\dfrac{1}{2}x + 3 - 3\right) = x$

9. **a.** $f(x) = \dfrac{3}{4}x - \dfrac{2}{3} = y$
$$y = \frac{3}{4}x - \frac{2}{3} \quad \text{(Exchange } x \text{ and } y)$$
$$\frac{3}{4}y - \frac{2}{3} = x$$
$$y = \frac{4}{3}x + \frac{8}{9}$$
$$\boxed{f^{-1}(x) = \frac{4}{3}x + \frac{8}{9}}$$

$$f[f^{-1}(x)] = \frac{3}{4}\left[\frac{1}{9}(12x+8)\right] - \frac{2}{3} = x + \frac{2}{3} - \frac{2}{3} = x$$
$$f^{-1}[f(x)] = \frac{1}{9}\left[12\left(\frac{3}{4}x - \frac{2}{3}\right) + 8\right]$$
$$= \frac{1}{9}[9x - 8 + 8] = x$$

11. a. $f(x) = \sqrt{2x+3}$

$$y = \sqrt{2x+3}$$
$$y^2 = 2x+3$$
$$x^2 = 2y+3 \quad \text{(Exchange } x \text{ and } y\text{)}$$
$$x^2 - 3 = 2y$$
$$\frac{x^2-3}{2} = y$$

$$\boxed{f^{-1}(x) = \frac{x^2-3}{2}}$$

b. $f[f^{-1}(x)] = \sqrt{2\left(\frac{x^2-3}{2}\right)+3} = \sqrt{x^2} = x$

$f^{-1}[f(x)] = \dfrac{(\sqrt{2x+3})^2 - 3}{2}$

13. a. $f(x) = \sqrt[3]{x+1}$

$$y = \sqrt[3]{x+1}$$
$$y^3 = x+1$$
$$x^3 = y+1 \quad \text{(Exchange } x \text{ and } y\text{)}$$
$$y = x^3 - 1$$

$$\boxed{f^{-1}(x) = x^3 - 1}$$

b. $f[f^{-1}(x)] = \sqrt[3]{x^3-1+1} = x$

$f^{-1}[f(x)] = (\sqrt[3]{x+1})^3 - 1 = x+1-1 = x$

15. a. $f(x) = x^3$

$$y = x^3$$
$$x = y^3 \quad \text{(Exchange } x \text{ and } y\text{)}$$
$$y = x^{1/3}$$

$$\boxed{f^{-1}(x) = x^{1/3}}$$

b. $f[f^{-1}(x)] = (x^{1/3})^3 = x$
$f^{-1}[f(x)] = (x^3)^{1/3} = x$

17. a. $f(x) = x^5 - 1$

$$y = x^5 - 1$$
$$\text{Inverse: } x = y^5 - 1$$
$$x + 1 = y^5$$
$$\sqrt[5]{(x+1)} = y$$

$$\boxed{f^{-1}(x) = \sqrt[5]{(x+1)}}$$

b. $\begin{aligned} f[f^{-1}(x)] &= [f^{-1}(x)]^5 - 1 = [\sqrt[5]{(x+1)}]^5 - 1 \\ &= x+1-1 = x \end{aligned}$

$\begin{aligned} f^{-1}[f(x)] &= \sqrt[5]{f(x)+1} \\ &= \sqrt[5]{(x^5-1)+1} \\ &= \sqrt[5]{(x^5)} = x \end{aligned}$

19. a. $f(x) = mx + b$

$y = mx + b$

Inverse: $x = my + b$

$x - b = my$

$\dfrac{x - b}{m} = y$

$\boxed{f^{-1}(x) = \dfrac{x - b}{m}}$

b. $f[f^{-1}(x)] = m[f^{-1}(x)] + b = m\left(\dfrac{x - b}{m}\right) + b$

$= x - b + b = x$

$f^{-1}[f(x)] = \dfrac{f(x) - b}{m} = \dfrac{(mx + b) - b}{m}$

$= \dfrac{mx}{m} = x$

21. $f(x) = 4x + 5$ and $g(x) = \dfrac{1}{4x + 5}$

$f(g(x)) = 4[g(x)] + 5 = 4\left(\dfrac{1}{4x + 5}\right) + 5 = \dfrac{4 + 20x + 25}{4x + 5} = \dfrac{20x + 29}{4x + 5} \neq x$

Since $f[g(x)] \neq x$, g is not the inverse of f.

$(g(x) \neq f^{-1}(x))$

23. If $f(x) = x, f[f(x)] = x$

Furthermore: $y = x$

inverse: $x = y$

$f^{-1}(x) = x$

25. The graph represents a function, but the inverse is not a function (horizontal line test).

$\boxed{\text{function; inverse not a function}}$

27. The graph does not represent a function (vertical line test).

$\boxed{\text{not a funciton}}$

29. The graph does not represent a function (vertical line test).

$\boxed{\text{not a function}}$

31. The graph represents a function, but the inverse is not a function (horizontal line test).

$\boxed{\text{function; inverse is a function}}$

33. The graph represents a function, and the inverse is also a funcion. Function; All vertical lines intersect the graph no more than once.

Inverse is a function (one-to-one); All horizontal lines intersect the graph more than once.

$\boxed{\text{function; inverse is a function}}$

35. $y = x^2 - 1$ Inverse: $x = y^2 - 1$.
$y^2 = x + 1$
$y = \pm\sqrt{(x + 1)}$ (not a function)

x	0	3
y	± 1	± 2

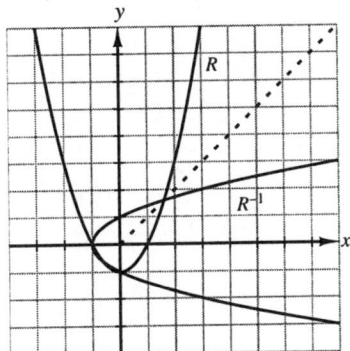

37. $y = 1$
Inverse: $x = 1$ (not a function)

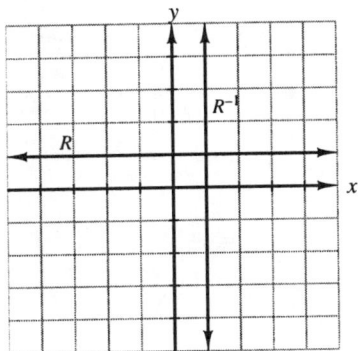

39. $y = x^3$

x	-2	-1	0	1	2
y	-8	-1	0	1	8

Inverse: $x = y^3$

$y = \sqrt[3]{x}$ (function)
$f^{-1}(x) = x^{1/3}$

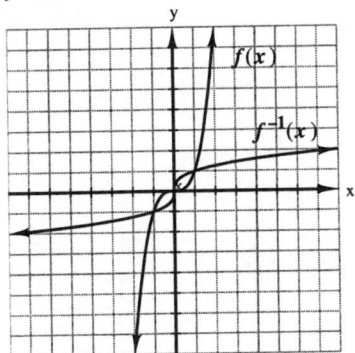

41. $x^2 + y^2 = 4$
Inverse: $y^2 + x^2 = 4$ (not a function)

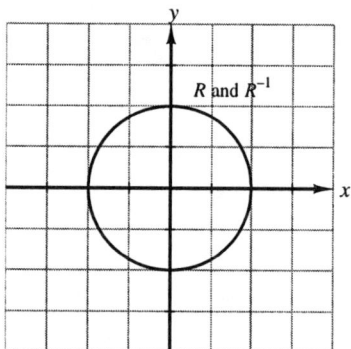

43. Ordered pairs on original function
    $(-3, 1), (-1, 0), (1, 1)$
Ordered pairs on inverse:
    $(1, -3), (0, -1), (1, 1)$
(*Note*: the inverse is not a function.)

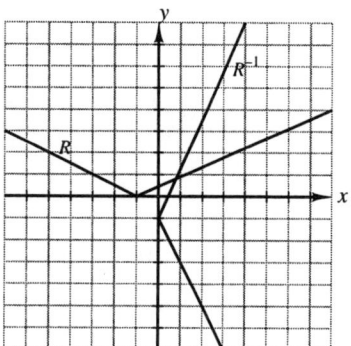

45. Ordered pairs on original function:
    $(0, 1), (1, 2), (2, 4)$
Ordered pairs on inverse:
    $(1, 0), (2, 1), (4, 2)$

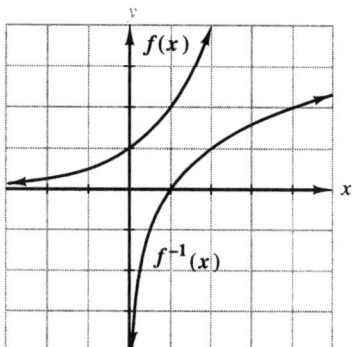

47. (*Note*: The inverse of f is not a function)

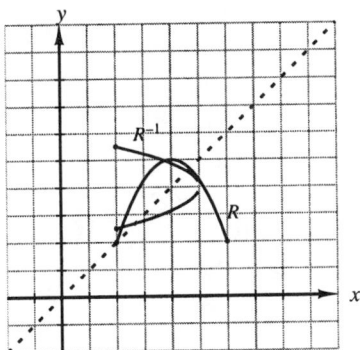

49. $f(x) = (x + 2)^2$

Restrict domain of f to $x \geq -2$. Then $f^{-1}(x) = \sqrt{x} - 2, \, x \geq -2$

51. $\boxed{C}$ is true;

53. $f(x) = \sqrt{x^2 - 1} = y$

inverse:

$$x = \sqrt{y^2 - 1}$$
$$x^2 = y^2 - 1$$
$$y^2 = x^2 + 1$$

$$\boxed{f^{-1}(x) = \sqrt{x^2 + 1}}$$

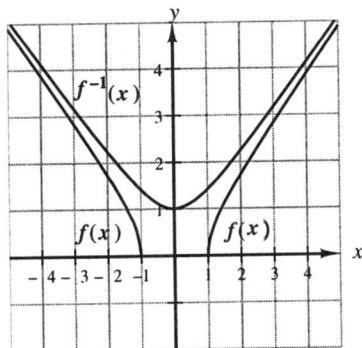

55. $f(x) = 6 - 5x$

$(f^{-1})^{-1}(x) = f(x) = \boxed{6 - 5x}$

57. $f(x) = 3 - \dfrac{2\sqrt[3]{3x + 2}}{3} = y$

$$2\sqrt[3]{3x + 2} = 9 - 3y$$
$$3x + 2 = \left(\frac{9 - 3y}{2}\right)^3$$
$$x = \frac{1}{3}\left[\left(\frac{9 - 3y}{2}\right)^3 - 2\right]$$

$$\boxed{f^{-1}(x) = \frac{1}{3}\left[\left(\frac{9 - 3x}{2}\right)^3 - 2\right]}$$

59. $f(5) = 13$

$$1 + f^{-1}(2x + 3) = 6$$
$$f^{-1}(2x + 3) = 6 - 1 = 5$$
$$2x + 3 = f(5) = 13$$
$$2x = 10$$
$$x = \boxed{5}$$

61. $f(x) = \dfrac{x+1}{x-2}$

$g(x) = \dfrac{2x+1}{x-1}$

$g[f(x)] = \dfrac{2[(x+1)/(x-2)]+1}{(x+1)/(x-2)-1}$

$= \dfrac{2(x+1)+(x-2)}{(x+1)-(x-2)}$

$= \dfrac{3x}{3} = x$

Review Problems

66. $y^{2/3} + 7y^{1/3} + 12 = 0$

$t^2 + 7t + 12 = 0$ (Let $t = y^{1/3}$)

$(t+4)(t+3) = 0$

$t+4 = 0$	or	$t+3 = 0$
$t = -4$	or	$t = -3$
$y^{1/3} = -4$		$y^{1/3} = -3$
$(y^{1/3})^3 = (-4)^3$		$(y^{1/3})^3 = (-3)^3$

$\boxed{\{-64, -27\}}$

67. $y = x^2 - 4x + 3$

x-intercept (set $y = 0$):

$0 = x^2 - 4x + 3$

$0 = (x-3)(x-1)$

$x = 3$ and $x = 1$

y-intercept (set $x = 0$): $y = 3$

vertex: $x = -\dfrac{b}{2a} = -\dfrac{(-4)}{2(1)} = \dfrac{4}{2} = 2$

$y = 2^2 - (4)(2) + 3 = -1$

Vertex: $(2, -1)$

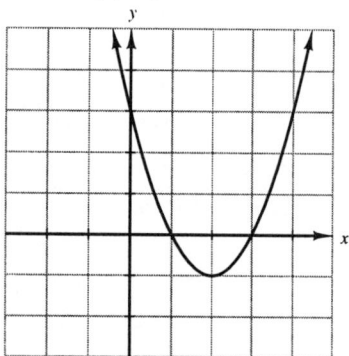

68. $x^2 + y^2 - 4x + 6y - 8 = 0$

$x^2 - 4x + 4 + y^2 + 6y + 9 - 8 = 4 + 9$

$(x-2)^2 + (y+3)^2 = 21$

center of circle $= \boxed{(2, -3)}$

Section 11.2   Exponential Functions

Problem Set 11.2, pp. 766-769

1. $f(x) = 3^x$

x	$f(x)$
-2	$\dfrac{1}{9}$
-1	$\dfrac{1}{3}$
0	1
1	3
2	9

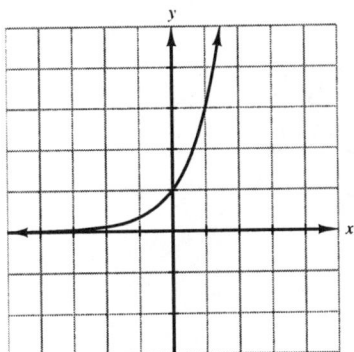

3. $f(x) = \left(\dfrac{1}{3}\right)^x$

x	$f(x)$
-2	9
-1	3
0	1
1	$\dfrac{1}{3}$
2	$\dfrac{1}{9}$

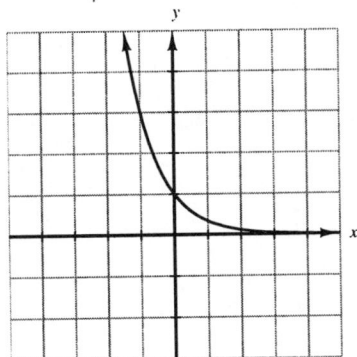

5. $f(x) = 5^x$

x	f(x)
−2	$\frac{1}{25}$
−1	$\frac{1}{5}$
0	1
1	5
2	25

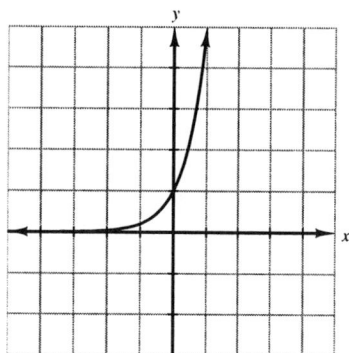

7. $f(x) = \left(\frac{1}{5}\right)^x$

x	f(x)
−2	25
−1	5
0	1
1	$\frac{1}{5}$
2	$\frac{1}{25}$

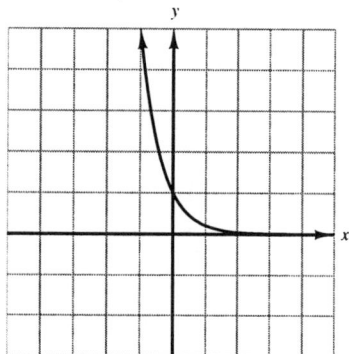

9. $f(x) = 2^{-x}$

x	$f(x)$
-3	8
-2	4
-1	2
0	1
1	$\frac{1}{2}$
2	$\frac{1}{4}$

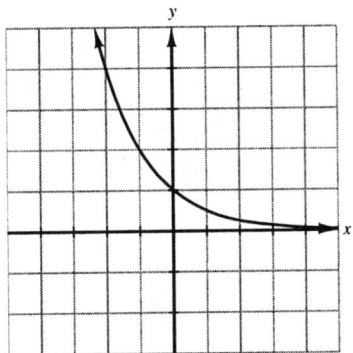

11. $f(x) = 2^{x+1}$

x	$f(x)$
-3	$\frac{1}{4}$
-2	$\frac{1}{2}$
-1	1
0	2
1	4
2	8

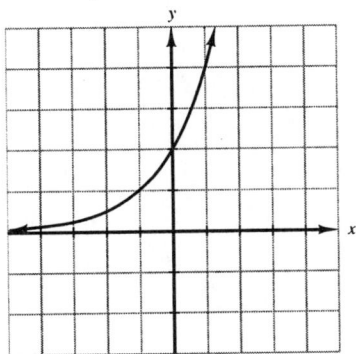

13. $f(x) = 2^x + 1$

x	$f(x)$
-3	$1\frac{1}{8}$
-2	$1\frac{1}{4}$
-1	$1\frac{1}{2}$
0	2
1	3
2	5

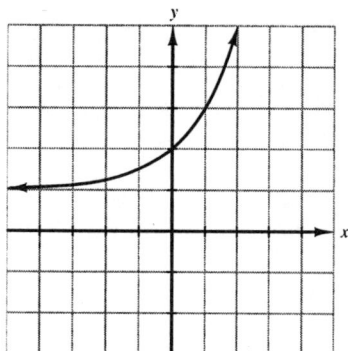

15. $f(x) = 2^{2x}$

x	$f(x)$
-2	$\frac{1}{16}$
-1	$\frac{1}{4}$
0	1
1	4
2	16

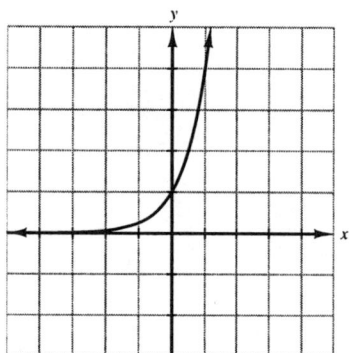

17. $f(x) = 2^{x-3}$

x	$f(x)$
-2	$\dfrac{1}{16}$
-1	$\dfrac{1}{8}$
0	$\dfrac{1}{4}$
1	$\dfrac{1}{2}$
2	1
3	2
4	4

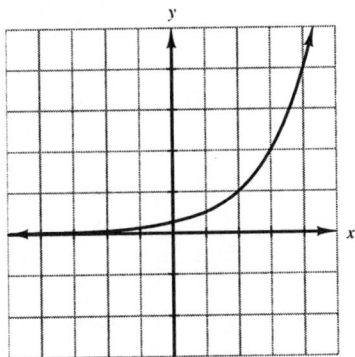

19. $f(x) = \left(\dfrac{3}{4}\right)^{x}$

x	$f(x)$
-3	$\dfrac{1}{(3/4)^3} = \dfrac{64}{27} \approx 2.4$
-2	$\dfrac{16}{9} = 1\dfrac{7}{9}$
-1	$\dfrac{4}{3} = 1\dfrac{1}{3}$
0	1
1	$\dfrac{3}{4}$
2	$\dfrac{9}{16}$

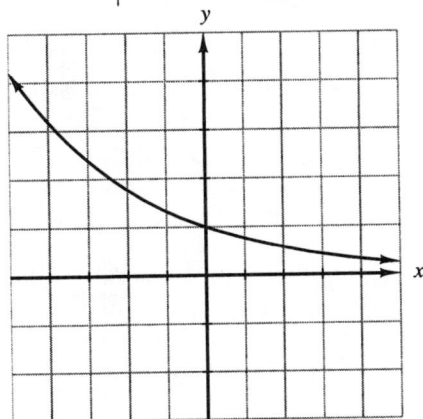

21. $f(x) = 2^{x/2}$

x	$f(x)$
−4	$\frac{1}{4}$
−2	$\frac{1}{2}$
0	1
2	2
4	4
6	8

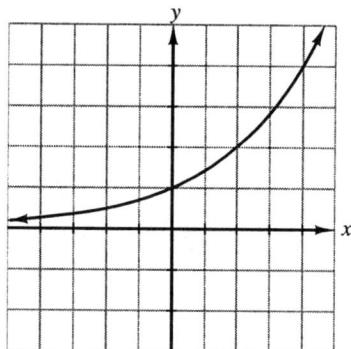

23. $f(x) = \left(\dfrac{1}{2}\right)^{-x}$

x	$f(x)$
−3	$(1/2)^3 = 1/8$
−2	1/4
−1	1/2
0	1
1	$(1/2)^{-1} = 2$
2	4
3	8

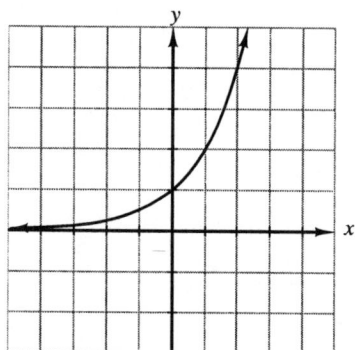

25. $f(x) = 2^{-x+1}$

x	$f(x)$
-3	16
-2	8
-1	4
0	2
1	1
2	1/2
3	1/4

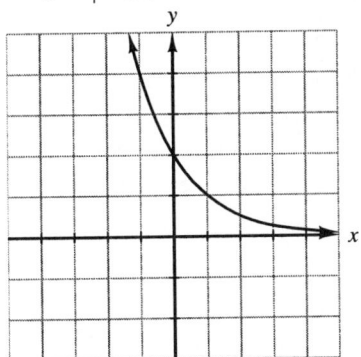

27. $f(x) = 2^{|x|}$

x	$f(x)$
-3	8
-2	4
-1	2
0	1
1	2
2	4
3	8

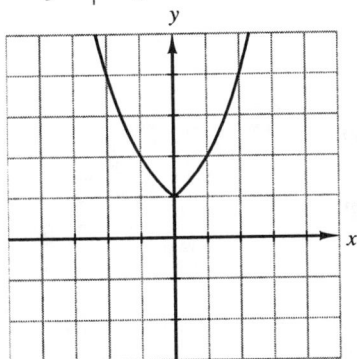

29.
$$2^x = 32$$
$$2^x = 2^5$$
$$x = 5$$
$$\{5\}$$

31.
$$\left(\frac{1}{3}\right)^x = \frac{1}{27}$$
$$(3^{-1})^x = 3^{-3}$$
$$-x = -3$$
$$x = 3$$
$$\{3\}$$

33. $4^{x+1} = 16$

$4^{x+1} = 4^2$

$x + 1 = 2$

$x = 1$

$\boxed{\{1\}}$

35. $7^{-x} = \dfrac{1}{7}$

$7^{-x} = 7^{-1}$

$-x = -1$

$x = 1$

$\boxed{\{1\}}$

37. $4^{2x-1} = 64$

$4^{2x-1} = 4^3$

$2x - 1 = 3$

$2x = 4$

$x = 2$

$\boxed{\{2\}}$

39. $16^x = 32$

$(2^4)^x = 2^5$

$2^{4x} = 2^5$

$4x = 5$

$x = \dfrac{5}{4}$

$\boxed{\left\{\dfrac{5}{4}\right\}}$

41. $\left(\dfrac{3}{4}\right)^x = \dfrac{9}{16}$

$\left(\dfrac{3}{4}\right)^x = \left(\dfrac{3}{4}\right)^2$

$x = 2$

$\boxed{\{2\}}$

43. $2^x + 6 = 38$

$2^x = 32$

$2^x = 2^5$

$x = 5$

$\boxed{\{5\}}$

45. $32^x = \dfrac{1}{8}$

$(2^5)^x = 2^{-3}$

$2^{5x} = 2^{-3}$

$5x = -3$

$x = -\dfrac{3}{5}$

$\boxed{\left\{-\dfrac{3}{5}\right\}}$

47. $10^x = 0.00001$

$10^x = 10^{-5}$

$x = -5$

$\boxed{\{-5\}}$

49. $3^{x+5} = \dfrac{1}{81}$

$3^{x+5} = 3^{-4}$

$x + 5 = -4$

$x = -9$

$\boxed{\{-9\}}$

51. $(5^{2x})(5^{4x}) = 125$

$5^{6x} = 5^3$

$6x = 3$

$x = \dfrac{3}{6} = \dfrac{1}{2}$

$\boxed{\left\{\dfrac{1}{2}\right\}}$

53. $\left(\dfrac{1}{5}\right)^{2x} = 125$

$(5^{-1})^{2x} = 5^3$

$5^{-2x} = 5^3$

$-2x = 3$

$x = -\dfrac{3}{2}$

$$\left\{-\dfrac{3}{2}\right\}$$

55. $6^3 = (2x-1)^3$

$6 = 2x - 1$

$7 = 2x$

$\dfrac{7}{2} = x$

$$\left\{\dfrac{7}{2}\right\}$$

57. $5^{x^2-12} = 25^{2x}$

$5^{x^2-12} = (5^2)^{2x}$

$5^{x^2-12} = 5^{4x}$

$x^2 - 12 = 4x$

$x^2 - 4x - 12 = 0$

$(x-6)(x+2) = 0$

$x - 6 = 0$ or $x + 2 = 0$

$x = 6$ $x = -2$

$$\{-2, 6\}$$

59. $2^{3x-1} = (1/4)^{x+2}$

$2^{3x-1} = (2^{-2})^{x+2}$

$2^{3x-1} = 2^{-2x-4}$

$3x - 1 = -2x - 4$

$5x = -3$

$x = -\dfrac{3}{5}$

$$\left\{-\dfrac{3}{5}\right\}$$

61. $f(x) = (10^6)(2^x)$

a.

x	$f(x)$
-3	$(1/8)(10^6)$
-2	$(1/4)(10^6)$
	Note: Since x represents time, the graph only shows the portion of the function for $x \geq 0$.
-1	$(1/2)(10^6)$
0	$(1)(10^6)$
1	$(2)(10^6)$
2	$(4)(10^6)$
3	$(8)(10^6)$

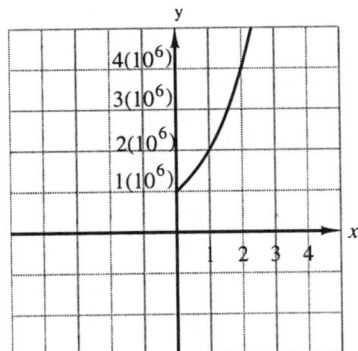

b. $\dfrac{f(5)}{f(2)} = \dfrac{(10^6)(2^5)}{(10^6)(2^2)} = 2^3 = 8$

The count is $\boxed{8}$ times as great.

63. $\boxed{C}$ is true; $f(x) = \left(\dfrac{1}{3}\right)^x = (3^{-1})^x = 3^{-x}$

65. $f(x) = 10,000e^{x/4.6}$

    a. $f(0) = 10,000\,e^0 = 10,000$

      $\boxed{10,000 \text{ cells}}$ were present initially

    b. $f(11.5) = 10,000e^{11.5/4.6} \approx 121,825$

      (Calculator: 11.5 $\boxed{\div}$ 4.6 $\boxed{=}$ $\boxed{e^x}$ $\boxed{\times}$ $10,000$ $\boxed{=}$)

      $\boxed{\text{Approximately } 121,825 \text{ cells}}$

67. $f(x) = 6,164e^{0.00667x}$

    a.

year	estimated population	actual population
1650	371,000,000	470,000,000
1950	2,745,000,000	2,501,000,000
1970	3,137,000,000	3,610,000,000

      For 1650:

      $f(1650) = 6,146\,e^{0.00667(1650)} \approx 371,000,000$

      (Calculator: 0.00667 $\boxed{\times}$ 1650 $\boxed{=}$ $\boxed{e}$ $\boxed{\times}$ $6,164$ $\boxed{=}$)

    b. The year 2000 is 20 years after 1980.

      $f(20) = 4.2\,e^{0.02(20)} = 4.2e^{0.4} \approx \boxed{6.3 \text{ billion}}$

69. $A = Pe^{rx}$

    $A(200) = 100e^{0.05(200)} = 100\,e^{10} \approx \$2,202,646$

    $\boxed{\text{Apporoximately } \$2,202,646}$

71. **a.** $A = 1(1 + 0.03)^{(1)300} \approx \boxed{\$7098.51}$

    b. $A = 1\left(1 + \dfrac{0.03}{4}\right)^{4(300)} \approx \boxed{\$7835.48}$

    c. $A = 1\left(1 + \dfrac{0.03}{52}\right)^{52(300)} \approx \boxed{\$8082.08}$

    d. $A = 1(e^{0.03(300)}) = \boxed{\$8103.09}$

73. $f(x) = \dfrac{0.8}{1 + e^{-0.2x}}$

    a. $\begin{aligned} f(0) &= \frac{0.8}{1 + e^{-0.2(0)}} \\ &= \frac{0.8}{1 + 1} \\ &= \frac{0.8}{2} \\ &= 0.4 \end{aligned}$

    0.4 (or $\boxed{40\%}$) of the responses are correct prior to learning.

    b. $\begin{aligned} f(10) &= \frac{0.8}{1 + e^{-0.2(10)}} \\ &= \frac{0.8}{1 + e^{-2}} \cdot \\ &= \frac{0.8}{1 + (0.1353)} \\ &\approx 0.7 \end{aligned}$

    0.7 (or approximately $\boxed{70\%}$) of the responses are correct after 10 learning trials.

c. As x gets larger and larger,

$$e^{-0.2x} = \frac{1}{e^{0.2x}}$$

gets very close to zero. Thus,

$$f(x) = \frac{0.8}{1 + e^{-0.2ix}}$$

gets closer to $\dfrac{0.8}{1 + 0} = 0.8$.

As continued learning takes place, 0.8 (or $\boxed{80\%}$) of the responses will be correct.

75. $f(x) = \dfrac{1}{\sqrt{2\pi}} e^{-x^2/2} \approx 0.4\, e^{-x^2/2}$

  a. $\begin{aligned}
  f(0) &\approx (0.4)e^0 \approx \boxed{0.4} \\
  f(1) &\approx (0.4)e^{-1/2} \approx \boxed{0.24} \\
  f(2) &\approx (0.4)e^{-4/2} \approx \boxed{0.05} \\
  f(-1) &\approx (0.4)e^{-1/2} \approx \boxed{0.24} \\
  f(-2) &\approx (0.4)e^{-4/2} \approx \boxed{0.05}
  \end{aligned}$

  b. $\dfrac{1}{\sqrt{2\pi}\, e^{x^2/2}}$ has a denominator which gets extremely large. Since the numerator stays the same size (1), $\boxed{\text{the expression approaches 0. This is shown by the graph getting closer and closer to the } x\text{-axis as}}$ $\boxed{x \text{ increases in size.}}$

  c. Again, the denominator gets large and $\boxed{\text{the expression approaches 0.}}$ $\boxed{\text{The graph gets closer to the}}$ $\boxed{x\text{-axis as } x \text{ decreases in size.}}$

77. $f(x) = \dfrac{100}{1 + 100{,}000e^{-0.4x}}$

  a. $\begin{aligned}
  f(10) &= \frac{100}{1 + 100{,}000e^{-0.4(10)}} \\
  &= \frac{100}{1 + 100{,}000e^{-4}} \\
  &= \frac{100}{1 + 100{,}000(0.018315)} \\
  &\approx .05 \\
  &\approx \boxed{5\%}
  \end{aligned}$

  b. As x increases, $100{,}000\,e^{-0.4x}$ gets very close to zero. Thus,

  $$y \approx \frac{100}{1 + 0} = 100 \quad (\text{or } \boxed{100\%})$$

  $\boxed{\text{At a high enough altitude, every pilot will suffer.}}$

79. $y = 5^{-x/2}$

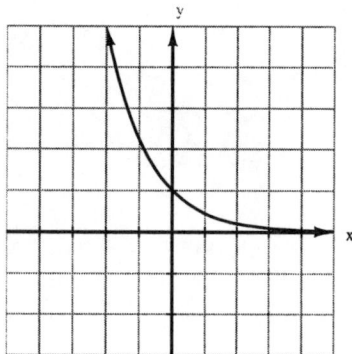

81. $y = 5^{x/2} + 2$

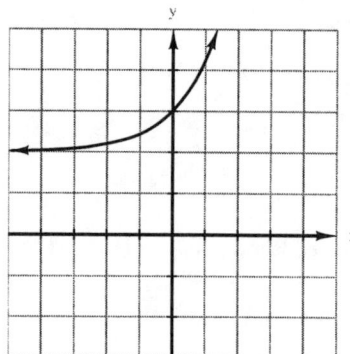

83. $y = 4.7^{x+2}$

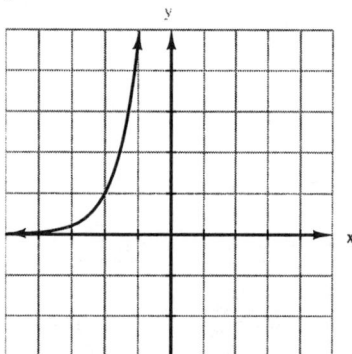

85. $y = 1 - 2^{-x/2}$

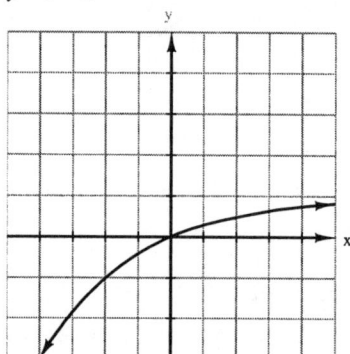

87. $f(x) = 2^x + 2^{-x}$
$g(x) = 2^x - 2^{-x}$

a. $f(x) + g(x) \;=\; (2^x + 2^{-x}) + (2^x - 2^{-x})$
$\qquad\qquad\qquad\; =\; (2)(2^x) = \boxed{2^{x+1}}$

b. $[f(x)][g(x)] \;=\; [(2^x + 2^{-x})][(2^x - 2^{-x})]$
$\qquad\qquad\qquad\quad =\; (2^x)(2^x) - (2^{-x})(2^{-x})$ (Outside and inside terms cancel.)
$\qquad\qquad\qquad\quad =\; \boxed{2^{2x} - 2^{-2x}}$

c. $\quad [f(x)]^2 \;=\; (2^x + 2^{-x})(2^x + 2^{-x})$
$\qquad\qquad\quad =\; 2^{2x} + 2^0 + 2^0 + 2^{-2x}$
$\qquad\qquad\quad =\; 2^{2x} + 2^{-2x} + 2$
$\qquad [g(x)]^2 \;=\; (2^x - 2^{-x})(2^x - 2^{-x})$
$\qquad\qquad\quad =\; 2^{2x} - 2^0 - 2^0 + 2^{-2x}$
$\qquad\qquad\quad =\; 2^{2x} + 2^{-2x} - 2$
$[f(x)]^2 - [g(x)]^2 \;=\; (2^{2x} + 2^{-2x} + 2) - (2^{2x} + 2^{-2x} - 2)$
$\qquad\qquad\qquad\quad =\; 2 - (-2)$
$\qquad\qquad\qquad\quad =\; \boxed{4}$

89. $f(x) = b^x$

$f(2) = b^2 = \dfrac{1}{4}$

$b^2 = \left(\dfrac{1}{2}\right)^2$

$b = \dfrac{1}{2}$

$f(x) = \left(\dfrac{1}{2}\right)^x$

$f(5) = \left(\dfrac{1}{2}\right)^5$

$= \boxed{\dfrac{1}{32}}$

Review Problems

96. $f(x) = x^2 + 4x - 3$

$\dfrac{f(b+h) - f(b)}{h}$

$= \dfrac{(b+h)^2 + 4(b+h) - 3 - (b^2 + 4b - 3)}{h}$

$= \dfrac{b^2 + 2nh + h^2 + 4b + 4h - 3 - b^2 - 4b + 3}{h}$

$= \dfrac{2bh + h^2 + 4h}{h}$

$= \boxed{2b + h + 4}$

97.

$y = \dfrac{10x - x^2}{2}$

$50 = \dfrac{10x - x^2}{2}$

$100 = 20x - x^2$

$x^2 - 20x + 100 = 0$

$(x - 10)^2 = 0$

$x = 10$

Density: $\boxed{10}$

98. Let x = Rider's speed

Time against the wind + Time with the wind = 3   (the total time of the top)

$$\dfrac{40}{x - 10} + \dfrac{40}{x + 10} = 3$$

$$(x - 10)(x + 10)\left[\dfrac{40}{x - 10} + \dfrac{40}{x + 10}\right] = (x - 10(x + 10)(3)$$

$$40(x + 10) + 40(x - 10) = 3x^2 - 300$$

$$40x + 400 + 40x - 400 = 3x^2 - 300$$

$$0 = 3x^2 - 80x - 300$$

$$0 = (x - 30)(3x + 10)$$

$x - 30 = 0$ or $3x + 10 = 0$

$x = 30$ $x = -\dfrac{10}{3}$ (rejct)

Rider's speed: $\boxed{30 \text{ mph}}$

Section 11.3   Inverses of Exponential Functions and Logarithms as Exponents

Problem Set 11.3, pp. 777-780

1.
$$2^3 = 8$$
$$\boxed{\log_2 8 = 3}$$

3.
$$3^2 = 9$$
$$\boxed{\log_3 9 = 2}$$

5.
$$10^3 = 1000$$
$$\boxed{\log_{10} 1000 = 3}$$

7.
$$8^1 = 8$$
$$\boxed{\log_8 8 = 1}$$

9.
$$5^{-3} = \frac{1}{125}$$
$$\boxed{\log_5 \frac{1}{125} = -3}$$

11.
$$3^0 = 1$$
$$\boxed{\log_3 1 = 0}$$

13.
$$\sqrt{100} = 100$$
$$100^{1/2} = 10$$
$$\boxed{\log_{100} 10 = \frac{1}{2}}$$

15.
$$\sqrt[3]{64} = 4$$
$$(64)^{1/3} = 4$$
$$\boxed{\log_{64} 4 = \frac{1}{3}}$$

17.
$$\left(\sqrt{81}\right)^3 = 729$$
$$[(81^{1/2})]^3 = 729$$
$$81^{3/2} = 729$$
$$\boxed{\log_{81} 729 = \frac{3}{2}}$$

19.
$$\left(\sqrt[5]{32}\right)^2 = 4$$
$$(32)^{2/5} = 4$$
$$\boxed{\log_{32} 4 = \frac{2}{5}}$$

21.
$$32^{-3/5} = \frac{1}{8}$$
$$\boxed{\log_{32} \frac{1}{8} = -\frac{3}{5}}$$

23.
$$\sqrt{\sqrt{16}} = 2$$
$$(16^{1/2})^{1/2} = 2$$
$$16^{1/4} = 2$$
$$\boxed{\log_{16} 2 = \frac{1}{4}}$$

25.
$$\log_2 64 = 6$$
$$\boxed{2^6 = 64}$$

27.
$$\log_{10} 1 = 0$$
$$\boxed{10^0 = 1}$$

29.
$$\log_e e^3 = 3$$
$$\boxed{e^3 = e^3}$$

31.
$$\log_5 \frac{1}{125} = -3$$
$$\boxed{5^{-3} = \frac{1}{125}}$$

33.
$$\log_b 1 = 0$$
$$\boxed{b^0 = 1}$$

35.
$$\log_{25} 5 = \frac{1}{2}$$
$$\boxed{25^{1/2} = 5}$$

37. $\log_8\left(\dfrac{1}{2}\right) = -\dfrac{1}{3}$

$$\boxed{8^{-1/3} = \dfrac{1}{2}}$$

39. $\log_3 9 = x$

$3^x = 9$

$x = \boxed{2}$

41. $\log_2 32 = x$

$2^x = 32$

$x = \boxed{5}$

43. $\log_7 \sqrt{7} = x$

$7^x = \sqrt{7}$

$7^x = 7^{1/2}$

$x = \boxed{\dfrac{1}{2}}$

45. $\log_7\left(\dfrac{1}{7}\right) = x$

$7^x = \dfrac{1}{7}$

$x = \boxed{-1}$

47. $\log_2\left(\dfrac{1}{32}\right) = x$

$2^x = \dfrac{1}{32}$

$2^x = 2^{-5}$

$x = \boxed{-5}$

49. $\log_{10} 10 = x$

$10^x = 10$

$x = \boxed{1}$

51. $\log_8 8^4 = x$

$8^x = 8^4$

$x = \boxed{4}$

53. $\log_{10} 1 = x$

$10^x = 1$

$x = \boxed{0}$

55. $\log_{10}(0.0001) = x$

$10^x = 0.0001$

$10^x = 10^{-4}$

$x = \boxed{-4}$

57. $\log_{0.01} 0.001 = x$

$0.01^x = 0.001$

$(10^{-2})^x = 10^{-3}$

$10^{-2x} = 10^{-3}$

$-2x = -3$

$x = \boxed{\dfrac{3}{2}}$

59. $\log_{0.5} 16 = x$

$(0.5)^x = 16$

$\left(\dfrac{1}{2}\right)^x = 16$

$(2^{-1})^x = 2^4$

$2^{-x} = 2^4$

$-x = 4$

$x = \boxed{-4}$

61. $\log_{81} 27 = x$

$81^x = 27$

$(3^4)^x = 3^3$

$3^{4x} = 3^3$

$4x = 3$

$x = \boxed{\dfrac{3}{4}}$

63. $10^{\log_{10} 8} = x$

In logarithmic form:

$\log_{10} x = \log_{10} 8$

$x = \boxed{8}$

65. $\log_3 (\log_7 7)$
   Consider: $\log_7 7 = x$
$$7^x = 7$$
$$x = 1$$
$$\log_3 (\log_7 7) = \log_3 1 = y$$
$$3^y = 1$$
$$y = \boxed{0}$$

67. $\log_2 (\log_3 81)$
$$\log_3 81 = x$$
$$3^x = 81$$
$$x = 4$$
$$\log_2 (\log_3 81) = \log_2 4 = y$$
$$2^y = 4$$
$$y = \boxed{2}$$

69.
$$\log_5 x = 3$$
$$5^3 = x$$
$$125 = x$$
$$\boxed{\{125\}}$$

71.
$$\log_{16} x = \frac{1}{2}$$
$$16^{1/2} = x$$
$$4 = x$$
$$\boxed{\{4\}}$$

73.
$$\log_3 x = \frac{1}{2}$$
$$3^{1/2} = x$$
$$\sqrt{3} = x$$
$$\boxed{\{\sqrt{3}\}}$$

75.
$$\log_8 x = \frac{2}{3}$$
$$8^{2/3} = x$$
$$\left(\sqrt[3]{8}\right)^2 = x$$
$$2^2 = x$$
$$4 = x$$
$$\boxed{\{4\}}$$

77.
$$\log_{125} x = -\frac{2}{3}$$
$$125^{-2/3} = x$$
$$\frac{1}{125^{2/3}} = x$$
$$\frac{1}{\left(\sqrt[3]{125}\right)^2} = x$$
$$\frac{1}{5^2} = x$$
$$\frac{1}{25} = x$$
$$\boxed{\left\{\frac{1}{25}\right\}}$$

79.
$$\log_b 25 = 2$$
$$b^2 = 25$$
$$b = 5$$
$$\boxed{\{5\}}$$

81.
$$\log_b 36 = \frac{1}{2}$$
$$b^{1/2} = 36$$
$$\sqrt{b} = 36$$
$$\left(\sqrt{b}\right)^2 = (36)^2$$
$$b = 1296$$
$$\boxed{\{1296\}}$$

83.
$$\log_{10} 27 = 3$$
$$b^3 = 27$$
$$b = 3$$
$$\boxed{\{3\}}$$

85.
$$\log_b 27 = 3$$
$$b^3 = 27$$
$$b = 3$$
$$\boxed{\{3\}}$$

87.
$$\log_3 (x-1) = 2$$
$$3^2 = x-1$$
$$9 = x-1$$
$$10 = x$$
$$\boxed{\{10\}}$$

89. $\log_{10}(x^2 + 9x) = 1$

$$10^1 = x^2 + 9x$$
$$0 = x^2 + 9x - 10$$
$$0 = (x + 10)(x - 1)$$
$$x + 10 = 0 \quad \text{or} \quad x - 1 = 0$$

$$x = -10 \quad \text{or} \quad x = 1$$

$$\boxed{\{-10, 1\}}$$

91. $\log_4\left(\dfrac{1}{64}\right) = -x^2 + x$

$$4^{-x^2+x} = \frac{1}{64}$$
$$4^{-x^2+x} = 4^{-3}$$
$$-x^2 + x = -3$$
$$0 = x^2 - x - 3$$
$$x = \frac{-b \pm \sqrt{b^2 - 4ac}}{2a}$$
$$= \frac{1 \pm \sqrt{(-1)^2 - 4(1)(-3)}}{2(1)}$$
$$= \frac{1 \pm \sqrt{1 - (-12)}}{2}$$
$$= \frac{1 \pm \sqrt{13}}{2}$$

$$\boxed{\left\{\frac{1 + \sqrt{13}}{2}, \frac{1 - \sqrt{13}}{2}\right\}}$$

93. $\log_b(b + 2) = 2$

$$b^2 = b + 2$$
$$b^2 - b - 2 = 0$$
$$(b - 2)(b + 1) = 0$$
$$b = 2 \quad \text{or} \quad b = -1 \quad \text{(reject since } b > 0\text{)}$$
$$\boxed{\{2\}}$$

95. a. (see part c) $f(x) = 4^x$

b.
$$f(x) = 4^x$$
$$y = 4^x$$
$$x = 4^y \quad \text{(Exchange } x \text{ and } y\text{)}$$
$$y = \log_4 x$$
$$f^{-1}(x) = \log_4 x$$

c. $f(x) = 4^x$ $f^{-1}(x) = \log_4 x$

x	$f(x)$
-2	$\dfrac{1}{16}$
-1	$\dfrac{1}{4}$
0	1
1	4
2	16

x	$f^{-1}(x)$
$\dfrac{1}{16}$	-2
$\dfrac{1}{4}$	-1
1	0
4	1
16	2

97. a. (see part c) $f(x) = \left(\dfrac{1}{3}\right)x$

b.
$$f(x) = \left(\frac{1}{3}\right)^x$$
$$y = \left(\frac{1}{3}\right)^x$$
$$x = \left(\frac{1}{3}\right)^y \quad \text{(Exchange } x \text{ and } y\text{)}$$
$$y = \log_{1/3} x$$
$$f^{-1}(x) = \log_{1/3} x$$

c. $f(x) = \left(\dfrac{1}{3}\right)^x$ $f^{-1}(x) = \log_{1/3} x$

x	$f(x)$
-2	9
-1	3
0	1
1	$\dfrac{1}{3}$
2	$\dfrac{1}{9}$

x	$f^{-1}(x)$
9	-2
3	-1
1	0
$\dfrac{1}{3}$	1
$\dfrac{1}{9}$	2

95. c.

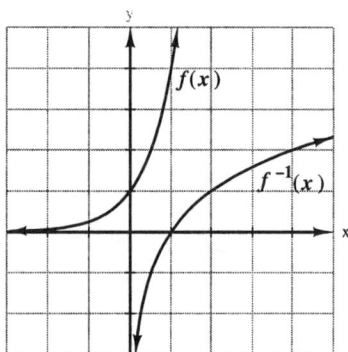

97. c.

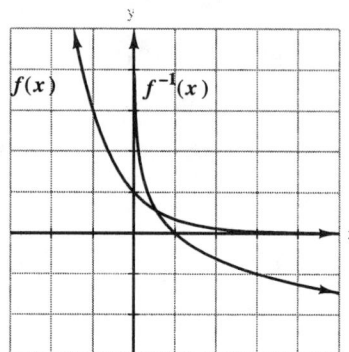

99.

$$f(x) = 12 \log_5 (2x - 5)$$

a.
$$f(15) = 12 \log_5 [(2)(15) - 5]$$
$$= 12 \log_5 25 \quad [\log_5 25 = y;\ 5^y = 25;\ y = 2]$$
$$= 12(2)$$
$$= \boxed{24}$$

b.
$$f(65) = 12 \log_5 [(2)(65) - 5]$$
$$= 12 \log_5 125 \quad [\log_5 125 = y;\ 5^y = 125;\ y = 3]$$
$$= 12(3)$$
$$= \boxed{36}$$

c.
$$f(315) = 12 \log_5 [(2)(315) - 5]$$
$$= 12 \log_5 625 \quad [\log_5 625 = y;\ 5^y = 625;\ y = 4]$$
$$= 12(4)$$
$$= \boxed{48}$$

101.

$$D = 10 \log_{10} \left(\frac{I}{10^{-12}} \right)$$

a.
$$D = 10 \log_{10} \left(\frac{10^{-8}}{10^{-12}} \right)$$
$$= 10 \log_{10} [10^{-8-(-12)}]$$
$$= 10 \log_{10} 10^4$$
$$= 10(4) \quad [\log_{10} 10^4 = y;\ 10^y = 10^4;\ y = 4]$$
$$= \boxed{40}$$

b.
$$D = 10 \log_{10} \left(\frac{10^{-6}}{10^{-12}} \right)$$
$$= 10 \log_{10} 10^6$$
$$= 10(6)$$
$$= \boxed{60}$$

c.
$$D = 10 \log_{10} \left(\frac{10^0}{10^{-12}} \right)$$
$$= 10 \log_{10} (10^{12})$$
$$= 10(12)$$
$$= \boxed{120}$$

103. $t = -18000 \log N_0$
$(N_0 = 0.01)$
$$t = -18000 \log 0.01 = -18000(-2) = \boxed{36,000 \text{ years}}$$

105. $\boxed{D}$ is true;

$$\log_5 5^7 \;=\; x$$
$$\text{Since } 5^x \;=\; 5^7$$
$$x \;=\; 7$$

107. $y = 4 \log_{10} x$

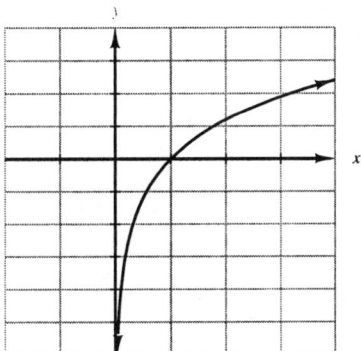

109. $y = 3 + 4 \log_{10} x$

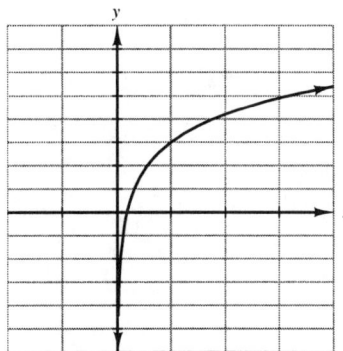

111.

$$\log_{25} 5 \;=\; x$$
$$25^x \;=\; 5$$
$$(5^2)^x \;=\; 5$$
$$5^{2x} \;=\; 5^1$$
$$2x \;=\; 1$$
$$x \;=\; \frac{1}{2}$$

$$\log_9 \frac{1}{27} \;=\; z$$
$$9^z \;=\; \frac{1}{27}$$
$$(3^2)^z \;=\; 3^{-3}$$
$$3^{2z} \;=\; 3^{-3}$$
$$2z \;=\; -3$$
$$z \;=\; -\frac{3}{2}$$

$$\log_{1/16} 8 \;=\; y$$
$$\left(\frac{1}{16}\right)^y \;=\; 8$$
$$(2^{-4})^y \;=\; 2^3$$
$$2^{-4y} \;=\; 2^3$$
$$-4y \;=\; 3$$
$$y \;=\; -\frac{3}{4}$$

$$\log_4 1 \;=\; w$$
$$4^w \;=\; 1$$
$$w \;=\; 0$$

Thus,

$$\frac{\log_{25} 5 - \log_{1/16} 8}{\log_9 \frac{1}{27} + \log_4 1} \;=\; \frac{\frac{1}{2} - \left(-\frac{3}{4}\right)}{-\frac{3}{2} + 0}$$

$$=\; \frac{\frac{5}{4}}{-\frac{3}{2}}$$

$$=\; \boxed{-\frac{5}{6}}$$

113.

$$\log_5 1 = x$$
$$5^x = 1$$
$$5^x = 5^0$$
$$x = 0$$

$$\log_8 \left[(4) \left(\sqrt[5]{16} \right) \right] = y$$

$$8^y = (4) \left(\sqrt[5]{16} \right)$$

$$(2^3)^y = (2^2) \sqrt[5]{2^4}$$

$$(2^3)^y = (2^2) \left(\sqrt[5]{2^4} \right)$$

$$2^{3y} = 2^{2+4/5}$$
$$2^{3y} = 2^{14/5}$$

$$3y = \frac{14}{5}$$

$$y = \frac{14}{15}$$

Thus,

$$\log_5 1 + \log_8 (4) \left(\sqrt[5]{16} \right) = 0 + \frac{14}{15}$$

$$= \boxed{\frac{14}{15}}$$

115.

Point	Coordinates	
A	$(0, 1)$	
B	$(1, 0)$	
C	$(4, 2)$	$(y = \log_2 4 = 2)$
D	$(2, 4)$	$(y = 2^2 = 4)$

117. $y = -\log_2 x$

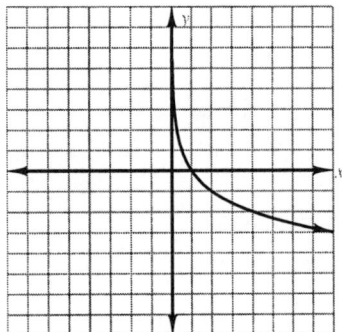

119.

$$\log_{1/25} 25\sqrt[3]{25} = x \quad \text{(Set the expression equal to } x.)$$
$$\log_{1/25} 25^{1+1/3} = x$$
$$\log_{1/25} 25^{4/3} = x$$
$$\left(\frac{1}{25}\right)^x = 25^{4/3} \quad \text{(Write in exponential form.)}$$
$$(25^{-1})^x = 25^{4/3} \quad \text{(Express with a common base.)}$$
$$25^{-x} = 25^{4/3}$$
$$-x = \frac{4}{3} \quad \text{(If } b^x = b^y, \text{ then } x = y.)$$
$$x = -\frac{4}{3}$$

Thus,

$$\log_{1/25} 25\sqrt[3]{25} = \boxed{-\frac{4}{3}}$$

Review Problems

123.

$$d = \sqrt{(x_2 - x_1)^2 + (y_2 - y_1)^2}$$
$$= \sqrt{(1 + 2)^2 + (-3 + 9)^2}$$
$$= \sqrt{3^2 + 6^2}$$
$$= \sqrt{9 + 36}$$
$$= \sqrt{45}$$
$$= \sqrt{(9)(5)}$$
$$= \boxed{3\sqrt{5}}$$

124. a.

$$m = \frac{y_2 - y_1}{x_2 - x_1}$$
$$= \frac{104 - 98}{323 - 129}$$
$$= \frac{6}{194}$$
$$= \frac{3}{97}$$
$$(x_1, y_1) = (129, 98)$$
$$y - y_1 = m(x - x_1)$$
$$y - 98 = \frac{3}{97}(x - 129)$$
$$y - 98 = \frac{3}{97}x - \frac{387}{97}$$
$$y = \frac{3}{97}x - \frac{387}{79} + 98 \quad \left[98 = \frac{9506}{97}\right]$$
$$y = \boxed{\frac{3}{97}x + \frac{9119}{97}}$$

b.

$$y = \frac{3}{97}(800) + \frac{9119}{97}$$
$$y = \frac{11519}{97} \approx 119$$

$\boxed{\text{Approximately 119}}$ people will die.

125.

$$2x = 11 - 5y \quad (\times 2) \rightarrow \quad 4x + 10y = 22$$
$$3x - 2y = 12 \quad (\times 5) \rightarrow \quad \underline{15x - 10y = -60}$$
$$19x = -38 \quad \text{(Add)}$$
$$x = -2$$

$$2x + 5y = 11$$
$$2(-2) + 5y = 11$$
$$-4 + 5y = 11$$
$$5y = 15$$
$$y = 3$$

$$\boxed{\{(-2, 3)\}}$$

Section 11.4   Properties of Logarithms; Common and Natural Logarithms

Problem Set 11.4, pp. 793-796

1. $\boxed{C}$ log 0 is undefined

3.
$$\log(10 \cdot 100) = \log 1000 = x$$
$$10^x = 1000$$
$$x = 3$$

$$\log 10 = y$$
$$10^y = 10$$
$$y = 1$$
$$\log 1000 = z$$
$$10^z = 100$$
$$z = 2$$

$$\log (10)(100) = \log 10 + \log 100$$
$$3 = 1 + 2$$
$$3 = 3 \quad \text{(verified)}$$

5.
$$\log_3(9)(1/3) = \log_3 3 = x$$
$$3^x = 3$$
$$x = 1$$

$$\log_3 9 = y$$
$$3^y = 9$$
$$y = 2$$
$$\log_3 1/3 = z$$
$$3^z = 1/3$$
$$z = -1$$

$$\log_3(9)(1/3) = \log_3 9 + \log_3 1/3$$
$$1 = 2 + (-1)$$
$$1 = 1 \quad \text{(verified)}$$

7.
$$\log_3 \frac{81}{3} = \log_3 27 = 3$$
$$\log_3 81 - \log_3 3 = 4 - 1 = 3$$
$$3 = 3 \quad \text{(verified)}$$

9.
$$\ln(e^{17}/e^4) = \ln e^{13} = x$$
$$e^x = e^{13}$$
$$x = 13$$

$$\ln e^{17} = y$$
$$e^y = e^{17}$$
$$y = 17$$

$$\ln e^4 = z$$
$$e^z = e^4$$
$$z = 4$$

$$\ln (e^{17}/e^4) = \ln e^{17} - \ln e^4$$
$$13 = 17 - 4$$
$$13 = 13 \quad \text{(verified)}$$

11.
$$\log_5 5^3 = \log_5 125 = x$$
$$5^x = 125$$
$$x = 3$$

$$\log_5 5 = y$$
$$5^y = 5$$
$$y = 1$$

$$\log_5 5^3 = 3\log_5 5$$
$$3 = 3(1)$$
$$3 = 3 \quad \text{(verified)}$$

13.
$$\log_5 25^{1/2} = \log_5 5 = 1$$
$$\frac{1}{2}\log_5 25 = \frac{1}{2}(2) = 1$$
$$1 = 1 \quad \text{(verified)}$$

15.
$$\log_3 3x = \log_3 3 + \log_3 x \quad (x > 0)$$
$$= \boxed{1 + \log_3 x}$$

17.
$$\log_b x^2 y = \log_b x^2 + \log_b y$$
$$= \boxed{2\log_b x + \log_b y} \quad (x > 0)$$

19.
$$\log x/100 = \log x - \log 100$$
$$= \log x - \log 10^2$$
$$= \boxed{\log x - 2} \quad (x > 0)$$

21.
$$\log_5 \sqrt[3]{x} = \log_5 x^{1/3}$$
$$= \boxed{\frac{1}{3} \log_5 x} \quad (x > 0)$$

23.
$$\log_4 \frac{\sqrt{x}}{16} = \log_4 \sqrt{x} - \log_4 16$$
$$= \log_4 x^{1/2} - \log_4 4^2$$
$$= \boxed{1/2 \log_4 x - 2}$$

25.
$$\log_b \frac{8}{\sqrt{3x-2}} = \log_b 2^3 - \log_b (3x-2)^{1/2}$$
$$= \boxed{3 \log_b 2 - \frac{1}{2} \log_b (3x-2)}$$

27.
$$\log_b x \sqrt{y}(\sqrt[3]{z}) = \log_b x + \log_b \sqrt{y} + \log_b \sqrt[3]{z}$$
$$= \log_b x + \log_b y^{1/2} + \log_b z^{1/3}$$
$$= \boxed{\log_b x + \frac{1}{2} \log_b y + \frac{1}{3} \log_b z}$$

29.
$$\ln 18e^3 = \ln 18 + \ln e^3$$
$$= \ln 18 + 3 \ln e$$
$$= \boxed{\ln 18 + 3}$$

31.
$$\log_b \sqrt{\frac{x}{y}} = \log_b \left(\frac{x}{y}\right)^{1/2}$$
$$= 1/2[\log_b(x/y)]$$
$$= \boxed{\frac{1}{2}(\log_b x - \log_b y)}$$

33.
$$\log_5 \sqrt{\frac{x^2 y}{25}} = \log_5 \left(\frac{x^2 y}{25}\right)^{1/3}$$
$$= \frac{1}{3} \log_5 \frac{x^2 y}{25}$$
$$= \frac{1}{3}[\log_5(x^2 y) - \log_5 25]$$
$$= \frac{1}{3}(\log_5 x^2 + \log_5 y - \log_5 5^2)$$
$$= \frac{1}{3}(2 \log_5 x + \log_5 y - 2)$$
$$= \boxed{\frac{2}{3} \log_5 x + \frac{2}{3} \log_5 y - \frac{2}{3}}$$

35.
$$\log_b \sqrt{\sqrt{x} \, y^3} = \log_b (x^{1/2})^{1/2} y^3$$
$$= \log_b x^{1/4} + \log_b y^3$$
$$= \boxed{\frac{1}{4} \log_b x + 3 \log_b y}$$

37.
$$\log_b x + \log_b (x-1) = \log_b x(x-1)$$
$$= \boxed{\log_b (x^2 - x)}$$

39.
$$3 \log_2 x + \frac{1}{2} \log_2 (x+3) = \log_2 x^3 + \log_2 (x+3)^{1/2}$$
$$= \log_2 [x^3 (x+3)^{1/2}]$$
$$= \boxed{\log_2 x^3 \sqrt{x+3}}$$

41.
$$\log_3 (x^2 - 9) - \log_3 (x-3) = \log_3 \frac{x^2 - 9}{x-3}$$
$$= \log_3 \frac{(x+3)(x-3)}{x-3}$$
$$= \boxed{\log_3 (x+3)} \quad (x \neq 3, x > -3)$$

43. $\dfrac{1}{2}\log_b x + \dfrac{1}{2}\log_b y = \log_b x^{1/2} + \log_b y^{1/2}$

$\qquad\qquad\qquad\qquad = \log_b x^{1/2} y^{1/2}$

$\qquad\qquad\qquad\qquad = \boxed{\log_b \sqrt{xy}}$

45. $\log x + \log y + 2\log z$

$\qquad = \log x + \log y + \log z^2$

$\qquad = \boxed{\log (xyz^2)}$

47. $\dfrac{1}{2}\log x - \dfrac{1}{2}\log y$

$\qquad = \dfrac{1}{2}(\log x - \log y)$

$\qquad = \dfrac{1}{2}\log \dfrac{x}{y}$

$\qquad = \boxed{\log \sqrt{\dfrac{x}{y}}}$

49. $\dfrac{1}{3}\log_4 x + 2\log_4 (3x + 2)$

$\qquad = \log_4 x^{1/3} + \log_4 (3x + 2)^2$

$\qquad = \boxed{\log_4 \sqrt[3]{x}(3x + 2)^2}$

51. $\dfrac{1}{2}(\log x + \log y) = \dfrac{1}{2}(\log xy)$

$\qquad\qquad\qquad\qquad = \log (xy)^{1/2}$

$\qquad\qquad\qquad\qquad = \boxed{\log \sqrt{xy}}$

53. $\dfrac{1}{2}(\log_5 x + \log_5 y) - 2\log_5 (x + 1)$

$\qquad = \dfrac{1}{2}\log_5 xy - \log_5 (x + 1)^5$

$\qquad = \boxed{\log_5 \dfrac{\sqrt{xy}}{(x + 1)^2}}$

55. $y = \log_3 (x + 1) = f(x)$

$\qquad\quad x + 1 \;>\; 0$

$\qquad\qquad\quad x \;>\; -1$

$\boxed{\text{domain } f = \{\, x \mid x > -1 \,\}}$ $\boxed{(-1, \infty)}$

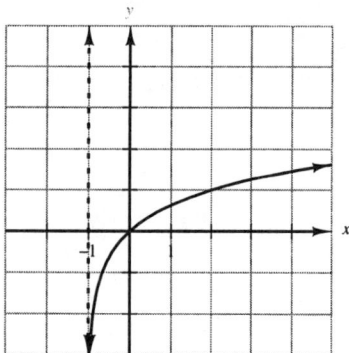

57. $y = \log_4 (2x - 4) = f(x)$

$\qquad\quad 2x - 4 \;>\; 0$

$\qquad\qquad\quad x \;>\; 2$

$\boxed{\text{domain } f = \{\, x \mid x > 2 \,\}}$ $\boxed{(2, \infty)}$

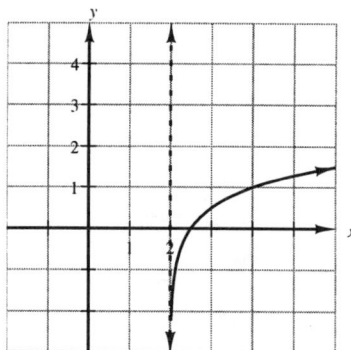

59. $y = \log(x - 10) = f(x)$

$$x - 10 > 0$$
$$x > 10$$

$\boxed{\text{domain } f = \{ x \mid x > 10 \}}$ $\boxed{(10, \infty)}$

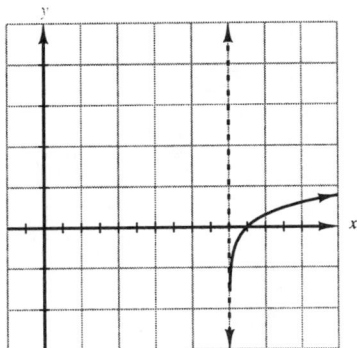

61. $\boxed{D}$ is true

$$\ln \sqrt{2} = \ln 2^{1/2}$$
$$= \frac{1}{2} \ln 2$$
$$= \frac{\ln 2}{2}$$

63. $\boxed{B}$ is true.

$$\ln e^x = \log_e e^x$$
$$= x$$

65. $\boxed{B}$ is true

$$\log_b \sqrt{\frac{xy}{z}} = \log_b \left(\frac{xy}{z}\right)^{1/2}$$
$$= \frac{1}{2} \log_b \left(\frac{xy}{z}\right)$$
$$= \frac{1}{2} (\log_b x + \log_b y - \log_b z)$$

$$(x > 0, y > 0, z > 0)$$

67.

x	Number of prime numbers that are less than x	$\dfrac{x}{\ln x}$	$\dfrac{x}{\ln x - 1.08366}$
100	25	22	28
10,000	1229	1086	1231
10^6	78,498	72,382	78,543
10^8	5,761,455	5,428,681	5,768,004
10^9	50,847,534	48,254,942	50,917,519
10^{10}	455,052,512	434,294,482	455,743,004

69. $\quad R = \log\left(\dfrac{a}{T}\right) + b$

$$= \log\left(\frac{300}{2.25}\right) + 4.11$$
$$= \boxed{6.2}$$

71. $N(r) = -5000 \ln r$

$N(0.6) = -5000 \ln (0.6) = \boxed{2554}$

$\boxed{\text{Approxmiately 2554 years have elapsed since the two languages having 60\% of their words from a}}$
$\boxed{\text{common ancestral language evolved from that common ancestral language.}}$

73. $\log_5 13 = \dfrac{\ln 13}{\ln 5} = \boxed{1.5937}$

75. $\log_3 1.87 = \dfrac{\ln 1.87}{\ln 3} = \boxed{0.5698}$

77. $\log_9 9.63 = \dfrac{\ln 9.63}{\ln 9} = \boxed{1.0308}$

79. $\log_{14} 87.5 = \dfrac{\ln 87.5}{\ln 14} = \boxed{1.6944}$

81. $\log_{50} 89.6 = \dfrac{\ln 89.6}{\ln 50} = \boxed{1.1491}$

83. a. $\log x = 0.6$

$ x \approx \boxed{4}$

b. $\log x = 0.7$

$ x \approx \boxed{5}$

c. $\log x = 0.8$

$ x \approx \boxed{6.3}$

d. $\log x = -1$

$ x \approx \boxed{0.1}$

e. $\log x = 0$

$ x \approx \boxed{1}$

f. $\log x = -1.2$

$ x \approx \boxed{0.06}$

85. Let

$$\log_b M = R$$

$$M = b^R$$
$$M^p = (b^R)^p$$
$$M^p = b^{pR}$$

In logarithmic form:
$$\log_b M^p = pR$$

Substituting $\log_b M$ for R:
$$\log_b M^p = p \log_b M$$

87. a. $\log_b a = \dfrac{\log_a a}{\log_a b}$

$ = \dfrac{1}{\log_a b}$

b. $\log_2 8 = x$

$ 2^x = 8$

$ x = 3$

$\log_8 2 = y$

$ 8^y = 2$

$ (2^3)^y = 2^1$

$ 2^{3y} = 2^1$

$ 3y = 1$

$ y = \dfrac{1}{3}$

Thus, $\log_2 8 = 3$ and $\log_8 2 = \dfrac{1}{3}$.

$$\boxed{\log_2 8 = \left(\dfrac{1}{\log_8 2}\right)}$$

$ 3 = \dfrac{1}{\frac{1}{3}}$

$ 3 = 3 \quad$ True

87. c.
$$\log_3 81 = x$$
$$3^x = 81$$
$$x = 4$$

$$\log_{81} 3 = y$$
$$81^y = 3$$
$$(3^4)^y = 3$$
$$3^{4y} = 3^1$$
$$4y = 1$$
$$y = \frac{1}{4}$$

Thus, $\log_3 81 = 4$ and $\log_{81} 3 = \frac{1}{4}$.

Substituting these values,

$$\log_3 81 = \frac{1}{\log_{81} 3} \quad \text{True}$$

89.
$$\log_{b^n} x = \frac{\log_b x}{\log_b b^n}$$
$$= \frac{\log_b x}{n}$$

Since $\log_{b^n} x = \dfrac{\log_b x}{n}$, then $\dfrac{1}{n}\log_b x = \log_{b^n} x$.

Multiplying the range values of $\log_b x$ by $\dfrac{1}{n}$ yields the range values of $\log_{b^n} x$.

91.
$$\log 36 = \log \frac{3600}{100}$$
$$= \log \frac{(80)(45)}{100}$$
$$= \log 80 + \log 45 - \log 100$$
$$= \boxed{A + B - 2}$$

93. $\log \dfrac{1}{2} + \log \dfrac{2}{3} + \log \dfrac{3}{4} + \log \dfrac{4}{5} + \ldots + \log \dfrac{96}{97} + \log \dfrac{97}{98} + \log \dfrac{98}{99} + \log \dfrac{99}{100}$

$= \log \left(\dfrac{1}{2} \cdot \dfrac{2}{3} \cdot \dfrac{3}{4} \cdot \dfrac{4}{5} \cdot \ldots \cdot \dfrac{96}{97} \cdot \dfrac{97}{98} \cdot \dfrac{99}{100} \right)$

$= \log \dfrac{1}{100}$

$= \log 1 - \log 100$

$= 0 - 2$

$= -2$

$\boxed{A \text{ represents } -2.}$

95. $\log_4 9 \; > \; \log_4 8 = \dfrac{3}{2}$

$$\left(\log_4 8 = x, \text{ so } 4^x = 8. \ (2^2)^x = 2^3, \text{ so } 2x = 3 \text{ and } x = \frac{3}{2} \right)$$

$\log_9 28 \; > \; \log_9 27 = \dfrac{3}{2}$

$$\left(\log_9 27 = x, \text{ so } 9^x = 27. \ (3^2)^x = 3^3, \text{ so } 2x = 3 \text{ and } x = \frac{3}{2} \right)$$

This means that $\log_4 9 + \log_4 28 > 3$.

Thus, $\boxed{\text{3 is the greatest integer}}$ less than the given number.

97. When $\qquad\qquad x \;=\; -\dfrac{1}{2},\ ax + c = 0$

$\qquad\qquad -\dfrac{1}{2}a + c \;=\; 0$

When $\qquad\qquad x \;=\; 0,\ ax + c = 1$

$\qquad\qquad\qquad \boxed{c \;=\; 1}$

$\qquad\qquad -\dfrac{1}{2}a \;=\; -1$

$\qquad\qquad\qquad \boxed{a \;=\; 2}$

Review Problems

104. $\qquad f(x) \;=\; 3x + 17$

$\qquad\quad\; y \;=\; 3x + 17$

$\qquad\quad\; x \;=\; 3y + 17 \qquad \text{(Exchange } x \text{ and } y\text{)}$

$\qquad x - 17 \;=\; 3y$

$\qquad \dfrac{x - 17}{3} \;=\; y$

$\qquad f^{-1}(x) \;=\; \dfrac{x - 17}{3}$

$f[f^{-1}(x)] \;=\; 3[f^{-1}(x)] + 17$

$\qquad\qquad\;\; = \; 3\left(\dfrac{x - 17}{3} \right) + 17$

$\qquad\qquad\;\; = \; x - 17 + 17$

$\qquad\qquad\;\; = \; x$

$f^{-1}[f(x)] \;=\; \dfrac{f(x) - 17}{3}$

$\qquad\qquad\;\; = \; \dfrac{3x + 17 - 17}{3}$

$\qquad\qquad\;\; = \; \dfrac{3x}{3}$

$\qquad\qquad\;\; = \; x \qquad \text{(verified)}$

105.
$$x - 5y - 2z = 6 \quad (1)$$
$$2x - 3y + z = 13 \quad (2)$$
$$3x - 2y + 4z = 22 \quad (3)$$

(Equations 1 and 2):
$$x - 5y - 2z = 6$$
$$(\times 2) \quad \underline{4x - 6y + 2z = 26}$$
$$5x - 11y = 32 \quad \text{(Equation 4)}$$

(Equations 1 and 3):
$$(\times 2) \quad 2x - 10y - 4z = 12$$
$$\underline{3x - 2y + 4z = 22}$$
$$5x - 12y = 34 \quad \text{(Equation 5)}$$

(Equations 4 and 5):
$$(\times -1) \quad -5x + 11y = -32$$
$$\underline{5x - 12y = 34}$$
$$-y = 2$$
$$y = -2$$

(Equation 4):
$$5x - 11(-2) = 32$$
$$5x = 10$$
$$x = 2$$

(Equation 2):
$$2(2) - 3(-2) + 2 = 13$$
$$10 + z = 13$$
$$z = 3$$

$$\boxed{\{(2, -2, 3)\}}$$

106. $\dfrac{x^2}{25} - \dfrac{y^2}{4} = 1 \qquad \left[\dfrac{x^2}{a^2} - \dfrac{y^2}{b^2} = 1\right]$

x-intercepts:
$$\frac{x^2}{25} = 1$$
$$x^2 = 25$$
$$x = \pm 5$$

$$a^2 = 25, \ a = 5$$
$$b^2 = 4, \ b = 2$$

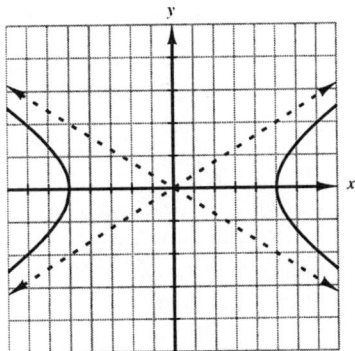

Section 11.5   Using Inverse Properties to Solve Exponential Equations

Problem Set 11.5, pp. 808-811

1.
$$10^x = 2.91$$
$$\log 10^x = \log 2.91$$
$$x = \log 2.91$$
$$\boxed{x = \log 2.91 \approx 0.464}$$

3.
$$10^x = 7823$$
$$\log 10^x = \log 7823$$
$$x = \log 7823$$
$$\boxed{x = \log 7823 \approx 3.893}$$

5.
$$e^x = 2.7$$
$$\ln e^x = \ln 2.7$$
$$x = \ln 2.7$$
$$\boxed{x = \ln 2.7 \approx 0.993}$$

7.
$$e^x = 0.8$$
$$\ln e^x = \ln(0.8)$$
$$\boxed{x = \ln 0.8 \approx -0.223}$$

9.
$$2e^{1.4x} = 26$$
$$e^{1.4x} = 13$$
$$\ln e^{1.4x} = \ln 13$$
$$1.4x = \ln 13$$
$$\boxed{x = \frac{\ln 13}{1.4} \approx 1.832}$$

11.
$$3.4(10^{1.8x}) = 68$$
$$10^{1.8x} = \frac{68}{3.4}$$
$$10^{1.8x} = 20$$
$$\log 10^{1.8x} = \log 20$$
$$1.8x = \log 20$$
$$\boxed{x = \frac{\log 20}{1.8} \approx 0.723}$$

13.
$$3.1e^{1.2x} - 28.3 = 49.2$$
$$3.1e^{1.2x} = 77.5$$
$$e^{1.2x} = 25$$
$$\ln e^{1.2x} = \ln 25$$
$$1.2x = \ln 25$$
$$\boxed{x = \frac{\ln 25}{1.2} \approx 2.682}$$

15.
$$0.7(10^{-1.3x}) - 21.7 = 16.24$$
$$0.7(10^{-1.3x}) = 16.24 + 21.7$$
$$0.7(10^{-1.3x}) = 37.94$$
$$10^{-1.3x} = \frac{37.94}{0.7}$$
$$10^{-1.3x} = 54.2$$
$$\log 10^{-1.3x} = \log 54.2$$
$$-1.3x = \log 54.2$$
$$\boxed{x = \frac{\log 54.2}{-1.3} \approx -1.334}$$

17.
$$800 - 500e^{-0.5x} = 733$$
$$-500e^{-0.5x} = -67$$
$$e^{-0.5x} = \frac{-67}{-500}$$
$$e^{-0.5x} = 0.134$$
$$\ln e^{-0.5x} = \ln 0.134$$
$$-0.5x = \ln 0.134$$
$$\boxed{x = \frac{\ln 0.134}{-0.5} \approx 4.020}$$

19.
$$3^x = 8$$
$$\log 3^x = \log 8$$
$$x \log 3 = \log 8$$
$$\boxed{x = \frac{\log 8}{\log 3} \approx 1.893}$$

21.
$$2^{x+1} = 5$$
$$\log 2^{x+1} = \log 5$$
$$(x + 1) \log 2 = \log 5$$
$$x \log 2 + \log 2 = \log 5$$
$$x \log 2 = \log 5 - \log 2$$
$$\boxed{x = \frac{\log 5 - \log 2}{\log 2} \approx 1.322}$$

23.
$$4^{x^2} = 9$$
$$\log 4^{x^2} = \log 9$$
$$x^2 = \frac{\log 9}{\log 4}$$
$$\boxed{x = \pm\sqrt{\frac{\log 9}{\log 4}} \approx \pm 1.259}$$

25.
$$3^{-x} = 7$$
$$\log 3^{-x} = \log 7$$
$$-x \log 3 = \log 7$$
$$-x = \frac{\log 7}{\log 3}$$
$$\boxed{x = \frac{-\log 7}{\log 3} \approx -1.771}$$

27.
$$\log C \cdot 10^{8t} = \log C + \log 10^{8t}$$
$$= \boxed{\log C + 8t}$$

29.
$$\ln A \cdot e^{9y^2} = \ln A + \ln e^{9y^2}$$
$$= \boxed{\ln A + 9y^2}$$

31.
$$10^{\log 3x^2 + \log 5x} = 10^{\log(3x^2)(5x)}$$
$$= 10^{\log 15x^3}$$
$$= \boxed{15x^3 \quad x > 0}$$

33.
$$10^{\log 12y^7 - \log 6y^2} = 10^{\log(12y^7/6y^2)}$$
$$= 10^{\log 2y^5}$$
$$= \boxed{2y^5 \quad (y > 0)}$$

35.
$$A = Pe^{rx}$$
$$16000 = 8000e^{0.08x}$$
$$e^{0.08x} = \frac{16000}{8000}$$
$$e^{0.08x} = 2$$
$$\ln e^{0.08x} = \ln 2$$
$$0.08x = \ln 2$$
$$x = \frac{\ln 2}{0.08} \approx 8.7$$

It will take $\boxed{8.7 \text{ years}}$.

37.
$$f(x) = 800 - 500e^{-0.5x}$$
$$800 - 500e^{-0.5x} = 733$$
$$-500e^{-0.5x} = -67$$
$$e^{-0.5x} = 0.134$$
$$\ln e^{-0.5x} = \ln 0.134$$
$$-0.5x = \ln 0.134$$
$$x = \frac{\ln 0.134}{-0.5} \approx 4$$

Approximately $\boxed{4 \text{ months}}$ of training.

39.
$$f(x) = 150e^{-0.1x} + 60$$
$(f(x) = 90)$:
$$150e^{-0.1x} + 60 = 90$$
$$150e^{-0.1x} = 30$$
$$e^{-0.1x} = 0.2$$
$$\ln e^{-0.1x} = \ln 0.2$$
$$-0.1x = \ln 0.2$$
$$x = \frac{\ln 0.2}{-0.1} \approx 16.1$$

After $\boxed{16.1 \text{ minutes}}$.

41. a.
$$f(x) = Ae^{kx}$$
$$Ae^{kx} = 2A$$
$$e^{kx} = 2$$
$$\ln e^{kx} = \ln 2$$
$$kx = \ln 2$$
$$x = \boxed{\dfrac{\ln 2}{k}}$$

b. $k = 0.02$

$$x = \dfrac{\ln 2}{0.02} \approx 34.7$$

It will take approximately $\boxed{35 \text{ years}}$ for the world's population to double. This implies that the population of the world in $1969 + 35 = 2004$ will be approximately $2(3.9) = 7.8$ billion. This is nearing the estimated capacity of the world.

43.
$$A = P\left(1 + \dfrac{r}{N}\right)^{Nx}$$
$(A = 2600,\ P = 1000,\ N = 4,\ r = 6\% = 0.06)$:
$$2600 = 1000\left(1 + \dfrac{0.06}{4}\right)^{4x}$$
$$(1 + 0.015)^{4x} = 2.6$$
$$(1.015)^{4x} = 2.6$$
$$\log (1.015)^{4x} = \log 2.6$$
$$4x \log (1.015) = \log 2.6$$
$$x \approx \dfrac{\log 2.6}{4 \log (1.015)} \approx 16.0$$

$\boxed{16 \text{ years}}$

45.
$$f(x) = Ae^{kx}$$
$(A = 60,\ f(12) = 30)$:
$$f(x) = 60e^{kx}$$

a.
$$60e^{k(12)} = 30$$
$$e^{12k} = 0.5$$
$$\ln e^{12k} = \ln 0.5$$
$$12k = \ln 0.5$$
$$\boxed{k} = \dfrac{\ln 0.5}{12} \boxed{\approx -0.058}$$

b.
$$f(x) = 60e^{-0.058x}$$
$(f(x) = 10)$:
$$10 = 60e^{-0.058x}$$
$$e^{-0.058x} = \dfrac{1}{6}$$
$$\ln e^{-0.058x} = \ln \dfrac{1}{6} \qquad \left[\ln \dfrac{1}{6} = \ln 1 - \ln 6 = -\ln 6\right]$$
$$-0.058x = -\ln 6$$
$$x = \dfrac{\ln 6}{0.058} \approx 30.9$$

$\boxed{\text{Approximately 31 years}}$

47.
$$y = Ae^{kx}$$
$$6.37 = 5.321\, e^{0.018x}$$
$$x = \dfrac{\ln\left(\dfrac{6.37}{5.321}\right)}{0.018} \approx 10 \text{ years}$$

or in $1990 + 10 = \boxed{2000}$

For less developed regions, population $= 4.107\, e^{0.021(10)} \approx 5.067$ billion

Percent of world's population $= \dfrac{5.067}{6.37} \cdot 100 = \boxed{79.5\%}$

49. a.
$$y = Ae^{kx}$$

$x = 0$:
$$y(0) = 6{,}907{,}387$$
$$Ae^{k \cdot 0} = 6{,}907{,}387$$
$$A = 6{,}907{,}387$$

$x = 10$:
$$y(10) = 10{,}586{,}223$$
$$10{,}586{,}223 = 6{,}907{,}387e^{10k}$$
$$\frac{10{,}586{,}223}{6{,}907{,}387} = e^{10k}$$
$$\ln\left(\frac{10{,}586{,}223}{6{,}907{,}387}\right) = \ln e^{10k}$$
$$\ln\left(\frac{10{,}586{,}223}{6{,}907{,}387}\right) = 10k$$
$$k = \frac{\ln\left(\dfrac{10{,}586{,}223}{6{,}907{,}387}\right)}{10} \approx 0.0427$$

$$\boxed{A = 6{,}907{,}387, \quad k \approx 0.0427}$$

b. In 1990, $x = 1990 - 1940 = 50$

$\boxed{\text{Estimated}}$ population $= 6{,}907{,}387\, e^{50(0.0427)} \approx \boxed{58{,}400{,}000}$

Actual population $= 29{,}279{,}000$

51. half-life: $y = \log 0.5 A_0$
$$y = Ae^{kx}$$
$$0.5A_0 = A_0 e^{8k}$$
$$0.5 = e^{8k}$$
$$\ln 0.5 = 9k$$
$$k = \frac{\ln 0.5}{8} \approx -0.0866$$
$$y = A_0 e^{-0.0866 x}$$

$(A_0 = 3000)$:
$$y = 3000e^{-0.0866 x}$$

$(y = 0.1 \text{ gram})$:
$$0.1 = 3000e^{-0.0866 x}$$
$$\ln \frac{0.1}{3000} = -0.0866x$$
$$x = \frac{1}{-0.0866} \ln \frac{0.1}{3000} \approx \boxed{119 \text{ days}}$$

53.
$$A = \frac{A_0}{k}(e^{kx} - 1)$$
$$1661 = \frac{21.7}{0.03}(e^{0.03x} - 1)$$
$$1661\left(\frac{0.03}{21.7}\right) = e^{0.03x} - 1$$
$$e^{0.03x} = 1 + 2.296$$
$$= 3.296$$
$$0.03x = \ln 3.296$$
$$= 1.1928$$

$x = \dfrac{1.1928}{0.03} = 39.8$ years, corresponding to the year $\boxed{2016}$.

55. a.
$$y = \frac{2000}{1 + 2e^{-x/2}}$$

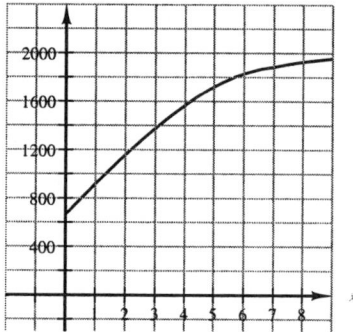

b. The population after 4 years is about $\boxed{1575}$.

c. Population = 1819 $\boxed{\text{after 6 years}}$.

d. The population will approach a maximum level of $\boxed{2000}$.

57. Iraq: \$8.8 million (1990 population); growth rate $(k) = 3.9\% = 0.039$
United Kingdom: 57.4 million (1990 population); growth rate $(k) = 0.2\% = 0.002$
Iraq: $p = 57.4e^{0.002x}$
United Kingdom: $P = 18.8\,e^{0.039x}$

$$57.4\,e^{0.002x} = 18.8\,e^{0.039x}$$
$$\frac{57.4}{18.8} = \frac{e^{0.039x}}{e^{0.002x}}$$
$$\frac{57.4}{18.8} = e^{(0.039 - 0.002)x}$$
$$\ln 3.053 = 0.037x$$
$$x \approx 30.2 \text{ years,}$$

or the year $1990 + 30 = \boxed{2020}$

Review Problems

61.
$$\frac{(x^{-1}y^2)^3}{(xy)^{1/2}} = \frac{x^{-3}y^6}{x^{1/2}y^{1/2}}$$
$$= x^{-3-(1/2)}y^{6-(1/2)}$$
$$= x^{-7/2}y^{11/2}$$
$$= \boxed{\dfrac{y^{11/2}}{x^{7/2}}}$$

62.
$$x^2 + y^2 = 17$$
$$x - y = 5 \quad \text{(solve for } y) \rightarrow \quad y = x - 5$$

(Substitute for y):
$$x^2 + (x-5)^2 = 17$$
$$x^2 + x^2 - 10x + 25 = 17$$
$$2x^2 - 10x + 8 = 0$$
$$x^2 - 5x + 4 = 0$$
$$(x-4)(x-1) = 0$$
$$x = 1 \quad \text{or} \quad x = 4$$

If $x = 1$, $y = 1 - 5 = -4$
If $x = 4$, $y = 4 - 5 = -1$
$$\boxed{\{(1, -4), (4, -1)\}}$$

Check: (1, −4): $1 + 16 = 17$ $1 − (−4) = 5$
          (4, −1): $16 + 1 = 17$ $4 − (−1) = 5$

Graphic verification:

$x^2 + y^2 = 17$ Circle: center (0, 0)

radius $\sqrt{17} \approx 4.1$

$x − y = 5$ Line: x–intercept $= 5$
                       y–intercept $= −5$

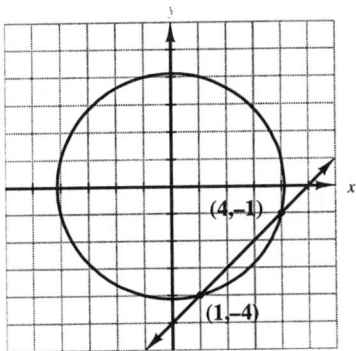

63.

$$4x^2 + 25y^2 = 100$$

$$\frac{4x^2}{100} + \frac{25y^2}{100} = 1$$

$$\frac{x^2}{25} + \frac{y^2}{4} = 1$$

x–intercepts: $\dfrac{x^2}{25} = 1$

$$x^2 = 25$$

$$x = \pm 5$$

y–intercepts: $\dfrac{y^2}{4} = 1$

$$y^2 = 4$$

$$y = \pm 2$$

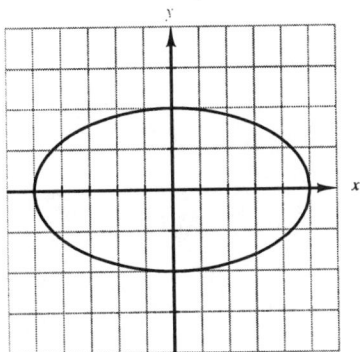

Section 11.6 Logarithmic Equations

Problem Set 11.6, pp. 822-824

1.
$$\log(x+4) = \log x + \log 4$$
$$\log(x+4) = \log(x)(4)$$
$$x+4 = 4x$$
$$4 = 3x$$
$$\frac{4}{3} = x$$

$$\boxed{\left\{\frac{4}{3}\right\}}$$

3.
$$\log(3x-3) = \log(x+1) + \log 4$$
$$\log(3x-3) = \log 4(x+1)$$
$$3x-3 = 4(x+1)$$
$$3x-3 = 4x+4$$
$$-7 = x$$

$x = -7$ causes the log of a negative number.

$$\boxed{\varnothing}$$

5.
$$2\log x = \log 25$$
$$\log x^2 = \log 25$$
$$x^2 = 25$$
$$x = 5 \quad \text{or} \quad x = -5 \quad \text{(reject)}$$

$$\boxed{\{5\}}$$

7.
$$\log(x+4) - \log 2 = \log(5x+1)$$
$$\log\frac{x+4}{2} = \log(5x+1)$$
$$\frac{x+4}{2} = 5x+1$$
$$x+4 = 10x+2$$
$$2 = 9x$$
$$\frac{2}{9} = x$$

$$\boxed{\left\{\frac{2}{9}\right\}}$$

9.
$$2\log x - \log 7 = \log 112$$
$$\log x^2 - \log 7 = \log 112$$
$$\log\frac{x^2}{7} = \log 112$$
$$\frac{x^2}{7} = 112$$
$$x^2 = 784$$
$$x = \pm\sqrt{784}$$

$x = 28$ or $x = -28$ (reject; produces the log of a negative number)

$$\boxed{\{28\}}$$

11.
$$\log x + \log(x+3) = \log 10$$
$$\log x(x+3) = \log 10$$
$$x(x+3) = 10$$
$$x^2 + 3x - 10 = 0$$
$$(x+5)(x-2) = 0$$
$$x+5 = 0 \qquad \text{or} \qquad x-2 = 0$$
$$x = -5 \quad \text{(reject)} \qquad\qquad x = 2$$

$$\boxed{\{2\}}$$

13.
$$\log_3(x+1) + \log_3(x+3) = \log_3 3$$
$$\log_3(x+1)(x+3) = \log_3 3$$
$$(x+1)(x+3) = 3$$
$$x^2 + 4x + 3 = 3$$
$$x^2 + 4x = 0$$
$$x(x+4) = 0$$

$x = 0$ or $x = -4$ (reject)

$$\boxed{\{0\}}$$

15.
$$\log_3(x+4) + \log_3(x+1) = 2\log_3(x+3)$$
$$\log_3(x+4)(x+1) = \log_3(x+3)^2$$
$$(x+4)(x+1) = (x+3)^2$$
$$x^2 + 5x + 4 = x^2 + 6x + 9$$
$$5x + 4 = 6x + 9$$
$$-5 = x \quad \text{(reject since } -5 \text{ produces the log of a negative number)}$$

$$\boxed{\varnothing}$$

17.
$$\ln(1-x) - \ln(1+x) = \ln e$$
$$\ln \frac{1-x}{1+x} = \ln e$$
$$\frac{1-x}{1+x} = e$$
$$1-x = e(1+x)$$
$$1-x = e + ex$$
$$1-e = x + ex$$
$$1-e = x(1+e)$$
$$\frac{1-e}{1+e} = x$$

Since $e \approx 2.72$, $\dfrac{1-e}{1+e} \approx -0.46$ and this value does not cause the natural log of a negative number in the equation.

$$\boxed{\left\{ \frac{1-e}{1+e} \right\}}$$

19.
$$\log(2x+1) + \log(x-2) = 2\log x$$
$$\log(2x+1)(x-2) = \log x^2$$
$$(2x+1)(x-2) = x^2$$
$$2x^2 - 3x - 2 = x^2$$
$$x^2 - 3x - 2 = 0$$
$$x = \frac{-b \pm \sqrt{b^2 - 4ac}}{2a} = \frac{3 \pm \sqrt{9 - 4(1)(-2)}}{2}$$
$$= \frac{3 \pm \sqrt{17}}{2}$$

$x = \dfrac{3+\sqrt{17}}{2}$ or $x = \dfrac{3-\sqrt{17}}{2} \approx -.56$ (reject)

$$\boxed{\left\{ \frac{3+\sqrt{17}}{2} \right\}}$$

21.

$$\log_b(2x-1) = \log_b(4x-3) - \log_b x$$

$$\log_b(2x-1) = \log_b \frac{4x-3}{x}$$

$$2x-1 = \frac{4x-3}{x}$$

$$2x^2 - x = 4x - 3$$

$$2x^2 - 5x + 3 = 0$$

$$(2x-3)(x-1) = 0$$

$$x = \frac{3}{2} \quad \text{or} \quad x = 1$$

$$\boxed{\left\{ 1, \frac{3}{2} \right\}}$$

23.

$$\log x + \log 50 = 2$$

$$\log(x)(50) = 2$$

Exponential form: $10^2 = 50x$

$$100 = 50x$$

$$2 = x$$

$$\boxed{\{2\}}$$

25.

$$\log_4 50 - \log_4 x = 2$$

$$\log_4 \frac{50}{x} = 2$$

$$4^2 = \frac{50}{x}$$

$$16 = \frac{50}{x}$$

$$16x = 50$$

$$x = \frac{50}{16} = \frac{25}{8}$$

$$\boxed{\left\{ \frac{25}{8} \right\}}$$

27.

$$\log_2(x+2) - \log_2(x-5) = 3$$

$$\log_2 \frac{x+2}{x-5} = 3$$

Exponential form:

$$2^3 = \frac{x+2}{x-5}$$

$$8 = \frac{x+2}{x-5}$$

$$8(x-5) = x+2$$

$$8x - 40 = x+2$$

$$7x = 42$$

$$x = 6$$

$$\boxed{\{6\}}$$

29. $\log_3(x-5) + \log_3(x+3) = 2$

$\log_3(x-5)(x+3) = 2$

Exponential form:

$$3^2 = (x-5)(x+3)$$
$$9 = x^2 - 2x - 15$$
$$0 = x^2 - 2x - 24$$
$$0 = (x-6)(x+4)$$

$x - 6 = 0$ or $x + 4 = 0$

$x = 6$ $x = -4$ (reject)

$\boxed{\{6\}}$

31. $\log_6 x + \log_6(x-12) = 2$

$\log_6 x(x-12) = 2$

Exponential form:

$$6^2 = x(x-12)$$
$$36 = x^2 - 12x$$
$$0 = x^2 - 12x - 36$$

$$x = \frac{-b \pm \sqrt{b^2 - 4ac}}{2a}$$

$$= \frac{12 \pm \sqrt{144 - 4(1)(-36)}}{2(1)}$$

$$= \frac{12 \pm \sqrt{288}}{2}$$

$$= \frac{12 \pm \sqrt{(144)(2)}}{2}$$

$$= \frac{12 \pm 12\sqrt{2}}{2}$$

$$= 6 \pm 6\sqrt{2}$$

(Reject $6 - 6\sqrt{2} \approx -2.5$, since this produces the log of a negative number.)

$\boxed{\{6 + 6\sqrt{2}\}}$

33. $\log_2(x-6) - \log_2 x + 5 = 7 - \log_2(x-4)$

$\log_2(x-6) - \log_2 x + \log_2(x-4) = 2$

$\log_2 \dfrac{(x-6)(x-4)}{x} = 2$

Exponential form:

$$2^2 = \frac{(x-6)(x-4)}{x}$$
$$4x = x^2 - 10x + 24$$
$$0 = x^2 - 14x + 24$$
$$0 = (x-12)(x-2)$$

$x - 12 = 0$ or $x - 2 = 0$

$x = 12$ $x = 2$ (reject; produces log of a negative number in equation)

$\boxed{\{12\}}$

35.
$$R = \log \frac{I}{I_0}$$
$$\log \frac{I}{I_0} = 8$$
$$\frac{I}{I_0} = 1(10^8)$$
$$I = 10^8 I_0$$

$\boxed{10^8 \text{ times more intense}}$

37.
$$R = \log \frac{I}{I_0}$$
(1906 earthquake: 8.25 on the Richter scale):
$$8.25 = \log \frac{I}{I_0}$$
$$10^{8.25} I_0 = I$$
(1989 earthquake: 7.1 on the Richter scale):
$$7.1 = \log \frac{I}{I_0}$$
$$10^{7.1} I_0 = I$$
$$\frac{I_{1906}}{I_{1989}} = \frac{10^{8.25}}{10^{7.1}}$$
$$= 10^{1.15}$$
$$\approx 14.1$$

$\boxed{10^{1.15} \approx 14.1 \text{ times greater in intensity}}$

39.
$$\log(1-r) = \frac{1}{T} \log \frac{W}{P}$$
$$\log(1-r) = \frac{1}{6} \log \frac{3{,}000}{12{,}000}$$
$$\log(1-r) = \frac{1}{6} \log (0.25)$$
$$\log(1-r) = \frac{1}{6} (-0.60206)$$
$$\log(1-r) = -0.10034$$
$$10^{-0.10034} = 1-r$$
$$1-r = 0.7937$$
$$1 - 0.7937 = r$$
$$0.2063 = r$$

Rate of depreciation: $\boxed{20.63\%}$

41. $\boxed{\text{C}}$ is true;
$$\log x + \log(x-1) = \log(8x-12) - \log 2$$
$$\log x(x-1) = \log \frac{8x-12}{2}$$
$$x(x-1) = \frac{8x-12}{2}$$
$$x^2 - x = \frac{8x-12}{2}$$

43. $P(x) = 95 - 30 \log_2 x$ $P(x) = 50$

$50 = 95 - 30 \log_2 x$

$30 \log_2 x = 95 - 50 = 45$

$\log_2 x = \dfrac{45}{30} = 1.5$

$x = 2^{1.5} \approx 2.83$ days

After $\boxed{\text{approximately 2.83 days}}$ only half the students will recall the important features.

The value can be located approximately on the graph.

45. $y = 302 - 50 \log_2(65 - x)$

$x = 62$ steps

$y = 302 - 50 \log_2(65 - 62)$

$= 302 - 50 \log_2 3$

$\approx 302 - 50(1.585)$

≈ 223 hours

$\boxed{\text{Approximately 223 hours}}$

47. $t = \dfrac{5600 \log R}{\log 0.5}$

$\dfrac{t}{5600} \log 0.5 = \log R$

$\log 0.5^{\,(t/5600)} = \log R$

$\boxed{R = \left(\dfrac{1}{2}\right)^{t/5600}}$

49. $f(x) = \log_3 x$

$f\{f[f(x)]\} = 3$

$\log_3\{\log_3[\log_3 x]\} = 3$

$\log_3[\log_3 x] = 3^3 = 27$

$\log_3 x = 3^{27} = 3^{3^3}$

$\boxed{x = 3^{3^{3^3}}}$

If $f(x) = \log_b x$ $(b > 0, b \neq 1)$

$f\{f[f(x)]\} = b$

$\boxed{x = b^{b^{b^b}}}$

51. $f(x) = x + 1$ and $g(x) = x$ $(x > -2)$

$3^{g(x) \log_3 f(x)} = f(x)$ (Given equation)

$3^{x \log_3(x+2)} = x + 2$ (substitute the given functions)

$3^{\log_3(x+2)^x} = x + 2$

$(x+2)^x = x + 2$ $(b^{\log_b x} = x)$

This equation is true if either the exponent (x) is 1 or the base $(x + 2)$ is 1.

$x = 1$ or $x + 2 = 1$

$x = -1$

(Although the equation $(x+2)^x = x + 2$ is true if $x = -2$, the domain of $\log_3(x + 2)$ is $\{x \mid x > -2\}$.)

Thus,

$\boxed{x = \pm 1}$

53. $\log_2 x + \dfrac{1}{\log_x 2} = 4$

$\log_2 x + \log_2 x = 4 \qquad \left(\log_b x = \dfrac{1}{\log_x b} \right)$

$\qquad\qquad 2\log_2 x = 4$

$\qquad\qquad\ \ \log_2 x = 2$

$\qquad\qquad\qquad 2^2 = x$

$\qquad\qquad\qquad\ \ 4 = x$

The solution set is $\boxed{\{4\}}$.

Review Problems

55. line passing through $(4, 2)$ and $(-2, 3)$ has slope

$\qquad \dfrac{3-2}{-2-4} = -\dfrac{1}{6}$

slope of line perpendicular $= -\dfrac{1}{-1/6} = 6$

line passing through $(1, 5)$ with slope 6 is

$\qquad y - 5 = 6(x - 1)$

$\qquad y - 5 = 6x - 6$

$\qquad \boxed{y = 6x - 1}$

56. $\dfrac{7\sqrt{2}}{2\sqrt{2}-1} + \dfrac{\sqrt{2}}{\sqrt{2}-1} \qquad = \dfrac{7\sqrt{2}}{2(1)} + \dfrac{\sqrt{2}}{(\sqrt{2}-1)} \cdot \dfrac{(\sqrt{2}+1)}{(\sqrt{2}+1)}$

$\qquad\qquad\qquad\qquad = \dfrac{7\sqrt{2}}{2} + \dfrac{\sqrt{2}(\sqrt{2}+1)}{1}$

$\qquad\qquad\qquad\qquad = \dfrac{7\sqrt{2} + 2(2 + \sqrt{2})}{2}$

$\qquad\qquad\qquad\qquad = \boxed{\dfrac{4 + 9\sqrt{2}}{2}}$

57. $\qquad \dfrac{3}{y+1} - \dfrac{5}{y} = \dfrac{19}{y^2 + y}$

$\qquad\qquad \dfrac{3}{y+1} - \dfrac{5}{y} = \dfrac{19}{y(y+1)}$

$y(y+1)\left[\dfrac{3}{y+1} - \dfrac{5}{y} \right] = y(y+1)\left[\dfrac{19}{y(y+1)} \right]$

$\qquad\qquad 3y - 5(y+1) = 19$

$\qquad\qquad 3y - 5(y+1) = 19$

$\qquad\qquad 3y - 5y - 5 = 19 \qquad (y \neq 0, -1)$

$\qquad\qquad\qquad\quad -2y = 24$

$\qquad\qquad\qquad\qquad y = -12$

$\boxed{\{-12\}}$

Chapter 11   Review Problems

Review Problems, pp. 827-830

1.

$$f(x) = 4x - 3$$
$$y = 4x - 3$$
$$x = 4y - 3 \quad \text{(Exchange } x \text{ and } y\text{)}$$
$$x + 3 = 4y$$
$$\frac{x + 3}{4} = y$$
$$\boxed{f^{-1}(x) = \frac{x + 3}{4}}$$

$$f[f^{-1}(x)] = 4[f^{-1}(x)] - 3$$
$$= 4\left(\frac{x + 3}{4}\right) - 4$$
$$= x + 3 - 3$$
$$= x$$

$$f^{-1}[f(x)] = \frac{f(x) + 3}{4}$$
$$= \frac{(4x - 3) + 3}{4}$$
$$= \frac{4x}{4}$$
$$= x$$

2.

$$f(x) = \sqrt[3]{2x + 1}$$
$$y = \sqrt[3]{2x + 1}$$
$$x = \sqrt[3]{2y + 1} \quad \text{(Exchange } x \text{ and } y\text{)}$$
$$x^3 = \left(\sqrt[3]{2y + 1}\right)^3$$
$$x^3 = 2y + 1$$
$$x^3 - 1 = 2y$$
$$\frac{x^3 - 1}{2} = y$$
$$\boxed{f^{-1}(x) = \frac{x^3 - 1}{2}}$$

$$f[f^{-1}(x)] = \sqrt[3]{2[f^{-1}(x)] + 1}$$
$$= \sqrt[3]{2\frac{(x^3 - 1)}{2} + 1}$$
$$= \sqrt[3]{x^3 - 1 + 1} = \sqrt[3]{x^3} = x$$

$$f^{-1}[f(x)] = \frac{[f(x)]^3 - 1}{2}$$
$$= \frac{(\sqrt[3]{2x + 1})^3 - 1}{2}$$
$$= \frac{2x + 1 - 1}{2} = \frac{2x}{2} = x$$

3. a. Domain = $(-\infty, \infty)$
Range = $[0, \infty)$
no inverse function

b. Domain = $(-6, -2) \cup (-2, 2) \cup (2, 6)$
Range = $(-\infty, \infty)$
no inverse function

c. Domain = $[-3, 3]$
Range = $[-3, 3]$
inverse function exists

d. Domain = $[-4, 4]$
Range = $[0, 3]$
no inverse function

Graph **C** because all horizontal lines intersect the graph only once. **C** has an inverse that is a function and **C** is one to one.

4.
$$2^{4x-2} = 64$$
$$2^{4x-2} = 2^6$$
$$4x - 2 = 6$$
$$4x = 8$$
$$x = 2$$
$$\boxed{\{2\}}$$

5.
$$3^{x^2+4x} = \frac{1}{27}$$
$$3^{x^2+4x} = 3^{-3}$$
$$x^2 + 4x = -3$$
$$x^2 + 4x + 3 = 0$$
$$(x+3)(x+1) = 0$$
$$x+3 = 0 \quad \text{or} \quad x+1 = 1$$
$$x = -3 \quad \text{or} \quad x = -1$$
$$\boxed{\{-3, -1\}}$$

6.
$$5^{2x^2} = 25^{2-x}$$
$$5^{2x^2} = 5^{2(2-x)}$$
$$2x^2 = 2(2-x)$$
$$x^2 = 2-x$$
$$x^2 + x - 2 = 0$$
$$(x+2)(x-1) = 0$$
$$x = 1 \quad \text{or} \quad x = -2$$
$$\boxed{\{1, -2\}}$$

7.
$$6^{x^2+2x} = 1$$
$$6^{x^2+2x} = 6^0$$
$$x^2 + 2x = 0$$
$$x(x+2) = 0$$
$$x = 0 \quad \text{or} \quad x = -2$$
$$\boxed{\{0, -2\}}$$

8.
$$8^x = 2^{2x^2+1}$$
$$2^{3x} = 2^{2x^2+1}$$
$$3x = 2x^2 + 1$$
$$2x^2 - 3x + 1 = 0$$
$$(2x-1)(x-1) = 0$$
$$x = \frac{1}{2} \text{ or } x = 1$$
$$\boxed{\left\{\frac{1}{2}, 1\right\}}$$

9.
$$27^{1-2x} \cdot 3^{x+1} = 9^{x-3}$$
$$3^{3(1-2x)} \cdot 3^{x+1} = 3^{2(x-3)}$$
$$3^{3(1-2x)+x+1} = 3^{2(x-3)}$$
$$3 - 6x + x + 1 = 2x - 6$$
$$7x = 10$$
$$x = \frac{10}{7}$$
$$\boxed{\left\{\frac{10}{7}\right\}}$$

10.
$$A = P\left(1 + \frac{r}{N}\right)^{Nx}$$

($P = \$7500$, $r = 6\frac{1}{4}\% = 0.0625$, $N = 12$ (compounded monthly)):

$$A = 7500\left(1 + \frac{0.0625}{12}\right)^{12(3)}$$
$$= 7500(1.0052083)^{36}$$
$$= \boxed{\$9042.33}$$

11.
$$A = Pe^{rx}$$
$$= 7500e^{0.0625(3)}$$
$$= \boxed{\$9046.73}$$
$$\text{Additional money} = 9046.73 - 9042.33$$
$$= \boxed{\$4.40}$$

12. $f(x) = 16 \cdot 2^{-x}$

a.

x	0	1	2	3
$f(x)$	16	8	4	2

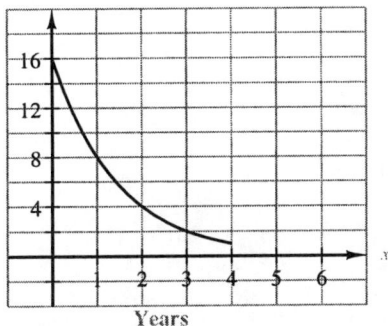

Years

b. $\dfrac{f(3)}{f(0)} = \dfrac{16 \cdot 2^{-3}}{16 \cdot 2^{0}}$

$\qquad = 2^{-3}$

$\qquad = \dfrac{1}{8}$

$\dfrac{f(3)}{f(0)} = \dfrac{1}{8}$

$f(3) = \dfrac{1}{8}f(0)$

$\boxed{\dfrac{1}{8} \text{ initial value.}}$

$\boxed{\text{At the end of 3 years the automobile is worth only } \dfrac{1}{8} \text{ of what it was worth when it was new.}}$

13. $f(x) = \dfrac{5000}{1 + 9\,e^{-0.2x}}$

a. $f(0) = \dfrac{5000}{1 + 9\,e^{0}}$

$\qquad = \dfrac{5000}{1 + 9}$

$\qquad = 500$

$\boxed{500}$ fish were stocked in the lake.

b. $f(10) = \dfrac{5000}{1 + 9\,e^{-0.2(10)}}$

$\qquad = \dfrac{5000}{1 + 9\,e^{-2}}$

$\qquad \approx \boxed{2254}$

c. $f(20) = \dfrac{5000}{1 + 9\,e^{-0.2(20)}}$

$\qquad = \dfrac{5000}{1 + 9\,e^{-4}}$

$\qquad \approx \boxed{4292}$

d. Consider $9e^{-0.2x} = \dfrac{9}{e^{0.2x}}$. As x gets larger, the denominator gets larger, but the numerator (9)

stays the same size. Thus, $\dfrac{9}{e^{0.2x}}$ gets closer and closer to 0. Consequently, the fish population

approaches $\dfrac{5,000}{1+0} = 5,000$ fish. $\boxed{5,000}$ fish will eventually inhabit the lake.

e.

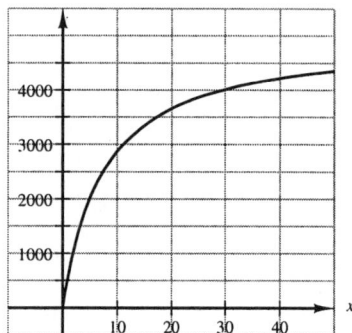

14. $\boxed{A = 100\left(\dfrac{1}{2}\right)^{t/5600}}$

t	0	2800	5600	11,200	15,000	16,800
A	100	70.7	50.0	25.0	15.6	12.5

The graph represents the exponential decay of the carbon–14.

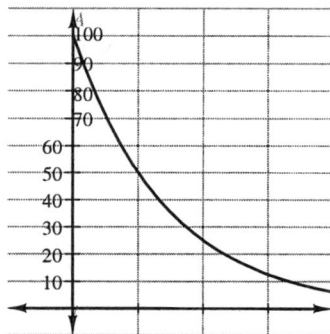

15.
$$A = 1000\left(\dfrac{1}{2}\right)^{t/30}$$
$$= 1000\left(\dfrac{1}{2}\right)^{100/30}$$
$$= 99.2 \text{ kg} < 100 \text{ kg}$$

$\boxed{\text{Yes}}$, the area will be safe after 100 years.

16. $f(x) = 3^x$

a.

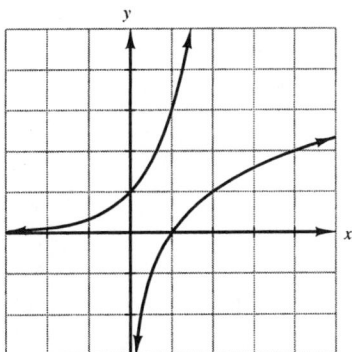

b.

$$
\begin{aligned}
f(x) &= 3^x \\
y &= 3^x \\
x &= 3^y \quad \text{(Exchange } x \text{ and } y.\text{)} \\
y &= \log_3 x \quad \text{(Rewrite in exponential form.)} \\
f^{-1}(x) &= \log_3 x
\end{aligned}
$$

c. See graph in part a.

17. $y = \log_2 (x + 1)$

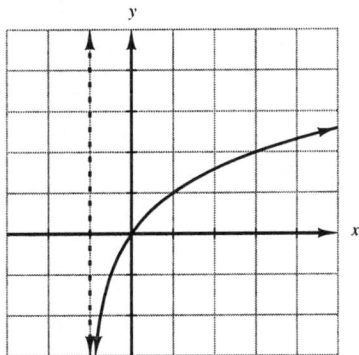

18. $y = (\log_2 x) + 1$

x	y
1	1
2	2
4	3
8	4
$\dfrac{1}{2}$	0
$\dfrac{1}{4}$	-1

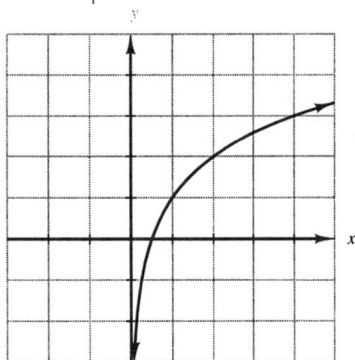

19.
$$\log_3 \frac{1}{27} = x$$
$$\boxed{3^x = \frac{1}{27}}$$
$$3^x = 3^{-3}$$
$$x = -3$$
$$\boxed{\{-3\}}$$

20. $\log_6 (3x + 4) = 2$
$$\boxed{6^2 = 3x + 4}$$
$$36 = 3x + 4$$
$$32 = 3x$$
$$\frac{32}{3} = x$$
$$\boxed{\left\{ \frac{32}{3} \right\}}$$

21.
$$\log_{25} x = -\frac{3}{2}$$
$$\boxed{(25)^{-3/2} = x}$$
$$\frac{1}{(25)^{3/2}} = x$$
$$\frac{1}{\left(\sqrt{25} \right)^3} = x$$
$$\frac{1}{5^3} = x$$
$$\frac{1}{125} = x$$
$$\boxed{\left\{ \frac{1}{125} \right\}}$$

22.
$$\log_4 8 = 2x + 1$$
$$\boxed{4^{2x+1} = 8}$$
$$(2^2)^{2x+1} = 2^3$$
$$2^{4x+2} = 2^3$$
$$4x + 2 = 3$$
$$4x = 1$$
$$x = \frac{1}{4}$$
$$\boxed{\left\{ \frac{1}{4} \right\}}$$

23.
$$\log_x 9 = 2$$

$$\boxed{x^2 = 9}$$

$$x = 3 \quad \text{(reject } x = -3\text{)}$$

$$\boxed{\{3\}}$$

24.
$$\log_x \frac{8}{27} = -3$$

$$\boxed{x^{-3} = \frac{8}{27}}$$

$$\frac{1}{x^3} = \frac{8}{27}$$

$$x^3 = \frac{27}{8}$$

$$x = \frac{3}{2}$$

$$\boxed{\left\{\frac{3}{2}\right\}}$$

25.
$$\log_6 \frac{x^3 y^2}{36} = \log_6 x^3 + \log_6 y^2 - \log_6 36$$
$$= \log_6 x^3 + \log_6 y^2 - \log_6 6^2$$
$$= 3\log_6 x + 2\log_6 y - 2\log_6 6$$
$$= \boxed{3\log_6 x + 2\log_6 y - 2}$$

26.
$$\log_2 \sqrt[5]{\frac{64}{x^3}} = \log_2 \left(\frac{64}{x^3}\right)^{1/5}$$
$$= \frac{1}{5}\log_2 \left(\frac{64}{x^3}\right)$$
$$= \frac{1}{5}[\log_2 64 - \log_2 x^3]$$
$$= \frac{1}{5}[\log_2 64 - \log_2 x^3]$$
$$= \frac{1}{5}(6 - 3\log_2 x)$$
$$= \boxed{\frac{6}{5} - \frac{3}{5}\log_2 x}$$

27.
$$\log \frac{B}{10^{7a}} = \log B - \log 10^{7a}$$
$$= \boxed{\log B - 7a} \quad (\log 10^x = x)$$

28.
$$\ln Ae^{9y^2} = \ln A + \ln e^{9y^2}$$
$$= \boxed{\ln A + 9y^2} \quad (\ln e^x = x)$$

29.
$$\log \frac{100x}{y^2 \sqrt{z}} = \log 100 + \log x - \log y^2 - \log \sqrt{z}$$
$$= 2 + \log x - 2\log y - \frac{1}{2}\log z$$
$$= \boxed{2 + \log x - 2\log y - \frac{1}{2}\log z}$$

30.
$$\log_5 \frac{\sqrt[3]{x}\sqrt[4]{y}}{625z} = \log_5 \sqrt[3]{x} + \log_5 \sqrt[4]{y} - \log_5 625 - \log_5 z$$
$$= \frac{1}{3}\log_5 x + \frac{1}{4}\log_5 y - \log_5 5^4 - \log_5 z$$
$$= \boxed{\frac{1}{3}\log_5 x + \frac{1}{4}\log_5 y - 4 - \log_5 z}$$

31. $\log_3 \sqrt{81}\sqrt[3]{x^2}\left(\sqrt[4]{y}\right)^3 = \log_3 \sqrt{81} + \log_3 \sqrt[3]{x^2} + \log_3 \left(\sqrt[4]{y}\right)$

$\qquad\qquad\qquad\qquad = \dfrac{1}{2}(4) + \dfrac{2}{3}\log_3 x + \dfrac{3}{4}\log_3 y$

$\qquad\qquad\qquad\qquad = \boxed{2 + \dfrac{2}{3}\log_3 x + \dfrac{3}{4}\log_3 y}$

32. $\log_b x + 2\log_b y = \log_b x + \log_b y^2$

$\qquad\qquad\qquad\quad = \boxed{\log_b xy^2}$

33. $3\log_b x - \dfrac{1}{2}\log_b y = 3\log_b x - \log_b \sqrt{y}$

$\qquad\qquad\qquad\qquad\quad = \boxed{\log_b \dfrac{x^3}{\sqrt{y}}}$

34. $\log_5 (x-2) - \log_5 (x^2-4) = \log_5 \dfrac{x-2}{x^2-4} \quad (x>2)$

$\qquad\qquad\qquad\qquad\qquad\quad = \log_5 \dfrac{x-2}{(x+2)(x-2)}$

$\qquad\qquad\qquad\qquad\qquad\quad = \boxed{\log_5 \dfrac{1}{x+2}}$

35. $\dfrac{1}{2}\log x + \dfrac{1}{2}\log y - 4\log z = \dfrac{1}{2}(\log x + \log y) - \log z^4$

$\qquad\qquad\qquad\qquad\qquad\qquad = \dfrac{1}{2}\log xy - \log z^4$

$\qquad\qquad\qquad\qquad\qquad\qquad = \log (xy)^{1/2} - \log z^4$

$\qquad\qquad\qquad\qquad\qquad\qquad = \boxed{\log \dfrac{\sqrt{xy}}{z^4}}$

36. $\dfrac{1}{3}(\ln x + 2\ln y - 3\ln z) = \dfrac{1}{3}(\ln x + \ln y^2) - \ln z$

$\qquad\qquad\qquad\qquad\qquad\qquad = \dfrac{1}{3}\ln xy^2 - \ln z$

$\qquad\qquad\qquad\qquad\qquad\qquad = \ln \sqrt[3]{xy^2} - \ln z$

$\qquad\qquad\qquad\qquad\qquad\qquad = \boxed{\ln \dfrac{\sqrt[3]{xy^2}}{z}}$

37. $\ln 2 + \ln \pi + \dfrac{1}{2}\ln (x+1) - \dfrac{2}{3}\ln x = \ln 2\pi + \ln \sqrt{x+1} - \ln x^{2/3}$

$\qquad\qquad\qquad\qquad\qquad\qquad\qquad = \boxed{\ln \dfrac{2\pi\sqrt{x+1}}{x^{2/3}}}$

38. $5 \log x + 3 \log (x + 1) - \dfrac{1}{2} \log (x - 1) = \log x^5 (x + 1)^3 - \dfrac{1}{2} \log (x - 1)$

$= \log x^5 (x + 1)^3 - \log \sqrt{x - 1}$

$= \boxed{\log \dfrac{x^5 (x + 1)^3}{\sqrt{x - 1}}}$

39. $\dfrac{1}{4} (\log_b x + \log_b y) - 2 \log_b (2x + 1) = \dfrac{1}{4} \log_b xy - 2 \log_b (2x + 1)$

$= \log_b \sqrt[4]{xy} - \log_b (2x + 1)^2$

$= \boxed{\log_b \dfrac{\sqrt[4]{xy}}{(2x + 1)^2}}$

40. $\log (1 + k) = \dfrac{0.3}{T}$

($T = 56$ years):

$\log (1 + k) = \dfrac{0.3}{56}$

$1 + k = 10^{0.3/56}$

$1 + k = 1.0124$

$k = \boxed{0.0124 \text{ or } 1.24\%}$

41. $t = \dfrac{1}{c} \ln \dfrac{A}{A - N}$

($A = 12$ mph, $c = 0.06$, $N = 5$ mph):

$t = \dfrac{1}{0.06} \ln \dfrac{12}{12 - 5}$

$t = \dfrac{1}{0.06} \ln \dfrac{12}{12 - 5}$

$= \dfrac{1}{0.06} \ln \dfrac{12}{7}$

≈ 9

$\boxed{\text{Approximately 9 weeks.}}$

42. a. $\boxed{\log_3 7 = \dfrac{\log 7}{\log 3}}$

b. $\boxed{\log_3 7 = \dfrac{\ln 7}{\ln 3}}$

43. $\boxed{B}$ is true;

$\ln (\log_5 60 - \log_5 12) = \ln \left(\log_5 \dfrac{60}{12} \right)$

$= \ln (\log_5 5)$

$= \ln 1$

$= 0$

44. $\log_5 97 - \log_{14} 7 = \dfrac{\log 97}{\log 5} - \dfrac{\log 7}{\log 14}$

$= 2.84243 - 0.73735$

$= \boxed{2.1051}$

45. $10^{\log 7x + \log 4x} = 10^{\log(7x)(4x)}$

$= 10^{\log 28 x^2}$

$= \boxed{28x^2}$ $[10^{\log y} = y]$

46. $e^{\ln 8x^3 - \ln 2x} = e^{\ln (8x^3/2x)}$

$= e^{\ln 4x^2}$

$= \boxed{4x^2}$ $[e^{\ln y} = y]$

47. $10^x = 72.3$

$\log 10^x = \log 72.3$

$\boxed{x = \log 72.3 \approx 1.86}$

48.
$$0.5 \cdot 10^{-0.2x} + 4.6 = 11.1$$
$$0.5 \cdot 10^{-0.2x} = 6.5$$
$$10^{-0.2x} = 13$$
$$\log 10^{-0.2x} = \log 13$$
$$-0.2x = \log 13$$
$$x = \frac{\log 13}{-0.2}$$
$$= \boxed{-5 \log 13 \approx -5.57}$$

(Scientific Calculator sequence:

13 $\boxed{\log}$ $\boxed{\div}$ 0.2 $\boxed{+/-}$ $\boxed{=}$)

49.
$$e^x = 47$$
$$\ln e^x = \ln 47$$
$$\boxed{x = \ln 47 \approx 3.85}$$

50.
$$1.2 e^{2.5x} - 8.1 = 21.3$$
$$1.2 e^{2.5x} = 29.4$$
$$e^{2.5x} = 24.5$$
$$\ln e^{2.5x} = \ln 24.5$$
$$2.5x = \ln 24.5$$
$$\boxed{x = \frac{\ln 24.5}{2.5} \approx 1.28}$$

51.
$$2^{x+1} = 9$$
$$\ln 2^{x+1} = \ln 9$$
$$(x+1)\ln 2 = \ln 9$$
$$x \ln 2 + \ln 2 = \ln 9$$
$$x \ln 2 = \ln 9 - \ln 2$$
$$x = \frac{\ln 9 - \ln 2}{\ln 2}$$
$$= \boxed{\frac{\ln 9}{\ln 2} - 1 \approx 2.17}$$

52.
$$y = 20,000 e^{0.03x}$$
$$(y = 50,000):$$
$$50,000 = 20,000 e^{0.03x}$$
$$2.5 = e^{0.03x}$$
$$0.03x = \ln 2.5$$
$$x = \frac{\ln 2.5}{0.03}$$
$$= \boxed{30.54 \text{ yr}}$$
$$\text{year} = 1980 + 31 = \boxed{2011}$$

53. a.
$$y = Ae^{kx}$$
If $x = 0$, $y = 226,549488$:
$$Ae^{k \cdot 0} = 226,549,488$$
$$\boxed{A = 226,549,488}$$
$$248,709,873 = 226,549,448 e^{10k}$$
$$1.0978 = e^{10k}$$
$$\ln 1.0978 = 10k$$
$$k = \frac{\ln 1.0978}{10}$$
$$k = \boxed{0.00933, \text{ or } 0.933\%}$$

b.
$$50,000,000 = 226,549,448 e^{0.00933x}$$
$$2.20702 = e^{0.00933x}$$
$$x = \frac{\ln 2.20702}{0.00933}$$
$$\approx 84.8$$
$$\text{year} = 1980 + 85 = \boxed{2065}$$

c.   Answers will vary.

54. $f(x) = 70 - 30e^{-kx}$

a. $f(0) = 70 - 30 = \boxed{40 \text{ °F}}$

b. $50 = 70 - 30e^{-k(1)}$

$\dfrac{-20}{-30} = e^{-k}$

$\ln \dfrac{2}{3} = -k$

$-k \approx -0.4055$

$\boxed{k \approx 0.4055}$

c. $65 = 70 - 30e^{0-0.4055x}$

$\dfrac{-5}{-30} = e^{-0.4055x}$

$\ln \dfrac{1}{6} = -0.4055x$

$x = \dfrac{-\ln 6}{-0.4055} \approx \boxed{4.42 \text{ hours}}$

55. $A = P\left(1 + \dfrac{r}{N}\right)^{Nx}$

($P = \$1000$, $A = \$4500$, $N = 4$ (compounded quarterly)):

$4500 = 1000\left(1 + \dfrac{0.04}{4}\right)^{4x}$

$4.5 = 1.01^{4x}$

$\ln 4.5 = 4x \ln 1.01$

$x = \dfrac{\ln 4.5}{4 \ln 1.01}$

$= \boxed{37.8 \text{ years}}$

56. $50,000 = P\left(1 + \dfrac{0.035}{12}\right)^{12(20)}$

$50,000 = P(1.002917)^{240}$

$50,000 = 2.0117P$

$P = \dfrac{50,000}{2.0117}$

$= \boxed{\$24,855}$

57. $A = \dfrac{A_0}{k}(e^{kx} - 1)$

$161,241 \times 10^{15} = \dfrac{250 \times 10^{15}}{0.02}(e^{0.02x} - 1)$

$15.73 = e^{0.02x} - 1$

$0.02x = \ln 16.73$

$x = \dfrac{\ln 16.73}{0.02} \approx 141 \text{ years}$

year $= 1976 + 141 = \boxed{2117}$

58. $f(x) = Ae^{kx}$
$3A = Ae^{kx}$
$e^{kx} = 3$
$\ln e^{kx} = \ln 3$
$kx = \ln 3$
$\boxed{x = \dfrac{\ln 3}{k}}$

59. $f(x) = 12 \cdot 2^{1.5x}$

a. $f(0) = 12 \cdot 2^{1.5(0)}$
$= 12 \cdot 2^0$
$= 12(1)$

$= 12$

$\boxed{12 \text{ crocodiles}}$

b.   The population would reach $12(2) = 24$ crocodiles.

$12 \cdot 2^{1.5x} = 24$
$2^{1.5x} = 2$
$1.5x = 1$
$x = \dfrac{1}{1.5}$

$= \dfrac{2}{3}$

$\boxed{\dfrac{2}{3} \text{ of a year.}}$

60.

$$\log_4(2x+1) = \log_4(x-3) + \log_4(x+5)$$

$$\log_4(2x+1) = \log_4(x-3)(x+5)$$

$$2x+1 = (x-3)(x+5)$$

$$2x+1 = x^2 + 2x - 15$$

$$16 = x^2$$

$$x = \pm\sqrt{16}$$

$$x = \pm 4$$

Reject –4 (produces log of a negative number, which is undefined.)

$$\boxed{\{4\}}$$

61.

$$\ln(x+1) - \ln 3 = \ln(1-2x)$$

$$\ln\frac{x+1}{3} = \ln(1-2x)$$

$$\frac{x+1}{3} = 1-2x$$

$$x+1 = 3(1-2x)$$

$$x+1 = 3-6x$$

$$7x = 2$$

$$x = \frac{2}{7}$$

$$\boxed{\left\{\frac{2}{7}\right\}}$$

62.

$$\log_2(x-5) - 5 = \log_2(3x+2) - 7$$

$$\log_2(x-5) - \log_2(3x+2) = -2$$

$$\log_2\frac{x-5}{3x+2} = -2$$

Exponential form:

$$2^{-2} = \frac{x-5}{3x+2}$$

$$\frac{1}{4} = \frac{x-5}{3x+2}$$

$$3x+2 = 4(x-5)$$

$$3x+2 = 4x-20$$

$$22 = x$$

$$\boxed{\{22\}}$$

63.

$$2\log_2 x - \log_2(x+4) = 1$$

$$\log_2\frac{x^2}{x+4} = 1$$

$$\frac{x^2}{x+4} = 2^1$$

$$x^2 = 2x+8$$

$$x^2 - 2x - 8 = 0$$

$$(x-4)(x+2) = 0$$

$$x = 4 \quad \text{or} \quad x = -2$$

$$x = 4 \quad (\text{reject } x = -2)$$

$$\boxed{\{4\}}$$

64.

$$\log_b(2x+1) + \log_b(x+2) = \log_b x$$

$$\log_b(2x+1)(x+2) = \log_b x$$

$$(2x+1)(x+2) = x$$

$$2x^2 + 5x + 2 = x$$

$$2x^2 + 4x + 2 = 0$$

$$x^2 + 2x + 1 = 0$$

$$(x+1)^2 = 0$$

$$x = -1 \text{ is not permitted (causes the log of a negative number) there is no solution.}$$

$$\boxed{\varnothing}$$

65.
$$\log(x+1) - \log(x+2) = \log\frac{1}{x}$$
$$\frac{x+1}{x+2} = \frac{1}{x}$$
$$x^2 + x = x + 2$$
$$x^2 - 2 = 0$$
$$x = \sqrt{2} \quad \left(\text{reject } x = -\sqrt{2}\right)$$

$$\boxed{\left\{\sqrt{2}\right\}}$$

66.
$$\log_\pi \pi^{2x-3} = 4$$
$$2x - 3 = 4$$
$$x = \frac{7}{2}$$

$$\boxed{\left\{\frac{7}{2}\right\}}$$

67.
$$\log(2x+3) = 0.73 + \log(x-1)$$
$$\log(2x+3) - \log(x-1) = 0.73$$
$$\log\frac{2x+3}{x-1} = 0.73$$
$$\frac{2x+3}{x-1} = 10^{0.73} \approx 5.3703$$
$$2x + 3 = 5.3703x - 5.3703$$
$$3.3703x = 8.3703$$
$$x = 2.4836$$

$$\boxed{\{2.4836\}}$$

68.
$$R = \log\frac{I}{I_0}$$
$$8.4 = \log\frac{I}{I_0}$$
$$I = I_0 \cdot 10^{8.4}$$
$$6.1 = \log\frac{I}{I_0}$$
$$I = I_0 \cdot 10^{6.1}$$
$$\frac{I_{8.4}}{I_{6.1}} = \frac{10^{8.4}}{10^{6.1}}$$
$$= 10^{2.3}$$
$$\approx 200$$

$\therefore$ the intensity is about $\boxed{200 \text{ times greater}}$ for the $R = 8.4$ quake as for the $R = 6.1$ quake.

69.
$$\log(1-r) = \frac{1}{T}\log\frac{W}{P}$$
$$\log(1-r) = \frac{1}{4}\log\frac{10{,}000}{20{,}000}$$
$$\log(1-r) = \frac{1}{4}\log\frac{1}{2}$$
$$\log(1-r) = \log\left(\frac{1}{2}\right)^{1/4}$$
$$1-r = \left(\frac{1}{2}\right)^{1/4} \approx 0.8409$$
$$r = 0.159, \text{ or } \boxed{15.9\%}$$

Cumulative Review Problems (Chapters 1-11)

Cumulative Review Problems, pp. 830-831

1.
$$\frac{2x}{x^2-1} - \frac{1}{x+3} = 0$$

$$(x+1)(x-1)(x+3)\left[\frac{2x}{(x+1)(x-1)} - \frac{1}{x+3}\right] = (x+1)(x-1)(x+3) = 0$$

$$x^2 + 6x + 1 = 0$$

$$x = \frac{-6 \pm \sqrt{36-4}}{2}$$

$$x = \frac{-6 \pm 4\sqrt{2}}{2} = -3 \pm 2\sqrt{2}$$

$$\boxed{\{-3 - 2\sqrt{2}, -2 + 2\sqrt{2}\}}$$

2.
$$(x^2 + 2x)^2 - 5(x^2 + 2x) + 6 = 0$$
$$t^2 - 5t + 6 = 0 \quad \text{(Let } t = x^2 + 2x)$$
$$(t-3)(t-2) = 0$$

$$t = 3 \quad \text{or} \quad t = 2$$
$$x^2 + 2x = 3 \qquad\qquad x^2 + 2x = 2$$
$$x^2 + 2x - 3 = 0 \qquad\qquad x^2 + 2x - 2 = 0$$

$$(x+3)(x-1) = 0 \qquad\qquad x = \frac{-2 \pm \sqrt{4+8}}{2}$$

$$x = -3 \quad \text{or} \quad x = 1 \qquad\qquad x = -1 \pm \sqrt{3}$$

$$\boxed{\{-3, 1, -1 - \sqrt{3}, -1 + \sqrt{3}\}}$$

3.
$$\frac{1}{x-1} < 2$$

$$\frac{1}{x-1} - 2 \cdot \frac{x-1}{x-1} < 0$$

$$\frac{1 - 2x + 2}{x-1} < 0$$

$$\frac{3 - 2x}{x-1} < 0$$

$$x = \frac{3}{2} \quad \text{or} \quad x = 1$$

T	F	T
1	3/2	

Test 0:　　　$\dfrac{3-0}{0-1} \; < \; 0$

　　　　　　　$-3 \; < \; 0$　　True

Test $1\frac{1}{4}$:　　$\dfrac{3-2\frac{1}{2}}{1\frac{1}{4}-1} \; < \; 0$

　　　　　　　$\dfrac{\frac{1}{2}}{\frac{1}{4}} \; < \; 0$

　　　　　　　$2 \; < \; 0$　　False

Test 2:　　　$\dfrac{3-4}{2-1} \; < \; 0$

　　　　　　　$\dfrac{-1}{1} \; < \; 0$

　　　　　　　$-1 \; < \; 0$　　True

$$\boxed{\left\{\, x \mid x < 1 \ \text{ or } \ x > \frac{3}{2} \right\} = (-\infty, 1) \cup \left(\frac{3}{2}, \infty\right)}$$

4.
$$\frac{y-2}{y^2+4y+3} + \frac{y-1}{y^2+y-6} \; = \; \frac{y+1}{y^2-y-2}$$

$$(y+3)(y+1)(y-2)\left[\frac{y-2}{(y+3)(y+1)} + \frac{y-1}{(y+3)(y-2)}\right] \; = \; \left[\frac{y+1}{(y-2)(y+1)}\right](y+3)(y+1)(y-2)$$

$$(y-2)^2 + (y+1)(y-1) \; = \; (y+1)(y+3)$$

$$y^2 - 4y + 4 + y^2 - 1 - y^2 - 4y - 3 \; = \; 0$$

$$y^2 - 8y \; = \; 0$$

$$y(y-8) \; = \; 0$$

$$y = 0 \quad \text{or} \quad y = 8$$

$$\boxed{\{0, 8\}}$$

5.
$$\frac{x}{3} + \frac{y}{3} \; = \; 1 \qquad (\times -3) \rightarrow \qquad -x - y \; = \; -3$$

$$\frac{x}{4} - y \; = \; 1 \qquad (\times 4) \qquad\qquad \underline{x - 4y \; = \; 4}$$

$$-5y \; = \; 1$$

$$x + y \; = \; 3 \qquad\qquad\qquad y \; = \; -\frac{1}{5}$$

$$x - \frac{1}{5} \; = \; 3$$

$$x \; = \; 3 + \frac{1}{5} = \frac{16}{5}$$

$$\boxed{\left\{\left(\frac{16}{5}, -\frac{1}{5}\right)\right\}}$$

6. $3x - 2y + z = 7$
$2x + 3y - z = 13$
$x - y + 2z = -6$

$$\begin{bmatrix} 3 & -2 & 1 & | & 7 \\ 2 & 3 & -1 & | & 13 \\ 1 & -1 & 2 & | & -6 \end{bmatrix} \quad \begin{matrix} R_2 - 2R_3 \to \\ \\ R_1 - 3R_3 \to \end{matrix} \quad \begin{bmatrix} 3 & -2 & 1 & | & 7 \\ 0 & 5 & -5 & | & 25 \\ 0 & 1 & -5 & | & 25 \end{bmatrix}$$

$$\tfrac{1}{5}R_2 \to \begin{bmatrix} 3 & -2 & 1 & | & 7 \\ 0 & 1 & -1 & | & 5 \\ 0 & 1 & -5 & | & 25 \end{bmatrix} \quad R_2 - R_3 \to \begin{bmatrix} 3 & -2 & 1 & | & 7 \\ 0 & 1 & -1 & | & 5 \\ 0 & 0 & 4 & | & -20 \end{bmatrix}$$

$$\begin{matrix} \tfrac{1}{3}R_1 \to \\ \\ \tfrac{1}{4}R_3 \to \end{matrix} \begin{bmatrix} 1 & -2/3 & 1/3 & | & 7/3 \\ 0 & 1 & -1 & | & 5 \\ 0 & 0 & 1 & | & -5 \end{bmatrix}$$

$x - \tfrac{2}{3}y + \tfrac{1}{3}z = \tfrac{7}{3}$
$y - z = 5$
$z = -5$ $y - (-5) = 5$
$y = 0$ $x - \tfrac{2}{3}(0) + \tfrac{1}{3}(-5) = \tfrac{7}{3}$
$x = \tfrac{12}{3} = 4$

$\boxed{\{(4, 0, -5)\}}$

7. $2x^2 + y^2 = 6$ $(\times -3) \to$ $-6x^2 - 3y^2 = -18$
$4x^2 + 3y^2 = 16$ $\underline{4x^2 + 3y^2 = 16}$
$-2x^2 = -2$
$x^2 = 1$
$x = \pm 1$
$x = \pm 1, \ 2(\pm 1)^2 + y^2 = 6$
$y^2 = 4$
$y = \pm 2$

If $x = 1, y = \pm 2$
If $x = -1, y = \pm 2$
$\boxed{\{(-1, -2), (-1, 2), (1, -2), (1, 2)\}}$

8. $4^{2x+1} = 32^{x+5}$
$2^{2(2x+1)} = 2^{5(x+5)}$
$2(2x+1) = 5(x+5)$
$4x + 2 = 5x + 25$
$-23 = x$
$\boxed{\{-23\}}$

9.

$$\log_5 2x + \log_5(x+1) = \log_5(3x+10)$$
$$\log_5 2x(x+1) = \log_5(3x+10)$$
$$2x^2 + 2x = 3x + 10$$
$$2x^2 - x - 10 = 0$$
$$(2x-5)(x+2) = 0$$
$$x = \frac{5}{2} \quad \text{or} \quad x = -2 \quad \text{(reject)}$$

$$\boxed{\left\{\frac{5}{2}\right\}}$$

10.

$$\sqrt{2y+5} - \sqrt{y+3} = 2$$
$$\sqrt{2y+5} = 2 + \sqrt{y+3}$$
$$2y+5 = 4 + 4\sqrt{y+3} + y + 3$$
$$y - 2 = 4\sqrt{y+3}$$
$$y^2 - 4y + 4 = 16y + 48$$
$$y^2 - 20y - 44 = 0$$
$$(y-22)(y+2) = 0$$
$$y = 22 \quad \text{or} \quad y = -2 \quad \text{(extraneous)}$$

$$\boxed{\{22\}}$$

11.

$$\{x \mid 3x+2 < 4\} \quad \cap \quad \{x \mid 4-x > 1\}$$
$$3x < 2 \qquad\qquad\quad -x > -3$$
$$x < \frac{2}{3} \qquad\qquad\quad x < 3$$

$$\boxed{\left\{x \mid x < \frac{2}{3}\right\} = \left(-\infty, \frac{2}{3}\right)}$$

12.

$$f(x) = 2x - 3$$
$$y = 2x - 3$$
$$x = 2y - 3 \quad \text{(Exchange } x \text{ and } y\text{)}$$
$$x + 3 = 2y$$
$$\frac{x+3}{2} = y$$

$$\boxed{f^{-1}(x) = \frac{x+3}{2}}$$

$$f[f^{-1}(x)] = f\left(\frac{x+3}{2}\right) = 2\left(\frac{x+3}{2}\right) - 3 = x + 3 - 3 = x$$

$$f^{-1}[f(x)] = f^{-1}(2x-3) = \frac{2x-3+3}{2} = \frac{2x}{2} = x$$

13. $3^{\log_3 7} - \log_6 36 + \log_4 32$

$$= 7 - \log_6 6^2 + \log_4 2^5$$

$$= 7 - 2 + \frac{5}{2} \qquad\qquad \left(\text{Note:} \quad \log_2 2^5 = x,\ 2^{2x} = 2^5,\ 2x = 5,\ x = \frac{5}{2}\right)$$

$$= 5 + 2\frac{1}{2}$$

$$= 7\frac{1}{2}$$

$$= \boxed{\frac{15}{2}}$$

14. $8a^4 + 16a^3b + ab^3 + 2b^4$
$= 8a^3(a + 2b) + b^3(a + 2b)$
$= (a + 2b)(8a^3 + b^3)$
$= \boxed{(a + 2b)(2a + b)(4a^2 - 2ab + b^2)}$

15. $25x^2 - 4y^2 = 100$
$\dfrac{x^2}{4} - \dfrac{y^2}{25} = 0$

hyperbola with center at (0, 0)
vertices at (±2, 0) and no y-intercept

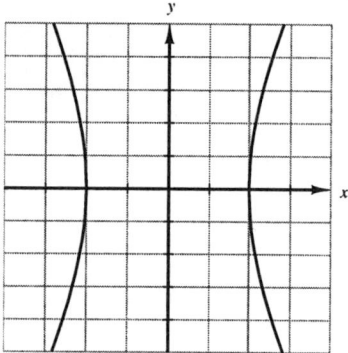

16. $f(x) = 2^{x+1}$
$y = 2^{x+1}$

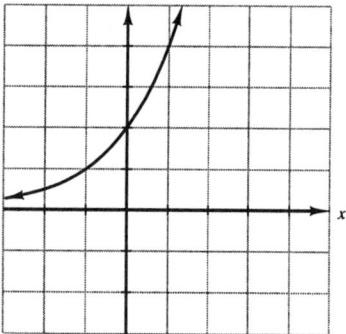

17. $f(x) = \log_3(x - 2)$
$y = \log_3(x - 2)$

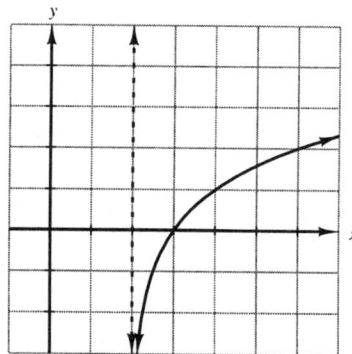

18. $y > x^2 - 4x + 1$

vertex: $x = -\dfrac{b}{2a} = \dfrac{4}{2} = 2$

$\quad\quad y = 2^2 - 4 \cdot 2 + 1 = -3$

Vertex $(2, -3)$

parabola with vertex at $(2, -3)$ and opens upward

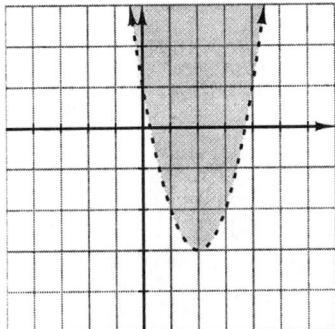

19. $\dfrac{(x-2)^2}{16} + \dfrac{(y+3)^2}{4} = 1$

ellipse with center at $(2, -3)$ and

vertices $(2 \pm 4, -2) = (6, -3)$ and $(-2, -3)$

$\quad\quad\quad\quad (2, -3 \pm 2) = (2, -1)$ and $(2, -5)$

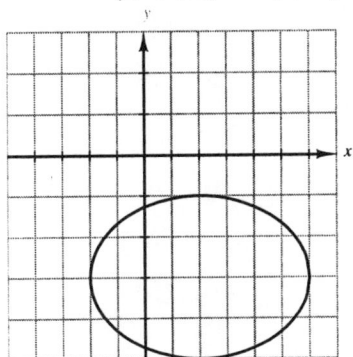

20. $x^2 + y^2 + 6x + 2y + 9 = 0$

$\quad\quad\quad\quad x^2 + 6x + y^2 + 2y = -9$

$(x^2 + 6x + 9) + (y^2 + 2y + 1) = -9 + 9 + 1$

$\quad\quad\quad (x + 3)^2 + (y + 1)^2 = 1$

circle with center at $(-3, -1)$ and radius 1

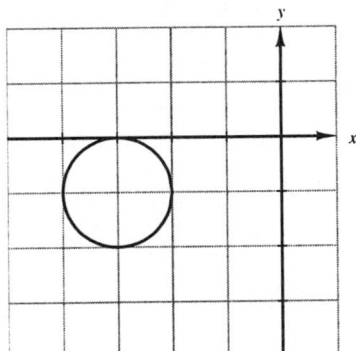

21. Let x = length of the ladder.

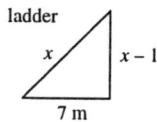

$$\begin{aligned} x^2 &= 7^2 + (x-1)^2 \\ x^2 &= 49 + x^2 - 2x + 1 \\ 2x &= 50 \\ x &= 25 \end{aligned}$$

$\boxed{25 \text{ meters}}$

22. Let x and y equal the dimensions of the rectangle.

$$\begin{aligned} 2x + 2y &= 10 \quad \text{(solve for } y) \rightarrow \\ x^2 + y^2 &= (\sqrt{13})^2 = 13 \quad \text{(substitute for } y) \rightarrow \end{aligned}$$

$$\begin{aligned} y &= 5 - x \\ x^2 + (5-x)^2 &= 13 \\ x^2 + 25 - 10x + x^2 &= 13 \\ 2x^2 - 10x + 12 &= 0 \\ x^2 - 5x + 6 &= 0 \\ (x-3)(x-2) &= 0 \end{aligned}$$

$$\begin{aligned} x &= 3 \\ y &= 5 - 3 = 2 \end{aligned} \quad \text{or} \quad \begin{aligned} x &= 2 \\ y &= 5 - 2 = 3 \end{aligned}$$

The dimensions are 2 ft by 3 ft.

$\boxed{\text{Width: 2 ft; Length: 3 ft}}$

23. $f = \dfrac{Kmv^2}{r}$

Given: $m = 2500$ pounds, $v = 40$ mph, $r = 800$ ft, $f = 1500$ pounds

$$1500 = \frac{K\,2500(40)^2}{800}$$

$$\frac{1500(800)}{2500(1600)} = K$$

$$\frac{3}{10} = K$$

$$f = \frac{3mv^2}{10r}$$

If $m = 4000$ pounds, $r = 600$ ft, $v = 30$ mph, $f = ?$

$$f = \frac{3(4000)(30)^2}{10(600)}$$

$$f = 1800$$

$\boxed{1800 \text{ lb}}$

24. $f(x) = x^3 - 2x - 5$

$$\dfrac{f(a+h) - f(a)}{h}$$

$$= \frac{[(a+h)^3 - 2(a+h) - 5] - (a^3 - 2a - 5)}{h}$$

$$= \frac{a^3 + 3a^2h + 3ah^2 + h^3 - 2a - 2h - 5 - a^3 + 2a + 5}{h}$$

$$= \frac{3a^2h + 3ah^2 + h^3 - 2h}{h}$$

$$= \boxed{3a^2 + 3ah + h^2 - 2}$$

25. parabola: $\quad y = x^2 - 4x - 1$

$$x = -\frac{b}{2a} = \frac{4}{2(1)} = 2$$

$$y = 2^2 - 4 \cdot 2 - 1 = -5$$

Vertex $(2, -5)$

circle: $\quad x^2 + y^2 - 12x - 6y + 33 = 0$

$$x^2 - 12x + y^2 - 6y = -33$$

$$x^2 - 12x + 36 + y^2 - 6y + 9 = -33 + 36 + 9$$

$$(x - 6)^2 + (y - 3)^2 = 12$$

center $(6, 3)$

distance between $(2, -5)$ and $(6, 3)$

$$\begin{aligned}
d &= \sqrt{(2 - 6)^2 + (-5 - 3)^2} \\
&= \sqrt{16 + 64} \\
&= \sqrt{80} \\
&= \boxed{4\sqrt{5} \approx 8.94}
\end{aligned}$$

26. line passing through the points $(-2, 3)$ and $(2, -3)$:

$$\text{slope} = \frac{-3 - 3}{2 - (-2)} = \frac{-6}{4} = \frac{-3}{2}$$

$$y - 3 = -\frac{3}{2}(x + 2)$$

$$y - 3 = -\frac{3}{2}x - 3$$

$$y = -\frac{3}{2}x$$

$$2y = -3x$$

$$\boxed{3x + 2y = 0}$$

27. $\quad \log_5(x + 2) - \log_5 x = 1$

$$\log_5 \frac{x + 2}{x} = 1$$

$$\frac{x + 2}{x} = 5^1$$

$$x + 2 = 5x$$

$$2 = 4x$$

$$\frac{1}{2} = x$$

$$x \text{ in the box} = \frac{1}{\frac{1}{2}} = 2$$

4	9	2
a	b	7
8	c	d

$$\text{Sum} = 4 + 9 + 2 = 15$$
$$4 + a + 8 = 15$$
$$a = 3$$
$$a + b + 7 = 3$$
$$3 + b + 7 = 15$$
$$b = 5$$
$$9 + b + c = 15$$
$$9 + 5 + c = 15$$
$$c = 1$$
$$2 + 7 + d = 15$$
$$d = 6$$

4	9	2
3	5	7
8	1	6

28. $\dfrac{2}{y+3} - \dfrac{1}{y^2 - 3y + 9} - \dfrac{54}{y^3 + 27}$

$= \dfrac{2}{y+3} \cdot \dfrac{y^2 - 3y + 9}{y^2 - 3y + 9} - \dfrac{1}{y^2 - 3y + 9} \cdot \dfrac{y+3}{y+3} - \dfrac{54}{(y+3)(y^2 - 3y + 9)}$

$= \dfrac{2(y^2 - 3y + 9) - (y + 3) - 54}{(y+3)(y^2 - 3y + 9)}$

$= \dfrac{2y^2 - 6y + 18 - y - 3 - 54}{(y+3)(y^2 - 3y + 9)}$

$= \dfrac{2y^2 - 7y - 39}{(y+3)(y^2 - 3y + 9)}$

$= \dfrac{(2y - 13)(y + 3)}{(y+3)(y^2 - 3y + 9)}$

$= \boxed{\dfrac{2y - 13}{y^2 - 3y + 9}}$

29. $\dfrac{\dfrac{2}{y+4} - \dfrac{1}{y-4}}{1 - \dfrac{128}{y^2 - 16}}$

$= \dfrac{\dfrac{2}{y+4} \cdot \dfrac{y-4}{y-4} - \dfrac{1}{y-4} \cdot \dfrac{y+4}{y+4}}{\dfrac{y^2 - 16}{y^2 - 16} - \dfrac{128}{y^2 - 16}}$

$= \dfrac{2(y-4) - (y+4)}{y^2 - 16} \div \dfrac{y^2 - 16 - 128}{y^2 - 16}$

$= \dfrac{2y - 8 - y - 4}{y^2 - 16} \cdot \dfrac{y^2 - 16}{y^2 - 144}$

$= \dfrac{y - 12}{(y-12)(y+12)}$

$= \boxed{\dfrac{1}{y + 12}}$

30. $\dfrac{\sqrt{6} + \sqrt{2}}{\sqrt{6} - \sqrt{2}}$

$= \dfrac{\sqrt{6} + \sqrt{2}}{\sqrt{6} - \sqrt{2}} \cdot \dfrac{\sqrt{6} + \sqrt{2}}{\sqrt{6} + \sqrt{2}}$

$= \dfrac{6 + 2\sqrt{12} + 2}{6 - 2}$

$= \dfrac{8 + 4\sqrt{3}}{4}$

$= \boxed{2 + \sqrt{3}}$

Algebra for College Students

Chapter 12   Sequences and Series

Section 12.1   Sequences

Problem Set 12.1, pp. 840-842

1.
$$a_n = 2n - 1$$
$$a_1 = 2(1) - 1 = 1$$
$$a_2 = 2(2) - 1 = 3$$
$$a_3 = 2(3) - 1 = 5$$
$$a_4 = 2(4) - 1 = 7$$
$$\boxed{1, 3, 5, 7, \ldots}$$

3.
$$a_n = n^2 + 2$$
$$a_1 = 1 + 2 = 3$$
$$a_2 = 4 + 2 = 6$$
$$a_3 = 9 + 1 = 11$$
$$a_4 = 16 + 2 = 18$$
$$\boxed{3, 6, 11, 18, \ldots}$$

5.
$$a_n = \frac{n}{n + 1}$$
$$a_1 = \frac{1}{1 + 1} = \frac{1}{2}$$
$$a_2 = \frac{2}{2 + 1} = \frac{2}{3}$$
$$a_3 = \frac{3}{4}$$
$$a_4 = \frac{4}{5}$$
$$\boxed{\frac{1}{2}, \frac{2}{3}, \frac{3}{4}, \frac{4}{5}, \ldots}$$

7.
$$a_n = \frac{1}{n^3}$$
$$a_1 = \frac{1}{1} = 1$$
$$a_2 = \frac{1}{8}$$
$$a_3 = \frac{1}{27}$$
$$a_4 = \frac{1}{64}$$
$$\boxed{1, \frac{1}{8}, \frac{1}{27}, \frac{1}{64}, \ldots}$$

9.
$$a_n = 2^n$$
$$a_1 = 2^1 = 2$$
$$a_2 = 2 = 4$$
$$a_3 = 2^3 = 8$$
$$a_4 = 2^4 = 16$$
$$\boxed{2, 4, 8, 16, \ldots}$$

11.
$$a_n = 3^{-n}$$
$$a_1 = 3^{-1} = \frac{1}{3}$$
$$a_2 = 3^{-2} = \frac{1}{3^2} = \frac{1}{9}$$
$$a_3 = 3^{-3} = \frac{1}{3^3} = \frac{1}{27}$$
$$a_4 = 3^{-4} = \frac{1}{3^4} = \frac{1}{81}$$
$$\boxed{\frac{1}{3}, \frac{1}{9}, \frac{1}{27}, \frac{1}{81}, \ldots}$$

13. $a_n = \dfrac{n-1}{n+2}$

$a_1 = \dfrac{1-1}{1+2} = 0$

$a_2 = \dfrac{1}{4}$

$a_3 = \dfrac{2}{5}$

$a_4 = \dfrac{3}{6}$

$\boxed{0, \dfrac{1}{4}, \dfrac{2}{5}, \dfrac{1}{2}, \cdots}$

15. $a_n = 1 + \dfrac{1}{n}$

$a_1 = 1 + \dfrac{1}{1} = 2$

$a_2 = 1 + \dfrac{1}{2} = \dfrac{3}{2}$

$a_3 = 1 + \dfrac{1}{3} = \dfrac{4}{3}$

$a_4 = 1 + \dfrac{1}{4} = \dfrac{5}{4}$

$\boxed{2, \dfrac{3}{2}, \dfrac{4}{3}, \dfrac{5}{4}, \cdots}$

17. $a_n = \dfrac{n(n+1)}{2}$

$a_1 = \dfrac{1(1+1)}{2} = 1$

$a_2 = \dfrac{2(2+1)}{2} = 3$

$a_3 = \dfrac{3(3+1)}{2} = 6$

$a_4 = \dfrac{4(4+1)}{2} = 10$

$\boxed{1, 3, 6, 10, ..}$

19. $a_n = (-1)^n n$

$a_1 = (-1)^1 \cdot 1 = -1$

$a_2 = (-1)^2 \cdot 2 = 2$

$a_3 = (-1)^3 \cdot 3 = -3$

$a_4 = (-1)^4 \cdot 4 = 4$

$\boxed{-1, 2, -3, 4, ..}$

21. $a_n = \dfrac{10}{2^{n-1}}$

$a_1 = \dfrac{10}{2^0} = 10$

$a_2 = \dfrac{10}{2^{2-1}} = \dfrac{10}{2}$

$a_3 = \dfrac{10}{2^{3-1}} = \dfrac{10}{4}$

$a_4 = \dfrac{10}{2^{4-1}} = \dfrac{10}{8}$

$\boxed{10, 5, \dfrac{5}{2}, \dfrac{5}{3}, \cdots}$

23. $a_n = 3n - 4$

$a_{12} = 3(12) - 4 = \boxed{32}$

25. $a_n = 2 - \dfrac{1}{n}$

$a_{20} = 2 - \dfrac{1}{20} = \boxed{\dfrac{39}{20}}$

27. $a_n = 3(2)^{2-n}$

$a_6 = 3(2)^{2-6}$

$= 3(2)^{-4} = \dfrac{3}{2^4} = \boxed{\dfrac{3}{16}}$

29. $2, 1, \frac{2}{3}, \frac{1}{2}, \ldots$

$$a_1 = 2 = \frac{2}{1}$$

$$a_2 = 1 = \frac{2}{2}$$

$$a_3 = \frac{2}{3}$$

$$a_4 = \frac{1}{2} = \frac{2}{4}$$

Thus, $\boxed{a_n = \frac{2}{n}}$.

31. $-1, 1, -1$

$$a_1 = -1 = (-1)^1$$

$$a^2 = 1 = (-1)^2$$

$$a_3 = -1 = (-1)^3$$

$$a_4 = 1 = (-1)^4$$

$$\boxed{a_n = (-1)^n}$$

33. $1, \frac{1}{4}, \frac{1}{9}, \frac{1}{16}, \ldots$

$$a_1 = 1 = \frac{1}{1^2}$$

$$a_2 = \frac{1}{4} = \frac{1}{2^2}$$

$$a_3 = \frac{1}{9} = \frac{1}{3^2}$$

$$a_4 = \frac{1}{16} = \frac{1}{4^2}$$

$$\boxed{a_n = \frac{1}{n^2}}$$

35.

100,	200,	400,
↑	↑	↑
end of	end of	end of
first hour	second hour	third hour

800,	1600,	3200
↑	↑	↑
a_4	a_5	a_6

After 6 hours: $\boxed{3200 \text{ bacteria}}$

After n hours:
$$a_1 = 100 = 100 \cdot 2^{1-1}$$
$$a_2 = 200 = 100 \cdot 2^{2-1}$$
$$a_3 = 400 = 100 \cdot 2^{3-1}$$
$$a_4 = 800 = 100 \cdot 2^{4-1}$$
$$a_n = 100 \cdot 2^{n-1} \quad \text{or} \quad 50(2^n)$$

After n hours: $\boxed{100 \cdot 2^{n-1} \text{ bacteria}}$

37. End of first year: 100,000
$$+0.05(100,000) = 105,000$$
End of 2 years: 105,000
$$+0.05(105,000) = 110,250$$
End of 3 years: 110,250
$$+0.05(110,250) = 115,762.50$$
End of 4 years: 115,762.50
$$+0.05(115,762.59) = 121,550.63$$

$\boxed{\$105,000; \$110,250; \$115,762.50; \$121,550.63, \ldots}$

39. a. $\boxed{1, 1, 2, 3, 5, 8, 13, 21, 34}$

b. $a_5 = a_4 + a_3 = 3 + 2 = 5$
$a_6 = a_5 + a_4 = 5 + 3 = 8$
$a_7 = a_6 + a_5 = 8 + 5 = 13$
$a_8 = a_7 + a_6 = 13 + 8 = 21$
$a_9 = a_8 + a_7 = 21 + 13 = 34$
$a_{10} = a_9 + a_8 = 34 + 21 = 55$
$a_{n+1} = a_n + a_{n-1}$
$a_1 = a_2 = 1$
$\boxed{1, 1, 2, 3, 5, 8, 13, 21, 34, 55}$

41. a.
$$a_1 = \frac{1}{2}$$
$$a_2 = \frac{1}{2} + \frac{1}{4} = \frac{3}{4}$$
$$a_3 = \frac{3}{4} + \frac{1}{8} = \frac{7}{8}$$
$$a_4 = \frac{7}{8} + \frac{1}{16} = \frac{15}{16}$$
$$a_5 = \frac{15}{16} + \frac{1}{32} = \frac{31}{32}$$
$$\boxed{\frac{1}{2}, \frac{3}{4}, \frac{7}{8}, \frac{15}{16}, \frac{31}{32}}$$

b. $\boxed{\text{The partial sums appear to be approaching 1. The number 1 will never be reached}}$; the value of the n-th partial sum is $1 - \frac{1}{2^n}$. The value approaches 1.

43. $\boxed{B}$ is true.

45. $\frac{4}{1}, \frac{9}{2}, \frac{16}{3}, \frac{25}{4}, \ldots$

$\frac{(1+1)^2}{1}, \frac{(2+1)^2}{2}, \frac{(3+1)^2}{3}, \frac{(4+1)^2}{4}, \ldots, \boxed{\frac{(n+1)^2}{n}}$

47. $a_n = \frac{1 + (-1)^{n+1}}{2i^{n-1}}$ $(i = \sqrt{-1})$

$$a_1 = \frac{1 + (-1)^2}{2i^0} = \frac{1+1}{2} = 1$$
$$a_2 = \frac{1 + (-1)^3}{2i} = 0$$
$$a_3 = \frac{1+1}{2i^2} = \frac{2}{-2} = -1$$
$$a_4 = 0$$
$$a_5 = \frac{2}{2i^4} = \frac{2}{2} = 1$$
$$a_6 = 0$$
$$a_7 \; \frac{2}{2i^6} = \frac{2}{-2} = -1$$
$$a_8 = 0$$
$$\boxed{1, 0, -1, 0, 1, 0, -1, 0}$$

Review Problems

51. $2y^6 + 16$
$= 2(y^6 + 8)$
$= 2[(y^2)^3 + 2^3]$
$2(y^2 + 2)[(y^2)^2 - y^2 \cdot 2 + 2^3]$
$= \boxed{2(y^2 + 2)(y^4 - 2y^2 + 4)}$

52. $\dfrac{x^2}{16} + \dfrac{y^2}{9} = 1$
Ellipse

x-intercepts: $\dfrac{x^2}{16} = 1$
$x^2 = 16$
$x = \pm 4$

y-intercepts: $\dfrac{y^2}{9} = 1$
$y^2 = 9$
$y = \pm 3$

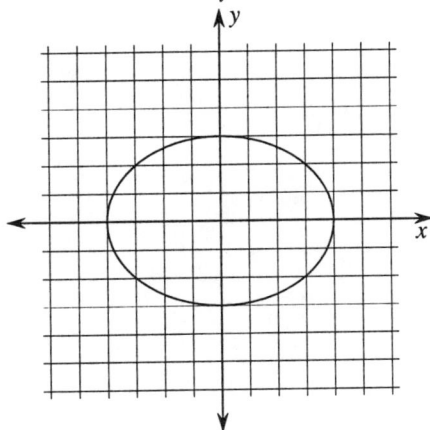

53. $\log_4 x + \log_4(x - 6) = 2$
$\log_4 x(x - 6) = 2$
Exponential form: $4^2 = x(x - 6)$
$16 = x^2 - 6x$
$0 = x^2 - 6x - 16$
$0 = (x - 8)(x + 2)$

$x - 8 = 0$ or $x + 2 = 0$
$x = 8$ $x = -2$ (reject; causes the log of a negative number)
$\boxed{\{8\}}$

Section 12.2 Series and Summation

Problem Set 12.2, pp. 847-849

1. $\displaystyle\sum_{i=1}^{4} 3i = 3(1) + 3(2) + 3(3) + 3(4)$
$= 3 + 6 + 9 + 12$
$= \boxed{30}$

3. $\displaystyle\sum_{i=2}^{6}(i^2 + 3)$ $= (2^2 + 3) + (3^2 + 3) + (4^2 + 3) + (5^2 + 3) + (6^2 + 3)$

$\qquad\qquad\qquad = 7 + 12 + 19 + 28 + 39$

$\qquad\qquad\qquad = \boxed{105}$

5. $\displaystyle\sum_{i=1}^{5}i(i + 4)$ $= 1(1 + 4) + 2(2 + 4) + 3(3 + 4) + 4(4 + 4) + 5(5 + 4)$

$\qquad\qquad\qquad = 5 + 12 + 21 + 32 + 45$

$\qquad\qquad\qquad = \boxed{115}$

7. $\displaystyle\sum_{i=1}^{4}(-1)^i$ $= (-1)^1 + (-1)^2 + (-1)^3 + (-1)^4$

$\qquad\qquad\qquad = -1 + 1 - 1 + 1$

$\qquad\qquad\qquad = \boxed{0}$

9. $\displaystyle\sum_{i=1}^{4}\left(-\frac{1}{2}\right)^i$ $= \left(-\frac{1}{2}\right)^1 + \left(-\frac{1}{2}\right)^2 + \left(-\frac{1}{2}\right)^3 + \left(-\frac{1}{2}\right)^4$

$\qquad\qquad\qquad = -\frac{1}{2} + \frac{1}{4} - \frac{1}{8} + \frac{1}{16}$

$\qquad\qquad\qquad = -\frac{8}{16} + \frac{4}{16} - \frac{2}{16} + \frac{1}{16}$

$\qquad\qquad\qquad = \boxed{-\frac{5}{16}}$

11. $\displaystyle\sum_{i=2}^{4}(-i)^i$ $= (-2)^2 + (-3)^3 + (-4)^4$

$\qquad\qquad\qquad = 4 - 27 + 256$

$\qquad\qquad\qquad = \boxed{233}$

13. $\displaystyle\sum_{i=3}^{5}\frac{2i - 1}{i - 1}$ $= \frac{2(3) - 1}{3 - 1} + \frac{2(4) - 1}{4 - 1} + \frac{2(5) - 1}{5 - 1}$

$\qquad\qquad\qquad = \frac{5}{2} + \frac{7}{3} + \frac{9}{4}$

$\qquad\qquad\qquad = \frac{30 + 28 + 27}{12}$

$\qquad\qquad\qquad = \boxed{\frac{85}{12}}$

15. $\displaystyle\sum_{i=1}^{4}x^i$ $= \boxed{x + x^2 + x^3 + x^4}$

17. $\displaystyle\sum_{i=4}^{7}x^{-i}$ $= \boxed{\frac{1}{x^4} + \frac{1}{x^5} + \frac{1}{x^6} + \frac{1}{x^7}}$

19. $\displaystyle\sum_{i=3}^{6}(x + i)$ $= (x + 3) + (x + 4) + (x + 5) + (x + 6)$

$\qquad\qquad\qquad = \boxed{4x + 18}$

21. $\displaystyle\sum_{i=1}^{4} i x^{i-1} = 1x^{1-1} + 2x^{2-1} + 3x^{3-1} + 4x^{4-1}$

$\phantom{\displaystyle\sum_{i=1}^{4} i x^{i-1}} = \boxed{1 + 2x + 3x^2 + 4x^3}$

23. $\displaystyle\sum_{i=3}^{6} \frac{x^i}{i^2} = \boxed{\dfrac{x^3}{9} + \dfrac{x^4}{16} + \dfrac{x^5}{25} + \dfrac{x^6}{36}}$

25. $\displaystyle\sum_{i=1}^{5} f(x_i)(x_i - x_{i-1}) = \boxed{f(x_1)(x_1 - x_0) + f(x_2)(x_2 - x_1) + f(x_3)(x_3 - x_2) + f(x_4)(x_4 - x_3) + f(x_5)(x_5 - x_4)}$

27. $\overline{x} = \dfrac{\displaystyle\sum_{i=1}^{4} x_i}{4}$

$\phantom{\overline{x}} = \dfrac{x_1 + x_2 + x_3 + x_4}{4}$

$\phantom{\overline{x}} = \dfrac{7.2 + 2.3 + 4.9 + 1.1}{4}$

$\phantom{\overline{x}} = \dfrac{15.5}{4}$

$\phantom{\overline{x}} = \boxed{3.875}$

29. $\overline{x} = \dfrac{\displaystyle\sum_{i=1}^{4} x_i}{4}$

$\phantom{\overline{x}} = \dfrac{0.01 + 0.01 + 0.1 + 0.4}{4}$

$\phantom{\overline{x}} = \dfrac{0.52}{4}$

$\phantom{\overline{x}} = 0.13$

31. $2 + 3 + 4 + 5 + 6 = \boxed{\displaystyle\sum_{i=2}^{6} i}$

$ = \boxed{\displaystyle\sum_{i=1}^{5} (i + 1)}$

33. $2 + 4 + 6 + 8 + 10 = 2(1) + 2(2) + 2(3) + 2(4) + 2(5)$

$ = \boxed{\displaystyle\sum_{i=1}^{5} 2i}$

35. $5 + 10 + 17 + 26 = (2^2 + 1) + (3^2 + 1) + (4^2 + 1) + (5^2 + 1)$

$ = \boxed{\displaystyle\sum_{i=1}^{4} (i^2 + 1)}$

37. $3 + 5 + 7 + 9 = [2(1) + 1] + [2(2) + 1] + [2(3) + 1] + [2(4) + 1]$

$ = \boxed{\displaystyle\sum_{i=1}^{4} (2i + 1)}$

39. $\dfrac{2}{3} + \dfrac{3}{4} + \dfrac{4}{5} + \dfrac{5}{6} + \dfrac{6}{7} + \dfrac{7}{8} = \boxed{\displaystyle\sum_{i=2}^{7} \dfrac{i}{i+1} \quad \text{or} \quad \displaystyle\sum_{i=1}^{6} \dfrac{i+1}{i+2}}$

41. $1 + x + x^2 + x^3 + x^4 + x^5 = x^0 + x^1 + x^2 + x^3 + x^4 + x^5$

$ = \boxed{\displaystyle\sum_{i=0}^{5} x^i \quad \text{or} \quad \displaystyle\sum_{i=1}^{6} x^{i-1}}$

43. $\dfrac{x}{x+3} + \dfrac{x}{x+4} + \dfrac{x}{x+5} + \dfrac{x}{x+6} = \boxed{\displaystyle\sum_{i=3}^{6} \dfrac{x}{x+i}}$

45. $x^3(x-3) + x^4(x-4) + x^5(x-5)$ $\boxed{\displaystyle\sum_{i=3}^{5} x^i(x-i)}$

47. $x_2 + x_3 + x_4 + x_5 = \boxed{\displaystyle\sum_{i=2}^{5} x_i}$

49. $x - \dfrac{x^2}{2} + \dfrac{x^3}{3} - \dfrac{x^4}{4} + \dfrac{x^5}{5}$

Consider:

$$x + \dfrac{x^2}{2} + \dfrac{x^3}{3} + \dfrac{x^4}{4} + \dfrac{x^5}{5} = \sum_{i=1}^{5} \dfrac{x^i}{i}$$

Now we must introduce the alternating sign for the terms.

Since $\displaystyle\sum_{i=1}^{5}(-1)^i = -1 + 1 - 1 + 1 - 1$, we see the signs are not alternating correctly. However,

$$\sum_{i=1}^{5}(-1)^{i+1} = (-1)^2 + (-1)^3 + (-1)^4 + (-1)^5 + (-1)^6$$
$$= 1 - 1 + 1 - 1 + 1,$$

and this follows the $+ - + - +$ pattern of the original series. Thus,

$$x - \dfrac{x^2}{2} + \dfrac{x^3}{3} - \dfrac{x^4}{4} + \dfrac{x^5}{5} = \boxed{\sum_{i=1}^{5}(-1)^{i+1} \cdot \dfrac{x^i}{i}}$$

51. $\boxed{D}$ is true;

$$\sum_{i=1}^{6} \dfrac{(-1)^i}{i^2} = \dfrac{(-1)^1}{1^2} + \dfrac{(-1)^2}{2^2} + \dfrac{(-1)^3}{3^2} + \dfrac{(-1)^4}{4^2} + \dfrac{(-1)^5}{5^2} + \dfrac{(-1)^6}{6^2}$$

$$\sum_{j=0}^{5} \dfrac{(-1)^{j+1}}{(j+1)^2} = \dfrac{(-1)^1}{1^2} + \dfrac{(-1)^2}{2^2} + \dfrac{(-1)^3}{3^2} + \dfrac{(-1)^5}{5^2} + \dfrac{(-1)^6}{6^2}$$

Thus,

$$\sum_{i=1}^{6} \dfrac{(-1)^i}{i^2} = \sum_{j=0}^{5} \dfrac{(-1)^{j+1}}{(j+1)^2}$$

53. $\boxed{C}$ is true;

$$\sum_{i=1}^{4} 3i + \sum_{i=1}^{4} 4i = 3\sum_{i=1}^{4} i + 4\sum_{i=1}^{4} i$$
$$= 7 \sum_{i=1}^{4} i$$
$$= \sum_{i=1}^{4} 7i$$

55. $$\sum_{i=-3}^{2}\frac{i}{i+4} = \frac{-3}{1}+\frac{-2}{2}+\frac{-1}{3}+0+\frac{1}{5}+\frac{2}{6}$$

$$\sum_{j=-20}^{-15}\frac{j+17}{j+21} = \frac{-3}{1}+\frac{-2}{2}+\frac{-1}{3}+0+\frac{1}{5}+\frac{2}{6}$$

$$\sum_{k=14}^{19}\frac{k-17}{k-13} = \frac{-3}{1}+\frac{-2}{2}+\frac{-1}{3}+0+\frac{1}{5}+\frac{2}{6}$$

Thus, $\boxed{\text{the sums are the same}}$:

$$\boxed{\frac{-3}{1}+\frac{-2}{2}+\frac{-1}{3}+0+\frac{1}{5}+\frac{2}{6}}$$

57. $$\sum_{i=1}^{4}\log 2i = \log 2 + \log 4 + \log 6 + \log 8$$

$$= \log 2 \cdot 4 \cdot 6 \cdot 8$$

$$= \boxed{\log 384}$$

59. $$\sum_{i=2}^{4}2i \log x = 4\log x + 6\log x + 8\log x$$

$$= \log x^4 + \log x^6 + \log x^8$$

$$= \log x^4 \cdot x^6 \cdot x^8$$

$$= \boxed{\log x^{18}}$$

61. $$\sum_{i=1}^{n}ka_i = ka_1 + ka_2 + ka_3 + ka_4 + \ldots + ka_n$$

$$= k(a_1 + a_2 + a_3 + a_4 + \ldots + a_n)$$

$$= \boxed{k\sum_{i=1}^{n}a_i}$$

63. $\displaystyle\sum_{i=1}^{6}a_i$ if $a_1 = 1$, $a_2 = 2$, and $a_n = a^2_{n-1} + a^2_{n-2}$ for $n \geq 3$

$$\sum_{i=1}^{6}a_i = 1 + 2 + (4+1) + (25+4) + (29^2 + 25) + (866^2 + 29^2)$$

$$= 1 + 2 + 5 + 29 + 866 + 750{,}797$$

$$= \boxed{751{,}700}$$

65. $$\sum_{i=1}^{n}a_i = \frac{1}{a_1}+\frac{1}{a_2}+\frac{1}{a_3}+\ldots+\frac{1}{a_n}$$

$$\frac{1}{\displaystyle\sum_{i=1}^{n}a_i} = \frac{1}{a_1 + a_2 + a_3 + \ldots + a_n}$$

$$\frac{1}{1}+\frac{1}{2}+\frac{1}{3}\ldots \neq \frac{1}{1+2+3+\ldots}$$

$$\frac{1}{a_1}+\frac{1}{a_2}+\frac{1}{a_3}+\ldots+\frac{1}{a_n} \neq \frac{1}{a_1 + a_2 + a_3 + \ldots + a_n}$$

The given statement is not true. $\boxed{\text{No.}}$

Review Problems

70. $\log_3 xy\sqrt[3]{z}$ $=$ $\log_3 x + \log_3 y + \log_3 z^{1/3}$

$= \boxed{\log_3 x + \log_3 y + \dfrac{1}{3}\log_3 z}$

71. $2x - 4y < 8$

Consider: $\qquad\qquad 2x - 4y = 8$

x–intercept: $\qquad\qquad 2x = 8$

$\qquad\qquad\qquad\qquad x = 4$

y–intercept: $\qquad\qquad -4y = 8$

$\qquad\qquad\qquad\qquad y = -2$

Test $(0, 0)$: $\qquad 2(0) - 4(0) < 8$

$\qquad\qquad\qquad\qquad\quad 0 < 8 \quad$ True

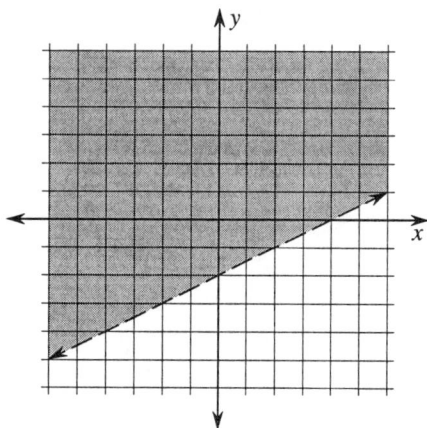

72.

$$5 - \sqrt{y + 5} = \sqrt{y}$$

$$\left(5 - \sqrt{y + 5}\right)^2 = \left(\sqrt{y}\right)^2$$

$$25 - 10\sqrt{y + 5} + y + 5 = y$$

$$30 + y - 10\sqrt{y + 5} = y$$

$$-10\sqrt{y + 5} = -30$$

$$\sqrt{y + 5} = 3$$

$$\left(\sqrt{y + 5}\right)^3 = 3^2$$

$$y + 5 = 9$$

$$y = 4$$

Check 4: $\qquad 5 - \sqrt{y + 5} = \sqrt{y}$

$\qquad\qquad\qquad 5 - \sqrt{9} = \sqrt{4}$

$\qquad\qquad\qquad\quad 5 - 3 = 2$

$\qquad\qquad\qquad\qquad\quad 2 = 2$

$\boxed{\{4\}}$

Section 12.3   Arithmetic Sequences

Problem Set 12.3, pp. 853-854

1. 2, 6, 10, 14, ...
$d = 6 - 2 = \boxed{4}$

3. −7, −2, 3, 8, ...
$d = -2 - (-7) = -2 + 7 = \boxed{5}$

5. $a, a + 5b, a + 10d, \ldots$
Common difference
$= a + 5d - a = \boxed{5d}$

7. $x - 8b, x - 5b, x - 2b, \ldots$
$d = x - 5b - (x - 8b) = -5b + 8b = \boxed{3b}$

9. 4, 7, 10, ...
$$\begin{aligned} d &= 7 - 4 = 3 \\ a_n &= a_1 + (n-1)d \\ a_{26} &= 4 + (26-1) \cdot 3 \\ a_{26} &= 4 + 25 \cdot 3 \\ a_{26} &= 4 + 75 \\ a_{26} &= \boxed{79} \end{aligned}$$

11. 3, 6, 9, ...
$$\begin{aligned} d &= 6 - 3 = 3 \\ a_n &= a_1 + (n-1)d \\ a_{15} &= 3 + (15-3) \cdot 3 \\ a_{15} &= 3 + (14)(3) \\ a_{15} &= 3 + 42 \\ a_{15} &= \boxed{45} \end{aligned}$$

13.
$$\begin{aligned} a_1 &= 9, d = 2 \\ a_n &= a_1 + (n-1)d \\ a_{16} &= 9 + (16-1)(2) \\ a_{16} &= 9 + (15)(2) \\ a_{16} &= 9 + 30 = \boxed{39} \end{aligned}$$

15.
$$\begin{aligned} a_1 &= 6, d = -\frac{1}{4} \\ a_n &= a_1 + (n-1)d \\ a_{10} &= 6 + (10-1)\left(-\frac{1}{4}\right) \\ a_{10} &= 6 - \frac{9}{4} = \boxed{\frac{15}{4}} \end{aligned}$$

17. 7, 10, 13, 16, ...
$$\begin{aligned} d &= 10 - 7 = 3 \\ a_n &= a_1 + (n-1)d \\ a_n &= -7 + (n-1) \cdot 3 \\ a_n &= 7 + 3n - 3 \\ a_n &= \boxed{3n + 4} \end{aligned}$$

19. 0, 5, 10, ...
$$\begin{aligned} d &= 5 - 0 = 5 \\ a_n &= a_1 + (n-1)d \\ a_n &= 0 + (n-1)5 \\ a_n &= \boxed{5n - 5} \end{aligned}$$

21. 14, −2, −18, ...
$$\begin{aligned} d &= -2 - 14 = -16 \\ a_n &= a_1 + (n-1)d \\ a_n &= 14 + (n-1)(-16) \\ a_n &= 14 - 16n + 16 \\ a_n &= \boxed{-16n + 30} \end{aligned}$$

23.
$$\begin{aligned} a_1 &= 2, d = 5 \\ a_n &= a_1 + (n-1)d \\ a_n &= 2 + (n-1)5 \\ a_n &= 2 + 5n - 5 \\ a_n &= \boxed{5n - 3} \end{aligned}$$

25.
$$\begin{aligned} a_1 &= 26, d = -10 \\ a_n &= a_1 + (n-1)d \\ a_n &= 26 + (n-1)(-10) \\ a_n &= \boxed{-10n + 36} \end{aligned}$$

27.
$$\begin{aligned} a_1 &= 60{,}000 \text{ (value during the first year)} \\ d &= -4{,}500 \\ a_n &= a_1 + (n-1)d \\ a_n &= 60{,}000 + (n-1)(-4{,}500) \\ a_n &= 60{,}000 - 4{,}500n + 4{,}500 \\ a_n &= \boxed{-4{,}500n + 64{,}500} \end{aligned}$$

29.
$$a_1 = 26{,}000, d = 3{,}500$$
$$a_n = a_1 + (n-1)d$$
Problem: What value of n will
result in $a_n = 99{,}500$?
$$99{,}500 = 26{,}000 + (n-1)(3{,}500)$$
$$99{,}500 = 26{,}000 + 3{,}500n - 3{,}500$$
$$99{,}500 = 22{,}500 + 3{,}500n$$
$$77{,}000 = 3{,}500n$$
$$n = \frac{77{,}000}{3{,}500} = 22$$

It will take $\boxed{22 \text{ years}}$ to reach
the maximum salary.

31.
$$a_1 = 22, d = 14.5$$
$$a_n = a_1 + (n-1)d$$
$$a_{30} = 22 + (30-1)(14.5)$$
$$a_{30} = 22 + (29)(14.5)$$
$$a_{30} = 442.5$$
$\boxed{442.5 \text{ cm}}$ above ground level

33. Company A:
$$a_1 = 12{,}000, d = 800$$
$$a_n = a_1 + (n-1)d$$
Year ten:
$$a_{10} = 12{,}000 + (10-1)(800)$$
$$= 19{,}200$$
Company B:
$$a_1 = 14{,}000, d = 500$$
Year ten:
$$a_{10} = 14{,}000 + (10-1)(500)$$
$$= 18{,}500$$
Year ten difference:
$$19{,}200 - 18{,}500 = 700$$
$\boxed{\text{Company } A \text{ pays \$700 more than Company } B \text{ in year 10.}}$

35.
$$a_n = a_1 + (n-1)d$$
$a_3 = 7$: $a_3 = a_1 + (3-1)d = 7$
$$a_1 + 2d = 7$$
$a_8 = 17$: $a_8 = a_1 + (8-1)d = 17$
$$a_1 + 7d = 17$$
$$a_1 + 2d = 7$$
$$\underline{a_1 + 7d = 17}$$
$$-5d = -10$$
$$d = 2$$
$$a_1 + 2d = 7$$
$$a_1 + 2(2) = 7$$
$$a_1 = 3$$
with $a_1 = 3$ and $d = 2$:
$$a_n = a_1 + (n-1)d$$
$$a_n = 3 + (n-1)(2)$$
$$\boxed{a_n = 2n+1}$$

37. $a_2 = 0$:

$$a_2 = a_1 + (2-1)d = 0$$
$$a_1 + d = 0$$

$a_9 = 35$:

$$a_9 = a_1 + (9-1)d = 35$$
$$a_1 + 8d = 35$$

$$a_1 + d = 0$$
$$\underline{a_1 + 8d = 35}$$
$$-7d = -35$$
$$d = 5$$
$$a_1 + d = 0$$
$$a_1 + 5 = 0$$
$$a_1 = -5$$

$$a_n = a_1 + (n-1)d$$
$$a_n = -5 + (n-1)(5)$$
$$\boxed{a_n = 5n - 10}$$

39. $\boxed{D}$ is true

$$a_1 = 5, a_3 = -3$$
$$a_3 = a_1 + (n-1)d$$
$$-3 = 5 + (3-1)d$$
$$-8 = 2d$$

$$-4 = d$$
$$a_4 = 5 + (4-1)(-4)$$
$$= 5 + 3(-4)$$
$$= -7 \text{ True}$$

41. $\dfrac{1}{1+\sqrt{2}}, \dfrac{3+2\sqrt{2}}{1+\sqrt{2}}, \dfrac{5+4\sqrt{2}}{1+\sqrt{2}}, \ldots$

$$a_1 = \frac{1}{1+\sqrt{2}}$$

$$d = \frac{3+2\sqrt{2}}{1+\sqrt{2}} - \frac{1}{1+\sqrt{2}}$$
$$= \frac{2+2\sqrt{2}}{1+\sqrt{2}}$$
$$= \frac{2\left(1+\sqrt{2}\right)}{1+\sqrt{2}}$$
$$= 2$$

$$a_n = a_1 + (n-1)d$$
$$a_n = \frac{1}{1+\sqrt{2}} + (2n-2) \cdot \frac{1+\sqrt{2}}{1+\sqrt{2}}$$
$$a_n = \frac{1 + 2n + 2n\sqrt{2} - 2 - 2\sqrt{2}}{1+\sqrt{2}}$$
$$\boxed{a_n = \frac{(2n-1) + (2n-2)\sqrt{2}}{1+\sqrt{2}}}$$

43. $3 - x, x, \sqrt{9-2x}$

The common difference is obtained by subtracting consecutive terms.

Common difference: $x - (3-x) = 2x - 3$

Common difference: $\sqrt{9-2x} - x$

Thus:

$$2x - 3 = \sqrt{9-2x} - x$$
$$3x - 3 = \sqrt{9-2x}$$
$$(3x-3)^2 = \left(\sqrt{9-2x}\right)^2$$
$$9x^2 - 18x + 9 = 9 - 2x$$
$$9x^2 - 16x = 0$$
$$x(9x - 16) = 0$$
$$x = 0 \text{ or } 9x - 16 = 0$$
$$x = \frac{16}{9}$$

0 is extraneous. Thus $\boxed{x = \dfrac{16}{9}}$.

Review Problems

47.

$$\begin{array}{rcl} x^2 + 4y^2 &=& 13 \\ x^2 - y^2 &=& 8 \end{array}$$

$\longrightarrow$

$(\times -1) \longrightarrow$

$$\begin{array}{rcl} x^2 + 4y^2 &=& 13 \\ \underline{-x^2 + y^2} &=& \underline{-8} \\ 5y^2 &=& 5 \quad \text{(add)} \\ y^2 &=& 1 \\ y &=& \pm 1 \end{array}$$

If $y = \pm 1$:

$$\begin{array}{rcl} x^2 - y^2 &=& 8 \\ x^2 - 1 &=& 8 \\ x^2 &=& 9 \\ x &=& \pm 3 \end{array}$$

$$\boxed{\{(3, 1), (3, -1), (-3, 1), (-3, -1)\}}$$

48. Resistance: R

Current: I

$$R = \frac{k}{I^2}$$

Given $I = 0.8$ ampere, $R = 50$ ohms

$$50 = \frac{k}{(0.8)^2}$$

$$50 = \frac{k}{0.64}$$

$$(50)(0.64) = k$$

$$k = 32$$

$$R = \frac{32}{I^2}$$

If $I = 0.5$ ampere, $R = ?$

$$R = \frac{32}{(0.5)^2}$$

$$R = \frac{32}{0.25}$$

$$R = 128$$

$$\boxed{128 \text{ ohms}}$$

49.

$$\frac{4x^2}{(x+y)(x-y)} + \frac{x+y}{x-y} \cdot \frac{(x+y)}{(x+y)} - \frac{x-y}{x+y} \cdot \frac{(x-y)}{(x-y)}$$

$$= \frac{4x^2 + (x+y)(x+y) - (x-y)(x-y)}{(x+y)(x-y)}$$

$$= \frac{4x^2 + x^2 + 2xy + y^2 - x^2 + 2xy - y^2}{(x+y)(x-y)}$$

$$= \frac{4x^2 + 4xy}{(x+y)(x-y)}$$

$$= \frac{4x(x+y)}{(x+y)(x-y)}$$

$$= \boxed{\frac{4x}{x-y}} \quad x \neq y$$

Section 12.4 Arithmetic Series

Problem Set 12.4, pp. 858-860

1. $4, 10, 16, 22, \ldots$ $d = 6$

$$a_n = a_1 + (n-1)d$$
$$a_{20} = 4 + (20-1)6$$
$$a_{20} = 118$$

$$S_n = \frac{n}{2}(a_1 + a_n)$$

$$S_{20} = \frac{n}{2}(a_1 + a_{20})$$

$$= \frac{20}{2}(4 + 118)$$

$$= \boxed{1{,}220}$$

3. $-15, -7, 1, 9, \ldots;$ $d = 8$

$$a_n = a_1 + (n-1)d$$
$$a_{10} = -15 + (10-1)(8)$$
$$= 57$$

$$S_n = \frac{n}{2}(a_1 + a_n)$$

$$S_{10} = \frac{10}{2}(-15 + 57)$$

$$= \boxed{210}$$

5. $100 + 95 + 90 + \ldots + 10$

$$S_n = \frac{n}{2}(a_1 + a_n)$$

$a_1 = 100$ and $a_n = 10$
We must find n (the number of terms we are adding).

$$a_n = a_1 + (n-1)d$$
$100, 95, 90, \ldots, 10 (d = -5)$

$$10 = 100 + (n-1)(-5)$$

$$10 = 100 - 5n + 5$$
$$10 = 105 - 5n$$
$$5n = 95$$
$$n = 19$$

$$S_{19} = \frac{19}{2}(100 + 10)$$

$$S_{19} = 9.5(110)$$

$$S_{19} = \boxed{1045}$$

7.

$$S_n = \frac{n}{2}(a_1 + a_n)$$

$$S_{25} = \frac{25}{2}(a_1 + a_{25})$$

$$a_1 = -9, \quad d = 5$$

We must find a_{25}.

$$a_n = a_1 + (n-1)d$$
$$a_{25} = -9 + (25-1)(5)$$
$$a_{25} = 111$$

$$S_{25} = \frac{25}{2}(-9 + 111)$$

$$S_{25} = \boxed{1275}$$

9.

$$S_n = \frac{n}{2}(a_1 + a_n)$$

$(n = 40, a_1 = 50, d = -3)$

$$S_{40} = \frac{40}{2}(50 + a_{40})$$

Find a_{40}.

$$a_n = a_1 + (n-1)d$$

$$a_{40} = 50 + (40-1)(-3)$$

$$a_{40} = -67$$

$$S_{40} = \frac{40}{2}(50 - 67)$$

$$S_{40} = \boxed{-340}$$

11.

$$S_n = \frac{n}{2}(a_1 + a_n)$$

$$\left(n = 12, a_1 = \frac{1}{2}, d = -\frac{1}{2}\right)$$

$$S_{12} = \frac{12}{2}\left(\frac{1}{2} + a_{12}\right)$$

Find a_{12}.

$$a_n = a_1 + (n-1)d$$

$$a_{12} = \frac{1}{2} + (12-1)\left(-\frac{1}{2}\right)$$

$$a_{12} = \frac{1}{2} - \frac{11}{2} = -5$$

$$S_{12} = \frac{12}{2}\left(\frac{1}{2} - 5\right)$$

$$S_{12} = 6(-4.5)$$

$$S_{12} = \boxed{-27}$$

13. Company A: $a_1 = 15,000 \quad d = 500$

$$S_{10} = \frac{n}{2}(a_1 + a_n)$$
$$= \frac{10}{2}(15,500 + a_{10})$$
$$a_{10} = a_1 + (10 - 1)d$$
$$a_{10} = 15,000 + 9(500)$$
$$= 19,500$$
$$S_{10} = \frac{10}{2}(15,000 + 19,500)$$
$$= 172,500$$

Company B: $a_1 = 16,000 \quad d = 400$

$$S_{10} = \frac{10}{2}(16,000 + a_{10})$$
$$a_{10} = 16,000 + 9(400)$$
$$= 19,600$$
$$S_{10} = \frac{10}{2}(16,000 + 19,600)$$
$$= 178,000$$

Company B will pay the greater amount.
($178,000, as opposed to $172,500).

15. $16, 48, 80, \ldots$

$a_1 = 16, \, d = 32$

$$a_6 = 16 + (6 - 1)(32)$$
$$= 176$$

176 feet during the sixth second

$$S_6 = \frac{6}{2}(a_1 + a_6)$$
$$= 3(16 + 176)$$
$$= 576$$

falls 576 feet during first six seconds

17. $70, 78, 86, \ldots$
$a_1 = 70, \, d = 8$

$$S_n = \frac{n}{2}(a_1 + a_n)$$
$$S_{24} = \frac{24}{2}(70 + a_{24})$$
$$a_{24} = a_1 + (24 - 1)d$$
$$a_{24} = 70 + (24 - 1)(8)$$
$$= 254$$
$$S_{24} = \frac{24}{2}(70 + 254)$$
$$= 3,888$$

3,888 seats

19. $\displaystyle\sum_{i=1}^{17}(5i + 3) = [5(1) + 3] + [5(2) + 3] + [5(3) + 3] + \ldots + [5(17) + 3]$

$$= 8 + 13 + 18 + \ldots + 88$$
$$S_n = \frac{n}{2}(a_1 + a_n)$$
$$S_{17} = \frac{17}{2}(a_1 + a_{17})$$
$$S_{17} = \frac{17}{2}(8 + 88)$$
$$= \boxed{816}$$

21. $\displaystyle\sum_{i=1}^{30}(-3i+5) = 2+(-1)+(-4)+\ldots+(-85)$

$$S_{30} = \frac{30}{2}(a_1+a_{30})$$

$$S_{30} = \frac{30}{2}(2-85)$$

$$= \boxed{-1{,}245}$$

23. $\displaystyle\sum_{i=1}^{100}4i = 4+8+12+\ldots+400$

$$S_{100} = \frac{100}{2}(a_1+a_{100})$$

$$S_{100} = \frac{100}{2}(4+400)$$

$$= \boxed{20{,}200}$$

25. Find S_{12} given that $a_1=3$ and $a_{10}=30$.

$$S_{10} = \frac{n}{2}(a_1+a_n)$$

$$S_{12} = \frac{12}{2}(3+a_{12})$$

Find a_{12}.

Given: $\begin{aligned} a_{10} &= 30 \\ a_n &= a_1+(n-1)d \\ a_{10} &= 3+(10-1)d=30 \\ 3+9d &= 30 \\ 9d &= 27 \\ d &= 3 \end{aligned}$

$$\begin{aligned} a_{12} &= a_1+(12-1)d \\ a_{12} &= 3+11(3)=36 \\ S_{12} &= \frac{12}{2}(3+36) \\ &= 6(39) \\ &= \boxed{234} \end{aligned}$$

27. Find S_{18} given that $a_1=-4$ and $a_8=-39$.

$$S_{18} = \frac{18}{2}(-4+a_{18})$$

Find a_{18}.

Given: $\begin{aligned} a_8 &= -39 \\ a_n &= a_1+(n-1)d \\ a_8 &= -4+(8-1)d \\ &= -39 \\ 7d &= -35 \\ d &= -5 \end{aligned}$

$$\begin{aligned} a_{18} &= a_1+(18-1)d \\ &= -4+17(-5) \\ &= -89 \\ S_{18} &= \frac{18}{2}(-4-89) \\ &= \boxed{-837} \end{aligned}$$

29. $2+4+6+\ldots+60$

$$S_n = \frac{n}{2}(a_1+a_n)$$

$$S_{30} = \frac{30}{2}(a_1+a_{30}) \quad a_1=2,\ d=2$$

Find a_{30}.

$$\begin{aligned} a_n &= a_1+(n-1)d \\ a_{30} &= 2+(30-1)(2) \\ a_{30} &= 60 \\ S_{30} &= \frac{30}{2}(2+60) \\ &= \boxed{930} \end{aligned}$$

31. $1 + 3 + 5 + \ldots + (18 \cdot 2 - 1) = 1 + 3 + 5 + \ldots + 35$

$$S_n = \frac{n}{2}(a_1 + a_n)$$

$$S_{18} = \frac{18}{2}(a_1 + a_{18}) \quad a_1 = 1,\, d = 2$$

Find a_{18}.

$$
\begin{aligned}
a_n &= a_1 + (n-1)d \\
a_{18} &= 1 + (18 - 1)(2) \\
&= 35 \\
S_{18} &= \frac{18}{2}(1 + 35) \\
&= \boxed{324}
\end{aligned}
$$

33. $22 + 24 + 26 + \ldots + 44 \quad d = 22$

$$S_n = \frac{n}{2}(a_1 + a_n) \quad a_1 = 22,\, a_n = 44$$

Find n.

$$
\begin{aligned}
a_n &= a_1 + (n-1)d \\
44 &= 22 + (n-1)2 \\
44 &= 22 + 2n - 2 \\
44 &= 20 + 2n \\
24 &= 2n \\
12 &= n
\end{aligned}
$$

$$
\begin{aligned}
S_n &= \frac{12}{2}(22 + 44) \\
S_n &= 6(66) \\
&= \boxed{396}
\end{aligned}
$$

35. $73 + 79 + 85 + \ldots$
$a_1 = 73,\, d = 6$

$$S_n = \frac{n}{2}(a_1 + a_n)$$

$$4{,}077 = \frac{n}{2}(73 + a_n)$$

Find n.

$$
\begin{aligned}
a_n &= a_1 + (n-1)d \\
a_n &= 73 + (n-1)6 \\
a_n &= 73 + 6n - 6 \\
a_n &= 6n + 67
\end{aligned}
$$

$$
\begin{aligned}
4{,}077 &= \frac{n}{2}(73 + 6n + 67) \\
8154 &= n(6n + 140) \\
8154 &= 6n^2 + 140n \\
0 &= 6n^2 + 140n - 8154 \\
0 &= 3n^2 + 70n - 4077
\end{aligned}
$$

$$n = \frac{-b \pm \sqrt{b^2 - 4ac}}{2a}$$

$$= \frac{-70 \pm \sqrt{(70)^2 - 4(3)(-4077)}}{2(3)}$$

$$= \frac{-70 \pm \sqrt{53,824}}{6}$$

$$= \frac{-70 \pm 232}{6}$$

$$n = \frac{-70 + 232}{6} \quad \text{or} \quad n = \frac{-70 - 232}{6}$$

$$= 27 \qquad\qquad\qquad = \frac{-302}{6} \quad \text{(reject, the number of rows must be a natural number)}$$

$\boxed{27 \text{ rows}}$

37. Let d = the fixed sum.

$700, 700 + d, 700 + 2d, \ldots$

$$S_n = \frac{n}{2}(a_1 + a_n)$$

$$S_8 = \frac{8}{2}(a_1 + a_n)$$

$$S_8 = 4(700 + a_8) = 6580$$

$$2800 + 4a_8 = 6580$$
$$4a_8 = 3780$$
$$a_8 = 945$$

$$a_n = a_1 + (n - 1)d$$
$$a_8 = a_1 + (8 - 1)d$$
$$945 = 700 + 7d$$
$$7d = 245$$
$$d = 35$$

Fixed sum: $\boxed{\$35}$.

39. $\boxed{D}$ is true:

$$\sum_{i=1}^{n} i = 1 + 2 + 3 + \ldots n$$

$$a_1 = 1, \, a_n = n$$

$$S_n = \frac{n}{2}(1 + n)$$

$$= \frac{n(n + 1)}{2} \quad \text{True}$$

41. $1 + 2 + 3 + 4 + \ldots = 4950$
Given: $a_1 = 1, \, d = 1, \, S_n = 4950$
Find n.

$$S_n = \frac{n}{2}(a_1 + a_n)$$

$$4950 = \frac{n}{2}(1 + a_n)$$

$$a_n = a_1 + (n - 1)d$$
$$a_n = 1 + (n - 1)(1)$$
$$a_n = n$$

$$4950 = \frac{n}{2}(1 + n)$$
$$9900 = n + n^2$$
$$n^2 + n - 9900 = 0$$
$$(n - 99)(n + 100) = 0$$
$$n = 99 \quad \text{or} \quad n = -100 \quad \text{(reject)}$$

The first $\boxed{99 \text{ numbers}}$ must be added.

43. $\displaystyle\sum_{i=1}^{n}(ai + b) = (a + b) + (2a + b) + (3a + b) + \ldots + (na + b)$

$$S_n = \frac{n}{2}(a_1 + a_n)$$

$$a_1 = a + b$$
$$a_n = na + b$$
$$n = n$$

$$S_n = \frac{n}{2}(a + b + na + b)$$

$$S_n = \frac{n}{2}(a + na + 2b)$$

$$S_n = \frac{na}{2} + \frac{n^2a}{2} + nb$$

$$S_n = \boxed{\frac{n(n+1)a}{2} + nb \quad \text{or} \quad \frac{n}{2}(a + na + 2b)}$$

45. $\displaystyle\sum_{i=1}^{2n+1} i = 1 + 2 + 3 + \ldots + (2n + 1)$

$$S_n = \frac{n}{2}(a_1 + a_n) \quad a_1 = 1, \, a_n = 2n + 1, \, n = 2n + 1$$

$$= \frac{2n+1}{2}(1 + 2n + 1)$$

$$= \frac{2n+1}{2} \cdot (2n + 2)$$

$$= \frac{2n+1}{2} \cdot \frac{2(n+1)}{1}$$

$$= (2n + 1)(n + 1)$$

$$\frac{1}{2n+1}\sum_{i=1}^{2n+1} i = \frac{1}{2n+1}(2n + 1)(n + 1)$$

$$= \boxed{n + 1}$$

47. $\dfrac{1}{1 + \sqrt{c}}, \dfrac{1}{1 - c}, \dfrac{1}{1 - \sqrt{c}}, \ldots$

Rationalize the denominator of the first term.

$$\frac{1}{1 + \sqrt{c}} \cdot \frac{1 - \sqrt{c}}{1 - \sqrt{c}} = \frac{1 - \sqrt{c}}{1 - c}$$

Now find the common difference.

$$d = a_2 - a_1$$

$$= \frac{1}{1 - c} - \left(\frac{1 - \sqrt{c}}{1 - c}\right)$$

$$= \frac{\sqrt{c}}{1 - c}$$

Now find a_n.

$$\begin{aligned}
a_n &= a_1 + (n-1)d \\
&= \frac{1-\sqrt{c}}{1-c} + (n-1)\frac{\sqrt{c}}{1-c} \\
&= \frac{1-\sqrt{c} + n\sqrt{c} - \sqrt{c}}{1-c} \\
&= \frac{1+(n-2)\sqrt{c}}{1-c}
\end{aligned}$$

Now find S_n.

$$\begin{aligned}
S_n &= \frac{n}{2}(a_1 + a_n) \\
&= \frac{n}{2}\left[\frac{1-\sqrt{c}}{1-c} + \frac{1+(n-2)\sqrt{c}}{1-c}\right] \\
&= \boxed{\frac{n}{2}\left[\frac{2+(n-3)\sqrt{c}}{1-c}\right]}
\end{aligned}$$

49. $\sqrt{3}, \sqrt{12}, \sqrt{27}, \sqrt{48}, \ldots$

$\sqrt{3}, 2\sqrt{3}, 3\sqrt{3}, 4\sqrt{3}, \ldots$

First, find the common difference.

$$\begin{aligned}
d &= a_2 - a_1 \\
&= \sqrt{12} - \sqrt{3} \\
&= 2\sqrt{3} - \sqrt{3} \\
&= \sqrt{3}
\end{aligned}$$

Now find a_{14}.

$$\begin{aligned}
a_n &= a_1 + (n-1)d \\
a_{14} &= \sqrt{3} + 13\sqrt{3} = 14\sqrt{3}
\end{aligned}$$

Now find S_{14}.

$$\begin{aligned}
S_n &= \frac{n}{2}(a_1 + a_n) \\
S_{14} &= \frac{14}{2}(a_1 + a_{14}) \\
&= 7\left(\sqrt{3} + 14\sqrt{3}\right) \\
&= 7\left(15\sqrt{3}\right) \\
&= \boxed{105\sqrt{3}}
\end{aligned}$$

51. Represent the integers by $x, x+1, x+2, \ldots, x+11$.

$$\begin{aligned}
S_n &= \frac{n}{2}(a_1 + a_n) \\
S_{12} &= \frac{12}{2}(a_1 + a_{12}) \\
&= 6(x + x + 11) \\
&= 6(2x + 11) \\
&= 12x + 66
\end{aligned}$$

$$\begin{aligned}
\frac{12x + 66}{4} &= \frac{12x}{4} + \frac{66}{4} \\
&= 3x + 16\frac{2}{4}
\end{aligned}$$

The remainder is $\boxed{2}$.

Review Problems

52.

$$\begin{aligned}
(y+1)(2y+3) - 3(y+2)(y+1) &= -3(y+5) \\
2y^2 + 5y + 3 - 3y^2 - 9y - 6 &= -3y - 15 \\
-y^2 - 4y - 3 &= -3y - 15 \\
0 &= y^2 + y - 12 \\
0 &= (y+4)(y-3)
\end{aligned}$$

$$\begin{array}{lll}
y+4 = 0 & \text{or} & y-3 = 0 \\
y = -4 & & y = 3 \quad \text{(both check)}
\end{array}$$

$\boxed{\{-4, 3\}}$

53.

$$\begin{vmatrix} 5 & 2 & 34 \\ -1 & 3 & 22 \\ 0 & 0 & 4 \end{vmatrix} = 0\begin{vmatrix} 2 & 34 \\ 3 & 22 \end{vmatrix} - 0\begin{vmatrix} 5 & 34 \\ -1 & 22 \end{vmatrix} + 4\begin{vmatrix} 5 & 2 \\ -1 & 3 \end{vmatrix}$$

$$= 4\begin{vmatrix} 5 & 2 \\ -1 & 3 \end{vmatrix}$$

$$= 4(15 + 2)$$

$$= 4(17)$$

$$= \boxed{68}$$

54.

$$\begin{aligned} 4x + y - 2z &= 8 \quad (1) \\ 3x + 2y - z &= 5 \quad (2) \\ -3y + z &= 9 \quad (3) \end{aligned}$$

(Equations 1 and 2):

$$\begin{aligned} (\times -3) \rightarrow \quad -12x - 3y + 6z &= -24 \\ (\times 4) \rightarrow \quad \underline{12x + 8y - 4z = 20} \\ 5y + 2z &= -4 \quad \text{(Equation 4)} \end{aligned}$$

(Equations 4 and 3):

$$\begin{aligned} 5y + 2z &= -4 \\ (\times -2) \rightarrow \quad \underline{6y - 2z = -18} \\ 11y &= -22 \quad \text{(Add)} \\ y &= -2 \end{aligned}$$

(Equation 4):

$$\begin{aligned} 5y + 2z &= -4 \\ 5(-2) + 2z &= -4 \\ -10 + 2z &= -4 \\ 2z &= 6 \\ z &= 3 \end{aligned}$$

(Equation 2):

$$\begin{aligned} 3x + 2y - z &= 5 \\ 3x + 2(-2) - 3 &= 5 \\ 3x - 7 &= 5 \\ 3x &= 12 \\ x &= 4 \end{aligned}$$

$$\boxed{\{(4, -2, 3)\}}$$

Section 12.5   Geometric Sequences

Problem Set 12.5, pp. 865-867

1. 81, 54, 36, . . .

$$\frac{a_2}{a_1} = \frac{54}{81} = 0.\bar{6} = \frac{2}{3}$$

$$\frac{a_3}{a_2} = \frac{36}{54} = 0.\bar{6} = \frac{2}{3}$$

There is a common ratio. The sequence is $\boxed{\text{geometric}}$.

$$\boxed{r = \frac{2}{3}}$$

3. 1, 4, 9, 16, . . .

$$\frac{4}{1} = 4$$

$$\frac{9}{4} = 2.25$$

No common ratio, not geometric

$$4 - 1 = 3$$

$$9 - 4 = 5$$

No common difference; not arithmetic

$\boxed{\text{Neither arithmetic nor geometric}}$

5. 1, −3, 9, −27, . . .

$$-\frac{3}{1} = -3$$

$$\frac{9}{-3} = -3$$

$$\frac{-27}{9} = -3$$

$\boxed{\text{Geometric}}$ with $\boxed{r = -3}$

7. 3, 3, 3, . . .

The sequence is $\boxed{\text{either arithmetic with } d = 0}$

or geometric with $r = 1$.

9. −2, 4, −2, 4, . . .

$$\frac{4}{-2} = -2$$

$$\frac{-2}{4} = -\frac{1}{2}$$

No common ratio; not geometric

$$4 - (-2) = 6$$

$$-2 - 4 = -6$$

No common difference; not arithmetic

$\boxed{\text{Neither arithmetic nor geometric}}$

11. 0.34, 0.33, 0.333, . . .

$$\frac{0.33}{0.3} = 1.1$$

$$\frac{0.333}{0.33} = 1.11$$

No common ratio; not geometric

$$0.33 - 0.3 = 0.03$$

$$0.333 - 0.33 = 0.003$$

No common difference; not arithmetic

$\boxed{\text{neither arithmetic nor geometric}}$

13. −2, 6, −18, . . .

$$\frac{6}{-2} = -3$$

$$\frac{-18}{6} = -3$$

$\boxed{\text{Geometric}}$; $\boxed{r = -3}$

15. $\sqrt{3}, 3, 3\sqrt{3}, \ldots$

$$\frac{3}{\sqrt{3}}$$

$$\frac{3\sqrt{3}}{3} = \sqrt{3}$$

Since $\dfrac{3}{\sqrt{3}} = \dfrac{3}{\sqrt{3}} \cdot \dfrac{\sqrt{3}}{\sqrt{3}} = \dfrac{3\sqrt{3}}{3} = \sqrt{3}$

the sequence is $\boxed{\text{geometric}}$ with $\boxed{r = \sqrt{3}}$

17. $\frac{a}{b}, \frac{a}{b^2}, \frac{a}{b^3}, \frac{a}{b^4}, \cdots$

$$\frac{\frac{a}{b^2}}{\frac{a}{b}} = \frac{1}{b}$$

$$\frac{\frac{a}{b^3}}{\frac{a}{b^2}} = \frac{1}{b}$$

The sequence is $\boxed{\text{geometric}}$ with $\boxed{r = \frac{1}{b}}$.

19.

$$a_1 = 10, r = \frac{1}{2}$$

$$a_1 = 10$$

$$a_2 = 10\left(\frac{1}{2}\right) = 5$$

$$a_3 = 5\left(\frac{1}{2}\right) = \frac{5}{2}$$

$$a_4 = \frac{5}{2}\left(\frac{1}{2}\right) = \frac{5}{4}$$

$$a_5 = \frac{5}{4}\left(\frac{1}{2}\right) = \frac{5}{8}$$

$$\boxed{10, 5, \frac{5}{2}, \frac{5}{4}, \frac{5}{8}}$$

21.

$$a_1 = -\frac{1}{4}, r = -2$$

$$a_2 = -\frac{1}{4}(-2) = \frac{1}{2}$$

$$a_3 = \frac{1}{2}(-2) = -1$$

$$a_4 = -1(-2) = 2$$

$$a_5 = 2(-2) = -4$$

$$\boxed{-\frac{1}{4}, \frac{1}{2}, -1, 2, -4}$$

23.

$$a_1 = 3, r = -3$$

$$a_2 = 3(-3) = -9$$

$$a_3 = -9(-3) = 27$$

$$a_4 = 27(-3) = -81$$

$$a_5 = -81(-3) = 243$$

$$\boxed{3, -9, 27, -81, 243}$$

25.

$$a_1 = \frac{a^2}{b}, r = \frac{2b}{a}$$

$$a_1 = \frac{a^2}{b}$$

$$a_2 = \frac{a^2}{b} \cdot \frac{2b}{a} = 2a$$

$$a_3 = 2a \cdot \frac{2b}{a} = 4b$$

$$a_4 = 4b \cdot \frac{2b}{a} = \frac{8b^2}{a}$$

$$a_5 = \frac{8b^2}{a} \cdot \frac{2b}{a} = \frac{16b^3}{a^2}$$

$$\boxed{\frac{a^2}{b}, 2a, 4b, \frac{8b^2}{a}, \frac{16b^3}{a^2}}$$

27. $-3, -15, -75, \ldots; a_6$

$$r = \frac{-15}{-3} = 5$$

$$a_n = a_1 r^{n-1}$$

$$a_6 = (-3)(5)^{6-1}$$

$$a_6 = (-3)(5)^5 = \boxed{-9,375}$$

29. $18, -6, 2, \ldots ; a_5$

$$r = -\frac{6}{18} = -\frac{1}{3}$$
$$a_n = a_1 r^{n-1}$$
$$a_5 = 18\left(-\frac{1}{3}\right)^{5-1}$$
$$= 18\left(-\frac{1}{3}\right)^4$$
$$= 18\left(\frac{1}{81}\right)$$
$$a_5 = \boxed{\frac{2}{9}}$$

31. $250, 50, 10, \ldots ; a_6$

$$r = \frac{50}{250} = \frac{1}{5}$$
$$a_n = a_1 r^{n-1}$$
$$a_6 = (250)\left(\frac{1}{5}\right)^{6-1}$$
$$a_6 = (250)\left(\frac{1}{5}\right)^5$$
$$a_6 = \frac{250}{3125} = \boxed{\frac{2}{25}}$$

33. $\sqrt{2}, 2, 2\sqrt{2}, \ldots ; a_7$

$$r = \frac{2}{\sqrt{2}} \cdot \frac{\sqrt{2}}{\sqrt{2}} = \frac{2\sqrt{2}}{2} = \sqrt{2}$$

$$a_n = a_1 r^{n-1}$$

$$a_7 = \sqrt{2}\left(\sqrt{2}\right)^{7-1}$$

$$a_7 = \sqrt{2}(2^{1/2})^6$$

$$a_7 = \sqrt{2} \cdot 2^3 = \boxed{8\sqrt{2}}$$

35. $222\frac{2}{9}, 22\frac{2}{9}, 2\frac{2}{9}, \ldots$

$$r = \frac{22\frac{2}{9}}{222\frac{2}{9}}$$
$$= \frac{\frac{200}{9}}{\frac{2000}{9}}$$
$$= \frac{200}{2000}$$
$$= \frac{1}{10}$$

$$a_n = a_1 r^{n-1}$$
$$a_6 = \left(222\frac{2}{9}\right)\left(\frac{1}{10}\right)^{6-1}$$
$$a_6 = \left(\frac{2000}{9}\right)\left(\frac{1}{10}\right)^5$$
$$a_6 = \left(\frac{2000}{9}\right)\left(\frac{1}{100,000}\right)$$
$$a_6 = \boxed{\frac{1}{450}}$$

37. $c^7 d^6, c^6 d^4, c^5 d^2, \ldots ; a_7$

$$r = \frac{c^6 d^4}{c^7 d^6} = \frac{1}{cd^2}$$
$$a_n = a_1 r^{n-1}$$
$$a_7 = (c^7 d^6)\left(\frac{1}{cd^2}\right)^{7-1}$$
$$a_7 = (c^7 d^6)\left(\frac{1}{cd^2}\right)^6$$
$$a_7 = \frac{c^7 d^6}{c^6 d^{12}} = cd^{-6} = \boxed{\frac{c}{d^6}}$$

39. $48, 12, 3, \ldots$

$$r = \frac{12}{48} = \frac{1}{4}$$
$$a_n = a_1 r^{n-1}$$
$$\boxed{a_n = (48)\left(\frac{1}{4}\right)^{n-1} \text{ or } \frac{3}{4^{n-3}}}$$

41. 3, –6, 12, . . .

$$r = -\frac{6}{3} = -2$$
$$a_n = a_1 r^{n-1}$$
$$\boxed{a_n = 3(-2)^{n-1}}$$

43. $3, \frac{3}{2}, \frac{3}{4}, \ldots$

$$r = \frac{\frac{3}{2}}{3} = \frac{1}{2}$$
$$a_n = a_1 r^{n-1}$$
$$\boxed{a_n = 3\left(\frac{1}{2}\right)^{n-1}}$$

45. a.
$$w_1 = 200, r = 0.99$$
$$w_1 = 200$$
$$w_2 = 200(0.99) = 198$$
$$w_3 = 198(0.99) = 196.02$$
$$w_4 = 196.02(0.99) = 194.0598$$
$$\boxed{200, 198, 196.02, 194.0598}$$

b.
$$w_n = w_1 r^{n-1}$$
$$\boxed{w_n = 200(0.99)^{n-1}}$$

47.
$$a_n = a_1 r^{n-1}$$
$$a_1 = 12{,}000$$
$$r = 0.75$$

$$a_n = 12{,}000(0.75)^{n-1}$$
$$a_4 = (12{,}000)(0.75)^{4-1}$$
$$= (12{,}000)(0.75)^3$$
$$a_4 = 5062.5$$

Value: $\boxed{5062.50}$

49. 100000,   120000,   144000 . . .
   ↑   in 5 years   in 10 years
            ↑   ↑
a_1 a_2 a_3
$$a_n = a_1 r^{n-1}$$
$$a_n = (100{,}000)(1.2)^{n-1}$$
$$r = \frac{120{,}000}{100{,}000} = 1.2$$
In twenty years, $n = 5$.
$$a_5 = (100{,}000)(1.2)^{5-1}$$
$$a_5 = 207{,}360$$
Population in 20 years: $\boxed{207{,}360}$

51. 10000,   20000,   40000, . . .
   ↑   in 6 hours   in 12 hours
            ↑   ↑
a_1 a_2 a_3
$$r = 2$$
$$a_n = a_1 r^{n-1}$$
$$a_n = (10{,}000)(2)^{n-1}$$

In 24 hours, $n = 5$
$$a_5 = (10{,}000)(2)^{5-1} = 160{,}000$$
Bacteria: $\boxed{160{,}000}$

53. Given: $a_1 = 6$, $a_5 = 96$

Find r.
$$a_n = a_1 r^{n-1}$$
$$a_5 = 6r^{5-1}$$
$$a_6 = 6r^4$$
$$16 = r^4$$
$$\boxed{r = \pm 2}$$

Two possible sequences:
   6, 12, 24, 48, 96, . . .
or   6, –12, 24, –48, 96, . . .
In both,
   $a_1 = 6$ and $a_5 = 96$

55. Given: $a_1 = 3$, $a_5 = \frac{1}{27}$

Find r.
$$a_n = a_1 r^{n-1}$$
$$a_5 = 3r^{5-1}$$
$$\frac{1}{27} = 3r^4$$
$$\frac{1}{81} = r^4$$
$$\boxed{r = \pm \frac{1}{3}}$$

57. Given: $a_3 = 28$, $a_5 = 112$

Find a_1.

$$a_n = a_1 r^{n-1}$$
$$a_3 = a_1 r^{3-1} = 28$$
$$a_1 r^2 = 28$$

$$a_5 = a_1 r^{5-1} = 112$$
$$a_1 r^4 = 112$$
$$\frac{a_1 r^4}{a_1 r^2} = \frac{112}{28}$$
$$r^2 = 4$$
$$r = \pm 2$$

$$a_1 (\pm 2)^2 = 28$$
$$4a_1 = 28$$
$$a_1 = \boxed{7}$$

59. Given: $a_3 = 4$, $a_6 = \frac{1}{2}$

Find a_1.

$$a_n = a_1 r^{n-1}$$
$$a_3 = a_1 r^2 = 4$$
$$a_6 = a_1 r^5 = \frac{1}{2}$$

$$\frac{a_1 r^5}{a_1 r^2} = \frac{\frac{1}{2}}{4}$$
$$r^3 = \frac{1}{8}$$
$$r = \frac{1}{2}$$

$$a_1 r^2 = 4$$
$$a_1 \left(\frac{1}{2}\right)^2 = 4$$
$$a_1 \left(\frac{1}{4}\right) = 4$$
$$a_1 = \boxed{16}$$

61. Given: $a_4 = 24$ and $r = 2$

Find a_8.

$$a_n = a_1 r^{n-1}$$
$$a_4 = a_1 (2)^{4-1}$$
$$24 = a_1 8$$
$$3 = a_1$$
$$a_8 = a_1 r^{8-1}$$
$$a_8 = 3(2)^7$$
$$a_8 = \boxed{384}$$

63. Company A:

20000,   21000,   22000, . . .
  ↑   ↑   ↑
year 1   year 2   year 3

Arithmetic sequence:
$$a_n = a_1 + (n-1)d$$

Year 6:
$$a_6 = 20{,}000 + (6-1)1000$$
$$= 20{,}000 + 5{,}000$$
$$= \$25{,}000$$

Company B:

20000,   21000,   22050, . . .
  ↑   ↑   ↑
year 1   year 2   year 3

Geometric Sequence:
$$a_n = a_1 r^{n-1}$$
$$r = \frac{21000}{20000} = 1.05$$

Year 6:
$$a_6 = 20{,}000(1.05)^{6-1}$$
$$a_6 = \$25{,}525.60$$

Company B will pay more in the sixth

year (approximately \$526 more).

65. $\boxed{D}$ is true.

67. $\dfrac{1}{a_1}, \dfrac{1}{a_2}, \dfrac{1}{a_3}, \ldots, \dfrac{1}{a_n}$ where $a_n = a_1 r^{n-1}$

Since $a_1, a_2, a_3, \ldots, a_n$ is geometric with

common ratio r: $\dfrac{a_2}{a_1} = r$

Consider: $\dfrac{1}{a_1}, \dfrac{1}{a_2}, \dfrac{1}{a_3}, \ldots, \dfrac{1}{a_n}$

Common ratio $= \dfrac{\frac{1}{a_2}}{\frac{1}{a_1}} = \dfrac{a_1}{a_2} = \boxed{\dfrac{1}{r}}$

69. Given: $\dfrac{a_2}{a_1} = r$

Consider: $3a_1, 3a_2, 3a_3, \ldots, 3a_n$

Common ratio $= \dfrac{3a_2}{3a_1} = \dfrac{a_2}{a_1} = \boxed{r}$

71. Given: $\dfrac{a_2}{a_1} = r$ and $\dfrac{a_n}{a_{n-1}} = r$

Consider: $a_n, a_{n-1}, a_{n-2}, \ldots, a_1$

Common ratio $= \dfrac{a_{n-1}}{a_n} = \boxed{\dfrac{1}{r}}$

73. Since $x, y, z, \ldots$ is geometric, then $\dfrac{z}{y} = \dfrac{y}{x}$.

$$\dfrac{z}{y} = \dfrac{y}{x}$$
$$xz = y^2$$
$$z = \dfrac{y^2}{x}$$

We must show that $\dfrac{1}{z+y} - \dfrac{1}{2y} = \dfrac{1}{2y} - \dfrac{1}{x+y}$.

Begin with

$$\dfrac{1}{z+y} - \dfrac{1}{2y}$$

$$= \dfrac{1}{\frac{y^2}{x}+y} - \dfrac{1}{2y} \quad \text{(Substitute the expression we found for } z.)$$

$$= \dfrac{x}{y^2+xy} - \dfrac{1}{2y} \quad \text{(Multiply numerator and denominator of the first fraction by } x.)$$

$$= \dfrac{x}{y(y+x)} - \dfrac{1}{2y}$$

$$= \dfrac{2x-(x+y)}{2y(x+y)} \quad \text{(The LCD is } 2y(x+y).)$$

$$= \dfrac{x-y}{2y(x+y)}$$

Now consider

$$\dfrac{1}{2y} - \dfrac{1}{x+y}$$

$$= \dfrac{(x+y)-2y}{2y(x+y)} \quad \text{(The LCD is } 2y(x+y).)$$

$$= \dfrac{x-y}{2y(x+y)}$$

Since $a_3 - a_2 = a_2 - a_1$, this shows that

$\dfrac{1}{x+y}, \dfrac{1}{2y}, \dfrac{1}{z+y}$ are three consecutive terms of an arithmetic sequence.

Review Problems

78. $\log 5x + 2\log x$
$= \log 5x + \log x^2$
$= \log (5x \cdot x^2)$
$= \boxed{\log 5x^3}$

79.
$$8^{y-1} = 4^{y+2}$$
$$(2^3)^{y-1} = (2^2)^{y+2}$$
$$2^{3y-3} = 2^{2y+4}$$
$$3y - 3 = 2y + 4$$
$$y = 7$$
$$\boxed{\{7\}}$$

80.
$$\frac{\sqrt{5}+\sqrt{3}}{\sqrt{5}-\sqrt{3}} = \frac{\sqrt{5}+\sqrt{3}}{\sqrt{5}-\sqrt{3}} \cdot \frac{\sqrt{5}+\sqrt{3}}{\sqrt{5}+\sqrt{3}}$$
$$= \frac{5 + 2\sqrt{15} + 3}{5-3}$$
$$= \frac{8 + 2\sqrt{15}}{2}$$
$$= \boxed{4 + \sqrt{15}}$$

Section 12.6 Finite Geometric Series

Problem Set 12.6, pp. 871-873

1. $2, 6, 18, \ldots$
$$S_n = \frac{a_1 - a_1 r^n}{1 - r}$$
$a_1 = 2, \ r = \frac{6}{2} = 3$
$$S_6 = \frac{2 - 2(3)^6}{1-3}$$
$$S_6 = \frac{2 - 1458}{-2}$$
$$S_6 = \frac{-1456}{-2}$$
$$= \boxed{728}$$

3. $3, -6, 12, \ldots$
$a_1 = 3, \ r = \frac{-6}{3} = -2$
$$S_n = \frac{a_1 - a_1 r^n}{1 - r}$$
$$S_6 = \frac{3 - 3(-2)^5}{1 - (-2)} = \frac{99}{3} = \boxed{33}$$

5. $-\dfrac{3}{2}, 3, -6, \ldots$

$a_1 = -\dfrac{3}{2}, \; r = \dfrac{3}{\left(-\dfrac{3}{2}\right)} = -2$

$\begin{aligned}
S_n &= \frac{a_1 - a_1 r^n}{1 - r} \\[6pt]
S_7 &= \frac{-\dfrac{3}{2} - \left(-\dfrac{3}{2}\right)(-2)^7}{1 - (-2)} \\[6pt]
S_7 &= \frac{-\dfrac{3}{2} + \dfrac{3}{2}(-128)}{3} \\[6pt]
S_7 &= \frac{-\dfrac{3}{2} - 192}{3} = \frac{-3 - 384}{6} \\[6pt]
S_7 &= \frac{-387}{6} \\[6pt]
&= \boxed{-\frac{129}{2}}
\end{aligned}$

7.

$\begin{aligned}
\sum_{i=0}^{6} 3^i &= 3^0 + 3^1 + 3^2 + \ldots + 3^6 \\[4pt]
&= 1 + 3 + 9 + \ldots + 729
\end{aligned}$

$a_1 = 1, \; r = 3, \; n = 7$ (There are seven terms.)

$\begin{aligned}
S_n &= \frac{a_1 - a_1 r^n}{1 - r} \\[6pt]
S_7 &= \frac{1 - 1(3)^7}{1 - 3} \\[6pt]
&= \frac{-2186}{-2} \\[6pt]
&= \boxed{1093}
\end{aligned}$

9. $\displaystyle\sum_{i=0}^{6}(-3)^i = 1 - 3 + 9 - 27 + \ldots + (-3)^6$

$a_1 = 1, \; r = -3, \; n = 7$

$\begin{aligned}
S_n &= \frac{a_1 - a_1 r^n}{1 - r} \\[6pt]
S_7 &= \frac{1 - 1(-3)^7}{1 - (-3)} \\[6pt]
&= \frac{2188}{4} \\[6pt]
&= \boxed{547}
\end{aligned}$

11. $\displaystyle\sum_{i=1}^{5} 2^{i-1} = 2^{1-1} + 2^{2-1} + 2^{3-1} + 2^{4-1} + 2^{5-1}$

$\qquad\qquad = 1 + 2 + 4 + 8 + 16$

$a_1 = 1, \; r = 2, \; n = 5$ (Since we begin at $i = 1$, not $i = 0$, we have five terms.)

$\begin{aligned}
S_n &= \frac{a_1 - a_1 r^n}{1 - r} \\[6pt]
S_5 &= \frac{1 - 1(2)^5}{1 - 2} \\[6pt]
&= \frac{-31}{-1} \\[6pt]
&= \boxed{31}
\end{aligned}$

13. $\displaystyle\sum_{i=1}^{4}\left(-\frac{2}{3}\right)^{i} = \left(-\frac{2}{3}\right)^{1} + \left(-\frac{2}{3}\right)^{2} + \left(-\frac{2}{3}\right)^{3} + \left(-\frac{2}{3}\right)^{4}$

$\qquad\qquad\quad = -\dfrac{2}{3} + \dfrac{4}{9} + \left(-\dfrac{2}{3}\right)^{3} + \left(-\dfrac{2}{3}\right)^{4}$

$a_1 = -\dfrac{2}{3},\ r = \dfrac{\frac{4}{9}}{-\frac{2}{3}} = -\dfrac{2}{3},\ n = 4$

$$S_n = \frac{a_1 - a_1 r^n}{1 - r}$$

$$S_4 = \frac{\left(-\frac{2}{3}\right) - \left(-\frac{2}{3}\right)\left(-\frac{2}{3}\right)^{4}}{1 - \left(-\frac{2}{3}\right)}$$

$$S_4 = \frac{-\frac{2}{3} + \frac{32}{243}}{\frac{5}{3}} \cdot \frac{243}{243}$$

$$S_4 = \frac{-162 + 32}{405}$$

$$\quad = \frac{-130}{405}$$

$$\quad = \boxed{\frac{-26}{81}}$$

15. $6,\ 5.4,\ 4.86,\ \ldots\ = \displaystyle\sum_{i=1}^{6} 6(0.9)^{i-1}$

$a_1 = 6,\ r = \dfrac{5.4}{6} = 0.9,\ n = 6$

$$S_n = \frac{a_1 - a_1 r^n}{1 - r}$$

$$S_6 = \frac{6 - 6(0.9)^{6}}{1 - 0.9}$$

$$S_6 = \frac{2.811354}{0.1}$$

$$S_6 = 28.11354$$

The bob will travel approximately $\boxed{28 \text{ inches}}$.

17. $\displaystyle\sum_{i=1}^{10} 25(3)^{i-1}$

$a_1 = 25,\ r = 3,\ n = 10$

$$S_n = \frac{a_1 - a_1 r^n}{1 - r}$$

$$S_{10} = \frac{25 - 25(3)^{10}}{1 - 3}$$

$$S_{10} = \frac{-1476200}{-2}$$

$$S_{10} = 738100¢$$

$$\quad = \boxed{\$7,381.00}$$

19. $\displaystyle\sum_{i=1}^{10} 10(2)^{i-1}$

$a_1 = 10,\ r = 2,\ n = 10$

$$S_n = \frac{a_1 - a_1 r^n}{1 - r}$$

$$S_{10} = \frac{10 - (10)(2)^{10}}{1 - 2}$$

$$\quad = 10230$$

Total: $\boxed{10,230 \text{ mg}}$ or $\boxed{10.23 \text{ grams}}$

21. Parents: 2
Grandparents: 4
Great grandparents: 8

$$\underset{a_1}{\underset{\uparrow}{2}} + \underset{a_2}{\underset{\uparrow}{4}} + \underset{a_3}{\underset{\uparrow}{8}} + \ldots + a_6 \quad = \quad \sum_{i=1}^{6} 2(2)^{i-1}$$

$a_1 = 2,\ r = 2,\ n = 6$

$$S_n \quad = \quad \frac{a_1 - a_1 r^n}{1 - r}$$

$$S_6 \quad = \quad \frac{2 - 2(2)^6}{1 - 2}$$

$$= \quad \frac{-126}{-1}$$

$$= \quad 126$$

$\boxed{126 \text{ ancestors}}$

(*Note:* there are 6 generations of ancestors represented.)

23. Given: $S_8 = -170,\ r = -2$

$$\sum_{i=1}^{8} a_1 r^{i-1} \quad = \quad -170$$

Find a_1.

$$S_n \quad = \quad \frac{a_1 - a_1 r^n}{1 - r}$$

$$S_8 \quad = \quad \frac{a_1 - a_1(-2)^8}{1 - (-2)}$$

$$= \quad -170$$

$$\frac{a_1 - 256a_1}{3} \quad = \quad -170$$

$$-255a_1 \quad = \quad -510$$

$$a_1 \quad = \quad \frac{-510}{-255} = \boxed{2}$$

25. Given: $S_7 = 547,\ r = -3$

$$\sum_{i=1}^{7} a_1 r^{i-1} \quad = \quad 547$$

Find a_1.

$$S_n \quad = \quad \frac{a_1 - a_1 r^n}{1 - r}$$

$$S_7 \quad = \quad \frac{a_1 - a_1(-3)^7}{1 - (-3)}$$

$$= \quad 547$$

$$2188a_1 \quad = \quad 2188$$

$$a_1 \quad = \quad \boxed{1}$$

27. Given: $S_5 = \dfrac{341}{8},\ r = \dfrac{1}{4}$

$$\sum_{i=1}^{5} a_1 r^{i-1} \quad = \quad \frac{341}{8}$$

Find a_1.

$$S_n \quad = \quad \frac{a_1 - a_1 r^n}{1 - r}$$

$$S_5 \quad = \quad \frac{a_1 - a_1\left(\frac{1}{4}\right)^5}{1 - \frac{1}{4}} = \frac{341}{8}$$

$$\frac{a_1 - \dfrac{a_1}{4^5}}{\dfrac{3}{4}} \cdot \frac{4^5}{4^5} \quad = \quad \frac{341}{8}$$

$$\frac{1024a_1 - a_1}{768} \quad = \quad \frac{341}{8}$$

$$\frac{1023a_1}{768} \quad = \quad \frac{341}{8}$$

$$8184a_1 \quad = \quad 261{,}888$$

$$a_1 \quad = \quad \boxed{32}$$

29. Given: $a_1 = 10$, $a_{10} = 30$

Find r and S_{10}.

$$\begin{aligned} a_n &= a_1 r^{n-1} \\ a_{10} &= a_1 r^9 \\ 30 &= 10 r^9 \\ 3 &= r^9 \\ \boxed{r &= \sqrt[9]{3}} \\ r &\approx 1.1298 \end{aligned}$$

$$\begin{aligned} S_n &= \frac{a_1 - a_1 r^n}{1 - r} \\ S_{10} &= \frac{10 - 10(3^{1/9})^{10}}{1 - \sqrt[9]{3}} \\ S_{10} &= \frac{10 - 10(3.38949)}{1 - 1.1298} \\ S_{10} &= \frac{-23.8949}{-0.1298} \\ &\approx \boxed{184} \end{aligned}$$

31. Given: $a_1 = 9$, $a_4 = \frac{8}{3}$

Find r and S_4.

$$\begin{aligned} a_n &= a_1 r^{n-1} \\ a_4 &= 9r^3 = \frac{8}{3} \\ r^3 &= \frac{8}{27} \\ r &= \sqrt[3]{\frac{8}{27}} = \boxed{\frac{2}{3}} \\ S_n &= \frac{a_1 - a_1 r^n}{1 - r} \\ S_4 &= \frac{9 - 9\left(\frac{2}{3}\right)^4}{1 - \frac{2}{3}} \\ S_4 &= \frac{9 - \frac{16}{9}}{\frac{1}{3}} \cdot \frac{9}{9} \\ S_4 &= \frac{81 - 16}{3} \\ &= \boxed{\frac{65}{3}} \end{aligned}$$

33. Given: $r = 2$, $S_6 = 20{,}000$

Find a_1.

$$\begin{aligned} S_n &= \frac{a_1 - a_1 r^n}{1 - r} \\ S_6 &= \frac{a_1 - a_1(2)^6}{1 - 2} \\ &= 20{,}000 \\ \frac{a_1 - 64a_1}{-1} &= 20{,}000 \\ -63a_1 &= -20{,}000 \\ a_1 &= \frac{-20{,}000}{-63} \\ &\approx 317.46 \end{aligned}$$

$$\boxed{\$317.46}$$

35. Company A:
Arithmetic sequence:

$$a_1 = 20,000$$

$$d = 1,000$$

$$S_n = \frac{n}{2}(a_1 + a_n)$$

$$S_6 = \frac{6}{2}(a_1 + a_6)$$

Find a_6:

$$a_n = a_1 + (n-1)d$$
$$a_6 = a_1 + (6-1)d$$
$$a_6 = 20,000 + 5(1,000)$$

$$a_6 = 25,000$$

$$S_6 = 3(a_1 + a_6)$$
$$S_6 = 3(20,000 + 25,000)$$
$$= 135,000$$

Company B:
Geometric sequence:

$$a_1 = 20,000$$
$$a_2 = 20,000(0.05) + 20,000$$
$$= 21,000$$

$$r = \frac{21,000}{20,000} = 1.05$$

$$S_n = \frac{a_1 - a_1 r^n}{1 - r}$$

$$S_6 = \frac{20,000 - 20,000(1.05)^6}{1 - 1.05}$$

$$S_6 = \frac{-6802}{-0.05} = 136,040$$

Over Six-Years: Company A: \$135,000
                 Company B: \$136,040

Company B produces the better total income.

37. Given: $a_1 = 1280$, $r = 1.25$, $S_n = 7380$
Find n.

$$S_n = \frac{a_1 - a_1 r^n}{1 - r}$$

$$7380 = \frac{1280 - 1280(1.25)^n}{1 - 1.25}$$

$$7380 = \frac{1280 - 1280(1.25)^n}{-0.25}$$

$$-1845 = 1280 - 1280(1.25)^n$$

$$-3125 = -1280 - 1280(1.25)^n$$

$$2.44 = (1.25)^n$$

$$\log 2.44 = \log(1.25)^n$$

$$\log 2.44 = n \log 1.25$$

$$\frac{\log 2.44}{\log 1.25} = n$$

$$n \approx 4$$

4 years

39. a.

$$a_1 = P$$
$$r = 1 + i$$

$$S_n = \frac{a_1 - a_1 r^n}{1 - r}$$

$$S_n = \frac{P - P(1 + i)^n}{1 - (1 + i)}$$

$$S_n = \frac{P - P(1 + i)^n}{-i}$$

$$= \boxed{\frac{P(1 + i)^n - P}{i}}$$

b.

$$S_{20} = \frac{3,000(1 + 0.08)^{20} - 3,000}{0.08}$$

$$S_{20} = \frac{10982.88}{0.08} \approx 137,286$$

Approximately \$137,286

41. $3 + 3^2 + 3^3 + \ldots + 3^n = 120$
$a_1 = 3$, $r = 3$, $S_n = 120$

Find n.

$$S_n = \frac{a_1 - a_1 r^n}{1 - r}$$

$$120 = \frac{3 - 3 \cdot 3^n}{1 - 3}$$

$$-240 = 3 - 3^{n+1}$$

$$3^{n+1} = 243$$

$$3^{n+1} = 3^5$$

$$n + 1 = 5$$

$$\boxed{n = 4}$$

43. $\displaystyle\sum_{i=1}^{n} 2^{2i-1} = 2 + 2^3 + 2^5 + 2^7 + \ldots + 2^{2n-1}$

$a_1 = 2,\ r = 2^2 = 4$

$$S_n = \frac{a_1 - a_1 r^n}{1-r}$$

$$S_n = \frac{2 - 2 \cdot 4^n}{1-4}$$

$$S_n = \frac{2 - 2 \cdot (2^2)^n}{1-4}$$

$$S_n = \frac{2 - 2^{2n+1}}{-3}$$

$$S_n = \boxed{\frac{2^{2n+1}-2}{3}}$$

45. $1, -\dfrac{1}{\sqrt{2}}, \dfrac{1}{2}, \dfrac{-\sqrt{2}}{4}, \ldots$

$a_1 = 1,\ r = -\dfrac{1}{\sqrt{2}}$

$$S_8 = \frac{a_1 - a_1 r^8}{1-r}$$

$$= \frac{1 - 1\left(-\dfrac{1}{\sqrt{2}}\right)^8}{1 - \left(-\dfrac{1}{\sqrt{2}}\right)}$$

$$= \frac{1 - 1\left(\dfrac{1}{16}\right)}{1 + \dfrac{1}{\sqrt{2}}}$$

$$= \frac{1 - \dfrac{1}{16}}{1 + \dfrac{1}{\sqrt{2}}} \cdot \frac{16\sqrt{2}}{16\sqrt{2}}$$

$$= \frac{16\sqrt{2} - \sqrt{2}}{16\sqrt{2} + 16}$$

$$= \frac{15\sqrt{2}}{16\left(\sqrt{2}+1\right)} \cdot \frac{\sqrt{2}-1}{\sqrt{2}-1}$$

$$= \frac{15\left(2-\sqrt{2}\right)}{16(2-1)}$$

$$= \boxed{\frac{15\left(2-\sqrt{2}\right)}{16}}$$

Review Problems

49.
$$\begin{aligned}
\log(x+2) + \log(x-1) &= \log 4 \\
\log(x+2)(x-1) &= \log 4 \\
(x+2)(x-1) &= 4 \\
x^2 + x - 6 &= 0 \\
(x+3)(x-2) &= 0
\end{aligned}$$

$$x + 3 = 0 \qquad \text{or} \qquad x - 2 = 0$$
$$x = -3 \quad \text{(reject)} \qquad\qquad x = 2$$

$$\boxed{\{2\}}$$

50. Let t = tens' digit, y = units' digit
$$\begin{aligned}
t &= u + 4 \\
\frac{tu}{t+u} &= \frac{3}{2} \\
2tu &= 3t + 3u
\end{aligned}$$

$$\begin{aligned}
2(u+4)u &= 3(u+4) + 3u \\
2u^2 + 8u &= 3u + 12 + 3u \\
2u^2 + 8u &= 6u + 12 \\
2u^2 + 2u - 12 &= 0 \\
u^2 + u - 6 &= 0 \\
(u+3)(u-2) &= 0
\end{aligned}$$

$$u = 2 \quad \text{or} \quad u = -3 \quad \text{(reject; digits can only be 0, 1, 2, \ldots, 9)}$$

$$\begin{aligned}
t &= u + 4 \\
t &= 2 + 4 \\
&= 6
\end{aligned}$$

$$\text{Number:} \qquad 10t + u = 10(6) + 2$$
$$= \boxed{62}$$

51.
$$\begin{aligned}
\left(\frac{y^{3/2} y^{-3/4}}{y^{-5/2}} \right)^{-8} &= \left(\frac{y^{(3/2)-(3/4)}}{y^{-5/2}} \right)^{-8} \\
&= \left(\frac{y^{3/4}}{y^{-5/2}} \right)^{-8} \\
&= \left(y^{(3/4)-(-5/2)} \right)^{-8} \\
&= \left(y^{(3/4)+(10/4)} \right)^{-8} \\
&= \left(y^{13/4} \right)^{-8} \\
&= y^{(13/4)(-8)} \\
&= y^{-26} \\
&= \boxed{\frac{1}{y^{26}}}
\end{aligned}$$

Section 12.7 Infinite Geometric Series

Problem Set 12.7, pp. 878-880

1. $1 + \dfrac{1}{4} + \dfrac{1}{16} + \dots$

$a_1 = 1,\ r = \dfrac{1}{4}$

$$S = \frac{a_1}{1-r}$$

$$S = \frac{1}{1 - \dfrac{1}{4}}$$

$$= \frac{1}{\dfrac{3}{4}} = \boxed{\dfrac{4}{3}}$$

3. $12 + 6 + 3 + \dots$

$a_1 = 12,\ r = \dfrac{6}{12} = \dfrac{1}{2}$

$$S = \frac{a_1}{1-r}$$

$$S = \frac{12}{1 - \dfrac{1}{2}}$$

$$= \frac{12}{\dfrac{1}{2}} = \boxed{24}$$

5. $27 - 18 + 12 - \dots$

$a_1 = 27,\ r = \dfrac{-18}{27} = -\dfrac{2}{3}$

$$S = \frac{a_1}{1-r}$$

$$= \frac{27}{1 - \left(-\dfrac{2}{3}\right)}$$

$$= \frac{27}{\dfrac{5}{3}} = \boxed{\dfrac{81}{5}}$$

7. $5 + 10 + 20 + \dots$

$a_1 = 5,\ r = \dfrac{10}{5} = 2$

Since r does not lie between -1 and 1 the infinite series has $\boxed{\text{no finite sum.}}$

9. $\dfrac{4}{3} + \dfrac{2}{9} + \dfrac{1}{27} + \dots$

$a_1 = \dfrac{4}{3},\ r = \dfrac{\dfrac{2}{9}}{\dfrac{4}{3}} = \dfrac{2}{9} \cdot \dfrac{3}{4} = \dfrac{1}{6}$

$$S = \frac{a_1}{1-r}$$

$$S = \frac{\dfrac{4}{3}}{1 - \dfrac{1}{6}}$$

$$= \frac{\dfrac{4}{3}}{\dfrac{5}{6}} = \frac{4}{3} \cdot \frac{6}{5} = \boxed{\dfrac{8}{5}}$$

11. $1 - \dfrac{1}{2} + \dfrac{1}{4} - \dfrac{1}{8} + \dots$

$a_1 = 1,\ r = \dfrac{-\dfrac{1}{2}}{1} = -\dfrac{1}{2}$

$$S = \frac{a_1}{1-r}$$

$$S = \frac{1}{1 - \left(-\dfrac{1}{2}\right)}$$

$$= \frac{1}{\dfrac{3}{2}} = \boxed{\dfrac{2}{3}}$$

13. $0.\overline{5} = 0.5555\ldots = 0.5 + 0.05 + 00.005 + \ldots$
$a_1 = 0.5, r = 0.1$

$$S = \frac{a_1}{1-r}$$

$$S = \frac{0.5}{1-0.1}$$

$$= \frac{0.5}{0.9} = \boxed{\frac{5}{9}}$$

15. $0.\overline{49} = 0.494949\ldots$
$\qquad\qquad = 0.49 + 0\,0049 + 0\,000049 + \ldots$
$a_1 = 0.49, \ r = 0.01$

$$S = \frac{a_1}{1-r}$$

$$S = \frac{0.49}{1-0.01} = \frac{0.49}{0.99} = \boxed{\frac{49}{99}}$$

17. $0.\overline{241} = 0.241241241\ldots$
$\qquad\qquad = 0.241 + 0.000241 + \ldots$
$a_1 = 0.241, r = 0.001$

$$S = \frac{a_1}{1-r}$$

$$S = \frac{0.241}{1-0.001}$$

$$S = \frac{0.241}{.999} = \boxed{\frac{241}{999}}$$

19. $5.\overline{47}$

$\qquad 0.\overline{47} = 0.474747\ldots$
$\qquad\qquad = 0.47 + 0.0047 + 0.000047 + \ldots$
$a_1 = 0.47, r = 0.01$

$$S = \frac{a_1}{1-r}$$

$$S = \frac{0.47}{1-0.01}$$

$$S = \frac{0.47}{0.99} = \frac{47}{99}$$

$$5.\overline{47} = 5\frac{47}{99} = \boxed{\frac{542}{99}}$$

21. $3.\overline{285}$

$\qquad 0.\overline{285} = 0.285285\ldots$

$$S = \frac{a_1}{1-r}$$

$$S = \frac{0.285}{1-0.001}$$

$$S = \frac{285}{999} = \frac{95}{333}$$

$$3.\overline{285} = 3\frac{95}{333} = \boxed{\frac{1094}{333}}$$

23. $0.1\overline{2} = 0.12222\ldots$
Consider $0.0\overline{2} = 0.02222\ldots$
$\qquad\qquad = 0.02 + 0.002 + 0.0002 + \ldots$
$a_1 = 0.02, r = 0.1$

$$S = \frac{a_1}{1-r}$$

$$S = \frac{0.02}{1-0.1}$$

$$S = \frac{0.02}{0.9} = \frac{2}{90} = \frac{1}{45}$$

$$0.1\overline{2} = 0.1 + \frac{1}{45} = \frac{1}{10} + \frac{1}{45}$$

$$= \frac{9}{90} + \frac{2}{90} = \boxed{\frac{11}{90}}$$

25. The amount spent is given by:
$0.9(40) + (0.9)^2(40) + (0.9)^3(40) + \ldots$

$a_1 = 0.9(40) = 36$

$r = 0.9$

$S = \dfrac{a_1}{1-r}$

$S = \dfrac{36}{1-0.9}$

$= \dfrac{36}{0.1} = 360$

Additional spending: $\boxed{\$360 \text{ billion}}$

27. $a_1 = 20$, $r = 0.9$

$S = \dfrac{a_1}{1-r}$

$S = \dfrac{20}{1-0.9}$

$S = \dfrac{20}{0.1} = 200$

$\boxed{\text{Distance traveled: 200 inches}}$

29. Perimeter of original square = 4(40) = 160
Perimeter of next – smaller square = 4(20) = 80
Perimeter of next – smaller square = 4(10) = 40
Series: $160 + 80 + 40 + 20 + \ldots$

$a_1 = 160$, $r = \dfrac{1}{2}$

$S = \dfrac{a_1}{1-r}$

$S = \dfrac{160}{1-\dfrac{1}{2}}$

$= \dfrac{160}{\dfrac{1}{2}} = 320$

$\boxed{\text{Sum of perimeters: 320 inches}}$

31. Given: $r = 0.9$, $S = 20{,}000$
a_1: Number of flies released each day

$S = \dfrac{a_1}{1-r}$

$20{,}000 = \dfrac{a_1}{1-0.9}$

$20{,}000 = \dfrac{a_1}{0.1}$

$2{,}000 = a_1$

$\boxed{\text{Release 2,000 flies each day.}}$

33. Employee is paid (after taxes):

$10,000 - $1,000 \quad = \quad $9,000

$+ \quad $1,000 - $100 \quad = \quad $900

$+ \qquad $100 - $10 \quad = \quad $90

$+ \qquad\quad $10 - $1 \quad = \quad $9

etc. ad. infinitum

Employee's salary:

$9,000 + 900 + 90 + 9 + \ldots$

$a_1 = 9,000, r = .01$

$$S = \frac{a_1}{1-r}$$

$$S = \frac{9,000}{1-0.01}$$

$$= \frac{9,000}{0.9} = \$10,000$$

Yes, the employee will get $\boxed{\$10,000}$

after taxes with this continued process.

35. Given: $r = 0.85, S = 200$

Find a_1.

$$S = \frac{a_1}{1-r}$$

$$200 = \frac{a_1}{1-0.85}$$

$$200 = \frac{a_1}{0.15}$$

$$a_1 = (200)(0.15) = 30$$

$\boxed{\text{first swing of the bob: 30 cm}}$

37.
$$\sum_{i=1}^{\infty} 0.3^i = \frac{0.3}{1-0.3}$$
$$= \frac{0.3}{0.7}$$
$$= \boxed{\frac{3}{7}} = 0.\overline{428571}$$

39.
$$\sum_{i=1}^{\infty} 3(0.4)^{i-1} = \frac{3}{1-0.4}$$
$$= \frac{3}{0.6}$$
$$= \boxed{5}$$

41.
$$\sum_{i=1}^{\infty} \frac{3}{5}\left(\frac{4}{7}\right)^i = \frac{\frac{12}{35}}{1-\frac{4}{7}}$$
$$= \frac{\frac{12}{35}}{\frac{3}{7}}$$
$$= \boxed{\frac{4}{5}}$$

43. $\left(1 + \frac{1}{2} + \cdot\frac{1}{4} + \frac{1}{8} + \ldots\right)\left(1 + \frac{1}{3} + \frac{1}{9} + \frac{1}{27} \ldots\right)\left(1 + \frac{1}{5} + \frac{1}{25} + \frac{1}{125} + \ldots\right)$

$$= \left(\frac{1}{1-\frac{1}{2}}\right)\left(\frac{1}{1-\frac{1}{3}}\right)\left(\frac{1}{1-\frac{1}{5}}\right)$$

$$= (2)\left(\frac{3}{2}\right)\left(\frac{5}{4}\right)$$

$$= \boxed{\frac{15}{4}}$$

45.

$$\sum_{i=1}^{\infty} \frac{7^{i-1}}{10^i} = \frac{1}{10} + \frac{7}{10^2} + \frac{7^2}{10^3} + \frac{7^3}{10^4} + \dots$$

$$= \frac{a_1}{1-r}$$

$$= \frac{\frac{1}{10}}{1 - \frac{7}{10}}$$

$$= \frac{\frac{1}{10}}{\frac{3}{10}}$$

$$= \boxed{\frac{1}{3}}$$

47. Sides are $4, \sqrt{2^2 + 2^2} = 2\sqrt{2}, \sqrt{2+2} = 2, \sqrt{1+1} = \sqrt{2}, \dots, \dfrac{4}{\left(\sqrt{2}\right)^{n-1}}$

$$\boxed{4, 2\sqrt{2}, 2, \sqrt{2}, \frac{\sqrt{2}}{2}, \dots}$$

49. Areas are $\boxed{16, 8, 4, 2\frac{1}{2}, \dots}$

$$16 + 8 + 4 + 2 + \dots = \frac{16}{1 - \frac{1}{2}} = 32$$

$$\boxed{32 \text{ in.}^2}$$

51. $x + x^2 + x^3 + \dots = \dfrac{1+x}{x}$

$a_1 = x, r = x$

$$\text{left-hand side} = \frac{x}{1-x} \quad (|x| < 1)$$

$$\frac{x}{1-x} = \frac{1+x}{x}$$

$$x^2 = 1 - x^2$$

$$2x^2 = 1$$

$$x = \pm \frac{1}{\sqrt{2}}$$

$$= \pm \frac{\sqrt{2}}{2}$$

$$\boxed{\left\{ -\frac{\sqrt{2}}{2}, \frac{\sqrt{2}}{2} \right\}}$$

53. $\dfrac{2}{1} + \dfrac{1}{3} + \dfrac{2}{9} + \dfrac{1}{27} + \dfrac{2}{81} + \dfrac{1}{243} + \cdots$

$= \dfrac{2}{1} + \dfrac{2}{9} + \dfrac{2}{81} + \cdots + \dfrac{1}{3} + \dfrac{1}{27} + \dfrac{1}{243} + \cdots$

$= \dfrac{a_1}{1-r} + \dfrac{a_1}{1-r}$

$= \dfrac{2}{1 - \dfrac{1}{9}} + \dfrac{\dfrac{1}{3}}{1 - \dfrac{1}{9}}$

$= \dfrac{2}{\dfrac{8}{9}} + \dfrac{\dfrac{1}{3}}{\dfrac{8}{9}}$

$= \dfrac{2}{1} \cdot \dfrac{9}{8} + \dfrac{1}{3} \cdot \dfrac{9}{8}$

$= \dfrac{9}{4} + \dfrac{3}{8}$

$= \dfrac{18 + 3}{8} = \boxed{\dfrac{21}{8}}$

Review Problems

56. $\sqrt[3]{54x^6 y^7} = \sqrt[3]{27 \cdot 2(x^2)^3 (y^2)^3 y}$

$= \boxed{3x^2 y^2 \sqrt[3]{2y}}$

57. Algebraically:

$$4x^2 + y^2 = 16$$
$$2x + y = 4 \quad \text{(solve for } y) \rightarrow$$

$$y = 4 - 2x$$
$$4x^2 + (4 - 2x)^2 = 16$$
$$\text{(substitute for } y) \rightarrow \quad 4x^2 + 16 - 16x + 4x^2 = 16$$
$$8x^2 - 16x = 0$$
$$8x(x - 2) = 0$$

$$8x = 0 \qquad \text{or} \qquad x - 2 = 0$$
$$x = 0 \qquad \text{or} \qquad x = 2$$

If $x = 0$: $\quad y = 4 - 2x \qquad$ If $x = 2$: $\quad y = 4 - 2(2)$
$$= 4 - 2(0) \qquad\qquad\qquad\quad = 4 - 4$$
$$= 4 \qquad\qquad\qquad\qquad\quad = 0$$

$\boxed{\{(0, 4), (2, 0)\}}$

Graphically:

$$4x^2 + y^2 = 16$$
$$\dfrac{4x^2}{16} + \dfrac{y^2}{16} = 1$$
$$\dfrac{x^2}{4} + \dfrac{y^2}{16} = 1$$

Ellipse: x-intercepts

$$\frac{x^2}{4} = 1$$
$$x^2 = 4$$
$$x = \pm 2$$

y-intercepts:

$$\frac{y^2}{16} = 1$$
$$y^2 = 16$$
$$y = \pm 4$$

$2x + y = 4$

Line: x-intercept: $2x = 4$
$$x = 2$$
y-intercept: $y = 4$

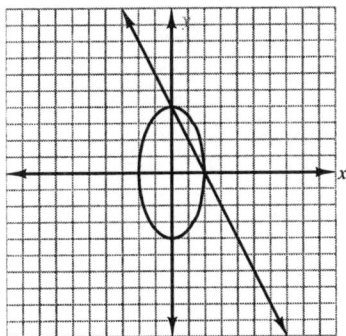

58.

$$f(x) = x^2 - 2x + 4$$
$$\frac{f(a + h) - f(a)}{h}$$
$$= \frac{(a + h)^2 - 2(a + h) + 4 - (a^2 - 2a + 4)}{h}$$
$$= \frac{a^2 + 2ah + h^2 - 2a - 2h + 4 - a^2 + 2a - 4}{h}$$
$$= \frac{2ah + h^2 - 2h}{h}$$
$$= \boxed{2a + h - 2}$$

Section 12.8 The Binomial Theorem

Problem Set 12.8, pp. 885-887

1. $3! = 3 \cdot 2 \cdot 1$
$= \boxed{6}$

3. $2! = 2 \cdot 1$
$= \boxed{2}$

5. $\dfrac{10!}{8!2!} = \dfrac{10 \cdot 9}{2 \cdot 1}$
$= \boxed{45}$
(Factor 8! from 10!)

7. $\dfrac{7!}{6!1!} = \dfrac{7}{1}$
$= \boxed{7}$

9. $\dbinom{6}{3} = \dfrac{6!}{3!(6-3)!}$

$= \dfrac{6!}{3!3!}$

$= \dfrac{6 \cdot 5 \cdot 4}{3 \cdot 2 \cdot 1}$

$= \boxed{20}$

11. $\dbinom{12}{1} = \dfrac{12!}{1!11!}$

$= \dfrac{12}{1}$

$= \boxed{12}$

13. $\dbinom{6}{6} = \dfrac{6!}{6!(6-6)!}$

$= \dfrac{6!}{6!0!}$

$= \dfrac{1}{0!}$

$= \dfrac{1}{1}$

$= \boxed{1}$

15. $(c+2)^5 = c^5 + \dbinom{5}{1}(2)c^4 + \dbinom{5}{2}(4)c^3 + \dbinom{5}{3}(8)c^2 + \dbinom{5}{4}(16)c + \dbinom{5}{5}(32)$

$= c^5 + 10c^4 + \dfrac{5 \cdot 4}{2 \cdot 1}(4)c^3 + \dfrac{5 \cdot 4 \cdot 3}{3 \cdot 2 \cdot 1}(8)c^2 + \dfrac{5 \cdot 4 \cdot 3 \cdot 2}{4 \cdot 3 \cdot 2 \cdot 1}(16)c + 32$

$= c^5 + 5c^4(2) + 10c^3(4) + 10c^2(8) + 5c(16) + 32$

$= \boxed{c^5 + 10c^4 + 40c^3 + 80c^2 + 80c + 32}$

17. $(a-2)^5 = [a+(-2)]^5$

$= \dbinom{5}{0}a^5 + \dbinom{5}{1}a^4(-2) + \dbinom{5}{2}a^3(-2)^2 + \dbinom{5}{3}a^2(-2)^3 + \dbinom{5}{4}a(-2)^4 + \dbinom{5}{5}(-2)^5$

$= 1a^2 + 5a^4(-2) + 10a^3(4) + 10a^2(-8) + 5a(16) + 1(-32)$

$= \boxed{a^5 - 10a^4 + 40a^3 - 80a^2 + 80a - 32}$

19. $\left(\dfrac{a}{2}+1\right)^4 = \dbinom{4}{0}\left(\dfrac{a}{2}\right)^4 + \dbinom{4}{1}\left(\dfrac{a}{2}\right)^3 \cdot 1 + \dbinom{4}{2}\left(\dfrac{a}{2}\right)^2 \cdot 1^2 + \dbinom{4}{3}\left(\dfrac{a}{2}\right)^1 \cdot 1^3 + \dbinom{4}{4} \cdot 1^4$

$= (1)\left(\dfrac{a^4}{16}\right) + 4\left(\dfrac{a^3}{8}\right) + 6\left(\dfrac{a^2}{4}\right) + 4\left(\dfrac{a}{2}\right) + 1$

$= \boxed{\dfrac{a^4}{16} + \dfrac{a^3}{2} + \dfrac{3a^2}{2} + 2a + 1}$

21. $(2x+3y)^4 = \dbinom{4}{0}(2x)^4 + \dbinom{4}{1}(2x)^3(3y) + \dbinom{4}{2}(2x)^2(3y)^2 + \dbinom{4}{3}(2x)^1(3y)^3 + \dbinom{4}{4}(3y)^4$

$= 1(16x^4) + 4(8x^3)(3y) + 6(4x^2)(9y^2) + 4(2x)(27y^3) + 1(81y^4)$

$= \boxed{16x^4 + 96x^3y + 216x^2y^2 + 216xy^3 + 81y^4}$

23. $(2x^2-y^2)^3 = [2x^2+(y^2)]^3$

$= \dbinom{3}{0}(2x^2)^3 + 8\dbinom{3}{1}(2x^2)^2(-y^2) + \dbinom{3}{2}(2x^2)(-y^2)^2 + \dbinom{3}{3}(-y^2)^3$

$= 1(8x^6) + 3(4x^4)(-y^2) + 3(2x^2)(y^4) + 1(-y^6)$

$= \boxed{8x^6 - 12x^4y^2 + 6x^2y^4 - y^6}$

25. $\left(\dfrac{a}{3}+2\right)^3 = \dbinom{3}{0}\left(\dfrac{a}{3}\right)^3 + \dbinom{3}{1}\left(\dfrac{a}{3}\right)^2(2) + \dbinom{3}{2}\left(\dfrac{a}{3}\right)^1(2)^2 + \dbinom{3}{3}(2)^3$

$\qquad\qquad = 1\left(\dfrac{a^3}{27}\right) + 3\left(\dfrac{a^2}{9}\right)(2) + 3\left(\dfrac{a}{3}\right)(4) + 1(8)$

$\qquad\qquad = \boxed{\dfrac{a^3}{27} + \dfrac{2a^2}{3} + 4a + 8}$

27. $(a^{1/2}+2)^4 = \dbinom{4}{0}(a^{1/2})^4 + \dbinom{4}{1}(a^{1/2})^3(2) + \dbinom{4}{2}(a^{1/2})^2(2)^2 + \dbinom{4}{3}(a^{1/2})(2)^3 + \dbinom{4}{4}(2)^4$

$\qquad\qquad = 1(a^2) + 4(a^{3/2})(2) + 6a(4) + 4a^{1/2}(8) + 1(16)$

$\qquad\qquad = \boxed{a^2 + 8a^{3/2} + 24a + 32a^{1/2} + 16}$

29. $(a^{-1}+b^{-1})^3 = \dbinom{3}{0}(a^{-1})^3 + \dbinom{3}{1}(a^{-1})^2(b^{-1}) + \dbinom{3}{2}(a^{-1})^1(b^{-1})^2 + \dbinom{3}{3}(b^{-1})^3$

$\qquad\qquad = 1a^{-3} + 3a^{-2}b^{-1} + 3a^{-1}b^{-2} + 1b^{-3}$

$\qquad\qquad = \boxed{\dfrac{1}{a^3} + \dfrac{3}{a^2b} + \dfrac{3}{ab^2} + \dfrac{1}{b^3}}$

31. $(x^2+x)^8 = \dbinom{8}{0}(x^2)^8 + \dbinom{8}{1}(x^2)^7 x + \dbinom{8}{2}(x^2)^6 x^2 + \ldots$

$\qquad\qquad = x^{16} + 8x^{14}(x) + \dfrac{8\cdot 7}{2\cdot 1}x^{12}x^2 + \ldots$

$\qquad\qquad = 1x^{16} + 8(x^{14})x + 28(x^{12})x^2 + \ldots$

$\qquad\qquad = \boxed{x^{16} + 8x^{15} + 28x^{14} + \ldots}$

33. $(2a+b)^8 = \dbinom{8}{0}(2a)^8 + \dbinom{8}{1}(2a)^7 b + \dbinom{8}{2}(2a)^6 b^2 + \ldots$

$\qquad\qquad = (2a)^8 + 8(2a)^7 b + \dfrac{8\cdot 7}{2\cdot 1}(2a)^6 b^2 + \ldots$

$\qquad\qquad = 1(256a^8) + 8(128a^7)b + 28(64a^6)b^2 + \ldots$

$\qquad\qquad = \boxed{256a^8 + 1024a^7 b + 1729a^6 b^2 + \ldots}$

35. $(a-2b)^8 = [a+(-2b)]^8$

$\qquad\qquad = \dbinom{8}{0}a^8 + \dbinom{8}{1}a^7(-2b) + \dbinom{8}{2}a^6(-2b)^2 + \ldots$

$\qquad\qquad = a^8 + 8a^7(-2b) + \dfrac{8\cdot 7}{2\cdot 1}a^6(-2b)^2 + \ldots$

$\qquad\qquad = 1a^8 + 8a^7(-2b) + 28a^6(4b^2) + \ldots$

$\qquad\qquad = \boxed{a^8 - 16a^7 b + 112a^6 b^2 + \ldots}$

37. $(a^2+b^2)^{10} = \dbinom{10}{0}(a^2)^{10} + \dbinom{10}{1}(a^2)^9 b^2 + \dbinom{10}{2}(a^2)^8(b^2)^2 + \ldots$

$\qquad\qquad = a^{20} + 10a^{18}b^2 + \dfrac{10\cdot 9}{2\cdot 1}a^{16}b^4 + \ldots$

$\qquad\qquad = \boxed{a^{20} + 10a^{18}b^2 + 45a^{16}b^4 + \ldots}$

39. $(a+b)^{42} = \dbinom{42}{0}a^{42} + \dbinom{42}{1}a^{41}b + \dbinom{42}{2}a^{40}b^2 + \ldots$

$\qquad\qquad = a^{42} + 42a^{41}b + \dfrac{42\cdot 41}{2\cdot 1}a^{40}b^2 + \ldots$

$\qquad\qquad = \boxed{a^{42} + 42a^{41}b + 861a^{40}b^2 + \ldots}$

41.
$$\left(y+\frac{1}{y}\right)^7 = \binom{7}{0}y^7 + \binom{7}{1}y^6\left(\frac{1}{y}\right) + \binom{7}{2}y^5\left(\frac{1}{y}\right)^2 + \ldots$$
$$= y^7 + 7y^6\left(\frac{1}{y}\right) + \frac{7\cdot 6}{2\cdot 1}y^5\left(\frac{1}{y}\right)^2 + \ldots$$
$$= 1y^7 + 7y^6\left(\frac{1}{y}\right) + 21y^5\left(\frac{1}{y^2}\right) + \ldots$$
$$= \boxed{y^7 + 7y^5 + 21y^3 + \ldots}$$

43. $(2a + b)^6$; 3rd term
$n = 6,\ r = 3$

$\binom{n}{r-1}a^{n-r+1}b^{r-1}$ becomes:

$$\binom{6}{3-1}(2a)^{6-3+1}b^{3-1} = \binom{6}{2}(2a)^4 b^2$$
$$= 15(16a^4)b^2$$
$$= \boxed{240a^4 b^2}$$

45. $(x + y)^{15}$; 7th term
$n = 15,\ r = 7$

$$\binom{n}{r-1}a^{n-r+1}b^{r-1} = \binom{15}{7-1}x^{15-7+1}y^{7-1}$$
$$= \binom{15}{6}x^9 y^6$$
$$= \frac{15\cdot 14\cdot 13\cdot 12\cdot 11\cdot 10}{6\cdot 5\cdot 4\cdot 3\cdot 2\cdot 1}x^9 y^6$$
$$= \boxed{5{,}005\,x^9 y^6}$$

47. $(c^5 + d^7)^9$; 3rd term
$n = 9,\ r = 3$

$$\binom{n}{r-1}a^{n-r+1}b^{r-1} = \binom{9}{3-1}(c^5)^{9-3+1}(d^7)^{3-1}$$
$$= \binom{9}{2}(c^5)^7(d^7)^2$$
$$= \frac{9\cdot 8}{2\cdot 1}(c^5)^7(d^7)^2$$
$$= \boxed{36c^{35}d^{14}}$$

49. a. $\boxed{1\quad 4\quad 6\quad 4\quad 1}$

If a coin is tossed four times, one may obtain four heads in one way, three heads and a tail in four ways, two heads and two tails in six ways, one head and three tails in four ways, and four tails in one way.

b. $\boxed{16\text{ outcomes}}$ when tossing a coin four times.

c. $P(3 \text{ heads, } 1 \text{ tail}) = 4$
$$\frac{4}{16} = \boxed{\frac{1}{4}}$$

51.
$$(x + h)^4 - (x^4 + h^4) = x^4 + 4x^3 h + 6x^2 h^2 + 4xh^3 + h^4 - x^4 - h^4$$
$$= \boxed{4x^3 h + 6x^2 h^2 + 4xh^3}$$

53. $\boxed{D}$ is true.

55.
$$f(x) = x^6 - x^4$$
$$\frac{f(a+h)-f(a)}{h} = \frac{(a+h)^6 - (a+h)^4 - a^6 + a^4}{h}$$
$$= \frac{a^6 + 6a^5h + 15a^4h^2 + 20a^3h^3 + 15a^2h^4 + 6ah^5 + h^6 - a^4 - 4a^3h - 6a^2h^2 - 4ah^3 - h^4 - a^6 + a^4}{h}$$
$$= \frac{6a^5h + 15a^4h^2 + 20a^3h^3 + 15a^2h^4 + 6ah^5 + h^6 - 4a^3h - 6a^2h^2 - 4ah^3 - h^4}{h}$$
$$= \boxed{6a^5 + 15a^4h + 20a^3h^2 + 15a^2h^3 + 6ah^4 + h^5 - 4a^3 - 6a^2h - 4ah^2 - h^3}$$

57.
$$(x^2 + x + 1)^3 = [x^2 + (x+1)]^3$$
$$= x^6 + 3x^4(x+1) + 3x^2(x+1)^2 + (x+1)^3$$
$$= x^6 + 3x^5 + 3x^4 + 3x^4 + 6x^3 + 3x^2 + x^3 + 3x^2 + 3x + 1$$
$$= \boxed{x^6 + 3x^5 + 6x^4 + 7x^3 + 6x^2 + 3x + 1}$$

59.
$$\frac{(n+1)!}{(n-1)!} = \frac{(n+1)n(n-1)!}{(n-1)!}$$
$$= n(n+1)$$
$$= \boxed{n^2 + n}$$

61.
$$(1+i)^6 = \binom{6}{0}1^6 + \binom{6}{1}1^5(i) + \binom{6}{2}1^4(i^2) + \binom{6}{3}1^3(i^3) + \binom{6}{4}1^2(i^4) + \binom{6}{5}1(i^5) + \binom{6}{6}(i^6)$$
$$= \boxed{1 + 6i + 15i^2 + 20i^3 + 15i^4 + 6i^5 + i^6}$$

$$i^2 = -1$$
$$i^3 = i^2 \cdot i = (-1)i = -i$$
$$i^4 = i^2 \cdot i^2 = (-1)(-1) = 1$$
$$i^5 = i^4 \cdot i = 1 \cdot i = 1$$
$$i^6 = i^5 \cdot i = i \cdot i = i^2 = -1$$

$$(1+i)^6 = 1 + 6i + 15(-1) + 20(-i) + 15(1) + 6i + (-1)$$
$$= 1 + 6i - 15 - 20i + 15 + 6i - 1$$
$$= \boxed{-8i}$$

63. $\left(\dfrac{1}{x} - x^2\right)^{12}$

There are 13 terms, so the middle term is the 7th term.
$n = 12, r = 7$
$$\binom{n}{r-1}a^{n-r+1}b^{r-1} = \binom{12}{7-1}\left(\frac{1}{x}\right)^{12-7+1}(-x^2)^{7-1}$$
$$= \binom{12}{6}\left(\frac{1}{x}\right)^6(-x^2)^6$$
$$= 924\left(\frac{1}{x^6}\right)(x^{12})$$
$$= \boxed{924x^6}$$

65. $(x^2 - y)^9$

Since the second term involves a factor of $(-y)$, the third term a factor of $(-y)^2$, the 4th term a factor of $(-y)^3$, etc., the term containing y^5 is the 6th term.

$n = 9$, $r = 6$

$$\binom{n}{r-1}a^{n-r+1}b^{r-1} = \binom{9}{6-1}(x^2)^{9-6+1}(-y)^{6-1}$$

$$= \binom{9}{5}(x^2)^4(-y)^5$$

$$= 126x^8(-y^5)$$

$$= \boxed{-126x^8y^5}$$

Review Problems

69. $f(x) = 2x - 1$ and $g(x) = x^2 - 3x + 2$

$$\begin{aligned}f[g(x)] &= 2[g(x)] - 1\\ &= 2(x^2 - 3x + 2) - 1\\ &= \boxed{2x^2 - 6x + 3}\end{aligned}$$

$$\begin{aligned}g[f(x)] &= [f(x)]^2 - 3[f(x)] + 2\\ &= (2x - 1)^2 - 3(2x - 1) + 2\\ &= 4x^2 - 4x + 1 - 6x + 3 + 2\\ &= \boxed{4x^2 - 10x + 6}\end{aligned}$$

70.

$$\begin{aligned}f(x) &= 3x + 5\\ y &= 3x + 5\\ x &= 3y + 5 \qquad \text{(Exchange } x \text{ and } y)\\ x - 5 &= 3y\\ \frac{x-5}{3} &= y\\ \boxed{f^{-1}(x) = \frac{x-5}{3}}\end{aligned}$$

$$\begin{aligned}f[f^{-1}(x)] &= 3[f^{-1}(x)] + 5\\ &= 3\left(\frac{x-5}{3}\right) + 5\\ &= x - 5 + 5\\ &= x\end{aligned}$$

$$\begin{aligned}f^{-1}[f(x)] &= \frac{f(x) - 5}{3}\\ &= \frac{3x + 5 - 5}{3}\\ &= \frac{3x}{3}\\ &= x\end{aligned}$$

71.
$$f(x) = Ae^{0.04x}$$
$$A = 1,000, f(x) = 4,000$$

$$
\begin{aligned}
1,000\, e^{0.04x} &= 4,000 \\
e^{0.04x} &= 4 \\
\ln e^{0.04x} &= \ln 4 \\
0.04x &= \ln 4 \\
x &= \frac{\ln 4}{0.04} \\
&\approx 35
\end{aligned}
$$

$\boxed{\text{Approximately 35 hours}}$

Chapter 12   Review Problems

Review Problems, pp. 890-891

1.
$$
\begin{aligned}
a_n &= n^2 + 1 \\
a_1 &= 1^2 + 1 = 2 \\
a_2 &= 2^2 + 1 = 5 \\
a_3 &= 3^2 + 1 = 10 \\
a_4 &= 4^2 + 1 = 17
\end{aligned}
$$

$\boxed{2, 5, 10, 17}$

2.
$$
\begin{aligned}
a_1 &= 1 = 1^2 \\
a_2 &= 4 = 2^2 \\
a_3 &= 9 = 3^2 \\
a_4 &= 16 = 4^2
\end{aligned}
$$

$\boxed{a_n = n^2}$

3. $\displaystyle\sum_{i=1}^{4} (2i^2 - 3)$

$= (2 \cdot 1^2 - 3) + (2 \cdot 2^2 - 3) + (2 \cdot 3^2 - 3) + (2 \cdot 4^2 - 3)$

$= (-1) + (5) + (15) + (29)$

$= \boxed{48}$

4. $\displaystyle\sum_{i=1}^{5} 6x^i$

$= \boxed{6x + 6x^2 + 6x^3 + 6x^4 + 6x^5}$

5. $x = \dfrac{\sum_{i}^{5} x}{n}$

$= \dfrac{3.8 + 2.3 + 1.1 + 7.2 + 8.1}{5}$

$= \dfrac{22.5}{5}$

$= \boxed{4.5}$

6. $1 + 8 + 27 + 64 + 125$

$= 1^3 + 2^3 + 3^3 + 4^3 + 5^3$

$= \boxed{\displaystyle\sum_{i=1}^{5} i^3}$

7. $\dfrac{x+1}{x} + \dfrac{x+2}{x} + \dfrac{x+3}{x} + \dfrac{x+4}{x}$

$= \boxed{\displaystyle\sum_{i=1}^{4} \dfrac{x+i}{x}}$

8. $-7, -3, 1, 5, \ldots$ $d = -3 - (-7) = 4$

$$a_n = a_1 + (n-1)d$$
$$a_{15} = -7 + (15-1)(4)$$
$$a_{15} = -7 + 14(4)$$
$$= -7 + 56$$
$$= \boxed{49}$$

9. $5, 2, -1, -4, \ldots$

$$d = 2 - 5 = -3$$
$$a_n = a_1 + (n-1)d$$
$$a_n = 5 + (n-1)(-3)$$
$$a_n = \boxed{8 - 3n}$$

10. $9, 15, 21, \ldots$

$$d = 15 - 9 = 6$$
$$a_n = a_1 + (n-1)d$$
$$a_{15} = 9 + (15-1)(6)$$
$$a_{15} = 9 + 84 = 93$$
$$\boxed{93 \text{ oranges}}$$

11. $18000, 18850, 19700, \ldots$

$a_1 = 18{,}000$ $d = 850$ $a_n = 25{,}650$
Find n.

$$a_n = a_1 + (n-1)d$$
$$25{,}650 = 18{,}000 + (n-1)(850)$$
$$25{,}650 = 18{,}000 + 850n - 850$$
$$25{,}650 = 17{,}150 + 850n$$
$$8{,}500 = 850n$$
$$10 = n$$
$$\boxed{10 \text{ years}}$$

12. $5, 12, 19, 26, \ldots$

$d = 12 - 5 = 7$

$$S_n = \frac{n}{2}(a_1 + a_n)$$
$$S_{22} = \frac{22}{2}(5 + a_{22})$$

Find a_{22}.

$$a_n = a_1 + (n-1)d$$
$$a_{22} = 5 + (22-1)(7)$$
$$a_{22} = 5 + (21)(7)$$
$$a_{22} = 152$$
$$S_{22} = \frac{22}{2}(5 + 152)$$
$$S_{22} = 11(157)$$
$$= \boxed{1{,}727}$$

13.   Given: $n = 16$, $a_1 = 3$, $d = 5$
Find S_{16}.

$$S_n = \frac{n}{2}(a_1 + a_n)$$
$$S_{16} = \frac{16}{2}(3 + a_{16})$$

Find a_{16}.

$$a_n = a_1 + (n-1)d$$
$$a_{16} = 3 + (16-1)(5)$$
$$a_{16} = 78$$

$$S_{16} = \frac{16}{2}(3 + 78)$$
$$S_{16} = \boxed{648}$$

14. $\displaystyle\sum_{i=1}^{16} (3i + 2)$

$$= (3 \cdot 1 + 2) + (3 \cdot 2 + 2) + (3 \cdot 3 + 2) + \ldots + (3 \cdot 16 + 2)$$
$$= 5 + 8 + 11 + \ldots + 50$$

$$a_1 = 5, n = 16 \quad \text{(There are 16 terms),}$$
$$a_{16} = 50$$
$$S_n = \frac{n}{2}(a_1 + a_n)$$
$$S_{16} = \frac{16}{2}(a_1 + a_{16})$$
$$S_{16} = \frac{16}{2}(5 + 50)$$
$$S_{16} = 8(55)$$
$$= \boxed{440}$$

15. Given: $a_1 = 68$, $a_5 = 59$
Find S_{17}.

$$S_n = \frac{n}{2}(a_1 + a_n)$$

$$S_{17} = \frac{17}{2}(68 + a_{17})$$

$$a_n = a_1 + (n-1)d$$

$$a_5 = 68 + (5-1)d = 59$$

$$68 + 4d = 59$$
$$4d = -9$$
$$d = -\frac{9}{4}$$

Find a_{17}.

$$a_{17} = a_1 + (17-1)d$$
$$a_{17} = 68 + 16\left(-\frac{9}{4}\right)$$
$$a_{17} = 68 + (-36) = 32$$
$$S_{17} = \frac{17}{2}(68 + 32)$$
$$S_{17} = \frac{17}{2}(100) = \boxed{850}$$

16. Let d = number of additional bushels each day
$35, 35 + d, 35 + 2d, \ldots$

$$a_1 = 35, S_{15} = 854$$

Find d.

$$S_n = \frac{n}{2}(a_1 + a_n)$$

$$S_{14} = \frac{14}{2}(35 + a_{14})$$

Find a_{14}.

$$a_n = a_1 + (n-1)d$$
$$a_{14} = 35 + (14-1)d$$
$$a_{14} = 35 + 13d$$
$$S_{14} = \frac{14}{2}(35 + 35 + 13d) = 854$$
$$7(70 + 13d) = 854$$
$$490 + 91d = 854$$
$$91d = 364$$
$$d = 4$$

$$\boxed{\text{4 additional bushels of fruit each day}}$$

17. $a_1 = \frac{9}{25}$

$$a_2 = \frac{9}{25}\left(-\frac{5}{3}\right) = -\frac{3}{5}$$

$$a_3 = -\frac{3}{5}\left(-\frac{5}{3}\right) = 1$$

$$a_4 = 1\left(-\frac{5}{3}\right) = -\frac{5}{3}$$

$$\boxed{\frac{9}{25}, -\frac{3}{5}, 1, -\frac{5}{3}}$$

18. $\frac{1}{3}, \frac{1}{2}, \frac{3}{4}, \ldots$

$$a_1 = \frac{1}{3} \qquad r = \frac{\frac{1}{2}}{\frac{1}{3}} = \frac{3}{2}$$

$$a_n = a_1 r^{n-1}$$

$$a_6 = \frac{1}{3}\left(\frac{3}{2}\right)^5 = \frac{3^4}{2^5} = \boxed{\frac{81}{32}}$$

19. $3, 3\sqrt{3}, 9, \ldots$

$$a_1 = 3 \qquad r = \frac{3\sqrt{3}}{3} = \sqrt{3}$$
$$a_n = a_1 r^{n-1}$$
$$a_{10} = 3\left(\sqrt{3}\right)^{10-1} = 3(3^{1/2})^9$$
$$= 3 \cdot 3^{9/2} = 3 \cdot 3^4 \cdot 3^{1/2}$$
$$a_{10} = 3^5 \cdot 3^{1/2} = \boxed{243\sqrt{3}}$$

20. $5, -10, 20, \ldots$

$$a_1 = 5, \qquad r = \frac{-10}{5} = -2$$
$$a_n = a_1 r^{n-1}$$
$$a_n = \boxed{5(-2)^{n-1}}$$

21. $a_1 = 10,000$ $r = 0.8$

$a_n = a_1 r^{n-1}$
$a_5 = 10,000(0.8)^4$

$a_5 = 10,000(0.4096) = 4096$

$\boxed{\$4,096}$

22. Given: $a_1 = 20$, $a_4 = \dfrac{5}{16}$

Find r.

$a_n = a_1 r^{n-1}$

$a_4 = 20 r^3 = \dfrac{5}{16}$

$r^3 = \dfrac{1}{64}$

$r = \sqrt[3]{\dfrac{1}{64}} = \boxed{\dfrac{1}{4}}$

23. $7, -14, 28, \ldots$

$a_1 = 7$ $r = \dfrac{-14}{7} = -2$

$S_n = \dfrac{a_1 - a_1 r^n}{1 - r}$

$S_6 = \dfrac{7 - 7(-2)^6}{1 - (-2)} = \dfrac{-441}{3} = \boxed{-147}$

24. $\displaystyle\sum_{i=1}^{5} \dfrac{1}{2}(6)^{i-1}$

$= \dfrac{1}{2}(6)^0 + \dfrac{1}{2}(6)^1 + \dfrac{1}{2}(6)^2 + \dfrac{1}{2}(6)^3 + \dfrac{1}{2}(6)^4$

$a_1 = \dfrac{1}{2}$, $r = 6$, $n = 5$ (There are five terms.)

$S_n = \dfrac{a_1 - a_1 r^n}{1 - r}$

$S_5 = \dfrac{\dfrac{1}{2} - \dfrac{1}{2}(6)^5}{1 - 6} = \dfrac{-3887.5}{-5} = \boxed{777.5}$

25. $4, 12, 36, \ldots$ $r = \dfrac{12}{4} = 3$

Find S_6.

$S_n = \dfrac{a_1 - a_1 r^n}{1 - r}$

$S_6 = \dfrac{4 - 4(3)^6}{1 - 3} = \dfrac{-2912}{-2} = 1456$

accumulated savings: $\boxed{\$1,456}$

26. Given: $S_4 = -100$, $r = -3$

Find a_1.

$S_n = \dfrac{a_1 - a_1 r_n}{1 - r}$

$S_4 = \dfrac{a_1 - a_1(-3)^4}{1 - (-3)} = -100$

$\dfrac{a_1 - 81 a_1}{4} = -100$

$-80 a_1 = -100$

$a_1 = \boxed{5}$

27. $36 + 12 + 4 + \ldots$

$r = \dfrac{12}{36} = \dfrac{1}{3}$

$S = \dfrac{a_1}{1 - r}$

$S = \dfrac{36}{1 - \dfrac{1}{3}} = \dfrac{36}{\dfrac{2}{3}} = 36 \cdot \dfrac{3}{2} = \boxed{54}$

28. $\dfrac{4}{3} - 1 + \dfrac{3}{4} - \ldots$

$a_1 = \dfrac{4}{3}$ $r = \dfrac{-1}{\dfrac{4}{3}} = -\dfrac{3}{4}$

$S = \dfrac{a_1}{1 - r}$

$S = \dfrac{\dfrac{4}{3}}{1 - \left(-\dfrac{3}{4}\right)} = \dfrac{\dfrac{4}{3}}{\dfrac{7}{4}} = \boxed{\dfrac{16}{21}}$

29. $0.36 = 0.363636\ldots$
$= 0.36 + 0.0036 + 0.000036 + \ldots$

$a_1 = 0.36 \qquad r = \dfrac{0.0036}{0.36} = 0.01$

$S = \dfrac{a_1}{1-r}$

$S = \dfrac{0.36}{1-0.01} = \dfrac{0.36}{0.99} = \dfrac{36}{99} = \boxed{\dfrac{4}{11}}$

30. $\displaystyle\sum_{i=1}^{12}(2i - 3)$

$a_1 = -1,\ d = 2,$

$a_{12} = -1 + 11(2) = 21$

$\begin{aligned}S_{12} &= \dfrac{12}{2}(-1 + 21)\\ &= \boxed{120}\end{aligned}$

31. $\displaystyle\sum_{i=1}^{10}3(10)^{-i}$

$a_1 = \dfrac{3}{10},\ r = \dfrac{1}{10}$

$\begin{aligned}S_{10} &= \dfrac{\left(\dfrac{3}{10}\right)\left(1 - \left(\dfrac{1}{10}\right)^{10}\right)}{1 - \dfrac{1}{10}}\\[2mm] &= \dfrac{3\left(1 - \dfrac{1}{10^{10}}\right)}{9}\\[2mm] &= \boxed{\dfrac{1}{3}\left(\dfrac{10^{10} - 1}{10^{10}}\right)}\end{aligned}$

32. $\displaystyle\sum_{i=1}^{\infty}18\left(\dfrac{2}{3}\right)^{i-1}$

$a_1 = 18,\ r = \dfrac{2}{3}$

$\begin{aligned}S &= \dfrac{18}{1 - \dfrac{2}{3}}\\[2mm] &= \boxed{54}\end{aligned}$

33. $\displaystyle\sum_{i=1}^{\infty}0.35(0.93)^{i-1}$

$a_1 = 0.35,\ r = 0.93$

$\begin{aligned}S &= \dfrac{0.35}{1 - 0.93}\\[2mm] &= \boxed{5}\end{aligned}$

34. $\displaystyle\sum_{i=1}^{25}(3i + 6)$

$a_1 = 9,\ d = 3,\ n = 25$

$\begin{aligned}a_{25} &= 9 + 24(3)\\ &= 81\\ S_{25} &= \dfrac{25}{2}(9 + 81)\\ &= \boxed{1125}\end{aligned}$

35. $\displaystyle\sum_{i=1}^{5} 5\left(\frac{1}{5}\right)^i$

$a_1 = 1,\ r = \frac{1}{5},\ n = 5$

$$S_5 = \frac{1\left(1 - \left(\frac{1}{5}\right)^5\right)}{1 - \frac{1}{5}}$$

$$= \frac{5}{4}\left(1 - \frac{1}{3125}\right)$$

$$= \frac{3905}{3125}$$

$$= \boxed{\frac{781}{625}}$$

36. $\displaystyle\sum_{i=1}^{\infty} \frac{2^i}{3^{i+1}} = \frac{1}{3}\sum_{i=1}^{\infty}\left(\frac{2}{3}\right)^i$

$$S = \frac{1}{3}\left(\frac{\frac{2}{3}}{1 - \frac{2}{3}}\right)$$

$$= \boxed{\frac{2}{3}}$$

37. $\sqrt{2} + \sqrt{8} + \sqrt{18} + \sqrt{32} + \ldots = \sqrt{2} + 2\sqrt{2} + 3\sqrt{2} + 4\sqrt{2} + \ldots$

$a_1 = \sqrt{2},\ d = \sqrt{2}$

$a_{30} = \sqrt{2} + (30 - 1)\sqrt{2} = \sqrt{2} + 29\sqrt{2} = 30\sqrt{2}$

$$S_{30} = \frac{30}{2}\left(\sqrt{2} + 30\sqrt{2}\right)$$

$$= 15(31\sqrt{2})$$

$$= \boxed{465\sqrt{2}}$$

38. $\qquad \sqrt{3} + \sqrt{12} + \sqrt{48} + \ldots = \sqrt{3} + 2\sqrt{3} + 4\sqrt{3} + \ldots$

$a_1 = \sqrt{3},\ r = 2,$

$$S_9 = \frac{\sqrt{3}(1 - 2^9)}{1 - 2}$$

$$= \sqrt{3}(512 - 1)$$

$$= \boxed{511\sqrt{3}}$$

39. $a_1 = 100,\ d = 2$

$a_{30} = 100 + 29(2) = 158$

$$S_{30} = \frac{30}{2}(100 + 158)$$

$$= 3870$$

Total number of seats: $\boxed{3870}$

40. $a_1 = 10,\ r = 3$

$$S_8 = \frac{10(1 - 3^8)}{1 - 3}$$

$$= 5(6561 - 1)$$

$$= 32{,}800$$

Remaining funds $= 100{,}000 - 32{,}000 = \boxed{\$67{,}200}$

41. $a_1 = 3(6) = 18,\ r = \dfrac{1}{3}$

$$S = \dfrac{18}{1 - \dfrac{1}{3}}$$

$$= 27\text{ m}$$

Sum of the perimeters: $\boxed{27\text{ m}}$

42. $a_1 = 30{,}000,\ r = 1.05$

$$a_{40} = 30{,}000(1.05)^{39}$$

$$\approx 201{,}142.50$$

Salary at the end of 40 years: $\boxed{\$201{,}142.50}$

$$S_{40} = \dfrac{30{,}000(1 - 1.05^{40})}{1 - 1.05}$$

$$\approx 3{,}623{,}993$$

Total salary over a 40 year period: $\boxed{\$3{,}623{,}993}$

43. $a_1 = 12,\ r = \dfrac{3}{4}$

$$S = \dfrac{12}{1 - \dfrac{3}{4}}$$

$$= 48$$

total distance covered: $\boxed{48\text{ m}}$

44.

$$A_1 = \pi r_1{}^2$$

$$= \pi \cdot 1^2$$

$$= \pi$$

$$A_2 = \pi r_2{}^2$$

$$= \pi\left(\dfrac{1}{2}\right)^2$$

$$= \dfrac{\pi}{4}$$

$$r = \dfrac{1}{4}$$

$$\text{total area} = \dfrac{\pi}{1 - \dfrac{1}{4}}$$

$$= \dfrac{4\pi}{3}$$

sum of the areas $= \boxed{\dfrac{4\pi}{3}\text{ square meters}}$

45.   number of levels $= 30 + 2 = 32$

$a_1 = 1,\ r = 2$

$$a_{32} = 2(2)^{32-1}$$

$$= \boxed{4{,}294{,}967{,}296}$$

46. $a_1 = 100,\ d = 10$

$$a_n = 100 + (n-1)(10)$$

$$= 10n + 90$$

$$S_n = \dfrac{n}{2}(a_1 + a_n)$$

$$= \dfrac{n}{2}(100 + 10n + 90)$$

$$= n(5n + 95)$$

$$5n^2 + 95n = 2550$$

$$n^2 + 19n - 510 = 0$$

$$(n + 34)(n - 15) = 0$$

$$n = 15\text{ months}$$

time to repay loan: $\boxed{15\text{ months}}$

47. $a_1 = 16{,}000;\ r = 0.80$

$$a_8 = 16{,}000(0.80)^{8-1}$$

$$\approx 3355.44$$

value of car at the start of the eighth year: $\boxed{\$3355.44}$

48. $\dbinom{9}{2} = \dfrac{9!}{2!(9-2)!}$

$\qquad\quad = \dfrac{9!}{2!7!}$

$\qquad\quad = \dfrac{9\cdot 8}{2\cdot 1}$

$\qquad\quad = \boxed{36}$

49. $(x^2+3y)^4 = \dbinom{4}{0}(x^2)^4 + \dbinom{4}{1}(x^2)^3(3y) + \dbinom{4}{2}(x^2)^2(3y)^2 + \dbinom{4}{3}(x^2)^1(3y)^3 + \dbinom{4}{4}(3y)^4$

$\qquad\qquad\quad = 1x^8 + 4x^6(3y) + 6x^4(9y^2) + 4x^2(27y^3) + 1(81y^4)$

$\qquad\qquad\quad = \boxed{x^8 + 12x^6y + 54x^4y^2 + 108x^2y^3 + 81y^4}$

50. $(x^3-2)^5 = [x^3+(-2)]^5$

$\qquad\qquad\quad = \dbinom{5}{0}(x^3)^5 + \dbinom{5}{1}(x^3)^4(-2) + \dbinom{5}{2}(x^3)^3(-2)^2 + \dbinom{5}{3}(x^3)^2(-2)^3 + \dbinom{5}{4}(x^3)^1(-2)^4 + \dbinom{5}{5}(-2)^5$

$\qquad\qquad\quad = 1x^{15} + 5x^{12}(-2) + 10x^9(4) + 10x^6(-8) + 5x^3(16) + 1(-32)$

$\qquad\qquad\quad = \boxed{x^{15} - 10x^2 + 40x^9 - 80x^6 + 80x^3 - 32}$

51. $(x+y)^{12} = \dbinom{12}{0}x^{12} + \dbinom{12}{1}x^{11}y + \dbinom{12}{2}x^{10}y^2 + \ldots$

$\qquad\qquad\quad = \boxed{x^{12} + 12x^{11}y + 66x^{10}y^2 + \ldots}$

52. $(2x-y)^6 = [2x+(-y)]^6$

$\qquad\qquad\quad = \dbinom{6}{0}(2x)^6 + \dbinom{6}{1}(2x)^5(-y) + \dbinom{6}{2}(2x)^4(-y)^2 + \ldots$

$\qquad\qquad\quad = 1(64x^6) + 6(32x^5)(-y) + 15(16x^4)(y^2) + \ldots$

$\qquad\qquad\quad = \boxed{64x^6 - 192x^5y + 240x^4y^2 + \ldots}$

53. $(3c+d)^9$

$\quad$ 7th term: $r=7,\ n=9$

$\qquad r^{\text{th}}\ \text{term} = \dbinom{n}{r-1}a^{n-r+1}b^{r-1}$

$\qquad 7^{\text{th}}\ \text{term} = \dbinom{9}{7-1}(3c)^{9-7+1}d^{7-1}$

$\qquad\qquad\quad = \dbinom{9}{6}(3c)^3d^6$

$\qquad\qquad\quad = 84(27c^3)d^6$

$\qquad\qquad\quad = \boxed{2268c^3d^6}$

54. $(a-2b)^7$

$\quad$ 4th term: $r=4,\ n=7$

$\qquad r^{\text{th}}\ \text{term} = \dbinom{n}{r-1}a^{n-r+1}b^{r-1}$

$\qquad 4^{\text{th}}\ \text{term} = \dbinom{7}{4-1}a^{7-4+1}(-2b)^{4-1}$

$\qquad\qquad\quad = \dbinom{7}{3}a^4(-2b)^3$

$\qquad\qquad\quad = \dfrac{7!}{3!4!}a^4(-8)b^3$

$\qquad\qquad\quad = 35a^4(-8b^3)$

$\qquad\qquad\quad = \boxed{-280a^4b^3}$

Algebra for College Students

Appendix B

Review Problems Covering the Entire Book

Appendix Review Problems, pp. A4-A8

1.
$$\frac{4y-2}{3} - \frac{y+2}{4} = \frac{7y-2}{12}$$
$$\frac{12(4y-2)}{3} - \frac{12(y+2)}{4} = \frac{12(7y-2)}{12}$$
$$4(4y-2) - 3(y+2) = 7y-2$$
$$16y - 8 - 3y - 6 = 7y - 2$$
$$6y = 12$$
$$y = 2$$

$\boxed{\{2\}}$

2. $\{x| -5(x-1)+3 > 3x - 4 - 4x\} \cup \{x| 3(x-2) - 5(2x-1) \geq 0\}$

$$-5x + 5 + 3 > -x - 4 \qquad\qquad 3x - 6 - 10x + 5 \geq 0$$
$$-4x > -12 \qquad\qquad\qquad -7x \geq 1$$
$$x < 3 \qquad\qquad\qquad\qquad x \leq -\frac{1}{7}$$

$\boxed{\{x| x < 3\} = (-\infty, 3)}$

3. $\{x| -7x > -14\} \cap \{x| 3x < 15\}$

$$x < 2 \cap x < 5$$

$\boxed{\{x| x < 2\} = (-\infty, 2)}$

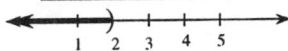

4.
$$1 \leq \frac{3-2x}{5} < 3$$
$$5 \leq 3 - 2x < 15$$
$$-1 \geq x > -6$$
$$-6 < x \leq -1$$

$\boxed{\{x| -6 < x \leq -1\} = (-6, -1]}$

5.
$$|3y + 1| \geq 16$$
$$3y + 1 \geq 16 \qquad\qquad \text{or} \qquad 3y + 1 \leq -16$$
$$3y \geq 15 \qquad\qquad\qquad\qquad 3y \leq -17$$
$$y \geq 5 \qquad\qquad\qquad\qquad y \leq -\frac{17}{3}$$

$\boxed{\{y| y \leq -\frac{17}{3} \text{ or } y \geq 5\} = (-\infty, -\frac{17}{3}] \cup [5, \infty)}$

6.

$$\frac{y}{y-3} - \frac{3y}{y^2-y-6} = \frac{4y^2-4y-18}{y^2-y-6}$$

$$(y-3)(y+2)\left[\frac{y}{y-3} - \frac{3y}{y^2-y-6}\right] = (y-3)(y+2)\left[\frac{4y^2-4y-18}{(y-3)(y+2)}\right]$$

$$y(y+2) - 3y = 4y^2-4y-18$$

$$3y^2 - 3y - 18 = 0$$

$$y^2 - y - 6 = 0$$

$$(y-3)(y+2) = 0$$

$$y = 3 \quad \text{or} \quad y = -3$$

(Reject both solutions; they cause division by 0.)

$$\boxed{\varnothing}$$

7.

$$\sqrt{5x+1} - \sqrt{3x} = 1$$

$$\sqrt{5x+1} = 1 + \sqrt{3x}$$

$$5x + 1 = 1 + 2\sqrt{3x} + 3x$$

$$2x = 2\sqrt{3x}$$

$$x = \sqrt{3x}$$

$$x^2 = 3x$$

$$x(x-3) = 0$$

$$x = 0 \quad \text{or} \quad x = 3 \quad \text{(both check)}$$

$$\boxed{\{0, 3\}}$$

8.

$$|x^2 + 2x - 4| = 4$$

$$x^2 + 2x - 4 = 4 \qquad \text{or} \qquad x^2 + 2x - 4 = -4$$

$$x^2 + 2x - 8 = 0 \qquad\qquad\qquad x^2 + 2x = 0$$

$$(x+4)(x-2) = 0 \qquad\qquad\qquad x(x+2) = 0$$

$$x = -4 \ \text{or}\ x = 2 \qquad\qquad x = 0 \ \text{or}\ x = -2$$

$$\boxed{\{-4, -2, 0, 2\}}$$

9.

$$\frac{1}{y+1} = 2 + \frac{2}{y-3}$$

$$(y+1)(y-3) \cdot \frac{1}{y+1} = (y+1)(y-3)\left[2 + \frac{2}{y-3}\right]$$

$$y - 3 = 2y^2 - 4y - 6 + 2y + 2$$

$$0 = 2y^2 - 3y - 1$$

$$y = \frac{3 \pm \sqrt{9+8}}{4}$$

$$y = \frac{3 \pm \sqrt{17}}{4}$$

$$\boxed{\left\{\frac{3-\sqrt{17}}{4}, \frac{3+\sqrt{17}}{4}\right\}}$$

10.

$$x^{2/3} - x^{1/3} - 6 = 0$$

$$t^2 - t - 6 = 0 \quad \text{(Let } t = x^{2/3}\text{)}$$

$$(t-3)(t+2) = 0$$

$$t = 3 \qquad\qquad \text{or} \qquad\qquad t = -2$$

$$x^{1/3} = 3 \qquad\qquad \text{or} \qquad\qquad x^{1/3} = -2$$

$$x = 3^3 \qquad\qquad\qquad\qquad x = (-2)^3$$

$$x = 27 \qquad\qquad\qquad\qquad x = -8$$

$$\boxed{\{-8, 27\}}$$

11.
$$\frac{1}{y+2} - \frac{1}{3} = \frac{1}{y}$$

$$3y(y+2)\left[\frac{1}{y+2} - \frac{1}{3}\right] = 3y(y+2)\cdot\frac{1}{y}$$

$$3y - y^2 - 2y = 3y + 6$$

$$-y^2 - 2y - 6 = 0$$

$$y^2 + 2y + 6 = 0$$

$$y = \frac{-2 \pm \sqrt{4-24}}{2}$$

$$= \frac{-2 \pm 2i\sqrt{5}}{2}$$

$$y = -1 \pm i\sqrt{5}$$

$$\boxed{\left\{-1 - i\sqrt{5}, -1 + i\sqrt{5}\right\}}$$

12.
$$(x^2 + x)^2 - 5(x^2 + x) = -6$$

$$(x^2 + x)^2 - 5(x^2 + x) + 6 = 0$$

$$t^2 - 5t + 6 = 0 \qquad\qquad \text{(Let } t = x^2 + x\text{)}$$

$$(t-3)(t-2) = 0$$

$$t = 3 \qquad\qquad \text{or} \qquad\qquad t = 2$$

$$x^2 + x = 3 \qquad\qquad\qquad\qquad\qquad x^2 + x = 2$$

$$x^2 + x - 3 = 0 \qquad\qquad\qquad\qquad x^2 + x - 2 = 0$$

$$x = \frac{-1 \pm \sqrt{1+12}}{2} \qquad\qquad\qquad (x+2)(x-1) = 0$$

$$x = \frac{-1 \pm \sqrt{13}}{2} \qquad\qquad\qquad\qquad\quad x = -2 \text{ or } x = 1$$

$$\boxed{\left\{\frac{-1-\sqrt{13}}{2}, \frac{-1+\sqrt{13}}{2}, -2, 1\right\}}$$

13.
$$3x^2 + 8x + 5 < 0$$

consider $\quad(3x+5)(x+1) = 0$

$$x = -\frac{5}{3} \text{ or } x = -1$$

F	T	F
−5/3	−1	

Test −2: $(-1)(-1) < 0$

$$1 < 0 \text{ False}$$

Test $-1\frac{1}{3}$: $(1)\left(-\frac{1}{3}\right) < 0$

$$-\frac{1}{3} < 0 \text{ True}$$

Test 0: $(5)(1) < 0$

$$5 < 0 \text{ False}$$

$$\boxed{\left\{x \mid -\frac{5}{3} < x < -1\right\} = \left(-\frac{5}{3}, -1\right)}$$

14.
$$\frac{x-1}{x+3} \le 0$$
$$x = 1, \quad x = -3$$

F	T	F
−3		1

Test −4:
$$\frac{-4-1}{-4+3} \le 0$$
$$5 \le 0 \text{ False}$$

Test 0:
$$\frac{-1}{3} \le 0 \text{ True}$$

Test 2:
$$\frac{2-1}{2+3} \le 0$$
$$\frac{1}{5} \le 0 \text{ False}$$

$$-3 < x \le 1$$

$$\boxed{\{x \mid -3 < x \le 1\} = (-3, 1]}$$

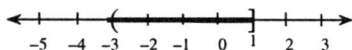

15.

$6x + 3y = -1$	$(\times -5) \rightarrow$	$-30x - 15y = 5$
$9x + 5y = 1$	$(\times 3) \rightarrow$	$27x + 15y = 3$

$$-3x = 8$$
$$x = -\frac{8}{3}$$

$$\boxed{\left\{\left(-\frac{8}{3}, 5\right)\right\}}$$

16.

$2x^2 + y^2 = 7$	$(\times 2) \rightarrow$	$4x^2 + 2y^2 = 14$
$x^2 - 2y^2 = -4$		$x^2 - 2y^2 = -4$

$$5x^2 = 10$$
$$x^2 = 2$$
$$x = \pm\sqrt{2}$$
$$2\left(\pm\sqrt{2}\right)^2 + y^2 = 7$$
$$4 + y^2 = 7$$
$$y^2 = 3$$
$$y = \pm\sqrt{3}$$

If $x = \sqrt{2}, y = \pm\sqrt{3}$
If $x = -\sqrt{2}, y = \pm\sqrt{3}$

$$\boxed{\left\{\left(-\sqrt{2}, -\sqrt{3}\right), \left(-\sqrt{2}, \sqrt{3}\right), \left(\sqrt{2}, -\sqrt{3}\right), \left(\sqrt{2}, \sqrt{3}\right)\right\}}$$

17.

$$\begin{aligned} 6x &= 10 + y \\ 3x^2 - xy &= 3 \end{aligned}$$

$\rightarrow$

(substitute for y) $\rightarrow$

$$\begin{aligned} y &= 6x - 10 \\ 3x^2 - x(6x - 10) &= 3 \\ 3x^2 - 6x^2 + 10x &= 3 \\ -3x^2 + 10x - 3 &= 0 \\ 3x^2 - 10x + 3 &= 0 \\ (3x - 1)(x - 3) &= 0 \end{aligned}$$

$$x = \frac{1}{3}$$

$$y = 6\left(\frac{1}{3}\right) - 10 = -8$$

or

$$x = 3$$

$$y = 6(3) - 10 = 8$$

$$\boxed{\left\{ \left(\frac{1}{3}, -8\right), (3, 8) \right\}}$$

18.

$$\begin{aligned} x - 2y + z &= -4 \\ 2x + 4y - 3z &= -1 \\ -3x - 6y + 7z &= 4 \end{aligned}$$

(Equations 1 and 2):

$$\begin{aligned} (\times 3) \quad 3x - 6y + 3z &= -12 \\ 2x + 4y - 3z &= -1 \\ \hline 5x - 2y &= -13 \text{ (Equation 4)} \end{aligned}$$

(Equations 1 and 3):

$$\begin{aligned} (\times -7) \quad -7x + 14y - 7z &= 28 \\ -3x - 6y + 7z &= 4 \\ \hline -10x + 8y &= 32 \text{ (Equation 5)} \end{aligned}$$

(Equations 4 and 5):

$$\begin{aligned} (\times 2) \quad 10x - 4y &= -26 \\ (\div 2) \quad -5x + 4y &= 16 \\ \hline 5x &= -10 \\ x &= -2 \end{aligned}$$

(Equation 4):

$$\begin{aligned} 5(-2) - 2y &= -13 \\ -2y &= -3 \\ y &= \frac{3}{2} \end{aligned}$$

(Equation 1):

$$\begin{aligned} -2 - 2\left(\frac{3}{2}\right) + z &= -4 \\ -2 - 3 + z &= -4 \\ z &= 1 \end{aligned}$$

$$\boxed{\left\{ \left(-2, \frac{3}{2}, 1\right) \right\}}$$

19.

$$\begin{aligned} 2^{2x+1} \cdot 4^x &= 8^{3-2x} \\ 2^{x+1} \cdot 2^{2x} &= 2^{3(3-2x)} \\ 2^{3x+1} &= 2^{9-6x} \\ 3x + 1 &= 9 - 6x \\ 9x &= 8 \\ x &= \frac{8}{9} \end{aligned}$$

$$\boxed{\left\{ \frac{8}{9} \right\}}$$

20.

$$\begin{aligned} \ln(x+2) + \ln(2x-1) &= \ln x \\ \ln(x+2)(2x-1) &= \ln x \\ (x+2)(2x-1) &= x \\ 2x^2 + 3x - 2 &= x \\ 2x^2 + 2x - 2 &= 0 \\ x^2 + x - 1 &= 0 \end{aligned}$$

$$\begin{aligned} x &= \frac{-1 \pm \sqrt{1+4}}{2} \\ &= \frac{-1 \pm \sqrt{5}}{2} \end{aligned}$$

$$\boxed{\left\{ \frac{-1 + \sqrt{5}}{2} \right\} \approx \{0.62\}}$$

21.
$$\log_2 x + \log_2 (2x-3) = 1$$
$$\log_2 x(2x-3) = 1$$
$$x(2x-3) = 2^1$$
$$2x^2 - 3x = 2$$
$$2x^2 - 3x - 2 = 0$$
$$(2x+1)(x-2) = 0$$
$$x = -\frac{1}{2} \text{ or } x = 2$$

$\left(\text{Reject } x = -\frac{1}{2} \text{ since } \log_2\left(-\frac{1}{2}\right) \text{ is not defined}\right)$

$\boxed{\{2\}}$

22.
$$2x^4 + x^3 - 9x^2 - 4x + 4 = 0$$

$$
\begin{array}{r|rrrrr}
-1 & 2 & 1 & -9 & -4 & 4 \\
   &   & -2 & 1 & 8 & -4 \\
\hline
   & 2 & -1 & -8 & 4 & 0
\end{array}
$$

$$(x+1)(2x^3 - x^2 - 8x + 4) = 0$$

$$
\begin{array}{r|rrrr}
2 & 2 & -1 & -8 & 4 \\
  &   & 4 & 6 & -4 \\
\hline
  & 2 & 3 & -2 & 0
\end{array}
$$

$$(x+1)(x-2)(2x^2 + 3x - 2) = 0$$
$$(x+1)(x-2)(2x-1)(x+2) = 0$$
$$x = -1, 2, \frac{1}{2}, -2$$

$\boxed{\left\{-1, 2, \frac{1}{2}, -2\right\}}$

23.
$$\frac{1}{B} = \frac{3}{C} - \frac{2}{A}$$
$$ABC\left(\frac{1}{B}\right) = ABC\left(\frac{3}{C} - \frac{2}{A}\right)$$
$$AC = 3AB - 2BC$$
$$AC + 2BC = 3AB$$
$$C(A+B) = 3AB$$
$$\boxed{C = \frac{3AB}{A+2B}}$$

24. $2x - z = 1$
$3y + 2z = 0$
$x - y = -3$

$$D = \begin{vmatrix} 2 & 0 & -1 \\ 0 & 3 & 2 \\ 1 & -1 & 0 \end{vmatrix} = 2\begin{vmatrix} 3 & 2 \\ -1 & 0 \end{vmatrix} - 0 + 1\begin{vmatrix} 0 & -1 \\ 3 & 2 \end{vmatrix}$$

$$= 2(0 + 2) - 0 + (0 + 3)$$
$$= 4 + 3 = 7$$

$$D_x = \begin{vmatrix} 1 & 0 & -1 \\ 0 & 3 & 2 \\ -3 & -1 & 0 \end{vmatrix} = 1\begin{vmatrix} 3 & 2 \\ -1 & 0 \end{vmatrix} - 0 + (-3)\begin{vmatrix} 0 & -1 \\ 3 & 2 \end{vmatrix}$$

$$= (0 + 2) - 3(0 + 3)$$
$$= 2 - 9 = -7$$

$x = \dfrac{D_x}{D}$

$\quad = \dfrac{-7}{7}$

$\quad = -1$

$\boxed{x = -1}$

$$D_y = \begin{vmatrix} 2 & 1 & -1 \\ 0 & 0 & 2 \\ 1 & -3 & 0 \end{vmatrix} = -0 + 0 - 2\begin{vmatrix} 2 & 1 \\ 1 & -3 \end{vmatrix}$$

$$= -2(-6 - 1) = -2(-7) = 14$$

$y = \dfrac{D_y}{D}$

$\quad = \dfrac{14}{7}$

$\quad = 2$

$$D_z = \begin{vmatrix} 2 & 0 & 1 \\ 0 & 3 & 0 \\ 1 & -1 & -3 \end{vmatrix} = -0 + 3\begin{vmatrix} 2 & 1 \\ 1 & -3 \end{vmatrix} - 0$$

$$= 3(-6 - 1) = 3(-7) = -21$$

$z = \dfrac{D_z}{D}$

$\quad = \dfrac{-21}{7}$

$\quad = -3$

$\boxed{\{(-1, 2, -3)\}}$

25.
$$x - 2y + z = 16$$
$$2x - y - z = 14$$
$$3x + 5y - 4z = -10$$

$$\begin{bmatrix} 1 & -2 & 1 & | & 16 \\ 2 & -1 & -1 & | & 14 \\ 3 & 5 & -4 & | & -10 \end{bmatrix}$$

$$\begin{bmatrix} 1 & -2 & 1 & | & 16 \\ 2 & -1 & -1 & | & 14 \\ 3 & 5 & -4 & | & -10 \end{bmatrix} \begin{matrix} R_2 - 2R_1 \to \\ R_3 - 3R_1 \to \end{matrix} \begin{bmatrix} 1 & -2 & 1 & | & 16 \\ 0 & 3 & -3 & | & -18 \\ 0 & 11 & -7 & | & -58 \end{bmatrix}$$

$$\to \tfrac{1}{3}R_2 \begin{bmatrix} 1 & -2 & 1 & | & 16 \\ 0 & 1 & -1 & | & -6 \\ 0 & 11 & -7 & | & -58 \end{bmatrix} R_3 - 11R_2 \to \begin{bmatrix} 1 & -2 & 1 & | & 16 \\ 0 & 1 & -1 & | & -6 \\ 0 & 0 & 4 & | & 8 \end{bmatrix}$$

$$\to \tfrac{1}{4}R_3 \begin{bmatrix} 1 & -2 & 1 & | & 16 \\ 0 & 1 & -1 & | & -6 \\ 0 & 0 & 1 & | & 2 \end{bmatrix}$$

$$x - 2y + z = 16$$
$$y - z = -6$$
$$z = 2$$

$$y - 2 = -6$$
$$y = -4$$

$$x - 2(-4) + 2 = 16$$
$$x = 6$$

$$\{(6, -4, 2)\}$$

26. $4y^4 + 4y$
$= 4y(y^3 + 1)$
$= \boxed{4y(y + 1)(y^2 - y + 1)}$

27. $x^3 - 2x^2 - 9x + 18$
$= x^2(x - 2) - 9(x - 2)$
$= \boxed{(x - 3)(x + 3)(x - 2)}$

28. $(2x + y)^2 + 15(2x + y) + 36$
$= t^2 + 15t + 36$ (Let $t = 2x + y$)
$= (t + 12)(t + 3)$
$= \boxed{(2x + y + 12)(2x + y + 3)}$

29. $(-2x^{-2}y)^2(-4x^2y^{-2})^{-2}$
$= (-2)^2(x^{-2})^2y^2(-4)^{-2}(x^2)^{-2}(y^{-2})^{-2}$
$= 4x^{-4}y^2 \cdot \dfrac{1}{(-4)^2}x^{-4}y^4$
$= \dfrac{4}{16}x^{-8}y^6 = \boxed{\dfrac{y^6}{4x^8}}$

30. $\left(\dfrac{-30x^{4n-3}y^{1-2n}}{10x^{4n-2}y^{-3-2n}}\right)^{-2}$
$= (-3x^{4n-3-(4n-2)}y^{1-2n-(-3-2n)})$
$= (-3x^{-1}y^4)^{-2}$
$= (-3)^{-2}(x^{-1})^{-2}(y^4)^{-2}$
$= \dfrac{1}{(-3)^2}x^2y^{-8}$
$= \boxed{\dfrac{x^2}{9y^8}}$

31. $a^{5/8}b^{1/2}(a^{5/2}b^{-5})^{-3/5}$
$= a^{5/8}b^{1/2}(a^{5/2})^{-3/5}(b^{-5})^{-3/5}$
$= a^{5/8}b^{1/2}a^{-3/2}b^3$
$= a^{5/8 - 3/2}b^{1/2 + 3}$
$= \boxed{\dfrac{b^{7/2}}{a^{7/8}}}$

32. $(2x - y)(x + 3y)(x - 2y)$

$= (2x - y)(x^2 + xy - 6y^2)$

$= 2x^3 + 2x^2y - 12xy^2$

$\quad\quad - x^2y - xy^2 + 6y^3$

$= \boxed{2x^3 + x^2y - 13xy^2 + 6y^3}$

33. $(3x^3 - 19x^2 + 17x + 4) \div (3x - 4) = \boxed{x^2 - 5x - 1}$

$$
\begin{array}{r}
x^2 - 5x - 1 \\
3x - 4 \overline{\smash{)}\, 3x^3 - 19x^2 + 17x + 4} \\
\underline{3x^3 - 4x^2} \\
-15x^2 + 17x \\
\underline{-15x^2 + 20x} \\
-3x + 4 \\
\underline{-3x + 4} \\
0
\end{array}
$$

34. $\dfrac{3x^3 - 5x^2 + 2x - 1}{x - 2} = \boxed{3x^2 + x + 4 + \dfrac{7}{x - 2}}$

$$
\begin{array}{r|rrrr}
2 & 3 & -5 & 2 & -1 \\
  &   & 6 & 2 & 8 \\
\hline
  & 3 & 1 & 4 & 7
\end{array}
$$

35. $\dfrac{y^3 - 7y^2 + 12y}{y^2 - y - 6} \div \dfrac{y^3 - 4y^2}{y^2 - 3y - 10}$

$= \dfrac{y(y^2 - 7y + 12)}{(y - 3)(y + 2)} \div \dfrac{y^2(y - 4)}{(y - 5)(y + 2)}$

$= \dfrac{y(y - 4)(y - 3)}{(y - 3)(y + 2)} \cdot \dfrac{(y - 5)(y + 2)}{y^2 + (y - 4)}$

$= \boxed{\dfrac{y - 5}{y}}$

36. $\dfrac{2y - 6}{3y^2 - 14y - 5} - \dfrac{y - 3}{y^2 - 5y}$

$= \dfrac{2y - 6}{(3y + 1)(y - 5)} - \dfrac{y - 3}{y(y - 5)}$ (LCD is $y(3y + 1)(y - 5)$)

$= \dfrac{2y - 6}{(3y + 1)(y - 5)} \cdot \dfrac{y}{y} - \dfrac{y - 3}{y(y - 5)} \cdot \dfrac{(3y + 1)}{(3y + 1)}$

$= \dfrac{y(2y - 6) - (y - 3)(3y + 1)}{y(3y + 1)(y - 5)}$

$= \dfrac{2y^2 - 6y - 3y^2 + 8y + 3}{y(3y + 1)(y - 5)}$

$= \boxed{\dfrac{-y^2 + 2y + 3}{y(3y + 1)(y - 5)}}$

37. $\dfrac{1 - \dfrac{14y - 45}{y^2}}{\dfrac{y}{9} - \dfrac{9}{y}} \cdot \dfrac{9y^2}{9y^2}$

$= \dfrac{9y^2 - 9(14y - 45)}{y^3 - 81y}$

$= \dfrac{9y^2 - 126y + 405}{y^3 - 81y}$

$= \dfrac{9(y^2 - 14y + 45)}{y(y^2 - 81)}$

$= \dfrac{9(y - 9)(y - 5)}{y(y + 9)(y - 9)}$

$= \boxed{\dfrac{9(y - 5)}{y(y + 9)}}$

38. $\sqrt[3]{4x^2y^5} \cdot \sqrt[3]{4xy^2z^2}$

$= \sqrt[3]{16x^3y^7z^2}$

$= \sqrt[3]{8 \cdot 2x^3(y^2)^3yz^2}$

$= \boxed{2xy^2\sqrt[3]{2yz^2}}$

39. $7\sqrt{18x^5} - 3x\sqrt{2x^3}$

$= 7\sqrt{9 \cdot 2(x^2)^2x} - 3x\sqrt{2x^2x}$

$= 7 \cdot 3x^2\sqrt{2x} - 3x(x)\sqrt{2x}$

$= 21x^2\sqrt{2x} - 3x^2\sqrt{2x}$

$= \boxed{18x^2\sqrt{2x}}$

40. $\dfrac{1 + \sqrt{3}}{3\sqrt{3} - 1}$

$= \dfrac{1 + \sqrt{3}}{3\sqrt{3} - 1} \cdot \dfrac{3\sqrt{3} + 1}{3\sqrt{3} + 1}$

$= \dfrac{3\sqrt{3} + 1 + 3(3) + \sqrt{3}}{9(3) - 1}$

$= \dfrac{4\sqrt{3} + 10}{26}$

$= \boxed{\dfrac{2\sqrt{3} + 5}{13}}$

41. $(2 + 3i)(4 - 5i)$

$= 8 - 10i + 12i - 15i^2$

$= 8 + 2i - 15(-1)$

$= \boxed{23 + 2i}$

42. $\dfrac{6}{3 + 5i}$

$= \dfrac{6}{3 + 5i} \cdot \dfrac{3 - 5i}{3 - 5i}$

$= \dfrac{6(3 - 5i)}{9 - 25i^2}$

$= \dfrac{6(3 - 5i)}{9 - 25(-1)}$

$= \dfrac{6(3 - 5i)}{34}$

$= \dfrac{3(3 - 5i)}{17}$

$= \dfrac{9 - 15i}{17}$

$= \boxed{\dfrac{9}{17} - \dfrac{15}{17}i}$

43. $\dfrac{5}{x^{1/2}} - 3x^{1/2} = \dfrac{5}{x^{1/2}} \cdot \dfrac{x^{1/2}}{x^{1/2}} - 3x^{1/2} \cdot \dfrac{x}{x}$

$\qquad\qquad\qquad = \dfrac{5x^{1/2} - 3x(x^{1/2})}{x}$

$\qquad\qquad\qquad = \boxed{\dfrac{5\sqrt{x} - 3x\sqrt{x}}{x}}$

44. $\dfrac{-3^2(5) - (-7)(2) - \left| 5 - 13 \right|}{-1 + 3(-4)} = \dfrac{-9(5) + 14 - 8}{-1 - 12}$

$\qquad\qquad\qquad\qquad\qquad = \dfrac{-45 + 6}{-13}$

$\qquad\qquad\qquad\qquad\qquad = \dfrac{-39}{-13}$

$\qquad\qquad\qquad\qquad\qquad = \boxed{3}$

45. Passing through $(1, -4)$ and $(-5, 8)$:

slope $= \dfrac{8 - (-4)}{-5 - 1} = \dfrac{12}{-6} = -2$

point–slope: $\boxed{y + 4 = -2(x - 1) \quad \text{or} \quad y - 8 = -2(x + 5)}$

$\qquad\qquad y + 4 = -2x + 2$

slope–intercept:

$\qquad\qquad \boxed{y = -2x - 2}$

46. Passing through $(3, -2)$ and perpendicular to the line whose euqation is:

$\qquad -\dfrac{1}{4}x + y = 5$

$\qquad\qquad\quad y = \dfrac{1}{4}x + 5$

$\qquad\quad \text{slope} = \dfrac{1}{4}$

slope of line perpendicular: $\dfrac{-1}{\frac{1}{4}} = -4$

point slope:

$\qquad \boxed{y + 2 = -4(x - 3)}$

slope–intercept:

$\qquad y + 2 = -4x + 12$

$\qquad \boxed{y = -4x + 10}$

47. $\qquad\qquad f(x) = 7x^2 - 5x - 3$

$\dfrac{f(a + h) - f(a)}{h} = \dfrac{[7(a + h)^2 - 5(a + h) - 3] - (7a^2 - 5a - 3)}{h}$

$\qquad\qquad\qquad = \dfrac{7a^2 + 14ah + 7h^2 - 5a - 5h - 3 - 7a^2 + 5a + 3}{h}$

$\qquad\qquad\qquad = \dfrac{14ah + 7h^2 - 5h}{h}$

$\qquad\qquad\qquad = \boxed{14a + 7h - 5}$

48. $f(x) = \dfrac{1}{x^2}$

$$\dfrac{f(a+h) - f(a)}{h} = \dfrac{\dfrac{1}{(a+h)^2} - \dfrac{1}{a^2}}{h}$$

$$= \dfrac{a^2 - (a+h)^2}{a^2(a+h)^2} \div h$$

$$= \dfrac{a^2 - a^2 - 2ah - h^2}{a^2(a+h)^2} \cdot \dfrac{1}{h}$$

$$= \dfrac{-2ah - h^2}{ha^2(a+h)^2}$$

$$= \boxed{\dfrac{-2a - h}{a^2(a^2 + 2ah + h^2)}}$$

49. $f(x) = \sqrt{x}$

$$\dfrac{f(a+h) - f(a)}{h} = \dfrac{\sqrt{a+h} - \sqrt{a}}{h}$$

$$= \dfrac{\sqrt{a+h} - \sqrt{a}}{h} \cdot \dfrac{\sqrt{a+h} + \sqrt{a}}{\sqrt{a+h} + \sqrt{a}}$$

$$= \dfrac{a + h - a}{h\left(\sqrt{a+h} + \sqrt{a}\right)}$$

$$= \dfrac{h}{h\left(\sqrt{a+h} + \sqrt{a}\right)}$$

$$= \boxed{\dfrac{1}{\sqrt{a+h} + \sqrt{a}}}$$

50. $f(x) = 3x^2 - 7x + 4$ and $g(x) = 2x - 1$

$$\begin{aligned}
f[g(x)] - g[f(x)] &= \{3[g(x)]^2 - 7[g(x)] + 4\} - \{2[f(x)] - 1\} \\
&= 3(2x-1)^2 - 7(2x-1) + 4 - 2(3x^2 - 7x + 4) + 1 \\
&= 12x^2 - 12x + 3 - 14x + 7 + 4 - 6x^2 + 14x - 8 + 1 \\
&= \boxed{6x^2 - 12x + 7}
\end{aligned}$$

51. $f(x) = \dfrac{2 + \dfrac{1}{x}}{2 - \dfrac{1}{x}}$ and $g(x) = x - 3$

$$\begin{aligned}
(f \circ g)(x) &= f[g(x)] \\
&= \dfrac{2 + \dfrac{1}{x-3}}{2 - \dfrac{1}{x-3}} \\
&= \dfrac{2x - 6 + 1}{x - 3} \div \dfrac{2x - 6 - 1}{x - 3} \\
&= \dfrac{2x - 5}{x - 3} \cdot \dfrac{x - 3}{2x - 7} \\
&= \boxed{\dfrac{2x - 5}{2x - 7} \quad \left(x \neq 3, \dfrac{7}{2}\right)}
\end{aligned}$$

$$\boxed{D_{f \circ g} = \left\{ x \mid x \neq 3, \dfrac{7}{2} \right\} = (-\infty, 3) \cup \left(3, \dfrac{7}{2}\right) \cup \left(\dfrac{7}{2}, \infty\right)}$$

52. $f(x) = 2x - 1$ and $g(x) = x^2 - 3x - 1$

$$f\left(\dfrac{1}{2}\right) = 2\left(\dfrac{1}{2}\right) - 1 = 1 - 1 = 0$$

$$g\left[f\left(\dfrac{1}{2}\right)\right] = g(0) = 0 - 0 - 1$$

$$= \boxed{-1}$$

53. $f(x) = \dfrac{1}{2}x - 5$

$y = \dfrac{1}{2}x - 5$

$x = \dfrac{1}{2}y - 5$ (Exchange x and y)

$x + 5 = \dfrac{1}{2}y$

$2x + 10 = y$

$\boxed{f^{-1}(x) = 2x + 10}$

$f[f^{-1}(x)] = \dfrac{1}{2}(2x + 10) - 5 = x + 5 - 5 = x$

$f^{-1}[f(x)] = 2\left(\dfrac{1}{2}x - 5\right) + 10 = x - 10 + 10 = x$

54. $f(x) = \sqrt{12 - 4x}$

$12 - 4x \geq 0$

$-4x \geq -12$

$x \leq 3$

$\boxed{D_f = \{x \mid x \leq 3\} = (-\infty, -3]}$

55. $f(3) = -2$

$f(2) = -3$

a. $\left|f(3) - f(2)\right| = \left|-2 - (-3)\right| = \left|-2 + 3\right| = \boxed{1}$

b. $\boxed{\begin{array}{l} \text{Domain} = [0, 4] \\ \text{Range} = [-3, 0] \end{array}}$

56. $y = -\dfrac{2}{3}x + 1$

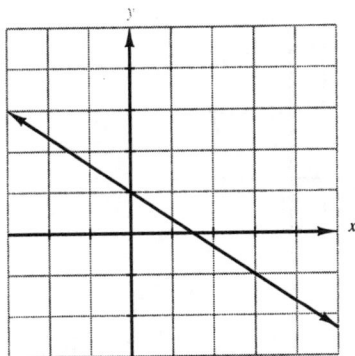

57. $2x - y \geq 4$ and $x \leq 2$

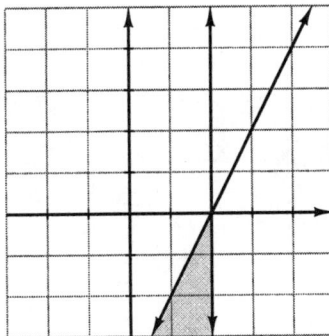

58. $f(x) = \dfrac{1}{x - 2}$

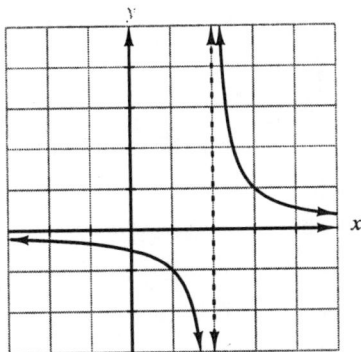

59. $f(x) = \begin{cases} \sqrt{x - 3} & \text{if } x \geq 3 \\ 3 - x & \text{if } x < 3 \end{cases}$

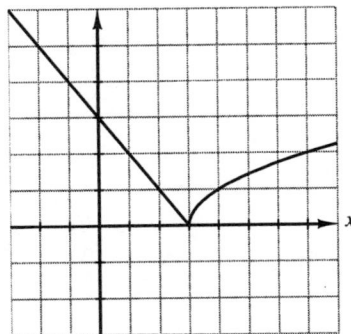

60. $f(x) = \dfrac{2x^2 - 5x - 3}{x - 3}$

$= \dfrac{(2x + 1)(x - 3)}{x - 3}$

$= 2x + 1 \quad (x \neq 3)$

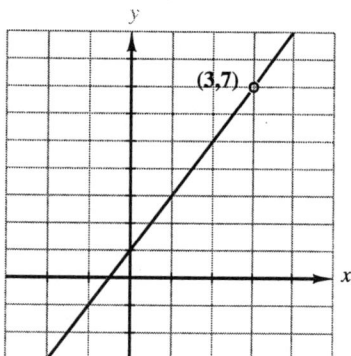

61. $x^2 + y^2 + 4x - 6y + 9 = 0$

$x^2 + 4x + y^2 - 6y = -9$

$x^2 + 4x + 4 + y^2 - 6y + 9 = -9 + 4 + 9$

$(x + 2)^2 + (y - 3)^2 = 4$

circle with center $(-2, 3)$ and radius 2.

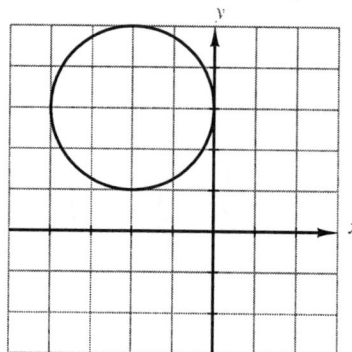

62. $\dfrac{x^2}{9} + \dfrac{y^2}{4} \leq 1$

ellipse with center $(0, 0)$ and vertices $(\pm 3, 0)$ and $(0, \pm 2)$

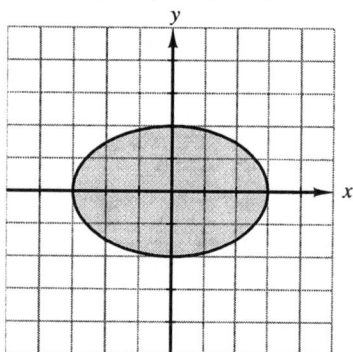

63. $\dfrac{x^2}{4} - \dfrac{y^2}{9} = 1$

hyperbola with center $(0, 0)$, vertices $(\pm 2, 0)$ and no y–intercept.

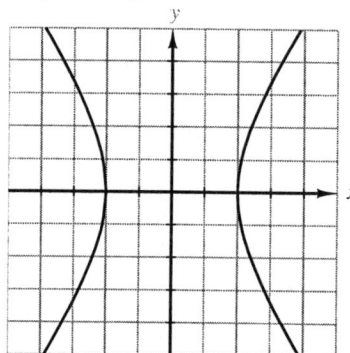

64. $y = x^2 - 4x - 5$
Vertex:
$$x = -\frac{b}{2a} = \frac{4}{2} = 2$$
$$y = 2^2 - 4(2) - 5 = -9$$
Vertex $(2, -9)$

x-intercepts:
$$x^2 - 4x - 5 = 0$$
$$(x - 5)(x + 1) = 0$$
$$x = 5 \qquad \text{or} \qquad x = -1$$

y-intercepts:
$$y = 0 - 0 - 5 = -5$$
parabola with vertex $(2, -5)$ opening upward

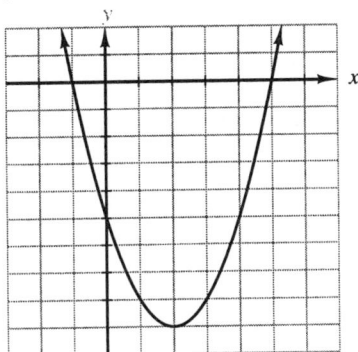

65. $\dfrac{(x-2)^2}{9} + \dfrac{(y+1)^2}{25} \geq 1$
ellipse with center $(2, -1)$ and vertices
$$(2 \pm 3, -1) = (5, -1) \text{ or } (-1, -1)$$
$$(2, -1 \pm 5) = (2, 4) \text{ or } (2, -6)$$

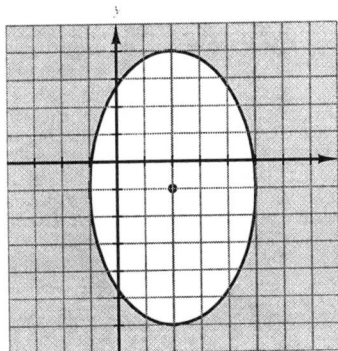

66. $f(x) = 2^{x+1}$

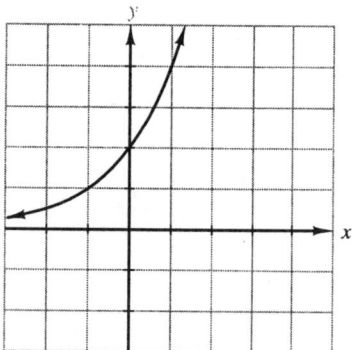

67. $f(x) = \log_2(x-1)$

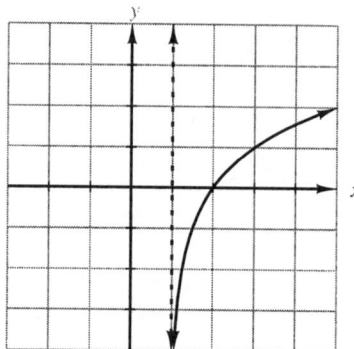

68. $x > (y-1)^2 - 1$
boundary (dashed curve)
$\qquad x = (y-1)^2 - 1$ is a parabola with vertex $(-1, 1)$ opening to the right.
y-intercepts (set $x = 0$):

$$
\begin{aligned}
0 &= (y-1)^2 - 1 \\
(y-1)^2 &= 1 \\
y - 1 &= \pm 1 \\
y &= 2 \quad \text{or} \quad y = 0
\end{aligned}
$$

x-intercepts (set $y = 0$):

$$x = (0-1)^2 - 1 = 0$$

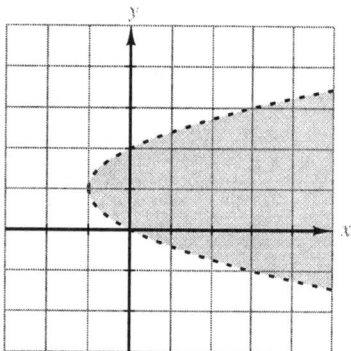

69. $f(x) = x^3 + 4x^2 - x - 4$

x–intercepts:

$$
\begin{aligned}
x^3 + 4x^2 - x - 4 &= 0 \\
x^2(x + 4) - (x + 4) &= 0 \\
(x + 4)(x^2 - 1) &= 0 \\
(x + 4)(x - 1)(x + 1) &= 0 \\
x &= -4, 1, -1
\end{aligned}
$$

intervals:

$x < -4$	$-4 < x < -1$	$-1 < x < 1$	$x > 1$
-4	-1	1	

interval	test value
$x < -4$	-5: $f(-5) = -24$
$-4 < x < -1$	-2: $f(-2) = 6$
$-1 < x < 1$	0: $f(0) = -4$
$x > 1$	2: $f(2) = 18$

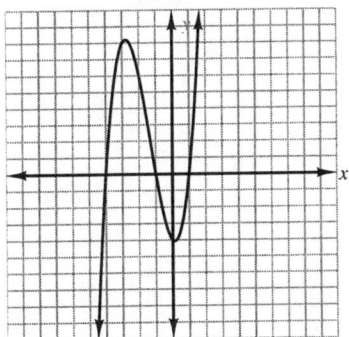

70. $f(x) = \dfrac{x^2 - 6x + 5}{x^2 - 6x + 9}$

vertical asymptote:

$$
\begin{aligned}
x^2 - 6x + 9 &= 0 \\
(x - 3)^2 &= 0 \\
x &= 3
\end{aligned}
$$

horizontal asymptote:

$$
f(x) = \frac{x^2 - 6x + 5}{x^2 - 6x + 9} = \frac{1 - \dfrac{6}{x} + \dfrac{5}{x^2}}{1 - \dfrac{6}{x} + \dfrac{9}{x^2}} \quad \rightarrow \quad \frac{1 - 0 + 0}{1 - 0 + 0} = 1
$$

$y = 1$ is a horizontal asymptote

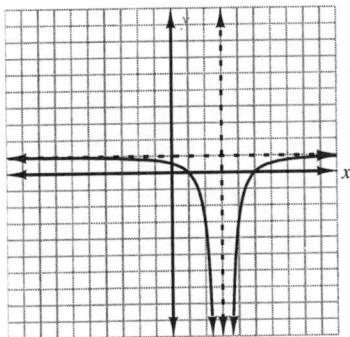

71. $3(7 + 4) = 3 \cdot 7 + 4 \cdot 3$

Distributive property

Commutative property of multiplication

72. $\{x \mid x$ is a prime number greater than 3 and less than 17 that yields a remainder of 3 when divided by 4$\}$

$3 <$ prime number < 17: 5, 7, 11, 13

$$\frac{5}{4} = 1 \text{ remainder } 1$$

$$\frac{7}{4} = 1 \text{ remainder } 3$$

$$\frac{11}{4} = 2 \text{ remainder } 3$$

$$\frac{13}{4} = 3 \text{ remainder } 1$$

$\{7, 11\}$

73. $\dfrac{(0.0045)(60,000)}{(1800)(0.00015)} = \dfrac{(4.5 \times 10^{-3})(6 \times 10^4)}{(1.8 \times 10^3)(1.5 \times 10^{-4})} = \dfrac{27 \times 10^1}{2.7 \times 10^{-1}} = 10 \times 10^2 = \boxed{10^3}$

74. $F = \dfrac{km_1 m_2}{d^2}$

Given: $m_1 = 64$ slugs, $m_2 = 108$ slugs, $d = 24$ feet, $F = 6$ pounds

$$6 = \frac{k(64)(108)}{(24)^2}$$

$$k = \frac{1}{2}$$

If $m_1 = m_2 = 30$ slugs and $d = 10$ feet, what is F?

$$F = \frac{\frac{1}{2}m_1 m_2}{d^2}$$

$$F = \frac{1}{2} \cdot \frac{(30)(30)}{10^2}$$

$$F = 4.5$$

4.5 pounds

75. $\begin{vmatrix} 2 & 4 & 0 \\ 5 & 0 & -1 \\ -2 & 1 & -1 \end{vmatrix} = 2 \begin{vmatrix} 0 & -1 \\ 1 & -1 \end{vmatrix} - 4 \begin{vmatrix} 5 & -1 \\ -2 & -1 \end{vmatrix} + 0$

$= 2(0 + 1) - 4(-5 - 2) + 0$

$= 2(1) - 4(-7) + 0$

$= 2 + 28$

$= \boxed{30}$

76. $(4, 3)$ to $(2, -1)$:

$$d = \sqrt{(4-2)^2 + (3+1)^2}$$

$$= \sqrt{2^2 + 4^2}$$

$$= \sqrt{4 + 16}$$

$\boxed{2\sqrt{5}}$

77. $\log_2\left(\dfrac{1}{64}\right) = y$

$$2^y = \frac{1}{64}$$

$$2^y = 2^{-6}$$

$$y = -6$$

$\{-6\}$

78. $\log_5 \dfrac{x^3 \sqrt{y}}{125} = \log_5 x^3 + \log_5 y^{1/2} - \log_5 5^3$

$= \boxed{3 \log_5 x + \dfrac{1}{2} \log_5 y - 3}$

79. $\dfrac{1}{4} \log_b x + 2 \log_b (x^2 - 3) - \dfrac{1}{3} \log_b z = \dfrac{1}{4} \log_b x + \log_b (x^2 - 3)^2 - \log_b z^{1/3}$

$= \boxed{\log_b \dfrac{x^{1/4} (x^2 - 3)^2}{\sqrt[3]{z}}}$

80. $f(x) = 0.015 x^2 - 4.5 x + 400$
minimum occurs when

$x = -\dfrac{b}{2a}$

$= \dfrac{-(-4.5)}{2(0.015)}$

$= 150$

$y = 0.015(150)^2 - 4.5(150) + 400$

$= 337.5 - 675 + 400$

$= 62.5$

minimum daily cost/person: $\boxed{\$62.50}$

81. $2, 6, 10, \ldots$
$a_1 = 2, n = 30, d = 6 - 2 = 4$

$S_n = \dfrac{n}{2}[2a_1 + (n - 1)d]$

$= 15[4 + 29(4)]$

$\boxed{1800}$

82. $a_{16} = -30, d = -3$

$a_n = a_1 + (n - 1)d$

$a_{16} = a_1 + (16 - 1)(-3)$

$-30 = a_1 - 45$

$15 = a_1$

$\boxed{15}$

83. $\dfrac{1}{2}, 2, 8, \ldots$

$a_1 = \dfrac{1}{2}, n = 8, r = \dfrac{a_2}{a_1} = \dfrac{2}{\frac{1}{2}} = 4$

$S_n = \dfrac{a_1 - a_1 r^n}{1 - r}$

$S_8 = \dfrac{\dfrac{1}{2} - \dfrac{1}{2}(4)^8}{1 - 4}$

$= \dfrac{\dfrac{1}{2}(1 - 65536)}{-3}$

$= -32767.5$

$= \boxed{10,922.5}$

84. $a_1 = 8, r = -\dfrac{1}{2}, n = 5$

$a_n = a_1 r^{n-1}$

$a_5 = 8 \left(-\dfrac{1}{2}\right)^{5-1} = 8 \left(-\dfrac{1}{2}\right)^4 = \dfrac{8}{16} = \boxed{\dfrac{1}{2}}$

85. $0.\overline{450} = 0.450\,450\ldots = 0.450 + 0.000450 + 0.000000450$

$a_1 = 0.450, r = \dfrac{0.000450}{0.450} = 0.001$

$$S = \frac{a^1}{1-r}$$
$$= \frac{0.450}{0-0.001}$$
$$= \frac{0.450}{0.999}$$
$$= \frac{450}{999}$$
$$= \boxed{\frac{50}{111}}$$

86. Distance ball travels:

$$90 + \frac{2}{3}(90) + \left(\frac{2}{3}\right)^2(90) + \left(\frac{2}{3}\right)^3(90) + \ldots$$

Infinite geometric serves 2 with $a = 90$ and $r = \dfrac{2}{3}$.

$$S = \frac{a}{1-r}$$
$$S = \frac{90}{1-\frac{2}{3}} = \frac{90}{\frac{1}{3}} = 270$$

$\boxed{270 \text{ meters}}$

87. $(2x - y^3)^5 = (2x)^5 + 5(2x)^4(-y^3) + 10(2x)^3(-y^3)^2 - 10(2x)^2(-y^3)^3 + 5(2x)(-y^3)^4 + (-y^3)^5$

$\qquad\qquad = \boxed{32x^5 - 80x^4y^3 + 80x^3y^6 - 40x^2y^9 + 10xy^{12} - y^{15}}$

88. 50¢ possibilities:

q	d	n
2	0	0
1	2	1
1	1	3
1	0	5
0	5	0
0	4	2
0	3	4
0	2	6
0	1	8
0	0	10

$\boxed{10 \text{ ways}}$

89. Odd number less than 100: 1, 3, 5, 7, . . ., 97, 99
A multiple of 5: 5, 15, 25, 35, 45, 55, 65, 75, 85, 95
Divisible by 3: 15, 45, 75
Sum of digitis is odd:
    15: $1 + 5 = 6$
    45: $4 + 5 = 9$
    75: $7 + 5 = 12$
$\boxed{45}$ is the only number

90.

	1st	2nd	3rd	4th	5th	6th	12th	nth
Pattern	$\dfrac{1(1)}{2}$	$\dfrac{2(2+1)}{2}$	$\dfrac{3(3+1)}{2}$	$\dfrac{4(4+1)}{2}$	$\dfrac{5(5+1)}{2}$	$\dfrac{6(6+1)}{2}$	$\dfrac{12(12+1)}{2}$	$\dfrac{n(n+1)}{2}$
Triangular number	1	3	6	10	15	21	78	$\dfrac{n(n+1)}{2}$

$$\boxed{15, 21, 78, \dfrac{n(n+1)}{2}}$$

91. $\quad \dfrac{10x+y}{w} = z \quad$ with $w, x, y,$ and z different, non–zero, one-digit positive number.

a. $\quad w = 3, y = 1$

$$\dfrac{10x+1}{3} = z$$

$$\boxed{x = 2 \text{ and } z = 7} \text{ since } \dfrac{20+1}{3} = 7$$

b. $\qquad x = y - 3$

$$\boxed{y = 4, x = 1, w = 7 \text{ so } z = 2}$$

$$\boxed{y = 4, x = 1, w = 2 \text{ so } z = 7}$$

since

$$\dfrac{10x+y}{w} = z$$

$$\dfrac{10+4}{7} = 2$$

and

$$\dfrac{10+4}{2} = 7$$

c. $\quad w = 2z$ and $x > y$

$$\boxed{x = 3, y = 2, z = 4, w = 8}$$

$$z = 4, \quad w = 2(4) = 8$$

$$\dfrac{10(3)+2}{8} = 4, \quad 3 > 2$$

d. $\qquad w + z = x + y$

$$\boxed{x = 1, y = 8 \text{ (with } w = 3 \text{ and } z = 6)}$$

$$3 + 6 = 1 + 8$$

$$\dfrac{10(1)+8}{3} = 6$$

92. Let $x, x + 2$ and $x + 4$ equal three consecutive even integers.

$$\boxed{\begin{aligned} (x+2) + (x+4) &= 2 + 3x \\ 2x + 6 &= 2 + 3x \end{aligned}}$$

93. $\qquad$
$$\begin{aligned} \text{Let } x &= \text{Lisa's present age} \\ 38 - x &= \text{Mark's present age} \\ 38 - x - 5 &= \text{Mark's age 5 years ago} \\ x + 7 &= \text{Lisa's age in 7 years} \end{aligned}$$

$$\boxed{x + 7 = 3(38 - x - 5)}$$

or $\qquad x + 7 = 3(33 - x)$

94. Let x = plane's speed during first 3 hours
distance: 2540 km
time: 5 hours
distance traveled in first 3 hours: $3x$
time remaining: $5 - 3 = 2$ hours
rate for last 2 hours: $x - 30$

(distance traveled first 3 hours) + (distance traveled last 2 hours) = (total distance)

$$\boxed{3x + (x - 30)(2) \ = \ 2540}$$

95. number of dimes: x
number of quarters: $2x - 1$
number of nickels: $2(2x - 1) - 1 = 4x - 3$
Value of the pennies = value of nickels: $5(4x - 3)$
value of dimes + value of quarters + value of nickels + value of pennies = 145

$$\boxed{10x + 25(2x - 1) + 5(4x - 3) + 5(4x - 3) \ = \ 145}$$

96. $\qquad\qquad 2Ax + By \ = \ -20$
(–2, 2): If $x = -2$, $y = 2$
$\qquad\qquad\qquad 2A(-2) + B(2) \ = \ -20$
$\qquad\qquad\qquad\quad -4A + 2B \ = \ -20$

$\qquad\qquad\quad Ax - 2By \ = \ 10$
(–2, 2): If $x = -2$, $y = 2$
$\qquad\quad A(-2) - 2B(2) \ = \ 10$
$\qquad\qquad\quad -2A - 4B \ = \ 10$

System:

$$\boxed{\begin{aligned} -4A + 2B &= -20 \\ -2A - 4B &= 10 \end{aligned}}$$

97. Let
$\qquad\quad x \ = \ $ Boat's rate of speed in still water
$\qquad\quad y \ = \ $ Rate of current
$\qquad$ rate $\times$ time $\ = \ $ distance
With current: 45 minutes $= \dfrac{45}{60}$ of an hour $= \dfrac{3}{4}$ h to cover 6 kilometers:

$$(x + y) \cdot \frac{3}{4} \ = \ 6$$

Against current: $1\dfrac{1}{2}$ hours to cover 6 kilometers: $(x - y) \cdot \dfrac{3}{2} = 6$

System:

$$\boxed{\begin{aligned} \frac{3}{4}(x + y) &= 6 \\ \frac{3}{2}(x - y) &= 6 \end{aligned}}$$

98. Let
$\qquad\qquad\quad t \ = \ $ tens' digit
$\qquad\qquad\quad u \ = \ $ units' digit
$\qquad\quad 10t + u \ = \ $ the number
$\qquad\quad 10u + t \ = \ $ the number with its digits reversed

$t + u = 8$	or	$t + u = 8$
$10u + t = 10t + u + 18$		$-9t + 9u = 18$

99. Let

$$x = \text{distance between edge of lot and edge of garden}$$
$$32 - 2x = \text{width of garden}$$
$$40 - 2x = \text{length of garden}$$

Area of garden = 560

$$\boxed{(32 - 2x)(40 - 2x) = 560}$$

100. Let x = amount invested at 6% for 3 years

$50,000 - x$ = amount invested at 7% for 2 years

$$\boxed{(0.06x)(3) + (0.07)(50,000 - x)(2) = 8800}$$

101. Let x = the greater number

$x - 4$ = the smaller number

sum of multiplicative inverses $= \dfrac{10}{21}$

$$\boxed{\dfrac{1}{x} + \dfrac{1}{x-4} = \dfrac{10}{21}}$$

102. Let x = number of hours to fill pool.

	Fractional Part of Job Completed in 1 Hour	Time Spent Working Together	Fractional Part of Job Completed in x hours
Inlet pipe (12 hours)	$\dfrac{1}{12}$	x	$\dfrac{x}{12}$
Outlet pipe (15 hours)	$\dfrac{1}{15}$	x	$\dfrac{x}{15}$

(Fractional part by inlet pipe) – (Fractional part by outlet pipe) = (one full pool)

$$\boxed{\dfrac{x}{12} - \dfrac{x}{15} = 1}$$

103. Let x = walking speed

$x + 4$ = running speed

	Distance (km)	Rate	Time $= \dfrac{\text{Distance}}{\text{Rate}}$
Running	17	$x + 4$	$\dfrac{17}{x+4}$
Walking	9	x	$\dfrac{9}{x}$

Time to run 17 km = Time to walk 9 km

$$\boxed{\dfrac{17}{x+4} = \dfrac{9}{x}}$$

104.

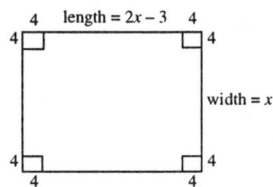

Let

$$x = \text{width of rectangular pices of metal}$$
$$2x - 3 = \text{length of rectangular piece of metal}$$

$$\text{height of box} = 4 \text{ meters}$$
$$x - 2(4) = x - 8 = \text{width of box}$$
$$2x - 3 - 2(4) = 2x - 3 - 8 = \text{length of box}$$

$$\text{Volume of box} = 532$$
$$\text{Volume} = \text{length} \times \text{width} \times \text{height}$$
$$(2x - 3 - 8)(x - 8)4 = 532$$
$$\boxed{4(2x - 11)(x - 8) = 532}$$

105.

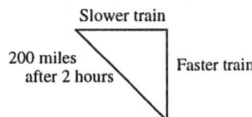

Let x = slower train's speed.
$x + 20$ = faster train's speed.
Distance of slower train in 2 hours: $2x$
Distance of faster train in 2 hours: $2(x + 20) = 2x + 40$

By the Pythagorean Theorem: $\boxed{(2x)^2 + (2x + 40)^2 = (200)^2}$

106. Let x = number of $8 tickets sold.
$250 - x$ = number of $12 tickets sold

Income from $8 tickets + Income from $12 tickets = $2200
$\boxed{8x + 12(250 - x) = 2200}$

107. Let x = number of grams of the 80% solution.

$$\begin{pmatrix} \text{Amount of} \\ \text{antifreeze} \\ \text{in the 80\%} \\ \text{solution} \end{pmatrix} + \begin{pmatrix} \text{Amount of} \\ \text{antifreeze} \\ \text{in the 12\%} \\ \text{solution} \end{pmatrix} = \begin{pmatrix} \text{Amount of} \\ \text{antifreeze} \\ \text{in the 60\%} \\ \text{mixture} \end{pmatrix}$$

$$\boxed{0.8x + (0.12)(175) = (0.6)(x + 175)}$$

108. Let x = number of people in the smaller group.
$x + 4$ = number of people in the larger group.

$$\left(\begin{array}{c}\text{Amount paid by}\\ \text{each person in the}\\ \text{smaller group}\end{array}\right) - \left(\begin{array}{c}\text{Amount paid by}\\ \text{each person in}\\ \text{the larger group}\end{array}\right) = \$20$$

$$\text{Amount paid by each person} = \frac{\text{Total cost}}{\text{Number of people}}$$

$$\boxed{\frac{19{,}200}{x} - \frac{19{,}200}{x+4} = 20}$$

or $\quad \dfrac{19{,}200}{x} - 20 = \dfrac{19{,}200}{x+4}$

109. Let x = edge of open box.
$x + 1$ = edge of closed box.
Surface area of open box (with five square faces) = $5x^2$
Surface area of closed box (with six square faces) = $6(x+1)^2$
$$\boxed{5x^2 = 6(x+1)^2 - 51}$$

110. Perimeter = 30

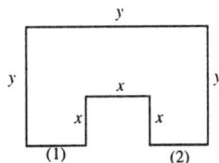

Length of (1) + Length of (2) = $y - x$

$$
\begin{aligned}
y - x &= 30 \\
\text{Perimeter} &= y + y + y + x + x + x + (y - x) \\
&= 4y + 2x \\
&= 30 \\
\text{Area} &= 27 \\
y^2 - x^2 &= 27
\end{aligned}
$$

System:
$$\boxed{\begin{aligned} 4y + 2x &= 30 \\ y^2 - x^2 &= 27 \end{aligned}}$$

111.
$$\text{Area of triangle} + \text{Area of square} \;=\; 37$$
$$\tfrac{1}{2}(2y)(4) + x^2 \;=\; 37$$
$$4y + x^2 \;=\; 37$$

By the Pythagorean Theorem:
$$4^2 + y^2 \;=\; x^2$$
$$16 + y^2 \;=\; x^2$$

System:

$$\boxed{\begin{aligned} 4y + x^2 &= 37 \\ 16 + y^2 &= x^2 \end{aligned}}$$

112.

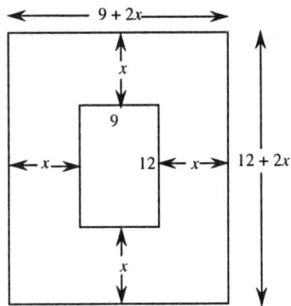

Let x = width of border.
$9 + 2x$ = width of pool plus border
$12 + 2x$ = length of pool plus border
$$\text{area of pool plus border} \;=\; \text{area of pool} + \text{area of border}$$
$$\boxed{(9 + 2x)(12 + 2x) \;=\; 9 \cdot 12 + 100}$$